H. Windrich

MECHANICS OF SOLIDS III

EDITOR

C. TRUESDELL

MECHANICS OF SOLIDS I–IV

EDITOR

C. TRUESDELL

VOLUME I

The Experimental Foundations of Solid Mechanics

VOLUME II

Linear Theories of Elasticity and Thermoelasticity
Linear and Nonlinear Theories of Rods, Plates, and Shells

VOLUME III

Theory of Viscoelasticity, Plasticity, Elastic Waves, and Elastic Stability

VOLUME IV

Waves in Elastic and Viscoelastic Solids
(Theory and Experiment)

Originally published as ENCYCLOPEDIA OF PHYSICS VIa/1–4

MECHANICS OF SOLIDS

VOLUME III

Theory of Viscoelasticity, Plasticity, Elastic Waves, and Elastic Stability

Editor
C. Truesdell

Contributions by
P. J. Chen G. M. C. Fisher H. Geiringer R. J. Knops
M. J. Leitman T. W. Ting E. W. Wilkes

With 56 Figures

Springer-Verlag
Berlin Heidelberg New York Tokyo 1984

Volume Editor

Professor Dr. C. Truesdell

The Johns Hopkins University, Baltimore, MD 21218, USA

Chief Editor of Encyclopedia of Physics

Professor Dr. S. Flügge

Fakultät für Physik, Universität Freiburg
D-7800 Freiburg, Fed. Rep. of Germany

This book originally appeared in hardcover as Volume VIa/3 of
Encyclopedia of Physics

ISBN 3-540-05536-3 Springer-Verlag Berlin Heidelberg New York
ISBN 0-387-05536-3 Springer-Verlag New York Heidelberg Berlin

ISBN 3-540-13162-0 Springer-Verlag Berlin Heidelberg New York Tokyo
ISBN 0-387-13162-0 Springer-Verlag New York Heidelberg Berlin Tokyo

Offset printing: Beltz Offsetdruck, 6944 Hemsbach
Bookbinding: J. Schäffer OHG, 6718 Grünstadt
2153/3130-543210

Preface.

Reissue of
Encyclopedia of Physics / Handbuch der Physik, Volume VIa

The mechanical response of solids was first reduced to an organized science of fairly general scope in the nineteenth century. The theory of small elastic deformations is in the main the creation of CAUCHY, who, correcting and simplifying the work of NAVIER and POISSON, through an astounding application of conjoined scholarship, originality, and labor greatly extended in breadth the shallowest aspects of the treatments of particular kinds of bodies by GALILEO, LEIBNIZ, JAMES BERNOULLI, PARENT, DANIEL BERNOULLI, EULER, and COULOMB. Linear elasticity became a branch of mathematics, cultivated wherever there were mathematicians. The magisterial treatise of LOVE in its second edition, 1906 – clear, compact, exhaustive, and learned – stands as the summary of the classical theory. It is one of the great "gaslight works" that in BOCHNER's words[1] "either do not have any adequate successor[s] ... or, at least, refuse to be superseded ...; and so they have to be reprinted, in ever increasing numbers, for active research and reference", as long as State and Society shall permit men to learn mathematics by, for, and of men's minds.

Abundant experimentation on solids was done during the same century. Usually the materials arising in nature, with which experiment most justly concerns itself, do not stoop easily to the limitations classical elasticity posits. It is no wonder that the investigations LOVE's treatise collects, condenses, and reduces to symmetry and system were in the main ill at ease with experiment and unconcerned with practical applications. In LOVE's words, they belong to "an abstract conceptual scheme of Rational Mechanics". He concluded thus his famous Historical Introduction:

The history of the mathematical theory of Elasticity shows clearly that the development of the theory has not been guided exclusively by considerations of its utility for technical Mechanics. Most of the men by whose researches it has been founded and shaped have been more interested in Natural Philosophy than in material progress, in trying to understand the world than in trying to make it more comfortable. From this attitude of mind it may possibly have resulted that the theory has contributed less to the material advance of mankind than it might otherwise have done. Be this as it may, the intellectual gain which has accrued from the work of these men must be estimated very highly. The discussions that have taken place concerning the number and meaning of the elastic constants have thrown light on most recondite questions concerning the nature of molecules and the mode of their interaction. The efforts that have been made to explain optical phenomena by means of the hypothesis of a medium having the same physical character as an elastic solid body led, in the first instance, to

[1] SALOMON BOCHNER: "Einstein between centuries", Rice Univ. Stud. 65 (3), 54 (1979).

the understanding of a concrete example of a medium which can transmit transverse vibrations, and, at a later stage, to the definite conclusion that the luminiferous medium has not the physical character assumed in the hypothesis. They have thus issued in an essential widening of our ideas concerning the nature of the aether and the nature of luminous vibrations. The methods that have been devised for solving the equations of equilibrium of an isotropic solid body form part of an analytical theory which is of great importance in pure mathematics. The application of these methods to the problem of the internal constitution of the Earth has led to results which must influence profoundly the course of speculative thought both in Geology and in cosmical Physics. Even in the more technical problems, such as the transmission of force and the resistance of bars and plates, attention has been directed, for the most part, rather to theoretical than to practical aspects of the questions. To get insight into what goes on in impact, to bring the theory of the behaviour of thin bars and plates into accord with the general equations - these and such-like aims have been more attractive to most of the men to whom we owe the theory than endeavours to devise means for effecting economies in engineering constructions or to ascertain the conditions in which structures become unsafe. The fact that much material progress is the indirect outcome of work done in this spirit is not without significance. The equally significant fact that most great advances in Natural Philosophy have been made by men who had a first-hand acquaintance with practical needs and experimental methods has often been emphasized; and, although the names of Green, Poisson, Cauchy show that the rule is not without important exceptions, yet it is exemplified well in the history of our science.

LOVE's treatise mentions experiment rarely and scantly. Its one passage concerning experiment in general, § 63, in effect warns its reader to have a care of experimental data because of their indirectness.

In an irony of history the ever-increasing use of mathematical notation in physical science, to the point that now often works on experiment are dominated by their authors' seemingly compulsive recourse to mathematical formulae interconnected by copied or adapted bits of old mathematical manipulation, LOVE's treatise is sometimes in reproaches upon modern "pure" or "abstract" researchers held up as a model of practical, applied theory.

Experiment on the mechanical properties of solids became in the later nineteenth century a science nearly divorced from theory. Nevertheless, no great treatise on experiment fit to be set beside LOVE's on theory ever appeared. Even such books of experiment as were published seem to have in the main taken positions either dominated by theory, usually crude, verbose, and ill presented, or flatly opposed to theory.

The modern reader will cite as objections against the foregoing coarse summary many individual masterpieces that do not support it: brilliant comparisons of theory with experiment by ST. VENANT, independent experiments of fundamental importance by WERTHEIM, CAUCHY's marvellously clear mathematical apparatus for conceiving stress and arbitrarily large strains and rotations, theories of internal friction and plasticity proposed by BOLTZMANN, ST. VENANT, and others. If he is searching for antecedents of what has happened in the second half of the twentieth century, he is abundantly right in citing these and other achievements of the nineteenth while passing over the work of the ruck, but in that century's gross product of solid mechanics they are exceptions that prove rules.

In planning this volume on the mechanics of solids for the *Encyclopedia of Physics* I designed

1) To provide a treatise on experimental mechanics of solids that, not dominated by mathematical theory and not neglecting the work of the eighteenth and nineteenth centuries in favor of recent, more popular, and more costly forays, should be comparable in authority, breadth, and scholarship with LOVE's.
2) To provide treatises on basic, mathematical theory that would stand at the level of LOVE's while in their own, narrower scope supplanting it by compact and efficient development of fundamentals, making use of modern, incisive, yet elementary mathematics to weave together old and recent insights and achievements.
3) To illustrate the power of modern mathematical theory and modern experiment by articles on selected topics recently developed for their intellectual and practical importance, these two qualities being closer to each other than to some they may seem.

I encouraged the authors to meet the standard established by LOVE in just citation and temperate respect for the discoverers.

The reader will be able to form his own judgment of such success and failure as did accrue.

On the first head, experiment in general, the reader will find the treatise by Mr. BELL, filling all of Part 1. While it is not primarily a historical work, the historian S. G. BRUSH pronounced it in 1975 "the most important new publication by a single author" on the history of physics.

On the second head, the reader should not expect to find the basic ideas of solids treated *ab novo* or in isolation. The general and unified mechanics of EULER and CAUCHY, in which fluids, solids, and materials of other kinds are but instances, has come into its own in our day. No wise scientist now can afford to shut out solids when studying fluids or to forget the nature and peculiarities of fluids when studying solids. The two are but extreme examples in the class of systems comprised by mechanics. Articles in Parts 1 and 3 of Volume III of the *Encyclopedia: The Classical Field Theories* and *The Non-Linear Field Theories of Mechanics,* are cited so often by the authors writing in Volume VIa as to make it fatuous to deny that they provide the basic concepts, structures, and mathematical apparatus for the articles on theoretical mechanics of solids. In particular *The Non-Linear Field Theories* goes into such detail regarding large mechanical deformation as to allow most of the text in Volume VIa to concentrate upon small strain.

This much understood, we see that while Mr. BELL's volume provides, at last, a monument of exposition and scholarship on experiment, the articles by Messrs. GURTIN, CARLSON, FICHERA, NAGHDI, ANTMAN, and FISHER & LEITMAN, by Mrs. GEIRINGER, and by Mr. TING together provide a modern treatise on mathematical theories of the classical kinds. The survey of theories of elastic stability by Messrs. KNOPS & WILKES, now justly regarded as the standard reference for its field, necessarily considers deformations that need not be small.

Coming finally to application, in which theory and experiment complement one another, the reader will find major examples in the articles by Messrs. CHEN; NUNZIATO, WALSH, SCHULER and BARKER; and THURSTON. Many more topics of application might have been included. I regret that I could not secure articles about them. The most serious want is a survey of applications of linear elasticity to problems of intrinsic or applied interest that have arisen in this century and that illustrate the power of new mathematical analysis in dealing with special problems. A long article of that kind, a veritable treatise, was twice contracted and twice defaulted. Fortunately the gap thus left has been abundantly and expertly filled by Mr. VILLAGGIO, *Qualitative Methods in Elasticity,* Leyden, Noordhoff, 1977.

Baltimore, December, 1983 C. TRUESDELL

Contents.

The Linear Theory of Viscoelasticity. By MARSHALL J. LEITMAN, Associate Professor of Mathematics, Case Western Reserve University, Cleveland, Ohio (USA) and Dr. GEORGE M. C. FISHER, Bell Telephone Laboratories, Holmdel, New Jersey (USA) . . . 1

A. Introduction . . . 1

1. Plan and scope of this article . . . 1
2. Notation. Vectors, tensors, and linear transformations . . . 3
3. Processes and histories . . . 4
4. Convolutions . . . 8
5. The Boltzmann operator . . . 10

B. Foundations of the linear theory . . . 11

6. Linear hereditary laws . . . 11
7. Boltzmann laws. Definitions . . . 12
8. Characterization of Boltzmann laws . . . 14
9. Constitutive relations. Linearly viscoelastic materials . . . 16
10. Constitutive equations. Boltzmann laws. Stress relaxation . . . 18
11. Relaxation and creep laws . . . 23
12. Isotropic materials . . . 29
13. Additional properties of Boltzmann laws. Mechanical forcing . . . 31
14. Differential operator laws . . . 34
15. Relaxation times and differential operator laws . . . 37
16. Special differential operators . . . 39
17. Field equations . . . 42

C. Quasi-static linear viscoelasticity . . . 42

18. The quasi-static assumption . . . 42
19. Quasi-static viscoelastic processes . . . 42
20. Displacement equations of equilibrium . . . 44
21. Equations of compatibility . . . 45
22. Boundary data . . . 46
23. The quasi-static boundary value problem . . . 47
24. Synchronous and separable boundary data . . . 48
25. Initial response. Elastic states . . . 49
26. The past-history problem . . . 50
27. Integral theorems . . . 53
28. Uniqueness of quasi-static viscoelastic processes . . . 55
29. Existence of quasi-static viscoelastic processes . . . 59
30. Quasi-static variational principles . . . 64
31. Elastic-viscoelastic correspondence . . . 68
32. Stress functions for quasi-static viscoelastic processes . . . 73
33. Singular solutions . . . 77
34. Green's processes and integral solutions . . . 81
35. Saint-Venant's principle . . . 85

D. Dynamic linear viscoelasticity . . . 89

36. Dynamic viscoelastic processes . . . 89
37. Field equations . . . 91
38. Complete dynamic displacement generating functions . . . 92
39. Power and energy . . . 95
40. Uniqueness. Boundary value problem . . . 95
41. Waves. Singular surfaces . . . 96
42. Initial value problem. Uniqueness and existence of solutions . . . 104

43. Oscillatory displacement processes. Free vibrations 106
44. Dynamic variational principles 112

References 115

Theory of Elastic Stability. By R. J. Knops, Professor of Mathematics, Heriot-Watt University, Edinburgh, and E. W. Wilkes, School of Mathematics, The University of Newcastle upon Tyne (Great Britain). (With 3 Figures) 125

A. Introduction 125

B. Abstract dynamical systems 126

1. Introduction 126
2. General features of a dynamical system 127
3. Definition of a dynamical system 128
4. The set of initial data and a related mapping 130

C. Definitions of stability 131

5. Introduction 131
6. Definition of Liapounov stability 133
7. Further definitions 141
8. Continuous dependence 145
9. Instability in the sense of Liapounov 147
10. Boundedness and Liapounov stability 148
11. Instability in the sense of Lagrange 150
12. Stability and uniqueness 151

D. Stability theorems for abstract dynamical systems 152

13. Introduction 152
14. Maximum principles 152
15. Liapounov's theorems on stability. (The second method) 153
16. Discussion of the theorems 158
17. Examples 160
18. Relation of stability to the calculus of variations 163
19. Theorems on instability 167
20. Boundedness and asymptotic stability 169

E. Eigenfunction analyses 170

F. Stability of elastic bodies 171

21. Introduction 171
22. Derivation of basic equations 173
23. The equations of perturbed motion. Isothermal linear elasticity 176
24. Incremental equations for thermoelasticity 183
25. Some causes of perturbations 186

G. Liapounov functions for finite thermoelasticity 187

26. Introduction 187
27. Liapounov functions from the energy balance equation 188
28. Liapounov functions from the entropy production inequality 189

H. Liapounov stability in the class of non-linear perturbations 191

29. Introduction 191
30. Sufficiency theorems 192

I. The energy criterion for stability 195

31. Statement and history of the criterion 195
32. Necessary conditions 199
33. Sufficient conditions 203
34. Criticism of the energy criterion 203

J. Stability for a fixed surface and under dead loads in the class of small incremental displacements 206

35. Stability of equilibrium 206
36. Stability of a body with time-dependent elasticities 209
37. Discussion of choice of measure 210
38. Stability with multipolar elasticity 213
39. Incompressible media 213

K. Stability under dead surface loads in the class of small incremental displacements 215
40. Introduction 215
41. Stability analysis I 217
42. Stability analysis II 219
43. Note on Korn's inequality 221

L. Instability under dead surface loads from the equations of linear incremental displacement 222
44. Instability from negative-definite total energy 222
45. Non-uniqueness and instability 224
46. The method of adjacent equilibrium 224
47. History and application of the test 226

M. Logarithmic convexity 228
48. Introduction 228
49. Convexity of the function $F(t; \alpha, t_0)$ 229
50. Applications 230

N. Extension of stability analysis for traction boundary conditions 239
51. Stability without an axis of equilibrium 239
52. Stability with an axis of equilibrium 242
53. Instability analysis 242
54. Incompressible media 244

O. Stability in special traction boundary value problems 244
55. Introduction 244
56. Isotropic compressible material under hydrostatic stress 244
57. Incompressible elastic material 248

P. Stability in the class of linear thermoelastic displacements under dead loads . . . 249
58. Introduction 249
59. Liapounov stability 250
60. Instability 251
61. Hölder stability 252
62. Asymptotic stability 254

Q. Classification of stability problems with non-dead loading 256
63. Introduction 256
64. Group (a): Persistent stability 256
65. Group (b): Motion-dependent data 257

R. Stability under weakly conservative loads 258
66. Definitions 258
67. Characterisations of weakly conservative forces 260
68. Stability analyses 263

S. Stability with time-dependent and position-dependent data 264
69. Introduction 264
70. Prescribed surface traction with zero initial data 264
71. Prescribed surface traction with non-zero initial data 266
72. Prescribed surface displacement 268
73. Prescribed body force 269
74. Variation in the elasticities 271
75. Change in initial data under dead loading 272
76. Convexity arguments 273
77. Further arguments 275

T. Stability under follower forces 278
78. Introduction 278
79. Examples using the Liapounov theory 280
80. Adjacent equilibrium method. Instability by divergence 282
81. Eigenfunction expansions. Analyses depending upon separation of variables . 284

U. Dissipative forces 285
82. Introduction 285
83. Snap-through 287

References 289

Growth and Decay of Waves in Solids. By Dr. PETER J. CHEN, Sandia Laboratories, Albuquerque, New Mexico (USA). (With 1 Figure) . . . 303

I. Introduction . . . 303
1. Nature of this article . . . 303
2. General scheme of notation . . . 304

II. Preliminaries . . . 304
3. Basic kinematical concepts . . . 304
4. Theory of singular surfaces . . . 305
5. Definition of shock waves and acceleration waves . . . 309

III. Acceleration waves in elastic bodies . . . 315
6. Longitudinal waves in anisotropic elastic bodies . . . 315
7. Transverse waves in anisotropic elastic bodies . . . 321
8. Thermodynamic influences on waves in anisotropic elastic bodies . . . 323
9. Waves in isotropic elastic bodies . . . 325
10. Waves of arbitrary shape in isotropic elastic bodies . . . 328
11. Thermodynamic influences on waves in isotropic elastic bodies . . . 333

IV. One dimensional waves in bodies of material with memory . . . 334
12. Acceleration waves in bodies of material with memory . . . 334
13. The local and global behavior of the amplitudes of acceleration waves . . . 339
14. Acceleration waves entering homogeneously deformed bodies of material with memory . . . 344
15. Thermodynamic influences on acceleration waves in bodies of material with memory . . . 348
16. Shock waves entering unstrained bodies of material with memory . . . 349
17. Consequences of the existence of steady shock waves . . . 354

V. One dimensional waves in elastic bodies . . . 356
18. Acceleration waves in elastic bodies . . . 356
19. Shock waves in elastic bodies . . . 357

VI. One dimensional waves in elastic non-conductors of heat . . . 362
20. Acceleration waves in elastic non-conductors of heat . . . 362
21. Acceleration waves entering deformed elastic non-conductors . . . 363
22. Shock waves in elastic non-conductors of heat . . . 369
23. Shock waves entering deformed elastic non-conductors . . . 376

VII. One dimensional waves in inhomogeneous elastic bodies . . . 378
24. Acceleration waves in inhomogeneous elastic bodies . . . 378
25. Acceleration waves in inhomogeneous elastic bodies at rest . . . 379
26. Shock waves in inhomogeneous elastic bodies . . . 382
27. Shock waves in inhomogeneous elastic bodies at rest . . . 384

Appendix . . . 385
1. Existence of the one dimensional kinematical condition of compatibility . . . 385
2. Proofs of Theorems 13.2, 13.3, 13.4 and 13.5 . . . 387
3. Derivation of (16.12) . . . 395

References . . . 398
List of works cited . . . 398
Additional references . . . 401

Ideal Plasticity. By HILDA GEIRINGER, Professor Emer. of Mathematics, Cambridge, Massachusetts (USA). (With 50 Figures) . . . 403

A. The basic equations . . . 403

I. The three-dimensional problem . . . 403
1. Quadratic yield condition . . . 403
2. Some basic formulas. Mohr circles . . . 407
3. Plastic potential . . . 408
4. Tresca's yield criterion. "Singular" yield conditions . . . 411

5. "Compatibility" relations 411
6. The flow equations of Prandtl and Reuss 412
7. Further stress strain laws 414
8. Remarks on some three-dimensional problems 415
8 bis. Remarks on uniqueness for rigid plastic solids 416

II. Discontinuous solutions 416

a) Characteristics. Application to the three-dimensional problem of the perfectly plastic body 416

9. Introduction 416
10. Examples 417
11. Systems of differential equations 420
12. Characteristics of the v. Mises plasticity equations 422
13. Further results and comments 424

b) Continuation 424

14. Characteristic surfaces. Characteristic condition 424
15. Compatibility conditions 426
16. Discontinuous solutions 426
17. Preliminary comments on discontinuous solutions in plasticity 427

c) Hadamard's theory 428

18. Moving surfaces 428
19. Geometrical and kinematical discontinuity conditions 429
19 bis. Continuation 430
20. Application to a system of equations 431
21. Compatibility conditions 432

d) Shock conditions. Stress discontinuities 433

22. "Shock conditions" 433
23. On the classification of discontinuities 434
24. Stress discontinuities 434

B. Plane problems 436

I. Plane strain, plane stress, and generalizations 436

25. Plane strain with v. Mises' or with Tresca's yield condition derived from three-dimensional problem 436
26. Plane strain under general yield condition 437
27. Plane stress with quadratic yield condition 438
27 bis. Plane stress. Continued 439
28. Generalized plane stress 440

II. The theory of plane strain 440

a) Differential relations 440

29. Basic equations 440
30. Continuation. Slip line field 445

b) Integration. Particular solutions 449

31. Integration 449
32. Examples of exact particular solutions 452
33. Discontinuities 456

C. The general plane problem 458

I. Basic theory 458

a) The equations 458

34. Linearization 458
35. Various yield conditions 461

b) Characteristics of the complete plane problem 464

36. Characteristic directions and compatibility relations 464
37. Continuation. Relation to O. Mohr's theory. Differential equations in characteristic coordinates 468
38. Examples for Sects. 36 and 37 471

c) Remarks on integration. Examples 474

39. On integration 474

II. Singular solutions and various remarks 477
a) Limit line singularities and branch line singularities 477
40. Limit line singularities 477
41. Limit line singularities. Continuation 480
42. Branch line singularities 483
b) Simple waves 484
43. Definition 484
44. Simple waves. Continuation 486
45. Simple waves for particular yield conditions 487
c) Various remarks 491
46. Remarks on the approximate solution of initial-value problems 491
47. Summary remarks on some further problems 494
D. Boundary-value problems 495
I. Some elastic-plastic problems 495
a) The torsion problem 495
48. Fully elastic and fully plastic torsion 495
49. Elastic-plastic torsion 499
50. Examples, further problems and concluding remarks 502
b) The thick walled tube 506
51. Expansion of a cylindrical tube 506
52. Partly plastic tube 508
53. Further solutions. Comments 511
c) Flat ring and flat sheet in plane stress. Further elastic-plastic problems . . 515
54. Flat ring radially stressed as a problem of plastic-elastic equilibrium . . 515
55. Continuation: Plastic-elastic equilibrium 516
56. Expansion of a circular hole in an infinite sheet 519
57. A few further elastic-plastic problems 521
II. Some plastic-rigid problems 523
a) Various remarks 523
58. The plastic-rigid body 523
58 bis. Axial symmetry. A few remarks 525
b) Wedge with pressure on one face 526
59. General discussion and velocity distribution 526
c) Plastic mass between rough rigid plates 529
60. Infinite slab 529
61. Slab of material with overhanging ends between rough plates 531
Some reference books 533

Topics in the Mathematical Theory of Plasticity. By Tsuan Wu Ting, Professor of Mathematics, University of Illinois, Urbana, Illinois (USA). (With 2 Figures) . . . 535
A. Introduction 535
B. Foundation of the theory 535
I. Thermodynamics for elastic-plastic materials 536
1. Motion and deformation history 536
2. Constitutive functionals 537
3. Elastic-plastic resolution of deformations 538
4. Consequences of the Clausius-Duhem inequality 539
5. Rate independent materials 542
6. Isotropy and principle of frame indifference 544
II. Prandtl-Reuss theory 545
7. Flow rule 546
8. Prandtl-Reuss equations 547
III. St. Venant-Lévy-v. Mises theory 548
IV. Theory of Hencky 549

C. General theorems 551
I. Intrinsic formulation of the variational principle 551
II. Examples 554
9. Elastic-plastic torsion in the sense of Hencky 555
10. Stationary creep deformation of a plate 555
11. Plane stress and plane deformation problems 555
III. Existence of certain steady plastic flows 556
12. Formulation of a weak problem 556
13. Existence of solution of certain weak problems 557

D. Torsion problems 560
I. Completely plastic torsion 560
14. Variational formulation of the problem 561
15. Solution of the problem 561
16. St. Venant's conjecture and its extension 562
II. Elastic-plastic torsion 562
17. Formal statement of the problem 563
18. Variational formulation of the problem 563
19. Existence and uniqueness of the extremal 564
20. Hölder continuity of the extremal 565
21. The existence of an elastic core 568
22. Continuity of stress 574
23. Natural partition of the cross section 578
24. Some properties of E and P 579
25. Elastic-plastic boundary 580
26. Dependence upon angle of twist 581
27. Dependence upon the yield strength 584
28. Dependence upon the angle of twist, continued 588

References 589

Namenverzeichnis — Author Index 591

Sachverzeichnis (Deutsch-Englisch) 597

Subject Index (English-German) 623

The Linear Theory of Viscoelasticity.

By

Marshall J. Leitman and George M. C. Fisher.

A. Introduction.

1. Plan and scope of this article. It has long been known that material behavior is not always elastic. Indeed, many substances exhibit the property of hereditary response. That is, the present state of stress depends not only upon the present state of deformation, but also upon previous states. This property is revealed experimentally in the phenomena of creep, stress relaxation, and the intrinsic attenuation of propagating waves. The nonlinear mechanical theories of materials with memory have been developed to characterize such behavior.[1] More specifically, the linearly viscoelastic material is a model which seeks to characterize hereditary effects within the context of an infinitesimal linearized theory. A theory based on this model should, in the absence of hereditary effects, reduce to the classical linearized theory of elasticity.

The linear theories of elasticity and viscoelasticity have much in common. Indeed, there is considerable formal similarity in their developments. This is not illusory since many results from the elastic theory have direct generalizations to the viscoelastic theory.

In this article we focus not only upon the formal similarities but also upon the intrinsic differences of the two theories. We attempt to isolate those conceptual and mathematical difficulties which arise over and above those inherent in elastic problems. The greatest of these arises directly from the inclusion of hereditary effects. In elastic theories the relationship between stress and strain is finite-dimensional, whereas, in a hereditary theory this relationship is generally infinite-dimensional. Even within the context of a linear theory, the additional complications may be formidable. It will be seen that the ramifications of this assertion permeate the entire body of the theory.

We develop our subject entirely within the context of an infinitesimal linearized theory. In this respect our presentation is phenomenological and follows

---

Acknowledgements. We are greatly indebted to our teacher Professor M. E. Gurtin for his friendship and guidance since our student days. To Professor C. Truesdell we express our gratitude for the opportunity to write this article and for his patience and help in its preparation. We should like to thank Professor A. J. Lohwater of Case Western Reserve University and J. S. Courtney-Pratt, F. E. Froehlich, F. T. Geyling, and R. N. Thurston, all of Bell Telephone Laboratories, for their encouragement, support, and valuable discussions. We also wish to acknowledge the cheerful and essential assistance given to us by Mrs. N. G. Roselli in the typing of the many versions of the manuscript. Last, but certainly not least, we thank our wives Carolyn Leitman and Ann Fisher for their help, encouragement, and great patience.

[1] Green and Rivlin [1957, *2*], Noll [1958, *16*], Coleman and Noll [1961, *2*], [1964, *5*], Coleman [1964, *4*], Coleman and Mizel [1966, *5*].

the ideas set down first by BOLTZMANN[2] and later by VOLTERRA.[3] Mathematically, our formulation is patterned after the more recent work of KÖNIG and MEIXNER[4] and GURTIN and STERNBERG.[5] We have been greatly influenced by the investigations of the non-linear theories of mechanics and the concept of fading memory. Most significant of these are the fundamental studies of GREEN and RIVLIN, NOLL, COLEMAN and NOLL, COLEMAN, and COLEMAN and MIZEL.[6]

We have attempted throughout to adopt a style which may be rendered precise, in the mathematical sense, without overburdening the reader with cumbersome notation. This article is intended as a sequel to the Linear Theory of Elasticity (LTE) by GURTIN, which appears in the preceding part of this volume. The completeness of that article enables us to use it as a primary reference.

Our survey is divided into four chapters. In Chap. A we provide some requisite technical background and notation. Since we adopt most of the notation and results developed in LTE, we are able to concentrate on those mathematical concepts which are central to the viscoelastic theory but absent in the elastic theory. A self-contained formulation of the foundations of the linear theory is presented in Chap. B. We do not approach this theory as a linear approximation to a non-linear theory but proceed directly within the phenomenological framework of a linear theory. In Chap. C we present the fundamental mathematical results of the quasi-static theory. By the term "quasi-static" we mean that all inertial effects are systematically neglected. We include the inertial effects in Chap. D, which concerns the full dynamic theory and the propagation of waves in viscoelastic media. Throughout, we concentrate on presentation of results without usually giving detailed proofs.

We are cognizant of certain omissions which might properly belong in an article of this type. However, our selection of topics has been based upon limitations of space coupled with a desire to give precedence to fundamental mathematical aspects of the theory.

We have made only casual reference to the vast experimental literature. For more information in this regard, the reader is referred to the reviews of FERRY and KOLSKY.[7]

We have paid little attention to the differential models for viscoelastic behavior—finite networks of springs and dashpots. These models are somewhat special and a separate treatment using them is unnecessary. We do, however, include a discussion of their place in the general linear theory. We take as fundamental the classical Boltzmann model and the notion of linear superposition.

Our treatment of material symmetry is brief. It is readily seen that the concept is the same for both elastic and viscoelastic materials. There is an extensive treatment of this topic in LTE which obviates a parallel treatment in this article. The only type of symmetry singled out for special consideration is that of isotropy. Furthermore, we do not separately consider the problems associated with constrained materials such as incompressible media. Our discussion is restricted to general compressible viscoelastic solids. The extension to constrained solids and fluids may be effected along lines of the general theories of mechanics.[8]

---

[2] BOLTZMANN [1874, *1*], [1878, *1*].
[3] VOLTERRA [1909, *1*], [1913, *1*].
[4] KÖNIG and MEIXNER [1958, *11*].
[5] GURTIN and STERNBERG [1962, *10*].
[6] Cited in footnote 1, p. 1.
[7] FERRY [1970, *4*], KOLSKY [1963, *10*], [1969, *3*].
[8] Cf. TRUESDELL and NOLL [1965, *29*].

Only the linearization based upon infinitesimal deformations is considered. It is possible, however, to develop a linear theory based upon finite deformations. This has been done, for example, by COLEMAN and NOLL.[9]

For brevity we have omitted completely the important discussion of thermodynamics for viscoelastic materials. A sound basis for this theory is presented in the work of COLEMAN.[10] Other work of interest includes that of BIOT and SCHAPERY[11] as well as the research concerning free energy, recoverable work, and related work bounds as exemplified by the articles of DAY, BREUER, MARTIN and PONTER, and BREUER and ONAT.[12] Furthermore, BIOT's[13] well known development of the concept of hidden variables has led VALANIS[14] to an interesting discussion of the viscoelastic potential function. COLEMAN and GURTIN[15] have generalized the internal state variables approach in the context of modern thermodynamics. THURSTON[16] has identified a set of hidden variables as ensemble-averaged occupation numbers which obey a relaxation type of differential equation and has subsequently derived, from microscopic concepts, the macroscopic theory of linear viscoelasticity as the appropriate linearization.

We do not mention the notion of thermorheologically simple materials in the sense of SCHWARZL and STAVERMAN.[17] Although this concept seems quite useful, especially for experimental investigations, the subject is still in a state of growth and the reader is referred to the works of LEADERMAN, FERRY, MUKI and STERNBERG, and MORLAND and LEE[18] for additional information.

Stability analysis for functional differential equations as applied to materials with memory is still a subject in its infancy. Even at this stage, however, we can refer to the works of COLEMAN and MIZEL and DAFERMOS,[19] who shed some light on the subject.

Throughout, we have not examined solutions to particular problems in linear viscoelasticity. This has meant the exclusion of, among others, the general boundary value problem, which includes the contact problem, the problem of ablating boundaries, numerical techniques, and the effect of boundaries in wave propagation.

Finally, we observe that no precise relationship between the quasi-static and dynamic theories has yet been revealed.

2. Notation. Vectors, tensors, and linear transformations.

The primary source for notation and basic mathematical notions is the article in the preceding part of this volume entitled The Linear Theory of Elasticity (LTE) by M. E. GURTIN. We use the notation in Subchaps. B.I and B.II of LTE virtually without change. However, we depart from this procedure for Subchap. B.III of LTE (Functions of position and time) since our requirements in this regard are somewhat different. *All definitions and notation not explicitly established in this article are established by* GURTIN *in LTE.*

---

[9] COLEMAN and NOLL [1961, *2*], [1964, *5*].

[10] COLEMAN [1964, *4*].

[11] BIOT [1958, *3*], [1965, *2*], SCHAPERY [1962, *20*], [1964, *23*], [1965, *26*].

[12] DAY [1970, *2*], BREUER [1969, *1*], MARTIN and PONTER [1966, *17*], BREUER and ONAT [1963, *1*], [1964, *1*].

[13] BIOT [1965, *2*].

[14] VALANIS [1968, *6*].

[15] COLEMAN and GURTIN [1967, *2*].

[16] THURSTON [1968, *5*].

[17] SCHWARZL and STAVERMAN [1952, *2*].

[18] LEADERMAN [1958, *12*], MUKI and STERNBERG [1961, *5*], MORLAND and LEE [1960, *13*].

[19] COLEMAN and MIZEL [1968, *2*], DAFERMOS [1970, *1*].

We write $\mathscr{E}$ for the three-dimensional Euclidean point space and $\mathscr{V}$ for the underlying translation space, so that $\mathscr{V}$ is a three-dimensional inner product space. Elements of $\mathscr{E}$ are called *points* and elements of $\mathscr{V}$ are called *vectors*.

The set of all linear transformations of $\mathscr{V}$ into $\mathscr{V}$ is a nine-dimensional inner product space. We denote this space by $\mathscr{T}$ and call its elements *tensors*. We denote by $\mathscr{T}_{\text{sym}}$ the six-dimensional subspace of symmetric tensors in $\mathscr{T}$, and by $\mathscr{T}_{\text{skw}}$ the three-dimensional subspace of skew-symmetric tensors in $\mathscr{T}$.

Let $\mathscr{X}$ and $\mathscr{Y}$ be a pair of finite-dimensional (inner product) spaces. Then the set of all linear transformations from $\mathscr{X}$ into $\mathscr{Y}$ is also a finite-dimensional (inner product) space. We denote this space by $[\mathscr{X}, \mathscr{Y}]$ and write $[\mathscr{X}]$ for $[\mathscr{X}, \mathscr{X}]$. The magnitude of an element ϕ in an inner product space $\mathscr{X}$ is the non-negative real number $|\phi|_{\mathscr{X}}$:

$$|\phi|_{\mathscr{X}} = \sqrt{\phi \cdot \phi}, \tag{2.1}$$

where the dot $(\cdot)$ denotes the appropriate inner product. If the space $\mathscr{X}$ is understood from the context we write $|\phi|$ for $|\phi|_{\mathscr{X}}$.

The most important examples in our physical context are the spaces generated by $\mathscr{V}$. Thus $[\mathscr{V}] \equiv \mathscr{T}$, the space of (second-order) tensors and $[\mathscr{T}]$ denotes the space of linear maps in $\mathscr{T}$ (the space of fourth-order tensors). If M is an element of $[\mathscr{T}]$, then its *magnitude* $|\mathsf{M}|_{[\mathscr{T}]}$ is given by

$$|\mathsf{M}|_{[\mathscr{T}]} = \sup\{|\mathsf{M}[\boldsymbol{A}]|_{\mathscr{T}} \colon \boldsymbol{A} \in \mathscr{T};\ |\boldsymbol{A}|_{\mathscr{T}} \leq 1\}. \tag{2.2}$$

For consistency and unity in presentation, we establish some general notational conventions. It should be understood that there are exceptions to these rules, but such deviations will be cited specifically or will be clear from the context.

Greek and Latin italic light face minuscules and majuscules are used to denote scalar quantities or elements in an abstract vector space. Points in $\mathscr{E}$ or vectors in $\mathscr{V}$ are indicated by Greek or Latin italic bold face minuscules while tensors in $\mathscr{T}$ are indicated by italic bold face Latin majuscules. Linear transformations defined in the space of tensors are indicated by italic sans-serif bold face Latin majuscules, while other linear transformations are denoted by italic sans-serif light face Latin majuscules. The latter may denote scalars when they are used in the sense of linear maps in the real line. Script Latin majuscules are reserved for point or vector spaces, subsets of such spaces, or other special sets or classes of objects.

3. Processes and histories.[20] We always denote the real line by $\mathscr{R}$, the non-negative real line by $\mathscr{R}^+$, and the positive real line by $\mathscr{R}^{++}$. The ***time*** t is always a real number.

Let $\mathscr{X}$ denote a point or vector space. By an $(\mathscr{X})$-***process*** or, simply, a ***process*** we mean a function[21] defined on $\mathscr{R}$ with values in $\mathscr{X}$. If ϕ is an $(\mathscr{X})$-process and $t \in \mathscr{R}$, we call $\phi(t) \in \mathscr{X}$ the ***value of*** ϕ ***at time*** t. By an $(\mathscr{X})$-***history*** or, simply, a ***history*** we mean a function defined on $\mathscr{R}^+$ with values in $\mathscr{X}$; and by an $(\mathscr{X})$-***past history*** or, simply, a ***past history***, we mean a function defined on $\mathscr{R}^{++}$ with values in $\mathscr{X}$. Where no confusion may occur, we suppress the dependence upon the range $\mathscr{X}$.

[20] That the definitions and conventions introduced in this section are appropriate to the study of hereditary phenomena has been amply demonstrated by the investigations of COLEMAN and MIZEL [1966, *5*], [1968, *2*]. Indeed, their work has significantly influenced the format and content of this, and other, sections. The notion of history as used here is due to NOLL [1958, *16*].

[21] We use the term *function* for *equivalence classes of functions modulo sets of Lebesgue measure zero*, whenever it is appropriate to do so, without explicit comment.

Given an $(\mathscr{X})$-process ϕ and a time $t\in\mathscr{R}$, define a history ϕ^t by

$$\phi^t(s) = \phi(t-s). \tag{3.1}$$

The function ϕ^t is called the ***history of ϕ up to time t***. The restriction of ϕ^t to $\mathscr{R}^{++}$ is called the ***past history of ϕ up to time*** t and denoted by ${}_r\phi^t$.

Let α be an element of $\mathscr{X}$. Then the $(\mathscr{X})$-process α^* defined by

$$\alpha^*(t) = \alpha, \qquad t\in\mathscr{R}, \tag{3.2}$$

is the ***constant process with value*** α.

If ψ is an $(\mathscr{X})$-history, then $\psi(0)\in\mathscr{X}$ is called the ***present value of*** ψ. For a process ϕ, observe that the present value of ϕ^t is the value of ϕ at time t:

$$\phi^t(0) = \phi(t). \tag{3.3}$$

We now proceed to develop some special terminology to distinguish classes of processes. Let $\mathscr{X}$ be a normed vector space. Then an $(\mathscr{X})$-process ϕ is said to have:

finite histories if, and only if, some (and hence every) history of ϕ has bounded support; that is, the set $\{s\in\mathscr{R}^+ \colon \phi^t(s) \neq 0\}$ is bounded for some (and hence every) time t,

infinite histories if, and only if, it does not have finite histories,

bounded histories if, and only if, every history of ϕ is bounded; that is, for each time t there is an $M(t) \geqq 0$ such that $\sup\{|\phi^t(s)| \colon s\in\mathscr{R}^+\} \leqq M(t)$, and

integrable[22] ***histories*** if, and only if, every history of ϕ is integrable on $\mathscr{R}^+$; that is, for each time t

$$\int_0^\infty |\phi^t(s)|\, ds < \infty. \tag{3.4}$$

We further say that an $(\mathscr{X})$-process ϕ is:

restricted if, and only if, some (and hence every) history of ϕ vanishes at infinity; that is, for some (and hence every) time t

$$\lim_{s\to\infty} \phi^t(s) = 0, \tag{3.5}$$

continuous at infinity if, and only if, there is an element $\overset{\infty}{\phi}$ in $\mathscr{X}$ such that

$$\lim_{t\to\infty} \phi(t) = \overset{\infty}{\phi}, \tag{3.6}$$

of ***Heaviside type*** if, and only if, the past history of ϕ up to time $t=0$ vanishes; that is,

$$\phi(t) = 0 \qquad \text{for } t<0,$$

and

locally integrable if, and only if, ϕ is integrable on every bounded subset of $\mathscr{R}$.

Let N be a non-negative integer. A process or history ϕ is said to be in ***class*** C^N if, and only if, ϕ is N times continuously differentiable on $\mathscr{R}$ or $\mathscr{R}^+$. We customarily write C for C^0. If a process or history ϕ is in class C^N for every N, we say that ϕ is in ***class*** C^∞.

---

[22] By *integrable* we mean integrable in the sense of LEBESGUE. Thus, a history ψ is integrable on $\mathscr{R}^+$ if, and only if, $|\psi|$ is integrable on $\mathscr{R}^+$.

For processes of Heaviside type we define the ***Heaviside classes.*** Let ϕ be a process of Heaviside type and let N be a non-negative integer. Then ϕ is said to be in ***class*** H^N if, and only if, ϕ is N times continuously differentiable on $\mathscr{R}^+$. We customarily write H for H^0. We note that if $\phi \in H$ and $\phi(0^+) \neq 0$, then, since $\phi(t) = 0$ for $t < 0$, ϕ has a jump discontinuity at $t = 0$ but nowhere else. If ϕ is in class H^N for *every* N, we say ϕ is in ***class*** H^∞. The process ϕ is said to be in ***class*** H^{ac} if, and only if, ϕ is absolutely continuous on $\mathscr{R}^+$. We have the following characterization of class H^{ac}.

An $(\mathscr{X})$*-process* ϕ *of Heaviside type is in class* H^{ac} *if, and only if, there is an element* $\overset{\circ}{\phi} \in \mathscr{X}$ *and a locally integrable function* $\dot{\phi}$: $\mathscr{R}^+ \to \mathscr{X}$ *such that*

$$\phi(t) = \overset{\circ}{\phi} + \int_0^t \dot{\phi}(\tau)\, d\tau, \qquad t \in \mathscr{R}^+. \tag{3.7}$$

The element $\overset{\circ}{\phi}$ *and the function* $\dot{\phi}$ *are uniquely*[23] *determined through*

$$\overset{\circ}{\phi} = \lim_{t \downarrow 0} \phi(t) \tag{3.8}$$

and

$$\dot{\phi}(t) = \frac{d}{dt} \phi(t) \tag{3.9}$$

for almost every t *in* $\mathscr{R}^+$[24].

A process ϕ in class H^{ac} will be called ***strong*** only if $\dot{\phi}$ is integrable on $\mathscr{R}^+$.[25] It then follows that:

A process ϕ *in class* H^{ac} *is continuous at infinity if (but not only if)*[26] *it is strong. If* ϕ *is continuous at infinity, then*

$$\overset{\infty}{\phi} = \overset{\circ}{\phi} + \int_0^\infty \dot{\phi}(s)\, ds. \tag{3.10}$$

We observe here that composite terms such as *continuous restricted process, locally integrable process of class H, class* C^∞ *process with finite histories*, etc. are easily interpreted.

By the ***Heaviside unit step process*** h we always mean the *scalar* process defined by

$$h(t) = \begin{Bmatrix} 0: & t < 0 \\ 1: & t \geq 0 \end{Bmatrix}. \tag{3.11}$$

Each of the classes of processes and histories thus far defined may be given the structure of a linear vector space under the pointwise operations of addition and multiplication by a scalar. We single out two for special attention.

---

[23] To be precise, $\dot{\phi}$ is uniquely determined on $\mathscr{R}^+$ up to a set of Lebesgue measure zero.

[24] The terms *almost every* and *almost everywhere* mean *except for a set of Lebesgue measure zero*. Cf. footnote 23, p. 6.

[25] Cf. footnote 22, p. 5. In particular, $\dot{\phi}$ is integrable if, and only if, $|\dot{\phi}|_{\mathscr{X}}$ is integrable.

[26] A process ϕ in H^{ac} may be continuous at infinity without being strong. Consider the scalar process ϕ in H^{ac} defined by

$$\phi(t) = \alpha + \pi - \int_0^t \frac{1}{s} \sin(s)\, ds, \qquad t \in \mathscr{R}^+.$$

Then ϕ is continuous at infinity ($\overset{\infty}{\phi} = \alpha$) but $|\dot{\phi}|$ is not integrable on $\mathscr{R}^+$.

First, consider the space $\mathscr{C}_0(\mathscr{R}^+; \mathscr{X})$ of *continuous* $(\mathscr{X})$*-histories which vanish at infinity*; that is $\mathscr{C}_0(\mathscr{R}^+; \mathscr{X})$ is the space of continuous functions $\phi\colon \mathscr{R}^+ \to \mathscr{X}$ such that $\lim_{s\to\infty} \phi(s) = 0$. $\mathscr{C}_0(\mathscr{R}^+; \mathscr{X})$ is a complete normed linear space (Banach space) under the norm $\|\phi\|$ given by

$$\|\phi\| = \sup\{|\phi(s)|_{\mathscr{X}} \colon s \in \mathscr{R}^+\}. \tag{3.12}$$

We call $\|\phi\|$ the ***amplitude of*** ϕ. Thus, if ϕ is a continuous restricted $(\mathscr{X})$-process, $\phi^t \in \mathscr{C}_0(\mathscr{R}^+; \mathscr{X})$ for each time t.

Second, consider the space $\mathscr{L}^1(\mathscr{R}^+; \mathscr{X})$ of *integrable* $(\mathscr{X})$*-histories*; that is, $\mathscr{L}^1(\mathscr{R}^+; \mathscr{X})$ is the space of $\phi\colon \mathscr{R}^+ \to \mathscr{X}$ such that the norm $\|\phi\|_1$, given by

$$\|\phi\|_1 = \int_0^\infty |\phi(s)|_{\mathscr{X}}\, ds \tag{3.13}$$

is finite. $\mathscr{L}^1(\mathscr{R}^+; \mathscr{X})$ is a complete normed linear space. Thus if ϕ is an $(\mathscr{X})$-process with integrable histories, then $\phi^t \in \mathscr{L}^1(\mathscr{R}^+; \mathscr{X})$ for each time t.

Let ϕ be a function defined in some neighborhood of a point $t \in \mathscr{R}$ with values in $\mathscr{X}$.[27] Let N be a positive integer. If the Nth derivative of ϕ at $t \in \mathscr{R}$ exists, we write

$$\phi^{(N)}(t) \quad \text{for} \quad \frac{d^N}{dt^N}\phi(t).$$

If $N=1$ or $N=2$, we usually deviate from this scheme and write $\dot\phi(t)$ for $\phi^{(1)}(t)$ and $\ddot\phi(t)$ for $\phi^{(2)}(t)$. Also, for consistency in notation, we sometimes write $\phi^{(0)}(t)$ for $\phi(t)$. If ϕ is an $(\mathscr{X})$-process of class C^N or class H^N $(N \geqq 0)$, then $\mathring{\phi}^{(n)}$ $(n=0, 1, 2, 3, \ldots, N)$ denotes the limit

$$\mathring{\phi}^{(n)} = \lim_{t\downarrow 0} \phi^{(n)}(t). \tag{3.14}$$

For any $(\mathscr{X})$-process ϕ we write $\mathring{\phi} \in \mathscr{X}$ for

$$\mathring{\phi} = \lim_{t\downarrow 0} \phi(t), \tag{3.15}$$

whenever the limit exists. In particular, if ϕ is in class H or class C, then $\mathring{\phi} \equiv \mathring{\phi}^{(0)}$.

We consistently use the notation established by GURTIN[28] regarding functions of position and time. However, for such functions we adopt some special terminology more suited to our purpose.

Let $\mathscr{D}$ be a domain (open set) in $\mathscr{E}$ and let $\mathscr{X}$ be a point or vector space. By an $(\mathscr{X})$***-field*** we mean a function $\phi\colon \mathscr{D} \to \mathscr{X}$. We use the terms *scalar field* if $\mathscr{X} = \mathscr{R}$, *point field* if $\mathscr{X} = \mathscr{E}$, *vector field* if $\mathscr{X} = \mathscr{V}$, *tensor field* if $\mathscr{X} = \mathscr{T}$, etc. The terminology and results established by GURTIN[29] regarding such fields are appropriate to our context.

A function on $\mathscr{D} \times \mathscr{R}$ with values in $\mathscr{X}$ will be called an $(\mathscr{X})$***-field process for*** $\mathscr{D}$. Thus if $\phi\colon \mathscr{D} \times \mathscr{R} \to \mathscr{X}$, we may consider ϕ as a *process* whose value at time t is the $(\mathscr{X})$-field $\phi(t)$ given by

$$\phi(t) = \phi(\cdot, t), \tag{3.16}$$

[27] Here, $\mathscr{X}$ may be any normed vector space or the point space $\mathscr{E}$.
[28] GURTIN [LTE] Sect. 9.
[29] LTE, Subchaps. B.I. and B.II.

the subsidiary map of $\mathscr{D}$ into $\mathscr{X}$ obtained by holding the time t fixed. Alternatively, if ϕ: $\mathscr{D}\times\mathscr{R}\to\mathscr{X}$, then $\phi_{\boldsymbol{x}}$ denotes the ($\mathscr{X}$)-process (at $\boldsymbol{x}\in\mathscr{D}$) given by

$$\phi_{\boldsymbol{x}}=\phi(\boldsymbol{x},\cdot), \tag{3.17}$$

the subsidiary map of $\mathscr{R}$ into $\mathscr{X}$ obtained by holding the position $\boldsymbol{x}$ fixed. We say that an ($\mathscr{X}$)-field process ϕ for $\mathscr{D}$ is *restricted*, is of *Heaviside type*, has *integrable histories*, is in *class* C^N or *class* H^N, etc. according as the ($\mathscr{X}$)-process $\phi_{\boldsymbol{x}}$ is of this type for each $\boldsymbol{x}\in\mathscr{D}$.

For functions ϕ: $\mathscr{D}\times\mathscr{R}^+\to\mathscr{X}$ or ϕ: $\mathscr{D}\times\mathscr{R}^{++}\to\mathscr{X}$ we adopt analogous terminology, replacing the term *process* by *history* or *past history* according as $\mathscr{R}$ is replaced by $\mathscr{R}^+$ or $\mathscr{R}^{++}$. For example, if ϕ is a field process (for $\mathscr{D}$), then ϕ^t denotes the field history (for $\mathscr{D}$) of ϕ up to time t.

We use the terms *process, history,* and *past history* without certain qualifying adjectives if to retain them is cumbersome and to omit them may cause no confusion.

We conclude this section with an example illustrating the use of the notation established by GURTIN in our context. Let $\boldsymbol{u}$ be a *displacement field process for a body* $\mathscr{B}$;[30] $\boldsymbol{u}$: $\mathscr{B}\times\mathscr{R}\to\mathscr{V}$. Then $\nabla\boldsymbol{u}$ is the *displacement gradient field process for* $\mathscr{B}$ whose value $\nabla\boldsymbol{u}(t)$ at time t is the displacement gradient field

$$\nabla\boldsymbol{u}(t)=\nabla\boldsymbol{u}(\cdot,t); \tag{3.18}$$

$\dot{\boldsymbol{u}}$ is the *velocity field process for* $\mathscr{B}$ whose value at each time t is the velocity field

$$\dot{\boldsymbol{u}}(t)=\dot{\boldsymbol{u}}(\cdot,t); \tag{3.19}$$

and, if $\boldsymbol{u}$ is of class $C^{M,N}$ on $\mathscr{B}\times\mathscr{R}$, then $\nabla^{(m)}\boldsymbol{u}^{(n)}$ $\big(m=0,1,2,\ldots,M;\ n=0,1,2,\ldots,N;\ (m+n)\leqq\max\{M,N\}\big)$ denotes the field process for $\mathscr{B}$ whose value at time t is the field

$$\nabla^{(m)}\boldsymbol{u}^{(n)}(t)=\nabla^{(m)}\boldsymbol{u}^{(n)}(\cdot,t). \tag{3.20}$$

Similarly, for a fixed point $\boldsymbol{x}\in\mathscr{B}$, $(\nabla\boldsymbol{u})_{\boldsymbol{x}}$ denotes the *displacement gradient process at* $\boldsymbol{x}$

$$(\nabla\boldsymbol{u})_{\boldsymbol{x}}=\nabla\boldsymbol{u}(\boldsymbol{x},\cdot), \tag{3.21}$$

$\dot{\boldsymbol{u}}_{\boldsymbol{x}}$ denotes the *velocity process at* $\boldsymbol{x}$

$$\dot{\boldsymbol{u}}_{\boldsymbol{x}}=\dot{\boldsymbol{u}}(\boldsymbol{x},\cdot), \tag{3.22}$$

and $(\nabla^{(m)}\boldsymbol{u}^{(n)})_{\boldsymbol{x}}$ denotes the process at $\boldsymbol{x}$

$$(\nabla^{(m)}\boldsymbol{u}^{(n)})_{\boldsymbol{x}}=\nabla^{(m)}\boldsymbol{u}^{(n)}(\boldsymbol{x},\cdot). \tag{3.23}$$

For processes which appear frequently in the sequel, such as the *displacement field process* $\boldsymbol{u}$ *(for* $\mathscr{B}$*)*, we will be considerably less formal. Thus we may refer to $\boldsymbol{u}$ as the *displacement process (for* $\mathscr{B}$*)* or, even more simply, the *displacement (for* $\mathscr{B}$*)*.

4. Convolutions. Let ϕ and ψ be scalar processes. *Formally* define a process $\phi*\psi$, called the ***convolution of*** ϕ ***with*** ψ, by

$$(\phi*\psi)(t)=\begin{Bmatrix} 0: t<0 \\ \int_0^t \phi(s)\,\psi(t-s)\,ds:\ t\geqq 0\end{Bmatrix}. \tag{4.1}$$

[30] For the definition of *body*, see LTE, Sect. 5.

To render the definition precise we observe[31] for scalar processes ϕ and ψ:

(i) *If* ϕ *and* ψ *are in class* H *(or in class* C *on* $\mathscr{R}^+$*), then* (4.1) *defines a scalar process* $\phi*\psi$ *in class* H.

(ii) *If* ϕ *is in class* $H^{N'}$ *(or in class* $C^{N'}$ *on* $\mathscr{R}^+$*) and* ψ *is in class* $H^{N''}$ *(or in class* $C^{N''}$ *on* $\mathscr{R}^+$*), where* N' *and* N'' *are non-negative integers, then* (4.1) *defines a scalar process* $\phi*\psi$ *in class* H^N, *where* $N=\min\{N', N''\}$.

(iii) *If* ϕ *and* ψ *are locally integrable and of Heaviside type (or processes which are locally integrable on* $\mathscr{R}^+$*), then* (4.1) *defines a scalar process* $\phi*\psi$ *which is locally integrable and of Heaviside type. In addition, if either* ϕ *or* ψ *is of class* H *(or class* C *on* $\mathscr{R}^+$*) then* $\phi*\psi$ *is of class* H.

Note that the definition of *convolution* applies to processes which are not of Heaviside type although the convolution itself is always of Heaviside type. In the sequel we will usually restrict the operation of convolution to the Heaviside classes. Exceptions to this rule will be specifically noted.

If ϕ and ψ are field processes for a domain $\mathscr{D}$, then the previous definition and results continue to apply. For example, if ϕ and ψ are scalar field processes for $\mathscr{D}$ of class H[32] then $\phi*\psi$ is the scalar field process for $\mathscr{D}$ of class H defined by (4.1) with the position in $\mathscr{D}$ held fixed:

$$(\phi*\psi)(\boldsymbol{x},t)=(\phi_{\boldsymbol{x}}*\psi_{\boldsymbol{x}})(t), \qquad (\boldsymbol{x},t)\in\mathscr{D}\times\mathscr{R}. \tag{4.2}$$

Thus far we have considered the convolution of scalar or scalar field processes. The definition of *convolution* and the results which follow may easily be extended to include processes with values in $\mathscr{R}$, $\mathscr{V}$, $\mathscr{T}$, and $[\mathscr{T}]$. For convenience we list those generalizations of (4.1) which appear in this article.

Let ϕ be a scalar process, $\boldsymbol{u}$ and $\boldsymbol{v}$ be vector processes, $\boldsymbol{A}$ and $\boldsymbol{B}$ be tensor processes, and M be a ($[\mathscr{T}]$)-process[33]. Then define the convolutions $\phi*\boldsymbol{u}$, $\phi*\boldsymbol{A}$, $\phi*\mathsf{M}$, $\boldsymbol{u}*\boldsymbol{v}$, $\boldsymbol{A}*\boldsymbol{B}$, $\boldsymbol{A}*\boldsymbol{u}$, and $\mathsf{M}*\boldsymbol{A}$, for $t\geqq 0$, by

$$(\phi*\boldsymbol{u})(t)=\int_0^t \phi(s)\,\boldsymbol{u}(t-s)\,ds, \tag{4.3}$$

$$(\phi*\boldsymbol{A})(t)=\int_0^t \phi(s)\,\boldsymbol{A}(t-s)\,ds, \tag{4.4}$$

$$(\phi*\mathsf{M})(t)=\int_0^t \phi(s)\,\mathsf{M}(t-s)\,ds, \tag{4.5}$$

$$(\boldsymbol{u}*\boldsymbol{v})(t)=\int_0^t \boldsymbol{u}(s)\cdot\boldsymbol{v}(t-s)\,ds, \tag{4.6}$$

$$(\boldsymbol{A}*\boldsymbol{B})(t)=\int_0^t \boldsymbol{A}(s)\cdot\boldsymbol{B}(t-s)\,ds, \tag{4.7}$$

$$(\boldsymbol{A}*\boldsymbol{u})(t)=\int_0^t \boldsymbol{A}(s)\,\boldsymbol{u}(t-s)\,ds, \tag{4.8}$$

$$(\mathsf{M}*\boldsymbol{A})(t)=\int_0^t \mathsf{M}(s)\,[\boldsymbol{A}(t-s)]\,ds. \tag{4.9}$$

[31] The verification of assertions (i) and (ii) is elementary. For a proof of (iii) see MIKUSINSKI [1959, 7].

[32] That is, for each $\boldsymbol{x}\in\mathscr{D}$, $\phi_{\boldsymbol{x}}$ and $\psi_{\boldsymbol{x}}$ are scalar processes in class H. See Sect. 3.

[33] That is, for all t, $\mathsf{M}(t)$ is a linear transformation from the space of all tensors into itself.

Note that $\phi * \boldsymbol{u}$ and $\boldsymbol{A} * \boldsymbol{u}$ are vector processes, $\phi * \boldsymbol{A}$ and $\mathbf{M} * \boldsymbol{A}$ are tensor processes, $\phi * \mathbf{M}$ is a $([\mathscr{T}])$-process, but $\boldsymbol{u} * \boldsymbol{v}$ and $\boldsymbol{A} * \boldsymbol{B}$ are scalar processes. Clearly the above definitions may be extended to include *field* processes following the pattern established for scalar processes.

Finally we observe that the properties of the convolution given by GURTIN[34] carry over to our context with only minor modifications in terminology and notation.

5. The Boltzmann operator. Let ϕ be a scalar process in class H^{ac}. Recall, from Sect. 3, that ϕ may be represented by

$$\phi(t) = \mathring{\phi} + \int_0^t \dot{\phi}(\tau)\, d\tau, \qquad t \in \mathscr{R}^+. \tag{5.1}$$

Let ψ be a (scalar) process and *formally* define a new scalar process $\phi \circledast \psi$ by

$$(\phi \circledast \psi)(t) = \mathring{\phi}\, \psi(t) + \int_0^\infty \dot{\phi}(s)\, \psi(t-s)\, ds, \qquad t \in \mathscr{R}. \tag{5.2}$$

By a ***Boltzmann operator*** $\phi \circledast$ ***with kernel*** ϕ we mean the map which assigns the process $\phi \circledast \psi$ to the process ψ.[35]

For scalar processes this definition is rendered precise through:

Properties of the Boltzmann operator. *Let ϕ be a process in class H^{ac}. Then the Boltzmann operator $\phi \circledast$ with kernel ϕ is a linear map in:*

(i) *the space of continuous processes with finite histories,*

(ii) *the space of locally integrable processes of Heaviside type, and*

(iii) *the processes of class H.*

If, in addition, the process ϕ is strong,[36] *then the Boltzmann operator $\phi \circledast$ with kernel ϕ is a linear map in:*

(iv) *the space of restricted continuous processes,*

(v) *the space of processes with bounded histories, and*

(vi) *the space of processes with integrable histories.*

Finally, if $\phi \circledast$ is the zero map in any of the spaces listed in (i)–(vi), *then the kernel ϕ is the zero process in H^{ac}; that is, the Boltzmann operator $\phi \circledast$ is uniquely determined by its kernel ϕ in H^{ac}.*

The verification of assertions (i), (iii), and (iv) depends upon the observation that the histories of the processes in question are uniformly continuous on their supports.[37] The verification of (v) is elementary. And the verification of (ii) and (vi) follows from Fubini's theorem.[38]

Bearing in mind the definition of convolution in Sect. 4, we see that (5.2) may be written in the form

$$(\phi \circledast \psi)(t) = \mathring{\phi}\, \psi(t) + (\dot{\phi} * \psi)(t) + \int_{-\infty}^0 \dot{\phi}(t-s)\, \psi(s)\, ds, \qquad t \in \mathscr{R}. \tag{5.3}$$

[34] LTE, Sect. 10.

[35] An operator in the form of $\phi \circledast$ was used by BOLTZMANN [1874, *1*] in his study of "elastic aftereffects." In the class of Heaviside processes, the operator $\phi \circledast$ defines a Volterra integral operator with kernel ϕ. See VOLTERRA [1909, *1*]. For functions in H^1, $\phi \circledast \psi$ is essentially the Stieltjes convolution as used by GURTIN and STERNBERG [1962, *10*].

[36] Cf. Sect. 3. In particular, recall that a strong process in class H^{ac} is always continuous at infinity.

[37] For a detailed proof of (iii), see GURTIN and STERNBERG [1962, *10*, Theorem 2.2]. The proofs of (i) and (iv) are similar.

[38] See, for example, SAKS [1964, *22*, Sect. 8].

If, in addition, ψ is of Heaviside type, then (5.3) reduces to[39]

$$(\phi \circledast \psi)(t) = \overset{\circ}{\phi}\, \psi(t) + (\dot{\phi} * \psi)(t), \qquad t \in \mathscr{R}. \tag{5.4}$$

In either case, if $\phi = h$, the Heaviside unit step process, then $h \circledast \psi = \psi$; that is, *the Boltzmann operator with kernel h is the identity operator.*

Let ϕ and ψ be two scalar processes in class H^{ac}; then the process $\phi \circledast \psi$ is also in class H^{ac}; in fact,

$$\overset{\circ}{\widehat{\phi \circledast \psi}} = \overset{\circ}{\phi}\, \overset{\circ}{\psi}, \tag{5.5a}$$

$$\dot{\widehat{\phi \circledast \psi}} = \overset{\circ}{\phi}\, \dot{\psi} + \dot{\phi}\, \overset{\circ}{\psi} + \dot{\phi} * \dot{\psi}. \tag{5.5b}$$

If, in addition, ϕ and ψ are strong, then $\phi \circledast \psi$ is strong. Furthermore, for an appropriate[40] process γ,

$$(\phi \circledast \psi) \circledast \gamma = \phi \circledast (\psi \circledast \gamma). \tag{5.6}$$

That is, the Boltzmann operator $(\phi \circledast \psi) \circledast$ is equivalent to the operator defined by the application of $\psi \circledast$ followed by the application of $\phi \circledast$.

We may easily extend the definition of the Boltzmann operator to include scalar field processes and processes whose values are not scalars or scalar fields. Indeed, the pattern established in Sect. 4 for convolutions is easily adapted to Boltzmann operators. For example, if the kernel $\mathbf{K}$ is a strong $([\mathscr{T}])$-process in class H^{ac}, then $\mathbf{K} \circledast$ maps continuous restricted $(\mathscr{T})$-processes into continuous restricted $(\mathscr{T})$-processes.

We close this section with the observation that (in the scalar case) the operator $\phi \circledast \psi$ is not generally symmetric. However, if ϕ and ψ are both in class H^{ac}, then $\phi \circledast \psi = \psi \circledast \phi$.

B. Foundations of the linear theory.

6. Linear hereditary laws. We begin our treatment of constitutive relations for the linear theory of viscoelasticity with the concept of a *linear hereditary law* in which the stress depends linearly and continuously not only upon the present state of strain, but also upon all past values of the strain. We do not here consider the linear theory within the more general framework of the non-linear theories of mechanics.[1]

In this section $\mathscr{X}$ and $\mathscr{Y}$ denote finite-dimensional inner product spaces.[2] Let E be an $(\mathscr{X})$-process and let S be a $(\mathscr{Y})$-process. By a ***linear hereditary law*** $\mathscr{L}$[3] we mean a relation $S = \mathscr{L}[E]$ between ordered pairs of processes (E, S) which is defined as follows: there exists a continuous linear functional $\boldsymbol{L}$: $\mathscr{C}_0(\mathscr{R}^+; \mathscr{X}) \to \mathscr{Y}$, called the ***response functional for*** $\mathscr{L}$, such that $S = \mathscr{L}[E]$ if, and only if, E is a

---

[39] Since $\overset{\circ}{\phi}\psi(t) = (\dot{\phi} * \psi)(t) = 0$ for $t < 0$, $\phi \circledast$ may properly be called a Volterra operator with kernel ϕ (VOLTERRA [1909, *1*], [1913, *1*]). To avoid confusion, we continue to use the term Boltzmann operator in all cases.

[40] Cf. the properties of the Boltzmann operator in this section.

[1] For the more general treatment in terms of the non-linear theories see the review by TRUESDELL and NOLL [1965, *29*] and the paper by COLEMAN and NOLL [1964, *5*].

[2] Cf. Sect. 1. Recall that the magnitude $|\phi|_{\mathscr{X}}$ of an element $\phi \in \mathscr{X}$ is given by $|\phi|_{\mathscr{X}} = \sqrt{\phi \cdot \phi}$. See also GURTIN, LTE, Subchap. B. I.

[3] The precise mathematical characterization of linear hereditary laws has been considered by KÖNIG and MEIXNER [1958, *11*], by GURTIN and STERNBERG [1962, *10*] and by LEITMAN and MIZEL [1970, *6*]. The format and terminology of this section follow closely that used by these authors.

continuous restricted ($\mathscr{X}$)-process[4] and

$$S(t) = \boldsymbol{L}[E^t], \quad t \in \mathscr{R}. \tag{6.1}$$

We next record four characteristic properties of linear hereditary laws. These properties are direct consequences of the above definition.

(i) ***Linearity*** or the ***superposition property:***[5] If $\hat{S} = \mathscr{L}[\hat{E}]$ and $S = \mathscr{L}[E]$, then, for any pair of real numbers $\hat{\alpha}$ and α, $\hat{\alpha}\hat{S} + \alpha S = \mathscr{L}[\hat{\alpha}\hat{E} + \alpha E]$.

(ii) ***Time-translation invariance:*** Let $S = \mathscr{L}[E]$. For any real number η, define the ***translate processes*** E_η and S_η by

$$E_\eta(t) = E(t-\eta), \quad S_\eta(t) = S(t-\eta), \quad t \in \mathscr{R}. \tag{6.2}$$

Then, for all real numbers η, $S_\eta = \mathscr{L}[E_\eta]$.

(iii) ***Determinism*** or ***causality:***[6] If $\hat{S} = \mathscr{L}[\hat{E}]$ and $S = \mathscr{L}[E]$ and if $\hat{E}^t = E^t$, for some time t, then $\hat{S}(t) = S(t)$.

(iv) ***Continuity:***[7] Let $S = \mathscr{L}[E]$. For each time t and any $\varepsilon > 0$ there exists a $\delta > 0$ such that $\|E^t\| < \delta$ implies $|S(t)|_{\mathscr{Y}} < \varepsilon$.[8]

Properties (i)–(iv) characterize linear hereditary laws. Consider the set of ordered pairs (E, S), where E is a continuous restricted ($\mathscr{X}$)-process and S is a ($\mathscr{Y}$)-process. Suppose this set of pairs is related through $\mathscr{L}$, which relation has the properties (i)–(iv). Then it is possible to infer the existence of a continuous linear response functional $\boldsymbol{L}\colon \mathscr{C}_0(\mathscr{R}^+; \mathscr{X}) \to \mathscr{Y}$ such that $S = \mathscr{L}[E]$ implies (6.1). This is the approach taken by KÖNIG and MEIXNER and expanded by GURTIN and STERNBERG.[9]

Whenever we write $S = \mathscr{L}[E]$, E is a restricted continuous process. The next result shows that the same is true for S.

A linear hereditary law $\mathscr{L}$ may be regarded as a linear map from the space of restricted continuous ($\mathscr{X}$)-processes into the space of restricted continuous ($\mathscr{Y}$)-processes. That is, if E is a restricted continuous process and $S = \mathscr{L}[E]$, then S is a restricted continuous process.[10]

7. Boltzmann laws. Definitions. In this section we consider a special class of linear hereditary laws which is particularly appropriate to the study of linear viscoelastic response.

We use $\mathscr{X}$ and $\mathscr{Y}$ to denote finite-dimensional inner product spaces and $[\mathscr{X}, \mathscr{Y}]$ to denote the set of linear maps on $\mathscr{X}$ into $\mathscr{Y}$.[11]

By a ***response function*** we will always mean an *($[\mathscr{X}, \mathscr{Y}]$)-process* K in class H^{ac}. Thus, if K is a response function, there is an element $\mathring{\mathsf{K}} \in [\mathscr{X}, \mathscr{Y}]$ and a

---

4 Cf. Sect. 3.

5 The superposition property was used by BOLTZMANN [1874, *1*] in his early work describing "elastic aftereffects ..."

6 Cf. KÖNIG and MEIXNER [1958, *11*], GURTIN and STERNBERG [1962, *10*], TRUESDELL and NOLL [1965, *29*].

7 Ibid.

8 Cf. Sect. 3. Recall that $\|E^t\| = \sup\{|E^t(s)|_{\mathscr{X}}\colon s \in \mathscr{R}^+\}$.

9 KÖNIG and MEIXNER [1958, *11*] and GURTIN and STERNBERG [1962, *10*].

10 This result is due essentially to GURTIN and STERNBERG [1962, *10*, Theorem 2.2]. However, these authors consider only processes in Heaviside class H. Compare this result with the properties of the Boltzmann operator in Sect. 5 of this article.

11 In the sequel $\mathscr{X}$ and $\mathscr{Y}$ will denote spaces of tensors—strain tensors, stress tensors, etc.

function $\dot{\mathsf{K}}\colon \mathscr{R}^+ \to [\mathscr{X}, \mathscr{Y}]$ such that[12]

$$\mathsf{K}(t) = \overset{\circ}{\mathsf{K}} + \int_0^t \dot{\mathsf{K}}(s)\,ds, \qquad t \in \mathscr{R}^+. \tag{7.1}$$

The element $\overset{\circ}{\mathsf{K}} \in [\mathscr{X}, \mathscr{Y}]$ is called the ***initial response.*** If K is continuous at infinity, the element $\overset{\infty}{\mathsf{K}} \in [\mathscr{X}, \mathscr{Y}]$ given by

$$\overset{\infty}{\mathsf{K}} = \lim_{t \to \infty} \mathsf{K}(t) \tag{7.2}$$

is called the ***equilibrium response.*** We say that a response function K is a ***strong response function*** only if K is a strong process in class H^{ac}; that is, $\dot{\mathsf{K}}$ is integrable on $\mathscr{R}^+$. If K is strong it always admits the equilibrium response $\overset{\infty}{\mathsf{K}}$, but not conversely.[13]

Let K be a strong response function. Define a map $\boldsymbol{L}_{\mathsf{K}}\colon \mathscr{C}_0(\mathscr{R}^+; \mathscr{X}) \to \mathscr{Y}$ by

$$\boldsymbol{L}_{\mathsf{K}}[H] = \overset{\circ}{\mathsf{K}}[H(0)] + \int_0^\infty \dot{\mathsf{K}}(s)\,[H(s)]\,ds. \tag{7.3}$$

Then $\boldsymbol{L}_{\mathsf{K}}$ is a linear and continuous (bounded) map with norm[14]

$$\|\boldsymbol{L}_{\mathsf{K}}\| = |\overset{\circ}{\mathsf{K}}|_{[\mathscr{X},\mathscr{Y}]} + \int_0^\infty |\dot{\mathsf{K}}(s)|_{[\mathscr{X},\mathscr{Y}]}\,ds < \infty. \tag{7.4}$$

A linear hereditary law $\mathscr{L}_{\mathsf{K}}$ whose response functional $\boldsymbol{L}_{\mathsf{K}}$ is given by (7.3), for some strong response function K, is called a ***Boltzmann law***[15] ***with (strong) response function*** K.

If E is a restricted continuous $(\mathscr{X})$-process, then $E^t \in \mathscr{C}_0(\mathscr{R}^+; \mathscr{X})$ for each time t. From (7.3) it follows that a Boltzmann law $\mathscr{L}_{\mathsf{K}}$, with strong response function K, is just a Boltzmann operator with strong kernel K:[16]

$$S = \mathscr{L}_{\mathsf{K}}[E] = \mathsf{K} \circledast E, \tag{7.5}$$

for every restricted continuous $(\mathscr{X})$-process E. Indeed

$$\begin{aligned} S(t) = \boldsymbol{L}_{\mathsf{K}}[E^t] &= \overset{\circ}{\mathsf{K}}[E(t)] + \int_0^\infty \dot{\mathsf{K}}(s)\,[E(t-s)]\,ds \\ &= (\mathsf{K} \circledast E)(t), \qquad t \in \mathscr{R}. \end{aligned} \tag{7.6}$$

It follows from the properties of the Boltzmann operator that the Boltzmann law $\mathscr{L}_{\mathsf{K}}$ is uniquely characterized by its response function K.

Notice the particular role played by the present time t in (7.6). That is, the value $S(t)$ may depend specifically upon the present value $E^t(0) \equiv E(t)$ while any particular past value $E^t(s) \equiv E(t-s)$, $s > 0$, has essentially no contribution to $S(t)$. Further discussion of this feature will be explored later in this article.

---

[12] Cf. Sect. 3.

[13] Recall the definition and properties of a strong process in class H^{ac} given in Sect. 3, especially footnote 26, Chap. A, p. 6.

[14] If K is not required to be a strong response function, then the integral in (7.4) may diverge, in which case $\boldsymbol{L}_{\mathsf{K}}$ is not continuous on $\mathscr{C}_0(\mathscr{R}^+; \mathscr{X}) \to \mathscr{Y}$.

[15] A linear hereditary law of this form was first used to investigate "elastic aftereffects" by Boltzmann [1874, *1*].

[16] Cf. Sect. 5 for the definition and discussion of the Boltzmann operator.

Recall that a Boltzmann law, like any linear hereditary law, expresses a linear relationship between continuous restricted processes. However, the expression in (7.6) remains meaningful for other classes of processes.[17] Suppose E were the constant process with value $\hat{E}\in\mathscr{X}$; i.e., $E=\hat{E}^*$. Then it follows from (7.6) that

$$S(t)=\overset{\circ}{\mathsf{K}}[\hat{E}]+\int_0^\infty \dot{\mathsf{K}}(s)[\hat{E}]\,ds=\overset{\infty}{\mathsf{K}}[\hat{E}], \qquad t\in\mathscr{R}; \tag{7.7}$$

i.e., S is $(\overset{\infty}{\mathsf{K}}[\hat{E}])^*$, the constant process with value $\overset{\infty}{\mathsf{K}}[\hat{E}]$. Thus if K is strong, $\mathscr{L}_{\mathsf{K}}$ may be regarded as a linear map between spaces of constant processes, which map is completely determined by the equilibrium modulus $\overset{\infty}{\mathsf{K}}$.[18] Now let h denote the scalar Heaviside unit step process given by (3.11), and consider the Heaviside step process $E=h\hat{E}$, where $\hat{E}\in\mathscr{X}$. From (7.6) it then follows that

$$S(t)=\mathsf{K}(t)[\hat{E}], \qquad t\in\mathscr{R}. \tag{7.8}$$

In fact, as was seen in Sect. 5, the Boltzmann law induces a linear map in the Heaviside class H. Continuing this argument, we see that if E were a process which was piecewise continuous (right continuous at the jumps) with bounded or integrable histories, then S determined through (7.6) will also be piecewise continuous (right continuous at the jumps) with bounded or integrable histories and, more importantly, S *may have a jump discontinuity only at those times where E has a jump discontinuity*. This latter property is sometimes called ***preservation of regularity;***[19] it is a direct consequence of the specific dependence upon the present value.[20]

Motivated by the above argument and the properties of the Boltzmann operator[21] we lay down the following:

Convention. *We always assume that the domain of a Boltzmann law, with a strong response function, has been extended to include the processes with bounded histories and the processes with integrable histories.*

These classes certainly include the piecewise continuous processes with bounded or integrable histories, the Heaviside classes, the locally integrable processes with finite histories, etc.

8. Characterization of Boltzmann laws. Not every linear hereditary law is a Boltzmann law. Indeed, consider a linear hereditary law $\mathscr{L}$ whose response functional $\boldsymbol{L}$ is defined on $\mathscr{C}_0(\mathscr{R}^+;\mathscr{X})$ by

$$\boldsymbol{L}[H]=\hat{\mathsf{K}}[H(\lambda)], \tag{8.1}$$

where $\hat{\mathsf{K}}$ is a linear map in $[\mathscr{X},\mathscr{Y}]$ and λ is some fixed *positive* real number. Thus if $S=\mathscr{L}[E]$ we have

$$S(t)=\hat{\mathsf{K}}[E(t-\lambda)], \qquad t\in\mathscr{R}. \tag{8.2}$$

Clearly, there is no response function K such that (7.6) reduces to (8.2).

Since Boltzmann laws are of great importance in mechanical theories including hereditary response,[22] it is reasonable to seek a characterization of Boltzmann laws within our context of a linear theory.

---

[17] Recall the properties of the Boltzmann operator in Sect. 5.

[18] This assertion remains true if K is not strong provided the equilibrium response $\overset{\infty}{\mathsf{K}}$ exists.

[19] See COLEMAN and MIZEL [1966, *5*].

[20] This fact is particularly important in our discussions of propagating singular surfaces in Sect. 41.

[21] Cf. Sect. 5.

[22] COLEMAN and MIZEL [1966, *5*], COLEMAN and NOLL [1964, *5*], TRUESDELL and NOLL [1965, *29*, Sect. 41].

We begin by observing that the continuity and time-translation invariance of any linear hereditary law $\mathscr{L}$ imply that for each restricted continuous process E, the map $S_\eta(t)$, defined through $S_\eta = \mathscr{L}[E_\eta]$, is continuous (at zero) for each time t.[23] It turns out that a stronger type of translation continuity characterizes Boltzmann laws.

Let $\mathscr{A}$ be any open set in $\mathscr{R}$ and define an open set $\mathscr{A}_\eta$ in $\mathscr{R}$ for each real η by

$$\mathscr{A}_\eta = \{\tau + \eta : \tau \in \mathscr{A}\}. \tag{8.3}$$

If $\mathscr{A}$ lies in $\mathscr{R}^+$, let $\mathscr{C}_0(\mathscr{A}; \mathscr{X})$ denote the set of all functions in $\mathscr{C}_0(\mathscr{R}^+; \mathscr{X})$ which vanish outside $\mathscr{A}$. For each such $\mathscr{A}$ define $\|\boldsymbol{L}\|(\mathscr{A})$, the norm of the response functional $\boldsymbol{L}$ restricted to $\mathscr{C}_0(\mathscr{A}; \mathscr{X})$, by

$$\|\boldsymbol{L}\|(\mathscr{A}) = \sup\{|\boldsymbol{L}(H)|_{\mathscr{Y}} : H \in \mathscr{C}_0(\mathscr{A}; \mathscr{X});\ \|H\|_{\mathscr{X}} \leqq 1\}. \tag{8.4}$$

Of course $\|\boldsymbol{L}\|(\mathscr{A}) < \infty$ for each open set $\mathscr{A}$ in $\mathscr{R}^+$ since $\boldsymbol{L}$ is continuous on $\mathscr{C}_0(\mathscr{R}^+; \mathscr{X})$.

We now may state the following

Theorem[24] ***(Characterization of Boltzmann laws).*** *A linear hereditary law $\mathscr{L}$ is a Boltzmann law if, and only if, its response functional $\boldsymbol{L}$ has the following translation property: the map*

$$\eta \to \|\boldsymbol{L}\|(\mathscr{A}_\eta)$$

is continuous (at zero) for every bounded open set $\mathscr{A}$ whose closure lies in $\mathscr{R}^{++}$.

Let t be a fixed time and write $S(t) = \boldsymbol{L}[E^t]$. Choose some interval $\mathscr{A}$ of past times and consider only those processes E whose histories up to time t vanish outside $\mathscr{A}$ and have unit amplitude. Then the above characterization roughly asserts that the Boltzmann laws are just those linear hereditary laws for which the maximum value of $|S(t)|$, over the considered class of processes E, varies continuously under small translations of the interval $\mathscr{A}$. Note that the intervals $\mathscr{A}$ in this characterization never contain the present time t. In this way, the special emphasis upon the present time is preserved.

Throughout our discussion of linear hereditary laws we have based our analysis upon the class of continuous restricted processes. Our characterization of Boltzmann laws was greatly dependent upon this class. Indeed, the requirement that $\boldsymbol{L}_K$, defined in (7.3), be a *continuous linear map on* $\mathscr{C}_0(\mathscr{R}^+; \mathscr{X})$ into $\mathscr{Y}$ is satisfied if, and only if, the response function K is strong. Therefore, by considering the class of restricted continuous processes we are led naturally to laws of Boltzmann type defined by *strong* response functions.

Suppose that at the outset we had considered the subspace of restricted continuous processes all of which vanish up to some fixed time t_0. Then, by virtue of the time-translation invariance of linear hereditary laws, we should need consider only restricted continuous processes of Heaviside type.[25] In this case, virtually all previous arguments would remain valid, but we could drop the restriction that the response function of a Boltzmann law be strong. Note that if K does not admit the equilibrium response $\overset{\infty}{K}$, then $\boldsymbol{L}_K$, given in (7.3), is *not* defined for the class of constant processes.

[23] Recall that S_η and E_η are the translate processes given, for all real η, by (6.2).

[24] LEITMAN and MIZEL [1970, *6*].

[25] To be consistent we should here follow KÖNIG and MEIXNER [1958, *11*] or GURTIN and STERNBERG [1962, *10*] and consider properties (i)–(iv) of our Sect. 6 as defining linear hereditary laws. See the comments following property (iv) of Sect. 6.

In summary, when the fundamental class of processes to be considered is the class of restricted continuous processes, it is necessary to associate strong response functions with Boltzmann laws; however, if only subspaces of Heaviside type are considered, the restriction that response functions be strong may be dropped. *When it is essential that a response function* K *be strong, we state so explicitly; otherwise* K *will simply be called a response function.*

It is often convenient to approximate processes of Heaviside type by smooth processes. More precisely we have the following

Proposition.[26] *Let* ϕ *be a process of Heaviside class* H^N. *Then there is a sequence of processes* $\{\phi_n\}$ *of Heaviside type in class* C^N *such that*

(i) *the sequence* $\{\phi_n\}$ *is uniformly bounded on closed and bounded sets containing zero and*

(ii) $\phi_n \to \phi$ *as* $n \to \infty$ *uniformly on closed sets not containing zero.*

Using this result it is possible to give a direct interpretation of Boltzmann laws for Heaviside processes.

Theorem.[27] *Consider a Boltzmann law with response function* K. *Suppose that* E *and* S *are processes of class* H *defined as follows: there exist sequences* $\{E_n\}$ *and* $\{S_n\}$ *of Heaviside processes in class* C *such that*

(i) $\{E_n\}$ *and* $\{S_n\}$ *are uniformly bounded on all compact sets,*

(ii) $E_n \to E$ *and* $S_n \to S$ *as* $n \to \infty$ *uniformly on closed sets not containing zero, and*

(iii) *for each* n

$$S_n = \mathsf{K} \circledast E_n. \tag{8.5}$$

Then

$$S = \mathsf{K} \circledast E. \tag{8.6}$$

Conversely, suppose (8.6) *holds for some processes* E *and* S *of class* H *(in* $\mathscr{X}$ *and* $\mathscr{Y}$*). Then there exist sequences of Heaviside processes* $\{E_n\}$ *and* $\{S_n\}$ *of class* C *such that* (i)–(iii) *hold.*

9. Constitutive relations. Linearly viscoelastic materials. In this section, and in the sequel, $\mathscr{B}$ is a *body* in $\mathscr{E}$, $\boldsymbol{x} \in \mathscr{B}$ is a *material point* in $\mathscr{B}$, and $t \in \mathscr{R}$ is a *time*. We consistently use $\boldsymbol{u}$ to denote a *displacement field process for* $\mathscr{B}$, $\boldsymbol{E}$ to denote a *strain field process for* $\mathscr{B}$, and $\boldsymbol{S}$ to denote a *stress*[28] *field process for* $\mathscr{B}$. Recall that, for each $(\boldsymbol{x}, t) \in \mathscr{B} \times \mathscr{R}$, $\boldsymbol{u}(\boldsymbol{x}, t) \in \mathscr{V}$, $\boldsymbol{E}(\boldsymbol{x}, t) \in \mathscr{T}_{\mathrm{sym}}$, and $\boldsymbol{S}(\boldsymbol{x}, t) \in \mathscr{T}_{\mathrm{sym}}$.[29]

The behavior of a body $\mathscr{B}$ is said to be ***linearly viscoelastic of relaxation type at a material point*** $\boldsymbol{x}$ if, and only if, there is a linear hereditary law $\mathscr{L}^{\boldsymbol{x}}$ such that the stress process $\boldsymbol{S}_{\boldsymbol{x}}$ and the displacement process $\boldsymbol{u}_{\boldsymbol{x}}$, at the material point $\boldsymbol{x}$, are related through $\mathscr{L}^{\boldsymbol{x}}$, by

$$\boldsymbol{S}_{\boldsymbol{x}} = \mathscr{L}^{\boldsymbol{x}}[\nabla \boldsymbol{u}_{\boldsymbol{x}}], \qquad \boldsymbol{x} \in \mathscr{B}, \tag{9.1}$$

or, equivalently, by

$$\boldsymbol{S}(\boldsymbol{x}, t) = \boldsymbol{L}^{\boldsymbol{x}}[\nabla \boldsymbol{u}^t(\boldsymbol{x})], \qquad (\boldsymbol{x}, t) \in \mathscr{B} \times \mathscr{R}. \tag{9.2}$$

---

[26] GURTIN and STERNBERG [1962, *10*, Lemma 3.1].

[27] This result is essentially that of GURTIN and STERNBERG [1962, *10*, Theorem 3.1]. We note that the proof requires only that K be of bounded variation on compact subsets of $\mathscr{R}^+$; therefore, continuity or integrability of $\dot{\mathsf{K}}$ is not essential.

[28] We identify material points of the body by their positions in a fixed reference configuration. (By the stress $\boldsymbol{S}$ we always mean the Cauchy stress.) In order to be consistent we assume that the residual stress in the reference configuration vanishes. See TRUESDELL and NOLL [1965, *29*].

[29] Cf. Sect. 3.

If the body $\mathscr{B}$ is linearly viscoelastic of relaxation type at *every* material point $\boldsymbol{x}$, then we say that $\mathscr{B}$ is a ***linearly viscoelastic body of relaxation type*** or, when no confusion may occur, $\mathscr{B}$ is a ***viscoelastic body.*** We call $\mathscr{L}^{\boldsymbol{x}}$ a ***relaxation law (at $\boldsymbol{x}$)*** and $\boldsymbol{L}^{\boldsymbol{x}}$ the corresponding ***relaxation functional (at $\boldsymbol{x}$).***

A viscoelastic body $\mathscr{B}$ is a *simple material*[30] in the sense that the stress at a material point $\boldsymbol{x}$ and time t is completely determined by the history up to time t of the displacement gradient at $\boldsymbol{x}$.

We make the following fundamental

Postulate on rigid motions.[31] *An (infinitesimal) rigid motion of the body $\mathscr{B}$ up to time t results in zero stress at time t at every material point $\boldsymbol{x}$ in $\mathscr{B}$.*

Let $\boldsymbol{u}$ be an (infinitesimal) rigid rotation process for $\mathscr{B}$. Then $\nabla \boldsymbol{u}_{\boldsymbol{x}} = \boldsymbol{W}$, where $\boldsymbol{W}$ is a $(\mathscr{T}_{\text{skw}})$-process which is independent of the material point $\boldsymbol{x}$. As an immediate consequence of the postulate we conclude that

$$\mathscr{L}^{\boldsymbol{x}}[\boldsymbol{W}] = \boldsymbol{0}, \qquad \boldsymbol{x} \in \mathscr{B}. \tag{9.3}$$

Thus, for each material point $\boldsymbol{x} \in \mathscr{B}$, the linear hereditary law $\mathscr{L}^{\boldsymbol{x}}$ vanishes on the class of $(\mathscr{T}_{\text{skw}})$-processes. Now $\nabla \boldsymbol{u}$ has the unique decomposition

$$\nabla \boldsymbol{u} = \boldsymbol{E} + \boldsymbol{W}, \tag{9.4}$$

where $\boldsymbol{E}$ is the symmetric (infinitesimal) strain field process and $\boldsymbol{W}$ is the skew (infinitesimal) rotation field process. Hence it follows from the linearity of $\mathscr{L}^{\boldsymbol{x}}$, and (9.1)–(9.4) that[32]

$$\boldsymbol{S}_{\boldsymbol{x}} = \mathscr{L}^{\boldsymbol{x}}[\hat{\nabla} \boldsymbol{u}_{\boldsymbol{x}}] \equiv \mathscr{L}^{\boldsymbol{x}}[\boldsymbol{E}_{\boldsymbol{x}}], \qquad \boldsymbol{x} \in \mathscr{B} \tag{9.5}$$

or, equivalently, that

$$\boldsymbol{S}(\boldsymbol{x}, t) = \boldsymbol{L}^{\boldsymbol{x}}[\hat{\nabla} \boldsymbol{u}^t(\boldsymbol{x})] \equiv \boldsymbol{L}^{\boldsymbol{x}}[\boldsymbol{E}^t(\boldsymbol{x})], \qquad (\boldsymbol{x}, t) \in \mathscr{B} \times \mathscr{R}. \tag{9.6}$$

By virtue of the postulate and the comments which follow it we see that for each material point $\boldsymbol{x}$ in $\mathscr{B}$, $\mathscr{L}^{\boldsymbol{x}}$ may be regarded as a linear map in the class of restricted continuous $(\mathscr{T}_{\text{sym}})$-processes or, equivalently, that $\boldsymbol{L}^{\boldsymbol{x}}$ may be regarded as a continuous linear map $\boldsymbol{L}^{\boldsymbol{x}}$: $\mathscr{C}_0(\mathscr{R}^+; \mathscr{T}_{\text{sym}}) \to \mathscr{T}_{\text{sym}}$. *We make no notational distinction between the maps $\mathscr{L}^{\boldsymbol{x}}$ or $\boldsymbol{L}^{\boldsymbol{x}}$ and their restriction to processes or histories whose values are symmetric tensors.* The distinction should be clear from the context. The constitutive relation expressed by (9.5) or (9.6) does not satisfy the principle of material frame indifference[33] since the strain $\boldsymbol{E}$ is not properly invariant under changes-in-frame. However, the infinitesimal theory does satisfy the principle of material frame in difference for infinitesimal rotations as can be shown from the full finite theory.

A viscoelastic body $\mathscr{B}$ is said to be ***homogeneous*** if, and only if, its relaxation law $\mathscr{L}^{\boldsymbol{x}}$ is independent of the material point $\boldsymbol{x} \in \mathscr{B}$. In this case we write $\mathscr{L}$ for $\mathscr{L}^{\boldsymbol{x}}$ and call it the ***relaxation law for*** $\mathscr{B}$. Of course the corresponding relaxation functional $\boldsymbol{L}^{\boldsymbol{x}}$ is also independent of $\boldsymbol{x}$, so that we also write $\boldsymbol{L}$ for $\boldsymbol{L}^{\boldsymbol{x}}$, the ***relaxation functional for*** $\mathscr{B}$. If the body $\mathscr{B}$ is not homogeneous, it is said to be ***inhomogeneous.***

---

[30] For a complete discussion of the concept of *simple materials* and collateral bibliography see TRUESDELL and NOLL [1965, *29*, Sect. 28] and the definitive work of NOLL [1958, *16*].

[31] Cf. LTE, Sect. 20.

[32] We write $\hat{\nabla} \boldsymbol{u}$ for sym $\nabla \boldsymbol{u}$ so that $\hat{\nabla} \boldsymbol{u} \equiv \boldsymbol{E}$.

[33] For an extensive treatment of the principle of material frame indifference see TRUESDELL and NOLL [1965, *29*, Sect. 19].

An orthogonal tensor Q is said to be a ***symmetry transformation***[34] ***at*** $x \in \mathscr{B}$ if, and only if,

$$QL^x[H]Q^T = L^x[QHQ^T] \tag{9.7}$$

for every history $H \in \mathscr{C}_0(\mathscr{R}^+; \mathscr{T})$. The collection of all such tensors forms a group $\mathscr{G}_x$ called the ***symmetry group at*** $x \in \mathscr{B}$. Since the set $\{\mathbf{1}, -\mathbf{1}\}$ is always a subgroup of $\mathscr{G}_x$, it follows that $\mathscr{G}_x$ is generated by $\{\mathbf{1}, -\mathbf{1}\}$ and $\mathscr{G}_x^+$, the subgroup of $\mathscr{G}_x$ consisting of all proper orthogonal tensors (rotations) in $\mathscr{G}_x$. The body $\mathscr{B}$ is said to be ***isotropic at*** $x \in \mathscr{B}$ if, and only if, $\mathscr{G}_x$ is the full orthogonal group; otherwise $\mathscr{B}$ is said to be ***anisotropic at*** $x \in \mathscr{B}$.

10. Constitutive equations. Boltzmann laws. Stress relaxation. In this section, and those which follow, we consider linear hereditary laws which are Boltzmann laws. Let $\mathscr{B}$ be a linearly viscoelastic body of relaxation type such that, at each point $x \in \mathscr{B}$, the relaxation law $\mathscr{L}^x$ is a Boltzmann law. This means that for each $x \in \mathscr{B}$, there is a response function $\mathbf{G}_x = \mathbf{G}(x, \cdot)$, called the ***relaxation function at*** $x \in \mathscr{B}$, such that, for each $(x, t) \in \mathscr{B} \times \mathscr{R}$, $\mathbf{G}(x, t) \in [\mathscr{T}, \mathscr{T}_{\text{sym}}]$ and[35]

$$\mathbf{G}(x, t) = \overset{\circ}{\mathbf{G}}(x) + \int_0^t \dot{\mathbf{G}}(x, s)\, ds, \qquad (x, t) \in \mathscr{B} \times \mathscr{R}^+. \tag{10.1}$$

The map $\overset{\circ}{\mathbf{G}}(x)$ in $[\mathscr{T}, \mathscr{T}_{\text{sym}}]$ is called the ***initial elasticity at*** $x \in \mathscr{B}$ and governs the response to instantaneous changes in strain. The map $\overset{\infty}{\mathbf{G}}(x)$ in $[\mathscr{T}, \mathscr{T}_{\text{sym}}]$, given by

$$\overset{\infty}{\mathbf{G}}(x) = \lim_{t \to \infty} \mathbf{G}(x, t), \qquad x \in \mathscr{B}, \tag{10.2}$$

which determines the material response at equilibrium, is called the ***equilibrium modulus at*** $x \in \mathscr{B}$. Furthermore, (9.2) becomes

$$\begin{aligned} S(x, t) &= (\mathbf{G} \circledast \nabla u)(x, t) \\ &= \overset{\circ}{\mathbf{G}}(x)[\nabla u(x, t)] + \int_0^\infty \dot{\mathbf{G}}(x, s)[\nabla u(x, t-s)]\, ds, \qquad (x, t) \in \mathscr{B} \times \mathscr{R}. \end{aligned} \tag{10.3}$$

Note that the value of the relaxation function, at each time t, is the $([\mathscr{T}, \mathscr{T}_{\text{sym}}])$-field on $\mathscr{B}$ denoted by $\mathbf{G}(t) = \mathbf{G}(\cdot, t)$. *We will always assume that the field* $\mathbf{G}(t) = \mathbf{G}(\cdot, t)$ *may be continuously extended to* $\overline{\mathscr{B}}$ *and is smooth on* $\mathscr{B}$. If the viscoelastic body $\mathscr{B}$ is homogeneous, the relaxation function $\mathbf{G}$ is independent of the material point $x \in \mathscr{B}$.

In this section, we consider a fixed point $x \in \mathscr{B}$ and suppress dependence upon it. Thus, we write $u(t)$ for $u(x, t)$, $S(t)$ for $S(x, t)$, $E(t)$ for $E(x, t)$, $\mathbf{G}(t)$ for $\mathbf{G}(x, t)$, etc., and (10.3) becomes

$$S(t) = (\mathbf{G} \circledast \nabla u)(t) = \overset{\circ}{\mathbf{G}}[\nabla u(t)] + \int_0^\infty \dot{\mathbf{G}}(s)[\nabla u(t-s)]\, ds, \qquad t \in \mathscr{R}. \tag{10.4}$$

By virtue of the *postulate on rigid motions*, we have

$$S(t) = (\mathbf{G} \circledast \widehat{\nabla} u)(t) = \overset{\circ}{\mathbf{G}}[\widehat{\nabla} u(t)] + \int_0^\infty \dot{\mathbf{G}}(s)[\widehat{\nabla} u(t-s)]\, ds, \qquad t \in \mathscr{R}, \tag{10.5}$$

---

[34] The concept of material symmetry is the same for viscoelastic materials as for elastic materials. Therefore, with minor modifications, the discussion and results of LTE, Sect. 21 apply here as well. For a more complete discussion of material symmetry see TRUESDELL and NOLL [1965, *29*] and TRUESDELL and TOUPIN [1960, *15*].

[35] For each $x \in \mathscr{B}$, $\mathbf{G}_x = \mathbf{G}(x, \cdot)$ is a process in class H^{ac}. Cf. the notational convention of Sect. 3.

where $\hat{\nabla} \boldsymbol{u} \equiv \operatorname{sym} \nabla \boldsymbol{u} = \boldsymbol{E}$. Usually (10.5) is written in the abbreviated form.

$$\boldsymbol{S} = \mathbf{G} \circledast \boldsymbol{E}. \tag{10.6}$$

The postulate also implies that

$$\mathbf{G}(t)[\boldsymbol{T}] = \mathbf{G}(t)[\operatorname{sym} \boldsymbol{T}], \quad t \in \mathscr{R}, \tag{10.7}$$

for every tensor $\boldsymbol{T} \in \mathscr{T}$. *Henceforth, the values of the relaxation function* $\mathbf{G}$ *will always be regarded as* $([\mathscr{T}_{\mathrm{sym}}])$*-fields on* $\mathscr{B}$; that is $\mathbf{G}(\boldsymbol{x}, t) \in [\mathscr{T}_{\mathrm{sym}}]$ for all $(\boldsymbol{x}, t) \in \mathscr{B} \times \mathscr{R}$ and hence is a mapping of symmetric tensors into symmetric tensors. Of course, similar assertions must then hold for $\overset{\circ}{\mathbf{G}}$, $\dot{\mathbf{G}}$, and $\overset{\infty}{\mathbf{G}}$.

We remark at this point, that the relaxation function $\mathbf{G}$ may be constant on $\mathscr{R}^+$. Thus, $\dot{\mathbf{G}}$ vanishes and $\mathbf{G}$ is of the form

$$\mathbf{G} = h \overset{\circ}{\mathbf{G}}, \tag{10.8}$$

where h is the scalar Heaviside unit step process defined in (3.11). The viscoelastic body $\mathscr{B}$ is then ***elastic*** with elasticity $\overset{\circ}{\mathbf{G}}$ and (10.5) reduces to

$$\boldsymbol{S}(t) = \overset{\circ}{\mathbf{G}}[\boldsymbol{E}(t)], \quad t \in \mathscr{R}. \tag{10.9}$$

Therefore, *an elastic material is just a viscoelastic material for which there is no hereditary response.*

The restriction of material symmetry, at a material point, may be characterized in terms of restrictions on the relaxation function $\mathbf{G}$: *An orthogonal tensor* $\boldsymbol{Q}$ *belongs to the symmetry group* $\mathscr{G}$ *if, and only if,* $\boldsymbol{Q}$ *belongs to the symmetry group of the relaxation function* $\mathbf{G}$; *that is,*

$$\boldsymbol{Q}\mathbf{G}(t)[\boldsymbol{T}]\boldsymbol{Q}^T = \mathbf{G}(t)[\boldsymbol{Q}\boldsymbol{T}\boldsymbol{Q}^T], \quad t \in \mathscr{R}, \tag{10.10}$$

for every symmetric tensor $\boldsymbol{T}$. The sufficiency of this condition is immediate while the necessity follows from the fact that $\mathscr{L}_{\mathbf{G}}$ is uniquely characterized by $\mathbf{G}$. This assertion makes it possible to adapt the results of linear elasticity directly to the study of linear viscoelasticity.[36]

Let $\mathbf{G}$ be a strong relaxation function with equilibrium modulus $\overset{\infty}{\mathbf{G}}$. Then if $\boldsymbol{E}$ is the constant process with value $\hat{\boldsymbol{E}} \in \mathscr{T}_{\mathrm{sym}}$ (i.e., $\boldsymbol{E} = \hat{\boldsymbol{E}}^*$) and if $\boldsymbol{S} = \mathbf{G} \circledast \boldsymbol{E}$, it follows[37] that $\boldsymbol{S}$ is a constant process with value $\overset{\infty}{\mathbf{G}}[\hat{\boldsymbol{E}}]$. That is

$$\boldsymbol{S} = \mathbf{G} \circledast (\hat{\boldsymbol{E}}^*) = (\overset{\infty}{\mathbf{G}}[\hat{\boldsymbol{E}}])^*. \tag{10.11}$$

Similarly, if $\boldsymbol{E}$ is the Heaviside step strain process $\boldsymbol{E} = h\hat{\boldsymbol{E}}$ and if $\boldsymbol{S} = \mathbf{G} \circledast \boldsymbol{E}$, it follows[38] that $\boldsymbol{S}$ is the Heaviside process

$$\boldsymbol{S}(t) = \mathbf{G}(t)[\hat{\boldsymbol{E}}], \quad t \in \mathscr{R}. \tag{10.12}$$

This relation shows that the relaxation function $\mathbf{G}(t)$ is simply the stress which results from a unit Heaviside strain process and hence provides the basis for the experimental determination of $\mathbf{G}(t)$. Moreover, the limit $\hat{\boldsymbol{S}} \in \mathscr{T}_{\mathrm{sym}}$ given by

$$\hat{\boldsymbol{S}} = \lim_{t \to \infty} \boldsymbol{S}(t) = \lim_{t \to \infty} \mathbf{G}(t)[\hat{\boldsymbol{E}}] = \overset{\infty}{\mathbf{G}}[\hat{\boldsymbol{E}}] \tag{10.13}$$

[36] Cf. LTE, Sects. 21, 22, and 26 for a detailed discussion of symmetry in the elastic case. Virtually all the results contained there may be adapted directly to our discussion. We consider the isotropic case in Sect. 12.

[37] Cf. Sect. 7, Eq. (7.7).

[38] Cf. Sect. 7, Eq. (7.7).

exists. The limiting stress $\hat{\boldsymbol{S}}$ is precisely that which would obtain if the strain process $\boldsymbol{E}$ had been constant with value $\hat{\boldsymbol{E}}$, and the value of $\hat{\boldsymbol{S}}$ determined by $\hat{\boldsymbol{E}}$ through the *elastic law*

$$\hat{\boldsymbol{S}} = \overset{\infty}{\mathbf{G}}[\hat{\boldsymbol{E}}], \tag{10.14}$$

where the elasticity is the equilibrium modulus $\overset{\infty}{\mathbf{G}}$.

The phenomenon described above is usually called *stress relaxation*. In fact, we have shown that for all Heaviside step processes $\boldsymbol{E} = h\hat{\boldsymbol{E}}$, $\hat{\boldsymbol{E}} \in \mathscr{T}_{\text{sym}}$, the stress $\boldsymbol{S} = \mathbf{G} \circledast \boldsymbol{E}$ satisfies

$$\hat{\boldsymbol{S}} = \lim_{t\to\infty} \boldsymbol{S}(t) = \overset{\infty}{\mathbf{G}}\left[\lim_{t\to\infty} \boldsymbol{E}(t)\right] = \overset{\infty}{\mathbf{G}}[\hat{\boldsymbol{E}}]. \tag{10.15}$$

More generally, we say that a strong relaxation function $\mathbf{G}$ has the ***stress relaxation property***[39] ***over a space of strain processes*** if, and only if, for each strain-process $\boldsymbol{E}$ in that space which is continuous at infinity, the stress process $\boldsymbol{S} = \mathbf{G} \circledast \boldsymbol{E}$ is also continuous at infinity and

$$\lim_{t\to\infty} \boldsymbol{S}(t) = \overset{\infty}{\mathbf{G}}\left[\lim_{t\to\infty} \boldsymbol{E}(t)\right]. \tag{10.16}$$

In this case the limiting stress, say, $\hat{\boldsymbol{S}}$ is determined by the limiting strain, say, $\hat{\boldsymbol{E}}$ and the equilibrium modulus $\overset{\infty}{\mathbf{G}}$ through the elastic law (10.14).

The next result characterizes some important spaces of strain processes for which there is a relaxation property.

A strong relaxation function $\mathbf{G}$ *has the relaxation property over the spaces of strain processes with bounded histories.* Included among these are the piecewise continuous processes with bounded histories which, in turn, includes the continuous restricted processes, the Heaviside classes, etc.

Note that the condition that the processes have bounded histories is essential. For there are strain processes $\boldsymbol{E}$ whose histories are integrable, but unbounded, such that $\lim_{t\to\infty} \boldsymbol{E}(t) = \mathbf{0}$ but $\lim_{t\to\infty} (\mathbf{G} \circledast \boldsymbol{E})(t)$ does not even exist. Thus a strong relaxation function does not have the relaxation property over the space of strain processes with integrable histories. However, if $\mathbf{G}$ is a strong relaxation function with the additional property that $|\dot{\mathbf{G}}|$ is eventually monotone,[40] then $\mathbf{G}$ will have the relaxation property over the space of strain processes with integrable histories.

If G is a strong scalar relaxation function such that G decreases monotonically from $\overset{\circ}{G}$ to $\overset{\infty}{G}$ then the condition that $|\dot{G}|$ be monotone decreasing is equivalent to the condition that G be concave upward on $\mathscr{R}^+$. In this case $\overset{\infty}{G} \leqq \overset{\circ}{G}$ and the stress corresponding to a unit Heaviside step strain "relaxes" from $\overset{\circ}{G}$ to $\overset{\infty}{G}$.[41]

The results outlined above have further implications regarding the regular limiting behavior of Boltzmann laws. Observe that since the spaces of strain processes for which $\mathbf{G}$ has the relaxation property are *linear*, we have the following statement regarding asymptotic behavior.

---

[39] Cf. COLEMAN and MIZEL [1966, *5*].

[40] A positive function k on $\mathscr{R}$ is ***eventually monotone*** if, and only if, it is monotone on some interval of the form $[T, \infty)$. In our context, all properties hold except, possibly, on a set of Lebesgue measure zero. However, we suppress such technical details.

[41] For scalar relaxation functions, the conditions $\overset{\circ}{G} \geqq \overset{\infty}{G} > 0$, $\dot{G} \leqq 0$, and G concave upward on $\mathscr{R}^+$ are just those used by DAFERMOS [1970, *1*] to guarantee the existence, uniqueness, and asymptotic stability of solutions to the dynamic initial past-history problem. Cf. Chap. D of this article, Sect. 42.

Let $\mathbf{G}$ *have the relaxation property over a space of strain processes. Let* $\boldsymbol{E}_1$ *and* $\boldsymbol{E}_2$ *be a pair of strain processes in that space which approach each other asymptotically:*

$$\lim_{t\to\infty} \left(\boldsymbol{E}_1(t) - \boldsymbol{E}_2(t)\right) = \mathbf{0}. \tag{10.17}$$

Then, if $\boldsymbol{S}_1 = \mathbf{G} \circledast \boldsymbol{E}_1$ *and* $\boldsymbol{S}_2 = \mathbf{G} \circledast \boldsymbol{E}_2$, *it follows that* $\boldsymbol{S}_1$ *and* $\boldsymbol{S}_2$ *also approach each other asymptotically:*

$$\lim_{t\to\infty} \left(\boldsymbol{S}_1(t) - \boldsymbol{S}_2(t)\right) = \mathbf{0}. \tag{10.18}$$

Consider any bounded strain processes $\boldsymbol{E}$ which is *periodic in time with period* $T \geqq 0$:

$$\boldsymbol{E}(t+T) = \boldsymbol{E}(t), \qquad t \in \mathscr{R}. \tag{10.19}$$

If $\boldsymbol{S} = \mathbf{G} \circledast \boldsymbol{E}$, a simple computation shows that $\boldsymbol{S}$ is also periodic in time with period $T \geqq 0$:

$$\boldsymbol{S}(t+T) = \boldsymbol{S}(t), \qquad t \in \mathscr{R}. \tag{10.20}$$

Now the periodic and bounded strain processes are certainly in the class of strain processes having bounded histories. Let $\boldsymbol{E}$ be such a periodic process, with period $T \geqq 0$, and consider the associated Heaviside process $h\boldsymbol{E}$. Then the process $\boldsymbol{E} - h\boldsymbol{E} \equiv (1-h)\,\boldsymbol{E}$ has bounded histories and

$$\lim_{t\to\infty} (1-h)\,\boldsymbol{E}(t) = \mathbf{0}.$$

Thus, since $\mathbf{G}$ has the stress relaxation property for such processes, it follows that

$$\lim_{t\to\infty} \left((\mathbf{G} \circledast \boldsymbol{E})(t) - (\mathbf{G} \circledast h\boldsymbol{E})(t)\right) = \mathbf{0}. \tag{10.21}$$

That is, the Heaviside stress process associated with the Heaviside strain process $h\boldsymbol{E}$ is asymptotically the same as the periodic stress associated with the periodic strain process $\boldsymbol{E}$.

Consider a periodic strain process $\boldsymbol{E}$ with amplitude $\hat{\boldsymbol{E}} \in \mathscr{T}_{\mathrm{sym}}$ and circular frequency $\omega \neq 0 \left(T = (2\pi/\omega) > 0\right)$

$$\boldsymbol{E}(t) = \sin(\omega t)\,\hat{\boldsymbol{E}}, \qquad t \in \mathscr{R}. \tag{10.22}$$

Then $\boldsymbol{S} = \mathbf{G} \circledast \boldsymbol{E}$ is given by the formula

$$\boldsymbol{S}(t) = \sin(\omega t)\,[\mathring{\mathbf{G}} + \hat{\dot{\mathbf{G}}}_c(\omega)]\,[\hat{\boldsymbol{E}}] - \cos(\omega t)\,[\hat{\dot{\mathbf{G}}}_s(\omega)]\,[\hat{\boldsymbol{E}}], \qquad t \in \mathscr{R}, \tag{10.23}$$

where $\hat{\dot{\mathbf{G}}}_c$ and $\hat{\dot{\mathbf{G}}}_s$ are the half-range Fourier cosine and sine transforms of $\dot{\mathbf{G}}$ defined for real ω by[42]

$$\hat{\dot{\mathbf{G}}}_c(\omega) = \int_0^\infty \cos(\omega s)\,\dot{\mathbf{G}}(s)\,ds, \tag{10.24a}$$

$$\hat{\dot{\mathbf{G}}}_s(\omega) = \int_0^\infty \sin(\omega s)\,\dot{\mathbf{G}}(s)\,ds. \tag{10.24b}$$

Now it follows from the Riemann-Lebesgue lemma that

$$\lim_{|\omega|\to\infty} [\mathring{\mathbf{G}} + \hat{\dot{\mathbf{G}}}_c(\omega)] = \mathring{\mathbf{G}}, \tag{10.25a}$$

$$\lim_{|\omega|\to\infty} [\hat{\dot{\mathbf{G}}}_s(\omega)] = \mathbf{0}. \tag{10.25b}$$

[42] The transforms $\hat{\dot{\mathbf{G}}}_c$ and $\hat{\dot{\mathbf{G}}}_s$ are well defined since $\dot{\mathbf{G}}$ is integrable on $\mathscr{R}^+$.

Hence, in the limit as the circular frequency ω becomes unbounded, there is no "phase lag" and the stress response becomes *elastic;* that is,

$$\lim_{|\omega|\to\infty} \left(\boldsymbol{S}(t) - \overset{\circ}{\boldsymbol{G}}[\widehat{\boldsymbol{E}}] \sin(\omega t)\right) = \boldsymbol{0}, \qquad t \in \mathscr{R}. \tag{10.26}$$

To exhibit clearly the nature of the phase lag consider a *scalar* relaxation function G:

$$\boldsymbol{S} = G \circledast \boldsymbol{E}. \tag{10.27}$$

Define the ***phase lag*** $\delta(\omega)$, for each circular frequency ω, through

$$\tan\delta(\omega) = \frac{\hat{\dot{G}}_s(\omega)}{\overset{\circ}{G} + \hat{\dot{G}}_c(\omega)}. \tag{10.28}$$

If the viscoelastic material is elastic, $\dot{G} \equiv 0$ and hence $\delta(\omega) = 0$ for all frequencies. For an arbitrary viscoelastic material

$$\lim_{|\omega|\to\infty} \delta(\omega) = 0$$

for any (integrable) $\dot{G}$. Thus, the phase lag is a reflection of the hereditary nature of viscoelastic materials, and $\delta(\omega)$ vanishes in the high frequency limit. In this sense, *viscoelastic materials behave elastically at high frequencies*. Finally, (10.23), (10.24) and (10.28) together imply, for scalar G,

$$\boldsymbol{S}(t) = \{[\overset{\circ}{G} + \hat{\dot{G}}_c(\omega)]^2 + [\hat{\dot{G}}_s(\omega)]^2\}^{\frac{1}{2}} \sin\left(\omega t - \delta(\omega)\right) \widehat{\boldsymbol{E}}, \qquad t \in \mathscr{R}. \tag{10.29}$$

We note that the property of stress relaxation expresses for hereditary laws a certain continuity property known as fading memory. This is a concept which is not limited to linear laws. However, a full treatment of this subject is beyond the scope of this article.[43]

We close this section with a few remarks on the symmetry of relaxation functions. GURTIN and HERRERA[44] have shown that if the material is ***dissipative*** in the sense that

$$\int_0^t \boldsymbol{S}(\tau) \cdot \dot{\boldsymbol{E}}(\tau)\, d\tau \geqq 0, \qquad t \in \mathscr{R}^+, \tag{10.30}$$

for all sufficiently smooth $\boldsymbol{E}$ with $\boldsymbol{E}(0) = \boldsymbol{0}$, then both the initial elasticity and the equilibrium modulus are positive semi-definite and symmetric. COLEMAN,[45] has shown using thermodynamic arguments that, not only are $\overset{\circ}{\boldsymbol{G}}$ and $\overset{\infty}{\boldsymbol{G}}$ symmetric, but also $\overset{\circ}{\boldsymbol{G}} - \overset{\infty}{\boldsymbol{G}}$ is positive semi-definite. SHU and ONAT[46] have also independently derived the symmetry of $\overset{\circ}{\boldsymbol{G}}$ based on the dissipation inequality (10.30). The assumption that equality holds in (10.30) for all times if and only if $\boldsymbol{E} \equiv 0$ defines, in the terminology of GURTIN and HERRERA,[47] a ***strongly dissipative material.*** Such materials can be shown to have positive definite and symmetric initial elasticity. It is important to note that neither the notion of dissipativity nor strong dissipativity implies the monotonicity often associated with relaxation functions. Further, one might wonder about the sym-

[43] For a precise and elegant discussion of the concept of fading memory see the essay by COLEMAN and MIZEL [1966, *5*].

[44] GURTIN and HERRERA [1965, *16*].

[45] COLEMAN [1964, *4*].

[46] SHU and ONAT [1965, *27*].

[47] GURTIN and HERRERA [1965, *16*].

metry constraints imposed on $\mathbf{G}(t)$ for $t>0$. Arguments for such symmetry usually appeal to Onsager type principles, but ROGERS and PIPKIN[48] rightfully question the adequacy of such arguments and suggest some experiments which might clarify the situation.

DAY[49] has greatly clarified this subject. He defines the ***time reversal*** $\tilde{\boldsymbol{E}}$ of a strain process $\boldsymbol{E}$ as $\tilde{\boldsymbol{E}}(-t)=\boldsymbol{E}(t)$ for every t. He then shows that $\mathbf{G}(t)$ is symmetric for every t if, and only if, the *work* $W(\boldsymbol{E})$:

$$W(\boldsymbol{E})=\int_{-\infty}^{\infty}\boldsymbol{S}(\tau)\cdot\dot{\boldsymbol{E}}(\tau)\,d\tau, \tag{10.31}$$

for every strain process $\boldsymbol{E}$ of compact support, is invariant with respect to time reversal; i.e., $W(\boldsymbol{E})=W(\tilde{\boldsymbol{E}})$.

GURTIN and STERNBERG[50] have proved a reciprocal theorem the validity of which requires the symmetry of $\mathbf{G}(t)$, for all times t, as a necessary and sufficient condition.

11. Relaxation and creep laws. Here we continue to consider a fixed point $\boldsymbol{x}$ in a body $\mathscr{B}$ and suppress all specific dependence upon it. If $\mathscr{B}$ is a viscoelastic body of relaxation type, there is a linear hereditary relaxation law $\mathscr{L}$ for $\mathscr{B}$ such that the stress process $\boldsymbol{S}$ is determined by the displacement process $\boldsymbol{u}$ through $\boldsymbol{S}=\mathscr{L}[\nabla\boldsymbol{u}]$ or, by virtue of the *postulate on rigid motions*, by the strain process $\boldsymbol{E}$ through $\boldsymbol{S}=\mathscr{L}[\boldsymbol{E}]$. Suppose there is a linear hereditary law $\overset{*}{\mathscr{L}}$ for $\mathscr{B}$ such that the strain process $\boldsymbol{E}$ is determined by the stress process $\boldsymbol{S}$ through $\boldsymbol{E}=\overset{*}{\mathscr{L}}[\boldsymbol{S}]$. In this event, the behavior of the body $\mathscr{B}$ is said to be ***linearly viscoelastic of creep type*** and $\overset{*}{\mathscr{L}}$ is called a ***creep law for*** $\mathscr{B}$. This section is devoted to an examination of the following question: when is a viscoelastic body of relaxation type also of creep type?

Let $\mathscr{L}$ be a relaxation law for $\mathscr{B}$ and let $\boldsymbol{S}$ be any restricted continuous stress process for $\mathscr{B}$; then the linear hereditary law $\overset{*}{\mathscr{L}}$ is a creep law for $\mathscr{B}$ if, and only if, the strain process $\boldsymbol{E}$ given by $\boldsymbol{E}=\overset{*}{\mathscr{L}}[\boldsymbol{S}]$ satisfies $\mathscr{L}[\boldsymbol{E}]=\boldsymbol{S}$. In other words: a viscoelastic body $\mathscr{B}$ of relaxation type with relaxation law $\mathscr{L}$ is also of creep type with creep law $\overset{*}{\mathscr{L}}$ if, and only if, there exists a linear hereditary law $\overset{*}{\mathscr{L}}$ such that the composition $\mathscr{L}\overset{*}{\mathscr{L}}$ is the identity map in the class of restricted continuous processes. In this case, we say that $\overset{*}{\mathscr{L}}$ ***is a creep law for*** $\mathscr{L}$.

Observe that "$\overset{*}{\mathscr{L}}$ is a creep law for $\mathscr{L}$" means "$\overset{*}{\mathscr{L}}$ is a *right inverse* of $\mathscr{L}$". This property is not necessarily symmetric. That is, if $\overset{*}{\mathscr{L}}$ is a creep law for $\mathscr{L}$, it does not generally follow that $\mathscr{L}$ is a relaxation law for $\overset{*}{\mathscr{L}}$ or, equivalently, that $\overset{*}{\mathscr{L}}$ is also a *left inverse* of $\mathscr{L}$. If $\overset{*}{\mathscr{L}}$ is both a right and left inverse of $\mathscr{L}$, then $\overset{*}{\mathscr{L}}$ is the *inverse* of $\mathscr{L}$ and we write $\mathscr{L}^{-1}$ for $\overset{*}{\mathscr{L}}$.

If $\overset{*}{\mathscr{L}}$ is a right inverse of $\mathscr{L}$, then $\mathscr{L}$ maps the class of restricted continuous processes *onto* itself. It follows that $\overset{*}{\mathscr{L}}$ is also a left inverse of $\mathscr{L}$ (hence $\overset{*}{\mathscr{L}}=\mathscr{L}^{-1}$) if, and only if, the map $\mathscr{L}$ is *one-to-one;* that is $\mathscr{L}[\boldsymbol{E}]=\mathbf{0}$ implies $\boldsymbol{E}=\mathbf{0}$.

---

[48] ROGERS and PIPKIN [1963, *20*]. See also ONSAGER [1931, *1*, *2*].
[49] DAY [1971, *1*].
[50] GURTIN and STERNBERG [1963, *7*].

We consider the problem of finding a right inverse $\overset{*}{\mathscr{L}}$ of $\mathscr{L}$ in the restricted context of Boltzmann laws. Very roughly speaking, we will show that if $\mathscr{L}$ is a Boltzmann relaxation law with nonsingular initial elasticity, then there is a Boltzmann creep law $\overset{*}{\mathscr{L}}$ for $\mathscr{L}$.

Let $\mathscr{L}_{\boldsymbol{G}}$ be a Boltzmann relaxation law for $\mathscr{B}$ with strong relaxation function $\boldsymbol{G}$. Thus, for each time t,

$$\boldsymbol{S}(t) = (\boldsymbol{G} \circledast \boldsymbol{E})(t) = \overset{\circ}{\boldsymbol{G}}[\boldsymbol{E}(t)] + \int_0^\infty \dot{\boldsymbol{G}}(s)[\boldsymbol{E}(t-s)]\,ds. \tag{11.1}$$

Suppose, in addition, that $\mathscr{L}_{\boldsymbol{J}}$ is a Boltzmann creep law for $\mathscr{B}$ with strong response function $\boldsymbol{J}$ called the ***creep compliance.*** Hence, for each time t,

$$\boldsymbol{E}(t) = (\boldsymbol{J} \circledast \boldsymbol{S})(t) = \overset{\circ}{\boldsymbol{J}}[\boldsymbol{S}(t)] + \int_0^\infty \dot{\boldsymbol{J}}(s)[\boldsymbol{S}(t-s)]\,ds. \tag{11.2}$$

We thus have the following assertion: given $\mathscr{L}_{\boldsymbol{G}}$, $\mathscr{L}_{\boldsymbol{J}}$ is a creep law for $\mathscr{L}_{\boldsymbol{G}}$ if, and only if, $\boldsymbol{J}$ satisfies the system[51]

$$\overset{\circ}{\boldsymbol{G}}\overset{\circ}{\boldsymbol{J}} = \boldsymbol{1} \tag{11.3a}$$

and, for almost every s in $\mathscr{R}^+$,

$$\overset{\circ}{\boldsymbol{G}}\dot{\boldsymbol{J}}(s) + \dot{\boldsymbol{G}}(s)\overset{\circ}{\boldsymbol{J}} + (\dot{\boldsymbol{G}} * \dot{\boldsymbol{J}})(s) = \boldsymbol{0}. \tag{11.3b}$$

Clearly (11.3a) holds if, and only if, the initial elasticity $\overset{\circ}{\boldsymbol{G}}$ is invertible in $[\mathscr{T}_{\text{sym}}]$; the inverse $\overset{\circ}{\boldsymbol{J}}$ is called the ***initial elastic compliance.*** Therefore the invertibility of $\overset{\circ}{\boldsymbol{G}}$ is a necessary condition that there be a creep law $\mathscr{L}_{\boldsymbol{J}}$ for the relaxation law $\mathscr{L}_{\boldsymbol{G}}$.

Set $\overset{\circ}{\boldsymbol{J}} = \overset{\circ}{\boldsymbol{G}}{}^{-1}$ or, equivalently, $\overset{\circ}{\boldsymbol{J}}{}^{-1} = \overset{\circ}{\boldsymbol{G}}$ and define functions $\boldsymbol{U}$ and $\boldsymbol{V}$ on $\mathscr{R}^+$ into $[\mathscr{T}_{\text{sym}}]$ by the alternatives

$$\boldsymbol{U}(s) = \dot{\boldsymbol{J}}(s)\,\overset{\circ}{\boldsymbol{J}}{}^{-1} \quad \text{or} \quad \overset{\circ}{\boldsymbol{J}}{}^{-1}\dot{\boldsymbol{J}}(s), \tag{11.4a}$$

$$\boldsymbol{V}(s) = -\overset{\circ}{\boldsymbol{G}}{}^{-1}\dot{\boldsymbol{G}}(s) \quad \text{or} \quad -\dot{\boldsymbol{G}}(s)\,\overset{\circ}{\boldsymbol{G}}{}^{-1}. \tag{11.4b}$$

Then (11.3) and (11.4) together imply that

$$\boldsymbol{U}(s) - (\boldsymbol{V} * \boldsymbol{U})(s) = \boldsymbol{V}(s) \tag{11.5}$$

for almost every s in $\mathscr{R}^+$ (for either alternative). Now since $\boldsymbol{G}$ is a strong response function, $\boldsymbol{V}$ is integrable on $\mathscr{R}^+$. We seek conditions on $\boldsymbol{V}$ in order that (11.5) have an integrable solution $\boldsymbol{U}$ on $\mathscr{R}^+$, in which case $\boldsymbol{J}$ will be a strong response function.

It is easily seen that (11.5) has the *formal* solution

$$\boldsymbol{U} = \sum_{n=0}^{\infty} \underbrace{\boldsymbol{V} * \boldsymbol{V} * \cdots * \boldsymbol{V}}_{n\text{-convolutions}}. \tag{11.6}$$

[51] The system (11.3) is equivalent to the statement:

$$\boldsymbol{G} \circledast \boldsymbol{J} = h\boldsymbol{1}.$$

where we formally use the $\circledast$ notation for the response functions $\boldsymbol{G}$ and $\boldsymbol{J}$. This follows at once from the properties of the Boltzmann operator discussed in Sect. 5. Cf. GURTIN and STERNBERG [1962, *10*, Eqs. (3.24), (3.25).

Now write $\|\mathbf{U}\|_1$ and $\|\mathbf{V}\|_1$ for

$$\|\mathbf{U}\|_1 = \int_0^\infty |\mathbf{U}(s)|_{[\mathscr{T}_{\text{sym}}]}\, ds, \tag{11.7a}$$

$$\|\mathbf{V}\|_1 = \int_0^\infty |\mathbf{V}(s)|_{[\mathscr{T}_{\text{sym}}]}\, ds. \tag{11.7b}$$

Then Fubini's theorem,[52] (11.6) and (11.7) imply

$$\|\mathbf{U}\|_1 \leqq \|\mathbf{V}\|_1 \sum_{n=0}^{\infty} \|\mathbf{V}\|_1^n. \tag{11.8}$$

Hence (11.6) defines an integrable function on $\mathscr{R}^+$ whenever the series in (11.8) converges. This series converges if, and only if,

$$\|\mathbf{V}\|_1 < 1 \tag{11.9}$$

in which case we obtain the estimate

$$\|\mathbf{U}\|_1 \leqq \left\{\frac{\|\mathbf{V}\|_1}{1-\|\mathbf{V}\|_1}\right\}. \tag{11.10}$$

Furthermore, the solution $\mathbf{U}$ is unique whenever (11.9) holds. For if there were two solutions, (11.5) implies their difference, say, $\hat{\mathbf{U}}$ must satisfy

$$\hat{\mathbf{U}}(s) = (\mathbf{V} * \hat{\mathbf{U}})(s), \tag{11.11}$$

for almost every s in $\mathscr{R}^+$. But (11.11) and Fubini's theorem imply the estimate

$$\|\hat{\mathbf{U}}\|_1 \leqq \|\mathbf{V}\|_1 \|\hat{\mathbf{U}}\|_1. \tag{11.12}$$

Finally, since (11.9) holds, (11.12) implies that $\hat{\mathbf{U}}$ vanishes almost everywhere on $\mathscr{R}^+$.

We have just outlined a proof of the

Inversion theorem for strong response functions.[53] *Let $\mathbf{G}$ be a strong relaxation function for the Boltzmann relaxation law $\mathscr{L}_{\mathbf{G}}$, and suppose that*

(i) *the initial elasticity $\mathring{\mathbf{G}}$ is invertible on $\mathscr{T}_{\text{sym}}$ and*

(ii) *either*

$$\int_0^\infty |\mathring{\mathbf{G}}^{-1}\dot{\mathbf{G}}(s)|_{[\mathscr{T}_{\text{sym}}]}\, ds < 1 \tag{11.13}$$

or

$$\int_0^\infty |\dot{\mathbf{G}}(s)\,\mathring{\mathbf{G}}^{-1}|_{[\mathscr{T}_{\text{sym}}]}\, ds < 1 \tag{11.14}$$

(or both). Then there exists a unique Boltzmann creep law $\mathscr{L}_{\mathbf{J}}$ for $\mathscr{L}_{\mathbf{G}}$ with strong creep compliance $\mathbf{J}$. Moreover, the creep compliance $\mathbf{J}$ is uniquely defined through

(iii)
$$\mathring{\mathbf{J}} = \mathring{\mathbf{G}}^{-1} \quad \textit{and} \tag{11.15}$$

(iv) *either*

$$\dot{\mathbf{J}}\mathring{\mathbf{J}}^{-1} = \sum_{n=0}^{\infty} (-1)^{n+1} \underbrace{\{(\mathring{\mathbf{G}}^{-1}\dot{\mathbf{G}}) * (\mathring{\mathbf{G}}^{-1}\dot{\mathbf{G}}) * \cdots * (\mathring{\mathbf{G}}^{-1}\dot{\mathbf{G}})\}}_{n\text{-convolutions}} \tag{11.16}$$

[52] See, for example, Riesz and Sz.-Nagy [1955, *6*].
[53] Cf. Gurtin and Sternberg [1962, *10*, Theorem 3.3].

or

$$\overset{\circ}{\mathbf{J}}^{-1}\mathbf{J} = \sum_{n=0}^{\infty} (-1)^{n+1} \underbrace{\{(\dot{\mathbf{G}}\overset{\circ}{\mathbf{G}}^{-1}) * (\dot{\mathbf{G}}\overset{\circ}{\mathbf{G}}^{-1}) * \cdots * (\dot{\mathbf{G}}\overset{\circ}{\mathbf{G}}^{-1})\}}_{n\text{-convolutions}}. \tag{11.17}$$

If $\mathscr{L}_{\mathbf{J}}$ is the creep law for $\mathscr{L}_{\mathbf{G}}$, we say that $\mathbf{J}$ is the ***creep compliance for*** $\mathbf{G}$. If $\mathscr{L}_{\mathbf{G}}$ is also a relaxation law for $\mathscr{L}_{\mathbf{J}}$, then we say $\mathbf{G}$ is the ***relaxation function for*** $\mathbf{J}$. Finally, if $\mathscr{L}_{\mathbf{G}}$ and $\mathscr{L}_{\mathbf{J}}$ are mutually inverse, we say that $\mathbf{G}$ *and* $\mathbf{J}$ *are mutually inverse.*

The inversion theorem for strong response functions has the following

Corollary. *If* $\mathbf{G}$ *satisfies the hypotheses of the theorem, then* $\mathscr{L}_{\mathbf{G}}$ *is invertible. That is,* $\mathscr{L}_{\mathbf{J}} = \mathscr{L}_{\mathbf{G}}^{-1}$ *and* $\mathbf{J}$ *is inverse to* $\mathbf{G}$.

If $\mathscr{L}_{\mathbf{J}} = \mathscr{L}_{\mathbf{G}}^{-1}$, we sometimes write $\mathbf{G}^{-1}$ for $\mathbf{J}$ so that $\mathbf{G}^{-1}\circledast = \mathscr{L}_{\mathbf{G}^{-1}} = \mathscr{L}_{\mathbf{G}}^{-1}$. Observe that[54]

$$\mathbf{G}^{-1} \circledast \mathbf{G} = \mathbf{G} \circledast \mathbf{G}^{-1} = h\mathbf{1}. \tag{11.18}$$

Recall that by virtue of (11.3a), condition (i) of the inversion theorem is *necessary* for the relaxation function $\mathbf{G}$ to have a creep compliance $\mathbf{J}$. Now if $\mathbf{G}$ and $\mathbf{J}$ are strong response functions and (11.3b) holds, it follows from Fubini's theorem that

$$\overset{\infty}{\mathbf{G}}\overset{\infty}{\mathbf{J}} = \mathbf{1}. \tag{11.19}$$

Hence, for a strong relaxation function $\mathbf{G}$ to have a strong creep compliance $\mathbf{J}$, it is also *necessary* that $\overset{\infty}{\mathbf{G}}$ be invertible in $[\mathscr{T}_{\text{sym}}]$, in which case $\overset{\infty}{\mathbf{J}} = \overset{\infty}{\mathbf{G}}^{-1}$. In summary, *for a viscoelastic body of relaxation type with a strong relaxation function* $\mathbf{G}$ *to be of creep with a strong creep compliance* $\mathbf{J}$, *it is necessary that both the initial and equilibrium response of* $\mathbf{G}$ *be non-singular.*[55]

As a simple example consider the relaxation function $\mathbf{G}$ given by

$$\mathbf{G} = G\mathbf{1} \tag{11.20}$$

where G is a scalar valued response function and $\mathbf{1}$ is the identity map in $[\mathscr{T}_{\text{sym}}]$. Condition (i) of the inversion theorem is then equivalent to

$$\overset{\circ}{G} \neq 0; \tag{11.21}$$

while condition (ii) of that theorem is equivalent to

$$\int_0^\infty \left| \frac{\dot{G}(s)}{\overset{\circ}{G}} \right| ds < 1. \tag{11.22}$$

Now conditions (11.21) and (11.22) together imply

$$\frac{\overset{\infty}{G}}{\overset{\circ}{G}} > 0; \tag{11.23}$$

that is, the scalar parts of the initial elasticity and the equilibrium modulus are either both positive or both negative. But (11.21) and (11.23) do not alone imply

---

[54] Here $\mathbf{G}^{-1}$ denotes the *inverse kernel* for $\mathbf{G}$ in the sense of relations (5.5) and (5.6). It should not be confused with the inverse function or the reciprocal of $\mathbf{G}$, which may not even be defined.

[55] Cf. the discussion at the end of Sect. 10 regarding dissipation inequalities.

(11.22). However, (11.21) and (11.23) together with the additional condition

$$\frac{\dot{G}(t)}{\overset{\circ}{G}} \leqq 0, \qquad t \in \mathscr{R}^+, \tag{11.24}$$

do imply (11.22). Thus, if the scalar relaxation function G is initially positive and decreases monotonically on $\mathscr{R}^+$ to a positive equilibrium value or if G is initially negative and increases monotonically on $\mathscr{R}^+$ to a negative equilibrium modulus, then (11.22) holds and G possesses a strong scalar creep compliance J. Furthermore, it follows from (iv) of the inversion theorem that (11.24) implies

$$\frac{\dot{J}(t)}{\overset{\circ}{J}} \geqq 0, \qquad t \in \mathscr{R}^+. \tag{11.25}$$

Thus, for example, if G is strictly positive and monotone decreasing on $\mathscr{R}^+$, then J is strictly positive and monotone increasing on $\mathscr{R}^+$.

By specializing this example we can see that condition (ii) of the inversion theorem cannot easily be relaxed. Let $\boldsymbol{G}$ be given by (11.20) with

$$G(t) = \exp(-t), \qquad t \in \mathscr{R}^+. \tag{11.26}$$

Then (11.21) is certainly satisfied but (11.22) is not:

$$\int_0^\infty \left| \frac{\dot{G}(s)}{\overset{\circ}{G}} \right| ds = \int_0^\infty e^{-t}\, dt = 1. \tag{11.27}$$

However, the formulae in (iv) of the inversion theorem remain meaningful in the sense that they define the Heaviside process

$$J(t) = 1 + t, \qquad t \in \mathscr{R}^+, \tag{11.28}$$

which is not a strong response function. Thus, the strong relaxation process $\boldsymbol{G} = G\boldsymbol{1}$, where G is given by (11.26), has a creep compliance $\boldsymbol{J} = J\boldsymbol{1}$, where J is given by (11.28), which does not admit an equilibrium response.

The latter example suggests that for response functions which do not necessarily admit an equilibrium response there is a counterpart to the inversion theorem for strong response functions.

Inversion theorem for response functions. *Let $\boldsymbol{G}$ be the relaxation function for the Boltzmann relaxation law $\mathscr{L}_{\boldsymbol{G}}$. Then there exists a unique Boltzmann creep law $\mathscr{L}_{\boldsymbol{J}}$, with creep compliance $\boldsymbol{J}$, for $\mathscr{L}_{\boldsymbol{G}}$ if, and only if, the initial elasticity $\overset{\circ}{\boldsymbol{G}}$ is invertible on $\mathscr{T}_{\text{sym}}$. Moreover, the creep compliance $\boldsymbol{J}$ is determined through* (iii) *and* (iv) *of the inversion theorem for strong response functions. If, in addition, $\boldsymbol{G}$ is of class H^N ($N \geqq 1$), then $\boldsymbol{J}$ is also of class H^N.*

The condition is clearly necessary. It is also sufficient. For the formulae of (iv) in the inversion theorem for strong response functions define a locally integrable function $\dot{\boldsymbol{J}}$ whenever $\dot{\boldsymbol{G}}$ is locally integrable.[56] That $\boldsymbol{J}$ is of class C^N ($N > 1$) whenever $\boldsymbol{G}$ is of class C^N follows from a classical result of VOLTERRA.[57]

As shown by a previous example, $\boldsymbol{J}$ may not exhibit an equilibrium response even if $\boldsymbol{G}$ does. Now the existence of the equilibrium creep response $\overset{\infty}{\boldsymbol{J}}$ is physically

[56] See MIKUSINSKI [1959, *7*].

[57] See GURTIN and STERNBERG [1962, *10*].

reasonable for solids. For if not, as the example above shows, (11.28) implies that a bounded stress process may produce an unbounded strain process, in which case our linear theory may not apply.

It is easily seen that $\mathbf{G}$ and $\mathbf{J}$ are mutually inverse whenever $\mathbf{J}$ is a creep compliance for $\mathbf{G}$. Thus, the inversion theorem for response functions just stated has precisely the same corollary as the inversion theorem for strong response functions; of course, the interpretations differ.

We have made no mention of the numerical techniques which might be employed in actual realization of the creep functions from experimental data on relaxation or conversely. For such a discussion the reader is referred to the work of HOPKINS and HAMMING.[58]

Let $\mathbf{K}$ denote the response function for a Boltzmann law. Consider the integral

$$\bar{\mathbf{K}}(\lambda) = \int_0^\infty e^{-\lambda s}\,\mathbf{K}(s)\,ds \tag{11.29}$$

which always converges in the complex λ-plane for all λ such that $\operatorname{Re}\lambda > \sigma_0$, where σ_0 is a real number or $+\infty$ depending on $\mathbf{K}$. If σ_0 is finite the function $\bar{\mathbf{K}}$, is called the ***Laplace transform of*** $\mathbf{K}$.

Now let $\mathbf{G}$ be a relaxation function with inverse creep compliance $\mathbf{J}$. Suppose further that $\bar{\mathbf{G}}$ and $\bar{\mathbf{J}}$ have a common half plane of convergence, say, $\operatorname{Re}\lambda > \sigma_0$. Then

$$\bar{\mathbf{G}}(\lambda)\,\bar{\mathbf{J}}(\lambda) = \frac{1}{\lambda^2}\,\mathbf{1} \tag{11.30}$$

in $[\mathscr{T}_{\mathrm{sym}}]$ for all λ such that $\operatorname{Re}\lambda > \sigma_0$. If $\mathbf{G}$ and $\mathbf{J}$ are strong response functions, then $\bar{\mathbf{G}}(\lambda)$ and $\bar{\mathbf{J}}(\lambda)$ are well defined for $\operatorname{Re}\lambda \geqq \sigma_0 > 0$.

Relations such as (11.30) indicate that the full power of the theories of Fourier and Laplace transforms may be used to augment the simple *sufficient* conditions for the inversion of response functions provided in this section. We do not imply, in the case of strong response functions, that our conditions are *necessary*. A summary of relationships between $\mathbf{G}$ and $\mathbf{J}$ which may be obtained through the Laplace transform has been given by GROSS.[59]

We close this section with a comment on Heaviside processes and convolutions. Let $\mathbf{G}$ be a relaxation function and let $\mathbf{J}$ be a creep compliance.

Then, for processes $\boldsymbol{E}$ and $\boldsymbol{S}$ in class H, the Boltzmann laws $\mathscr{L}_{\mathbf{G}}$ and $\mathscr{L}_{\mathbf{J}}$ have the form[60]

$$\boldsymbol{S} = \mathbf{G} \circledast \boldsymbol{E} = \mathring{\mathbf{G}}[\boldsymbol{E}] + \dot{\mathbf{G}} * \boldsymbol{E}, \tag{11.31a}$$

$$\boldsymbol{E} = \mathbf{J} \circledast \boldsymbol{S} = \mathring{\mathbf{J}}[\boldsymbol{S}] + \dot{\mathbf{J}} * \boldsymbol{S}. \tag{11.31b}$$

If $\mathscr{L}_{\mathbf{J}}$ is a creep law for the relaxation law $\mathscr{L}_{\mathbf{G}}$, then

$$\mathbf{G} \circledast (\mathbf{J} \circledast \boldsymbol{E}) = \boldsymbol{E} \tag{11.32}$$

for every strain process $\boldsymbol{E}$ of class H or, equivalently,

$$\mathring{\mathbf{G}}\mathring{\mathbf{J}}[\boldsymbol{E}] + (\mathring{\mathbf{G}}\dot{\mathbf{J}} + \dot{\mathbf{G}}\mathring{\mathbf{J}} + \dot{\mathbf{G}} * \dot{\mathbf{J}}) * \boldsymbol{E} = \boldsymbol{E}, \tag{11.33}$$

for every strain-process $\boldsymbol{E}$ in class H. This, in turn, is equivalent to the system (11.3).[61] Thus, we can obtain the properties of the creep compliance $\mathbf{J}$ for the

[58] HOPKINS and HAMMING [1957, *3*], [1958, *8*].

[59] GROSS [1953, *1*].

[60] Cf. Sects. 4 and 5. Especially Eq. (5.4).

[61] Compare with (5.5) and (5.6) of Sect. 5 and footnote 51, p. 24.

relaxation function G strictly within the context of Heaviside processes. In this case, of course, questions of strong response functions or equilibrium response never arise.

Thus, in the context of Heaviside processes, it is usually not essential to distinguish response functions which admit an equilibrium response from those which do not.

12. Isotropic materials. In this section we consider only viscoelastic bodies $\mathscr{B}$ which are isotropic. Recall the definition of material symmetry in Sect. 4 and its specialization to Boltzmann laws in Sect. 5. If the material symmetry group $\mathscr{G}$ is the full orthogonal group, then (10.10) holds for every symmetric tensor. It then follows[62] that there are *scalar* response functions μ and λ such that

$$\mathsf{G}(t)[\boldsymbol{T}] = 2\mu(t)\boldsymbol{T} + \lambda(t)(\operatorname{tr}\boldsymbol{T})\,\mathbf{1}, \qquad t\in\mathscr{R}, \tag{12.1}$$

for every symmetric tensor $\boldsymbol{T}$. This the direct extension to linear viscoelasticity of the classical result from linear elasticity. The viscoelastic response functions μ and λ correspond to the elastic Lamé moduli.

For isotropic materials the Boltzmann law (10.5) reduces to

$$\begin{aligned}\boldsymbol{S}(t) &= (2\mu \circledast \hat{\nabla}\boldsymbol{u})(t) + (\lambda \circledast \operatorname{tr}\hat{\nabla}\boldsymbol{u})(t)\,\mathbf{1}\\ &= 2\mathring{\mu}\,\hat{\nabla}\boldsymbol{u}(t) + \int_0^\infty 2\dot{\mu}(s)\,\hat{\nabla}\boldsymbol{u}(t-s)\,ds\\ &\quad + \mathring{\lambda}(\operatorname{tr}\hat{\nabla}\boldsymbol{u}(t))\,\mathbf{1} + \int_0^\infty \dot{\lambda}(s)\,(\operatorname{tr}\hat{\nabla}\boldsymbol{u}(t-s))\,ds\,\mathbf{1}, \qquad t\in\mathscr{R},\end{aligned} \tag{12.2}$$

where $\hat{\nabla}\boldsymbol{u} \equiv \operatorname{sym}\nabla\boldsymbol{u} = \boldsymbol{E}$. Of course, (12.2) has the equivalent abbreviated form

$$\boldsymbol{S} = 2\mu \circledast \boldsymbol{E} + (\lambda \circledast \operatorname{tr}\boldsymbol{E})\,\mathbf{1}. \tag{12.3}$$

It is important to observe the formal similarity of (12.3) and its elastic counterpart, which is obtained by merely replacing the operators $\mu\circledast$ and $\lambda\circledast$ in (12.3) with the corresponding elastic constants. Each of these expressions reduces to the classical elastic law for isotropic materials if, and only if, the response functions μ and λ are constant on $\mathscr{R}^+$:

$$\mu = \mathring{\mu} h \quad\text{and}\quad \lambda = \mathring{\lambda} h, \tag{12.4}$$

where $\mathring{\mu}$ and $\mathring{\lambda}$ are the scalar Lamé moduli and h is the scalar Heaviside unit step process.

Now both the stress process $\boldsymbol{S}$ and the strain process $\boldsymbol{E}$ possess the unique decompositions

$$\boldsymbol{S} = \boldsymbol{S}_0 + (\tfrac{1}{3}\operatorname{tr}\boldsymbol{S})\,\mathbf{1}, \qquad \operatorname{tr}\boldsymbol{S}_0 = 0, \tag{12.5}$$

$$\boldsymbol{E} = \boldsymbol{E}_0 + (\tfrac{1}{3}\operatorname{tr}\boldsymbol{E})\,\mathbf{1}, \qquad \operatorname{tr}\boldsymbol{E}_0 = 0. \tag{12.6}$$

The traceless symmetric tensors $\boldsymbol{S}_0$ and $\boldsymbol{E}_0$ are called the ***deviatoric stress*** and ***deviatoric strain tensors.*** Define scalar valued response functions G_1 and G_2 in terms of μ and λ by

$$\mathsf{G}_1 = 2\mu, \tag{12.7}$$

$$\mathsf{G}_2 = 2\mu + 3\lambda. \tag{12.8}$$

[62] See, for example, LTE, Sect. 22 (2).

Then, (12.7) and (12.8) imply that (12.3) has the equivalent form

$$S = G_1 \circledast E + [\tfrac{1}{3}(G_2 - G_1)] \circledast (\operatorname{tr} E)\, \mathbf{1}. \tag{12.9}$$

or, using (12.5) and (12.6)

$$S_0 = G_1 \circledast E_0, \tag{12.10}$$

$$\operatorname{tr} S = G_2 \circledast \operatorname{tr} E. \tag{12.11}$$

Following the discussion of GURTIN[63] or GURTIN and STERNBERG[64] we see that the scalar response function G_1 characterizes the behavior of the material in pure shear: *the response to a pure shearing motion is a pure shear stress;* while the scalar response function G_2 characterizes the behavior of the material in compression: *the response to a pure dilatation is a uniform pressure.* Hence the response function G_1 will be called the ***shear modulus***[65] and the response function G_2 will be called the ***compression modulus***.[66]

We may apply the inversion theorems of Sect. 11 directly to the "scalar" laws (12.10) and (12.11). Write G_α for either G_1 or G_2. Conditions (i) and (ii) of the inversion theorem for strong response functions then have the form

(i)
$$\mathring{G}_\alpha \neq 0, \tag{12.12}$$

(ii)
$$\int_0^\infty \left| \frac{\dot{G}_\alpha(s)}{\mathring{G}_\alpha} \right| ds < 1. \tag{12.13}$$

Thus, if G_α $(\alpha = 1, 2)$ satisfies condition (i) and (ii) above, then (12.10) and (12.11) may be inverted. That is, there are strong scalar creep compliances J_α for G_α $(\alpha = 1, 2)$ such that

$$E_0 = J_1 \circledast S_0, \tag{12.14}$$

$$\operatorname{tr} E = J_2 \circledast \operatorname{tr} S. \tag{12.15}$$

Moreover, the creep law for the relaxation law (12.9) is given in terms of J_1 and J_2 by

$$E = J_1 \circledast S + [\tfrac{1}{3}(J_2 - J_1)] \circledast (\operatorname{tr} S)\, \mathbf{1}. \tag{12.16}$$

This last assertion follows by direct computation or by applying the theorem on the characteristic values of the elasticity tensor[67] to the "elasticities" $\mathbf{G}(t)$ for each fixed t.

Observe that the condition $G_\alpha \neq 0$ $(\alpha = 1, 2)$ is equivalent to

$$2\mathring{\mu} \neq 0, \tag{12.17}$$

$$3\mathring{\lambda} + 2\mathring{\mu} \neq 0, \tag{12.18}$$

which suffices to guarantee the existence of the initial elastic compliance $\mathring{\mathbf{J}}$: $\mathring{\mathbf{J}} = \mathring{\mathbf{G}}^{-1}$. Indeed

$$\mathring{\mathbf{J}}[S] = \frac{1}{2\mathring{\mu}} \left[S - \frac{\mathring{\lambda}}{(2\mathring{\mu} + 3\mathring{\lambda})} (\operatorname{tr} S)\, \mathbf{1} \right], \tag{12.19}$$

---

[63] LTE, Sect. 22, (1) and (2).

[64] GURTIN and STERNBERG [1962, *10*].

[65] Some authors call μ the shear modulus and write G for μ.

[66] Sometimes the compression modulus G_2 is denoted by k. This is especially so in the elastic case.

[67] LTE, Sect. 22, (3).

or, equivalently,

$$\overset{\circ}{\boldsymbol{J}}[\boldsymbol{S}] = \frac{1}{\overset{\circ}{G}_1}\left[\boldsymbol{S} - \left(1 - \frac{1}{3}\frac{\overset{\circ}{G}_1}{\overset{\circ}{G}_2}\right)(\operatorname{tr}\boldsymbol{S})\,\mathbf{1}\right], \tag{12.20}$$

for every symmetric tensor $\boldsymbol{S}$.[68] Furthermore, (12.12) and (12.13) together imply that

$$\frac{\overset{\infty}{G}_\alpha}{\overset{\circ}{G}_\alpha} > 0 \qquad (\alpha = 1, 2); \tag{12.21}$$

that is, the initial elasticity and the equilibrium modulus have the same algebraic sign. Now (12.12) and (12.21) do not alone imply (12.13). However, if (12.12) and (12.21) hold and

$$\frac{\dot{G}_\alpha(t)}{\overset{\circ}{G}_\alpha} \leqq 0, \qquad t \in \mathscr{R}, \tag{12.22}$$

then (12.13) also holds. That is, if G_α is initially positive and decreases monotonically to positive $\overset{\infty}{G}_\alpha$ or if G_α is initially negative and increases monotonically to negative $\overset{\infty}{G}_\alpha$, then (12.13) holds and G_α possesses a strong inverse J_α.

Finally, there are analogous comments which apply to scalar response functions which are not necessarily strong. In this case, of course, there may be no equilibrium modulus.[69]

13. Additional properties of Boltzmann laws. Mechanical forcing. Let $\hat{\boldsymbol{S}}$ be a symmetric tensor and consider the periodic stress process $\boldsymbol{S}$

$$\boldsymbol{S}(t) = \hat{\boldsymbol{S}} \sin(\omega t), \qquad t \in \mathscr{R}. \tag{13.1}$$

If $\boldsymbol{G}$ is a strong relaxation function with a strong creep compliance $\boldsymbol{J}$, then there is a unique periodic strain process $\boldsymbol{E}$ such that

$$(\boldsymbol{G} \circledast \boldsymbol{E})(t) = \hat{\boldsymbol{S}} \sin(\omega t), \qquad t \in \mathscr{R}. \tag{13.2}$$

Indeed, $\boldsymbol{E}$ is given by[70]

$$\begin{aligned}\boldsymbol{E}(t) = (\boldsymbol{J} \circledast \boldsymbol{S})(t) &= \sin(\omega t)\left(\overset{\circ}{\boldsymbol{J}} + \int_0^\infty \cos(\omega s)\,\dot{\boldsymbol{J}}(s)\,ds\right)[\hat{\boldsymbol{S}}] \\ &\quad - \cos(\omega t)\left(\int_0^\infty \sin(\omega s)\,\dot{\boldsymbol{J}}(s)\,ds\right)[\hat{\boldsymbol{S}}], \qquad t \in \mathscr{R}.\end{aligned} \tag{13.3}$$

Now let h be the scalar Heaviside unit step process and consider the new stress process $\boldsymbol{S}'$ given by

$$\boldsymbol{S}'(t) = \hat{\boldsymbol{S}}\, h(t) \sin(\omega t). \tag{13.4}$$

Then the equation

$$(\boldsymbol{G} \circledast \boldsymbol{E})(t) = \hat{\boldsymbol{S}}\, h(t) \sin(\omega t), \qquad t \in \mathscr{R}, \tag{13.5}$$

also has a unique solution $\boldsymbol{E}'$ of Heaviside type given by

$$\begin{aligned}\boldsymbol{E}'(t) &= (\boldsymbol{J} \circledast \boldsymbol{S}')(t) \\ &= \sin(\omega t)\left(\overset{\circ}{\boldsymbol{J}} + \int_0^t \cos(\omega s)\,\dot{\boldsymbol{J}}(s)\,ds\right)[\hat{\boldsymbol{S}}] \\ &\quad - \cos(\omega t)\left(\int_0^t \sin(\omega s)\,\dot{\boldsymbol{J}}(s)\,ds\right)[\hat{\boldsymbol{S}}], \qquad t \in \mathscr{R}^+.\end{aligned} \tag{13.6}$$

[68] LTE, Sect. 22, (4).

[69] Cf. the comments regarding scalar response functions in Sect. 11. We have repeated some of these comments here for completeness only.

[70] Cf. Sect. 10.

Thus

$$\boldsymbol{E}(t)-\boldsymbol{E}'(t)=\sin(\omega t)\left(\int_t^\infty \cos(\omega s)\,\dot{\boldsymbol{J}}(s)\,ds\right)[\hat{\boldsymbol{S}}]$$
$$-\cos(\omega t)\left(\int_t^\infty \sin(\omega s)\,\dot{\boldsymbol{J}}(s)\,ds\right)[\hat{\boldsymbol{S}}],\qquad t\in\mathcal{R}^+, \tag{13.7}$$

from which it follows that

$$\lim_{t\to\infty}\left(\boldsymbol{E}'(t)-\boldsymbol{E}(t)\right)=\boldsymbol{0}. \tag{13.8}$$

We conclude that if the material is "forced" by the Heaviside stress process $\boldsymbol{S}'$, then, as $t\to\infty$, the corresponding strain process $\boldsymbol{E}'$ approaches the unique periodic strain process $\boldsymbol{E}$ corresponding to the periodic stress process $\boldsymbol{S}$. Of course inertial effects have not been considered, so that this material stability under periodic forcing does not necessarily represent a true dynamic stability. Moreover, the conclusions above hold for any bounded periodic forcing stress process and are not limited to the sinusoidal case.

The essential constitutive assumption which provides these conclusions is that the Boltzmann law $\mathcal{L}_{\boldsymbol{G}}$ have an inverse $\mathcal{L}_{\boldsymbol{J}}$ where the relaxation function $\boldsymbol{G}$ and the corresponding creep compliance $\boldsymbol{J}$ are both *strong*. In particular, if G is a strong *scalar* response function which has a positive initial response and which decreases monotonically on $\mathcal{R}^+$ to a positive equilibrium response, then G possesses a strong scalar creep compliance J. Indeed, $\overset{\circ}{G}\geqq\overset{\infty}{G}>0$ and $\dot{G}\leqq 0$ on $\mathcal{R}^+$ together imply

$$\int_0^\infty |\overset{\circ}{G}{}^{-1}|\,|\dot{G}(s)|\,ds<1. \tag{13.9}$$

Hence the inversion theorem of Sect. 10 may be applied. In this case the creep compliance J has positive initial response and increases monotonically to its equilibrium response; that is, $\overset{\infty}{J}\geqq\overset{\circ}{J}>0$ and $\dot{J}\geqq 0$ on $\mathcal{R}^+$.[71]

Summarizing this discussion we have the following

Theorem on asymptotic behavior. *Let $\boldsymbol{G}$ be a strong relaxation function which has a strong creep compliance $\boldsymbol{J}$. If $\boldsymbol{S}$ is a prescribed bounded periodic stress process, with period $T\geqq 0$, then there is a unique periodic strain process $\boldsymbol{E}$, with period $T\geqq 0$, such that*

$$\mathcal{L}_{\boldsymbol{G}}[\boldsymbol{E}]\equiv\boldsymbol{G}\circledast\boldsymbol{E}=\boldsymbol{S}. \tag{13.10}$$

Furthermore, if $\boldsymbol{S}'$ is the Heaviside stress process determined through $\boldsymbol{S}$ by

$$\boldsymbol{S}'=h\boldsymbol{S}, \tag{13.11}$$

then there is a unique Heaviside strain process $\boldsymbol{E}'$ such that

$$\mathcal{L}_{\boldsymbol{G}}[\boldsymbol{E}']\equiv\boldsymbol{G}\circledast\boldsymbol{E}'=\boldsymbol{S}'. \tag{13.12}$$

Finally, and most importantly, $\boldsymbol{E}'$ approaches $\boldsymbol{E}$ asymptotically as $t\to\infty$; that is,

$$\lim_{t\to\infty}\left[\boldsymbol{E}(t)-\boldsymbol{E}'(t)\right]=\boldsymbol{0}. \tag{13.13}$$

The results of this section are closely related, but not equivalent, to those of Sect. 10 regarding stress relaxation. In that section we showed that strain processes which were asymptotically the same produce stress processes which

[71] Cf. Sect. 11, inversion theorems, especially the discussion of the scalar case.

are asymptotically the same. In the present section we considered the "converse" problem. That is, if a viscoelastic material is stressed periodically, then the *only* strain process compatible with the periodic stress is also periodic. We have further shown that if the body is initially unstressed and then given a periodic forcing, the *only* compatible strain process is asymptotically periodic.

The theorem on asymptotic behavior, stated here for bounded periodic processes, may further be extended to include non-periodic processes with bounded histories.[72]

If periodic strains rather than periodic stresses are imposed and if the Boltzmann laws are (strongly) invertible, then the previous results and discussion remain valid with the roles of stress and strain, $\boldsymbol{G}$ and $\boldsymbol{J}$, and relaxation and creep interchanged.

Consider, as an example, the case of a scalar law for which the strain process E',

$$E'(t) = h(t)\sin(\omega t)\,\hat{E}, \qquad t\in\mathscr{R}, \tag{13.14}$$

is imposed. Then the results just obtained imply that the corresponding stress process S' is asymptotic, as $t\to\infty$, to the periodic stress process S given by

$$S(t) = \hat{E}\{\sin(\omega t)[\mathring{G} + \hat{\dot{G}}_c(\omega)] - \cos(\omega t)\,\hat{\dot{G}}_s(\omega)\}, \qquad t\in\mathscr{R}; \tag{13.15}$$

here $\hat{\dot{G}}_c(\omega)$ and $\hat{\dot{G}}_s(\omega)$ are the half-range Fourier cosine and sine transforms of $\dot{G}$:[73]

$$\hat{\dot{G}}_c(\omega) = \int_0^\infty \cos(\omega s)\,\dot{G}(s)\,ds, \tag{13.16a}$$

$$\hat{\dot{G}}_s(\omega) = \int_0^\infty \sin(\omega s)\,\dot{G}(s)\,ds. \tag{13.16b}$$

By virtue of the Fourier integral theorem it is possible to recover the relaxation function G, provided $\mathring{G}$ and either $\hat{\dot{G}}_c$ or $\hat{\dot{G}}_s$ are known. Indeed, since G is strong, it follows that

$$G(t) = \mathring{G} + \frac{2}{\pi}\int_0^\infty \frac{\sin(\omega t)}{\omega}\,\hat{\dot{G}}_c(\omega)\,d\omega, \qquad t\in\mathscr{R}^+, \tag{13.17a}$$

or, alternatively,

$$G(t) = \mathring{G} - \frac{2}{\pi}\int_0^\infty \frac{\cos(\omega t)-1}{\omega}\,\hat{\dot{G}}_s(\omega)\,d\omega, \qquad t\in\mathscr{R}^+. \tag{13.17b}$$

The function $\hat{\dot{G}}_c$ is called the ***dynamic modulus*** and $-\hat{\dot{G}}_s$ is called the ***loss modulus.*** Also, $\mathring{G} + \hat{\dot{G}}_c$ is known as the ***storage modulus***, while the function η, given by

$$\eta(\omega) = -\frac{1}{\omega}\,\hat{\dot{G}}_s(\omega), \tag{13.18}$$

---

[72] If the imposed stress process has bounded histories, then so does the corresponding strain process. The result then follows from the fact that the strong creep compliance $\boldsymbol{J}$ has the creep dual of the stress relaxation process. On the other hand, if the imposed stress process has unbounded but integrable histories, then so may the corresponding strain process. The asymptotic result may not then follow unless $\boldsymbol{J}$ has the creep dual of the stress relaxation property for such processes. Cf. Sect. 10.

[73] The assumption that G be strong guarantees the existence of the transforms $\hat{\dot{G}}_c$ and $\hat{\dot{G}}_s$. However, G itself may have no such transform. See also Sect. 10.

is called the ***dynamic viscosity.*** Additional relations among these functions and their creep duals have been collected by GROSS.[74]

As we have already seen, the conditions $\overset{\circ}{G} \geqq \overset{\infty}{G} > 0$ and $\dot{G} \leqq 0$ on $\mathscr{R}^+$ guarantee that G has a strong inverse J. If, in addition, $\dot{G}$ is monotone, in which case G is positive, decreasing, and concave upward on $\mathscr{R}^+$, then $\hat{\dot{G}}_c(\omega)$ and $\omega\hat{\dot{G}}_s(\omega)$ are non-positive for *all* real ω. Thus, in particular, the power expended per cycle is never negative.[75] Indeed, the power expended in a time interval of length $T = 2\pi/\omega$ is given in terms of the loss modulus $-\hat{\dot{G}}_s$ by

$$\int_0^T S(t)\,\dot{E}(t)\,dt = -\pi \hat{E}^2 \hat{\dot{G}}_s(\omega) \geqq 0. \tag{13.19}$$

14. Differential operator laws. Linear hereditary laws are essentially integral laws, a special case of which is the Boltzmann law. However, for those bodies whose microstructure is imagined mechanically equivalent to finite networks of linearly elastic and viscous elements (springs and dashpots), the stress and strain processes are related through differential operator laws.[76] It will turn out that this hypothesis is somewhat restrictive, for, in a sense, every differential operator law is a Boltzmann law but not conversely. The presentation of this and the next two sections follows closely the work of GURTIN and STERNBERG.[77]

As usual we consider a fixed point $\boldsymbol{x} \in \mathscr{B}$ and suppress all dependence upon it. For simplicity in presentation and without essential loss in generality, we consider only scalar laws.

In this section, since we consider Heaviside processes, we use the convolution notation. It is therefore not essential that response functions admit an equilibrium response. Throughout this section E and S denote *scalar* valued strain and stress processes of class H^N $(N \geqq 0)$. Recall the proposition and theorem in Sect. 8 regarding the relation between restricted continuous processes and processes of Heaviside class.

For any nonnegative integer N, define autonomous differential operators P and Q on sufficiently smooth scalar functions ϕ by

$$P\phi = \sum_{k=0}^{N} p_k\, \phi^{(k)}, \tag{14.1a}$$

$$Q\phi = \sum_{k=0}^{N} q_k\, \phi^{(k)}, \tag{14.1b}$$

where p_k and q_k $(k = 0, 1, \ldots, N)$ are scalar constants and $\phi^{(k)}$ denotes the kth derivative of ϕ. The pair $\langle P, Q\rangle$ is said to be of ***order*** N if, and only if, either $p_N \neq 0$ or $q_N \neq 0$. In particular, if $p_N \neq 0$ or $q_N \neq 0$ the pair $\langle P, Q\rangle$ is said to be of order N and ***relaxation*** or ***creep type.***

Let $\langle P, Q\rangle$ be a pair of differential operators of order $N \geqq 1$ and let G be a process of class H^N. Then the pair $\langle P, Q\rangle$ ***belongs to the relaxation function***

---

[74] GROSS [1953, *1*] Observe that conditions which guarantee the existence of Fourier transforms for G itself may not be desirable. If, say, G were integrable, then its transform exists; however, the equilibrium modulus must then vanish.

[75] Cf. Sect. 10. Notice also that these conditions on G guarantee that G has the stress relaxation property for all classes of processes considered in this article.

[76] For a more complete discussion of theories based upon this hypothesis, see BLAND [1960, *2*].

[77] GURTIN and STERNBERG [1962, *10*, Sect. 4].

$\boldsymbol{G}$ if, and only if, every pair of processes E and S of class H^N related through the Boltzmann relaxation law

$$S = \boldsymbol{G} \circledast E = \mathring{\boldsymbol{G}}[E] + \dot{\boldsymbol{G}} * E \tag{14.2}$$

satisfies the ***differential operator law***

$$\sum_{k=0}^{N} p_k S^{(k)} = \sum_{k=0}^{N} q_k E^{(k)} \tag{14.3}$$

on $\mathscr{R}^{++}$ together with the ***initial conditions***[78]

$$\sum_{r=k}^{N} p_r \mathring{S}^{(r-k)} = \sum_{r=k}^{N} q_r \mathring{E}^{(r-k)}, \qquad k = 1, 2, \ldots, N. \tag{14.4}$$

Conversely, the relaxation function $\boldsymbol{G}$ ***belongs to the pair*** $\langle P, Q\rangle$ if, and only if, every pair of processes E and S of class H^N satisfying the differential operator law (14.3) on $\mathscr{R}^{++}$, and the initial conditions (14.4) is related through the Boltzmann relaxation law (14.2). Observe that if E and S were also of class C^N the initial conditions are always satisfied. Of course, such processes E and S are restricted.

To discuss the relationship between differential and integral laws it suffices to use "test" processes in class H. The connection between these processes and restricted continuous processes as well as the significance of the initial conditions is exhibited through the following

Theorem.[79] *Let the pair $\langle P, Q\rangle$ be of order $N \geqq 1$ and suppose E and S are processes of class H^N defined as follows: there are sequences of processes $\{E_n\}$ and $\{S_n\}$ such that*

(i) *for each $n = 1, 2, \ldots$, E_n and S_n are Heaviside processes of class C^N,*

(ii) *the sequences $\{E_n\}$ and $\{S_n\}$ are uniformly bounded on closed and bounded neighborhoods of zero,*

(iii) *for each $n = 1, 2, \ldots$, the differential operator law is satisfied;*

$$\sum_{k=0}^{N} p_k S_n^{(k)} = \sum_{k=0}^{N} q_k E_n^{(k)}, \tag{14.5}$$

(iv) *$E_n \to E$ and $S_n \to S$ as $n \to \infty$ uniformly on closed and bounded subsets not containing zero.*

Then the pair of processes E and S satisfies the differential operator law (14.3) *on $\mathscr{R}^{++}$ and the initial conditions* (14.4).

Conversely, suppose E and S are of class H^N and satisfy the differential operator law (14.3) *on $\mathscr{R}^{++}$ and the initial conditions* (14.4). *Then there exist sequences of processes $\{E_n\}$ and $\{S_n\}$ which satisfy* (i)–(iv).

A previous theorem (Sect. 8) shows that every pair of discontinuous processes E and S of class H related through a Boltzmann law is the limit of sequences of pairs of restricted continuous processes related through the same Boltzmann law, and conversely. The above theorem shows that this is not the case with differential operator laws. Indeed, *only* those pairs E and S of class H^N ($N \geqq 1$) satisfying the differential operator law (14.3) on $\mathscr{R}^{++}$ which meet the initial condition (14.4) may be approximated by sufficiently smooth restricted continuous processes satisfying the same differential operator law.

[78] Cf. Boley and Weiner [1960, *3*, Chap. 15] for a discussion of such initial conditions for differential operator laws.

[79] Gurtin and Sternberg [1962, *10*, Theorem 4.2].

The next result exhibits a sufficient condition that a Boltzmann relaxation law reduce to a differential operator law.

Theorem.[80] *Let* G *be a relaxation function of class* H^N $(N \geqq 1)$ *and suppose there exist constants* q_0 *and* p_k $(k=0, 1, 2, \ldots, N)$ *with* $p_N \neq 0$ *such that* G *satisfies the differential equation* ***(generating equation)***

$$\sum_{k=0}^{N} p_k \, \mathsf{G}^{(k)} = q_0 \tag{14.6}$$

on $\mathscr{R}^{++}$. *Then the pair* $\langle P, Q\rangle$ *of order* N *and relaxation type belongs to the relaxation function* G, *where* Q *is determined through*

$$q_k = \sum_{r=k}^{N} p_r \, \mathring{\mathsf{G}}^{(r-k)}, \qquad k = 1, 2, 3, \ldots, N. \tag{14.7}$$

The condition that G satisfy the generating equation (14.6) on $\mathscr{R}^{++}$ and meet the initial condition (14.7) is also necessary that a pair $\langle P, Q\rangle$ of order $N \geqq 1$ belongs to G. More precisely we have

Theorem.[81] *Let* G *be of class* H^N $(N \geqq 1)$ *and suppose that* $\langle P, Q\rangle$ *is a pair of differential operators of order* $N \geqq 1$ *which belongs to* G. *Then*

(i) G *is of class* H^∞ *and satisfies the generating equation* (14.6) *on* $\mathscr{R}^{++}$ *and meets the initial condition* (14.7) *and*

(ii) *the pair* $\langle P, Q\rangle$ *is of relaxation type.*

The previous two theorems provide necessary and sufficient conditions that a Boltzmann relaxation law reduces to a differential operator law. Clearly not every Boltzmann relaxation law reduces to a differential operator law of relaxation type. However, the next result asserts that every differential operator law of relaxation type corresponds to a Boltzmann relaxation law.

Theorem.[82] *Let* $\langle P, Q\rangle$ *be a pair of differential operators of order* $N \geqq 1$. *Then there exists a relaxation function* G *of class* H^N *that belongs to* $\langle P, Q\rangle$ *if, and only if,* $\langle P, Q\rangle$ *is of relaxation type. Moreover, if there be such a relaxation function* G *it is uniquely determined, of class* H^∞, *and satisfies the generating equation* (14.6) on $\mathscr{R}^{++}$ *together with the initial conditions:*

$$\mathring{\mathsf{G}} = \frac{q_1}{p_1} \quad \text{for } N = 1, \tag{14.8a}$$

$$\mathring{\mathsf{G}} = \frac{q_N}{p_N}, \quad \mathring{\mathsf{G}}^{(k)} = \frac{1}{p_N}\left[q_{N-k} - \sum_{r=0}^{k-1} p_{N-k+r}\, \mathring{\mathsf{G}}^{(r)}\right], \quad k = 1, 2, \ldots, (N-1), \quad \text{for } N > 1. \tag{14.8b}$$

The preceding results may be combined to yield the following

Theorem.[83] *Let* $\langle P, Q\rangle$ *be a pair of differential operators of order* $N \geqq 1$ *and let* G *be of class* H^N $(N \geqq 1)$. *Then* $\langle P, Q\rangle$ *belongs to the relaxation function* G *if, and only if,* G *belongs to* $\langle P, Q\rangle$.

---

[80] GURTIN and STERNBERG [1962, *10*, Theorem 4.1].
[81] GURTIN and STERNBERG [1962, *10*, Theorem 4.3].
[82] GURTIN and STERNBERG [1962, *10*, Theorem 4.4].
[83] GURTIN and STERNBERG [1962, *10*, Theorem 4.5].

This result implies that "belonging to" is a symmetric relationship. Suppose $\langle P, Q\rangle$ belongs to G. Then every pair of processes E and S of class H^N which satisfies the Boltzmann relaxation law (14.2) also satisfies the differential operator law (14.3) on $\mathscr{R}^{++}$ and meets the initial conditions (14.4). Conversely, every pair of processes E and S of class H^N which satisfies (14.3) on $\mathscr{R}^{++}$ and (14.4) also satisfies (14.2) on $\mathscr{R}^{++}$. A similar statement follows if we suppose that G belongs to $\langle P, Q\rangle$.

Standard results from the theory of ordinary differential equations guarantee that for every strain process E of class H^N there exists a unique stress process S such that E and S satisfy a differential operator law of relaxation type.

In fact we have the

Theorem.[84] *Let $\langle P, Q\rangle$ be a pair of differential operators of order $N \geqq 1$. If $\langle P, Q\rangle$ is of relaxation type then, for every process E of class H^N, there is a unique process S of class H^N such that the pair E and S satisfies the differential operator law* (14.3) *on $\mathscr{R}^{++}$ and meets the initial conditions* (14.4).

The preceding five theorems regarding relaxation laws of Boltzmann type and differential operator laws of relaxation type have their duals for laws of creep type. To obtain these dual results it is only necessary to reverse simultaneously the roles of strain E and stress S, the differential operators P and Q, the relaxation function G and the creep compliance J, and to replace the word "relaxation" by "creep" whenever it appears.

To extend the above results to include tensor laws, we observe that all assertions remain valid if the scalar constants p_k and q_k $(k=0, 1, 2, \ldots, N)$ are interpreted as linear maps in $\mathscr{T}_{\text{sym}}$. In this case, the restriction "$p_N \neq 0$ or $q_N \neq 0$" should be replaced by "p_N or q_N is invertible" and $1/p_N$ or $1/q_N$ should be replaced by p_N^{-1} or q_N^{-1} whenever they appear.

Observe that all results thus far concerned differential operator laws of order $N \geqq 1$. If $\langle P, Q\rangle$ is of order zero we see that (14.3) reduces to

$$p_0 S = q_0 E \tag{14.9}$$

on $\mathscr{R}^{++}$, where either $p_0 \neq 0$ or $q_0 \neq 0$. To avoid the obvious degeneracies which may result, we suppose that differential operator pairs $\langle P, Q\rangle$ of order zero are simultaneously of relaxation and creep type. In this case, for example, p_0 and q_0 are not zero, and the "differential" operator law (14.9) on $\mathscr{R}^{++}$ is equivalent to

$$S = G \circledast E \quad \text{or} \quad E = J \circledast S, \tag{14.10}$$

where G and J are given by

$$G = \frac{q_0}{p_0}\, h \quad \text{and} \quad J = \frac{p_0}{q_0}\, h. \tag{14.11}$$

15. Relaxation times and differential operator laws. Let $\langle P, Q\rangle$ be a differential operator pair of order $N \geqq 1$ which belongs to a relaxation function G of class C^N on $\mathscr{R}^{+}$. A previous result then shows that G satisfies the generating equation (14.6) on $\mathscr{R}^{++}$ and meets the initial conditions (14.7). Define N scalar functions G_k $(k=1, 2, \ldots, N)$ by

$$G_k = G^{(k-1)}, \qquad k = 1, 2, \ldots, N, \tag{15.1}$$

and define N constants f_k $(k=1, 2, \ldots, N)$ by

$$f_k = 0, \qquad k = 1, 2, \ldots, (N-1), \qquad f_N = \frac{q_0}{p_N}. \tag{15.2}$$

[84] Gurtin and Sternberg [1962, *10*, Theorem 4.6].

Let (A_{jk}) $(j, k=1, 2, \ldots, N)$ be the entries in an $N\times N$ matrix A defined by

$$A=(A_{jk})=\left\{\begin{array}{ll} 1:\ j=k-1, & 1\leqq j\leqq N-1 \\ 0:\ j\neq k-1, & 1\leqq j\leqq N-1 \\ -\dfrac{p_{k-1}}{p_N}:\ j=N, & 1\leqq k\leqq N \end{array}\right\}. \tag{15.3}$$

Then (14.6) is equivalent to the first order system

$$\dot{G}_j(s)=\sum_{k=1}^{N} A_{jk}\,G_k(s)+f_j, \qquad 1\leqq j\leqq N, \tag{15.4}$$

for $s\in\mathscr{R}^{++}$. The ***spectrum*** of the differential operator law $\langle P, Q\rangle$ consists of the N eigenvalues of the matrix A.

Recall that in this and the preceding section we have not insisted that the relaxation function G be strong and, hence, admit the equilibrium response $\overset{\infty}{G}$. Since solutions of (15.4) may be unbounded or oscillatory, we see that any condition implying the existence of $\overset{\infty}{G}$ would have placed additional restrictions on the results obtained in the preceding section. Moreover, in a theory based on processes whose histories have a common fixed support (such as Heaviside processes) such a condition does not appear without additional assumption. We point out that a sufficient, but not necessary, condition that G be a solution of (15.4) on $\mathscr{R}^{++}$ such that $\overset{\infty}{G}$ exists is that the elements in the spectrum of $\langle P, Q\rangle$ have negative real parts, in which case $\overset{\infty}{G}=0$. In this case, the negative reciprocals of the real parts of the eigenvalues of A are called ***relaxation times***. Further, if zero is a simple eigenvalue of A the corresponding relaxation time is said to be infinite.

Consider a law $\langle P, Q\rangle$ whose spectrum has the property that each member λ satisfies $\operatorname{Re}\lambda<\sigma_0<0$ for some real number σ_0. Then there is a constant $c>0$ such that

$$|G(t)|\leqq c\exp(\sigma_0 t), \qquad s\in\mathscr{R}^{+}, \tag{15.5}$$

where G is the relaxation function belonging to $\langle P, Q\rangle$. It follows that $-1/\sigma_0$ is greater than every relaxation time and, if $t\geqq -1/\sigma_0$, then (15.5) implies that

$$|G(t)|\leqq c\,\frac{1}{e} \qquad (e=\exp(1)). \tag{15.6}$$

Recall that if E is of class H^N with $E(t)=1$, for $t\geqq 1$, then the stress S corresponding to E satisfies $S(t)=G(t)$, for $t\geqq 0$. Hence if $t\geqq -1/\sigma_0$, then

$$|S(t)|\leqq c\,\frac{1}{e} \qquad (e=\exp(1)) \tag{15.7}$$

and, moreover,

$$\lim_{t\to\infty} S(t)=0. \tag{15.8}$$

If the differential operator pair belongs to a creep compliance J, then the above comments all have creep duals. To obtain these creep duals, simply replace the term *relaxation* by *creep* and interchange the roles of G and J and P and Q wherever they appear. In this context, the negative reciprocals of the eigenvalues of A are called ***retardation times.***

16. Special differential operators. In this section we focus upon some particular differential operator laws of classical interest.

The pair $\langle P, Q\rangle$ is said to be ***elastic*** if, and only if, it is of order $N \geqq 0$ and there is a nonzero constant c such that $P = cQ$; that is, $p_k = c\, q_k$, $k = 1, 2, 3, \ldots, N$. We then see that $\langle P, Q\rangle$ is of relaxation and creep type with relaxation function G and creep compliance J given by

$$\mathsf{G} = c\, h, \qquad \mathsf{J} = \frac{1}{c}\, h. \tag{16.1}$$

Thus G and J are constant on $\mathscr{R}^+$, which justifies the term *elastic* for such a pair.

If $N = 1$, we recover the classical one-dimensional viscoelastic materials. The differential operator law is said to be ***standard*** if the pair $\langle P, Q\rangle$ is of order $N = 1$. In particular, $\langle P, Q\rangle$ is a pair of ***Maxwell operators***[85] if, and only if,

$$p_0 \neq 0, \qquad p_1 = 1, \qquad q_0 = 0, \qquad q_1 \neq 0 \tag{16.2}$$

and $\langle P, Q\rangle$ is a pair of ***Kelvin*** or ***Kelvin-Voigt operators***[86] if, and only if,

$$p_0 = 1, \qquad p_1 = 0, \qquad q_0 \neq 0, \qquad q_1 \neq 0. \tag{16.3}$$

The results listed in Sect. 14 show that for standard differential operators of relaxation type the relaxation function G is given by

$$\mathsf{G}(t) = \begin{cases} \dfrac{q_0}{p_0} + \left(\dfrac{q_1}{p_1} - \dfrac{q_0}{p_0}\right) \exp\left(-\dfrac{p_0}{p_1}\, t\right) : p_0 \neq 0 \\ \dfrac{q_0}{p_1}\, t + \dfrac{q_1}{p_1} : p_0 = 0 \end{cases}, \qquad t \in \mathscr{R}^+. \tag{16.4}$$

The dual statement of (16.4) provides the creep compliance J belonging to a pair of standard differential operators.

From (16.4) and its dual we obtain, for a pair of Maxwell operators

$$\mathsf{G}(t) = q_1 \exp(-p_0\, t), \qquad t \in \mathscr{R}^+, \tag{16.5a}$$

$$\mathsf{J}(t) = \frac{1}{q_1}\,(1 + p_0\, t), \qquad t \in \mathscr{R}^+. \tag{16.5b}$$

Similarly, for a pair of Kelvin operators

$$\mathsf{J}(t) = \frac{1}{q_0}\left[1 - \exp\left(\frac{q_0}{q_1}\, t\right)\right], \qquad t \in \mathscr{R}^+, \tag{16.6}$$

Observe that the creep compliance (16.5b) for a pair of Maxwell operators is unbounded. This is consistent with the inversion theorems of Sect. 11. Observe also that a Kelvin body is of creep type *only* and has no corresponding relaxation function G. Indeed, the Kelvin body exhibits no initial elastic compliance ($\mathring{\mathsf{J}} = 0$) and, hence, fails to satisfy the necessary condition ($\mathring{\mathsf{J}} \neq 0$) for inversion.[87] Furthermore, the present values of the stress and strain do not depend specifically upon one another.

---

[85] Maxwell [1868, *1*].

[86] Meyer [1874, *2*], [1878, *2*], Thomson (Lord Kelvin) [1875, *1*], Voigt [1890, *1*], [1892, *2*]. For a more complete discussion of these special types of operators, see Bland [1960, *2*] and Gurtin and Sternberg [1962, *10*].

[87] Cf. Sect. 11 (inversion theorems).

Let 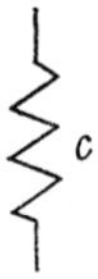 c be a linearly elastic element (spring) with elasticity $c > 0$ and let 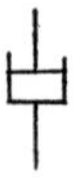 ν be a linearly viscous element (dashpot) with viscosity $\nu > 0$. Let S and E denote one-dimensional stress and strain processes. Then the differential operator laws for springs and dashpots are

$$S^{(0)} = c\,E^{(0)} \tag{16.7}$$

and

$$S^{(0)} = \nu\,E^{(1)} \tag{16.8}$$

on $\mathscr{R}^{++}$. A ***Maxwell element*** is given by

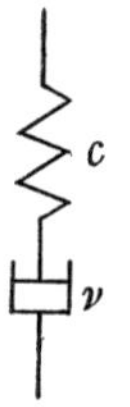

(series) (16.9)

and its differential operator law is

$$S^{(1)} + \frac{c}{\nu} S^{(0)} = c\,E^{(1)} \tag{16.10}$$

on $\mathscr{R}^{++}$, which conforms with (16.2) with $q_1 = c$ and $p_0 = c/\nu$. A ***Kelvin*** or ***Kelvin-Voigt element*** is given by

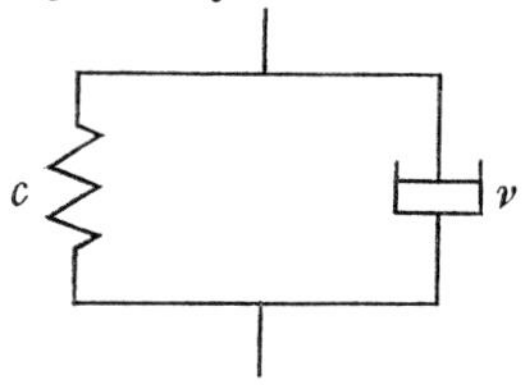

(parallel) (16.11)

and its differential operator law is

$$S^{(0)} = \nu\,E^{(1)} + c\,E^{(0)} \tag{16.12}$$

on $\mathscr{R}^{++}$, which conforms with (16.3) with $q_0 = c$ and $q_1 = \nu$. Notice that an element such as (16.11) can permit no (finite) stress corresponding to the discontinuity of a Heaviside (step) strain process.

Two other standard models are[88]

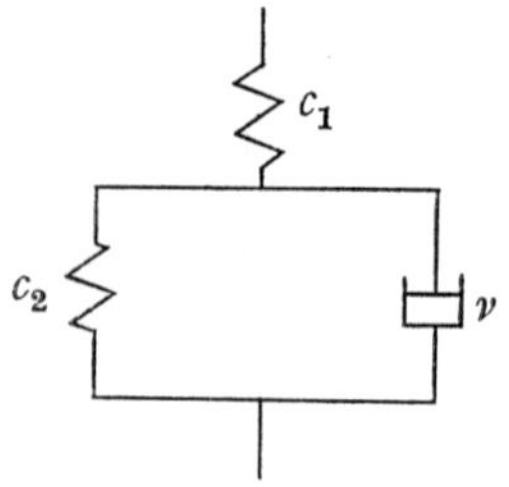

and

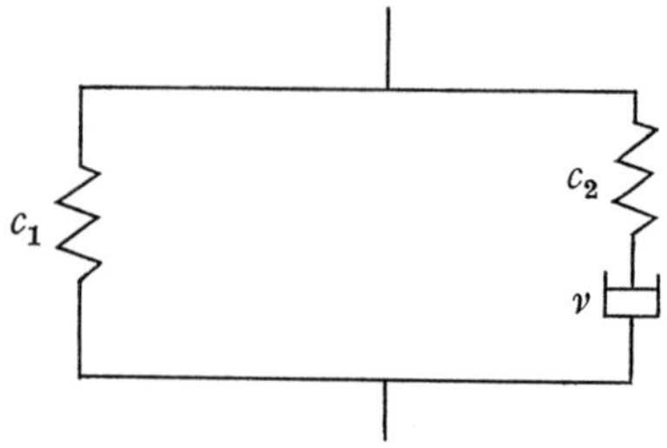

(16.13)

[88] It should be noted that the mathematical simplicity of the differential operator formulation often leaves much to be desired from a physical viewpoint. In particular the "discrete element" models discussed here generally represent real material dynamic behavior for only rather limited ranges of frequencies and hence should be used with caution. See KOLSKY and SHI [1958, *10*].

and their standard differential operator laws are

$$S^{(1)} + \frac{c_1 + c_2}{\nu} S^{(0)} = c_1 E^{(1)} + \frac{c_1 c_2}{\nu} E^{(0)} \tag{16.14}$$

and

$$S^{(1)} + \frac{c_2}{\nu} S^{(0)} = (c_1 + c_2) E^{(1)} + \frac{c_1 c_2}{\nu} E^{(0)}. \tag{16.15}$$

Returning to the Maxwell element, we see from (15.3), (16.2), (16.9), and (16.10) that the relaxation time $\tau = 1/p_0 = \nu/c_0$, where c is the elasticity of the spring and ν is the viscosity of the dashpot. Then the relaxation function G is

$$G(t) = c\, e^{-(c/\nu)t} = c\, e^{-t/\tau}, \qquad t \in \mathscr{R}^+. \tag{16.16}$$

Now consider N such elements in parallel with elasticities c_n and viscosities ν_n ($n = 1, 2, 3, \ldots, N$). Then the relaxation function G_N is given by

$$G_N(t) = \sum_{n=1}^{N} c_n\, e^{-t/\tau_n}, \qquad t \in \mathscr{R}^+, \tag{16.17}$$

where $\tau_n = c_n/\nu_n$ ($n = 1, 2, 3, \ldots, N$). If this model is further generalized in such a way that $N \to \infty$, then the relaxation function G_∞ is given by

$$G_\infty(t) = \sum_{n=1}^{\infty} c_n\, e^{-t/\tau_n}, \qquad t \in \mathscr{R}^+. \tag{16.18}$$

A necessary and sufficient condition that G_∞ be well defined on $\mathscr{R}^+$ by (16.18) is that the series $\sum_{n=1}^{\infty} c_n$ converge. Of course $\tau_n > 0$ for all $n = 1, 2, 3, \ldots$. We always assume, without loss in generality, that $\tau_1 < \tau_2 < \cdots < \tau_n < \cdots$.

Finally, it is possible to consider a distributed system of Maxwell elements in parallel with the distribution of elasticities determined by a function α which is increasing and bounded on $\mathscr{R}^+$. In this case, the generalization of G_∞ in (16.18) is given by the Stieltjes integral[89]

$$G(t) = \int_0^\infty e^{-t/\tau}\, d\alpha(\tau), \qquad t \in \mathscr{R}^+. \tag{16.19}$$

There is a dual argument for the generalization of a sequence of Kelvin-Voigt elements in series. Simply replace *relaxation function* by *creep compliance, elasticity* by *compliance,* and *relaxation time* by *retardation time* in the above statements. Furthermore, for Kelvin-Voigt elements in series, Eqs. (16.16)–(16.19) have the creep duals

$$\begin{aligned} J(t) &= \frac{1}{c}\left[1 - e^{-(c/\nu)\,t}\right] \\ &= \frac{1}{c}\left[1 - e^{-t/\tau}\right], \qquad t \in \mathscr{R}^+. \end{aligned} \tag{16.20}$$

$$J_N(t) = \sum_{n=1}^{N} \frac{1}{c_n}\left[1 - e^{-t/\tau_n}\right], \qquad t \in \mathscr{R}^+, \tag{16.21}$$

$$J_\infty(t) = \sum_{n=1}^{\infty} \frac{1}{c_n}\left[1 - e^{-t/\tau_n}\right], \qquad t \in \mathscr{R}^+, \tag{16.22}$$

$$J(t) = \int_0^\infty \left[1 - e^{-t/\tau}\right] d\beta(\tau), \qquad t \in \mathscr{R}^+, \tag{16.23}$$

[89] A positive, increasing, and bounded function on $\mathscr{R}^+$ is of bounded total variation on $\mathscr{R}^+$. Hence the Stieltjes integral (16.19) is well defined for $t \in \mathscr{R}^+$.

where $1/c_n$ and τ_n $(n=1,2,3,\ldots)$ are the elastic compliances and retardation times and β is the increasing and bounded function on $\mathscr{R}^+$ which determines the distribution of elastic compliances.

17. Field equations. In this section we record the basic field equations of the linear theory of viscoelasticity. Throughout this section $\mathscr{B}$ denotes a viscoelastic body with closure $\bar{\mathscr{B}}$ and boundary $\partial\mathscr{B}$. Dependence of quantities upon material points in $\mathscr{B}$ will be suppressed unless confusion may occur.

The fundamental laws of balance of linear and angular momentum are expressed through the following result.

Theorem*[90] *(Cauchy-Poisson). *Balance of linear and angular momentum hold for $\mathscr{B}$ if, and only if there exists a process $\boldsymbol{S}$ for $\mathscr{B}$, called the* ***stress process*** *such that:*

(S-1) *the process $\boldsymbol{S}$ is continuous (class C) and the values of $\boldsymbol{S}$ are smooth (class C^1) tensor fields on $\mathscr{B}$;*

(S-2) *for each unit vector $\boldsymbol{n}$ in $\mathscr{V}$*

$$\boldsymbol{s}_{(\boldsymbol{n})} = \boldsymbol{S}\,\boldsymbol{n}. \tag{17.1}$$

where $\boldsymbol{s}_{(\boldsymbol{n})}$ is the surface traction on a surface in $\mathscr{B}$ oriented by the unit normal vector $\boldsymbol{n}$;

(S-3) *the equations of motion are satisfied on $\mathscr{B}$:*

$$\operatorname{div}\boldsymbol{S} + \boldsymbol{b} = \varrho\ddot{\boldsymbol{u}}, \tag{17.2}$$

$$\boldsymbol{S} = \boldsymbol{S}^T, \tag{17.3}$$

where ϱ is the density of $\mathscr{B}$ and $\boldsymbol{b}$ and $\boldsymbol{u}$ are the body force and displacement processes for $\mathscr{B}$.

A stress process $\boldsymbol{S}$ for $\mathscr{B}$ is said to be ***quasi-static*** if, and only if, for a given body force process $\boldsymbol{b}$, the ***equations of equilibrium*** are satisfied on $\mathscr{B}$:

$$\operatorname{div}\boldsymbol{S} + \boldsymbol{b} = \boldsymbol{0}, \tag{17.4}$$

$$\boldsymbol{S} = \boldsymbol{S}^T. \tag{17.5}$$

C. Quasi-static linear viscoelasticity.

18. The quasi-static assumption. In this chapter we consider the quasi-static linearized theory of viscoelasticity. By the term ***quasi-static*** we mean that all inertial effects are systematically neglected. We consider a fixed viscoelastic body $\mathscr{B}$ in $\mathscr{E}$ with boundary $\partial\mathscr{B}$. Material points in $\mathscr{B}$ are denoted by $\boldsymbol{x}, \boldsymbol{y}, \boldsymbol{z}, \ldots$ and the number $t \in \mathscr{R}$ is the time.

19. Quasi-static viscoelastic processes. Let $\mathscr{B}$ be a body in $\mathscr{E}$ with boundary $\partial\mathscr{B}$ and closure $\bar{\mathscr{B}}$. By an ***admissible state for*** $\mathscr{B}$ we mean an ordered triplet $[\hat{\boldsymbol{u}}, \hat{\boldsymbol{E}}, \hat{\boldsymbol{S}}]$ consisting of a vector displacement field $\hat{\boldsymbol{u}}$ and symmetric tensor strain and stress fields $\hat{\boldsymbol{E}}$ and $\hat{\boldsymbol{S}}$ defined on $\bar{\mathscr{B}}$. An ***admissible process for*** $\mathscr{B}$ is an ordered triplet $[\boldsymbol{u}, \boldsymbol{E}, \boldsymbol{S}]$ whose value $[\boldsymbol{u}, \boldsymbol{E}, \boldsymbol{S}](t) = [\boldsymbol{u}(t), \boldsymbol{E}(t), \boldsymbol{S}(t)]$ is an admissible state at each time t. We say that an admissible process $[\boldsymbol{u}, \boldsymbol{E}, \boldsymbol{S}]$ is restricted, continuous, of Heaviside type, etc., according as each of its elements is a process of this type.

[90] See, for example, LTE, Sect. 15.

Observe that the set of admissible states for $\mathscr{B}$ can be given the structure of a linear vector space with addition and multiplication by a scalar defined by

$$[\hat{\boldsymbol{u}}, \hat{\boldsymbol{E}}, \hat{\boldsymbol{S}}] + [\boldsymbol{u}, \boldsymbol{E}, \boldsymbol{S}] = [\hat{\boldsymbol{u}} + \boldsymbol{u}, \hat{\boldsymbol{E}} + \boldsymbol{E}, \hat{\boldsymbol{S}} + \boldsymbol{S}] \tag{19.1}$$

and

$$\alpha [\hat{\boldsymbol{u}}, \hat{\boldsymbol{E}}, \hat{\boldsymbol{S}}] = [\alpha \hat{\boldsymbol{u}}, \alpha \hat{\boldsymbol{E}}, \alpha \hat{\boldsymbol{S}}]. \tag{19.2}$$

Similarly, the set of all admissible processes may be given the structure of a linear vector space by requiring that (19.1) and (19.2) hold at each time t.

Let $\mathscr{B}$ be a viscoelastic body with relaxation function $\mathbf{G}$ and body force process $\boldsymbol{b}$. Then by a ***quasi-static viscoelastic process for $\mathscr{B}$ corresponding to the relaxation function $\mathbf{G}$ and body force $\boldsymbol{b}$*** we mean an admissible process $[\boldsymbol{u}, \boldsymbol{E}, \boldsymbol{S}]$ such that:

(R-1) the values of $\boldsymbol{u}$, $\boldsymbol{E}$, and $\boldsymbol{S}$ are class C (continuous) fields on $\bar{\mathscr{B}}$;

(R-2) the values of $\boldsymbol{u}$ are class C^2 vector fields on $\bar{\mathscr{B}}$; and

(R-3) on $\mathscr{B}$, $\boldsymbol{u}$, $\boldsymbol{E}$, and $\boldsymbol{S}$ satisfy the *equation of equilibrium*[1]

$$\operatorname{div} \boldsymbol{S} + \boldsymbol{b} = \boldsymbol{0}, \tag{19.3}$$

the *strain-displacement equation*

$$\boldsymbol{E} = \tfrac{1}{2}[\nabla \boldsymbol{u} + \nabla \boldsymbol{u}^T], \tag{19.4}$$

and the *Boltzmann relaxation law*

$$\boldsymbol{S} = \mathscr{L}_{\mathbf{G}}[\boldsymbol{E}] = \mathbf{G} \circledast \boldsymbol{E}. \tag{19.5}$$

The set of all quasi-static viscoelastic processes for $\mathscr{B}$ corresponding to $\mathbf{G}$ and $\boldsymbol{b}$ is a vector subspace of the space of admissible processes denoted by QSVP$(\mathscr{B}, \mathbf{G}, \boldsymbol{b})$ or, if the dependence on $\mathscr{B}$ is understood, by QSVP$(\mathbf{G}, \boldsymbol{b})$.

Let $\mathscr{B}$ be a viscoelastic body with creep compliance $\mathbf{J}$ and body force process $\boldsymbol{b}$. Then by a ***quasi-static viscoelastic process for $\mathscr{B}$ corresponding to the creep function $\mathbf{J}$ and body force $\boldsymbol{b}$*** we mean an admissible process $[\boldsymbol{u}, \boldsymbol{E}, \boldsymbol{S}]$ such that:

(C-1) the values of $\boldsymbol{u}$, $\boldsymbol{E}$, and $\boldsymbol{S}$ are class C (continuous) fields on $\bar{\mathscr{B}}$;

(C-2) the values of $\boldsymbol{u}$ are class C^2 vector fields on $\bar{\mathscr{B}}$;

(C-3) the values of $\boldsymbol{S}$ are class C^1 (smooth) symmetric tensor fields on $\bar{\mathscr{B}}$;

and

(C-4) on $\mathscr{B}$, $\boldsymbol{u}$, $\boldsymbol{E}$, and $\boldsymbol{S}$ satisfy the equation of equilibrium (19.3),[2] the strain-displacement equation (19.4) and the *Boltzmann creep law*

$$\boldsymbol{E} = \mathscr{L}_{\mathbf{J}}[\boldsymbol{S}] = \mathbf{J} \circledast \boldsymbol{S}. \tag{19.6}$$

The set of all quasi-static viscoelastic processes for $\mathscr{B}$ corresponding to $\mathbf{J}$ and $\boldsymbol{b}$ is a vector subspace of the space of admissible processes denoted by QSVP$(\mathscr{B}, \mathbf{J}, \boldsymbol{b})$ or, if the dependence on $\mathscr{B}$ is understood, by QSVP$(\mathbf{J}, \boldsymbol{b})$.

If $[\boldsymbol{u}, \boldsymbol{E}, \boldsymbol{S}]$ belongs to QSVP$(\mathbf{G}, \boldsymbol{b})$ or QSVP$(\mathbf{J}, \boldsymbol{b})$,[3] we assume that the process is sufficiently regular in time to render the definitions meaningful. For example, if $\mathbf{G}$ and $\mathbf{J}$ are *strong* response functions for $\mathscr{B}$, then we may suppose

[1] The requirement that angular momentum be balanced is automatically satisfied since $\boldsymbol{S} = \boldsymbol{S}^T$.

[2] Cf. footnote 1.

[3] We shall continue to use the notation QSVP$(\mathbf{G}, \boldsymbol{b})$ and QSVP$(\mathbf{J}, \boldsymbol{b})$ loosely since clearly the association indicated is different for creep functions from that for relaxation functions.

that $[\boldsymbol{u}, \boldsymbol{E}, \boldsymbol{S}]$ has integrable histories or is piecewise right-continuous with bounded histories. If $\mathbf{G}$ and $\mathbf{J}$ are not required to be strong response functions for $\mathscr{B}$, then it is usually appropriate to consider processes $[\boldsymbol{u}, \boldsymbol{E}, \boldsymbol{S}]$ of Heaviside type or which have finite histories.

Suppose the relaxation function $\mathbf{G}$ is *constant* on $\mathscr{R}^+$, in which case $\mathbf{G} = h\overset{\circ}{\mathbf{G}}$, then $[\boldsymbol{u}, \boldsymbol{E}, \boldsymbol{S}]$ belongs to QSVP$(\mathbf{G}, \boldsymbol{b})$ if, and only if, $[\boldsymbol{u}, \boldsymbol{E}, \boldsymbol{S}](t)$ is an elastic state corresponding to elasticity $\overset{\circ}{\mathbf{G}}$ and body force $\boldsymbol{b}(t)$, at each time t. A similar assertion is true if the relaxation function $\mathbf{G}$ is replaced by the creep compliance $\mathbf{J}$.

It is also easy to see that if $\mathscr{L}_{\mathbf{G}}$ and $\mathscr{L}_{\mathbf{J}}$ are mutually inverse; that is, $\mathbf{G}$ is a relaxation function for $\mathbf{J}$ and $\mathbf{J}$ is a creep function for $\mathbf{G}$, then $[\boldsymbol{u}, \boldsymbol{E}, \boldsymbol{S}]$ belongs to QSVP$(\mathbf{G}, \boldsymbol{b})$ if, and only if, $[\boldsymbol{u}, \boldsymbol{E}, \boldsymbol{S}]$ belongs to QSVP$(\mathbf{J}, \boldsymbol{b})$.

For differential operator laws we say that $[\boldsymbol{u}, \boldsymbol{E}, \boldsymbol{S}]$ is a ***weak quasi-static viscoelastic process for $\mathscr{B}$ corresponding to the differential operator pair $\langle P, Q\rangle$ of order N and body force process $\boldsymbol{b}$*** if, and only if,

(i) $\langle P, Q\rangle$ is of relaxation/creep type and

(ii) $[\boldsymbol{u}, \boldsymbol{E}, \boldsymbol{S}]$ is of Heaviside type in QSVP$(\mathbf{G}, \boldsymbol{b})$/QSVP$(\mathbf{J}, \boldsymbol{b})$ where $\mathbf{G}/\mathbf{J}$ belongs to $\langle P, Q\rangle$ in the sense of Sect. 14.

We say that $[\boldsymbol{u}, \boldsymbol{E}, \boldsymbol{S}]$ is a ***strong quasi-static viscoelastic process for $\mathscr{B}$ corresponding to the differential operator pair $\langle P, Q\rangle$ of order N and body force process $\boldsymbol{b}$*** if, and only if, $[\boldsymbol{u}, \boldsymbol{E}, \boldsymbol{S}]$ satisfies (i) and (ii) above and is in Heaviside class H^N. Clearly a strong quasi-static viscoelastic process is necessarily weak. It can be shown that weak processes satisfy constitutive relations of Boltzmann type while strong processes, by virtue of their increased smoothness, satisfy the differential operator laws and initial conditions. By use of the concepts of weak and strong processes, the various boundary and initial value problems may be formulated for viscoelastic bodies which obey differential operator laws.[4] In the sequel, we consider only Boltzmann laws in integral form.

20. Displacement equations of equilibrium. Let $\mathscr{B}$ be a viscoelastic body with a Boltzmann relaxation law $\mathscr{L}_{\mathbf{G}}$, and let $\boldsymbol{b}$ be a body force process for $\mathscr{B}$. Then if $[\boldsymbol{u}, \boldsymbol{E}, \boldsymbol{S}]$ is in QSVP$(\mathbf{G}, \boldsymbol{b})$ the displacement process $\boldsymbol{u}$ satisfies the *displacement equation of equilibrium*[5]

$$\operatorname{div} \mathscr{L}_{\mathbf{G}}[\hat{\nabla} \boldsymbol{u}] + \boldsymbol{b} = \mathbf{0} \tag{20.1 a}$$

or, equivalently,

$$\operatorname{div} (\mathbf{G} \circledast \hat{\nabla} \boldsymbol{u}) + \boldsymbol{b} = \mathbf{0}. \tag{20.1 b}$$

Conversely, suppose $\boldsymbol{u}$ is a process whose values are class C^2 vector fields on $\overline{\mathscr{B}}$ and which satisfies the displacement equation of equilibrium (20.1). If the strain process $\boldsymbol{E}$ is *defined* through the strain-displacement equation (20.4) and the stress process $\boldsymbol{S}$ is *defined* through the Boltzmann relaxation law (20.5), then $[\boldsymbol{u}, \boldsymbol{E}, \boldsymbol{S}]$ is in QSVP$(\mathbf{G}, \boldsymbol{b})$.

In case the body $\mathscr{B}$ is *isotropic*, we see from (12.1) that (20.1) becomes[6]

$$\operatorname{div}\{\mathscr{L}_{2\mu}[\hat{\nabla}\boldsymbol{u}] + \mathscr{L}_{\lambda}[(\operatorname{tr} \hat{\nabla}\boldsymbol{u})\, \mathbf{1}]\} + \boldsymbol{b} = \mathbf{0} \tag{20.2a}$$

[4] These definitions and results concerning differential operator laws are those given by GURTIN and STERNBERG [1962, *10*, Sect. 5].

[5] The relaxation function has values in $[\mathscr{T}, \mathscr{T}_{\mathrm{sym}}]$, and $\mathbf{G}(s)$ operating on $\nabla\boldsymbol{u}$ is identically equal to $\mathbf{G}(s)$ operating on the symmetric part $\hat{\nabla}\boldsymbol{u}$ of $\nabla\boldsymbol{u}$ for all $\boldsymbol{u}$. Cf. Sect. 9.

[6] Here $\hat{\nabla}\boldsymbol{u}$ denotes sym $\nabla\boldsymbol{u}$. If $\varkappa$ is a *scalar* response function $\mathscr{L}_{\varkappa}$ means $\mathscr{L}_{\varkappa\mathbf{1}}$, where $\mathbf{1}$ is the identity map in $[\mathscr{T}_{\mathrm{sym}}]$.

or, equivalently,

$$\operatorname{div}\{2\mu \circledast \hat{\nabla} \boldsymbol{u} + \lambda \circledast (\operatorname{tr} \hat{\nabla} \boldsymbol{u})\, \mathbf{1}\} + \boldsymbol{b} = \mathbf{0}. \tag{20.2b}$$

If, in addition, the body $\mathscr{B}$ is *homogeneous*, Eq. (20.2) reduces to

$$\mathscr{L}_{\mu}[\Delta \boldsymbol{u}] + \mathscr{L}_{\lambda+\mu}[\nabla(\nabla \cdot \boldsymbol{u})] + \boldsymbol{b} = \mathbf{0} \tag{20.3a}$$

or, equivalently,

$$\mu \circledast \Delta \boldsymbol{u} + (\lambda + \mu) \circledast [\nabla(\nabla \cdot \boldsymbol{u})] + \boldsymbol{b} = \mathbf{0}. \tag{20.3b}$$

Observe that if the scalar valued response functions λ and μ are *constant* on $\mathscr{R}^+$ — $\lambda = \mathring{\lambda} h$, $\mu = \mathring{\mu} h$ — then (20.3) reduces to the displacement equation of equilibrium for an isotropic and homogeneous elastic body with Lamé constants $\lambda \equiv \mathring{\lambda}$ and $\mu \equiv \mathring{\mu}$.

Within the class of Heaviside processes, (20.3) becomes

$$\mathring{\mu} \Delta \boldsymbol{u} + \dot{\mu} * \Delta \boldsymbol{u} + (\mathring{\lambda} + \mathring{\mu})\, \nabla(\nabla \cdot \boldsymbol{u}) + (\dot{\lambda} + \dot{\mu}) * \nabla(\nabla \cdot \boldsymbol{u}) + \boldsymbol{b} = \mathbf{0}. \tag{20.4}$$

21. Equations of compatibility. Let $\mathscr{B}$ be a viscoelastic body with creep compliance $\boldsymbol{J}$ and let $\boldsymbol{b}$ be a body force process for $\mathscr{B}$. For completeness we record the

Compatibility theorem.[7] *The strain process* $\boldsymbol{E}$ *corresponding to a displacement process* $\boldsymbol{u}$ *whose values are class* C^3 *vector-fields on* $\mathscr{B}$ *satisfies the* ***equation of compatibility***

$$\operatorname{curl} \operatorname{curl} \boldsymbol{E} = \mathbf{0} \tag{21.1a}$$

or, equivalently,

$$\Delta \boldsymbol{E} + \nabla\nabla(\operatorname{tr} \boldsymbol{E}) - 2\hat{\nabla} \operatorname{div} \boldsymbol{E} = \mathbf{0} \tag{21.1b}$$

on $\mathscr{B}$. *Conversely, let* $\mathscr{B}$ *be simply connected and let* $\boldsymbol{E}$ *be a process whose values are class* C^N $(N \geqq 2)$ *symmetric-tensor fields which satisfy* (21.1) *on* $\mathscr{B}$. *Then there is a displacement process* $\boldsymbol{u}$ *whose values are class* C^{N+1} *vector-fields on* $\mathscr{B}$ *such that* $\boldsymbol{E}$ *and* $\boldsymbol{u}$ *satisfy the strain-displacement equation* (19.4). *The process* $\boldsymbol{u}$ *is uniquely determined by* $\boldsymbol{E}$ *to within a superposed (infinitesimal) rigid motion.*

Let $[\boldsymbol{u}, \boldsymbol{E}, \boldsymbol{S}]$ be in QSVP$(\boldsymbol{J}, \boldsymbol{b})$. Suppose the values of $\boldsymbol{u}$ are class C^3 vector fields on $\bar{\mathscr{B}}$, the values of $\boldsymbol{J}$ are class C^2 fields on $\bar{\mathscr{B}}$, and $\boldsymbol{J}$ belongs to a relaxation function $\boldsymbol{G}$ whose values are also class C^2 fields on $\bar{\mathscr{B}}$. Then $[\boldsymbol{u}, \boldsymbol{E}, \boldsymbol{S}]$ belongs to QSVP$(\boldsymbol{G}, \boldsymbol{b})$ and the values of $\boldsymbol{S}$ are symmetric tensor processes of class C^2 on $\bar{\mathscr{B}}$. Furthermore, the stress process $\boldsymbol{S}$ satisfies the ***stress equation of compatibility***[8]

$$\operatorname{curl} \operatorname{curl} \mathscr{L}_{\boldsymbol{J}}[\boldsymbol{S}] = \mathbf{0} \tag{21.2a}$$

or, equivalently,

$$\Delta \boldsymbol{J} \circledast \boldsymbol{S} + \nabla\nabla \operatorname{tr}(\boldsymbol{J} \circledast \boldsymbol{S}) - 2\hat{\nabla} \operatorname{div}(\boldsymbol{J} \circledast \boldsymbol{S}) = \mathbf{0} \tag{21.2b}$$

on $\bar{\mathscr{B}}$. Eq. (21.2) and the stress equation of equilibrium (19.3) together constitute the generalization to the linear theory of anisotropic viscoelastic solids of the classical Beltrami-Donati-Michell equations of isotropic linear elasticity theory.[9]

The stress equations of equilibrium (19.3) and compatibility (21.2) may be used to construct processes in QSVP$(\boldsymbol{J}, \boldsymbol{b})$. Let $\mathscr{B}$ be a simply connected region in $\mathscr{E}$. Suppose the values of $\boldsymbol{J}$ are class C^2 fields on $\bar{\mathscr{B}}$. Let $\boldsymbol{S}$ be a symmetric-tensor field of class C^2 on $\bar{\mathscr{B}}$ which satisfies (19.3) and (21.2) on $\mathscr{B}$. Define $\boldsymbol{E}$ through (19.6). Then $\boldsymbol{E}$ is a process whose values are symmetric-tensor fields of

[7] Cf. LTE, Sect. 14.

[8] Gurtin and Sternberg [1962, *10*, Theorem 5.7]; cf. LTE, Sect. 14.

[9] See, e.g., LTE, Sect. 27.

class C^2 on $\bar{\mathscr{B}}$. Furthermore, $\boldsymbol{E}$ satisfies the compatibility equation (21.1) on $\mathscr{B}$. Hence, by the compatibility theorem, there is a process $\boldsymbol{u}$, whose values are class C^3 vector fields on $\mathscr{B}$, which satisfies the strain-displacement equation (19.4). Thus $[\boldsymbol{u}, \boldsymbol{E}, \boldsymbol{S}]$ lies in QSVP$(\boldsymbol{J}, \boldsymbol{b})$. Moreover, $\boldsymbol{u}$ is determined (to within infinitesimal rigid motions) by the line integral[10]

$$\boldsymbol{u}(\boldsymbol{x}, t) = \int_{\boldsymbol{x}_0}^{\boldsymbol{x}} \boldsymbol{E}(\boldsymbol{y}, t)\, d\boldsymbol{y} + \int_{\boldsymbol{x}_0}^{\boldsymbol{x}} \{2 \operatorname{skw} \nabla[\boldsymbol{E}(\boldsymbol{y}, t)\, d\boldsymbol{y}](\boldsymbol{x} - \boldsymbol{y})\}, \qquad (\boldsymbol{x}, t) \in \mathscr{B} \times \mathscr{R}, \tag{21.3}$$

where $\boldsymbol{x}_0$ is some fixed point in $\mathscr{B}$. For each time t, the integral in (21.3) is independent of the path provided $\mathscr{B}$ is simply connected, and

$$\nabla \boldsymbol{u}(\boldsymbol{x}, t) = \boldsymbol{E}(\boldsymbol{x}, t) + \int_{\boldsymbol{x}_0}^{\boldsymbol{x}} 2 \operatorname{skw} \nabla[\boldsymbol{E}(\boldsymbol{y}, t)\, d\boldsymbol{y}], \qquad (\boldsymbol{x}, t) \in \mathscr{B} \times \mathscr{R}. \tag{21.4}$$

Since $\boldsymbol{E}$ is symmetric, it follows that sym $\nabla \boldsymbol{u} = \boldsymbol{E}$.

For a homogeneous, isotropic viscoelastic body $\mathscr{B}$ of relaxation type, the stress equations of equilibrium (19.3) and compatibility (21.2) may be combined to yield

$$J_1 \circledast \Delta \boldsymbol{S} + \tfrac{1}{3}(2J_1 + J_2) \circledast \nabla\nabla(\operatorname{tr} \boldsymbol{S}) - \tfrac{1}{3}(J_1 - J_2) \circledast \Delta(\operatorname{tr} \boldsymbol{S})\, \mathbf{1} + 2J_1 \circledast \widehat{\nabla \boldsymbol{b}} = \mathbf{0}. \tag{21.5}$$

Taking the trace of (21.5), we obtain

$$\tfrac{1}{3}(J_1 + 2J_2) \circledast \Delta(\operatorname{tr} \boldsymbol{S}) + J_1 \circledast \operatorname{div} \boldsymbol{b} = \mathbf{0}. \tag{21.6}$$

If the Boltzmann laws $\mathscr{L}_{J_1}$ and $\mathscr{L}_{J_1+2J_2}$ are invertible, then (21.5) and (21.6) may be combined to yield the viscoelastic counterpart of the Beltrami-Donati-Michell equations of classical elastodynamics:[11]

$$\begin{aligned} &\Delta \boldsymbol{S} + \tfrac{1}{3} J_1^{-1} \circledast (2J_1 + J_2) \circledast \nabla\nabla(\operatorname{tr} \boldsymbol{S}) \\ &\quad + J_1^{-1} \circledast (J_1 - J_2) \circledast (J_1 + 2J_2)^{-1} \circledast J_1 \circledast \operatorname{div} \boldsymbol{b} + 2\widehat{\nabla \boldsymbol{b}} = \mathbf{0}. \end{aligned} \tag{21.7}$$

22. Boundary data. Let $\mathscr{B}$ be a regular (open) region in $\mathscr{E}$ with boundary ∂B and closure $\bar{\mathscr{B}}$:

$$\bar{\mathscr{B}} = \mathscr{B} \cup \partial\mathscr{B}. \tag{22.1}$$

We assume that the boundary $\partial\mathscr{B}$ is decomposed into two ***separate surface elements*** $\mathscr{S}_{\hat{\boldsymbol{u}}}$ and $\mathscr{S}_{\hat{\boldsymbol{s}}}$; that is,

$$\partial\mathscr{B} = \mathscr{S}_{\hat{\boldsymbol{u}}} \cup \mathscr{S}_{\hat{\boldsymbol{s}}} \tag{22.2}$$

and[12]

$$\mathring{\mathscr{S}}_{\hat{\boldsymbol{u}}} \cap \mathring{\mathscr{S}}_{\hat{\boldsymbol{s}}} = \emptyset. \tag{22.3}$$

The Eq. (22.2) expresses the requirement that points in $\partial\mathscr{B}$ belong either to $\mathscr{S}_{\hat{\boldsymbol{u}}}$ or $\mathscr{S}_{\hat{\boldsymbol{s}}}$ (or both); while Eq. (22.3) expresses the requirement that the points

[10] See, e.g., BOLEY and WEINER [1960, 3] and LTE, Sect. 14.

[11] Here we write $J^{-1}\circledast$ for $\mathscr{L}_J^{-1}$ (cf. Sects. 5 and 11). Note that if J_1 and J_2 are *formally* replaced by

$$\frac{1}{2\mu} \quad \text{and} \quad \frac{1}{2\mu} \cdot \frac{1-2\nu}{1+\nu},$$

where μ is the Lamé shear modulus and ν is POISSON's ratio $(\nu = \lambda/2(\lambda+\mu))$, then (21.7) becomes the classical Beltrami-Donati-Michell equation of elastostatics. See LTE, Sect. 27. Here we must take $\nu \neq -1$ to obtain a meaningful expression.

[12] Here $\mathring{\mathscr{S}}_{\hat{\boldsymbol{u}}}$ and $\mathring{\mathscr{S}}_{\hat{\boldsymbol{s}}}$ denote the interiors relative to $\partial\mathscr{B}$ of $\mathscr{S}_{\hat{\boldsymbol{u}}}$ and $\mathscr{S}_{\hat{\boldsymbol{s}}}$.

common to $\mathscr{S}_{\hat{\boldsymbol{u}}}$ and $\mathscr{S}_{\hat{\boldsymbol{s}}}$ are boundary points of both $\mathscr{S}_{\hat{\boldsymbol{u}}}$ and $\mathscr{S}_{\hat{\boldsymbol{s}}}$ (relative to $\partial\mathscr{B}$). We always assume that, except for a set of area measure zero, the points of $\partial\mathscr{B}$ are regular and, hence, have a well defined unit outward normal vector $\boldsymbol{n}\in\mathscr{V}$. Throughout this article the body $\mathscr{B}$ and, hence, the boundary $\partial\mathscr{B}$ are fixed in time. However, many arguments in the sequel remain valid if the sets $\mathscr{S}_{\hat{\boldsymbol{u}}}$ and $\mathscr{S}_{\hat{\boldsymbol{s}}}$ vary with time provided (22.2) and (22.3) hold at each time.

By ***displacement boundary data*** we mean a process $\hat{\boldsymbol{u}}$ whose values are vector fields defined and continuous at the regular points of $\mathscr{S}_{\hat{\boldsymbol{u}}}$. By ***stress*** or ***traction boundary data*** we mean a process $\hat{\boldsymbol{s}}$ whose values are vector fields defined and continuous at the regular points of $\mathscr{S}_{\hat{\boldsymbol{s}}}$. The pair $[\hat{\boldsymbol{u}},\hat{\boldsymbol{s}}]$ consisting of displacement and traction boundary data will be called, simply, ***boundary data for*** $\partial\mathscr{B}$. If neither $\mathscr{S}_{\hat{\boldsymbol{s}}}$ nor $\mathscr{S}_{\hat{\boldsymbol{u}}}$ is empty, $[\hat{\boldsymbol{u}},\hat{\boldsymbol{s}}]$ is called ***mixed boundary data.***

In the linear theory of viscoelasticity, as in the linear theory of elasticity, it is possible to consider other types of boundary data such as *mixed-mixed boundary data*[13] and cases where $\mathscr{B}$ and, hence, $\partial\mathscr{B}$ vary with time. For simplicity in presentation, we do not do so in this article.

23. The quasi-static boundary value problem. In this section we suppose that we are given a viscoelastic body with relaxation function $\mathbf{G}$ or creep compliance $\mathbf{J}$. Furthermore, we have prescribed body forces $\boldsymbol{b}$ on $\mathscr{B}$ and boundary data $[\hat{\boldsymbol{u}},\hat{\boldsymbol{s}}]$ on $\partial\mathscr{B}$. By a ***solution to the quasi-static boundary value problem for*** $\mathscr{B}$ ***(corresponding to the boundary data*** $[\hat{\boldsymbol{u}},\hat{\boldsymbol{s}}]$***, the body force b, and either the relaxation function*** $\mathbf{G}$ ***or the creep compliance*** $\mathbf{J}$***)*** we mean a process $[\boldsymbol{u},\boldsymbol{E},\boldsymbol{S}]$ in either QSVP$(\mathbf{G},\boldsymbol{b})$ or QSVP$(\mathbf{J},\boldsymbol{b})$ which satisfies the ***boundary conditions on*** $\partial\mathscr{B}$

$$\boldsymbol{u}=\hat{\boldsymbol{u}} \qquad \text{on } \mathscr{S}_{\hat{\boldsymbol{u}}}, \tag{23.1a}$$

$$\boldsymbol{S}\boldsymbol{n}\equiv\boldsymbol{s}_{(\boldsymbol{n})}=\hat{\boldsymbol{s}} \qquad \text{on } \mathscr{S}_{\hat{\boldsymbol{s}}}, \tag{23.1b}$$

where $\boldsymbol{n}$ denotes the unit outward normal vector field on $\partial\mathscr{B}$.[14] If neither $\mathscr{S}_{\hat{\boldsymbol{u}}}$ nor $\mathscr{S}_{\hat{\boldsymbol{s}}}$ is empty, we refer to the ***mixed boundary value problem;*** if $\mathscr{S}_{\hat{\boldsymbol{s}}}$ is empty, we refer to the ***displacement boundary value problem;*** and if $\mathscr{S}_{\hat{\boldsymbol{u}}}$ is empty, we refer to the ***stress*** or ***traction boundary value problem.***

A solution to the quasi-static boundary value problem may be characterized purely in terms of displacements. Suppose we are given boundary data $[\hat{\boldsymbol{u}},\hat{\boldsymbol{s}}]$ on $\partial\mathscr{B}$, a body force process $\boldsymbol{b}$ for $\mathscr{B}$, and a relaxation function $\mathbf{G}$ for $\mathscr{B}$. First determine a displacement process $\boldsymbol{u}$ on $\overline{\mathscr{B}}$ which satisfies the displacement equation of equilibrium (20.1) on $\mathscr{B}$ and which meets the boundary conditions on $\partial\mathscr{B}$

$$\boldsymbol{u}=\hat{\boldsymbol{u}} \qquad \text{on } \mathscr{S}_{\hat{\boldsymbol{u}}}, \tag{23.2a}$$

$$\{\mathbf{G}\circledast\widehat{\nabla\boldsymbol{u}}\}\,\boldsymbol{n}=\hat{\boldsymbol{s}} \qquad \text{on } \mathscr{S}_{\hat{\boldsymbol{s}}}. \tag{23.2b}$$

Then *define* the strain process $\boldsymbol{E}$ on $\overline{\mathscr{B}}$ through the strain-displacement relation (19.4) and the stress process $\boldsymbol{S}$ on $\overline{\mathscr{B}}$ through the Boltzmann relaxation law (19.5). The process $[\boldsymbol{u},\boldsymbol{E},\boldsymbol{S}]$ thus computed is a solution to the quasi-static boundary value problem corresponding to $[\hat{\boldsymbol{u}},\hat{\boldsymbol{s}}]$, $\boldsymbol{b}$, and $\mathbf{G}$.

---

[13] The extension of most results to include mixed-mixed boundary data may be carried out following the pattern established by GURTIN, LTE, Subchap. D.V.

[14] The normal vector field $\boldsymbol{n}$ is defined only at the regular points of $\partial\mathscr{B}$. Strictly speaking, therefore, (23.1b) is meaningful only at the regular points of $\partial\mathscr{B}$. Since the set of nonregular points has area measure zero, this distinction is unimportant.

On the other hand, a solution to the quasi-static *traction* boundary value problem may be characterized purely in terms of the stresses. That is, suppose we are given traction boundary data $\hat{\mathbf{s}}$ on $\partial\mathscr{B}$ ($\mathscr{S}_{\hat{\boldsymbol{u}}}$ is empty), a body force process $\boldsymbol{b}$ for $\mathscr{B}$, and a creep compliance $\boldsymbol{J}$ for $\mathscr{B}$. First determine a stress process $\boldsymbol{S}$ on $\overline{\mathscr{B}}$ which satisfies the stress equation of compatibility (21.2) and the stress equation of equilibrium (19.3) on $\mathscr{B}$, and which satisfies the traction boundary data (23.1 b) on $\mathscr{S}_{\hat{\mathbf{s}}}=\partial\mathscr{B}$. Then compute a strain process $\boldsymbol{E}$ on $\overline{\mathscr{B}}$ through the Boltzmann creep law (19.6) and a displacement process $\boldsymbol{u}$ on $\overline{\mathscr{B}}$ through the line integral (21.3).[15] The process $[\boldsymbol{u}, \boldsymbol{E}, \boldsymbol{S}]$ thus determined is a solution to the quasi-static traction boundary value problem corresponding to $\hat{\mathbf{s}}$, $\boldsymbol{b}$, and $\boldsymbol{J}$. Note that it is not possible to express displacement boundary data in terms of the stresses in a simple way.

24. Synchronous and separable boundary data.[16] Solutions to the quasi-static boundary value problem are particularly simple to characterize if the boundary data are synchronous and separable (as defined below).

Let ϕ be a *scalar-valued* process in the Heaviside class H^{ac}. Let $\boldsymbol{v}$ and $\boldsymbol{r}$ be vector fields defined and continuous at the regular points of $\mathscr{S}_{\hat{\boldsymbol{u}}}$ and $\mathscr{S}_{\hat{\mathbf{s}}}$. The boundary data $[\hat{\boldsymbol{u}}, \hat{\mathbf{s}}]$ is said to be ***synchronous and separable*** if, and only if, for all t, $\hat{\boldsymbol{u}}(\boldsymbol{x}, t)$ and $\hat{\mathbf{s}}(\boldsymbol{x}, t)$ are of the form

$$\hat{\boldsymbol{u}}(\boldsymbol{x}, t)=\boldsymbol{v}(\boldsymbol{x})\,\phi(t) \quad \text{for} \quad \boldsymbol{x}\in\mathscr{S}_{\hat{\boldsymbol{u}}}, \tag{24.1 a}$$

$$\hat{\mathbf{s}}(\boldsymbol{x}, t)=\boldsymbol{r}(\boldsymbol{x})\,\phi(t) \quad \text{for} \quad \boldsymbol{x}\in\mathscr{S}_{\hat{\mathbf{s}}} \tag{24.1 b}$$

or, more concisely,

$$\hat{\boldsymbol{u}}=\boldsymbol{v}\,\phi \quad \text{on } \mathscr{S}_{\hat{\boldsymbol{u}}}, \tag{24.2 a}$$

$$\hat{\mathbf{s}}=\boldsymbol{r}\,\phi \quad \text{on } \mathscr{S}_{\hat{\mathbf{s}}}. \tag{24.2 b}$$

Using the scalar valued Heaviside unit process h:

$$h(t)=\begin{Bmatrix} 0: & t<0 \\ 1: & t\geqq 0 \end{Bmatrix}, \tag{24.3}$$

define boundary data $[\hat{\boldsymbol{u}}', \hat{\mathbf{s}}']$ on $\partial\mathscr{B}$ by

$$\hat{\boldsymbol{u}}'\equiv\boldsymbol{v}\,h \quad \text{on } \mathscr{S}_{\hat{\boldsymbol{u}}}, \tag{24.4}$$

$$\hat{\mathbf{s}}'\equiv\boldsymbol{r}\,h \quad \text{on } \mathscr{S}_{\hat{\mathbf{s}}}, \tag{24.5}$$

where $\boldsymbol{v}$ and $\boldsymbol{r}$ are the same vector fields used in (24.1) and (24.2).

Let $[\boldsymbol{u}', \boldsymbol{E}', \boldsymbol{S}']$ be a solution to the quasi-static boundary value problem corresponding to the boundary data, $[\hat{\boldsymbol{u}}', \hat{\mathbf{s}}']$, a body force $\boldsymbol{b}'$ of Heaviside class H, and the relaxation function $\boldsymbol{G}$. Then the process $[\boldsymbol{u}, \boldsymbol{E}, \boldsymbol{S}]$ of Heaviside type defined by[17]

$$\boldsymbol{u}=\phi\circledast\boldsymbol{u}', \tag{24.6}$$

$$\boldsymbol{E}=\phi\circledast\boldsymbol{E}', \tag{24.7}$$

$$\boldsymbol{S}=\phi\circledast\boldsymbol{S}' \tag{24.8}$$

---

[15] Here we must assume that $\mathscr{B}$ is simply connected in order to obtain a unique displacement process $\boldsymbol{u}$.

[16] The presentation of this section follows that of GURTIN and STERNBERG [1962, *10*, Sect. 5]. The results contained constitute an application of "Duhamel's principle" to the viscoelastic problem.

[17] For any process of Heaviside class H, say ψ, the function $\phi\circledast\psi$ of Heaviside class is defined by $\phi\circledast\psi=\overset{\circ}{\phi}\psi+\dot{\phi}*\psi$. Cf. Sect. 5.

is a solution to the quasi-static boundary value problem corresponding to the boundary data $[\hat{\boldsymbol{u}}, \hat{\mathbf{s}}]$, the body force $\boldsymbol{b}$ defined by

$$\boldsymbol{b} = \phi \circledast \boldsymbol{b}', \tag{24.9}$$

and the relaxation function $\mathbf{G}$. This result follows from the properties of the Boltzmann operator; in particular,

$$\gamma \circledast h = h \circledast \gamma = \gamma, \tag{24.10}$$

for any scalar valued process γ in H^{ac}.[18]

Thus, we see that if the boundary data $[\hat{\boldsymbol{u}}, \hat{\mathbf{s}}]$ is synchronous and separable and the body force $\boldsymbol{b}$ is given by (24.9), where $\boldsymbol{b}'$ is some process in Heaviside class H,[19] then it suffices to solve the quasi-static boundary value problem corresponding to the simpler boundary data $[\hat{\boldsymbol{u}}', \hat{\mathbf{s}}']$, body force $\boldsymbol{b}'$, and relaxation function $\mathbf{G}$.

25. Initial response. Elastic states. If $\mathscr{B}$ is a viscoelastic body having a Boltzmann relaxation law with relaxation function $\mathbf{G}$, then its initial mechanical response is elastic. Let t_0 be some arbitrary initial time which, without loss in generality, we take to be $t_0 = 0$. Suppose $[\boldsymbol{u}, \boldsymbol{E}, \boldsymbol{S}]$ is in $\mathrm{QSVP}(\mathbf{G}, \boldsymbol{b})$ and of Heaviside type. Then the initial quasi-static viscoelastic state $[\mathring{\boldsymbol{u}}, \mathring{\boldsymbol{E}}, \mathring{\boldsymbol{S}}]$ is an *elastic state* corresponding to the elasticity $\mathring{\mathbf{G}}$ and body force $\mathring{\boldsymbol{b}}$.[20] This follows from the definition of $\mathrm{QSVP}(\mathbf{G}, \boldsymbol{b})$ and the observation that $\mathring{\boldsymbol{S}} = \mathring{\mathbf{G}}[\mathring{\boldsymbol{E}}]$ whenever $\boldsymbol{E}$ is a process of Heaviside type.

Now let $[\boldsymbol{u}, \boldsymbol{E}, \boldsymbol{S}]$ be a process of Heaviside type in $\mathrm{QSVP}(\mathbf{G}, \boldsymbol{b})$. For $N \geqq 1$, assume that $[\boldsymbol{u}, \boldsymbol{E}, \boldsymbol{S}]$, $\mathbf{G}$, and $\boldsymbol{b}$ are processes of class H^N.[21] Define a symmetric-tensor field $\boldsymbol{S}_N$ and a vector field $\boldsymbol{b}_N$ by[22]

$$\boldsymbol{S}_N = \mathring{\mathbf{G}}[\mathring{\boldsymbol{E}}^{(N)}], \tag{25.1}$$

$$\boldsymbol{b}_N = \mathring{\boldsymbol{b}}^{(N)} + \operatorname{div} \sum_{n=0}^{N-1} \mathring{\mathbf{G}}^{(N-n)}[\mathring{\boldsymbol{E}}^{(n)}]. \tag{25.2}$$

It then follows, by elementary computations,[23] that $[\mathring{\boldsymbol{u}}^{(N)}, \mathring{\boldsymbol{E}}^{(N)}, \boldsymbol{S}_N]$ is an elastic state corresponding to elasticity $\mathring{\mathbf{G}}$ and body force $\boldsymbol{b}_N$. Moreover

$$\mathring{\boldsymbol{S}}^{(N)} = \boldsymbol{S}_N + \sum_{n=0}^{N-1} \mathring{\mathbf{G}}^{(N-n)}[\mathring{\boldsymbol{E}}^{(n)}]. \tag{25.3}$$

The classical uniqueness theorem for elastic states[24] may be used to guarantee the uniqueness of the initial viscoelastic states. In particular, suppose that

---

[18] This form of the result is due to GURTIN and STERNBERG [1962, *10*].

[19] If $\mathring{\phi} \neq 0$, then ϕ is invertible in the sense that there is a scalar process ψ in H^{ac} such that $\psi \circledast (\phi \circledast \boldsymbol{b}') = \boldsymbol{b}'$ for any $\boldsymbol{b}'$ in Heaviside class H. Hence, if the body force $\boldsymbol{b}$ is given, $\boldsymbol{b}'$ can be computed through $\boldsymbol{b}' = \psi \circledast \boldsymbol{b}$. See the inversion theorem for response functions in Sect. 11 and related comments in Sect. 5.

[20] See GURTIN and STERNBERG [1962, *10*, Theorem 6.1]. As previously mentioned the superposed circle (°) denotes the value of the process at zero. For a complete discussion of elastic states see LTE, Sect. 28.

[21] If $[\boldsymbol{u}, \boldsymbol{E}, \boldsymbol{S}]$ and $\mathbf{G}$ are in class H^N and $[\boldsymbol{u}, \boldsymbol{E}, \boldsymbol{S}]$ is in $\mathrm{QSVP}(\mathbf{G}, \boldsymbol{b})$, then $\boldsymbol{b}$ is necessarily of class H^N.

[22] $\mathring{\boldsymbol{b}}^{(N)}$ is equal to $\left[\dfrac{d^N}{dt^N}\boldsymbol{b}(t)\right]_{t=0^+}$.

[23] GURTIN and STERNBERG [1962, *10*, Theorem 6.2].

[24] LTE, Sect. 32, (1).

$[\boldsymbol{u}, \boldsymbol{E}, \boldsymbol{S}]$ is in QSVP$(\mathbf{G}, \boldsymbol{b})$ and is in class H^N with all the smoothness properties described above. Suppose, in addition, that $\mathscr{B}$ is a regular region in $\mathscr{E}$, the initial elasticity $\mathring{\mathbf{G}}$ is positive definite on $\mathscr{B}$, and for each $n=0, 1, \ldots, N$

$$\mathring{\boldsymbol{b}}^{(n)} = \mathbf{0} \quad \text{on } \mathscr{B}, \tag{25.4a}$$

$$\mathring{\boldsymbol{u}}^{(n)} = \mathbf{0} \quad \text{on } \mathscr{S}_{\hat{\boldsymbol{u}}}, \tag{25.4b}$$

$$\mathring{\boldsymbol{s}}^{(n)} = \mathbf{0} \quad \text{on } \mathscr{S}_{\hat{\boldsymbol{s}}}. \tag{25.4c}$$

Then, for $n=0, 1, 2, \ldots, N$,

$$[\mathring{\boldsymbol{u}}^{(n)}, \mathring{\boldsymbol{E}}^{(n)}, \mathring{\boldsymbol{S}}^{(n)}] = [\boldsymbol{w}_n, \mathbf{0}, \mathbf{0}], \tag{25.5}$$

where $\boldsymbol{w}_n$ is an (infinitesimal) rigid displacement.[25]

The above results show that for Heaviside processes it is possible to compute the initial viscoelastic state and its existing initial time derivatives directly from the boundary data and the body force process by solving a sequence of *elastic* problems for the region $\mathscr{B}$.

26. The past-history problem. The definition of *quasi-static viscoelastic process* given in Sect. 19 did not include specific restrictions upon the histories of such processes.[26] Thus the values of a quasi-static viscoelastic process need not vanish at any time. Recall the earlier observation that the material characterization (Boltzmann law) depends upon the class of processes under consideration: *for the class of restricted continuous processes, response functions must admit an equilibrium response; while for the class of restricted continuous processes which vanish up to some initial time t_0, response functions need not admit an equilibrium response.*[27]

In this section we consider quasi-static viscoelastic processes whose past history[28] up to some initial time t_0 is prescribed. By an ***initial past history*** we mean a past history $[\boldsymbol{v}, \boldsymbol{R}, \boldsymbol{T}]$ whose values are admissible states for $\mathscr{B}$.[29]

Henceforth, t_0 denotes some fixed initial time. By virtue of the time-translation invariance of linear hereditary laws, we assume, without loss in generality, that $t_0=0$. If $[\boldsymbol{v}, \boldsymbol{R}, \boldsymbol{T}]$ is a prescribed initial past history and $[\boldsymbol{u}, \boldsymbol{E}, \boldsymbol{S}]$ is an admissible process for $\mathscr{B}$, we say that $[\boldsymbol{u}, \boldsymbol{E}, \boldsymbol{S}]$ ***has initial past history*** $[\boldsymbol{v}, \boldsymbol{R}, \boldsymbol{T}]$ ***up to time*** t_0 if, and only if, $[\boldsymbol{u}, \boldsymbol{E}, \boldsymbol{S}]$ satisfies the ***initial past-history condition***

$${}_r[\boldsymbol{u}, \boldsymbol{E}, \boldsymbol{S}]^{t_0} = [\boldsymbol{v}, \boldsymbol{R}, \boldsymbol{T}]. \tag{26.1}$$

We observe that if $[\boldsymbol{u}, \boldsymbol{E}, \boldsymbol{S}]$ has the initial past history $[\boldsymbol{v}, \boldsymbol{R}, \boldsymbol{T}]$ up to time $t_0=0$ and is in either QSVP$(\mathbf{G}, \boldsymbol{b})$ or QSVP$(\mathbf{J}, \boldsymbol{b})$, then

$$\operatorname{div} \boldsymbol{T} + {}_r\boldsymbol{b}^{t_0} = \mathbf{0}, \tag{26.2}$$

$$\boldsymbol{R} = \tfrac{1}{2}[\nabla \boldsymbol{v} + \nabla \boldsymbol{v}^T], \tag{26.3}$$

---

[25] See GURTIN and STERNBERG [1962, *10*, Theorem 6.4]. Note that Eq. (25.3) implies that S_N is equal to $\mathring{\boldsymbol{S}}^{(N)}$ when $\mathring{\boldsymbol{E}}^{(n)} = \mathbf{0}$ for $n=0, 1, 2, \ldots, N-1$.

[26] Of course, we always assume that a process is sufficiently regular to render our operations meaningful.

[27] Cf. Sect. 8.

[28] Recall that if p is some process, then the *past history of p up to time t_0*, denoted by ${}_rp^{t_0}$, is defined on $\mathscr{R}^{++}$ by

$${}_rp^{t_0}(s) = p(t_0 - s), \quad s \in \mathscr{R}^{++}.$$

[29] For *relaxation* problems we assume that the values of $\boldsymbol{R}$ and $\boldsymbol{T}$ are class C symmetric tensor fields on $\bar{\mathscr{B}}$ and the values of $\boldsymbol{v}$ are class C^2 vector fields on $\bar{\mathscr{B}}$. For *creep* problems we assume that the values of $\boldsymbol{R}$ are class C symmetric tensor fields on $\bar{\mathscr{B}}$, the values of $\boldsymbol{T}$ are class C^1 symmetric tensor fields on $\bar{\mathscr{B}}$, and the values of $\boldsymbol{v}$ are class C^2 vector fields on $\bar{\mathscr{B}}$.

and, for $s \in \mathscr{R}^{++}$, either

$$\boldsymbol{T}(s) = \mathring{\mathbf{G}}[\boldsymbol{R}(s)] + \int_0^\infty \dot{\mathbf{G}}(\xi)[\boldsymbol{R}(s+\xi)]\,d\xi \tag{26.4}$$

or

$$\boldsymbol{R}(s) = \mathring{\mathbf{J}}[\boldsymbol{T}(s)] + \int_0^\infty \dot{\mathbf{J}}(\xi)[\boldsymbol{T}(s+\xi)]\,d\xi \tag{26.5}$$

according as $[\boldsymbol{u}, \boldsymbol{E}, \boldsymbol{S}]$ lies in QSVP$(\mathbf{G}, \boldsymbol{b})$ or QSVP$(\mathbf{J}, \boldsymbol{b})$. It follows that if $[\boldsymbol{v}, \boldsymbol{R}, \boldsymbol{T}]$ is the initial past history up to time t_0 of a quasi-static viscoelastic process (corresponding to body force $\boldsymbol{b}$), then the values of $\boldsymbol{v}$, $\boldsymbol{R}$, $\boldsymbol{T}$, and ${}_r\boldsymbol{b}^{t_0}$ may not be specified independently on $\mathscr{R}^{++}$.

Let $[\hat{\boldsymbol{u}}, \hat{\mathbf{s}}]$ be prescribed boundary data on $\partial\mathscr{B}$. Suppose $[\boldsymbol{u}, \boldsymbol{E}, \boldsymbol{S}]$ is a solution to the viscoelastic boundary value problem corresponding to the boundary data $[\hat{\boldsymbol{u}}, \hat{\mathbf{s}}]$, the body force $\boldsymbol{b}$, and either the relaxation function $\mathbf{G}$ or the creep compliance $\mathbf{J}$. Then, if $[\boldsymbol{u}, \boldsymbol{E}, \boldsymbol{S}]$ has the initial past history $[\boldsymbol{v}, \boldsymbol{R}, \boldsymbol{T}]$ up to time t_0, it follows that the values of $\boldsymbol{v}$, $\boldsymbol{R}$, and $\boldsymbol{T}$ must satisfy (26.2)–(26.5) on $\mathscr{B}$ together with the conditions on $\partial\mathscr{B}$

$$\boldsymbol{v} = {}_r\hat{\boldsymbol{u}}^{t_0} \quad \text{on } \mathscr{S}_{\hat{\boldsymbol{u}}}, \tag{26.6}$$

$$\boldsymbol{T}\boldsymbol{n} = {}_r\hat{\mathbf{s}}^{t_0} \quad \text{on } \mathscr{S}_{\hat{\mathbf{s}}}. \tag{26.7}$$

Consider the viscoelastic boundary value problem determined by $[\hat{\boldsymbol{u}}, \hat{\mathbf{s}}]$, $\boldsymbol{b}$, and either $\mathbf{G}$ or $\mathbf{J}$. By a ***solution to the viscoelastic initial past-history problem*** we mean a solution to the viscoelastic mixed boundary value problem which has prescribed initial past history up to time t_0. The prescribed initial past history must be ***consistent*** with the given boundary data, body force, and response function in the sense that the relationship expressed by (26.2)–(26.7) must hold.

Consider now the special case in which displacement boundary data are prescribed on $\partial\mathscr{B}$ ($\mathscr{S}_{\hat{\mathbf{s}}}$ is empty). Suppose further that the body $\mathscr{B}$ is of relaxation type with relaxation function $\mathbf{G}$. Then we may characterize solutions to the displacement problem purely in terms of displacements.[30] This formulation may be extended to include initial past-history problems for *arbitrary* prescribed initial past histories of the displacement.

Let $\boldsymbol{v}$ be an arbitrary prescribed past history whose values are class C^2 vector fields on $\mathscr{B}$. Further prescribe a function $\hat{\hat{\boldsymbol{u}}}$ on $\mathscr{R}^+$ whose values are continuous vector fields on $\partial\mathscr{B}$ and a function $\hat{\hat{\boldsymbol{b}}}$ on $\mathscr{R}^+$ whose values are continuous vector fields on $\overline{\mathscr{B}}$. Define displacement boundary data $\hat{\boldsymbol{u}}$ on $\partial\mathscr{B}$ by

$$\hat{\boldsymbol{u}}(t) = \begin{Bmatrix} \boldsymbol{v}(-t): & t < 0 \\ \hat{\hat{\boldsymbol{u}}}(t): & t \geqq 0 \end{Bmatrix} \tag{26.8}$$

and define a body force process $\boldsymbol{b}$ on $\overline{\mathscr{B}}$ by

$$\boldsymbol{b}(t) = \begin{Bmatrix} -\operatorname{div}\left\{\mathring{\mathbf{G}}[\hat{\nabla}\boldsymbol{v}(-t)] + \int_0^\infty \dot{\mathbf{G}}(s)[\hat{\nabla}\boldsymbol{v}(-t+s)]\,ds\right\}: & t < 0 \\ \hat{\hat{\boldsymbol{b}}}(t) & : t \geqq 0 \end{Bmatrix}. \tag{26.9}$$

Now let $\boldsymbol{u}$ be a process whose values are class C^2 vector fields on $\overline{\mathscr{B}}$ such that $\boldsymbol{u}$ satisfies the displacement equation of equilibrium (20.1) on $\mathscr{B}$ with the body

[30] Cf. Sects. 20, 23.

force $\boldsymbol{b}$ defined in (26.9). Further suppose that $\boldsymbol{u}$ satisfies the boundary condition

$$\boldsymbol{u} = \ddot{\boldsymbol{u}} \tag{26.10}$$

on $\partial\mathscr{B}$ and the initial past-history condition

$${}_r\boldsymbol{u}^0 = \boldsymbol{v} \tag{26.11}$$

on $\mathscr{B}$. Then the values of $\boldsymbol{u}$ on $\mathscr{B}$ for $t \geqq 0$ satisfy[31]

$$\operatorname{div}\{\overset{\circ}{\mathsf{G}}[\hat{\nabla}\boldsymbol{u}] + \dot{\mathsf{G}} * \hat{\nabla}\boldsymbol{u}\}(t) + \tilde{\boldsymbol{b}}(t) = \boldsymbol{0}, \tag{26.12}$$

where $\tilde{\boldsymbol{b}}(t)$ is given by

$$\tilde{\boldsymbol{b}}(t) = \hat{\hat{\boldsymbol{b}}}(t) + \operatorname{div} \int_0^\infty \dot{\mathsf{G}}(t+s)[\hat{\nabla}\boldsymbol{v}(s)]\, ds. \tag{26.13}$$

Also, on $\partial\mathscr{B}$, the values of $\boldsymbol{u}$ for $t \geqq 0$ satisfy

$$\boldsymbol{u}(t) = \hat{\boldsymbol{u}}(t). \tag{26.14}$$

Conversely, suppose $\boldsymbol{v}$, $\hat{\hat{\boldsymbol{u}}}$, and $\hat{\hat{\boldsymbol{b}}}$ are prescribed. Let $\tilde{\boldsymbol{u}}$ be a function on $\mathscr{R}^+$, whose values are class C^2 vector fields on $\overline{\mathscr{B}}$, such that $\tilde{\boldsymbol{u}}$ satisfies (26.12)–(26.14). Define a process $\boldsymbol{u}$ by

$$\boldsymbol{u}(t) = \begin{Bmatrix} \boldsymbol{v}(-t): & t < 0 \\ \tilde{\boldsymbol{u}}(t): & t \geqq 0 \end{Bmatrix}, \tag{26.15}$$

a process $\boldsymbol{E}$ through the strain-displacement relation (19.4), and a process $\boldsymbol{S}$ through the Boltzmann relaxation law (19.5). Then the process $[\boldsymbol{u}, \boldsymbol{E}, \boldsymbol{S}]$ is a solution to the initial past-history problem with displacement boundary data $\hat{\boldsymbol{u}}$ given by (26.8), body force $\boldsymbol{b}$ given by (26.9), and the relaxation function G. The initial past-history condition (26.1) becomes

$${}_r[\boldsymbol{u}, \boldsymbol{E}, \boldsymbol{S}]^0 = [\boldsymbol{v}, \boldsymbol{R}, \boldsymbol{T}] \tag{26.16}$$

where the past histories $\boldsymbol{R}$ and $\boldsymbol{T}$ are defined through $\boldsymbol{v}$ by (26.3) and (26.4).

In summary, if the values of the displacement process are known up to (but not including) the initial time $t_0 = 0$ and the body forces and surface displacements are prescribed for times greater than or equal to time $t_0 = 0$, then a solution to the initial past-history problem may be characterized purely in terms of displacements through solutions of (26.12)–(26.14). Observe that prescribed past histories enter the field equation (26.12) only through a modified body force.

If the prescribed initial past-history vanishes, a particularly simple formulation of the viscoelastic problem is obtained. Let the boundary data $[\hat{\boldsymbol{u}}, \hat{\boldsymbol{s}}]$ and body forces $\boldsymbol{b}$ be of Heaviside type. Then a vanishing initial past history is certainly consistent in the sense that (26.2)–(26.7) are identically satisfied. Hence, to construct a solution to the initial past-history problem for vanishing initial past history it suffices to find an admissible process $[\boldsymbol{u}, \boldsymbol{E}, \boldsymbol{S}]$ of Heaviside type whose values on $\mathscr{R}^+$ satisfy the equation of equilibrium (19.3), the strain-displacement relation (19.4) and either

$$\boldsymbol{S} = \mathsf{G} \circledast \boldsymbol{E} = \overset{\circ}{\mathsf{G}}[\boldsymbol{E}] + \dot{\mathsf{G}} * \boldsymbol{E} \tag{26.17}$$

or

$$\boldsymbol{E} = \mathsf{J} \circledast \boldsymbol{S} = \overset{\circ}{\mathsf{J}}[\boldsymbol{S}] + \dot{\mathsf{J}} * \boldsymbol{S} \tag{26.18}$$

[31] Recall the definitions and conventions regarding the convolution operator ($*$) established in Sect. 4.

on $\mathscr{B}$ together with the boundary conditions

$$\boldsymbol{u} = \hat{\boldsymbol{u}} \quad \text{on } \mathscr{S}_{\hat{\boldsymbol{u}}}, \tag{26.19}$$

$$\boldsymbol{S}\boldsymbol{n} \equiv \boldsymbol{s}_{(\boldsymbol{n})} = \hat{\boldsymbol{s}} \quad \text{on } \mathscr{S}_{\hat{\boldsymbol{s}}} \tag{26.20}$$

on $\partial\mathscr{B}$.

Finally suppose $[\boldsymbol{u}', \boldsymbol{E}', \boldsymbol{S}']$ and $[\boldsymbol{u}'', \boldsymbol{E}'', \boldsymbol{S}'']$ are two processes in QSVP$(\boldsymbol{G}, \boldsymbol{b})$ or QSVP$(\boldsymbol{J}, \boldsymbol{b})$ having the *same* initial past history up to time $t_0 = 0$. Then the process

$$[\boldsymbol{u}, \boldsymbol{E}, \boldsymbol{S}] = [\boldsymbol{u}', \boldsymbol{E}', \boldsymbol{S}'] - [\boldsymbol{u}'', \boldsymbol{E}'', \boldsymbol{S}'']$$

is in QSVP$(\boldsymbol{G}, \boldsymbol{0})$ or QSVP$(\boldsymbol{J}, \boldsymbol{0})$ and has vanishing initial past history; that is, $[\boldsymbol{u}, \boldsymbol{E}, \boldsymbol{S}]$ is of Heaviside type. If, in addition, the processes $[\boldsymbol{u}', \boldsymbol{E}', \boldsymbol{S}']$ and $[\boldsymbol{u}'', \boldsymbol{E}'', \boldsymbol{S}'']$ meet the *same* boundary conditions, then $[\boldsymbol{u}, \boldsymbol{E}, \boldsymbol{S}]$ has null boundary data.

Recalling the remark at the beginning of this section, we see that the important class of processes to be considered in the initial past-history problem is the class of Heaviside processes. For this reason, it is usually appropriate to omit the condition that response functions be strong and, hence, admit an equilibrium response. However, in its place we must require that the initial past history be such as to render all operations meaningful.

27. Integral theorems. If $\mathscr{B}$ is of relaxation type and $\boldsymbol{E}$ is a strain process for $\mathscr{B}$, define a scalar process $W_{\boldsymbol{E}}$ by[32]

$$W_{\boldsymbol{E}} = \int_{\mathscr{B}} \{\boldsymbol{E} \cdot (\boldsymbol{G} \circledast \boldsymbol{E})\}\, dV. \tag{27.1}$$

Similarly, if $\mathscr{B}$ is of creep type and $\boldsymbol{S}$ is a stress process for $\mathscr{B}$, define a scalar process $W_{\boldsymbol{S}}$ by[33]

$$W_{\boldsymbol{S}} = \int_{\mathscr{B}} \{\boldsymbol{S} \cdot (\boldsymbol{J} \circledast \boldsymbol{S})\}\, dV. \tag{27.2}$$

We now exhibit the viscoelastic analogs of the classical work-energy identities of elastostatics.[34]

If $[\boldsymbol{u}, \boldsymbol{E}, \boldsymbol{S}]$ is in QSVP$(\boldsymbol{G}, \boldsymbol{b})$, then[35]

$$\int_{\partial\mathscr{B}} \boldsymbol{s} \cdot \boldsymbol{u}\, dA + \int_{\mathscr{B}} \boldsymbol{b} \cdot \boldsymbol{u}\, dV = W_{\boldsymbol{E}}. \tag{27.3}$$

And if $[\boldsymbol{u}, \boldsymbol{E}, \boldsymbol{S}]$ is in QSVP$(\boldsymbol{J}, \boldsymbol{b})$ then[36]

$$\int_{\partial\mathscr{B}} \boldsymbol{s} \cdot \boldsymbol{u}\, dA + \int_{\mathscr{B}} \boldsymbol{b} \cdot \boldsymbol{u}\, dV = W_{\boldsymbol{S}}. \tag{27.4}$$

To verify the identities expressed by (27.3) and (27.4) we observe that the symmetry of the stress, the equation of equilibrium (19.3), and the divergence theorem together imply

$$\int_{\partial\mathscr{B}} \boldsymbol{s} \cdot \boldsymbol{u}\, dA + \int_{\mathscr{B}} \boldsymbol{b} \cdot \boldsymbol{u}\, dV = \int_{\mathscr{B}} \boldsymbol{E} \cdot \boldsymbol{S}\, dV. \tag{27.5}$$

Finally, the right-hand side of (27.5) is equal to $W_{\boldsymbol{E}}$ or $W_{\boldsymbol{S}}$ according as $[\boldsymbol{u}, \boldsymbol{E}, \boldsymbol{S}]$ is in QSVP$(\boldsymbol{G}, \boldsymbol{b})$ or QSVP$(\boldsymbol{J}, \boldsymbol{b})$.

---

32 The process $W_{\boldsymbol{E}}$ given in (27.1) is well defined whenever $\boldsymbol{E}$ has finite integrable histories or, if $\boldsymbol{G}$ is a strong relaxation function, whenever $\boldsymbol{E}$ has bounded or integrable histories.

33 Cf. footnote 32 with $\boldsymbol{E}$ and $\boldsymbol{G}$ replaced by $\boldsymbol{S}$ and $\boldsymbol{J}$.

34 See, e.g., LTE, Sect. 28.

35 Volterra [1909, *1*].

36 Breuer and Onat [1962, *3*].

Suppose $\mathbf{J}$ and $\mathbf{G}$ are strong response functions in class H^1 and that $\boldsymbol{u}$, $\boldsymbol{E}$, and $\boldsymbol{S}$ are restricted processes of class C^1 whose time derivatives have integrable histories.[37] If $\mathscr{B}$ is of relaxation type, define a scalar process $\mathscr{E}_{\boldsymbol{E}}$ by

$$\mathscr{E}_{\boldsymbol{E}}(t)=\int_{\mathscr{B}}\int_{-\infty}^{t}\int_{-\infty}^{t}\dot{\boldsymbol{E}}(\eta)\cdot\mathbf{G}(\eta-\eta')[\dot{\boldsymbol{E}}(\eta')]\,d\eta'\,d\eta\,dV. \tag{27.6}$$

Similarly, if $\mathscr{B}$ is of creep type, define a scalar process $\mathscr{E}_{\boldsymbol{S}}$ by

$$\mathscr{E}_{\boldsymbol{S}}(t)=\int_{\mathscr{B}}\int_{-\infty}^{t}\int_{-\infty}^{t}\dot{\boldsymbol{S}}(\eta)\cdot\mathbf{J}(\eta-\eta')[\dot{\boldsymbol{S}}(\eta')]\,d\eta'\,d\eta\,dV. \tag{27.7}$$

We now have the following

Theorem.[38] *Let* $[\boldsymbol{u},\boldsymbol{E},\boldsymbol{S}]$ *be a restricted admissible process for* $\mathscr{B}$.

(i) *If* $[\boldsymbol{u},\boldsymbol{E},\boldsymbol{S}]$ *is in* QSVP$(\mathbf{G},\boldsymbol{b})$ *and* $\boldsymbol{u}$ *is in class* C^1, *then for each time* t

$$\int_{-\infty}^{t}\left\{\int_{\partial\mathscr{B}}\mathbf{s}(\tau)\cdot\dot{\boldsymbol{u}}(\tau)\,dA+\int_{\mathscr{B}}\boldsymbol{b}(\tau)\cdot\dot{\boldsymbol{u}}(\tau)\,dV\right\}d\tau=\mathscr{E}_{\boldsymbol{E}}(t). \tag{27.8}$$

(ii) *If* $[\boldsymbol{u},\boldsymbol{E},\boldsymbol{S}]$ *is in* QSVP$(\mathbf{J},\boldsymbol{b})$ *and* $\boldsymbol{S}$ *is in class* C^1, *then for each time* t

$$\int_{-\infty}^{t}\left\{\int_{\partial\mathscr{B}}\dot{\mathbf{s}}(\tau)\cdot\boldsymbol{u}(\tau)\,dA+\int_{\mathscr{B}}\dot{\boldsymbol{b}}(\tau)\cdot\boldsymbol{u}(\tau)\,dV\right\}d\tau=\mathscr{E}_{\boldsymbol{S}}(t). \tag{27.9}$$

(iii) *If* $[\boldsymbol{u},\boldsymbol{E},\boldsymbol{S}]$ *satisfies the hypotheses of both* (i) *and* (ii), *then*

$$\int_{\partial\mathscr{B}}\mathbf{s}\cdot\boldsymbol{u}\,dA+\int_{\mathscr{B}}\boldsymbol{b}\cdot\boldsymbol{u}\,dV=\mathscr{E}_{\boldsymbol{E}}+\mathscr{E}_{\boldsymbol{S}}. \tag{27.10}$$

The identities (27.8)–(27.10) follow by standard manipulations from the *power identity:*

$$\int_{\partial\mathscr{B}}\mathbf{s}\cdot\dot{\boldsymbol{u}}\,dA+\int_{\mathscr{B}}\boldsymbol{b}\cdot\dot{\boldsymbol{u}}\,dV=\int_{\mathscr{B}}\boldsymbol{S}\cdot\dot{\boldsymbol{E}}\,dV. \tag{27.11}$$

The process on the left in (27.11) is the *rate of working on* $\mathscr{B}$ *of the surface tractions* $\mathbf{s}$ *and the body forces* $\boldsymbol{b}$. Hence $\mathscr{E}_{\boldsymbol{E}}$ represents the work done on $\mathscr{B}$ by the tractions $\mathbf{s}$ and the body forces $\boldsymbol{b}$. We see that, under sufficient smoothness hypotheses, part (iii) of the theorem implies,

$$\mathscr{E}_{\boldsymbol{E}}+\mathscr{E}_{\boldsymbol{S}}=W_{\boldsymbol{E}}=W_{\boldsymbol{S}}. \tag{27.12}$$

If we consider only processes of Heaviside type, we may remove the restriction that $\mathbf{G}$ and $\mathbf{J}$ be strong response functions. In this case the previous theorem remains true. In particular, if $t\geqq 0$, the integrals in (27.8)–(27.10) over $(-\infty, t)$ may be replaced by integrals over $[0, t)$, while if $t<0$, all expressions vanish identically. Note that a process of Heaviside type in class C^N $(N\geqq 0)$ must vanish at time $t=0$. The conclusions of the theorem do not obtain if the Heaviside processes have jump discontinuities at time $t=0$.

---

[37] These conditions guarantee that there is a non-negative constant c such that (27.6) implies

$$|\mathscr{E}_{\boldsymbol{E}}(t)|\leqq c\left(\int_0^\infty|\dot{\boldsymbol{E}}^t(s)|\,ds\right)^2,\quad t\in\mathscr{R}.$$

In fact, $c=\max\{|\mathbf{G}(s)|: s\in\mathscr{R}^+\}<\infty$. The same inequality is implied by (27.7) with $\boldsymbol{E}$ and $\mathbf{G}$ replaced by $\boldsymbol{S}$ and $\mathbf{J}$.

[38] This is a minor variation of a result of GURTIN and STERNBERG [1962, *10*, Theorem 7.2].

Recall Betti's theorem[39] in the theory of elastostatics. This result has the following viscoelastic counterpart, which result reduces to Betti's theorem if the response functions are constant on $\mathscr{R}^+$.

Reciprocal theorem.[40] *Let* $[\boldsymbol{u}, \boldsymbol{E}, \boldsymbol{S}]$ *and* $[\boldsymbol{u}', \boldsymbol{E}', \boldsymbol{S}']$ *be in* QSVP$(\mathbf{G}, \boldsymbol{b})$ *and* QSVP$(\mathbf{G}, \boldsymbol{b}')$, *or* QSVP$(\mathbf{J}, \boldsymbol{b})$ *and* QSVP$(\mathbf{J}, \boldsymbol{b}')$, *where the values of the relaxation function* $\mathbf{G}$, *or creep compliance* $\mathbf{J}$, *are symmetric (on* $\mathscr{T}_{\text{sym}}$*). Further, suppose the displacement processes* $\boldsymbol{u}$ *and* $\boldsymbol{u}'$, *or stress processes* $\boldsymbol{S}$ *and* $\boldsymbol{S}'$, *are of class* H^1. *Then, for each time* $t \geqq 0$,

$$\begin{aligned}\int_{\partial\mathscr{B}} \boldsymbol{u}' \circledast \boldsymbol{s}\, dA + \int_{\mathscr{B}} \boldsymbol{u}' \circledast \boldsymbol{b}\, dV &= \int_{\partial\mathscr{B}} \boldsymbol{u} \circledast \boldsymbol{s}'\, dA + \int_{\mathscr{B}} \boldsymbol{u} \circledast \boldsymbol{b}'\, dV \\ &= \int_{\mathscr{B}} \boldsymbol{E}' \circledast \boldsymbol{S}\, dV \\ &= \int_{\mathscr{B}} \boldsymbol{E} \circledast \boldsymbol{S}'\, dV.\end{aligned} \tag{27.13}$$

Suppose that the boundary data and the body force are *synchronous and separable* in the sense of Sect. 24; that is, there is a scalar process ϕ in Heaviside class H^1 such that[41]

$$\begin{aligned}\boldsymbol{b} &= \tilde{\boldsymbol{b}}\,\phi, & \boldsymbol{b}' &= \tilde{\boldsymbol{b}}'\,\phi & &\text{on } \mathscr{B}, \\ \boldsymbol{u} &= \tilde{\boldsymbol{u}}\,\phi, & \boldsymbol{u}' &= \tilde{\boldsymbol{u}}'\,\phi & &\text{on } \mathscr{S}_{\hat{\boldsymbol{u}}}, \\ \boldsymbol{s} &= \tilde{\boldsymbol{s}}\,\phi, & \boldsymbol{s}' &= \tilde{\boldsymbol{s}}'\,\phi & &\text{on } \mathscr{S}_{\hat{\boldsymbol{s}}}.\end{aligned} \tag{27.14}$$

Then Betti's theorem holds for the viscoelastic body; that is,[42]

$$\begin{aligned}\int_{\partial\mathscr{B}} \boldsymbol{s} \cdot \boldsymbol{u}'\, dA + \int_{\mathscr{B}} \boldsymbol{b} \cdot \boldsymbol{u}'\, dV &= \int_{\partial\mathscr{B}} \boldsymbol{s}' \cdot \boldsymbol{u}\, dA + \int_{\mathscr{B}} \boldsymbol{b}' \cdot \boldsymbol{u}\, dV \\ &= \int_{\mathscr{B}} \boldsymbol{S} \cdot \boldsymbol{E}'\, dV \\ &= \int_{\mathscr{B}} \boldsymbol{S}' \cdot \boldsymbol{E}\, dV.\end{aligned} \tag{27.15}$$

The symmetry of the response functions together with the restriction to processes of Heaviside type is necessary for the validity of these reciprocal relations.[43] If the body is isotropic, the condition of symmetry is always satisfied.

28. Uniqueness of quasi-static viscoelastic processes. Let $\mathscr{B}$ be a viscoelastic body which occupies a bounded regular region in $\mathscr{E}$. As usual, explicit dependence upon material points in $\mathscr{B}$ is suppressed.

Let $\mathbf{K}$ denote either a relaxation function $\mathbf{G}$ or a creep compliance $\mathbf{J}$ for $\mathscr{B}$. Suppose $[\boldsymbol{u}', \boldsymbol{E}', \boldsymbol{S}']$ and $[\boldsymbol{u}'', \boldsymbol{E}'', \boldsymbol{S}'']$ are processes in QSVP$(\mathbf{K}, \boldsymbol{b})$ which satisfy the same boundary data on $\partial\mathscr{B}$

$$\boldsymbol{u}' = \boldsymbol{u}'' \quad \text{on } \mathscr{S}_{\hat{\boldsymbol{u}}}, \tag{28.1}$$

$$\boldsymbol{s}' = \boldsymbol{s}'' \quad \text{on } \mathscr{S}_{\hat{\boldsymbol{s}}}. \tag{28.2}$$

Then the process $[\boldsymbol{u}, \boldsymbol{E}, \boldsymbol{S}]$ given by

$$[\boldsymbol{u}, \boldsymbol{E}, \boldsymbol{S}] = [\boldsymbol{u}', \boldsymbol{E}', \boldsymbol{S}'] - [\boldsymbol{u}'', \boldsymbol{E}'', \boldsymbol{S}''] \tag{28.3}$$

---

[39] See LTE, Sect. 30 (1).

[40] This result is given for isotropic materials by GURTIN and STERNBERG [1962, *10*, Theorem 7.3], and for anisotropic materials by GURTIN and STERNBERG [1963, *7*].

[41] We replace H^{ac} by the stronger condition H^1 for consistency within this section.

[42] That Betti's theorem holds for synchronous and separable loadings of a viscoelastic body was observed by GURTIN and STERNBERG [1962, *10*, Theorem 7.5].

[43] GURTIN and STERNBERG [1963, *7*].

is in QSVP$(\mathbf{K}, \mathbf{0})$ and satisfies *null boundary data on* $\partial\mathscr{B}$:

$$\boldsymbol{u} = \mathbf{0} \quad \text{on } \mathscr{S}_{\hat{\boldsymbol{u}}}, \tag{28.4}$$

$$\mathbf{s} = \mathbf{0} \quad \text{on } \mathscr{S}_{\hat{\mathbf{s}}}. \tag{28.5}$$

If the body $\mathscr{B}$ is of relaxation type, so that $\mathbf{K} = \mathbf{G}$, then the identity (27.3) implies that the scalar process $W_{\boldsymbol{E}}$ vanishes, where $W_{\boldsymbol{E}}$ is defined in (27.1). Similarly, if the body $\mathscr{B}$ is of creep type, so that $\mathbf{K} = \mathbf{J}$, then the identity (27.4) implies that the scalar process $W_{\boldsymbol{S}}$ vanishes, where $W_{\boldsymbol{S}}$ is defined in (27.2). Therefore, either $W_{\boldsymbol{E}}$ or $W_{\boldsymbol{S}}$ vanishes for sufficiently regular $[\boldsymbol{u}, \boldsymbol{E}, \boldsymbol{S}]$.

If $W_{\boldsymbol{E}} = 0$ implies that the strain process $\boldsymbol{E} = \mathbf{0}$, then the corresponding stress process $\boldsymbol{S} = 0$. Moreover, it can be shown that the displacement process $\boldsymbol{u}$ corresponding to $\boldsymbol{E} = \mathbf{0}$ is an arbitrary (infinitesimal) rigid motion of $\mathscr{B}$.[44] Similarly, if $W_{\boldsymbol{S}} = 0$ implies that the stress process $\boldsymbol{S} = 0$, then the corresponding strain process $\boldsymbol{E} = \mathbf{0}$. Hence, the displacement process $\boldsymbol{u}$ corresponding to $\boldsymbol{E} = \mathbf{0}$ is an arbitrary (infinitesimal) rigid motion of $\mathscr{B}$. In summary, *if the body $\mathscr{B}$ is of relaxation type and $W_{\boldsymbol{E}} = 0$ implies $\boldsymbol{E} = \mathbf{0}$ or if the body $\mathscr{B}$ is of creep type and $W_{\boldsymbol{S}} = 0$ implies $\boldsymbol{S} = \mathbf{0}$, then $[\boldsymbol{u}, \boldsymbol{E}, \boldsymbol{S}] = [\boldsymbol{w}, \mathbf{0}, \mathbf{0}]$, where $\boldsymbol{w}$ is an (infinitesimal) rigid motion of $\mathscr{B}$.*

We are thus led to seek conditions on the response functions $\mathbf{G}$ and $\mathbf{J}$ such that the conditions of the preceding paragraph are satisfied for a suitable class of processes $[\boldsymbol{u}, \boldsymbol{E}, \boldsymbol{S}]$. In this case, quasi-static viscoelastic states will be unique to within arbitrary (infinitesimal) rigid motions.

Let $\boldsymbol{P}$ denote a process whose values are continuous symmetric-tensor fields on $\bar{\mathscr{B}}$. Consider the scalar process $Q_{\boldsymbol{P}}$ given by:

$$Q_{\boldsymbol{P}} = \int_{\mathscr{B}} \{\boldsymbol{P} \cdot (\mathbf{K} \circledast \boldsymbol{P})\}\, dV. \tag{28.6}$$

Observe that $Q_{\boldsymbol{P}} = W_{\boldsymbol{E}}$ if $\mathbf{K} = \mathbf{G}$ and $\boldsymbol{P} = \boldsymbol{E}$ and $Q_{\boldsymbol{P}} = W_{\boldsymbol{S}}$ if $\mathbf{K} = \mathbf{J}$ and $\boldsymbol{P} = \boldsymbol{S}$.

Volterra's lemma.[45] *Let $\mathbf{K}$ be a response function for $\mathscr{B}$ of class H^1 whose initial response $\overset{\circ}{\mathbf{K}}$ is symmetric and definite (in $\mathscr{T}_{\text{sym}}$).*[46] *Suppose $\boldsymbol{P}$ is a process of Heaviside class H whose values are symmetric-tensor fields on $\mathscr{B}$. Then the process $Q_{\boldsymbol{P}}$ given in* (28.6) *is well defined, and $Q_{\boldsymbol{P}}$ vanishes if, and only if, $\boldsymbol{P}$ vanishes.*

By virtue of the time-translation invariance of linear hereditary laws, it follows that Volterra's lemma may be extended to include processes $\boldsymbol{P}$ which vanish up to some time t_0. Volterra's lemma and the discussion preceding it together yield the following

Uniqueness theorem for quasi-static viscoelastic processes (Volterra). *Let $[\boldsymbol{u}, \boldsymbol{E}, \boldsymbol{S}]$ and $[\boldsymbol{u}', \boldsymbol{E}', \boldsymbol{S}']$ be processes of Heaviside class H in* QSVP$(\mathbf{K}, \boldsymbol{b})$, *where $\mathbf{K}$ denotes either the relaxation function $\mathbf{G}$ or the creep compliance $\mathbf{J}$. Further suppose that these processes satisfy the same boundary data on $\partial\mathscr{B}$:*

$$\boldsymbol{u} = \boldsymbol{u}' \quad \text{on } \mathscr{S}_{\hat{\boldsymbol{u}}}, \tag{28.7}$$

$$\mathbf{s} = \mathbf{s}' \quad \text{on } \mathscr{S}_{\hat{\mathbf{s}}}. \tag{28.8}$$

---

[44] See the discussion of infinitesimal rigid motions in LTE, Sects. 12–13.

[45] VOLTERRA [1909, *1*], [1913, *1*], the version stated here for tensor processes is a straightforward modification of Volterra's result. Also see GURTIN and STERNBERG [1962, *10*, Sect. 8].

[46] For Volterra's lemma to hold it is not necessary that $\mathbf{K}$ be a strong response function. Cf. Sect. 8.

If K *is of class* H^1 *and its initial response* $\overset{\circ}{\mathsf{K}}$ *is symmetric and definite (in* $\mathscr{T}_{\text{sym}}$*),*[47] *then*

$$[\boldsymbol{u}, \boldsymbol{E}, \boldsymbol{S}] = [\boldsymbol{u}', \boldsymbol{E}', \boldsymbol{S}'] + [\boldsymbol{w}, \boldsymbol{0}, \boldsymbol{0}], \tag{28.9}$$

where $\boldsymbol{w}$ *is an (infinitesimal) rigid motion of* $\mathscr{B}$ *in class* H.

By virtue of the remarks following the statement of Volterra's lemma, it follows that the uniqueness theorem above also applies to processes in QSVP$(\mathsf{K}, \boldsymbol{b})$ which vanish up to some time t_0. Moreover, recalling the definition of weak viscoelastic process corresponding to a differential operator law,[48] we see that the uniqueness theorem above also holds for such processes provided the associated response function satisfies the hypotheses of the theorem.

The uniqueness theorem is easily extended to include mixed-mixed boundary data as well as those cases in which $\overline{\mathscr{B}}$ is fixed but $\mathscr{S}_{\hat{\boldsymbol{u}}}$ and $\mathscr{S}_{\hat{\boldsymbol{s}}}$ vary with time.[49] For the special case of an isotropic material, the conditions on the response function K may be replaced by: K_1 and K_2 are of class H^1 and $\overset{\circ}{\mathsf{K}}_1 \overset{\circ}{\mathsf{K}}_2 > 0$, where K_1 and K_2 denote the (scalar) shear and compression relaxation moduli or their corresponding creep compliances[50].

The above results and comments reduce to their counterparts in the classical theory of elastostatics whenever the response function K is constant on $\mathscr{R}^+$.

Concerning the initial past-history problem[51] we have the following result, corollary to the uniqueness theorem: *Let* $[\boldsymbol{u}, \boldsymbol{E}, \boldsymbol{S}]$ *and* $[\boldsymbol{u}', \boldsymbol{E}', \boldsymbol{S}']$ *be continuous processes in* QSVP$(\mathsf{K}, \boldsymbol{b})$ *whose past histories up to some time* t_0*, say,* $t_0 = 0$*, agree:*

$$_r[\boldsymbol{u}, \boldsymbol{E}, \boldsymbol{S}]^0 = {}_r[\boldsymbol{u}', \boldsymbol{E}', \boldsymbol{S}']^0. \tag{28.10}$$

Furthermore, on $\partial\mathscr{B}$ *let*

$$\boldsymbol{u} = \boldsymbol{u}' \quad \text{on } \mathscr{S}_{\hat{\boldsymbol{u}}}, \tag{28.11}$$

$$\boldsymbol{s} = \boldsymbol{s}' \quad \text{on } \mathscr{S}_{\hat{\boldsymbol{s}}}. \tag{28.12}$$

Then, if K *satisfies the hypotheses of the uniqueness theorem,*

$$[\boldsymbol{u}, \boldsymbol{E}, \boldsymbol{S}] = [\boldsymbol{u}', \boldsymbol{E}', \boldsymbol{S}'] + [\boldsymbol{w}, \boldsymbol{0}, \boldsymbol{0}], \tag{28.13}$$

where $\boldsymbol{w}$ *is a continuous (infinitesimal) rigid motion of* $\mathscr{B}$ *of class* H.

Suppose now that the initial response $\overset{\circ}{\mathsf{K}}$ is not definite (in $\mathscr{T}_{\text{sym}}$). This would be the case, for example, if the material were of Kelvin-Voigt type,[52] in which case $\overset{\circ}{\mathsf{K}} = \boldsymbol{0}$. Then Volterra's uniqueness theorem does not apply. If the initial response $\overset{\circ}{\mathsf{K}}$ vanishes, the conclusion of Volterra's theorem holds, provided that the condition that $\overset{\circ}{\mathsf{K}}$ be symmetric and definite (in $\mathscr{T}_{\text{sym}}$) is replaced by the con-

[47] Recall the variant of the inversion theorem for response functions (of class H^1) in Sect. 11. If the initial response $\overset{\circ}{\mathsf{K}}$ is definite (on $\mathscr{T}_{\text{sym}}$), it is certainly invertible in $\mathscr{T}_{\text{sym}}$. Hence the conditions on K in VOLTERRA's uniqueness theorem guarantee that the Boltzmann law $\mathscr{L}_{\mathsf{K}}$ is invertible in the Heaviside classes.

[48] Cf. Sect. 19.

[49] Cf. Sect. 22.

[50] Cf. Sect. 13. For the special case of an isotropic material of relaxation type, if displacement boundary data are given, then $\overset{\circ}{G}_1 \overset{\circ}{G}_2 > 0$ may be replaced by the weaker condition $\overset{\circ}{G}_1 (2\overset{\circ}{G}_1 + \overset{\circ}{G}_2) > 0$. See GURTIN and STERNBERG [1960, *5*], [1962, *10*].

[51] Cf. Sect. 26.

[52] Cf. Sect. 16 for the scalar versions.

dition that $\dot{\mathbf{K}}$ be of definite type.[53] That is, the scalar process $\Omega_{\boldsymbol{P}}$, given by[54]

$$\Omega_{\boldsymbol{P}}(t) = \int_{\mathscr{B}} \int_{-\infty}^{t} \int_{-\infty}^{t} \boldsymbol{P}(\tau) \cdot \dot{\mathbf{K}}(\tau - s)\,[\boldsymbol{P}(s)]\, ds\, d\tau\, dV, \qquad t \in \mathscr{R} \tag{28.14}$$

where $\boldsymbol{P}$ is a process in class H whose values are continuous symmetric-tensor fields on $\overline{\mathscr{B}}$, is of constant sign for non-zero $\boldsymbol{P}$.[55]

The problem of establishing uniqueness (and existence) of viscoelastic processes has been studied with precision and elegance by BABUŠKA, HLAVAČEK, and PREDELEANU.[56] These authors restrict themselves to processes of Heaviside type and consider several different types of boundary data. The work of HLAVAČEK and PREDELEANU concerns the linear problem and includes non-time-translation invariant (aging) effects. BABUŠKA and HLAVAČEK include nonlinear hereditary response in their study. For linear hereditary laws, the results obtained by these authors regarding uniqueness reduce essentially to those considered in this section.

Thus far we have considered only processes $[\boldsymbol{u}, \boldsymbol{E}, \boldsymbol{S}]$ of Heaviside type or which have finite histories. To include processes where histories are not finite, we provide the following

Lemma. *Let* $\mathbf{K}$ *be a strong response function whose initial response* $\overset{\circ}{\mathbf{K}}$ *is symmetric and positive-definite on* $\mathscr{B}$.

(i) *Let* $\boldsymbol{P}$ *be a process, whose values are symmetric tensor fields on* $\mathscr{B}$, *such that the scalar process*

$$\int_{\mathscr{B}} |\boldsymbol{P}|\, dV \tag{28.15}$$

has integrable histories. If[57]

$$\int_0^{\infty} \sup_{\mathscr{B}} \{|\overset{\circ}{\mathbf{K}}{}^{\frac{1}{2}}\, \dot{\mathbf{K}}(s)\, \overset{\circ}{\mathbf{K}}{}^{\frac{1}{2}}|_{[\mathscr{T}_{\text{sym}}]}\}\, ds < 1, \tag{28.16}$$

then $Q_{\boldsymbol{P}} = 0$ *if, and only if,* $\boldsymbol{P} = \mathbf{0}$.

(ii) *Let* $\boldsymbol{P}$ *be a process, whose values are symmetric tensor fields on* $\mathscr{B}$, *such that the scalar processes*

$$\int_{\mathscr{B}} |\boldsymbol{P}|\, dV \quad \text{and} \quad \int_{\mathscr{B}} |\boldsymbol{P}|^2\, dV \tag{28.17}$$

---

[53] For a discussion of the scalar version of this result see GURTIN and STERNBERG [1962, *10*, Sect. 8].

[54] Here we consider $\mathbf{K}$ as a process in class H^1 so that $\dot{\mathbf{K}}$ may be considered as a process of class H.

[55] GURTIN and STERNBERG [1962, *10*] provide characterizations for scalar functions on $\mathscr{R}^+$ of positive definite type. In particular the Kelvin-Voigt solid of Sect. 16 is of positive definite type. BREUER and ONAT [1962, *3*] provide a uniqueness theorem using scalar response functions of positive definite type. They show that if the scalar response function K has positive initial response ($\overset{\circ}{K} > 0$), decreases monotonically to $\overset{\infty}{K} > 0$, and is convex from below, then K is of positive definite type and viscoelastic processes are unique within a rigid motion of $\mathscr{B}$. It has been observed by GURTIN and STERNBERG [1962, *10*, Sect. 8] that these assumptions are more restrictive than those of VOLTERRA [1909, *1*], [1913, *1*].

[56] BABUŠKA and HLAVAČEK [1966, *3*], HLAVAČEK and PREDELEANU [1964, *9*], [1965, *19*], [1966, *11*].

[57] Recall the conditions given in Sect. 11 on a response function $\mathbf{K}$ which guarantee the invertibility of $\mathscr{L}_{\mathbf{K}}$.

have integrable histories. If[58]

$$\sup_{\mathscr{B}} \left\{ \int_0^\infty |\overset{\circ}{\mathbf{K}}{}^{\frac{1}{2}} \dot{\mathbf{K}}(s) \overset{\circ}{\mathbf{K}}{}^{\frac{1}{2}}|_{[\mathscr{T}_{\text{sym}}]} \, ds \right\} < 1 \tag{28.18}$$

then $Q_{\boldsymbol{P}} = 0$ if, and only if, $\boldsymbol{P} = \mathbf{0}$.

Notice that (28.16) implies (28.18), but not conversely, so that the hypotheses of assertion (i) do not imply those of assertion (ii), or conversely.[59]

The lemma and previous arguments provide the following

Uniqueness theorem for processes with infinite histories. *Let $[\boldsymbol{u}, \boldsymbol{E}, \boldsymbol{S}]$ and $[\boldsymbol{u}', \boldsymbol{E}', \boldsymbol{S}']$ be processes for $\mathscr{B}$ in* QSVP$(\mathbf{K}, \boldsymbol{b})$ *such that on $\partial\mathscr{B}$*

$$\boldsymbol{u} = \boldsymbol{u}' \quad \text{on } \mathscr{S}_{\hat{\boldsymbol{u}}}, \tag{28.19}$$

$$\mathbf{s} = \mathbf{s}' \quad \text{on } \mathscr{S}_{\hat{\mathbf{s}}}. \tag{28.20}$$

If $\mathbf{K} = \mathbf{G}$ and the hypotheses of the lemma are satisfied for $\boldsymbol{P} \equiv \boldsymbol{E} - \boldsymbol{E}'$ or if $\mathbf{K} = \mathbf{J}$ and the hypotheses of the lemma are satisfied for $\boldsymbol{P} \equiv \boldsymbol{S} - \boldsymbol{S}'$, then

$$[\boldsymbol{u}, \boldsymbol{E}, \boldsymbol{S}] = [\boldsymbol{u}', \boldsymbol{E}', \boldsymbol{S}'] + [\boldsymbol{w}, \mathbf{0}, \mathbf{0}],$$

where $\boldsymbol{w}$ is an (infinitesimal) rigid motion of $\mathscr{B}$.

29. Existence of quasi-static viscoelastic processes. In this section we show that solutions to the viscoelastic problem may be determined through the solutions to a family of elastic problems together with the solution to an integral equation. Our method is essentially that presented by Babuška, Hlavaček, Predeleanu, and Fichera.[60] To avoid the extensive technical considerations necessary for a general treatment of this problem, we proceed formally and specifically remark those places where questions of a technical nature must be addressed.

Let $\mathscr{B}$ denote a body which occupies a bounded regular region in $\mathscr{E}$. As usual, explicit dependence upon material points in $\mathscr{B}$ will be shown only when confusion may otherwise occur. Throughout this section $\boldsymbol{b}$ denotes a prescribed body force process for $\mathscr{B}$ and $[\hat{\boldsymbol{u}}, \hat{\mathbf{s}}]$ denotes prescribed boundary data for $\partial\mathscr{B}$. For definiteness, let $\mathscr{B}$ be of relaxation type, with relaxation function $\mathbf{G}$ of class H^1. This assumption is not too restrictive, for it will be seen in the sequel that, *roughly speaking*, invertibility of the Boltzmann law is sufficient to guarantee the existence, and uniqueness, of quasi-static viscoelastic processes.

Consider the viscoelastic mixed boundary-value problem for $\mathscr{B}$ corresponding to the relaxation function $\mathbf{G}$, the body force $\boldsymbol{b}$, and the boundary data $[\hat{\boldsymbol{u}}, \hat{\mathbf{s}}]$.[61] That is, we seek an admissible process $[\boldsymbol{u}, \boldsymbol{E}, \boldsymbol{S}]$ for $\mathscr{B}$ whose value at each time t satisfies the field equations

$$\operatorname{div} \boldsymbol{S}(t) + \boldsymbol{b}(t) = \mathbf{0}, \qquad \boldsymbol{S}(t) = \boldsymbol{S}^T(t), \tag{29.1}$$

$$\boldsymbol{E}(t) = \tfrac{1}{2}[\nabla \boldsymbol{u}(t) + \nabla \boldsymbol{u}^T(t)] \equiv \hat{\nabla} \boldsymbol{u}(t), \tag{29.2}$$

$$\boldsymbol{S}(t) = (\mathbf{G} \circledast \boldsymbol{E})(t) = \overset{\circ}{\mathbf{G}}[\boldsymbol{E}(t)] + \int_0^\infty \dot{\mathbf{G}}(s)[\boldsymbol{E}(t-s)] \, ds \tag{29.3}$$

---

[58] Cf. footnote 57.

[59] Recall that a square integrable function need not be integrable on an *unbounded* interval.

[60] Babuška and Hlavaček [1966, *3*]; Hlavaček and Predeleanu [1964, *9*], [1965, *19*], [1966, *11*]; Fichera [1965, *12*]. See also Edelstein [1966, *6*].

[61] Cf. Sects. 22, 23, 26.

on $\mathscr{B}$, and the boundary conditions

$$\boldsymbol{u}(t) = \hat{\boldsymbol{u}}(t) \qquad \text{on } \mathscr{S}_{\hat{\boldsymbol{u}}}, \tag{29.4}$$

$$\boldsymbol{S}(t)\,\boldsymbol{n} = \mathbf{s}_{(\boldsymbol{n})}(t) = \hat{\mathbf{s}}(t) \qquad \text{on } \mathscr{S}_{\hat{\mathbf{s}}} \tag{29.5}$$

on $\partial\mathscr{B}$. Such a process will be called a ***solution to the viscoelastic problem*** or, simply, a ***viscoelastic solution.***

We have supposed the body $\mathscr{B}$ to be viscoelastic with relaxation function $\mathbf{G}$. If $\mathring{\mathbf{G}}$ denotes the initial elasticity, we call $\mathscr{B}$ the ***associated elastic body*** if, and only if, $\mathscr{B}$ is elastic with elasticity $\mathring{\mathbf{G}}$. Of course, the associated elastic body is viscoelastic with the relaxation function $h\mathring{\mathbf{G}}$.[62] By a ***solution to the associated elastic problem*** or, simply, an ***associated elastic solution*** we mean an admissible process $[\tilde{\boldsymbol{u}}, \tilde{\boldsymbol{E}}, \tilde{\boldsymbol{S}}]$ for $\mathscr{B}$ whose value at each time t satisfies the field equations

$$\operatorname{div} \tilde{\boldsymbol{S}}(t) + \boldsymbol{b}(t) = \boldsymbol{0}, \qquad \tilde{\boldsymbol{S}}(t) = \tilde{\boldsymbol{S}}^T(t), \tag{29.6}$$

$$\tilde{\boldsymbol{E}}(t) = \tfrac{1}{2}[\nabla\tilde{\boldsymbol{u}}(t) + \nabla\tilde{\boldsymbol{u}}^T(t)] \equiv \hat{\nabla}\tilde{\boldsymbol{u}}(t), \tag{29.7}$$

$$\tilde{\boldsymbol{S}}(t) = \mathring{\mathbf{G}}[\tilde{\boldsymbol{E}}(t)] \tag{29.8}$$

on $\mathscr{B}$, and the boundary conditions

$$\tilde{\boldsymbol{u}}(t) = \hat{\boldsymbol{u}}(t) \qquad \text{on } \mathscr{S}_{\hat{\boldsymbol{u}}}, \tag{29.9}$$

$$\tilde{\boldsymbol{S}}(t)\,\boldsymbol{n} = \tilde{\mathbf{s}}_{(\boldsymbol{n})}(t) = \hat{\mathbf{s}}(t) \qquad \text{on } \mathscr{S}_{\hat{\mathbf{s}}} \tag{29.10}$$

on $\partial\mathscr{B}$. Clearly, an associated elastic solution is a solution to the viscoelastic boundary-value problem for $\mathscr{B}$ corresponding to the constant relaxation function $h\mathring{\mathbf{G}}$, the body force $\boldsymbol{b}$, and the boundary data $[\hat{\boldsymbol{u}}, \hat{\mathbf{s}}]$. In this case the viscoelastic states $[\tilde{\boldsymbol{u}}, \tilde{\boldsymbol{E}}, \tilde{\boldsymbol{S}}](t)$ determined for each time t through (29.6)–(29.10) are elastic states.

Let $[\boldsymbol{u}, \boldsymbol{E}, \boldsymbol{S}]$ be a viscoelastic solution and let $[\tilde{\boldsymbol{u}}, \tilde{\boldsymbol{E}}, \tilde{\boldsymbol{S}}]$ be an associated elastic solution. Define an admissible process $[\check{\boldsymbol{u}}, \check{\boldsymbol{E}}, \check{\boldsymbol{S}}]$ for $\mathscr{B}$ by

$$[\check{\boldsymbol{u}}, \check{\boldsymbol{E}}, \check{\boldsymbol{S}}] = [\boldsymbol{u}, \boldsymbol{E}, \boldsymbol{S}] - [\tilde{\boldsymbol{u}}, \tilde{\boldsymbol{E}}, \tilde{\boldsymbol{S}}]. \tag{29.11}$$

Then (29.1)–(29.11) imply that the value of $[\check{\boldsymbol{u}}, \check{\boldsymbol{E}}, \check{\boldsymbol{S}}]$ at each time t satisfies the field equations

$$\operatorname{div} \check{\boldsymbol{S}}(t) = \boldsymbol{0}, \qquad \check{\boldsymbol{S}}(t) = \check{\boldsymbol{S}}^T(t), \tag{29.12}$$

$$\check{\boldsymbol{E}}(t) = \tfrac{1}{2}[\nabla\check{\boldsymbol{u}}(t) + \nabla\check{\boldsymbol{u}}^T(t)] \equiv \hat{\nabla}\check{\boldsymbol{u}}(t), \tag{29.13}$$

$$\check{\boldsymbol{S}}(t) = \mathring{\mathbf{G}}[\check{\mathbf{E}}(t) - \boldsymbol{E}^*(t)] \tag{29.14}$$

on $\mathscr{B}$, where $\boldsymbol{E}^*(t)$ is defined on $\mathscr{B}$ by[63]

$$\boldsymbol{E}^*(t) = -\int_0^\infty \mathring{\mathbf{G}}^{-1}\,\dot{\mathbf{G}}(s)\,[\boldsymbol{E}(t-s)]\,ds. \tag{29.15}$$

[62] Recall that h denotes the scalar Heaviside unit step process so that $\mathbf{G} = h\mathring{\mathbf{G}}$ is constant on $\mathscr{R}^+$.

[63] Since we are proceeding formally, we suppose that the initial elasticity $\mathring{\mathbf{G}}$ is invertible in $\mathscr{T}_{\text{sym}}$.

Moreover, the value of $[\check{\boldsymbol{u}}, \check{\boldsymbol{E}}, \check{\boldsymbol{S}}]$ at each time t satisfies the null boundary data on $\partial\mathscr{B}$

$$\check{\boldsymbol{u}}(t) = \boldsymbol{0} \qquad \text{on } \mathscr{S}_{\hat{\boldsymbol{u}}}, \tag{29.16}$$

$$\check{\boldsymbol{S}}(t)\,\boldsymbol{n} = \check{\boldsymbol{s}}_{(\boldsymbol{n})}(t) = \boldsymbol{0} \qquad \text{on } \mathscr{S}_{\hat{\boldsymbol{s}}}. \tag{29.17}$$

Let $\boldsymbol{E}^*$ denote a *prescribed* process for $\mathscr{B}$ whose values are symmetric tensor fields on $\mathscr{B}$. Then by a ***solution to the auxiliary problem for*** $\boldsymbol{E}^*$ or, simply, an ***auxiliary solution*** we mean an admissible process $[\check{\boldsymbol{u}}, \check{\boldsymbol{E}}, \check{\boldsymbol{S}}](\boldsymbol{E}^*)$ whose value at each time t satisfies the field equations (29.12)–(29.14) on $\mathscr{B}$ and the null boundary conditions (29.16) and (29.17) on $\partial\mathscr{B}$.

Assuming that the associated elastic problem has a solution $[\tilde{\boldsymbol{u}}, \tilde{\boldsymbol{E}}, \tilde{\boldsymbol{S}}]$, we can summarize the previous discussion as follows: *if* $[\boldsymbol{u}, \boldsymbol{E}, \boldsymbol{S}]$ *is a viscoelastic solution and* $\boldsymbol{E}^*$ *satisfies* (29.15), *then* $[\check{\boldsymbol{u}}, \check{\boldsymbol{E}}, \check{\boldsymbol{S}}]$ *defined through* (29.11) *is a solution to the auxiliary problem for* $\boldsymbol{E}^*$: $[\check{\boldsymbol{u}}, \check{\boldsymbol{E}}, \check{\boldsymbol{S}}] = [\check{\boldsymbol{u}}, \check{\boldsymbol{E}}, \check{\boldsymbol{S}}](\boldsymbol{E}^*)$. If $[\tilde{\boldsymbol{u}}, \tilde{\boldsymbol{E}}, \tilde{\boldsymbol{S}}]$ is a known associated elastic solution, then a viscoelastic solution $[\boldsymbol{u}, \boldsymbol{E}, \boldsymbol{S}]$ is completely determined by the process $[\check{\boldsymbol{u}}, \check{\boldsymbol{E}}, \check{\boldsymbol{S}}]$ through (29.11).

Observe that if $\boldsymbol{E}^*$ were known and satisfied (29.15), then $[\check{\boldsymbol{u}}, \check{\boldsymbol{E}}, \check{\boldsymbol{S}}] = [\check{\boldsymbol{u}}, \check{\boldsymbol{E}}, \check{\boldsymbol{S}}](\boldsymbol{E}^*)$ could be computed and, by virtue of the above comments, the process $[\boldsymbol{u}, \boldsymbol{E}, \boldsymbol{S}]$ defined through (29.11) is a viscoelastic solution. However $\boldsymbol{E}^*$ depends upon $\boldsymbol{E}$ through (29.15), which process is generally not known in advance. The next result shows that the auxiliary problem may still be used to determine a viscoelastic solution.

Suppose the auxiliary problem has a *unique* solution $[\check{\boldsymbol{u}}, \check{\boldsymbol{E}}, \check{\boldsymbol{S}}](\boldsymbol{E}^*)$ for a prescribed $\boldsymbol{E}^*$. Let $\check{\mathscr{E}}$ denote the function which assigns $\check{\boldsymbol{E}}$ to $\boldsymbol{E}^*$:

$$\check{\boldsymbol{E}} = \check{\mathscr{E}}(\boldsymbol{E}^*). \tag{29.18}$$

Clearly the function $\check{\mathscr{E}}$ is linear.

Theorem (Babuška and Hlavaček).[64] *In order to determine the process* $[\check{\boldsymbol{u}}, \check{\boldsymbol{E}}, \check{\boldsymbol{S}}]$ *by means of the auxiliary problem, that is,* $[\check{\boldsymbol{u}}, \check{\boldsymbol{E}}, \check{\boldsymbol{S}}] = [\check{\boldsymbol{u}}, \check{\boldsymbol{E}}, \check{\boldsymbol{S}}](\boldsymbol{E}^*)$, *it is necessary and sufficient that the process* $\boldsymbol{E}^*$ *satisfy the following* ***hereditary integral equation*** *on* $\mathscr{B}$:

$$\boldsymbol{E}^*(t) = -\int_0^\infty \mathring{\mathsf{G}}^{-1}\,\dot{\mathsf{G}}(s)\,[\tilde{\boldsymbol{E}}(t-s) + \check{\mathscr{E}}(\boldsymbol{E}^*)(t-s)]\,ds, \qquad t \in \mathscr{R}. \tag{29.19}$$

From the above discussion we conclude that if the associated elastic problem, the auxiliary problem, and the hereditary integral equation each have (sufficiently regular) unique solutions, then the viscoelastic problem will also have a (sufficiently regular) unique solution.

Observe that the time appears only as a parameter in the associated elastic and auxiliary problems. Thus, the solution of the viscoelastic problem reduces to the solution of two families of elastic problems (parameterized by time) and a single hereditary integral equation. The hereditary integral equation characterizes the history dependence of the solutions to the viscoelastic problem.

In the particular case of an initial past-history problem the following modifications are required. Choose an initial time t_0, say, $t_0 = 0$. Let $\boldsymbol{R}$ be the prescribed

---

[64] Babuška and Hlavaček [1966, *3*, Theorem I.3]. Only the formal contents of their result are presented here in order to avoid extensive technical preliminaries.

initial history of $\boldsymbol{E}$ up to time $t_0=0$. Then the hereditary integral equation (29.19) becomes

$$\boldsymbol{E}^*(t)+\int_0^t \overset{\circ}{\mathbf{G}}{}^{-1}\dot{\mathbf{G}}(t-s)\,[\check{\mathscr{E}}(\boldsymbol{E}^*)(s)]\,ds=\boldsymbol{F}(t),\qquad t\in\mathscr{R}^+, \tag{29.20}$$

where

$$\boldsymbol{F}(t)=-\int_{-\infty}^0 \overset{\circ}{\mathbf{G}}{}^{-1}\dot{\mathbf{G}}(t-s)\,[\boldsymbol{R}(-s)]\,ds-\int_0^t \overset{\circ}{\mathbf{G}}{}^{-1}\dot{\mathbf{G}}(t-s)\,[\widetilde{\boldsymbol{E}}(s)]\,ds,\qquad t\in\mathscr{R}^+. \tag{29.21}$$

Since $\boldsymbol{F}$ is known on $\mathscr{R}^+$, Eq. (29.20) is a Volterra equation of classical type. Thus, to determine the viscoelastic solution for the initial past-history problem, it suffices to solve the integral equation (29.20), the associated elastic problem, and the auxiliary problem, for times $t\geqq 0$.[65]

By a change in variable, the hereditary integral equation (29.19) has the form

$$\boldsymbol{E}^*(t)+\int_{-\infty}^t \overset{\circ}{\mathbf{G}}{}^{-1}\dot{\mathbf{G}}(t-s)\,[\check{\mathscr{E}}(\boldsymbol{E}^*)(s)]\,ds=\widehat{\boldsymbol{F}}(t),\qquad t\in\mathscr{R}, \tag{29.22}$$

where

$$\widehat{\boldsymbol{F}}(t)=-\int_{-\infty}^t \overset{\circ}{\mathbf{G}}{}^{-1}\dot{\mathbf{G}}(t-s)\,[\widetilde{\boldsymbol{E}}(s)]\,ds,\qquad t\in\mathscr{R}. \tag{29.23}$$

Here $\widehat{\boldsymbol{F}}$ is known on $\mathscr{R}$, but (29.22) is not a Volterra equation of classical type, since the interval of integration is unbounded for each time t. Mathematically, the integral equations (29.20) and (29.22) require separate consideration.

The initial elasticity $\overset{\circ}{\mathbf{G}}$ is said to be ***uniformly positive definite* (*on* $\mathscr{B}$)** if, and only if, there is a positive number γ such that for every material point in $\overline{\mathscr{B}}$ and any symmetric tensor $\boldsymbol{A}$

$$\boldsymbol{A}\cdot\overset{\circ}{\mathbf{G}}[\boldsymbol{A}]\geqq\gamma\boldsymbol{A}\cdot\boldsymbol{A}. \tag{29.24}$$

The initial elasticity $\overset{\circ}{\mathbf{G}}$ is said to be ***symmetric* (*on* $\mathscr{B}$)** in $[\mathscr{T}_{\text{sym}}]$ if, and only if, for every material point in $\overline{\mathscr{B}}$ and any pair of symmetric tensors $\boldsymbol{A}$ and $\boldsymbol{B}$.[66]

$$\boldsymbol{A}\cdot\overset{\circ}{\mathbf{G}}[\boldsymbol{B}]=\boldsymbol{B}\cdot\overset{\circ}{\mathbf{G}}[\boldsymbol{A}]. \tag{29.25}$$

If the initial elasticity $\overset{\circ}{\mathbf{G}}$ is uniformly positive definite on $\mathscr{B}$, then the associated elastic and auxiliary problems are strongly elliptic boundary-value problems for each time t.[67] Henceforth we assume that this is the case. Furthermore we assume that $\overset{\circ}{\mathbf{G}}$ is symmetric on $\mathscr{B}$.[68]

---

[65] Most authors consider the initial past-history problem with the special assumption that the initial past history vanishes. Cf. footnote 60, p. 59.

[66] The initial symmetry condition expressed by (29.25) is sometimes referred to as a type of Onsager's relation which we do not attempt to justify on physical grounds. Mathematically, the condition is necessary for the associated elastic problem to have an equivalent variational formulation.

[67] AGMON [1965, *1*] or FICHERA [1965, *12*]. The initial elasticity $\overset{\circ}{\mathbf{G}}$ determines a *uniformly strongly elliptic operator* if, and only if, there is a positive constant γ such that

$$\boldsymbol{a}\cdot\overset{\circ}{\mathbf{G}}[(\boldsymbol{a}\otimes\boldsymbol{b})]\,\boldsymbol{b}\equiv(\boldsymbol{a}\otimes\boldsymbol{b})\cdot\overset{\circ}{\mathbf{G}}[(\boldsymbol{a}\otimes\boldsymbol{b})]\geqq\gamma|\boldsymbol{a}|^2|\boldsymbol{b}|^2$$

for arbitrary pairs of vectors $\boldsymbol{a}$ and $\boldsymbol{b}$.

[68] Mathematically, the assumption that the initial elasticity $\overset{\circ}{\mathbf{G}}$ be symmetric on $\mathscr{B}$ is needed to insure variational formulations of the same problem and is not essential for existence theorems. See Sect. 30.

In order to present some specific results without extensive technical preparation, we will make some simplifying regularity and smoothness assumptions. Those which we present may be considerably relaxed.[69]

We assume:

1. The body $\mathscr{B}$ is a bounded regular region in $\mathscr{E}$ whose boundary $\partial\mathscr{B}$ is a class C^∞ manifold in $\mathscr{E}$;[70]

2. the values of the boundary data $[\hat{\boldsymbol{u}}, \hat{\mathbf{s}}]$ are class C^∞ fields on $\mathscr{S}_{\hat{u}}$ and $\mathscr{S}_{\hat{s}}$;

3. the values of the body force process $\boldsymbol{b}$ are class C^∞ fields on $\overline{\mathscr{B}}$; and

4. the values of the relaxation function $\mathbf{G}$ are class C^∞ fields on $\overline{\mathscr{B}}$.

We further restrict our attention to data of Heaviside type and suppose that the body force $\boldsymbol{b}$ and the boundary data $[\hat{\boldsymbol{u}}, \hat{\mathbf{s}}]$ are processes of Heaviside class H.[71]

We may now guarantee the existence of a solution to the viscoelastic boundary value problem determined by $\mathbf{G}$, $\boldsymbol{b}$, and $[\hat{\boldsymbol{u}}, \hat{\mathbf{s}}]$.

Theorem. *Let the initial elasticity $\overset{\circ}{\mathbf{G}}$ be uniformly positive definite and symmetric on $\overline{\mathscr{B}}$.*[72] *If the boundary data are of mixed or displacement type ($\mathscr{S}_{\hat{s}}$ is empty), there exists a unique solution $[\boldsymbol{u}, \boldsymbol{E}, \boldsymbol{S}]$ to the viscoelastic problem. Moreover, the process $[\boldsymbol{u}, \boldsymbol{E}, \boldsymbol{S}]$ is of Heaviside class H and has values which are class C^∞ fields on $\overline{\mathscr{B}}$. If the boundary data are of traction type ($\mathscr{S}_{\hat{u}}$ is empty) and the body forces and surface tractions are equilibrated,*[73] *there exists a solution $[\boldsymbol{u}, \boldsymbol{E}, \boldsymbol{S}]$ to the viscoelastic problem. Moreover, the process $[\boldsymbol{u}, \boldsymbol{E}, \boldsymbol{S}]$ is in Heaviside class H and has values which are class C^∞ fields on $\overline{\mathscr{B}}$. The solution $[\boldsymbol{u}, \boldsymbol{E}, \boldsymbol{S}]$ is unique to within infinitesimal rigid motions. The condition that the body forces and surface tractions be equilibrated is also necessary for the traction problem to have a solution.*[74]

The conclusions of the theorem follow from the classical existence theory in the linear theory of elasticity and the fact that the hereditary integral equation (29.19) reduces to a Volterra equation of standard type.[75]

In fact, the result just obtained provides a solution to the initial past-history problem with vanishing initial past history. Furthermore, by virtue of the above observation, we see that the hypotheses of the theorem guarantee the existence

---

[69] For an extensive treatment of this problem under weaker hypotheses, see the analyses of HLAVAČEK, BABUŠKA, and PREDELEANU cited in footnote 60, p. 59. For general reference to the technical aspects of the problem, see the works of AGMON and FICHERA cited in footnote 67, p. 62.

[70] Roughly speaking, $\mathscr{B}$ should be regular enough that the Green-Gauss identity and the Sobolev embedding theorem hold. See AGMON [1965, *1*]. Moreover, anticipating the prescription of boundary data on $\partial\mathscr{B}$, we suppose the sets $\mathscr{S}_{\hat{u}}$ and $\mathscr{S}_{\hat{s}}$ in $\partial\mathscr{B}$ have boundaries which are class C^∞ manifolds in $\partial\mathscr{B}$.

[71] We also suppose that $\nabla\boldsymbol{b}$ and all higher gradients of $\boldsymbol{b}$ are in the Heaviside class H.

[72] Recall that at the outset we have assumed the relaxation function $\mathbf{G}$ to be of class H^1.

[73] Recall that the body forces $\boldsymbol{b}$ and the surface tractions $\hat{\mathbf{s}}$ are *equilibrated* if, and only if,

$$\int_{\mathscr{B}} \boldsymbol{b}\, dV + \int_{\partial\mathscr{B}} \hat{\mathbf{s}}\, dA = \mathbf{0}$$

and

$$\int_{\mathscr{B}} \boldsymbol{p}_0 \times \boldsymbol{b}\, dV + \int_{\partial\mathscr{B}} \boldsymbol{p}_0 \times \hat{\mathbf{s}}\, dA = \mathbf{0}$$

where $\boldsymbol{p}_0$ is defined on $\overline{\mathscr{B}}$ for some fixed $\boldsymbol{x}_0 \in \mathscr{E}$ by $\boldsymbol{p}_0(\boldsymbol{x}) = \boldsymbol{x} - \boldsymbol{x}_0$.

[74] For the displacement problem a similar result has been proved by EDELSTEIN [1966, *6*]. EDELSTEIN shows that existence is guaranteed if *either* $\overset{\circ}{\mathbf{G}}$ is uniformly positive definite *or* $\overset{\circ}{\mathbf{G}}$ is strongly elliptic and independent of material points in $\overline{\mathscr{B}}$.

[75] FICHERA [1965, *12*].

of solutions to the general initial past-history problem provided the prescribed initial past histories are sufficiently regular. In particular, the initial past history should be such as to render the function $\boldsymbol{F}$, defined in (29.21), continuous on $\mathscr{R}^+$.

The uniform positive definiteness of $\overset{\circ}{\mathbf{G}}$ guarantees that $\overset{\circ}{\mathbf{G}}$ is invertible at each material point in $\overline{\mathscr{B}}$. Hence, the *inversion theorem for response functions* of Sect. 11 may be applied to conclude that there is a creep compliance $\mathbf{J}$ of class H^1 wich belongs to $\mathbf{G}$. Thus, the condition which provides existence of viscoelastic solutions (for initial history problems) also guarantees that the Boltzmann law is invertible in the Heaviside classes. In this case, the response functions $\mathbf{G}$ and $\mathbf{J}$ need not be strong.[76]

If the viscoelastic problem is not of initial past-history type,[77] the problem of existence is somewhat different. In this case, if the initial elasticity $\overset{\circ}{\mathbf{G}}$ is required to be uniformly positive definite and symmetric, the associated elastic and auxiliary problems remain strongly elliptic boundary value problems (for each time t). However, the hereditary integral equation (29.19) or, equivalently, (29.22) and (29.23) is no longer of classical Volterra type.

We will not pursue this problem further since considerable technical detail must be provided. We do point out, however, that it is appropriate to assume that $\mathbf{G}$ be a strong relaxation function. Moreover, conditions which guarantee the existence of some type of viscoelastic solution should also guarantee the (strong) invertibility of the Boltzmann law.[78]

30. Quasi-static variational principles. In this section we seek a variational characterization of solutions to the quasi-static boundary value problem with prescribed initial past history (up to time $t_0=0$). To begin with we consider the special case in which the initial past history vanishes. Therefore unless stated otherwise, all processes in this section are assumed to be of Heaviside type.[79] Recall that the Boltzmann laws $\mathscr{L}_{\mathbf{G}}$ and $\mathscr{L}_{\mathbf{J}}$ now have the form

$$\boldsymbol{S} = \mathbf{G} \circledast \boldsymbol{E} = \overset{\circ}{\mathbf{G}}[\boldsymbol{E}] + \dot{\mathbf{G}} * \boldsymbol{E}, \tag{30.1}$$

$$\boldsymbol{E} = \mathbf{J} \circledast \boldsymbol{S} = \overset{\circ}{\mathbf{J}}[\boldsymbol{S}] + \dot{\mathbf{J}} * \boldsymbol{S} \tag{30.2}$$

for processes $\boldsymbol{E}$ and $\boldsymbol{S}$ in Heaviside class H.[80]

Denote by $\mathscr{A}$ the class of all admissible processes $[\boldsymbol{u}, \boldsymbol{E}, \boldsymbol{S}]$ for $\mathscr{B}$ of Heaviside class H whose values are smooth (class C^1) fields on $\overline{\mathscr{B}}$. For simplicity, write $\boldsymbol{P} \equiv [\boldsymbol{u}, \boldsymbol{E}, \boldsymbol{S}]$ for an element of $\mathscr{A}$. Let $\mathscr{K}$ be a subset of $\mathscr{A}$ and let $\Psi: \mathscr{K} \to \mathscr{R}$ be a functional on $\mathscr{K}$.[81] If $\boldsymbol{P} \in \mathscr{K}$ and if $\check{\boldsymbol{P}} \in \mathscr{A}$ is such that $\boldsymbol{P} + \alpha \check{\boldsymbol{P}} \in \mathscr{K}$ for all sufficiently small real numbers α, then the ***variation of Ψ at $\boldsymbol{P}$ in the direction $\check{\boldsymbol{P}}$*** is the real number $\delta_{\check{\boldsymbol{P}}} \Psi\{\boldsymbol{P}\}$ defined by[82]

$$\delta_{\check{\boldsymbol{P}}} \Psi\{\boldsymbol{P}\} = \frac{d}{d\alpha} \Psi\{\boldsymbol{P} + \alpha \check{\boldsymbol{P}}\}\big|_{\alpha=0}. \tag{30.3}$$

---

[76] Cf. Sects. 10, 11.

[77] That is, if the body forces and boundary data were prescribed and nonvanishing for all times and the initial past history were unknown.

[78] We are not aware, at this time, of any general existence theory for these problems. However, it should be possible to construct such a theory using the outline presented here together with classical methods from the study of integro-partial differential equations.

[79] Cf. Sect. 3.

[80] Cf. Sects. 5, 7, 11.

[81] The term *functional* here denotes a real-valued function defined on a subset of $\mathscr{A}$. Such a function need not be linear.

[82] By writing $\delta_{\check{\boldsymbol{P}}} \Psi\{\boldsymbol{P}\}$, we assume that it is well defined by the operation in Eq. (30.3).

We say ***the variation of*** Ψ ***vanishes at*** $\boldsymbol{P}\in\mathscr{K}$ and write

$$\delta\Psi\{\boldsymbol{P}\}=0 \qquad \textbf{\textit{over}}\ \mathscr{K} \tag{30.4}$$

if, and only if, $\delta_{\check{\boldsymbol{P}}}\Psi\{\boldsymbol{P}\}=0$ for all $\check{\boldsymbol{P}}\in\mathscr{A}$ such that $\boldsymbol{P}+\alpha\check{\boldsymbol{P}}\in\mathscr{K}$ for all α sufficiently small.

Let $[\hat{\boldsymbol{u}}, \hat{\mathbf{s}}]$ be boundary data for $\partial\mathscr{B}$ in Heaviside class H whose values are uniformly continuous vector fields on the regular points of $\mathscr{S}_{\hat{\boldsymbol{u}}}$ and $\mathscr{S}_{\hat{\mathbf{s}}}$, and let $\boldsymbol{b}$ be a body force process for $\mathscr{B}$ in Heaviside class H whose values are uniformly continuous vector fields on $\mathscr{B}$. Finally, suppose that the values of the response functions $\mathbf{G}$ and $\boldsymbol{J}$ are symmetric (on $\mathscr{T}_{\text{sym}}$).[83]

For each time t define a functional Λ_t on $\mathscr{A}$ by[84]

$$\begin{aligned}\Lambda_t\{[\boldsymbol{u}, \boldsymbol{E}, \boldsymbol{S}]\} = \tfrac{1}{2}\int_{\mathscr{B}} \big((\mathbf{G}\circledast\boldsymbol{E})*\boldsymbol{E}\big)(t)\,dV \\ -\int_{\mathscr{B}} (\boldsymbol{S}*\boldsymbol{E})(t)\,dV \\ -\int_{\mathscr{B}} \big((\operatorname{div}\boldsymbol{S}+\boldsymbol{b})*\boldsymbol{u}\big)(t)\,dV \\ +\int_{\mathscr{S}_{\hat{\boldsymbol{u}}}} (\mathbf{s}*\hat{\boldsymbol{u}})(t)\,dA \\ +\int_{\mathscr{S}_{\hat{\mathbf{s}}}} \big((\mathbf{s}-\hat{\mathbf{s}})*\boldsymbol{u}\big)(t)\,dA\,.\end{aligned} \tag{30.5}$$

We may now state the

First variational principle for relaxation problems.[85] *Let* $\mathscr{K}=\mathscr{A}$. *Then the process* $[\boldsymbol{u}, \boldsymbol{E}, \boldsymbol{S}]$ *in* $\mathscr{K}$ *is a solution to the quasi-static boundary value problem for* $\mathscr{B}$ *corresponding to the boundary data* $[\hat{\boldsymbol{u}}, \hat{\mathbf{s}}]$, *the body force* $\boldsymbol{b}$, *and the symmetric relaxation function* $\mathbf{G}$ *if, and only if,*

$$\delta\Lambda_t\{[\boldsymbol{u}, \boldsymbol{E}, \boldsymbol{S}]\}=0 \qquad \textit{over}\ \mathscr{K} \tag{30.6}$$

for all times $t\geqq 0$.[86]

Now let $\mathscr{K}$ be that subset of $\mathscr{A}$ consisting of all those processes $[\boldsymbol{u}, \boldsymbol{E}, \boldsymbol{S}]$ which satisfy the strain-displacement relation (19.4), the Boltzmann relaxation law (19.5) (in the form given by (30.1)), and the displacement boundary conditions on $\mathscr{S}_{\hat{\boldsymbol{u}}}$ given in (23.1 a).[87] Let Φ_t denote the *restriction* to $\mathscr{K}\subset\mathscr{A}$ of the functional Λ_t defined by (30.5). Then Φ_t may be written in the form

$$\begin{aligned}\Phi_t\{[\boldsymbol{u}, \boldsymbol{E}, \boldsymbol{S}]\} = \tfrac{1}{2}\int_{\mathscr{B}} \big((\mathbf{G}\circledast\boldsymbol{E})*\boldsymbol{E}\big)(t)\,dV \\ -\int_{\mathscr{B}} (\boldsymbol{b}*\boldsymbol{u})(t)\,dV \\ -\int_{\mathscr{S}_{\hat{\mathbf{s}}}} (\hat{\mathbf{s}}*\boldsymbol{u})(t)\,dA\,, \qquad t\in\mathscr{R}.\end{aligned} \tag{30.7}$$

---

[83] Recall that a response function $\mathbf{K}$ is *symmetric* (on $\mathscr{T}_{\text{sym}}$) if, and only if, for each time t and any pair of symmetric tensors $\boldsymbol{A}$ and $\boldsymbol{B}$

$$\boldsymbol{A}\cdot\mathbf{K}(t)[\boldsymbol{B}]=\boldsymbol{B}\cdot\mathbf{K}(t)[\boldsymbol{A}].$$

[84] Recall the definition and properties of convolutions in Sect. 4.

[85] GURTIN [1963, *5*]. For the corresponding theorem in elasticity see HU [1955, *3*], WASHIZU [1955, *10*]. See also LTE, Sect. 38.

[86] Observe that Λ_t vanishes identically whenever $t<0$.

[87] Note that $\mathscr{K}\subset\mathscr{A}$ is not a vector subspace of $\mathscr{A}$ unless $\hat{\boldsymbol{u}}\equiv\boldsymbol{0}$ on $\mathscr{S}_{\hat{\boldsymbol{u}}}$. However, if $[\boldsymbol{u}, \boldsymbol{E}, \boldsymbol{S}]$ is in $\mathscr{K}$ and $[\check{\boldsymbol{u}}, \check{\boldsymbol{E}}, \check{\boldsymbol{S}}]$ is chosen in $\mathscr{A}$ such that (19.4) and (30.1) are satisfied and $\check{\boldsymbol{u}}\equiv\boldsymbol{0}$ on $\mathscr{S}_{\hat{\boldsymbol{u}}}$, then the process $[\boldsymbol{u}, \boldsymbol{E}, \boldsymbol{S}]+\alpha[\check{\boldsymbol{u}}, \check{\boldsymbol{E}}, \check{\boldsymbol{S}}]$ lies in $\mathscr{K}$ for all real α.

Using the functionals Φ_t we have the

Second variational principle for relaxation problems[88]. *Let $\mathscr{K}$ consist in all processes in $\mathscr{A}$ which satisfy the strain-displacement relation* (19.4), *the Boltzmann relaxation law* (30.1), *and the displacement boundary condition* (23.1 a) *on $\mathscr{S}_{\hat{\mathbf{u}}}$. Then the process $[\boldsymbol{u}, \boldsymbol{E}, \boldsymbol{S}]$ in $\mathscr{K}$ is a solution to the quasi-static boundary value problem for $\mathscr{B}$ corresponding to the boundary data $[\hat{\mathbf{u}}, \hat{\mathbf{s}}]$, the body force $\boldsymbol{b}$, and the symmetric relaxation function $\mathbf{G}$ if, and only if,*

$$\delta\,\Phi_t\{[\boldsymbol{u}, \boldsymbol{E}, \boldsymbol{S}]\} = 0 \qquad \text{over } \mathscr{K} \tag{30.8}$$

for all times $t \geqq 0$.

The proofs of the two variational principles stated above are given by GURTIN.[89] They depend upon the properties of convolutions and the fundamental lemma of the calculus of variations. It should be emphasized that the symmetry of the relaxation function $\mathbf{G}$ is essential for a variational characterization of the quasi-static viscoelastic problem.

Observe that a solution to the quasi-static viscoelastic problem considered here is characterized as the solution to a one-parameter family of stationary value problems. Contrast this with the characterization of solutions to problems in elastostatics by a single stationary value problem.[90]

With a view toward developing variational principles for problems of creep type, let $\mathscr{K}$ be that subset of $\mathscr{A}$ consisting of all processes $[\boldsymbol{u}, \boldsymbol{E}, \boldsymbol{S}]$ which satisfy the strain-displacement relation (19.4). Define a functional θ_t on $\mathscr{K}$ for each time t by

$$\begin{aligned}\theta_t\{[\boldsymbol{u}, \boldsymbol{E}, \boldsymbol{S}]\} = &\int_{\mathscr{B}} (\boldsymbol{S} * \boldsymbol{E})(t)\, dV \\ &- \tfrac{1}{2}\int_{\mathscr{B}} \big((\mathbf{J} \circledast \boldsymbol{S}) * \boldsymbol{S}\big)(t)\, dV \\ &- \int_{\mathscr{B}} (\boldsymbol{b} * \boldsymbol{u})(t)\, dV \\ &- \int_{\mathscr{S}_{\hat{\boldsymbol{u}}}} \big(\mathbf{s} * (\boldsymbol{u} - \hat{\boldsymbol{u}})\big)(t)\, dA \\ &- \int_{\mathscr{S}_{\hat{\mathbf{s}}}} (\hat{\mathbf{s}} * \boldsymbol{u})(t)\, dA\,.\end{aligned} \tag{30.9}$$

Using the functionals θ_t we state the

First variational principle for creep problems.[91] *Let $\mathscr{K}$ be the set of all processes in $\mathscr{A}$ which meet the strain-displacement relation* (19.4). *Then the process $[\boldsymbol{u}, \boldsymbol{E}, \boldsymbol{S}]$ in $\mathscr{K}$ is a solution to the quasi-static boundary value problem for $\mathscr{B}$ corresponding to the boundary data $[\hat{\mathbf{u}}, \hat{\mathbf{s}}]$, the body force $\mathbf{b}$, and the symmetric creep compliance $\mathbf{J}$ if, and only if,*

$$\delta\,\theta_t\{[\boldsymbol{u}, \boldsymbol{E}, \boldsymbol{S}]\} = 0 \qquad \text{over } \mathscr{K} \tag{30.10}$$

for all times $t \geqq 0$.

The next variational principle is given in terms of the stress process only. We denote by $\mathscr{F}$ the vector space of all stress processes for $\mathscr{B}$ in Heaviside class H whose values are smooth (class C^1) symmetric tensor fields on $\overline{\mathscr{B}}$. Let $\mathscr{K}$ be that

[88] GURTIN [1963, *5*]. This result generalizes the principle of stationary potential energy of classical elastostatics. See e.g., LTE, Sect. 34.

[89] GURTIN [1963, *5*].

[90] Cf. LTE, Sects. 34–39.

[91] GURTIN [1963, *5*].

subset of $\mathscr{F}$ consisting of all stress processes $\boldsymbol{S}$ in $\mathscr{F}$ which satisfy the stress equation of equilibrium (19.3) and the traction boundary condition (23.1b) on $\mathscr{S}_{\hat{\mathbf{s}}}$. Define the functional Ψ_t on $\mathscr{K}$ for each time t by

$$\Psi_t\{\boldsymbol{S}\} = \tfrac{1}{2}\int_{\mathscr{B}} \big((\mathbf{J} \circledast \boldsymbol{S}) * \boldsymbol{S}\big)(t)\, dV - \int_{\mathscr{S}_{\hat{\boldsymbol{u}}}} (\mathbf{s} * \hat{\boldsymbol{u}})(t)\, dA. \tag{30.11}$$

Then we have the

Second variational principle for creep problems.[92] *Let $\mathscr{K}$ be the subset of all stress processes in $\mathscr{F}$ which satisfy the stress equation of equilibrium* (19.3) *and the traction boundary condition* (23.1b). *Then*

$$\delta\Psi_t\{\boldsymbol{S}\} = 0 \qquad \text{over } \mathscr{K} \tag{30.12}$$

for all times $t \geqq 0$ if there exist processes $\boldsymbol{u}$ and $\boldsymbol{E}$ for $\mathscr{B}$ such that $[\boldsymbol{u}, \boldsymbol{E}, \boldsymbol{S}]$ lies in $\mathscr{A}$ and is a solution to the quasi-static boundary value problem corresponding to $[\hat{\boldsymbol{u}}, \hat{\mathbf{s}}]$, $\boldsymbol{b}$, and symmetric $\mathbf{J}$.

Conversely, suppose

(i) *$\mathscr{B}$ is convex with respect to $\mathscr{S}_{\hat{\boldsymbol{u}}}$;*[93]

(ii) *$\mathscr{B}$ is simply connected;*

(iii) *the values of $\mathbf{J}$ and $\boldsymbol{S}$ are class C^2 fields on $\overline{\mathscr{B}}$;*

(iv) *$\hat{\boldsymbol{u}}$ is in Heaviside class H^1 on $\mathscr{S}_{\hat{\boldsymbol{u}}}$; and*

(v) (30.12) *holds for all times $t \geqq 0$.*

Then there exist processes $\boldsymbol{u}$ and $\boldsymbol{E}$ such that $[\boldsymbol{u}, \boldsymbol{E}, \boldsymbol{S}]$ lies in $\mathscr{A}$ and is a solution to the quasi-static boundary value problem corresponding to $[\hat{\boldsymbol{u}}, \hat{\mathbf{s}}]$, $\boldsymbol{b}$, and symmetric $\mathbf{J}$.

The four variational principles cited above are appropriate for initial past-history problems with vanishing initial past histories (up to time $t_0 = 0$). To illustrate how they may be extended to include nonvanishing initial past-history problems, we modify the *first variational principle for relaxation problems.*

Let $[\boldsymbol{v}, \boldsymbol{R}, \boldsymbol{T}]$ be a prescribed initial past history and suppose the values of the boundary data $[\hat{\boldsymbol{u}}, \hat{\mathbf{s}}]$ and the body force $\boldsymbol{b}$, for times $t < 0$, are such that the initial past history $[\boldsymbol{v}, \boldsymbol{R}, \boldsymbol{T}]$ is consistent in the sense of Sect. 14. Define a function $\boldsymbol{S}^+$ on $\mathscr{R}^+$ by[94]

$$\boldsymbol{S}^+(t) = \int_t^\infty \dot{\mathbf{G}}(s)\,[\boldsymbol{R}(s-t)]\, ds, \qquad t \in \mathscr{R}^+. \tag{30.13}$$

Let $\mathscr{A}'$ denote the set of all admissible processes $[\boldsymbol{u}, \boldsymbol{E}, \boldsymbol{S}]$ for $\mathscr{B}$ such that the values of $[\boldsymbol{u}, \boldsymbol{E}, \boldsymbol{S}]$ are smooth (class C^1) fields on $\overline{\mathscr{B}}$, the restriction of $[\boldsymbol{u}, \boldsymbol{E}, \boldsymbol{S}]$ to $\mathscr{R}^+$ is continuous, and $[\boldsymbol{u}, \boldsymbol{E}, \boldsymbol{S}]$ has the given initial past history $[\boldsymbol{v}, \boldsymbol{R}, \boldsymbol{T}]$. Thus the difference of two processes in $\mathscr{A}'$ is a process in $\mathscr{A}$. For each time $t \geqq 0$, define a functional Λ'_t on $\mathscr{A}'$ by

$$\Lambda'_t\{[\boldsymbol{u}, \boldsymbol{E}, \boldsymbol{S}]\} = \Lambda_t\{[\boldsymbol{u}, \boldsymbol{E}, \boldsymbol{S}]\} + \int_{\mathscr{B}} (\boldsymbol{S}^+ * \boldsymbol{E})(t)\, dV, \tag{30.14}$$

where Λ_t denotes the functional defined by the expression (30.5).[95]

---

[92] Gurtin [1963, 5]. This result generalizes the principle of stationary complementary energy in the classical theory of elastostatics. See LTE, Sect. 36.

[93] ***$\mathscr{B}$ is convex with respect to $\mathscr{S}_{\hat{\boldsymbol{u}}}$*** if, and only if, the line $\boldsymbol{x}(\lambda) = \boldsymbol{x}' + (\boldsymbol{x}'' - \boldsymbol{x}')\lambda$, $\lambda \in \mathscr{R}$, intersects $\mathscr{S}_{\hat{\boldsymbol{u}}}$ only at $\boldsymbol{x}'$ and $\boldsymbol{x}''$ for any pair of points $\boldsymbol{x}'$ and $\boldsymbol{x}''$ in $\mathscr{S}_{\hat{\boldsymbol{u}}}$.

[94] The function $\boldsymbol{S}^+$ is well defined on $\mathscr{R}^+$ if, say, $\mathbf{G}$ is a strong relaxation function and the history $\boldsymbol{R}$ is integrable or bounded on $\mathscr{R}^+$.

[95] Recall the convention established for convolutions in Sect. 4 for functions not of Heaviside type.

Using the functionals Λ_t' we have the

First variational principle for relaxation problems with non-vanishing initial past histories. *Let $\mathcal{K} = \mathcal{A}'$. Then the process $[\boldsymbol{u}, \boldsymbol{E}, \boldsymbol{S}]$ in $\mathcal{K}$ is a solution to the quasi-static boundary value problem with prescribed initial past history corresponding to the boundary data $[\hat{\boldsymbol{u}}, \hat{\boldsymbol{s}}]$, the body forces $\boldsymbol{b}$, the symmetric relaxation function G, and the past history $[\boldsymbol{v}, \boldsymbol{R}, \boldsymbol{T}]$ if, and only if*

$$\delta\Lambda_t'\{[\boldsymbol{u}, \boldsymbol{E}, \boldsymbol{S}]\} = 0 \quad \text{over } \mathcal{K} \tag{30.15}$$

for all times $t \geqq 0$.

The proof of this result is a straightforward adaptation of the proof of the first variational principle for relaxation problems.[96]

31. Elastic-viscoelastic correspondence. In this section we exhibit relations between solutions to quasi-static viscoelastic problems and their counterparts in elastostatics.

These relations are explored through the mechanism of the operational calculus, particularly the Laplace transform. The formalism of these methods is especially useful for viscoelastic stress analysis. The observations recorded here follow closely the work of ALFREY, TSIEN, LEE, and RADOK.[97]

We assume that the viscoelastic body $\mathcal{B}$ and, hence, its boundary $\partial\mathcal{B}$ are fixed in time. Furthermore, when considering mixed boundary data on $\partial\mathcal{B} = \mathcal{S}_{\hat{\boldsymbol{u}}} \cup \mathcal{S}_{\hat{\boldsymbol{s}}}$, we assume that the sets $\mathcal{S}_{\hat{\boldsymbol{u}}}$ and $\mathcal{S}_{\hat{\boldsymbol{s}}}$ are also fixed in time. That the methods we present here are not generally restricted to this case has been shown by RADOK, LEE, WOODWARD, and HUNTER.[98]

For simplicity, we consider only processes of Heaviside type. Thus the Boltzmann relaxation and creep laws $\mathcal{L}_{\mathsf{G}}$ and $\mathcal{L}_{\mathsf{J}}$ have the form

$$\boldsymbol{S} = \mathring{\mathsf{G}}[\boldsymbol{E}] + \dot{\mathsf{G}} * \boldsymbol{E}, \tag{31.1}$$

$$\boldsymbol{E} = \mathring{\mathsf{J}}[\boldsymbol{S}] + \dot{\mathsf{J}} * \boldsymbol{S}, \tag{31.2}$$

where $\boldsymbol{E}$ and $\boldsymbol{S}$ are in class H.

Let a superscribed bar $(^{-})$ denote the Laplace transform of a process in class H. Observe that the response functions G and J have Laplace transforms given by

$$\lambda\bar{\mathsf{G}}(\lambda) = \mathring{\mathsf{G}} + \bar{\dot{\mathsf{G}}}(\lambda), \tag{31.3}$$

$$\lambda\bar{\mathsf{J}}(\lambda) = \mathring{\mathsf{J}} + \bar{\dot{\mathsf{J}}}(\lambda) \tag{31.4}$$

provided, say, G and J are strong and $\operatorname{Re}\lambda > 0$. If $\boldsymbol{S}$ and $\boldsymbol{E}$ are in class H and are related through either (31.1) or (31.2), then

$$\bar{\boldsymbol{S}}(\lambda) = (\mathring{\mathsf{G}} + \bar{\dot{\mathsf{G}}}(\lambda))\,[\boldsymbol{E}(\lambda)], \tag{31.5}$$

$$\bar{\boldsymbol{E}}(\lambda) = (\mathring{\mathsf{J}} + \bar{\dot{\mathsf{J}}}(\lambda))\,[\bar{\boldsymbol{S}}(\lambda)] \tag{31.6}$$

---

[96] See GURTIN [1963, *5*].

[97] The methods discussed here were first set down by ALFREY [1944, *1*] and TSIEN [1950, *4*]. An extension of these methods was carried out by LEE [1955, *4*], [1958, *13*]. RADOK [1957, *8*], and LEE and RADOK [1957, *5*]. Other authors using operational methods which exploit the separability of space and time in quasi-static problems are JEFFRIES [1932, *1*], READ [1950, *2*], MINDLIN [1949, *1*], GRAFFI [1951, *1*], [1952, *1*], LEE, RADOK, and WOODWARD [1959, *5*]. For general references on viscoelastic stress analysis see ROGERS [1963, *17*], LEE [1964, *12*].

[98] RADOK [1957, *8*], LEE and RADOK [1957, *5*], LEE, RADOK, and WOODWARD [1959, *5*], HUNTER [1960, *8*].

in some right half-plane where the transforms $\bar{S}$, $\bar{E}$, $\bar{\dot{G}}$, and $\bar{\dot{J}}$ all exist. If, in addition, (31.3) and (31.4) hold, then (31.5) and (31.6) have the form of elastic laws for each fixed λ

$$\bar{S}(\lambda) = \lambda\bar{G}(\lambda)[\bar{E}(\lambda)], \tag{31.7}$$

$$\bar{E}(\lambda) = \lambda\bar{J}(\lambda)[\bar{S}(\lambda)]. \tag{31.8}$$

If the boundary data $[\hat{u}, \hat{s}]$ and the body force b also have Laplace transforms $[\bar{\hat{u}}, \bar{\hat{s}}]$ and $\bar{b}$ and $[u, E, S]$ is a solution to the quasi-static boundary value problem corresponding to G, b, and $[\hat{u}, \hat{s}]$, then the triplet $[\bar{u}(\lambda), \bar{E}(\lambda), \bar{S}(\lambda)]$ satisfies the field equations (31.5) or (31.6) and

$$\operatorname{div}\bar{S}(\lambda) + \bar{b}(\lambda) = 0, \tag{31.9}$$

$$\bar{E}(\lambda) = \tfrac{1}{2}[\nabla\bar{u}(\lambda) + \nabla\bar{u}^T(\lambda)], \tag{31.10}$$

and the boundary conditions:

$$\bar{u}(\lambda) = \bar{\hat{u}}(\lambda) \qquad \text{on } \mathscr{S}_{\hat{u}}, \tag{31.11}$$

$$\bar{S}(\lambda)\, n = \bar{s}_{(n)}(\lambda) = \bar{\hat{s}}(\lambda) \qquad \text{on } \mathscr{S}_{\hat{s}}. \tag{31.12}$$

Thus, the triplet $[\bar{u}(\lambda), \bar{E}(\lambda), \bar{S}(\lambda)]$ is a solution to the *elastic* problem defined, for each λ, by Eqs. (31.5), (31.6), and (31.9)–(31.12). The elasticities in this case are $(\overset{\circ}{G} + \bar{\dot{G}}(\lambda))$ or $(\overset{\circ}{J} + \bar{\dot{J}}(\lambda))$.[99]

The above arguments may be reversed in the following sense: if, for each λ, $[\bar{u}(\lambda), \bar{E}(\lambda), \bar{S}(\lambda)]$ is an elastic solution to (31.9)–(31.12) and either (31.5) or (31.6) and $\bar{u}$, $\bar{E}$, and $\bar{S}$ are the Laplace transforms of the Heaviside processes u, E, and S, then $[u, E, S]$ is a solution to the given viscoelastic problem.

Consider the special case of a homogeneous and isotropic viscoelastic body of, say, relaxation type. Then the Boltzmann relaxation law (31.1) may be written in the form[100]

$$S_0 = \overset{\circ}{G}_1 E_0 + \dot{G}_1 * E_0, \tag{31.13}$$

$$\operatorname{tr} S = \overset{\circ}{G}_2 \operatorname{tr} E + \dot{G}_2 * \operatorname{tr} E \tag{31.14}$$

where E_0 and S_0 are the deviatoric strain and stress processes. Similarly (31.7) becomes

$$\bar{S}_0(\lambda) = \lambda\bar{G}_1(\lambda)\bar{E}_0(\lambda), \tag{31.15}$$

$$\operatorname{tr}\bar{S}(\lambda) = \lambda\bar{G}_2(\lambda) \operatorname{tr}\bar{E}(\lambda). \tag{31.16}$$

Thus, if $[u, E, S]$ $(\hat{G}_1, \hat{G}_2, \hat{b})$ is an elastic state corresponding to (constant) shear and bulk moduli $\hat{G}_1$ and $\hat{G}_2$ and the body force $\hat{b}$, then replacing $\hat{G}_1$, $\hat{G}_2$, and $\hat{b}$ by $\lambda\bar{G}_1(\lambda)$, $\lambda\bar{G}_2(\lambda)$, and $\bar{b}(\lambda)$, for each fixed λ, yields the Laplace transform of a process $[u, E, S]$ in QSVP(G_1, G_2, b) provided the inverse transformation is possible.

The case of a scalar differential operator law is particularly simple.[101] Let $\langle P, Q\rangle$ be a differential operator pair of order $N \geqq 0$ and suppose the processes S and E are in class H^N and satisfy Eq. (14.3) on $\mathscr{R}^{++}$:

$$\sum_{k=0}^{N} p_k S^{(k)} = \sum_{k=0}^{N} q_k E^{(k)}. \tag{31.17}$$

---

[99] If G and J have Laplace transforms, we can use (31.3) and (31.4) to write the elasticities as $\lambda\bar{G}(\lambda)$ and $\lambda\bar{J}(\lambda)$.

[100] Cf. Sect. 12, Eqs. (12.10) and (12.11).

[101] Cf. Sects. 14–16.

If the Laplace transforms $\bar{S}$ and $\bar{E}$ exist in some right half-plane, it follows that if $N=0$,

$$P(\lambda)\,\bar{S}(\lambda)=Q(\lambda)\,\bar{E}(\lambda) \tag{31.18}$$

while if $N>0$,

$$P(\lambda)\bar{S}(\lambda)-\frac{1}{\lambda}\sum_{k=1}^{N}p_k\sum_{r=1}^{k}\lambda^r\,\overset{\circ}{S}^{(k-r)}=Q(\lambda)\,\bar{E}(\lambda)-\frac{1}{\lambda}\sum_{k=1}^{N}q_k\sum_{r=1}^{k}\lambda^r\overset{\circ}{E}^{(k-r)} \tag{31.19}$$

where $P(\lambda)$ and $Q(\lambda)$ are polynomials defined for all complex λ by

$$P(\lambda)=\sum_{k=0}^{N}p_k\,\lambda^k, \tag{31.20}$$

$$Q(\lambda)=\sum_{k=0}^{N}q_k\,\lambda^k. \tag{31.21}$$

For $N>0$, (31.19) may be rewritten in the form

$$P(\lambda)\bar{S}(\lambda)-\frac{1}{\lambda}\sum_{k=1}^{N}\lambda^k\sum_{r=k}^{N}p_r\,\overset{\circ}{S}^{(r-k)}=Q\,(\lambda)\,\bar{E}(\lambda)-\frac{1}{\lambda}\sum_{k=1}^{N}\lambda^k\sum_{r=k}^{N}q_r\,\overset{\circ}{E}^{(r-k)}. \tag{31.22}$$

Now suppose that the pair $\langle P, Q\rangle$ *belongs* to a scalar relaxation function G in the sense of Sect. 14. Then, if $N>0$, the initial conditions (14.4) are satisfied:

$$\sum_{r=k}^{N}p_r\,\overset{\circ}{S}^{(r-k)}=\sum_{r=k}^{N}q_r\,\overset{\circ}{E}^{(r-k)},\qquad k=1,2,\ldots,N. \tag{31.23}$$

We have verified the following assertion.[102] *Let $\langle P, Q\rangle$ be a differential operator pair of order $N\geqq 1$ which belongs to a relaxation function G. If S and E are in H^N, have Laplace transforms $\bar{S}$ and $\bar{E}$ in some right half-plane, and*

$$S=\overset{\circ}{\mathsf{G}}E+\dot{\mathsf{G}}*E, \tag{31.24}$$

then

$$P(\lambda)\,\bar{S}(\lambda)=Q(\lambda)\,\bar{E}\,(\lambda) \tag{31.25}$$

for all λ in the half-plane where $\bar{S}$ and $\bar{E}$ are defined.

If $\langle P, Q\rangle$ is of order $N\geqq 1$ and belongs to a relaxation function, then $\langle P, Q\rangle$ is of relaxation type and $P(\lambda)$ is a polynomial of degree N in λ while $Q(\lambda)$ is a polynomial of degree $\leqq N$ in λ. It then follows from (31.7) (for scalar processes), (31.24), and (31.25) that if $\langle P, Q\rangle$ belongs to G[103] then there is a right half-plane such that[104]

$$\lambda\bar{\mathsf{G}}(\lambda)=\frac{Q\,(\lambda)}{P\,(\lambda)}, \tag{31.26}$$

for all complex λ in that half-plane. In fact,

$$\bar{\mathsf{G}}\,(\lambda)=O\left(\frac{1}{\lambda}\right)\quad\text{as}\quad \operatorname{Re}\lambda\to\infty. \tag{31.27}$$

---

[102] The precise formulation of this result is due to GURTIN and STERNBERG, [1962, *10*, Theorem 4.7]. The same assertion is used by LEE [1955, *4*], [1958, *13*], RADOK [1957, *8*], and ALFREY [1944, *1*]. These authors assume that the processes S and E are so smooth at time $t=0$ that the initial values $\overset{\circ}{S},\ldots,\overset{\circ}{S}^{(N-1)}$, and $\overset{\circ}{E},\ldots,\overset{\circ}{E}^{(N-1)}$ all vanish.

[103] From the definition of *belongs* in Sect. 14 we have $\langle P, Q\rangle$ of order $N\geqq 1$ and the process G in H^N.

[104] See, for example, GURTIN and STERNBERG [1962, *10*, Theorem 4.8], for a detailed proof of this assertion.

We observe here that the consideration of relaxation laws was not essential. Indeed, all assertions have duals for creep laws. We have also avoided the degenerate case $N=0$ in which the differential operator laws are elastic. We merely point out that if $N=0$ and p_0 and q_0 do not vanish, then

$$\lambda \bar{G}(\lambda) = q_0/p_0 \tag{31.28}$$

which implies that

$$G = \frac{q_0}{p_0} h, \tag{31.29}$$

where h is the scalar Heaviside unit process.

Consider now a homogeneous and isotropic *incompressible* viscoelastic body $\mathscr{B}$. Let $\boldsymbol{S}_0$ and $\boldsymbol{E}_0$ be the deviatoric stress and strain and let S and E be the mean normal stress and strain:

$$\boldsymbol{S} = \boldsymbol{S}_0 + S\boldsymbol{1}, \quad \operatorname{tr} \boldsymbol{S}_0 = 0, \tag{31.30}$$

$$\boldsymbol{E} = \boldsymbol{E}_0 + E\boldsymbol{1}, \quad \operatorname{tr} \boldsymbol{E}_0 = 0. \tag{31.31}$$

The constraint of incompressibility is expressed by the requirement that the displacement $\boldsymbol{u}$ satisfies

$$\operatorname{div} \boldsymbol{u} = 0. \tag{31.32}$$

For such a displacement, the strain-displacement equation (19.4) and (31.31) together imply

$$\boldsymbol{E} = \boldsymbol{E}_0, \quad E = 0. \tag{31.33}$$

The deviatoric stress and strain are related through the Boltzmann relaxation law

$$\boldsymbol{S}_0 = G \circledast \boldsymbol{E}_0, \tag{31.34}$$

where G is the relaxation function for shear response.[105] Mechanically, the constraint of incompressibility is reflected by the fact that the mean normal stress S remains undetermined by the strain.[106] In the absence of body forces, the stress equation of equilibrium has the form

$$\operatorname{div} \boldsymbol{S}_0 + \nabla S = \boldsymbol{0}. \tag{31.35}$$

The equation of compatibility (31.1b) for a traceless symmetric tensor process $\boldsymbol{E}_0$ reduces to

$$\Delta \boldsymbol{E}_0 - 2\hat{\nabla} \operatorname{div} \boldsymbol{E}_0 = \boldsymbol{0}. \tag{31.36}$$

Since the material is assumed homogeneous, it follows from the relaxation law (31.34) that if $\boldsymbol{E}_0$ satisfies the compatibility equation (31.36) then so does $\boldsymbol{S}_0$:

$$\Delta \boldsymbol{S}_0 - 2\hat{\nabla} \operatorname{div} \boldsymbol{S}_0 = \boldsymbol{0}. \tag{31.37}$$

This result may be combined with the equation of equilibrium (31.35) to yield

$$\Delta \boldsymbol{S}_0 + 2\nabla(\nabla S) = \boldsymbol{0}. \tag{31.38}$$

By taking the trace of each side of Eq. (31.38) we conclude that the mean normal stress S must be harmonic:

$$\Delta S = 0. \tag{31.39}$$

In this case, $\Delta \boldsymbol{S} = \Delta \boldsymbol{S}_0$ so that (31.38) has the alternate form

$$\Delta \boldsymbol{S} + 2\nabla(\nabla S) = \boldsymbol{0}. \tag{31.40}$$

---

[105] In Sect. 12, G was denoted by G_1.

[106] Often, the mean normal stress is written $S = -p$, where p is the *hydrostatic pressure*.

Hence, for the traction problem, we have the set of equations

$$\operatorname{div} \boldsymbol{S}_0 + \nabla S = \boldsymbol{0}, \tag{31.35}$$

$$\Delta \boldsymbol{S} + 2\nabla(\nabla S) = \boldsymbol{0}, \tag{31.40}$$

and

$$\boldsymbol{S}\boldsymbol{n} = \hat{\boldsymbol{s}} \quad \text{on } \partial\mathscr{B} \tag{31.41}$$

which, for a simply connected body, completely characterize the solution to the viscoelastic problem. Further if $\boldsymbol{S}$ is a solution to (31.35) and (31.40), then it follows that there exists an incompressible displacement field $\boldsymbol{u}$ such that $\boldsymbol{u}$ and S satisfy Eqs. (31.32), (31.34), and (19.4).

The Eqs. (31.35), (31.40), and (31.41) also characterize the solution to the corresponding elastic problem. Thus the stress field associated with the elastic problem is also a viscoelastic stress field. In fact, for the traction problem in the incompressible theory *the stresses are independent of the material*. The displacements, however, depend on the material and are found by solving (31.32), (31.34) and the strain displacement relation (19.4).

The displacement problem of incompressible viscoelasticity consists of finding the displacement $\boldsymbol{u}$ and a scalar field ϕ such that

$$\Delta \boldsymbol{u} + \nabla\phi = \boldsymbol{0}, \tag{31.42}$$

$$\operatorname{div} \boldsymbol{u} = 0 \tag{31.43}$$

and

$$\boldsymbol{u} = \hat{\boldsymbol{u}} \quad \text{on } \partial\mathscr{B}, \tag{31.44}$$

where, for invertible Boltzmann laws[107],

$$\phi = 2\mathsf{G}^{-1} \circledast S. \tag{31.45}$$

These equations completely characterize the viscoelastic problem. If $\boldsymbol{u}$ and ϕ are consistent with the above equations and if S and $\boldsymbol{S}$ are defined through (31.34), (31.45) and the strain displacement relation (19.4), then $\boldsymbol{u}$ and $\boldsymbol{S}$ constitute a solution of the viscoelastic problem. But (31.42), (31.43), and (31.44) are the equations for the corresponding incompressible elastic problem (with hydrostatic pressure $p = (-1/2\mu)\,\phi$). Thus the viscoelastic displacement field is exactly the same as the corresponding elastic field. In fact for the displacement problem in the theory of incompressible materials *the displacement field is independent of the material.*

The traction and displacement problems discussed above can be shown to have unique solutions as in classical elasticity.[108]

From these observations we obtain the following

Theorem (Alfrey-Graffi).[109] *Let $\mathscr{B}$ be a homogeneous and isotropic incompressible viscoelastic body for which the body forces vanish.*

(i) *For a simply connected body $\mathscr{B}$, let* $\boldsymbol{S}$ *be a process whose values are class C^2 symmetric-tensor fields on $\mathscr{B}$. Then* $\boldsymbol{S}$ *is a stress process corresponding to a solution of the viscoelastic traction boundary value problem for $\mathscr{B}$ if, and only if,* $\boldsymbol{S}$ *is a stress process corresponding to a solution of the elastic traction boundary value problem for $\mathscr{B}$. Thus, for this problem the stress process is independent of the material.*

---

[107] Cf. Sect. 11.

[108] For a complete discussion of the incompressible theory of linear elasticity, see LTE, Sect. 58.

[109] ALFREY [1944, *1*]. In this paper, differential operator rather than Boltzmann laws are used. As noted by GRAFFI [1952, *1*], the conclusions are, in either case, the same.

(ii) *Let $\boldsymbol{u}$ be a process for $\mathscr{B}$ whose values are class C^4 vector fields on $\mathscr{B}$. Suppose further that the Boltzmann relaxation law* (31.34) *is invertible. Then $\boldsymbol{u}$ is the displacement process corresponding to a solution of the viscoelastic displacement boundary value problem for $\mathscr{B}$ if, and only if, $\boldsymbol{u}$ is the displacement process corresponding to a solution of the elastic displacement boundary value problem for $\mathscr{B}$. Thus, for this problem the displacement field is independent of the material.*

We observe here that if the Boltzmann law (31.34) is not invertible, then the "if" but not the "only if" part of assertion (ii) of the theorem holds. More importantly, while both the stress process $\boldsymbol{S}$, in part (i), and the displacement process $\boldsymbol{u}$, in part (ii), are simultaneously elastic and viscoelastic, the same is not true of the remaining elements of the processes $[\boldsymbol{u}, \boldsymbol{E}, \boldsymbol{S}]$.

Finally, if the body $\mathscr{B}$ were of creep type, then the Boltzmann relaxation law (31.34) must be replaced by

$$\boldsymbol{E}_0 = J \circledast \boldsymbol{S}_0. \tag{31.46}$$

In this case, all arguments proceed as before and the theorem holds with the condition regarding the invertibility of the relaxation law in part (ii) replaced by the same condition on the invertibility of the creep law in part (i). Of course the comments following the statement of the theorem are also applicable with the obvious modifications.

TSIEN[110] has extended ALFREY'S result to include a class of homogeneous and isotropic viscoelastic bodies. Specifically, the isotropic scalar response functions are assumed to be proportional:

$$G_1 = \alpha G_2 \quad \text{or} \quad J_1 = \beta J_2 \tag{31.47}$$

where α and β are real constants. Under this hypothesis the conclusions of ALFREY'S theorem hold for nonvanishing body force, provided the boundary data and body forces are separable and synchronous.

32. Stress functions for quasi-static viscoelastic processes. Let $\mathscr{B}$ be an isotropic and homogeneous viscoelastic body. Recall that $\mathscr{B}$ is an open region in $\mathscr{E}$ whose boundary $\partial\mathscr{B}$ consists of a finite number of regular surface elements; $\mathscr{B}$ may be bounded or unbounded. All processes in this section are assumed to be of Heaviside type and, at the very least, integrable on bounded subsets of $\mathscr{R}$.

A vector field process $\boldsymbol{f}$ on $\mathscr{B}$ is said to be ***simple*** if, and only if, its values are class C^2 fields and either

$$\operatorname{div} \boldsymbol{f} = 0 \tag{32.1}$$

or

$$\operatorname{curl} \boldsymbol{f} = \boldsymbol{0}. \tag{32.2}$$

If $\boldsymbol{f}$ satisfies (32.1) it is said to be ***solenoidal***, while if $\boldsymbol{f}$ satisfies (32.2) it is said to be ***irrotational***. If $\boldsymbol{f}$ is irrotational and $\mathscr{B}$ is simply connected, there exists a *scalar* field process ϕ such that

$$\boldsymbol{f} = \nabla \phi. \tag{32.3}$$

[110] TSIEN [1950, *4*] uses differential operator rather than Boltzmann laws. Also the conditions given by TSIEN for isotropy imply that the time dependences of shear and bulk behavior are proportional, which condition is not necessary for the material to be isotropic. (See also Sect. 43.) The results of ALFREY and TSIEN presented for differential operator laws have been related to similar results for integral laws (Boltzmann laws) by GRAFFI [1952, *1*]. He too considers the case where the time dependences of shear and bulk behavior are proportional.

If in addition $\boldsymbol{f}$ is solenoidal, the values of ϕ are harmonic on $\mathscr{B}$:

$$\Delta\phi = 0. \tag{32.4}$$

Finally, since the values of ϕ are class C^∞ scalar fields on $\mathscr{B}$, it follows that

$$\Delta \boldsymbol{f} = \Delta(\nabla\phi) = \nabla(\Delta\phi) = \boldsymbol{0}. \tag{32.5}$$

Hence the values of $\boldsymbol{f}$ are harmonic vector fields on $\mathscr{B}$ whenever $\boldsymbol{f}$ is both solenoidal and irrotational.

We now have the

First regularity theorem.[111] *Let $[\boldsymbol{u}, \boldsymbol{E}, \boldsymbol{S}]$ be in* QSVP$(G_1, G_2, \boldsymbol{b})$[112] *where $\boldsymbol{b}$ is both solenoidal and irrotational. Further suppose that the values of $\boldsymbol{u}$ are class C^5 vector fields on $\mathscr{B}$ and that the response functions G_1 and $2G_1 + G_2$ do not vanish. Then:*

(i) *the values $\boldsymbol{u}$, $\boldsymbol{E}$, and $\boldsymbol{S}$ are class C^∞ fields on $\mathscr{B}$;*

(ii) $\Delta(\operatorname{div}\boldsymbol{u}) = 0$, $\Delta(\operatorname{curl}\boldsymbol{u}) = \boldsymbol{0}$, $\Delta(\operatorname{tr}\boldsymbol{E}) = 0$, *and* $\Delta(\operatorname{tr}\boldsymbol{S}) = 0$ *on $\mathscr{B}$; and*

(iii) $\Delta\Delta\boldsymbol{u} = \boldsymbol{0}$, $\Delta\Delta\boldsymbol{E} = \boldsymbol{0}$, *and* $\Delta\Delta\boldsymbol{S} = \boldsymbol{0}$ *on $\mathscr{B}$.*

This result follows from the displacement equation of equilibrium, in the form[113]

$$G_1 \circledast \Delta\boldsymbol{u} + \tfrac{1}{3}(G_1 + 2G_2) \circledast \nabla\operatorname{div}\boldsymbol{u} + 2\boldsymbol{b} = \boldsymbol{0} \tag{32.6}$$

together with the properties of $\boldsymbol{b}$ and Titchmarsh's theorem.

By using arguments of DUFFIN and FRIEDRICHS[114] the results of the first regularity theorem may be strengthened.

Second regularity theorem.[115] *Let $[\boldsymbol{u}, \boldsymbol{E}, \boldsymbol{S}]$ be in* QSVP$(G_1, G_2, \boldsymbol{b})$ *where $\boldsymbol{b}$ is both solenoidal and irrotational. Suppose the processes G_1 and $2G_1 + G_2$ do not vanish. Then:*

(i) *the values of $\boldsymbol{u}$ and $\boldsymbol{E}$ are class C^∞ fields on $\mathscr{B}$;*

(ii) $\Delta(\operatorname{div}\boldsymbol{u}) = \boldsymbol{0}$, $\Delta(\operatorname{curl}\boldsymbol{u}) = \boldsymbol{0}$ *on $\mathscr{B}$; and*

(iii) $\Delta\Delta\boldsymbol{u} = \boldsymbol{0}$, $\Delta\Delta\boldsymbol{E} = \boldsymbol{0}$ *on $\mathscr{B}$.*

In addition, if the values of $\boldsymbol{u}$ are class C^3 vector fields on $\mathscr{B}$, then:

(iv) *the values of $\boldsymbol{S}$ are class C^∞ symmetric tensor fields on $\mathscr{B}$; and*

(v) $\Delta(\operatorname{tr}\boldsymbol{S}) = 0$, $\Delta\Delta\boldsymbol{S} = \boldsymbol{0}$ *on $\mathscr{B}$.*

Now we turn to the problem of complete sets of solutions to viscoelastic problems. In particular, we exhibit the viscoelastic generalizations of the Boussinesq-Somigliana-Galerkin and Boussinesq-Papkovich-Neuber solutions of classical elastostatics.[116]

Let $\boldsymbol{u}$ be the displacement process of $[\boldsymbol{u}, \boldsymbol{E}, \boldsymbol{S}]$ in QSVP$(G_1, G_2, \boldsymbol{b})$. Then the values of $\boldsymbol{u}$ are class C^2 vector fields on $\mathscr{B}$ and $\boldsymbol{u}$ satisfies the displacement equation of equilibrium (32.6). Conversely, if the values of $\boldsymbol{u}$ are class C^2 vector fields on $\mathscr{B}$ and $\boldsymbol{u}$ satisfies (32.6), then there exist processes $\boldsymbol{E}$ and $\boldsymbol{S}$ such that the

[111] GURTIN and STERNBERG [1962, *10*, Theorem 6.5]. A similar regularity result has been obtained by ELDER [1964, *7*], in the absence of body forces.

[112] Since $\mathscr{B}$ is isotropic, its relaxation function $\boldsymbol{G}$ is completely determined by the scalar relaxation functions G_1 and G_2. We therefore write QSVP$(G_1, G_2, \boldsymbol{b})$ for QSVP$(\boldsymbol{G}, \boldsymbol{b})$.

[113] The equation of equilibrium in the form (32.6) follows from (20.3) and the comments in Sect. 12.

[114] DUFFIN [1956, *2*]; FRIEDRICHS [1947, *1*].

[115] GURTIN and STERNBERG [1962, *10*, Theorem 6.6].

[116] For the elastic counterparts refer to LTE, Sect. 44.

process $[\boldsymbol{u}, \boldsymbol{E}, \boldsymbol{S}]$ lies in QSVP$(G_1, G_2, \boldsymbol{b})$. We call such a process $\boldsymbol{u}$ a ***displacement process for $\mathscr{B}$ corresponding to G_1, G_2, and $\boldsymbol{b}$.***

Theorem (Generalized Boussinesq-Somigliana-Galerkin solution).[117] *Let $\boldsymbol{b}$ be a body force process whose values are continuous vector fields on $\mathscr{B}$ and suppose the isotropic scalar relaxation functions G_1 and G_2 are such that the relaxation laws $\mathscr{L}_{G_1}$ and $\mathscr{L}_{2G_1+G_2}$ are invertible (in the Heaviside classes).*[118] *Let $\boldsymbol{g}$ be a process whose values are class C^4 vector fields on $\mathscr{B}$ such that*

$$\Delta\Delta \boldsymbol{g} = -\boldsymbol{f}, \tag{32.7}$$

where $\boldsymbol{f}$ is defined by[119]

$$\boldsymbol{f} = (2G_1 + G_2)^{-1} \circledast G_1^{-1} \circledast \boldsymbol{b}. \tag{32.8}$$

Then the process $\boldsymbol{u}$, defined by

$$\boldsymbol{u} = 2(2G_1 + G_2) \circledast \Delta \boldsymbol{g} - (G_1 + 2G_2) \circledast \nabla \operatorname{div} \boldsymbol{g}, \tag{32.9}$$

is a displacement process for $\mathscr{B}$ corresponding to G_1, G_2, and $\boldsymbol{b}$.

The proof of this result is direct and uses Eq. (32.6). Observe that, with additional regularity, the restriction to Heaviside type processes may be omitted.

Theorem (Generalized Boussinesq-Papkovich-Neuber solution).[120] *Let $\boldsymbol{b}$ be a body force process whose values are continuous vector fields on $\mathscr{B}$ and suppose the isotropic scalar relaxation functions G_1 and G_2 are such that the relaxation laws $\mathscr{L}_{G_1}$ and $\mathscr{L}_{2G_1+G_2}$ are invertible (in the Heaviside classes). Let ϕ and $\boldsymbol{\psi}$ be processes whose values are class C^3 scalar and vector fields on $\mathscr{B}$ such that*

$$\Delta\phi = -\tfrac{1}{2}\boldsymbol{p}_0 \cdot \boldsymbol{f}, \tag{32.10}$$

and

$$\Delta\boldsymbol{\psi} = \tfrac{1}{2}\boldsymbol{f}, \tag{32.11}$$

where $\boldsymbol{f}$ is defined in (32.8) and $\boldsymbol{p}_0$ is the position function defined in $\mathscr{E}$ for any fixed point $\boldsymbol{x}_0 \in \mathscr{E}$ by

$$\boldsymbol{p}_0(\boldsymbol{x}) = \boldsymbol{x} - \boldsymbol{x}_0. \tag{32.12}$$

Then the process $\boldsymbol{u}$ defined by

$$\boldsymbol{u} = (G_1 + 2G_2) \circledast \nabla(\phi + \boldsymbol{p}_0 \cdot \boldsymbol{\psi}) - 4(2G_1 + G_2) \circledast \boldsymbol{\psi} \tag{32.13}$$

is a displacement process for $\mathscr{B}$ corresponding to G_1, G_2, and $\boldsymbol{b}$.

The comments following the previous theorem apply here as well. Of course the solutions (32.9) and (32.13) reduce to the Boussinesq-Somigliana-Galerkin and Boussinesq-Papkovich-Neuber solution of elastostatics whenever the solid is elastic; that is, whenever G_1 and G_2 are constant on $\mathscr{R}^+$.[121]

We see from the previous two results that the generation of a displacement process $\boldsymbol{u}$ for $\mathscr{B}$ corresponding to G_1, G_2 and $\boldsymbol{b}$ reduces to the solution of either the inhomogeneous *biharmonic* equation (32.7) or the inhomogeneous *harmonic* equa-

---

[117] GURTIN and STERNBERG [1962, *10*, Theorem 9.1]. Essentially the same result has been given by ELDER [1964, *7*] for the case where the body forces vanish.

[118] This will be the case if, for example, $\mathring{G}_1$ and $\mathring{G}_2$ are positive (cf. Sect. 11).

[119] If a response function K is such that $\mathscr{L}_K$ is invertible, we write $K^{-1}\circledast$ for $\mathscr{L}_K^{-1}$ (cf. Sects. 5 and 11).

[120] GURTIN and STERNBERG [1962, *10*, Theorem 9.2]. See also ELDER [1964, *7*] for a version with vanishing body forces. A generalization of LOVE's solution for axially symmetric problems in terms of a single scalar bi-harmonic function is also carried out in this article. See also the historical note in LTE, Sect. 44.

[121] See LTE, Sect. 44.

tions (32.10) and (32.11). Solutions to these problems may be generated through a generalization of the Newtonian potential.[122]

Let ϱ be a process for $\mathscr{B}$ such that the values of ϱ are scalar fields which are of class C on $\overline{\mathscr{B}}$ and of class C^N on $\mathscr{B}$ ($N \geqq 1$). Define the ***Newtonian potential*** U for an integrable region $\mathscr{B}$ by

$$U(\boldsymbol{x}, t) = -\frac{1}{4\pi}\int\limits_{\mathscr{B}} \frac{\varrho(\boldsymbol{\xi}, t)}{|\boldsymbol{x}-\boldsymbol{\xi}|}\, dV(\boldsymbol{\xi}), \qquad (\boldsymbol{x}, t) \in \mathscr{B}\times\mathscr{R}. \tag{32.14}$$

Note that U defined in (32.14) may be considered a process (of Heaviside type) whose values are scalar fields on $\mathscr{B}$. It can be shown that the values of U are scalar fields of class $N+1$ on $\mathscr{B}$ and

$$\Delta U = \varrho \tag{32.15}$$

on $\mathscr{B}$[123].

Using the Newtonian potential and the second of the previous two theorems we see that if the region $\mathscr{B}$ is integrable, the hypotheses of the theorem regarding G_1, G_2, and $\boldsymbol{b}$ are satisfied, and the values of $\boldsymbol{b}$ are of class C on $\overline{\mathscr{B}}$ and of class C^N on $\mathscr{B}$ ($N \geqq 2$), then the processes ϕ and $\boldsymbol{\psi}$ given on $\mathscr{B}$ by

$$\phi(\boldsymbol{x}, t) = \frac{1}{8\pi}\int\limits_{\mathscr{B}} \frac{\boldsymbol{\xi}\cdot\boldsymbol{f}(\boldsymbol{\xi}, t)}{|\boldsymbol{x}-\boldsymbol{\xi}|}\, dV(\boldsymbol{\xi}), \qquad (\boldsymbol{x}, t) \in \mathscr{B}\times\mathscr{R}, \tag{32.16}$$

and

$$\boldsymbol{\psi}(\boldsymbol{x}, t) = \frac{-1}{8\pi}\int\limits_{\mathscr{B}} \frac{\boldsymbol{f}(\boldsymbol{\xi}, t)}{|\boldsymbol{x}-\boldsymbol{\xi}|}\, dV(\boldsymbol{\xi}), \qquad (\boldsymbol{x}, t) \in \mathscr{B}\times\mathscr{R}, \tag{32.17}$$

where $\boldsymbol{f}$ is defined in (32.8), generate a displacement process $\boldsymbol{u}$ corresponding to G_1, G_2, and $\boldsymbol{b}$ through (32.13).[124] In this fashion, it is possible to generate particular solutions to the displacement equation of equilibrium (32.6).

The generalized Boussinesq-Somigliana-Galerkin and Boussinesq-Papkovich-Neuber solutions discussed above are complete as are their elastic counterparts.[125] By *complete* we mean that every sufficiently regular displacement process $\boldsymbol{u}$ corresponding to G_1, G_2, and $\boldsymbol{b}$ admits such generalized representations.

Theorem[126] ***(Completeness).*** *Let $\mathscr{B}$ be an integrable region and suppose that $\boldsymbol{u}$ is a displacement process for $\mathscr{B}$ corresponding to G_1, G_2, and $\boldsymbol{b}$, where $\mathscr{L}_{G_1}$ and $\mathscr{L}_{2G_1+G_2}$ are invertible (in the Heaviside classes). Further suppose that the values of $\boldsymbol{u}$ are of class C on $\overline{\mathscr{B}}$ and of class C^4 on $\mathscr{B}$. Then there exists a process $\boldsymbol{g}$, whose values are class C^4 vector fields on $\mathscr{B}$ and which satisfies* (32.7), *such that* (32.9) *holds. Moreover, if the values of $\boldsymbol{u}$ are class C^5 vector fields on $\mathscr{B}$, then there exist processes ϕ and $\boldsymbol{\psi}$, whose values are class C^3 scalar and vector fields on $\mathscr{B}$ and which satisfy* (32.10) *and* (32.11), *such that* (32.13) *holds.*

Observe that the method of generating functions described here reduces the problem of finding processes $[\boldsymbol{u}, \boldsymbol{E}, \boldsymbol{S}]$ in QSVP$(G_1, G_2, \boldsymbol{b})$ to the solution of time-dependent Poisson equations,[127] provided there is sufficient regularity. However, in order to determine the boundary conditions which must be satisfied, it

[122] See, for example, COURANT and HILBERT [1962, *8*]. Here we follow the exposition of GURTIN and STERNBERG [1962, *10*, Sect. 9].
[123] GURTIN and STERNBERG [1962, *10*, Lemma 9.1].
[124] For a formal proof see GURTIN and STERNBERG [1962, *10*, Theorem 9.3].
[125] See LTE, Sect. 44.
[126] GURTIN and STERNBERG [1962, *10*, Theorem 9.4].
[127] Cf. Sect. 29.

is generally necessary to solve linear integral equations of the type considered in Sect. 29.

33. Singular solutions. Consider a viscoelastic body which occupies the entire space $\mathscr{E}$. Suppose the body $\mathscr{E}$ is homogeneous, isotropic, and of relaxation type with scalar response functions G_1 and G_2. Unless stated otherwise, all processes for $\mathscr{E}$ are assumed to be in Heaviside class H. We further assume that G_1 and G_2, are in Heaviside class H^1 and are *initially positive:*

$$\mathring{G}_1 > 0, \quad \mathring{G}_2 > 0. \tag{33.1}$$

Thus it follows that the Boltzmann laws $\mathscr{L}_{G_1}$, $\mathscr{L}_{G_2}$, and $\mathscr{L}_{2G_1+G_2}$ are all invertible (in the Heaviside classes).[128]

Since the body $\mathscr{E}$ is unbounded, we amend the definition of quasi-static viscoelastic process to include unbounded bodies $\mathscr{B}$: an admissible process $\boldsymbol{P} \equiv [\boldsymbol{u}, \boldsymbol{E}, \boldsymbol{S}]$ for a body $\mathscr{B}$ is said to be a ***quasi-static viscoelastic process for $\mathscr{B}$ corresponding to the relaxation function G and body force b*** if, and only if, the processes $\boldsymbol{u}$, $\boldsymbol{E}$, and $\boldsymbol{S}$ satisfy the conditions (R-1)–(R-3) listed in Sect. 19, and, if $\mathscr{B}$ is unbounded,

(R-4) $\boldsymbol{u}_{\boldsymbol{x}} = O(|\boldsymbol{x}|^{-1})$, $\boldsymbol{S}_{\boldsymbol{x}} = O(|\boldsymbol{x}|^{-2})$, and $\boldsymbol{b}_{\boldsymbol{x}} = O(|\boldsymbol{x}|^{-3})$ as $|\boldsymbol{x}| \to \infty$, uniformly on closed, bounded subsets of $\mathscr{R}$.[129]

The class of all quasi-static viscoelastic processes for $\mathscr{B}$ corresponding to $\mathbf{G}$ and $\boldsymbol{b}$ is denoted by QSVP$(\mathscr{B}, \mathbf{G}, \boldsymbol{b})$. If the body forces vanish, we write QSVP$(\mathscr{B}, \mathbf{G})$ for QSVP$(\mathscr{B}, \mathbf{G}, \mathbf{0})$ and if $\mathbf{G}$ is isotropic (as in this section) we replace $\mathbf{G}$ by the pair (G_1, G_2).[130]

If $\boldsymbol{x}_0$ is a fixed point in $\mathscr{E}$ and $\varrho \geqq 0$, $\mathscr{B}(\boldsymbol{x}_0, \varrho)$ denotes the open ball in $\mathscr{E}$ with center $\boldsymbol{x}_0$ and radius ϱ. We also write $\mathscr{E}_{\boldsymbol{x}_0}$ for the open set $\mathscr{E} - \{\boldsymbol{x}_0\}$.

In this section, we consider the viscoelastic problem associated with a concentrated load applied at a point in a viscoelastic body. This is the viscoelastic counterpart of Kelvin's problem in elastostatics.[131] The formulation presented here is that of STERNBERG and AL-KHOZAIE.[132]

Let $\boldsymbol{l}$ be a process of class H whose values are vectors in $\mathscr{V}$. Then, following STERNBERG and AL-KHOZAIE, we say that ***a sequence $\{\boldsymbol{b}^n\}$ of body force processes tends to a concentrated load $\boldsymbol{l}$ applied at $\boldsymbol{x}_0 \in \mathscr{E}$*** if, and only if:

(C.L.-1) For each n ($n = 1, 2, 3, \ldots$), $\boldsymbol{b}^n$ is a process for $\mathscr{E}$ in class H and the values of $\boldsymbol{b}^n$ are class C^2 vector fields on $\mathscr{E}$ which vanish outside $\mathscr{B}(\boldsymbol{x}_0, \varrho_n)$, where $\varrho_n \to 0$ as $n \to \infty$;

(C.L.-2)
$$\int_{\mathscr{B}(\boldsymbol{x}_0, \varrho_n)} \boldsymbol{b}^n \, dV \to \boldsymbol{l} \quad \text{as} \quad n \to \infty,$$

uniformly on closed and bounded subsets of $\mathscr{R}$; and

---

[128] Cf. Sect. 11.

[129] Let ϕ be a field process for a body $\mathscr{B}$. Then $\phi_{\boldsymbol{x}} = O(|\boldsymbol{x}|^n)$ ($n = 0, \pm 1, \pm 2, \ldots$) *as* $|\boldsymbol{x}| \to \infty$ *uniformly on closed and bounded subsets* $\mathscr{J}$ *of* $\mathscr{R}$ means that there exist numbers α and γ, which may depend upon $\mathscr{J}$, such that $|\boldsymbol{x}| > \alpha$ implies $|\phi(\boldsymbol{x}, t)| < \gamma$ for all times $t \in \mathscr{J}$.

[130] Observe that in the case of a *bounded* body $\mathscr{B}$ the definition given here agrees with that of Sect. 19. The notation has been modified to make the dependence upon $\mathscr{B}$, G_1, and G_2 explicit.

[131] See LTE, Sect. 51.

[132] STERNBERG and AL-KHOZAIE [1964, *24*]. These authors follow the development by STERNBERG and EUBANKS [1955, *9*] for the theory of elastostatics. The essential method for describing concentrated loads was deduced by THOMSON and TAIT [1879, *1*], [1883, *1*] and refined by LOVE [1944, *2*]. The first elastic solution appeared in a note by THOMSON (Lord KELVIN) [1884, *1*]. See also STERNBERG [1954, *2*] and LTE, Sect. 51.

(C.L.-3) The sequence of scalar processes

$$\left\{ \int_{\mathscr{B}(\boldsymbol{x}_0,\, \varrho_n)} |\boldsymbol{b}^n| \, dV \right\}$$

is uniformly bounded on closed and bounded subsets of $\mathscr{R}$.

Using this definition of *concentrated load* we have the following result concerning the basic singular solution.

Theorem[133] ***(Viscoelastic Kelvin processes).*** *Let $\{\boldsymbol{b}^n\}$ be a sequence of body force processes for $\mathscr{E}$ which tends to the concentrated load $\boldsymbol{l}$ applied at $\boldsymbol{x}_0 \in \mathscr{E}$. Then:*

(i) *there exists a sequence of processes $\boldsymbol{P}^n \equiv [\boldsymbol{u}^n, \boldsymbol{E}^n, \boldsymbol{S}^n]$ for $\mathscr{E}$ such that, for each n $(n = 1, 2, 3, \ldots)$, $\boldsymbol{P}^n$ lies in* QSVP$(\mathscr{E}, G_1, G_2, \boldsymbol{b}^n)$;

(ii) *$\boldsymbol{P}^n$ converges to a limit process $\boldsymbol{P}$ as $n \to \infty$ in such a way that the convergence is uniform on sets of the form $\overline{\mathscr{B}}_{\boldsymbol{x}_0} \times [0, T]$, where $\overline{\mathscr{B}}_{\boldsymbol{x}_0}$ is any closed and bounded set in $\mathscr{E}$ not containing $\boldsymbol{x}_0 \in \mathscr{E}$ and $T \geqq 0$;*[134] *and*

(iii) *the limit process $\boldsymbol{P}$ is independent of the choice of the sequence $\{\boldsymbol{b}^n\}$ and is generated by the Boussinesq-Papkovich-Neuber stress processes ϕ and $\boldsymbol{\psi}$ defined for $(\boldsymbol{x}, t) \in \mathscr{E}_{\boldsymbol{x}_0} \times \mathscr{R}$ through*[135]

$$\phi(\boldsymbol{x}, t) = 0, \tag{33.2}$$

$$\boldsymbol{\psi}(\boldsymbol{x}, t) = -\frac{\boldsymbol{f}(t)}{8\pi |\boldsymbol{x} - \boldsymbol{x}_0|}, \tag{33.3}$$

where $\boldsymbol{f}$ is the vector process

$$\boldsymbol{f} = (2G_1 + G_2)^{-1} \circledast G_1^{-1} \circledast \boldsymbol{l}. \tag{33.4}$$

The process $\boldsymbol{P}$ is called ***the viscoelastic Kelvin process corresponding to the concentrated load $\boldsymbol{l}$ applied at $\boldsymbol{x}_0 \in \mathscr{E}$ and to the relaxation functions G_1 and G_2.***

Observe that conditions (C.L.-1) and (C.L.-3) in the definition of concentrated load imply:

(C.L.-3)′

$$\int_{\mathscr{B}(\boldsymbol{x}_0,\, \varrho_n)} \boldsymbol{p}_0 \times \boldsymbol{b}^n \, dV \to \boldsymbol{0} \quad \text{as} \quad n \to \infty,$$

uniformly on closed and bounded subsets of $\mathscr{R}$, where $\boldsymbol{p}_0$ is the position map defined in $\mathscr{E}$ for fixed $\boldsymbol{x}_0 \in \mathscr{E}$ by

$$\boldsymbol{p}_0(\boldsymbol{x}) = \boldsymbol{x} - \boldsymbol{x}_0. \tag{33.5}$$

It can be shown that condition (C.L.-3) of the definition cannot be replaced by the physically more natural condition (C.L.-3)′ without rendering the conclusion of the theorem invalid.[136] Indeed, for the special case where $\mathscr{E}$ is elastic, a sequence of body force processes $\{\boldsymbol{b}^n\}$ can be found which tends to the concentrated load $\boldsymbol{l} \equiv \boldsymbol{0}$ at $\boldsymbol{x}_0 \in \mathscr{E}$ but which generates, according to the theorem, a nontrivial process $\boldsymbol{P}$.[137] However, conditions (C.L.-1) and (C.L.-2) of the definition imply condition (C.L.-3) if the body force processes $\{\boldsymbol{b}^n\}$ are parallel and of the same sense.

Let $\{\boldsymbol{e}_i\}$ $(i = 1, 2, 3)$ be a fixed right-handed orthonormal basis in $\mathscr{V}$ and let h denote the scalar Heaviside unit process. Write $\boldsymbol{P}^i_{(\boldsymbol{x}_0)}$ for the ***(normalized) Kelvin process corresponding to the concentrated load $\boldsymbol{l} = h\boldsymbol{e}_i$ applied at $\boldsymbol{x}_0$ and the relaxation functions G_1 and G_2.*** The symbol $\boldsymbol{P}^i_{(\boldsymbol{x}_0)}$ represents the

---

[133] STERNBERG and AL-KHOZAIE [1964, *24*, Theorem 2.1].

[134] Here, for each n $(n = 1, 2, 3, \ldots)$, we consider $\boldsymbol{P}^n$ as a function on $\mathscr{E} \times \mathscr{R}$.

[135] Cf. Sect. 32.

[136] STERNBERG and AL-KHOZAIE [1964, *24*, Sect. 2].

[137] STERNBERG and EUBANKS [1955, *9*].

function on $\mathscr{E}_{x_0} \times \mathscr{R}$ whose values are $P^i_{(x_0)}(x, t)$ $(i = 1, 2, 3)$. It is easily seen from part (iii) of the theorem that for fixed $x_0 \in \mathscr{E}$

$$P^i_{(x_0)}(x, t) = P^i_{(0)}(x - x_0, t), \qquad (x, t) \in \mathscr{E}_{x_0} \times \mathscr{R}. \tag{33.6}$$

Hence it suffices, in most cases, to consider the concentrated load applied at the origin in $\mathscr{E}$. If $x_0 = 0$, we will write P^i for $P^i_{(0)}$.

For reference we record the specific representation of $P^i \equiv [u^i, E^i, S^i]$. For $x \neq 0$ write $x = re$, so that $r = |x| > 0$ and $|e| = 1$. Then, for $(x, t) \in \mathscr{E}_0 \times \mathscr{R}$,

$$u^i(x, t) = \frac{1}{8\pi r}\{2J_1(t)\,[e_i + (e \cdot e_i)\,e] + 3\,Q_1(t)\,[e_i - (e \cdot e_i)\,e]\}, \tag{33.7}$$

$$\begin{aligned} E^i(x, t) = \frac{-3}{8\pi r^2}\{&[2J_1(t) - 3\,Q_1(t)]\,(e \cdot e_i)\,(e \otimes e) + Q_1(t) \\ &\cdot [2\,\mathrm{sym}\,(e \otimes e_i) - (e \cdot e_i)\,\mathbf{1}]\}, \end{aligned} \tag{33.8}$$

$$\begin{aligned} S^i(x, t) = \frac{-3}{8\pi r^2}\{&[2h(t) - 3\,Q_2(t)]\,(e \cdot e_i)\,(e \otimes e) + Q_2(t) \\ &\cdot [2\,\mathrm{sym}\,(e \otimes e_i) - (e \cdot e_i)\,\mathbf{1}]\}, \end{aligned} \tag{33.9}$$

where J_1, Q_1, and Q_2 are the auxiliary response functions given by

$$J_1 = G_1^{-1}, \qquad Q_1 = (2G_1 + G_2)^{-1}, \qquad Q_2 = G_1 \circledast Q_1. \tag{33.10}$$

Theorem[138] ***(Properties of the Kelvin process).*** *The Kelvin process* P^i $(i = 1, 2, 3)$ *has the properties:*

(i) *the process* $P^i \equiv [u^i, E^i, S^i]$ *for the body* $\mathscr{E}_0$ *lies in* $\mathrm{QSVP}(\mathscr{E}_0, G_1, G_2, 0)$ *and its values are class* C^∞ *fields on* $\mathscr{E}_0$; *if we let* $\mathscr{S}$ *be any closed regular surface surrounding the origin in* $\mathscr{E}$ *and let* $s^i_{(n)} = S^i n$, *where* n *is the inward unit normal vector field on* $\mathscr{S}$, *then*

(ii) $\int_{\mathscr{S}} s^i_{(n)}\, dA = h\, e_i$;

(iii) $\int_{\mathscr{S}} p_0 \times s^i_{(n)}\, dA = 0 \quad (x_0 = 0)$; *and*

(iv) $u^i_x = O(|x|^{-1})$, $S^i_x = O(|x|^{-2})$ *as* $|x| \to \infty$ *uniformly on closed and bounded subsets of* $\mathscr{R}$.

The properties (i), (ii), and (iii) of the Kelvin processes P^i do not determine P^i uniquely. Indeed, there are viscoelastic processes which are not null and yet possess self-equilibrated singularities at the origin.[139]

Observe that the form of the Kelvin processes P^i given in (33.7)–(33.9) may be *formally* obtained from their elastic counterparts[140] through the correspondence principle discussed in Sect. 31.

For any pair of indices i, j $(i, j = 1, 2, 3)$ define the ***Kelvin (normalized) doublet process*** $P^{ij}_{(x_0)}$ ***at*** x_0 by

$$P^{ij}_{(x_0)}(x, t) = \lim_{\alpha \to 0}\left\{\frac{1}{\alpha}\,[P^i_{(x_0)}(x + \alpha e_j, t) - P^i_{(x_0)}(x, t)]\right\}, \quad \text{on} \quad \mathscr{E}_{x_0} \times \mathscr{R}, \tag{33.11}$$

or, equivalently,

$$P^{ij}_{(x_0)}(x, t) = \lim_{\alpha \to 0}\left\{\frac{1}{\alpha}\,[P^i_{(x_0 - \alpha e_j)}(x, t) - P^i_{(x_0)}(x, t)]\right\}, \quad \text{on} \quad \mathscr{E}_{x_0} \times \mathscr{R}. \tag{33.12}$$

---

[138] STERNBERG and AL-KHOZAIE [1964, *24*, Theorem 2.2].

[139] See STERNBERG and EUBANKS [1955, *9*] for the elastic version of this statement. In the sequel to this section, we will also consider such processes.

[140] See LTE, Sect. 51.

It follows that

$$\boldsymbol{P}^{ij}_{(\boldsymbol{x}_0)}(\boldsymbol{x},t) = \boldsymbol{P}^{ij}_{(0)}(\boldsymbol{x}-\boldsymbol{x}_0,t), \quad \text{on} \quad \mathscr{E}_{\boldsymbol{x}_0}\times\mathscr{R}. \tag{33.13}$$

As before, it suffices to consider only Kelvin doublets at the origin in $\mathscr{E}$. Thus we write $\boldsymbol{P}^{ij}$ for $\boldsymbol{P}^{ij}_{(0)}$.

For reference, we record the specific form of $\boldsymbol{P}^{ij} \equiv [\boldsymbol{u}^{ij}, \boldsymbol{E}^{ij}, \boldsymbol{S}^{ij}]$. For $\boldsymbol{x} \neq \boldsymbol{0}$ write $\boldsymbol{x} = r\boldsymbol{e}$ so that $r = |\boldsymbol{x}| > 0$ and $|\boldsymbol{e}| = 1$. Then, for $(\boldsymbol{x}, t) \in \mathscr{E}_0 \times \mathscr{R}$,

$$\begin{aligned}\boldsymbol{u}^{ij}(\boldsymbol{x},t) = \frac{-1}{8\pi r^2}\{&[2J_1(t)+3Q_1(t)](\boldsymbol{e}\cdot\boldsymbol{e}_j)\boldsymbol{e}_i \\ &+[2J_1(t)-3Q_1(t)][3(\boldsymbol{e}\cdot\boldsymbol{e}_i)(\boldsymbol{e}\cdot\boldsymbol{e}_j)\boldsymbol{e} \\ &-(\boldsymbol{e}_i\cdot\boldsymbol{e}_j)\boldsymbol{e}-(\boldsymbol{e}\cdot\boldsymbol{e}_i)\boldsymbol{e}_j]\},\end{aligned} \tag{33.14}$$

$$\begin{aligned}\boldsymbol{E}^{ij}(\boldsymbol{x},t) = \frac{3}{8\pi r^3}\{&[2J_1(t)-3Q_1(t)][(5(\boldsymbol{e}\cdot\boldsymbol{e}_i)(\boldsymbol{e}\cdot\boldsymbol{e}_j) \\ &-(\boldsymbol{e}_i\cdot\boldsymbol{e}_j))(\boldsymbol{e}\otimes\boldsymbol{e})-2(\boldsymbol{e}\cdot\boldsymbol{e}_i)\operatorname{sym}(\boldsymbol{e}\otimes\boldsymbol{e}_j)] \\ &+Q_1(t)[6(\boldsymbol{e}\cdot\boldsymbol{e}_j)\operatorname{sym}(\boldsymbol{e}\otimes\boldsymbol{e}_i)-2\operatorname{sym}(\boldsymbol{e}_i\otimes\boldsymbol{e}_j) \\ &-(3(\boldsymbol{e}\cdot\boldsymbol{e}_i)(\boldsymbol{e}\cdot\boldsymbol{e}_j)-(\boldsymbol{e}_i\cdot\boldsymbol{e}_j))\boldsymbol{1}]\},\end{aligned} \tag{33.15}$$

$$\begin{aligned}\boldsymbol{S}^{ij}(\boldsymbol{x},t) = \frac{3}{8\pi r^3}\{&[2h(t)-3Q_2(t)][(5(\boldsymbol{e}\cdot\boldsymbol{e}_i)(\boldsymbol{e}\cdot\boldsymbol{e}_j) \\ &-(\boldsymbol{e}_i\cdot\boldsymbol{e}_j))(\boldsymbol{e}\otimes\boldsymbol{e})-2(\boldsymbol{e}\cdot\boldsymbol{e}_i)\operatorname{sym}(\boldsymbol{e}\otimes\boldsymbol{e}_j)] \\ &+Q_2(t)[6(\boldsymbol{e}\cdot\boldsymbol{e}_j)\operatorname{sym}(\boldsymbol{e}\otimes\boldsymbol{e}_i)-2\operatorname{sym}(\boldsymbol{e}_i\otimes\boldsymbol{e}_j) \\ &-(3(\boldsymbol{e}\cdot\boldsymbol{e}_i)(\boldsymbol{e}\cdot\boldsymbol{e}_j)-(\boldsymbol{e}_i\cdot\boldsymbol{e}_j))\boldsymbol{1}]\},\end{aligned} \tag{33.16}$$

where J_1, Q_1, and Q_2 are defined in (33.10).

From (33.14)–(33.16) the properties of the Kelvin doublets may be deduced

Theorem[141] ***(Properties of the Kelvin doublets).*** *The Kelvin doublet process $\boldsymbol{P}^{ij}$ has the properties:*

(i) *the process $\boldsymbol{P}^{ij} \equiv [\boldsymbol{u}^{ij}, \boldsymbol{E}^{ij}, \boldsymbol{S}^{ij}]$ for the body $\mathscr{E}_0$ lies in* QSVP$(\mathscr{E}_0, G_1, G_2, \boldsymbol{0})$ *and its values are class C^∞ fields on $\mathscr{E}_0$; if we let $\mathscr{S}$ be any closed regular surface surrounding the origin in $\mathscr{E}$ and let $\boldsymbol{s}^{ij}_{(\boldsymbol{n})} = \boldsymbol{S}^{ij}\boldsymbol{n}$, where $\boldsymbol{n}$ is the unit inward normal vector field on $\mathscr{S}$, then*

(ii) $\int_{\mathscr{S}} \boldsymbol{s}^{ij}_{(\boldsymbol{n})}\, dA = \boldsymbol{0}$;

(iii) $\int_{\mathscr{S}} \boldsymbol{p}_0 \times \boldsymbol{s}^{ij}_{(\boldsymbol{n})}\, dA = \varepsilon_{ijk}\boldsymbol{e}_k h$, $(\boldsymbol{x}_0 = \boldsymbol{0})$; *and*

(iv) $\boldsymbol{u}^{ij}_{\boldsymbol{x}} = O(|\boldsymbol{x}|^{-2})$, $\boldsymbol{S}^{ij}_{\boldsymbol{x}} = O(|\boldsymbol{x}|^{-3})$ *as* $|\boldsymbol{x}| \to \infty$ *uniformly on closed and bounded subsets of* $\mathscr{R}$.

Observe that $\boldsymbol{P}^{ij}$ is statically equivalent to a force couple if $i \neq j$ and statically null if $i = j$. Therefore, $\boldsymbol{P}^{ij}_{(\boldsymbol{x}_0)}$ is called a ***Kelvin doublet process at $\boldsymbol{x}_0$, with moment*** or ***without moment*** according as $i \neq j$ or $i = j$. Using the doublet processes $\boldsymbol{P}^{ij}_{(\boldsymbol{x}_0)}$, two other processes of some interest may be defined. Define a ***viscoelastic center of compression $\boldsymbol{P}^{0}_{(\boldsymbol{x}_0)}$ at $\boldsymbol{x}_0$*** by

$$\boldsymbol{P}^{0}_{(\boldsymbol{x}_0)} = \boldsymbol{P}^{ii}_{(\boldsymbol{x}_0)}, \tag{33.17}$$

141 Op. cit. footnote 136, Theorem 2.3; ε_{ijk} is the alternator tensor. As usual, summation is implied by repeated indices. Cf. LTE, Sect. 3.

and a ***viscoelastic center of rotation*** $\overline{\boldsymbol{P}}^i_{(\boldsymbol{x}_0)}$ ***at*** $\boldsymbol{x}_0$ ***in the direction*** $\boldsymbol{e}_i$ by

$$\overline{\boldsymbol{P}}^i_{(\boldsymbol{x}_0)} = \tfrac{1}{2}\varepsilon_{ijk}\boldsymbol{P}^{ik}_{(\boldsymbol{x}_0)}. \tag{33.18}$$

From (33.14)–(33.18) we have, for each $(\boldsymbol{x}, t) \in \mathscr{E}_0 \times \mathscr{R}$ $(\boldsymbol{x}_0 = \boldsymbol{0})$

$$\boldsymbol{u}^0(\boldsymbol{x}, t) = \frac{-3}{4\pi r^2} Q_1(t)\, \boldsymbol{e}, \tag{33.19}$$

$$\boldsymbol{E}^0(\boldsymbol{x}, t) = \frac{3Q_1(t)}{4\pi r^2}\left[3(\boldsymbol{e}\otimes\boldsymbol{e}) - \boldsymbol{1}\right], \tag{33.20}$$

$$\boldsymbol{S}^0(\boldsymbol{x}, t) = \frac{3Q_2(t)}{4\pi r^2}\left[3(\boldsymbol{e}\otimes\boldsymbol{e}) - \boldsymbol{1}\right], \tag{33.21}$$

$$\overline{\boldsymbol{u}}^i(\boldsymbol{x}, t) = \frac{J_1(t)}{4\pi r^2}\varepsilon_{ijk}(\boldsymbol{e}\cdot\boldsymbol{e}_j)\,\boldsymbol{e}_k, \tag{33.22}$$

$$\overline{\boldsymbol{E}}^i(\boldsymbol{x}, t) = \frac{-3J_1(t)}{8\pi r^3}\varepsilon_{ijk}(\boldsymbol{e}\cdot\boldsymbol{e}_j)(\boldsymbol{e}_k\otimes\boldsymbol{e}), \tag{33.23}$$

$$\overline{\boldsymbol{S}}^i(\boldsymbol{x}, t) = \frac{-3h(t)}{8\pi r^3}\varepsilon_{ijk}(\boldsymbol{e}\cdot\boldsymbol{e}_j)(\boldsymbol{e}_k\otimes\boldsymbol{e}). \tag{33.24}$$

It is easily seen that $\boldsymbol{P}^0$ is a self-equilibrated Kelvin process at the origin and that $\overline{\boldsymbol{P}}^i$ is statically equivalent to a couple at the origin with axis $\boldsymbol{e}_i$. Specifically,

$$\int_{\mathscr{S}} \overline{\boldsymbol{s}}^i_{(\boldsymbol{n})}\, dA = \boldsymbol{0}, \tag{33.25}$$

$$\int_{\mathscr{S}} \boldsymbol{p}_0 \times \overline{\boldsymbol{s}}^i_{(\boldsymbol{n})}\, dA = h\,\boldsymbol{e}_i \qquad (\boldsymbol{x}_0 = \boldsymbol{0}). \tag{33.26}$$

Higher order Kelvin processes can be constructed following the pattern established above. These processes will also have values which are class C^∞ fields on $\mathscr{E}_{\boldsymbol{x}_0}$, but the singularities at $\boldsymbol{x}_0 \in \mathscr{E}$ will be of higher order.

34. Green's processes and integral solutions. The singular solutions (Kelvin processes) constructed in Sect. 33 may be used to obtain integral formula solutions of viscoelastic problems. The method is analogous to that used for the construction of Green's functions and integral formulas in classical elastostatics.[142] This section is a continuation of Sect. 33, so that the formulation presented here is that of STERNBERG and AL-KHOZAIE.[143]

Let $\mathscr{B}$ be a homogeneous and isotropic body which occupies a regular region in $\mathscr{E}$. Now $\mathscr{B}$ may be bounded or unbounded. Since the boundary $\partial\mathscr{B}$ of $\mathscr{B}$ consists of a finite number of regular closed surfaces, if $\mathscr{B}$ is unbounded it is necessarily an exterior region. We continue to assume that all processes are of Heaviside class H and that the class QSVP$(\mathscr{B}, G_1, G_2, \boldsymbol{b})$ is defined as in Sect. 33. We further assume that if $[\boldsymbol{u}, \boldsymbol{E}, \boldsymbol{S}]$ lies in QSVP$(\mathscr{B}, G_1, G_2, \boldsymbol{b})$ then $\boldsymbol{u}$ is of Heaviside class H^1 and, if $\mathscr{B}$ is unbounded, $\boldsymbol{u}_{x} = O(|\boldsymbol{x}|^{-1})$ as $|\boldsymbol{x}| \to \infty$, uniformly on closed bounded subsets of $\mathscr{R}$.

First we seek integral formula solutions to the viscoelastic *displacement* boundary value problem. The processes

$$\widehat{\boldsymbol{P}}^i_{(\boldsymbol{x})} = [\hat{\boldsymbol{u}}^i_{(\boldsymbol{x})}, \widehat{\boldsymbol{E}}^i_{(\boldsymbol{x})}, \widehat{\boldsymbol{S}}^i_{(\boldsymbol{x})}] \quad \text{and} \quad \widehat{\boldsymbol{P}}^{ij}_{(\boldsymbol{x})} = [\hat{\boldsymbol{u}}^{ij}_{(\boldsymbol{x})}, \widehat{\boldsymbol{E}}^{ij}_{(\boldsymbol{x})}, \widehat{\boldsymbol{S}}^{ij}_{(\boldsymbol{x})}] \qquad (i, j = 1, 2, 3)$$

[142] See LTE, Sects. 51–53.

[143] STERNBERG and AL-KHOZAIE [1964, *24*, Sect. 3]. Their formulation parallels that of STERNBERG and EUBANKS [1955, *9*] for the elastic theory. See also LTE, Sects. 51–53.

are called ***Green's processes at x of the first kind for $\mathscr{B}$ corresponding to the relaxation functions G_1 and G_2*** if, and only if: for each $\boldsymbol{x} \in \mathscr{B}$ and all $(\boldsymbol{\xi}, t) \in \overline{\mathscr{B}} \times \mathscr{R}$ such that $\boldsymbol{\xi} \neq \boldsymbol{x}$

$$\hat{\boldsymbol{P}}^i_{(\boldsymbol{x})}(\boldsymbol{\xi}, t) = \boldsymbol{P}^i_{(\boldsymbol{x})}(\boldsymbol{\xi}, t) + \tilde{\boldsymbol{P}}^i_{(\boldsymbol{x})}(\boldsymbol{\xi}, t), \tag{34.1}$$

$$\hat{\boldsymbol{P}}^{ij}_{(\boldsymbol{x})}(\boldsymbol{\xi}, t) = -\tfrac{1}{2}[\boldsymbol{P}^{ij}_{(\boldsymbol{x})}(\boldsymbol{\xi}, t) + \boldsymbol{P}^{ji}_{(\boldsymbol{x})}(\boldsymbol{\xi}, t)] + \tilde{\boldsymbol{P}}^{ij}_{(\boldsymbol{x})}(\boldsymbol{\xi}, t), \tag{34.2}$$

where $\boldsymbol{P}^i_{(\boldsymbol{x})}$ and $\boldsymbol{P}^{ij}_{(\boldsymbol{x})}$ are the (normalized) Kelvin and Kelvin doublet processes (applied at x) of Sect. 33 corresponding to G_1 and G_2; further, for each $\boldsymbol{x} \in \mathscr{B}$

$$\tilde{\boldsymbol{P}}^i_{(\boldsymbol{x})} \equiv [\tilde{\boldsymbol{u}}^i_{(\boldsymbol{x})}, \tilde{\boldsymbol{E}}^i_{(\boldsymbol{x})}, \tilde{\boldsymbol{S}}^i_{(\boldsymbol{x})}] \quad \text{and} \quad \tilde{\boldsymbol{P}}^{ij}_{(\boldsymbol{x})} \equiv [\tilde{\boldsymbol{u}}^{ij}_{(\boldsymbol{x})}, \tilde{\boldsymbol{E}}^{ij}_{(\boldsymbol{x})}, \tilde{\boldsymbol{S}}^{ij}_{(\boldsymbol{x})}]$$

are field processes such that

(i) $\tilde{\boldsymbol{P}}^i_{(\boldsymbol{x})}$ and $\tilde{\boldsymbol{P}}^{ij}_{(\boldsymbol{x})}$ are in QSVP$(\mathscr{B}, G_1, G_2)$; and

(ii) $\hat{\boldsymbol{u}}^i_{(\boldsymbol{x})} = \boldsymbol{0}$, and $\hat{\boldsymbol{u}}^{ij}_{(\boldsymbol{x})} = \boldsymbol{0}$ on $\partial\mathscr{B} \times \mathscr{R}$.

When dependence upon the distinguished point $\boldsymbol{x}$ is unimportant, or when no confusion may occur, we write $\boldsymbol{P}^i$ for $\boldsymbol{P}^i_{(\boldsymbol{x})}$, $\boldsymbol{P}^{ij}$ for $\boldsymbol{P}^{ij}_{(\boldsymbol{x})}$, etc.

Now the uniqueness theorem for quasi-static viscoelastic processes of Sect. 28 holds for unbounded regions under our present hypotheses, hence the conditions (i) and (ii) serve to determine uniquely the processes $\tilde{\boldsymbol{P}}^i$ and $\tilde{\boldsymbol{P}}^{ij}$. It therefore follows that the Green's processes $\hat{\boldsymbol{P}}^i$ and $\hat{\boldsymbol{P}}^{ij}$ are uniquely determined.

Theorem*[144] *(Integral representation of the solution to the displacement boundary value problem). *Let* $\boldsymbol{P} \equiv [\boldsymbol{u}, \boldsymbol{E}, \boldsymbol{S}]$ *be in* QSVP$(\mathscr{B}, G_1, G_2, \boldsymbol{b})$. *Let* $\hat{\boldsymbol{P}}^i$ *and* $\hat{\boldsymbol{P}}^{ij}$ *be the Green's processes of the first kind for* $\mathscr{B}$ *corresponding to* G_1 *and* G_2. *Then, the components of the displacement and strain processes of* $\boldsymbol{P} \equiv [\boldsymbol{u}, \boldsymbol{E}, \boldsymbol{S}]$ *with respect to the basis* $\{\mathbf{e}_i\}$ *are given by*

$$\begin{aligned} u_i(\boldsymbol{x}, t) = &\int_{\mathscr{B}} [\hat{\boldsymbol{u}}^i_{(\boldsymbol{x})} \circledast \boldsymbol{b}](\boldsymbol{\xi}, t)\, dV(\boldsymbol{\xi}) \\ &- \int_{\partial\mathscr{B}} [\boldsymbol{u}_{(\boldsymbol{x})} \circledast \hat{\mathbf{s}}^i_{(\boldsymbol{n})}](\boldsymbol{\xi}, t)\, dA(\boldsymbol{\xi}), \qquad (\boldsymbol{x}, t) \in \mathscr{B} \times \mathscr{R}, \end{aligned} \tag{34.3}$$

$$\begin{aligned} E_{ij}(\boldsymbol{x}, t) = &\int_{\mathscr{B}} [\boldsymbol{u}^{ij}_{(\boldsymbol{x})} \circledast \boldsymbol{b}](\boldsymbol{\xi}, t)\, dV(\boldsymbol{\xi}) \\ &- \int_{\partial\mathscr{B}} [\boldsymbol{u}_{(\boldsymbol{x})} \circledast \hat{\mathbf{s}}^{ij}_{(\boldsymbol{n})}](\boldsymbol{\xi}, t)\, dA(\boldsymbol{\xi}), \qquad (\boldsymbol{x}, t) \in \mathscr{B} \times \mathscr{R}. \end{aligned} \tag{34.4}$$

Observe that the integral formulas (34.3) and (34.4), with the exception of the Green's processes of the first kind, involve *only the values of the displacement* $\boldsymbol{u}$ of $\boldsymbol{P} = [\boldsymbol{u}, \boldsymbol{E}, \boldsymbol{S}]$ *on the surface* $\partial\mathscr{B}$ *and the body forces* $\boldsymbol{b}$ *on* $\mathscr{B}$. The stress process $\boldsymbol{S}$ can, of course, be computed through the Boltzmann law $\mathscr{L}_{\boldsymbol{G}}$. Thus $\boldsymbol{P} = [\boldsymbol{u}, \boldsymbol{E}, \boldsymbol{S}]$ may be computed through (34.3) and (34.4) provided the Green's processes of the first kind for $\mathscr{B}$ are known and the surface displacements and body forces are prescribed. Also, if $\mathscr{B} = \mathscr{E}$, then $\hat{\boldsymbol{P}}^i \equiv \boldsymbol{P}^i$ and $\hat{\boldsymbol{P}}^{ij} \equiv \boldsymbol{P}^{ij}$, since the processes $\tilde{\boldsymbol{P}}^i$ and $\tilde{\boldsymbol{P}}^{ij}$ vanish.

Before turning to the traction boundary value problem, we observe that the Green's processes are *symmetric*: for each $(\boldsymbol{\xi}, t) \in \mathscr{B} \times \mathscr{R}$ and $\boldsymbol{x} \in \mathscr{B}$ such that $\boldsymbol{\xi} \neq \boldsymbol{x}$,

$$\hat{\boldsymbol{P}}^i_{(\boldsymbol{x})}(\boldsymbol{\xi}, t) = \hat{\boldsymbol{P}}^i_{(\boldsymbol{\xi})}(\boldsymbol{x}, t), \tag{34.5}$$

$$\hat{\boldsymbol{P}}^{ij}_{(\boldsymbol{x})}(\boldsymbol{\xi}, t) = \hat{\boldsymbol{P}}^{ij}_{(\boldsymbol{\xi})}(\boldsymbol{x}, t). \tag{34.6}$$

---

[144] STERNBERG and AL-KHOZAIE [1964, *24*, Theorem 3.1].

The proof of this assertion depends upon the reciprocal theorem of Betti type of Sect. 27, which holds under our present hypotheses even if $\mathscr{B}$ is unbounded.[145]

For the *traction* boundary value problem, the following definition is crucial. The processes $\hat{\boldsymbol{P}}^i_{(\boldsymbol{x})} = [\hat{\boldsymbol{u}}^i_{(\boldsymbol{x})}, \hat{\boldsymbol{E}}^i_{(\boldsymbol{x})}, \hat{\boldsymbol{S}}^i_{(\boldsymbol{x})}]$ and $\hat{\boldsymbol{P}}^{ij}_{(\boldsymbol{x})} = [\hat{\boldsymbol{u}}^{ij}_{(\boldsymbol{x})}, \hat{\boldsymbol{E}}^{ij}_{(\boldsymbol{x})}, \hat{\boldsymbol{S}}^{ij}_{(\boldsymbol{x})}]$ $(i, j = 1, 2, 3)$ are called ***Green's processes at $\boldsymbol{x}$ of the second kind for $\mathscr{B}$ corresponding to the relaxation functions G_1 and G_2 and, if $\mathscr{B}$ is bounded, to the fixed point $\boldsymbol{x}_0 \in \mathscr{B}$*** if, and only if: for each $\boldsymbol{x} \in \mathscr{B}$ and all $(\boldsymbol{\xi}, t) \in \mathscr{B} \times \mathscr{R}$ such that $\boldsymbol{\xi} \neq \boldsymbol{x}$ and $\boldsymbol{\xi} \neq \boldsymbol{x}_0$,

$$\begin{aligned}\hat{\boldsymbol{P}}^i_{(\boldsymbol{x})}(\boldsymbol{\xi}, t) = \boldsymbol{P}^i_{(\boldsymbol{x})}(\boldsymbol{\xi}, t) + \tilde{\boldsymbol{P}}^i_{(\boldsymbol{x})}(\boldsymbol{\xi}, t)\\ + c\left[-\boldsymbol{P}^i_{(\boldsymbol{x}_0)}(\boldsymbol{\xi}, t) + \varepsilon_{ijk}\left((\boldsymbol{x} - \boldsymbol{x}_0)\cdot \boldsymbol{e}_j\right) \overline{\boldsymbol{P}}^k_{(\boldsymbol{x}_0)}(\boldsymbol{\xi}, t)\right],\end{aligned} \tag{34.7}$$

$$\hat{\boldsymbol{P}}^{ij}_{(\boldsymbol{x})}(\boldsymbol{\xi}, t) = -\tfrac{1}{2}\left[\boldsymbol{P}^{ij}_{(\boldsymbol{x})}(\boldsymbol{\xi}, t) + \boldsymbol{P}^{ji}_{(\boldsymbol{x})}(\boldsymbol{\xi}, t)\right] + \tilde{\boldsymbol{P}}^{ij}_{(\boldsymbol{x})}(\boldsymbol{\xi}, t), \tag{34.8}$$

where $\boldsymbol{P}^i$, $\boldsymbol{P}^{ij}$, and $\overline{\boldsymbol{P}}^i$ are the Kelvin, Kelvin doublet, and center of rotation (with axis $\boldsymbol{e}_i$) processes of Sect. 33 corresponding to G_1 and G_2, while $c = 1$ or $c = 0$ according as $\mathscr{B}$ is bounded or unbounded; further for each $\boldsymbol{x} \in \mathscr{B}$

$$\tilde{\boldsymbol{P}}^i_{(\boldsymbol{x})} = [\tilde{\boldsymbol{u}}^i_{(\boldsymbol{x})}, \tilde{\boldsymbol{E}}^i_{(\boldsymbol{x})}, \tilde{\boldsymbol{S}}^i_{(\boldsymbol{x})}] \quad \text{and} \quad \tilde{\boldsymbol{P}}^{ij}_{(\boldsymbol{x})} = [\tilde{\boldsymbol{u}}^{ij}_{(\boldsymbol{x})}, \tilde{\boldsymbol{E}}^{ij}_{(\boldsymbol{x})}, \tilde{\boldsymbol{S}}^{ij}_{(\boldsymbol{x})}]$$

are field processes such that

(i) $\tilde{\boldsymbol{P}}^i_{(\boldsymbol{x})}$ and $\tilde{\boldsymbol{P}}^{ij}_{(\boldsymbol{x})}$ are in $\mathrm{QSVP}(\overline{\mathscr{B}}, G_1, G_2)$;

(ii) $\hat{\boldsymbol{s}}^i_{(\boldsymbol{x})} = \boldsymbol{0}$ and $\hat{\boldsymbol{s}}^{ij}_{(\boldsymbol{x})} = \boldsymbol{0}$ on $\partial\mathscr{B} \times \mathscr{R}$; and

(iii) when $\mathscr{B}$ is unbounded,

$$\tilde{\boldsymbol{u}}^i_{(\boldsymbol{x})}(\boldsymbol{x}_0, t) = \tilde{\boldsymbol{\omega}}^i_{(\boldsymbol{x})}(\boldsymbol{x}_0, t) = \boldsymbol{0}, \qquad t \in \mathscr{R}, \tag{34.9}$$

$$\tilde{\boldsymbol{u}}^{ij}_{(\boldsymbol{x})}(\boldsymbol{x}_0, t) = \tilde{\boldsymbol{\omega}}^{ij}_{(\boldsymbol{x})}(\boldsymbol{x}_0, t) = \boldsymbol{0}, \qquad t \in \mathscr{R}, \tag{34.10}$$

where $\tilde{\boldsymbol{\omega}}^i$ and $\tilde{\boldsymbol{\omega}}^{ij}$ are the rotation processes associated with the processes $\tilde{\boldsymbol{P}}_i$ and $\tilde{\boldsymbol{P}}^{ij}$.[146]

For Green's processes of the first kind, the nonsingular field processes $\tilde{\boldsymbol{P}}^i_{(\boldsymbol{x})}$ and $\tilde{\boldsymbol{P}}^{ij}_{(\boldsymbol{x})}$ were, for fixed $\boldsymbol{x} \in \mathscr{B}$, solutions to a *displacement* boundary value problem. For Green's processes of the second kind, however, the nonsingular field processes $\tilde{\boldsymbol{P}}^i_{(\boldsymbol{x})}$ and $\tilde{\boldsymbol{P}}^{ij}_{(\boldsymbol{x})}$ are, for fixed $\boldsymbol{x} \in \mathscr{B}$, solutions for a *traction* boundary value problem. This solution is unique if $\mathscr{B}$ is unbounded[147] and unique to within superposed (infinitesimal) rigid motions if $\mathscr{B}$ is bounded. Now condition (iii) rules out the possibility of (infinitesimal) rigid motions in case $\mathscr{B}$ is bounded. Hence the processes $\tilde{\boldsymbol{P}}^i$ and $\tilde{\boldsymbol{P}}^{ij}$ are uniquely determined. Moreover, the Green's processes $\hat{\boldsymbol{P}}^i$ and $\hat{\boldsymbol{P}}^{ij}$ are also uniquely determined and, if $\mathscr{B}$ is unbounded ($c = 0$), independent of the point $\boldsymbol{x}_0 \in \mathscr{B}$.

Observe that if $\mathscr{B}$ is bounded the traction boundary value problem determining $\tilde{\boldsymbol{P}}^i$ has a solution only if the associated surface tractions $\tilde{\boldsymbol{s}}^i$ on $\partial\mathscr{B}$ are self-equilibrated (in the absence of body forces). Hence condition (iii) implies

[145] For details, see STERNBERG and AL-KHOZAIE [1964, *24*, Theorem 3.2]. Note that if $\mathscr{B}$ is isotropic, then $\mathbf{G}$ is always symmetric.

[146] The *rotation process* $\boldsymbol{\omega}$ associated with $\boldsymbol{P} = [\boldsymbol{u}, \boldsymbol{E}, \boldsymbol{S}]$ is twice the axial vector of the gradient of $\boldsymbol{u}$; that is, $\boldsymbol{\omega} = \operatorname{curl} \boldsymbol{u}$.

[147] The uniqueness for unbounded $\mathscr{B}$ follows under our hypotheses by a straight forward modification of the theorem of Sect. 28. The possibility of rigid motions is ruled out since the displacements must vanish at infinity.

that the singularities of $\hat{\boldsymbol{P}}^i$ must be self-equilibrating. The term in (34.7) with coefficient c ($c=1$) and singularity at $\boldsymbol{x}_0$ serves to balance the singularity of the term $\boldsymbol{P}^i_{(\boldsymbol{x})}$ at $\boldsymbol{x}$. Thus the complete system of singularities of $\hat{\boldsymbol{P}}^i_{(\boldsymbol{x})}$ at $\boldsymbol{x}$ and $\boldsymbol{x}_0$ is self-equilibrated.

If $\mathscr{B}$ is unbounded ($c=0$), the singularity at $\boldsymbol{x}$ for $\hat{\boldsymbol{P}}^i_{(\boldsymbol{x})}$ is merely that of the (normalized) Kelvin process $\boldsymbol{P}^i_{(\boldsymbol{x})}$ (concentrated load $\boldsymbol{l}=h\boldsymbol{e}_i$). Finally, the singularities of $\hat{\boldsymbol{P}}^{ij}_{(\boldsymbol{x})}$ at $\boldsymbol{x}$ are always self-equilibrated whether $\mathscr{B}$ is bounded or unbounded.

Theorem[148] ***(Integral representation of the solution of the traction boundary value problem).*** *Let* $\boldsymbol{P}=[\boldsymbol{u}, \boldsymbol{E}, \boldsymbol{S}]$ *be in* QSVP$(\mathscr{B}, G_1, G_2, \boldsymbol{b})$ *and, if* $\mathscr{B}$ *is bounded, the processes* $\boldsymbol{u}$ *and* $\boldsymbol{\omega}$ *satisfy*[149]

$$\boldsymbol{u}_{\boldsymbol{x}_0}=\boldsymbol{\omega}_{\boldsymbol{x}_0}=\boldsymbol{0}, \tag{34.11}$$

where $\boldsymbol{\omega}$ *is the rotation process associated with* $\boldsymbol{P}$ *and* $\boldsymbol{x}_0 \in \mathscr{B}$. *Let* $\hat{\boldsymbol{P}}^i$ *and* $\hat{\boldsymbol{P}}^{ij}$ *be the Green's processes of the second kind for* $\mathscr{B}$ *corresponding to* G_1 *and* G_2 *and, if* $\mathscr{B}$ *is bounded, to* $\boldsymbol{x}_0$. *Then, the components of the displacement and strain processes of* $\boldsymbol{P}=[\boldsymbol{u}, \boldsymbol{E}, \boldsymbol{S}]$ *with respect to the basis* $\{\boldsymbol{e}_i\}$ *are given by*

$$\begin{aligned} u_i(\boldsymbol{x}, t) &= \int_{\mathscr{B}} [\hat{\boldsymbol{u}}^i_{(\boldsymbol{x})} \circledast \boldsymbol{b}]\,(\boldsymbol{\xi}\, t)\, dV(\boldsymbol{\xi}) \\ &\quad + \int_{\partial\mathscr{B}} [\hat{\boldsymbol{u}}^i_{(\boldsymbol{x})} \circledast \boldsymbol{s}_{(\boldsymbol{n})}]\,(\boldsymbol{\xi}, t)\, dA(\boldsymbol{\xi}), \quad (\boldsymbol{x}, t) \in \mathscr{B}\times\mathscr{R}, \end{aligned} \tag{34.12}$$

$$\begin{aligned} E_{ij}(\boldsymbol{x}, t) &= \int_{\mathscr{B}} [\hat{\boldsymbol{u}}^{ij}_{(\boldsymbol{x})} \circledast \boldsymbol{b}]\,(\boldsymbol{\xi}, t)\, dV(\boldsymbol{\xi}) \\ &\quad + \int_{\partial\mathscr{B}} [\hat{\boldsymbol{u}}^{ij}_{(\boldsymbol{x})} \circledast \boldsymbol{s}_{(\boldsymbol{n})}]\,(\boldsymbol{\xi}, t)\, dA(\boldsymbol{\xi}), \quad (\boldsymbol{x}, t) \in \mathscr{B}\times\mathscr{R}. \end{aligned} \tag{34.13}$$

The integral formulas (34.12) and (34.13) with the exception of the Green's processes of the second kind, involve *only the values of the surface tractions* $\boldsymbol{s}$ *associated with* $\boldsymbol{P}=[\boldsymbol{u}, \boldsymbol{E}, \boldsymbol{S}]$ *on the surface* $\partial\mathscr{B}$ *and the body forces* $\boldsymbol{b}$ *on* $\mathscr{B}$. Of course, the stress process $\boldsymbol{S}$ can be computed through the Boltzmann law $\mathscr{L}_{\boldsymbol{G}}$. Thus $\boldsymbol{P}=[\boldsymbol{u}, \boldsymbol{E}, \boldsymbol{S}]$ may be computed through (34.12) and (34.13) provided the Green's processes of the second kind for $\mathscr{B}$ are known and the surface tractions and body forces are prescribed. If $\mathscr{B}$ is bounded, a change in the position of the fixed point $\boldsymbol{x}_0$ results in a rigid displacement added to $\boldsymbol{u}$ but leaves $\boldsymbol{E}$ unaltered. This is seen to be consistent with the corresponding changes in the Green's processes of the second kind.

The Green's processes of the second kind for $\mathscr{B}$ satisfy the same symmetry relations—(34.5) and (34.6)—as the Green's processes of the first kind for $\mathscr{B}$. This assertion is established in the same way.[150]

The formulation of this section may be extended to include integral representations for the solution of *mixed* boundary value problems, which formulation would contain the displacement and traction boundary value problems as special cases. Such a formulation would be similar to that presented but more complicated in aspect.[151]

---

[148] STERNBERG and AL-KHOZAIE [1964, *24*, Theorem 3.3].

[149] Here recall the convention established in Sect. 3 for field processes. Thus $\boldsymbol{u}_{\boldsymbol{x}_0}$ denotes the process $\boldsymbol{u}_{\boldsymbol{x}_0}=\boldsymbol{u}(\boldsymbol{x}_0, \cdot)$ and $\boldsymbol{\omega}_{\boldsymbol{x}_0}$ denotes the process $\boldsymbol{\omega}_{\boldsymbol{x}_0}=\boldsymbol{\omega}(\boldsymbol{x}_0, \cdot)$.

[150] Cf. footnote [145], p. 83.

[151] STERNBERG and AL-KHOZAIE [1964, *24*] assert that the analysis of concentrated *surface* loads given by STERNBERG and EUBANKS [1955, *9*, Sect. 7], may be extended to the viscoelastic case through the use of the second theorem (integral representation of the traction boundary value problem). See also LTE, Sects. 51–53.

35. Saint-Venant's principle. In this section we consider some formulations of the classical principle introduced by SAINT-VENANT[152] for the equilibrium theory of elasticity. In the classical theory of elastostatics there are various formulations of this principle[153] which admit generalization to viscoelasticity. We first present a version of the principle formulated by STERNBERG and AL-KHOZAIE[154]. Their formulation utilizes the integral representations of Sect. 34 and constitutes a generalization to quasi-static viscoelasticity of the work of STERNBERG[155] for the elastostatic theory.

Let $\mathscr{B}(\boldsymbol{x}, \varrho)$ denote the open ball in $\mathscr{E}$ with center $\boldsymbol{x}\in\mathscr{E}$ and radius $\varrho \geqq 0$. Let $\mathscr{B}$ be a body in $\mathscr{E}$ which occupies a regular region in $\mathscr{E}$. Let $\{\boldsymbol{\xi}^n\}$ $(n=1, 2, 3, \ldots, N)$ be a set of N regular points of $\partial\mathscr{B}$ with the following property: for each n $(n=1, 2, \ldots, N)$ there is a positive number ϱ_n such that the ball $\mathscr{B}(\boldsymbol{\xi}^n, \varrho_n)$ may be mapped, in a twice continuously differentiable way, onto the ball $\mathscr{B}(\mathbf{0}, 1)$; furthermore, the image of $\mathscr{B}(\boldsymbol{\xi}^n, \varrho_n)\cap\partial\mathscr{B}$ under this map is a plane disc. Let $\varrho_0=\min\{\varrho_n: n=1, 2, 3, \ldots, N\}$. The sets Λ_ϱ^n $(0<\varrho<\varrho_0;\ n=1, 2, 3, \ldots, N)$ are said to constitute ***a collection of** N **families of load regions which contract to the** N **points*** $\{\boldsymbol{\xi}^n\}$ ***in*** $\partial\mathscr{B}$ if, and only if, each Λ_ϱ^n is of the form

$$\Lambda_\varrho^n=\overline{\mathscr{B}}(\boldsymbol{\xi}^n, \varrho)\cap\partial\mathscr{B} \qquad (0<\varrho<\varrho_0;\ n=1, 2, 3, \ldots, N), \tag{35.1}$$

and

$$\Lambda_\varrho^n\cap\Lambda_\varrho^m=\emptyset \qquad (0<\varrho<\varrho_0;\ m, n=1, 2, 3, \ldots, N) \tag{35.2}$$

whenever $m\neq n$.

Let $\mathscr{B}$ be a homogeneous and isotropic viscoelastic body of relaxation type with isotropic relaxation functions G_1 and G_2. Further suppose that the body force process for $\mathscr{B}$ vanishes. We continue to assume that all processes are in Heaviside class H and that the class QSVP$(\mathscr{B}, G_1, G_2)$ is defined as in Sects. 33 and 34.

Suppose $\{\Lambda_\varrho^n\}$ $(0<\varrho<\varrho_0;\ n=1, 2, 3, \ldots, N)$ is a collection of N families of load regions in $\partial\mathscr{B}$ contracting to the N points $\{\boldsymbol{\xi}^n\}$ in $\partial\mathscr{B}$. Then $\boldsymbol{P}_\varrho\equiv[\boldsymbol{u}_\varrho, \boldsymbol{E}_\varrho, \boldsymbol{S}_\varrho]$ is a ***family of quasi-static viscoelastic processes for*** $\mathscr{B}$ ***corresponding to loads on*** $\{\Lambda_\varrho^n\}$ if, and only if,

(i) the process $\boldsymbol{P}_\varrho$, $0<\varrho<\varrho_0$, lies in QSVP$(\mathscr{B}, G_1, G_2)$;

(ii) the surface traction process $\boldsymbol{s}_\varrho$, $0<\varrho<\varrho_0$, corresponding to $\boldsymbol{P}_\varrho$ vanishes on

$$\partial\mathscr{B}-\bigcup_n \Lambda_\varrho^n;$$

(iii) if $\mathscr{B}$ is bounded, for some $\boldsymbol{x}_0\in\mathscr{B}$

$$\boldsymbol{u}_\varrho(\boldsymbol{x}_0, t)=\boldsymbol{\omega}_\varrho(\boldsymbol{x}_0, t)=\mathbf{0}, \qquad t\in\mathscr{R},\ \varrho\in(0, \varrho_0), \tag{35.3}$$

where $\boldsymbol{\omega}_\varrho$ is the rotation process associated with $\boldsymbol{P}_\varrho$; and

(iv) the surface tractions $\boldsymbol{s}_\varrho$ corresponding to $\boldsymbol{P}_\varrho$ are uniformly bounded as a function on $\partial\mathscr{B}\times[0, T]\times(0, \varrho_0)$ for all $T\geqq 0$.

---

[152] SAINT-VENANT [1855, *1*]. In his memoir, only extension, torsion, and flexure of cylindrical bodies were considered.

[153] For early statements of Saint-Venant's principle see BOUSSINESQ [1885, *1*], LOVE [1944, *2*] and v. MISES [1945, *2*]. For a complete discussion of the principle in elastostatics, see STERNBERG [1954, *2*], TOUPIN [1965, *28*], or LTE, Subchap. D.*X*.

[154] STERNBERG and AL-KHOZAIE [1964, *24*, Sect. 4].

[155] STERNBERG [1954, *2*].

For each ϱ and n $(0<\varrho<\varrho_0;\ n=1, 2, \ldots, N)$ define vector processes $\boldsymbol{l}_\varrho^n$ and $\boldsymbol{m}_\varrho^n$ by

$$\boldsymbol{l}_\varrho^n = \int_{\Lambda_\varrho^n} \mathbf{s}_\varrho \, dA, \tag{35.4}$$

$$\boldsymbol{m}_\varrho^n = \int_{\Lambda_\varrho^n} \boldsymbol{p}_0 \times \mathbf{s}_\varrho \, dA \qquad (\boldsymbol{x}_0 = \mathbf{0}). \tag{35.5}$$

Thus, $\boldsymbol{l}_\varrho^n$ and $\boldsymbol{m}_\varrho^n$ denote the resultant force and moment (about the origin) of the surface tractions $\mathbf{s}_\varrho$ acting on Λ_ϱ^n.

We may now state a

Saint-Venant principle for viscoelastic bodies.[156] *Let* $\{\Lambda_\varrho^n\}$ $(0<\varrho<\varrho_0;$ $n=1, 2, 3, \ldots, N)$ *be a collection of* N *families of load regions for the boundary* $\partial\mathscr{B}$ *of a (regular) viscoelastic body* $\mathscr{B}$ *which contract to a sequence of* N *points* $\{\boldsymbol{\xi}^n\}$ *in* $\partial\mathscr{B}$. *(Assume* $N \geqq 2$ *if* $\mathscr{B}$ *is bounded.) Let* $\mathbf{P}_\varrho$ $(0<\varrho<\varrho_0)$ *be a family of quasi-static viscoelastic processes for* $\mathscr{B}$ *corresponding to loads on* Λ_ϱ^n. *Let* $\boldsymbol{x} \in \mathscr{B}$ *be fixed. Assume that there exist Green's processes of the second kind for* $\mathscr{B}$.[157] *Then, uniformly in time* $t \in [0, T]$, $T \geqq 0$,

$$\boldsymbol{u}_\varrho(\boldsymbol{x}, t) = O(\varrho^\delta), \quad \mathbf{E}_\varrho(\boldsymbol{x}, t) = O(\varrho^\delta), \quad \mathbf{S}_\varrho(\boldsymbol{x}, t) = O(\varrho^\delta), \quad \text{as} \quad \varrho \to 0, \tag{35.6}$$

where $\delta = 2$. *Moreover:*

(i) $\delta = 3$ *if, for* $t \in [0, T]$,

$$\boldsymbol{l}_\varrho^n(t) = \mathbf{0} \qquad (0<\varrho<\varrho_0;\ n=1, 2, 3, \ldots, N); \tag{35.7}$$

(ii) $\delta = 4$ *if, for* $t \in [0, T]$,

$$\boldsymbol{l}_\varrho^n(t) = \mathbf{0}, \quad \int_{\Lambda_\varrho^n} (\boldsymbol{p} \cdot \boldsymbol{e}_i)\, \mathbf{s}_\varrho(t) \, dA = \mathbf{0} \qquad (0<\varrho<\varrho_0;\ n=1, 2, 3, \ldots, N); \tag{35.8}$$

(iii) $\delta = 4$ *if, for* $t \in [0, T]$,

$$\boldsymbol{l}_\varrho^n(t) = \mathbf{0}, \qquad \boldsymbol{m}_\varrho^n(t) = \mathbf{0} \qquad (0<\varrho<\varrho_0;\ n=1, 2, 3, \ldots, N), \tag{35.9}$$

provided the surface tractions on Λ_ϱ^n *for* $t \in [0, T]$ *and* $\varrho \in (0, \varrho_0)$ *are of the form*

$$\mathbf{s}_\varrho(\boldsymbol{\xi}, t) = \phi_\varrho^n(\boldsymbol{\xi}, t)\, \boldsymbol{k}^n(t) \qquad (n=1, 2, 3, \ldots, N), \tag{35.10}$$

where each $\boldsymbol{k}^n$ *is a vector valued process in class* H *whose values on* $[0, T]$ *are such that*

$$\boldsymbol{k}^n(t) \cdot \mathbf{v}^n \neq 0 \qquad (n=1, 2, 3, \ldots, N), \tag{35.11}$$

$\mathbf{v}^n$ *being the unit normal vector to* $\partial\mathscr{B}$ *at* $\boldsymbol{\xi}^n$.

From the first conclusion of the above result, we see that the displacements, strains, and stresses, at a point, always vanish at least to order $O(\varrho^2)$ as $\varrho \to 0$. This is true under the essentially unrestrictive conditions of the basic definition and, in particular, whether the loads on each family of load regions are self-equilibrated or not. However, conclusion (i) asserts that $O(\varrho^2)$ may be replaced by $O(\varrho^3)$ if, but not only if, the loads on each load region vanish. Similar comments hold for conclusions (ii) and (iii).

Observe that the restrictions (35.8) of (ii) imply, but are not implied by, the equilibrium conditions (35.9) of (iii). A counterexample due to v. MISES[158] shows

[156] STERNBERG and AL-KHOZAIE [1964, *24*, Theorem 4.1].

[157] Cf. Sect. 34.

[158] v. MISES [1945, 2].

(in the elastic case) that replacing (35.7) in condition (i) by (35.9) in condition (iii), without the provision that the tractions on each load region be parallel and not tangent to $\partial\mathscr{B}$ (provisions (35.10) and (35.11)), does not permit the replacement of $O(\varrho^3)$ by $O(\varrho^4)$.

The conclusions of this formulation of Saint-Venant's principle are of "potential theory" type; that is, they reflect the basic character of the singular solutions used to construct the Green's processes. To see this, let $\boldsymbol{P}^i$ and $\boldsymbol{P}^{ij}$ be the Kelvin and Kelvin doublet processes (with singularities at the origin) of Sect. 33. Let $\mathscr{B}$ be any regular region containing the origin and write $\boldsymbol{l}^i$, $\boldsymbol{m}^i$ and $\boldsymbol{l}^{ij}$, $\boldsymbol{m}^{ij}$ for the resulting forces and moments (with respect to the origin) induced on $\partial(\mathscr{E}-\mathscr{B})$ by the stresses of $\boldsymbol{P}^i$ and $\boldsymbol{P}^{ij}$. Then it follows from the properties of Kelvin and Kelvin doublet processes stated in Sect. 33 that

$$\boldsymbol{l}^i \neq \boldsymbol{0}, \qquad \boldsymbol{m}^i = \boldsymbol{0}, \tag{35.12}$$

$$\boldsymbol{l}^{ij} = \boldsymbol{0}, \qquad \boldsymbol{m}^{ij} \neq \boldsymbol{0} \qquad (i \neq j), \tag{35.13}$$

$$\boldsymbol{l}^{ij} = \boldsymbol{0}, \qquad \boldsymbol{m}^{ij} = \boldsymbol{0} \qquad (i = j). \tag{35.14}$$

However, uniformly for a closed and bounded subset of $\mathscr{R}$, as $|\boldsymbol{x}| \to \infty$

$$\boldsymbol{s}^i_{\boldsymbol{x}} = O(|\boldsymbol{x}|^{-2}), \qquad \boldsymbol{s}^i_{\boldsymbol{x}} \neq O(|\boldsymbol{x}|^{-\delta})\,(\delta > 2), \tag{35.15}$$

$$\boldsymbol{s}^{ij}_{\boldsymbol{x}} = O(|\boldsymbol{x}|^{-3}), \qquad \boldsymbol{s}^{ij}_{\boldsymbol{x}} \neq O(|\boldsymbol{x}|^{-\delta})\,(\delta > 3) \qquad i \neq j, \tag{35.16}$$

$$\boldsymbol{s}^{ij}_{\boldsymbol{x}} = O(|\boldsymbol{x}|^{-3}), \qquad \boldsymbol{s}^{ij}_{\boldsymbol{x}} \neq O(|\boldsymbol{x}|^{-\delta})\,(\delta > 3) \qquad i = j. \tag{35.17}$$

Thus, the stresses decay (at infinity) more rapidly when the load on $\partial(\mathscr{E}-\mathscr{B})$ vanishes than when it does not. However, the vanishing of both the force *and* moment on $\partial(\mathscr{E}-\mathscr{B})$ does not further increase the rate of decay.

The results stated here for bodies of general shape certainly apply to cylindrical or prismatic bodies. However, there is no preferred direction in the sense that these results reflect a comparison between a linear dimension of the surface load region with a distance from that region in the body. For a cylinder loaded at one end there is a preferred direction, namely the distance along a generator from the end. For elastic bodies, TOUPIN[159] has shown that for a cylinder with a self-equilibrated load at one end the strain energy stored in segments beyond a fixed point along the axis decreases *exponentially* with the distance from that point to the loaded end. This result obtains in a modified form, for regions other than cylindrical bodies.

A generalization of TOUPIN's result to linear viscoelasticity has been supplied by EDELSTEIN.[160] Here we state EDELSTEIN's result for a right cylindrical body loaded at one end. We do this for simplicity in presentation; the result is formulated and proved by EDELSTEIN for bodies of general shape.

Let $\mathscr{B}$ be a homogeneous and isotropic viscoelastic body in the shape of a right cylinder of length $L > 0$. Denote the loaded end of $\mathscr{B}$ by $\mathscr{C}_0$ and the terminal end of $\mathscr{B}$ by $\mathscr{C}_L$. Let π_s denote a plane normal to the generators of $\mathscr{B}$ (parallel to $\mathscr{C}_0$ and $\mathscr{C}_L$) at a distance s $(0 \leqq s \leqq L)$ from $\mathscr{C}_0$, and write $\mathscr{C}_s$ for the intersection $\pi_s \cap \mathscr{B}$. Also let $\mathscr{B}_s$ denote the segment of $\mathscr{B}$ bounded at the ends by $\mathscr{C}_s$ and $\mathscr{C}_L$; of course, $\mathscr{B}_0 \equiv \mathscr{B}$. Finally, we assume that the body forces for $\mathscr{B}$ vanish.

By a ***Saint-Venant solution for*** $\mathscr{B}$ we mean a process $\boldsymbol{P} \equiv [\boldsymbol{u}, \boldsymbol{E}, \boldsymbol{S}]$ for $\mathscr{B}$ in QSVP$(\mathscr{B}, G_1, G_2)$ (of Heaviside class H) such that the displacement is in

[159] TOUPIN [1965, *28*]. See also LTE, Sect. 55.
[160] EDELSTEIN [1970, *3*].

class H^1 and the values of $\boldsymbol{u}$ are class C^4 vector fields on $\mathscr{B}$;[161] further, the surface traction process $\boldsymbol{s}$ associated with $\boldsymbol{P}$ must satisfy:

$$\boldsymbol{s}=\boldsymbol{0} \quad \text{on } \partial\mathscr{B}-\mathscr{C}_0. \tag{35.18}$$

Notice that no specific boundary conditions are imposed upon the loaded end $\mathscr{C}_0$. Of course, if traction boundary data is given on $\mathscr{C}_0$, a necessary condition that there exist a Saint-Venant solution for $\mathscr{B}$ is that these tractions be self-equilibrated.[162]

Let $\boldsymbol{E}$ be a process of class H whose values are continuous (class C) symmetric tensor fields on $\overline{\mathscr{B}}$. For each s, $0\leqq s\leqq L$, define a scalar process W_s by

$$W_s(t)=\int_{\mathscr{B}_s}\int_0^t \{\boldsymbol{E}(\boldsymbol{x},\tau)\cdot\boldsymbol{E}(\boldsymbol{x},\tau)+[\operatorname{tr}\boldsymbol{E}(\boldsymbol{x},\tau)]^2\}\,d\tau\,dV(\boldsymbol{x}), \qquad t\in\mathscr{R}. \tag{35.19}$$

For the cylinder $\mathscr{B}$ we have the following

Theorem.[163] *Let the response functions G_1 and (G_2-G_1) both be positive, non-decreasing, and bounded away from zero.*[164] *Then for $l>0$, there exists a constant c (depending only upon $\mathscr{B}$, G_1, G_2, and l) such that, for any Saint-Venant solution for $\mathscr{B}$,*

$$W_s\leqq\exp\left[-\left(\frac{s-l}{c}\right)\right]W_0, \quad l\leqq s\leqq L-l, \tag{35.20}$$

where W_s is the process defined for $0\leqq s\leqq L$ by (35.19).

Note that the decay estimate depends upon the time t only through the processes W_s $(l\leqq s\leqq L-l)$ and W_0. Furthermore, this is not a direct extension of TOUPIN's result in elastostatics. TOUPIN's estimate is of the same form as (35.20) with the stored elastic energy in $\mathscr{B}_s$ replacing the process W_s; however, W_s does not represent the stored energy processes for $\mathscr{B}_s$.

EDELSTEIN[165] also provides a *pointwise estimate*, which we state in words for cylindrical bodies $\mathscr{B}$. *For any line segment in $\mathscr{B}$ parallel to a generator of $\mathscr{B}$ and for each time t, the magnitude of the strain corresponding to a Saint-Venant solution for $\mathscr{B}$ decreases exponentially with the distance along the line segment from the loaded end $\mathscr{C}_0$.* This estimate depends upon the time t and the distance of the line segment from the lateral boundary of $\mathscr{B}$.

The generalization of the results of this section to include anisotropic and inhomogeneous viscoelastic bodies may be carried out with some technical difficulties.

Finally we observe that, in some special cases, the Saint-Venant principles of elastostatics may carry over directly to the viscoelastic case. In particular, if there is a correspondence principle such as that of TSIEN's[166], where G_1 and G_2 are proportional, or if the body $\mathscr{B}$ is essentially one-dimensional (filament), then TOUPIN's result carries over directly to the viscoelastic case.[167]

---

[161] Recall our initial assumption that all processes are in class H. If $[\boldsymbol{u},\boldsymbol{E},\boldsymbol{S}]$ lies in $\mathrm{QSVP}(\mathscr{B},G_1,G_2)$, then the values of $\boldsymbol{u}$ are automatically class C^2 vector fields on $\overline{\mathscr{B}}$. Also, G_1 and G_2 are processes in class H^1.

[162] Recall that the body forces for $\mathscr{B}$ are assumed to vanish.

[163] EDELSTEIN [1970, *3*, Theorem 1]. This is EDELSTEIN's generalization of TOUPIN's classical result in elastostatics restricted to the case of cylindrical bodies. See also TOUPIN [1965, *28*].

[164] In particular, these conditions imply that G_1, (G_2-G_1) and (hence) G_2 are strong response functions which possess strong inverses (cf. Sect. 11).

[165] EDELSTEIN [1970, *3*, Theorem 2].

[166] TSIEN [1950, *4*]. Cf. footnote [110], p. 73.

[167] See EDELSTEIN [1970, *3*], TOUPIN [1965, *28*], and KNOWLES and STERNBERG [1966, *14*].

D. Dynamic linear viscoelasticity.

36. Dynamic viscoelastic processes. In this section we formulate the dynamic boundary value problem. For the case where the initial past history is prescribed, an alternative formulation is given.

Let $\mathscr{B}$ be a viscoelastic body with relaxation function G, body force $\boldsymbol{b}$, and density ϱ. Then by a ***dynamic viscoelastic process for*** $\mathscr{B}$ ***corresponding to*** G, $\boldsymbol{b}$, ***and*** ϱ we mean an admissible process $[\boldsymbol{u}, \boldsymbol{E}, \boldsymbol{S}]$ for $\mathscr{B}$ such that:

(R-1) the values of $\boldsymbol{u}$, $\boldsymbol{E}$, and $\boldsymbol{S}$ are continuous (class C) fields on $\overline{\mathscr{B}}$;

(R-2) the displacement $\boldsymbol{u}$ is a process of class C^2 whose values are class C^2 vector fields on $\overline{\mathscr{B}}$; and

(R-3) on $\mathscr{B}$, the processes $\boldsymbol{u}$, $\boldsymbol{E}$, and $\boldsymbol{S}$ satisfy the equation of motion

$$\operatorname{div} \boldsymbol{S} + \boldsymbol{b} = \varrho \ddot{\boldsymbol{u}}, \tag{36.1}$$

the strain-displacement relation

$$\boldsymbol{E} = \tfrac{1}{2}[\nabla \boldsymbol{u} + \nabla \boldsymbol{u}^T], \tag{36.2}$$

and the Boltzmann relaxation law

$$\boldsymbol{S} = \mathscr{L}_{\mathsf{G}}[\boldsymbol{E}] \equiv \mathsf{G} \circledast \boldsymbol{E}. \tag{36.3}$$

The set of all dynamic viscoelastic processes corresponding to G, $\boldsymbol{b}$, and ϱ is a vector subspace of the space of all admissible processes for $\mathscr{B}$. Henceforth, it is denoted by $\mathrm{DVP}(\mathsf{G}, \boldsymbol{b}, \varrho)$.

Similarly, for a viscoelastic body $\mathscr{B}$ with creep compliance J, body force $\boldsymbol{b}$, and density ϱ, we define a ***dynamic viscoelastic process for*** $\mathscr{B}$ ***corresponding to*** J, $\boldsymbol{b}$, ***and*** ϱ to be an admissible process $[\boldsymbol{u}, \boldsymbol{E}, \boldsymbol{S}]$ for $\mathscr{B}$ such that:

(C-1) the values of $\boldsymbol{u}$, $\boldsymbol{E}$, and $\boldsymbol{S}$ are continuous (class C) fields on $\overline{\mathscr{B}}$;

(C-2) the displacement $\boldsymbol{u}$ is a process of class C^2 whose values are class C^2 vector fields on $\overline{\mathscr{B}}$;

(C-3) the values of $\boldsymbol{S}$ are smooth (class C^1) symmetric-tensor valued fields on $\overline{\mathscr{B}}$; and

(C-4) on $\mathscr{B}$, the processes $\boldsymbol{u}$, $\boldsymbol{E}$, and $\boldsymbol{S}$ satisfy the equation of motion (36.1), the strain-displacement relation (36.2), and the Boltzmann creep law

$$\boldsymbol{E} = \mathscr{L}_{\mathsf{J}}[\boldsymbol{S}] \equiv \mathsf{J} \circledast \boldsymbol{S}. \tag{36.4}$$

The set of all dynamic viscoelastic processes corresponding to J, $\boldsymbol{b}$, and ϱ is also a vector subspace of the space of all admissible processes for $\mathscr{B}$. Henceforth it is denoted by $\mathrm{DVP}(\mathsf{J}, \boldsymbol{b}, \varrho)$.

If $[\hat{\boldsymbol{u}}, \hat{\mathbf{s}}]$ denotes boundary data for $\partial\mathscr{B}$, then a process $[\boldsymbol{u}, \boldsymbol{E}, \boldsymbol{S}]$ in $\mathrm{DVP}(\mathsf{G}, \boldsymbol{b}, \varrho)$ or $\mathrm{DVP}(\mathsf{J}, \boldsymbol{b}, \varrho)$ is said to be a ***solution to the dynamic viscoelastic boundary value problem corresponding to*** $[\hat{\boldsymbol{u}}, \hat{\mathbf{s}}]$ if, and only if, $[\boldsymbol{u}, \boldsymbol{E}, \boldsymbol{S}]$ satisfies the boundary conditions on $\partial\mathscr{B}$[1]

$$\boldsymbol{u} = \hat{\boldsymbol{u}} \qquad \text{on } \mathscr{S}_{\hat{\boldsymbol{u}}}, \tag{36.5}$$

$$\boldsymbol{S}\boldsymbol{n} \equiv \mathbf{s}_{(\boldsymbol{n})} = \hat{\mathbf{s}} \qquad \text{on } \mathscr{S}_{\hat{\mathbf{s}}}. \tag{36.6}$$

Following the pattern established in Sect. 26, we say that an initial past history $[\boldsymbol{v}, \boldsymbol{R}, \boldsymbol{T}]$ for $\mathscr{B}$ is ***consistent*** with $[\hat{\boldsymbol{u}}, \hat{\mathbf{s}}]$, $\boldsymbol{b}$, ϱ, and either G or J if, and

[1] Cf. Sect. 23, Eqs. (36.1a) and (36.1b).

only if, $\boldsymbol{v}$, $\boldsymbol{R}$, and $\boldsymbol{T}$ satisfy (26.3)–(26.7) and on $\mathscr{B}$[2]

$$\operatorname{div} \boldsymbol{T} + {}_r\boldsymbol{b}^0 = \varrho \ddot{\boldsymbol{v}}. \tag{36.7}$$

Finally, we recall that an admissible process $[\boldsymbol{u}, \boldsymbol{E}, \mathbf{S}]$ *has the initial past history* $[\boldsymbol{v}, \boldsymbol{R}, \boldsymbol{T}]$ if, and only if,[3]

$$_r[\boldsymbol{u}, \boldsymbol{E}, \mathbf{S}]^0 = [\boldsymbol{v}, \boldsymbol{R}, \boldsymbol{T}]. \tag{36.8}$$

Let h denote the scalar Heaviside unit process and define a scalar process g of Heaviside type by

$$g = h * h. \tag{36.9}$$

Thus, for $t \geqq 0$ we have

$$h(t) = 1, \tag{36.10}$$

$$g(t) = t. \tag{36.11}$$

Let $\boldsymbol{a}$ and $\boldsymbol{c}$ denote prescribed continuous vector fields on $\mathscr{B}$. Using the fields $\boldsymbol{a}$ and $\boldsymbol{c}$ together with the body force process $\boldsymbol{b}$, define a function $\boldsymbol{f}$ on $\mathscr{R}^+$ by[4]

$$\boldsymbol{f}(t) = (g * \boldsymbol{b})(t) + \varrho(\boldsymbol{a} + t\boldsymbol{c}), \qquad t \in \mathscr{R}^+. \tag{36.12}$$

The values of $\boldsymbol{f}$ are continuous vector fields on $\mathscr{B}$.

The following result, due to IGNACZAK,[5] is useful in obtaining other formulations of dynamic problems:

The stress process $\mathbf{S}$ *and the displacement process* $\boldsymbol{u}$ *satisfy the equation of motion* (36.1) *for times* $t > 0$ *and meet the initial conditions (at time* $t = 0$)[6]

$$\mathring{\boldsymbol{u}}^{(0)} = \boldsymbol{a}, \tag{36.13}$$

$$\mathring{\boldsymbol{u}}^{(1)} = \boldsymbol{c} \tag{36.14}$$

if, and only if, for $t > 0$

$$(g * \operatorname{div} \mathbf{S})(t) + \boldsymbol{f}(t) = \varrho\, \boldsymbol{u}(t). \tag{36.15}$$

Let $[\boldsymbol{v}, \boldsymbol{R}, \boldsymbol{T}]$ be a prescribed initial past history for $\mathscr{B}$ which is consistent with $[\hat{\boldsymbol{u}}, \hat{\mathbf{s}}]$, $\boldsymbol{b}$, ϱ, and either $\mathbf{G}$ or $\mathbf{J}$. Define vector fields $\boldsymbol{a}$ and $\boldsymbol{c}$ on $\mathscr{B}$ through the limits[7]

$$\lim_{s \downarrow 0} \boldsymbol{v}^{(0)}(s) = \boldsymbol{a}, \tag{36.16}$$

$$\lim_{s \downarrow 0} \boldsymbol{v}^{(1)}(s) = \boldsymbol{c}. \tag{36.17}$$

Then we have the following

---

[2] Recall that ${}_r\boldsymbol{b}^0$ is defined on $\mathscr{R}^{++}$ by

$${}_r\boldsymbol{b}^0(s) = \boldsymbol{b}(-s), \qquad s \in \mathscr{R}^{++}.$$

Also, if $\boldsymbol{v} = {}_r\boldsymbol{u}^0$, then

$$\frac{d^2}{ds^2}\boldsymbol{v}(s) = \ddot{\boldsymbol{u}}(-s), \qquad s \in \mathscr{R}^{++}.$$

[3] Cf. Sect. 26, Eq. (26.1) with $t_0 = 0$.

[4] Recall the convention for the convolution of functions which are not processes of Heaviside type established in Sect. 1.

[5] IGNACZAK [1963, *8*].

[6] Here, of course, $\mathring{\boldsymbol{u}}^{(0)} = \lim_{t \downarrow 0} \boldsymbol{u}(t)$ and $\mathring{\boldsymbol{u}}^{(1)} = \lim_{t \downarrow 0} \boldsymbol{u}^{(1)}(t)$.

[7] Here we assume that the limits (36.16) and (36.17) exist and define continuous vector fields on $\mathscr{B}$.

Theorem.[8] *Let* $[\boldsymbol{u}, \boldsymbol{E}, \boldsymbol{S}]$ *be an admissible process for* $\mathscr{B}$ *which satisfies the conditions* (R-1) *and* (R-2) (*or* (C-1), (C-2), *and* (C-3)) *and which has the consistent initial past history* $[\boldsymbol{v}, \boldsymbol{R}, \boldsymbol{T}]$. *Then* $[\boldsymbol{u}, \boldsymbol{E}, \boldsymbol{S}]$ *is a solution of the dynamic boundary value problem corresponding to* $[\hat{\boldsymbol{u}}, \hat{\mathbf{s}}]$, $\mathbf{G}$ *(or* $\mathbf{J}$*),* $\boldsymbol{b}$*, and* ϱ *if, and only if, for times* $t > 0$*, the processes* $\boldsymbol{u}$*,* $\boldsymbol{E}$*, and* $\boldsymbol{S}$ *satisfy the strain-displacement relation* (36.2), *the Boltzmann relaxation law* (36.3) (*or the Boltzmann creep law* (36.4)), *the boundary conditions* (36.5) *and* (36.6), *and the Eq.* (36.15).

37. Field equations. We recall the governing field equation for the dynamic theory:

(i) the strain displacement relation

$$\boldsymbol{E} = \tfrac{1}{2}\left[\nabla \boldsymbol{u} + \nabla \boldsymbol{u}^T\right] = \operatorname{sym} \nabla \boldsymbol{u}\,; \tag{37.1}$$

(ii) the equations of motion

$$\operatorname{div} \boldsymbol{S} + \boldsymbol{b} = \varrho\, \ddot{\boldsymbol{u}}, \tag{37.2}$$

$$\boldsymbol{S} = \boldsymbol{S}^T; \tag{37.3}$$

(iii) the Boltzmann relxation law

$$\boldsymbol{S} = \mathscr{L}_{\mathbf{G}}[\boldsymbol{E}] = \mathbf{G} \circledast \boldsymbol{E}. \tag{37.4}$$

We may combine the above relations to derive the displacement equations of motion

$$\operatorname{div} \mathscr{L}_{\mathbf{G}}[\operatorname{sym} \nabla \boldsymbol{u}] + \boldsymbol{b} = \varrho\, \ddot{\boldsymbol{u}}. \tag{37.5}$$

For the isotropic case we recall from (12.1), (12.7), and (12.8) that the relaxation function $\mathbf{G}$ may be written in terms of the two isotropic scalar relaxation functions as:

$$\mathbf{G}[\boldsymbol{E}] = G_1 \boldsymbol{E} + \tfrac{1}{3}(G_2 - G_1)(\operatorname{tr} \boldsymbol{E})\, \mathbf{1}, \tag{37.6}$$

for every symmetric tensor $\boldsymbol{E}$. Hence, we may write the isotropic displacement equation of motion as:

$$\mathscr{L}_{K_2}[\nabla^2 \boldsymbol{u}] + \mathscr{L}_{(K_1 - K_2)}[\nabla(\nabla \cdot \boldsymbol{u})] + \frac{1}{\varrho}\boldsymbol{b} = \ddot{\boldsymbol{u}} \tag{37.7}$$

where the scalar response functions K_1 and K_2 are given by:

$$K_1 \equiv \frac{2G_1 + G_2}{3\varrho}, \tag{37.8}$$

$$K_2 \equiv \frac{G_1}{2\varrho}. \tag{37.9}$$

In (37.7) we continue to use the operator notation with scalar response functions realizing that the context obviates the necessity for explicitly stating the range and domain of the relevant operators. It is, however, worthwhile pointing out[9] some useful properties of $\mathscr{L}_K$ for the case when K is scalar valued. We restrict our attention to Heaviside processes, bearing in mind the definitions of *Boltzmann*

---

[8] LEITMAN [1966, *16*, Theorem 3.1]. In the work cited the author also uses IGNACZAK's result to obtain an alternate characterization of the stress process of a solution to the dynamic problem.

[9] EDELSTEIN and GURTIN [1965, *11*].

operator in Sect. 5 and of *response function* in Sect. 7. Thus, for any scalar field process f

$$\mathscr{L}_K[f](t) = \overset{\circ}{K} f(t) + \int_0^t \dot{K}(t-\tau) f(\tau)\, d\tau = (K \circledast f)(t). \tag{37.10}$$

It is easy to verify that, for any pair (K, M) of scalar response functions,

$$\mathscr{L}_K \mathscr{L}_M = \mathscr{L}_M \mathscr{L}_K, \tag{37.11a}$$

$$\mathscr{L}_M + \mathscr{L}_K = \mathscr{L}_{K+M}. \tag{37.11b}$$

Furthermore, if we assume $\overset{\circ}{K} \neq 0$ then there exists a scalar inverse response function K^{-1} such that[10]

$$\mathscr{L}_{K^{-1}} \mathscr{L}_K = \mathscr{L}_K \mathscr{L}_{K^{-1}} = \mathscr{I} \tag{37.12}$$

where $\mathscr{I}$ denotes the identity operator. Through $\mathscr{L}_K$ we define a ***hereditary wave operator*** $\Box_K$ by

$$\Box_K f = \mathscr{L}_K[\nabla^2 f] - \ddot{f}. \tag{37.13}$$

If $K = \overset{\circ}{K} h$, where $\overset{\circ}{K}$ is a constant, then $\Box_K$ reduces to the usual linear wave operator with "wave speed" $\sqrt{\overset{\circ}{K}}$. EDELSTEIN and GURTIN[11] show that formal manipulations of $\mathscr{L}_K$ as though it were a constant and $\Box_K$ as an ordinary wave operator are justified in the sense of the following remarks:

(i) if $\overset{\circ}{f}{}^{(k)} = 0$ $(k = 0, 1, 2, \ldots, n-1)$, then

$$(\mathscr{L}_K[f])^{(n)} = \mathscr{L}_K[f^{(n)}] \tag{37.14}$$

and

(ii) if $\overset{\circ}{f} = \overset{\circ}{f}{}^{(1)} = 0$, then

$$\mathscr{L}_M[\Box_K f] = \Box_K \mathscr{L}_M[f], \tag{37.15}$$

$$\Box_M \Box_K f = \Box_K \Box_M f, \tag{37.16}$$

$$\int_0^t \int_0^\tau \Box_K f(s)\, ds\, d\tau = \Box_K \int_0^t \int_0^\tau \dot{f}(s)\, ds\, d\tau, \quad t \geqq 0, \tag{37.17}$$

$$(\mathscr{L}_K[f])(0) = (\mathscr{L}_K[f])^{(1)}(0) = 0. \tag{37.18}$$

38. Complete dynamic displacement generating functions. If we restrict our attention to the isotropic dynamic theory, we find there are viscoelastic generalizations to some of the stress function solutions of classical elastodynamics[12]. For the present discussion we consider only Heaviside processes[13] and vanishing body force. We then have the following three equivalent forms of the isotropic displacement equations of motion (37.7):

(i) $$\mathscr{L}_{K_2}[\nabla^2 \boldsymbol{u}] + \mathscr{L}_{(K_1-K_2)}[\nabla(\nabla\cdot\boldsymbol{u})] = \ddot{\boldsymbol{u}}, \tag{38.1a}$$

(ii) $$\mathscr{L}_{K_1}[\nabla(\nabla\cdot\boldsymbol{u})] - \mathscr{L}_{K_2}[\nabla\times(\nabla\times\boldsymbol{u})] = \ddot{\boldsymbol{u}}, \tag{38.1b}$$

(iii) $$\Box_{K_2}\boldsymbol{u} + \mathscr{L}_{(K_1-K_2)}[\nabla(\nabla\cdot\boldsymbol{u})] = \boldsymbol{0}, \tag{38.1c}$$

where K_1 and K_2 are given by (38.8) and (38.9).

[10] Cf. Sect. 11, Eq. (11.18). Here we apply the inversion theorem for response functions, for which K and K^{-1} need not be strong.

[11] EDELSTEIN and GURTIN [1965, *11*].

[12] See, for example, LTE, Sect. 67.

[13] Cf. Sect. 3 for the definition of Heaviside process.

From the governing field equations we shall seek general solutions which are *complete* in the sense that the derived representation in terms of partial differential equations for the displacement generating functions hold for *every* dynamic viscoelastic process corresponding to $\boldsymbol{G}$, $\boldsymbol{b}$ and ϱ. We shall look for such complete solutions of the field equations under the material constraints that

$$\varrho > 0, \quad \mathring{G}_1 > 0, \quad \mathring{G}_2 > 0 \tag{38.2}$$

and the initial conditions

$$\mathring{\boldsymbol{u}} = \mathring{\boldsymbol{u}}^{(1)} = \boldsymbol{0}. \tag{38.3}$$

By an application of BIOT's correspondence principle,[14] through which the squares of the wave speeds of linear isotropic elastodynamics may be replaced in dynamic isotropic viscoelasticity by the operators $\mathscr{L}_{K_1}$ and $\mathscr{L}_{K_2}$, EDELSTEIN and GURTIN[15] prove the following viscoelastic generalizations of the elastodynamic Lamé and Somigliana stress function solutions.[16]

Theorem 1. Viscoelastic Lamé stress functions. *If α and $\boldsymbol{\beta}$ are scalar and vector processes which satisfy*

$$\mathring{\alpha} = \mathring{\alpha}^{(1)} = 0, \quad \mathring{\boldsymbol{\beta}} = \mathring{\boldsymbol{\beta}}^{(1)} = \boldsymbol{0} \tag{38.4}$$

and

$$\Box_{K_1} \alpha = 0, \tag{38.5}$$

$$\Box_{K_2} \boldsymbol{\beta} = \boldsymbol{0}, \tag{38.6}$$

and we define

$$\boldsymbol{u} = \nabla\alpha + \nabla \times \boldsymbol{\beta}, \tag{38.7}$$

then $\boldsymbol{u}$ satisfies the displacement equations of motion (38.1) *and the initial conditions* (38.3).

The proof of this theorem is provided by substitution of (38.7) into (38.1b) using (38.4), (38.5), and (38.6). The important completeness property is stated as

Theorem 2. Completeness of viscoelastic Lamé stress functions. *For every displacement process $\boldsymbol{u}$ which satisfies the displacement equation of motion* (38.1) *and the initial conditions* (38.3), *there exist processes α and $\boldsymbol{\beta}$ (whose values at any time t are scalar and vector fields on $\mathscr{B}$) such that* (38.4), (38.5), (38.6), *and* (38.7) *hold. Moreover*

$$\nabla \cdot \boldsymbol{\beta} = 0. \tag{38.8}$$

We shall give the proof of EDELSTEIN and GURTIN[17] since it is typical of the utility of the present notation. This proof is an adaptation of a method of SOMIGLIANA.[18]

Proof. Since $\mathring{\boldsymbol{u}} = \mathring{\boldsymbol{u}}^{(1)} = \boldsymbol{0}$ by (38.3), the displacement equations of motion (38.1b) may be integrated to obtain

$$\begin{aligned} \boldsymbol{u}(\boldsymbol{x}, t) = {} & \nabla \int_0^t \int_0^\tau \mathscr{L}_{K_1}[\nabla \cdot \boldsymbol{u}](\boldsymbol{x}, s)\, ds\, d\tau \\ & - \nabla \times \int_0^t \int_0^\tau \mathscr{L}_{K_2}[\nabla \times \boldsymbol{u}](\boldsymbol{x}, s)\, ds\, d\tau, \quad (\boldsymbol{x}, t) \in \mathscr{B} \times \mathscr{R}^+. \end{aligned} \tag{38.9}$$

---

[14] BIOT [1955, *1*].
[15] EDELSTEIN and GURTIN [1965, *11*].
[16] LTE, Sect. 67.
[17] EDELSTEIN and GURTIN [1965, *11*].
[18] SOMIGLIANA [1892, *1*].

Hence we need only identify α and $\boldsymbol{\beta}$ as

$$\alpha(\boldsymbol{x},t)=\int_0^t\int_0^\tau \mathscr{L}_{K_1}[\nabla\cdot\boldsymbol{u}](\boldsymbol{x},s)\,ds\,d\tau,\qquad (\boldsymbol{x},t)\in\mathscr{B}\times\mathscr{R}^+, \tag{38.10}$$

$$\boldsymbol{\beta}(\boldsymbol{x},t)=-\int_0^t\int_0^\tau \mathscr{L}_{K_2}[\nabla\times\boldsymbol{u}](\boldsymbol{x},s)\,ds\,d\tau,\qquad (\boldsymbol{x},t)\in\mathscr{B}\times\mathscr{R}^+. \tag{38.11}$$

We see immediately that

$$\boldsymbol{u}=\nabla\alpha+\nabla\times\boldsymbol{\beta}, \tag{38.12}$$

$$\nabla\cdot\boldsymbol{\beta}=0, \tag{38.13}$$

and (38.4) hold. To show that (38.5) and (38.6) are satisfied by α and $\boldsymbol{\beta}$, we merely take the divergence and curl of (38.9). This completes the proof.

As a corollary EDELSTEIN and GURTIN state:

Corollary. *If $\boldsymbol{u}$ satisfies* (38.1) and (38.3), *then there exist processes $\boldsymbol{u}_1$ and $\boldsymbol{u}_2$ such that*

$$\boldsymbol{u}=\boldsymbol{u}_1+\boldsymbol{u}_2 \tag{38.14}$$

and

$$\nabla\times\boldsymbol{u}_1=\boldsymbol{0},\qquad \square_{K_1}\boldsymbol{u}_1=\boldsymbol{0}, \tag{38.15}$$

$$\nabla\cdot\boldsymbol{u}_2=0,\qquad \square_{K_2}\boldsymbol{u}_2=\boldsymbol{0}. \tag{38.16}$$

The next theorem of EDELSTEIN and GURTIN[19] provides the generalization of the Somigliana stress function solution of elastodynamics.[20]

Theorem 3. Viscoelastic Somigliana stress function. *If $\boldsymbol{g}$ is a vector-field process which satisfies*

$$\mathring{\boldsymbol{g}}=\mathring{\boldsymbol{g}}^{(1)}=\mathring{\boldsymbol{g}}^{(2)}=\mathring{\boldsymbol{g}}^{(3)}=\boldsymbol{0} \tag{38.17}$$

and

$$\square_{K_2}\square_{K_1}\boldsymbol{g}=\boldsymbol{0}, \tag{38.18}$$

then $\boldsymbol{u}$ defined by

$$\boldsymbol{u}=\square_{K_1}\boldsymbol{g}-\mathscr{L}_{(K_1-K_2)}[\nabla(\nabla\cdot\boldsymbol{g})] \tag{38.19}$$

satisfies the displacement equations of motion (38.1) *and the initial conditions* (38.3). *Furthermore, this solution is complete in the usual sense: for every $\boldsymbol{u}$ which satisfies the displacement equation of motion and the initial conditions, there exists a process $\boldsymbol{g}$ such that* (38.17), (38.18), *and* (38.19) *hold.*

EDELSTEIN and GURTIN[21] also point out that from the Somigliana representation there is a viscoelastic analog to Boggio's theorem of elastodynamics.[22] This is the content of

Theorem 4. Viscoelastic counterpart to Boggio's theorem. *If $\boldsymbol{g}$ is a vector field process which satisfies conditions* (38.17) *and* (38.18) *of Theorem 3, then there exist processes $\boldsymbol{g}_1$ and $\boldsymbol{g}_2$ such that*

$$\boldsymbol{g}=\boldsymbol{g}_1+\boldsymbol{g}_2 \tag{38.20}$$

and

$$\square_{K_1}\boldsymbol{g}_1=\square_{K_2}\boldsymbol{g}_2=\boldsymbol{0}. \tag{38.21}$$

---

[19] EDELSTEIN and GURTIN [1965, *11*].
[20] LTE, Sect. 67.
[21] EDELSTEIN and GURTIN [1965, *11*].
[22] See LTE, Sect. 67.

39. Power and energy. For the dynamic theory we can, for sufficiently smooth processes, write the rate at which work is done.[23]

Theorem of power expended. *The rate at which work is done on a body $\mathscr{B}$ by the surface forces* $\mathbf{s}$ *and body forces* $\boldsymbol{b}$ *equals the stress power plus the time rate of change of the kinetic energy:*

$$\int_{\partial\mathscr{B}} \mathbf{s}\cdot\dot{\boldsymbol{u}}\,dA+\int_{\mathscr{B}}\boldsymbol{b}\cdot\dot{\boldsymbol{u}}\,dV=\int_{\mathscr{B}}\boldsymbol{S}\cdot\dot{\boldsymbol{E}}\,dV+\frac{d}{dt}\frac{1}{2}\int_{\mathscr{B}}\varrho|\dot{\boldsymbol{u}}|^2\,dV \tag{39.1}$$

where

$$\textit{stress power}\equiv\int_{\mathscr{B}}\boldsymbol{S}\cdot\dot{\boldsymbol{E}}\,dV, \tag{39.2a}$$

$$\equiv\int_{\mathscr{B}}\mathscr{L}_{\mathbf{G}}[\boldsymbol{E}]\cdot\dot{\boldsymbol{E}}\,dV \tag{39.2b}$$

and

$$\textit{kinetic energy}\equiv\tfrac{1}{2}\int_{\mathscr{B}}\varrho|\dot{\boldsymbol{u}}|^2\,dV. \tag{39.3}$$

Hence, for a dynamic viscoelastic process (in DVP$(\mathbf{G}, \boldsymbol{b}, \varrho)$), we have

$$\int_{\partial\mathscr{B}} \mathbf{s}\cdot\dot{\boldsymbol{u}}\,dA+\int_{\mathscr{B}}\boldsymbol{b}\cdot\dot{\boldsymbol{u}}\,dV=\int_{\mathscr{B}}(\mathscr{L}_{\mathbf{G}}[\boldsymbol{E}])\cdot\dot{\boldsymbol{E}}\,dV+\frac{d}{dt}\frac{1}{2}\int_{\mathscr{B}}\varrho|\dot{\boldsymbol{u}}|^2\,dV \tag{39.4}$$

or, from the definition of $\mathscr{L}_{\mathbf{G}}$,

$$\begin{aligned}\int_{\partial\mathscr{B}} &\mathbf{s}(\boldsymbol{x},t)\cdot\dot{\boldsymbol{u}}(\boldsymbol{x},t)\,dA(\boldsymbol{x})+\int_{\mathscr{B}}\boldsymbol{b}(\boldsymbol{x},t)\cdot\dot{\boldsymbol{u}}(\boldsymbol{x},t)\,dV(\boldsymbol{x})\\ &=\int_{\mathscr{B}}\mathring{\mathbf{G}}\boldsymbol{E}(\boldsymbol{x},t)\cdot\dot{\boldsymbol{E}}(\boldsymbol{x},t)\,dV(\boldsymbol{x})+\int_{\mathscr{B}}\dot{\boldsymbol{E}}(\boldsymbol{x},t)\cdot\int_0^\infty\dot{\mathbf{G}}(\boldsymbol{x},\tau)\,\boldsymbol{E}(\boldsymbol{x},t-\tau)\,d\tau\,dV\\ &\quad+\frac{d}{dt}\frac{1}{2}\int_{\mathscr{B}}\varrho|\dot{\boldsymbol{u}}|^2(\boldsymbol{x},t)\,dV(\boldsymbol{x}),\qquad t\in\mathscr{R}.\end{aligned} \tag{39.5}$$

This relation, applied to the case of null boundary data, is basic to the proofs of uniqueness theorems in dynamic linear viscoelasticity.

40. Uniqueness. Boundary value problem. In this section we consider the question of uniqueness of a dynamic viscoelastic process for $\mathbf{G}$, $\boldsymbol{b}$, ϱ and:[24]

(i) the initial past history on $\mathscr{B}$,

$$_r\boldsymbol{u}^0\equiv\mathbf{0} \tag{40.1}$$

and

(ii) the boundary conditions on $\partial\mathscr{B}$,

$$\boldsymbol{u}=\hat{\boldsymbol{u}}\qquad\text{on } \mathscr{S}_{\hat{\boldsymbol{u}}}, \tag{40.2}$$

$$\boldsymbol{S}\boldsymbol{n}=\mathbf{s}_{(\boldsymbol{n})}=\hat{\mathbf{s}}\quad\text{on } \mathscr{S}_{\hat{\mathbf{s}}}. \tag{40.3}$$

Assume that the density ϱ is continuous and positive on $\overline{\mathscr{B}}$, the body force $\boldsymbol{b}$ vanishes, and the relaxation process $\mathbf{G}$ is in Heaviside class H^1 with values which are class C^2 fields on $\overline{\mathscr{B}}$. We can then state some important unique-

[23] Cf. LTE, Sect. 60 (2).
[24] Cf. Sect. 36 for the precise meaning of *dynamic viscoelastic process.*

ness theorems for solutions to the dynamic viscoelastic boundary value problem satisfying the system of Eqs. (37.1), (37.2), (37.3), (37.4), (40.1), (40.2), and (40.3). We state without proof the following three theorems of EDELSTEIN and GURTIN.[25]

Theorem 1. Mixed problem.[26] *If the initial elasticity* $\overset{\circ}{\mathbf{G}}$ *is symmetric and positive definite, then there is at most one solution to the mixed boundary value problem.*

Theorem 2. Mixed problem (Laplace transform formulation). *If* $\mathbf{G}$ *is independent of position* $\boldsymbol{x}$ *(i.e., homogeneous), initially positive definite, and possesses a Laplace transform with respect to time t, then there is at most one solution to the mixed problem which has a Laplace transform with respect to t.*

Theorem 3. Displacement boundary data ($\mathscr{S}_{\hat{\boldsymbol{s}}} \equiv \emptyset$). *If* $\mathbf{G}$ *is independent of position* $\boldsymbol{x}$*, and the initial elasticity* $\overset{\circ}{\mathbf{G}}$ *is symmetric and strongly elliptic,*[27] *then there is at most one solution to the displacement problem.*

The second of the above theorems removes the initial symmetry requirement on $\mathbf{G}$ at the expense of requiring that the solution have a Laplace transform and that the material be homogeneous. The application of such Laplace transform techniques in a discussion of uniqueness in linear viscoelasticity was first made by ONAT and BREUER,[28] whose techniques were used by EDELSTEIN and GURTIN[29] to prove Theorem 2. EDELSTEIN and FOSDICK[30] have shown that the homogeneity assumption is essential and provide a counterexample to some previous work which claimed to remove this requirement.[31] BARBERÁN and HERRERA[32] have removed the restriction of translation invariance of the Boltzmann law in giving an alternate proof of Theorem 1.

41. Waves. Singular surfaces. In this section we shall examine necessary conditions under which finite jump discontinuities in displacement derivatives (e.g., at a pulse front) may propagate as waves through a viscoelastic material. We shall consider both the speed of propagation and, for plane waves,[33] the variation of amplitude with time.

Let $\mathscr{S}_t$ denote a one parameter (time) family of surfaces in three-dimensional Euclidean space and Σ denote the hypersurface

$$\Sigma = \{(\boldsymbol{x}, t) \mid \boldsymbol{x} \in \mathscr{S}_t,\ 0 \leqq t < \infty\}. \tag{41.1}$$

Then the process $[\boldsymbol{u}, \boldsymbol{E}, \boldsymbol{S}]$ is a ***viscoelastic wave of order*** $N > 1$ if, and only if, $[\boldsymbol{u}, \boldsymbol{E}, \boldsymbol{S}]$ is a solution of the governing field equations (37.1), (37.2),[34] (37.3), (37.4), and furthermore satisfies the following constraints:[35]

[25] EDELSTEIN and GURTIN [1964, *6*].
[26] We refer to the "mixed problem" if neither $\mathscr{S}_{\hat{\boldsymbol{u}}}$ nor $\mathscr{S}_{\hat{\boldsymbol{s}}}$ is empty.
[27] That is, for any pair $\boldsymbol{x}$, $\boldsymbol{y}$ of nonzero vectors, $\boldsymbol{x} \cdot \overset{\circ}{\mathbf{G}}(\boldsymbol{x} \otimes \boldsymbol{y})\, \boldsymbol{y} > 0$. Cf. footnote 67, Chap. C, p. 62.
[28] ONAT and BREUER [1963, *14*].
[29] EDELSTEIN and GURTIN [1964, *6*].
[30] EDELSTEIN and FOSDICK [1968, *4*].
[31] ODEH and TADJBAKHSH [1965, *22*].
[32] BARBERÁN and HERRERA [1966, *4*, Theorem 1].
[33] Effects of other than plane wave geometry in our context have been considered to various degrees by CHEN [1965, *6*], VARLEY [1965, *31*], and VALANIS [1965, *30*].
[34] Actually the consideration of shock waves ($N = 1$) requires the use of the integral form of the balance of linear momentum.
[35] These constraints characterize a singular surface of order N. See the historical notes in LTE, Subchap. E.V., and TRUESDELL and TOUPIN [1960, *15*, Sect. 187].

(A.1) The displacement $\boldsymbol{u}$, considered as a function defined on $\mathscr{B}\times\mathscr{R}$, is $N-1$ times continuously differentiable; $\boldsymbol{u}$ possesses continuous N-th derivatives at every point of $\mathscr{B}\times\mathscr{R}$ except, possibly, on Σ where the N-th derivatives may possess jump discontinuities.

(A.2) The response function $\boldsymbol{G}$ is N times continuously differentiable on $\mathscr{B}\times\mathscr{R}^+$, while the body force $\boldsymbol{b}$ and the density ϱ are continuous and, if $N\geqq 3$, $N-2$ times continuously differentiable on their respective domains.

(A.3) The hypersurface Σ is smooth and orientable.

(A.4) For each $\boldsymbol{x}$, the set $\omega_{\boldsymbol{x}}=\{t\,|\,\boldsymbol{x}\in\mathscr{S}_t\}$ has measure zero (roughly, the wave does not cross the same point $\boldsymbol{x}$ too many times).

Condition (A.1) is essentially the classical definition of a singular surface in the sense of DUHEM and HADAMARD.[36] Condition (A.3) guarantees that there exists a normal $\boldsymbol{v}$ to Σ at $(\boldsymbol{x}, t)\in\Sigma$. If $\boldsymbol{n}$ is a unit normal to $\mathscr{S}_t$ at $\boldsymbol{x}$, then such a $\boldsymbol{v}$ is given by

$$\boldsymbol{v}\equiv(\boldsymbol{n},\ -U)\equiv(n_1, n_2, n_3,\ -U) \tag{41.2}$$

where $U\geqq 0$. The vector $\boldsymbol{n}$ is called the ***direction of propagation***, and the number U is called the ***speed of propagation.***

For the ***jump*** of a function $g(\boldsymbol{x}, t)$ across Σ at $(\boldsymbol{x}, t)$ we shall use the bracket notation:

$$[g(\boldsymbol{x}, t)]=\lim_{(\boldsymbol{x}^+,t^+)\to(\boldsymbol{x},t)} g(\boldsymbol{x}^+, t^+)-\lim_{(\boldsymbol{x}^-,t^-)\to(\boldsymbol{x},t)} g(\boldsymbol{x}^-, t^-) \tag{41.3}$$

where $(\boldsymbol{x}^+, t^+)$ and $(\boldsymbol{x}^-, t^-)$ are points on "opposite" sides of Σ. Waves of order 1 shall be referred to as ***shock waves*** and waves or order 2 as ***acceleration waves.*** We further define a ***wave axis*** or ***polarization*** to be any nonzero vector which is parallel to the jump in the N-th time derivative of $\boldsymbol{u}(\boldsymbol{x}, t)$.

The mathematical existence of viscoelastic waves of order N has been demonstrated in particular cases by SIPS, LEE and KANTER, BERRY, and CHU.[37] They show that shocks and higher order waves may exist in the context of linear viscoelasticity even though the *spontaneous* generation of shock-waves is inherently due to nonlinear processes. A systematic approach to the more general question of finding *necessary* conditions for the propagation of viscoelastic waves was initiated by HERRERA and GURTIN.[38] In that work necessary conditions were established for the propagation of acceleration waves in anisotropic linear viscoelasticity. Subsequently, there followed complete treatments of waves, not only in linear viscoelasticity, but also in the more general nonlinear theories of materials with memory.[39] While we do not consider the nonlinear theories in this article, it suffices to say that much of the following discussion of the linear theory, when restricted to one space dimension, arises as a special case in the aforementioned papers on the nonlinear theories.

We shall divide the study of waves into two parts, the first pertaining to the determination of wave speeds and wave axes and the second regarding the time dependence of the jump discontinuities for plane waves.[40] The basic result concerning wave speeds is

---

[36] See TRUESDELL and TOUPIN [1960, *15*, Sect. 187].

[37] SIPS [1951, *3*], LEE and KANTER [1953, *2*], BERRY [1958, *1*, *2*], CHU [1962, *4*].

[38] HERRERA and GURTIN [1965, *17*].

[39] See, for example, the contributions by COLEMAN, GURTIN and HERRERA [1965, *9*], COLEMAN and GURTIN [1965, *7*, *8*], and VARLEY [1965, *31*].

[40] VALANIS [1965, *30*] has derived results for the nonplane wave case in the linear theory.

Theorem 1. *The speed of propagation U of a viscoelastic wave at a point $(\boldsymbol{x}, t) \in \Sigma$ is independent of the order N of the wave and satisfies the eigenvalue problem*

$$\mathring{\boldsymbol{Q}}(\boldsymbol{n})\, \boldsymbol{q} - \varrho\, U^2 \boldsymbol{q} = \boldsymbol{0} \quad \begin{pmatrix}\text{Fresnel-Hadamard} \\ \text{Propagation Condition}\end{pmatrix} \tag{41.4}$$

where $\mathring{\boldsymbol{Q}}(\boldsymbol{n})$ is the ***acoustic tensor***[41] *defined on $\mathscr{V}$, for each unit vector* $\boldsymbol{n}$, *through*

$$\mathring{\boldsymbol{Q}}(\boldsymbol{n})\, \boldsymbol{a} = \mathring{\mathbf{G}}[(\boldsymbol{a} \otimes \boldsymbol{n})]\, \boldsymbol{n}, \qquad \boldsymbol{a} \in \mathscr{V}. \tag{41.5}$$

The eigenvector $\boldsymbol{q}$ is a ***wave axis.***

The essential point of this theorem is that the viscoelastic wave speeds are dictated by the "elastic" response of the material. That is, only the initial value $\mathring{\mathbf{G}}$ of the relaxation function $\mathbf{G}$ enters into the determination of the wave speeds and wave axes. Further, since we have the same propagation condition as for linear elasticity, we can adopt the numerous results for elastic wave propagation reviewed by GURTIN[42] by merely identifying the elastic acoustic tensor with the viscoelastic acoustic tensor. Of course, the linear elastic case does not have the important wave decay feature which is characteristic of viscoelastic waves.

Theorem 1 was first stated in a general anisotropic viscoelastic context for $N=2$ by HERRERA and GURTIN[43] and, soon after, for shocks and waves of arbitrary order by FISHER and GURTIN.[44] For simplicity, we shall confine our arguments here to the case of acceleration waves, $N=2$. Those for other orders are similar. For shocks, however, it is necessary to utilize the integral form of the balance of momentum because of the discontinuous velocity field.[45]

To verify the assertion of Theorem 1 for $N=2$, we take the jump in the displacement equations of motion:

$$[\operatorname{div}\{\mathring{\mathbf{G}} \operatorname{sym} \nabla \boldsymbol{u}(t)\}] + \left[\operatorname{div} \int_0^t \dot{\mathbf{G}}(t-\tau) \operatorname{sym} \nabla \boldsymbol{u}(\tau)\, d\tau\right] + [\varrho\, \boldsymbol{b}(t)] = [\varrho\, \ddot{\boldsymbol{u}}(t)], \qquad t \in \mathscr{R}. \tag{41.6}$$

Then, noting conditions (A. 1) and (A.2) and the smoothness of integrals possessing jump discontinuities in their integrands, we apply a slight variant of Maxwell's theorem[46] to obtain

$$\mathring{\mathbf{G}}[(\boldsymbol{q} \otimes \boldsymbol{n})]\, \boldsymbol{n} = \varrho\, U^2 \boldsymbol{q} \tag{41.7}$$

where

$$U \boldsymbol{q} = -[\ddot{\boldsymbol{u}}]. \tag{41.8}$$

This provides the desired result.

---

[41] LTE, Sect. 70, lists some of the more important properties of the acoustic tensor. We associate the term "acoustic tensor" with the initial value of the relaxation function although we shall also have use for the term "*complex* acoustic tensor", which is a frequency dependent operator of related form.

[42] LTE, Subchap. E.V. Cf. Sect. 25 for additional relations between elastic states and the initial response of viscoelastic bodies.

[43] HERRERA and GURTIN [1965, *17*].

[44] FISHER and GURTIN [1965, *14*].

[45] Essentially one develops Hugoniot type shock relations.

[46] See LTE, Sect. 72, for historical notes and derivation.

Clearly, if $\overset{\circ}{\mathsf{G}}$ is symmetric, then there exist three, not necessarily distinct, wave speeds. Further, if $\overset{\circ}{\mathsf{G}}$ is strongly elliptic, i.e., if

$$\boldsymbol{m}\cdot\overset{\circ}{\mathsf{G}}[(\boldsymbol{m}\otimes\boldsymbol{n})]\,\boldsymbol{n}>0, \quad \text{when} \quad \boldsymbol{m}\neq\boldsymbol{0},\ \boldsymbol{n}\neq\boldsymbol{0} \tag{41.9}$$

then the wave speeds are real and nonzero.

We characterize ***longitudinal waves*** as those obeying

$$\boldsymbol{n}\times\boldsymbol{q}=\boldsymbol{0} \tag{41.10}$$

(i.e., wave axis and propagation direction are parallel) and ***transverse waves*** as those for which the inner product

$$\boldsymbol{n}\cdot\boldsymbol{q}=0 \tag{41.11}$$

(i.e., wave axis and propagation direction are perpendicular). Then, for a visco-elastic wave of order N, the following three statements are equivalent:

$$\begin{aligned} &\text{(i)} \quad \text{the wave is longitudinal,}\\ &\text{(ii)} \quad [\nabla\times\boldsymbol{u}^{(N-1)}]=\boldsymbol{0},\\ &\text{(iii)} \quad \boldsymbol{q}=\boldsymbol{n}\,[\nabla\cdot\boldsymbol{u}^{(N-1)}]. \end{aligned} \tag{41.12}$$

Further, for transverse waves we have the corresponding set of equivalent statements:

$$\begin{aligned} &\text{(i)} \quad \text{the wave is transverse,}\\ &\text{(ii)} \quad [\nabla\cdot\boldsymbol{u}^{(N-1)}]=0,\\ &\text{(iii)} \quad \boldsymbol{q}=\boldsymbol{n}\times[\nabla\times\boldsymbol{u}^{(N-1)}]\times\boldsymbol{n}. \end{aligned} \tag{41.13}$$

These remarks are readily confirmed if we observe from Maxwell's theorem[47] that

$$\begin{aligned} \boldsymbol{n}\cdot\boldsymbol{q}&=[\nabla\cdot\boldsymbol{u}^{(N-1)}],\\ \boldsymbol{n}\times\boldsymbol{q}&=[\nabla\times\boldsymbol{u}^{(N-1)}]; \end{aligned}$$

and hence, the identity

$$\boldsymbol{q}=\boldsymbol{n}\,(\boldsymbol{n}\cdot\boldsymbol{q})-\boldsymbol{n}\times(\boldsymbol{n}\times\boldsymbol{q}), \qquad (|\boldsymbol{n}|=1)$$

gives[48]

$$\boldsymbol{q}=\boldsymbol{n}\,[\nabla\cdot\boldsymbol{u}^{(N-1)}]-\boldsymbol{n}\times[\nabla\times\boldsymbol{u}^{(N-1)}]. \tag{41.14}$$

From this point the remark is easily verified.

We have remarked previously that symmetry of the acoustic tensor gives rise to three, not necessarily distinct, wave speeds. We now restrict our attention to the even stronger symmetry assumption that the relaxation function be initially isotropic; then

$$\overset{\circ}{\mathsf{G}}[\boldsymbol{E}]=\overset{\circ}{G}_1\,\boldsymbol{E}+\tfrac{1}{3}\,(\overset{\circ}{G}_2-\overset{\circ}{G}_1)\,(\operatorname{tr}\boldsymbol{E})\,\boldsymbol{1}, \tag{41.15}$$

for all symmetric $\boldsymbol{E}$, and we have the following well known corollary to Theorem 1.

[47] See LTE, Sect. 72.

[48] This is the main statement of Weingarten's theorem; see LTE, Sect. 72, or Truesdell and Toupin [1960, *15*].

Corollary.[49] *If the relaxation function is initially isotropic, then there are two, and only two, possible speeds of propagation (U_L and U_T) of a viscoelastic wave of any order. These speeds are given by*

$$\text{(i)}\quad U_L^2 = (\mathring{G}_2 + 2\mathring{G}_1)/3\varrho, \text{ in which case the wave is longitudinal,} \tag{41.16}$$

and

$$\text{(ii)}\quad U_T^2 = \mathring{G}_1/2\varrho, \text{ in which case the wave is transverse.} \tag{41.17}$$

Conversely: if the wave is longitudinal, then $U=U_L$; if the wave is transverse, then $U=U_T$.

The results stated thus far give information only about wave speeds and wave axes. The magnitude of a propagating viscoelastic jump discontinuity may, however, vary with time. We shall now consider the nature of such variations for some simple wave shapes. Historically, it is interesting to note that many of the results to be discussed here in the context of the linear theory were actually obtained first in the context of more general nonlinear theories.[50]

For the present we shall restrict our discussion to a consideration of ***plane wave*** propagation[51] in the sense that we admit only those displacement fields $\boldsymbol{u}$ which have the form

$$\boldsymbol{u}(\boldsymbol{x}, t) = \tilde{\boldsymbol{u}}(\boldsymbol{x}\cdot\boldsymbol{n}, t) \tag{41.18}$$

where $\tilde{\boldsymbol{u}}$ is defined on $\mathscr{R}\times\mathscr{R}$, and $\boldsymbol{n}$ is a fixed direction of propagation ($|\boldsymbol{n}|=1$). For an N-th order wave we shall refer to the ***strength*** $\boldsymbol{q}$

$$\boldsymbol{q} \equiv \left[\frac{\partial^N \tilde{\boldsymbol{u}}(z,t)}{\partial z^N}\right](-U)^{N-1} \tag{41.19}$$

and the ***induced jump*** γ

$$\gamma \equiv -\left[\frac{\partial^{N+1} \tilde{\boldsymbol{u}}(z,t)}{\partial z^{N+1}}\right](-U)^{N}. \tag{41.20}$$

Our prime interest is to show the time dependence of the wave strength $\boldsymbol{q}(\boldsymbol{x},t)$. It should be noted that the corresponding nonlinear theories show that $\boldsymbol{q}$ may either decay or grow, but plane wave strengths in the linear theory never grow under reasonable physical restrictions.[52] The approach presented here is basically that given by FISHER[53] for the linear theory, but some of the results have been deduced in the forementioned nonlinear theories.

We shall henceforth restrict our attention to homogeneous materials. Central to our discussion is the following

Lemma. *The strength $\boldsymbol{q}(t)$[54] of a plane viscoelastic wave of order N and wave speed U propagating in a homogeneous medium is governed by the vector differential equation*

$$\dot{\boldsymbol{Q}}(\boldsymbol{n})\,\boldsymbol{q} - 2\varrho U^2 \frac{d\boldsymbol{q}}{dt} = \mathring{\boldsymbol{Q}}(\boldsymbol{n})\,\gamma - \varrho U^2 \gamma, \tag{41.21}$$

[49] TRUESDELL and TOUPIN [1960, *15*], FISHER and GURTIN [1965, *14*].

[50] Particularly in the works of COLEMAN and GURTIN [1965, *7, 8*] and VARLEY [1965, *31*].

[51] Some aspects of non-plane wave propagation have been treated by VARLEY [1965, *31*. and VALANIS [1965, *30*].

[52] There can, of course, be geometric growth effects for non-plane waves. See VALANIS [1965, *30*].

[53] FISHER [1965, *13*].

[54] Henceforth we shall suppress the spatial dependence, writing $\boldsymbol{q}(z,t)=\boldsymbol{q}(t)$, etc.

where the ***attenuation tensor*** $\dot{Q}(n)$ *is defined on* $\mathscr{V}$, *for each unit vector* $\boldsymbol{n}$, *by*

$$\dot{Q}(\boldsymbol{n})\,\boldsymbol{a} = \dot{\mathsf{G}}(0)\,[(\boldsymbol{a}\otimes\boldsymbol{n})]\,\boldsymbol{n}, \qquad \boldsymbol{a}\in\mathscr{V}. \tag{41.22}$$

It is important to note that the time dependence of $\boldsymbol{q}$ is governed, independent of order N, by the same set of equations. Furthermore, Eq. (41.2) shows the connection between the wave strength $\boldsymbol{q}$ and the induced jump γ.

We shall outline the proof of the lemma.[55] If we differentiate the displacement equations of motion, given by Eq. (37.5), $N-2$ times with respect to $\boldsymbol{x}$ and once with respect to t, we arrive at a set of $(N+1)$-th order differential equations in $\boldsymbol{u}(\boldsymbol{x}, t)$ for $(\boldsymbol{x}, t)\notin\Sigma$. We then formally find the jump in these equations across Σ noting the smoothness of the integrals possessing integrands with jump discontinuities. To complete the proof we need only appeal to a result of THOMAS,[56] which can be adapted to say that if $f(\boldsymbol{x}, t)$ is a function with the jump $[f]$ at $(\boldsymbol{x}, t)\in\Sigma$, then the time rate of change of the jump in f is given by

$$\frac{d}{dt}[f] = \left[\frac{\partial f}{\partial t}\right] + U\boldsymbol{n}\cdot[\nabla f]. \tag{41.23}$$

Hence, applying this in conjunction with Maxwell's theorem,[57] and requiring the wave to obey the propagation condition of Theorem 1, we recover (41.21), where, for the case of $N=2$,

$$\boldsymbol{q} = -\left[\frac{\partial^2\tilde{\boldsymbol{u}}}{\partial z^2}\right] U \tag{41.24}$$

and

$$\gamma = -\left[\frac{\partial^3\tilde{\boldsymbol{u}}}{\partial z^3}\right] U^2. \tag{41.25}$$

Given the propagation condition of Eqs. (41.4) and (41.21), we see that the time dependence of $\boldsymbol{q}$ is intimately associated with the relation of γ to $\boldsymbol{q}$ and the form of the operators $\mathring{Q}$ and $\dot{Q}$. Henceforth, when no confusion may arise, the dependence of the tensors $\mathring{Q}$ and $\dot{Q}$ upon the unit vector $\boldsymbol{n}$ will be suppressed.

We shall consider two cases which are naturally suggested by Eq. (41.21). If γ is also an eigenvector of $\mathring{Q}$ for the direction $\boldsymbol{n}$ and wave speed U, then we shall speak of ***axially similar waves.***[58] Since the left-hand side of Eq. (41.21) then vanishes identically, we have

Theorem 2. *If a plane wave of order N propagating in a homogeneous visco-elastic medium, in the direction* $\boldsymbol{n}$ *with wave speed U, is axially similar, then the wave strength* $\boldsymbol{q}(t)$ *is given by*

$$\boldsymbol{q}(t) = \exp\left[\left(\frac{t}{2\varrho U^2}\right)\dot{Q}\right]\boldsymbol{q}(0) \tag{41.26a}$$

or, for symmetric $\dot{Q}$, *by*

$$\boldsymbol{q}(t) = \sum_{i=1}^{3}\alpha_i\,\boldsymbol{a}_i\,e^{\mu_i t}, \tag{41.26b}$$

where $2\mu_i\varrho U^2$ *is the i-th eigenvalue and* $\boldsymbol{a}_i$ *the corresponding eigenvector of the attenuation operator* $\dot{Q}$ *and the* α_i *are scalar constants.*

---

55 FISHER [1965, *13*].

56 THOMAS [1961, *10*].

57 See LTE, Sect. 72.

58 Note that in one dimensional theories all waves are necessarily axially similar.

The multiplicity of the eigenvalues ϱU^2 of $\overset{\circ}{\underset{\sim}{Q}}$ may be used to reduce the number of terms in Eq. (41.26b). In fact, it is easily seen that if ϱU^2 is a distinct eigenvalue, $\boldsymbol{q}$ must be of constant direction and only a single value of the μ_i may enter into Eq. (41.26b). The situation is complicated somewhat by the fact that the μ_i may be multiple eigenvalues of $\dot{\underset{\sim}{Q}}$. The proof of Theorem 2 follows immediately from the general solution of the homogeneous vector differential equation

$$\dot{\underset{\sim}{Q}}\,\boldsymbol{q} - 2\varrho\, U^2 \frac{d\boldsymbol{q}}{dt} = \boldsymbol{0}. \tag{41.27}$$

In the case when $\overset{\circ}{\underset{\sim}{Q}}\,\gamma - \varrho U^2 \gamma \neq \boldsymbol{0}$, we call the wave ***axially dissimilar.*** We can then apply the Fredholm alternative for symmetric matrices to obtain the time dependence of $\boldsymbol{q}(t)$. Thus, if $\overset{\circ}{\underset{\sim}{Q}}$ is symmetric and $\boldsymbol{b}$ any eigenvector of $\overset{\circ}{\underset{\sim}{Q}}$ for the direction $\boldsymbol{n}$ and wave speed U then

$$\{\overset{\circ}{\underset{\sim}{Q}}\,\gamma - \varrho\, U^2 \gamma\}\cdot\boldsymbol{b} = 0. \tag{41.28}$$

Hence, we have the orthogonality condition

$$\left\{\dot{\underset{\sim}{Q}}\,\boldsymbol{q} - 2\varrho\, U^2 \frac{d\boldsymbol{q}}{dt}\right\}\cdot\boldsymbol{b} = 0 \tag{41.29}$$

which suffices to give us the following information for axially dissimilar waves.[59]

Theorem 3. *The strength $\boldsymbol{q}(t)$ of a plane axially dissimilar wave of order N propagating in a homogeneous viscoelastic solid with symmetric acoustic tensor $\varrho^{-1}\overset{\circ}{\underset{\sim}{Q}}$ is given by:*

(i) *if ϱU^2 is a distinct eigenvalue of $\overset{\circ}{\underset{\sim}{Q}}$, then*

$$\boldsymbol{q}(t) = \alpha\,\boldsymbol{b}\, e^{\mu t} \qquad (\alpha \ldots \text{constant}) \tag{41.30}$$

where

$$\mu \equiv \frac{\boldsymbol{b}\cdot\dot{\underset{\sim}{Q}}\,\boldsymbol{b}}{2\varrho\, U^2} \tag{41.31}$$

and $\boldsymbol{b}$ is a unit wave axis.[60]

(ii) *if ϱU^2 is a double eigenvalue of $\overset{\circ}{\underset{\sim}{Q}}$, then*

$$\boldsymbol{q}(t) = \sum_{\gamma=1}^{2} \alpha_\gamma\, \boldsymbol{a}_\gamma\, e^{\mu_\gamma t}, \tag{41.32}$$

where $2\mu_\gamma \varrho U^2$ is the eigenvalue corresponding to eigenvector $\boldsymbol{a}_\gamma$ of the ***reduced attenuation matrix*** *$\dot{\underset{\sim}{Q}}_{\alpha\beta} \equiv \boldsymbol{e}_\alpha\cdot\dot{\underset{\sim}{Q}}\,\boldsymbol{e}_\beta$, with $\boldsymbol{e}_\alpha$, $\boldsymbol{e}_\beta$ two orthonormal basis vectors in the principle plane of $\overset{\circ}{\underset{\sim}{Q}}$.*[61]

We note that in the case that ϱU^2 is a triple eigenvalue of $\overset{\circ}{\underset{\sim}{Q}}$, then any γ is also an eigenvector of $\overset{\circ}{\underset{\sim}{Q}}$, and all such waves are axially similar.

Proof of Theorem 3.

(i) If ϱU^2 is a distinct eigenvalue of $\overset{\circ}{\underset{\sim}{Q}}$ then there is only one unit vector $\boldsymbol{b}$ which is a wave axis with $\overset{\circ}{\underset{\sim}{Q}}\,\boldsymbol{b} - \varrho U^2 \boldsymbol{b} = \boldsymbol{0}$; hence, using the orthogonality con-

[59] FISHER [1965, *13*].
[60] We note that Eq. (41.31), for distinct ϱU^2, also provides an explicit form for the single decay parameter μ mentioned after Theorem 2 for axially similar waves.
[61] Here the indices α, β, and γ range over 1, 2.

dition (41.29) and writing $\boldsymbol{q}(t) = \beta(t)\,\boldsymbol{b}$, where β is the scalar magnitude of $\boldsymbol{q}$, gives (41.30) and (41.31).

(ii) If ϱU^2 is a double eigenvalue of $\overset{\circ}{\boldsymbol{Q}}$ corresponding to the principal plane $\mathscr{P}$, then every solution $\boldsymbol{u}$ of the homogeneous equation

$$\overset{\circ}{\boldsymbol{Q}}\,\boldsymbol{u} - \varrho U^2 \boldsymbol{u} = \boldsymbol{0} \tag{41.33}$$

may be written as

$$\boldsymbol{u} = u_\gamma \boldsymbol{e}_\gamma, \tag{41.34}$$

where $\boldsymbol{e}_1$, $\boldsymbol{e}_2$ denote orthonormal basis vectors of $\mathscr{P}$. Similarly $\boldsymbol{b}$ may be written as

$$\boldsymbol{b}(t) = b_\gamma(t)\,\boldsymbol{e}_\gamma. \tag{41.35}$$

Hence, the orthogonality condition (41.29) gives

$$\left\{\dot{\boldsymbol{Q}}\, b_\alpha \boldsymbol{e}_\alpha - 2\varrho U^2 \boldsymbol{e}_\alpha \frac{d}{dt}(b_\alpha)\right\} \cdot u_\gamma \boldsymbol{e}_\gamma = 0. \tag{41.36}$$

Therefore, if we define the ***reduced attenuation matrix*** $\dot{Q}_{\gamma\alpha}$:

$$\boldsymbol{e}_\gamma \cdot \dot{\boldsymbol{Q}}\, \boldsymbol{e}_\alpha \equiv \dot{Q}_{\gamma\alpha} \tag{41.37}$$

then

$$\left\{\dot{Q}_{\gamma\alpha}\, b_\alpha - 2\varrho\, U^2 \frac{d}{dt} b_\gamma\right\} u_\gamma = 0 \tag{41.38}$$

for every $\boldsymbol{u}$. Hence

$$\dot{Q}_{\alpha\beta}\, b_\alpha - 2\varrho\, U^2 \frac{d}{dt} b_\beta = 0, \tag{41.39}$$

and the result in (ii) follows.

We note that if ϱU^2 is a distinct eigenvalue of $\overset{\circ}{\boldsymbol{Q}}$, and if $\overset{\circ}{\boldsymbol{Q}}$ and $\dot{\boldsymbol{Q}}$ commute, then $\boldsymbol{q}$ is also an eigenvector of $\dot{\boldsymbol{Q}}$ and all $\boldsymbol{q}$ are hence axially similar. Further, if ϱU^2 is a double eigenvalue of $\overset{\circ}{\boldsymbol{Q}}$, and if $\overset{\circ}{\boldsymbol{Q}}$ and $\dot{\boldsymbol{Q}}$ commute, then we must require that the characteristic manifolds of $\overset{\circ}{\boldsymbol{Q}}$, which are by necessity invariant manifolds of $\dot{\boldsymbol{Q}}$, also be characteristic manifolds of $\dot{\boldsymbol{Q}}$ in order to assure that all such waves be axially similar. It is of practical interest to note that both the commutativity of $\overset{\circ}{\boldsymbol{Q}}$ and $\dot{\boldsymbol{Q}}$ and the stronger requirement of identical characteristic manifolds hold for isotropic materials with relaxation function given by Eq. (41.15). This leads immediately to

Theorem 4. *A plane wave of order N propagating in a homogenous, isotropic viscoelastic material is either*

(i) *longitudinal with*

$$\boldsymbol{q}(t) = b_0\, \boldsymbol{n}\, e^{\mu_L t}, \tag{41.40}$$

where

$$\mu_L = \frac{\overset{\circ}{G}{}_2^{(1)} + 2\overset{\circ}{G}{}_1^{(1)}}{2(\overset{\circ}{G}_2 + 2\overset{\circ}{G}_1)} \tag{41.41}$$

and $\boldsymbol{q}$ *and* γ *are linearly dependent; or*

(ii) *transverse with*

$$\boldsymbol{q}(t) = b_0\, \boldsymbol{\tau}\, e^{\mu_T t}, \tag{41.42}$$

where

$$\mu_T = \frac{\overset{\circ}{G}{}_1^{(1)}}{2\overset{\circ}{G}_1}, \tag{41.43}$$

$\boldsymbol{\tau}$ *is a vector normal to* $\boldsymbol{n}$, *and* $\boldsymbol{q}$ *and* γ *are both in the transverse plane normal to* $\boldsymbol{n}$.

From these simple cases one can derive the essential aspects of the time dependence of propagating plane singular surfaces in the linear theory. That is, the magnitude of the discontinuity is governed by the ratio of the initial slope to the initial value of the relevant relaxation function.

One might wonder about the time dependence of the induced jump γ. Basically nothing can be learned in this respect for the axially similar waves. For axially dissimilar waves one can give explicit formulae[62] for those components of γ which are not in the characteristic manifold of $\overset{\circ}{\underset{\sim}{Q}}$ for ϱU^2. Furthermore, if $\overset{\circ}{\underset{\sim}{Q}}$ and $\dot{\underset{\sim}{Q}}$ share the same characteristic manifolds, as in the isotropic case, then these components vanish identically.

42. Initial value problem. Uniqueness and existence of solutions. BARBERÁN and HERRERA[63] have formulated the initial value problem (Cauchy's problem) for dynamic viscoelasticity. They have proved a uniqueness theorem for three dimensional problems in the anisotropic theory and, based on HERRERA's[64] development of the Riemann representation method for viscoelasticity, they have proved existence and uniqueness theorems for the Cauchy problem in the one-dimensional theory.[65] We shall, in this section, summarize the results of some of this work. Without giving all the details of the development of BARBERÁN and HERRERA[66] we shall sketch enough of their formulation of the Cauchy initial value problem for dynamic viscoelasticity to enable us to state the uniqueness theorem of interest.

For some $T>0$ we have for each time $t\in[0,T]$ a compact region $\mathscr{R}(t)\subset\mathscr{E}\times\{t\}$, a copy of the three-dimensional Euclidean subspace embedded in four-dimensional space-time. Further, the compact region $V(t)$ may be defined through

$$V(t)=\bigcup_{\tau=0}^{\tau=t}\mathscr{R}(\tau) \tag{42.1}$$

where we associate $\mathscr{R}(0)$ with $\mathscr{B}\times\{0\}$ and adopt the following notation for the boundary ∂V of the hypervolume $V(t)$:

$$S_1(t)=\mathscr{R}(0), \tag{42.2a}$$

$$S_2(t)=\overline{\partial V(t)\cap[\mathscr{E}\times(0,t)]}, \tag{42.2b}$$

$$S_3(t)=\mathscr{R}(t). \tag{42.2c}$$

The four-dimensional unit normal $\boldsymbol{n}$ to $\partial V(t)$ is assumed to be defined and continuous on $S_2(T)$. By convention, the first three components of $\boldsymbol{n}$ are spatial and the fourth, n_4, is the time component. Hence,

$$\boldsymbol{n}=(0,0,0,-1) \quad \text{on } S_1, \tag{42.3a}$$

$$\boldsymbol{n}=(0,0,0,1) \quad \text{on } S_3. \tag{42.3b}$$

Our purpose in this section is to find those space-time volumes $V(t)$ for which there is at most one solution to a broad class of initial-value—initial-history problems. For this purpose we shall say that $V(T)$ is a *conoid* when $\mathscr{R}(t')\subset\mathscr{R}(t)$ for

[62] FISHER [1965, *13*].
[63] BARBERÁN and HERRERA [1966, *4*].
[64] HERRERA [1966, *9*].
[65] BARBERÁN and HERRERA [1967, *1*].
[66] BARBERÁN and HERRERA [1966, *4*].

all $t' > t$. Moreover we require that the time component of the normal to the "sides" of the conoid be strictly positive and less than unit magnitude for $(\boldsymbol{x}, t) \in S_2(T)$. We then say that $\boldsymbol{u}(\boldsymbol{x}, t)$ is a ***solution to the Cauchy initial value problem*** for dynamic viscoelasticity if, for the conoid $V(T)$, $\boldsymbol{u}$ is defined and continuous on $(\mathscr{B} \times \mathscr{R}^+) \cap V(T)$ and satisfies[67]

a) $\boldsymbol{u}(\boldsymbol{x}, t) \in C^2$ in $V(T)$;

b) the basic field equations (37.1)–(37.4) in $V(T)$;[68]

c) the initial past-history condition

$$ {}_r\boldsymbol{u}^0 = \boldsymbol{v} \quad \text{on } \mathscr{B} \times \mathscr{R}^+; $$

d) the initial conditions

$$ \left.\begin{aligned} \mathring{\boldsymbol{u}} &= \boldsymbol{a} \\ \mathring{\boldsymbol{u}}^{(1)} &= \boldsymbol{c} \end{aligned}\right\} \quad \text{on } \mathscr{B}. $$

Thus, if the past history $\boldsymbol{v}$ and fields $\boldsymbol{a}$ and $\boldsymbol{c}$ are prescribed independently, the displacement and velocity may have jump discontinuities at time $t = 0$. We are now in a position to state the main results of BARBERÁN and HERRERA:[69]

Uniqueness—Cauchy initial value problem. *If the initial elastic response $\mathring{\boldsymbol{G}}$ is symmetric and positive definite, and if the normal unit vector is given on S_2 by*

$$ \boldsymbol{n} = \lambda(e_1, e_2, e_3, c) $$

for the unit vector $\mathbf{e}$ *and normalizing factor* λ, *Cauchy's initial value problem has at most one solution in* $(\mathscr{B} \times \mathscr{R}^+) \cap V(T)$ *providing*

$$ c > U_3 \geqq U_2 \geqq U_1 > 0 $$

for the three, not necessarily distinct, wave speeds U_1, U_2, U_3 *satisfying the Fresnel-Hadamard propagation condition* (41.4).

The essence of this theorem is that if solutions exist to the Cauchy problem, then they are unique in a conoid $V(T)$ which, in the sense of the theorem, "shrinks" from $\mathscr{B} \times \{0\}$ at a rate faster than the fastest singular surface wave speed U_3.

The question of existence, in the generality of the above theorem (i.e., anisotropic and three space dimensions), has not been treated. However, HERRERA[70] and BARBERÁN and HERRERA[71] have developed and used the Riemann representation method in one space dimension for the dynamic theory of viscoelasticity. In that context they demonstrate both existence and uniqueness for the Cauchy problem.

Recently, DAFERMOS[72] has proved an existence, uniqueness, and asymptotic stability theorem for the dynamic initial past-history problem. These results are quite general, but here we mention only the version applicable to homogenous one-dimensional bodies (filaments).

---

[67] The definition is actually stated under certain mild smoothness conditions on ϱ, $\boldsymbol{u}$ and $\boldsymbol{G}$; see BARBERÁN and HERRERA [1966, *4*].

[68] Actually the proof of BARBERÁN and HERRERA does not require the relaxation law to be translation invariant and hence is more general than stated here.

[69] BARBERÁN and HERRERA [1966, *4*].

[70] HERRERA [1966, *9*].

[71] BARBERÁN and HERRERA [1967, *1*].

[72] DAFERMOS [1970, *1*].

Dafermos' theorem.[73] *For the dynamic initial past-history problem, existence and uniqueness are guaranteed provided that the strong (scalar) relaxation function* G *is initially positive and decreases monotonically on* $\mathscr{R}^+$ *to a positive equilibrium modulus. Furthermore, if* G *is concave upward on* $\mathscr{R}^+$, *the displacement decreases asymptotically to zero whatever be the initial past-history.*

43. Oscillatory displacement processes. Free vibrations. The subject of dynamic viscoelastic processes for which the displacement has sinusoidally varying time dependence is well treated in the literature. It is particularly important because of the availability of simple analytical expressions which can be easily correlated with experiment. Obviously the linearity of the system of equations gives added importance, through superposition and Fourier analysis, to the single frequency response of the material. The interested reader is referred to an excellent review by HUNTER[74] which covers this material, especially for one dimensional motion, in great depth and, in addition, cites useful experimental work.

In this section we shall confine our attention to developing some of the formalisms associated with the "complex modulus" formulation and deduce some results which relate the high frequency limits of oscillatory viscoelastic processes with the wavespeed and attenuation of the damped, propagating singular surfaces of Sect. 41.

If we restrict our attention to oscillatory displacement processes[75] over $\mathscr{B}$ at the real frequency ω,

$$\boldsymbol{u}(\boldsymbol{x},t)=\boldsymbol{a}(\boldsymbol{x})\,e^{i\omega t},\qquad(\boldsymbol{x},t)\in\mathscr{B}\times\mathscr{R},\tag{43.1}$$

we find that the constitutive relation for strong response functions G

$$\boldsymbol{S}=\mathscr{L}_{\mathsf{G}}[\boldsymbol{E}]=\mathsf{G}\circledast\boldsymbol{E}\tag{43.2}$$

becomes

$$\boldsymbol{S}(\boldsymbol{x},t)=\Big\{\mathring{\mathsf{G}}(\boldsymbol{x})+\int_0^\infty\dot{\mathsf{G}}(\boldsymbol{x},\tau)\,e^{-i\omega\tau}\,d\tau\Big\}\,\tfrac{1}{2}\big(\nabla\boldsymbol{a}(\boldsymbol{x})+\nabla\boldsymbol{a}(\boldsymbol{x})^T\big)\,e^{i\omega t},\qquad(\boldsymbol{x},t)\in\mathscr{B}\times\mathscr{R}.\tag{43.3}$$

Thus

$$\boldsymbol{S}(\boldsymbol{x},t)=\mathsf{C}^*(\boldsymbol{x},\omega)\,\boldsymbol{E}(\boldsymbol{x},t),\qquad(\boldsymbol{x},t)\in\mathscr{B}\times\mathscr{R},\tag{43.4}$$

where

$$\mathsf{C}^*(\boldsymbol{x},\omega)\equiv\mathring{\mathsf{G}}(\boldsymbol{x})+\int_0^\infty\dot{\mathsf{G}}(\boldsymbol{x},\tau)\,e^{-i\omega\tau}\,d\tau,\qquad(\boldsymbol{x},\omega)\in\mathscr{B}\times\mathscr{R}.\tag{43.5}$$

$\mathsf{C}^*(\boldsymbol{x},\omega)$ is the well known and experimentally useful ***complex modulus.*** We see that for motions of the form of Eq. (43.1) the linear viscoelastic constitutive relation is identical to the relation for linear elasticity except that the real elasticity tensor is replaced by a *complex valued, frequency dependent, fourth order tensor* C^*. We shall have occasion to refer to the real and imaginary parts of C^*; thus

$$\mathsf{C}^*=\mathsf{C}_1+i\,\mathsf{C}_2,\tag{43.6}$$

[73] DAFERMOS [1970, *1*, Theorems 2.1 and 3.1]. The conditions for the general case are Hilbert space generalizations of those given here. For example, the condition of positive initial response must be replaced by a uniform strong ellipticity condition. Compare with other aspects of viscoelastic behavior discussed in this article in Sects. 11, 13, 29, and 35.

[74] HUNTER [1960, *8*].

[75] In this section we shall be considering processes which are usually not of Heaviside type and shall consequently require the use of strong response functions. See Sects. 10 and 13 where such processes were considered.

where[76]

$$\mathbf{C}_1(\omega) = \overset{\circ}{\mathbf{G}} + \operatorname{Re} \int_0^\infty \dot{\mathbf{G}}(\tau)\, e^{-i\omega\tau}\, d\tau, \tag{43.7}$$

$$\mathbf{C}_2(\omega) = \operatorname{Im} \int_0^\infty \dot{\mathbf{G}}(\tau)\, e^{-i\omega\tau}\, d\tau. \tag{43.8}$$

If we further restrict our attention to ***plane travelling wave*** displacement fields propagating in the direction $\boldsymbol{n}$

$$\boldsymbol{a}(\boldsymbol{x}) = \boldsymbol{p}\, e^{-ik(\boldsymbol{x}\cdot\boldsymbol{n})}, \qquad \boldsymbol{x} \in \mathscr{B} \tag{43.9}$$

then

$$\boldsymbol{u}(\boldsymbol{x}, t) = \boldsymbol{p}\, e^{i(\omega t - k(\boldsymbol{x}\cdot\boldsymbol{n}))}, \qquad (\boldsymbol{x}, t) \in \mathscr{B} \times \mathscr{R} \tag{43.10}$$

where k is a ***complex wave number***

$$k \equiv \frac{\omega}{c} - i\alpha \tag{43.11}$$

and

$$\boldsymbol{p} = \boldsymbol{p}_1 + i\,\boldsymbol{p}_2 \tag{43.12}$$

is the complex valued "polarization vector" for which $\boldsymbol{p}_1$ and $\boldsymbol{p}_2$ are independent of $\boldsymbol{x}$ and t. Following the usual notation, $c(\omega)$ is the ***phase velocity*** and $\alpha(\omega)$ is the ***attenuation*** at frequency ω. It should be noted that in the elastic case we can find travelling plane waves for which there is no attenuation ($\alpha = 0$), no intrinsic phase velocity dispersion ($dc/d\omega = 0$), and no need to consider complex polarization $\boldsymbol{p}$. It is worth recalling that phase velocity dispersion in a general wave propagation problem in a viscoelastic material includes not only effects created by physical boundaries (geometric dispersion) but in addition includes inherent dependence of phase velocity on frequency resulting from the frequency dependence of the complex modulus $\mathbf{C}^*$.[77]

To examine this problem briefly we look for solutions of the form of Eq. (43.10) for which the displacement equation of motion

$$\operatorname{div}\{\mathbf{C}^*[\nabla\boldsymbol{a} + \nabla\boldsymbol{a}^T]\} = -2\varrho\,\omega^2\,\boldsymbol{a} \tag{43.13}$$

yields the complex propagation condition for a homogeneous body:

$$\mathbf{C}^*[(\boldsymbol{p}\otimes\boldsymbol{n})]\,\boldsymbol{n} - \frac{\varrho\omega^2}{k^2}\,\boldsymbol{p} = \boldsymbol{0}. \tag{43.14}$$

For brevity we shall define the frequency dependent linear operator $\boldsymbol{Q}^*(\boldsymbol{n})$, the ***complex acoustic tensor***, which is defined on the *complex* three-dimensional Euclidean vector space $\mathscr{V}^*$ for each unit vector $\boldsymbol{n}$ by

$$\boldsymbol{Q}^*(\boldsymbol{n})\,\boldsymbol{a} = \mathbf{C}^*[(\boldsymbol{a}\otimes\boldsymbol{n})]\,\boldsymbol{n}, \qquad \boldsymbol{a} \in \mathscr{V}^*. \tag{43.15}$$

---

[76] Henceforth, we shall suppress the dependence of the complex modulus upon position unless explicitly needed. The integral in (43.5) defines the half-range Fourier transform $\hat{\dot{\mathbf{G}}}$ of $\dot{\mathbf{G}}$. Thus, $\mathbf{C}_1$ and $\mathbf{C}_2$, defined through $\mathbf{C}^*$ by (43.6)–(43.8), have the form:

$$\mathbf{C}_1 = \overset{\circ}{\mathbf{G}} + \hat{\dot{\mathbf{G}}}_c$$

$$\mathbf{C}_2 = \hat{\dot{\mathbf{G}}}_s$$

in terms of the half-range Fourier cosine and sine transforms $\hat{\dot{\mathbf{G}}}_c$ and $\hat{\dot{\mathbf{G}}}_s$. Cf. Sect. 10, Eqs. (10.24) or Sect. 13, Eqs. (13.16).

[77] Cf. Sect. 10. Also see the reviews of HUNTER [1960, *8*], KOLSKY [1963, *10*] and FERRY [1970, *4*], for excellent references to the related experimental literature.

Hence, suppressing the dependence upon $\boldsymbol{n}$, we rewrite Eq. (43.14) as

$$\boldsymbol{Q}^* \boldsymbol{p} - \lambda \boldsymbol{p} = \boldsymbol{0}, \tag{43.16}$$

where the frequency dependent function λ is given by[78]

$$\lambda = \frac{\varrho \omega^2}{k^2}. \tag{43.17}$$

We are now in a position to find the complex polarization vectors $\boldsymbol{p}$ and the associated function $\lambda(\omega)$. We note that, depending on the direction $\boldsymbol{n}$ and the symmetry of $\boldsymbol{Q}^*$, there may be up to three of these complex functions $\lambda(\omega)$, each giving, in general, different frequency dependence for the attenuation and phase velocity.

As an important example we can look at an isotropic viscoelastic material, for which

$$\varrho^{-1} \boldsymbol{Q}^*(\boldsymbol{n})\, \boldsymbol{a} = \varrho^{-1} \mathbf{C}^*[(\boldsymbol{a} \otimes \boldsymbol{n})]\, \boldsymbol{n} \tag{43.18}$$

$$= \left\{\frac{G_2^* + 2G_1^*}{3\varrho}\right\} (\boldsymbol{n} \otimes \boldsymbol{n})\, \boldsymbol{a} + \left\{\frac{G_1^*}{2\varrho}\right\} \{\mathbf{1} - (\boldsymbol{n} \otimes \boldsymbol{n})\}\, \boldsymbol{a} \tag{43.19}$$

$$= K_1^* (\boldsymbol{n} \otimes \boldsymbol{n})\, \boldsymbol{a} + K_2^* \{\mathbf{1} - (\boldsymbol{n} \otimes \boldsymbol{n})\}\, \boldsymbol{a}, \quad \boldsymbol{a} \in \mathscr{V}^*, \tag{43.20}$$

where K_1 and K_2 are given by (37.8), (37.9), and the star (*) notation denotes the transform operation in (37.5): for any (strong) scalar process M in H^{ac}

$$M^*(\omega) = \overset{\circ}{M} + \int_0^\infty \dot{M}(s)\, e^{-i\omega s} ds = \overset{\circ}{M} + \hat{\dot{M}}(\omega). \tag{43.21}$$

We can easily show that for the isotropic case there are two distinct functions $\lambda(\omega)$ in (43.16) such that

(i) $$\lambda(\omega) = \frac{G_1^*(\omega)}{2} \tag{43.22}$$

if and only if the wave is transverse; i.e., $\boldsymbol{p} \cdot \boldsymbol{n} = 0$.

(ii) $$\lambda(\omega) = \tfrac{1}{3}\left(G_2^*(\omega) + 2G_1^*(\omega)\right) \tag{43.23}$$

if and only if the wave is longitudinal; i.e., $\boldsymbol{p} \times \boldsymbol{n} = \boldsymbol{0}$.[79]

Thus, for the isotropic symmetry, we have (recalling that $\lambda(\omega) = \varrho\, \omega^2/((\omega/c) - i\,\alpha))$ a simple way of calculating the phase velocity dispersion and the frequency dependence of the attenuation if we know the two isotropic scalar response functions G_1 and G_2.

Returning to the general problem, we remark that the vectors $\boldsymbol{p}$ of the complex propagation condition (43.14) corresponding to the function λ will, in general, be complex, thus allowing solutions for which the direction of $\boldsymbol{u}$ varies with time in the plane determined by $\boldsymbol{p}_1$ and $\boldsymbol{p}_2$. We shall see in succeeding paragraphs that the assumption that $\boldsymbol{p}$ in Eq. (43.10) be real from the outset (i.e., one then seeks "linearly polarized" waves), as is quite appropriate in linear elasticity, is in fact a restriction on the operator $\boldsymbol{Q}^*$ in linear viscoelasticity. Furthermore, the assumption that $\boldsymbol{Q}^*$ be self adjoint is certainly seen to be inappropriate in the viscoelastic case as this would immediately imply that λ is real and hence $\alpha = 0$ (i.e., no attenuation).

---

[78] In linear elasticity the corresponding relation for the real acoustic tensor and real vector $\boldsymbol{p}$ is $\lambda = \varrho c^2$, and thus c is independent of frequency.

[79] In these expressions we formally consider $\boldsymbol{n}$ to be a complex valued vector with zero imaginary part.

We decompose λ and $\boldsymbol{Q}^*$ into real and imaginary parts[80]

$$\boldsymbol{Q}^* = \boldsymbol{Q}_1 + i\boldsymbol{Q}_2, \tag{43.24}$$

$$\lambda = \lambda_1 + i\lambda_2 \tag{43.25}$$

defined in the natural way through Eqs. (43.6), (43.15) and (43.17). Then we see that the real and imaginary parts of (43.16) yield

$$(\boldsymbol{Q}_1 - \lambda_1 \mathbf{1})\,\boldsymbol{p}_1 - (\boldsymbol{Q}_2 - \lambda_2 \mathbf{1})\,\boldsymbol{p}_2 = \mathbf{0}, \tag{43.26a}$$

$$(\boldsymbol{Q}_1 - \lambda_1 \mathbf{1})\,\boldsymbol{p}_2 + (\boldsymbol{Q}_2 - \lambda_2 \mathbf{1})\,\boldsymbol{p}_1 = \mathbf{0} \tag{43.26b}$$

where explicitly

$$\lambda_1(\omega) = \frac{\varrho\,\omega^2 c^2 (\omega^2 - \alpha^2 c^2)}{(\omega^2 + \alpha^2 c^2)^2}, \tag{43.27a}$$

$$\lambda_2(\omega) = \frac{2\varrho\,\omega^2 c^2 (\omega c \alpha)}{(\omega^2 + \alpha^2 c^2)^2}. \tag{43.27b}$$

For conciseness we shall define the linear operators $\boldsymbol{L}_1$ and $\boldsymbol{L}_2$ by

$$\boldsymbol{L}_1 \equiv \boldsymbol{Q}_1 - \lambda_1 \mathbf{1}, \tag{43.28a}$$

$$\boldsymbol{L}_2 \equiv \boldsymbol{Q}_2 - \lambda_2 \mathbf{1}; \tag{43.28b}$$

hence, (43.26) may be written as

$$\boldsymbol{L}_1 \boldsymbol{p}_1 - \boldsymbol{L}_2 \boldsymbol{p}_2 = \mathbf{0}, \tag{43.29a}$$

$$\boldsymbol{L}_1 \boldsymbol{p}_2 + \boldsymbol{L}_2 \boldsymbol{p}_1 = \mathbf{0}. \tag{43.29b}$$

Thus, we see that the assumption of linearly polarized solutions (i.e., $\boldsymbol{p}_2 = \mathbf{0}$) implies that $\boldsymbol{p}_1$ must be a homogeneous solution of both $\boldsymbol{L}_1$ and $\boldsymbol{L}_2$. Since this amounts to a symmetry constraint on the real and imaginary parts of the complex modulus, we prefer to consider the less restrictive case with $\boldsymbol{p}_2 \neq \mathbf{0}$.

We now proceed to demonstrate the relation between the high frequency behavior for plane traveling waves of Eq. (43.10) and the speed and decay of propagating singular surfaces of Sect. 41. From classical Fourier transform theory we note that when c and α are finite as $\omega \to \infty$

$$\lim_{\omega\to\infty} \boldsymbol{Q}_2(\omega) = \mathbf{0}, \tag{43.30}$$

$$\lim_{\omega\to\infty} \lambda_2(\omega) = 0, \tag{43.31}$$

$$\lim_{\omega\to\infty} \boldsymbol{Q}_1(\omega) = \mathring{\boldsymbol{Q}}, \tag{43.32}$$

$$\lim_{\omega\to\infty} \lambda_1(\omega) = \varrho\, c_\infty^2, \tag{43.33}$$

where, for each unit vector $\boldsymbol{n}$, $\varrho^{-1}\mathring{\boldsymbol{Q}} \equiv \varrho^{-1}\mathring{\boldsymbol{Q}}(\boldsymbol{n})$ is the acoustic tensor defined on $\mathscr{V}$ by

$$\mathring{\boldsymbol{Q}}(\boldsymbol{n})\,\boldsymbol{a} = \mathring{\boldsymbol{G}}[(\boldsymbol{a}\otimes\boldsymbol{n})]\,\boldsymbol{n}, \qquad \boldsymbol{a}\in\mathscr{V}, \tag{43.34}$$

and $c_\infty \equiv \lim_{\omega\to\infty} c(\omega)$. Then Eqs. (43.26) become in the limit as $\omega\to\infty$

$$\mathring{\boldsymbol{G}}(\boldsymbol{p}_1 \otimes \boldsymbol{n})\,\boldsymbol{n} - \varrho\, c_\infty^2\, \boldsymbol{p}_1 = \mathbf{0}, \tag{43.35a}$$

$$\mathring{\boldsymbol{G}}(\boldsymbol{p}_2 \otimes \boldsymbol{n})\,\boldsymbol{n} - \varrho\, c_\infty^2\, \boldsymbol{p}_2 = \mathbf{0}. \tag{43.35b}$$

[80] This is merely a decomposition of $\boldsymbol{Q}^*$, component by component, into real and imaginary parts giving the two real linear operators $\boldsymbol{Q}_1$ and $\boldsymbol{Q}_2$.

Thus, we see that in the infinite frequency limit both $\boldsymbol{p}_1$ and $\boldsymbol{p}_2$ are solutions of the same Fresnel-Hadamard propagation equation. Further, if we allow $\overset{\circ}{\mathbf{G}}$ to be symmetric, we see that if $\varrho\, c_\infty^2$ is a distinct eigenvalue of $\overset{\circ}{\boldsymbol{Q}}$, then $\boldsymbol{p}_1$ and $\boldsymbol{p}_2$ are in the same direction. Also, if $\varrho\, c_\infty^2$ is a double eigenvalue, then $\boldsymbol{p}_1$ and $\boldsymbol{p}_2$ are in the associated principal plane of $\overset{\circ}{\boldsymbol{Q}}$. From these remarks and Eqs. (43.35) we have the following

Equivalence principle[81] ***for wave speeds and wave axes between infinite frequency damped progressive waves and propagating singular surfaces.*** *If U and $\boldsymbol{b}$ are the wave speed and wave axis of a propagating singular surface in a linear viscoelastic homogeneous medium, then $U = c_\infty$ and both $\boldsymbol{p}_1$ and $\boldsymbol{p}_2$ are wave axes.*

It is this statement which leads to the frequent assertion that in the infinite frequency limit, damped progressive waves behave as propagating singular surfaces. In this case, the associated wave speeds are identical and dictated by the elastic response of the material; i.e., by the "acoustic tensor" $\overset{\circ}{\boldsymbol{Q}}$ given by (43.34).

We now wish to compare the high frequency limit of the attenuation (i.e., $\alpha_\infty \equiv \lim_{\omega\to\infty} \alpha(\omega)$) with the decay of the strength of viscoelastic waves (propagating singular surfaces). If we take the inner product of $\boldsymbol{p}_2$ with (43.29a) and subtract the inner product of $\boldsymbol{p}_1$ with (43.29b), we have

$$\boldsymbol{p}_2 \cdot \boldsymbol{L}_2 \boldsymbol{p}_2 + \boldsymbol{p}_1 \cdot \boldsymbol{L}_2 \boldsymbol{p}_1 = 0, \tag{43.36}$$

where we have assumed that $\boldsymbol{L}_1$ is symmetric:

$$\boldsymbol{p}_2 \cdot \boldsymbol{L}_1 \boldsymbol{p}_1 = \boldsymbol{p}_1 \cdot \boldsymbol{L}_1 \boldsymbol{p}_2. \tag{43.37}$$

If we multiply (43.36) by ω, we note from classical Fourier transform theory[82] that

$$\lim_{\omega\to\infty} \omega\, \boldsymbol{Q}_2(\omega)\, \boldsymbol{a} = -\,\dot{\mathbf{G}}(0)\,[(\boldsymbol{a}\otimes\boldsymbol{n})]\,\boldsymbol{n} = -\,\dot{\boldsymbol{Q}}(\boldsymbol{n})\,\boldsymbol{a}, \qquad \boldsymbol{a}\in\mathscr{V} \tag{43.38}$$

(i.e., this limit gives the attenuation operator $\dot{\boldsymbol{Q}}$, familiar in the discussion of singular surfaces).[83] Furthermore,

$$\lim_{\omega\to\infty} \omega\, \boldsymbol{L}_2(\omega) = \dot{\boldsymbol{Q}} - 2\varrho\, c^2 (\alpha_\infty\, c_\infty)\, \mathbf{1}. \tag{43.39}$$

From Eqs. (43.35), (43.36), and (43.39) we thus see that if we let $\boldsymbol{v}$ be any vector in the direction of $(\boldsymbol{p}_1 + \boldsymbol{p}_2)$, then

$$\alpha_\infty\, c_\infty = \frac{\boldsymbol{v}\cdot\dot{\boldsymbol{Q}}\boldsymbol{v}}{2\boldsymbol{v}\cdot\overset{\circ}{\boldsymbol{Q}}\boldsymbol{v}} = \frac{-\boldsymbol{v}\cdot\dot{\mathbf{G}}(0)\,[(\boldsymbol{v}\otimes\boldsymbol{n})]\boldsymbol{n}}{2\boldsymbol{v}\cdot\overset{\circ}{\mathbf{G}}[(\boldsymbol{v}\otimes\boldsymbol{n})]\boldsymbol{n}}. \tag{43.40}$$

From Eq. (43.40) and Theorem 3 of Sect. 41 we can now state a form of the

Equivalence principle for attenuation between infinite frequency damped progressive waves and propagating singular surfaces: *If U is a distinct wave speed of a plane singular surface propagating in a homogeneous*

---

[81] This is a slight modification of Remark 7.1 of COLEMAN and GURTIN [1965, *7*], which relates closely to TRUESDELL's "Second Theorem of Equivalence for Elastic Materials," TRUESDELL and TOUPIN [1960, *15*].

[82] Formally it suffices to assume that $\dot{\mathbf{G}}$ and $\ddot{\mathbf{G}}$ are both continuous and integrable on $\mathscr{R}^+$.

[83] Cf. Sect. 41.

viscoelastic medium with symmetric acoustic tensor $\varrho^{-1}\overset{\circ}{\boldsymbol{Q}}$ *in the direction* $\boldsymbol{n}$ *with fixed wave axis* $\boldsymbol{b}$, *then*

$$\boldsymbol{q}(t) = \boldsymbol{b}\, e^{\mu t}, \tag{43.41}$$

where μ *is identical to the infinite frequency limit of* $\alpha(\omega)\, c(\omega)$ *for a damped progressive wave;* i.e.,

$$\mu = \alpha_\infty\, c_\infty \tag{43.42}$$

if we associate $\boldsymbol{p}_1 + \boldsymbol{p}_2$ *(in this case they are collinear) with the wave axis* $\boldsymbol{b}$.

A similar statement results for the case where $\boldsymbol{b}$ is an eigenvector of $\dot{\boldsymbol{Q}}$ as well as $\overset{\circ}{\boldsymbol{Q}}$.

We conclude this section on oscillatory processes with the statement of a correspondence principle[84] which relates the solutions of a wide class of isotropic viscoelastic vibration problems with their elastic counterparts.

We have seen that for the isotropic material there are two distinct "dispersion" relations $\lambda(\omega)$ depending on whether the wave is transverse or longitudinal. General motions are, of course, considerably more complex. But when a particular displacement process is of one of these simple types, or more generally, when the motion is governed by a single response function, then there is a direct correspondence between the mode shapes and those which may occur in an "associated elastic problem." To analyze this notion precisely we consider the isotropic displacement equation of motion (38.1 b).

$$\mathscr{L}_{K_1}[\nabla\nabla\cdot\boldsymbol{u}] - \mathscr{L}_{K_2}[\nabla\times\nabla\times\boldsymbol{u}] = \ddot{\boldsymbol{u}}. \tag{43.43}$$

As usual, we restrict K_1 and K_2 through

$$\overset{\circ}{G}_1 > 0, \qquad \overset{\circ}{G}_2 > 0. \tag{43.44}$$

We shall need the following

Lemma. *Every free vibration* $\boldsymbol{u} = \boldsymbol{f}(\boldsymbol{x})\, e^{i\omega t}$ *of a homogeneous, isotropic viscoelastic solid has the form*

$$\boldsymbol{u}(\boldsymbol{x}, t) = \boldsymbol{u}_1(\boldsymbol{x}, t) + \boldsymbol{u}_2(\boldsymbol{x}, t), \tag{43.45}$$

where

$$\boldsymbol{u}_1(\boldsymbol{x}, t) = \boldsymbol{f}_1(\boldsymbol{x})\, e^{i\omega t}, \qquad \nabla\times\boldsymbol{f}_1 = \boldsymbol{0}, \tag{43.46a}$$

$$\boldsymbol{u}_2(\boldsymbol{x}, t) = \boldsymbol{f}_2(\boldsymbol{x})\, e^{i\omega t}, \qquad \nabla\cdot\boldsymbol{f}_2 = 0. \tag{43.46b}$$

Moreover,

$$\nabla^2\boldsymbol{f}_1 + \left(\frac{\omega^2}{K_1^*(\omega)}\right)\boldsymbol{f}_1 = \boldsymbol{0}, \tag{43.47a}$$

$$\nabla^2\boldsymbol{f}_2 + \left(\frac{\omega^2}{K_2^*(\omega)}\right)\boldsymbol{f}_2 = \boldsymbol{0}, \tag{43.47b}$$

where we now allow ω *to be a complex valued frequency.*

The proof of this is immediate if we define $\boldsymbol{f}_1$ and $\boldsymbol{f}_2$ through

$$\boldsymbol{f}_1(\boldsymbol{x}) = -\left(\frac{K_1^*}{\omega^2}\right)\nabla(\nabla\cdot\boldsymbol{f})(\boldsymbol{x}), \tag{43.48a}$$

$$\boldsymbol{f}_2(\boldsymbol{x}) = \left(\frac{K_2^*}{\omega^2}\right)\nabla\times(\nabla\times\boldsymbol{f})(\boldsymbol{x}), \tag{43.48b}$$

[84] Fisher and Leitman [1966, *8*].

with the complex isotropic moduli K_i^* given by the transform operation (43.21) with K_i in place of M:

$$K_i^*(\omega) \equiv \overset{\circ}{K}_i + \int_0^\infty \dot{K}_i(\tau)\, e^{-i\omega\tau}\, d\tau \equiv \overset{\circ}{K}_i + \hat{\dot{K}}_i(\omega) \qquad (i=1,2). \tag{43.49}$$

Let the ***associated elastic material*** be that for which $K_1 = h\overset{\circ}{K}_1$ and $K_2 = h\overset{\circ}{K}_2$. Then, if we call a vector field ***simple*** whenever its divergence or its curl vanishes, we can state the following

Correspondence principle for simple free vibrations. *The displacement field $\boldsymbol{f}(\boldsymbol{x})\, e^{i\omega_v t}$ describes a simple free vibration of a homogeneous, isotropic viscoelastic solid if, and only if, $\boldsymbol{f}(\boldsymbol{x})\, e^{i\omega_e t}$ describes a simple free vibration of the associated elastic material, and the frequencies ω_v and ω_e are related by*

$$\left(\frac{\omega_v}{\omega_e}\right)^2 = 1 + \frac{\hat{\dot{K}}_i(\omega_v)}{\overset{\circ}{K}_i}, \tag{43.50}$$

where

$$K_i = K_1 \quad \text{for} \quad \nabla\times\boldsymbol{f} = \boldsymbol{0}; \qquad K_i = K_2 \quad \text{for} \quad \nabla\cdot\boldsymbol{f} = 0. \tag{43.51}$$

In other words: For these simple motions, the viscoelastic mode shapes $\boldsymbol{f}$ are the same as those for the associated elastic case provided the frequencies are related through Eq. (43.50). The proof relies upon the fact that the operator $-\nabla^2$ is selfadjoint and positive definite when taken together with its natural boundary conditions.

One can prove similar correspondence principles for motions, such as those in the classical one-dimensional theory, which are governed by only one material response function. *For nonsimple motions, however, no such correspondence exists in general.*[85]

44. Dynamic variational principles. Variational principles in the dynamic linear theory of viscoelasticity may be deduced in much the same way as they are in linear elastodynamics.[86] The definition of the *variation of a functional* over some subset of a linear space has been given in Sect. 30 (Quasi-static variational principles).

Throughout this section, $[\boldsymbol{v}, \boldsymbol{R}, \boldsymbol{T}]$ denotes a fixed initial past history for $\mathscr{B}$ which is always assumed to be consistent with $[\hat{\boldsymbol{u}}, \hat{\mathbf{s}}]$, $\boldsymbol{b}$, ϱ, and either $\boldsymbol{G}$ or $\boldsymbol{J}$. Furthermore we assume that the limits

$$\lim_{s\downarrow 0} \boldsymbol{v}^{(0)}(s) = \boldsymbol{a} \tag{44.1}$$

and

$$\lim_{s\downarrow 0} \boldsymbol{v}^{(1)}(s) = \boldsymbol{c} \tag{44.2}$$

exist and define continuous vector fields $\boldsymbol{a}$ and $\boldsymbol{c}$ on $\mathscr{B}$. Finally, and most importantly, we assume that the relaxation function $\boldsymbol{G}$ and the creep compliance $\boldsymbol{J}$ are symmetric (in $[\mathscr{T}_{\text{sym}}]$).

---

[85] FISHER and LEITMAN [1966, *8*].

[86] For a discussion of variational principles in linear elastodynamics see LTE, Subchap. E.III. The extension to linear dynamic viscoelasticity, presented here, is due to LEITMAN [1966, *16*].

The Boltzmann laws (36.3) and (36.4) may be written in the form

$$\boldsymbol{S}(t) = \frac{d}{dt}(\mathbf{G} * \boldsymbol{E})(t) + \int_{-\infty}^{0} \dot{\mathbf{G}}(t-s)\,\boldsymbol{E}(s)\,ds, \tag{44.3a}$$

$$\boldsymbol{E}(t) = \frac{d}{dt}(\mathbf{J} * \boldsymbol{S})(t) + \int_{-\infty}^{0} \dot{\mathbf{J}}(t-s)\,\boldsymbol{S}(s)\,ds, \tag{44.3b}$$

for times $t \geqq 0$. Thus, if ${}_r\boldsymbol{E}^0 = \boldsymbol{R}$ and ${}_r\boldsymbol{S}^0 = \boldsymbol{T}$ and (44.3) holds, it follows that for times $t \geqq 0$

$$\boldsymbol{S}(t) = \frac{d}{dt}(\mathbf{G} * \boldsymbol{E})(t) + \boldsymbol{S}^+(t), \tag{44.4a}$$

$$\boldsymbol{E}(t) = \frac{d}{dt}(\mathbf{J} * \boldsymbol{S})(t) + \boldsymbol{E}^+(t), \tag{44.4b}$$

where $\boldsymbol{S}^+$ and $\boldsymbol{E}^+$ are defined on $\mathscr{R}^+$ by

$$\boldsymbol{S}^+(t) = \int_0^\infty \dot{\mathbf{G}}(\xi + t)\,\boldsymbol{R}(\xi)\,d\xi, \qquad t \in \mathscr{R}^+, \tag{44.5a}$$

$$\boldsymbol{E}^+(t) = \int_0^\infty \dot{\mathbf{J}}(\xi + t)\,\boldsymbol{T}(\xi)\,d\xi, \qquad t \in \mathscr{R}^+. \tag{44.5b}$$

Let $\mathscr{A}$ denote the space of all continuous admissible processes $[\boldsymbol{u}, \boldsymbol{E}, \boldsymbol{S}]$ for $\mathscr{B}$ which satisfy the conditions (R-1) and (R-2) of Sect. 36. For each time $t \geqq 0$ define the functional Λ_t on $\mathscr{A}$ by

$$\begin{aligned}
\Lambda_t\{[\boldsymbol{u}, \boldsymbol{E}, \boldsymbol{S}]\} = {} & \tfrac{1}{2}\int_{\mathscr{B}} (h * \boldsymbol{E} * \mathbf{G} * \boldsymbol{E})(t)\,dV \\
& + \tfrac{1}{2}\int_{\mathscr{B}} \varrho\,(\boldsymbol{u} * \boldsymbol{u})(t)\,dV \\
& - \int_{\mathscr{B}} \big(g * (\boldsymbol{S} - \boldsymbol{S}^+) * \boldsymbol{E}\big)(t)\,dV \\
& - \int_{\mathscr{B}} \big((g * \operatorname{div} \boldsymbol{S} + \boldsymbol{f}) * \boldsymbol{u}\big)(t)\,dV \\
& + \int_{\mathscr{S}_{\hat{\mathbf{s}}}} (g * \mathbf{s} * \hat{\boldsymbol{u}})(t)\,dA \\
& + \int_{\mathscr{S}_{\hat{\boldsymbol{u}}}} \big(g * (\mathbf{s} - \hat{\mathbf{s}}) * \boldsymbol{u}\big)(t)\,dA,
\end{aligned} \tag{44.6}$$

where h, g, and $\boldsymbol{f}$ are defined through (36.9)–(36.12) and $\boldsymbol{S}^+$ is defined by (44.5a).

Variational principle for dynamic relaxation problems.[87] *Let $\mathscr{K}$ be the subset of $\mathscr{A}$ consisting of processes which have the initial past history $[\boldsymbol{v}, \boldsymbol{R}, \boldsymbol{T}]$. Then*

$$\delta\Lambda_t\{[\boldsymbol{u}, \boldsymbol{E}, \boldsymbol{S}]\} = 0 \qquad \textit{over } \mathscr{K} \tag{44.7}$$

for each time $t \geqq 0$ if, and only if, $[\boldsymbol{u}, \boldsymbol{E}, \boldsymbol{S}] \in \mathscr{K}$ is a solution to the dynamic boundary value problem corresponding to $[\hat{\boldsymbol{u}}, \hat{\mathbf{s}}]$, $\boldsymbol{b}$, ϱ, and symmetric $\mathbf{G}$.

Let $\tilde{\mathscr{A}}$ be the space of all continuous admissible processes for $\mathscr{B}$ which satisfy the conditions (C-1), (C-2), and (C-3) of Sect. 36. For each time $t \geqq 0$ define

[87] Leitman [1966, *16*, Theorem 4.1].

the functional θ_t on $\tilde{\mathscr{A}}$ by

$$\begin{aligned}\theta_t\{[\boldsymbol{u}, \boldsymbol{E}, \boldsymbol{S}]\} = & \int_{\mathscr{B}} \left(g * \boldsymbol{S} * (\boldsymbol{E} - \boldsymbol{E}^+)\right)(t)\, dV \\ & - \tfrac{1}{2}\int_{\mathscr{B}} (h * \boldsymbol{S} * \boldsymbol{J} * \boldsymbol{S})(t)\, dV \\ & + \tfrac{1}{2}\int_{\mathscr{B}} \varrho (\boldsymbol{u} * \boldsymbol{u})(t)\, dV \\ & - \int_{\mathscr{B}} (\boldsymbol{f} * \boldsymbol{u})(t)\, dV \\ & - \int_{\mathscr{S}_{\hat{\boldsymbol{u}}}} \left(g * \boldsymbol{S} * (\boldsymbol{u} - \hat{\boldsymbol{u}})\right)(t)\, dA \\ & - \int_{\mathscr{S}_{\hat{\mathbf{s}}}} (g * \hat{\mathbf{s}} * \boldsymbol{u})(t)\, dA,\end{aligned} \tag{44.8}$$

where h, g, and $\boldsymbol{f}$ are defined in (36.9)–(36.12) and $\boldsymbol{E}^+$ is defined in (44.5b).

Variational principle for dynamic creep problems.[88] *Let $\tilde{\mathscr{K}}$ be the subset of $\tilde{\mathscr{A}}$ consisting of all processes which satisfy the strain-displacement relation* (36.2) *and which have the initial past history $[\boldsymbol{v}, \boldsymbol{R}, \boldsymbol{T}]$. Then*

$$\delta\theta_t\{[\boldsymbol{u}, \boldsymbol{E}, \boldsymbol{S}]\} = 0 \qquad \textit{over } \tilde{\mathscr{K}} \tag{44.9}$$

for each time $t \geqq 0$ if, and only if, $[\boldsymbol{u}, \boldsymbol{E}, \boldsymbol{S}] \in \tilde{\mathscr{K}}$ is a solution to the dynamic boundary value problem corresponding to $[\hat{\boldsymbol{u}}, \hat{\mathbf{s}}]$ $\boldsymbol{b}$, ϱ, and symmetric $\boldsymbol{J}$.

Let $\mathscr{F}$ denote the space of all continuous admissible stress processes $\boldsymbol{S}$ for $\mathscr{B}$ such that the values of $\boldsymbol{S}$ are class C^2 symmetric-tensor fields on $\overline{\mathscr{B}}$. Define a functional Γ_t on $\mathscr{F}$ for each time $t \geqq 0$ by

$$\begin{aligned}\Gamma_t\{\boldsymbol{S}\} = & \tfrac{1}{2}\int_{\mathscr{B}} (g * \operatorname{div}\boldsymbol{S} * \operatorname{div}\boldsymbol{S})(t)\, dV \\ & + \frac{1}{2}\int_{\mathscr{B}} \varrho \frac{d}{dt}(\boldsymbol{S} * \boldsymbol{J} * \boldsymbol{S})(t)\, dV \\ & - \int_{\mathscr{B}} \left((\operatorname{sym} \nabla\boldsymbol{f} - \varrho\, \boldsymbol{E}^+) * \boldsymbol{S}\right)(t)\, dV \\ & + \int_{\mathscr{S}_{\hat{\boldsymbol{u}}}} \left((\boldsymbol{f} - \varrho\, \hat{\boldsymbol{u}}) * \mathbf{s}\right)(t)\, dA \\ & + \int_{\mathscr{S}_{\hat{\mathbf{s}}}} \left(g * (\hat{\mathbf{s}} - \mathbf{s}) * \operatorname{div}\boldsymbol{S}\right)(t)\, dA.\end{aligned} \tag{44.10}$$

Variational principle for dynamic creep problems in terms of stresses.[89] *Let $\mathscr{K}$ be the subset of all stress processes $\boldsymbol{S}$ in $\mathscr{F}$ which satisfy the initial past-history condition*

$$_r\boldsymbol{S}^0 = \boldsymbol{T} \tag{44.11}$$

on $\mathscr{B}$. Then

$$\delta\Gamma_t\{\boldsymbol{S}\} = 0 \qquad \textit{over } \mathscr{K} \tag{44.12}$$

for all times $t \geqq 0$ if, and only if, $\boldsymbol{S}$ is the stress process of a solution of the dynamic boundary value problem corresponding to $[\hat{\boldsymbol{u}}, \hat{\mathbf{s}}]$, $\boldsymbol{b}$, ϱ, and symmetric $\boldsymbol{J}$.

---

[88] LEITMAN [1966, *16*, Theorem 4.2].

[89] LEITMAN [1966, *16*, Theorem 5.1].

References.

This list contains mainly works cited in the article. For each of these, the number of the section in which it is cited is given in parentheses at the end of the reference. The list includes also a few other articles and books not specifically mentioned in the text but of relevance to the subject matter discussed.

1855 *1.* SAINT-VENANT, A.-J.-C. B. DE: Mémoire sur la torsion de prismes (1853). Mém. Divers Savants Acad. Sci. Paris **14**, 233–560. (*35*)

1868 *1.* MAXWELL, J. C.: On the dynamical theory of gases. Phil. Mag. **35**, No. 235, 129–145. (*16*)

1874 *1.* BOLTZMANN, L.: Zur Theorie der elastischen Nachwirkung. Sitzber. Kaiserl. Akad. Wiss. Wien, Math.-Naturw. Kl. **70**, Sect. II, 275–306. (*1, 5, 6, 7*)

2. MEYER, O. E.: Theorie der elastischen Nachwirkung. Ann. Physik u. Chemie (6), **1**, 108–118. (*16*)

1875 *1.* THOMSON, W. (Lord KELVIN): Math. Phys. Papers **3**, 27. (*16*)

1878 *1.* BOLTZMANN, L.: Zur Theorie der elastischen Nachwirkung. Ann. Physik u. Chemie **5**, 430–432. (*1*)

2. MEYER, O. E.: Über die elastische Nachwirkung. Ann. Physik u. Chemie **4**, 249–267. (*16*)

1879 *1.* THOMSON, W. (Lord KELVIN), and P. G. TAIT: Treatise on natural philosophy, part I. Cambridge. (*33*)

1883 *1.* THOMSON, W. (Lord KELVIN), and P. G. TAIT: Treatise on natural philosophy, part II. Cambridge. (*33*)

1884 *1.* THOMSON, W. (Lord KELVIN): Note on the integration of the equations of equilibrium of an elastic solid. Cambridge and Dublin Math. J. **3**, 87. (*33*)

1885 *1.* BOUSSINESQ, J.: Application des potentiels à l'étude de l'équilibre et des mouvements des solides élastiques. Paris: Gauthier-Villars. (*35*)

1890 *1.* VOIGT, W.: Über die innere Reibung der festen Körper, insbesondere der Krystalle. Göttinger Abh. **36**, No. 1. (*16*)

1892 *1.* SOMIGLIANA, C.: Sulle espressioni analitiche generali dei movimenti oscillatori. Atti Reale Accad. Lincei Roma, Ser. 5, **1**, 111–119. (*38*)

2. VOIGT, W.: Über innere Reibung fester Körper, insbesondere der Metalle. Ann. Physik u. Chemie **47**, 671–693. (*16*)

1894 *1.* VOIGT, W.: Über eine anscheinend nothwendige Erweiterung der Elastizitätstheorie. K. Ges. d. W. Nachrichten, Math.-Phys. Kl., No. 1, 33–43.

1909 *1.* VOLTERRA, V.: Sulle equazioni integrodifferenziali della teoria dell'elasticità. Atti Reale Accad. Lincei **18**, No. 2, 295. (*1, 5, 27, 28*)

1913 *1.* VOLTERRA, V.: Leçons sur les fonctions des lignes. Paris: Gauthier-Villars. (*1, 5, 28*)

1931 *1.* ONSAGER, L.: Reciprocal relations in irreversible processes I. Phys. Rev. **37**, No. 4, 405–426. (*10*)

2. — Reciprocal relations in irreversible processes II. Phys. Rev. **38**, No. 12, 2265–2279. (*10*)

1932 *1.* JEFFRIES, H.: On plasticity and creep in solids. Proc. Roy. Soc. (London), Ser. A **138**, 283–297. (*31*)

1942 *1.* SIMHA, R.: On relaxation effects in amorphous media. J. Appl. Phys. **13**, 201–207.

1944 *1.* ALFREY, T.: Non-homogeneous stresses in viscoelastic media. Quart. Appl. Math. **2**, 113–119. (*31*)

2. LOVE, A. E. H.: The mathematical theory of elasticity. New York: Dover Publications [This is a reprint of the fourth edition, 1934.] (*33, 35*)

1945 *1.* ALFREY, T., and P. DOTY: The methods of specifying the properties of viscoelastic materials. J. Appl. Phys. **16**, 700–713.

2. MISES, R. v.: On Saint-Venant's principle. Bull. Amer. Math. Soc. **51**, 555. (*35*)

1947 *1.* FRIEDRICHS, K. O.: On the first boundary value problem of the theory of elasticity and Korn's inequality. Ann. of Math. **48**, No. 2, 441. (*32*)

2. GROSS, B.: On creep and relaxation. J. Appl. Phys. **18**, 212–221.

3. SCHOLTE, J. G.: On Rayleigh waves in viscoelastic media. Physica **13**, 245–250.

1949 *1.* MINDLIN, R. D.: A mathematical theory of photoviscoelasticity. J. Appl. Phys. **20**, 206–216. (*31*)

1950 *1.* OLDROYD, J. G.: On the formulation of rheological equations of state. Proc. Roy. Soc. (London), Ser. A **200**, 523–541.

2. READ, W. T.: Stress analysis for compressible viscoelastic materials. J. Appl. Phys. **21**, No. 7, 671–674. (*31*)

3. TER HAAR, D.: A phenomenological theory of viscoelastic behaviour, I, II, III. Physica **16**, 719–737, 738–752, 839–850.

4. TSIEN, H. S.: A generalization of Alfrey's theorem for viscoelastic media. Quart. Appl. Math. **8**, 104–106. (*31, 35*)

1951 *1.* GRAFFI, D.: Su alcune questioni di elasticità ereditaria. Atti Accad. Naz. Lincei, Rend. Classe Sci. Fìs. Mat. Nat. **10**, No. 8, 25–30. (*31*)

2. OESTREICHER, H. L.: Field and impedance of an oscillating sphere in a viscoelastic medium with an application to biophysics. J. Acoust. Soc. Am. **23**, No. 6.

3. SIPS, R.: General theory of deformation of viscoelastic substances. J. Polymer Sci. **7**, 191. (*41*)

4. VOLTERRA, E.: On elastic continua with hereditary characteristics. J. Appl. Mech. **18**, 273–279.

1952 *1.* GRAFFI, D.: Sulla teoria dei materiali elastico-viscosi. Att. Accad. Ligure **9**, 77–83. (*31*)

2. SCHWARZL, F., and A. J. STAVERMAN: Time-temperature dependence of linear viscoelastic behavior. J. Appl. Phys. **23**, 838. (*1*)

1953 *1.* GROSS, B.: Mathematical structure of the theories of viscoelasticity. Paris: Hermann & Co. (*11, 13*)

2. LEE, E. H., and I. KANTER: Wave propagation in finite rods of viscoelastic material. J. Appl. Phys. **24**, No. 9, 1115–1122. (*41*)

1954 *1.* BIOT, M. A.: Theory of stress-strain relations in anisotropic viscoelasticity and relaxation phenomenon. J. Appl. Phys. **25**, 1385.

2. STERNBERG, E.: On Saint-Venant's principle. Quart. Appl. Math. **11**, No. 4, 393. (*33, 35*)

1955 *1.* BIOT, M. A.: Dynamics of viscoelastic anisotropic media. Proc. Fourth Midwestern Conf. on Solid Mechanics, Publ. No. 129, Eng. Exp. Sta., Purdue, U., Lafayette, Ind., 94–108. (*38*)

2. COTTER, B. A., and R. S. RIVLIN: Tensors associated with time-dependent stress. Quart. Appl. Math. **13**, No. 2, 177–182.

3. HU, H.: On some variational principles in the theory of elasticity and theory of plasticity. Sci. Sinica (Peking) **4**, 33–54. (*30*)

4. LEE, E. H.: Stress analysis in viscoelastic bodies. Quart. Appl. Math. **13**, No. 2, 183. (*31*)

5. LORSCH, H. G., and A. M. FREUDENTHAL: On mixed boundary value problems of linear viscoelastic solids. Proc. of Second U.S. Nat. Congr. of Applied Mechanics, Ann Arbor, 1954, p. 539–545. New York: ASME.

6. RIESZ, F., and B. SZ-NAGY: Functional analysis. New York: F. Ungar Publishing Co. (*11*)

7. RIVLIN, R. S.: Further remarks on the stress-deformation relations for isotropic materials. J. Rational Mech. Anal. **4**, No. 5, 681–702.

8. SEMYAKIN, E. I.: Propagation of unsteady disturbances in a visco-elastic medium. Dokl. Akad. Nauk SSSR (N.S.) **104**, 34–37.

9. STERNBERG, E., and R. A. EUBANKS: On the concept of concentrated loads and an extension of the uniqueness theorem in the linear theory of elasticity. J. Rational Mech. Anal. **4**, 135–168. (*33, 34*)

10. WASHIZU, K.: On the variational principles of elasticity and plasticity. Rept. No. 25–18, Cont. Nsori-07833, MIT, March. (*30*)

1956 *1.* BERRY, D. S., and S. C. HUNTER: The propagation of dynamic stresses in viscoelastic rods. J. Mech. Phys. Solids **4**, 72–95.

2. DUFFIN, R. J.: Analytic continuation in elasticity. J. Rational Mech. Anal. **5**, No. 6, 939. (*32*)

3. KOLSKY, H.: The propagation of stress pulses in viscoelastic solids. Phil. Mag. **8**, 693.

4. LEE, E. H.: Stress analysis in viscoelastic materials. J. Appl. Phys. **27**, 665–672.

5. LOVE, E. R.: Linear superposition in viscoelasticity and theories of delayed effects. Australian J. Phys. **9**, No. 1, 1–12.

6. MIYAKE, A.: General theory of relaxation phenomena. Nat. Sci. Rep. Lib. Arts Fac. Shizuoka Univ., No. 9, 3–8.

7. MORRISON, J. A.: Wave propagation in rods of Voigt material and visco-elastic materials with three parameter models. Quart. Appl. Math. **14**, 153–169.

1957 *1.* EWING, W. M., JARDETSKY, W. S., and F. PRESS: Elastic waves in layered media. New York: McGraw Hill Book Co.

2. GREEN, A. E., and R. S. RIVLIN: The mechanics of nonlinear materials with memory. Arch. Rational Mech. Anal. **1**, 1. Reprinted in Continuum Mechanics II, Rational Mechanics of Materials, ed. C. TRUESDELL. New York: Gordon and Breach 1965. (*1*)

3. Hopkins, I. L., and R. W. Hamming: On creep and relaxation. J. Appl. Phys. **28**, 906. (*11*)
4. Leaderman, H.: Proposed nomenclature for linear viscoelastic behavior. Trans. Soc. Rheol. **1**, 213–222.
5. Lee, E. H., and J. R. M. Radok: Stress analysis in linearly viscoelastic materials. Proc. Ninth Int. Congr. Appl. Mech. **5**, 321. (*31*)
6. Mandel, J.: Sur les vibrations des corps viscoélastiques à comportement linéaire. C. R. Acad. Sci. Paris **245**, 2176–2178.
7. Miyake, A.: Notes on the general theory of relaxation. Nat. Sci. Res. Lib. Arts Fac. Shizuoka Univ., No. 10, 1–5.
8. Radok, J. R. M.: Viscoelastic stress analysis. Quart. Appl. Math. **15**, 198. (*31*)
9. Rivlin, R. S.: Preliminary notes to a course of lectures on solid mechanics. Seminar in Applied Mathematics, Amer. Math. Soc. and U. of Cal.
10. Smith, G. F., and R. S. Rivlin: Stress-deformation relations for anisotropic solids. Arch. Rational. Mech. Anal. **1**, No. 2, 107–112.
1958 *1.* Berry, D. S.: A note on stress pulses in viscoelastic rods. Phil. Mag. **25**, 100. (*41*)
2. — Stress propagation in viscoelastic bodies. J. Mech. Phys. Solids **6**, 177–185. (*41*)
3. Biot, M. A.: Linear thermodynamics and the mechanics of solids. Proc. Third U.S. Nat. Cong. Appl. Mech., p. 1–18. New York: ASME. (*1*)
4. Borodacov, M. M.: Longitudinal vibrations of viscoelastic rods. Akad. Nauk. Ukr., RSR Prikl. Meh. **4**, 176–181.
5. Freudenthal, A. M., and H. Geiringer: The mathematical theories of the inelastic continuum. Encyclopedia of physics, ed. S. Flügge, vol. 6. Berlin-Göttingen-Heidelberg: Springer.
6. Green, A. E., Radok, J. R. M., and R. S. Rivlin: Thermoelastic similarity laws. Quart. Appl. Math. **15**, No. 4, 381–393.
7. Gross, B.: Single lines in retardation and relaxation spectra. Rheol. Acta **1**, No. 2–3, 131–133.
8. Hopkins, I. L., and R. W. Hamming: Note on "On creep and relaxation". J. Appl. Phys. **29**, 742. (*11*)
9. Jung, H.: Zur Theorie der Maxwellschen Flüssigkeiten. Rheol. Acta **1**, No. 2–3, 280–285.
10. Kolsky, H., and Y. Y. Shi: The validity of model representations for linear viscoelastic behaviour. Brown Univ. Rep. NONR 562 (14)/5. (*16*)
11. König, H., and J. Meixner: Lineare Systeme und lineare Transformationen. Math. Nachrichten **19**, 256. (*1, 6, 8*)
12. Leaderman, H.: Viscoelasticity phenomenon in amorphous high polymeric systems. In: Rheology: Theory and applications, vol. 2, ed. by F. R. Eirich, p. 1–61 New York: Academic Press Inc. (*1*)
13. Lee, E. H.: Viscoelastic stress analysis. Brown U., Div. of Appl. Math., Tech. Report No. C11–42, NONR 562 (10). (*31*)
14. Mandel, J.: Les corps viscoélastiques Boltzmanniens. Rheol. Acta **1**, No. 2–3.
15. Meixner, J., u. H. König: Zur Theorie der linearen dissipativen Systeme. Rheol. Acta **1**, No. 2–3.
16. Noll, W.: A mathematical theory of the mechanical behavior of continuous media. Arch. Rational Mech. Anal. **2**, 197. Reprinted in Continuum Mechanics II, Rational Mechanics of Materials, ed. C. Truesdell, New York: Gordon and Breach, 1965, and in Continuum Theory of Inhomogeneities in Simple Bodies, New York: Springer 1968. (*1, 3, 9*)
17. Sternberg, E.: On transient thermal stresses in linear viscoelasticity. Proc. Third U.S. Nat. Cong. Appl. Mech. 673.
18. Terry, N. B.: The behavior of a vibrating viscoelastic cylinder. Proc. Phys. Soc. (London) **71**, 973–978.
1959 *1.* Biot, M. A.: On the instability and folding deformation of a layered viscoelastic medium in compression. J. Appl. Mech. **26**, 393–400.
2. Bland, D. R.: On the foundations of linear isotropic viscoelasticity. Proc. Roy. Soc. (London) **250**, 524–549.
3. Distéfano, J. N.: Sulla stabilità in regime viscoelastico a comportamento lineare, I, II. Atti Accad. Naz. Lincei, Rend. Classe Sci. Fìs. Mat. Nat. **27**, No. 8, 205–211, 356–361.
4. Green, A. E., Rivlin, R. S., and A. J. M. Spencer: The mechanics of non-linear materials with memory, Part II. Arch. Rational Mech. Anal. **3**, No. 1, 82–90.
5. Lee, E. H., Radok, J. R. M., and W. B. Woodward: Stress analysis for linear viscoelastic materials. Trans. Soc. Rheol. **3**, 41–59. (*31*)

6. MAY, W. D., MORRIS, E. L., and D. ATOCK: Rolling friction of a hard cylinder over a viscoelastic material. J. Appl. Phys. **30**, 1713–1724.
7. MIKUSINSKI, J.: Operational calculus. New York: Pergamon Press. (*4, 11*)
8. MOVCHAN, A. A.: The direct method of Liapunov in stability problems of elastic systems. PMM **23**, No. 3, 1959, 483–493. (Moscow)
9. NOWACKI, W.: Thermal stresses due to the action of heat sources in a viscoelastic space. Arch. Mech. Stos. **11**, 111–125.
10. OLSZAK, W., and P. PERZYNA: Variational theorems in general viscoelasticity. Ing. Arch. **28**, 241–250.
11. PIPKIN, A. C., and R. S. RIVLIN: The formulation of constitutive equations in continuum physics I. Arch. Rational Mech. Anal. **4**, No. 2, 129–144.
12. SPENCER, A. J. M., and R. S. RIVLIN: The theory of matrix polynomials and its application to the mechanics of isotropic continua. Arch. Rational Mech. Anal. **2**, No. 4, 309–336.
13. THURSTON, G. B.: Theory of oscillation of a viscoelastic medium between parallel planes. J. Appl. Phys. **30**, 1855–1860.
14. VOLTERRA, V.: Theory of functionals and of integral and integro-differential equations. New York: Dover. [This is a reprint of a volume first published in 1930.]

1960 *1.* BERGEN, J. T. (editor): Viscoelasticity: Phenomenological aspects. New York-London: Academic Press.
2. BLAND, D. R.: The theory of linear viscoelasticity. New York: Pergamon Press. (*14, 16*)
3. BOLEY, B. A., and J. H. WEINER: Theory of thermal stresses. New York: John Wiley & Sons. (*14, 21*)
4. ELDER, A. S.: Stress function theory for linearly viscoelastic solids. Memorandum Rep. No. 1282, Ballistic Research Lab., Aberdeen Proving Ground, Maryland.
5. GURTIN, M. E., and E. STERNBERG: On the first boundary-value problem of linear elastostatics. Arch. Rational. Mech. Anal. **6**, No. 3, 177. (*28*)
6. GREEN, A. E., and R. S. RIVLIN: The mechanics of non-linear materials with memory, Part. III. Arch. Rational Mech. Anal. **4**, No. 5, 387–404.
7. HUNTER, S. C.: The Hertz problem for a rigid spherical indenter and a viscoelastic half space. J. Mech. Phys. Solids **8**, 219–234.
8. — Viscoelastic waves. Progress in solid mechanics, vol. 1, p. 1–57. Amsterdam: North-Holland Publishing Co. (*31, 43*)
9. KOLSKY, H.: Experimental wave-propagation in solids. Proc. First Symp. on Naval Structural Mech. New York: Pergamon Press.
10. — Viscoelastic waves. Proc. Conf. Stress Propagation 1959, Penn. State Univ. New York: Interscience.
11. LEE, E. H.: Viscoelastic stress analysis. Proc. First Symp. on Naval Structural Mech., p. 456–482. New York: Pergamon Press.
12. —, and J. R. M. RADOK: The contact problem for viscoelastic bodies. J. Appl. Mech. **27**, 438–444.
13. MORLAND, L. W., and E. H. LEE: Stress analysis for linear viscoelastic material with temperature variation. Trans. Soc. Rheol. **4**, 233. (*1*)
14. SNEDDON, I. N., and R. HILL: Progress in solid mechanics, I. Amsterdam: North-Holland Publishing Co.
15. TRUESDELL, C., and R. A. TOUPIN: The classical field theories, In: Encyclopedia of physics, vol. III/1, ed. S. FLÜGGE. (*9, 41, 43*)

1961 *1.* CHU, B. T.: Stability criteria for isotropic linear viscoelastic materials. Tech. Rep. No. 26, Brown U., NONR. 562. (*20*)
2. COLEMAN, B. D., and W. NOLL: Foundations of linear viscoelasticity. Rev. Mod. Phys. **33**, No. 2, 239–249. (*1*)
3. HUNTER, S. C.: Tentative equations for the propagation of stress, strain, and temperature fields in viscoelastic solids. J. Mech. Phys. Solids **9**, 39–51.
4. LOCKETT, F. J.: Interpretation of mathematical solutions in viscoelasticity theory illustrated by a dynamic spherical cavity problem. J. Mech. Phys. Solids **9**, 215–229.
5. MUKI, R., and E. STERNBERG: On transient thermal stresses in viscoelastic materials with temperature dependent properties. J. Appl. Mech. **28**, 193. (*1*)
6. PARKS, J. R., and L. COOPER: The application of numerical integration to a problem in viscoelasticity. Soc. Ind. Appl. Math. Rev. **3**, 315–321.
7. PIPKIN, A. C., and R. S. RIVLIN: Small deformations superposed on large deformations in materials with fading memory. Arch. Rational Mech. Anal. **8**, No. 4, 297–308.

8. Radok, J. R. M.: Plane problems of linear viscoelasticity, In: Problems of continuum mechanics, p. 359–361. Soc. Ind. Appl. Math.

9. Reiss, E. L.: On the quasistatic theory of viscoelasticity. Arch. Rational Mech. Anal. 7, 402–411.

10. Thomas, T. Y.: Plastic flow and fracture in solids. New York: Academic Press. (*41*)

1962 *1.* Achenbach, J. D., and C. C. Chao: A three-parameter viscoelastic model particularly suited for dynamic problems. J. Mech. Phys. Solids **10**, 245.

2. Al Khozaie, S., and E. H. Lee: Influence of material compressibility in the viscoelastic contact problem. Brown U. Report, Armstrong Cork Co., May.

3. Breuer, S., and E. T. Onat: On uniqueness in linear viscoelasticity. Quart. Appl. Math. **19**, No. 4, 355. (*27, 28*)

4. Chu, B. T.: Stress waves in isotropic linear viscoelastic materials I. J. Méc. **1**, 439–462. (*41*)

5. Coleman, B. D.: Kinematical concepts with applications in the mechanics and thermodynamics of incompressible viscoelastic fluids. Arch. Rational Mech. Anal. **9**, 4, 273–300.

6. — Mechanical and thermodynamical admissibility of stress-strain functions. Arch. Rational Mech. Anal. **9**, No. 2, 172–186.

7. Corneliussen, A. H., and E. H. Lee: Stress distribution analysis for linear viscoelastic materials. Proc. IUTAM Colloq. Creep in Structures, Stanford Univ. 1960, ed. N. J. Hoff, p. 1–20. Berlin-Göttingen-Heidelberg: Springer.

8. Courant, R., and D. Hilbert: Methods of mathematical physics, vol. 2. New York: Interscience. (*32*)

9. Gurtin, M. E.: A note on the principle of minimum potential energy for linear anisotropic elastic solids. Brown U. NONR 562 (25)/15.

10. –, and E. Sternberg: On the linear theory of viscoelasticity. Arch. Rational Mech. Anal. **11**, No. 4, 291–356. (*1, 5, 6, 8, 11, 12, 14, 16, 19, 21, 24, 25, 27, 28, 31, 32*)

11. Hertelendy, P.: Displacement and strain energy distribution in a longitudinally vibrating cylindrical rod with a viscoelastic coating. J. Appl. Mech. **29**, 47–52.

12. Kolsky, H., and E. H. Lee: The propagation and reflection of stress pulses in linear viscoelastic media. Brown U. Rep. NONR 562 (30)/5.

13. Lockett, F. J.: The reflection and refraction of waves at an n terface between viscoelastic materials. J. Mech. Phys. Solids **10**, 53.

14. Lubliner, J.: Cylindrical waves in viscoelastic solids. J. Acoust. Soc. Am. **34**, 1706–1710.

15. Man, F.: Stress distributions in semi-infinite viscoelastic media. Acta Mech. Sinica **5**, 117–126.

16. Nowacki, W.: Theory of creep. Ankady, Warsaw, Poland [in Polish], also see Appl. Mech. Rev. **18**, 3310.

17. Onat, E. T.: On a variational principle in linear viscoelasticity. J. Méc. **1**, 135–140.

18. Sackman, J. L.: Uniformly progressing surface pressure on a viscoelastic half plane. Proc. Fourth U.S. Nat. Cong. Appl. Mech. **2**, p. 1067–1074. New York: ASME.

19. Schapery, R. A.: Approximate methods for transform inversion for viscoelastic stress analysis. Proc. Fourth U.S. Nat. Cong. Appl. Mech. **2**, **p.** 1075–1085. New York: ASME.

20. — Irreversible thermodynamics and variational principles with applications to viscoelasticity. Aeronautical Research Lab, Wright-Patterson A.F.B. ARL 62-418. (*1*)

1963 *1.* Breuer, S., and E. T. Onat: On the determination of free energy in linear viscoelastic solids. Brown U. Rep. NONR 562 (10), C 11-87. (*1*)

2. Corneliussen, A. H., Kamowitz, E. F., Lee, E. H., and J. R. M. Radok: Viscoelastic stress analysis of a spinning hollow circular cylinder with an ablating pressurized cavity. Trans. Soc. Rheol. **7**, 357.

3. Ericksen, J. L.: Non-existence theorems in linear elasticity theory. Arch. Rational Mech. Anal. **14**, No. 3, 180–183.

4. Gottenberg, W. G., and R. M. Christensen: Some interesting aspects of general linear viscoelastic deformation. Trans. Soc. Rheol. **7**, 171–180.

5. Gurtin, M. E.: Variational principles in the linear theory of viscoelasticity. Arch. Rational Mech. Anal. **13**, No. 3, 179–191. (*30*)

6. — A generalization of the Beltrami stress functions in continuum mechanics. Div. of Appl. Math., Brown U. Tech. Report No. 20, NONR 562 (25)/20. (*30*)

7. GURTIN, M., and E. STERNBERG: A reciprocal theorem in the linear theory of anisotropic viscoelastic solids. J. Soc. Ind. Appl. Math. **11**, 607–613. (*10, 27*)
8. IGNACZAK, J.: A completeness problem for stress equations of motion in the linear theory of elasticity. Arch. Mech. Stos. **15**, No. 2, 225–234. (*36*)
9. KALISKI, S.: Absorption of magneto-viscoelastic surface waves in a real conductor in a magnetic field. Proc. Vibration Prob. **4**, 319–330.
10. KOLSKY, H.: Stress waves in solids. New York: Dover Publ. Inc. (*1, 43*)
11. LEE, E. H., and T. G. ROGERS: Solution of viscoelastic stress analysis problems using measured creep or relaxation properties. J. Appl. Mech. **30**, 127–133.
12. MISICU, M.: Theory of viscoelasticity with couple stresses and some reductions to two dimensional problems. Rev. Méc. Appl. **8**, 921–952.
13. MORLAND, L. W.: Dynamic stress analysis for a viscoelastic half-plane subjected to moving surface tractions. Proc. London Math. Soc. (3), **13**, 471–492.
14. ONAT, E. T., and S. BREUER: On uniqueness in linear viscoelasticity. Progress in appl. mechanics, p. 349–353. New York: Macmillan. (*40*)
15. PIPKIN, A. C.: Small finite deformations of viscoelastic solids. Proc. Princeton U. Conf. on Solid Mech., p. 75–100.
16. PREDELÉANU, M.: Mécanique des milieux continus. C. R. Acad. Sci. Paris **256**, 1, 71.
17. ROGERS, T. G.: Viscoelastic stress analysis. Proc. Princeton U. Conf. on Solid Mech., p. 49–74. (*31*)
18. —, and E. H. LEE: The cylinder problem in viscoelastic stress analysis. Tech. Rep. No. 138, Div. Eng. Mech., Stanford University.
19. — — Thermoviscoelastic stresses in a sphere with an ablating cavity. Progress in appl. mechanics, p. 355. London: Macmillan.
20. —, and A. C. PIPKIN: Asymmetric relaxation and compliance matrices in linear viscoelasticity. Z. Angew. Math. Phys. **14**, 334–343. (*10*)
21. SCHAPERY, R. A.: Heating of thermorheologically simple viscoelastic media due to cyclic loading. Purdue U. Rep. A&ES 63-4.
22. SHINOZUKA, M.: Stresses in a linear incompressible viscoelastic cylinder with moving inner boundary. J. Appl. Mech. **30**, 335.
23. STERNBERG, E., and M. E. GURTIN: Uniqueness in the theory of thermorheologically simple ablating solids. Progress in appl. mechanics, p. 373–384. New York: Macmillan.
24. TAO, L. N.: The associated elastic problem in dynamic viscoelasticity. Quart. Appl. Math. **21**, 215–222.
25. WARD, I. M., and E. T. ONAT: Non-linear mechanical behaviour of oriented polypropylene. J. Mech. Phys. Solids **11**, 217.
1964 *1.* BREUER, S., and E. T. ONAT: On reversible work in linear viscoelasticity. Z. Angew. Math. Phys. **15**, 12–21. (*1*)
2. CHACON, R. V. S., and R. S. RIVLIN: Representation theorems in the mechanics of materials with memory. Z. Angew. Math. Phys. **15**, 444.
3. CHAO, C., and J. D. ACHENBACH: A simple viscoelastic analogy for stress waves. In: Stress waves in anelastic solids (H. KOLSKY and W. PRAGER, ed.), p. 222–238. Berlin-Göttingen-Heidelberg-New York: Springer.
4. COLEMAN, B. D.: On thermodynamics, strain impulses, and viscoelasticity. Arch. Rational Mech. Anal. **17**, 230–254. (*1, 10*)
5. —, and W. NOLL: Foundations of linear viscoelasticity. Rev. Mod. Phys. **36** (Erratum. for [1961, *2*]) 1103–1104. (*1, 6, 8*)
6. EDELSTEIN, W., and M. E. GURTIN: Uniqueness theorems in the linear dynamic theory of anisotropic viscoelastic solids. Arch. Rational Mech. Anal. **17**, No. 1, 47–60. (*40*)
7. ELDER, A. S.: Biharmonic solution of certain integro-differential equations of linear viscoelasticity. Trans Soc. Rheol. **8**, 101–115. (*32*)
8. GURTIN, M. E.: Variational principles for linear elastodynamics. Arch. Rational Mech. Anal. **16**, 1, 34–50.
9. HLAVAČEK, I., et M. PREDELÉANU: Sur l'existence et l'unicité de la solution dans la théorie du fluage linéaire, I. Aplikace Matematiky **9**, No. 5. (*28,29*)
10. HUANG, N. C., LEE, E. H., and T. G. ROGERS: On the influence of viscoelastic compressibility in stress analysis. Proc. Fourth Internat. Congr. on Rheology, part 2, p. 213. New York: John Wiley & Sons, Inc.
11. KALISKI, S., and W. NOWACKI: Propagation of magneto-elastic disturbances in viscoelastic bodies. In: Stress waves in anelastic solids (H. KOLSKY and W. PRAGER, ed.), p. 43–53. Berlin-Göttingen-Heidelberg-New York: Springer.
12. LEE, E. H.: Recent developments of linear viscoelastic stress analysis. Stanford U. Div. of Eng. Mech., Tech. Rep. No. 150. (*31*)

13. Lee, E. H., and T. G. Rogers: Non-linear effects of temperature variation in stress analysis of isothermally linear viscoelastic materials. IUTAM Symp. — Second order effects in elasticity, plasticity and fluid dynamics, Haifa, 1962, ed. M. Reiner and D. Abir. New York: Macmillan.
14. Lee, T. M.: Spherical waves in viscoelastic media, J. Acoust. Soc. Am. **36**, 2402–2407.
15. — Dilatation constants and complex ratio from forced vibration of a free viscoelastic sphere. J. Acoust. Soc. Am. **36**, 458–462.
16. Meyer, M. L.: On spherical near fields and far fields in elastic and viscoelastic solids. J. Mech. Phys. Solids **12**, 77–111.
17. Miklowitz, J.: Pulse propagation in a viscoelastic solid with geometric dispersion. In: Stress waves in anelastic solids (H. Kolsky and W. Prager, ed.), p. 255–276. Berlin-Göttingen-Heidelberg-New York: Springer.
18. Petrof, R. C., and S. Gratch: Wave propagation in a viscoelastic material with time dependent properties and thermomechanical coupling. J. Appl. Mech. **31**, 423–429.
19. Pipkin, A. C.: Small finite deformations of viscoelastic solids. Rev. Mod. Phys. **36**, 1034–1041.
20. Rogers, T. G., and E. H. Lee: The cylinder problem in viscoelastic stress analysis. Quart. Appl. Math. **22**, 117–131.
21. Sakar, S. K.: Disturbances in a semi-infinite viscoelastic medium due to transient tangential forces on a plane boundary. Indian J. Mech. Math. **2**, 28–29.
22. Saks, S.: Theory of the integral. New York: Dover. (*5*)
23. Schapery, R. A.: Applications of thermodynamics to thermomechanical fracture and birefringent phenomenon in viscoelastic media. J. Appl. Phys. **35**, 1451–1465. (*1*)
24. Sternberg, E., and S. Al-Khozaie: On Green's functions and Saint-Venant's principle in the linear theory of viscoelasticity. Arch. Rational Mech. Anal. **15**, 112–146. (*33, 34, 35*)
25. —, and M. E. Gurtin: Further study of thermal stresses in viscoelastic materials with temperature dependent properties. IUTAM Symp., Second order effects in elasticity, plasticity, and fluid dynamics, Haifa, 1962, ed. M. Reiner and D. Abir, p. 51–76. New York: Macmillan.
26. Taylor, R. L.: Creep and relaxation. AIAAJ **2**, 1659–1660.
27. Valanis, K. C., and G. Lianis: Thermal stresses in a viscoelastic cylinder with temperature dependent properties. AIAAJ **2**, 1642–1644.

1965 *1.* Agmon, S.: Lectures on elliptic boundary value problems. Princeton: Van Nostrand. (*29*)
2. Biot, M. A.: Mechanics of incremental deformations. New York: John Wiley & Sons, Inc. (*1*)
3. — Internal stability of anisotropic viscous and viscoelastic media under initial stress. J. Franklin Inst. **279**, 65–82.
4. Breuer, S., and E. T. Onat: On uniqueness in linear viscoelasticity. Quart. Appl. Math. **19**, 355–359.
5. Christensen, R. M., and R. N. Schreiner: Response to pressurization of a viscoelastic cylinder with an eroding internal boundary. AIAAJ **3**, 1451–1455.
6. Chen, P.: Acceleration waves in rheological materials. Ph. D. thesis, U. of Washington, Dept. of Aeronautics and Astronautics. (*41*)
7. Coleman, B. D., and M. E. Gurtin: Waves in materials with memory II, III. Arch. Rational Mech. Anal. **19**, No. 4, 239–265, 266–298. [These two papers and the two cited just following are reprinted in Wave Propagation in Dissipative Materials. New York: Springer 1965.] (*41, 43*)
8. — — Waves in materials with memory IV. Arch. Rational Mech. Anal. **19**, No. 5, 317–338. (*41*)
9. — —, and I. Herrera: Waves in materials with memory, I: The velocity of one-dimensional shock and acceleration waves. Arch. Rational Mech. Anal. **19**, No. 1, 1–19. (*41*)
10. Cooper, H. F., and E. L. Reiss: Propagation and reflection of viscoelastic waves. J. Acoust. Soc. Am. **38**, 24–34.
11. Edelstein, W. S., and M. E. Gurtin: A generalization of the Lamé and Somigliana stress functions for the dynamic linear theory of viscoelastic solids. Int. J. Eng. Sci. **3**, 109–117. (*37, 38*)
12. Fichera, G.: Linear elliptic differential systems and eigenvalue problems. Berlin-Heidelberg-New York: Springer. (*29*)
13. Fisher, G. M. C.: The decay of plane waves in the linear theory of viscoelasticity. Brown U. Report NONR 562 (40)/2. (*41*)

14. FISCHER, G. M. C., and M. E. GURTIN: Wave propagation in the linear theory of viscoelasticity. Quart. Appl. Math. **23**, No. 3, 257–263. (*41*)
15. GRAHAM, G. A. C.: The contact problem in the linear theory of viscoelasticity. Int. J. Eng. Sci. **3**, 27–46.
16. GURTIN, M. E., and I. HERRERA: On dissipation inequalities and linear viscoelasticity. Quart. Appl. Math. **23**, No. 3, 235–245. (*10*)
17. HERRERA, I., and M. E. GURTIN: A correspondence principle for viscoelastic wave propagation. Quart. Appl. Math. **22**, 360–364. (*41*)
18. HERRMANN, L. R.: On a general theory of viscoelasticity. J. Franklin Inst. **280**, 244–255.
19. HLAVAČEK, I., et M. PREDELÉANU: Sur l'existence et l'unicité de la solution dans la théorie du fluage linéaire II. Aplikace Matematiky, Praha **10**, 5. (*28, 29*)
20. JOSEPH, D. D.: On the stability of the Boussinesq equations. Arch. Rational Mech. Anal. **20**, No. 1, 59–71.
21. NOWACKI, W.: Theory of creep, linear viscoelasticity. Vienna: Franz Deuticke [in German], also see Appl. Mech. Rev. **19**, (6), 3455.
22. ODEH, F., and I. TADJBAKHSH: Uniqueness in the linear theory of viscoelasticity. Arch. Rational Mech. Anal. **18**, 244–250. (*40*)
23. PIPKIN, A. C.: Approximate constitutive equations for viscoelastic flows. Brown U. DA-G 484/9.
24. RIVLIN, R. S.: Non-linear viscoelastic solids. Soc. Indust. Appl. Math. Rev. **7**, 323–340.
25. SCHAPERY, R. A.: A method of viscoelastic stress analysis using elastic solutions. J. Franklin Inst. **279**, 268–289.
26. — Thermomechanical behavior of a viscoelastic media with variable properties subject to cyclic loading. J. Appl. Mech. **32**, 611–619. (*1*)
27. SHU, L. S., and E. T. ONAT: On anisotropic viscoelastic solids. Proc. Fourth Symp. Naval Structural Mech., Purdue U. (*10*)
28. TOUPIN, R. A.: Saint-Venant's principle. Arch. Rational Mech. Anal. **18**, No. 2, 83. (*35*)
29. TRUESDELL, C., and W. NOLL: The non-linear field theories of mechanics. In: Encyclopedia of physics, ed. S. FLÜGGE, vol. III/3. Berlin-Heidelberg-New York: Springer. (*1, 6, 8, 9*)
30. VALANIS, K. C.: Propagation and attenuation of waves in linear viscoelastic solids. J. Math. Phys. **44**, 227–239. (*41*)
31. VARLEY, E.: Acceleration fronts in viscoelastic materials. Arch. Rational Mech. Anal. **19**, No. 4, 215–225. (*41*)

1966 *1.* APPLEBY, E. J., and E. H. LEE: Linearization in nonlinear viscoelasticity and its use in the determination of relaxation moduli. J. Mech. Phys. Solids **14**, 375–391.
2. AXELRAD, D. R.: On the longitudinal wave propagation in a viscoelastic cylinder. Z. Angew. Math. Phys. **17**, 511–518.
3. BABUŠKA, I., and I. HLAVAČEK: On the existence and uniqueness of solutions in the theory of viscoelasticity. Arch. Mech. Stos. **18**, 47–84. (*28, 29*)
4. BARBERÁN, J., and I. HERRERA: Uniqueness theorems and speed of propagation of signals in viscoelastic materials. Arch. Rational Mech. Anal. **23**, 173–190. (*40, 42*)
5. COLEMAN, B. D., and V. J. MIZEL: Norms and semigroups in the theory of fading memory. Arch. Rational Mech. Anal. **23**, 87–123. (*1, 3, 7, 8, 10*)
6. EDELSTEIN, W. S.: Existence of solutions to the displacement problem for quasistatic viscoelasticity. Arch. Rational Mech. Anal. **22**, 121–128. (*29*)
7. EFIMOV, A. B.: An axially symmetric contact problem for linear viscoelastic bodies. Vestn. Mosk. Univ., Ser. I Mat. Meh. **21**, No. 2, 120–127.
8. FISHER, G. M. C., and M. J. LEITMAN: A correspondence principle for free vibrations of viscoelastic solids. J. Appl. Mech. **33**, 924–926. (*43*)
9. HERRERA, I.: Riemann representation method in viscoelasticity. I. Characterization and construction of Riemann function—Solution of problems with prescribed body forces. Arch. Rational Mech. Anal. **22**, 270–291. (*42*)
10. HLAVAČEK, I.: Sur quelques théorèmes variationnels dans la théorie du fluage linéaire. Aplikace Matematiky, Praha **11**, 4.
11. —, et M. PREDELÉANU: Sur l'existence et l'unicité de la solution dans la théorie du fluage linéaire III. Aplikace Matematiky, Praha **11**, 3. (*28, 29*)
12. HUANG, N. C., and E. H. LEE: Nonlinear viscoelasticity for short time ranges. J. Appl. Mech. **33**, No. 2, 313–321.
13. KNOWLES, J. K.: On Saint-Venant's principle in the two-dimensional linear theory of elasticity. Arch. Rational Mech. Anal. **21**, No. 1, 1.

14. KNOWLES, J. K., and E. STERNBERG: On Saint-Venant's problem and the torsion of solids of revolution. Arch. Rational Mech. Anal. **22**, No. 2, 100. (*35*)
15. LEE, E. H.: Some recent developments in linear viscoelastic stress analysis. Proc. 11th Internat. Congr. of Appl. Mech., Munich 1964, p. 396–402.
16. LEITMAN, M. J.: Variational principles in the linear dynamic theory of viscoelasticity. Quart. Appl. Math. **24**, No. 1, 37–46. (*36, 44*)
17. MARTIN, J. B., and A. R. S. PONTER: A note on a work inequality in linear viscoelasticity. Quart. Appl. Math. **24**, 161–165. (*1*)
18. ODQVIST, F. K. G.: Mathematical theory of creep and rupture. Oxford: Clarendon Press.
19. STERNBERG, E., and J. K. KNOWLES: Minimum energy characterization of Saint-Venant's solution to the related Saint-Venant problem. Arch. Rational Mech. Anal. **21**, No. 2, 89.
20. TING, T. Y.: The contact stresses between a rigid indenter and a viscoelastic half space. J. Appl. Mech. **4**, 845–854.
1967 *1.* BARBERÁN, I., and I. HERRERA: Riemann representation method in viscoelasticity. II: Cauchy's initial value problem. Arch. Rational Mech. Anal. **25**, 178–187. (*42*)
2. COLEMAN, B. D., and M. E. GURTIN: Thermodynamics with internal state variables. J. Chem. Phys. **47**, No. 2, 597–613. (*1*)
3. FLÜGGE, W.: Viscoelasticity. Waltham, Mass.: Blaisdell Publ. Co.
4. MORLAND, L. W.: Exact solutions for rolling contact between viscoelastic cylinders. Quart. J. Mech. Appl. Math. **20**, 73–106.
5. NEIS, V. V., and J. L. SACKMAN: An experimental study of a nonlinear material with memory. Trans. Soc. Rheol. **11**, 3, 307–333.
1968 *1.* BARTENEV, G. M., and Y. S. ZUYEV: Strength and failure of viscoelastic materials. London: Pergamon Press.
2. COLEMAN, B. D., and V. J. MIZEL: On the stability of solutions of functional differential equations. Arch. Rational Mech. Anal. **30**, 173–196. (*1, 3*)
3. DAFERMOS, C. M.: On the existence and the asymptotic stability of solutions to the equations of linear thermoelasticity. Arch. Rational Mech. Anal. **29**, No. 4, 241–355.
4. EDELSTEIN, W. S., and R. L. FOSDICK: A note on nonuniqueness in linear elasticity theory. Z. Angew. Math. Phys. **19**, 906–912. (*40*)
5. THURSTON, R. N.: A connection between nonlinear elasticity and ultrasonic attenuation. 6th Internat. Congr. on Acoustics, Tokyo, Japan, p. 177–180. (*1*)
6. VALANIS, K. C.: The viscoelastic potential and its thermodynamic foundations. J. Math. Phys. **47**, No. 3, 262–275. (*1*)
1969 *1.* BREUER, S.: Lower bounds on work in linear viscoelasticity. Quart. Appl. Math. **27**, 139–146. (*1*)
2. DAY, W. A.: Useful strain histories in linear viscoelasticity. Quart. Appl. Math. **27**, 255–259.
3. KOLSKY, H.: Recent experimental studies of the mechanical response of inelastic solids to rapidly changing stresses. Brown University, Div. Appl. Math., Tech. Rep. No. 13. (*1*)
4. TSAI, Y. M.: Stress distribution in elastic and viscoelastic plates subjected to symmetric rigid indentations. Quart. Appl. Math. **27**, No. 3, 271.
1970 *1.* DAFERMOS, C. M.: Asymptotic stability in viscoelasticity. Arch. Rational Mech. Anal. **37**, No. 4, 297–308. (*1, 10, 42*)
2. DAY, W. A.: Reversibility, recoverable work and free energy in linear viscoelasticity. Quart. J. Mech. Appl. Math. **23**, 1, 1–15. (*1*)
3. EDELSTEIN, W. S.: On Saint-Venant's principle in linear viscoelasticity. Arch. Rational Mech. Anal. **36**, No. 5, 366–380. (*35*)
4. FERRY, J. D.: Viscoelastic properties of polymers, 2nd ed. New York: Wiley Interscience. (*1, 43*)
5. GURTIN, M. E.: The linear theory of elasticity [LTE]. In the preceding part of this volume, Encyclopedia of Physics. (*1, 2, 3, 4, 6, 9, 10, 12, 17, 21, 25, 27, 28, 29, 30, 32, 33, 34, 35, 38, 39, 40, 41, 44*)
6. LEITMAN, M. J., and V. J. MIZEL: On linear hereditary laws. Arch. Rational Mech. Anal. **38**, No. 1, 46–58. (*6, 8*)
1971 *1.* DAY, W. A.: Time-reversal and the symmetry of the relaxation function of a linear viscoelastic material. Arch. Rational Mech. Anal. **40**, No. 3, 155–159. (*10*)

Theory of Elastic Stability.

By

R. J. Knops and E. W. Wilkes.

With 3 Figures.

A. Introduction.

The major concern throughout the history of elastic stability has been with applications and particular problems dictated by the demands of engineering. Comparatively little progress has been made in the general theory, which, with few exceptions, has scarcely developed beyond that proposed by Kirchhoff. It is only in the last ten years, as a result of the adaptation of Liapounov's ideas to continuous systems, that a measure of understanding of basic principles has been reached and new results added to the fundamental theory of elastic stability.

Our objective in this treatise is to explain these principles and results and for this purpose we consider it necessary to refer first to elements of the corresponding theory of dynamical systems. The generality of this setting lends perspective to elastic stability theory and indicates its limitations. In this context we are able not only to form a better judgment on the place occupied in the general theory by the energy criterion and other accepted classical tests, but also on the relation between linear and non-linear theories. The text has been divided into two main parts each of which forms a nearly independent unit. Chaps. A to E concern the abstract discussion, while the remainder concerns elasticity.

To review the vast literature on applications and special problems is not our intention. Those illustrations included are examples considered relevant to the purpose of the text. The selection is subjective and the omission of an example should not be interpreted as a reflection on its lack of importance. In contrast to this perfunctory treatment of problems, the literature has been searched for contributions to the general theory of stability in both linear and non-linear elasticity and it is hoped that the most important articles have been referenced. A history and account of modern derivations is thus intended to be represented. Despite our attempt to cover every area of current active interest in stability, lack of space has naturally compelled the treatment of several interesting topics to be superficial. For example, little attention has been paid to post-buckling behaviour, bifurcation and the treatment of elastic stability problems by means of discrete approximations.

Standard mathematical notation is employed, such as the comma to denote partial differentiation and the repeated suffix to denote summation. Other notation, mainly in conformity with the relevant articles in this Encyclopedia, is explained as introduced.[1]

[1] *Acknowledgement.* The authors are grateful to Dr. R. N. Hills for reading parts of the manuscript and making critical comments on an early draft.

B. Abstract dynamical systems.

1. Introduction. A full appreciation of the concepts of stability rests in an understanding of abstract dynamical systems. In this and the next few sections we propose to outline in sufficient generality all the features of these systems which will be required for an adequate discussion of elastic stability. Comprehensive studies of dynamical systems may be found in standard treatises,[1] although for present purposes it is convenient to use a definition differing slightly from that usually put forward by other writers.[2]

The concept of a dynamical system was probably first introduced by POINCARÉ[3] in connexion with a finite real system of ordinary first order differential equations of the form

$$\frac{dx}{dt} = f(x). \tag{1.1}$$

POINCARÉ managed to derive certain qualitative facts about the global behaviour of the system without actually solving the Eqs. (1.1). Following closely on POINCARÉ's work and inspired by it, came LIAPOUNOV's[4] first publications in this field of investigation. LIAPOUNOV, however, soon produced new and profound ideas, supplying stimuli to this study and stability that even now has hardly begun to abate. We shall see that much of the theory we discuss in this treatise is attributable to him. In spite of its importance, LIAPOUNOV's work remained unknown outside Russia for a long period, and development of POINCARÉ's ideas on dynamical systems was made by BENDIXSON[5] and then at a later date by BIRKHOFF[6] independently of the work by LIAPOUNOV and his co-workers. Among BIRKHOFF's contributions to the subject was the introduction of many new notions, including those of limit sets. Although various generalisations of these early ideas were subsequently made by other authors, the first definitions of an abstract system appear to have been formulated by MARKOV,[7] WHITNEY,[8] and BARBASHIN.[9] They retained in their definitions all those properties of a system of ordinary differential equations which enabled the previously obtained qualitative results to hold. At the present time, the spaces used in establishing abstract dynamical systems are far more general than the euclidean spaces of the original ordinary differential equations and enable a wider concept of dynamical system to be employed. Indeed, in the definition introduced below an attempt is made to dispense altogether with an underlying space and to use instead an underlying set possessing hardly any structure. This is done for the express purpose of emphasising that (Liapounov) stability is a primitive concept valid in very general circumstances. What is more significant, we shall see that several of the fundamental theorems are also valid for equally general circumstances. Naturally, refinements of both concepts and theorems call for a corresponding sophistication in the structure of the underlying set on which the system is defined.

---

[1] See, e.g., NEMYTSKII and STEPANOV [1947, *3*], GOTTSCHALK and HEDLUND [1955, *4*], BHATIA and SZEGÖ [1967, 5; 1970, *5*], ROSENBROCK and STOREY [1970, *27*], and especially ZUBOV [1964, *17*].

[2] Cp., for instance, ZUBOV [1964, *17*], WANG [1966, *32*], HALE [1969, *10*].

[3] POINCARÉ [1881, *2*], [1882, *1*].

[4] LIAPOUNOV [1892, *1*].

[5] BENDIXSON [1901, *1*].

[6] BIRKHOFF [1927, *1*].

[7] MARKOV [1931, *2*].

[8] WHITNEY [1932, *1*].

[9] BARBASHIN [1948, *1*].

2. General features of a dynamical system. In order first to give an introductory idea of the nature of a dynamical system, we describe the basic ingredients required in the precise formulation of the next section. These ingredients are:

(i) An independent time variable, t, which takes values in a real interval T, finite or infinite.

(ii) A set X whose elements are of a certain class (to be designated when the specific system is prescribed). At any instant of time t in the given interval T, each element of X usually lies in the range of a mapping of a subset B of $\mathbb{R}^3$. When a physical system is being considered, X, for example, might contain at any instant position vectors, displacements, velocities, stresses, temperature, etc.[10]

(iii) A function φ which maps T into[11] X (i.e., $\varphi: T \to X$) *subject to certain restrictions* to be specified. The images of t in X may thus be designated $\varphi(t)$.

We may then loosely say that the *dynamical system* comprising these ingredients is considered to be the *set* $\mathscr{B}(T, X)$ *of all functions* φ *conforming to the imposed restrictions.*

The set $\mathscr{B}(T, X)$ defines a function space with X as the set of *values* of its functions. A *motion* in $\mathscr{B}(T, X)$ is said to be the function whose values are given by $\varphi(\tau+t)$ for some fixed instant $\tau \in T$ and with $t \in T$, $t \geqq 0$, wherever $\tau+t$ is defined. The basic requirement on φ in order that $\mathscr{B}(T, X)$ should define a dynamical system is that the *associated motion should be encompassed in the system for all* $t \in T$. This is the primitive restriction on φ implied in (iii) above. The specific system with given set X will be known once the rule φ has been particularised. The function φ may be thought of as a *solution* in the system.

Examples of dynamical systems are to be found in the system of ordinary differential Eqs. (1.1) (also in the corresponding non-autonomous system), in functional differential equations and certain partial differential equations, including systems of such equations.[12] When the dynamical system concerns differential equations, the elements of X will be chosen to satisfy certain smoothness conditions and the boundary conditions to the problem, while the mapping φ is determined, in part, by the equation. In particular, X could be a euclidean, Banach or metric space.

A simple example of a dynamical system is given by the set of solutions to the wave equation

$$u_{,xx} = \ddot{u} \qquad 0 \leqq x \leqq 1, \qquad 0 \leqq t < \infty \tag{2.1}$$

with displacement $u(t, x)$ subject to the boundary conditions

$$u(t, 0) = u(t, 1) = 0. \tag{2.2}$$

The space X is now the subset of $C[0,1]$ limited to functions ψ such that

(i) they are real-valued and have continuous derivatives up to the second order;

(ii) they satisfy the conditions $\psi(0) = \psi(1) = 0$.

The system defined by (2.1) and (2.2) is therefore the set of all functions $u(t, \cdot)$ defined on $[0, \infty)$ taking values in X such that (2.1) is satisfied. X could have been chosen alternatively as a Sobolev space.

---

[10] As usual, these variables are labelled by the names of the physical quantities, the values of whose measures they assume at any given instant.

[11] Neither surjective nor injective.

[12] Hale [1969, *10*]. Other examples are considered, for instance, by Zubov [1964, *17*].

Another example is furnished by the solutions to the linear equations of thermoelasticity, discussed in Sect. 24. SLEMROD and INFANTE[13] have demonstrated that the generalised solutions may be regarded as a dynamical system in which the set X is an appropriate Banach space, provided that the coefficients in the equations are smooth functions of position, and that certain symmetry and positive-definiteness conditions are satisfied.

3. Definition of a dynamical system.[14] We now proceed to formalise these brief introductory remarks by defining the abstract dynamical system on which the subsequent developments of this treatise will be based.

Thus, let X be an arbitrary set containing all the relevant dependent variables and let T be a locally compact semi-group with identity 0, which for most purposes can be taken to be the set of non-negative real numbers. Let $\mathscr{B}(T, X)$ denote a set of functions φ defined on T with values in X, i.e. $\varphi: T \to X$. Now suppose that $\mathscr{T}$ is a subset of T. Then for each $\tau \in \mathscr{T}$ and $\varphi \in \mathscr{B}(T, X)$ we introduce a notation φ_τ for the *translate* of φ defined on T by

$$\varphi_\tau(t) = \varphi(\tau + t) \qquad t \in T. \tag{3.1}$$

In general the translate need not belong to $\mathscr{B}(T, X)$.

The definition of a dynamical system is now as follows:

Definition[15] *3.1.* A dynamical system corresponding to a triple $(T, \mathscr{T}, X)$ is a set $\mathscr{B}(T,X)$ of functions defined on T taking values in X such that

(i) $\varphi_\tau \in \mathscr{B}(T, X)$ whenever $\varphi \in \mathscr{B}(T,X)$, $\tau \in \mathscr{T}$,

(ii) $\lim_{t \to 0} \varphi_\tau(t) = \varphi(\tau)$, $\varphi \in \mathscr{B}(T,X)$, $\tau \in \mathscr{T}$.

Although condition (ii) of the definition is a natural requirement of a dynamical system, it is unnecessary in certain theorems on stability given later.

In Sect. 2 certain traditional expressions (motion, solution) with geometrical connotations were used heuristically. Precise definitions will now be given of terms which play a useful role in describing the behaviour of a dynamical system and in assisting with the fixing of ideas where geometrical analogies are relevant.

Definition 3.2. The function φ is the *solution*.

Definition 3.3. The *motion* is the translate, φ_τ.

Definition 3.4. The *trajectory* is the set of all pairs $(t, \varphi_\tau(t))$ for $t \in T$. That is, it is the graph of *the motion*.

Definition 3.5. The *orbit* is the projection of the trajectory onto X. That is, it is the set of values, $\gamma(\varphi(\tau))$, of the *motion*:

$$\gamma(\varphi(\tau)) = \{\varphi_\tau(t) : t \in T\}.$$

It should be observed that the *motion* is a function belonging to $\mathscr{B}(T, X)$, whereas the *trajectory* and *orbit* are sets derived from the values of this function. For $X = E^2$, the euclidean two-space, the geometrical representation of these definitions is given in Fig. 1.

[13] SLEMROD and INFANTE [1971, *12*]. Their work is based, in part, on that of DAFERMOS [1968, *7*], and employs a dynamical system defined in the sense of HALE. See Sect. 4.

[14] This chapter is based largely on the relevant portions of the article by GILBERT and KNOPS [1967, *10*].

[15] GILBERT and KNOPS [1967, *10*].

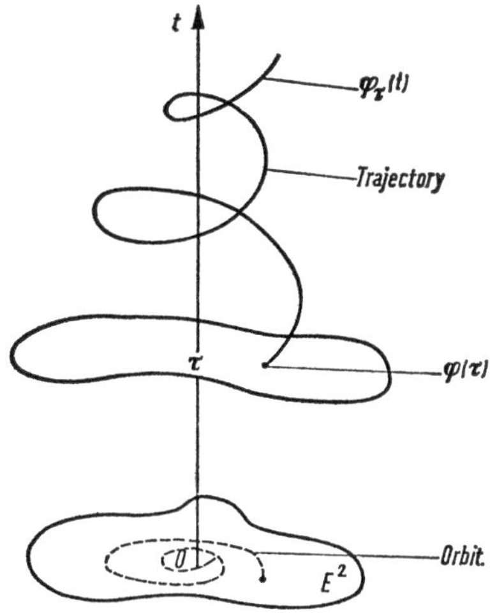

Fig. 1. Geometrical representation of the trajectory and orbit.

Three further definitions are also set down here. Although these are not immediately required in our treatment of stability they will be needed later in connexion with invariance principles and the concept of asymptotic stability.[16]

Definition 3.6. A set $M \subset X$ is said to be an *invariant set* of the dynamical system $\mathscr{B}(T, X)$ if for each $\varphi(\tau) \in M$ and $\varphi \in \mathscr{B}(T, X)$ it follows that $\gamma(\varphi(\tau)) \subset M$.

In the literature of dynamical systems it is usual to distinguish between a positively invariant set, which corresponds to what we have here called an invariant set, and an invariant set, which is defined more restrictively.[17] With these definitions, every invariant set is a positively invariant set, but not *vice versa*. In elasticity, moreover, little meaning can be attached to an invariant set in the more restricted sense, whereas a positively invariant set is meaningful. Also, when the definition of a positively invariant set is used none of the relevant theorems is invalidated. For the purposes of this treatise, therefore, there is little point in modifying the idea given in Definition 3.6.

Examples of invariant sets, most commonly used here, are the values of equilibrium solutions, defined below, and the values of *periodic* motions, defined by

$$\varphi_\tau(t) = \varphi_\tau(t+s),$$

where s is the *period* of the motion.

Definition 3.7. The solution φ is an *equilibrium solution* when $\varphi_\tau(t) = \varphi(\tau)$, $\varphi \in \mathscr{B}(T, X)$, $\tau \in \mathscr{T}$, $t \in T$.

Definition 3.8. The identically vanishing equilibrium solution $\varphi_\tau(t) = 0$, $t \in T$, $\tau \in \mathscr{T}$ is the *null solution*.

Definition 3.9. An *equilibrium point* $\varphi(\tau) \in X$ is such that $\varphi(\tau) = \varphi_\tau(t)$, $\varphi \in \mathscr{B}(T, X)$, $\tau \in \mathscr{T}$, $t \in T$.

An equilibrium solution is alternatively designated *critical*, *rest* or *stationary*. It is clearly independent of time and is a solution of a static problem in the

[16] Invariance principle is used here in the sense of LASALLE [1967, *20*] for ordinary differential equations. For generalisation to abstract dynamical systems on a Banach space, see HALE [1969, *10*], SLEMROD [1969, *21*], SLEMROD and INFANTE [1971, *12*].

[17] HALE [1969, *10*], KERSHAW [1970, *17*].

system. It may be non-unique in the sense that more than one equilibrium solution may exist under given boundary conditions. It should be kept in mind, however, that when an equilibrium solution is regarded as belonging to a dynamic problem, not only must it satisfy certain boundary data but also *initial data.* Uniqueness of the equilibrium solution then depends upon the properties of the dynamic problem. It should also be remarked, here, that for non-linear problems, the uniqueness of one solution does not imply that of any other solution in the system. The preceding observations are particularly significant when the so-called "adjacent equilibrium" criteria for stability have to be handled, and also when the relation between stability and buckling is considered.

4. The set of initial data and a related mapping. In order to investigate the behaviour of a dynamical system it is necessary to introduce further concepts. In most previous discussions of stability it has been customary to deal with the set X explicitly and formulate theorems in terms of neighbourhoods defined in this set. We here take a different approach and work with a subset $Y(\tau)$ of X and a mapping T_τ from $Y(\tau)$ to $\mathscr{B}(T,X)$ for each $\tau\in\mathscr{T}$. This procedure allows a greater simplicity in the statement of stability theorems and in presenting the ideas underlying the proofs of some of these theorems. Alternative formulations of these proofs relying upon geometrical notions are however possible and a brief comparison with this approach is given in Chaps. C and D, where definitions and concepts of other authors are also contrasted.

Let $Y(\tau)$ be the subset of X defined as follows:

$$Y(\tau)=\{\varphi(\tau):\ \varphi\in\mathscr{B}(T,X),\ \tau\in\mathscr{T}\subseteq T\}.$$

This may be regarded as the set of "initial" values of the motions φ_τ which in many problems will be a proper subset of X. For example, it is well-known that in the system defined by the wave equation of spatial dimension higher than one under Cauchy data, the solution for $t>0$ is not always accommodated in the space of initial data, but in some larger space.[18]

The mapping T_τ is now the map from the "initial" values $Y(\tau)$ to $\mathscr{B}(T,X)$ given by:

$$T_\tau:\ \varphi(\tau)\to\varphi_\tau,\qquad \varphi\in\mathscr{B}(T,X),\qquad \tau\in\mathscr{T}.$$

The map T_τ need not be single-valued unless the solutions of the system are unique, but at no stage in the proof of stability theorems is uniqueness required. In fact, with the definition of stability given in this treatise, *uniqueness will be shown to be implied by stability.*

Among the dynamical systems that have features in common with the preceding definition, mention should be made of the "general" system of ZUBOV[19] and the system defined by HALE.[20] ZUBOV defines his system on a metric space, and in this comes closest to our definition. On the other hand, HALE's *definition,* on which are based some of the theorems used later, specifies the system to be the function $u:\mathbb{R}^+\times B\to B$, where B is a Banach space. In addition, it stipulates

(a) a homomorphism axiom equivalent to

$$u(t+t_1,\psi)=u(t,u(t_1,\psi)),\qquad t,t_1\in\mathbb{R}^+,\qquad \psi\in B,$$

[18] See, e.g. COURANT and HILBERT [1962, *3*, p. 669], SOBOLEV [1964, *12*, p. 201], HELLWIG [1960, *4*, p. 23].

[19] ZUBOV [1964, *17*, p. 199]. On p. 20, ZUBOV gives another definition similar to that of NEMYTSKII and STEPANOV [1947, *3*].

[20] HALE [1969, *10*]. References to definitions by some other authors are contained in the footnotes to Sect. 1.

(b) $u(0,\psi)=\psi, \quad \psi\in B$,

(c) the assumption that u is continuous in a suitable sense.

The connexion with our definition of a dynamical system may be seen from the equivalence relations

$$\left.\begin{aligned} u(t,\psi)&=\varphi_\tau(t),\\ \psi&=\varphi(\tau), \end{aligned}\right\} \quad \tau, t\in \mathbb{R}^+,$$

with the set X becoming the Banach space B. But, in general, our definition omits both (a) and (c) as neither is required for the analysis of stability, and it is only in later problems involving asymptotic stability and instability that HALE's definition must be called upon.

Instead of our mapping T_τ, some authors[21] use a *phase map*

$$\pi\colon\ T\times Y(\tau)\to X,$$

defined by

$$\pi(t,\varphi(\tau))=\varphi(\tau+t), \quad \varphi(\tau)\in Y(\tau), \quad \varphi\in\mathscr{B}(T,X), \quad \tau\in\mathscr{T}, \quad t\in T.$$

The motion then becomes the induced map:

$$\pi_{\varphi(\tau)}\colon\ T\to X.$$

The phase map, as well as inducing the motion, also induces the *action*, which is the map $\pi_t\colon X\to X$ defined by

$$\pi_t(\varphi(\tau))=\varphi(\tau+t), \quad \tau\in\mathscr{T}, \quad t\in T.$$

This last map allows us to interpret equilibrium points (Definition 3.9) as fixed points in a *transition map*, defined to be the mapping of the subset Y of initial values into itself,

$$\pi_t\colon\ Y(\tau)\to Y(\tau).$$

C. Definitions of stability.

5. Introduction. In the "dynamical system" we have the means of establishing mathematical models for a wide range of mechanical and other physical systems. When creating a model we attempt to characterise properties of the physical system by the structure of the dynamical system, that is by the set X and the mapping $\varphi\colon T\to X$. A qualitative description of the map, or solution, φ is furnished by the notion of stability. The purpose of this chapter is therefore to consider stability in relation to dynamical systems and to provide abstract definitions of stability in their different forms we follow this by discussion of the associated concepts of instability, boundedness and uniqueness. In later chapters these are related to the particular study of elastic systems.

Historically, ideas on stability of *discrete* systems culminated with LAGRANGE's definition[1] in a version modified by DIRICHLET.[2] As it is the Lagrange-Dirichlet concept which has been the basis of most subsequent definitions of stability we quote here DIRICHLET's modified version which is more appropriate in the present context. This refers to the stability of equilibrium: "... the equilibrium is stable if, in displacing the points of the system from their equilibrium positions by an infinitesimal amount and giving each one a small initial velocity, the displacements of different points relative to the equilibrium positions of the system, remain,

[21] E.g., BHATIA and SZEGÖ [1967, *5*], [1970, *5*].

[1] LAGRANGE [1788, *1*].

[2] DIRICHLET [1846, *1*].

throughout the course of the motion, contained between certain small prescribed limits."[3]

In order to analyse the fundamental elements of this definition prior to giving abstract definitions, it is instructive to reword it in an alternative form, (presupposing a suitable definition of neighbourhood): "The given (equilibrium) solution is stable if and only if all solutions starting in the neighbourhood of the given solution remain in that neighbourhood for all times $t \in T$."

The essential ingredients of this definition, when critically analysed, are as follows:

(a) Stability is a quality of one solution—an equilibrium solution of the system. LAGRANGE, DIRICHLET and other writers treat this solution only, but this restriction may be removed and stability regarded as a property of *any particular solution.*

(b) Initial and subsequent perturbations have been described by certain measures tacitly taken to be the same in each case and to involve both displacement and velocity in some undefined way. When generalising, these measures may be made distinct from one another and may involve variables other than displacement and velocity.

(c) The problem of ascertaining the stability of a solution concerns the "neighbourhood" of the particular solution and is therefore a *local one.* When stability is formulated in terms of "neighbourhoods", a connexion with the concept of "continuity" is suggested. That stability and continuity in suitable contexts are equivalent concepts will be shown later.

In the light of this analysis it is seen that progress towards formulating an abstract definition based on the ideas set out by LAGRANGE and DIRICHLET can only be made if the notions of *measure*[4] and *neighbourhood* are first made precise. The additional concept of *equivalence* is also of importance and will be introduced in the course of this chapter. We therefore proceed to give definitions of these and associated terms.

Definition 5.1. Given a set X with elements $x_1, x_2 \in X$, a real scalar-valued function $\varrho(x_1, x_2)$ defined on $X \times X$ is said to be *positive-definite* if

(i) $\varrho(x_1, x_2) > 0 \qquad x_1, x_2 \in X, \qquad x_1 \neq x_2.$

(ii) $\varrho(x_1, x_2) = 0$ if and only if $x_1 = x_2.$

Any such positive-definite function can be used to define a neighbourhood of a solution $\varphi \in \mathscr{B}(T, X)$. In order to achieve this let $\varrho(\varphi, \psi)$ be given by

$$\varrho(\varphi, \psi) = \sup_{t \in T} \varrho(\varphi(t), \psi(t)) \qquad \varphi, \psi \in \mathscr{B}(T, X),$$

[3] The above transcription is taken from the French note by LEJEUNE DIRICHLET appearing as an appendix to LAGRANGE's Mécanique Analytique [1853, *1*]. The passage in question exhibits slight differences from that appearing in the original paper of LEJEUNE DIRICHLET [1846, *1*]. We quote this passage so that the reader may form his own interpretation of its meaning: "Findet wirklich das Maximum Statt, so hat das Gleichgewicht den Character der Stabilität, d.h. das System wird sich, wenn die Puncte aus einer solchen Lage nur wenig verrückt werden und kleine Anfangsgeschwindigkeiten erhalten, im Laufe der Zeit nie über gewisse enge Grenzen hinaus von derselben entfernen." Other early definitions of stability are due, for instance, to MAXWELL [1856, *1*, p. 295], ROUTH [1877, *2*], KLEIN and SOMMERFELD [1897, *2*]. A discussion of some of these is contained in the articles by STOKER [1955, *7*], [1967, *32*], while a fuller description is provided by MOISEEV [1949, *2*].

[4] Throughout the treatise, the meaning of the term measure is slightly different from that used in analysis.

and define the open ball with centre φ and radius r to be the set in $\mathscr{B}(T,X)$:

$$S(\varphi, r)=\{\psi\in\mathscr{B}(T,X): \varrho(\varphi,\psi)<r\} \qquad \varphi\in\mathscr{B}(T,X),$$

where r is a non-negative real number.

Definition 5.2. A subset A of $\mathscr{B}(T,X)$ is said to be a *neighbourhood of* $\varphi\in\mathscr{B}(T,X)$ if and only if there exists a positive number r such that $S(\varphi, r)\subseteq A$.[5]

Examples of positive-definite functions that will be used subsequently are the norms on the usual linear functional spaces, such as those of HILBERT and SOBOLEV, and also the uniform (or supremum) metric on the set of continuous bounded real valued functions defined on a euclidean space. In fact, whenever X is a normable linear space we shall always assume that the positive-definite function arises from a norm $\|(\cdot)\|$ defined on X.

The reader should note, however, that the class of positive-definite functions is larger than the class of metrics.[6]

In some circumstances it is necessary to indicate both the set and the positive-definite function defined on it. In this case we shall write $Y_{\varrho_\tau}(\tau)$ or X_ϱ for sets $Y(\tau)$ and X respectively. Where two different functions ϱ_1, ϱ_2 are defined on X the notation for X will be X_{ϱ_1} or X_{ϱ_2}. From the definition of $\mathscr{B}(T,X)$ it also follows that if a positive-definite function ϱ is defined on $X\times X$ then a positive-definite function is induced on $\mathscr{B}(T,X)$. This will be denoted by $\mathscr{B}_\varrho(T,X)$.

6. Definition of Liapounov stability. With these preliminaries it is now possible to give a definition of stability. Of all the definitions that have been proposed and are possible in an abstract framework, LIAPOUNOV'S[7] extension of LAGRANGE'S statement is perhaps the most natural. Although many writers, especially on continuum mechanics, fail to be explicit about their notion of stability, in most cases it is clear that the Liapounov definition underlies their work. This definition is therefore given first, formulated however, in terms of the map T_τ defined in Sect. 4.

Discussion of modifications of the definition and also of associated concepts and their relation to stability follow later in Sect. 7.

Definition 6.1.[8] A solution $\varphi\in\mathscr{B}(T,X)$ is said to be *Liapounov stable* if and only if, for each $\tau\in\mathscr{T}$, the mapping T_τ from $Y(\tau)$ to $\mathscr{B}(T,X)$ is continuous at φ. That is, for each $\tau\in\mathscr{T}$ and for each positive real number ε, there exists a positive real number δ which depends on τ and ε such that

$$\varrho_\tau(\varphi(\tau),\psi(\tau))<\delta$$

---

[5] If it is desired to equip $\mathscr{B}(T, X)$ with a topology, one method is to make X into a metric space. This requires that the positive-definite functions satisfy the triangle inequality and symmetry condition:

(iii) $\varrho(x_1, x_2)\leqq\varrho(x_1, x_3)+\varrho(x_3, x_2), \quad x_1, x_2, x_3\in X,$

(iv) $\varrho(x_1, x_2)=\varrho(x_2, x_1), \quad x_1, x_2\in X.$

Another, but perhaps less obvious method, which does not require conditions (iii) and (iv) is to form a base from the sets

$$S(\varphi, r), \quad S(\varphi_1, r_1)\cup S(\varphi_2, r_2), \quad \varphi_1\neq\varphi_2$$

for all φ, φ_1, $\varphi_2\in\mathscr{B}(T, X)$ and positive r, r_1, r_2, and use TAYLOR'S construction: TAYLOR [1958, *2*, p. 58].

[6] For further comments on positive-definite functions, see BHATIA and SZEGÖ [1967, 5, p. 4].

[7] LIAPOUNOV [1892, *1*].

[8] The form of the definition presented here is due to GILBERT and KNOPS [1967, *10*] and MOVCHAN [1960, *7*]. MOVCHAN'S definition corresponds to the second form of the statement above with additional continuity assumptions on ϱ_τ, ϱ.

implies

$$\varrho(\varphi_\tau, \psi_\tau) < \varepsilon, \qquad \varphi, \psi \in \mathscr{B}(T, X).$$

Here, $\varrho(\varphi_\tau, \psi_\tau)$ is a positive-definite function defined by

$$\varrho(\varphi_\tau, \psi_\tau) = \sup_{t \in T} \varrho\big(\varphi_\tau(t), \psi_\tau(t)\big),$$

in agreement with Definition 5.1.

As has already been mentioned in Sect. 3, in most applications the set T will be the semi-infinite interval of non-negative real numbers. When stability is limited to closed bounded intervals the definition is equivalent to that of continuous dependence on the data. This is discussed in Sect. 8.

We now make the following observations on this definition:

(i) This is a definition of *stability of a given motion*,[9] free from any reference to an equilibrium solution.

(ii) The definition makes it clear that for a *given* dynamical system a solution may be stable for one choice of positive-definite function but not for another.[10]

(iii) When the mapping T_τ is a homeomorphism at each τ, it can be shown that, if φ is stable at any instant $\tau \in \mathscr{T}$, then it is stable for all instants.[11]

In connexion with (i) it should be observed that when X is a normed *linear* space, and when the set $\mathscr{B}(T, X)$ is also linear, by means of the substitution $\Phi = \varphi - \psi$, it is always possible to express the stability of a given φ in terms of the stability of the null solution. Thus, the stability of every solution belonging to a linear system is determined by the stability of the null solution. For this reason in most researches into linear elastic systems it is only the stability of null equilibrium solutions that needs investigation.

When the system is non-linear the behaviour of the null solution does not describe the behaviour of all solutions. It is therefore natural to introduce the following definition:

Definition 6.2. The dynamical system defined by the triplet $(T, \mathscr{T}, X)$ is said to be *stable* if and only if every solution $\varphi \in \mathscr{B}(T, X)$ is stable.

We now turn to observation (ii) concerning the choice of measures. As this feature of the theory is one on which stability is so finely hinged, we now comment on the way in which the measures influence the estimate of stability and illustrate this with several examples.

It has been seen in (b) of Sect. 5 that the Lagrange-Dirichlet definition is imprecise about the manner in which the measures of perturbation of a system are prescribed. It also tacitly assumes the use of the same measure for initial and subsequent perturbations. That different measures should be adopted for these disturbances was apparently first suggested by MOVCHAN[12] who used positive-definite functions for these quantities. In the present analysis it is clear

[9] Stability of motions was probably first considered by ROUTH [1860, *1*], [1877, *2*].

[10] This dependence on the characterisation of a neighbourhood is similar to the well-known property of the calculus of variations in which a functional may have a minimum in one space but not in another. Cf. SAGAN [1969, *19*, p. 82–89], COURANT and HILBERT [1953, *1*, p. 173], HADAMARD [1907, *1*], and the remarks by DUHEM [1902, *1*].

[11] See HALE [1969, *11*, p. 27], CESARI [1963, *1*, p. 5].

[12] MOVCHAN [1959, *5*], [1960, *7*]. The positive-definite functions used by MOVCHAN were required to obey certain other conditions (see discussion of Theorem 15.1).

that positive-definite functions define neighbourhoods whose purpose is to accommodate the perturbations. *It is thus necessary for us to assume that on the set $X \times X$ of our dynamical system, a positive-definite function ϱ can be defined which is suitable to serve as a measure for the general perturbation.* Further, for each $\tau \in \mathcal{T}$ we assume the existence of a positive-definite function ϱ_τ defined on $Y(\tau) \times Y(\tau)$ so that neighbourhoods on $Y(\tau)$ are defined by ϱ_τ. This item serves as a measure of the initial disturbance.

We have just emphasised that the existence of a positive-definite measure is assumed in order to define neighbourhoods on the set X. There are circumstances, however, in which measures of the type required may fail to exist for a particular X. The failure might result from one of a number of different factors including the unboundedness of the measures. This last mentioned case naturally contravenes definitions of stability set up in terms of finite measures and this aspect of the study is discussed further in Sect. 11.

Hence, from Definition 6.1 it is clear that our aim is to construct neighbourhoods of a given solution on the set $\mathcal{B}(T, X)$. Measures in this set are induced by those in $X \times X$ so that failure to define neighbourhoods in $\mathcal{B}(T, X)$ might arise as a consequence of the definition of X or its measures. In particular, as the measure on $\mathcal{B}(T, X)$ is by definition $\sup\limits_{t \in T} \varrho$, unboundedness of ϱ on $X \times X$ means unboundedness of the measure on $\mathcal{B}(T, X)$. This could occur at a finite time $t \in T$ or in the limit as $t \to \infty$. In the trivial case in which $\mathcal{B}(T, X)$ happens to be a singleton or null, then, although measures can be induced on $X \times X$, no neighbourhoods of $\mathcal{B}(T, X)$ exist and no perturbations are possible. In all cases in which Definition 6.1 cannot be applied, we shall say that the solution is *Liapounov unstable*. There are situations, however, like that of the singleton above for which the measures may become undefined without becoming unbounded. In these circumstances some may prefer to regard the situation as vacuous rather than as one yielding instability. Whenever the measures admit definition on the class of solutions in $\mathcal{B}(T, X)$, a formal definition of Liapounov instability is given in Sect. 9.

The final choice of ϱ_τ and ϱ depends on the particular investigation. Provided that the above assumptions on existence and boundedness of these measures are not violated, Liapounov stability may be investigated with arbitrarily selected ϱ_τ and ϱ. If, for example, the *overall* value of displacement is judged to give a reliable description of stability then the mean-value integrals

$$\int_\Omega u^2 \, d\Omega \quad \text{or} \quad \int_\Omega |u| \, d\Omega$$

may be used.[13] On the other hand, where *local* behaviour of the body is of importance, as in fracture, a uniform measure such as $\sup\limits_\Omega |u|$ must be employed. In general, where there are several measures consistent with the specification of the system their choice, like that of safety factors, must be left to the engineer. The proper function of the theorist, in this regard, should be that of exhaustively exploring the consequences of choosing all possible measures for a given system.

We now illustrate how the estimate of stability can be influenced by choice of measures in the following examples. These examples also show how the prescription of the dynamical system itself influences stability of a given solution.

---

[13] Here, Ω represents either a surface or volume measure.

A) Laplace equation.[14]

Let T be the semi-group of all non-negative real numbers with $\mathcal{T}=T$. We consider the solution $u(t, x)$ of the two-dimensional Laplace equation

$$u_{,xx}+\ddot{u}=0, \quad 0\leqq x\leqq 1, \quad t>0, \tag{i}$$

which vanishes at $x=0$, and is subject to the initial data

$$u(0, x)=0, \quad \dot{u}(0, x)=\frac{1}{n}\sin n\pi x, \quad n=0, 1, \ldots \tag{ii}$$

at $\tau=0$. The set $Y(\tau)$ thus becomes the set $Y(0)$ and is defined by this data, while the set X is the subset of functions of class $C^2[0,1]$ which vanish at $x=0,1$. The dynamical system representing this problem consists of all functions defined on bounded sub-sets of T and taking values in X such that (i) is satisfied.

A solution to (i), (ii) is

$$u(t, x)=\frac{1}{n^2\pi}\sin n\pi x \sinh n\pi t.$$

Hence on choosing the positive-definite functions to be

$$\varrho_\tau(u(\xi))=\varrho(u(\xi))=\sup_{x\in[0,1]}|u(\xi, x)|,$$

we see that the null solution is unstable. However, on taking for ϱ the form

$$\varrho(u(t))=\tfrac{1}{2}\int_0^1[\dot{u}^2-u_{,x}^2]\,dx,$$

we again recover stability. When the solutions are restricted so as to be bounded, stability of the null solution may be recovered on compact sub-intervals of the time interval,[15] allowed to be infinite in the above example. Stability obtained in this sense may not necessarily correspond to Liapounov stability. The latter method of regaining stability illustrates one case in which altering the specification of the system may change the estimate of stability.

B) Simply supported elastic plate.[16]

We consider next the problem of the bending of an elastic thin plate of infinite length simply supported along two edges and subject to a constant force in its plane. The transverse displacement $u(t, x)$ then satisfies the equation

$$u_{,xxxx}-\alpha u_{,xx}+\ddot{u}=0, \quad 0\leqq x\leqq 1, \quad t>0 \tag{i}$$

with constant $\alpha>-\pi^2$, and is subject to the boundary conditions

$$u(t, x)=u_{,xx}(t, x)=0 \quad \text{at } x=0,1 \quad t>0. \tag{ii}$$

For the set $Y(\tau)$ of initial data we take

$$u(0, x)=c\sin n\pi x, \quad \dot{u}(0, x)=0, \quad 0\leqq x\leqq 1 \quad t>0 \tag{iii}$$

[14] This example is well-known, being due to HADAMARD, who apparently first announced it at a meeting of the Swiss Mathematical Society in 1917. (See HADAMARD [1923, *1*].) It helps to reveal the relation between the notions of "well-posedness" and stability. See Sect. 8.

[15] LAVRENTIEV [1962, *10*], [1967, *21*] see also Chap. L and the discussion by F. JOHN [1970, *15*, p. 275].

[16] Elements of this discussion may be found in the articles of MOVCHAN [1959, *5*] and WANG [1966, *32*].

where c is arbitrary and n is an integer. A solution under these conditions is

$$u(t, x) = c \sin n\pi x \cos \beta_n t, \quad \beta_n^2 = n^2 \pi^2 (n^2 \pi^2 + \alpha). \tag{iv}$$

We now choose T, $\mathscr{T}$, τ as in the previous example and for X take the set of all functions $C^4[0, 1]$ which together with their second spatial derivatives vanish at $x=0$, $x=1$. The dynamical system corresponding to the problem given by (i), (ii), (iii) is the set of all functions defined on T and taking values in X such that (i) is satisfied.

Let us now examine the stability of the null solution of (i) with respect to the positive-definite functions ϱ_τ, ϱ given by

$$\varrho_\tau(u(\xi)) = \varrho(u(\xi)) = \sup_{x \in [0,1]} [|u(\xi, x)| + |\dot{u}(\xi, x)|] \tag{v}$$

in the class of perturbed displacements $u(t, x)$ in X. Then again at $\tau=0$, we have

$$\varrho_0(u(0)) = c$$

$$\varrho(u) \equiv \sup_{t>0} \varrho(u(t)) = |c| (1 + \beta_n^2)^{\frac{1}{2}},$$

so that no matter how small c is chosen, we can always select n of such a value that $\varrho(u)$ is arbitrarily large. The null solution is thus unstable for choice (v) of positive-definite functions. The same conclusion also applies when ϱ_τ takes the value given in (v) but ϱ is the function given by the total energy:

$$\varrho(u(t)) = \tfrac{1}{2} \int_0^1 [u_{,xx}^2 + \alpha u_{,x}^2 + \dot{u}^2]\, dx. \tag{vi}$$

The total energy may be made arbitrarily large with increasing n.

On the other hand there is stability of the null solution in $\mathscr{B}(T, X)$ when ϱ_τ is chosen to be either the total energy (vi) or the measure

$$\varrho_\tau(u(\tau)) = \sup_{x\in[0,1]} \left| \frac{\partial^4 u}{\partial x^4} \right|$$

with either (v) or (vi) as choice of ϱ.

The foregoing provides an example of the manner in which stability can be influenced by the choice of measure. The previous general discussion shows, however, that stability can also depend on the choice of the set X or $Y(\tau)$. In this example stability may be recovered with respect to the measures (v) if the set $Y(\tau)$ is changed to

$$u(0, x) = \frac{c}{n^2} \sin n x, \quad \dot{u}(0, x) = 0.$$

C) The three-dimensional wave equation.[17]

We now turn to the wave equation

$$u_{,ii} = \ddot{u} \quad \text{in} \quad D, \quad t > 0 \tag{i}$$

[17] See e.g., SOBOLEV [1964, *12*, Chap. III, § 18] for the n-dimensional wave equation. The example in the text is based upon the discussion by SHIELD [1965, *21*], but see also PETROVSKY [1954, *1*, § 13], JOHN [1964, *7*, pp. 13–14], and other standard text books.

where D is a domain of three-dimensional euclidean space, with boundary conditions prescribed by

$$u=0 \quad \text{on } \overline{\partial D_1}, \quad t>0, \tag{ii}$$

$$\frac{\partial u}{\partial n}=0 \quad \text{on } \partial D_2, \quad t>0 \tag{iii}$$

where ∂D_1, ∂D_2 are disjoint parts of the surface ∂D of D, such that $\partial D=\overline{\partial D_1}\cup\partial D_2$ and $\boldsymbol{n}$ is the unit outward normal on ∂D_2. A solution of (i) may be expressed in the form

$$u=\frac{1}{r}\,[f(t+r)-f(t-r)], \qquad r^2=x_i\,x_i \tag{iv}$$

where f is three times continuously differentiable. This corresponds to the initial data

$$u(0, x)=2f', \qquad \dot{u}=0,$$

where f' denotes the derivative of f.

For sufficiently small times t and a suitably large domain D we choose f to be

$$f=\frac{1}{\varepsilon^7\,r_0^7}\,(r-r_0)^4\,(r-r_0-2\varepsilon r_0)^4, \qquad r_0\leqq r\leqq r_0+2\varepsilon r_0, \tag{v}$$

with $f=0$ elsewhere.

It then follows that

$$f(r)=0(\varepsilon), \qquad f'(r)=0(1) \qquad \text{as } \varepsilon\to 0.$$

The total energy defined by

$$E(t)=\tfrac{1}{2}\int_D [\dot{u}^2+u_{,i}\,u_{,i}]\,dx$$

is thus of order ε, but the solution $u(t, \boldsymbol{x})$ is of order unity at the origin for the times $r_0<t<r_0+2\varepsilon r_0$.

We take T to be the set of non-negative real numbers and specify that the dynamical system consists of all functions (iv) with f given by (v). It is understood that the limits of these expressions are included in the system so that it consists of certain classical and certain generalised solutions to (i).

Let us now examine the consequences of these results for the stability of the null solution. On choosing the positive-definite functions ϱ_τ and ϱ to be

$$\varrho_0\big(u(0)\big)=E(0)$$

$$\varrho\big(u(t)\big)=\sup_{\boldsymbol{x}\in D}|u(t, \boldsymbol{x})|$$

we see that the null solution cannot be stable in the sense of LIAPOUNOV. On taking

$$\varrho\big(u(t)\big)=E(t)$$

for the same choice of ϱ_τ, however, we recover Liapounov stability of this solution on recalling that (i), (ii), (iii), imply the conservation of total energy $E(t)$. Stability thus depends again on the choice of positive-definite functions.

This phenomenon, when related to a physical system, may be interpreted as the focussing of the waves at the origin. The local peak behaviour is suppressed in a stability measure taken to be the total energy. This behaviour is common to solutions of the wave equation in a space with any odd number of dimensions greater than one. SHIELD and GREEN[18] have applied this example to linear elasticity and ACHENBACH[19] has related it to thermoelasticity. In the elastic model it is shown that a homogeneous isotropic sphere with zero displacement and stress when given an infinitesimal perturbation can develop perturbed displacements that are not infinitesimal for all time at all places. It has been argued that this example of SHIELD and GREEN shows that any elastic body at rest must therefore be unstable, a conclusion which is asserted to be physically unreasonable.

That this assertion may be misleading is indicated by the following facts: (a) the instability appears in a *particular* problem, that of spherical waves in a linear isotropic elastic medium, (b) the spherical waves can be produced by the application of physically realisable loads applied prior to the time $t=0$ (as SHIELD and GREEN have shown), (c) Liapounov instability is measured in this problem against the uniform norm. We thus see that this apparently unusual behaviour may not be unnatural when related to this particular mathematical system and the measure chosen for stability. Some authors adhere to the view that conclusions of this kind are unacceptable and they circumvent the difficulty either by changing the measure[20] or by restricting the class of perturbations.[21] Both methods are discussed in Sects. 35, 37, 50. The above analysis considers only mathematical models irrespective of whether or not they are suitable for representing a physical system.

The above examples serve, among other things, to underline the topological nature of stability. The topological properties concerned are not intrinsic to the dynamical system and must be supplied as additional specifications relevant to the problem in hand. To add further emphasis to this statement, we note that, although positive-definiteness of the stability measures ϱ_τ, ϱ has been specified (in common with MOVCHAN) in Definition 6.1, this is not an essential requirement. Non-negativeness would still be adequate. With this requirement only we have the artificial but illustrative example for which $\varrho_\tau=\varrho=0$. This then makes *every* solution of *any* system stable. There is also a further example, this time consistent with the original definition which does not allow *any* solution of *any* system to be stable. This is the choice

$$\begin{aligned} \varrho&=0, & x_1&=x_2,\\ \varrho&=1, & x_1&\neq x_2. \end{aligned}$$

It is now a natural question to ask in connexion with the topological character of stability whether a solution, stable under one pair of measures ϱ_τ, ϱ is stable under any other pair $\hat{\varrho}_\tau$, $\hat{\varrho}$. This situation can in fact occur when the following equivalence relations hold between the respective pairs of measures:

$$\left.\begin{aligned} \varrho_\tau(x,y)&\leqq\alpha\,\hat{\varrho}_\tau(x,y)\\ \hat{\varrho}(x,y)&\leqq\beta\,\varrho(x,y) \end{aligned}\right\}\quad x,\ y\in X$$

where α, β are positive real numbers. It should be emphasised that analytically these relations correspond to semi-equivalence but despite this we adhere to the term equivalence in this context.

[18] SHIELD and GREEN [1963, *18*]. See also KOITER [1963, *7*], HSU [1966, *12*] and NEMAT-NASSER and HERRMANN [1966, *22*].

[19] ACHENBACH [1967, *1*].

[20] This is the approach favoured, for instance, by HSU [1966, *12*] and KOITER [1963, *7*], [1965, *12*], [1967, *18*], who use, for example, L_2-norms of the displacements.

[21] Cf. SHIELD [1965, *21*].

When X is an appropriate normed linear space with suitable norms as measures the well-known Sobolev embedding theorems may be applied to establish equivalence in certain circumstances.[22]

Equivalent formulations of Definition 6.1 may be given in terms either of neighbourhoods or in terms of equicontinuity.[23] The first reformulation is effectively a paraphrase of the main definition, while the second is enunciated in terms of the action maps defined in Sect. 4. Each is easily seen to be mathematically equivalent to Definition 6.1.

Definition 6.1A. The solution $\varphi \in \mathcal{B}(T, X)$ is said to be *stable relative to the set* $(X, \varrho_\tau, \varrho)$ if and only if the map T_τ from $Y(\tau)$ to $\mathcal{B}(T, X)$ is continuous at φ with the neighbourhoods of φ on $Y(\tau)$ and $\mathcal{B}(T, X)$ defined respectively by ϱ_τ and ϱ.

Definition 6.1B. A solution $\varphi \in \mathcal{B}(T, X)$ is said to be *stable* if and only if for each $\tau \in \mathcal{T}$ the family of maps π_t from $Y(\tau)$ to X are equicontinuous at φ.

The accompanying diagrams give a representation of the geometrical concepts underlying Liapounov stability. Fig. 2 employs a representation of stability in terms of trajectories of motions rather than in terms of the set $\mathcal{B}(T, X)$, as depicting the former is naturally easier. Fig. 3 on the other hand is intended to represent the continuity of the map T_τ at φ.[24]

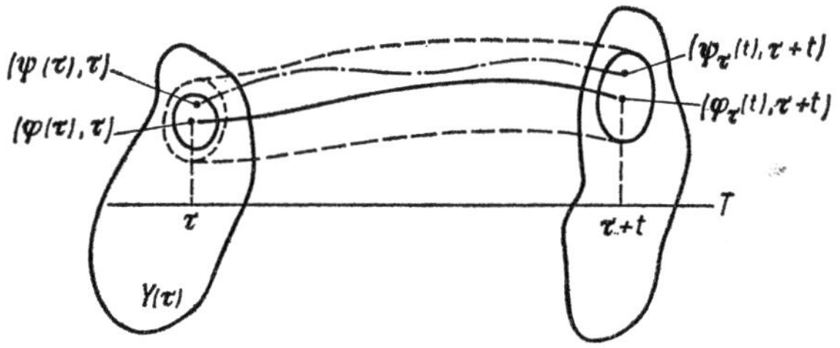

Fig. 2. Liapounor stability illustrated by trajectories.

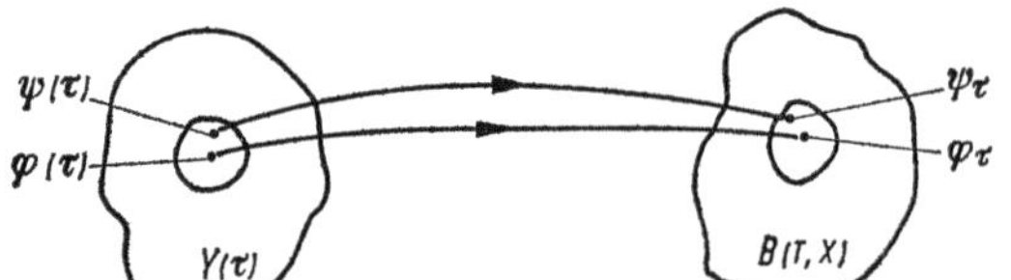

Fig. 3. Continuity of the map T_τ and φ.

The essentially topological character of stability contained in the definition of this section has already been emphasised. It is this dependence on certain selected measures ϱ, making it possible to have stability for some measures while giving instability with respect to others that distinguishes *stability in continuous systems from that in discrete systems.*

For a discrete system, in which X is a normable finite-dimensional vector space and measures ϱ_1, ϱ_2 are chosen to be norms on X, the following well-known

[22] See any modern introduction to partial differential equations.

[23] Cf. BHATIA and SZEGÖ [1967, *5*, Definition 2.10: 2.12.1], LEFSCHETZ [1957, *3*, IV, § 2.2, § 6] and ZUBOV [1964, *17*].

[24] For recent discussions of certain stability topics which emphasise its geometric aspects see ROSENBERG [1966, *30*] and TAKAHASHI [1966, *31*].

theorem will apply:[25] for any two norms $\|(\cdot)\|_1$, $\|(\cdot)\|_2$ defined on X there must exist positive real numbers α, β such that for every $x \in X$ the following inequalities are true

$$\|x\|_1 \leqq \alpha \|x\|_2,$$

$$\|x\|_2 \leqq \beta \|x\|_1.$$

Note that for stability only semi-equivalence is required. Thus in a discrete system with norms as measures on X our earlier question about stability under different pairs of measures is always answered in the affirmative. That is, if stability is established for one pair ϱ_τ, ϱ, it is automatically established with respect to any other pair $\hat{\varrho}_\tau$, $\hat{\varrho}$. This fact helps to expose the inadequacy of investigations of stability in continuous systems based on the model of the discrete system.

In general, a disregard for the topological nature of stability in continuous systems is a characteristic of classical investigations. These have always relied on LAGRANGE'S original analysis, correct in its context but inapplicable when topological properties are an issue. One of the most glaring errors resulting from this misunderstanding has been the conclusion that the wave equation is stable for all measures, a statement just shown to be false. Confusion can only be avoided in any analysis of continuous systems if their topological aspects are correctly considered, whether the method of analysis involves the use of eigenfunctions, Liapounov methods or *ad hoc* procedures.

7. Further definitions. There have been several modifications, extensions and variations on the definition of Liapounov stability. We now proceed to give the most important of these.

Definition 7.1.[26] A solution $\varphi \in \mathscr{B}(T, X)$ is said to be *uniformly stable* if and only if the map T_τ from $Y(\tau)$ to $\mathscr{B}(T, X)$ is continuous at φ uniformly in τ. That is, if for each $\tau \in \mathscr{T}$ and each positive real number ε, there exists a positive real number $\delta(\varepsilon)$, depending only on ε, such that

$$\varrho_\tau(\varphi(\tau), \psi(\tau)) < \delta$$

implies

$$\varrho(\varphi_\tau, \psi_\tau) < \varepsilon, \qquad \varphi, \psi \in \mathscr{B}(T, X).$$

As an example of a situation in which this concept is relevant we may consider a body subject to a time-dependent finite deformation. In linear elasticity, however, the only applications appear to be those in which a material has time-dependent properties, e.g., visco-elasticity.

The next Definition 7.2 concerns the dynamical system $\mathscr{B}(T, X)$ and introduces the idea of a limit. This is preliminary to the Definition 7.3 that follows in sequence.

Definition 7.2. If φ, ψ belong to $\mathscr{B}(T, X)$ then φ is said to tend to ψ at ∞ if, given any $\varepsilon > 0$, there is a compact set G in T such that

$$\varrho(\varphi(t), \psi(t)) < \varepsilon, \qquad t \in T, \quad t \notin G.$$

When T is the semi-group of non-negative real numbers this definition states that φ tends to ψ at ∞ if

$$\lim_{t \to \infty} \varrho(\varphi(t), \psi(t)) = 0.$$

---

[25] See, for example, GOFFMAN [1965, *9*], BROWN and PAGE [1970, *7*].

[26] This formulation follows that of GILBERT and KNOPS [1967, *10*].

Definition 7.3. The solution $\varphi \in \mathscr{B}(T, X)$ is said to be *asymptotically stable* if and only if (i) φ is stable, (ii) for each $\tau \in \mathscr{T}$ and all $\psi(\tau)$ in some neighbourhood of $\varphi(\tau)$ (defined by the function ϱ_τ), $T_\tau(\psi(\tau))$ tends to φ_τ at ∞.

In mathematical detail this definition reads:

The solution $\varphi \in \mathscr{B}(T, X)$ is said to be asymptotically stable if and only if

(i) for each $\tau \in \mathscr{T}$ and for each positive real number ε there exists a positive real number $\delta(\varepsilon, \tau)$ such that

$$\varrho_\tau(\varphi(\tau), \psi(\tau)) < \delta$$

implies

$$\varrho(\varphi_\tau, \psi_\tau) < \varepsilon,$$

(ii) for each $\tau \in \mathscr{T}$, there exists a positive real number $\gamma(\tau)$ such that for every $\varphi(\tau)$ satisfying $\varrho_\tau(\varphi(\tau), \psi(\tau)) < \gamma(\tau)$, ψ_τ tends to φ_τ at ∞ in the sense of Definition 7.2.

Definition 7.4. The solution φ is said to be *weakly asymptotically stable* if condition (ii) only of Definition 7.3 obtains.

Definition 7.5. The solution of $\varphi \in \mathscr{B}(T, X)$ is said to be *uniformly asymptotically stable* if and only if the numbers δ and γ in the previous Definition 7.4 are independent of τ.

Definition 7.6. For a solution φ that is asymptotically stable, the *region of stability* or *region of attraction* is said to be the neighbourhood of $Y(\tau)$ defined by

$$\varrho_\tau(\varphi(\tau), \psi(\tau)) < \gamma.$$

This region may be a proper subset of X or the whole set X.

Definition 7.7. The solution φ is said to be *globally asymptotically stable* or *completely stable* when the region of stability coincides with the whole of the set X.

A trivial consequence of Definition 7.3 is that asymptotic stability implies stability but not conversely.

In general, Liapounov stability and weak asymptotic stability are distinct.[27] When X is a normed linear space, however, weak asymptotic stability of the solution implies that $\varrho(\psi_\tau, \varphi_\tau)$ remains bounded for $t \in [0, \infty)$.[28] In this case the solution is stable as shown at the end of Sect. 10.

Definition 7.8.[29] The solution $\varphi \in \mathscr{B}(T, X)$ is said to be *exponentially asymptotically stable* if and only if for each $\tau \in \mathscr{T}$ and or each positive real number ε there are positive real numbers $\delta(\varepsilon, \tau)$, t_0 and α such that for $t \geqq t_0$

$$\varrho_\tau(\varphi(\tau), \psi(\tau)) < \delta$$

implies

$$\varrho(\varphi(t), \psi(t)) \leqq \varepsilon \exp[-\alpha(t - t_0)].$$

In the remainder of this section we record for completeness a few definitions relating to special aspects of stability. These include *practical stability* and some variations, *neutral stability* and *orbital stability*.

It has already been mentioned in Sect. 3 and again in connexion with Definition 6.1 that the semi-group T has usually to be regarded as the set of non-negative real numbers. If stability is considered over any compact sub-intervals

[27] CESARI [1963, *1*, p. 96] gives an example from ordinary differential equations which illustrates the difference.

[28] CESARI [1963, *1*, pp. 95–96].

[29] Cp. LEFSCHETZ [1957, *3*, p. 85], HALE [1969, *11*, p. 84].

of T then in physical systems having solutions which depend continuously on the initial data it is possible to choose δ and ε such that the conditions of Definition 6.1 are trivially satisfied. In order to overcome this difficulty for situations in which a physical system has to be considered over a finite interval of time two further definitions of *local stability* and *practical* or *finite time stability* have been proposed respectively by KRASOVSKII[30] and LASALLE and LEFSCHETZ.[31]

Definition 7.9. The solution $\varphi \in \mathscr{B}(T, X)$ is said to be *locally stable* on intervals of length t_1, for approximations $\Delta(\varepsilon) > 0$ if and only if for each $\tau \in \mathscr{T}$ and each positive real number ε there exists a positive real number $\delta(\tau, \varepsilon) > \Delta(\varepsilon)$ such that

$$\varrho_\tau(\varphi(\tau), \psi(\tau)) < \delta$$

implies

$$\varrho_{t_1}(\varphi_\tau, \psi_\tau) < \varepsilon,$$

where

$$\varrho_{t_1}(\varphi_\tau, \psi_\tau) = \sup_{0 \leqq t \leqq t_1} \varrho(\varphi_\tau(t), \psi_\tau(t)),$$

and $\varphi, \psi \in \mathscr{B}(T, X)$; $\tau \in \mathscr{T}$; $t_1 \in T$.

Definition 7.10.[32] The solution $\varphi \in \mathscr{B}(T, X)$ is said to be *practically stable* if and only if, given positive real numbers A, B, t_1 (all chosen arbitrarily with $A \leqq B$), it follows that

$$\varrho_\tau(\varphi(\tau), \psi(\tau)) < A$$

implies

$$\varrho_{t_1}(\varphi_\tau, \psi_\tau) < B.$$

This definition provides an independent notion of stability in an oscillating system which is unstable by the Liapounov definition. Perturbations of oscillatory but bounded nature originating in a given state of the system might give rise to oscillations which remain within acceptable limits throughout the interval of length t_1. This form of stability is closely associated with general considerations of boundedness and will be taken up again later in Sect. 10.

It should be noted that a solution may be stable in any one of the senses, LIAPOUNOV, local (KRASOVSKII) or practical (LASALLE and LEFSCHETZ) and yet not in the other two. Emphasis, too, should also be laid on the fact that, in practical stability, the numbers A, B, ε, t are all independent (subject to $A \leqq B$), a feature which distinguishes it from most other definitions.

Two further variations on practical stability have been proposed:

Definition 7.11.[33] The solution $\varphi \in \mathscr{B}(T, X)$ is said to be *quasi-contractively stable* with respect to A, B, C, t_1 ($A \leqq B < C$) if and only if φ is practically stable and

$$\varrho_\tau(\varphi(\tau), \psi(\tau)) < A < \varrho(\varphi_\tau(t), \psi_\tau(t)) \leqq C, \quad 0 \leqq t \leqq t_1$$

implies

$$\varrho(\varphi_\tau(t), \psi_\tau(t)) < B, \quad t_2 < t < t_1,$$

for at least one t_2.

---

[30] KRASOVSKII [1963, *11*, p. 3]. ROUTH'S [1877, *2*] definition was possibly motivated by this idea. See also WILLKE [1967, *34*], CESARI [1963, *1*] and HAHN [1959, *4*].

[31] LASALLE and LEFSCHETZ [1961, *5*, p. 121].

[32] Compare LASALLE and LEFSCHETZ [1961, *5*, p. 121], INFANTE and WEISS [1965, *11*], [1967, *17*]. See also the definition attributed to KAMENKOV by KUZMIN [1967, *19*] in which a motion φ_τ is stable if, starting in some set G, it remains in that set.

[33] INFANTE and WEISS [1965, *11*], [1967, *17*].

Definition 7.12.[34] The solution $\varphi \in \mathscr{B}(T, X)$ is said to be *contractively stable* if and only if Definition 7.11 is repeated with the exception that the condition on A, B, C is replaced by $B < A \leqq C$.

A condition of "neutral stability" may occur in a system in equilibrium and is defined as follows:

Definition 7.13. The equilibrium solution $\varphi \in \mathscr{B}(T, X)$ is said to be *neutrally stable* (or in neutral equilibrium), in the class of initial data $Y(\tau)$, if and only if ψ_τ is an equilibrium motion for all $\psi(\tau) \in Y(\tau)$ and $\psi_\tau \in \mathscr{B}(T, X)$.

Examples of neutrally stable solutions are the non-unique equilibrium solutions to a boundary value problem with $Y(\tau)$ as the class of all such solutions.[35] We have another example in a physical system subject to fixed forces if the system remains in equilibrium for all possible forces.[36]

This definition is important in connexion with non-uniqueness of equilibrium solutions discussed in Sect. 12 and will be examined again in Sects. 33, 45 in relation to classical criteria used in investigating elastic stability. It is particularly relevant to the so-called adjacent equilibrium technique usually associated with EULER and with what is called *latent instability* and discussed in Sects. 46, 47.

Finally, definitions of *orbital stability* and *asymptotically orbital stability* are given. These ideas had their origins in the study of the paths of planetary bodies and were formalised by POINCARÉ.[37] A perturbation of a closed orbit gives rise to a trajectory of the motion which remains close to the orbit for all subsequent instants of time. This is given formally as follows:

Definition 7.14. The solution $\varphi \in \mathscr{B}(T, X)$ is said to be *orbitally stable* if and only if the mapping T_τ is continuous at φ, the neighbourhoods on $\mathscr{B}(T, X)$ being defined by

$$U_\varepsilon = \{\psi\colon \sup_{t\in T} \inf_{s\in T} \varrho(\varphi_\tau(t), \psi_\tau(s)) < \varepsilon\} \qquad \varphi, \psi \in \mathscr{B}(T, X).$$

Related to this is the definition of asymptotic orbital stability given below:

Definition 7.15.[38] The solution $\varphi \in \mathscr{B}(T, X)$ is said to be *asymptotically orbitally stable* if and only if it is orbitally stable and for some $\gamma > 0$

$$\inf_{s\in T} \varrho_\tau(\varphi(\tau), \psi_\tau(s)) < \gamma,$$

implies

$$\lim_{t\to\infty} \inf_{s\in T} \varrho(\varphi_\tau(t), \psi_\tau(s)) = 0.$$

While the Liapounov definition of stability, as we have shown, may be easily applied to the stability of a motion this may lead to conclusions not altogether satisfactory from a practical viewpoint. Thus, in illustration we consider the vibrating string[39]

$$u_{,xx} = c^2 \ddot{u}, \quad 0 \leqq x \leqq \pi, \quad t > 0, \quad u(t, 0) = u(t, \pi) = 0$$

---

[34] INFANTE and WEISS [1965, *11*], [1967, *17*].

[35] See Sect. 12.

[36] THOMSON and TAIT [1879, *1*, vol. I, Art. 291], THOMSON [1888, *4*].

[37] POINCARÉ [1881, *2*], [1882, *1*]. The idea has been generalised to "stability of sets". See BHATIA and SZEGÖ [1967, *5*, p. 63].

[38] Cf. CESARI [1963, *1*, § 1.8].

[39] This example is analogous to that of the simple pendulum commonly produced to demonstrate the deficiency of Liapounov stability in such cases. Several other examples are given by CESARI [1963, *1*, § 1.9].

and examine the stability of the motion (for suitable α)

$$u(t, x) = \sin c\alpha x \sin(\alpha t + d),$$

where d is constant. For an appropriate choice of the constant d_1, a solution with neighbouring initial values is (for suitable β)

$$v(t, x) = \sin c\beta x \sin(\beta t + d_1).$$

Clearly, however, we have that $\varrho(\varphi_\tau, \psi_\tau)$, chosen in accordance with Definition 6.1, cannot remain arbitrarily small and hence the original motion is unstable.[40] This could be considered as an incorrect classification, and such situations where Liapounov stability is too stringent may be redeemed merely by introducing Definition 7.14 and the new measure implied there. Other situations may demand measures different from any of those given hith reto, e.g.,

$$\varrho(\psi_\tau, \varphi_\tau) \equiv \int_0^\infty \varrho(\psi_\tau(t), \varphi_\tau(t))\, dt.$$

Since the criteria developed later for Liapounov stability are essentially topological in nature, the precise form of the measures adopted is immaterial for the validity of those criteria.

8. Continuous dependence. Stability as defined in Sect. 6 is closely related to the concept of "continuous dependence of the solution on its data", well studied in the field of partial differential equations.[41] As with Liapounov stability, where the issue is dependence of the solution on the initial data, continuous dependence in general can only be defined after the ideas of "smallness" and "continuity" have been made precise. When this is done both concepts may be equated with the idea of continuity of the mapping from the set of data to the set of solutions. Liapounov stability is defined, however, in terms of $\varrho(\varphi_\tau, \psi_\tau)$, that is the supremum of $\varrho(\varphi_\tau(t), \psi_\tau(t))$ taken over the whole interval T[42] and thus concerns the mapping in the entire interval. Continuous dependence on the other hand is concerned with mappings in compact sub-intervals of T.[43]

We thus introduce the notation $\varrho_{t_1}(\varphi_\tau, \psi_\tau)$ as follows

$$\varrho_{t_1}(\varphi_\tau, \psi_\tau) = \sup_{0 \leq t \leq t_1} \varrho(\varphi_\tau(t), \psi_\tau(t))$$

where $\varphi, \psi \in \mathscr{B}(T, X)$, $\tau \in \mathscr{T}$, and $t, t_1 \in T$, so that $\mathscr{B}(T, X)$ is equipped with a positive-definite function ϱ_{t_1}. The definition of continuous dependence then runs:

Definition 8.1. The solution φ is said to *depend continuously on the initial data* $\varphi(\tau)$ if and only if the map T_τ from $Y(\tau)$ to $\mathscr{B}_{\varrho_{t_1}}(T, X)$ is continuous at φ. Alternatively, the definition may be expressed as: the solution φ is said to depend continuously on the data $\varphi(\tau)$ if and only if, for each $\tau \in \mathscr{T}$ and each positive real number ε there exists a positive real number $\delta(\tau, \varepsilon, t_1)$ such that

$$\varrho_\tau(\varphi(\tau), \psi(\tau)) < \delta$$

implies

$$\varrho_{t_1}(\varphi_\tau, \psi_\tau) < \varepsilon.$$

[40] That is, not stable. A more explicit definition of instability is given later.

[41] See, for example, Sobolev [1964, *12*, Lecture 2, § 2], who describes the problem as "correctly formulated" if its solution depends continuously on the data. Other authors use the terms "well-posed" (Hadamard [1923, *1*, pp. 23–24], John [1970, *15*, p. 275]), and "reasonable" (Petrovsky [1954, *1*, § 8]).

[42] We now regard T as the set of non-negative real numbers.

[43] Cf. Hale [1969, *11*, p. 25], Sobolev [1964, *12*, Lecture 2, § 2].

Liapounov stability thus implies continuous dependence of the solution on the initial data, whereas continuous dependence of the solution on the initial data does not imply Liapounov stability. The latter contingency is readily illustrated by considering that

$$\inf_{t_1 \in T} \delta(\tau, \varepsilon, t_1)$$

may be zero. This is demonstrated in an interesting example given by HALE.[44]

For a dynamical system to serve as a useful model of a physical system the continuous dependence of the solution on the data has usually been regarded as a necessary requirement. That is to say that a small error in the data must not lead to a large error in the solution. This idea is reflected in PETROVSKY's epithet "reasonableness" and SOBOLEV's "correctly formulated" for models simulating physical systems and these probably originated from HADAMARD's concept of a "well-posed problem".[45] In this he demands that the solution (a) exist, (b) be unique, and (c) depend continuously on the data. Mathematical models not conforming to (b) or (c) do, however, in some problems give information about certain physical systems as, for example, in the processes concerning buckling. Uniqueness of the solution ceases at the critical load and the equilibrium solution fails to depend continuously on the load parameter. Moreover, instability, in some sense, of the original state must occur in order for one of the buckled states to be attained. HADAMARD's requirements are thus too restrictive for the investigation of certain systems and a critical study of systems failing to be unique or continuous becomes necessary.

In some problems in which continuous dependence of the solution on the data is known to fail, it is necessary to know the conditions under which continuous dependence can be recovered. An investigation by F. JOHN[46] has shown that in certain situations it is sufficient to impose a uniform bound on the solution to achieve this. In these cases continuity must be modified to the corresponding concept of Hölder continuity.

As a class of problems involving this phenomenon will be investigated later, we now give another definition of stability which incorporates this modification and which is termed Hölder stability.

Definition 8.2.[47] A solution $\varphi \in \mathscr{B}(T, X)$ is said to be *Hölder stable* on the interval $[\tau, \tau + t_1]$ for $\tau \in \mathscr{T}$, $t_1 \in T$, if and only if for each positive real number ε we have that

$$\varrho_\tau(\varphi(\tau), \psi(\tau)) < \varepsilon$$

implies

$$\varrho_{t_2}(\varphi_\tau, \psi_\tau) < A\,\varepsilon^\alpha, \qquad 0 < \alpha \leq 1,$$

where A is a positive constant independent of ε and

$$\varrho_{t_2}(\varphi_\tau, \psi_\tau) = \sup_{0 \leq t \leq t_2} \varrho(\varphi_\tau(t), \psi_\tau(t)), \qquad t_2 < t_1.$$

---

[44] HALE [1969, *11*, p. 26].

[45] HADAMARD [1923, *1*, pp. 23–24].

[46] F. JOHN [1955, *5*], [1960, *5*], [1970, *15*].

[47] Cp. KNOPS and PAYNE [1971, *9*]. Definition 8.2, in a more general context, was proposed by F. JOHN [1960, *5*], [1970, *15*]. He also considered logarithmic continuity, which is weaker than Hölder continuity, and which could also be made the basis of a stability definition. However, such a definition appears to have little significance for elasticity. Similar ideas are briefly discussed by CESARI [1963, *1*, p. 9]. We also mention here BIRKHOFF's definitions of "perturbative stability" and "trigonometric stability" (BIRKHOFF [1927, *1*], CESARI [1963, *1*, p. 13]), but their significance for elasticity again appears to be limited.

To provide an intuitive idea of Hölder stability and the situations in which it might be applicable, we suppose there is a linear system admitting solutions whose time dependence is of the type e^{nt}. The solutions bounded at $t=t_1$ are then of the form $e^{-n(t_1-t)}$, and thus remain arbitrarily near the null solution for $0 \leqq t < t_1$. However, these solutions cannot remain arbitrarily close to the null solution for $0 \leqq t \leqq t_1$, but must increase to their bounded value of one, no matter how large we take n. Thus, there is continuous dependence on the initial data in all compact sub-intervals of $[0, t_1]$, but not on the interval $[0, t_1]$ itself.

9. Instability in the sense of Liapounov. Any solution which fails to satisfy the requirement of Definition 6.1 may be said to be unstable, and we have taken the term to have this meaning so far in the treatise. Many authors, including LIAPOUNOV, define instability in this manner, but it has been convenient, however, to formulate independent definitions[48] to meet the cases of some of the forms of stability treated in Sect. 7.

Definition 9.1.[49] A solution $\varphi \in \mathscr{B}(T, X)$ is said to be *Liapounov unstable*[50] if and only if for some $\tau \in \mathscr{T}$, the mapping T_τ from $Y(\tau)$ to $\mathscr{B}(T, X)$ is discontinuous at φ with respect to the neighbourhoods of φ induced on $Y(\tau)$ and $\mathscr{B}(T, X)$ by ϱ_τ and ϱ respectively.

Expressed explicitly, the definition becomes: φ is unstable if and only if, for at least one instant $\tau \in \mathscr{T}$, there exists a positive real number ε such that for all positive real numbers δ, the condition

$$\varrho_\tau(\varphi(\tau), \psi(\tau)) < \delta$$

implies

$$\varrho(\varphi_\tau, \psi_\tau) \geqq \varepsilon.$$

Definition 9.2.[51] A solution $\varphi \in \mathscr{B}(T, X)$ is said to be *conditionally stable* if and only if it is unstable according to Definition 9.1, but is stable in a proper sub-family of perturbations.

For linear systems with the set T chosen to be the non-negative real numbers, conditional stability of an equilibrium solution φ is always implied by its neutral stability. Perturbations of the form $t\varphi k$, where k is a constant, demonstrate instability; whereas within the class of perturbations of type $k\varphi$, for sufficiently small constant k, there is stability.

Another simple example of conditional stability occurs in the linear system which consists of solutions possessing exponential functions, with two exponents, one generating an increasing, and the other a decreasing, time factor.

In the two examples just quoted, instability arises because some of the perturbations become unbounded in time. It must be emphasised, however, that

[48] One of the earliest definitions of instability is due to LAGRANGE [1788, *1*]. It states that "a state of equilibrium is unstable if, on being slightly disturbed, the system can perform oscillations which will not be small, and which will become larger and larger". Of the first writers to consider instability directly in elasticity, we mention GREEN [1839, *1*], KIRCHHOFF [1850, *1*], [1859, *1*], [1877, *1*, 4th lecture], CLEBSCH [1862, *1*, § 20, § 21], and LIPSCHITZ [1874, *1*], the last two being entirely concerned with it. THOMSON [1888, *4*] makes some interesting comments on GREEN's work.

[49] Cp. MOVCHAN [1960, *7*], GILBERT and KNOPS [1967, *10*].

[50] The term "labil" is sometimes used in the German literature; see, for example, KLEIN and SOMMERFELD [1897, *2*, p. 343], PFLÜGER [1964, *10*, p. 1]. This should not be confused with the term "labile", found in the English literature of the late nineteenth century, whose modern equivalent corresponds to "neutral equilibrium".

[51] Cp. LEFSCHETZ [1957, *3*, pp. 78, 83].

instability in the sense of Definition 9.1 is not necessarily related to unboundedness, and may occur for perturbations that remain bounded. The example of SHIELD, described in Sect. 6, provides one illustration and another occurs in the non-linear treatment of "snap-through" problems.[52] Nevertheless, it is true that situations exist in which boundedness and stability, when appropriately defined, are related and in the next section this question is developed further.

10. Boundedness and Liapounov stability. The ideas of stability examined so far rest on the concept of continuity and reflect our intuitive expectations. These can be formalised in the definition of Liapounov stability and its variations. Of these variations, *practical stability* (Definition 7.10) has taken us farthest from the original idea which demanded that ϱ_τ and ϱ should be arbitrarily small on the perturbations of the solution φ. However, in some contexts it is altogether inappropriate to use the Liapounov definition of stability, and instead it becomes preferable to use the idea of boundedness. The need for bounded, but not necessarily arbitrarily small, measures motivated early attempts at definitions of stability such as those arising from the study of planetary motions. The same need is encountered in elasticity when dealing with problems such as buckling and snap-through from one equilibrium position to another. The relevant ideas can mostly be classed under the heading of *Lagrange stability*[53] and they express fundamentally the condition that solutions should remain bounded.

Before defining Lagrange stability, we make precise some ideas concerning boundedness.

Definition 10.1. The solution $\psi\in\mathscr{B}(T, X)$ is said to be *bounded* with respect to the given solution $\varphi\in\mathscr{B}(T, X)$ and the measure ϱ if and only if

$$\varrho(\varphi, \psi) = \sup_{t\in T} \varrho(\varphi(t), \psi(t)) < \infty .$$

When φ is the null solution we merely say that ψ is a bounded solution. This definition then agrees with that given by YOSHIZAWA.[54] When X is a normable linear space, with norm ϱ, any solution ψ that is unbounded with respect to φ may be trivially converted into a bounded solution by using the norm $\varrho/(1+\varrho)$.

Definition 10.2. The mapping T_τ, where $\tau\in\mathscr{T}$, is said to be *bounded* with respect to ϱ_τ and ϱ at $\varphi\in\mathscr{B}(T, X)$ if there exist constants A and B (depending on φ and τ) such that

$$\varrho_\tau(\varphi(\tau), \psi(\tau)) \leqq A$$

implies

$$\varrho(\varphi_\tau, \psi_\tau) \leqq B .$$

As with Liapounov stability the boundedness of the map T_τ is estimated at φ by the behaviour of solutions which at the instant τ are in a (finite) neighbourhood of $\varphi(\tau)$, and not by the behaviour of φ alone. Despite this similarity, we prefer not to define Lagrange stability by Definition 10.2, but rather, in conformity with modern usage, we reserve this concept to describe the behaviour of a single solution that remains bounded. We therefore propose the following definition:

---

[52] See, for instance, HSU [1966, *12*, *13*], and also Sect. 83.

[53] There is some confusion in the literature as to whether the concept is due to LAGRANGE or to LAPLACE.

[54] YOSHIZAWA [1955, *9*].

Definition 10.3.[55] The solution $\psi \in \mathscr{B}(T, X)$ is said to be *Lagrange stable*[56] at the solution $\varphi \in \mathscr{B}(T, X)$ if and only if, for each $\tau \in \mathscr{T}$ there exist positive real numbers A and B (depending on φ, ψ and τ) such that

$$\varrho_\tau(\varphi(\tau), \psi(\tau)) \leqq A$$

implies

$$\varrho(\varphi_\tau, \psi_\tau) \leqq B.$$

This definition is similar to that of *practical stability* (Definition 7.10) when the time interval T is the set of non-negative real numbers.

Lagrange stability is not a special case of the Definition 10.2 for the boundedness of the map T_τ. The latter definition involves *all* solutions emanating from a finite neighbourhood, whereas Lagrange stability applies to just *one* such solution. However, relations do exist between these two definitions. For example, it may be easily seen that the map T_τ is bounded at φ if and only if every solution, at the instant $\tau \in \mathscr{T}$, is Lagrange stable at φ. Relations also exist with Liapounov stability. Obviously, when φ is Liapounov stable then T_τ is bounded at φ for every $\tau \in \mathscr{T}$. But when the set X is a normed linear space and each T_τ is homogeneous we may prove that these concepts are exactly equivalent.[57] The following theorems embodying this result are merely formal restatements of the theorem in analysis that a linear functional is bounded if and only if it is continuous.

Theorem 10.1.[58] *Let X be a normed linear space and let each map T_τ satisfy $T_\tau(\alpha\varphi) = \alpha T_\tau(\varphi)$ for $\varphi \in X$ and each positive number α. Then, if ϱ_τ and ϱ are norms, a solution $\varphi \in \mathscr{B}(T, X)$ is Liapounov stable if and only if each T_τ is bounded at φ: more precisely, if and only if there is a constant B_τ such that*

$$\varrho_\tau(\psi(\tau) - \varphi(\tau)) \leqq 1$$

implies

$$\varrho(\psi_\tau - \varphi_\tau) \leqq B_\tau, \quad \psi, \varphi \in \mathscr{B}(T, x).$$

Even when X is not a linear space the equivalence of Liapounov stability of the *null solution* and the boundedness of the dynamical system at the null solution can be proved. The proof of Theorem 10.1 can be easily modified to yield

Theorem 10.1a. Let X be a set with the property that $\alpha\varphi \in X$ for each positive number α whenever $\varphi \in X$ and let each T_τ satisfy $T_\tau(\alpha\varphi) = \alpha T_\tau(\varphi)$. Then, if ϱ_τ and ϱ are positive-definite functions satisfying

$$\varrho_\tau(\alpha\psi, 0) \leqq \alpha\varrho_\tau(\psi, 0), \qquad \varrho(\alpha\psi, 0) = \alpha\varrho(\psi, 0)$$

---

[55] Cp. NEMYTSKII and STEPANOV [1947, *3*, p. 340], BHATIA and SZEGÖ [1967, *5*, p. 129], [1970, *5*]. These authors define Lagrange stability of the solution ψ at the null solution and for a metric space X. Their definition then states that ψ is Lagrange stable if and only if the *orbit* through $\psi(\tau)$ is compact. This stronger form is required as they wish to study the connexion between Lagrange stability and the limit of the orbit as $t \to \infty$. A consequence of their definition is the boundedness of ψ at the null solution.

[56] When φ is the null solution, YOSHIZAWA [1955, *9*] calls this *equiboundedness*; when, in addition A and B are independent of τ, he calls it *uniform boundedness*. A further definition used by YOSHIZAWA [1955, *9*] and LASALLE and LEFSCHETZ [1961, *5*, p. 117] is quoted here: The solution $\varphi \in \mathscr{B}(T, X)$ is said to be *ultimately bounded* if and only if there exists $t_0 > 0$ and a positive real number N such that

$$\sup_{t \geqq t_0} \varrho(\varphi_\tau(t), 0) < N.$$

[57] In general, these concepts are distinct. See, for instance, examples discussed by CESARI [1963, *1*, p. 7].

[58] GILBERT and KNOPS [1967, *10*]. A special case of the theorem is reported by CESARI [1963, *1*, p. 7].

the null solution of $\mathscr{B}(T, X)$ *is Liapounov stable if and only if each* T_τ *is bounded at* 0.

Theorem 10.1 is proved using the following well-known result whose proof is not given.

Lemma. Suppose X *is a set with the property of Theorem 10.1a. Suppose further that* ϱ_1 *and* ϱ_2 *are positive-definite functions on* X *satisfying*

$$\varrho_1(\alpha\varphi, 0) \leqq \alpha\varrho_1(\varphi, 0), \quad \varrho_2(\alpha\varphi, 0) = \alpha\varrho_2(\varphi, 0), \quad \varphi\in X, \quad \alpha > 0.$$

Then, if T *is a mapping from* X *to* X *such that* $T(\alpha\varphi) = \alpha T(\varphi)$, $\varphi\in X$, *the following assertions are equivalent.*[59]

(i) T *is continuous from* X_{ϱ_1} *to* X_{ϱ_2} *at* 0,

(ii) *the set* $\{T(\varphi)\colon \varphi\in X_{\varrho_1}, \varrho_1(\varphi)\leqq 1\}$ is bounded in X_{ϱ_2},

(iii) *for some positive* A_T,

$$\varrho_2(T(\varphi)) \leqq A_T\,\varrho_1(\varphi), \quad \varphi\in X.$$

To prove Theorem 10.1 we note that because X is linear it suffices to apply the Lemma. Then the continuity of T_τ at φ is equivalent to the boundedness of T_τ at φ. The second statement of the theorem follows from the equivalence of assertions (i) and (ii).

11. Instability in the sense of Lagrange. Coupled with Lagrange stability is the idea of Lagrange instability which we define as follows:

Definition 11.1. The solution $\psi\in\mathscr{B}(T, X)$ is said to be *Lagrange unstable* at the solution $\varphi\in\mathscr{B}(T, X)$ if and only if for every $\tau\in\mathscr{T}$, there exists a positive real number A such that as a consequence of the inequality

$$\varrho_\tau(\varphi(\tau), \psi(\tau)) \leqq A,$$

no positive real number B can be found to satisfy the inequality

$$\varrho(\varphi_\tau, \psi_\tau) \leqq B. \tag{11.1}$$

It follows immediately from the respective definitions that the Lagrange instability of a solution ψ at φ implies the Liapounov instability of the solution φ.

When $\varrho(\varphi_\tau, \psi_\tau)$ remains bounded on compact sub-intervals of T, but the conditions of Definition 11.1 continue to hold on the whole set T, we say that the solution φ is *asymptotically unstable.* In particular, when T is the set of non-negative real numbers and the inequality

$$\varrho(\varphi_\tau(t), \psi_\tau(t)) \geqq \lambda\, e^{\alpha(t-t_1)}, \quad t\geqq t_1$$

holds for sufficiently large t_1, with λ, α positive constants, we say that the solution φ is *exponentially asymptotically unstable.* Solutions ψ with exponentially increasing time factors provide an illustration of this last concept.

Since all physical systems remain bounded the concept of Lagrange instability is a theoretical one. It would thus seem that Liapounov instability, which can occur in a system without its solutions' being unbounded, has the greater practical significance. Against this must be weighed the fact that mathematical models of physical systems are imperfect. The idealisations made in creating the model can lead to greater misrepresentations than those introduced by representing

[59] X_{ϱ_i} stands for the set X provided with the positive-definite function ϱ_i.

large values of solutions as unbounded ones. There are obvious examples of inadequate models in linear systems and in buckling problems. In the former, the inadequacy of the model compels instability, when it occurs, to be of Lagrange type (this follows from theorems proved at the end of the previous section). The use of a more adequate model, in this instance, would probably enable Liapounov stability to have a meaningful application.

No matter under what circumstances Lagrange instability occurs, whether this be due to an inadequate model or not, it is always an indication of a breakdown in the existence of the solution. In these circumstances, the measure ϱ becomes unbounded and therefore the solution ψ cannot remain in the class of functions yielding finite values of ϱ. This class will often be a Banach or Sobolev space for appropriate choice of ϱ. Situations illustrating failures of the foregoing type are provided by the first and second examples treated in Sect. 6 and the following discussion. However, it must be recognised that the Lagrange instability of a solution ψ *at the solution* φ does not necessarily imply the non-existence of the solution φ itself. This applies especially when φ is an equilibrium solution.

It is still largely an open question as to what is the relation between the existence of a solution and its stability.

12. Stability and uniqueness.

Definition 12.1. The solution φ of the dynamical system $\mathscr{B}(T, X)$ is said to be *unique* if, for each $\tau\in\mathscr{T}$

$$\psi(\tau)=\varphi(\tau)$$

implies

$$\psi_\tau(t)=\varphi_\tau(t), \qquad \tau\in\mathscr{T}, \qquad t\in T.$$

In the context of partial differential equations it should be noted that this definition refers to the *initial* boundary value problem and not to the *equilibrium* boundary value problem.

From this definition the connexion between Liapounov stability and uniqueness is given immediately by the following theorem:

Theorem 12.1.[60] *If* $\varphi\in\mathscr{B}(T, X)$ *is Liapounov stable it is unique.*

To prove the theorem, we suppose φ is not unique and let $\psi\in\mathscr{B}(T, X)$ be chosen so that $\psi(\tau)=\varphi(\tau), \psi_\tau(t)\neq\varphi_\tau(t)$ for some $\tau\in\mathscr{T}$, $t\in T$. Then, since ϱ is positive-definite,

$$\varrho\left(\varphi_\tau(t), \psi_\tau(t)\right)\neq 0,$$

or, equivalently, $\varrho(\varphi_\tau, \psi_\tau)\neq 0$. Since φ is stable, given $\varepsilon>0$, there exists $\delta>0$ such that for $\chi\in\mathscr{B}(T, X)$ we have

$$\varrho_\tau\left(\varphi(\tau), \chi(\tau)\right)<\delta$$

implies

$$\varrho(\varphi_\tau, \chi_\tau)<\varepsilon.$$

When $\varepsilon=\frac{1}{2}\varrho(\varphi_\tau, \psi_\tau)$ the last statement gives an obvious contradiction.

An immediate consequence of the uniqueness theorem is that non-uniqueness of a solution automatically implies its instability. The boundedness results of Sect. 10 further show that for a *linear system*, a solution is Liapounov stable if and only if the perturbed trajectories remain bounded. It thus follows that

[60] Movchan [1960, *7*], Gilbert and Knops [1967, *10*]. The theorem also applies to a solution which depends continuously on the data.

linear systems have unique solutions in the class of solutions bounded in the sense of Definition 10.1.

In the special case of linear elasticity, a relationship exists between uniqueness of solutions to the *static* problem and the stability of solutions to an associated dynamic problem. It may be shown by means of examples[61] that non-uniqueness of a static solution implies Lagrange instability at an equilibrium solution in the dynamic problem with the same boundary conditions. Inversely, the stability of an equilibrium solution in elastodynamic problems implies the uniqueness of solutions to the associated elastostatic problem.

For non-linear systems it should be noted that the uniqueness of one solution does not necessarily imply the uniqueness of any other solution in the system.

D. Stability theorems for abstract dynamical systems.

13. Introduction. In the last section Liapounov stability was defined abstractly for a dynamical system and emphasis was laid on the fact that stability is a topological concept associated with continuity, that is with neighbourhoods of a solution and its initial values. The neighbourhoods are defined from arbitrarily chosen positive-definite measures ϱ and ϱ_τ.

We now turn our attention to the methods devised for ascertaining the stability or instability of a solution for a given choice of measure in the framework of a particular definition. These methods fall into three classes, each employing a distinct technique, whose success or failure depends on the specification of the particular problem. These are:

(i) application of a maximum principle,

(ii) application of the direct (or second) method of Liapounov,

(iii) application of eigenfunction (or spectrum) decomposition.

In the present chapter we consider (i) and (ii), treating (ii) exhaustively. The technique (iii) of eigenfunction decomposition is deferred until Chap. E where only a brief discussion is given as the method is well-known. The development given is intended to familiarise and illustrate all the procedures so that their range, content and uniqueness will be immediately apparent in Chaps. F–U where extensive applications are made to elasticity.

14. Maximum principles. Maximum principles can be used to determine the stability of a given solution as a consequence of the following theorem:

Theorem 14.1. A sufficient condition for the solution $\varphi\in\mathscr{B}(T,X)$ to be Liapounov stable is that ϱ_τ and ϱ satisfy the inequality

$$\varrho(\varphi_\tau(t),\psi_\tau(t))\leq M_\tau(t)\,\varrho_\tau(\varphi(\tau),\psi(\tau)),\qquad t\in T,$$

for each $\tau\in\mathscr{T}$ and $\psi\in\mathscr{B}(T,X)$, where $M_\tau(t)$ is a bounded real function on T.

Corollary 14.2. If $M_\tau(t)$ is independent of $\tau\in\mathscr{T}$ then φ is uniformly stable.

Corollary 14.3. If $M_\tau(t)$ tends to zero as $t\to\infty$ then φ is asymptotically stable.

The theorem and its corollaries follow easily from the definitions and provide a practical test for stability in those systems demanding no finer condition.[1] It

[61] See Sect. 45.

[1] In functional analysis, the application of a device similar to that of the theorem often proves the most practical means of establishing continuity.

(i) T_τ from $Y(\tau)$ to $\mathscr{B}_{F_\tau}(T, X)$,

(ii) the identity mapping i from $\mathscr{B}_{F_\tau}(T, X)$ to $\mathscr{B}_\varrho(T, X)$.

It is thus clear if T_τ is continuous at φ from $Y(\tau)$ to $\mathscr{B}_{F_\tau}(T, X)$ and i is continous at φ from $\mathscr{B}_{F_\tau}(T, X)$ to $\mathscr{B}_\varrho(T, X)$ then T_τ is continuous at φ from $Y(\tau)$ to $\mathscr{B}_\varrho(T, X)$. When this holds for each $\tau \in \mathscr{T}$, then φ is obviously stable. Conversely, if φ is stable one such function $F_{\tau,t}$ always exists as can be seen by setting $F_{\tau,t}$ equal to ϱ on $X \times X$. Therefore, based upon the standard law of composition of continuous maps, the following theorem is proved:[29]

Theorem 15.1. The solution $\varphi \in \mathscr{B}(T, X)$ is stable if and only if there exist positive-definite functions $F_{\tau,t}$, where $t \in T$, $\tau \in \mathscr{T}$, defined on $X \times X$ for which

(i) *given a real positive number ε there exists a real positive number $\delta(\varepsilon, \tau)$ such that*

$$\varrho_\tau(\varphi(\tau), \psi(\tau)) < \delta \tag{a}$$

implies

$$F_\tau(\varphi_\tau, \psi_\tau) < \varepsilon, \quad \psi \in \mathscr{B}(T, X), \tag{b}$$

(ii) *given a real positive number η there exists a real positive number $\xi(\eta, \tau)$ such that*

$$F_\tau(\varphi_\tau, \psi_\tau) < \xi \tag{c}$$

implies

$$\varrho(\varphi_\tau, \psi_\tau) < \eta, \quad \psi \in \mathscr{B}(T, X). \tag{d}$$

The pattern for the foregoing and for other formulations of the theorem is set by LIAPOUNOV's original theorem enunciated for discrete systems.[30] The real nature of the results required for continuous systems, where infinite-dimensional spaces are needed, is obscured by the use of the finite-dimensional theory. This fact has already made its appearance in Sect. 6 where the equivalence of norms was discussed. So much confusion has arisen and so many erroneous conclusions have been drawn in applying the theorem inadequately that an analysis of basic ideas in the two forms of the theorem seems appropriate. We therefore quote the finite-dimensional form of the theorem in the framework of a "system" used throughout this article. The proof will be taken up later in Sect. 16 during the course of discussion.

Theorem 15.2. Let $X = \mathbb{R}^n$, the n-dimensional euclidean space, and let ϱ_τ, ϱ both be (equivalent to) the euclidean norm. The solution φ is stable if and only if there exist continuous positive-definite functions $F_{\tau,t}$ where $\tau \in \mathscr{T}$, $t \in T$, defined on $\mathbb{R}^n \times \mathbb{R}^n$ such that $F_{\tau,t}(\varphi_\tau(t), \psi_\tau(t))$, $\varphi, \psi \in \mathscr{B}(T, X)$, are non-increasing with respect to t.

Many variations on Theorem 15.1 are possible and we find the position summed up in a perspicacious comment by LIAPOUNOV after establishing his own results on the second method. In his own words:[31]

"By varying the conditions satisfied by the unknown functions, one could, of course, propose many other theorems similar to those demonstrated. For the applications which we have in view, however, these theorems are perfectly sufficient and therefore we limit ourselves to them."

We may thus, with justification, name after LIAPOUNOV all theorems which employ auxiliary functions to establish stability. Examples of other Liapounov

---

[29] GILBERT and KNOPS [1967, *10*]. The theorem, under slightly more restrictive assumption has also been established by ZUBOV [1964, *17*, p. 199], and MOVCHAN [1960, *7*]. A similar theorem for the stability of closed sets is given by BHATIA and SZEGÖ [1967, *5*, p. 203].

[30] LIAPOUNOV [1892, *1*].

[31] Quoted by V. P. BASOV in the introduction to the book by LIAPOUNOV [1966, *20*]. See also vol. 2 of LIAPOUNOV's Collected Works.

theorems are given in Sects. 19, 20. The original theorem is still that most frequently evoked for analyses of elastic stability and tests relying on the energy criterion are basically an application of the theorem: see Chap. I.

The form of the theorem used in this article and presented as Theorem 15.1 is set in a more general framework than that used for the classical theory[32] or by most recent writers on the subject. Comparison between Theorems 15.1 and 15.2 shows that X has been generalised from a euclidean space to an arbitrary set. Yet again, the continuity requirement of F_τ with respect to ϱ_τ on $\tau\in\mathcal{T}$ makes the current expression of Theorem 15.1 particularly mild compared with the versions of ZUBOV[33] and MOVCHAN[34] who have pioneered generalisations in this field. ZUBOV discusses stability of invariant sets and both writers prescribe the exact manner in which condition (i) of Theorem 15.1 is to be satisfied.[35] We name these conditions (iii) and (iv):

(iii) *given a real positive number* ε *there exists a real positive number* $\delta(\varepsilon,\tau)$ *such that*

$$\varrho_\tau(\varphi(\tau),\psi(\tau))<\delta$$

implies

$$F_{\tau,0}(\varphi(\tau),\psi(\tau))<\varepsilon,$$

(iv) $F_{\tau,t}(\varphi_\tau(t),\psi_\tau(t))$ *are non-increasing with respect to t.*

This exposition of the theorem is clearly prompted by the LIAPOUNOV version for discrete systems (Theorem 15.2) and the function $F_{\tau,t}$ is required to satisfy conditions similar to those prescribed in the original version. It is this function therefore that serves the role of *Liapounov function.*

In selecting Liapounov functions we have $\varrho(\varphi_\tau,\psi_\tau)$ or $F_\tau(\varphi_\tau,\psi_\tau)$ as obvious choices whenever the system is stable. With the Zubov-Movchan condition (iv), extra boundedness conditions are thus imposed:

$$\varrho(\varphi_\tau,\psi_\tau)<\infty,$$

or

$$F_\tau(\varphi_\tau,\psi_\tau)<\infty,$$

for all $\psi\in\mathcal{B}(T,X)$ i.e., every $\psi\in\mathcal{B}(T,X)$ is bounded with respect to φ in the "topology" defined by either ϱ or $F_{\tau,t}$. Put otherwise, every $\psi\in\mathcal{B}(T,X)$ is Lagrange stable at φ (Definition 10.3) so that the existence of a Liapounov function satisfying MOVCHAN's theorem enables us to establish Lagrange and Liapounov stability simultaneously. Another common choice of Liapounov function is the total energy which must therefore be finite under the Zubov-Movchan conditions.

By way of comparison we note that the boundedness properties which feature in ZUBOV's and MOVCHAN's versions of the stability theorem are entirely absent in the present statement of Theorem 15.1. We emphasise that in this theorem the functions ϱ_τ, ϱ and $F_{\tau,t}$ are defined on the set $X\times X$ and the stability is concerned only with the behaviour of solutions of the system at or near the solution whose stability is under investigation.[36] It is therefore unnatural to

[32] LIAPOUNOV [1892, *1*].
[33] ZUBOV [1964, *17*, p. 202].
[34] MOVCHAN [1960, *7*].
[35] ZUBOV differs from MOVCHAN in working with a single measure only.
[36] When sufficiency only is required, the measures need be defined only in the respective neighbourhoods of the solution.

introduce global boundedness properties into a stability analysis. We also note that, unlike MOVCHAN's theorem, no demands are made on the continuity of ϱ with respect to t nor the continuity on T of ϱ with respect to ϱ_τ.

A slightly more practical expression of condition (ii) of Theorem 15.1 is that[37] there exists a continuously strictly increasing function $\lambda(x)$, $x \in \mathbb{R}$, with $\lambda(0)=0$ and $\lambda(x) \to \infty$ as $x \to \infty$, such that[38]

$$F_\tau(\varphi_\tau, \psi_\tau) \geqq \lambda(\varrho(\varphi_\tau, \psi_\tau)). \tag{1}$$

We may similarly show that the Zubov-Movchan condition (iii) is equivalent to the statement that *there exists a continuous strictly increasing function* $\mu(x)$, $x \in \mathbb{R}$, *with* $\mu(0)=0$, *such that*

$$F_{\tau,0}(\varphi(\tau), \psi(\tau)) \leqq \mu(\varrho_\tau(\varphi(\tau), \psi(\tau))). \tag{2}$$

The Zubov-Movchan conditions (iii) and (iv) are not the only ones ensuring condition (i) of Theorem 15.1. LASALLE and LEFSCHETZ[39] have suggested taking for auxiliary function $F_{\tau,t}$ any function satisfying a relation of the type $\dot{x}=f(t, x)$. More generally, any function satisfying an ordinary differential equation with stable solutions would serve. The idea is further exploited in papers by CORDUNEANU[40] and ANTOSIEWICZ[41] for ordinary differential equations. Similar techniques involving differential inequalities have been applied to general dynamical systems by LAKSHMIKANTHAM[42] and co-workers. An application to elasticity is made in Sect. 77.

In the same manner that equivalence relations were introduced in Sect. 6 for the measures ϱ_τ and ϱ, so equivalence relations may be introduced between the Liapounov functions $F_{\tau,t}$. The method for doing so is obvious, and no further allusion to this topic will be made.

Irrespective of the form in which the Liapounov theorem is presented its basic importance, as we have already stressed, centres round the fact that stability of a solution may be ascertained without the explicit knowledge of that or any other solution. In application, nevertheless, the theorem must be couched in a form which is relevant to the problem. Use of the discrete (finite-dimensional) form may obscure the behaviour of a continuous system and lead to erroneous conclusions. The discussion and comparison of the forms of the theorem are taken up in Sect. 16.

Before investigating the relation between the theorems for the discrete and continuous systems we conclude this section with the corresponding theorems for uniform stability and asymptotic stability. These may be considered as corollaries to Theorem 15.1.

Corollary 15.3. The solution is uniformly stable if and only if conditions (i) *and* (ii) *of Theorem* 15.1 *hold uniformly in* τ.

---

[37] The statement follows from results of BHATIA and SZEGÖ [1967, *5*, § 0.3.3].

[38] In applications, λ frequently takes the form: $\lambda(x) = c\,x$, for positive bounded constant c. MOVCHAN [1960, *7*] then calls "positive-definite", functions F_τ satisfying (1), while MIKHLIN [1965, *17*] describes as "positive-definite bounded below", any operator whose "energy", F_τ, satisfies the modified form (1).

[39] LASALLE and LEFSCHETZ [1961, *5*, p. 107].

[40] CORDUNEANU [1960, *2*].

[41] ANTOSIEWICZ [1962, *1*].

[42] See, for instance, LAKSHMIKANTHAM and LEELA [1969, *15*], where fuller references may be found.

Corollary 15.4. The solution φ *is asymptotically stable if and only if condition* (i) *holds together with the condition*

$$\lim_{t\to\infty} F_{\tau,t}\big(\varphi_\tau(t), \psi_\tau(t)\big) = 0 \quad \text{[43]} \tag{i}$$

for all $\varphi(\tau)$ *in the "region of stability".*[44]

We note that in asymptotic stability, condition (i) of Corollary 15.4 is ensured by imposing the condition that $F_{\tau,t}$ is a *strictly decreasing function* of t, except at the solution φ where it remains constant with respect to $t\in T$. This condition, on its own, suffices for the *weak asymptotic stability* of φ.

It is unlikely that the solutions of a conservative system (for example, elasticity) will be asymptotically stable. Once damping is introduced, however, as in thermoelasticity or in certain theories of mixtures, then we may expect this kind of stability. The problem of determining the region of asymptotic stability then immediately arises, and some general results are provided in Sect. 20 where HALE's invariance principle is described.

16. Discussion of the theorems. We now look at the nature of conditions (i) and (ii) of Theorem 15.1. Condition (i) involves the intrinsic properties of the system together with the topological characteristics imposed by the choice of ϱ_τ and $F_{\tau,t}$. In contrast to this, condition (ii) is concerned with the topological aspects only. In certain problems condition (i) may fail outside specified neighbourhoods and this is particularly likely to happen where linearisation has been used to modify the system in the vicinity of φ. Implicit in LIAPOUNOV's theorem, therefore, is the fact that application is restricted to certain prescribed regions of X. In the case of condition (ii) there may be certain choices of $F_{\tau,t}$ and ϱ for which it may never be satisfied at all. Some examples are given in Sect. 17, a number being taken from the more familiar field of the calculus of variations.

The topological character of the second condition can be noted in another way by recognising that it is equivalent to the continuous embedding of the set $\mathscr{B}_{F_{\tau,t}}(T,X)$ in $\mathscr{B}_\varrho(T,X)$, an operation dependent on the choice of $F_{\tau,t}$ and ϱ and not on the system. When X is a Sobolev space we can naturally achieve this embedding by means of the well-known embedding theorems and inequalities. Examples in Sect. 17 demonstrate the technique, but in some circumstances other types of inequality can also usefully be employed for effecting embedding. Some of these are used in the treatment of elastic stability contained in later sections of this treatise. There are cases in which continuous embedding of $\mathscr{B}_{F_{\tau,t}}(T,X)$ in $\mathscr{B}_\varrho(T,X)$ is impossible with certain choices of $F_{\tau,t}$ and ϱ and it is thus clear that in these cases stability fails for a topological reason.[45]

The above comments should be read in conjunction with those in Sect. 6 referring to the measures ϱ_τ and ϱ and possible breakdowns that may occur in their definition.

The characters of conditions (i) and (ii) have so far been discussed separately. The Liapounov theorem (Theorem 15.1) however, requires that both these conditions be satisfied. Expressed loosely, condition (i) imposes a restriction on the way in which the trajectories of the system may evolve while condition (ii) ensures that these trajectories are close to the solution φ with respect to the

---

[43] We return in Sect. 20 to give a further discussion of asymptotic stability in connexion with HALE's work.

[44] "Region of stability" is sometimes used in connexion with stability diagrams in the plane of parameters. This should not be confused with the term as used here which refers to dynamical systems. In the corollary, T is the set of non-negative real numbers.

[45] DUHEM [1902, *1*] was one of the first to note this reason for failure of stability.

measure chosen for stability. Both conditions are essential. That the second cannot be omitted because of its apparently non-intrinsic role may be seen by examining the topological nature of the condition.

In order to do this we now study the proof of Liapounov's theorem for the discrete case i.e., when X is a set in $\mathbb{R}^n$. The result is stated in Theorem 15.2 and it is noted that condition (ii) does not occur explicitly there.

Proof of Theorem 15.2.[46] Since ϱ_τ, ϱ are the euclidean norm, denote each by $\|(\cdot)\|$. For any positive real number ε, the set

$$\|x-y\|=\varepsilon \qquad x, y\in \mathbb{R}^n$$

is compact,[47] and the continuous function $F_{\tau,t}(x, y)$ achieves its minimum on the set. We may thus put

$$0<k\equiv \min_{\|x-y\|=\varepsilon} F_{\tau,t}(x, y), \qquad x, y\in \mathbb{R}^n. \tag{a}$$

Now let us consider the solution φ_τ which may conveniently be regarded as held fixed in the following. By continuity we may choose a positive real number $\delta(k, \tau)$ such that $F_{\tau,0}(\varphi(\tau), \psi(\tau))<k$ whenever $\|\varphi(\tau)-\psi(\tau)\|<\delta$. Then, from the monotonicity of $F_{\tau,t}$ we have $F_{\tau,t}(\varphi_\tau(t), \psi_\tau(t))<k$ for $t\in T$. Let us next consider

$$g(t)=\sup_{0\leq s\leq t} \|\varphi_\tau(s)-\psi_\tau(s)\|, \quad \text{with} \quad \|\varphi(\tau)-\psi(\tau)\|<\delta.$$

Since $g(s)$ is a non-decreasing function we suppose that t_1 is the first value of t for which $g(t_1)\geqq\varepsilon$. But then by (a), $F_{\tau,t_1}(\varphi_\tau(t_1), \psi_\tau(t_1))\geqq k$ and we arrive at a contradiction. We therefore conclude that

$$\sup_{t\in T} \|\varphi_\tau(t)-\psi_\tau(t)\|<\varepsilon,$$

and sufficiency is proved.

Necessity now follows by taking

$$F_{\tau,t}(\varphi_\tau(t), \psi_\tau(t))=\inf_{0\leq s\leq t}\|\varphi_\tau(s)-\psi_\tau(s)\|.$$

The crucial feature of our discussion is now brought to light in the proof of sufficiency. We observe that the proof rests on the fact that in a euclidean space $F_{\tau,t}$ assumes its extreme values on the frontier of balls, so that (a) may always be satisfied without the need for imposing any additional condition on $F_{\tau,t}$. But (a) satisfies condition (ii) of Theorem 15.1, which can therefore always be omitted from this theorem in its finite-dimensional form.

On the other hand, for arbitrary sets, even infinite-dimensional Banach spaces, F_τ does not necessarily attain its extremes on the frontier of balls so that condition (ii) of Theorem 15.1 is not automatically satisfied and must be postulated independently.[48] If this is not done, there is no guarantee that $\inf_{\varrho=\varepsilon} F_\tau$ is non-zero, or, alternatively, that ϱ is bounded when F_τ is sufficiently small. This now discloses how the issue of stability is obscured in finite-dimensional systems

[46] Continuity and monotonicity with respect to t of $F_{\tau,t}$ may be replaced by the more general condition (i) of the Theorem 15.1.

[47] For metric spaces, a necessary and sufficient condition that a continuous function achieves its upper and lower bounds on a set is that the set be compact. The unit ball and its frontier is compact in a normed linear space (see, for instance, Brown and Page [1970, 7, Theorem 4.4.6]) *if and only if*, the space is finite-dimensional.

[48] Note that Theorem 15.1 allows $F_{\tau,t}$ to be discontinuous, so that even if $X\times X$ were compact, the point in the text would still hold.

by the very limitation of the system. Many errors in elastic stability have arisen from applying this limited theory in the context of more general sets. Examples in which condition (ii) of Theorem 15.1 alone fails to guarantee stability are to be found in those given by SHIELD and SHIELD and GREEN (see Sect. 6) and by HELLINGER[49] and HAMEL.[50]

Of those working independently of LIAPOUNOV and ZUBOV and who avoided the error, we mention VOLTERRA,[51] KOITER[52] and SHIELD.[53]

17. Examples. We next discuss two simple examples with a view to providing straightforward illustrations of the theorems just discussed. No attempt is made to present the most general results, since the present object is merely that of familiarising the reader with the applications of the foregoing techniques.

(i) *The one-dimensional wave equation.*[54] The displacement $u(t, x)$ satisfies the wave equation

$$u_{,xx}=\ddot{u}, \qquad 0\leqq x\leqq 1, \qquad t>0 \tag{17.1}$$

and the boundary conditions

$$u(t,x)=0, \qquad x=0,1. \tag{17.2}$$

We investigate the stability of the null solution $u(t,x)=0$ and for the set X take the subset of $C^2[0,1]$ consisting of real-valued functions ψ defined on $[0,1]$ satisfying the conditions

$$\psi(0)=\psi(1)=0.$$

Since X is linear we select as positive-definite function the norm

$$\varrho(0,\psi)=\sup_{0\leqq x\leqq 1}|\psi(x)|, \qquad \psi\in C^2[0,1].$$

When T is the semi-group of all non-negative real numbers and $\mathscr{T}=T$, the dynamical system representing the one-dimensional wave equation consists of all functions u defined on $[0,\infty)$ taking values in X and such that (17.1) is satisfied. The norm induced by ϱ on $\mathscr{B}(T,X)$ is therefore

$$\varrho(u)=\sup_{0\leqq t<\infty}\Big(\sup_{0\leqq x\leqq 1}|u(t,x)|\Big), \qquad u\in\mathscr{B}(T,X).$$

In order to avoid questions of existence, we assume that $Y(\tau)$, $\tau\in\mathscr{T}$ is the subset of X consisting of functions of the form $u(\tau,x)$.

By virtue of Eq. (17.1) and the boundary conditions (17.2) the total energy is conserved:

$$E(t)\equiv\tfrac{1}{2}\int_0^1(\dot{u}^2+u_{,x}^2)\,dx=E(\tau). \tag{17.3}$$

We may choose τ arbitrarily so let us take it to be zero, and let us choose

$$\varrho_0(u)=E(0), \tag{17.4}$$

while for the Liapounov function, we take,

$$F_{0,t}(u)=E(t), \qquad t\geqq 0. \tag{17.5}$$

---

[49] HELLINGER [1914, *1*, pp. 653, 654].
[50] HAMEL [1949, *1*, p. 269].
[51] VOLTERRA [1899, *1*].
[52] KOITER [1963, *7*], [1965, *12*], [1967, *18*].
[53] SHIELD [1965, *21*].
[54] See, for instance, SLOBODKIN [1962, *13*], SHIELD [1965, *21*], KNOPS and WILKES [1966, *15*], GILBERT and KNOPS [1967, *10*], MOVCHAN [1959, *5*], [1960, *6*, *7*].

Then, because of the inequality,

$$\varrho(u) \leqq \sup_{0 \leqq t < \infty} \left(\int_0^1 u_{,x}^2 \, dx \right)^{\frac{1}{2}} \leqq 2^{\frac{1}{2}} \sup_{0 \leqq t < \infty} F_{0,t}^{\frac{1}{2}} \tag{17.6}$$

and definition (17.5) both conditions (i) and (ii) of Theorem 15.1 are satisfied, and consequently the null solution is stable with respect to the measures ϱ_0 and ϱ.

Stability of the null solution also exists with respect to the measures

$$\varrho_{01}(u) = \sup_{0 \leqq x \leqq 1} |u_{,x}(0, x)| + \sup_{0 \leqq x \leqq 1} |\dot{u}(0, x)|,$$

and

$$\varrho_1(u) = \sup_{0 \leqq t < \infty} \int_0^1 u^2 \, dx,$$

since ϱ_{01} and ϱ_1 are obviously equivalent (in the sense of Sect. 6) to ϱ_0 and ϱ, respectively. For either pair of positive-definite functions, it follows that an equivalent Liapounov function is the potential energy

$$V(t) = \tfrac{1}{2} \int_0^1 u_{,x}^2 \, dx,$$

but $V(t)$ does not have a sign-definite time-derivative, and hence cannot be used in the Zubov-Movchan version of Theorem 15.1.

To obtain the stability of the null solution with respect to positive-definite functions defined on the first derivatives of u, we must introduce the first order energy[55] $E_1(t)$ defined by

$$E_1(t) \equiv \tfrac{1}{2} \int_0^1 (\ddot{u}^2 + \dot{u}_{,x}^2) \, dx.$$

This function may be proved to be constant in time by means of Eq. (17.1) and boundary conditions (17.2). On taking $E_1(t)$ as Liapounov function and repeating the previous argument, we obtain stability with respect to

$$\varrho_{02}(u) = E_1(0),$$

and

$$\varrho_2(u) = \sup_{0 \leqq t < \infty} \left(\sup_{0 \leqq x \leqq 1} |\dot{u}| \right).$$

In addition, the inequalities

$$E(t) \leqq 2 E_1(t),$$

and[56]

$$\sup_{0 \leqq x \leqq 1} |u_{,x}| \leqq \left(\int_0^1 u_{,xx}^2 \, dx \right)^{\frac{1}{2}} \leqq (2 E_1(t))^{\frac{1}{2}}$$

prove, in a similar way, that the null solution is stable with respect to any function chosen from

$$\varrho_{02}(u) = E_1(0)$$

$$\varrho_{03}(u) = \sup_{0 \leqq x \leqq 1} |\dot{u}_{,x}(0, x)| + \sup_{0 \leqq x \leqq 1} |\ddot{u}(0, x)|,$$

and from

$$\varrho_3(u) = \sup_{0 \leqq t < \infty} \left(\sup_{0 \leqq x \leqq 1} |u| + \sup_{0 \leqq x \leqq 1} |u_{,x}| + \sup_{0 \leqq x \leqq 1} |\dot{u}| \right)$$

$$\varrho_4(u) = \sup_{0 \leqq t < \infty} \left(\int_0^1 (u^2 + u_{,x}^2 + \dot{u}^2, \, dx \right).$$

---

[55] Cp. SHIELD [1965, *21*].

[56] This inequality follows on recalling that $u_{,x}(\xi) = 0$ for some $\xi \in [0, 1]$.

It is interesting to note that sharper stability results may be derived directly from the complete solution when expressed in d'ALEMBERT's form:[57]

$$u(t,x)=\tfrac{1}{2}\left[f(x-t)+f(x+t)+\int_{x-t}^{x+t} g(\xi)\,d\xi\right],$$

where f and g are the odd periodic continuations to the whole line of the initial value of u and its time-derivative. It follows immediately that the null solution is stable with respect to

$$\varrho_{04}=\sup_{0\le x\le 1}\left(|u(0,x)|+|\dot{u}(0,x)|\right)$$

$$\varrho_{05}=\sup\left(|u_{,x}(0,x)|+|\dot{u}(0,x)|\right)$$

and $\varrho(u)$, $\varrho_3(u)$, respectively. Thus, the energy could be unbounded, but stability is still possible with respect to ϱ_{04}, $\varrho(u)$.

(ii) *The three-dimensional wave equation.*[58] We next consider the displacement $u_i(t,\boldsymbol{x})$ which in a bounded region B satisfies the three-dimensional wave equation

$$u_{,ii}=\ddot{u},\qquad \boldsymbol{x}\in B,\qquad t>0 \tag{17.7}$$

and on the surface satisfies the homogeneous boundary conditions

$$u(t,\boldsymbol{x})=0,\qquad \boldsymbol{x}\in\overline{\partial B_1},\qquad t>0, \tag{17.8}$$

$$\frac{\partial u}{\partial n}(t,\boldsymbol{x})=0,\qquad \boldsymbol{x}\in\partial B_2,\qquad t>0 \tag{17.9}$$

where $\partial B=\overline{\partial B_1}\cup\partial B_2$ and $\boldsymbol{n}$ is the unit outward normal on ∂B_2. We shall examine the null solution of (17.7)–(17.9) and for X take the subset of real-valued functions which belong to $C^2(B)$ and meet the boundary conditions (17.8), (17.9). We again suppose that T is the set of real non-negative numbers and that $\tau=0$. Then, the dynamical system consists of all functions defined on $[0,\infty)$, taking values in X and satisfying (17.7).

We have already seen in Sect. 6 that the null solution is not stable with respect to the spatial uniform norm. We shall return to the question of this norm shortly, but meanwhile let us consider the stability of the null solution with respect to the measure

$$\varrho(u(t))=\int_{B(t)} u^2\,dx \tag{17.10}$$

where $\varrho(u)=\varrho(0,u)$ and $B(t)$ denotes integration over B at time t. From standard eigenvalue inequalities we then obtain

$$\varrho(u(t))\le\lambda\int_{B(t)} u_{,i}\,u_{,i}\,dx \tag{17.11}$$

where λ is the lowest eigenvalue of the corresponding membrane problem. The expression on the right side of inequality (17.11) is 2λ times the potential energy and therefore on recalling that the kinetic energy is non-negative we may obtain

$$\varrho(u(t))\le 2\lambda\,E(t) \tag{17.12}$$

in which $E(t)$ denotes the total energy at time t. On setting $F_{0,t}=E(t)$, we see that (17.12) satisfies condition (ii) of Theorem 15.1. Condition (i) of that theorem is satisfied by taking $\varrho_0(u(t))=E(0)$, and noting that $E(0)=E(t)$ by virtue of (17.7)–(17.9). The null solution is therefore stable with respect to the appropriate measures. Other measures could be taken for ϱ_0 and clearly the analysis may be extended to include weak solutions.

[57] Cp. SLOBODKIN [1962, *13*].

[58] SLOBODKIN [1962, *13*], SHIELD [1965, *21*], KNOPS and WILKES [1966, *15*].

Let us next consider stability with respect to the uniform measure

$$\varrho(u(t)) = \sup_{\boldsymbol{x}\in\bar{B}} u^2(t, \boldsymbol{x}). \tag{17.13}$$

The second example of Sect. 6 shows that an inequality of type (17.12) cannot hold and in fact the null solution is unstable with respect to the measure (17.13). We shall now prove that suitable restriction of initial data enables stability to be recovered.[59] From the Sobolev inequality, we have

$$\varrho(u(t)) \leqq c_1 \|u\|^2_{W^2_2} = c_1 \int\limits_{B(t)} (u^2 + u_{,i}\, u_{,i} + u_{,ij}\, u_{,ij})\, dx,$$

for some positive constant c_1. Provided the region B is sufficiently regular, it may then be proved that, for positive constant c_2,[60]

$$\|u\|^2_{W^2_2} \leqq c_2 \int\limits_{B(t)} u_{,ii}\, u_{,jj}\, dx$$

and hence

$$\varrho(u(t)) \leqq c_1 c_2 \int\limits_{B(t)} (\dot{u}_{,i}\, \dot{u}_{,i} + \ddot{u}^2)\, dx = 2 c_1 c_2 E_1(t),$$

where (17.8) has been used and $E_1(t)$ is the total energy of the first order. But from (17.7)–(17.9) we may prove that $E_1(t) = E_1(0)$. Hence, on setting $\varrho_0(u(0)) = E_1(0)$, $F_{0,t} = E_1(t)$, Theorem 15.1 tells us that the null solution is stable with respect to these measures. The initial data must now have derivatives such that $E_1(0)$ can be made arbitrarily small. The previous analysis only required the initial data to make $E(0)$ small.

Further discussion of this problem is provided by SHIELD who, in particular, considers other kinds of measures besides those treated here.

18. Relation of stability to the calculus of variations. The discussion concerning condition (ii) of Theorem 15.1 in Sect. 16 led us to observe that on the frontier of the ball $\|x-y\| = \varepsilon$ in a finite-dimensional euclidean space X the function $F_{\tau,t}$ attains its extreme values. In other spaces, condition (ii) has to be applied to $F_{\tau,t}$ to ensure the appropriate behaviour for stability. These observations suggest a direct connexion between stability and the calculus of variations. This connexion finds partial expression in LAGRANGE'S theorem, mentioned at the beginning of Sect. 15, which is the simple criterion for determining stability of an equilibrium position in a discrete conservative system. According to this theorem it is necessary to examine a given functional[61] to find its minimum. This criterion, while valid for discrete conservative systems, is often applied directly to continuous systems and in practice sometimes yields useful results. When this procedure is viewed in the light of the Liapounov theorem it is realised that a number of tacit assumptions lie behind its use. That these assumptions are fulfilled in some applications seems to have been a matter of chance. We therefore recapitulate briefly the content of the Liapounov theorem to bring out the true connexion of the procedure with the calculus of variations.

(a) Before a meaning can be assigned to the idea of stability, a choice must be made of the measures ϱ_τ and ϱ which are positive-definite in the sense that

$$\varrho(\psi_\tau(t), \varphi_\tau(t)) > 0, \qquad \varphi \neq \psi$$

$$\varrho(\varphi_\tau(t), \varphi_\tau(t)) = 0;$$

[59] SHIELD [1965, *21*]. This author provides a direct proof.
[60] HLAVÁČEK and NEČAS [1970, *12*].
[61] In elasticity, the functional is usually the potential energy.

(b) a class of functions $F_{\tau,t}$ must be chosen which are positive-definite in the sense of (a);

(c) for stability, conditions (i) and (ii) of Theorem 15.1 must be satisfied by

$$F_\tau = \sup_{\tau \in T} F_{\tau,t},$$

or $F_{\tau,t}$ must obey conditions (iii) and (iv) in the ZUBOV-MOVCHAN version of the theorem.

We thus see that the link with the calculus of variations is provided by (a) and (b). The positive-definite function introduced into the Liapounov theorem is, by definition, a function with a strict (proper) minimum at $\varphi \in \mathscr{B}(T, X)$ for all $\psi \in \mathscr{B}(T, X)$.[62] Since stability of φ depends only on the behaviour of the "perturbed" solutions ψ in the neighbourhood of φ, the corresponding minimum problem need clearly only be in the *relative* or *local* sense. In addition, if the minimum is weak, stability of φ is conditional, since the perturbed solutions ψ must now be restricted to the class in which the weak minimum is attained.[63]

The important parallel between the calculus of variations and stability lies in the fact that, in each case, certain conclusions can only be drawn if a specified function actually attains its minimum. In the calculus of variations a condition similar to (ii) of Theorem 15.1 is required to ensure the convergence of minimising sequences in the so-called direct method and associated approximation procedures.[64] There is an abundance of examples[65] in which a condition equivalent to (ii) is violated and these are counterparts to the examples of HELLINGER, HAMEL and SHIELD and GREEN quoted earlier.

We thus see that it is important to know under what conditions a function attains its minimum. In stability analysis, as our concern is with a positive-definite function we require this function to assume the value zero at the minimum, this being the analogy with the corresponding problem of the calculus of variations. In this context it is most important to note that functions may achieve their minima *in one space but not in another*.[66,67]

---

[62] In fact some writers actually state LIAPOUNOV's theorem in terms of a strict minimum of a function. Cp. TIKHONOV, VASIL'EVA and VOLOSOV [1970, *29*, p. 208].

[63] KOITER [1965, *12*] gives an example showing that even a weak proper minimum of the energy function is insufficient for stability of the equilibrium solution under L_2-norms as measures.

[64] See, for example, FUNK [1962, *4*, pp. 452–497], MORREY [1966, *21*, p. 15].

[65] See HADAMARD [1907, *1*], COURANT and HILBERT [1953, *1*, p. 183], KANTOROVICH and KRYLOV [1958, *1*, p. 338], GELFAND and FOMIN [1963, *4*], etc.

[66] The study of the subject originates from WEIERSTRASS' objection to the Dirichlet problem. See the account by COURANT [1950, *1*] or MORREY [1966, *21*, p. 5].

[67] We insert here a preliminary examination of the minimum potential energy criterion for ascertaining stable equilibrium in the light of the above earlier remarks. To state that the potential energy has a minimum is nothing more than to say it is an admissible positive-definite Liapounov function. This requirement of itself restricts the form of the potential energy. Of the other conditions of the Liapounov theorem nothing is generally said. Measures are chosen and silently incorporated in an unspecified way, condition (i) of Theorem 15.1 may be automatically satisfied (as in most conservative systems), while tacit expediency in the choice of measures may also succeed in satisfying conditions (ii). This last condition cannot be expected to be satisfied universally, however, so that there will be examples in which the criterion fails and erroneous conclusions made. Another source of error in the criterion arises in taking the minimum of the potential energy with respect to the set X_1 of functions satisfying the kinematic boundary conditions of the equilibrium problem (the so-called "virtual displacements"), whereas, even in conservative systems, stability at the minimum point should be evaluated on the set X_2 of solutions to the dynamic problem. In some examples, X_1 and X_2 may be such that the respective minima are different. It is thus clear that the minimum energy criterion is insufficient *by itself* and that the method as usually presented may give rise to results that are mathematically suspect. The above remarks are amplified in Sects. 31–34.

In view of the importance of knowing how a function behaves at its minimum, we now discuss some theorems giving necessary and sufficient conditions for a function $F(x, y)$ to reach its strict minimum, and we assume familiarity with concept of the Gateau and Fréchet derivatives. The set X is limited to a normed linear space and throughout the discussion we shall assume that *x is held fixed*, so in fact we shall treat a function of the variable y alone. We write $F_x(y)=F(x, y)$ and we suppose that $F_x(y)$ attains its minimum at $y=x$ with $F_x(x)=0$. The conditions for necessity and those for sufficiency are supplied by separate theorems. Proofs may be found in standard references and are omitted unless required to illustrate a particular aspect of the theory.[68]

These theorems are later used in the discussion of elastic stability.

Theorem 18.1. Let the function $F_x(y)$ be defined on X, where X is a normed linear space, and suppose that F_x has first and second Gateau differentials at x. Then in order that $y=x$ be a proper minimum of $F_x(y)$, $y\in X$, it is necessary that[69]

$$\text{(a)}\quad DF_x(x, h)=0, \qquad \forall h\in X,$$

$$\text{(b)}\quad D^2 F_x(x, h, h)>0, \qquad \forall h\in X$$

where $DF_x(x, h)$ is the Gateau differential of $F_x(y)$ at $y=x$ in the direction h.

For a relative minimum of $F_x(y)$ condition (b) becomes

$$\text{(c)}\quad D^2 F_x(x, h, h)\geqq 0 \quad \forall h\in X.$$

When equality applies for some $h\neq\emptyset$ further necessary conditions are required on the third and fourth Gateau derivatives of F_x at x.[70]

For the next theorem we assume that the function $F_x(y)$ possesses Fréchet derivatives.

Theorem 18.2.[71] *Let the function $F_x(y)$ be defined on an open set Ω of X where X is a normed linear space with norm $\|(\cdot)\|$, and suppose that F_x has first and second Fréchet derivatives at x which satisfy the relations*

$$\text{(d)}\quad dF_x(x, h)=0,$$

$$\text{(e)}\quad d^2 F_x(x, h, h)\geqq\alpha\|h\|^2,$$

where $h=y-x$, $x, y\in\Omega$, and α is a positive real number. Then the point x is a proper minimum of $F_x(y)$ for sufficiently small $\|h\|$.

It is important to note that the condition $d^2 F_x(x, h, h)>0$ is *not* sufficient for a minimum as can be demonstrated by examples.[72]

---

[68] For completeness, we include here a theorem giving general conditions under which a functional achieves its minimum; see VAINBERG [1964, *14*, Theorem 9.2]. *A weakly lower semi-continuous functional on a bounded weakly closed non-empty subset of a reflexive Banach space assumes its minimum there.* We make no direct use of this theorem.

[69] Condition (a) is usually satisfied for equilibrium solutions derived from a variational process. For hyperelastic bodies see TRUESDELL and NOLL [1965, *22*, § 88], and WASHIZU [1968, *27*]; in these studies, however, it seems reasonable to suppose that the variations are Fréchet derivatives.

[70] This aspect is the subject of extensive study by KOITER [1945, *1*], [1962, *9*], [1963, *8*], [1965, *12*] who in particular examines its significance for post-buckling behaviour.

[71] A more general theorem in terms of Gateau derivatives is proved by VAINBERG [1964, *14*, Theorem 9.4], but it is not needed in the treatise. KOITER [1945, *1*], [1962, *9*], [1965, *12*], deals with related results in which $\|(\cdot)\|$ is a homogeneous positive-definite quadratic function.

[72] See, for example, BOLZA [1931, *1*, p. 73ff.], KOITER [1965, *12*], SAGAN [1969, *19*, p. 39].

The proof of the above theorem depends upon the expansion

$$F_x(y) - F_x(x) = dF_x(x, h) + \tfrac{1}{2} d^2 F_x(x, h, h) + A(h),$$

where

$$\lim_{\|h\| \to 0} \frac{A(h)}{\|h\|^2} = 0$$

which also leads immediately to:

Corollary 18.3. Under the conditions of Theorem 18.2, there exists a real positive number α *such that*

$$F_x(y) - F_x(x) \geqq \alpha \|h\|^2, \qquad h = y - x, \qquad x, y \in \Omega,$$

provided $\|h\|$ *is sufficiently small.*

We shall take up the case for a semi-positive-definite second Fréchet derivative in Sect. 32,[73] while the case of an indefinite or negative-definite second derivative is considered under instability in Sect. 44.

The above theorems may be applied to special forms of the function F_x. A common representation, required particularly in elasticity, is

$$F_x(y) = \int_B U(\operatorname{grad}(y - x))\, d\Omega, \qquad x, y \in X$$

where B is a bounded smooth region of $\mathbb{R}^n$ with volume element $d\Omega$, X is the set of vector-valued functions defined on $\mathbb{R}^n$ and the scalar-valued function U is twice continuously differentiable in its arguments. Then condition (c) of Theorem 18.1 becomes[74]

$$\text{(f)} \qquad \int_B \left(\frac{\partial^2 U}{\partial h_{i,j}\, \partial h_{k,l}}\right)_{h=0} \eta_{ij}\, \eta_{kl}\, d\Omega \geqq 0, \qquad h = y - x$$

where η_{ij} is an arbitrary second order tensor. The following theorem incorporates the well-known *Legendre-Hadamard* condition:

Theorem 18.4.[75] *A necessary condition for*

$$F_x(y) = \int_B U(\operatorname{grad}(y - x))\, d\Omega$$

to attain a proper minimum at x *under the condition that* h *vanishes*[76] *on the boundary* ∂B *of* B *is that*

$$\left(\frac{\partial^2 U}{\partial h_{i,j}\, \partial h_{k,l}}\right)_{h=0} \lambda_i\, \lambda_k\, \mu_j\, \mu_l > 0, \qquad h = y - x,$$

holds at every point of B, *where* λ_i *and* μ_j *are arbitrary non-zero vectors. For* $F_x(y)$ *to achieve a relative minimum at* x *the condition is*

$$\left(\frac{\partial^2 U}{\partial h_{i,j}\, \partial h_{k,l}}\right)_{h=0} \lambda_i\, \lambda_k\, \mu_j\, \mu_l \geqq 0, \qquad h = y - x.$$

---

[73] See also KOITER [1945, *1*], [1963, *8*], [1965, *12*].

[74] See, for example, MORREY [1966, *21*, p. 10], SAGAN [1969, *19*, p. 393].

[75] HADAMARD [1902, *2*], [1903, *1*, § 270] apparently gave the first proof in connexion with stability of elastic bodies, for which he took (f) as the condition. TRUESDELL and NOLL [1965, *22*, Sect. 68 bis] record that DUHEM [1906, *2*, Chap. III, § IV] criticised HADAMARD's proof but did not succeed in supplying a rigorous substitute. Correct proofs have since been given by CATTANEO [1946, *1*], NOLL (see TRUESDELL and NOLL [1965, *22*, Sect. 68 bis]) and MORREY [1966, *21*, p. 11]. The theorem of the text is taken from MORREY. Like HADAMARD's proof, NOLL's arose in the derivation of a necessary condition for stability from condition (f), again taken as the stability definition for boundary conditions of displacement and traction.

[76] In the version of the theorem due to CATTANEO and NOLL (see previous footnote) the condition that $y - x$ vanishes on ∂B may be relaxed.

From Theorem 18.3, a sufficient condition for the function F_x of Theorem 18.4 to achieve a minimum at x is

$$\int_B \left(\frac{\partial^2 U}{\partial h_{i,j}\,\partial h_{k,l}}\right)_{h=0} \eta_{ij}\,\eta_{kl}\,d\Omega \geqq \gamma \|\eta\|^2, \qquad h=y-x.$$

However, when $h=y-x$ vanishes on the boundary ∂B, we have the well-known theorem of VAN HOVE:

Theorem 18.5.[77] *A sufficient condition for the function $F_x(y)$ of Theorem 18.4 to achieve a weak relative minimum at $y=x$, when $h=y-x$ vanishes on ∂B, is that there exists a real positive number β such that*

$$\left(\frac{\partial^2 U}{\partial h_{i,j}\,\partial h_{k,l}}\right)_{h=0} \lambda_i\,\lambda_k\,\mu_j\,\mu_l \geqq \beta\,\lambda_i\,\lambda_i\,\mu_j\,\mu_j,$$

holds at every point of B for arbitrary vectors λ_i and μ_i.

19. Theorems on instability. To prove that a solution φ is unstable it is sufficient to demonstrate that at least one solution ψ beginning in every neighbourhood of $\varphi(\tau)$ has a motion ψ_τ whose values do not lie in the neighbourhood of $\varphi_\tau(t)$ for at least one instant t. One of the earliest criteria used for ascertaining instability was that proposed by LAGRANGE[78] relating to the equilibrium solution in a discrete conservative system. This states that the solution is unstable if the potential energy achieves a maximum value at this solution.[79] It fails, however, to provide an exact converse to LAGRANGE's stability theorem (see Sect. 15) since for instability this converse requires that the potential energy *does not have a minimum* at the equilibrium solution. The proof of the converse was partially achieved by CHETAEV,[80] using a generalisation of two theorems established by LIAPOUNOV. One of our later tasks is to establish a similar instability result in linear elasticity, but meanwhile we describe a general theorem on instability applicable to the abstract dynamical systems defined earlier. This theorem is likewise related to those proved by LIAPOUNOV. At the end of this section and in the next we mention two other theorems which this time are applicable to HALE's dynamical system, and which may also be used to obtain conditions for instability.

Conditions for instability could, of course, be deduced from those already given for stability. As results derived in this way may be difficult to apply in practice, a theorem providing more convenient conditions is now given. In Sect. 12 it was shown that non-uniqueness of a solution φ implied its instability. Moreover, if every neighbourhood of $\varphi(\tau)$ contains solutions which become unbounded with respect to φ_τ then φ cannot be stable. A relevant theorem on instability may therefore exclude reference to non-unique solutions or solutions which are known to be unbounded irrespective of the neighbourhood of φ from which they originate.

---

[77] VAN HOVE [1947, *2*].

[78] LAGRANGE [1788, *1*], [1853, *1*, Pt. I, Sect. III, § V].

[79] JACOBI [1842, *1*] established instability for discrete systems in the case in which the potential energy is a maximum at equilibrium by a method employing LAGRANGE's identity. This work was later extended by LIPSCHITZ [1866, *5*], SUNDMAN [1913, *1*] and BIRKHOFF [1927, *1*]. See also HILL [1964, *5*] who gives an account of the work and extends the results. THOMSON and TAIT [1879, *1*, § 345 (ii)] still treating discrete systems, gave a geometrical proof under the condition of maximum potential energy.

[80] CHETAEV [1961, *2*, p. 33]. KOITER [1965, *13*] provides a critical survey.

Theorem 19.1.[81] *The solution* $\varphi\in\mathscr{B}(T,X)$ *is unstable if and only if there exist positive-definite functions* $F_{\tau,t}$, $\tau\in\mathscr{T}$, $t\in T$, defined on $X\times X$ *such that*

(i) *the mapping* T_τ *from* $Y(\tau)$ *to* $\mathscr{B}_{F_\tau}(T,X)$ *is discontinuous at* φ *for some* $\tau\in\mathscr{T}$;

(ii) *the identity mapping* i *is continuous from* $\mathscr{B}(T,X)$ *to* $\mathscr{B}_{F_\tau}(T,X)$.

Mathematically, these conditions take the form:

(ia) *for at least one* $\tau\in\mathscr{T}$ *there is a positive real number* ε *such that for all positive real numbers* δ

$$\varrho_\tau(\varphi(\tau),\psi(\tau))<\delta\Rightarrow F_\tau(\varphi_\tau,\psi_\tau)\geqq\varepsilon;$$

(iia) *given any positive real number* η *there exists a positive real number* $\xi(\eta,\tau)$ *for which*

$$\varrho(\varphi_\tau,\psi_\tau)<\xi\Rightarrow F_\tau(\varphi_\tau,\psi_\tau)<\eta.$$

The proof of the above theorem is straightforward. Sufficiency of the conditions is obvious on inverting the implication in (iia), and necessity is established by setting

$$F_{\tau,t}(\varphi(t),\psi(t))=\varrho(\varphi(t),\psi(t)).$$

It is also possible to give conditions under which the solution φ is not uniformly stable. The relevant theorem is given by Movchan.[82]

The motivation for Theorem 19.1 is to be found, as with that for stability, in the theorems given by Liapounov[83] for discrete systems. The more general form of Liapounov's work given by Chetaev[84] forms the basis of direct extensions to continuous systems represented in the theorems by Zubov,[85] Movchan[86] and Hale.[87] As with the stability theorem, the essential difference between the theorems expounded by Zubov and Movchan and the current presentation, lies in the fact that these authors prescribe the precise manner in which condition (ia) of Theorem 19.1 is to be achieved. That is, they require $F_{\tau,t}$ to possess a derivative with respect to t that is bounded below by some positive quantity, and further they demand the condition that solutions arising from all neighbourhoods of $\varphi(\tau)$ must remain bounded with respect to φ_τ.

Another theorem on instability which is of later use is that due to Hale. It applies, however, only to equilibrium solutions belonging to the dynamical system defined by Hale[88] (Sect. 4) but it relaxes the requirement of positive-definiteness of the Liapounov function. Before quoting the theorem, we therefore present a revised definition of Liapounov function, noting that, since X in this instance must be a Banach space, the equilibrium solution φ may be taken as the null solution without loss.

Definition 19.1. The function $F_{\tau,t}$ *is a Liapounov function in the sense of Lasalle on a set* $G\subset X$, *if* $F_{\tau,t}$ *is continuous on* $\bar{G}$, *the closure of* G, *and if*

$$\dot{F}_{\tau,t}(\psi_\tau(t))\equiv\overline{\lim_{\eta\to 0}}\frac{1}{\eta}\left[F_{\tau,t+\eta}(\psi_\tau(t+\eta))-F_{\tau,t}(\psi_\tau(t))\right]\leqq 0,\qquad \psi\in\mathscr{B}(T,X),$$

where $\tau\in\mathscr{T}$, $t\in T$, *and* $F_{\tau,t}(\psi_\tau(t))=F_{\tau,t}(\psi_\tau(t),0)$.

[81] This statement is taken from Gilbert and Knops [1967, *10*].
[82] Movchan [1960, *7*].
[83] Liapounov [1892, *1*, § 16].
[84] Chetaev [1961, *2*, p. 28].
[85] Zubov [1964, *17*].
[86] Movchan [1960, *7*], [1963, *16*].
[87] Hale [1969, *10*].
[88] The discussion in the remainder of this section is based on the article by Hale [1969, *10*].

Let us recall Definition 3.6 for an invariant set and then define the set S by

$$S=\{\psi_\tau(t)\in\bar{G}: \dot{F}_{\tau,t}(\psi_\tau(t))=0\},$$

and let M be the largest invariant set in S of the given dynamical system.

Theorem 19.2.[89] *Let a dynamical system $\mathscr{B}(T, X)$ in the sense of* HALE *be defined by $(\mathscr{T}, T, X)$, where X is a Banach space. Let $0\in\mathscr{B}(T, X)$ be an equilibrium point and suppose 0 belongs to the closure of some open set $U\subset X$ and let $N\subset X$ be an open neighbourhood of 0. We suppose that*

(i) *$F_{\tau,t}$ is a Liapounov function in the sense of Lasalle on $G=N\cap U$,*
(ii) *$M\cap G$ is either the empty set or zero,*
(iii) *$F_{\tau,t}(x)<\eta$ on G for $x\neq 0$, and some positive real number η,*
(iv) *$F_{\tau,t}(0)=\eta$ and $F_{\tau,t}(x)=\eta$ when x is in that part of the boundary of G in N.*

If N_0 is a bounded neighbourhood of zero properly contained in N, then $\psi(\tau)\neq 0$ in $G\cap N_0$ implies either there exists a $t_1>0$ such that $\psi_\tau(t_1)$ belongs to the boundary of N_0 or $\psi_\tau(t)$ remains in $G\cap N_0$ but does not belong to a compact set of $G\cap N_0$.

20. Boundedness and asymptotic stability. Stability and boundedness have long been recognised as intimately related, not only in linear systems as shown in Sect. 10, but also in general. The connexion exhibits itself through the properties of the Liapounov function and several of the preceding theorems on stability may be modified in an obvious fashion to yield conditions for boundedness. The ideas involved have been exploited by several authors[90] although there has been little application of these to elasticity.

A somewhat different kind of theorem on boundedness, an example of which is cited below, may assist in determining the regions of asymptotic stability for particular sets X.[91] The theorem on asymptotic stability given in Sect. 15 gives no indication of the manner of determining this region, and in general more structure must be introduced into X in order to obtain any theorems of practical value. The theorem we quote below is applicable only to equilibrium solutions of the dynamical system defined by HALE (Sect. 4) and employs the Liapounov function defined in the sense of LASALLE (Definition 19.1). We quote the theorem, deducing in a corollary its application to determining the regions of asymptotic stability.[92] There is no loss in taking the solution φ to be *the null solution*, and we note the set M is defined as in the previous section.

Theorem. Let a dynamical system in the sense of HALE *be defined by $(\mathscr{T}, T, X)$, where X is a Banach space. If $F_{\tau,t}$ is a Liapounov function in the sense of Lasalle on a set $G\subset X$ and an orbit $\gamma(\psi(\tau))$ belongs to G and is in a compact set of X then $\psi_\tau(t)\to M$ as $t\to\infty$.*

The theorem may clearly be used to obtain information about the boundedness of ψ and is used as such in the discussion of snap-through buckling given in Sect. 83. But it also provides information on asymptotic stability.

Corollary. When $M=\{0\}$, under the conditions of the theorem the null (equilibrium) solution is asymptotically stable with region of stability G.

[89] HALE's original version [1969, *10*, Theorem 2] is more general than the one quoted in the text.

[90] See, for example, YOSHIZAWA [1955, *9*], LASALLE and LEFSCHETZ [1961, *5*, pp. 113, 118], BHATIA and LAKSHMIKANTHAM [1965, *2*].

[91] It is evident that when X is linear and ϱ_τ, ϱ are linear maps, then the whole set $Y(\tau)$ becomes the region of asymptotic stability provided that the null solution is the only equilibrium solution and is asymptotically stable.

[92] This work is based on that by LASALLE [1967, *20*], HALE [1969, *10*], SLEMROD and INFANTE [1971, *12*].

E. Eigenfunction analyses.

In this chapter we very briefly mention some points in connexion with the well-known method of examining stability based on eigenfunction expansions.[1] Our remarks are intended to draw attention to the fact that in eigenfunction studies where sufficient conditions are derived for instability it is often the practice to regard, without adequate cause, these same conditions as also being necessary. In certain circumstances therefore it must be expected that this procedure will lead to erroneous stability estimates, emphasing the need for care.

It is an assumption intrinsic to the entire eigenfunction method that the governing operator possesses the separability property[2] and that solutions exist to the associated eigenfunction problem. That these assumptions are not always applicable is demonstrated by several counter-examples. The existence of at least one separable solution that is unbounded in time is taken as the condition of (Lagrange) instability at the null solution, while the existence of separable solutions, bounded for all time, is taken as the condition for stability. Separable solutions remaining constant in time give rise to neutral equilibrium which is not discussed until Sect. 32.

This technique for determining instability is perfectly justified, but for stability certain other conditions must be verified. For instance, it must be ensured that the eigenvalues form a denumerable ordered sequence of finite multiplicity[3] and that the eigenfunctions combine with the null solution to form a set which after completion results in a Hilbert space $\mathscr{H}$ with norm $\|(\cdot)\|_{\mathscr{H}}$. Correctly investigated, stability of the null solution then requires the initial data to lie in $\mathscr{H}$ and the related eigenfunction expansion to converge in $\mathscr{H}$ for all t. This, coupled with the boundedness of separable solutions (that is, terms in the expansion) means that solutions ψ with arbitrary initial data small in $\|(\cdot)\|_{\mathscr{H}}$ remain in $\mathscr{H}$ and are small in $\|(\cdot)\|_{\mathscr{H}}$ for all time. Establishing this convergence in the norm $\|(\cdot)\|_{\mathscr{H}}$ and dependence on the initial data can be achieved by demonstrating the existence of a time-dependent function which bounds $\|(\cdot)\|_{\mathscr{H}}$ from above. In many cases the result can be accomplished by a method equivalent to that of LIAPOUNOV.[4] This latter approach is the more general method as it does not require existence of eigenfunctions.

An essential feature of the method outlined in the preceding paragraph is that the eigenfunctions form a complete space, or a space that can be completed. We may, in fact, use any set of complete functions in the expansion of a solution ψ and stability will follow in the same way. For example, in non-linear analysis, eigenfunctions of the associated linear operator have been used,[5] while in non-self adjoint problems, those of the adjoint operator have been used.[6] Also falling into this category are the standard approximation procedures such as the Galerkin, Rayleigh-Ritz and finite-element methods.

[1] A recent general study of this method has been carried out by COTSAFTIS [1969, *4*], [1971, *1*]. Applications to elasticity are numerous, with a comprehensive account being given by BOLOTIN [1961, *1*]. Some further references are contained in Sect. 81.

[2] For separability of the operator M, defined on $\mathbb{R} \times \mathbb{R}^+$ say, a function $\psi(t, x)$ must exist such that $MS(t)\,Q(x)/\psi(t, x)\,S(t)\,Q(x)$ is of the form $P(x) + R(t)$, where S, Q, P, R are functions of the indicated variables.

[3] Operators with continuous spectra have been treated especially in the fluid mechanics literature.

[4] See, for example, SOBOLEV [1964, *12*], MIKHLIN [1964, *9*], LEIPHOLZ [1962, *11*, *12*], [1963, *14*], LEVINSON [1966, *19*].

[5] ECKHAUS [1965, *7*], [1967, *9*].

[6] PRASAD and HERRMANN [1967, *29*], DIPRIMA and HABETLER [1969, *6*].

F. Stability of elastic bodies.

21. Introduction. We come now to a consideration of the stability and instability of motions and equilibrium configurations of a deformed or undeformed elastic body. Non-linear and linear deformations are treated. When dealing with the class of linear deformations, stability or instability will be assigned according to the concepts of LIAPOUNOV, LAGRANGE or HÖLDER. On the other hand, when stability in the class of non-linear deformations is under examination the Liapounov notion will be used exclusively. For the most part, however, we shall be concerned with equilibrium configurations in the former class.

A variety of techniques are described for analysing stability properties. In the case of the more general non-linear elastic perturbations, direct appeal is necessarily made to fundamental theorems developed in Chap. D. For the particular theory relating to small linear perturbations we have recourse, in addition, to a number of specialised methods which permit greater analytical refinement in the analysis.

According to the overall scheme laid down in Chaps. B, C, D the formulation of problems in elastic stability will involve the following steps:

(i) Identification of the dynamical system (if any) associated with the given system of partial differential equations.

(ii) Choice of type of stability or instability to be investigated, i.e. whether Liapounov, Hölder, Lagrange etc.

(iii) Selection of positive-definite measures ϱ_τ and ϱ.

(iv) Choosing an appropriate Liapounov function in those cases for which the Liapounov method is applicable.

These will now be discussed in order.

(i) *Identification of the dynamical system.* This requires the specification of the set X and the sets T and $\mathscr{T}$. Throughout the remainder of this work T and $\mathscr{T}$ are both taken to be the semi-group of non-negative real numbers, while in the set of initial values $\mathscr{T}$, τ will be either zero or t_0. For X we take the set of vector-valued functions defined on either the spatial region B_0 of the body in the reference configuration, or the region B of the body in its primary state (see Sect. 23), and satisfying, in either case, certain boundary conditions at the surface of the region. At this stage there is still some freedom of choice as to which classes of vector-valued functions are to be admitted in X. The dynamical system $\mathscr{B}(T, X)$ associated with our stability problem consists, in general, of the incremental displacement $u_i(\cdot, \cdot)$[1] defined on $[0, \infty)$, satisfying the governing equations and taking values in X. The introduction of further regularity or smoothness requirements on the functions defining X will delimit the classes of solution admissible as perturbations. In this work X will in general be restricted to admit only classical solutions. There are examples where the stability analysis for classical solutions can be immediately made valid for various weak solutions, in which case such phenomena as shock waves could figure in the perturbations. It will usually be clear for which problems this can be effected and the extension to weak solutions is left to the reader.

(ii) *Choice of definition.* This matter has already been discussed. The choice of stability definition like the choice of measure is largely in the hands of the analyst. We note, for example, that although Liapounov stability is generally

[1] In some problems the dynamical system may consist of the set of all stresses $\sigma_{ij}(\cdot, \cdot)$ satisfying the equations of motion and taking values in X. Dynamical systems can similarly be defined for other variables.

the form used in testing the stability of equilibrium configurations, it might be advantageous to employ the criterion for *orbital stability* in many cases of motions.

(iii) *Selection of Measures* ϱ_τ, ϱ. We reiterate here what has already been enlarged upon earlier, that stability assessment depends upon the choice of measures and the set X. Choice of measure is not entirely arbitrary, however, as there will be an interrelationship between the set X and the various positive-definite functions that are to be defined on it. Certain functions will be incompatible with a given choice of X, for example, we cannot impose a Sobolev norm on a non-Sobolev space. In these cases our definitions give us trivial instabilities. A general programme is therefore one of listing the more obvious measures, pointing out any equivalence between them and examining the consequences of each choice for stability. This programme has been fairly completely carried out in the case of the linearised problem in which conservative loads are involved. Here the problem is simplified to some extent since with X a normed linear space and natural norms as measures, stability is entirely controlled by the choice of X. In the case of the non-linear problem the situation is very different and only a few preliminary results can be stated.

(iv) *Choice of Liapounov function.* As the Liapounov function enjoys a role similar in many ways to that of the positive-definite measures ϱ_τ, ϱ in defining a neighbourhood in $\mathscr{B}(T, X)$, it is also partially dependent on the properties of the set X. It is, however, time-dependent in our problems and its existence implies something about the stability of solutions which comprise the set $\mathscr{B}(T, X)$. As these solutions are governed by the differential equations, the selection of the function must be guided by these equations. In certain types of problem there exist integral invariants or certain functional forms derivable from the equations which become obvious candidates for Liapounov functions. Examples of these are to be found in the total energy in isothermal elasticity and the forms given in Chap. G for general thermoelasticity. In circumstances in which the procedures used to obtain these functions are not applicable the search for Liapounov functions is more difficult. In such cases the method may fail and other means of stability analysis may have to be employed.

In what follows, we shall study consequences of the above steps. Attention is focussed mainly on the three-dimensional hyperelastic theory in order to exhibit the basic methodology. It must be emphasised that, because the problems are formulated in a euclidean 3-space, conclusions obtained for these are not automatically valid for other elastic systems such as the rod, plate or shell, or even valid in two dimensions. In these theories a similar *general* approach may be adopted, but the conclusions naturally differ as a result of the different circumstances. This remark is also true of other systems in continuum mechanics:[2] their stability may be discussed in exactly the same *general* manner as that of three-dimensional elasticity.

We begin by deriving the basic equations of the theory and then proceed to select appropriate Liapounov functions apposite to the discussion of stability. After this, stability of an *equilibrium* position in the class of *non-small* incremental thermoelastic displacements is examined. The so-called "energy criterion" is next critically discussed and the circumstances in which it is justified are fully explained. Conservative, non-conservative and dissipative systems are also treated in the light of theorems of Liapounov and Hale. Our approach is based on the usual thermodynamic postulates which allow variations in both temperature and entropy. At this stage, the analysis is not restricted to small defor-

[2] E.g. multipolars, mixtures, materials with memory.

mations.[3] We then turn to the major contents of the treatise in which we consider in much greater detail stability in the class of *infinitesimally small* incremental displacements.

Isothermal and thermal elastic perturbations are considered separately. Both stability and instability analyses are treated.

22. Derivation of basic equations.[4] In this section we derive equations for the motion of an elastic body on which the whole of the subsequent stability analysis will be based. This will be conducted for hyperelastic bodies where the existence of a strain-energy (or stored-energy) function is assumed. Some of the calculations are valid, however, under less restrictive assumptions and it will be clear from the results themselves that some can be generalised to Cauchy elastic bodies or even to hypoelastic materials. As we shall deal mainly with *elastic bodies*, dissipation will not appear explicitly in our equations.[5] Should it be necessary to consider a theory involving dissipation, however, this will be treated separately in its own right as, for example, when dissipative surface and body forces are incorporated. This is also the case where thermoelasticity is concerned and this topic will be dealt with as a separate subject.

As usual the motion of the body is referred to a reference configuration and a fixed set of rectangular cartesian axes. The body in the reference configuration is taken to be homogeneous and the coordinates of a particle in the reference configuration are X_A with respect to these axes. In the subsequent motion of the body this particle has coordinates x_i given by

$$x_i = x_i(X_1, X_2, X_3, t), \tag{22.1}$$

with the condition that

$$\det\left[\frac{\partial x_i}{\partial X_A}\right] > 0. \tag{22.2}$$

Although an inverse of (22.1) exists whenever $\det[\partial x_i/\partial X_A]$ is non-zero, we have assumed in the usual way that this determinant is positive so that the density remains positive and deformations correspond to those of a real material.

For the greater part of the work classical solutions of the equations are used so that we assume the continuous differentiability of x_i with respect to each of the variables X_A, t as many times as required. x_i is thus usually of class $C^2[B_0 \times [0, t_1)]$, where B_0 is the reference configuration of the body and $[0, t_1)$ a time interval. Exceptions to this may occur at singular points, curves or surfaces. Whenever a weak solution of the equations of motion or equilibrium is required this restriction is naturally weakend.

The components of velocity at the point x_i are given by v_i where

$$v_i = \dot{x}_i, \tag{22.3}$$

and the superposed dot denotes differentiation with respect to t holding X_A constant in (22.1). That is, the dot denotes the material time derivative. Second order material time derivatives of x_i are similarly defined and are denoted by

[3] For general treatments of non-linear dynamic stability problems see ERICKSEN [1966, *7*, *8*], HSU [1966, *12*, *13*], KNOPS and WILKES [1966, *15*], KOITER [1965, *12*], [1966, *16*], [1967, *18*], and NEMAT-NASSER [1970, *24*].

[4] For further details see, for example, GREEN and ZERNA [1968, *10*, Chap. VII], TRUESDELL and NOLL [1965, *22*, § 79ff.], RIVLIN [1966, *29*], or any of the recent text books on modern continuum mechanics.

[5] See, for example, KOITER [1966, *16*], [1967, *18*], NEMAT-NASSER [1970, *24*]. These authors arbitrarily postulate the existence of dissipative stresses. Such theories are non-elastic by definition.

two superposed dots. Partial derivatives will be denoted by a comma preceding a subscript, differentiation with respect to x_i being indicated by small roman type and differentiation with respect to X_A by roman capitals, thus

$$\frac{\partial x_i}{\partial X_A} = x_{i,A}, \quad \frac{\partial q_i}{\partial x_i} = q_{i,i}.$$

Summation is implied by repeated subscripts of either kind.

Now consider a body, which, to begin with may be either elastic or not, and which at time t_0 occupies the region B_0 of euclidean three space and is bounded by the piece-wise smooth surface ∂B_0. The configuration of the body at time t_0 is taken as the reference configuration.

We postulate that an energy balance exists at time t in the form

$$\int_{\tilde{B}_0} \varrho_0 \dot{x}_i \ddot{x}_i \, dX + \int_{\tilde{B}_0} \varrho_0 \dot{U} \, dX = \int_{\tilde{B}_0} \varrho_0 [r + F_i \dot{x}_i] \, dX + \int_{\partial \tilde{B}_0} (T_i \dot{x}_i - H) \, dA. \tag{22.4}$$

Here, dX and dA denote the elements of volume and surface area respectively, ϱ_0 is the mass density of the material at time t_0, assumed bounded and positive[6] and U is the internal energy per unit mass. The function r represents a heat supply distributed throughout the volume and is measured per unit mass and unit time, while H, the heat flux across the surface, is measured per unit area of the surface at time t_0. The body force is denoted by F_i per unit mass, and T_i is the surface force per unit area of the surface at time t_0. Conservation of mass is assumed and integration is taken over any arbitrary sub-domain $\tilde{B}_0$ of the body in its reference configuration.

We also postulate an entropy production inequality for the material. At time t this assumes the form

$$\int_{\tilde{B}_0} \varrho_0 \dot{S} \, dX - \int_{\tilde{B}_0} \varrho_0 \frac{r}{T} \, dX + \int_{\partial \tilde{B}_0} \frac{H}{T} \, dA \geqq 0, \tag{22.5}$$

where S is the entropy per unit mass and $T (> 0)$ is the absolute temperature.

On introducing the Piola-Kirchhoff stress tensor T_{Ai}[7] and the heat flux Q_A[8] and requiring that (22.4) should be invariant under rigid-body motions in the usual way, we obtain

$$T_{Ai,A} + \varrho_0 F_i = \varrho_0 \ddot{x}_i, \tag{22.6}$$

with the relations

$$\left. \begin{aligned} T_i &= T_{Ai} N_A, \quad (22.7) \\ H &= Q_A N_A, \quad (22.8) \end{aligned} \right\} \quad \text{on } \partial B_0$$

where N_A are the components of the unit outward normal on the surface ∂B_0 of the body in its configuration at time t_0.

Use of these relations in (22.4) then leads to the local residual equation of energy balance

$$T_{Ai} v_{i,A} = \varrho_0 \dot{U} + Q_{A,A} - \varrho_0 r, \tag{22.9}$$

which represents an expression for the *stress power* $T_{Ai} v_{i,A}$.

[6] I.e. $\varrho_0 (X_A) \geqq \bar{\varrho}_0 > 0$, where $\bar{\varrho}_0$ is constant. It is always clear from the context whether the symbol ϱ designates density or a measure.

[7] I.e. stress at time t measured per unit area of surface in the body at time t_0. Note that in general $T_{iA} \neq T_{Ai}$.

[8] Q_A is the heat flux over the coordinate surfaces in the reference configuration per unit area in this configuration per unit time.

In addition, the entropy production inequality (22.5) may be written as

$$\varrho_0\, T\dot{S} \geqq \varrho_0\, r - Q_{A,A} + \frac{Q_A}{T}\, T_{,A}, \tag{22.10}$$

which, on using (22.9) becomes,

$$\varrho_0 (\dot{U} - T\dot{S}) \leqq T_{Ai}\, v_{i,A} - \frac{Q_A}{T}\, T_{,A}. \tag{22.11}$$

It is sometimes useful to have the above relations expressed in terms of quantities referred to the current configuration B. These quantities are the Cauchy (or true) stress σ_{ij}, the heat flux h per unit time per unit surface area of the deformed body and the heat flux q_i across the coordinate planes measured per unit area of the deformed body. We then have in place of Eq. (22.6)

$$\sigma_{ij,j} + \varrho\, F_i = \varrho\, \ddot{x}_i, \tag{22.12}$$

together with the boundary condition corresponding to (22.7)

$$t_i = \sigma_{ji}\, n_j \quad \text{on } \partial B,$$

where t_i is the surface traction per unit surface area of the deformed body and n_j is the unit outward normal on the deformed surface ∂B.

In the same way Eq. (22.9) has a counterpart

$$\sigma_{ij}\, v_{i,j} = \varrho\, \dot{U} + q_{i,i} - \varrho\, r, \tag{22.13}$$

where

$$h = q_i\, n_i \quad \text{on } \partial B \tag{22.14}$$

Eq. (22.11) likewise becomes

$$\varrho (\dot{U} - T\dot{S}) \leqq \sigma_{ij}\, v_{j,i} - \frac{q_i}{T}\, T_{,i}. \tag{22.15}$$

With the introduction of the Helmholtz free energy function $\mathscr{A} = U - TS$, an elastic material is defined by means of the constitutive equations

$$\begin{gathered} \mathscr{A} = \mathscr{A}(x_{i,A}, T), \qquad S = S(x_{i,A}, T), \qquad Q_A = Q_A(x_{i,A}, T, T_{,i}), \\ T_{Ai} = T_{Ai}(x_{i,A}, T). \end{gathered} \tag{22.16}$$

Because the inequality (22.5) holds for arbitrary volumes, arbitrary motions and arbitrary values of T it may be deduced in the usual way that

$$T_{Ai} = \varrho_0 \frac{\partial \mathscr{A}}{\partial x_{i,A}}, \qquad S = -\mathscr{A}_{,T}. \tag{22.17}$$

From the Eqs. (22.16) and (22.17) and the definition of $\mathscr{A}$, the residual energy balance (22.9) becomes

$$\varrho_0\, T\dot{S} = \varrho_0\, r - Q_{A,A}, \tag{22.18}$$

and the entropy production inequality (22.11) becomes

$$Q_A\, T_{,A} \leqq 0. \tag{22.19}$$

With respect to the current configuration Eqs. (22.17)–(22.19) are

$$\sigma_{ji} = \sigma_{ij} = \frac{\varrho}{2}\, x_{i,A}\, x_{j,B} \left(\frac{\partial \mathscr{A}}{\partial e_{AB}} + \frac{\partial \mathscr{A}}{\partial e_{BA}} \right), \tag{22.17a}$$

where

$$e_{AB}=\tfrac{1}{2}(x_{i,A}\,x_{i,B}-\delta_{AB}),$$

and

$$\varrho\, T\dot{S}=\varrho\, r-q_{i,i}, \tag{22.18a}$$

$$q_i T_{,i}\leqq 0. \tag{22.19a}$$

Note that either (22.19) or (22.19a) imply that if

$$T_{,i}\equiv 0$$

then

$$q_i=0.$$

An appropriate integral form of the energy balance equation is

$$\int_{\tilde{B}_0}\varrho_0\,\dot{x}_i\ddot{x}_i\,dX+\int_{\tilde{B}_0}\varrho_0(\dot{\mathscr{A}}+\dot{\overline{TS}})\,dX =\int_{\tilde{B}_0}\varrho_0\,T\dot{S}\,dX+\int_{\tilde{B}_0}\varrho_0\,F_i\dot{x}_i\,dX+\int_{\partial\tilde{B}_0}T_i\dot{x}_i\,dA, \tag{22.20}$$

in which all integrals are with respect to quantities measured in the reference configuration.

If the material is incompressible then

$$\det\,[x_{i,A}]=1 \tag{22.21}$$

and

$$\varrho=\varrho_0. \tag{22.22}$$

On letting p be a scalar function, it follows in the usual way that (22.17a) is replaced by

$$\sigma_{ij}=\sigma_{ji}=-p\,\delta_{ij}+\frac{\varrho_0}{2}\,x_{i,A}\,x_{j,B}\left(\frac{\partial\mathscr{A}}{\partial e_{AB}}+\frac{\partial\mathscr{A}}{\partial e_{BA}}\right). \tag{22.23}$$

23. The equations of perturbed motion. Isothermal linear elasticity. In the last section we set down the equations governing the motion of an elastic body, and we now wish to look at perturbations of this motion. In order to distinguish clearly between the unperturbed and perturbed motions of the body, the terms "primary" state and "secondary" state will be employed and the corresponding configurations of the body will be denoted by B and B^* respectively. Moreover, to avoid a possible confusion that can arise in terminology, we shall call "incremental" those quantities associated with the difference of motions in the primary and secondary states. For example, if the point X_A in the reference configuration B_0 moves to x_i in the primary state and to y_i in the secondary, then $u_i=y_i-x_i$ is the incremental displacement.

The equations governing the secondary state are of the same form as those for the primary state. In order to distinguish those variables representing these two states, thermodynamic quantities and forces associated with the secondary state will be denoted with an asterisk and the position coordinates of the particle X_A at time t will be denoted by y_i. We thus have

$$y_i=y_i(X_1,X_2,X_3,t),$$

with

$$\det\,[y_{i,A}]>0.$$

Of particular interest for the generation of Liapounov functions is Eq. (22.20) which becomes

$$\int_{\tilde{B}_0} \varrho_0 \dot{y}_i \ddot{y}_i \, dX + \int_{\tilde{B}_0} \varrho_0 (\dot{\mathscr{A}}^* + \overline{T^* S^*}^{\bullet}) \, dX = \int_{\tilde{B}_0} \varrho_0 T^* \dot{S}^* dX + \int_{\tilde{B}_0} \varrho_0 F_i^* \dot{y}_i \, dX + \int_{\partial \tilde{B}_0} T_i^* \dot{y}_i \, dA. \tag{23.1}$$

Now let

$$u_i = y_i - x_i.$$

On subtracting (23.1) from (22.20) we obtain

$$\begin{aligned} \int_{\tilde{B}_0} \varrho_0 (\dot{y}_i \ddot{y}_i - \dot{x}_i \ddot{x}_i) \, dX + \int_{\tilde{B}_0} \varrho_0 [\dot{\mathscr{A}}^* - \dot{\mathscr{A}} + \overline{T^* S^*}^{\bullet} - \overline{TS}^{\bullet}] \, dX \\ = \int_{\tilde{B}_0} \varrho_0 (T^* \dot{S}^* - T\dot{S}) \, dX + \int_{\tilde{B}_0} \varrho_0 [F_i^* \dot{u}_i + (F_i^* - F_i) \dot{x}_i] \, dX \\ + \int_{\partial \tilde{B}_0} [T_i^* \dot{u}_i + (T_i^* - T_i) \dot{x}_i] \, dA. \end{aligned} \tag{23.2}$$

This equation may also be put in another useful form

$$\begin{aligned} \int_{\tilde{B}_0} \varrho_0 (\dot{y}_i \ddot{y}_i - \dot{x}_i \ddot{x}_i) \, dX + \int_{\tilde{B}_0} \varrho_0 (\dot{U}^* - \dot{U}) \, dX \\ = \int_{\tilde{B}_0} \varrho_0 (r^* - r) \, dX + \int_{\tilde{B}_0} \varrho_0 [F_i^* \dot{u}_i + (F_i^* - F_i) \dot{x}_i] \, dX \\ + \int_{\partial \tilde{B}_0} N_A (Q_A - Q_A^*) \, dA + \int_{\partial \tilde{B}_0} [T_i^* \dot{u}_i + (T_i^* - T_i) \dot{x}_i] \, dA. \end{aligned} \tag{23.2a}$$

Either of the forms (23.2) or (23.2a) may be used for a stability analysis in cases for which the displacement is of finite magnitude. As Liapounov stability of a given primary state is governed by the behaviour of secondary states in its neighbourhood, some bound on the relative displacements is obviously implied. On the other hand, when the displacement u_i may safely be assumed small, a further form of (23.2) proves convenient in certain linearisation procedures. This is

$$\int_{\tilde{B}_0} \varrho_0 \dot{u}_i \ddot{u}_i \, dX + \int_{\tilde{B}_0} (T_{Aj}^* - T_{Aj}) \dot{u}_{j,A} \, dX = \int_{\tilde{B}_0} \varrho_0 (F_i^* - F_i) \dot{u}_i \, dX + \int_{\partial \tilde{B}_0} (T_i^* - T_i) \dot{u}_i \, dA, \tag{23.3}$$

where a substitution from Eq. (22.17) has been used. We may then employ the usual Taylor expansion of T_{Aj}^* about T_{Aj} when u_i is sufficiently small. A quadratic functional in $u_{i,j}$ will result when terms of order higher than the second are neglected. It should be noted that (23.3) can also be applied in analyses of stability with finite displacements but then the second integrand is not a perfect differential, a fact that has repercussion when it is desired to construct a Liapounov functional with the help of this equation.

We now wish to analyse more fully systems in which u_i is small. In this respect we will be making assumptions about the order of smallness of u_i similar to those made in the usual approximate theory of small deformations superposed on large. The fact that we do this does not mean that we believe the stability of a given solution in the class of non-linear perturbations may be judged from the stability of the solution in the class of the corresponding linearised perturbations.[9] We prefer, in studying linear systems, to regard them, not as approximations to non-linear ones, but as systems in their own right. Any conclusions drawn will then not be automatically applicable to non-linear stability.

Let us therefore suppose that the secondary motion is such that the displacement u_i enables terms of order higher than the second to be neglected in the expansion of T_{Ai}^* about T_{Ai}. Then for isothermal problems in which $Q_A = Q_A^* = 0$ we obtain from (23.3) in the usual way

$$\int_{\tilde{B}_0} \varrho_0 \dot{u}_i \ddot{u}_i \, dX + \int_{\tilde{B}_0} A_{iAkB} u_{k,B} \dot{u}_{i,A} \, dX = \int_{\tilde{B}_0} \varrho_0 (F_i^* - F_i) \dot{u}_i \, dX + \int_{\partial \tilde{B}_0} (T_i^* - T_i) \dot{u}_i \, dA, \tag{23.4}$$

where

$$A_{iAkB} = \frac{\partial T_{Ai}}{\partial x_{k,B}} = \varrho_0 \frac{\partial^2 \mathscr{A}}{\partial x_{k,B} \, \partial x_{i,A}} = A_{kBiA}. \tag{23.5}$$

The "incremental" rate of work Eq. (23.4) yields the equations of linearised elasticity

$$(A_{iAkB} u_{k,B})_{,A} + \varrho_0 (F_i^* - F_i) = \varrho_0 \ddot{u}_i \quad \text{in } B_0 \times (0, t_1), \tag{23.6}$$

where t_1 is some time instant that may be infinite. On the boundary ∂B_0 of B_0 we have

$$\begin{aligned} u_i &= h_i && \text{on } \overline{\partial B_{01}} \times [0, t_1) \\ N_A A_{iAkB} u_{k,B} &= T_i^* - T_i && \text{on } \partial B_{02} \times [0, t_1) \end{aligned}, \tag{23.7}$$

where h_i is a specified function of X_A and t, and $\partial B_0 = \overline{\partial B_{01}} \cup \partial B_{02}$. We also assu me that the body force and surface traction F_i, F_i^*, T_i, T_i^* are given functions of position and time, and also of displacement, velocity, etc. Their differences, $F_i^* - F_i$, $T_i^* - T_i$ will be expanded and expressed in terms of incremental quantities at a later stage when particular problems with special loads are discussed.

The coefficients A_{iAkB} of (23.5) are evaluated in the primary state and are functions of the state of strain, and hence of the deformation, of the body in its primary state. Thus, in general, A_{iAkB} are functions of both position and time. When the primary configuration is in equilibrium, these quantities are time-independent and when the primary configuration is homogeneously deformed these quantities are position-independent.

To complete the specification of the problem given by (23.6) and (23.7) we must adjoin the initial data

$$u_i = g_i, \quad \dot{u}_i = f_i \quad \text{on } B_0 \times 0, \tag{23.8}$$

where f_i and g_i are given functions. The problem thus formulated gives rise to the notion of a classical solution, by which is meant a vector-valued function $u_i(t, \boldsymbol{X}) \in C^2(B_0 \times (0, t_1))$ satisfying (23.6) and the boundary and initial data (23.7) and (23.8). While most subsequent calculations are worked under the assumption that solutions are classical, several can in fact be extended to include weak solutions. To define such solutions,[10] we let $\boldsymbol{W}$ denote the set of all vector

[9] Several examples are known which illustrate the contrary. See for example, HALE and LASALLE [1963, *6*], where various situations are described and also CESARI [1963, *1*]. As we have already noted (Sect. 15), DIRICHLET [1846, *1*] pointed out that linearisation often conceals in itself a tautology when stability is being estimated. KLEIN and SOMMERFELD [1897, *2*] reemphasised this when discussing the stability of a top. See also STOKER [1955, *7*], [1967, *32*].

[10] See KNOPS and PAYNE [1968, *18*], DAFERMOS [1968, *7*]. The latter author treats thermoelasticity and shows that suitably defined weak solutions belong to a Banach space provided certain positive-definiteness conditions are fulfilled. See also SLEMROD and INFANTE [1971, *12*].

functions $\boldsymbol{w}(t, \boldsymbol{x})$ that are continuous in $\bar{B}_0 \times [0, t_1]$ and have square integrable derivatives in $B_0 \times [0, t_1]$. Further, for almost all $\boldsymbol{X} \in B_0$, the derivatives $\dot{w}_i$ and $w_{i,A}$ are assumed to exist and to be continuous functions of $t \in (0, t_1]$. We let Φ denote the set of vector-valued functions $\boldsymbol{\varphi}(t, \boldsymbol{X})$ that are continuously differentiable in $B_0 \times [0, t_1]$ and that vanish on the sets $B_0 \times 0$ and $\overline{\partial B_{10}} \times [0, t_1)$, and we impose on the data the following conditions:

(a) $\varrho_0(\boldsymbol{X})$ and $A_{iAkB}(t, \boldsymbol{X})$ are bounded measurable functions in B_0,

(b) g_i and h_i are continuous functions on their respective regions of definition,

(c)

$$\int_{\partial B_{02} \times [0, t_1]} l_i l_i \, dA + \int_0^{t_1} \int_{B_0(\eta)} (F_i^* - F_i)(F_i^* - F_i) \, dX \, d\eta + \int_{B_0} (g_{i,A} g_{i,A} + f_i f_i) \, dX < \infty, \tag{23.9}$$

where

$$l_i = T_i^* - T_i, \tag{23.10}$$

and $B_0(t)$ denotes that the integration is over B_0 at time t. The weak solution is then defined to be a vector-valued function $\boldsymbol{u}(t, \boldsymbol{x}) \in \boldsymbol{W}$ which for each $\boldsymbol{\varphi} \in \Phi$ and for each arbitrary $t \in [0, t_1]$ satisfies

$$\int_{B_0(t)} \varrho_0 \varphi_i \dot{u}_i \, dX - \int_0^t \int_{B_0(\eta)} (\varrho_0 \varphi_{i,\eta} u_{i,\eta} - A_{iAkB} \varphi_{i,A} u_{k,B}) \, dX \, d\eta = \int_{\partial B_{02}(t)} l_i \varphi_i \, dA + \int_0^t \int_{B_0(\eta)} \varrho_0 (F_i^* - F_i) \varphi_i \, dX \, d\eta, \tag{23.11}$$

$$u_i = h_i \quad \text{on } \overline{\partial B_{01}} \times [0, t_1]. \tag{23.12}$$

In addition, the weak solution satisfies the initial condition

$$\begin{aligned} u_i &= g_i && B_0 \times 0 \\ \dot{u}_i = f_i, \quad u_{i,A} &= g_{i,A} && \text{a.e. in } B_0 \times 0, \end{aligned} \tag{23.13}$$

and the conservation law

$$E(t) \equiv \tfrac{1}{2} \int_{B_0(t)} (\varrho_0 \dot{v}_i \dot{v}_i + A_{iAkB} v_{i,A} v_{k,B}) \, dX = E(0), \tag{23.14}$$

where $E(0)$ is constant and v_i is the difference of any two solutions of (23.11).

Weak solutions are required in problems for which the data and geometry are insufficiently smooth to allow the existence of classical solutions, and therefore in view of the comments of Sect. 6, they have great importance in stability analyses.

It is sometimes convenient to refer all quantities to the configuration of the body in its primary motion at time t instead of the reference configuration at time t_0. This is especially useful whenever the primary deformation happens to be a configuration of equilibrium so that it is independent of time. KOITER[11] has treated the isothermal equilibrium condition of the primary state in this way, taking the free energy $\mathscr{A}$ and entropy S to be zero for this configuration. If we do this, however, it must always be remembered that the energy and entropy of the secondary motions depend not only on the displacement gradients, the temperature and its gradients, but also on the position of a material particle in

[11] KOITER [1967, *18*].

the primary state. This latter state is deformed and hence in general is not homogeneous.

When referred to the primary configuration B the rate of work equation (23.4) can be transformed in the conventional manner to the form

$$\int_{\tilde{B}} \varrho \dot{u}_i \ddot{u}_i \, dx + \int_{\tilde{B}} d_{ijkl} u_{i,j} \dot{u}_{k,l} \, dx = \int_{\tilde{B}} \varrho (F_i^* - F_i) \dot{u}_i \, dx + \int_{\partial \tilde{B}} (t_i^* - t_i) \dot{u}_i \, da, \tag{23.15}$$

where[11a]

$$d_{ijkl} = \sigma_{jl} \delta_{ik} + c_{ijkl} = d_{klij}, \tag{23.16}$$

and c_{ijkl} is defined by

$$\begin{aligned} c_{ijkl} &= 4\varrho \, x_{i,N} x_{k,M} x_{l,B} x_{j,A} \frac{\partial^2 \mathscr{A}}{\partial C_{AN} \partial C_{BM}} \\ &= c_{jikl} = c_{klij}, \end{aligned} \tag{23.17}$$

and C_{AB} is the Green deformation tensor $x_{i,A} x_{i,B}$. The integrals are taken over the respective volumes $\tilde{B}$ and surfaces $\partial\tilde{B}$ of the body in its *primary* configuration B.

The corresponding differential equations of motion are

$$(d_{ijkl} u_{k,l})_{,j} + \varrho (F_i^* - F_i) = \varrho \ddot{u}_i, \quad \text{in } B \times (0, t_1) \tag{23.18}$$

while on the bounding surface ∂B there holds

$$\begin{aligned} u_i &= h_i && \text{on } \overline{\partial B_1} \times [0, t_1) \\ n_j d_{ijkl} u_{k,l} &= t_i^* - t_i && \text{on } \partial B_2 \times [0, t_1), \end{aligned} \tag{23.19}$$

where h_i is an assigned function of position and time on $\overline{\partial B_1}$, $\partial B = \overline{\partial B_1} \cup \partial B_2$, and t_1 is some time instant, either finite or infinite.

Classical and weak solutions to this problem may be defined as in the previous case. We also remark that the differences in body force and surface tractions, which may depend, for instance, on the respective displacements and velocities, will be considered in greater detail as the need arises.

We next look at the case of an isotropic homogeneous elastic medium and examine the expressions assumed by the elasticities.[12]

For an isotropic body the Cauchy stress is given by[13]

$$\sigma_{ji} = \alpha_1 b_{ij} + \alpha_2 b_{ik} b_{kj} + \alpha_0 \delta_{ij}, \tag{23.20}$$

where

$$b_{ij} = x_{i,A} x_{j,A} \tag{23.21}$$

and α_0, α_1, α_2 are scalar invariant functions of the three invariants I_1, I_2, I_3, where

$$\begin{aligned} I_1 &= \operatorname{tr} C \\ I_2 &= \tfrac{1}{2}[(\operatorname{tr} C)^2 - \operatorname{tr} C^2] \\ I_3 &= \det [x_{i,A}]^2 \end{aligned} \tag{23.22}$$

and $C = [C_{AB}]$, and $\operatorname{tr} C = C_{AA}$. Moreover, the free energy $\mathscr{A}$ is now an invariant function of I_1, I_2, I_3:

$$\mathscr{A} = \mathscr{A}(I_1, I_2, I_3).$$

[11a] The major symmetry on the elasticities is intimately related to the assumption of a strain energy function; see TRUESDELL [1963, *19*].

[12] Other writers have fully considered crystal classes. See, for example, GREEN and ADKINS [1960, *3*, Chap. V].

[13] See, for example, TRUESDELL and NOLL [1965, *22*, §§ 47–49], GREEN and ADKINS [1960, *3*] and RIVLIN [1966, *29*, § 18].

Expressed explicitly,

$$\alpha_0=2\varrho\, I_3\frac{\partial\mathscr{A}}{\partial I_3},\qquad \alpha_1=2\varrho\left(\frac{\partial\mathscr{A}}{\partial I_1}+I_1\frac{\partial\mathscr{A}}{\partial I_2}\right),\qquad \alpha_2=-2\varrho\frac{\partial\mathscr{A}}{\partial I_2}. \tag{23.23}$$

On the other hand, the Piola-Kirchhoff stress is given by

$$\begin{aligned} T_{A\,i}=\beta_0\, x_{i,A}+\beta_1\, C_{AB}\, x_{i,B}+\beta_2\, e_{iqr}(\delta_{A1}\, x_{q,2}\, x_{r,3}\\ +\delta_{A2}\, x_{q,3}\, x_{r,1}+\delta_{A3}\, x_{q,1}\, x_{r,2})\end{aligned} \tag{23.24}$$

where

$$\beta_0=2\varrho_0\left(\frac{\partial\mathscr{A}}{\partial I_1}+I_1\frac{\partial\mathscr{A}}{\partial I_2}\right),\qquad \beta_1=-2\varrho_0\frac{\partial\mathscr{A}}{\partial I_2},\qquad \beta_2=2\varrho_0\, I_3^{\frac{1}{2}}\frac{\partial\mathscr{A}}{\partial I_3}. \tag{23.25}$$

Moreover, for isotropic materials, the coefficients of (23.17) become

$$\begin{aligned} c_{ijkl}=\sigma_{ij}\,\delta_{kl}-\alpha_0(\delta_{ik}\,\delta_{jl}+\delta_{il}\,\delta_{jk})+\alpha_2(b_{ik}\,b_{jl}+b_{il}\,b_{jk})\\ +2\delta_{ij}\,A^0_{kl}+2b_{ij}\,A^1_{kl}+2b_{ir}\,b_{rj}\,A^2_{kl},\end{aligned} \tag{23.26}$$

where

$$A^\gamma_{ij}=\frac{\partial\alpha_\gamma}{\partial I_1}\,b_{ij}+\frac{\partial\alpha_\gamma}{\partial I_2}\,(I_1\,b_{ij}-b_{ik}\,b_{kj})+I_3\frac{\partial\alpha_\gamma}{\partial I_3}\,\delta_{ij}\qquad \gamma=0,1,2. \tag{23.27}$$

Finally, when the primary state represents a small deformation from a stress-free isotropic reference state, then expression (23.26) may be reduced to[14]

$$\begin{aligned} c_{ijkl}=\lambda\delta_{ij}\,\delta_{kl}+2\mu\,\delta_{ik}\,\delta_{jl}+\frac{1}{(3\lambda+2\mu)}\left[2l-2m-\lambda-\frac{4\lambda(\lambda+m)+n\lambda}{2\mu}\right]\sigma_{mm}\,\delta_{ij}\,\delta_{kl}\\ +\frac{1}{(3\lambda+2\mu)}\left[2(m-\mu)-4\lambda+\frac{n\lambda}{2\mu}\right]\sigma_{mm}\,\delta_{ik}\,\delta_{jl}+\frac{\lambda+m}{\mu}\,(\delta_{ij}\,\sigma_{kl}+\sigma_{ij}\,\delta_{kl})\\ +2(\sigma_{ik}\,\delta_{jl}+\sigma_{jl}\,\delta_{ik})+\frac{n}{2\mu}\,e_{ils}\,e_{jkt}\,\sigma_{st}.\end{aligned} \tag{23.28}$$

In this, λ, μ are the Lamé constants associated with the initial small deformation, l, m, n are Murnaghan's second order elastic constants, and e_{ijk} is the permutation symbol.

The displacements considered in the preceding sections must always be subject to the constraints of the problem. As an example of constraint we consider the case of an incompressible solid.[15] Examples of more general constraints will be found in GREEN and ADKINS.[16] For incompressibility we have

$$\det[x_{i,A}]=1,$$

so that

$$\varrho=\varrho_0,$$

and therefore $\dot\varrho=0$. The equation of continuity is

$$\dot\varrho+\varrho\,\dot x_{i,i}=0,$$

and hence

$$\dot x_{i,i}=\dot x_{i,A}\,X_{A,i}=0. \tag{23.29}$$

This is then the required condition on the velocity gradients $\dot x_{i,A}$. From Eq. (22.17), on introducing a scalar function p as Lagrange multiplier to incorporate this

[14] See HOLDEN [1964, *6*] who gives this result and also remarks that the usual engineering practice ignores the terms in (23.28) which are linear in σ_{ij}. This presupposes that these linear terms are not of the same order of magnitude as the first term, a fact that cannot be assumed without further investigation. See also MURNAGHAN [1951, *2*, p. 65].

[15] See HILL [1957, *1*, § 6], GREEN and ADKINS [1960, *3*, Chap. V], BEATTY [1967, *4*], [1968, *2*].

[16] GREEN and ADKINS [1960, *3*, § 9.6].

constraint, we obtain now, in place of (22.17)

$$T_{Aj} = \varrho_0 \frac{\partial \mathscr{A}}{\partial x_{j,A}} - p X_{A,j}. \tag{23.30}$$

The scalar function p has the nature of pressure at each point. As it can change its value with position it does not qualify as a hydrostatic pressure in the usual sense.

The forms of the exact rate of work equations (23.2), (23.2a), (23.3) remain unaltered. Under the new assumptions, however, Eq. (23.5) is modified to

$$\frac{\partial T_{Ai}}{\partial x_{k,B}} = A_{iAkB} + p X_{A,k} X_{B,i}, \tag{23.31}$$

and Eq. (23.4) changes to

$$\begin{aligned} \int_{\tilde{B}_0} \varrho_0 \dot{u}_i \ddot{u}_i \, dX + \int_{\tilde{B}_0} (A_{iAkB} u_{k,B} \dot{u}_{i,A} + p u_{k,i} \dot{u}_{i,k}) \, dX \\ = \int_{\tilde{B}_0} \varrho_0 (F_i^* - F_i) \dot{u}_i \, dX + \int_{\partial \tilde{B}_0} (T_i^* - T_i) \dot{u}_i \, dA \end{aligned} \tag{23.32}$$

when second order terms are excluded. A_{iAkB} is still given by

$$A_{iAkB} = \varrho_0 \frac{\partial^2 \mathscr{A}}{\partial x_{k,B} \partial x_{i,A}}.$$

Quantities may again be referred to the body in the primary state instead of the reference state. Hence on using the corresponding stress relations and employing the manipulation to be found in any of the standard references,[17] we have

$$\begin{aligned} \sigma_{ji} &= \varrho_0 x_{j,A} \frac{\partial \mathscr{A}}{\partial x_{i,A}} - p \delta_{ij} \\ &= \varrho_0 x_{j,A} x_{i,B} \frac{\partial \mathscr{A}}{\partial e_{AB}} - p \delta_{ij} \end{aligned} \tag{23.33}$$

for the stresses and for (23.32)

$$\begin{aligned} \int_{\tilde{B}} \varrho_0 \dot{u}_i \ddot{u}_i \, dx + \int_{\tilde{B}} [\sigma_{jl} \delta_{ik} + p(\delta_{jl} \delta_{ik} + \delta_{jk} \delta_{il}) + b_{ijkl}] u_{k,l} \dot{u}_{i,j} \, dx \\ = \int_{\tilde{B}} \varrho_0 (F_i^* - F_i) \dot{u}_i \, dx + \int_{\partial \tilde{B}} (T_i^* - T_i) \dot{u}_i \, da, \end{aligned} \tag{23.34}$$

where

$$\begin{aligned} b_{ijkl} &= 4 \varrho_0 x_{i,N} x_{k,M} x_{l,B} x_{j,A} \frac{\partial^2 \mathscr{A}}{\partial C_{AN} \partial C_{BM}} \\ &= b_{jikl} = b_{klij} = b_{ijlk}. \end{aligned} \tag{23.35}$$

It must be emphasised that both (23.32) and (23.34) are restricted to the incremental displacements satisfying the incompressibility condition

$$\dot{u}_{i,i} = 0.$$

The corresponding equations satisfied by u_i are then

$$(D_{ijkl} u_{k,l})_{,j} + [p(u_{i,j} + u_{j,i})]_{,j} + \varrho_0 (F_i^* - F_i) + p'_{,i} = \varrho_0 \ddot{u}_i, \tag{23.36}$$

[17] See footnote 13 on p. 180.

with

$$\left.\begin{aligned} &u_i = h_i \quad \text{on } \overline{\partial B_1} \times [0, t_1) \\ &n_j D_{ijkl} u_{k,l} + n_j p (u_{i,j} + u_{j,i}) + n_i p' = t_i^* - t_i \quad \text{on } \partial B_2 \times [0, t_1) \\ &D_{ijkl} = \sigma_{jl} \delta_{ik} + b_{ijkl} \end{aligned}\right\} \tag{23.37}$$

and ... where

and p' is an arbitrary scalar function of position and time.

When the material is isotropic, incompressible and homogeneous, the third invariant I_3 is simply unity, and hence the free energy is a function of the invariants I_1, I_2 alone. The constitutive relation for the Piola-Kirchhoff stress is then given by

$$T_{Ai} = -p X_{A,i} + \tilde{\beta}_0 x_{i,A} + \tilde{\beta}_1 C_{AB} x_{i,B}, \qquad \tilde{\beta}_\gamma = \beta_\gamma |_{I_3=1}, \qquad \gamma = 0, 1, \tag{23.38}$$

while the Cauchy stress is given by

$$\sigma_{ij} = -p \delta_{ij} + \tilde{\alpha}_1 b_{ij} + \tilde{\alpha}_2 b_{ik} b_{kj}, \qquad \tilde{\alpha}_\gamma = \alpha_\gamma |_{I_3=1}, \qquad \gamma = 1, 2. \tag{23.39}$$

The coefficients b_{ijkl} become[18]

$$b_{ijkl} = \tilde{\alpha}_2 (b_{ik} b_{jl} + b_{il} b_{jk}) + 2 b_{ji} \tilde{A}^1_{kl} + 2 b_{ir} b_{rj} \tilde{A}^2_{kl}, \tag{23.40}$$

where

$$\tilde{A}^\gamma_{ij} = \frac{\partial \tilde{\alpha}_\gamma}{\partial I_1} b_{ij} + \frac{\partial \tilde{\alpha}_\gamma}{\partial I_2} (I_1 b_{ij} - b_{ik} b_{kj}), \qquad \gamma = 1, 2. \tag{23.41}$$

24. Incremental equations for thermoelasticity. We now once again introduce thermal effects into our considerations. The exact rate of work equations are given by (23.2), (23.2a) or (23.3), and to these must be adjoined the residual rate of work Eq. (22.9) and the residual entropy production inequality (22.10). These exact equations will be employed later in a discussion of thermoelastic stability in the class of finite displacements, but here we treat the reduction of the respective equations when the incremental displacement u_i is small. We are therefore concerned with the equations of small thermoelastic deformations superposed upon large. These have been developed by ENGLAND and GREEN[19] and by GREEN,[20] and we rely on their work in the following discussion.

The derivations are conveniently based on Eq. (23.3).

We recall that the Piola-Kirchhoff stress T_{Ai} and the free energy $\mathscr{A}$ are now both functions of the absolute temperature T, in addition to being functions of the deformation gradients. For simplicity the body is assumed homogeneous in its reference configuration at time t_0 so that

$$T^*_{Ai} = T_{Ai} + \frac{\partial T_{Ai}}{\partial x_{k,B}} u_{k,B} + \frac{\partial T_{Ai}}{\partial T} (T^* - T) + \cdots \tag{24.1}$$

where dots indicate higher order terms. Let us put

$$\theta = T^* - T, \tag{24.2}$$

which is the incremental temperature variable. Then to a second order approximation, (23.3) becomes

$$\begin{aligned} &\int_{\tilde{B}_0} \varrho_0 \dot{u}_i \ddot{u}_i \, dX + \int_{\tilde{B}_0} (A_{iAkB} u_{k,B} \dot{u}_{i,A} + F_{Ai} \theta \dot{u}_{i,A}) \, dX \\ &\quad = \int_{\tilde{B}_0} \varrho_0 (F_i^* - F_i) \dot{u}_i \, dX + \int_{\partial \tilde{B}_0} (T_i^* - T_i) \dot{u}_i \, dA, \end{aligned} \tag{24.3}$$

---

[18] TRUESDELL and NOLL [1965, *22*, § 45–47], BEATTY [1968, *2*].

[19] ENGLAND and GREEN [1961, *3*].

[20] GREEN [1962, *6*].

with the coefficients A_{iAkB} being given by (23.5), and

$$F_{Ai} = T_{Ai,T}.$$

F_{Ai} are thus, in general, functions of position and time.

The equation of motion corresponding to (24.3) is

$$\left.\begin{aligned} [A_{iAkB}\, u_{k,B} + F_{Ai}\,\theta]_{,A} + \varrho_0 (F_i^* - F_i) &= \varrho_0\, \ddot{u}_i &&\text{in } B\times(0, t_1)\\ T_i^* - T_i &= N_A (A_{iAkB}\, u_{k,B} + F_{Ai}\,\theta) &&\text{on } \partial B_2\times[0, t_1)\\ u_i &= h_i &&\text{on } \overline{\partial B_1}\times[0, t_1), \end{aligned}\right\} \tag{24.4}$$

where, as before h_i and $T_i^* - T_i$ are supposed given on $\partial B = \overline{\partial B_1} \cup \partial B_2$ and t_1 is an instant of time. The equation of heat conduction must be adjoined to (24.4) and this will be derived later.

To reveal explicit dependence upon the stress and temperature of the primary state, we refer the above expressions to the primary configuration B of the body. The derivation, similar to that in the previous section, leads to

$$\begin{aligned} &\int_{\tilde{B}} \varrho\, \dot{u}_i\, \ddot{u}_i\, dx + \int_{\tilde{B}} (d_{ijkl}\, u_{i,j}\, \dot{u}_{k,l} + f_{ij}\,\theta\, \dot{u}_{i,j})\, dx\\ &\qquad = \int_{\tilde{B}} \varrho (F_i^* - F_i)\, \dot{u}_i\, dx + \int_{\partial\tilde{B}} (t_i^* - t_i)\, \dot{u}_i\, da, \end{aligned} \tag{24.5}$$

where

$$f_{ij} = 2\frac{\varrho}{\varrho_0}\, \frac{\partial^2 \mathcal{A}}{\partial T\, \partial C_{AB}}\, x_{i,B}\, x_{j,A}, \tag{24.6}$$

and d_{ijkl} is given by (23.16). Corresponding to (24.5) we have the equations of motion

$$(d_{ijkl}\, u_{k,l} + f_{ij}\,\theta)_{,j} + \varrho (F_i^* - F_i) = \varrho \ddot{u}_i \quad \text{in } B\times(0, t_1), \tag{24.7}$$

$$\begin{aligned} n_j (d_{ijkl}\, u_{k,l} + f_{ij}\,\theta) &= t_i^* - t_i &&\text{on } \partial B_2\times[0, t_1),\\ u_i &= h_i &&\text{on } \overline{\partial B_1}\times[0, t_1), \end{aligned} \tag{24.8}$$

where the notation is the same as before. Again, the equation of heat conduction must be added to (24.7) and (24.8).

We now give the heat conduction equation for the case in which the primary state is in equilibrium so that $\dot{S} = 0$. On letting $\tilde{q}_i = q_i^* - q_i$, under the further assumptions $r^* = 0 = r$, this equation is[21]

$$\begin{aligned} &-T\left[-\frac{\partial^2 \mathcal{A}}{\partial T^2}\,\dot{\theta} + f_{ij}\, \dot{u}_{i,j}\right]\\ &\quad = \frac{\varrho_0}{\varrho}\left[\tilde{q}_i + \frac{1}{2}\, q_i\, X_{M,k}\, X_{N,k} (x_{j,N}\, u_{j,M} + x_{j,M}\, u_{j,N})\right]_{,i} \quad \text{in } B\times(0, t_1). \end{aligned} \tag{24.9}$$

Related boundary conditions are

$$\begin{aligned} \theta &= \hat{\theta} &&\text{on } \partial B_{\mathrm{I}}\times[0, t_1)\\ n_i\, q_i &= q &&\text{on } \overline{\partial B_{\mathrm{II}}}\times[0, t_1), \end{aligned} \tag{24.10}$$

where $\hat{\theta}$, q are given functions, and $\partial B = \partial B_{\mathrm{I}} \cup \overline{\partial B_{\mathrm{II}}}$. Referred to the reference configuration these equations become:

$$-T\left[-\frac{\partial^2 \mathcal{A}}{\partial T^2}\,\dot{\theta} + F_{iA}\, \dot{u}_{i,A}\right] = \tilde{Q}_{A,A} \quad \text{in } B_0\times(0, t_1), \tag{24.11}$$

$$\begin{aligned} \theta &= \hat{\theta} &&\text{on } \partial B_{0\mathrm{I}}\times[0, t_1)\\ N_A\, \tilde{Q}_A &= Q &&\text{on } \overline{\partial B_{0\mathrm{II}}}\times[0, t_1), \end{aligned} \tag{24.12}$$

where $\hat{\theta}$, Q are given functions and $\partial B_0 = \partial B_{0\mathrm{I}} \cup \overline{\partial B_{0\mathrm{II}}}$.

[21] GREEN [1962, *6*].

Two special cases are now considered: these are the case of the isotropic homogeneous elastic body and the case in which the primary state corresponds to a small equilibrium deformation from the reference configuration.

Firstly, when the body is initially isotropic and homogeneous, the Cauchy and Piola-Kirchhoff stresses are still given respectively by (23.20) and (23.24) except that the free energy function now has the form

$$\mathscr{A}=\mathscr{A}(I_1, I_2, I_3, T).$$

The scalar invariant functions $\alpha_i, \beta_i (i=0, 1, 2)$ continue to be given by their previous expressions (23.23), (23.25) so that the coefficients c_{ijkl} retain their form (23.26). On the other hand, f_{ij} assumes the expression

$$f_{ij}=\frac{2\varrho}{\varrho_0} x_{i,B}\, x_{j,A} \frac{\partial}{\partial T}\left[\frac{\partial \mathscr{A}}{\partial I_k} \frac{\partial I_k}{\partial C_{AB}}\right]. \tag{24.13}$$

In the case of *incompressible material* the same modifications apply as in the isothermal case.[22]

In order to obtain expressions for $\tilde{q}_i$ and $\widetilde{Q}_A$, we restrict the primary state to be one in equilibrium at constant temperature. Then $\tilde{q}_i$ can be written

$$-\tilde{q}_i=(\gamma_0\, \delta_{ij}+\gamma_1\, b_{ij}+\gamma_2\, b_{im}\, b_{mj})\, \theta_{,i}=a_{ij}\, \theta_{,i}, \tag{24.14}$$

where $\gamma_{,0}\, \gamma_{,1}\, \gamma_{,2}$ are polynomials (scalar invariant functions) of I_1, I_2, I_3 and T. Similarly for the reference state

$$\begin{aligned}-\widetilde{Q}_A&=[\varepsilon_0\, \delta_{AB}+\varepsilon_1\, C_{AB}+\varepsilon_2\, C_{AN}\, C_{N,B}]\, X_{B,k}\, X_{C,k}\, \theta_{,C}\\ &=A_{AB}\, \theta_{,B},\end{aligned}$$

where again $\varepsilon_0, \varepsilon_1, \varepsilon_2$ are polynomials in I_1, I_2, I_3. It follows from the residual entropy production inequalities (22.19), (22.19a) that the quadratic forms created from the coefficients a_{ij} and A_{AB}, are positive semi-definite, that is

$$\begin{aligned}a_{ij}\, \xi_i\, \xi_j &\geqq 0,\\ A_{AB}\, \xi_A\, \xi_B &\geqq 0,\end{aligned}$$

for arbitrary vectors ξ_i, ξ_A.

Secondly,[23] when the primary state is a small deformation from the reference state, the coordinates x_i may be replaced by X_A to the degree of approximation involved. Then provided initial stress σ_{ij} is present, the equations are

$$(c_{ijkl}\, u_{k,l}+\sigma_{ir}\, u_{j,r}-f_{ij}\, \theta)_{,j}+\varrho\, (F_i^*-F_i)=\varrho\, \ddot{u}_i \quad \text{in } B\times(0, t_1), \tag{24.15}$$

$$\begin{aligned}t_i^*-t_i&=(c_{ijkl}\, u_{k,l}+\sigma_{ir}\, u_{j,r}-f_{ij}\, \theta)\, n_j && \text{on } \partial B_2\times[0, t_1)\\ u_i&=h_i && \text{on } \overline{\partial B_1}\times[0, t_1).\end{aligned} \tag{24.16}$$

The heat conduction equation is now

$$+T_0\left[-\frac{\partial^2 \mathscr{A}}{\partial T^2}\, \dot{\theta}+f_{ij}\, \dot{u}_{i,j}\right]=(h_{ij}\, \theta_{,j})_{,i} \quad \text{in } B\times(0, t_1), \tag{24.17}$$

where h_{ij} are functions of the coordinates x_i, and satisfy

$$h_{ij}\, \xi_i\, \xi_j \geqq 0, \tag{24.18}$$

for arbitrary vectors ξ_i.

---

[22] See ENGLAND and GREEN [1961, *3*].

[23] GREEN [1962, *6*], HUANG and SHIEH [1970, *13*].

25. Some causes of perturbations. In order to ascertain the Liapounov stability of a motion of the body we must consider, as mentioned earlier, other motions of the body which at time $t=t_0$ commence in some neighbourhood of the given primary motion. If the perturbed motion (which in our notation is the secondary motion) remains in a neighbourhood of the given primary motion, then, in accordance with the previous definitions, this is described as stable. Correspondingly, if the secondary motion leaves the neighbourhood the primary motion is unstable. Equations governing the secondary motion, and in particular the "incremental" motion, have been discussed in the last two sections, and in this section we comment on the variety of factors producing such motions. In the problems which we consider, the disturbed (or secondary) state is initiated by alteration in some of the data to the problem. As a result there are three broad classes of disturbance which we now briefly consider in turn: (i) those arising from changes in initial data, (ii) those obtained by varying the different forms of the loading, for example, body forces, surface tractions, surface displacements or heat data, (iii) those arising from a variation in the material properties of the body.

(i) *Initial data.* In this first group, the *displacement*, *velocity* and *temperature* in the secondary motion at time $t=t_0$ differ from those in the primary motion. These initial data are usually required to satisfy certain smoothness hypotheses and are also required to be compatible with the prescribed boundary conditions. Otherwise shocks, or other types of disturbance, representable only through weak solutions, may occur in the secondary motion and possibly induce instability due to failure in the definition of the positive-definite functions ϱ_τ, ϱ or $F_{\tau,t}$. Of course, instability of this kind may be avoided by different choice of positive-definite functions. In fact, in obtaining a non-trivial assessment of stability with a given choice of measure we are verifying that the solution is regular enough and the data smooth enough for us to apply our definition. Part of our aim, not completely fulfilled, will be to classify the initial data both as regards smoothness and magnitude in order that the primary motion may be stable or unstable. The magnitude of the data is clearly of great importance in applications of stability theory to engineering practice. It is natural to suppose that anything "kicked" sufficiently hard will become unstable.

In practical terms, the prescribed initial displacements and velocities may be produced in a number of ways. For instance,[24] the required alterations in these quantities may be brought about by changing the surface tractions and displacements, body forces and heat boundary conditions for a period $t_2 \leqq t \leqq t_0$ *immediately prior to* t_0.[25] In this way a total energy may be prescribed, but displacements and velocities so produced are not in general compatible with the given boundary data, as Ericksen has pointed out.[26] An alternative way of obtaining velocities, at least,[27] is by applying an impulsive load at $t=t_0$, but again incompatibility with the boundary data may be expected.

(ii) *Loading.* In this second class we contemplate a perturbed motion which is obtained by varying the body force, surface traction, surface displacement or heat-data terms. To adopt a general standpoint, we shall assume that the load-data terms producing the secondary motion are independent of those for the primary motion and may depend upon position, time, temperature and the state of deformation of the body and may vary in direction and magnitude as the

[24] See Shield and Green [1963, *18*], Ericksen [1966, *7*, *8*], Shield [1965, *21*].
[25] See Shield and Green [1963, *18*].
[26] Ericksen [1966, *7*, *8*].
[27] Hsu [1966, *12*, *13*].

motion of the body proceeds. We admit arbitrary variations and enquire which classes of "loads" produce stability and which instability. This classification will identify the load not only by its *smoothness* but also by its *magnitude*.

In the most general analysis the loads will be time-varying. If stability is maintained under a system of time-varying loads it is referred to as "persistent" or "parametric" stability and such loads may be either continuous or discrete. When the loads are applied impulsively, or are suddenly applied for a finite period of time before being instantaneously released, shocks may develop which, as noted in (i), clearly introduce another type of effect which may influence stability.

Examples of different types of load variation are given in detail in Sects. 64, 65.

(iii) *Material properties.* The third class of perturbations treats those problems in which we change the material properties of the body. Again we wish to examine the stability of a given motion but this time with respect to neighbouring motions which occur in a body of different constitution, conceivably time-dependent. This group will include the consideration of stability of an elastic body in the class of small perturbations arising from changes in the so-called elasticities. These changes may be produced by a change in the loading of the fundamental equilibrium state, which itself may initiate the secondary motion.

A full investigation of stability must therefore examine all the perturbations produced by factors occurring under one or all of the preceding headings. It must aspire to finding the restrictions to be placed on their magnitude and regularity in order that stability should occur for any one of a set of arbitrarily chosen measures. Even leaving aside mention of regularity conditions that must be satisfied for the validity of the basic thermodynamic postulates, this is an ambitious programme and has so far not been fully carried out for any one type of body or structure. In the sequel, we describe how far success has been achieved in the case of elastic bodies, but our investigations (reflecting the current level of advance in the subject) will be performed only when certain of the factors are held fixed. For example, we shall deal with the stability of an equilibrium configuration in the class of small incremental isothermal displacements for which the initial data is zero and the body force and surface displacement (where prescribed) are the same as those of the equilibrium configuration. Conditions on the incremental surface tractions for the primary equilibrium condition to be stable will be obtained.

Our principal concern in the first place will be, however, with perturbations in the initial data alone, all other factors being held fixed. We shall see that in this group of problems it is possible to supply fairly complete answers in the stability programme outlined above.

G. Liapounov functions for finite thermoelasticity.

26. Introduction. In order to apply the stability theorems of LIAPOUNOV it is self-evident that a knowledge of suitable Liapounov functions is required. In this section we determine such functions for problems concerned with thermoelastic bodies and in particular we consider a number of possibilities for the different primary, secondary or incremental states. For example, they may be variously thermoelastic, isothermal, isentropic, equilibrium and dynamic. Liapounov functions will be defined, generally speaking, on quantities belonging to the incremental state and will fall into two distinct classes according to the equations used to generate them:

(A) Those which make no explicit use of the residual entropy production inequality $Q_A\, T_{,A} \leqq 0$. These functions are suitable for problems of the following types:

(a) The primary and incremental states both *thermoelastic* and *nonlinear.*

(b) The primary and incremental states both *isentropic* and *nonlinear.*

(c) The primary and incremental states both *isothermal* and *nonlinear.*

(d) The primary state isothermal but the incremental state *isothermal* and *linear.*

(B) Those functions which for their construction require the residual entropy production inequality $Q_A\, T_{,A} \leqq 0$ and in addition the equations stemming from energy balance. In deriving this second group of functions recourse is made directly to the basic postulates (22.4), (22.5) and the constitutive equations of the material are ignored entirely. Generality of this kind can only be achieved, however, at the expense of imposing the following conditions:

(a) The primary state is one of isothermal deformation for which

$$\frac{dT}{dt}=0, \quad Q_{A,A}-\varrho_0\, r=0 \quad \text{in } B_0$$
$$T=\text{const} \quad \text{on } \partial B_0. \tag{26.1}$$

(b) The condition

$$N_A\, Q_A^*\,(T^*-T) \geqq 0 \tag{26.2}$$

holds at each point on the boundary ∂B_0. Thus, an outward increase in heat flow across the boundary must be accompanied by a local increase in the incremental temperature.

Special cases of (26.2) occur for the following conditions:

(i) $N_A Q_A^*=0$, corresponding to the case in which the boundary ∂B_0 is thermally insulated throughout its deformation from B_0 and in the incremental perturbations.

(ii) $T^*=T$, in which case the boundary ∂B_0 is maintained at constant temperature throughout the deformations and perturbations.

(c) The condition $(T-T^*)\, r^* \geqq 0$ at each point of the domain B_0. This means that a positive incremental heat supply must be accompanied by a local decrease in the incremental temperature. The retention of heat sources and body force produces an artificial theory, however, and can rarely be realised in practice.

Liapounov functions generated in this class require the primary state to be one of isothermal deformation while the incremental perturbations are in general fully thermoelastic.

The derivations of functions in classes A and B are now dealt with in Sects. 27 and 28 respectively.

27. Liapounov functions from the energy balance equation. In this section we assume that both primary and incremental displacements satisfy only the equations of thermoelasticity and neglect the residual entropy production inequality. The incremental motion is assumed subject to a general set of data. Then from Eq. (23.2a) we may define a function

$$V_1(t)=\int_{B_0(t)} \varrho_0 (U^*-U)\, dX - \int_0^t \int_{B_0(\eta)} \varrho_0 (T^*\dot{S}^* - T\dot{S})\, dX\, d\eta + H(t), \tag{27.1}$$

where

$$H(t)=\tfrac{1}{2}\int_{B_0(t)}\varrho_0(\dot y_i\,\dot y_i-\dot x_i\,\dot x_i)\,dX-\int_0^t\int_{B_0(\eta)}\varrho_0[F_i^*\,\dot u_i+(F_i^*-F_i)\,\dot x_i]\,dX\,d\eta$$
$$-\int_0^t\int_{\partial B_0(\eta)}[T_i^*\,\dot u_i+(T_i^*-T_i)\,\dot x_i]\,dA\,d\eta. \tag{27.2}$$

This is readily seen to satisfy the condition

$$\frac{dV_1}{dt}=0. \tag{27.3}$$

Alternatively, we may take account of (23.2) to find that the function[1] V_2, given by

$$V_2(t)=\int_{B_0(t)}\varrho_0(\mathscr{A}^*-\mathscr{A})\,dX+\int_0^t\int_{B_0(\eta)}\varrho_0(\dot T^*S^*-\dot T S)\,dX\,d\eta+H(t) \tag{27.4}$$

also satisfies the equation

$$\frac{dV_2}{dt}=0. \tag{27.5}$$

From (27.1) and (27.4) we obtain the relation,

$$V_1(t)=V_2(t)+\int_{B_0(0)}\varrho_0(T^*S^*-TS)\,dX, \tag{27.6}$$

which shows that V_1 and V_2 are equal to within an additive constant.

Thus both V_1 and V_2 qualify as Liapounov functions for problems under the class A (a) of Sect. 26.

Special forms of these functions could be considered, as, for example, when the primary and incremental states are either both *isothermal* or both *isentropic*. We consider the former in which $\dot T=0$ so that $T=T^*$. The function $V_2(t)$ is now given by

$$V_2^T(t)=\int_{B_0(t)}\varrho_0(\mathscr{A}^{T^*}-\mathscr{A})\,dX+H(t), \tag{27.7}$$

where $\mathscr{A}^{T^*}=\mathscr{A}(e_{ij}^*, T)$.

An important point concerning the class of displacement gradients entering into (27.4) and (27.7) may be conveniently made here. Insofar as isothermal elasticity may be regarded as a special case of thermoelasticity, the displacement gradients in (27.7) belong to a sub-class of those in (27.4). However, by replacing the thermodynamical postulates by equations of balance we may for isothermal elasticity arrive at an expression formally similar to (27.7) but with one notable difference. This difference lies in the fact that the displacement gradients in the two separate derivations may belong to distinct classes. This has obvious consequences for the stability assessment.

28. Liapounov functions from the entropy production inequality. We next turn to Liapounov functions in Group B and accordingly assume that the conditions (a)–(c) of this Group are satisfied. Thus, we do not necessarily deal with thermoelastic bodies. We first consider an important Liapounov function discovered by COLEMAN[2] during the course of an investigation into fluid bodies, and named by

[1] KNOPS and WILKES [1966, *15*].

[2] COLEMAN [1970, *8*]. NEMAT-NASSER [1970, *24*] also derived the function for a general class of materials which obey FOURIER's heat conduction law, and KOITER [1967, *18*] determined the function in the context of thermoelasticity for which heat sources were absent and the loads conservative.

him the *canonical free energy*. In order to construct this function we observe that inequality (26.2) may be written in the form

$$-\frac{N_A Q_A^*}{T} \leqq -\frac{N_A Q_A^*}{T^*}. \tag{28.1}$$

Hence, on inserting (28.1) into (23.2a) we obtain

$$\int_{B_0(t)} \varrho_0 (\dot{U}^* - \dot{U})\, dX + \dot{H}(t) \leqq \int_{B_0(t)} \varrho_0 (r^* - r)\, dX + T \int_{\partial B_0(t)} \left(\frac{N_A Q_A}{T} - \frac{N_A Q_A^*}{T^*} \right) dA. \tag{28.2}$$

Then from (22.5), on using (22.8), the divergence theorem and hypotheses B(a), B(c) of Sect. 26, we see that the function $V_3(t)$, defined by

$$V_3(t) = \int_{B_0(t)} \varrho_0 [U^* - U - TS^*]\, dX + H(t), \tag{28.3}$$

has a time-derivative satisfying the condition

$$\frac{dV_3}{dt} \leqq 0. \tag{28.4}$$

$V_3(t)$ is thus a Liapounov function and is COLEMAN'S canonical free energy. It may be alternatively expressed as

$$V_3(t) = \int_{B_0(t)} \varrho_0 [\mathscr{A}^* - \mathscr{A} + (T^* - T)\, S^*]\, dX - \int_{B_0(t)} TS\, dX + H(t). \tag{28.5}$$

When we assume that the *primary and the reference states are coincident and* $S=0$, the expression (28.5) formally coincides with one obtained by ERICKSEN[3] within the context of thermoelasticity. This function which we call V_4 is given by

$$V_4(t) = \int_{B_0(t)} \varrho_0 [\mathscr{A}^* - \mathscr{A} + (T^* - T)\, S^*]\, dX + H(t). \tag{28.6}$$

ERICKSEN, however, did not make the assumption that $S=0$ and his result in this respect is therefore more general. The derivation adopted by ERICKSEN is essentially as follows: the function $V_2(t)$ given by (27.4) may be decomposed to give

$$V_2(t) = V_4(t) - \int_0^t \int_{B_0(\eta)} (T^* - T)\, \dot{S}^* dX\, d\eta. \tag{28.7}$$

It then follows, on recalling that now we are confining ourselves to *thermoelastic deformations*, that

$$\begin{aligned} \int_{B_0(t)} \left(\frac{T}{T^*} - 1 \right) T^* \dot{S}^* dX = & \int_{B_0(t)} \left[\left(\frac{T}{T^*} - 1 \right) \varrho_0 r^* - \frac{Q_A^* T T_{,A}^*}{T^{*2}} \right] dX \\ & - \int_{\partial B_0(t)} \left(\frac{T}{T^*} - 1 \right) N_A Q_A^*\, dA, \end{aligned} \tag{28.8}$$

where the divergence theorem has been applied. Hence, because we are supposing that hypotheses B(a), (b), (c) of Sect. 26 still hold the right side is non-negative and so we finally deduce the relation

$$V_2 \geqq V_4. \tag{28.9}$$

[3] ERICKSEN [1966, *7*, *8*] obtained his result under the separate boundary data $N_A Q_A = 0$ or $T^* = T$, assuming also that no heat sources are present. See also NEMAT-NASSER [1970, *24*].

On noting that $dV_2/dt=0$ from, (28.7), (28.8), we have

$$\frac{dV_4}{dt} \leqq 0. \tag{28.10}$$

The function $V_4(t)$ may be further reduced on applying the second mean value theorem to express $\mathscr{A}(e_{ij}^*, T)$ in terms of T^*-T, thus

$$\mathscr{A}(e_{ij}^*, T) = \mathscr{A}(e_{ij}^*, T^*) + (T-T^*) \left.\frac{\partial \mathscr{A}}{\partial T}\right|_{e_{ij}^*, T^*} + \frac{1}{2}(T^*-T)^2 \left.\frac{\partial^2 \mathscr{A}}{\partial T^2}\right|_{e_{ij}^*, \hat{T}}, \tag{28.11}$$

where $\hat{T}$ lies between T and T^*. If we now define $\hat{\mathsf{K}}$ by the equation

$$\hat{\mathsf{K}} = -\left.\frac{\partial^2 \mathscr{A}}{\partial T^2}\right|_{e_{ij}^*, \hat{T}},$$

and use the constitutive relation $S = -\dfrac{\partial \mathscr{A}}{\partial T}$, we see that $V_4(t)$ may be alternatively expressed as[4]

$$V_4(t) = \int_{B_0(t)} \varrho_0 [\mathscr{A}^{T^*} - \mathscr{A} + \tfrac{1}{2}(T^*-T)^2 \hat{\mathsf{K}}]\, dX + H(t), \tag{28.12}$$

where $\mathscr{A}^{T^*} = \mathscr{A}(e_{ij}^*, T)$. The function $\mathscr{A}^{T^*}$ thus corresponds to the Helmholtz free energy for isothermal deformations at constant temperature T. It follows therefore that $\mathscr{A}^{T^*} - \mathscr{A}$ is the increase in the free energy for isothermal deformations at the constant temperature of the primary state.

Finally, it may be noted that further Liapounov functions may be generated from some linear combinations of functions produced in this section and Sect. 27. One such example is $V_3 + \alpha V_1$ where α is time-independent.

In this section no mention has been made of the positive-definiteness condition required for the Liapounov functions. When the above derived functions are used in the Liapounov test this condition will have to be imposed on the functions selected. In HALE's invariance principle, however, this condition is not required and the fact that the time-derivatives of the function are non-positive suffices. This principle is used mainly in a treatment of asymptotic stability and instability.

H. Liapounov stability in the class of non-linear perturbations.

29. Introduction. In any non-linear theory the detailed study of Liapounov stability is limited by the knowledge of the classes of solutions that can arise from the governing equations and the data. In general it is not possible to define measures until the set X is determined which in turn needs the knowledge of the classes of solutions. For the same reasons it is not possible to specify Liapounov functions unless these explicitly involve the thermodynamic variables which are by hypothesis well-defined. In those circumstances, however, in which a function of this kind can be found and which can be linearised with respect to small incremental displacements it will be possible to introduce specific measures relevant to this function.

Because of these difficulties, inherent in any non-linear analysis, we shall not in this section be able to specify the measures and we shall in general assume that the various embedding inequalities needed in the stability theorems are satisfied.

[4] ERICKSEN [1966, 7, 8].

In this respect our analysis is incomplete and must be restricted to a study of properties derivable from the chosen Liapounov function.[1]

Of the several examples of Liapounov functions given in the previous Chap. G the function $V_4(t)$, proposed by ERICKSEN, proves to be the most suitable for our investigations in the class of non-linear perturbations of an isothermally deformed primary state. In making this choice we assume that $V_4(t)$ is well-defined. As $V_4(t)$ contains terms corresponding to the free energy of the body, however, making this assumption is equivalent to assuming that the solutions are bounded in energy. If this condition is not satisfied then, by definition, we shall have instability of the primary state in the sense of LAGRANGE or LIAPOUNOV depending on whether the primary or secondary solutions have unbounded energies respectively. Stability or instability is then measured by $\varrho=V_4(t)$. For any other measures it may happen that the primary state could be proved stable with a different choice of Liapounov function.

The imprecision resulting from inability to specify the measures is removed, as mentioned above, under certain circumstances. These occur in investigations for which stability in the class of non-linear perturbations can be assessed from stability in the class of linear perturbations. In some cases it is only necessary for us to specify one pair of the measures and the stability investigation with these is automatically valid for several different measures which now serve for both linear and non-linear problems. The feasible choices will be apparent from the subsequent sections.

In all problems the primary state is assumed to satisfy the conditions of hypothesis B(a) in Sect. 26.

30. Sufficiency theorems. The first result is based directly on the Liapounov Theorem 15.1 and is therefore stated without proof.

Theorem 30.1. For a given measure ϱ let the function $V_4(t)$ satisfy the inequality[2]

$$V_4(t) \geqq c\varrho, \quad c>0. \tag{30.1}$$

Then the primary state is stable in the class of non-linear thermoelastic perturbations with respect to the measures $V_4(0)$ and ϱ.

An immediate deduction concerning the stability of an equilibrium position is the following

Corollary 30.2.[3] *For given measure ϱ let the primary state be an* equilibrium position *and let the free energy $\mathscr{A}$ satisfy the inequality*[4]

$$\int_{B_0(t)} \varrho_0 [\mathscr{A}(e^*_{ij}, T) - \mathscr{A}(e_{ij}, T)]\, dX + H_0(t) \geqq c\varrho, \tag{30.2}$$

where

$$H_0(t) = -\int_0^t \int_{B_0(\eta)} \varrho_0 F_i^* \dot{u}_i\, dX\, d\eta - \int_0^t \int_{\partial B_0(\eta)} T_i^* \dot{u}_i\, dA\, d\eta. \tag{30.3}$$

---

[1] Examples of a complete non-linear analysis are given by COLEMAN and DILL [1968, *5*] who study the problems of inflation of a circular tube and spherical shell.

[2] We have chosen inequality (30.1) rather than the more abstract condition (ii) of Theorem 15.1 because of its more practical application.

[3] The greater part of this result is due to ERICKSEN [1966, *7*, *8*]. KOITER [1967, *18*] also derived the result for L_2-norms still under condition (30.2). See also NEMAT-NASSER [1970, *24*].

[4] On the basis of (30.2), KOITER [1967, *18*] claims, contrary to PEARSON [1956, *2*], that elastic stability is an *isothermal* phenomenon. But we note that (30.2) is not strictly a statement about isothermal quantities since the strains e^*_{ij}, e_{ij} must still satisfy the equations of thermoelasticity, and hence KOITER's conclusion does not appear to follow.

Then, provided

$$\widehat{\mathsf{K}} = -\left.\frac{\partial^2 \mathscr{A}}{\partial T^2}\right|_{e_{ij}^*, \hat{T}} \geqq 0,$$

the equilibrium state is stable in the class of non-linear thermoelastic perturbations with respect to the measures ϱ_0 *and* ϱ, *where*

$$\varrho_0 = \tfrac{1}{2} \int_{B_0(0)} \varrho_0 [\dot{u}_i \dot{u}_i + \mathscr{A}(e_{ij}^*, T) - \mathscr{A}(e_{ij}, T) + \tfrac{1}{2}(T^* - T)^2 \widehat{\mathsf{K}}]\, dX. \tag{30.4}$$

To prove this corollary we note that because $\widehat{\mathsf{K}} \geqq 0$ and the kinetic energy is non-negative, the condition (30.2) implies (30.1). Moreover, when $\dot{x}_i = 0$, we have $\varrho_0 = V_4(0)$ and the corollary is proved.[5]

If $\widetilde{C}_\gamma$ is the specific heat at the constant deformation e_{ij}^* and at the temperature $\widetilde{T}$, then by the definition of $\widehat{\mathsf{K}}$, we have

$$\widetilde{C}_\gamma = \widetilde{T}\, \widehat{\mathsf{K}}. \tag{30.5}$$

Thus postulating the condition $\widehat{\mathsf{K}} \geqq 0$ is equivalent to postulating that the specific heat at constant deformation is positive for all temperatures.

We note further that Corollary 30.2 is a test valid in the class of *thermal perturbations* of the primary state although this itself is obtained by isothermal deformation.[6]

Inequality (30.2) must hold for *all* thermoelastic displacements compatible with the mechanical boundary conditions of the perturbed problem. Now, because the loads F_i^* and T_i^* may themselves be path-dependent,[7] the function H_0 will in general, vary in sign according to which particular incremental deformation is being followed. The task of verifying the conditions under which inequality (30.2) is valid is thus by no means straightforward.

The problem is simplified however when the loads possess a potential in the following sense[8]

$$\int_{\dot{B}_0(t)} \varrho_0 F_i^* \dot{u}_i\, dX + \int_{\partial \dot{B}_0(t)} T_i^* \dot{u}_i\, dA = \frac{d\Psi}{dt} + \frac{d\Phi}{dt}, \tag{30.6}$$

where

$$\Phi = \int_{\partial B_0(t)} \varphi\, dA, \qquad \Psi = \int_{B_0(t)} \psi\, dX, \tag{30.7}$$

are functions depending upon the displacement and possibly other variables so that (30.2) becomes

$$\int_{B_0(t)} \varrho_0 [\mathscr{A}^*(e_{ij}^*, T) - \mathscr{A}(e_{ij}, T)]\, dX - \Phi(t) - \Psi(t) + \Phi(0) + \Psi(0) \geqq c\varrho. \tag{30.8}$$

---

[5] Nemat-Nasser [1970, *24*] uses the function

$$V_5(t) = \int_{B_0(t)} \varrho_0 [\mathscr{A}(e_{ij}^*, T) - \mathscr{A}(e_{ij}, T)]\, dX$$

as Liapounov function and then requires the condition $\dot{H}_0 \leqq 0$. He argues further that, in the case of motion-dependent forces $\dot{H}_0$ could exceed the rate of entropy production thereby vitiating the present test. He concludes that entropy production may therefore critically affect stability of a body subjected to non-conservative loads.

[6] Attention has already been drawn in Sect. 27 to the difference between classes of functions that can arise when isothermal perturbations are regarded as originating from a completely non-thermal theory and from a thermal theory with $\dot{T} = 0$.

[7] A discussion of the loads is given in Chap. Q.

[8] Such loads, said to be *weakly conservative*, are discussed in Chap. R.

When Φ and Ψ depend only upon the displacement u_i, then the above Corollary 30.2 with inequality (30.2) replaced by (30.8) and with the specific measure

$$\varrho = \int_{B_0(t)} \varrho_0 \, u_i \, u_i \, dX$$

was first established by Koiter.[9] It is again emphasised that (30.8) is derived from thermal considerations with temperature a constant so that our previous remarks on classes of functions apply.[10]

We are now able to specify alternative forms of sufficient conditions for the stability of an *equilibrium solution* under conditions of *(weakly) conservative loading* in terms of quantities governed by the linearised equations only.[11] For this we return to Theorem 18.2.

Corollary 30.3. Let the displacements take values in the set X, which is now a normed linear space with norm $\varrho(\cdot)$. Let the function

$$V(t) = \int_{B_0(t)} \varrho_0 [\mathscr{A}(e_{ij}^*, T) - \mathscr{A}(e_{ij}, T)] \, dX - \Phi - \Psi$$

be well-defined on X and possess first and second Fréchet derivatives. Then, provided $\hat{\mathsf{K}} \geqq 0$ and the second Fréchet derivative satisfies the condition

$$d^2 V \geqq c \varrho^2 \tag{30.10}$$

for some positive constant c, the equilibrium position is Liapounov stable in the class of non-linear perturbations with respect to the measures ϱ_0 and ϱ, where

$$\varrho_0 = V(0) + \tfrac{1}{2} \int_{B_0(0)} \varrho_0 \, \dot{u}_i \, \dot{u}_i \, dX. \tag{30.11}$$

For the proof of this corollary, we note that since u_i satisfies the dynamic equations and the unperturbed state is in equilibrium, then by well-known variational principles $dV = 0$.[12] Thus from Corollary 18.3 and (30.10) we verify that inequality (30.8) holds. We may now, without loss, put $\Phi(0) = \Psi(0) = 0$ so that the corollary is proved.

The importance of this corollary lies in the fact that the problem of stability in the class of non-linear perturbations is reduced *by means of inequality* (30.10) to the *problem of stability in the corresponding linear* problem. Thus, provided that the linear stability problem rests on an inequality of type (30.10) the non-linear problem is automatically solved. As a discussion of inequalities of the type (30.10) and their relation to stability is the content of subsequent sections, we do

[9] Koiter [1967, *18*].

[10] Other notions of thermoelastic stability are not discussed here. For further references see Truesdell and Noll [1965, *22*, § 89] and Ericksen [1966, *8*].

[11] This is essentially the result of Koiter [1965, *12*] except that he employed an L_2-norm for his measures. See also Koiter [1945, *1*] and Trefftz [1933, *1*].

[12] On assuming that $\mathscr{A}$ has sufficient smoothness the Fréchet and Gateau derivatives coincide. Consequently dV and d^2V are the ordinary first and second variations of V in the sense of the variational calculus. We then have, apart from terms in Φ, Ψ,

$$dV = \frac{1}{2} \int_{B_0} \varrho_0 \left(\frac{\partial \mathscr{A}}{\partial e_{ij}} + \frac{\partial \mathscr{A}}{\partial e_{ji}} \right) \tilde{e}_{ij} \, dX,$$

$$d^2 V = \frac{1}{8} \int_{B_0} \varrho_0 \left[\left(\frac{\partial}{\partial e_{pq}} + \frac{\partial}{\partial e_{qp}} \right) \left(\frac{\partial \mathscr{A}}{\partial e_{ij}} + \frac{\partial \mathscr{A}}{\partial e_{ji}} \right) \tilde{e}_{pq} \tilde{e}_{ij} + 2 \left(\frac{\partial \mathscr{A}}{\partial e_{ij}} + \frac{\partial \mathscr{A}}{\partial e_{ji}} \right) u_{r,j} u_{r,i} \right] dX,$$

where e_{ij}, $\tilde{e}_{ij}$ are the non-linear and linear strains respectively.

not dwell further on this matter here. It should be mentioned in passing, nevertheless, that another condition sufficient for stability may be based on Theorem 18.4 and this is treated in Sect. 37. We also note that when $d^2V=0$ for some incremental displacements, the previous sufficient condition for stability breaks down. By pursuing the analogy with the variational calculus, we may examine higher Fréchet derivatives and thus continue to obtain information about stability from these. This aspect of the subject has been extensively studied by KOITER[13] whose ideas in this connection are of importance in non-linear post-buckling behaviour.

The isothermal analysis also includes the sufficient conditions for stability given by MOVCHAN.[14] His Liapounov function is equivalent to V_2^T with the terms in kinetic energy subtracted and the loads restricted to being conservative. However, the equivalence between V_2^T and the functional used by MOVCHAN is only formal since MOVCHAN derived his solely for an isothermal theory in the strict "non-thermal" sense. As previously noted, the functions involved may belong to different classes for the two functionals. MOVCHAN'S conditions, with their still wider interpretation given here in terms of the thermal theory, are the natural generalisations of the criteria propounded by LAGRANGE and DIRICHLET.

To conclude this section we investigate sufficient conditions for the stability of the *natural state* of a body composed of thermoelastic material. We postulate that the natural state is an unloaded equilibrium state without stress, displacements and heat sources and is maintained at uniform temperature T. Moreover it will be assumed that $\mathscr{A}(0, 0)=0$. An initial displacement and velocity are given to the body. Then, since there are no perturbing forces or thermal effects, we see from Corollary 30.2 that sufficient conditions for stability with respect to measure ϱ are $\hat{\mathsf{K}}\geqq 0$ and

$$\int_{B_0(t)} \varrho_0 \mathscr{A}(e_{ij}, T)\, dX \geqq c\varrho, \tag{30.12}$$

where u_i is any actual (thermoelastic) displacement and e_{ij} are the associated non-linear strains. Because of the previous Corollary 30.3, we may assume in (30.12) that $\mathscr{A}$ corresponds to the linearised free energy, in which case e_{ij} become the linear strains $\tilde{e}_{ij}$.

Inequality (30.12) therefore represents a restriction on the free energy, and hence on the constitutive relations, in order that the natural state of a thermoelastic body should be stable in the measure ϱ.

The approach followed in this section has been based on thermodynamical postulates and the use of Liapounov theorems. In Sect. 32 we shall develop necessary conditions for stability from the same viewpoint. Nowhere in the Treatise, however, do we deal with another important aspect of the theory in which relations are sought with postulates such as the inequality proposed by COLEMAN and NOLL[15] and DUHEM'S "loi du déplacement isothermique de l'équilibre".[16]

I. The energy criterion for stability.

31. Statement and history of the criterion. Early ideas concerning stability of discrete systems found in the works of LAGRANGE and DIRICHLET have already

[13] KOITER [1945, *1*], [1962, *9*], [1963, *8*], [1965, *12*].
[14] MOVCHAN [1963, *16*].
[15] COLEMAN and NOLL [1959, *1*]. See also TRUESDELL and NOLL [1965, *22*] and the discussions by HILL [1968, *11*], [1970, *11*].
[16] DUHEM [1906, *2*, p. 133].

been generalised to the *Liapounov definition* for abstract systems (see Definition 6.1). LAGRANGE also developed a test, known later as the energy criterion,[1] for discrete conservative systems which corresponds to the criterion for the minimum of the potential energy at the equilibrium position in the class of virtual displacements satisfying the kinematical constraints of the problem.[2] As such it has been incorporated into elasticity theory. The following statement of the principle is selected as being representative of the form usually assumed or implied in this context.

For an elastic body under conservative loads and subject to isothermal conditions, a necessary and sufficient condition for stability of an equilibrium position is that the potential energy of the system assumes at the equilibrium position a weak relative minimum in the class of virtual displacements satisfying the kinematical constraints.

Mathematical statements of the principle may be found at the end of the following section but meanwhile we notice that because only virtual displacements, the potential energy and loads are involved in the test it is *static* in nature.[3] Further, the precise notion of stability to which it refers is left undefined, although most writers using the test usually tacitly adopt the Liapounov definition, taking, by implication, the measures to be the uniform norm over the body of the displacement and velocity.

In order to make an appraisal of the test and comment on its validity it is helpful to give a brief history of development and ideas fundamental to the criterion. This now follows.

Historical development.

The history of the endeavour to apply the energy criterion as a test suitable for assessing stability in elastic bodies can be roughly categorised by five phases. These do not represent clearcut stages of chronological development (as exemplified by the rather confused contemporary scene) but indicate instead a sequence in the modes of thinking on the whole subject of elastic (and thermoelastic) stability.

The first phase, covering the middle of the nineteenth century, includes the ideas of GREEN, CLEBSCH, KIRCHHOFF, THOMSON and LIPSCHITZ and is concerned with applying the test to null solutions or special problems. The second era begins with BRYAN, who attempted to bring the test into use for more general states, and includes the works of HADAMARD and DUHEM. Thirdly, the work of DUHEM began a new phase when, in the early part of the present century, he treated correctly the theory of instability in non-linear elasticity. The fourth development originates with TREFFTZ who extended the ideas of the energy test to include

[1] Sometimes also known as the criterion for static stability.

[2] LAGRANGE's proof [1788, *1*], based on the growth properties of eigenfunctions, and DIRICHLET's criticism are mentioned in Sect. 15. DIRICHLET's proof for conservative systems which does not require any linear approximation is the one usually employed for sufficiency. See also WHITTAKER [1937, *1*], ROUTH [1860, *1*], [1877, *3*], THOMSON and TAIT [1879, *1*, § 292], ZHUKOVSKII [1882, *2*]. The proof of sufficiency follows most directly, however, from LIAPOUNOV's stability theorem as demonstrated, for instance, by MALKIN [1952, *3*], CHETAEV [1961, *2*] and HALE [1969, *11*]. Necessity has been treated by LAGRANGE [1788, *1*], [1853, *1*, Sect. VI], ROUTH [1877, *3*, § 462], MINCHIN [1888, *3*] and LIAPOUNOV [1892, *1*], who proved instability when the potential energy acquires a maximum at the equilibrium point, and by LAURENT [1878, *1*] and THOMSON and TAIT [1879, *1*, §§ 292, 345], who introduce dissipation. CHETAEV [1938, *1*], [1961, *2*] (see also MALKIN [1952, *3*]) proved instability in certain circumstances when the potential energy does not have a minimum at equilibrium; see footnote 80, p. 167.

[3] One of the earliest static tests for stability was formulated by ARCHIMEDES in his study of floating bodies.

bodies with initial stress. Finally, the fifth stage comes with the realisation that the energy test has only a limited application to elasticity. This results from the better understanding of stability given by Liapounov theory for continuous systems and may be said to have originated with ZUBOV and MOVCHAN about 1960.

Although KIRCHHOFF'S[4] name stands to the fore in the matter of extending the energy test to elasticity, both GREEN[5] and CLEBSCH[6] had made statements about its application at earler dates. GREEN'S references were oblique but CLEBSCH, without referring explicitly to LAGRANGE or DIRICHLET, provided elements of a proof for the test in linear elasticity in the cases for which the potential energy is either a minimum or a maximum. His discussion, based on the time-evolution of eigenfunctions, used solutions expanded in infinite series whose convergence was not established. KIRCHHOFF, however, put forward the claim that a positive-definite strain energy is necessary for stability and hence that the solutions to the standard boundary-value problems are unique. For justification he made direct appeal to DIRICHLET'S[7] paper. The rigorous proof that a maximum potential energy is sufficient for instability was left to LIPSCHITZ,[8] while THOMSON[9] clearly relying on the validity of the test, used it to obtain conclusions relating to the positive-definite character of the linear strain energy.

BRYAN,[10] apparently breaking new ground, considered the stability of a stressed elastic body. He cited KIRCHHOFF, and thus indirectly LAGRANGE and DIRICHLET, for justification of the assumption that a *necessary and sufficient condition for stability is that there is a positive increase in the potential energy in any virtual displacement compatible with the geometrical constraints from the given equilibrium position.* BRYAN'S work is limited firstly, by the fact that the initially stressed state is subject to only an infinitesimally small deformation and secondly his supposition that the work done by the forces in the incremental displacement is positive. Thus, only the strain energy is restricted by the criterion.

An application more successful in being free from the faults in BRYAN'S work, just described, was that due to HADAMARD.[11] He based his work on the form of the criterion quoted in the previous paragraph and applied it to the displacement boundary-value problem. Unlike BRYAN, however, he did not restrict the incremental displacement to be infinitesimal but obtained instead necessary and sufficient conditions for the validity of the criterion in terms of the second variation of the strain energy.[12] He related these conditions to the reality of the speeds of waves propagating in the body.

HADAMARD'S criterion of stability required the potential energy V to be a weak relative minimum. A condition *necessary* for this therefore[13] is

$$D^2 V \geqq 0, \tag{31.1}$$

---

[4] KIRCHHOFF [1877, *1*, pp. 392–395].

[5] GREEN [1839, *1*].

[6] CLEBSCH [1862, *1*].

[7] DIRICHLET [1846, *1*].

[8] LIPSCHITZ [1874, *1*].

[9] THOMSON [1863, *1*], THOMSON and TAIT [1879, *1*, § 673, Appendix C], THOMSON [1888, *4*].

[10] BRYAN [1888, *1*, *2*]. In (*1*) BRYAN'S major concern is with instability, which he assumes to occur for a negative-definite second variation of the potential energy.

[11] HADAMARD [1903, *1*, § 269].

[12] TRUESDELL and NOLL [1965, *22*, § 89] have extended this test to the case in which mixed boundary conditions are allowed.

[13] See MORREY [1966, *21*, p. 11] and Theorem 18.4.

where D^2V is the Gateau derivative of V. In terms of the incremental virtual displacement $v_i(\boldsymbol{x})$, condition (31.1) is strictly equivalent to

$$\int_B d_{ijkl} v_{i,j} v_{k,l} \, dx \geqq 0, \tag{31.2}$$

which in turn necessarily implies that the condition[14]

$$d_{ijkl} \xi_i \xi_k \eta_j \eta_l \geqq 0 \tag{31.3}$$

holds at each point of B for arbitrary real vectors ξ_i and η_i.

We shall adopt the term *Hadamard stability* for the notion of stability defined by means of the energy criterion. More specifically, we shall say the equilibrium solution is *Hadamard stable* if and only if the potential energy acquires a minimum value at the equilibrium solution in the class of finite virtual displacements meeting the displacement boundary data. The equilibrium solution is *Hadamard infinitesimally stable*[15] if and only if (31.2) is satisfied in the class of infinitesimal virtual displacements meeting the displacement boundary data.

Our reason for introducing this new definition is not only to accommodate the practice of some authors, but also for convenience in the subsequent assessment of the energy criterion. Further discussion of Hadamard stability may be found in the Treatise by TRUESDELL and NOLL.

Shortly after the appearance of HADAMARD'S work, DUHEM[16] published a unified account of his studies on stability begun prior to 1897.[17] His results, again seeking vindication in LAGRANGE'S, were valid for boundary conditions of place and pressure, and included conditions either necessary or sufficient for stability, mainly for isothermal conditions, although temperature effects were taken into account. In instability, DUHEM'S work is based on the full non-linear equations of elastodynamics and is therefore exact. The usual criticisms of the energy test, applicable to DUHEM'S work on stability, do not therefore carry over to his work on instability. The presentations of both HADAMARD and DUHEM were fairly comprehensive and later contributions were in general elaborations of their work. VON MISES, SOUTHWELL and BIEZENO and HENCKY were concerned more with neutral equilibrium and only tacitly admitted their belief in the energy criterion.[18]

TREFFTZ'S[19] new approach, on the other hand, consisted in displaying for the first time the role of primary stress. His analysis was limited, however, by his continuing to assume that HOOKE'S law held for the incremental displacement.

Of later authors concerned exclusively with the energy criterion mention should be made of KOITER,[20] PEARSON, HILL, HOLDEN, BEATTY, TRUESDELL and NOLL, KRALL and WEBER.[21] The analyses of all these authors are applied to finite elasticity and amplify the ideas of HADAMARD, DUHEM and TREFFTZ: they refer mainly to problems in which the loading is conservative. However, in some cases there is confusion over interpretation as a *proper* minimum of the energy is

---

[14] Condition (31.3) holds also for boundary conditions of displacement and traction, a fact proved by CATTANEO [1946, 1] and NOLL [TRUESDELL and NOLL 1965, *22*, § 69bis].

[15] Compare TRUESDELL and NOLL [1965, *22*, § 69bis].

[16] DUHEM [1906, *2*].

[17] See DUHEM [1897, *1*, Chap. X]. The *Traité Elémentaire de Mécanique Chimique* published that year contained an account of thermoelastic stability.

[18] See Chap. L for an account of their work.

[19] TREFFTZ [1930, *2*], [1933, *1*].

[20] KOITER [1945, *1*], [1963, *7*, *8*], [1965, *12*, *13*].

[21] PEARSON [1956, *2*], HILL [1957, *1*], HOLDEN [1964, *6*], BEATTY [1965, *1*], [1967, *4*], [1968, *2*], TRUESDELL and NOLL [1965, *22*, § 69bis, § 89], KRALL [1966, *17*], WEBER [1967, *33*], [1969, *24*].

implied. BEATTY'S *criterion* is perhaps the most general and is given in Sect. 32 (32.8). The *criterion of* TRUESDELL *and* NOLL is obtained from BEATTY'S criterion when the body force is assumed derivable from a potential and the surface tractions are dead. This is also given in Sect. 32 (32.9).

This history indicates that belief in the energy test as a necessary and sufficient condition for stability had in most cases become a matter of faith. This attitude had arisen partly because, at the earlier stages of development, extending the theorems formulated for discrete systems to continuous materials seemed a natural procedure to be taken for granted and partly because the concept of stability, against which the test could be gauged, had not been fully developed. This lack of suitable definition of stability is reflected in the fact that the *criterion* itself has sometimes been adopted as the *definition*. Further support for the use of the test comes from the fact that it is a necessary condition under some circumstances. These circumstances require, however, an appropriate definition of stability, a suitable interpretation of the potential energy in terms of strain energy and *conservative* loading and a clear definition of the class of incremental displacements involved. These points are discussed in the next section. As a sufficient condition the test is not adequate and thus, on its own, cannot guarantee stability.

Adequate criteria in the form of the Liapounov theorems have, on the other hand, been proved. In order, therefore, to investigate the efficacy of the traditional test, we examine it in the light of some necessary and sufficient conditions derivable from Liapounov theory. These are given in the next two sections and a final appraisal of the energy criterion is made in Sect. 34. Of course, our comments do not apply to Hadamard stability—that is to stability *defined* by means of the energy criterion—where the physical relevance of the definition is difficult to understand. The last mentioned criterion has the form used by PEARSON and HILL, with direct antecedents in the work of KOITER, TREFFTZ, HADAMARD, DUHEM, BRYAN, KIRCHHOFF and LAGRANGE.

Contemporary applications of the energy criterion include those of DUPUIS and KERR and EL-BAYOUMY,[22] with extensions to Cosserat media being made by DIKMAN and BEATTY.[23] Much valuable work has also been done on properties of the potential energy in non-linear elastic systems with a finite number of degrees of freedom,[24] but since the relation of this work with continuous systems is not altogether certain we do not include an account.

32. Necessary conditions. In order to establish criteria necessary for Liapounov stability we first examine sufficient conditions for instability. In this respect, the function V_4, introduced in Eq. 28.6, forms a convenient Liapounov function which, however, has only a non-positive time-derivative as distinct from a negative-definite one required by conditions of the instability Theorem 19.1. Therefore, instead of this theorem, we must turn to HALE'S Theorem 19.2 to derive sufficient conditions for instability.

Theorem 32.1.[25] *Let X be the space of vector-valued functions equipped with a suitable norm and completed if necessary. Let the solutions of the given thermoelastic problem form the dynamical system $\mathcal{B}(T, X)$ and suppose that the function V_4 is*

---

[22] DUPUIS [1968, *8*], [1969, *7*], KERR and EL-BAYOUMY [1970, *16*].

[23] BEATTY [1970, *3*], DIKMAN [1971, *2*].

[24] A comprehensive bibliography is provided by THOMPSON [1969, *22*]; see also SEWELL [1970, *28*].

[25] Other versions of this theorem are possible. The present form is adequate for our purposes, however.

defined and continuous on X *while an equilibrium solution* $w_i(\boldsymbol{x})$ *is contained in the closure of an open subset* U *of* X. *Let* N *be an open neighbourhood of* $w_i(\boldsymbol{x})$ *and in* $\overline{N \cap U}$ *let* $w_i(\boldsymbol{x})$ *be the only invariant set on which* $\dot{V}_4 = 0$.[26] *Then, provided that* V_4 *satisfies the condition*

$$V_4 < 0 \quad \text{for } e^*_{ij} \neq e_{ij}, \qquad T^* \neq T \tag{32.1}$$

the equilibrium solution is unstable in the sense of LIAPOUNOV.

The proof of this theorem follows at once from Theorem 19.2.[27] Because $\dot{V}_4$ is assumed to vanish only at an equilibrium solution w_i, condition (32.1) may be replaced to give the theorem an alternative formulation which we present as a corollary. The notation is that of Corollary 30.2.

Corollary 32.2.[28] *Let the conditions of Theorem 32.1 be satisfied with the exception that inequality* (32.1) *is replaced by*

$$\begin{aligned} \int_{B_0(t)} \varrho_0 [\mathscr{A}(e^*_{ij}, T) - \mathscr{A}(e_{ij}, T) + \tfrac{1}{2}(T^* - T)^2 \hat{\mathsf{K}}]\, dX + H_0(t) < 0, \\ e^*_{ij} \neq e_{ij}, \qquad T^* \neq T. \end{aligned} \tag{32.2}$$

Then Theorem 32.1 is still true provided that the initial incremental velocities are sufficiently small.

To prove this we note that in any perturbed deformation the condition $\dot{V}_4 \leqq 0$ implies $V_4(t) < V_4(0)$, since $\dot{V}_4$ vanishes only in the primary state. Then, as we may choose the initial incremental velocity so small that $V_4(0) \leqq 0$ for the secondary motion, condition (32.1) is satisfied and the corollary proved.

By inverting these conditions, sufficient for instability, some conditions necessary for stability may be produced. It must always be remembered, however, that the conditions so obtained are valid only for the situations allowed by the corollary. In particular, it is assumed that *no equilibrium solutions, other than the primary state itself, exist in the neighbourhood for which* $\dot{V}_4 = 0$,[29] that the class of deformations is that specified by the theorem and that the measures are supplied by norms possible on X.

We present these results as a theorem followed by a number of corollaries.[30]

Theorem 32.3. Under the conditions of Corollary 32.2 a necessary condition for stability is the following inequality

$$\int_{B_0(t)} \varrho_0 [\mathscr{A}(e^*_{ij}, T) - \mathscr{A}(e_{ij}, T) + \tfrac{1}{2}(T^* - T)^2 \hat{\mathsf{K}}]\, dX + H_0(t) \geqq 0. \tag{32.3}$$

---

[26] This condition states that $\overline{N \cup U}$ can contain no isothermal motions nor other equilibrium states.

[27] Without loss, we may ignore an alternative which is strictly possible. In this the perturbations instead of reaching the boundary of $N \cap U$, and thus demonstrating the required instability, might remain in a non-compact set of $\overline{N \cup U}$.

[28] This form of the theorem was first established by KOITER [1967, *18*] with certain restrictions and for weakly conservative loads. CLEBSCH [1862, *1*] and LIPSCHITZ [1874, *1*] obtained results for *isothermal linear* deformations while DUHEM [1906, *2*, p. 97] established it for *isothermal* deformations under boundary conditions of displacement and pressure. The general form (32.2), but applied only to *isothermal deformations*, was first given by MOVCHAN [1963, *16*] with the strain energy a homogeneous polynomial.

[29] If this condition is omitted all the following necessary conditions are invalidated as the result of CAUGHEY and SHIELD [1968, *4*] shows. Their examples, however, do demonstrate necessity under the strict inequality.

[30] The basic idea is due to ERICKSEN [1966, *7*, *8*] who uses a direct analysis and assumes the primary equilibrium state to be asymptotically stable. This approach possibly allows a wider class of incremental deformations than those in the Banach space X of the present theorem but restricts the necessary conditions to asymptotic rather than Liapounov stability.

Corollary 32.4. Under the conditions of Corollary 32.2 a necessary condition for stability is that the specific heat in the primary state should be positive, that is

$$C_\gamma = T\hat{\mathsf{K}} \geqq 0. \tag{32.4}$$

This follows as (32.3) holds for all incremental deformations so that we may set $u_i = \dot{u}_i = 0$ to obtain the result.

Corollary 32.5. Under the conditions of Corollary 32.2, when the temperature T is constant a necessary condition for stability is given by the following inequality

$$\int_{B_0(t)} \varrho_0 [\mathscr{A}(e_{ij}^*, T) - \mathscr{A}(e_{ij}, T)]\, dX + H_0(t) \geqq 0. \tag{32.5}$$

For conservative loading when the forces possess potentials Φ, Ψ in the notation of (30.6), (30.7), another result follows.

Corollary 32.6. Under the conditions of Corollary 32.5, when the loads are conservative, with potentials Φ, Ψ, a necessary condition for stability is given by the following inequality

$$\int_{B_0(t)} \varrho_0 [\mathscr{A}(e_{ij}^*, T) - \mathscr{A}(e_{ij}, T)]\, dX \geqq \Phi + \Psi - (\Phi(0) + \Psi(0)). \tag{32.6}$$

Corollary 32.7. Under the conditions of Corollary 32.6, for sufficiently small displacements, the necessary condition for stability (32.6) becomes

$$\int_{B_0(t)} \varrho_0 D^2 \mathscr{A}(e_{ij})\, dX \geqq \Phi + \Psi - (\Phi(0) + \Psi(0)), \tag{32.7}$$

where $\Phi(0)$, $\Psi(0)$ are the respective work terms of the forces in the primary state.[31]

The derivation of necessary conditions is now complete and we take up again the discussion of the classical energy criterion in its various formulations. It is supposed that the incremental displacement is held fixed on the part ∂B_{01} of the surface ∂B_0. Other constraints, such as incompressibility may also be present.

The test proposed by BEATTY[32] may formally be obtained from Corollary 32.5, when the time-dependence of the displacement is replaced by dependence upon a parameter.

Corollary 32.8. Under the conditions of Corollary 32.2 when the conditions are also isothermal, so that the free energy becomes the strain energy and the displacement $v_i(\eta, \boldsymbol{x})$ is restricted so as to be time-independent, a necessary condition for stability is given by the following inequality

$$\int_{B_0(t)} \varrho_0 [\mathscr{A}(e_{ij}^*) - \mathscr{A}(e_{ij})]\, dX \geqq \int_0^t \left[\int_{B_0(\eta)} \varrho_0 F_i^* \frac{\partial v_i}{\partial \eta}\, dX + \int_{\partial B_0(\eta)} T_i^* \frac{\partial v_i}{\partial \eta}\, dA \right] d\eta, \tag{32.8}$$

where η and t are parameters.

BEATTY's criterion, supposed necessary and sufficient for stability, is formally represented by inequality (32.8) in which the displacement v_i is regarded as virtual[33] and subject to the kinematical constraints. Unlike the versions which

[31] DUHEM [1906, *2*] obtained this result as a sufficient condition for instability with boundary conditions of displacement and pressure in the isothermal case. His measures were the uniform norm of the displacement and its first spatial and time derivatives.

[32] BEATTY [1965, *1*], [1967, *4*], [1968, *2*].

[33] Note, however, that these restricted displacements are still a subclass of those considered in the theorem. In all classical tests listed below the displacements are virtual and may not comply with this requirement, a matter further discussed in Sect. 34.

follow, the body forces and surface tractions are not required to be conservative. The criterion of Truesdell and Noll[34] follows by restricting further the body force to be conservative, and derivable from the potential φ, and the surface tractions to be dead. The virtual displacement $v_i(\boldsymbol{x})$ depends only on position and satisfies the boundary conditions of displacement. Their criterion, likewise asserted necessary and sufficient for stability, therefore reads

$$\int_{B_0} \varrho_0 [\mathscr{A}(e^*_{ij}) - \mathscr{A}(e_{ij})]\, dX \geqq \int_{B_0} \varrho_0 \varphi\, dX + \int_{\partial B_0} T_i v_i\, dA. \tag{32.9}$$

Pearson's formulation[35] is obtained by requiring the virtual displacement to be in addition infinitesimal. The remaining classical tests may likewise be associated with necessary conditions for stability. The classical definiteness conditions (31.2) and (31.3) are obtained by restricting the (virtual) displacement to be both time-independent and infinitesimal for the case of dead loading and given surface displacements.

We once again emphasise that the above necessary conditions were derived under the assumption that no other equilibrium solution exists in the neighbourhood of the primary configuration. This assumption is equivalent to supposing that the primary configuration is not one of neutral equilibrium and when false sometimes leads to (Lagrange) instability. Caughey and Shield[36] have shown that a position of neutral equilibrium is unstable in the class of linearised incremental displacements or for the class of materials with homogeneous polynomial strain energies. In the class of finite incremental displacements, stability of a position of neutral equilibrium is undecided, although by examining the class of displacements "close" to those of the neutral state, Koiter[37] has succeeded in obtaining further stability results. (His work has application to post-buckling behaviour and is beyond the scope of the present article.) Hence, whenever the primary configuration is in neutral equilibrium the necessity of the above conditions is no longer certain.

Most writers who use the energy criterion in fact exclude the primary configuration from being in neutral equilibrium or, what amounts to the same thing, from being neutrally stable[38] (Definition 7.13). It is tacitly supposed that positions of neutral equilibrium are always unstable and this has given rise to the "method of adjacent equilibrium" for assessing stability. In this other classic test, fully discussed in Chap. L, the "limit of stability" is attained when the equilibrium of the primary configuration first becomes neutral, stability having been assumed beforehand. As we have remarked in the last paragraph, in only a few circumstances has neutral equilibrium been demonstrated to imply instability, and therefore the test lacks any widespread substantiation. On the contrary, there are several instances where it is known to fail.[39]

---

[34] Truesdell and Noll [1965, *22*, § 89].

[35] Pearson [1956, *2*]. See also Hill [1957, *1*] and Truesdell and Noll [1965, *22*, § 68 bis].

[36] Caughey and Shield [1968, *4*]. An account is given in Chap. L.

[37] Koiter [1945, *1*], [1963, *8*], [1965, *12*], [1967, *18*]; see also Budiansky [1967, *6*].

[38] Confusion has arisen over this as some authors use proper weak relative minimum in the criterion thus requiring $D^2V > 0$ and specifying that rigid body rotations are excluded. See, for example, Hill [1957, *1*]; Ericksen and Toupin [1956, *1*]. Truesdell and Noll [1965, *22*, § 68 bis] call this *infinitesimal superstability* in contradistinction to stability defined from $D^2V \geqq 0$ which they define to be *infinitesimal stability*. Superstability implies the uniqueness of the equilibrium boundary value problems for *infinitesimal displacements* and hence precludes neutral equilibrium.

[39] An example of failure is the cantilevered beam under a tangential follower force at its free end. See, for instance, Pflüger [1964, *10*]. Further discussion may be found in Chap. Q and Sect. 80.

33. Sufficient conditions. We have just seen in the last section that, if the "potential" energy V of a body in equilibrium under conservative loading can be suitably defined, as, for example by the left-side of inequality (32.5) (with work terms in H_0 relating to conservative forces), then a *necessary* condition for Liapounov stability takes the form

$$V \geqq 0 \tag{33.1}$$

for all perturbations in a suitably defined class of displacements. This condition is equivalent to the necessary condition

$$D^2 V \geqq 0, \tag{33.2}$$

where D is the Gateau derivative of V. Condition (33.1) is applicable to both non-linear and linear perturbations.

Let us suppose now that the potential function V is smooth enough for both Gateau and Fréchet derivatives to reduce to the simple differential δ. Now under suitable restrictions the function V_4 of Chap. H may be identified with the potential energy V so that the condition sufficient for stability derived in that section becomes

$$V \geqq c\varrho, \tag{33.3}$$

where ϱ is a suitably chosen measure, and c is a positive constant. This, in turn, is implied by

$$\delta^2 V \geqq c\varrho. \tag{33.4}$$

The case of neutral equilibrium, for which a non-trivial solution exists to the equation

$$\delta^2 V = 0 \tag{33.5}$$

has been discussed in the previous Sect. 32.

While sets of necessary and sufficient conditions have been found for stability, the above summary indicates that the sets are not the same. A single complete set of necessary and sufficient conditions has not yet been found.

34. Criticism of the energy criterion. We now survey the classical energy criterion as a test for stability in its own right and also from the viewpoint of the Liapounov theory developed in the preceding sections. In Sect. 31, we saw that the energy test is used to ascertain the stability of an equilibrium position under isothermal conditions with conservative or dead loading and in the class of virtual displacements subject to the kinematical constraints. Now, any stability test must assume a definition of stability against which it either affirms or negates stability in prescribed circumstances. The Lagrange-Dirichlet criterion (the seeking of a minimum of potential energy in a discrete system under conservative loading) relied on a concept of stability concerned with the behaviour of displacements and velocities governed by *dynamical equations* near an equilibrium configuration. There is a definition of stability clearly implied in this work which, as we have seen, can be regarded as the Liapounov definition with uniform measures assumed. On the other hand, the energy test as usually applied is a static test, likewise dealing with conservative forces, which seeks a minimum of potential energy in the class of virtual displacements, again subject to kinematical constraints being satisfied.[40] In those works relating the test to continuous

[40] It is important to observe that a distinction is being drawn between the Lagrange-Dirichlet criterion and the energy test. This distinction was made by DIRICHLET [1846, *1*] in the paper containing the statement and proof of his theorem and turned on the fact that in the theorem, as first stated, the minimum of potential energy is taken in the class of actual dynamic displacements and not in the class of virtual displacements. Thus, the Dirichlet criterion, as originally formulated, is quite distinct from the energy criterion. Only later in the same paper did DIRICHLET *postulate* that both criteria were the same, and it is apparently only from DIRICHLET's second formulation of his result that we must reckon the origin of the energy criterion.

systems attention has usually been focussed on the *test itself* and the idea of stability which it was intended to measure has receded into the background. It thus no longer gives a statement about stability, that is about Liapounov stability with DIRICHLET'S concept as a special case, but merely a statement about the minimum of a functional.[41] Consequently, inconsistencies have become inevitable. Logical compatibility can be restored in this state of affairs by *defining* stability to relate in some way to the minimum of potential energy. This can be accomplished, for example, by adopting the mathematical definition of what we have called Hadamard stability. It may be argued that this is the course taken by some authors, but, in restoring consistency, stability is entirely divorced from the original physical grounding that it had in the hands of LAGRANGE and DIRICHLET.

Setting aside Hadamard stability, inconsistencies encountered in the applications of the energy test fall into two groups. Those in the first group arise from applications in which an underlying definition other than that of HADAMARD (most probably LIAPOUNOV) is tacitly assumed. Because this definition has not been explicitly stated or clarified, unacceptable results have sometimes been obtained as, for example, in the counter-examples of HELLINGER, HAMEL and SHIELD and GREEN, in which the uniform norm of displacements and velocities are effectively employed as measures. The energy test predicts stability but in fact there is instability.[42] The second group concerns applications of the test, inappropriately used even when the underlying definition of stability can be clarified. Its use has often been indiscriminate as, for instance, when it has been applied to non-conservative loading. It is therefore important to distinguish between cases in which this has occurred and those in which the test has failed on account of its intrinsic weakness. ZIEGLER[43] has pointed out many circumstances of the first kind and further critical comments on this aspect have been given by NEMAT-NASSER and ROORDA.[44]

Yet a third type of misuse arises from wrongly applied conditions for a minimum under the Hadamard (or equivalent) definition. Sets of necessary and sufficient conditions for a minimum may be found in Sect. 18 and no further discussion of their misapplication in this context is given here.

These remarks indicate briefly the confusion that surrounds the energy criterion and the ideas of stability on which it is reputed to rest. Some of the reasons why we prefer the Liapounov approach to stability are now apparent. This theory starts with a physically transparent definition of stability and then proceeds to formulate mathematically relevant necessary and sufficient criteria. In the light of these criteria, we may view in better perspective the classical energy test. We may identify the potential energy of the test with the function given in inequality (30.2) or (32.5), the free energy $\mathscr{A}$ now being the strain energy and the function H_0, the work done by body and surface forces, being given by (30.3) with the restriction that the loads are conservative. Then, as shown in the last section, the classical test emerges as formally *necessary* for Liapounov stability of the equilibrium solution under isothermal conditions, provided the equilibrium is not neutral. However, the classical test is stipulated to hold in the class of virtual displacements, whereas the necessary conditions are valid for

[41] However, as we have already seen, the test as applied to discrete systems provides mathematically proven necessary and sufficient conditions for stability in the sense of DIRICHLET.

[42] Compare footnote 67, p. 164, of Sect. 18.

[43] ZIEGLER [1952, *7*], [1953, *6*], [1956, *3*], [1968, *29*].

[44] NEMAT-NASSER and ROORDA [1967, *25*]. See also HSU [1968, *12*].

actual displacements occurring in the dynamic deformation. Thus, there may be circumstances in which the energy criterion is applicable and the Liapounov theory is not, so that the resemblance between the two, just noted, would then be nothing but formal.

Bearing somewhat on the last point is the suggestion, and its consequences, that the static nature of the energy criterion makes it simpler to use, since a large class of materials are elastic in static behaviour. An analysis similar to that already conducted here for elastodynamics would likewise show that the energy criterion is necessary for Liapounov stability of (non-neutral) equilibrium positions in these materials. However, the energy criterion, insofar as it involves the elastic strain energy, cannot, even suitably qualified, be sufficient for the same stability since it would then ignore dynamic behaviour, different in each material. Further, it is to be expected that the same material loaded by, say, conservative and non-conservative forces, will have different stability characteristics.[45] The energy test, by virtue of the very simplicity claimed in its favour, fails to take into account the rates of working of these loads and so cannot distinguish between them. The significance of a test like BEATTY'S, which admits general loads together with virtual displacements, is therefore doubly uncertain.

It should be stated here that the adjacent equilibrium method, which gives likewise a static criterion, is similarly open to criticism.

The position occupied by the classical test in relation to sufficiency for Liapounov stability is even more tenuous than its relation to necessity. The weak relative minimum of the potential energy V which we have taken in the statement of the criterion requires that the potential energy should satisfy the condition

$$\delta^2 V \geqq 0,^{46} \tag{34.1}$$

where V is now assumed to be sufficiently smooth for both Gateau and Fréchet derivatives to reduce to the simple differential (variation) δ. Now this condition is not sufficient in general to ensure the condition $V \geqq 0$ and certainly not $V \geqq c\varrho$, where ϱ is the measure of stability (now related to a norm on the set X) and c some positive constant. It is thus clear that the conditions usually applied in the energy tests to obtain stability are not sufficient. A sufficient condition, on the other hand, is that the second Fréchet derivative satisfies the condition

$$d^2 V \geqq c\varrho. \tag{34.2}$$

This does imply $V \geqq c\varrho$ and consequently Liapounov stability. That $d^2 V > 0$ is insufficient to ensure stability is seen from the fact that $d^2 V > 0$ may always be true yet the condition (34.2) violated for *every* ϱ.

Condition (34.2) is also the condition for a proper minimum of V at the primary configuration. If the set X is *finite-dimensional* then the condition $d^2 V > 0$ does imply $d^2 V > c\varrho$, however, and stability follows.[47] But this result is then generally truc for discrete systems only.

We may thus summarise by remarking that although under certain conditions the sufficient requirements for Liapounov stability, using V as the Liapounov function, and those for a minimum of V, do coincide,[48] this is not so in general.

[45] These and other points, are made by ERICKSEN [1966, *8*] and NEMAT-NASSER and ROORDA [1967, *25*].

[46] With neutral equilibrium, as usual, excluded.

[47] Sect. 14. ϱ is here associated with the norm on the space.

[48] The conditions for which coincidence obtains occur, for example, when inequalities $V \geqq c\,\varrho$ or $\delta^2 V \geqq c\,\varrho$ hold, for suitable ϱ, by virtue of an embedding theorem. Then $V > 0$ or $\delta^2 V > 0$ clearly suffice for Liapounov stability with respect to the stipulated measure ϱ. This observation is the basis of various justifications of the energy criterion which have appeared in the literature.

The energy criterion is only one of the sufficient conditions for Liapounov stability. The classical test is, however, always formally a necessary condition subject to the reservations noted earlier.

From these conclusions we are able to say that there is apparently only one logical manner of relating the energy criterion with a physical notion of stability and that, in the light of the present theory, the procedure of attempting to establish stability by seeking a minimum of V is not the most efficacious and can lead to spurious results. The better method is that of applying the Liapounov theorem whenever possible, for which dynamic perturbations are considered and in which the full knowledge of whether a condition is necessary or sufficient is available.

In spite of these criticisms and objections there are numerous circumstances in which the energy test has been successfully applied.[49] In general, the circumstances relating to the successful application must be capable of justification by appeal to more sophisticated theory. This means that when the test is applied it must either be checked in each individual circumstance or classes of problems must be created for which the test is known to be true. It is only in the latter case that the test will be of value. However, it is not our present task to investigate all these classes.

We may thus conclude by remarking that the inadequacies of the test have gradually become realised by the majority of writers. ZIEGLER's objections have been met by applying the test only to conservative systems; the criticisms of HELLINGER, HAMEL and SHIELD and GREEN, being more fundamental and proving more difficult to dispose of, have only been fully accommodated by using the work of LIAPOUNOV as generalised by ZUBOV and MOVCHAN.[50]

J. Stability for a fixed surface and under dead loads in the class of small incremental displacements.

35. Stability of equilibrium. The system we study in this section is that of a finitely deformed elastic body whose primary state is an equilibrium configuration. The body occupies a bounded region of three-space B with displacements prescribed on $\overline{\partial B_1}$ and dead loads on ∂B_2 where the whole boundary ∂B is $\overline{\partial B_1} \cup \partial B_2$. The incremental displacements u_i are taken to be small so that the analysis is valid only for the *class of linear perturbations*. The body force is taken to be zero.

This system is determined by the equations:

$$(d_{ijkl}\,u_{k,l})_{,j} = \varrho\,\ddot{u}_i \quad \text{in } B\times(0, t_1), \tag{35.1}$$

$$u_i = 0 \quad \text{on } \overline{\partial B_1}\times[0, t_1), \tag{35.2}$$

$$n_j\,d_{ijkl}\,u_{k,l} = 0 \quad \text{on } \partial B_2\times[0, t_1). \tag{35.3}$$

Special cases of the boundary conditions can be taken, for example, $\partial B_2 = \emptyset$, so that $\partial B_1 = \partial B$. The latter case is considered separately below. We also recall that the "elasticities" d_{ijkl} are functions of position alone and are determined explicitly by the finite deformation of the body whose stability is under investigation.

[49] Reference may be made to the book by TIMOSHENKO and GERE [1961, 7] and the use of the test by engineers in the design of structures.

[50] It is only fair to point out that a little before the writings of ZUBOV and MOVCHAN became known in the west, both KOITER [1963, 7, 8], [1965, 12] and SHIELD [1965, 21] had suggested possible solutions which in fact embodied elements of the general approach.

The configuration of the body whose stability is to be investigated corresponds to the null solution of (35.1)–(35.3), that is

$$u_i(t, \boldsymbol{x}) = 0. \tag{35.4}$$

For the set X we take the subset of $C^2(B)$ consisting of real vector-valued functions ψ defined on B and satisfying the boundary data (35.2), (35.3). Since X is linear we take for positive-definite function ϱ the square of the L_2-norm:[1]

$$\varrho(\psi, 0) = \varrho(\psi) = \int_B \varrho \psi_i \psi_i \, dx, \qquad \psi_i \in X.$$

When T is the semi-group of all non-negative real numbers i.e. $T = [0, \infty)$, and $T = \mathscr{T}$, the dynamical system $\mathscr{B}(T, X)$, representing the equations of dynamical linear elasticity, consists of all vector-valued functions u_i defined on $[0, \infty)$, taking values in X and such that (35.1) is satisfied.[2] The norm which ϱ induces on $\mathscr{B}(T, X)$ is therefore given by its square:

$$\varrho(u) = \sup_{0 \leq t < \infty} \int_{B(t)} \varrho u_i u_i \, dx. \tag{35.5}$$

It is clearly immaterial which instant τ is selected as the initial one and hence we arbitrarily choose $\tau = 0$. We are thus dealing with the concept of *uniform Liapounov stability*.

As measure of initial perturbations we select

$$\varrho_\tau(u) = \sup_{\substack{\boldsymbol{x} \in B \\ t=0}} (|u_{i,j}|^2 + |\dot{u}_i|^2) \tag{35.6}$$

and as Liapounov functions $F_{\tau,t}$ take twice the strain energy

$$F_{\tau,t} = \int_{B(t)} d_{ijkl} u_{i,j} u_{k,l} \, dx. \tag{35.7}$$

Because of the boundary data (35.2), (35.3), we see from (35.1) or by integrating Eq. (23.15) that the total energy is conserved, that is

$$2E(t) = \int_{B(t)} \varrho \dot{u}_i \dot{u}_i \, dx + \int_{B(t)} d_{ijkl} u_{i,j} u_{k,l} \, dx = 2E(0). \tag{35.8}$$

Let us now impose the following conditions on the elasticities:

(i) d_{ijkl} are of class $C^2(B)$;

(ii) d_{ijkl} are positive-definite in the sense that there exists a positive constant d such that

$$\int_B d_{ijkl} \xi_{ij} \xi_{kl} \, dx \geqq d \int_B \xi_{ij} \xi_{ij} \, dx, \tag{35.9}$$

for all (symmetric) tensors ξ_{ij}.[3]

Now, by virtue of condition (i) above we have, for some positive constant k the inequality

$$2E(0) \leqq k \varrho_\tau(u), \tag{35.10}$$

[1] It is always apparent from the context when ϱ denotes a positive-definite function, and when it denotes the density in the primary state B.

[2] The functions u_i are clearly classical solutions. However, it is a simple matter to extend the analysis to the weak solutions defined in Sect. 23. See, for example, KNOPS and PAYNE [1968, *18*], DAFERMOS [1968, *7*], SLEMROD and INFANTE [1971, *12*].

[3] Symmetry is essential when regarding the quantitites d_{ijkl} as the elasticities of the classical linear theory for which $d_{ijkl} = c_{ijkl}$, with $c_{ijkl} = c_{jikl} = c_{ijlk} = c_{lkij}$.

so that by (35.8)

$$2E(t) \leqq k\,\varrho_\tau(u) \tag{35.11}$$

and hence, on using the property of non-negativeness of the kinetic energy, we obtain the inequality

$$\int_{B(t)} d_{ijkl}\, u_{i,j}\, u_{k,l}\, dx \leqq k\,\varrho_\tau(u). \tag{35.12}$$

Thus condition (i) of Liapounov's theorem is satisfied.

Next, by (35.9) and Poincaré's inequality,[4] we have

$$\frac{\lambda_1 d}{\bar{\varrho}} \int_B \varrho\, u_i\, u_i\, dx \leqq \int_B d_{ijkl}\, u_{i,j}\, u_{k,l}\, dx, \tag{35.13}$$

where $\bar{\varrho} = \sup_{x \in B} \varrho$, and λ_1 is the lowest eigenvalue of the corresponding membrane problem,

$$\left.\begin{aligned} u_{,ii} + \lambda^2 u &= 0, \\ u &= 0 \quad \text{on } \partial B_1 \\ n_i\, u_{,i} &= 0 \quad \text{on } \partial B_2. \end{aligned}\right\} \tag{35.14}$$

Inequality (35.13) shows that condition (ii) of the Liapounov theorem is also satisfied so that stability is established.[5]

We see that (35.10), (35.12), (35.13) give the continued inequalities

$$\varrho(u) = \int_{B(t)} \varrho\, u_i\, u_i\, dx \leqq \frac{\bar{\varrho}}{\lambda_1 d} \int_{B(t)} d_{ijkl}\, u_{i,j}\, u_{k,l}\, dx \leqq \frac{k\,\bar{\varrho}}{\lambda_1 d}\, \varrho_\tau(u),$$

illustrating precisely the various steps required in the stability analysis.

It can be seen that the stability proof for the present problem depends on establishing inequality (35.13) which is itself equivalent to ascertaining that the Rayleigh quotient

$$\frac{\int_B d_{ijkl}\, u_{i,j}\, u_{k,l}\, dx}{\int_B \varrho\, u_i\, u_i\, dx}$$

has a positive lower bound. It is known that the lower bound of this quotient is the smallest minus eigenvalue of Eq. (35.1). The sufficient condition for stability is thus equivalent to having an imaginary smallest eigenvalue. In a more general context inequality (35.13) has formed the basis of Koiter's[6] method of treating stability, which in turn is a generalisation of a procedure used by Trefftz.[7] The same idea may apparently be found in the work of Kneser[8] and Born[9] on the stability of elastic strings and rods. These authors use Osgood's[10] theorem on the functional of a function of a single variable. Whether an inequality such as (35.13) is valid in other problems depends on the operator defined by equations corresponding to (35.1)–(35.3).

[4] We assume that B satisfies appropriate conditions.

[5] A similar result is obtained by Slobodkin [1962, *13*] and Knops and Wilkes [1966, *15*].

[6] Koiter [1965, *12*].

[7] Trefftz [1933, *1*].

[8] Kneser [1903, *2*].

[9] Born [1906, *1*]. See also Funk [1925, *2*]. Similar use of the above inequalities arises in the related work of Friedrichs [1947, *1*], Mikhlin [1964, *9*], [1965, *17*] and Zorski [1962, *16*].

[10] Osgood [1901, *2*], Hadamard [1902, *2*].

When $\partial B_1 = \partial B$ and the elasticities d_{ijkl} are *constant*, inequality (35.9) is guaranteed by the condition

$$d_{ijkl}\,\xi_i\,\xi_k\,\eta_j\,\eta_l \geqq d\,\xi_i\,\xi_i\,\eta_j\,\eta_j \tag{35.15}$$

for arbitrary vectors ξ_i, η_i. The result is a trivial consequence of GÅRDING's inequality,[11] and shows that in this special case (35.15) is an alternative condition for stability.[12] Instead of inequality (35.15), we could postulate

$$d_{ijkl}\,\xi_i\,\xi_k\,\eta_j\,\eta_l \geqq d_0\,\xi_i\,\xi_i, \tag{35.16}$$

where d_0 is a positive constant and η_i is non-zero. When ξ_i form a compact set, (35.16) is a direct consequence of (35.15). To prove that stability in the present problem is implied by (35.16), we extend u_i to the whole space E_3 by putting $u_i \equiv 0$ in $E_3 - \bar{B}$ and introduce the Fourier transform

$$W_i(\boldsymbol{z}) = (2\pi)^{-\frac{3}{2}} \int_{E_3} e^{i x_k z_k}\, w_i(\boldsymbol{x})\, dx.$$

A well-known theorem in Fourier analysis[13] then shows that

$$\int_B d_{ijkl}\, u_{i,j}\, u_{k,l}\, dx = \int_{E_3} d_{ijkl}\, z_j\, z_l\, \mathrm{Re}\,[\bar{U}_i\, U_k]\, dx,$$

a bar denoting the complex conjugate, so that the conclusion immediately follows on using (35.16) and PARSEVAL's theorem.

An inequality of type (35.16) has been discussed by HAYES in connexion with certain uniqueness results in linearised elasticity.

36. Stability of a body with time-dependent elasticities. When the configuration under examination is varying with time, the coefficients d_{ijkl} are functions of time as well as position.[14] We shall investigate the stability of the null state in the class of small perturbations measured by

$$\varrho(u) = \int_0^t \int_{B(\eta)} \varrho\, u_i\, u_i\, dx\, d\eta. \tag{36.1}$$

The equations governing the small incremental perturbations are then given by (35.1)–(35.3). We denote the total energy of the body at time t by $E(t)$ and then establish the following theorem:

Theorem 36.1. The null solution is stable in the class of small twice continuously differentiable displacements with respect to the measures $E(0)$, ϱ provided that

(a) *the elasticities are positive-definite:*

$$\int_B d_{ijkl}\,\xi_{ij}\,\xi_{kl}\, dx \geqq d_0 \int_B \xi_{ij}\,\xi_{ij}\, dx, \tag{36.2}$$

for d_0 a positive constant and ξ_{ij} an arbitrary second order tensor, and

[11] See, for instance, VAN HOVE [1947, *2*] and FICHERA [1965, *8*, Lect. 14], who also proves that (35.9) implies (35.15) under the stated conditions.

[12] SHIELD [1965, *21*] gives a direct proof of stability. See also KNOPS and WILKES [1966, *15*].

[13] See, for example, GOLDBERG [1961, *4*].

[14] See Sect. 23.

(b) *the time-derivatives of d_{ijkl} are negative-definite in the sense given by the inequality*:

$$-\int_B \dot{d}_{ijkl}\,\xi_{ij}\,\xi_{kl}\,dx \geqq d_1 \int_B \xi_{ij}\,\xi_{ij}\,dx, \tag{36.3}$$

for d_1 a positive constant and ξ_{ij} an arbitrary second order tensor.

To prove this theorem we note that Eq. (35.1) when multiplied by $\dot{u}_i$ and integrated over the volume B of the primary state yields, on applying the boundary conditions,

$$\int_{B(t)} (d_{ijkl}\,\dot{u}_{i,j}\,u_{k,l} + \varrho\,\dot{u}_i\,\ddot{u}_i)\,dx = 0. \tag{36.4}$$

Hence we have

$$2E(t) - \int_0^t \int_{B(\eta)} \dot{d}_{ijkl}\,u_{i,j}\,u_{k,l}\,dx\,d\eta = 2E(0). \tag{36.5}$$

We now apply the inequalities (36.2), (36.3) postulated in the theorem together with the Poincaré inequality so that we have

$$\alpha_1 \int_{B(t)} \varrho\,u_i\,u_i\,dx + \alpha_2 \int_0^t \int_{B(\eta)} \varrho\,u_i\,u_i\,dx\,d\eta \leqq 2E(0), \tag{36.6}$$

where α_1, α_2 are certain positive constants. This equation can be integrated to give

$$\int_0^t \int_{B(\eta)} \varrho\,u_i\,u_i\,dx\,d\eta \leqq 2E(0)\,(1 - e^{-\alpha_2 t/\alpha_1}) \tag{36.7}$$

and the result follows.

37. Discussion of choice of measure. In Sect. 35 the initial measure $\varrho_\tau(u)$ could alternatively have been chosen to be

$$\varrho_\tau(u) = \int_{B(0)} (u_{i,j}\,u_{i,j} + \varrho\,\dot{u}_i\,\dot{u}_i)\,dx \qquad u(t,\cdot) \in X. \tag{37.1}$$

On the other hand, it is not possible in general to choose as initial measure the positive-definite function

$$\varrho_\tau(u) = \sup_{\substack{x\in B\\ t=0}} (u_i\,u_i), \tag{37.2}$$

since the displacement, itself a continuous function, may not possess bounded derivatives.[15] If, however, it is known that u_i does possess bounded second spatial derivatives in B then the measure (37.2) may be employed as may be verified by considering the inequality[16]

$$\sup_{x\in \mathring{B}} (|u_i| + |u_{i,j}|) \leqq \alpha \big(\sup_{x\in B} |u_i|\big)^{\frac{1}{2}} \big[\sup_{x\in B} (|u_i| + |u_{i,j}| + |u_{i,jk}|\big]^{\frac{1}{2}},$$

where $\mathring{B}$ is a compact subset of B and α a positive constant.

With regard to alternative choices for the positive-definite function ϱ we know, from the example of SHIELD and GREEN referred to in Sect. 6, that

$$\varrho(u, 0) \equiv \varrho(\psi_i\,\psi_i) = \sup_{\substack{x\in B\\ t\geqq 0}} (\psi_i\,\psi_i), \qquad \psi_i \in X \tag{37.3}$$

[15] For example, in the case of a "saw-tooth" function.

[16] F. JOHN [1960, *5*].

is not a suitable measure for assessment of stability of solutions in the *dynamical system defined at the beginning of this section*. This example shows that initial data in the class of bounded total energy can lead to a displacement that is not small at all places and for all time. In these circumstances, even though the elasticities d_{ijkl} may be positive-definite in the sense of (35.9), an inequality of the kind

$$\int_{B(t)} d_{ijkl}\, u_{i,j}\, u_{k,l}\, dx \geqq d_0 \sup_{\substack{x\in B\\ t\geqq 0}} (u_i\, u_i) \tag{37.4}$$

cannot be satisfied.[17]

We now consider how the *dynamical system* may be modified so that stability can be recovered with the measure (37.3). We first note that, even when the regularity of functions defining set X is relaxed so that X now includes incremental displacements which are *weak solutions*,[18] the null solution is still unstable with respect to measures $E(0)$ and (37.3). This follows from an argument similar to that used by SHIELD and GREEN. On the other hand if we restrict appropriately the coefficients d_{ijkl} and the initial data we may still be able to establish stability using classical solutions and the uniform norm (37.3). With this end in view we supplement the positive-definiteness inequality (35.9) with the conditions

(iii) d_{ijkl} belongs to the class $C^3(B)$,

(iv) u_i belongs to the class which has initially bounded first order total energy[19] $E_1(0)$, where

$$E_1(t) = \tfrac{1}{2} \int_{B(t)} (\varrho\, \ddot{u}_i\, \ddot{u}_i + d_{ijkl}\, \dot{u}_{i,j}\, \dot{u}_{k,l})\, dx. \tag{37.5}$$

By virtue of (35.1)–(35.3) this quantity is conserved, that is,

$$E_1(t) = E_1(0). \tag{37.6}$$

We assume that the primary configuration is in equilibrium and establish stability of the null solution. From inequality (35.9) we see that the following holds

$$\int_{B(t)} (d_{ijkl}\, v_{i,j}\, v_{k,l} - d\, v_{i,j}\, v_{i,j})\, dx \geqq 0, \tag{37.7}$$

for any vector-valued function $v_i \in C^1(B)$ satisfying the displacement boundary condition (35.2). Equality in (37.7) is achieved for $v_i = 0$, and therefore Theorem 18.4 shows that (37.7) implies

$$d_{ijkl}\, \xi_i\, \xi_k\, \eta_j\, \eta_l \geqq d\, \xi_i\, \xi_i\, \eta_j\, \eta_j \tag{37.8}$$

uniformly in B as d is constant. The operator on the left-side of (35.1) is consequently strongly elliptic and thus elliptic, so that we have for positive constant d_2,[20]

$$\int_{B(t)} (u_i\, u_i + u_{i,j}\, u_{i,j} + u_{i,jk}\, u_{i,jk})\, dx \leqq d_2 \int_{B(t)} \varrho\, \ddot{u}_i\, \ddot{u}_i\, dx. \tag{37.9}$$

[17] Counterexamples disproving (37.4) also exist in the calculus of variations. Demonstrating their validity usually amounts to showing that the measure (37.3) can become unbounded so that the solution fails to exists in $C(B)$. See, for instance, HADAMARD [1907, *1*]. For other examples in continuum mechanics see references entered on p. 60.

[18] See Sect. 23.

[19] The development in the text follows closely that of SHIELD [1965, *21*] who seems to have been the first to introduce higher order energies into elastic stability analyses of this kind. See also KNOPS and WILKES [1966, *15*], who, however, consider only constant d_{ijkl}.

[20] Inequality (37.9) arises in proofs concerning regularity of solutions in the Dirichlet problem in elliptic equations: see, for instance, FICHERA [1965, *8*]. We conjecture that it holds also in the mixed problem.

The left-side of (37.9) is the norm associated with the Sobolev space $W_2^{(2)}$,[21] and so by the Sobolev embedding theorems,

$$\sup_{\substack{x\in B\\ t\geqq 0}} (u_i u_i) \leqq d_2 c \int_{B(t)} \varrho \ddot{u}_i \ddot{u}_i \, dx, \tag{37.10}$$

for some positive constant c. We now appeal to the Liapounov stability theorem, taking as Liapounov functions

$$F_{\tau,t}(u) = \int_{B(t)} \varrho \ddot{u}_i \ddot{u}_i \, dx,$$

which, from inequality (35.9). satisfies the inequality

$$F_{\tau,t}(u) \leqq 2E_1(t).$$

Then Eqs. (37.6) and (37.10) show that we have established stability with respect to the measures

$$\varrho_\tau(u) = E_1(0)$$

and (37.3). An equivalent measure for the initial data is

$$\varrho_\tau(u) = \sup_{\substack{x\in B\\ t=0}} (\ddot{u}_i \ddot{u}_i + \dot{u}_{i,j} \dot{u}_{i,j}).$$

In a similar way, this time using $E_2(t)$ instead of $E_1(t)$, where

$$E_2(t) = \tfrac{1}{2} \int_{B(t)} (\varrho \dddot{u}_i \dddot{u}_i + d_{ijkl} \ddot{u}_{i,j} \ddot{u}_{k,l}) \, dx,$$

and imposing extra smoothness on the coefficients d_{ijkl}, stability may be established with respect to[22]

$$\varrho_\tau(u) = \begin{cases} E_2(0), \\ \sup\limits_{\substack{x\in B\\ t=0}} (\dddot{u}_i \dddot{u}_i + \dot{u}_{i,jk} \dot{u}_{i,jk}) \end{cases}$$

and

$$\varrho(u) = \sup_{\substack{x\in B\\ t\geqq 0}} (\dot{u}_i \dot{u}_i + u_{i,j} u_{i,j}).$$

The above deductions may be expressed in a single theorem as the result of a final calculation. On taking the strain energy, performing an integration by parts and applying SCHWARZ's inequality we obtain

$$\left(\int_{B(t)} d_{ijkl} v_{i,j} v_{k,l} \, dx\right)^2 \leqq \int_{B(t)} v_i v_i \, dx \int_{B(t)} (d_{ijkl} v_{k,l})_{,j} (d_{ipmn} v_{m,n})_{,p} \, dx.$$

After using (35.13) we find that

$$\int_{B(t)} d_{ijkl} u_{i,j} u_{k,l} \, dx \leqq \frac{\overline{\varrho}}{\lambda_1 d} \int_{B(t)} (d_{ijkl} u_{k,l})_{,j} (d_{ipmn} u_{m,n})_{,p} \, dx. \tag{37.11}$$

Inequality (35.13) together with (37.11) and the equations of motion (35.1) then show that

$$E(t) \leqq \frac{\overline{\varrho}}{\lambda_1 d} E_1(t), \tag{37.12}$$

---

[21] The space $W_2^{(k)}$ consists of (generalised) functions whose (generalised) derivatives up to and including the k-th order are in L_2.

[22] SHIELD [1965, *21*]. The set X must now consist of vector-valued functions in $C^3(B)$.

and a similar inequality relates $E_1(t)$ and $E_2(t)$, where $\bar{\varrho}$ is an upper bound for the density. On combining (37.12) with the two previous stability results we obtain SHIELD's theorem.[23]

Theorem 37.1. When the coefficients d_{ijkl} are at least three times continuously differentiable on the open region B and satisfy the relation (35.9), *then the solution u_i of* (35.1) *subject to the boundary conditions* (35.2), (35.3) *is such that u_i, $\dot{u}_i$, $u_{i,j}$ all tend to zero when the second order energy $E_2(0)$ tends to zero.*

SHIELD also proves that *u_i tends to zero as $E_1(0)$ tends to zero, and $\dot{u}_i$ tends to zero as $E_2(0)$ tends to zero,* when the coefficients d_{ijkl} satisfy the strong ellipticity condition (37.8) and the inequality

$$\int_{B(t)} d_{ijkl}\, v_{i,j}\, v_{k,l}\, dx > 0,$$

where v_i is a non-zero vector satisfying (35.1).

38. Stability with multipolar elasticity. From the general theory underlying the Sobolev embedding theorems it is clear that if we are able to use an energy integral for the Liapounov function to establish stability with respect to a uniform norm for measure, this function must contain second-order partial derivatives at least. We have seen in Sect. 37 that SHIELD introduced such derivatives by considering the higher-order energies. Another technique which takes us to models beyond those normally encompassed by classical elasticity theory, is that of considering multipolar elasticity, in particular the theory of simple force and stress multipoles.[24] As the higher spatial derivatives of displacement are also present in the internal and free energies, these too should prove suitable for Liapounov functions when stability in the uniform norm is to be established. This is confirmed by the work of KOITER,[25] who, following a series of steps similar to those in the previous section succeeds in justifying the sufficiency of the energy criterion[26] for this model. This conclusion is invalidated, however, when classical elasticity is considered. The proof of necessity requires the introduction of dissipation which again excludes its application in theories of elasticity.

In our approach we allow the mathematical theory to predict that the null solution may not be stable with respect to all common measures in the class of linear displacements. SHIELD's theorem shows in which sense stability can be recovered and under what conditions.[27] If better agreement with physical fact can be obtained by appealing to another model then it may be necessary to abandon classical theory completely in favour of this new model. This does not invalidate classical theory as a *mathematical model,* however, neither does it justify the use of a new model to establish specific arbitrarily selected results in the classical theory which may turn out not to be limiting cases of the new theory.

39. Incompressible media. When the elastic body is incompressible, the preceding analysis requires modification. The equations of motion and boundary

[23] SHIELD [1965, *21*].
[24] GREEN and RIVLIN [1964, *2, 3*]. Other treatments relating to multipolar mechanics are due notably to TOUPIN [1964, *13*], and ERINGEN and SUHUBI [1964, *1*].
[25] KOITER [1965, *12*].
[26] See Sect. 34.
[27] Another technique is demonstrated in Sect. 50.

conditions are now (see Eqs. (23.36), (23.37))

$$(D_{ijkl}u_{k,l})_{,j}+[p(u_{i,j}+u_{j,i})]_{,j}+p'_{,i}=\varrho_0\ddot{u}_i \quad \text{in } B\times(0,t_1), \tag{39.1}$$

$$u_{i,i}=0 \quad \text{in } B\times[0,t_1], \tag{39.2}$$

$$u_i=0 \quad \text{on } \overline{\partial B_1}\times[0,t_1), \tag{39.3}$$

$$n_j D_{ijkl}u_{k,l}+n_j p(u_{i,j}+u_{j,i})+n_i p'=0 \quad \text{on } \partial B_2\times[0,t_1), \tag{39.4}$$

where p is the hydrostatic pressure in the primary state, and p' is an undetermined scalar function of position and time. Also,

$$D_{ijkl}=b_{ijkl}+\sigma_{jl}\delta_{ik},$$

with b_{ijkl} defined in Eq. (23.35). As before, we select T and $\mathcal{T}$ to be the set of non-negative real numbers, with $\tau=0$, and for X take the subset of $C^2(B)$ consisting of real-valued vector functions satisfying (39.2) and (39.3). The dynamical system then consists of the set of vector-valued functions u_i defined on $[0,\infty)$ with values in X and such that (39.1) is satisfied.

We shall discuss stability only in the case in which ϱ is the square of the weighted L_2-norm of the incremental displacement u_i.

From (39.1), (39.3) and (39.4), or more directly from the rate of work equation (23.34), we find that

$$E(t)\equiv\tfrac{1}{2}\int_{B(t)}[\varrho_0\dot{u}_i\dot{u}_i+D_{ijkl}u_{i,j}u_{k,l}+\tfrac{1}{2}p(u_{i,j}+u_{j,i})(u_{i,j}+u_{j,i})]\,dx=E(0). \tag{39.5}$$

Corresponding to (35.9), we now impose the following conditions on the elasticities:

(i) D_{ijkl}, p, ϱ_0 are uniformly bounded in B,

(ii) there exists a positive constant D such that

$$\int_B D_{ijkl}\xi_{ij}\xi_{kl}\,dx\geqq D\int_B \xi_{ij}\xi_{ij}\,dx, \tag{39.6}$$

for all tensors ξ_{ij}.[28]

For ϱ_τ we choose (35.6) and then obtain, for positive constant k,

$$\int_{B(t)}[D_{ijkl}u_{i,j}u_{k,l}+\tfrac{1}{2}p(u_{i,j}+u_{j,i})(u_{i,j}+u_{j,i})]\,dx\leqq 2k\,\varrho_\tau(u). \tag{39.7}$$

We consider separately the cases $p\geqq 0$, $p<0$. In the former, we may omit the second term on the left in (39.7) without affecting the inequality and then use (39.6) together with POINCARÉ's inequality to obtain

$$\frac{\lambda_1 D}{\bar{\varrho}_0}\int_{B(t)}\varrho_0 u_i u_i\,dx\leqq\int_{B(t)}D_{ijkl}u_{i,j}u_{k,l}\,dx\leqq 2k\,\varrho_\tau(u),$$

from which stability follows in an obvious manner. When $p<0$, we note that

$$p(u_{i,j}+u_{j,i})(u_{i,j}+u_{j,i})=p\,u_{i,j}u_{i,j}-p(u_{i,j}-u_{j,i})(u_{i,j}-u_{j,i})\geqq p\,u_{i,j}u_{i,j}.$$

Hence,

$$\int_{B(t)}(D_{ijkl}u_{i,j}u_{k,l}+\tfrac{1}{2}p(u_{i,j}+u_{j,i})(u_{i,j}+u_{j,i})\,dx\geqq\int_{B(t)}(D+p)\,u_{i,j}u_{i,j}\,dx.$$

Thus, when $D+p>0$, stability follows as before.

---

[28] When $\sigma_{ij}\equiv 0$, ξ_{ij} must be restricted to be a symmetric tensor.

K. Stability under dead surface loads in the class of small incremental displacements.

40. Introduction. We consider now the isothermal traction boundary value problem again under the assumption of an equilibrium primary configuration in which the body forces are assumed absent for simplicity of treatment. Stability will be examined in the class of small incremental displacements which satisfy the following equations

$$(\sigma_{jl} u_{i,l} + c_{ijkl} u_{k,l})_{,j} = \varrho \ddot{u}_i \qquad \text{in } B\times(0, t_1), \tag{40.1}$$

$$n_j(\sigma_{jl} u_{i,l} + c_{ijkl} u_{k,l}) = 0 \qquad \text{on } \partial B\times[0, t_1), \tag{40.2}$$

where the expressions for c_{ijkl} are given explicitly in (23.17) and

$$c_{ijkl} = c_{jikl} = c_{klij}. \tag{40.3}$$

Since the primary state is in equilibrium we have

$$\int_{\partial B} \sigma_{ij} n_j \, da = 0, \tag{40.4}$$

and

$$e_{ijk} \int_{\partial B} x_j \sigma_{kp} n_p \, da = 0, \tag{40.5}$$

or

$$e_{ijk} \int_{\partial B_0} x_j T_{Ak} N_A \, dA = 0, \tag{40.5a}$$

where e_{ijk} is the permutation symbol.

As in Chap. J we examine the stability of the null solution to Eqs. (40.1)–(40.3) subject to the equilibrium conditions (40.4), (40.5). We choose these equations which refer the motion to the primary configuration rather than the formally equivalent set (23.6), (23.7) as in the following the properties of the primary stress field are crucial in deciding the existence of certain instabilities. The set X is defined as in Chap. J with the exception that the functions belonging to X satisfy the boundary conditions (40.2) instead of (35.2), (35.3). We again arbitrarily take the initial instant to be zero and note that the analysis in several parts of this section is valid in fact for weak solutions defined in the sense of Sect. 23. The elements of $\mathscr{B}(T, X)$ are unique, however, only to within rigid body motions and it is the instabilities just mentioned that may correspond to these motions.[1]

Let us consider first rigid body translations. The resultant incremental force on the body remains zero during the incremental deformation. The body can thus undergo both a rigid body translational displacement and a uniform rigid body translational velocity. The velocity clearly produces an instability in which the displacement grows with time and can be excluded by ensuring that its initial value is zero. That is, we shall always assume that

$$\int_{B(0)} \dot{u}_i \, dx = 0. \tag{40.6}$$

The rigid body translational displacement, on the other hand, then remains constant in time and can thus be controlled by including it in the measure of the initial disturbance imparted to the system.

Similar stipulations apply to the rigid body rotational displacement and velocity, with however, one important proviso, namely, that the equilibrium of the

[1] Other instabilities are discussed in Chap. N.

body is not disturbed. When the rotational balance is broken by, say, an initial incremental displacement, the rigid body part of the motion is governed by the usual laws of classical mechanics. This part of the motion may or may not lead to instability. When the full symmetries of the elasticities given in (41.1) are present the rigid body motions may be separated from the elastic deformations and their stability investigated separately, although these instabilities will also appear in the normal course of an analysis conducted without separation. Rotational velocities of rigid body type not disturbing equilibrium may be excluded by putting their initial values equal to zero, that is by setting

$$e_{ijk} \int_{B(0)} x_k \dot{u}_j \, dx = 0, \tag{40.7}$$

where the i-th axis is the axis of equilibrium. The associated rigid body rotational displacement may be accounted for by including it in the measure of initial perturbation.

In order, therefore, to proceed with the analysis we need to determine the rigid body rotational motion that maintains equilibrium. Let the particle originally at X_A be at y_i in such a motion.[2] Then for rotational balance

$$e_{ijk} \int_{\partial B_0(t)} y_j T_{Ak} N_A \, dA = 0 \tag{40.8}$$

as a result of the dead loading. Hence from Eqs. (40.5 a)–(40.8) we obtain

$$e_{ijk} \int_{\partial B_0(t)} u_j T_{Ak} N_A \, dA = 0,$$

so that on integration by parts we have

$$e_{ijk} \int_{B_0(t)} T_{Ak} u_{j,A} \, dX = 0, \tag{40.9}$$

or

$$e_{ijk} \int_{B_0(t)} T_{Ak} x_{l,A} u_{j,l} \, dX = 0,$$

or

$$e_{ijk} \int_{B(t)} \sigma_{lk} u_{j,l} \, dx = 0.^{3} \tag{40.10}$$

We now insert the condition that u_i is a small rigid body rotational displacement, that is,

$$u_i = e_{ijk} d_j x_k, \tag{40.11}$$

for some d_j independent of x_i. This gives, from Eq. (40.10)

$$(\bar{\sigma}_{kk} \delta_{ij} - \bar{\sigma}_{ij}) d_i = 0, \tag{40.12}$$

where $\bar{\sigma}_{ij}$ denotes mean stress over the volume. On transferring to axes of principal mean stress, Eq. (40.12) becomes

$$(\bar{\sigma}_2 + \bar{\sigma}_3) d_1 = 0, \quad (\bar{\sigma}_3 + \bar{\sigma}_1) d_2 = 0, \quad (\bar{\sigma}_1 + \bar{\sigma}_2) d_3 = 0.$$

It therefore follows that $d_i = 0$, unless some or all of the quantities

$$\bar{\sigma}_i + \bar{\sigma}_j \, (i \neq j;\; i, j = 1, 2, 3)$$

[2] The development is due mainly to HILL [1967, *15*].

[3] This equation has been obtained by BEATTY [1965, *1*], [1967, *4*] and also by HILL [1967, *15*].

are zero. In the case in which some of these quantities are zero the tractions have an *axis of equilibrium*, that is d_i is arbitrary in the following sense. If $\bar{\sigma}_2+\bar{\sigma}_3=0$, then d_1 is undetermined; if, in addition, $\bar{\sigma}_3+\bar{\sigma}_1=0$ then d_2 is also undetermined; and, if further, $\bar{\sigma}_1+\bar{\sigma}_2=0$, then all three components of d_i are undetermined.

Clearly, when the primary state is stress-free d_i is again entirely arbitrary and any rigid body rotational motion is possible. We begin the analysis in Sect. 41 with a body assumed to be in this condition and exclude the rigid body motions. We then proceed in the following section to consider the case of a body in a stressed primary state.

Beatty, Holden and Truesdell and Noll[4] propose dealing with the problem of rigid body rotations by admitting in their stability analysis only those incremental displacements $u_i(t, x_j)$ which preserve the overall equilibrium of the body. That is, they require the admissible displacements to satisfy condition (40.10) which they call the *zero moment* condition. In particular, rigid body rotations not about an axis of equilibrium are excluded by this condition. Rotations about an axis of equilibrium are however allowed.

Their device has thus the effect opposite to the one adopted here. We reject all rotations about an axis of equilibrium, since these lead to clear instabilities. On the other hand, rotations not about these axes are retained and examined for stability in the subsequent analysis. Their presence normally also indicates instability, but it becomes obvious how they may be excluded from the displacement. Evidently, we are here dealing with an example of *conditional stability* (see Definition 9.2).

All authors cited above exclude rigid body translational motions.

41. Stability analysis I.[5] In this section we treat the problem in which the elasticities possess the symmetries

$$d_{ijkl}=d_{jikl}=d_{ijlk}. \tag{41.1}$$

These conditions are both necessary and sufficient for the *primary state to be stress-free. In this analysis we shall consider only those deformations which are free from rigid body motions*. In order to exclude these motions, both translational and rotational, we express the displacement u_i in terms of the function $w_i(t, x_j)$ by the following relation[6]

$$w_i(t, x_j)=u_i(t, x_j)+a_i(t)+e_{ijk}\,d_j(t)\,x_k, \tag{41.2}$$

where w_i satisfies the following conditions

$$\int_{B(t)} w_i\,dx=0, \tag{41.3}$$

$$\int_{B(t)} (w_{i,j}-w_{j,i})\,dx=0=\int_{B(t)} (w_i x_j-w_j x_i)\,dx. \tag{41.4}$$

It then follows from (41.2) and (41.3) with the origin at the mass centre that

$$a_i(t)=-\frac{1}{B}\int_{B(t)} u_i\,dx, \tag{41.5}$$

[4] Beatty [1965, *1*], [1967, *4*], [1968, *2*], [1970, *2*], Holden [1964, *6*], Truesdell and Noll [1965, *22*, § 68 bis].

[5] See Knops and Payne [1968, *17*].

[6] This decomposition, although convenient, is not essential to the analysis.

where B is the volume of the body at time t. Now, by hypothesis, the incremental resultant force acting on the body is zero and therefore both the integrals

$$\int_{B(t)} \dot{u}_i\,dx \quad \text{and} \quad \int_{B(t)} (\dot{u}_{i,j}-\dot{u}_{j,i})\,dx$$

remain constant. We therefore put each equal to zero,

$$\int_{B(t)} \dot{u}_i\,dx=0=\int_{B(t)} (\dot{u}_{i,j}-\dot{u}_{j,i})\,dx \tag{41.6}$$

to avoid the trivial instabilities that arise from the constant rigid body velocities. These instabilities are generated for example by displacements of the type $u_i=\varepsilon_i t$ or $u_i=t\,R_{ij}x_j$ with $R_{ij}=-R_{ji}$, where ε_i and R_{ij} are constant. It is easily seen that these are solutions of (40.1) and, no matter how small ε_i and R_{ij} are chosen, the displacement eventually becomes unbounded in time.

Now condition (41.6) implies the condition

$$\int_{B(t)} u_i\,dx=\int_{B(0)} u_i\,dx, \tag{41.6a}$$

and hence by Eq. (41.5), $a_i(t)$ is an absolute constant. If we now treat condition (41.4) in a similar manner we obtain in conjunction with (41.1) the result

$$B\,d_i=-\tfrac{1}{2}\,e_{ijk}\int_{B(t)} u_{k,j}\,dx. \tag{41.6b}$$

From the second equation in (41.6) it then follows that d_i is also constant. Let us now consider the L_2-norm of u_i. We obtain

$$\begin{aligned}\int_{B(t)} w_i\,w_i\,dx &= \int_{B(t)} u_i\,u_i\,dx+B\,a_i\,a_i+2a_i\int_{B(t)} u_i\,dx\\ &\quad+I_{mn}\,d_n\,d_m+2e_{ijk}\,d_j\int_{B(t)} u_i\,x_k\,dx+2a_i\,e_{ijk}\,d_j\int_{B(t)} x_k\,dx\\ &=\int_{B(t)} u_i\,u_i\,dx-B\,a_i\,a_i-I_{mn}\,d_n\,d_m,\end{aligned} \tag{41.7}$$

in which the origin has been taken at the mass centre and the identity

$$\int_{B(t)} e_{ijk}\,u_j\,x_k\,dx=I_{im}\,d_m \tag{41.8}$$

has been used, where

$$I_{ij}=\int_{B(t)} (x_k\,x_k\,\delta_{ij}-x_i\,x_j)\,dx. \tag{41.9}$$

Now, we suppose the existence of a KORN's constant K such that, when (41.4) is satisfied, we have KORN's inequality[7]

$$\int_B c_{ijkl}\,w_{i,j}\,w_{k,l}\,dx \geqq K\int_B w_{i,j}\,w_{i,j}\,dx. \tag{41.10}$$

Then by the Poincaré inequality which we here put in the form

$$\int_B v_{,i}\,v_{,i}\,dx \geqq c_2\left[\int_B v^2\,dx-\frac{1}{B}\left(\int_B v\,dx\right)^2\right], \tag{41.11}$$

where c_2 is a positive constant, we get from Eqs. (41.10), (41.11) and (41.3), the inequality

$$\int_{B(t)} w_i\,w_i\,dx \leqq c_2 K\int_{B(t)} c_{ijkl}\,w_{i,j}\,w_{k,l}\,dx. \tag{41.12}$$

---

[7] See Sect. 43.

Further on recalling Eqs. (41.1), (41.2) we see that

$$c_{ijkl}\,w_{i,j}\,w_{k,l}=c_{ijkl}\,u_{i,j}\,u_{k,l}, \tag{41.13}$$

so that we obtain

$$\int_{B(t)} w_i\,w_i\,dx \leqq 2c_2\,KE(t)=2c_2\,KE(0). \tag{41.14}$$

We thus obtain from inequality (41.14) and the relation (41.7) the condition

$$\int_{B(t)} u_i\,u_i\,dx \leqq 2c_2\,KE(0)+B\,a_i\,a_i+I_{mn}\,d_n\,d_m. \tag{41.15}$$

Thus stability has been established according to the following theorem.

Theorem 41.1. The null solution of (40.1), *subject to the conditions* (40.2) *(both with* $\sigma_{ij}=0$*), is stable in the class of displacements satisfying* (41.6), *with respect to the measures*

$$\varrho_\tau[u_i(0)]\equiv 2c_2\,KE(0)+B\,a_i\,a_i+I_{mn}\,d_n\,d_m,$$

and

$$\varrho[u_i(t)]\equiv \int_{B(t)} u_i\,u_i\,dx.$$

This is evidently an example of conditional stability.

The above theorem may also be established if the inequality (41.12) is replaced by the following

$$\int_{B(t)} c_{ijkl}\,\tilde{e}_{ij}\,\tilde{e}_{kl}\,dx \geqq c_0 \int_{B(t)} \tilde{e}_{ij}\,\tilde{e}_{ij}\,dx \tag{41.16}$$

for some positive constant c_0, where $\tilde{e}_{ij}$ are the linear strains

$$\tilde{e}_{ij}=\tfrac{1}{2}(u_{i,j}+u_{j,i}).$$

The left side introduces an expression proportional to the classical strain energy. The theorem is proved under this condition by using a result of Bramble and Payne for the traction boundary value problem,[8] that is

$$\int_{B(t)} w_i\,w_i\,dx \leqq c_3 \int_{B(t)} (w_{i,j}+w_{j,i})\,(w_{i,j}+w_{j,i})\,dx \tag{41.17}$$

where w_i satisfy the conditions (41.3), (41.4) and c_3 is a positive constant.

42. Stability analysis II. We now assume that the coefficients d_{ijkl} satisfy only the major symmetries, that is

$$d_{ijkl}=d_{klij}.$$

As has already been noted, the existence of these symmetries implies that the pre-stress σ_{ij} of the primary state is non-zero. Under these conditions it is not possible in general for the body to have arbitrary rigid body rotations, except when these are executed about an axis of equilibrium. It is thus unnecessary to exclude them *a priori*. Certain primary states will, however, be unstable in the class of rigid body rotations. We shall later identify such states and again exclude the rotations and examine the stability of the configuration with respect to the residual (otherwise arbitrary) displacements. Rigid body translational motions, on the other hand, are accounted for in the manner previously described.

We thus introduce, first of all, the function $v_i(t, x_j)$ related to the displacement $u_i(t, x_j)$ by the equation

$$v_i(t, x_j)=u_i(t, x_j)+a_i(t), \tag{42.1}$$

[8] Bramble and Payne [1962, *2*].

and subject to the condition

$$\int_{B(t)} v_i\, dx = 0, \tag{42.2}$$

so that rigid body translations are excluded from v_i. From (42.1), (42.2) we immediately obtain the result

$$a_i(t) = -\frac{1}{B}\int_{B(t)} u_i\, dx. \tag{42.3}$$

Since the incremental resultant force on the body remains zero, it follows from the conservation of momentum that

$$\int_{B(t)} \dot{u}_i\, dx = \text{const.} \tag{42.4}$$

Rigid body translational velocities may thus be excluded from u_i by taking

$$\int_{B(0)} \dot{u}_i\, dx = 0, \tag{42.5}$$

so that we have the relation

$$\int_{B(t)} u_i\, dx = \int_{B(0)} u_i\, dx, \tag{42.6}$$

and a_i is constant.

Hence from (42.1) and (42.2) we get

$$\int_{B(t)} u_i u_i\, dx = \int_{B(t)} v_i v_i\, dx + a_i a_i B. \tag{42.7}$$

We next require the coefficients d_{ijkl} to satisfy the inequality

$$\int_{B(t)} d_{ijkl}\xi_{ij}\xi_{kl}\, dx \geqq d_1 \int_{B(t)} \xi_{ij}\xi_{ij}\, dx \tag{42.8}$$

for all arbitrary tensors ξ_{ij} and positive constant d_1. Then, on using POINCARÉ'S inequality, we get for some positive constant d_2,

$$\int_{B(t)} v_i v_i\, dx \leqq d_2 \int_{B(t)} d_{ijkl} v_{i,j} v_{k,l}\, dx \leqq 2 d_2 E(0).$$

Hence, by (42.7) and (42.3),

$$\int_{B(t)} u_i u_i\, dx \leqq 2 d_2 E(0) + \frac{1}{B}\int_{B(0)} u_i\, dx \int_{B(0)} u_i\, dx$$

and stability is established according to the definition of LIAPOUNOV with the following choice of measures:

$$\varrho_\tau[u_i(0)] = 2 d_2 E(0) + \frac{1}{B}\int_{B(0)} u_i\, dx \int_{B(0)} u_i\, dx,$$

$$\varrho[u_i(t)] = \int_{B(t)} \varrho\, u_i u_i\, dx.$$

Clearly we have only established conditional stability within the class of displacements restricted by (42.5). That condition (42.5) is necessary for stability may be easily demonstrated by examining the particular displacement $u_i = \varepsilon_i t$, for constant ε_i.

43. Note on Korn's inequality. This inequality, named after A. KORN,[9] who originally proved it in the context of elastostatics,[10] takes the form

$$\int_B u_{i,j} u_{i,j} \, dx \leq \frac{K}{4} \int_B (u_{i,j}+u_{j,i})(u_{i,j}+u_{j,i}) \, dx, \tag{43.1}$$

with u_i restricted by the condition

$$\int_B (u_{i,j}-u_{j,i}) \, dx = 0. \tag{43.2}$$

The smallest value of K appearing in (43.1) is referred to as KORN'S *constant.* Although inequalities of the types (41.10), are sometimes associated with the name of KORN the inequality (43.1) is strictly that implied by the name. The following two inequalities, however, are strictly equivalent to (43.1):

$$\int_B (u_{i,j}-u_{j,i})(u_{i,j}-u_{j,i}) \, dx \leq \frac{4(K-1)}{K} \int_B u_{i,j} u_{i,j} \, dx, \tag{43.3}$$

and

$$\int_B (u_{i,j}-u_{j,i})(u_{i,j}-u_{j,i}) \, dx \leq (K-1) \int_B (u_{i,j}+u_{j,i})(u_{i,j}+u_{j,i}) \, dx, \tag{43.4}$$

both with $K \geqq 1$.

Proofs for the existence of a KORN'S constant depend crucially on the nature of the region B and one set of conditions is given in the treatment by FRIEDRICHS. The proof of PAYNE and WEINBERGER applies to any region that can be mapped into a sphere by means of a map with bounded first and second derivatives. They also show that if KORN'S inequality holds for several domains it also holds for their union. It can then be shown that B may have edges and corners but not cusps or cusp-like edges. DAFERMOS extends this result to include confocal ellipsoidal shells. His method gives, in principle, numerical values for the constant. Recently, HLAVÁČEK and NEČAS have established the Korn inequality for Lipschitz regions of euclidean N-space.

The best possible value of the Korn constant for a sphere is shown by PAYNE and WEINBERGER to be

$$K = \frac{56}{13},$$

while BERNSTEIN and TOUPIN obtain the simple lower bound

$$K \geqq 1 + \frac{4 I_1}{I_2 + I_3} \geqq 3,$$

where $I_1 \geqq I_2 \geqq I_3$ are the principal values of $I_{ij} = \int_B x_i x_j \, dx$. That a lower bound of 3 must always occur for this constant has been demonstrated by HOLDEN[11] who by a simple example, also shows that an upper bound must depend on the region.

[9] KORN [1908, *1*], [1909, *1*].

[10] Later proofs have been supplied by FRIEDRICHS [1947, *1*], BERNSTEIN and TOUPIN [1960, *1*], PAYNE and WEINBERGER [1961, *6*], DAFERMOS [1968, *6*], HLAVÁČEK and NEČAS [1970, *12*], with further studies by EYDUS [1951, *1*], MIKHLIN [1965, *17*]. See also GOBERT [1962, *5*] and HORGAN and KNOWLES [1971, *4*].

[11] HOLDEN [1964, *6*].

L. Instability under dead surface loads from the equations of linear incremental displacement.

44. Instability from negative-definite total energy. In Sect. 32 some conditions *necessary* for stability have been obtained by establishing *sufficient* conditions for instability using the function V_4 (equivalent to the total energy in certain circumstances) as Liapounov function. *Sufficient* conditions for stability have also been obtained from V_4 in Chap. H. These results apply in general to non-linear and linear perturbations. The failure of the sufficiency test with V_4 (total energy) does not imply instability, however, as another Liapounov function may be found still capable of proving stability.

In this section we turn to *linear elasticity*[1] and examine some further conditions which give sufficiency for *instability*. On inversion they therefore become conditions *necessary* for stability. Even in this case of small incremental displacements a complete set of necessary and sufficient conditions for stability is as yet unknown.

The analysis described is based on a method used by Jacobi[2] to investigate dispersion in the n-body problem. The idea involved was first applied to elasticity by Lipschitz[3] and more fully by Caughey and Shield.[4] We follow the latter presentation here.

Let us consider the measure

$$F(t) = \int_{B(t)} \varrho\, u_i\, u_i\, dx, \tag{44.1}$$

where u_i is the incremental displacement satisfying the equations

$$(d_{ijkl}\, u_{k,l})_{,j} = \varrho \ddot{u}_i \quad \text{in } B\times(0, t_1), \tag{44.2}$$

with

$$d_{ijkl} = d_{klij}. \tag{44.3}$$

The boundary conditions are taken to be

$$u_i = 0 \quad \text{on } \overline{\partial B_1}\times(0, t_1), \tag{44.4}$$

$$n_j\, d_{ijkl}\, u_{k,l} = 0 \quad \text{on } \partial B_2\times(0, t_1), \tag{44.5}$$

so that we are again concerned with the problem of *dead loading* on part or all of the boundary.

When ∂B_1 is empty, so that we are dealing with traction boundary conditions, we shall assume, for simplicity, that, in the appropriate cases, the various rigid body instabilities have been excluded by the method described in Chap. K.

We assume first that the initial value of the total energy is negative. That is, in the notation of Eq. (35.8),

$$E(0) \leqq -k < 0, \tag{44.6}$$

---

[1] This is understood to exclude dissipation.

[2] Jacobi [1842, *1*].

[3] Lipschitz [1874, *1*].

[4] Caughey and Shield [1968, *4*]. This analysis is actually executed for a class of materials whose internal energy is a homogeneous polynomial of the strains of which linear elasticity is a special case. Very nearly the same approach was used by Duhem [1906, *2*] for non-linear elasticity under boundary conditions of displacement (see Sect. 32). In the same work, Duhem specialises his result to the linearised equations as a result of which he is able to consider boundary conditions of displacement and pressure.

where k is a positive constant. Condition (44.6) implies that the strain energy is negative-definite:

$$W(t) = \delta^2 V(t) = \tfrac{1}{2} \int_{B(t)} d_{ijkl} u_{i,j} u_{k,l} \, dx < 0. \tag{44.7}$$

We do not require the strain energy, or the total energy, to be negative-definite bounded above.

By differentiating (44.1) we get

$$\dot{F}(t) = 2 \int_{B(t)} \varrho u_i \dot{u}_i \, dx, \tag{44.8}$$

and

$$\ddot{F}(t) = 2 \int_{B(t)} \varrho (\dot{u}_i \dot{u}_i + u_i \ddot{u}_i) \, dx. \tag{44.9}$$

On substituting for the inertia term from (44.2), integrating by parts, and using the boundary conditions (44.4), (44.5), the last expression becomes

$$\begin{aligned} \ddot{F}(t) &= 2 \int_{B(t)} (\varrho \dot{u}_i \dot{u}_i - d_{ijkl} u_{i,j} u_{k,l}) \, dx \\ &= 4K(t) - 4W(t) \\ &= 8K(t) - 4E(0), \end{aligned} \tag{44.10}$$ [5]

where $K(t)$ is the kinetic energy and the conservation equation (35.8) has been used. A similar calculation is clearly valid for the weak solution defined in Sect. 23. We now have from (44.6), (44.10) and non-negativeness of $K(t)$, the inequality

$$\ddot{F}(t) \geqq 4k > 0, \tag{44.11}$$

which on integration yields

$$F(t) \geqq F(0) + \dot{F}(0)\, t + 2k\, t^2. \tag{44.12}$$

Hence, for sufficiently large values of t, $F(t)$ is bounded below by a quadratic function of t, from which we deduce that there is instability provided that condition (44.6), and thus (44.7), are satisfied.[6] This is, in fact, Lagrange instability, although Liapounov instability is implied.

The above method tells us nothing about instability when $E(0) > 0$. For $E(0) = 0$, however, we still obtain instability from (44.12) provided that the additional condition

$$\dot{F}(0) > 0 \tag{44.13}$$

is satisfied. Stronger results of this kind are derived in Chap. M where arguments based on logarithmic convexity are employed.

---

[5] When the strain energy is a homogeneous polynomial of order $h\,(>2)$ in the displacement and its first and second-order derivatives, the expression (44.10) becomes $\ddot{F} = (2+h)\,K - hE$ and the analysis may be concluded as described in the text. See CAUGHEY and SHIELD [1968, *4*].

[6] This conclusion was recorded by LIPSCHITZ [1874, *1*] and DUHEM [1906, *2*], both authors adopting a proof similar to that described in the text. CLEBSCH [1862, *1*], using a method of decomposing the displacement into sinusoidal parts, derived a similar result. MOVCHAN [1963, *16*], employing the Liapounov function $-E(0)\,\dot{F}(t)$, obtained this result from the Liapounov instability theorem.

45. Non-uniqueness and instability. Now let us suppose that the strain energy W does not satisfy a sign-definiteness condition, so that there exists a non-trivial vector $v_i(x)$ independent of t which satisfies the boundary conditions (44.4), (44.5) and for which the strain energy vanishes, that is,

$$\int_B d_{ijkl} v_{i,j} v_{k,l} \, dx = 0, \quad v_i \not\equiv 0, \quad \text{in } B. \tag{45.1}$$

From Eq. (45.1), by definition $v_i(x)$ is a weak solution of the *equilibrium* equations corresponding to (44.2) under the boundary conditions (44.4), (44.5). Under the appropriate smoothness conditions $v_i(x)$ could also become a classical solution satisfying these equations. We assume this to be the case.

Let us now introduce the functions $w_i(t, x)$ defined from the relation

$$w_i(t, x) = (a + b t) \, v_i,$$

where a and b are real-valued constants. The vector w_i then satisfies the equations of motion (44.2) and the boundary conditions (44.4), (44.5) so that it represents a possible solution to the elastodynamic problem. It is thus evident that, no matter how small a, b are chosen, w_i always becomes unbounded with time. We therefore conclude that the null solution is unstable with respect to all the usual measures, for example, the uniform and Sobolev norms.[7] The total energy remains constant, however, so that there is not instability with respect to this measure.

The above result may be summarised by the following (equivalent) theorems, if we bear in mind that the analysis is based on linearised equations and that the conclusions are therefore valid only in the special context of linear elasticity.[8] The instability of the null solution may well be altered to stability when the full non-linear equations are taken into account.

Theorem 45.1a. The non-uniqueness of the equilibrium solution of Eqs. (44.2) *with* (44.4), (44.5) *implies the instability of the null solution of the corresponding dynamic problem.*

This may be inverted to give

Theorem 45.1b. The stability of the null solution of the dynamic problem specified by Eqs. (44.2) *with* (44.4), (44.5) *implies the uniqueness of the solution to the equilibrium problem with the same boundary conditions.*[9]

This result is thus complementary to the more general result, proved earlier, that stability of a dynamical system implies its uniqueness. Under a definition[10] of stability different from that used here, the result of Theorem 45.1b has been derived by ERICKSEN and TOUPIN[11] and also by HILL.[12] Theorem 45.1a may be used to give some justification for the "adjacent equilibrium" criterion for stability. This criterion will be discussed next.

46. The method of adjacent equilibrium. In the previous section, we have seen that the primary equilibrium state of a finitely deformed elastic body is stable under dead loading in the class of linearised incremental displacements if the

[7] CAUGHEY and SHIELD [1968, *4*]. Elements of this result were obtained in a special case by SOUTHWELL and SKAN [1924, *2*].

[8] This naturally includes the theory of small deformations superposed on large.

[9] We leave as understood the precise stability measure.

[10] It corresponds to Hadamard superstability, i.e., when strict inequality holds in the respective definition of Hadamard stability. See Sect. 31, and also footnote 38 of Sect. 32.

[11] ERICKSEN and TOUPIN [1956, *1*].

[12] HILL [1957, *1*].

following inequality (or some other coercive inequality) holds

$$\delta^2 V \equiv \int_{B(t)} d_{ijkl}\, u_{i,j}\, u_{k,l}\, dx \geqq d_0 \int_{B(t)} u_{i,j}\, u_{i,j}\, dx, \tag{46.1}$$

where d_0 is some positive constant, V is the incremental strain energy function and $\delta^2 V$ is its second variation. The primary equilibrium position becomes unstable in the class of linearised incremental displacements if $\delta^2 V$ is either indefinite or negative-definite. Let us suppose that the loads acting in the primary state depend on a monotonically increasing parameter λ such that $\delta^2 V$ vanishes for $\lambda = \lambda'$, but is positive for $\lambda < \lambda'$. For reasons that will become apparent later, when the displacement assumes the non-trivial value which makes $\delta^2 V$ vanish at $\lambda = \lambda'$ the system is often said to be in an eigenmode or buckling mode. Its determination is equivalent to finding the positions of equilibrium of the body that lie in the neighbourhood of the primary equilibrium position, and, as we saw in Chap. I, the occurrence of this neutral equilibrium suggested that the primary position was unstable. From this the following technique has been made the basis of examining stability and is known as the *method of adjacent equilibrium*.

Assume that an equilibrium position of an elastic body is initially stable and let the load parameter be progressively increased. Under the varying load, the body is presumed to remain in stable equilibrium until the value of the parameter is reached at which $\delta^2 V$ is indefinite. The value of λ for which this occurs is said to be *critical* and corresponds to the critical or *buckling load*. This load, or the corresponding value of the load parameter, is said to give the *limit of stability*.

In applying this test, it is supposed that for values of λ in excess of λ' the primary state is unstable. The justification put forward for its use is that λ' represents the first value of the load parameter at which the strain energy ceases to have a minimum at the primary state and therefore, by the energy criterion, this state becomes unstable at $\lambda = \lambda'$.

We infer at once from the manner in which the adjacent equilibrium test is used that stability is judged according to the energy criterion. We know however that the conditions required by this criterion are not sufficient to ensure stability in the Liapounov sense. The latter requires for its sufficiency condition the relation $\delta^2 V > c\varrho$ to be satisfied (under the assumption of dead loads so that potential energy reduces to strain energy), and in the case of the L_2-norm (and certain other norms controlled by Poincaré or Sobolev inequalities) this is satisfied if (46.1) holds. This sufficiency requirement may cease to be satisfied, however, at some value of λ (say λ'') less than the critical value λ' although $\delta^2 V$ still remains positive in the interval $\lambda'' \leqq \lambda < \lambda'$. In this case stability cannot be guaranteed up to the critical value. If, further, $\delta^2 V \leqq 0$ for $\lambda > \lambda'$ then we know that the post-critical condition will be unstable in the class of linearised incremental displacements, whereas if $\delta^2 V > 0$ in this region the behaviour is not determined by the test so that instability may or may not set in. The argument just stated may be extended from the case of dead loading to general conservative loading by replacing V by overall incremental potential energy which includes strain energy and work from loads. Hence, the test can only definitely tell us that at the critical value there is instability in the class of the linearised incremental displacements.

There are various ways in which the criterion has been applied. In one method the equations of the linearised incremental motion are used and a separable solution examined. The transition from sinusoidal to non-sinusoidal form corresponding to waves of zero frequency gives the critical value of the load parameter at which the strain energy $\delta^2 V$ (potential energy) becomes indefinite.[13] Alternatively, we may regard the non-trivial linearised incremental displacement which makes $\delta^2 V$ vanish as the eigenfunction corresponding to a zero eigenvalue of the associated equations, and this has given rise to the name "eigenmode". This aspect of the method is evidently related to the well-known method of "exchange of stabilities" in fluid mechanics.[14] In another interpretation, the critical value is obtained directly from conditions producing neutral equilibrium and is related to the problem of determining non-unique equilibrium solutions to the equations of small elastic deformations superposed upon large.[15] This version of the test is also used in conjunction with buckling problems, since bifurcation of the linearised equilibrium equations occurs at the critical value of the load parameter, and the non-trivial solutions may be regarded as the buckling modes. However, at the moment it appears uncertain whether buckling, an essentially non-linear phenomenon, can be predicted in this fashion from the linearised equations. Often buckling is used synonymously with instability but apart from the simple relation noted in Sect. 45 for the linear theory,[16] the relation is complicated and is still the subject of investigation.

We may thus summarise by saying that the critical value obtained from the adjacent equilibrium test when applied in any one of its forms to a body in equilibrium under dead or conservative loads is not sufficient on its own for ascertaining the true limit of stability of this body in the sense that the body ceases to be stable and becomes unstable after this point. In general the conditions of the test need to be checked in individual cases or supplementary information acquired.

In the next section a brief history of the test is given together with references to its application.

47. History and application of the test. The method of adjacent equilibrium is one of the oldest of the criteria used in elastic stability, having been employed by EULER in his analysis of the strut and the tube.[17] Other early applications were also made to particular problems[18] and a comprehensive account is presented by TIMOSHENKO.[19] The first general application was made by SOUTHWELL[20] who, unlike BRYAN in his treatment of the energy criterion, did not assume that the primary displacement was small. Nevertheless, SOUTHWELL did make certain simplifications, assuming for instance that the primary state was a homogeneous deformation and that the material obeyed HOOKE's law.

BIEZENO and HENCKY[21] repeated SOUTHWELL's calculation, but did not require the primary state to be elastic which they therefore took as the reference configuration. However, HOOKE's law was still taken as the constitutive relation in the secondary deformation.

---

[13] Chap. E.

[14] For a description see, for example, DAVIS [1969, *5*].

[15] KNOPS and PAYNE [1971, *8*]. Contrary to a sometimes expressed belief, the equilibrium solution to the classical linear equations may fail to be unique.

[16] For the corresponding problem in a discrete theory see POINCARÉ [1885, *1*] and the recent survey by HUSEYIN [1970, *14*].

[17] EULER [1744, *1*], [1757, *1*]. See also LAGRANGE [1853, *1*, p. 143]. For other early work on tubes see the discussion by SOUTHWELL [1914, *2*].

[18] We may mention here the work of LÉVY [1883, *2*], [1884, *2*], GREENHILL [1881, *1*], [1883, *1*], and HALPHEN [1884, *1*].

[19] TIMOSHENKO [1953, *4*].

[20] SOUTHWELL [1914, *2*].

[21] BIEZENO and HENCKY [1928, *1*, *2*], [1929, *1*]. See also BIEZENO and GRAMMEL [1955, *1*], and the related study by KRALL [1966, *17*].

A presentation of the equations of neutral equilibrium in general curvilinear coordinates, devoid of any simplifying assumptions on either the initial or superposed state (beyond that of smallness) was given by NEUBER[22] and BUFLER.[23] The method, however, also involved neglecting second and higher-order terms in the superposed displacements. This approach also forms the basis of the treatment to be found in the books by PEARSON[24] and NOVOZHILOV.[25]

The intention in the approaches listed thus far is to determine by direct means a non-trivial solution to the equilibrium equations. A departure from this line of thinking was initiated by VON MISES,[26] who in examining the stability in a special case, used the condition that the potential energy associated with the superposed strain should vanish for certain of the incremental displacements. While this method is basically identical with that of finding a non-trivial solution to the equations, it has certain computational advantages. It was extended to the general case by TREFFTZ,[27] and in a variant of this method the condition for neutral equilibrium is given by the variational formulation for which $\delta W = 0$, where W is the second variation of the potential energy V. This idea may be found, for instance in the work of REISSNER,[28] MARGUERRE[29] and, in rather fuller detail, in that of KAPPUS.[30] It is also the basis of BIOT'S[31] method for examining stability. Accounts are also provided by PRAGER,[32] KERR,[33] PFLÜGER,[34] BURGERMEISTER, STEUP and KRETZCHMAR,[35] and SHEN.[36] See also LEIBENSON[37] and NOVOZHILOV.[38]

The conditions for neutral equilibrium may also be obtained from a consideration of possible wave motions, and correspond to waves possessing zero frequency. This approach, used by SOUTHWELL and SKAN,[39] for the special problem of the rectangular plate, has been considered in a general setting by KOITER,[40] PEARSON[41] and HILL.[42]

Although these various formulations all coexist in the literature there is a good deal of similarity in the actual method of solving problems by the adjacent equilibrium test. This usually consists in firstly representing the displacement as a formal infinite expansion. Relations between the coefficients are obtained as a result of applying one form of the test to ascertain the conditions on the initial stress under which the incremental displacement is a non-trivial solution of the equilibrium solution. This is the technique of GREENHILL,[43] BRYAN,[44] SOUTHWELL,[45]

---

[22] NEUBER [1943, *1*], [1965, *18*].
[23] BUFLER [1965, *5*].
[24] PEARSON [1959, *6*].
[25] NOVOZHILOV [1953, *3*].
[26] VON MISES [1923, *2*].
[27] TREFFTZ [1930, *2*], [1933, *1*].
[28] REISSNER [1925, *3*].
[29] MARGUERRE [1938, *2*, *3*].
[30] KAPPUS [1939, *1*].
[31] BIOT [1965, *3*].
[32] PRAGER [1947, *4*].
[33] KERR [1962, *8*].
[34] PFLÜGER [1964, *10*, pp. 67–69].
[35] BURGERMEISTER, STEUP and KRETZCHMAR [1966, *2*, pp. 192–194].
[36] SHEN [1968, *25*].
[37] LEIBENSON [1922, *1*].
[38] NOVOZHILOV [1953, *3*].
[39] SOUTHWELL and SKAN [1924, *2*].
[40] KOITER [1945, *1*].
[41] PEARSON [1956, *2*].
[42] HILL [1957, *1*].
[43] GREENHILL [1881, *1*], [1883, *1*].
[44] BRYAN [1891, *1*], [1894, *1*].
[45] SOUTHWELL [1914, *2*].

DEAN[46] and BIOT.[47] The books by TIMOSHENKO[48] supply numerous examples using this method.

In some of the earlier work (and in some more recent ones) various approximations are introduced into the derivations, in many cases with little attempt at justification. For some more recent applications see the list of works given below in which reference is also made to studies on post-buckling and "snap-through".[49]

Experimental work has long been known to give results at variance with the predicted values.[50] In the case of thin tubes under compression the actual collapse occurs at lower values than predicted, whilst flat plates are known to support loads in their plane exceeding the critical load when calculated by the linear theory.

M. Logarithmic convexity.

48. Introduction. The arguments of CAUGHEY and SHIELD described in an earlier section show that whenever the potential energy associated with the incremental displacement is negative-definite, or semi-definite, initial data may be chosen for which the incremental displacements become unbounded in time. More precisely, these displacements, measured by a weighted L_2-norm, $F(t)$, are bounded below by a time-dependent quadratic function for large values of t. The null solution is thus unstable under the conditions on the potential energy just stated. This result of CAUGHEY and SHIELD merely states, however, that there are initial data producing unbounded displacements, and gives no indication of how to characterise this data. Further, while the inequality employed is adequate for establishing instability, it does not always supply the best lower bound for the growth of $F(t)$ in time. We find an immediate illustration of this in the example for which the solution admits increasing exponential time factors. In this chapter we set out to describe fully the classes of initial data for which there is, and those for which there is not, subsequent growth of the associated incremental displacements. Moreover, we attempt to determine the best possible lower bound when the displacement is measured by $F(t)$.

The argument we employ is based on proving the convexity with respect to time of a positive-definite function closely related to $F(t)$. The approach is not new as it has already been used in establishing the Hadamard three circle theorem, various interpolation formulae, the uniqueness theorem of CARLEMAN and more recently certain results on the continuous dependence of a solution upon its data.[1]

---

[46] DEAN [1925, *1*], [1952, *2*].

[47] BIOT [1965, *3*].

[48] See TIMOSHENKO [1936, *1*], TIMOSHENKO and GERE [1961, *7*].

[49] WILKES [1955, *8*], GREEN and SPENCER [1959, *2*], WESOŁOWSKI [1962, *14*], [1963, *20*, *21*], [1964, *16*], FOSDICK and SHIELD [1963, *3*], LEVINSON [1968, *21*], BROMBERG [1970, *6*], KERR [1962, *8*], BUFLER [1965, *5*], NEUBER [1943, *1*], [1965, *18*], KOITER [1963, *8*], SEWELL [1969, *20*], [1970, *28*], HSU [1966, *12*, *13*], [1967, *16*], [1968, *12*, *13*], WU and WIDERA [1969, *25*], NOWINSKI [1969, *17*] and NOWINSKI and SHAHINPOOR [1969, *18*]. A general survey of buckling problems is supplied in the books by WOLMIR [1962, *15*] and STOKER [1968, *26*]. See also WOLKOWISKY [1967, *35*].

[50] See, for instance, ARBOCZ and BABCOCK [1969, *1*]. An early paper by VON KARMAN and TSIEN [1939, *2*] is of importance, while other contributions are discussed by WOLMIR [1962, *15*] and STOKER [1968, *26*].

[1] See, for example, LAVRENTIEV [1957, *2*], [1962, *10*], [1967, *21*], PUCCI [1955, *6*], PAYNE [1966, *28*], AGMON [1966, *1*], AGMON and NIRENBERG [1967, *2*], OGAWA [1965, *19*], [1967, *26*], [1970, *25*], LEVINE [1970, *21*, *22*]. In addition to the references cited in PAYNE [1966, *28*], the convexity argument is applied to non-elastic problems in continuum mechanics by, for example, OGAWA [1968, *23*], KNOPS and PAYNE [1968, *16*] and KNOPS and STEEL [1969, *14*].

As well as using the method to investigate instability we also employ it to obtain results concerning uniqueness and Hölder stability.

Since the results of this chapter and Chaps N, O hold for conservative loads, we use the term potential rather than strain energy. Other notation remains unaltered.

As before, we assume that the elasticities d_{ijkl} and mass density ϱ are bounded measurable functions for $\boldsymbol{x}\in B$, and that

$$\varrho \geqq \varrho_0 > 0, \tag{48.1}$$

where ϱ_0 is constant. We take the initial instant to be zero, and treat the equations

$$(d_{ijkl}u_{k,l})_{,j} = \varrho\,\ddot{u}_i \quad \text{on} \quad B\times(0,t_1), \tag{48.2}$$

subject to the homogeneous boundary conditions

$$u_i = 0 \quad \text{on } \overline{\partial B_1}\times[0,t_1), \tag{48.3}$$

$$n_j\,d_{ijkl}u_{k,l} = 0 \quad \text{on } \partial B_2\times[0,t_1), \tag{48.4}$$

so that in the primary state the part $\overline{\partial B_1}$ of the surface is fixed, while the complementary part ∂B_2 is subject to dead loads. Other standard types of boundary conditions may be similarly treated provided that energy is not supplied to the system under the new conditions. We may, for instance, take $\partial B_1=\emptyset$, and hence recover the traction problem considered earlier; we still exclude trivial instabilities about axes of equilibrium, but treat this topic again in Chap. N.

In the following it is essential that the time interval $[0,t_1)$ on which the equations are defined is at least half-open. This requirement is needed not only in the discussion of Hölder stability, but also for instability when t_1 will be allowed to tend to infinity.

We again consider only classical solutions, but it may be easily shown that our techniques apply equally well to weak solutions defined in the sense of Sect. 23.[2] A remarkable feature of the entire method is that proofs for uniqueness and Hölder stability do not require any definiteness assumptions of the elasticities d_{ijkl}.

49. Convexity of the function $F(t;\alpha,t_0)$. The logarithm of the function $H(t)$ will be proved convex, where

$$H(t) \equiv F(t;\alpha,t_0) = F(t) + \alpha(t+t_0)^2, \qquad 0\leqq t<t_1, \tag{49.1}$$

$$F(t) = \int_{B(t)} \varrho\,u_i u_i\,dx, \tag{49.2}$$

and α, t_0 are positive constants whose values will be determined later. We thus obtain the following properties as immediate consequences of the definition. All are required subsequently.

(i) $F(t;0,0)=F(t), \quad t\in[0,t_1),$ (49.3)

(ii) $F(t;\alpha,t_0)\geqq 0, \quad \forall\,\alpha,t_0, \quad t\in[0,t_1),$ (49.4)

(iii) $F(t;\alpha,t_0)=0 \Leftrightarrow u_i(t,\boldsymbol{x})=0, \quad t\in[0,t_1).$ (49.5)

In order to establish the convexity of $\ln F(t;\alpha,t_0)$, for convenience we write $H(t)$ for $F(t;\alpha,t_0)$ and show that $H(t)$ satisfies the following inequality

$$H\ddot{H} - \dot{H}^2 \geqq -2\big(2E(0)+\alpha\big)H \qquad 0\leqq t<t_1. \tag{49.6}$$

---

[2] See Knops and Payne [1971, *10*].

To prove this we first differentiate (49.1) to give

$$\dot{H}=2\int_{B(t)}\varrho u_i \dot{u}_i\,dx+2\alpha(t+t_0), \tag{49.7}$$

$$\ddot{H}=2\int_{B(t)}\varrho(\dot{u}_i\dot{u}_i+u_i\ddot{u}_i)\,dx+2\alpha. \tag{49.8}$$

We now use Eq. (48.2) to eliminate the inertia term in (49.8) and after integrating by parts obtain

$$\ddot{H}=8K(t)-8E(0)+2\alpha, \tag{49.9}$$

where $E(0)$ as before is the total energy, and $K(t)$ is the kinetic energy:

$$2K(t)=\int_{B(t)}\varrho\dot{u}_i\dot{u}_i\,dx. \tag{49.10}$$

The left side of (49.6) may now be evaluated to give

$$H\ddot{H}-\dot{H}^2=S^2(t;\alpha,t_0)-2[2E(0)+\alpha]H \qquad 0\leqq t<t_1 \tag{49.11}$$

in which the function $S^2(t;\alpha,t_0)$ is defined by

$$\begin{aligned}S^2(t;\alpha,t_0)=4\Big[\int_{B(t)}\varrho u_i u_i\,dx+\alpha(t+t_0)^2\Big]\Big[\int_{B(t)}\varrho\dot{u}_i\dot{u}_i\,dx+\alpha\Big]\\ -4\Big[\int_{B(t)}\varrho u_i\dot{u}_i\,dx+\alpha(t+t_0)\Big]^2.\end{aligned} \tag{49.12}$$

Now by SCHWARZ's inequality,

$$S^2(t;\alpha,t_0)\geqq 0,^3 \tag{49.13}$$

and the result (49.6) follows.

The remainder of the chapter is devoted to using inequality (49.6) in conjunction with a number of elementary arguments to obtain various estimates from which our conclusions follow. The discussion cannot be conducted simultaneously for all types of initial data, however, so we separate accordingly the situations in which the initial total energy is respectively negative, zero or positive. We also distinguish where necessary within each class the cases corresponding to different initial values of $H(t)$ and its derivative.

50. Applications.

(i) We begin by considering the homogeneous data

$$u_i(0,\boldsymbol{x})=\dot{u}_i(0,\boldsymbol{x})=0 \tag{50.1}$$

and prove that $u_i(t,\boldsymbol{x})$ remains identically zero for $t>0$. For this, let $\alpha=t_0=0$ so that $H(t)=F(t)$ and $H(0)=E(0)=0$. By continuity, it follows that either $F(t)\equiv 0$ for $0\leqq t<t_1$, or there exists an open interval $(t_2,t_3)\in[0,t_1]$ for which $F(t)>0$. In the latter case inequality (49.6) may be immediately rewritten as

$$\frac{d^2}{dt^2}[\ln F(t)]\geqq 0. \tag{50.2}$$

By means of expansions in finite Taylor series, or alternatively by JENSEN's inequality, we obtain from (50.2) the inequality

$$F(t)\leqq F(t_2)^{\frac{t_3-t}{t_3-t_2}}F(t_3)^{\frac{t-t_2}{t_2-t_3}} \qquad 0\leqq t_2<t<t_3\leqq t_1. \tag{50.3}$$

[3] Circumstances exist in which $S^2(t;\alpha,t_0)$ itself possesses a convex logarithm.

Now either $t_2=0$ or by continuity, $F(t_2)=0$. When $F(t_2)=0$, (50.3) gives at once $F(t)=0$ for $0\leqq t<t_3\leqq t_1$,[4] and continuity then reveals that $F(t_3)=0$. Thus, $F(t)=0$ for $0\leqq t\leqq t_1$. When $t_2=0$ a repetition of the argument leads to the same conclusions. Finally, property (iii) Sect. 49 implies that $u_i(t, \boldsymbol{x})=0$ for $t\in[0, t_1]$, and the assertion is proved. The conclusion is equivalent to establishing the uniqueness of the classical solution to standard problems in linear elastodynamics, and as such was given by KNOPS and PAYNE.[5] The same result had previously been given by BRUN,[6] who did not, however, employ convexity arguments.

Because Eq. (48.2) is invariant under a time reversal, a simple modification of the previous argument tells us that if $F(\hat{t})=0$ for $\hat{t}\in(0, t_1)$, then $F(t)\equiv 0$ for $t\in[0, t_1]$. Without loss, therefore, we may assume that $F(t)>0$ for $t\in[0, t_1]$, *provided* $E(0)\leqq 0$.

(ii) We next prove the Hölder stability of the null solution. To do this, we assume first that $E(0)\leqq 0$ and again set $\alpha=t_0=0$, so that $H(t)=F(t)$. By hypothesis $F(t)>0$, $t\in[0, t_1]$, and therefore (50.2), and thus (50.3) hold, the latter in the form

$$F(t)\leqq F(0)^{\frac{t_1-t}{t_1}}\,F(t_1)^{\frac{t}{t_1}}\qquad t\in(0, t_1)\,. \tag{50.4}$$

We say that incremental displacements are of class $\mathscr{M}$ if at time t_1 they are uniformly bounded in the sense that

$$F(t_1)<M\,, \tag{50.5}$$

for positive finite constant M. For such displacements, we find from (50.4) that

$$F(t)\leqq M^{1-\delta}[F(0)]^{\delta},\qquad \delta=1-(t/t_1)\,,\qquad 0\leqq\delta<1\,, \tag{50.6}$$

which by virtue of Definition 8.2 establishes that for incremental displacements of class $\mathscr{M}$ *the null solution is Hölder stable on compact sub-intervals of* $[0, t_1)$ *in the norm* F *for initial data with non-positive total energy.*

When the initial data produces a positive total energy we use a similar argument applied to the function[7]

$$F(t)+2E(0)\,.$$

The previous manipulations then result in the inequality

$$F(t)+2E(0)\leqq e^{\delta(1-\delta)}\,M_1^{1-\delta}\left(F(0)+2E(0)\right)^{\delta},\qquad \delta=1-(t/t_1)\,,\qquad 0\leqq\delta<1\,, \tag{50.7}$$

for incremental displacements satisfying the uniform boundedness condition (50.5) with M replaced by M_1. Here $M_1=M+2E(0)$, so that now the *null solution is Hölder stable on compact subintervals of* $[0, t_1]$ *in the norm* $F(t)+2E(0)$ *for initial data with positive total energy.*

An alternative way of writing inequality (50.4), needed below, is

$$F(t)\leqq F(0)\exp\left[\frac{t}{t_1}\ln\left(\frac{F(t_1)}{F(0)}\right)\right],\qquad t\in[0, t_1)\,, \tag{50.8}$$

---

[4] Detailed justification of this step is provided by LEVINE [1970, *22*]. See also AGMON [1966, *1*].

[5] KNOPS and PAYNE [1968, *18*]. They proved the result for weak solutions. LEVINE [1970, *21*, *22*] has extended the argument to abstract operators in a Hilbert space.

[6] BRUN [1965, *4*], [1969, *2*]. His methods admit the conditions of footnote 6, p. 254.

[7] KNOPS and PAYNE [1968, *19*]. This paper also explains why Theorem 10.1 on the equivalence between boundedness and Liapounov stability in linear systems does not hold for Hölder stability.

which demonstrates that *when* $E(0) \leqq 0$, $F(t)$ *cannot be greater than some time increasing exponential function on* $[0, t_1)$.

Inequalities (50.6) and (50.7) may be made the basis of a method for establishing local stability in the sense of Definitions 7.9 and 7.10 respectively. The conclusions obtained in this way are naturally valid only in the respective classes $\mathcal{M}$ or $\mathcal{M}_1$.

(iii) We now obtain further estimates for the growth of $F(t)$ on the semi-infinite time interval[8] and begin with the case of negative total energy

$$E(0) \leqq -k < 0, \tag{50.9}$$

where k is a positive constant. Because the kinetic energy is positive-definite and total energy is conserved, condition (50.9) implies that $W(t) < 0$ for $t \in [0, t_1]$, which is precisely the circumstance in which the incremental displacement can possess an exponentially increasing time factor. We shall in fact now show that for all initial data admitting (50.9) the incremental displacement as measured by $F(t)$ is bounded below by an exponentially increasing function of time when t is sufficiently large.

By hypothesis, we have

$$F(t; \alpha, t_0) > 0 \qquad t \in [0, t_1],$$

and on choosing $\alpha = 2k$, we find from (49.6) that

$$\frac{d^2}{dt^2} \ln F(t; \alpha, t_0) \geqq 0 \qquad t \in [0, t_1), \tag{50.10}$$

so that $\ln F(t; \alpha, t_0)$ is a convex function of $t \in [0, t_1)$. Expansion of this function in a finite Taylor series about $t = 0$ then gives

$$F(t; \alpha, t_0) \geqq F(0; \alpha, t_0) \exp\left[\left\{\frac{dF}{dt}(0; \alpha, t_0)/F(0; \alpha, t_0)\right\} t\right] \qquad t \in [0, t_1). \tag{50.11}$$

We observe that for sufficiently large values of t_0, we always have

$$\frac{dF}{dt}(0; \alpha, t_0) > 0, \tag{50.12}$$

irrespective of the initial value of the derivative of $F(t)$, and so, on letting $t_1 \to \infty$ we arrive at the following conclusion: *provided the initial data is such that* $u_i(0, \boldsymbol{x}) \neq 0$ *and the total energy is strictly negative* $F(t)$ *is bounded below for sufficiently large values of* t *by an exponentially increasing function of* t.

Whenever the derivative of $F(t)$ is initially positive we may put $\alpha = 0$ and obtain directly from (50.11) the lower bound:

$$F(t) \geqq F(0) \exp\left[\left(\dot{F}(0)/F(0)\right) t\right] \qquad t \geqq 0. \tag{50.13}$$

When the derivative of $F(t)$ is initially non-positive $F(t)$ decays to a (non-zero) minimum and thereafter grows in accordance with (50.11). Estimates may be derived for the minimum of $F(t)$ and also the time at which it is attained. Details will be found in the work of KNOPS and PAYNE,[9] where examples are presented showing that the estimates are the best possible.

---

[8] All the subsequent results are specialisations of those established by KNOPS and PAYNE [1971, *10*] for abstract operators in a Hilbert space. These authors use generalisations of the techniques described in the text.

[9] KNOPS and PAYNE [1971, *10*].

(iv) We next consider initial data for which the total energy is zero, that is

$$E(0)=0, \tag{50.14}$$

which implies that the initial value of the potential energy must be non-positive. In this situation we find from inequality (49.6) that, on again setting $\alpha=0$ we have $\ln F(t)$ a convex function of t. Hence inequality (50.8) may be applied. Let us suppose that

$$\lim_{t_1\to\infty}\left(\frac{1}{t_1}\ln F(t_1)\right)=0, \tag{50.15}$$

a condition which is implied, for instance, if, for sufficiently large t_1, we have

$$F(t_1)\leqq\gamma\exp\left(\beta t_1^{1-k}\right)t_1^N, \tag{50.16}$$

where γ and β are finite positive constants, N is an arbitrary integer and $k>0$. On letting $t_1\to\infty$, we then see from (50.8) that

$$F(t)\leqq F(0)\qquad t\geqq 0. \tag{50.17}$$

Further results can be obtained only by separating the various cases corresponding to the different initial values of the derivative of $F(t)$, and this we proceed to do.

When

$$\frac{dF}{dt}(0)>0, \tag{50.18}$$

the convexity of $\ln F$ leads as before to the estimate (50.13) and hence $F(t)$ is bounded below by an increasing exponential function of t, for $t>0$. Should (50.18) hold at, say, $t=t_0$, then the estimate (50.13) must be modified to

$$F(t)\geqq F(0)\exp\left[\left(\dot{F}(t_0)/F(t_0)\right)(t-t_0)\right]\qquad t\geqq t_0. \tag{50.19}$$

When

$$\frac{dF}{dt}(0)=0,$$

the estimate (50.13) still holds and in fact yields

$$F(t)\geqq F(0)\qquad t\geqq 0. \tag{50.20}$$

Thus, when $F(t)$ has the limiting behaviour (50.15) or (50.16), we gather from (50.17) and (50.20) that $F(t)=F(0)$, $t\geqq 0$. We may prove more. Since (50.20) holds it follows by continuity that t_0 exists such that either $F(t)=F(0)$ for $t\in[0,t_0]$ or

$$\frac{dF}{dt}(t_0)>0. \tag{50.21}$$

When (50.21) is valid we know that $F(t)$ satisfies (50.19) for $t\geqq t_0$ and so we may immediately conclude that *for initial data with zero total energy and zero initial derivative of $F(t)$ either*

(a) *$F(t)$ is bounded below by an increasing exponential function of t for sufficiently large t*, or

(b) *$F(t)$ remains equal to its initial value for all values of t.*

On presupposing that the second alternative holds, we find from Eq. (49.9) that

$$0\equiv\frac{d^2F}{dt^2}(t)=8K(t)\qquad t\geqq 0. \tag{50.22}$$

The positive-definitiness of the kinetic energy then shows that the velocity vanishes for $t \geqq 0$ and hence, in particular, that (b) necessarily implies

$$u_i(0, \boldsymbol{x}) = v_i(\boldsymbol{x}), \qquad \dot{u}_i(0, \boldsymbol{x}) = 0, \tag{50.23}$$

where $v_i(\boldsymbol{x})$ is a non-trivial solution to the corresponding homogeneous equilibrium problem, that is, the equilibrium problem with boundary conditions (48.3) and (48.4).[10] On setting

$$w_i(t, \boldsymbol{x}) = u_i(t, \boldsymbol{x}) - v_i(\boldsymbol{x}),$$

we may use the uniqueness theorem established earlier in this section to show that (50.23) is also sufficient for condition (b) to hold.[11] Hence, conditions (a) and (b) may be restated as follows:

For initial data with zero total energy and zero derivative of $F(t)$, the function $F(t)$ remains equal to its initial value if, and only if, $\dot{u}_i(0, \boldsymbol{x}) = 0$ and $u_i(0, \boldsymbol{x}) = v_i(\boldsymbol{x})$, where $v_i(\boldsymbol{x})$ is a solution to the equilibrium problem with homogeneous boundary conditions corresponding to (48.3) *and* (48.4). *Otherwise for sufficiently large t, $F(t)$ is bounded below by an increasing exponential function of t.*

(v) We turn our attention to initial data for which $E(0) = 0$ and the initial value of the derivative of $F(t)$ is negative:

$$\frac{dF}{dt}(0) < 0. \tag{50.24}$$

By continuity there exists a finite value t_2 of t such that the derivative of $F(t)$ is negative for $t \in [0, t_2)$ but is such that at $t = t_2$ this derivative is either zero or negative. We consider first the possibility that $\dot{F}(t_2) = 0$. According to the previous result $F(t)$ either increases or remains equal to its value at t_2 for $t \geqq t_2$. Necessary and sufficient conditions for $F(t)$ to remain uniform are that a solution $v_i(\boldsymbol{x})$ exists to the equilibrium problem and moreover that at $t = t_2$ we have

$$u_i(t_2, \boldsymbol{x}) = v_i(\boldsymbol{x}), \qquad \dot{u}_i(t_2, \boldsymbol{x}) = 0. \tag{50.25}$$

The uniqueness theorem derived above may be applied backwards in time to show that conditions (50.25) imply that $\dot{u}_i(t, \boldsymbol{x}) = 0$ for $t \in [0, t_2]$, which in particular at $t = 0$ contradicts (50.24). Hence, we conclude that when $\dot{F}(t)$ vanishes at a finite value of t, $F(t)$ increases thereafter.

We next examine the consequences of the condition $\dot{F}(t_2) < 0$. This alternative is equivalent to assuming that $F(t)$ possesses a negative derivative for all finite values of t, which implies that

$$F(t) < F(0) \qquad t > 0. \tag{50.26}$$

On the other hand, we are still supposing that $E(0) = 0$, so that $\ln F(t)$ is convex and consequently inequality (50.13) continues to hold. The exponential index is now, however, negative and hence we infer from (50.13) that $F(t)$ can never decay faster than a certain exponential function, and so cannot vanish for finite values of t. We may hence conclude that *for initial data with zero total energy and negative derivative of $F(t)$ at $t = 0$, either*

(a) *$F(t)$ is bounded below by an increasing exponential function of t for sufficiently large values of time* or

---

[10] We observe that $v_i(\boldsymbol{x})$ must be non-trivial, since otherwise $u_i(0, \boldsymbol{x}) = 0$ contrary to the hypothesis that $F(0) > 0$.

[11] KNOPS and PAYNE [1971, *10*].

(b) *$F(t)$ decays for all (finite) values of time no faster than a certain decaying exponential function and may or may not tend to zero as t tends to infinity.*

It may be shown by means of examples that, without further restrictions on the initial data, both alternatives are possible.[12] Below we give further consideration to this ambiguity but meanwhile we treat the problem in which initial data gives rise to positive total energy.

(vi) We recall that when the total energy is positive the previous uniqueness theorem fails. The potential energy, however, may be positive, negative or zero and included in these cases are those for which it is sign-definite — either positive or negative. Although precise statements concerning the behaviour of $F(t)$ will be made for the sign-definite potential energies, it is more convenient to defer consideration of these until after the general discussion in which types of potential energy are not distinguished. In this analysis it is natural to expect that we can only make general statements, and to recover the precision of earlier results, extra assumptions on the initial data must be introduced.

Then with the single supposition that the total energy is initially positive we may proceed as follows. The fundamental inequality (49.6) remains valid, provided that α and t_0 are not zero, and it may be rewritten:

$$\frac{d^2}{dt^2}\left[\ln\frac{F(t;\alpha,t_0)}{(t+t_0)^{2+\varepsilon}}\right]\geqq 0,\qquad \varepsilon=\frac{4E(0)}{\alpha},\qquad t\in[0,t_1). \tag{50.27}$$

Thus, $(t+t_0)^{-2-\varepsilon}F(t;\alpha,t_0)$ has its logarithm convex with respect to time so that some of the previous arguments may be applied to this function. In particular under the condition

$$\lim_{t_1\to\infty}\left(\frac{1}{t_1}\ln F(t_1;\alpha,t_0)\right)=0 \tag{50.28}$$

it may be shown that *either*:

(a) *for sufficiently large t, $F(t)$ is bounded above by $(t+t_0)^{2+\varepsilon}$ for any $\varepsilon>0$ and $t_0>0$* or

(b) *the asymptotic behaviour indicated by* (50.28) *is not satisfied.*

Without further assumptions, not much further progress can be made and so we now require the initial data to be such that $\dot{F}(0)>0$. On taking $\alpha=t_0=0$ in inequality (49.6) it may be shown that the derivative of $F(t)$ is positive on the whole time interval so that (49.6) may be directly integrated. In this way, we find that, under the condition

$$\frac{dF}{dt}(0)>2[2F(0)\,E(0)]^{\frac{1}{2}}, \tag{50.29}$$

the following estimate holds for $t\geqq 0$:

$$F(t)\geqq[F(0)+4E(0)A^{-2}]\cosh At+[\dot{F}(0)A^{-1}]\sinh At-4E(0)A^{-2}, \tag{50.30}$$

where

$$A^2=[\dot{F}(0)/F(0)]^2-8E(0)\,F^{-1}(0). \tag{50.31}$$

We remark that (50.29) restricts the initial potential energy so that it is strictly negative and hence (50.29) cannot be satisfied by a positive-definite potential energy. This case is, however, mentioned below.

[12] Knops and Payne [1971, *10*].

When

$$\frac{dF}{dt}(0) = 2[2F(0)\,E(0)]^{\frac{1}{2}}, \tag{50.32}$$

we have, for $t \geqq 0$

$$F(t) \geqq F(0) + 2[2F(0)\,E(0)]^{\frac{1}{2}}\,t + 2E(0)\,t^2. \tag{50.33}$$

We note that (50.32) can be satisfied only for a potential energy with non-negative initial value. Inequality (50.33) is similar to that obtained by Caughey and Shield (see Sect. 44), but improves upon the latter when $E(0) > 0$. The reverse is true when $E(0) < 0$, but under this condition the prior estimates of this section operate and supply better bounds.

Particular examples may be constructed which show that the estimates (50.30) and (50.33) are the best possible.

Before continuing the argument based upon convexity we briefly mention the conclusions of another based upon the reciprocal theorem for linear elastodynamics. Details may be found in Knops and Payne,[13] where it is shown that *provided*

$$F(t) = o(t^2) \quad \text{as} \quad t \to \infty \tag{50.34}$$

and $\dot{u}_i(0, \boldsymbol{x}) = 0$ *then* $F(t) \leqq F(0)$, $t \geqq 0$, *without any restriction on the initial value of the total energy.*

(vii) So far in this discussion, we have produced estimates which are best possible when the total energy has negative initial value. When the total energy is initially zero or positive, the estimates are best possible when the derivative of $F(t)$ has an initial value which is, in this case, positive and satisfies either condition (50.29) or (50.32). Of the remaining cases covered in this section, some were left with indeterminacies which we now endeavour to remove. A number of different means are employed but in spite of this, completely determinate results cannot be found in all cases. In every subsequent computation, we assume that $u_i(t, \boldsymbol{x})$ possesses the necessary smoothness. Full details of the manipulations involved may be found in the above cited reference.

The first method we examine arises from the observation that Eq. (48.2) and boundary conditions (48.3), (48.4) still retain the same form after differentiation with respect to time to any order.

All the previous arguments thus apply to the *n-th order kinetic energy* $K_n(t)$, where

$$K_n(t) = \frac{1}{2}\int_{B(t)} \varrho \overset{(n+1)}{u_i} \overset{(n+1)}{u_i}\,dx, \qquad \overset{(n)}{u_i} = \frac{\partial^n u_i}{\partial t^n} \qquad n \geqq 0 \tag{50.35}$$

with the role of the total energy now be taken by

$$E_{n+1}(t) = \tfrac{1}{2}\int_{B(t)} (\varrho \overset{(n+2)}{u_i} \overset{(n+2)}{u_i} + d_{ijkl} \overset{(n+1)}{u_{i,j}} \overset{(n+1)}{u_{k,l}})\,dx \qquad n \geqq 0,$$

which is the $(n+1)$*-th order total energy*. Estimates on the behaviour of $K_n(t)$ may then be used to provide information about $F(t)$ on noting that, for $n \geqq 1$,

$$\frac{d^{2n+2}F}{dt^{2n+2}}(t) = 2^{2n+3} K_n(t) - 2^{2n+2} E_n(0), \tag{50.36}$$

obtained by repeated differentiation of (48.2) and use of the conservation laws

$$E_n(t) = E_n(0). \tag{50.37}$$

[13] Knops and Payne [1971, *10*].

Substitution of the estimates for $K_n(t)$ into (50.36) followed by an integration leads to the desired results.

Next we consider sign-definite potential energies. The case of positive-definiteness has already been considered, forming the content of Chaps. J and K on Liapounov stability. For a negative-definite potential energy, we suppose that $W(t)$ satisfies the relation

$$-W(t) = -\tfrac{1}{2}\int_{B(t)} d_{ijkl}\, u_{i,j}\, u_{k,l}\, dx > c_0 \int_{B(t)} u_i\, u_i\, dx \qquad t>0, \tag{50.38}$$

for positive constant c_0, and introduce the function

$$G(t;\alpha,t_0) = -2\,W(t) + \alpha(t+t_0)^2 \qquad t\in[0,t_1). \tag{50.39}$$

Differentiation of the function G and use of Eq. (48.2) together with conditions (48.3), (48.4) enables us to reach the equation

$$G\ddot{G} - \dot{G}^2 = 4R^2(t) + 2G[2E_1(0) - \alpha] \tag{50.40}$$

where G is written for $G(t;\alpha,t_0)$ and

$$\begin{aligned} R^2(t) = {} & [-2W(t) + \alpha(t+t_0)^2]\,[-\int_{B(t)} d_{ijkl}\,\dot{u}_{i,j}\dot{u}_{k,l}\,dx + \alpha] \\ & -[-\int_{B(t)} d_{ijkl}\, u_{i,j}\,\dot{u}_{k,l}\,dx + \alpha(t+t_0)]^2 . \end{aligned} \tag{50.41}$$

We find from SCHWARZ's inequality that $R^2(t) \geqq 0$, and so (50.40) may then be reduced to an inequality analogous to (49.6). Hence the convexity arguments described in the first part of this section may be applied to the function G.

Various conclusions may now be obtained in a manner similar to that of the previous treatment. On combining appropriate results for the potential energy with those for the kinetic energy we may conclude, for instance, *that when the potential energy is negative-definite and the initial data is such that* $E_1(0) \neq 0$, $F(t)$ *must grow exponentially for sufficiently large values of time.*

Clearly, higher order negative-definite potential energies may be similarly treated to produce results supplementing those for the higher order kinetic energies.

(viii) Finally, in the discussion of methods which remove the indeterminacies noted above, we return to the function $S^2(t)$, defined in Eq. (49.12) and which we know, by SCHWARZ's inequality, is non-negative. We now prove that $\ln S^2(t)$ is a convex function of t under suitable conditions. It easily follows, in fact, by differentiation, use of (48.2), (48.3), (48.4) and the conservation laws (50.37) that

$$\begin{aligned} S^2(t)\,\frac{d^2S^2(t)}{dt^2} - \left[\frac{dS^2(t)}{dt}\right]^2 = {} & Q^2(t) - 16\,S^2(t)\,[F(t)\,E_1(0) + 2K(t)\,E(0) \\ & -2E^2(0)] \qquad t\in[0,t_1), \end{aligned} \tag{50.42}$$

where

$$\begin{aligned} Q^2(t) = {} & 64\Big[\int_{B(t)} \varrho u_i\, u_i\, dx \int_{B(t)} \varrho\dot{u}_i\,\dot{u}_i\, dx - \Big(\int_{B(t)} \varrho u_i \dot{u}_i\Big)^2\Big]\Big[\int_{B(t)} \varrho u_i\, u_i\, dx \int_{B(t)} \varrho\ddot{u}_i\,\ddot{u}_i\, dx \\ & -\Big(\int_{B(t)} \varrho u_i\,\ddot{u}_i\, dx\Big)^2\Big] \\ & -64\Big[\int_{B(t)} \varrho u_i\, u_i\, dx \int_{B(t)} \varrho\dot{u}_i\,\ddot{u}_i\, dx - \int_{B(t)} \varrho u_i\,\dot{u}_i\, dx \int_{B(t)} \varrho u_i\,\ddot{u}_i\, dx\Big]^2 . \end{aligned} \tag{50.43}$$

By SCHWARZ's inequality, $Q^2(t) \geqq 0$, and so (50.42) becomes

$$S^2(t)\frac{d^2 S^2(t)}{dt^2} - \left[\frac{d\,S^2(t)}{dt}\right]^2 \geqq -16\,S^2(t)\,[F(t)\,E_1(0) + 2K(t)\,E(0) - 2E^2(0)]. \quad (50.44)$$

We suppose the initial data is such as to produce one or other of the conditions

(a) $E(0)=0, \quad E_1(0)=0,$ (50.45)

(b) $E(0)>0, \quad E_1(0)=0, \quad \text{with} \quad W(t)\geqq 0,$ (50.46)

these being some of the circumstances under which ambiguity in the behaviour of $F(t)$ has not so far been eliminated. In (b) we note that the potential energy is allowed to be positive semi-definite and not merely definite as hitherto sometimes supposed. Either condition (a) or (b) taken with (50.44) establishes the convexity of $\ln S^2(t)$ and we now describe briefly some consequences of this property.

The previous argument for uniqueness, applied in the present context, shows that when $S^2(t)$ vanishes initially it vanishes for all time $t \geqq 0$. This conclusion inserted into (49.11) with $\alpha=0$ leads to

$$\frac{d}{dt}[\dot F(t)/F(t)] = -\frac{4E(0)}{F(t)} \quad (50.47)$$

which may now be integrated. In case (a) we then obtain

$$F(t) = F(0)\exp[\dot F(0)/F(0)]\,t, \quad (50.48)$$

and because the condition $S^2(0)=0$ is satisfied if, and only if,

$$\dot u_i(0,\boldsymbol{x}) = \beta u_i(0,\boldsymbol{x}), \quad (50.49)$$

for some arbitrary (real) constant β, we see that (50.48) demonstrates that $F(t)$ has exact exponential growth or decay depending upon whether β is positive or negative. A more refined calculation would tell us that the last property is shared also by the incremental displacement u_i. We have, in fact,

$$u_i(t,\boldsymbol{x}) = u_i(0,\boldsymbol{x})\,e^{\beta t}. \quad (50.50)$$

We next consider case (b), when we have $W(t)\geqq 0$. Firstly then suppose that $W(0)=0$, so that we may integrate (50.47) to obtain

$$F(t) = F(0) + 2[2F(0)\,E(0)]^{\frac{1}{2}}\,t + 2E(0)\,t^2. \quad (50.51)$$

Secondly, let $W(0)>0$. Integration of (50.47) then gives

$$F(t) = [F(0) - 4E(0)\,D^{-2}]\cos Dt + (\dot F(0)\,D^{-1})\sin Dt + 4E(0)\,D^{-2}, \quad (50.52)$$

where

$$D^2 = 8\,W(0)\,F^{-1}(0). \quad (50.53)$$

Other conditions, different from those described in either case (a) or (b), may also be considered; for instance, initial data may be taken for which $E(0)=0$ and $E_1(0)>0$. Or again, by suitable restriction of the higher order total energies, it may be shown that the function $Q^2(t)$ and other higher order ones analogous to $S^2(t)$ have convex logarithms. Moreover, the kinetic energies of all orders together with the negative-definite potential energies of all orders may be treated in a manner similar to their first order counterparts. It is thus obvious that, by substitution, we may determine the corresponding behaviour of $F(t)$.

The conclusions obtained in this section are capable of various interpretations. Firstly, they indicate strong dependence of growth properties on the initial data and provide a comprehensive description of initial data for which the null solution is either stable or unstable. In this sense, therefore, we find that together with the results of Chaps. J and K on stability in the displacement, mixed and traction initial boundary value problems, we have virtually completed the programme of investigation which we stated in Sect. 6 as being the desirable end to be accomplished in any stability analysis.

Secondly, they reveal conditions under which certain types of solution cannot exist. For example, when $E(0) < 0$ no solution can exist subject to the requirement of polynomial growth at infinity. Similar inferences may be made from the other conclusions, and this aspect of the subject is analogous to the Phragmén-Lindelöf principle in the theory of elliptic equations.

Thirdly, it is evident from our results that whenever a non-trivial solution to the equilibrium problem with homogeneous boundary data exists, there is Lagrange instability at the null solution. However, when the equilibrium problem possesses only the null solution, no definite observation may be made about the stability of the null solution to the dynamic problem. The reason for this lies in the fact that uniqueness of the equilibrium solution is secured by a sign-definite potential energy. For positive-definiteness we have demonstrated stability, while for negative-definiteness we have shown instability.

N. Extension of stability analysis for traction boundary conditions.

51. Stability without an axis of equilibrium. Separate sufficient and necessary conditions have been established in earlier sections for the stability of an equilibrium configuration of an elastic body in the class of linear incremental displacements under dead loading. The cases of a body with prescribed mixed boundary data and that of a body under traction surface loading are discussed in Chaps. J and K respectively. In this chapter we look further into the restrictions that must be imposed on rigid body rotations and primary stress to ensure stability in the problem where dead loads are prescribed on the whole surface. When the primary stress is not zero (the special case dealt with in Sect. 42) complications arise not only from the destabilising influence of the rigid body rotations but also from their interaction with the primary stress. To indicate how to eradicate some of these interacting rotations is part of the programme of this chapter.[1]

When the displacement field $v_i(t, \boldsymbol{x})$ *excludes rigid body translations* it was shown in Sect. 42 that a condition sufficient for stability of the body under the boundary conditions already specified is

$$\int_{B(t)} d_{ijkl}\, v_{i,j}\, v_{k,l}\, dx \geqq c_0 \int_{B(t)} v_{i,j}\, v_{i,j}\, dx, \tag{51.1}$$

where c_0 is some positive constant. This result may also be expressed in terms of the linearised strains $\tilde{e}_{ij}$.

In order to exclude further the rigid body rotations we proceed as in Chap. K to express $v_i(t, \boldsymbol{x})$ in terms of $w_i(t, \boldsymbol{x})$ by the relation

$$w_i(t, \boldsymbol{x}) = v_i(t, \boldsymbol{x}) + e_{ijk}\, d_j(t)\, x_k, \tag{51.2}$$

where w_i satisfies the equations

$$\int_{B(t)} w_i\, dx = \int_{B(t)} e_{ijk}\, w_j\, x_k\, dx = \int_{B(t)} e_{ijk}\, w_{j,k}\, dx = 0. \tag{51.3}$$

[1] Some basic ideas used in this chapter are due to Holden [1964, *6*] and Hill [1967, *15*] who, however, deal with the related problem of the energy criterion for stability.

From (51.2), (51.3) we then have

$$d_i(t) = \frac{1}{2B} \int\limits_{B(t)} e_{ijk}\, v_{j,k}\, dx, \tag{51.4}$$

where B is the volume of the body in the primary configuration. The vector d_i is thus proportional to the rotational part of v_i and hence may be varied independently of w_i.[2] We now insert the expression for v_i from Eq. (51.2) into inequality (51.1). The left side of this inequality is the incremental potential energy of the body under the assumed dead loads and after substitution the form of the resulting expression indicates its explicit dependence on the initial stress σ_{ij} and on rigid body rotations[3]

$$\begin{aligned} W = \int\limits_{B(t)} (c_{ijkl}\, w_{i,j}\, w_{k,l} + \sigma_{ij}\, w_{k,i}\, w_{k,j} - 2 e_{kni}\, \sigma_{ij}\, w_{k,j}\, d_n)\, dx \\ + (\bar{\sigma}_{kk}\, \delta_{ij} - \bar{\sigma}_{ij})\, d_i\, d_j \geq c_0 \int\limits_{B(t)} w_{i,j}\, w_{i,j}\, dx + 2c_0\, B\, d_i\, d_i. \end{aligned} \tag{51.5}$$

The mean stress $\bar{\sigma}_{ij}$ is given by

$$\bar{\sigma}_{ij} = \int\limits_{B} \sigma_{ij}\, dx. \tag{51.6}$$

The energy W may also be usefully evaluated from displacements measured with respect to the principal axes of mean stress, in which case it becomes

$$\begin{aligned} W = \int\limits_{B(t)} (c_{ijkl}\, w_{i,j}\, w_{k,l} + \sigma_{ij}\, w_{k,i}\, w_{k,j} - 2 e_{kni}\, \sigma_{ij}\, w_{k,j}\, d_n)\, dx \\ + (\bar{\sigma}_2 + \bar{\sigma}_3)\, d_1^2 + (\bar{\sigma}_3 + \bar{\sigma}_1)\, d_2^2 + (\bar{\sigma}_1 + \bar{\sigma}_2)\, d_3^2, \end{aligned} \tag{51.7}$$

where $\bar{\sigma}_1, \bar{\sigma}_2, \bar{\sigma}_3$ are the principal mean stresses.

We now consider the conditions on the rotations and initial stress in order that *stability*, determined by (51.1) should be maintained.

(i) *Case with* $\mathbf{d} = 0$. This excludes rigid body rotations from the motion completely and inequality (51.5) is satisfied provided that the following holds

$$\int\limits_{B(t)} (c_{ijkl} + \sigma_{jl}\, \delta_{ik})\, w_{i,j}\, w_{k,l}\, dx \geq K_1 \int\limits_{B(t)} w_{i,j}\, w_{i,j}\, dx \tag{51.8}$$

for positive constant K_1. This is therefore equivalent to assuming the existence of a Korn's constant.

(ii) *Case for which* $\mathbf{d}$ *satisfies the conditions*

$$d_n \int\limits_{B(t)} e_{kni}\, \sigma_{ij}\, w_{k,j}\, dx = 0, \tag{51.9}$$

and

$$\bar{\sigma}_1 + \bar{\sigma}_2 > kB, \quad \bar{\sigma}_2 + \bar{\sigma}_3 > kB, \quad \bar{\sigma}_3 + \bar{\sigma}_1 > kB, \tag{51.10}$$

for some positive constant k.

It is seen by analogy with Eq. (40.10) that the integral in (51.9) is the resultant moment M_i of the loads acting in the primary state taken over the incremental displacement w_i. Condition (51.9) is thus satisfied whenever w_i preserves rotational balance, or when the axis of rigid body rotation is perpendicular to M_i. On transforming to the principal axes of mean stress, condition (51.5) is satisfied provided that conditions (51.9) together with (51.8) and (51.10) are satisfied.

[2] On setting $R_{ij} = -e_{ipj}\, d_p$ we obtain the quantity introduced by Holden [1964, *6*, Eq. (3.8)].

[3] In Sect. 46 we denoted W by $\delta^2 V$.

Under these conditions the constant c_0 of inequality (51.5) assumes the value

$$c_0 = \min\,(k/2,\, K_1)\,.$$

(iii) *Case of restriction of primary stress only.* The assumptions in this case are (a) the existence of a KORN'S constant K_2 such that the following inequality is satisfied

$$\int\limits_{B(t)} c_{ijkl}\, w_{i,j}\, w_{k,l}\, dx \geqq K_2 \int\limits_{B(t)} w_{i,j}\, w_{i,j}\, dx\,, \tag{51.11}$$

with the stipulation

$$\left[K_2 + \min_B(\sigma_{ij}) - \max_B(\sigma_{ij}\,\sigma_{ij})\right] \geqq c > 0\,. \tag{51.12}$$

(b) the conditions (51.10) *with* $k > 1$.

From the third term in the expression (51.5), on applying the arithmetic geometric mean inequality followed by the Lagrange inequality

$$e_{ijk}\, e_{irs}\, a_j\, a_r\, b_k\, b_s \leqq a_i\, a_i\, b_j\, b_j\,, \tag{51.13}$$

for arbitrary vectors a_i, b_i, we obtain the result

$$2\int\limits_{B(t)} e_{kni}\, \sigma_{ij}\, w_{k,j}\, d_n\, dx \leqq \max_B(\sigma_{ij}\,\sigma_{ij}) \int\limits_{B(t)} w_{i,j}\, w_{i,j}\, dx + B\, d_i\, d_i\,. \tag{51.14}$$

Condition (51.5) is thus satisfied provided that, referred to the principal axes of mean stress, (51.10), (51.11) and (51.12) hold. The constant c_0 in (51.5) becomes in this case

$$c_0 = \min\left(c,\, (k-1)/2\right). \tag{51.15}$$

In a similar manner, let us assume that conditions (51.10) hold with $k > 1$, KORN'S inequality is satisfied in the form of Eq. (43.1), with the linear strain $\tilde{e}_{ij}$ introduced into its right side, and that the following inequalities hold:

$$\begin{aligned} \int\limits_{B(t)} \left\{c_{ijkl}\, \tilde{e}_{ij}\, \tilde{e}_{kl} + \frac{K}{4}\left[\min_B(\sigma_{ij}) - \max_B(\sigma_{ij}\,\sigma_{ij})\right] \tilde{e}_{ij}\, \tilde{e}_{ij}\right\} dx \\ > c_1 \int\limits_{B(t)} \tilde{e}_{ij}\, \tilde{e}_{ij}\, dx\,, \end{aligned} \tag{51.16}$$

where c_1 is some positive constant, and

$$\min_B(\sigma_{ij}) - \max_B(\sigma_{ij}\,\sigma_{ij}) \leqq 0\,. \tag{51.17}$$

It then follows that inequality (51.16) is sufficient for stability with respect to measures $E(0)$ and the L_2-norm of the linearised incremental strains.[4]

[4] Related results are obtained by HOLDEN [1964, *6*] who, while placing no sign restriction on the primary stress, requires overall rotational balance to be maintained. Thus, (40.10) holds and HOLDEN writes

$$0 = \int\limits_B e_{knj}\, \sigma_{ij}\, v_{k,i}\, dx = \int\limits_B e_{knj}\, \sigma_{ij}\, w_{k,i}\, dx - \int\limits_B e_{knj}\, \sigma_{ij}\, e_{kpi}\, d_p\, dx\,,$$

so that

$$d_n \int\limits_B e_{knj}\, \sigma_{ij}\, w_{k,i}\, dx = \int\limits_B e_{knj}\, e_{kpi}\, \sigma_{ij}\, d_p\, d_n\, dx\,,$$

in our notation. The potential energy, by (51.5), becomes

$$\int\limits_B (c_{ijkl}\, w_{i,j}\, w_{k,l} + \sigma_{ij}\, w_{k,i}\, w_{k,j} - e_{kni}\, \sigma_{ij}\, w_{k,j}\, d_n)\, dx$$

and the discussion may be conducted as before to obtain stability with respect to $E(0)$ and the L_2-integral of e_{ij}. HOLDEN, being concerned with Hadamard stability, does not obtain these final estimates.

52. Stability with an axis of equilibrium. Now let us assume that an axis of equilibrium exists such that

$$\bar{\sigma}_2+\bar{\sigma}_3=0. \tag{52.1}$$

Then on using Eqs. (40.10) and (51.2), we obtain the relation

$$\int_{B(t)} e_{k1i}\,\sigma_{ij}\,w_{k,j}\,dx=0, \tag{52.2}$$

with d_1 arbitrary. Hence it follows from expression (51.7) that condition (51.1) can never be satisfied. This trivial instability resulting from the presence of d_1 may be removed, however, by setting to zero the initial spin about the equilibrium axis. On the other hand, the value of the strain energy (51.7) is independent of d_1 as long as rotational balance is maintained, so that the previous arguments show that stability exists for the classes of measures $E(0)$ and

$$\int_{B(t)} w_i\,w_i\,dx+d_2^2+d_3^2.$$

If rotational balance about the equilibrium axis is not maintained then Eq. (52.2) is no longer valid and from expression (51.7) the initial incremental strain energy may be made non-positive for appropriate choice of d_1. Instability therefore results.[5]

53. Instability analysis. In Chap. L it has been seen that a sufficient condition for instability of the null solutions is

$$\int_{B(t)} d_{ijkl}\,v_{i,j}\,v_{k,l}\,dx\leqq 0, \qquad t\geqq 0, \tag{53.1}$$

where the displacement v_i has been chosen in order to dispense with trivial translational instabilities. Inequality need only occur in (53.1) at one instant t_1 for instability to be produced for all $t>t_1$.

We now introduce the vector b_m defined by

$$b_m(\bar{\sigma}_{kk}\,\delta_{mn}-\bar{\sigma}_{nm})=-\int_{B(t)} e_{kni}\,\sigma_{ij}\,w_{k,j}\,dx \tag{53.2}$$

where the integral on the right represents the moment of external forces on B taken over incremental displacement. It is assumed that $(\bar{\sigma}_1+\bar{\sigma}_2)(\bar{\sigma}_2+\bar{\sigma}_3)(\bar{\sigma}_3+\bar{\sigma}_1)\neq 0$, so that b_i exists. When this assumption does not hold, the components b_i separately exist only when rotational balance is maintained about the appropriate axes of equilibrium. On using the Eq. (53.2) and the value of the incremental potential energy (51.7), we obtain, with components along principal axes of mean stress,

$$\begin{aligned} W=\int_{B(t)} &(c_{ijkl}\,w_{i,j}\,w_{k,l}+\sigma_{ij}\,w_{k,j}\,w_{k,i})\,dx+(\bar{\sigma}_2+\bar{\sigma}_3)(d_1+b_1)^2\\ &+(\bar{\sigma}_3+\bar{\sigma}_1)(d_2+b_2)^2+(\bar{\sigma}_1+\bar{\sigma}_2)(d_3+b_3)^2-(\bar{\sigma}_2+\bar{\sigma}_3)\,b_1^2\\ &-(\bar{\sigma}_3+\bar{\sigma}_1)\,b_2^2-(\bar{\sigma}_1+\bar{\sigma}_2)\,b_3^2. \end{aligned} \tag{53.3}$$

We then have *instability* when

$$\bar{\sigma}_2+\bar{\sigma}_3<0, \tag{53.4}$$

since d_i is always independent of w_i and hence, for given σ_{ij} and w_i, d_1 can always be chosen so as to make W negative.

[5] This discussion is essentially a reiteration of the results of HILL [1967, *15*] who treated the case for uniform σ_{ij} and $v_{i,j}$.

By way of contrast, however, if *rotational balance is restored* we may deduce from Eq. (53.2), for components along the principal axes of mean stress, the relation[6]

$$d_i + b_i = 0, \tag{53.5}$$

provided that $\bar{\sigma}_1, \bar{\sigma}_2, \bar{\sigma}_3$ satisfy the condition

$$(\bar{\sigma}_1 + \bar{\sigma}_2)(\bar{\sigma}_2 + \bar{\sigma}_3)(\bar{\sigma}_3 + \bar{\sigma}_1) \neq 0. \tag{53.6}$$

We may then recover stability under the conditions imposed by inequality (51.5) for $c_0 = \min(K_1, m)$ [provided that Eq. (51.8) holds] and the restrictions given by

$$\bar{\sigma}_1 + \bar{\sigma}_2 < -mB, \quad \bar{\sigma}_2 + \bar{\sigma}_3 < -mB, \quad \bar{\sigma}_3 + \bar{\sigma}_1 < -mB, \tag{53.7}$$

where m is a positive constant.

Alternatively stability may be recovered under the conditions

$$\bar{\sigma}_1 + \bar{\sigma}_2 > 0, \quad \bar{\sigma}_2 + \bar{\sigma}_3 > 0, \quad \bar{\sigma}_3 + \bar{\sigma}_1 > 0,$$

with $\bar{\sigma}_1 \geqq \bar{\sigma}_2 \geqq \bar{\sigma}_3$, provided that we have

$$\min_B (\sigma_{ij}) - 2B \max_B (\sigma_{ij}\, \sigma_{ij})/\bar{\sigma}_3 \leqq 0.$$

This is achieved by applying SCHWARZ's inequality and the Lagrange inequality (51.13) [as in (51.14)] to the component form of Eq. (53.2) to show that, together with KORN's inequality (43.1), the potential energy (51.7) bounds the L_2-norm of the linear strains from above. Stability then exists with respect to the measures $E(0)$ and this L_2-norm. Similar arguments show that condition (51.10) with $k \geqq 1$ is also sufficient for stability with respect to $E(0)$ and the L_2-norm of v_i.

From condition (53.4) it has been seen that instability can occur if

$$(\bar{\sigma}_{kk}\, \delta_{mn} - \bar{\sigma}_{mn}) \leqq 0, \tag{53.8}$$

although stability can be recovered if the rigid body rotations can be suitably restrained. Under the condition (51.10),

$$(\bar{\sigma}_{kk}\, \delta_{mn} - \bar{\sigma}_{mn}) \geqq c > 0, \tag{53.9}$$

it has been seen that with a further restriction on σ_{ij} stability is maintained. That *instability* can also occur under condition (53.9) is seen in the special case for which σ_{ij} and w_i are uniform.[7] The minimum of potential energy (53.3) under the condition of rotational balance $d_i + b_i = 0$ is

$$\begin{aligned} W_{\min} = \int_B c_{ijkl}\, \tilde{e}_{ij}\, \tilde{e}_{kl}\, dx + \sigma_1\, \tilde{e}_{11}^2 + \sigma_2\, \tilde{e}_{22}^2 + \sigma_3\, \tilde{e}_{33}^2 \\ + \frac{4\sigma_2\, \sigma_3}{\sigma_2 + \sigma_3} \tilde{e}_{23}^2 + \frac{4\sigma_3\, \sigma_1}{\sigma_3 + \sigma_1} \tilde{e}_{31}^2 + \frac{4\sigma_1\, \sigma_2}{\sigma_1 + \sigma_2} \tilde{e}_{12}^2, \end{aligned} \tag{53.10}$$

referred to the principal axes of σ_{ij} whose principal components are σ_i. It is thus possible to choose these components such that this expression is negative, indicating instability.

[6] See HOLDEN [1964, *6*, Eq. (3.16)], HILL [1967, *15*, Eq. (4.2)], whose equations correspond formally to those obtained in executing the above analysis. These are basically the "zero moment condition" used by HOLDEN [1964, *6*], BEATTY [1965, *1*], [1967, *4*], [1968, *2*], TRUESDELL and NOLL [1965, *22*] to restrict certain "trivial instabilities".

[7] This analysis is essentially that given by HILL [1967, *15*]. His criterion for stability of the primary state is $W_{\min} > 0$ and not (51.1) and for instability $W < 0$ as here.

54. Incompressible media. When the material is incompressible, we note that the modified form (23.37) of the elasticities must be used in the various inequalities of the previous section. The essential arguments for stability, however, remain unaltered so that it easily follows that (51.1) is replaced by[8]

$$\int_{B(t)} [b_{ijkl} v_{i,j} v_{k,l} + \sigma_{ij} v_{k,i} v_{k,j} + p (v_{i,j} v_{i,j} + v_{i,j} v_{j,i})] \, dx \geqq c_0 \int_{B(t)} v_{i,j} v_{i,j} \, dx \quad (54.1)$$

and the previous analysis applies. In particular, rigid body rotational motions may be treated according to the last section. Expression (54.1) is not the only one sufficient for stability, and others may be used as required.

O. Stability in special traction boundary value problems.

55. Introduction.[1] We now apply the preceding theory to some special problems in which either compressible or incompressible isotropic elastic materials are involved. Results (23.26), (23.40), which record the respective expressions for elasticities c_{ijkl} and b_{ijkl}, will thus be needed. Dead loading is prescribed everywhere on the surface ∂B of the body in its primary state for all cases. In several problems we shall assume further that the primary state is one of uniform dilatation (and possibly also one of rigid rotation from the reference configuration) and equilibrium under a uniform hydrostatic stress $\sigma_{ij} = \sigma \delta_{ij}$, so that the respective forms of Eqs. (23.20), (23.39) in compressible and incompressible cases become,

$$\sigma = \alpha_0 + \alpha_1 \lambda^2 + \alpha_3 \lambda^4, \quad (55.1)$$

and

$$\sigma = -p + \tilde{\alpha}_1 + \tilde{\alpha}_2. \quad (55.2)$$

In (55.1), $\lambda\ (>0)$ denotes the stretch and $\alpha_i = \alpha_i(3\lambda^2, 3\lambda^4, \lambda^6)$ with $i = 0, 1, 2$; while for (55.2), $\tilde{\alpha}_\beta = \tilde{\alpha}_\beta(3, 3)$. There is hydrostatic tension or pressure in the primary state according as $\sigma > 0$ or $\sigma < 0$.

56. Isotropic compressible material under hydrostatic stress.[2] Let us consider first the stability of a homogeneous isotropic compressible elastic material which in its primary state is subject to a uniform hydrostatic stress and is in a state of pure homogeneous deformation for which $x_i = \lambda X_i$, $\lambda > 0$. Hence from (23.21),

$$b_{ij} = \lambda^2 \delta_{ij} \quad (56.1)$$

and the elasticities (23.26) become

$$c_{ijkl} = (\tilde{\mu} - \sigma)(\delta_{ik}\delta_{jl} + \delta_{il}\delta_{jk}) + (\tilde{\lambda} + \sigma)\delta_{ij}\delta_{kl}, \quad (56.2)$$

where

$$\tilde{\mu} = \lambda^2(\alpha_1 + 2\lambda^2\alpha_2), \quad \tilde{\lambda} = \tilde{k} - 2\tilde{\mu}/3, \quad 3\tilde{k} = \frac{d\sigma}{d(\ln\lambda)}, \quad (56.3)$$

so that

$$\tilde{\lambda} = 2\lambda^2(A_0^* + \lambda^2 A_1^* + \lambda^4 A_2^*),$$

with

$$A_i^* = \frac{\partial \alpha_i}{\partial I_1} + 2\lambda^2 \frac{\partial \alpha_i}{\partial I_2} + \lambda^4 \frac{\partial \alpha_i}{\partial I_3}, \quad i = 0, 1, 2. \quad (56.4)$$

[8] Condition (54.1) with the right side replaced by zero, has been obtained by Hill [1967, *15*], Green and Adkins [1960, *3*], Beatty [1967, *4*], [1968, *2*]. In their expressions, these writers put $2(v_{i,j} v_{i,j} + v_{i,j} v_{j,i})$ for $(v_{i,j} + v_{j,i})(v_{i,j} + v_{j,i})$.

[1] Portions of this section rely on articles by Hill [1957, *1*], [1967, *15*], Beatty [1967, *4*], [1968, *2*] and Holden [1964, *6*].

[2] Beatty [1968, *2*].

The moduli $\tilde{\mu}$, $\tilde{k}$ correspond to the shear and bulk moduli evaluated in the primary state.

The potential energy is then

$$\int_{B(t)} [(\tilde{\lambda}+\sigma)\, v_{i,i}\, v_{k,k}+(\tilde{\mu}-\sigma)\,(v_{i,j}\, v_{i,j}+v_{i,j}\, v_{j,i})+\sigma v_{k,i}\, v_{k,i}]\, dx, \tag{56.5}$$

which may be written in the equivalent forms

$$\int_{B(t)} [(\tilde{\lambda}+\sigma)\, \tilde{e}_{ii}\, \tilde{e}_{kk}+(2\tilde{\mu}-\sigma)\, \tilde{e}_{ij}\, \tilde{e}_{ij}+\sigma\, \omega_{ij}\, \omega_{ij}]\, dx, \tag{56.6}$$

or

$$\int_{B(t)} [(\tilde{\lambda}+\sigma)\, \tilde{e}_{ii}\, \tilde{e}_{kk}+(2\tilde{\mu}-\sigma)\, \tilde{e}_{ij}\, \tilde{e}_{ij}+\sigma(\Omega_{ij}\,\Omega_{ij}-2\Omega_{ij}\, e_{jpi}\, d_p+2d_p\, d_p)]\, dx, \tag{56.7}$$

where

$$\omega_{ij}=\tfrac{1}{2}\,(v_{i,j}-v_{j,i}), \qquad \Omega_{ij}=\tfrac{1}{2}(w_{i,j}-w_{j,i}), \tag{56.8}$$

and we use the notation of (51.2). Since d_p and σ are spatially constant, the last expression for the potential energy may be written

$$\int_{B(t)} [(\tilde{\lambda}+\sigma)\, \tilde{e}_{ii}\, \tilde{e}_{kk}+(2\tilde{\mu}-\sigma)\, \tilde{e}_{ij}\, \tilde{e}_{ij}+\sigma\,\Omega_{ij}\,\Omega_{ij}+2\sigma\, d_p\, d_p]\, dx. \tag{56.9}$$

We consider separately the cases $\sigma>0$, $\sigma=0$, $\sigma<0$, and begin with $\sigma>0$.

(i) $\sigma>0$. On using the inequality

$$\tilde{e}_{ii}\, \tilde{e}_{jj} \leqq 3\, \tilde{e}_{ij}\, \tilde{e}_{ij}, \tag{56.10}$$

we see that, provided

$$(2\tilde{\mu}-\sigma-\nu)\geqq 0, \qquad (3\tilde{\lambda}+2\tilde{\mu}+2\sigma-\nu)\geqq 0, \tag{56.11}$$

or correspondingly

$$2\tilde{\mu}-\sigma\geqq\nu, \qquad \tilde{k}\geqq-\frac{2\sigma}{3}+\nu, \tag{56.12}$$

where ν is some positive constant, then

$$\begin{aligned}\int_{B(t)} [(\tilde{\lambda}+\sigma)\, \tilde{e}_{ii}\, \tilde{e}_{jj}+(2\tilde{\mu}-\sigma)\, \tilde{e}_{ij}\, \tilde{e}_{ij}+\sigma\Omega_{ij}\,\Omega_{ij}+2\sigma\, d_i\, d_i]\, dx \\ \geqq k \int_{B(t)} (\tilde{e}_{ij}\, \tilde{e}_{ij}+\Omega_{ij}\,\Omega_{ij}+2d_i\, d_i)\, dx,\end{aligned} \tag{56.13}$$

where

$$k=\min(\nu, \sigma). \tag{56.14}$$

Our previous analysis then shows that (56.13) establishes stability in the class of measures

$$\varrho_\tau(u)=E(0)+\frac{1}{B}\int_{B(0)} u_i\, dx \int_{B(0)} u_i\, dx, \tag{56.15}$$

$$\varrho(u)=\int_{B(t)} \varrho u_i\, u_i\, dx. \tag{56.16}$$

Inequality (56.13) also holds under the conditions

$$2\tilde{\mu}-\sigma>0, \qquad \tilde{k}\geqq\frac{2\tilde{\mu}}{3}\, q-\sigma(1-p/3), \tag{56.17}$$

where $p+q=1$, $p\geqq 0$, $q>0$.

When $d_i(0)=0$, $w_i(0, \boldsymbol{x})=\varphi_{,i}$, the potential energy may be given a non-positive *initial* value provided the quadratic form in $\tilde{e}_{ij}$ is negative semi-definite. That is, if

$$2\tilde{\mu}-\sigma\leqq 0, \quad \tilde{k}\leqq -2\sigma/3. \tag{56.18}$$

Then, we know from Chap. L that the null solution is unstable.

BEATTY has shown that sufficient conditions for the null solution to be *Hadamard stable*[3] are

$$2\tilde{\mu}\geqq\sigma, \quad \tilde{k}\geqq -2\sigma/3. \tag{56.19}$$

We see from (56.18) that these conditions with strict inequality are, in fact, necessary for the Liapounov stability of the null solution, but they are not sufficient. [This may be easily seen by taknig equality in (56.18), which still produces instability. The resulting values of the moduli give the "limit of stability".] The discrepancy between the conditions for the two respective notions of stability is in accordance with that already noted in Chap. I and Sect. 46.

The case $\sigma=0$ may be easily obtained from the foregoing and requires no further comment.

(ii) $\sigma<0$. For this case we find from expression (56.9) that the potential energy may always be given negative initial values by suitable choice of d_i, and therefore the null solution is unstable in this case. In order to discover how the rigid body rotation $d_i(t)$ may be excluded in this example, we must determine its evolution in time. To do this we recall the decomposition (51.2), in which the incremental displacement v_i is given by

$$w_i(t, x_j)=v_i(t, x_j)+e_{ijk}\, d_j(t)\, x_k, \tag{56.20}$$

with w_i having excluded from it rigid body motions:

$$\int_{B(t)} w_i\, dx=e_{jik}\int_{B(t)} w_j\, x_k\, dx=0. \tag{56.21}$$

We next impose the conditions

$$\int_{B(0)} \dot{v}_i\, dx=e_{ijk}\int_{B(0)} x_k\, \dot{v}_j\, dx=0, \tag{56.22}$$

and then by virtue of the identity

$$\int_{B(t)} \varrho x_k\, v_j\, dx=\int_{B(0)} \varrho x_k\, v_j\, dx+t\int_{B(0)} \varrho x_k\, \dot{v}_j\, dx+\int_0^t\int_{B(\eta)} (t-\eta)\, x_k\, \varrho\ddot{v}_j\, dx\, d\eta, \tag{56.23}$$

we obtain from (40.1), (40.2) the equality

$$\begin{aligned}\int_{B(t)} \varrho x_k\, v_j\, dx=&\int_{B(0)} \varrho x_k\, v_j\, dx\\ &+t\int_{B(0)} \varrho x_k\, \dot{v}_j\, dx-\int_0^t\int_{B(\eta)} (t-\eta)\,(c_{jkpl}+\sigma\delta_{kl}\,\delta_{jp})\, v_{p,l}\, dx\, d\eta.\end{aligned} \tag{56.24}$$

But $c_{ijkl}=c_{jikl}=c_{ijlk}$, and therefore from (56.22) and (56.24)

$$e_{ijk}\int_{B(t)} \varrho x_k\, v_j\, dx=e_{ijk}\int_{B(0)} \varrho x_k\, v_j\, dx-\int_0^t\int_{B(\eta)} (t-\eta)\, e_{ijk}\,\sigma v_{j,k}\, dx\, d\eta. \tag{56.25}$$

[3] BEATTY [1968, 2].

We already have one expression for $d_i(t)$ represented in (51.4). Another may be derived from (56.20) and (56.21) and is

$$I_{ii}\, d_i = e_{ijk} \int_{B(t)} v_j\, x_k\, dx \qquad \text{(no sum on } i\text{)}, \tag{56.26}$$

where

$$I_{ij} = \int_{B(t)} (x_k\, x_k\, \delta_{ij} - x_i\, x_j)\, dx \tag{56.27}$$

and the cartesian axes are selected to coincide with the principal axes of inertia with origin at the mass centre. Thus, we may write (56.25) as

$$d_i(t) = d_i(0) - \frac{2B\sigma}{\varrho I_{ii}} \int_0^t (t-\eta)\, d_i(\eta)\, d\eta, \tag{56.28}$$

and by integration obtain the result

$$d_i(t) = d_i(0) \cosh t(-2B\sigma/\varrho I_{ii})^{\frac{1}{2}}. \tag{56.29}$$

Expression (56.29) is true irrespective of the sign of σ, and shows that when (56.22) holds the initial vanishing of d_i implies its vanishing for all time. Moreover, when $d_i(0) \neq 0$, (56.29) shows that for $\sigma < 0$, $d_i(t)$ is an increasing exponential function and therefore the null solution is unstable in accordance with our first result.

Let us now assume that (56.22) holds and $d_i(0) = 0$. The potential energy is then from (56.9)

$$W \equiv \int_{B(t)} ((\tilde{\lambda}+\sigma)\, \tilde{e}_{ii}\, \tilde{e}_{kk} + (2\tilde{\mu}-\sigma)\, \tilde{e}_{ij}\, \tilde{e}_{ij} + \sigma \Omega_{ij}\, \Omega_{ij})\, dx. \tag{56.30}$$

By KORN's inequality (43.4), we have

$$\int_{B(t)} \Omega_{ij}\, \Omega_{ij}\, dx \leqq 4(K-1) \int_{B(t)} \tilde{e}_{ij}\, \tilde{e}_{ij}\, dx, \tag{56.31}$$

so that

$$W \geqq \int_{B(t)} [(\tilde{\lambda}+\sigma)\, \tilde{e}_{ii}\, \tilde{e}_{kk} + (2\tilde{\mu}-\sigma+4\sigma(K-1))\, \tilde{e}_{ij}\, \tilde{e}_{ij}]\, dx, \tag{56.32}$$

and hence

$$W \geqq \nu \int_{B(t)} \tilde{e}_{ij}\, \tilde{e}_{ij}\, dx, \tag{56.33}$$

provided that the following hold

$$2\tilde{\mu} + \sigma(4K-5) \geqq \nu, \tag{56.34}$$

$$3\tilde{k} + \sigma(4K-2) \geqq \nu, \tag{56.35}$$

for positive constant ν. We may now use inequality (41.17) to obtain the bound

$$W \geqq c \int_{B(t)} v_i\, v_i\, dx, \tag{56.36}$$

where we have used the fact that $w_i = v_i$ since $d_i = 0$. Hence stability is established again.

Instability under the condition $d_i = 0$ occurs whenever $W \leqq 0$ and from (56.30) we see that this is possible when, for instance, the quadratic form in the strains is negative semi-definite; that is, when

$$2\tilde{\mu} - \sigma \leqq 0, \qquad \tilde{k} \leqq -2\sigma/3. \tag{56.37}$$

On the other hand, when w_i is solenoidal, W is negative for any value of $(\tilde{\lambda}+2\tilde{\mu})$ when $2\tilde{\mu}=\sigma$. Other conditions may be easily found.

BEATTY[4] has shown that for the Hadamard stability of the null solution, sufficient conditions are provided by (56.34) and (56.35) with $\nu=0$. He remarks that these conditions are dependent, through the Korn constant K, on the geometry of the body in its primary configuration, and refers to some preliminary experimental work in support of this observation.

57. Incompressible elastic material. The same problem as in Sect. 56 is now considered but for an incompressible elastic material. Thus, b_{ij} is given by (56.1), but the elasticities b_{ijkl} now become[5]

$$b_{ijkl}=2(\tilde{A}_1+\tilde{A}_2)\,\delta_{ij}\,\delta_{kl}+\tilde{\alpha}_2(\delta_{ik}\,\delta_{jl}+\delta_{il}\,\delta_{jk}), \tag{57.1}$$

while the hydrostatic pressure p is given by

$$p=\tilde{\alpha}_1+\tilde{\alpha}_2-\sigma. \tag{57.2}$$

In these expressions

$$\tilde{A}_\beta=\left(\frac{\partial\tilde{\alpha}_\beta}{\partial I_1}+\frac{2\partial\tilde{\alpha}_\beta}{\partial I_2}\right)\Bigg|_{\lambda=1},\qquad \beta=1,2, \tag{57.3}$$

$$\tilde{\alpha}_\beta=\tilde{\alpha}_\beta(3,3). \tag{57.4}$$

On letting $\tilde{\mu}=(\tilde{\alpha}_1+2\tilde{\alpha}_2)_{\lambda=1}$, the potential energy becomes (recall $v_{i,i}=0$)

$$W\equiv\int_{B(t)}[(2\tilde{\mu}-\sigma)\,\tilde{e}_{ij}\,\tilde{e}_{ij}+\sigma\,\omega_{ij}\,\omega_{ij}]\,dx \tag{57.5}$$

$$=\int_{B(t)}[(2\tilde{\mu}-\sigma)\,\tilde{e}_{ij}\,\tilde{e}_{ij}+\sigma\Omega_{ij}\,\Omega_{ij}+2\sigma d_p\,d_p]\,dx, \tag{57.6}$$

in which we adhere to the notation of the previous section. Cases $\sigma>0$, $\sigma<0$ are again considered separately.

(i) $\sigma>0$. It follows immediately from (57.6) and KORN's inequality that, provided $2\tilde{\mu}-\sigma\geqq\nu>0$,

$$W\geqq k\int_{B(t)} v_i\,v_i\,dx,$$

where k is some positive constant. Thus, as before there is stability of the null solution with respect to the measures (56.15), (56.16).

The potential energy becomes zero when $\sigma=2\tilde{\mu}$ and the initial displacement v_i is chosen to be $v_1=x_1$, $v_2=a\,x_2$, $v_3=-(1+a)\;x_3$.[6] We then see that the null solution is unstable on using results from Chap. M. Furthermore, when $\sigma<2\tilde{\mu}$, the initial displacement can be chosen to give a negative potential energy, and hence there is again instability.

For stability of the null solution according to the energy test, HILL,[7] and GREEN and ADKINS[8] have shown that a sufficient condition is $2\tilde{\mu}-\sigma>0$. HILL also obtained the stability limit derived above. BEATTY,[9] adopting the Hadamard definition, proved that a sufficient condition was $2\tilde{\mu}-\sigma\geqq 0$.

[4] BEATTY [1968, *2*].

[5] Compare BEATTY [1967, *4*], GREEN and ADKINS [1960, *3*].

[6] HILL [1957, *1*].

[7] HILL [1957, *1*], [1967, *15*] considers in particular Mooney-Rivlin and Neo-Hookean elastic materials. Stability of the latter has also been treated by RIVLIN [1948, *2*].

[8] GREEN and ADKINS [1960, *3*].

[9] BEATTY [1967, *4*], [1968, *2*].

When $\sigma=0$, conditions for stability may be obtained from the foregoing.

(ii) $\sigma<0$. We may now choose the initial values of d_i and the displacement w_i so as to make the potential energy negative. The null solution is therefore unstable. However, it may be shown exactly as in Sect. 56 (ii) that $d_i(t)$ has the growth behaviour (56.29), provided that initial conditions (56.22) are imposed, so that setting $d_i(0)=0$ means that $d_i(t)=0$, $t\geqq 0$. Arguments similar to those used in deriving (56.36) now show that the null solution is Liapounov stable in the L_2-norm of v_i provided (56.34) holds. Moreover, there is instability whenever $2\tilde{\mu}-\sigma\leqq 0$.[10]

Hill[11] has studied the Neo-Hookean material when the primary state is subject to a homogeneous stress and he has detailed the situations for which there is stability (according to the energy criterion) and for which the potential energy may become zero. His discussion, which may easily be modified to cover Liapounov stability, is intended to illustrate a theory of eigenmodal deformations which he developed for general materials and on which parts of this chapter are based. He considers as a particular case the primary state in uniaxial tension (again for a Neo-Hookean material) which corresponds to the problem of the Euler column. Another treatment of the same problem is due to Beatty[12] but his results disagree with those of Hill. The Euler column composed of a compressible isotropic elastic material has also been considered by Holden,[13] while an account of experimental work for columns composed of rubber-like materials is given by Beatty and Hook.[14] It should be noted, however, that none of these last named authors is concerned with stability in the sense of Liapounov, but instead deal with either the energy criterion or the Hadamard definition.

P. Stability in the class of linear thermoelastic displacements under dead loads.

58. Introduction. In this chapter, we present some partial results for the stability of a thermoelastic body in equilibrium in the class of small thermoelastic incremental displacements. In the analysis we shall apply ideas developed in the previous sections for isothermal problems, with the important difference, however, that we must now take into account internal "dissipation" due to the presence of heat terms. This also means that asymptotic stability becomes of significance in the discussion of this chapter. We assume that body forces are dead and that in its primary state B the body has part of its surface $\overline{\partial B_1}$ held fixed while the remainder ∂B_2 is dead loaded. In addition, the temperature is prescribed on a portion $\overline{\Sigma}$ of the surface ∂B and the heat flux is prescribed on $\partial B-\Sigma$. We shall limit our attention to the special cases

(i) $\partial B-\overline{\Sigma}$ is empty, or

(ii) ∂B_2 is empty,

and we shall always suppose that $\partial B_1\neq\emptyset$. The relevant equations were obtained in Sect. 24 and have the general form

$$(d_{ijkl}u_{k,l})_{,j}+(f_{ij}\theta)_{,j}=\varrho\ddot{u}_i \qquad \text{in } B\times(0,t_1) \tag{58.1}$$

$$\dot{\theta}-cf_{ij}\dot{u}_{i,j}=(a_{ij}\theta_{,j})_{,i} \tag{58.2}$$

[10] Beatty [1967, *4*], [1968, *2*] obtains the sufficient condition $2\mu-\sigma\geqq 0$ for Hadamard stability.

[11] Hill [1967, *15*].

[12] Beatty [1967, *4*].

[13] Holden [1964, *6*].

[14] Beatty and Hook [1968, *3*].

where c is constant and the conductivity tensor $a_{ij}(\boldsymbol{x})$ satisfies

$$a_{ij}\xi_i\xi_j \geqq 0, \quad a_{ij}(\boldsymbol{x}) = a_{ji}(\boldsymbol{x}). \tag{58.3}$$

On the boundary we have

$$u_i = 0 \quad \text{on } \partial \overline{B}_1 \times (0, t_1), \tag{58.4}$$

$$n_j d_{ijkl} u_{k,l} + n_j f_{ij}\theta = 0 \quad \text{on } \partial B_2 \times (0, t_1), \tag{58.5}$$

$$\theta = 0 \quad \text{on } \overline{\Sigma} \times (0, t_1), \tag{58.6}$$

$$n_i \theta_{,i} = 0 \quad \text{on } B - \overline{\Sigma} \times (0, t_1). \tag{58.7}$$

As in Chap. J we assume that $d_{ijkl} = d_{klij}$, and that the density is positive.

We shall mainly consider only classical solutions to the above problem. Hence, let the set X be given by $Y \times Z$, where Y is the set of vector-valued functions satisfying the boundary conditions (58.4) and (58.5) and twice continuously differentiable on $\overline{B}$; Z is the set of scalar functions satisfying (58.6) and (58.7) and twice continuously differentiable on $\overline{B}$. Then the dynamical system $\mathscr{B}(T, X)$ consists of the pair of functions (u_i, θ) defined on $[0, t_1]$, taking values in X and satisfying (58.1), (58.2). In these equations we assume that d_{ijkl} and a_{ij} are twice continuously differentiable and f_{ij} once continuously differentiable on B. A weak solution will be defined in a similar manner to that described in Sect. 23,[1] and will be used later to examine asymptotic stability in connexion with HALE'S invariance principle.

Stability of the body in its equilibrium position B corresponds to the stability of the null solution to Eqs. (58.1), (58.2). We shall examine the stability in the light of different concepts and begin with that due to LIAPOUNOV.[2]

59. Liapounov stability. In order to ascertain whether the position B is stable in this sense we require an appropriate Liapounov function which could be obtained from those constructed in Chap. G. Instead, however, we shall construct our function directly from the equations stated above. From (58.2) we find that

$$\int_{B(t)} \theta^2\, dx + 2\int_0^t \int_{B(\eta)} a_{ij}\theta_{,i}\theta_{,j}\, dx\, d\eta = \int_{B(0)} \theta^2\, dx - 2c\int_0^t \int_{B(\eta)} \dot{u}_i (f_{ij}\theta)_{,j}\, dx\, d\eta, \tag{59.1}$$

while from (58.1) we obtain

$$E(t) = E(0) + \int_0^t \int_{B(\eta)} \dot{u}_i (f_{ij}\theta)_{,j}\, dx\, d\eta, \tag{59.2}$$

where $E(t)$ is the total energy

$$E(t) = \tfrac{1}{2}\int_{B(t)} (\varrho\, \dot{u}_i\, \dot{u}_i + d_{ijkl}\, u_{i,j}\, u_{k,l})\, dx. \tag{59.3}$$

From (59.1) and (59.2) it then follows that

$$J(t) \equiv \int_{B(t)} \theta^2\, dx + 2\int_0^t \int_{B(\eta)} a_{ij}\theta_{,i}\theta_{,j}\, dx\, d\eta + 2c\, E(t) = J(0). \tag{59.4}$$

---

[1] DAFERMOS [1968, *7*], SLEMROD and INFANTE [1971, *12*].

[2] For some applications of linear thermoelastic stability, not treated explicitly in the text, see WITTRICK [1953, *5*], SHAPOVALA [1964, *11*], AUGUSTI [1968, *1*], VAHIDI and HUANG [1969, *23*], HUANG and SHIEH [1970, *13*], AGGARWALA and SAIBEL [1970, *1*].

We now suppose that $c>0$ and that

$$\int_B d_{ijkl}\psi_{ij}\psi_{kl}\,dx \geqq d\int_B \psi_{ij}\psi_{ij}\,dx, \tag{59.5}$$

where d is a positive constant and ψ_{ij} any arbitrary tensor. Then from (59.4) and (59.5) and the arguments of Sect. 35 it may be easily shown that the *null solution* $u_i \equiv 0$ *of* (58.1) *and* (58.2) *is Liapounov stable with respect to the measures*

$$\varrho_\tau(u,\theta) = \int_{B(0)} \theta^2\,dx + E(0), \tag{59.6}$$

$$\varrho(u(t)) = \int_{B(t)} \varrho u_i u_i\,dx. \tag{59.7}$$

(58.3), (59.4), (59.5) give stability for spatial L_2-norms of θ. But when we also have:

$$\int_B a_{ij}\xi_i\xi_j\,dx \geqq a\int_B \xi_i\xi_i\,dx, \tag{59.8}$$

for positive constant a and arbitrary vectors ξ_i, we find from (59.4) that

$$\int_{B(t)} \theta^2\,dx + 2\int_0^t \int_{B(\eta)} a_{ij}\theta_{,i}\theta_{,j}\,dx\,d\eta \leqq J(0). \tag{59.9}$$

Application of POINCARÉ's inequality to the second term on the left of (59.9) then reduces this inequality to

$$\int_{B(t)} \theta^2\,dx + \lambda\int_0^t \int_{B(\eta)} \theta^2\,dx\,d\eta \leqq J(0), \tag{59.10}$$

for some computable positive constant λ. An integration of (59.10) then gives

$$\int_0^t \int_{B(\eta)} \theta^2\,dx\,d\eta \leqq J(0)\,(1-e^{-\lambda t}). \tag{59.11}$$

Inequality (59.11), taken together with (59.4) and (59.5), then shows that, *under the additional assumption* (59.8), *the null solution is Liapounov stable with respect to the measures* (59.6) *and*

$$\varrho(u,\theta) = \int_{B(t)} \varrho u_i u_i\,dx + \int_0^t \int_{B(\eta)} \theta^2\,dx\,d\eta. \tag{59.12}$$

60. Instability. We turn next to instability of the null solution and impose on the elasticities the negative-definiteness condition

$$-\int_B d_{ijkl}\psi_{ij}\psi_{kl}\,dx \geqq d\int_B \psi_{ij}\psi_{ij}\,dx, \tag{60.1}$$

for positive constant d and arbitrary second order tensors ψ_{ij}. On the quantity f_{ij} we impose the boundedness condition

$$f_{ij}f_{ij} \leqq M_1^2, \tag{60.2}$$

for finite constant M_1. The conductivity tensor a_{ij} is assumed still to be positive semi-definite in the sense of (58.3) and in accordance with thermodynamics. The technique used for establishing instability is similar to that of Chap. L, although we note that condition (60.1) is stronger than any requirement on the elasticities considered in the Chap. L.

Let us therefore introduce the function

$$F(t) = \int_{B(t)} \varrho u_i u_i\,dx. \tag{60.3}$$

On differentiating $F(t)$ twice, substituting for the inertial term from (58.1) and integrating by parts we obtain

$$\ddot{F}(t) = 2\int\limits_{B(t)} (\varrho \dot{u}_i \dot{u}_i - d_{ijkl} u_{i,j} u_{k,l} - f_{ij} \theta u_{i,j})\, dx. \tag{60.4}$$

We now apply the arithmetic-geometric mean inequality to the last term on the right of (60.4), followed by an application of inequalities (60.1), (60.2). This yields

$$-2\int\limits_{B(t)} f_{ij} \theta u_{i,j}\, dx \geqq -\alpha M_1^2 \int\limits_{B(t)} \theta^2\, dx + \frac{1}{\alpha d} \int\limits_{B(t)} d_{ijkl} u_{i,j} u_{k,l}\, dx, \tag{60.5}$$

where α is some positive constant. Insertion of (60.5) into (60.4) then produces the inequality

$$\ddot{F}(t) \geqq \left(4 - \frac{1}{\alpha d}\right) \int\limits_{B(t)} \varrho \dot{u}_i \dot{u}_i\, dx - \beta \left[c E(t) + \int\limits_{B(t)} \theta^2 dx\right], \tag{60.6}$$

where we have chosen $2d\alpha > 1$, and put

$$\beta = \max\left[\alpha M_1^2, \frac{4}{c}\left(1 - \frac{1}{2\alpha d}\right)\right]. \tag{60.7}$$

On using the fact that the kinetic energy is non-negative, together with (58.3) and the conservation law (59.4), we obtain from (60.6) the inequality

$$\ddot{F}(t) \geqq -\beta J(0).$$

By integration we thus finally reach the result

$$F(t) \geqq F(0) + t\dot{F}(0) - \beta \frac{t^2}{2} J(0). \tag{60.8}$$

Because of (60.1) we may choose $J(0)$ to be negative and so inequality (60.8) demonstrates instability of the null solution.

61. Hölder stability. We next consider the Hölder stability of the null solution[3] and for this purpose impose the following extra restrictions

(i) there exist finite constants M_1, M_2 such that

$$f_{ij} f_{ij} \leqq M_1^2, \quad f_{ij,j} f_{ik,k} \leqq M_2^2, \tag{61.1}$$

(ii) a_{ij} is positive-definite in the sense of (59.8).

We further confine our attention only to solutions u_i of Eqs. (58.1), (58.2) which satisfy the boundary conditions (58.4), (58.5), subject to (i), (ii) of Sect. 58, and which moreover satisfy the boundedness condition

$$\int\limits_0^{t_1} \int\limits_{B(\eta)} \varrho u_i u_i\, dx\, d\eta \leqq N^2, \tag{61.2}$$

where N is some prescribed finite constant. Such displacements we shall say belong to the class $\mathcal{N}$.

We make no assumptions concerning the definiteness of the elasticities d_{ijkl} nor concerning the positiveness of c.

Hölder stability in the class $\mathcal{N}$ is now a direct consequence of the following theorem:[4]

[3] KNOPS and PAYNE [1970, *19*].
[4] KNOPS and PAYNE [1970, *19*].

Theorem. If (u_i, θ) is a solution to (58.1), (58.2), (58.4), (58.5) and $u_i \in \mathcal{N}$ and if the initial temperature, displacement and velocity are square integrable and the initial energy is bounded, then for finite time t_1 it is possible to determine explicit positive constants K_i, N_1 such that

$$\int_0^t \int_{B(\eta)} \varrho u_i u_i \, dx \, d\eta \leqq K_1 N_1^{2\delta} \Big\{ K_2 \int_{B(0)} \varrho u_i u_i \, dx + K_3 \int_{B(0)} \varrho \dot{u}_i \dot{u}_i \, dx + K_4 \int_{B(0)} d_{ijkl} \dot{u}_{i,j} \dot{u}_{k,l} \, dx + K_5 \int_{B(0)} \theta^2 \, dx \Big\}^{1-\delta}, \quad t \in [0, t_1), \tag{61.3}$$

where

$$\delta = \frac{1 - \exp(-K_0 t)}{1 - \exp(-K_0 t_1)}. \tag{61.4}$$

The proof of this theorem, full details of which are supplied by Knops and Payne, relies on logarithmic convexity arguments of the type already used in Chap. M, together with various other estimates.

We remark that the theorem establishes both Hölder stability and continuous dependence of the solution on the initial data, with respect to the measure

$$\varrho(u) = \int_0^t \int_{B(\eta)} \varrho u_i u_i \, dx \, d\eta, \tag{61.5}$$

which is different from that adopted in the above discussion of Liapounov stability. However, the theorem does not establish either Hölder stability or continuous dependence on the initial data of θ in *any* appropriate norm. To obtain results of this kind, we must make use of the inequality[5]

$$\int_{B(t)} \theta^2 \, dx \leqq (1 + \nu t) \int_{B(0)} \theta^2 \, dx + [(1 + \nu t) \, d_1 + (\nu^{-1} + t^2) \, d_2] \times \int_0^t \int_{B(\eta)} \varrho \dot{u}_i \dot{u}_i \, dx \, d\eta, \quad t \in [0, t_1), \tag{61.6}$$

where ν is an arbitrary positive constant, $\varrho \geqq \varrho_m > 0$, and

$$d_1 = \frac{M_1^2 c^2}{\varrho_m a}, \qquad d_2 = \frac{2 M_2^2 c^2}{\varrho_m}, \tag{61.7}$$

with a given by (59.8). Hölder stability and continuous dependence now follow for θ, measured in an L_2-norm, from inequality (61.6) and inequality (61.3), but with u_i replaced by $\dot{u}_i$ in the latter. But now the initial data must be such that the displacement, its first and second derivatives, are all initially square-integrable.

We may also use the inequalities

$$\int_0^t \int_{B(\eta)} (t-\eta) \, \theta^2 \, dx \, d\eta + \int_0^t \int_{B(\eta)} (t-\eta)^2 \, a_{ij} \, \theta_{,i} \, \theta_{,j} \, dx \, d\eta \leqq t^2 \int_{B(0)} \theta^2 \, dx + (d_1 + d_2 t_1) \int_0^t \int_{B(\eta)} (t-\eta)^2 \varrho \dot{u}_i \dot{u}_i \, dx \, d\eta \tag{61.8}$$

and

$$\int_0^t \int_{B(\eta)} (t-\eta)^2 \varrho \dot{u}_i \dot{u}_i \, dx \, d\eta \leqq b_1 \int_0^t \int_{B(\eta)} \varrho u_i u_i \, dx \, d\eta + b_1 (t_1 - t) \int_{B(0)} \varrho u_i u_i \, dx + b_2 \gamma, \quad t \in [0, t_1), \tag{61.9}$$

[5] Knops and Payne [1970, *19*].

where b_1, b_2, γ are computable positive constants. Substitution of (61.9) into (61.8) and use of (61.3) then proves the Hölder stability and continuous dependence of θ in the measure

$$\varrho(\theta) = \int_0^t \int_{B(\eta)} (t-\eta)\, \theta^2\, dx\, d\eta + \int_0^t \int_{B(\eta)} (t-\eta)^2\, a_{ij}\, \theta_{,i}\, \theta_{,j}\, dx\, d\eta. \tag{61.10}$$

Let us note that the inequalities (61.3) and (61.6), for example, form the basis of a uniqueness theorem for the linear thermoelastic initial-boundary value problem[6] which, unlike classical theorems, does not impose any definiteness restrictions upon the elasticities.

62. Asymptotic stability. We finally study asymptotic stability of the null solution using the invariance principle of HALE (see Sect. 20) which, it will be recalled, applies only to dynamical systems defined in the sense of HALE. A major feature of these systems is that they are restricted to a set X which is a Banach space. The invariance principle then requires trajectories to remain in a compact Banach space so that with increasing time they will, loosely speaking, tend to an invariant set of the system. We must thus identify the Banach spaces associated with the solution to the thermoelastic equations (58.1)–(58.7) and then determine the invariant set. Although we treat the stability of the null solution it will be seen that this invariant set is not necessarily the null solution but may correspond instead to an isothermal vibration. The present treatment repeatedly refers to papers by DAFERMOS[7] and SLEMROD and INFANTE.[8] Detailed calculations are omitted as they can be found in the work of these authors.

The development is conducted with generalised solutions and it is always assumed that the elasticities and conductivity tensor are positive-definite in the sense of (59.5) and (59.8) respectively and that they are bounded and measurable in B. The last property must also be possessed by ϱ and f_{ij}. For the sake of simplicity we consider only boundary conditions of the form

$$u_i = 0 \quad \text{on } \partial B \times (0, t_1), \tag{62.1}$$

$$\theta = 0 \quad \text{on } \partial B \times (0, t_1). \tag{62.2}$$

In order to study the Banach spaces on which the dynamical system is defined, we introduce first the space $H_0(B) = \mathring{W}_2^{(1)}(B) \times L_2(B) \times L_2(B)$,[9] with norm

$$\|(v_i, w_i, \varphi)\|_0^2 = \int_B (\varrho\, w_i\, w_i + d_{ijkl}\, v_{i,j}\, v_{k,l} + \varphi^2)\, dx, \tag{62.3}$$

and also the space $H(B) = \mathring{W}_2^{(1)}(B) \times \mathring{W}_2^{(1)}(B) \times \mathring{W}_2^{(1)}(B)$. We define the map $P: H_0(B) \to H_1(B)$ sending $(v_i, w_i, \varphi) \in H_0(B)$ onto $(u_i, v_i, \theta) \in H_1(B) \subset H(B)$, where $(u_i, \theta) \in \mathring{W}_2^{(1)}(B) \times \mathring{W}_2^{(1)}(B)$ is defined by the solution of the system

$$\int_B d_{ijkl}\, u_{k,l}\, \psi_{i,j}\, dx = -\int_B (\varrho\, w_i\, \psi_i + f_{ij}\, \theta\, \psi_{i,j})\, dx, \tag{62.4}$$

$$\int_B a_{ij}\, \theta_{,j}\, \chi_{,i}\, dx = -\int_B (\varrho\, \varphi - c f_{ij}\, v_{i,j})\, \chi\, dx, \tag{62.5}$$

[6] See KNOPS and PAYNE [1970, *19*]. BRUN [1965, *4*], [1969, *2*], using different methods, obtains uniqueness for either symmetric, positive semi-definite or strongly elliptic elasticities. DAFERMOS [1968, *7*] treates generalised solutions with a positive-definite energy.

[7] DAFERMOS [1968, *7*].

[8] SLEMROD and INFANTE [1971, *12*].

[9] The space $\mathring{W}_2^{(1)}(B)$ is the set of functions in $W_2^{(1)}(B)$ with compact support in B.

for every $\chi, \psi \in \overset{\circ}{W}_2^{(1)}(B)$. The mapping is linear, well-defined on $H_0(B)$ and one to one. Let $H_m(B)$ denote the range of the map $P_m = P \circ P \circ \cdots \circ P$, where the composition is m-fold. Then P_m^{-1} is well-defined and maps $H_m(B)$ onto $H_0(B)$. Finally, let $\|h\|_m = \|P_m^{-1} h\|_0$, $h \in H_m(B)$. Then, DAFERMOS[10] has proved that *H_m is a Banach space with norm $\|(\cdot)\|_m$, $H_0(B) \supset H(B) \cdots \supset H_m(B)$ algebraically and topologically, and $H_m(B)$ is dense in $H_l(B)$ for $m > l$. Furthermore, the embedding $H_m(B) \to H_l(B)$ is compact.*

We shall call $(u_i, \dot{u}_i, \theta)$ a *generalised solution* of (58.1), (58.2) for the boundary conditions (62.1), (62.2) on $B \times (0, t)$, if for all smooth test functions (v_i, φ) with compact support on B and v_i vanishing on $B \times 0$ the following holds:

$$\begin{aligned} \int_0^t \int_{B(\eta)} [(\eta - t)\,(\varrho \dot{u}_i \ddot{v}_i - d_{ijkl} u_{k,l} \dot{v}_{i,j} - f_{ij} \theta \dot{v}_{i,j} + \varrho c^{-1} \theta \dot{\varphi} \\ - f_{ij} u_{i,j} \dot{\varphi}) + \varrho \dot{u}_i \dot{v}_i + \varrho c^{-1} \theta \varphi - f_{ij} u_{i,j} \varphi \\ - \int_0^\eta (a_{ij} \varphi_{,i})_{,j}\, \theta \, d\tau]\, dx\, d\eta \\ = -t \int_{B(0)} (\varrho g_i \dot{v}_i + \frac{\varrho}{c}\, \Theta \varphi - f_{ij} h_{i,j} \varphi)\, dx, \end{aligned} \tag{62.6}$$

where u_i, θ satisfy the prescribed initial conditions

$$u_i(0, \boldsymbol{x}) = h_i(\boldsymbol{x}), \qquad \dot{u}_i(0, \boldsymbol{x}) = g_i(\boldsymbol{x}), \qquad \theta(0, \boldsymbol{x}) = \Theta(\boldsymbol{x}). \tag{62.7}$$

It then follows that a *generalised solution $(u_i, \dot{u}_i, \theta)$ describes a dynamical system (in the sense of* HALE*) on $H_m(B)$*, $m = 0, 1, 2, \ldots$[11] and moreover satisfies the conservation law (59.4).

With these facts assembled, we are ready to start the stability analysis.

For the trajectory $(u_i, \dot{u}_i, \theta) \in H_m(B)$, it follows that $P \circ (u_i, \dot{u}_i, \theta) \equiv (\bar{u}_i, \dot{\bar{u}}_i, \bar{\theta})$ is a dynamical system on $H_{m+1}(B)$ with initial data $P \circ (h_i, g_i, \Theta) \in H_{m+1}(B)$. Because the embedding $H_m(B) \to H_l(B)$, $l < m$, is compact, it is then possible to show that the trajectory $(\bar{u}_i, \dot{\bar{u}}_i, \bar{\theta})$ lies in a compact set G of $H_l(B)$, $l \leqq m$, and so the conditions of HALE's invariance principle (Sect. 20) are met. On choosing for the Liapounov function $V = \|(\bar{u}_i, \dot{\bar{u}}_i, \bar{\theta})\|_m^2$ we are led from the above to the following result:[12]

For any initial data (h_i, g_i, Θ) in $H_m(B)$, $m \geqq 1$, provided the elasticities and the conductivity tensor are positive-definite, the density ϱ is positive and c is positive, the trajectory through (h_i, g_i, Θ) approaches the set

$$\begin{aligned} M = \{(w_i, \dot{w}_i, Y) \in H_0(B) \,|\, f_{ij} w_{i,j} = 0,\ Y = 0, \\ \int_0^t \int_{B(\eta)} [(\eta - t)\,(\varrho \dot{w}_i \ddot{v}_i - d_{ijkl} w_{k,l} \dot{v}_{i,j}) + \varrho \dot{w}_i \dot{v}_i]\, dx\, d\eta = -t \int_{B(0)} \varrho g_i \dot{v}_i\, dx, \end{aligned}$$

$\forall$ test functions v_i with compact support on B and vanishing on $B \times 0$},

in the norm of the space $H_0(B)$ as $t \to \infty$.

---

[10] DAFERMOS [1968, *7*].

[11] DAFERMOS [1968, *7*].

[12] SLEMROD and INFANTE [1971, *12*]. These authors construct in detail the invariant set.

Q. Classification of stability problems with non-dead loading.

63. Introduction. In this treatise attention so far has been concentrated mainly on problems in which the elastic body has part or all of its boundary either fixed or subject to dead loads. The body forces, when they exist, have also been assumed dead. We shall now consider other types of load and surface displacements, which do not remain unaltered during the deformation of the body, but which vary in some specified manner. A body's behaviour, once it has been disturbed from equilibrium, under, say, prescribed initial data, evidently depends, amongst other things, on the way the load varies in the subsequent motion. If, for instance, the load resists the motion, there will be a greater likelihood that the equilibrium position will be stable than if it favours the motion.

Some preliminary results have already been noted in Sect. 30 in which stability in the class of non-linear incremental displacements was discussed. In this earlier analysis the loads and surface displacements were left completely arbitrary and influenced the stability through the properties of the functions $H(t)$ and $H_0(t)$. This degree of generality could have been retained in the various linear stability analyses of Chaps. J–P and it is not difficult to see what restrictions must be imposed on the functions H and H_0 for the conclusions of these sections to remain valid. Usually however, H and H_0 do not have associated definiteness properties and it is thus unrealistic to expect that these restrictions will apply in general. As a special instance we note that when the loads depend on the motion of the body,[1] H_0 may assume opposite signs in different parts of the time interval.

Loads and surface displacements may be associated with reference, primary, secondary or incremental conditions. In the discussion of Chaps. R–U we shall be concerned entirely with the incremental values of these quantities and the remarks that we make there do not necessarily apply to the loads and surface displacements in the primary or secondary states. We shall consider two main groups of data: (a) that which is independent of the motion of the body and (b) that which depends on the motion.

64. Group (a): Persistent stability. In this group the loads and surface displacements are *given* functions of time and position in either reference or primary configuration. The notion of stability related to problems in this class will be termed "persistent". The loads maintaining the body in its secondary configuration satisfy the same law as those producing the primary motion. As the loads are time-dependent, both configurations are motions and application of our previous theory will lead to analysis of *Liapounov stability of motions*.

An important subset of this group concerns thin members and the time-dependence of the loads is reflected by the presence of time-dependent coefficients in the respective equations of motion. Such problems are sometimes associated with the name of "parametric excitation" and a classical example is that of the Euler column under pulsating end loads. Other examples include those of whirling shafts and different kinds of vibratory phenomena. Because of its engineering interest, the subject has a vast and thriving literature[2] in which stability analyses depend upon solutions obtained by separating the variables.

---

[1] These remarks are amplified, for example, in the work of NEMAT-NASSER and HERRMANN [1967, *14*], NEMAT-NASSER and ROORDA [1967, *25*]; and NEMAT-NASSER [1968, *22*], [1970, *24*].

[2] Of this literature, we content ourselves with merely citing the work of BOLOTIN [1961, *1*], EVAN-IWANOWSKI [1963, *2*], [1969, *8*], METTLER [1967, *24*], HERRMANN [1967, *12*], ZIEGLER [1968, *29*]. All these references contain extensive bibliographies.

This leads to the Mathieu equation and Strutt's diagram[3] with extensions in more complex systems to the generalised Mathieu-Hill equation. We do not repeat this well-known approach but present instead an application of LIAPOUNOV's theorem due to HSU and LEE.[4] For the purpose of illustration a one-dimensional system is considered which is treated in Sect. 77. This method gives a technique for finding a Liapounov function not actually time-decreasing but with a current value not greater than its initial value.

The second type of problem in this group contrasts with that just described in that the incremental data is time-dependent while the data maintaining the body in its primary configuration may be time-independent. *The primary configuration may thus be an equilibrium position.* Further, it is possible for the secondary motion to arise solely from the incremental data without the need for also imposing non-zero initial data. In these circumstances we are no longer using Liapounov stability in the sense of the preceding sections on elastic stability. The measure ϱ_τ of the initial disturbances must now be modified to measure the incremental time-dependent data and must thus be defined over space-time. With this extension, however, the basic interpretation of stability in terms of continuity continues to hold. The remaining problems in this group concern the general three-dimensional body and for them we develop a stability analysis relying upon a study of the total energy. In addition, we also introduce certain dual problems in Sect. 71 to obtain other stability estimates.

65. Group (b): Motion-dependent data. The second group in our discussion of loads and surface displacements concerns those that depend on the *motion of the body* in some specified way. In these problems our interest lies chiefly in the loads or more specifically with the surface stress vectors when these arise as contact forces representing the action between contiguous bodies and the given elastic body. The stress vector at a point of the surface between the bodies therefore depends on the relative motion of the bodies in some neighbourhood of the point. This vector could also be influenced by thermal or electromagnetic factors but these will not be considered in this analysis. General studies of these problems have been carried out by SEWELL,[5] NEMAT-NASSER[6] and BEATTY and LEIGH.[7] While the first mentioned author considers only stress vectors dependent on the relative positions of the bodies, referring to them as "configuration-dependent", the second mentioned writer allows dependence on the relative motion also and calls them "motion-dependent". We adopt their terminology. The last named authors develop a fully invariant constitutive theory in which the stress vector has a functional dependence on the history of the relative motion.

Within the second group, we may distinguish three categories of force for the loads. These are (i) (weakly) conservative force, (ii) dissipative force and (iii) so-called "follower force". Mention should perhaps also be made of gyroscopic force which does no work in a displacement. In the first two categories the names carry adequate description. Forces of the third kind have contemporary technological significance and arise in problems concerned, for instance, with rocket propulsion and plates in a gas flow. They are characterised by the property that they follow assigned directions, usually fixed in the body. While theoretical

[3] An early application of the technique is due to BELIAEV [1924, *1*], but for a more general account see ZIEGLER [1968, *29*].

[4] HSU and LEE [1971, *5*].

[5] SEWELL [1967, *31*].

[6] NEMAT-NASSER [1970, *24*].

[7] BEATTY and LEIGH [1970, *4*].

descriptions of follower forces have been given for a long time, it was thought that many had no physical significance, and it is only recently that practical models have been constructed in the laboratory for such theories.[8]

As far as possible each category is separately described and stability analyses are presented. Apart from the treatment of conservative forces, however, it is possible to obtain only fragmentary results, a fact that reflects the difficulty involved in dealing with non-conservative problems for which the corresponding equations are no longer self-adjoint. The majority of work for follower forces has been restricted either to one-dimensional models or to models with a finite number of degrees of freedom.[9] Because of these difficulties we leave a general description of follower forces until Chap. T.

In many follower force problems, inadequacies have long been known to arise from the classical criteria of energy or adjacent equilibrium, a fact probably first recognised by NIKOLAI and BECK.[10] Instead, it is essential that some form of analysis based on dynamic equations should be carried out and this has usually taken the form of an eigenfunction decomposition, although in a few instances the Liapounov approach has been used.

Because of its simplicity, we shall consider first, in Chap. R the case of weakly conservative forces and then deal with the time-dependent motion-independent forces and surface displacements in Chap. S. We shall next treat follower forces in Chap. T and finally finish in Chap. U with dissipative forces and their effect on stability in the presence of follower forces.

Our account is intended to outline methods and makes no claim to be an exhaustive treatment of problems. Solutions are restricted to being classical throughout, although it is evident where generalisation to weak solutions is possible.

We take care always to distinguish between the *stress vector* (or *surface traction*) and the *load* on a surface or volume element. The stress vector, denoted by T_i or t_i, is referred per unit area of the reference or primary configuration, respectively, while the corresponding loads are $T_i\,dA$ or $t_i\,da$. As usual, unstarred quantities belong to the primary configuration. In the secondary configuration the stress vectors are T_i^* and t_i^*, but now t_i^* is measured per unit area of surfaces in the primary configuration. The loads are therefore $T_i^*\,dA$ and $t_i^*\,da$.

R. Stability under weakly conservative loads.

66. Definitions. Of especial interest in elastic stability are incremental body forces and incremental stress vectors or loads that are conservative in a sense to be defined. In the ordinary way, if the incremental stress vector on the surface is conservative, there exists a single-valued function h of the incremental displacement u_i such that on the surface ∂B_2

$$t_i^* - t_i = \frac{\partial h}{\partial u_i}, \tag{66.1}$$

and hence

$$(t_i^* - t_i)\,\dot{u}_i = \dot{h}. \tag{66.2}$$

In particular, the last relation implies that the total work done by the tractions acting on ∂B_2 is zero around any closed path described in displacement space.

[8] HERRMANN, NEMAT-NASSER and PRASAD [1966, *11*].

[9] Surveys are given by ZIEGLER [1956, *3*], BOLOTIN [1961, *1*] and HERRMANN [1967, *12*].

[10] NIKOLAI [1928, *3*], [1929, *2*], [1930, *1*], BECK [1952, *1*]. A systematic account is given by ZIEGLER [1956, *3*]. Nevertheless it would be a mistake to suppose that the classical criteria always fail to produce results in these problems. See, for instance, PFLÜGER [1964, *10*].

For our purposes, the conventional idea of a conservative force is too restrictive. We do not require the local conditions (66.1) and (66.2) to hold since it is sufficient that the respective integrals of the incremental stress vector and body forces possess this property.[1] Moreover, there is no obvious reason why these quantities should depend only upon the local incremental displacement, and we shall extend their domain to include dependence on, for instance, displacement and velocity gradients. We thus propose the following definitions of conservativeness:

The incremental surface tractions are weakly conservative if and only if there exists a potential function $\Phi(t, \ldots)$ such that

$$\int_{\partial B_2(t)} (t_i^* - t_i)\, \dot{u}_i\, da = \dot{\Phi}, \tag{66.3}$$

where the potential function Φ is given by

$$\Phi = \int_{\Omega} \varphi\, d\chi. \tag{66.4}$$

The integration in (66.4) may be over either the volume B, in which case $\varphi(t, \ldots)$ is said to be *a volume potential density*, or the surface ∂B_2, when φ is said to be a *surface potential density*. Apart from time t, the potential densities may be functions of any of the kinematical variables. Similar definitions and comments apply to the incremental stress vector $T_i^* - T_i$ on the surface ∂B_{02}.

The incremental body force is weakly conservative if and only if there exists a potential function $\Psi(t)$ such that

$$\int_{B(t)} \varrho (F_i^* - F_i)\, \dot{u}_i\, dx = \dot{\Psi}, \tag{66.5}$$

where

$$\Psi = \int_{\Omega} \psi\, d\chi, \tag{66.6}$$

and $\psi(t, \ldots)$ is either a volume or surface potential density. As before, ψ is a function of t and the kinematical variables.

It immediately follows from the above definitions that the potential functions and their densities are not required to be single-valued. Nor do the definitions require that zero work be performed by the respective forces around closed paths in *any* space, which feature alone represents considerable relaxation of the conditions in the conventional definition.[2] For a stability analysis, we require only that the stress vector and body force are weakly conservative in the above sense, as then the function $H_0(t)$ appearing in Eq. (30.3) depends only upon the current incremental deformation and may be combined with the energy to form a new Liapounov function. The precise behaviour of the potential functions will, as components of the Liapounov function, influence the stability assessment and it remains to be seen whether multi-valued potentials, or potentials that do not give zero work around closed paths, always produce instability.

There are some obvious examples of weakly conservative forces. The first of these occurs when the whole surface ∂B of the body in its primary configuration

[1] That the property of "conservativeness" could be weakened in this way was apparently first suggested by HILL [1962, *7*].

[2] In this sense our definition differs from that of SEWELL [1965, *20*], [1967, *31*] and also from the original version of HILL [1962, *7*]. Both authors wish to preserve the equivalence of their definitions with a "zero work" principle. Note also that all our computations are conducted with actual and not virtual displacements.

is held fixed, so that $u_i = 0$ on ∂B, and therefore the incremental surface stress possesses constant potential. Again, by definition, gyroscopic forces do no work in a displacement, and hence also possess potentials that are identically zero. As a third, and final example, we consider *dead loading*, for which several stability analyses have already been carried out in the previous sections.

In so-called dead loading the *stress vectors* are retained fixed at their values in the primary configuration so that

$$T_i^* = T_i = T_i(t, X_j) \quad \text{and} \quad t_i^* = t_i = t_i(t, x_j).$$

From (66.3) we find that the potential Φ is a constant. Hence, even though the stress vectors are time-dependent, provided their increments are dead, then the incremental stress vector is conservative. In the situation where $t_i = T_i = 0$, and t_i^* or T_i^* are fixed in magnitude and direction, then the associated potential is given by

$$\dot{\Phi} = \int\limits_{\partial B_{02}(t)} T_i^* \dot{u}_i \, dA = \int\limits_{\partial B_2(t)} t_i^* \dot{u}_i \, da = \frac{d}{dt} \int\limits_{\partial B_{02}(t)} T_i^* u_i \, dA = \frac{d}{dt} \int\limits_{\partial B_2(t)} t_i^* u_i \, da.$$

Analogous statements hold for body forces.[3]

67. Characterisations of weakly conservative forces. A simple sufficient condition for this is provided by[4] the requirement

$$\int\limits_{\partial B_2(t)} (t_i^* - t_i) \dot{u}_i \, da = \int\limits_{\partial B_2(t)} \overline{(t_i^* - t_i)}^{\boldsymbol{\cdot}} u_i \, da, \tag{67.1}$$

for which the surface forces possess the surface potential density[5]

$$\varphi = \tfrac{1}{2} (t_i^* - t_i) u_i. \tag{67.2}$$

Examples may easily be constructed to show that condition (67.1) is not necessary for forces to be weakly conservative. In these examples the potential need not be given by (67.2).

As an application of (67.1) let us consider the incremental stress vector given by[6]

$$(t_i^* - t_i) = K_{ij} u_j + \pi_{ijk} D_k u_j, \tag{67.3}$$

[3] A related situation has been considered by the authors GUO ZHONG-HENG and URBANOWSKI [1963, *22*] and SEWELL [1967, *31*]. During the deformation of the body from its primary to its secondary configuration, the stress vector per unit area of current surface remains unaltered. On letting t_i^{**} denote this quantity in the secondary configuration, we have $t_i^{**} = t_i$, further $t_i^{**} \, da^* = t_i^* \, da$, so that

$$(t_i^* - t_i) \, da = t_i \left(\frac{da^*}{da} - 1 \right) da$$
$$\simeq t_i (u_{k,k} - n_i n_j u_{j,i}) \, da,$$

where the last expression is correct to first order. Here, n_i denotes the unit normal on ∂B_2. By considering the special case $t_i = -n_i p$, for constant p, it may easily be shown that the increment $(t_i^* - t_i) \, da$ is *not* conservative (WESOŁOWSKI [1964, *15*]).

[4] For simplicity the condition is stated only for stress vectors, but it holds equally for body force. When these are included, Eq. (67.1) bears a formal resemblance to the condition of self-adjointness. See HILL [1962, *7*] and SEWELL [1967, *31*].

[5] SEWELL [1965, *20*], [1967, *31*] considers the potential $\psi = (\alpha/2) \varrho (f_i^* - f_i) u_i$. Condition (67.1) is then replaced by $(1-\alpha) \int\limits_{\partial B} \varrho (f_i^* - f_i) \dot{u}_i \, dx = \alpha \int\limits_{\partial B} \varrho \overline{(f_i^* - f_i)}^{\boldsymbol{\cdot}} u_i \, dx$.

[6] SEWELL [1965, *20*], [1967, *31*], HILL [1962, *7*].

where K_{ij} and π_{ijk} are functions of position and D_k is the surface gradient operator defined by

$$D_k = (\delta_{kp} - n_k n_p) \frac{\partial}{\partial x_p}, \tag{67.4}$$

where n_i is the unit outward normal on ∂B_2. We note also the relations

$$\frac{\partial u_i}{\partial x_j} = n_j n_p \frac{\partial u_i}{\partial x_p} + D_j u_i, \tag{67.5}$$

and

$$\pi_{ijk}(D_k u_j)\, \dot{u}_i = D_p[\pi_{ijk}(\delta_{kp} - n_k n_p)(u_j \dot{u}_i - \dot{u}_j u_i)] - \pi_{ijk} u_j (D_k \dot{u}_i). \tag{67.6}$$

The insertion of (67.3) into (67.1) and use of (67.6) then shows that the incremental stress vector (67.3) has the potential density

$$\varphi = \tfrac{1}{2} K_{ij} u_j u_i + \tfrac{1}{2} \pi_{ijk} u_i (D_k u_j), \tag{67.7}$$

provided that

$$\pi_{ijk} + \pi_{jik} = 0, \tag{67.8}$$

$$K_{ij} - K_{ji} = D_p[\pi_{ijk}(\delta_{kp} - n_k n_p)], \tag{67.9}$$

$$\int_{\partial B_2(t)} D_p[\pi_{ijk}(\delta_{kp} - n_k n_p)\, u_j \dot{u}_i]\, da = 0. \tag{67.10}$$

By use of Stokes' theorem, it may be shown that (67.10) is identically satisfied when either $\partial B_1 = \emptyset$ or $u_i = 0$ on ∂B_1, and then (67.8) and (67.9) alone are sufficient conditions.

Another method of characterising weakly conservative loads results from seeking conditions under which a potential function of a given form exists. We may thus ask for conditions on the incremental stress vector such that the surface potential density in (66.4) is given by

$$\varphi = \varphi(x_i, u_i, \ldots, u_{i,jk}, \ldots, \dot{u}_i, \ldots, \overset{(n)}{u}_{i,jk} \ldots). \tag{67.11}$$

A technique, first described by Sewell,[7] may be extended, in principle, to the potential (67.11). We shall not present the full detail of this method but we shall consider instead the special case in which the loads correspond to a *hydrostatic* stress.[8] In this situation, the stress vectors are always normal to the deformed surface of the body and have the same magnitude per unit area of the deformed surface. They thus exemplify the notion of follower force introduced in Chap. Q. For given scalar function σ, the stress vectors are given by

$$t_i = \sigma n_i, \qquad t_i^{**} = \sigma n_i^*, \tag{67.12}$$

where n_i, n_i^* are the outward unit normals on ∂B_2, ∂B_2^* and t_i^{**} is the stress vector per unit surface area of ∂B_2^*. There is a tension or compression according as $\sigma > 0$ or $\sigma < 0$. We then find from (67.12) that

$$(t_i^* - t_i)\, da = \sigma \left(n_i^* \frac{da^*}{da} - n_i \right) da, \tag{67.13}$$

and with the help of standard results in differential geometry we then obtain

$$\int_{\partial B_2(t)} (t_i^* - t_i)\, \dot{u}_i\, da = \int_{\partial B_2(t)} \sigma (n_r \dot{u}_r u_{k,k} - \dot{u}_i u_{r,i} n_r + A_{ir} \dot{u}_i n_r)\, da, \tag{67.14}$$

[7] Sewell [1965, *20*], [1967, *31*]. He was apparently the first to consider this topic, but concerned himself, however, with virtual displacements and with potentials that gave zero work round closed paths. The latter requirement restricts the arguments of φ to the displacement and its surface derivatives.

[8] See Duhem [1906, *2*], Pearson [1956, *2*], Herbert [1966, *9*], Sewell [1967, *31*], Beatty [1970, *2*].

where

$$2A_{ir}=e_{ijk}\,e_{rpq}\,u_{j,p}\,u_{k,q}. \tag{67.15}$$

Let us first suppose that σ is constant and that either $u_i=0$ on ∂B_1 or $\partial B_1=\emptyset$. Then, by use of the divergence theorem, (67.14) becomes

$$\int\limits_{\partial B_2(t)} (t_i^*-t_i)\,\dot u_i\,da=\frac{\sigma}{2}\,\frac{d}{dt}\int\limits_{B(t)} (u_{k,k}\,u_{i,i}-u_{i,r}\,u_{r,i}+2\det[u_{i,j}])\,dx. \tag{67.16}$$

Hence, under these circumstances the hydrostatic load is weakly conservative in the sense of (66.3). For later use we observe that linearisation of (67.16) produces[9]

$$\int\limits_{\partial B_2(t)} (t_i^*-t_i)\,\dot u_i\,da=\frac{\sigma}{2}\,\frac{d}{dt}\int\limits_{B(t)} (u_{k,k}\,u_{i,i}-u_{i,r}\,u_{r,i})\,dx. \tag{67.17}$$

When σ is variable, and $u_i\neq 0$ on ∂B_1 or $\partial B_1\neq\emptyset$, the hydrostatic stress is not, in general, weakly conservative. However, it is still possible to derive conditions for which this is so. We may rearrange (67.14) to obtain

$$\int\limits_{\partial B_2(t)} (t_i^*-t_i)\,\dot u_i\,da=\tfrac{1}{2}\dot\Phi_1-\tfrac{1}{2}\int\limits_{\partial B_2(t)}\sigma(\boldsymbol{n}\times\nabla\cdot(\boldsymbol{u}\times\dot{\boldsymbol{u}}))\,da-\tfrac{1}{3}\int\limits_{\partial B_2(t)}\sigma(\boldsymbol{n}\times\nabla\cdot\boldsymbol{B})\,da, \tag{67.18}$$

where

$$\Phi_1=\int\limits_{\partial B_2(t)}\sigma(n_r\,u_r\,u_{k,k}-u_i\,u_{r,i}\,n_r+\tfrac{1}{3}A_{ir}\,u_i\,n_r)\,da, \tag{67.19}$$

$$B_p=e_{ijk}\,u_j\,u_{i,p}\,\dot u_k. \tag{67.20}$$

When STOKES' theorem is applied to the last two terms on the right of (67.18) we then get

$$\begin{aligned}\int\limits_{\partial B_2(t)} (t_i^*-t_i)\,\dot u_i\,da=\tfrac{1}{2}\dot\Phi_1&-\tfrac{1}{2}\oint\limits_C\sigma\boldsymbol{u}\times\dot{\boldsymbol{u}}\cdot d\boldsymbol{s}-\tfrac{1}{3}\oint\limits_C\sigma\boldsymbol{B}\cdot d\boldsymbol{s}\\&+\tfrac{1}{2}\int\limits_{\partial B_2(t)}\boldsymbol{u}\times\dot{\boldsymbol{u}}\cdot(\boldsymbol{n}\times\nabla\sigma)\,da+\tfrac{1}{3}\int\limits_{\partial B_2(t)}\boldsymbol{B}\cdot(\boldsymbol{n}\times\nabla\sigma)\,da,\end{aligned}$$

where C is the curve bounding ∂B_2 and s is the arc-length along C. Hence, sufficient conditions for the hydrostatic stress to be weakly conservative are

$$\boldsymbol{n}\times\nabla\sigma=0, \tag{67.21}$$

$$\oint\limits_C\sigma\boldsymbol{u}\times\dot{\boldsymbol{u}}\cdot d\boldsymbol{s}=0,\qquad \oint\limits_C\sigma\boldsymbol{B}\cdot d\boldsymbol{s}=0. \tag{67.22}$$

These results are essentially those obtained by BEATTY using some conclusions of SEWELL.[10] BEATTY also considers the effect of rigid body motions and shows that in order for the uniform hydrostatic stress to be weakly conservative with potential Φ then necessarily the moment of the hydrostatic tractions must vanish.

Let us observe that even for a uniform σ, when (67.21) is identically satisfied, the hydrostatic stress is not weakly conservative unless (67.22) are also satisfied. This is the case when, for example, $\partial B_1=\emptyset$, which has already been considered.

---

[9] PEARSON [1956, *2*], HERBERT [1966, *9*], SEWELL [1967, *31*], NEMAT-NASSER [1968, *22*] and BEATTY [1970, *2*].

[10] BEATTY [1970, *2*], SEWELL [1967, *31*]; see also HERBERT [1966, *9*].

In a manner similar to that just described, NEMAT-NASSER[11] has treated hydrostatic loads for which

$$t_i = \sigma n_i, \quad t_i^{**} = \sigma^* n_i^* \quad \text{and} \quad \sigma^* \, da^* = \sigma \, da.$$

68. Stability analyses. For a body with a prescribed weakly conservative body force and prescribed weakly conservative stress vector on a part ∂B_2 of the surface with the remainder of the surface ∂B_1 being held fixed, the function H_0 arising in the stability analyses of Chap. H is given by

$$H_0(t) = \Phi(t) + \Psi(t), \tag{68.1}$$

where we have arbitrarily set $\Phi(0) + \Psi(0) = 0$. Consequently, if we modify the internal energy by the addition of the potentials (68.1), then the stability analyses become identical to those for dead loading, and there is no need to repeat them in detail. As an alternative approach we could apply the conclusions of Sect. 30 exactly. These require $H_0(t)$ to satisfy certain inequalities which in the present circumstances are transferred to $\Phi(t) + \Psi(t)$. Naturally, we may still include in the analysis non-conservative forces, but then the inconclusiveness already noted in Sect. 63 continues to hold for the non-conservative part of $H_0(t)$. Where the forces are path-dependent, it cannot be expected, in general, that these conditions are fulfilled, but where dissipative forces are also present, then the non-conservative part of H_0 will be positive-definite when the dissipation is greater than the rate of working of the non-conservative forces.[12]

Similar comments apply to stability analyses of the linearised equations, provided only that the linearised incremental body forces and surface tractions are also weakly conservative.

Extensions of theorems and results obtained in Chaps. I–P can be obtained in an obvious manner and so there is no need to record them. In particular, the critiques of the classical tests of energy and adjacent equilibrium still hold.

Particular examples may also be discussed in a similar fashion.[13] For instance, for a body subject to uniform hydrostatic stress,[14] the expression (67.17) shows that the linearised conservation of energy equation becomes

$$\frac{1}{2}\int_B \varrho \dot{u}_i \dot{u}_i \, dx + \frac{1}{2}\int_B d_{ijkl} u_{i,j} u_{k,l} \, dx$$
$$- \frac{\sigma}{2}\int_B (\tilde{e}_{ii}\tilde{e}_{kk} - \tilde{e}_{ir}\tilde{e}_{ir} + \omega_{ir}\omega_{ir}) \, dx = \text{const},$$

where $\tilde{e}_{ij}$, ω_{ij} are components of the linearised strain and rotation tensors. The discussion may now be completed in exactly the same way as for dead loads, and this shows that isotropic homogeneous compressible elastic bodies have a stable configuration of uniform dilatation provided

$$\tilde{\mu} > 0, \quad \tilde{k} > 0,$$

---

[11] NEMAT-NASSER [1968, *22*].

[12] See also NEMAT-NASSER and HERRMANN [1966, *22*], NEMAT-NASSER [1968, *22*], [1970, *24*], NEMAT-NASSER and ROORDA [1967, *25*].

[13] PLAUT [1967, *28*] has described many applications of LIAPOUNOV's theorem to one and two dimensional models of conservative systems.

[14] BEATTY [1970, *2*] considers this problem for the energy criterion of stability. See also HERBERT [1966, *9*].

where $\tilde{\mu}$, $\tilde{k}$ are as previously defined. For incompressible bodies, the corresponding result is

$$\tilde{\mu} > 0.$$

S. Stability with time-dependent and position-dependent data.

69. Introduction. We turn now to problems in which the data depend upon time and position in the reference or primary state. At this stage motion- or configuration-dependent data are not considered. The data include surface displacement, surface tractions and body forces, and by means of the latter we shall also be able to examine initial data and elasticities which are time-dependent and position-dependent. The major difficulty in all problems encountered in the remainder of the Treatise is that the governing equations cease to be self-adjoint. The customary procedure is then to use eigenfunction expansions, but for the group of problems under immediate consideration it is possible to employ either energy arguments or convexity arguments of a kind previously used in Chaps. J, L, M. We shall describe two energy arguments, a novel feature in one of these being the introduction of a dual problem.

All our discussions concern linearised problems. Because stability is assessed with respect to square integrable measures, our results extend easily to generalised solutions in those cases where essential smoothness assumptions are not violated.

70. Prescribed surface traction with zero initial data. We consider the stability, in a sense to be made precise, of the null solution in the class of linearised incremental displacements when the displacement is held fixed on a part $\overline{\partial B_1}$ of the surface and time-dependent tractions are specified on the remainder, ∂B_2. Body forces are assumed to be absent. The equations for this problem are therefore

$$(d_{ijkl}\, u_{k,l})_{,j} = \varrho\, \ddot{u}_i \qquad \text{on } B \times (0, t_1), \tag{70.1}$$

$$u_i = 0 \qquad \text{on } \overline{\partial B_1} \times (0, t_1), \tag{70.2}$$

$$n_j\, d_{ijkl}\, u_{k,l} = t_i \qquad \text{on } \partial B_2 \times (0, t_1), \tag{70.3}$$

where $t_i = t_i(t, \boldsymbol{x})$ is a specified function of position and time. The initial displacement and velocity are also specified, and will be taken to be zero in present circumstances.

The elasticities are time-independent so that the primary state is an equilibrium configuration. They are also positive-definite in the sense given by

$$\int_B d_{ijkl}\, \xi_{ij}\, \xi_{kl}\, dx \geqq d_0 \int_B \xi_{ij}\, \xi_{ij}\, dx, \tag{70.4}$$

for positive constant d_0 and arbitrary second-order tensors ξ_{ij}. In addition, the elasticities satisfy the symmetry relations

$$d_{ijkl} = d_{klij}, \tag{70.5}$$

and are assumed to be continuously differentiable on B. The problem as defined is formally self-adjoint, but because of the non-homogeneous boundary condition (70.3) it fails in fact to be self-adjoint.

Now from (70.1)–(70.3) we find the equation[1]

$$\frac{dE(t)}{dt} = \int_{\partial B_2(t)} t_i \dot{u}_i \, da, \tag{70.6}$$

where $E(t)$ is the total energy. After a time-integration, we obtain

$$E(t) - E(0) = \int_{\partial B_2(t)} t_i u_i \, da - \int_{\partial B_2(0)} t_i u_i \, da - \int_0^t \int_{\partial B_2(\eta)} \frac{\partial t_i}{\partial \eta} u_i \, da \, d\eta, \tag{70.7}$$

where we have assumed that the order of integration may be reversed. An application of SCHWARZ's inequality to the last integral on the right then yields

$$\int_0^t \int_{\partial B_2(\eta)} \frac{\partial t_i}{\partial \eta} u_i \, da \, d\eta \leqq \int_0^t \left(\int_{\partial B_2(\eta)} \frac{\partial t_i}{\partial \eta} \frac{\partial t_i}{\partial \eta} \, da \right)^{\frac{1}{2}} \left(\int_{\partial B_2(\eta)} u_i u_i \, da \right)^{\frac{1}{2}} d\eta. \tag{70.8}$$

We next use the Stekloff inequality associated with the variational quotient:

$$\mu = \min_{w=0 \text{ on } \overline{\partial B_1}} \left(\frac{\int_B w_{,i} w_{,i} \, dx}{\int_{\partial B_2} w^2 \, da} \right), \tag{70.9}$$

where μ is the solution to the eigenvalue problem

$$\left.\begin{aligned} \Delta w &= 0 && \text{in } B, \\ w &= 0 && \text{on } \overline{\partial B_1}, \\ \frac{\partial w}{\partial n} &= \mu w && \text{on } \partial B_2. \end{aligned}\right\} \tag{70.10}$$

Substituting (70.9) into (70.8) and using (70.4) and treating the first integral on the right in similar fashion then enables the energy equation (70.7) to be written

$$\begin{aligned} E(t) - E(0) + \int_{\partial B_2(0)} t_i u_i \, da \leqq [2\tilde{E}(t) \, d_0 \mu]^{\frac{1}{2}} \Bigg[\Bigg(\int_{\partial B_2(t)} t_i t_i \, da \Bigg)^{\frac{1}{2}} \\ + \int_0^t \Bigg(\int_{\partial B_2(\eta)} \frac{\partial t_i}{\partial \eta} \frac{\partial t_i}{\partial \eta} \, da \Bigg)^{\frac{1}{2}} d\eta \Bigg], \end{aligned} \tag{70.11}$$

where $\tilde{E}(t)$ is the maximum value of the total energy on $[0, t)$. We assume $\tilde{E}(t)$ remains bounded on $[0, \infty)$.

Let us now suppose

$$u_i(0, \boldsymbol{x}) = \dot{u}_i(0, \boldsymbol{x}) = t_i(0, \boldsymbol{x}) = 0. \tag{70.12}$$

Then

$$E(0) = \int_{\partial B_2(0)} t_i u_i \, da = 0.$$

Moreover, we have

$$\frac{d}{dt} \int_{\partial B_2(t)} t_i t_i \, da = 2 \int_{\partial B_2(t)} t_i \dot{t}_i \, da \leqq 2 \left(\int_{\partial B_2(t)} t_i t_i \, da \right)^{\frac{1}{2}} \left(\int_{\partial B_2(t)} \dot{t}_i \dot{t}_i \, da \right)^{\frac{1}{2}}, \tag{70.13}$$

[1] The argument that follows is based essentially on that first proposed by SHIELD [1965, *21*].

which after an integration and use of (70.12) leads to

$$\Big(\int\limits_{\partial B_2(t)} t_i\, t_i\, da\Big)^{\frac{1}{2}} \leqq \int\limits_0^t \Big(\int\limits_{\partial B_2(\eta)} \frac{\partial t_i}{\partial \eta}\,\frac{\partial t_i}{\partial \eta}\, da\Big)^{\frac{1}{2}} d\eta. \tag{70.14}$$

Thus, with the help of (70.12) and (70.14) we may write (70.11) as

$$E(t) \leqq (8/d_0\,\mu)\left(\int\limits_0^t \Big(\int\limits_{\partial B_2(\eta)} \frac{\partial t_i}{\partial \eta}\,\frac{\partial t_i}{\partial \eta}\, da\Big)^{\frac{1}{2}} d\eta\right)^2. \tag{70.15}$$

A cruder inequality, but one with a more obvious physical interpretation, may be obtained by replacing (70.8) by

$$\int\limits_0^t \int\limits_{\partial B_2(\eta)} \frac{\partial t_i}{\partial \eta}\, u_i\, da\, d\eta \leqq \left(\int\limits_0^t \int\limits_{\partial B_2(\eta)} \frac{\partial t_i}{\partial \eta}\,\frac{\partial t_i}{\partial \eta}\, da\, d\eta\right)^{\frac{1}{2}} \left(\int\limits_0^t \int\limits_{\partial B_2(\eta)} u_i\, u_i\, da\, d\eta\right)^{\frac{1}{2}}.$$

This leads to

$$E(t) \leqq (8t/d_0\,\mu) \int\limits_0^t \int\limits_{\partial B_2(\eta)} \frac{\partial t_i}{\partial \eta}\,\frac{\partial t_i}{\partial \eta}\, da\, d\eta, \tag{70.16}$$

which corresponds to (70.15).

When $\partial B_1 = \emptyset$ certain normalisations must be imposed on the solution in order to exclude any instability that may arise from possible rigid body motions. The analysis, however, is similar to that conducted for dead loading and therefore is not repeated.

To complete the present analysis we observe that the elasticities are positive-definite, and therefore exactly as in Sect. 35, the strain energy is an upper bound for the weighted L_2-norm of the displacement. We have therefore proved that *under the initial conditions* (70.12) *the null solution of* (70.1) *is stable for the measure*

$$\varrho(u(t)) = \int\limits_{B(t)} \varrho\, u_i\, u_i\, dx, \tag{70.17}$$

provided that

$$\left(\int\limits_0^t \Big(\int\limits_{\partial B_2(\eta)} \frac{\partial t_i}{\partial \eta}\,\frac{\partial t_i}{\partial \eta}\, da\Big)^{\frac{1}{2}} d\eta\right)^2 \quad \text{or} \quad t \int\limits_0^t \int\limits_{\partial B_2(\eta)} \frac{\partial t_i}{\partial \eta}\,\frac{\partial t_i}{\partial \eta}\, da\, d\eta \tag{70.18}$$

remain sufficiently small, and the elasticities obey the positive-definiteness condition (70.4) *and the symmetry* (70.5).

We note that if $u_i^{(1)}$ and $u_i^{(2)}$ represent two solutions of (70.1)–(70.3) corresponding to the surface tractions $t_i^{(1)}, t_i^{(2)}$, then $w_i = u_i^{(1)} - u_i^{(2)}$ satisfies the same set of equations and boundary conditions corresponding to the surface traction $t_i^{(1)} - t_i^{(2)}$. Thus, inequalities (70.15), (70.16) hold for w_i and $t_i = t_i^{(1)} - t_i^{(2)}$, and hence we may prove that any solution of (70.1)–(70.3) is stable with respect to changes in surface tractions in the above measures. A similar statement holds for all subsequent results of like nature.

71. Prescribed surface traction with non-zero initial data. We next discuss a second method of determining stability for which the zero initial data (70.12) is not essential. The method depends on a certain dual problem which we now introduce.[2] Let $u_i(t, \boldsymbol{x})$ obey (70.1)–(70.3) and $v_i(t, \boldsymbol{x})$ satisfy the conditions

$$(d_{ijkl}\, v_{i,j})_{,l} = \varrho\, \ddot{v}_k \quad \text{in } B \times (0, t_1), \tag{71.1}$$

$$v_i = 0 \quad \text{on } \overline{\partial B_1} \times (0, t_1), \tag{71.2}$$

$$n_l\, d_{ijkl}\, v_{i,j} = 0 \quad \text{on } \partial B_2 \times (0, t_1), \tag{71.3}$$

$$v_i(t_1, \boldsymbol{x}) = 0, \quad \dot{v}_i(t_1, \boldsymbol{x}) = u_i(t_1, \boldsymbol{x}) \quad \text{on } B \times t_1. \tag{71.4}$$

We assume that a function v_i satisfying this problem always exists.[3] An integration by parts with respect to both space and time coordinates shows that

$$\begin{aligned} \int_{B(t_1)} \varrho\, u_i\, u_i\, dx = \int_{B(0)} \varrho\,(u_i\, \dot{v}_i - \dot{u}_i\, v_i)\, dx + \int_{B(t_1)} \varrho\, v_i\, \dot{u}_i\, dx \\ + \int_0^{t_1} \int_{\partial B(\eta)} d_{ijkl}(u_k\, n_l\, v_{i,j} - v_i\, n_j\, u_{k,l})\, da\, d\eta, \end{aligned} \tag{71.5}$$

and then using the assumed conditions on u_i and v_i we find that

$$\int_{B(t_1)} \varrho\, u_i\, u_i\, dx = -\int_0^{t_1} \int_{\partial B_2(\eta)} v_i\, t_i\, da\, d\eta + \int_{B(0)} \varrho\,(u_i\, \dot{v}_i - \dot{u}_i\, v_i)\, dx. \tag{71.6}$$

The first integral on the right in (71.6) is of the same form as integrals on the right of (70.7), and hence we may apply SCHWARZ's and STEKLOFF's inequalities in the same way to obtain

$$\int_0^{t_1} \int_{\partial B_2(\eta)} v_i\, t_i\, da\, d\eta \leq \alpha\, E_v^{\frac{1}{2}}(t_1) \int_0^{t_1} \Big(\int_{\partial B_2(\eta)} t_i\, t_i\, da \Big)^{\frac{1}{2}} d\eta, \tag{71.7}$$

where $\alpha^2 = 2\mu^{-1}\, d_0^{-1}$, and $E_v(t)$ is the total energy associated with the dual problem (71.1)–(71.4). We note at once that the homogeneous data (71.2) and (71.3) imply

$$E_v(t) = E_v(t_1), \quad t \in [0, t_1], \tag{71.8}$$

which also has been used in deriving (71.8). Next, we treat the second term on the right of (71.6) and by means of the arithmetic-geometric mean inequality and the inequalities of Sect. 35 obtain

$$\int_{B(0)} \varrho\,(u_i\, \dot{v}_i - v_i\, \dot{u}_i)\, dx \leq 2^{\frac{1}{2}} \beta\, E_v^{\frac{1}{2}}(t_1) \Big(\int_{B(0)} \varrho\,(u_i\, u_i + \dot{u}_i\, \dot{u}_i)\, dx \Big)^{\frac{1}{2}}, \tag{71.9}$$

where β^2 is a computable constant, and we have again used (71.7). Inserting (71.9) and (71.7) into (71.6), and recalling that by (71.4)

$$E_v(t_1) = \tfrac{1}{2} \int_{B(t_1)} \varrho\, u_i\, u_i\, dx, \tag{71.10}$$

(71.6) then produces

$$\begin{aligned} \int_{B(t)} \varrho\, u_i\, u_i\, dx \leq \alpha^2 \left[\int_0^t \Big(\int_{\partial B_2(\eta)} t_i\, t_i\, da \Big)^{\frac{1}{2}} d\eta \right]^2 \\ + 2\beta^2 \int_{B(0)} \varrho\,(u_i\, u_i + \dot{u}_i\, \dot{u}_i)\, dx \qquad t \geq 0. \end{aligned} \tag{71.11}$$

---

[2] This technique is due to L. E. PAYNE (private communication).

[3] Because of the symmetry condition (70.5), existence of v_i is equivalent to existence of the solution to problem (70.1)–(70.3), which is one of our basic assumptions.

We may alternatively complete the analysis by replacing (71.7) with

$$\int_0^{t_1}\int_{\partial B_2(\eta)} v_i t_i \, da \, d\eta \leqq \alpha t_1^{\frac{1}{2}} \left(\int_0^{t_1}\int_{\partial B_2(\eta)} t_i t_i \, da \, d\eta\right)^{\frac{1}{2}} E_v^{\frac{1}{2}}(t_1),$$

and then instead of (71.11) we obtain

$$\int_{B(t)} \varrho u_i u_i \, dx \leqq \alpha^2 t \int_0^t \int_{\partial B_2(\eta)} t_i t_i \, da \, d\eta + 2\beta^2 \int_{B(0)} \varrho (u_i u_i + \dot{u}_i \dot{u}_i) \, dx. \tag{71.12}$$

These inequalities then clearly establish that the *null solution of* (70.1) *is stable in the L_2-norm measure* (70.17) *provided that the data, as measured by the right sides of* (71.11) *or* (71.12), *remain small, and the elasticities satisfy the positive-definiteness condition* (70.4) *and the symmetry* (70.5).

The stability estimates (71.11) and (71.12) represent an important improvement on (70.15) and (70.16) since they do not contain time-rates of the surface traction, and they also admit non-zero initial data. Thus, for instance, (71.11) and (71.12) provide useful stability estimates even when severe oscillations occur in the surface tractions.

In the next section we use the same dual problem to consider stability in the corresponding case in which displacements are specified on the surface.

72. Prescribed surface displacement. In this section we consider the stability of an equilibrium position of an elastic body for which the incremental surface tractions, where prescribed, are zero, while the surface displacement is a specified function of position and time. The body force is either dead or zero. We shall use the technique based on the dual problem at the end of the previous section, although in Sect. 76 we shall mention, for completeness, an entirely different method.

Our problem is equivalent to examining the stability of $u_i \equiv 0$ in the class of displacements satisfying

$$(d_{ijkl} u_{k,l})_{,j} = \varrho \ddot{u}_i \quad \text{in } B \times (0, t_1), \tag{72.1}$$

$$u_i = h_i \quad \text{on } \overline{\partial B_1} \times (0, t_1), \tag{72.2}$$

$$n_j d_{ijkl} u_{k,l} = 0 \quad \text{on } \partial B_2 \times (0, t_1), \tag{72.3}$$

where $h_i(t, \boldsymbol{x})$ is a specified function of position and time. An obvious special case occurs when $\partial B_2 = \emptyset$. The elasticities are assumed positive-definite in the sense of (70.4) and to satisfy the symmetry condition (70.5).

We introduce the dual problem defined by (71.1)–(71.4), and observe that because of the conditions now assumed on u_i and v_i, Eq. (71.5) becomes

$$\int_{B(t_1)} \varrho u_i u_i \, dx = \int_{B(0)} \varrho (u_i \dot{v}_i - \dot{u}_i v_i) \, dx + \int_0^{t_1} \int_{\partial B_1(\eta)} d_{ijkl} h_k n_l v_{i,j} \, da \, d\eta. \tag{72.4}$$

Repeated application of SCHWARZ's inequality to the second term on the right of (72.4) then gives

$$\int_{\partial B_1(t)} d_{ijkl} h_k n_l v_{i,j} \, da \leqq M \left(\int_{\partial B_1(t)} h_k h_k \, da\right)^{\frac{1}{2}} \left(\int_{\partial B_1(t)} \frac{\partial v_i}{\partial n} \frac{\partial v_i}{\partial n} \, da\right)^{\frac{1}{2}}, \tag{72.5}$$

where

$$M^2 = \max_{\boldsymbol{x} \in B} (d_{ijkl} d_{ijkl}), \tag{72.6}$$

and we have used the fact that v_i vanishes on ∂B_1. We next use the inequality

$$\nu = \min_{w=0 \text{ on } \partial B_2} \left(\frac{\int\limits_{B} w_{,i}\, w_{,i}\, dx}{\int\limits_{\partial B_1} \left(\frac{\partial w}{\partial n}\right)^2 da} \right), \tag{72.7}$$

where ν is the solution to the eigenvalue problem

$$\left.\begin{aligned} \Delta w &= 0 && \text{in } B \\ w &= 0 && \text{on } \partial B_2 \\ \nu \frac{\partial w}{\partial n} &= w && \text{on } \partial B_1. \end{aligned}\right\} \tag{72.8}$$

Substituting (72.7) into (72.5) and using (70.4) yields

$$\int\limits_{\partial B_1(t)} d_{ijkl}\, h_k\, n_l\, v_{i,j}\, da \leqq 2^{\frac{1}{2}} E_v^{\frac{1}{2}}(t_1)\, M \Big(\int\limits_{\partial B_1(t)} h_k\, h_k\, da \Big)^{\frac{1}{2}} \tag{72.9}$$

where we have again used the conservation law (71.8). Thus, from (72.9), (72.4), (71.9), and (71.10) we finally obtain

$$\int\limits_{B(t)} \varrho u_i\, u_i\, dx \leqq 4 M^2 \left[\int\limits_0^t \Big(\int\limits_{\partial B_1(\eta)} h_k\, h_k\, da \Big)^{\frac{1}{2}} d\eta \right]^2 + 2\beta^2 \int\limits_{B(0)} \varrho (u_i\, u_i + \dot{u}_i\, \dot{u}_i)\, dx, \tag{72.10}$$

from which stability can easily be ascertained.

Alternatively, we may apply SCHWARZ's inequality over space-time to the second term on the right of (72.4) and obtain an estimate analogous to (71.12).

73. Prescribed body force. We consider the stability of an equilibrium position of an elastic body part of whose boundary is fixed and the remainder is subject to dead loading. The body force $f_i(t, \boldsymbol{x})$ (previously denoted by $F_i(t, \boldsymbol{x})$) is a prescribed function of position and time. While such problems have limited practical application, we shall use the results derived to establish further physically important conclusions in subsequent sections.

We therefore consider the problem

$$(d_{ijkl}\, u_{k,l})_{,j} + \varrho f_i = \varrho \ddot{u}_i \quad \text{in } B \times (0, t_1), \tag{73.1}$$

$$u_i = 0 \quad \text{on } \overline{\partial B_1} \times (0, t_1), \tag{73.2}$$

$$n_j\, d_{ijkl}\, u_{k,l} = 0 \quad \text{on } \partial B_2 \times (0, t_1), \tag{73.3}$$

in which the elasticities are positive-definite in the sense of (70.4) and satisfy the symmetry condition (70.5).

We shall again develop two methods for stability, one depending directly on a treatment of the total energy,[4] the other on the introduction of a dual problem. In the first method, we assume that there is homogeneous initial data, and we shall find that several estimates are possible.

From (73.1)–(73.3) we easily obtain

$$\dot{E}(t) = \int\limits_{B(t)} \varrho f_i\, \dot{u}_i\, dx, \tag{73.4}$$

[4] We adopt SHIELD's [1965, *21*] method for obtaining stability estimates.

where, as usual, $E(t)$ denotes the total energy. SCHWARZ's inequality together with the positive-definiteness of the elasticities next show that

$$\dot{E}(t) \leqq 2^{\frac{1}{2}} E^{\frac{1}{2}}(t) \Big(\int\limits_{B(t)} \varrho f_i f_i \, dx \Big)^{\frac{1}{2}}, \tag{73.5}$$

which upon integration becomes

$$2E(t) \leqq \Big(\int\limits_0^t \Big(\int\limits_{B(\eta)} \varrho f_i f_i \, dx \Big)^{\frac{1}{2}} d\eta \Big)^2, \tag{73.6}$$

since the initial displacement and velocity are zero by hypothesis. On recalling the computations of Sect. 35 showing that the strain energy is an upper bound for the weighted L_2-norm integral of the displacement we find that the following is proved: *for homogeneous mixed boundary data*[5] *and homogeneous initial data the null displacement* $u_i = 0$ *is stable in the class of solutions of* (73.1) *as measured by* (70.17) *provided the body force is small in measure*

$$\int\limits_0^t \Big(\int\limits_{B(\eta)} \varrho f_i f_i \, dx \Big)^{\frac{1}{2}} d\eta \tag{73.7}$$

and the elasticities obey the positive-definiteness condition (70.4) *and the symmetry* (70.5).

An alternative estimate to (73.6) may be obtained as follows. We integrate by parts the right side of (73.4) to obtain

$$E(t) = \int\limits_{B(t)} \varrho f_i u_i \, dx - \int\limits_0^t \int\limits_{B(\eta)} \varrho \frac{\partial f_i}{\partial \eta} u_i \, dx \, d\eta. \tag{73.8}$$

SCHWARZ's inequality and the estimates of Sect. 35 in the *mixed* boundary value problem applied to (73.8) then give

$$\tilde{E}^{\frac{1}{2}}(t) \leqq \alpha \Big(\int\limits_{B(t)} \varrho f_i f_i \, dx \Big)^{\frac{1}{2}} + \alpha \int\limits_0^t \Big(\int\limits_{B(\eta)} \varrho \frac{\partial f_i}{\partial \eta} \frac{\partial f_i}{\partial \eta} \, dx \Big)^{\frac{1}{2}} d\eta \tag{73.9}$$

for some computable positive constant α. An inequality similar to (70.14) then finally yields

$$E(t) \leqq 4\alpha^2 \Big(\int\limits_0^t \Big(\int\limits_{B(\eta)} \varrho f_i f_i \, dx \Big)^{\frac{1}{2}} d\eta \Big)^2. \tag{73.10}$$

It is clear from the above that for the validity of (73.6) we require only that the elasticities be non-negative. We shall now obtain a stability estimate when this less restrictive condition holds.

Let

$$F(t) = \int\limits_{B(t)} \varrho u_i u_i \, dx. \tag{73.11}$$

Then on differentiating and applying SCHWARZ's inequality, we find the inequality

$$\dot{F}(t) \leqq 2^{\frac{3}{2}} F^{\frac{1}{2}}(t) \, E^{\frac{1}{2}}(t), \tag{73.12}$$

in which the non-negativeness of the elasticities has also been used. Substitution for $E(t)$ from (73.6) and integration then produce

$$F(t) \leqq \Big(\int\limits_0^t (t-\eta) \Big(\int\limits_{B(\eta)} \varrho f_i f_i \, dx \Big)^{\frac{1}{2}} d\eta \Big)^2, \tag{73.13}$$

[5] We do not treat the case in which the traction is specified dead over all the surface.

which is the required estimate. It is valid also when the traction is specified to be dead everywhere on the surface.

Further estimates may be derived with the help of the dual problem defined by (71.1)–(71.4), where u_i is the solution to the problem (73.1)–(73.3). Eq. (71.5) therefore becomes

$$\int_{B(t_1)} \varrho u_i u_i \, dx = \int_{B(0)} (u_i \dot{v}_i - \dot{u}_i v_i) \, dx - \int_0^{t_1} \int_{B(\eta)} \varrho f_i v_i \, dx \, d\eta, \tag{73.14}$$

which after application of SCHWARZ's inequality and condition (71.8) and (71.10) yields

$$\begin{aligned} \int_{B(t)} \varrho u_i u_i \, dx \leqq 2\alpha^2 \left(\int_0^t \left(\int_{B(\eta)} \varrho f_i f_i \, dx \right)^{\frac{1}{2}} d\eta \right)^2 \\ + 2\beta^2 \int_{B(0)} \varrho (u_i u_i + \dot{u}_i \dot{u}_i) \, dx, \quad t \geqq 0, \end{aligned} \tag{73.15}$$

where α, β have the same values as in Eq. (71.11). Clearly, (73.15) provides an estimate from which stability may be deduced. We note that while the initial data is not required to be zero in (73.15), the elasticities are required to be positive-definite and to satisfy the symmetry condition (70.5).

An estimate containing integrals analogous to those in (71.12) may be obtained as before and reads

$$\int_{B(t)} \varrho u_i u_i \, dx \leqq 2\alpha^2 \int_0^t \int_{B(\eta)} \varrho f_i f_i \, dx \, d\eta + 2\beta^2 \int_{B(0)} \varrho (u_i u_i + \dot{u}_i \dot{u}_i) \, dx. \tag{73.16}$$

By means of the inequalities just derived for the L_2-measures of the displacement in the body force problem, we may derive several more estimates corresponding to a variation in other data.

74. Variation in the elasticities. We assume there are two sets of elasticities $d^{(1)}_{ijkl}$ and $d^{(2)}_{ijkl}$, both of which are taken to be continuous, together with their spatial derivatives in B. We further suppose that $d^{(1)}_{ijkl}$ are functions of position alone and satisfy the positive-definiteness condition (70.4) and the symmetry condition (70.5). On the other hand, $d^{(2)}_{ijkl}$, apart from the continuity conditions just imposed, may be arbitrary, so that Eq. (73.1) with d_{ijkl} replaced by $d^{(2)}_{ijkl}$ governs small incremental displacements from a *motion* of the elastic body for which the deformations are large.

We assume that there exist solutions $u^{(1)}_i$, $u^{(2)}_i$ to the equations of linearised elastodynamics corresponding to the different elasticities $d^{(1)}_{ijkl}$, $d^{(2)}_{ijkl}$ when the remaining data continue to be the same.

Attention is now restricted to those solutions $u^{(2)}_i$ satisfying the boundedness condition

$$\int_{B(t)} \varrho (u^{(2)}_{i,j} u^{(2)}_{i,j} + u^{(2)}_{i,jk} u^{(2)}_{i,jk}) \, dx \leqq M^4 \tag{74.1}$$

where M is some given positive real finite constant. If we set

$$w_i = u^{(1)}_i - u^{(2)}_i, \tag{74.2}$$

then w_i satisfies (73.1)–(73.3), but with d_{ijkl} replaced by $d^{(1)}_{ijkl}$ and with

$$\varrho f_i = (\mathscr{D}_{ijkl} u^{(2)}_{k,l})_{,j}, \tag{74.3}$$

where

$$\mathscr{D}_{ijkl} = d^{(1)}_{ijkl} - d^{(2)}_{ijkl}. \tag{74.4}$$

We next find from (73.15) that

$$\int_{B(t)} \varrho w_i w_i\, dx \leqq 2\alpha^2 M \int_0^t \Big(\int_{B(\eta)} \varrho (\mathcal{D}_{ijkl}\mathcal{D}_{ijkl} + \mathcal{D}_{ijkl,r}\mathcal{D}_{ijkl,r})\, dx \Big)^{\frac{1}{2}} d\eta + 2\beta^2 \int_{B(0)} \varrho (w_i w_i + \dot{w}_i \dot{w}_i)\, dx, \tag{74.5}$$

which is one form of the required stability estimate. From (74.5), we may read off a self-evident stability result which we do not record explicitly. Other stability estimates may likewise be derived from (73.6) or related inequalities.

We now show how (74.5) may be used in the assessment of stability of the motion of a finitely deformed body.

Let u_i satisfy

$$(d^{(2)}_{ijkl} u_{k,l})_{,j} + \varrho f_i = \varrho \ddot{u}_i \quad \text{in } B \times (0, t_1), \tag{74.6}$$

where $d^{(2)}_{ijkl}$ may be time-dependent but are otherwise arbitrary apart from the implied smoothness condition. On the boundary let

$$u_i = h_i \quad \text{on } \overline{\partial B_1} \times (0, t_1), \tag{74.7}$$

$$n_j d^{(2)}_{ijkl} u_{k,l} = l_i \quad \text{on } \partial B_2 \times (0, t_1), \tag{74.8}$$

where h_i, l_i are prescribed functions of position and time, while initially let

$$u_i(0, \boldsymbol{x}) = p_i(\boldsymbol{x}), \quad \dot{u}_i(0, \boldsymbol{x}) = q_i(\boldsymbol{x}) \tag{74.9}$$

where p_i and q_i are prescribed functions of position. Let $v_i(t, \boldsymbol{x})$ satisfy

$$(d^{(1)}_{ijkl} v_{k,l})_{,j} + \varrho f_i = \varrho \ddot{v}_i \quad \text{in } B \times (0, t_1), \tag{74.10}$$

where $d^{(1)}_{ijkl}$ are positive-definite and symmetric in the stated sense and also possess the implied smoothness. Further we suppose v_i satisfies the boundary data (74.7), (74.8) and the initial data

$$v_i(0, \boldsymbol{x}) = p^{(1)}_i(\boldsymbol{x}), \quad \dot{v}_i(0, \boldsymbol{x}) = q^{(1)}_i(\boldsymbol{x}). \tag{74.11}$$

The solutions u_i are the incremental displacements against which stability of the motion is to be assessed and we suppose these satisfy condition (74.1). We obtain the appropriate stability estimates from the triangle inequality

$$\int_{B(t)} \varrho u_i u_i\, dx \leqq \int_{B(t)} \varrho (u_i - v_i)(u_i - v_i)\, dx + \int_{B(t)} \varrho v_i v_i\, dx.$$

Estimate (74.5) applies to the first term on the right, while the earlier estimates apply in the respective cases to the second term on the right. Stability may now be obtained and shows that if the null solution of the second problem involving v_i is stable with respect to its data, and the coefficients $d^{(1)}_{ijkl}$, $d^{(2)}_{ijkl}$ are sufficiently close in norm, then the motion of the finitely deformed elastic body is stable in the class of linearised incremental displacements u_i satisfying (74.6)–(74.9) and the boundedness condition (74.1).

75. Change in initial data under dead loading. In Sect. 35 we obtained stability of the null solution with respect to changes in the initial data under dead loading by employing the total energy as Liapounov function. However, the method only established stability with respect to an initial measure equal to the total energy so that in particular $u_i(0, \boldsymbol{x})$ had to be, for example, once continuously differentiable on B.[6]

[6] It is actually sufficient for $u_i(0, \boldsymbol{x})$ to have square integrable derivatives on B.

The results we have just proved in connexion with the dual and body force problems show, in fact, that for the mixed problem, at least, we may obtain stability under less restrictive assumptions.

First, let us consider (71.12) with $t_i \equiv 0$. Careful examination of the proof leading to (71.12) reveals that no spatial differentiability is required of the initial value of u_i and thus we have established the *stability of the null solution* $u_i = 0$ *to the mixed problem under dead loads in the class of small incremental displacements with respect to the measures*

$$\varrho_\tau(u) = \int_{B(0)} \varrho(u_i\, u_i + \dot{u}_i\, \dot{u}_i)\, dx, \tag{75.1}$$

$$\varrho\big(u(t)\big) = \int_{B(t)} \varrho u_i\, u_i\, dx, \tag{75.2}$$

provided the elasticities are positive-definite and obey the symmetry condition (70.5).

We may also use the body force problem to obtain estimates.[7] Let us suppose that $u_i(t, \boldsymbol{x})$ satisfies the Eqs. (73.1)–(73.3) with $f_i = 0$ and non-zero initial data. If we set

$$v_i(t, \boldsymbol{x}) = \int_0 (t - \eta)\, u_i(\eta, \boldsymbol{x})\, d\eta, \tag{75.3}$$

then $v_i(t, \boldsymbol{x})$ satisfies (73.1)–(73.3) with homogeneous initial and boundary conditions and with

$$\varrho f_i = u_i(0, \boldsymbol{x}) + t\dot{u}_i(0, \boldsymbol{x}). \tag{75.4}$$

We may thus use (73.16) to prove that the *null solution of* (73.1)–(73.3) *with* $f_i = 0$ *is stable with respect to the measures*

$$\varrho_\tau(u) = t_1 \int_{B(0)} \varrho\left(\frac{t_1^2}{3}\, \dot{u}_i\, \dot{u}_i + t_1\, u_i\, \dot{u}_i + u_i\, u_i\right) dx, \tag{75.5}$$

$$\varrho\big(u(t)\big) = \int_{B(t)} \varrho v_i\, v_i\, dx, \tag{75.6}$$

where v_i *is given by* (75.3), *provided the elasticities are positive-definite and satisfy the symmetry condition* (70.5).

76. Convexity arguments.[8] As in the treatment of problems with dead loads, we may employ convexity arguments in the present group of problems to remove altogether any definiteness requirement on the elasticities. It may be proved that under certain mild hypotheses the solution to the equations of linearised elasticity depends continuously, in the sense of HÖLDER, on several data in a L_2-measure defined over space-time. The choice of this measure may perhaps appear artificial until it is realised that the analysis must accommodate all possible instabilities. It must therefore give estimates valid in the case of resonance for which a spatial measure of the displacement cannot be expected to remain small.

The precise measure used for the incremental displacements is defined by

$$\|w\|_t^2 = \int_0^t \int_{B(\eta)} \varrho w_i(\eta, \boldsymbol{x})\, w_i(\eta, \boldsymbol{x})\, dx\, d\eta, \tag{76.1}$$

[7] Cp. KNOPS and PAYNE [1969, *13*, § 4].

[8] All the results in this section are due to KNOPS and PAYNE [1969, *13*].

and we shall restrict attention to those solutions u_i which for some prescribed constant M satisfy

$$\|u\|_{t_1}^2 \leqq M^2, \tag{76.2}$$

where t_1 is some time instant. Such solutions we say belong to the *class* $\mathscr{M}$.

The fundamental result concerns the linearised equations in which all data, apart from the body force, is kept zero. Thus, the equations governing the incremental displacements are given by (73.1)–(73.3) supplemented with zero initial data. The elasticities, however, are no longer positive-definite but still satisfy the symmetry condition (70.5) and are time-independent. In addition, we shall require that they are twice continuously differentiable on B.

Hence, stability statements made in this Section refer to the equilibrium position of a finitely deformed elastic body on part of whose surface the displacement is held fixed while on the remainder the surface tractions are dead. The *incremental*, but not the *primary*, body force is a specified function of position and time which, in view of the zero incremental initial data, alone is responsible for the small incremental motion against which stability is assessed.

KNOPS and PAYNE have proved the following theorem:

Theorem 76.1. There exist constants K and M_1 and a function $\delta(t)$ $(0 \leqq \delta \leqq 1)$ independent of u_i such that any solution of (73.1)–(73.3) *in the class $\mathscr{M}$ with zero initial data satisfies the inequality*

$$\|u\|_t^2 \leqq K M_1^{2\delta} \|f\|_{t_1}^{2(1-\delta)}, \tag{76.3}$$

for $0 \leqq t \leqq t_2 < t_1$, provided the elasticities are twice continuously differentiable in B and obey the symmetry condition (70.5).

On the basis of inequality (76.3) and the substitutions of the previous section, it may be established that the solution also depends continuously on the elasticities and on the initial data. A similar technique, which however also requires the introduction of a certain dual problem, may also be used to prove the following corollary of Theorem 76.1, from which follows the continuous dependence of the solution on the prescribed surface displacement.

Corollary 76.2. In the initial displacement boundary value problem for zero initial data and surface displacement specified as a function h_i of position and time, the corresponding null solution $u_i \equiv 0$ depends Hölder-continuously in $\|(\cdot)\|_t$ on the surface displacement in $|||(\cdot)|||_{t_1}$ for twice continuously differentiable solutions $u_i(t, \boldsymbol{x})$ on $B \times [0, t_1]$ of class $\mathscr{M}$ and all $t \in [0, t_1)$. Here

$$|||u|||_{t_1}^2 \equiv \int_0^{t_1} \int_{\partial B(\eta)} (D_s^2 h_i D_s^2 h_i + D_s h_i D_s h_i + h_i h_i + \varrho \ddot{h}_i \ddot{h}_i)\, da\, d\eta,$$

where $D_s h_i D_s h_i$ and $D_s^2 h_i D_s^2 h_i$ denote the sum of squares of tangential derivatives of h_i on the boundary ∂B evaluated at the same given instant of time t.

Let us note that analogous results may be similarly obtained for the continuous dependence of the solution on the boundary data in the traction boundary value problem, i.e., the initial data and body force are zero and the traction is a prescribed function of position and time everywhere on the surface ∂B. The results hold provided that the elasticities are negative-definite and satisfy the symmetry condition (70.5) and provided also that rigid body motions are excluded.

Finally, convexity arguments may be used in the following problem. The initial data are not all taken at time $t = 0$ but at each point x_i of B the initial values of the displacement and velocities (as used in the mathematical problem) are actually measured at time $t=-f(\boldsymbol{x})$. We assume that

$$\sup_{\boldsymbol{x}\in B} |f(\boldsymbol{x})| \leqq \varepsilon,$$

and let $u_i^{(2)}$ be the solution for initial data given on the surface $t=-f(\boldsymbol{x})$ and $u_i^{(1)}$ denote the solution for initial data measured at $t=0$. Then on setting

$$v_i(t, \boldsymbol{x}) = \int_0^t (t-\eta)\left(u_i^{(1)}(\eta, \boldsymbol{x}) - u_i^{(2)}(\eta, \boldsymbol{x})\right) d\eta,$$

and restricting $u_i^{(1)}$ to lie in a certain class of bounded functions it may be proved that

$$\|v\|_t^2 \leqq K M_3^{2\delta} \varepsilon^{1-\delta}, \qquad t\in[0, t_1),$$

for perturbations v_i of class $\mathscr{M}$ and constants K, M_3, and variable $\delta(t)\in(0, 1)$, all independent of v_i. Details of the proof are described by Knops and Payne.[9]

77. Further arguments. In Sects. 36, 74 we discussed methods for investigating stability in problems with time-dependent coefficients in the governing differential equation. These problems correspond to an elastic body in motion and acted upon by time-dependent loads. Stability of the motion is then assessed in the class of linearised incremental displacements subject to non-zero initial data and homogeneous boundary data, and is therefore in the strict sense of Liapounov. As remarked earlier, an example of such problems is the Euler column with pulsating compressive end load and the usual method of analysis is that of decomposing the solution in terms of eigenfunctions.

We now discuss a method of studying asymptotic stability in these problems in which we do not decompose the solution but employ a Liapounov function. This method is thus applicable in those problems which do not allow eigenfunction decomposition. Unlike any of the Liapounov functions treated thus far, the Liapounov function in the following application is not required to have a sign-definite time-derivative. It is sufficient for it to possess a current value not greater than its initial value, in accordance with the general statement of Theorem 15.1 and Corollary 15.4. For the example considered below, we shall in fact prove that the Liapounov function satisfies a first-order differential inequality and as a consequence establish a simple exponential upper bound. The example and stability analysis, due primarily to Hsu and Lee,[10] have wide applicability and therefore, following the original presentation, they are discussed in an abstract setting. For completeness, dissipative terms are included in the differential equation.

Let $u(t, \boldsymbol{x})$ be a vector-valued function defined on $\bar{B}\times[0, \infty)$[11] taking values in a subset X_1 of a real Hilbert space X on which is defined the inner product

$$[u, v] = \int_B u_i v_i \, dx, \qquad u, v\in X. \tag{77.1}$$

The norm associated with this inner product is denoted by $\|(\cdot)\|$. Let M, N_1, N_2 be linear operators mapping X_1 into X, and let M be time-independent and

[9] Knops and Payne [1969, *13*].

[10] Hsu and Lee [1971, *5*]. The discussion of these authors differs slightly from that in the text. Similar techniques have been used by Kozin [1963, *10*], Caughey and Gray [1965, *6*], Caughey and Dickerson [1967, *8*], [1969, *3*], Wang [1965, *23*] and Infante [1968, *15*].

[11] As usual, B denotes the region occupied by the body in its primary configuration.

N_1, N_2 time-dependent. We suppose that u is a solution to the equation

$$\ddot{u}+M\dot{u}+N_1(t)\,\dot{u}+N_2(t)\,u=0, \quad \text{in } B, \quad t>0, \tag{77.2}$$

subject to the boundary condition

$$Gu=0, \quad \text{on } \partial B\times(0, t_1), \tag{77.3}$$

where G is a time-dependent operator from X_1 into X.

We now examine the asymptotic stability of the null solution to (77.1)–(77.3). As a measure for stability we take

$$\varrho(w, v)=\|w-v\|^2+\|A_1(w-v)\|^2+\cdots+\|A_n(w-v)\|^2, \quad w, v\in X_1, \tag{77.4}$$

in which $A_i (i=1, \ldots, n)$ are linear operators from X_1 into X. Let A_i^* be the formal adjoint to A_i. We then have the equation

$$\|A_i(w-v)\|^2=[w-v, A_i^* A_i(w-v)], \quad w, v\in X_1, \tag{77.5}$$

on supposing that boundary terms disappear in the implied integration by parts. Hence, we find from (77.5) and (77.4) the equalities

$$\varrho(w)\equiv\varrho(w, 0)=(w, Pw), \tag{77.6}$$

where

$$P=I+A_1^* A_1+\cdots+A_n^* A_n. \tag{77.7}$$

The Liapounov function is defined on X_1 by

$$V(t)=[\dot{u}+Mu, R(\dot{u}+Mu)], \tag{77.8}$$

where u is a solution to (77.2) and (77.3), and $R(t)$ is a symmetric time-dependent operator mapping X to X. We shall assume that there exists a positive constant c such that

$$V(t)\geqq c\,\varrho(\dot{u}+Mu), \tag{77.9}$$

or, equivalently, we shall assume that a solution exists to the variational inequality

$$[\dot{u}+Mu, (R-\alpha P)(\dot{u}+Mu)]\geqq 0. \tag{77.9a}$$

We now obtain an estimate for the behaviour of $V(t)$ over time. From (77.2) we have

$$\dot{V}=[\dot{u}+Mu, Q(\dot{u}+Mu)], \tag{77.10}$$

where the time-dependent operator $Q(t)$, assumed symmetric, is defined by

$$Q=N^* R+RN+\dot{R}, \tag{77.11}$$

and the operator N, with adjoint N^*, is defined by

$$N(\dot{u}+Mu)=-(N_1\dot{u}+N_2 u). \tag{77.12}$$

Let $\lambda(t)$ denote

$$\lambda(t)=\max_{u\in X_1}\frac{[\dot{u}+Mu, Q(\dot{u}+Mu)]}{[u+Mu, R(\dot{u}+Mu)]}, \tag{77.13}$$

so that $\lambda(t)$ is the maximum eigenvalue of the problem

$$\begin{aligned} Qw&=\lambda Rw \quad w\in X_1 \quad \text{in } B,\\ Gw&=0 \quad \text{on } \partial B. \end{aligned} \tag{77.14}$$

Then, on assuming that $R^{-1}Q$ is compact and λ is bounded, we clearly obtain from (77.8), (77.10) and (77.13) and an integration, the inequality

$$V(t) \leqq V(0) \exp[\gamma(t)\, t], \tag{77.15}$$

where

$$\gamma(t) = \frac{1}{t} \int_0^t \lambda(\eta)\, d\eta. \tag{77.16}$$

Because λ remains bounded, (77.15) and (77.9) demonstrate the continuous dependence of the null solution in $\|(\cdot)\|$ on the initial data in $V(0)$ for all finite time. However, (77.15) cannot be used to establish Liapounov stability of the null solution.

On imposing, for some $\varepsilon > 0$, the condition

$$\gamma = \lim_{t\to\infty} \gamma(t) \leqq -\varepsilon, \tag{77.17}$$

we have that $V(t) \to 0$ as $t \to \infty$, and therefore from (77.9) we see that the null solution is asymptotically stable.

We have thus proved the following:

Theorem 77.1. Suppose a functional $V(t)$ is defined on X_1 by (77.8) *and satisfies* (77.9). *Moreover, suppose that*

(i) *$N(t)$ given by* (77.12) *has an adjoint $N^*(t)$,*

(ii) *$Q(t)$, defined by* (77.11) *is symmetric,*

(iii) *the eigenvalue problem* (77.14) *is well-defined,*

(iv) *λ is bounded.*

Then the null solution of (77.2), (77.3) *is asymptotically stable with respect to $\|\dot{u} + M u\|$ provided that for some $\varepsilon > 0$ condition* (77.17) *holds.*

To illustrate this theorem, consider the simply supported column subjected to a periodic time-varying axial compressive load. The governing equation is

$$\ddot{w} + \beta \dot{w} + w_{,xxxx} + p(t)\, w_{,xx} = 0, \qquad 0 \leqq x \leqq 1, \qquad t > 0, \tag{77.18}$$

where the deflexion $w(t, x)$ must obey the boundary conditions

$$w(t, 0) = w_{,xx}(t, 0) = 0, \qquad x = 0, \qquad t > 0, \tag{77.19}$$

$$w(t, 1) = w_{,xx}(t, 1) = 0, \qquad x = 1, \qquad t > 0. \tag{77.20}$$

In (77.18) β is constant and the load is assumed given by

$$p(t) = \left(\frac{p_0}{p_{cr}} \pi^2\right) \cos \omega t, \tag{77.21}$$

where p_{cr} is the Euler critical load, corresponding to a constant force, and p_0 is the amplitude of the periodic applied load.

For the norm $\|(\cdot)\|$ we take

$$\begin{aligned} \|w\|^2 &= \int_0^1 [w_{,xx}^2 + w_{,x}^2 + w^2 + \dot{w}^2)\, dx \\ &= \|\dot{w}\|_{L_2}^2 + \|w\|_{W_2^2}^2, \end{aligned} \tag{77.22}$$

and also we have

$$P=\begin{bmatrix} \frac{\partial^4}{\partial x^4}-\frac{\partial^2}{\partial x^2}+1 & 0 \\ 0 & 1 \end{bmatrix}.$$

For the operator R, we select the matrix operator

$$R(t)=\begin{bmatrix} \frac{\partial^4}{\partial x^4}+g(t)\frac{\partial^2}{\partial x^2}+\alpha_2+\frac{\beta^2}{4} & \frac{\beta}{2} \\ \frac{\beta}{2} & 1 \end{bmatrix},$$

where α_2, a constant, and $g(t)$, a function of time, are to be chosen.

Various results may be obtained by taking different expressions for the Liapounov function V. For instance, to determine the effect of small damping and small excitation frequencies, we may take

$$V(t)=\int_0^1\left[w_{,xx}^2-g(t)\,w_{,x}^2+\left(\alpha_2+\frac{\beta^2}{4}\right)w^2+\beta\,w\dot{w}+\dot{w}^2\right]dx.$$

On choosing $g(t)$ to be

$$g(t)=\pi^2 A\cos c\omega t,$$

where c is equal to $1, 2, 3\ldots$ or $\frac{1}{2}, \frac{1}{3}, \ldots$ and A is a constant, it may be shown that for optimal values of the various parameters, the stability criteria for (77.18)–(77.21) approach the Euler critical value in the limit.

Using the same basic idea as that just described, CAUGHEY and DICKERSON[12] have succeeded in treating equation (77.2) when these are augmented by a term non-linear in u, and when the operators N_1 and N_2 are given by

$$N_1\equiv 0,$$
$$N_2=N_3+\sum_{i=1}^{n}a_i(t)\,l_i,$$

where N_3 and l_i are linear spatial operators independent of time and $a_i(t)$ are functions of time. They establish two general criteria and discuss several examples. For instance, they treat the string of constant tension with parametric excitation proportional to the slope and with fixed-fixed or fixed-free boundary conditions; the beam under axial load with damping proportional to the velocity and a restoring force proportional to the cube of the displacement for various boundary conditions; and a flat rectangular plate with damping proportional to the velocity and time varying in-plane loads.

Similar methods have also been employed by WANG[13] to study the stability of a non-uniform cantilevered beam in an airstream. Damping is assumed to be present.

T. Stability under follower forces.

78. Introduction. We come now to a study of the stability of an equilibrium position of an elastic body subject to surface and body force loads that alter their direction and magnitude in some prescribed manner depending on the deformation

[12] CAUGHEY and DICKERSON [1969, *3*].
[13] WANG [1965, *23*].

of the body. Such loads may be said to "follow" assigned directions and, in conformity with common practice, we have already termed them "follower forces". The magnitude of these loads usually varies because the surface and volume elements on which they act vary with the deformation.

Problems involving follower forces first arose in connexion with various types of loads on rods, bars and plates, the majority of these being of importance in technology and especially aeronautics. An early example of this kind concerns a cantilevered rod subject, at its free end, to either a force or torque which remains inclined at a fixed angle to the axis of the rod throughout its deformation. Other examples are concerned with gas flows over elastic panels or plates. Once phenomena such as these became more widely known,[1] several theoretical studies appeared whose applicability, however, to physical situations was somewhat dubious. This contrasts with other types of loads considered in this treatise, for which experiment and theory have progressed at approximately equal rates. In partial remedy of the discrepancy, HERRMANN, NEMAT-NASSER and PRASAD[2] have recently constructed laboratory models for many of these theories. These models are of one- and two-dimensional systems such as rods and plates and provide physical motivation for the development of a complete three-dimensional theory of follower forces. The derivation of such a theory is quite straightforward, being a simple exercise in differential geometry, and results are not recorded here.[3] However, let us remark that for most follower forces, expressions for their linearised increments depend, at least, on first-order displacement gradients and it is the presence of such terms in the data that lend an additional mathematical interest to the subject.[4]

As we have stated before, follower loads are not responsible for initiating the secondary motion. In problems under discussion here, this is done by perturbation of the displacement and velocity at some initial instant, the follower loads then influencing in some way the secondary motion once this has started. Thus, in a stability analysis concerned with these loads, the stability concept most natural to use is that of LIAPOUNOV. However, because, in general, follower forces are path-dependent and non-conservative,[5] equations governing the displacement are usually non-self-adjoint and it is not surprising therefore that Liapounov techniques have had little success in applications involving these forces. We shall in fact present two examples in which the Liapounov theory may be successfully applied, but in most problems of this kind it is customary to rely upon an expansion of the displacement in terms of eigenfunctions. Alternative methods use an

---

[1] Largely due to the book by BOLOTIN [1961, *1*].

[2] HERRMANN, NEMAT-NASSER and PRASAD [1966, *11*].

[3] Several cases have been examined by SEWELL [1967, *31*] and NEMAT-NASSER [1968, *22*], [1970, *24*]. The general method of obtaining expressions for incremental loads is, of course, well known; see, for instance, SOUTHWELL [1914, *2*], BIEZENO and HENCKY [1928, *1*, *2*], [1929, *1*], BIEZENO and GRAMMEL [1955, *1*] and GREEN and ZERNA [1968, *10*].

[4] For instance, a slight generalisation of an expression for the linearised incremental stress tensor mentioned by NEMAT-NASSER [1970, *24*] reads

$$t_i^* - t_i = c_i + L_{ij}^{(1)} u_j + L_{ijk}^{(2)} u_{j,k} + \cdots + K_{ij}^{(1)} \dot{u}_j + \cdots + K_{ijk}^{(2)} \dot{u}_{j,k} + \cdots$$

where c_i, $L_{ij}^{(1)} \ldots$ are tensor-valued functions defined over the primary (equilibrium) configuration of the body. When $L_{ijk}^{(2)} = \cdots K_{ij}^{(1)} = \cdots K_{ijk}^{(2)} = 0$ we have the case of "elastic support"; when $c_i = L_{ij}^{(1)} = \cdots 0$, we have various forms of dissipation, which are considered later.

[5] An example of a conservative follower *force* is the hydrostatic pressure considered in Chap. O; a conservative follower torque arises in NIKOLAI's problem when the torque bisects the angle between the axes of the straight and deformed positions of the rod. See ZIEGLER [1951, *3*].

expansion, either in terms of the eigenfunctions of the adjoint operator, or in terms of any complete set of functions. The latter approach naturally leads to a study of approximation methods. Even so, difficulties encountered in dealing with non-self-adjoint problems in general, combined with those encountered in dealing with the coupling of these follower loads and deformation in particular, have severely curtailed developments in this area, and attention has largely been confined to one- and two-dimensional systems, or to systems with a finite number of degrees of freedom. We shall content ourselves therefore with a brief description of general aspects of this work and not venture to give an exhaustive account of particular and specialised problems. For these and other topics the reader is referred to the comprehensive bibliographies contained in the writings of ZIEGLER, BOLOTIN, PFLÜGER, HERRMANN and DOWELL.[6]

The follower forces to be treated in this section are non-gyroscopic and non-dissipative. Dissipative follower forces are considered in Chap. U.[7]

79. Examples using the Liapounov theory. To indicate possible ways of applying the Liapounov theory to problems involving follower forces we consider three situations. The first problem deals with the case of tangential follower forces applied everywhere on the surface of a three-dimensional elastic body, while the remaining two are concerned with one-dimensional systems. In all three we consider stability of the null solution in the class of linearised incremental displacements, and only outline the analysis.[8]

The governing equations in the first problem are provided by[9]

$$(d_{ijkl}\, u_{k,l})_{,j} = \varrho\, \ddot{u}_i \quad \text{in } B\times(0, t_1), \tag{79.1}$$

with

$$n_j\, d_{ijkl}\, u_{k,l} = n_k\, \sigma_{jk}\, u_{i,j} - n_k\, \sigma_{ik}\, t_p\, u_{p,j}\, t_j \quad \text{on } \partial B\times(0, t_1), \tag{79.2}$$

where, in addition to the notation previously adopted, n_i and t_i denote the cartesian components respectively of the unit outward normal and unit tangent direction being followed on ∂B.

On multiplying (79.1) by $\dot{u}_i$ and integrating by parts we obtain with help of (79.2) the following, in which K is the kinetic energy,

$$\begin{aligned}\frac{dK}{dt} + \frac{1}{2}\frac{dW}{dt} &= \frac{1}{2}\int\limits_{B(t)} (\sigma_{jk}\, u_{i,jk}\, \dot{u}_i - \sigma_{jk}\, \dot{u}_{i,jk}\, u_i)\, dx \\ &\quad + \frac{1}{2}\int\limits_{\partial B(t)} (n_k\, \sigma_{ik}\, t_p\, \dot{u}_{p,j}\, t_j\, u_i - n_k\, \sigma_{ik}\, t_p\, u_{p,j}\, t_j\, \dot{u}_i)\, da,\end{aligned} \tag{79.3}$$

where

$$W(t) = \int\limits_{B(t)} c_{ijkl}\, u_{i,j}\, u_{k,l}\, dx + \int\limits_{\partial B(t)} (n_k\, \sigma_{ik}\, t_p\, u_{p,j}\, t_j\, u_i - n_k\, \sigma_{ik}\, u_{j,i}\, u_j)\, da, \tag{79.4}$$

and we have used the facts that σ_{ij} is the stress in the equilibrium primary configuration and $d_{ijkl} = c_{ijkl} + \sigma_{jl}\, \delta_{ik}$.

On taking $W(t)$ to be the Liapounov function it is easy to determine conditions which are sufficient and necessary for stability[10] in the L_2-norm. However, circumstances when the conditions hold are not clear.

[6] ZIEGLER [1956, *3*], BOLOTIN [1961, *1*], PFLÜGER [1964, *10*], HERRMANN [1967, *12*], DOWELL [1971, *3*].

[7] Retarded follower forces are considered by KIUSALASS and DAVIES [1970, *18*].

[8] See also LEIPHOLZ [1970, *20*].

[9] NEMAT-NASSER [1968, *22*].

[10] NEMAT-NASSER [1968, *22*] obtains the requisite inequalities from an eigenmodal analysis in the slightly more general case when dissipative stresses are present.

As a second example we consider the stability of a simply supported flat infinite two-dimensional plate loaded along its edges and placed in a supersonic air-stream.[11] The equation governing the displacement $w(t, x)$ normal to the plate is,

$$d w_{,xxxx} - f w_{,xx} + M w_{,x} + \dot{w} + \mu \ddot{w} = 0, \qquad 0 \leqq x \leqq 1, \qquad t > 0, \tag{79.5}$$

while the boundary conditions are

$$w(t, x) = w_{,xx}(t, x) = 0 \qquad x = 0, 1 \qquad t > 0. \tag{79.6}$$

In (79.5), the load parameter f is sign-indefinite, but the flexural stiffness d, the plate-air mass ratio μ and the "Mach number" M are all positive.

In choosing a Liapounov function, the obvious choice is the function

$$V_1 = \int_0^1 [\mu(\dot{w}^2 + f w_{,x}^2 + d w_{,xx}^2] \, dx, \tag{79.7}$$

but this has a sign-indefinite time-derivative. Hence, we consider the function $V = V_1 + (2\mu)^{-1} V_2$, where V_2, given by

$$V_2 = \int_0^1 (w^2 + 2\mu\, w \dot{w}) \, dx \tag{79.8}$$

is chosen to make the time-derivative of V negative-definite. The coefficient $(2\mu)^{-1}$ is introduced to optimise bounds on the parameter M in the subsequent analysis. Stability of the null solution to (79.5), (79.6) may now be established according to LIAPOUNOV's theorem for the measures

$$\varrho_\tau(u) = \varrho(u(t)) = \int_0^1 [\dot{w}^2 + w^2] \, dx, \tag{79.9}$$

provided that $(\pi^2 d + f) > 0$ and

$$\mu M^2 < (f + \pi^2 d). \tag{79.10}$$

When the displacements are restricted to those of the form

$$w(t, x) = \sum_{n=0}^{\infty} a_n(t) \sin n \pi x \tag{79.11}$$

where a_n are suitably smooth functions of time, writing down the conditions that V and its time-derivative are respectively positive- and negative-definite shows that the upper bound (79.10) may be improved by a factor of $(9\pi^2)/64$.

The preceding analysis has been extended by PLAUT[12] to the case of a shallow curved panel in an airstream. He proves that there is stability with respect to the measures (79.9) provided that

$$0 < M^2 < (f + \pi^2 d) - 2a \int_0^1 [y_{o,x}^2 - y_{s,x}^2]^2 \, dx, \tag{79.12}$$

where a is some constant, $y_o(x)$ and $y_s(x)$ denote the equilibrium positions of the plate when under the load alone and when exposed to the airstream and $w(t, x)$ now measures the deflection from $y_s(x)$.

[11] The solution of this problem described in the text is due to PARKS [1966, *27*], [1967, *27*]. See also PRITCHARD [1968, *24*].

[12] PLAUT [1967, *28*].

A third example of the Liapounov technique in problems concerning follower forces is provided by the cantilevered rod subjected to a tangential load distributed along its length.[13] We shall investigate the stability of the rod in its straight position of equilibrium in the class of linearised transverse deflections $w(t, x)$ which satisfy the equation

$$a w_{,xxxx} + q(1-x) w_{,xx} + \mu \ddot{w} = 0, \quad 0 \leqq x \leqq 1, \quad t > 0 \tag{79.13}$$

and the boundary conditions

$$\begin{aligned} w(t, x) = w_{,x}(t, x) &= 0, \quad x = 0, \quad t > 0 \\ w_{,xx}(t, x) = w_{,xxx}(t, x) &= 0, \quad x = 1, \quad t > 0 \end{aligned} \tag{79.14}$$

in which a, μ are constants and q is the uniform tangential load per unit length.

As Liapounov function we take

$$V = \frac{1}{2} \int_0^1 [\mu \dot{w}^2 + a w_{,xx}^2 - q(1-x) w_{,x}^2]\, dx + \int_0^t \int_0^1 \frac{\partial w}{\partial \eta} w_{,x}\, dx\, d\eta,$$

which, with the help of (79.13), (79.14), may easily be shown to have $\dot{V} = 0$. On using SCHWARZ's inequality, together with the mean value theorem

$$\int_0^1 (1-x) w_{,x}^2\, dx = \alpha \int_0^1 w_{,x}^2\, dx, \quad 0 < \alpha < 1,$$

and the assumption that $w(t, x) = X(x)\, T(t)$ for sufficiently small functions X, T and for X satisfying (79.13), we may prove that

$$V \geqq \alpha (a\lambda - q) \int_0^1 w^2\, dx, \tag{79.15}$$

where λ is the eigenvalue defined by

$$w_{,xxxx} + \lambda [(1-x) w_{,x}]_{,x} = 0, \tag{79.16}$$

and the boundary conditions (79.14). The dynamic counterpart of the last equation corresponds to a cantilevered rod subjected to a distributed vertical constant force q over its length, and it may be shown that λ is the critical load in this problem.

We conclude from (79.15) that provided $q < a\lambda$, the null solution of (79.13), (79.14) is stable with respect to the measures

$$\varrho_\tau(w) = V(0), \quad \varrho(w(t)) = \int_0^1 w^2\, dx.$$

80. Adjacent equilibrium method. Instability by divergence. It is common practice in the engineering literature of follower forces to borrow from aerodynamics terms which describe two broad ways in which instability may occur. The first, called *divergence*, is used when instability arises from non-uniqueness of the solution to the associated equilibrium problem. We have seen that under dead loading, this condition always implies instability within the class of linearised incremental displacements, and this method of determining instability was called the method of adjacent equilibrium. In the present context the method is

[13] LEIPHOLZ [1969, *16*], [1970, *20*]. These papers consider other boundary conditions and treat the corresponding problems by the same general method. The general conclusion is that the particular non-conservative systems considered are more stable (i.e. have lower critical load) than their conservative counterpart.

not necessarily valid, and in this section we briefly describe some known situations in which it does and in which it does not hold.[14]

The other way in which instability occurs is called *flutter* and occurs as a result of increasing amplitudes in the incremental displacement. Flutter is usually established by means of an eigenmodal analysis, and is dealt with in subsequent pages.

A simple case in which the adjacent equilibrium method holds and there is instability by divergence occurs when the body and surface follower forces are linear functions of the displacement and its spatial gradient and the surface displacement, where specified, is zero. Then obviously if $v_i(\boldsymbol{x})$ is a non-trivial solution to the associated equilibrium problem, $u_i(t, \boldsymbol{x}) = v_i(\boldsymbol{x})\,t$ is a solution to the dynamical problem and the null solution is consequently unstable.

Certain other situations in which the adjacent equilibrium method is true have been described by PFLÜGER, which we mention below, and LEIPHOLZ.[15]

The first to realise that the method of adjacent equilibrium is not possibly valid in problems of follower forces were NIKOLAI and PFLÜGER. NIKOLAI (see footnote 10, p. 258), considered an elastic rod (or bar) clamped at one end, the other end being free and acted upon by a compressive force and a torque which during the deformation of the rod remain inclined at a fixed angle to its axis. NIKOLAI found that when the torque always acted along the tangent to the axis there was only a single equilibrium position of the rod when it was straight. According to the method of adjacent equilibrium the straight position of the rod should therefore be stable for all values of the magnitude of the force and torque. However, by examining small oscillations of the rod about the straight position, NIKOLAI concluded that for a rod of circular cross section this position is unstable for any non-vanishing magnitude of the torque.

While the system of loads in NIKOLAI's problem is in general non-conservative, ZIEGLER[16] has shown that under zero compressive force the torque whose direction bisects the angle between the deformed and straight positions of the rod is, in fact, conservative. The adjacent equilibrium method, as we saw in Chap. L, is thus applicable in this case, a fact established by NIKOLAI in 1928.

PFLÜGER[17] considered a cantilevered rod under the action of a compressive force alone applied at the free end and which remains tangential to the axis. It was shown by both PFLÜGER and FEODOS'EV[18] that no equilibrium position exists in the neighbourhood of the straight position of the rod, and therefore the adjacent equilibrium test again predicts stability of this position for all values of the load. BECK,[19] on the other hand, using an eigenmodal decomposition of the linear displacement obtained a critical value of 20.05 for the load (in dimensionless quantities) above which there is instability.

However, as PFLÜGER[20] himself later proved, the method of adjacent equilibrium is valid for the problem of a rod simply supported at both ends and loaded axially by a uniformly distributed load tangential to the deformed length. The

[14] GUO ZHONG-HENG and URBANOWSKI [1963, *22*] indicate that the method of adjacent equilibrium is valid provided the equations are formally self-adjoint. See also HERBERT [1966, *9*] and NEMAT-NASSER and HERRMANN [1966, *24*].

[15] LEIPHOLZ [1963, *12*, *13*]. See also WESOŁOWSKI [1964, *15*] and CONTRI [1966, *4*].

[16] ZIEGLER [1951, *3*]. For other early work on this problem, see ZIEGLER [1951, *4*], [1952, *5*, *6*, *7*]; TROESCH [1952, *4*]. An account is given by BOLOTIN [1961, *1*].

[17] PFLÜGER [1964, *10*]. See also ZIEGLER [1952, *7*].

[18] FEODOS'EV [1953, *2*].

[19] BECK [1952, *1*]; see also DEINEKO and LEONOV [1955, *3*] who obtained the value 19.77.

[20] PFLÜGER [1964, *10*].

critical value of the load compares favourably with that predicted by LEIPHOLZ[21] from an eigenmodal analysis.

For special types of loading, stability may be lost either by divergence or by flutter, depending on which critical load is reached first. HERRMANN and BUNGAY[22] investigate a model with two degrees of freedom subjected to a follower force that for different values of the parameter α varies from the tangential ($\alpha=1$) to the fixed directional force ($\alpha=0$), corresponding to conservative loading. By means of an eigenmodal analysis, these authors show that for a given value of α there may be multiple ranges of stability and instability, and in the latter ranges instability may be caused by either divergence or flutter. CONTRI[23] has discussed conditions under which stability once lost cannot be regained.

81. Eigenfunction expansions. Analyses depending upon separation of variables. As we have just seen, the Liapounov method has had only limited success in solving stability problems of a non-conservative nature and we therefore consider next a more popular method which depends on separation of variables. The technique includes as special cases expansion of the incremental displacement in terms of eigenfunctions of the given operator or its adjoint, and also the Galerkin procedure of expansion in terms of any complete set of functions. In addition, mention should be made of a hybrid method combining separation of variables with the Liapounov approach which has been used, for example, by PRITCHARD[24] in extending the treatment of the second example of Sect. 79. No matter which particular method is being adopted, it must always be remembered that full (Liapounov) stability cannot be established without certain convergence and completeness questions being first satisfied. Otherwise, stability can at best be regarded as conditional.

Many investigators do not consider the adjoint operator, but express the incremental displacement in the form[25]

$$u(t, \boldsymbol{x}) = \Sigma a_n v_n(\boldsymbol{x}) e^{i\omega_n t}, \tag{81.1}$$

where $v_n(\boldsymbol{x})$ and ω_n are the eigenfunctions and eigenvalues respectively of the operator associated with the problem.[26] Provided non-trivial eigenfunctions exist, then there is stability with respect to suitable measures provided also that the eigenvalues have non-negative imaginary part. There is instability when at least one eigenvalue has negative imaginary part. As the load parameter increases, stability may be lost through either divergence or flutter. Divergence here is alleged to occur when the real part of ω_n becomes zero with a simultaneous change in sign of the imaginary part, and includes the case when the real part always is zero. If the real part does not remain zero or is not zero at the point of change then flutter is said to occur.[27] A unified method for the determination of eigenvalues and the significance of this method to stability is discussed by COTSAFTIS.[28]

[21] LEIPHOLZ [1962, *11*].

[22] HERRMANN and BUNGAY [1964, *4*].

[23] CONTRI [1966, *4*].

[24] PRITCHARD [1968, *24*].

[25] This approach is sometimes called the "kinetic stability criterion". This term and the analogous "static stability test" have very little to commend them.

[26] Several of these investigations are described in the book by BOLOTIN [1961, *1*]. A similar, but more general, approach is applied by DICKEY [1970, *9*] to the non-linear problem of the hinged extensible beam.

[27] Flutter and divergence are related to the principle of exchange of stabilities. See, e.g., DAVIS [1969, *5*].

[28] COTSAFTIS [1969, *4*], [1971, *1*].

Another common device, used mainly in problems for systems of one-dimension, is to assume that the incremental displacement may be represented by only a single term of the expansion (81.1). This is equivalent to assuming that stability can be reliably assessed in the class of such displacements and indeed conditions on material parameters are sought for existence of a non-trivial eigenfunction $v(x)$. This technique was used by BECK to determine the critical load in the problem considered by PFLÜGER (mentioned at the end of Sect. 80) and has also been used to study many other problems in one- and two-dimensions.[29]

Another standard method of treating non-self adjoint problems is to expand the solution in terms of eigenfunctions of the adjoint operator. This method is used in fluid mechanics by, for example, DIPRIMA and HABETLER.[30] In elasticity, the idea of considering the adjoint problem may be found in works by LEIPHOLZ and PRASAD and HERRMANN.[31] The latter authors treat a problem in which the non-conservativeness arises from the presence of an incremental follower surface force of the type $a_{ij} u_j + b_j u_{i,j}$, where, as usual, u_i denotes the incremental displacement. In addition, these authors treat the problems of PFLÜGER[32] by the method and obtain various estimates for the critical load.

As NEMAT-NASSER and HERRMANN[33] have pointed out, in particular cases the adjoint problem may be of physical interest. They showed that the problem adjoint to PFLÜGER'S is one first proposed by REUT.[34] In this problem, an elastic rod, cantilevered at one end, has its free end subjected to a compressive force acting in a fixed direction usually taken along the axis of the undeformed rod.[35]

Finally, mention should be made of the method of decomposing the incremental displacement in terms of the complete set of eigenfunctions of the corresponding problem under dead loading.[36] A closely allied technique is the Galerkin procedure and the related convergence question has been successfully studied in a number of one-dimensional problems by LEIPHOLZ and LEVINSON.[37] The procedure has also been applied to the two-dimensional problem of panel flutter by, DOWELL[38] amongst others.

U. Dissipative forces.

82. Introduction. Several authors regard the presence of dissipative forces as essential in any theory if it is to be a reasonable model for a physical process. Such forces imply a non-zero rate of change of energy and may be introduced

[29] Such problems abound in the engineering literature, typical examples being those considered by BOLOTIN [1961, *1*], COMO [1966, *3*], DUGUNDJI [1966, *6*], LIN, NEMAT-NASSER and HERRMANN [1967, *23*]. Bibliographies are contained in these papers and in the surveys by HERRMANN [1967, *12*] and DOWELL [1971, *3*]. It is beyond the scope of the present article to attempt an evaluation of these contributions as regards their engineering importance since in many of them approximations are made from a continuous system to one with a finite number of degrees of freedom.

[30] DIPRIMA and HABETLER [1969, *6*].

[31] LEIPHOLZ [1965, *14*, *15*], PRASAD and HERRMANN [1967, *29*].

[32] I.e., a cantilevered rod subject at its free end to a compressive force which remains tangential to the rod's axis. See Sect. 80.

[33] NEMAT-NASSER and HERRMANN [1966, *23*].

[34] REUT [1939, *4*]. A laboratory model is described by HERRMANN, NEMAT-NASSER and PRASAD [1966, *11*]. A stability analysis has been carried out by NIKOLAI [1939, *3*] and by FELDT, PRASAD, NEMAT-NASSER and HERRMANN [1969, *9*].

[35] Cp. KORDAS and ŻYCZKOWSKI [1963, *9*] who study this and similar problems.

[36] Cp., e.g., BOLOTIN [1961, *1*, p. 55f.], NEMAT-NASSER and HERRMANN [1966, *22*].

[37] LEIPHOLZ [1962, *11*], [1963, *14*], [1966, *18*], [1967, *22*] (see also LEIPHOLZ [1968, *20*]), LEVINSON [1966, *19*]. In particular, these authors considered PFLÜGER'S problem.

[38] DOWELL [1966, *5*].

by constitutive assumptions either on the usual variables or on the loads. Of the theories considered in this Treatise, only thermoelasticity falls into the first category but other examples concern elastic bodies with electro-magnetic effects and also the theory of elastic mixtures. The isothermal elasticity that we have been considering (that requires the existence of an internal energy function) does not fall into the first group and dissipation may be introduced into it only by postulating an appropriate dependence for the loads. In this chapter we study briefly the consequences of the presence of such dissipative loads in the theory of isothermal elasticity for the stability of the primary state.

Following the suggestion of Sect. 78, we assume that the incremental loads may be schematically expressed by

$$G_i^*(t, \boldsymbol{x}) - G_i(t, \boldsymbol{x}) = \alpha_{ij}^{(1)} \dot{u}_j + \cdots + \alpha_{ij}^{(n)} \overset{(n)}{u}_j + \cdots \tag{82.1}$$

where G_i is either the body force or surface loads, and $\alpha_{ij}^{(n)}$ may be functions of position $\boldsymbol{x}$.

For systems in which $\alpha_{ij}^{(1)}$ is symmetric and is the only non-zero coefficient in (82.1) and G_i corresponds to a body force, the boundary data being zero, it may easily be shown that, under appropriate definiteness conditions, many of the previous results for dead or weakly conservative loading continue to hold. For instance, when $\alpha_{ij}^{(1)}$ is negative semi-definite it follows that the null solution of the corresponding linearised elastodynamic problem is stable[1] or unstable[2] under the same conditions as found in Chaps. J and L respectively. The effect on these previous results of a different definiteness postulate may be studied in a straightforward manner suggested by earlier calculations and hence there is no need to present the results here.

While the effect of dissipative forces (82.1) on a (conservative) elastic system may be therefore easily resolved, the effect on a system under the action of non-conservative forces is not yet properly understood. Because of the difficulty in estimating the rate of work, most studies have employed an eigenmodal analysis and in addition attention has been restricted to systems with a finite number of degrees of freedom. The main result to emerge from these preliminary studies is that dissipative forces of the kind (82.1) can have a destabilising effect.[3] However, in at least one study of a continuous system,[4] dissipation, which in this case is represented by Coriolis forces, may have either a stabilising or destabilising effect depending on parameters describing material properties of the system. Thermoelastic dissipative effects in Pflüger's problem have been shown

---

[1] Movchan [1963, *16*], Shield [1965, *21*]. This result, analogous to Kelvin's theorem for discrete systems (cp. Chetaev [1961, *2*, p. *96*]) is proved by both authors to be true when gyroscopic forces are also present.

[2] Caughey and Shield [1968, *4*]. Koiter [1965, *12*, *13*], [1967, *18*] considers instability in the presence of dissipation and with finite incremental displacements. His conclusions, however, resting on the assumption that a decreasing function drops below a given level, are incomplete.

[3] Ziegler [1952, *7*], Bottema [1955, *2*], Bolotin [1961, *1*], Leipholz [1964, *8*], Leonov and Zovii [1963, *15*], Nemat-Nasser and Herrmann [1966, *25*], Herrmann and Nemat-Nasser [1967, *14*], Herrmann and I.-C. Jong [1965, *10*], [1966, *10*] and Zhinzher [1968, *30*]. See also the survey by Herrmann [1967, *12*]. Non-linear damping in a discrete system is considered by Hagedorn [1970, *10*].

[4] Herrmann and Nemat-Nasser [1967, *13*]. See also Nemat-Nasser, Prasad and Herrmann [1966, *26*] and Prasad and Herrmann [1967, *30*], who deal with the problem of small velocity-dependent forces in continuous systems, and Plaut and Infante [1970, *26*].

by HUANG and SHIEH[5] to cause significant reduction in the critical load and the same effect is recorded by WATHER and LEVINSON[6] in the case of rotary inertia.

The presence of dissipation in a system may be expected, under appropriate definiteness assumptions, to lead not only to the stability of the null solution but also to its asymptotic stability. We have already encountered in Chap. P an example in the field of thermoelasticity. More generally, we may say that dissipative forces damp out incremental motion so that even where incremental displacements do not tend to zero with increasing time, they may tend to some other equilibrium solution provided, of course, these exist. This suggests a possible mechanism whereby an elastic body can pass by means of an actual motion from one equilibrium position to another and is evidently important when dealing with buckling problems. In such problems, as well as determining non-unique equilibrium solutions, or buckling modes, we must also ensure that at least one of the buckled modes is accessible from the unbuckled one. Otherwise the mathematical model being used does not permit the actual occurrence of the buckling process. The analysis must be conducted on the non-linear equations and usually requires the presence of dissipation in some form or other. Most investigations into the non-linear problem are restricted to a determination of the equilibrium solutions and fail to investigate whether in any actual motion the body can reach one of its buckled modes.

An exception[7] may be found in the work of HSU[8] dealing with the particular instance of buckling known as "snap-through"[9] in which there is passage from one mode to another. We shall devote the remainder of this section to a short description and statement of the sufficient conditions for the occurrence and non-occurrence of snap-through given by HSU. We shall see that the proof of these criteria suggests a relation with HALE's invariance principle and is valid for non-linear elasticity. Since, strictly, the determining of equilibrium positions is not part of a stability analysis we shall suppose that these are always known.

83. Snap-through. We consider a bounded elastic body which initially is either undeformed or in some equilibrium position which may be a possible buckling mode. A secondary motion is imparted to the body by application of loads of various types[10] and dissipative forces are assumed to exist which cause the total energy in this motion to have a non-increasing time-derivative which vanishes only when the velocity is zero. The total energy $E(t)$, as before, consists of kinetic plus potential energy, the latter quantity being composed of the strain energy of the body together with the potential functions of any (weakly) conservative external forces.

For convenience of presentation, we now again introduce the notion of a dynamical system and let X be the set of vector-valued functions satisfying the boundary conditions to the problem and also certain smoothness requirements depending on the type of solution being required. We next suppose that the

[5] HUANG and SHIEH [1970, *13*].

[6] WATHER and LEVINSON [1968, *28*].

[7] Other exceptions are, for example, due to MATKOWSKY and REISS [1971, *11*] in elasticity, and LYNDEN-BELL [1965, *16*] in astrophysics. See also MATKOWSKY [1970, *23*].

[8] HSU [1966, *12, 13*], [1968, *13*]. The first two papers consist mainly of general studies.

[9] Also known as the "tin-canning" effect. It arises largely in buckling of shells, but we treat the three-dimensional problem. A fuller explanation of snap-through than given here is presented subsequently.

[10] These may be body forces or surface tractions either impulsively applied, step-wise increased or time-dependent and vanishing asymptotically with increasing time. The surface displacement, where prescribed, is supposed time-independent. See HSU [1966, *13*], [1968, *13*], HSU, KUO and PLAUT [1969, *12*], HSU, KUO and LEE [1968, *14*].

solutions to the non-linear elastic equations may be associated with the dynamical system $\mathscr{B}(T, X)$ and that the invariant sets are completely given by functions φ_i. The secondary motion is taken to start at the instant $\tau=0$ so that the translates are given by

$$\varphi_{i0}(t)=\varphi_i(0), \quad \varphi_i\in\mathscr{B}(T, X), \quad i=0, 1, 2\ldots.$$

The invariant sets are thus the buckling modes or critical points, in the earlier notation, of the system and correspond to the various possible positions of equilibrium that may be assumed by the body under the stipulated load conditions. *It is important to realise that the system is assumed to possess no other types of invariant sets.* The number and location of critical points may vary with the size of the load, but we shall always suppose that *at least one critical point always exists.* We assume moreover that the critical points are denumerable and are isolated in the sense that the neighbourhood of each, as defined by the measure ϱ, contains no other critical point.

The Liapounov stability of each buckling mode or critical point φ_i is examined by any of the linear analyses already described and one, φ_0 say, is selected that is stable. The preferred point may correspond to the equilibrium position first attained by the body from the reference configuration as the load parameter is increased, or it may correspond to an equilibrium position in which the body has been placed. For simplicity of description, we now suppose that apart from φ_0 there is just one other critical point φ_1. If, after the secondary motion has been imparted to the body, it settles down in the equilibrium position corresponding to φ_1 we say that *snap-through* has occurred. On the other hand, if the body returns to the equilibrium position corresponding to φ_0 then snap-through has not occurred.

To obtain a sufficiency criterion against snap-through we let $\varphi_0(0)$ be located at the origin in X and the critical point φ^* as follows. Let $G(\varphi_0)$ be the open subset of X specified by

$$G(\varphi_0)=\{x\in X \mid E(x)<U(\varphi^*(0))\}$$

where $E(x)$ is the total energy evaluated at the point x and $U(\varphi^*(0))$ is the potential energy of the body evaluated at the equilibrium solution $\varphi^*(0)$. Then φ^* is such that the closure of $G(\varphi_0)$ contains $\varphi_0(0)$ and $\varphi^*(0)$ (and possibly other equilibrium solutions) but that $G(\varphi_0)$ itself contains only $\varphi_0(0)$.

Hsu's criterion[11] *against snap-through is then that if initial values $\psi(\tau)$ of the secondary (disturbed) solutions lie in $G(\varphi_0)$ then φ_0 is asymptotically stable and snap-through will not occur.*

The proof of this criterion given by Hsu is not entirely satisfactory and it is better to use Hale's invariance principle.[12] To do this, however, we must ensure that the conditions of the theorem are met which in particular require X to be a Banach space and the paths $\psi_\tau(t)$ to lie in compact subsets of X. It is not clear for what problems these conditions can be established.

When φ_0 is the only critical point, so that the body has only a single equilibrium position, the body can never snap-through because now the whole space forms the region of attraction for φ_0.

Hsu's criterion *for snap-through* rests on the simple observation that as the load parameter increases, the equilibrium position corresponding to φ_0 may cease to exist and another appear corresponding to φ_1, say. Provided φ_1 is asymp-

[11] Hsu [1966, *12, 13*], Hsu, Kuo and Lee [1968, *14*].

[12] See Theorem 19.2.

totically stable and is the only critical point at the given value of the load parameter, the body must asymptotically adopt the corresponding equilibrium position. Under these conditions, snap-through must occur. When, after φ_0 ceases to exist, more than one new critical point appears, the final position adopted by the body will depend upon the region of attraction in which the initial data is located.

Again, to make this heuristic proof rigorous we must appeal, for instance, to HALE'S invariance principle. The proof in this case would then apply only to a restricted class of problems.

Several applications of HSU'S criteria have been made to the stability of shallow arches.[13] The various equilibrium positions have all been calculated and their stability against snap-through examined. The analysis is conducted on non-linear equations without approximations. A noteworthy feature of these researches is that the loads at which snap-through occurs for simply supported arches are different from those for arches which are clamped.[14]

References.

Italic numbers in parentheses following the reference indicate the sections in which the work is mentioned.

1744 *1.* EULER, L.: Methodus inveniendi lineas curvas maximi minimive proprietate gaudentes. Appendix: De curvis elasticis. Lausanne and Geneva. English transl. with appendix in Isis **20** (No. 58), 1 (1933). (*47*)

1757 *1.* EULER, L.: Sur la force des colonnes. Hist. Acad. Berlin **13**. (*47*)

1788 *1.* LAGRANGE, J.L.: Méchanique analitique. 1st Ed. Paris: Veuve Desaint. (*5, 9, 15, 19, 31*)

1838 *1.* MINDLING, F.: Handbuch der Differential- und Integral-Rechnung und ihrer Anwendungen auf Geometrie und Mechanik. Zweiter Theil, enthaltend die Mechanik. Handbuch der Theoretischen Mechanik. Berlin. (*15*)

1839 *1.* GREEN, G.: On the laws of the reflexion and rarefaction of light at the common surface of two non-crystallized media. (1837) Trans. Cambridge Phil. Soc. **7**, 1–24 (1839) = Collected Papers, pp. 245–269. London: MacMillan 1871. (*9, 15, 31*)

1842 *1.* JACOBI, C. G.: Vorlesungen über Dynamik. 1842–1843. Edited by A. CLEBSCH. Berlin: Reimer 1866. See also Gesammelte Werke. Berlin: Reimer 1884. (*19, 44*)

1846 *1.* DIRICHLET, G. LEJEUNE: Ueber die Stabilität des Gleichgewichts. J. reine angew. Math. **32**, 85–88 (1846) = Sur la stabilité de l'équilibre. J. Math. pures et appl. (Liouville) **12**, 474 (1867) = Mécanique Analytique by J.L. LAGRANGE: Note II, pp. 399–401. 3rd Ed. annotated by J. BERTRAND 1853. (*5, 15, 23, 31, 34*)

1850 *1.* KIRCHHOFF, G.: Ueber das Gleichgewicht und die Bewegung einer elastischen Scheibe. J. reine angew. Math. **40**, 51–58 = Ges. Abh. pp. 237–279. Leipzig: Barth 1882. (*9*)

1853 *1.* LAGRANGE, J.L.: Mécanique Analytique, Vol. 1, 3rd Ed. (annotated by J. BERTRAND). Paris: Mallet-Bachelier. (*5, 9, 15, 19, 31, 47*)

1856 *1.* MAXWELL, J.C.: Essay on the stability of the motion of Saturn's rings. Adams Prize = Scientific Papers I, p. 288. Cambridge: University Press 1890. (*5*)

1859 *1.* KIRCHHOFF, G.: Ueber das Gleichgewicht und die Bewegung eines unendlich dünnen elastischen Stabes. J. reine angew. Math. **56**, 285–313 = Ges. Abh. pp. 285–316. Leipzig: Barth 1882. (*9*)

1860 *1.* ROUTH, E. J.: The Advanced Part of a Treatise on the Dynamics of a System of Rigid Bodies. 1st Ed. London: Macmillan. 6th Ed. reprinted at New York: Dover 1955. (*6, 15, 31*)

1862 *1.* CLEBSCH, A.: Theorie der Elasticität der festen Körper. Leipzig. Teubner. French edition translated and annotated by B. DE ST. VENANT and A. FLAMANT. Paris: Dunod 1883. (*9, 15, 31, 32, 44*)

1863 *1.* THOMSON, W.: Dynamical problems regarding elastic spheroidal shells and spheroids of incompressible liquid. Appendix. Equations of equilibrium of an elastic solid

[13] HSU [1966, *12, 13*], [1967, *16*], [1968, *12, 13*], HSU, KUO and LEE [1968, *14*].

[14] HSU, KUO and PLAUT [1969, *12*].

deduced from the principle of energy. Phil. Trans. Roy. Soc. London **153**, 610–616. See also THOMSON and TAIT, Treatise on Natural Philosophy, Vol. 1, Part 2, Appendix C. Cambridge: University Press 1879. (*15, 31*)

1866 *1.* LIPSCHITZ, R.: Ueber einen algebraischen Typus der Bedingungen eines bewegten Massensystems. J. für Mathematik **66**, 363–374. (*19*)

1872 *1.* BOUSSINESQ, J.: Théorie des ondes et des remous qui se propagent le long d'une canal rectangulaire horizontal, en communiquant au liquide contenu dans ce canal des vitesses sensiblement pareilles de la surface au fond. J. Math. pures et appl., 2nd Ser. **17**, 55–108. (*15*)

1874 *1.* LIPSCHITZ, R.: Beweis eines Satzes der Elasticitätslehre. J. reine angew. Math. **78**, 329–337. (*9, 31, 32, 44*)

1877 *1.* KIRCHHOFF, G.: Vorlesungen über Mathematische Physik. Mechanik. 2nd Ed. Leipzig: Teubner. (*9, 15, 31*)

2. ROUTH, E. J.: A Treatise on the Stability of a Given State of Motion. Adams Prize. London: Macmillan. (*5, 6, 7*)

3. — Elementary Treatise on the Dynamics of a System of Rigid Bodies. 3rd Ed. London: Macmillan. (*31*)

1878 *1.* LAURENT, H.: Traité de Mécanique Rationnelle, Vol. 2. Paris. (*15, 31*)

1879 *1.* THOMSON, W., and P. G. TAIT: Treatise on Natural Philosophy. Cambridge. Last revised ed. (1912) reprinted at New York: Dover Publications (1962) as Principles of Mechanics. (*7, 15, 19, 31*)

1881 *1.* GREENHILL, A. G.: Determination of the greatest height consistent with stability that a vertical pole or mast can be made, and of the greatest height to which a tree of given proportions can grow. Proc. Cambridge Phil. Soc. **4**, 65–73. (*47*)

2. POINCARÉ, H.: Mémoire sur les courbes définies par les équations différentielles. J. Math. pures et appl. (3) **7**, 375–422 = Analyse des travaux scientifiques de H. Poincaré faite par lui-même. Acta Math. **38**, 1–135 (1921) = Œuvres de H. Poincaré **1**, pp. 3–84. See also pp. 85–222. Paris: Gauthier-Villars 1951. (*1, 7, 15*)

1882 *1.* POINCARÉ, H.: Mémoire sur les courbes définies par les équations différentielles. J. Math. pures et appl. (3) **8**, 251–296. See also previous reference. (*1, 7, 15*)

2. ZHUKOVSKII, N. E.: On strength of motion. Scientific notes of Moscow University. Physics-Mathematics Division No. 4 [in Russian]. See also Collected Works **1**, Gostekhizdat 1948. (*31*)

1883 *1.* GREENHILL, A. G.: On the strength of shafting when exposed both to torsion and to end thrust. Proc. Inst. Mech. Eng. 182–209. (*47*)

2. LÉVY, M.: Sur un nouveau cas intégrable du problème de l'élastique et l'une de ses applications. Comp. Rend. **97**, 694–697. (*47*)

1884 *1.* HALPHEN, G. H.: Sur une courbe élastique. Comp. Rend. **98**, 422–425. (*47*)

2. LÉVY, M.: Mémoire sur un nouveau cas intégrable de problème de l'élastique et l'une de ses applications. J. Math. pures et appl. 3rd Ser. **10**, 5–42. (*47*)

1885 *1.* POINCARÉ, H.: Sur l'équilibre d'une masse fluide animée d'un mouvement de rotation. Acta Math. **7**, 259–380. (*46*)

1888 *1.* BRYAN, G. H.: On the stability of elastic systems. Proc. Cambridge Phil. Soc. **6**, 199–210. (*15, 31*)

2. — Application of the energy test to the collapse of a long thin pipe under external pressure. Proc. Cambridge Phil. Soc. **6**, 287–292. (*15, 31*)

3. MINCHIN, G. M.: Treatise on Statics II. 4th Ed. Oxford: Clarendon Press. (*15, 31*)

4. THOMSON, W.: Reflexion and refraction of light. Phil. Mag., 5th Ser. **26**, 414–425. (*7, 9, 15, 31*)

1891 *1.* BRYAN, G. H.: On the stability of a plane plate under thrusts in its own plane, with applications to the "buckling" of the sides of a ship. Proc. London Math. Soc. **22**, 54–67. (*47*)

1892 *1.* LIAPOUNOV, A. M.: Problème général de la stabilité du mouvement. (Société Mathématique de Kharkow). Translated by DAVAUX, E., Ann. Fac. Sci. Toulouse, 2nd Ser. **9**, Reprinted at Princeton: Princeton University Press (Ann. of Math. **17**) 1949. (*1, 6, 15, 19, 31*)

1894 *1.* BRYAN, G. H.: On the buckling and wrinkling of plating when supported on parallel ribs or on a rectangular framework. Proc. London Math. Soc. **25**, 141–150. (*47*)

1895 *1.* REYNOLDS, O.: On the dynamical theory of incompressible viscous fluids and the determination of the criterion. Phil. Trans. Roy. Soc. London, Ser. A **186**, 123–164. (*15*)

1897 *1.* DUHEM, P.: Traité Élémentaire de Mécanique Chimique, Vol. 1. Paris: Hermann. (*31*)

2. KLEIN, F., u. A. SOMMERFELD: Über die Theorie des Kreisels. Leipzig: Teubner. (*5, 9, 23*)

1899 *1.* Volterra, V.: Sur la théorie des variations des latitudes. Acta Math. **22**, 201–357. (*15, 16*)

1901 *1.* Bendixson, I.: Sur les courbes définies par des équations différentielles. Acta Math. **24**, 1–88. (*1*)

2. Osgood, W. F.: On a fundamental property of a minimum in the calculus of variations and the proof of a theorem of Weierstrass's. Trans. Am. Math. Soc. **2**, 273–295. (*35*)

1902 *1.* Duhem, P.: Sur la stabilité, pour des perturbations quelconques, d'un système animé d'un mouvement de rotation uniforme. J. Math. pures et appl., 5th Sér. **8**, 5–18. (*6, 16*)

2. Hadamard, J.: Sur quelques questions du calcul des variations. Bull. Soc. Math. France **30**, 253–256. (*18, 35*)

1903 *1.* Hadamard, J.: Leçons sur la Propagation des Ondes et les Equations de l'Hydrodynamique. (Lectures of 1898–1900.) Paris. Reprinted at New York: Chelsea 1949. (*18, 31*)

2. Kneser, A.: Die Stabilität des Gleichgewichts hängender schwerer Fäden. J. reine angew. Math. **125**, (3). (*35*)

1906 *1.* Born, M.: Untersuchungen über die Stabilität der elastischen Linie in Ebene und Raum. Dissertation. Göttingen. (*35*)

2. Duhem, P.: Recherches sur l'Élasticité. Paris: Gauthier-Villars. (*18, 30, 31, 32, 44, 67*)

1907 *1.* Hadamard, J.: Sur quelques questions de calcul des variations. Ann. Sci. École Normale **24**, 203–231 = Œuvres **2**, pp. 485–513. Paris: Centre National de la Recherche Scientifique 1968. (*6, 18, 37*)

2. Orr, W. McF.: The stability or instability of the steady motions of a perfect liquid and of a viscous liquid. Proc. Roy. Irish Acad., Sect. A **27**, 9–68, 69–138. (*15*)

1908 *1.* Korn, A.: Solution générale du problème d'équilibre dans la théorie de l'élasticité dans le cas ou les efforts sont donnés à la surface. Ann. Fac. Sci. Toulouse, Ser. 2 **10**, 165–269. (*43*)

1909 *1.* Korn, A.: Über einige Ungleichungen, welche in der Theorie der elastischen und elektrischen Schwingungen eine Rolle spielen. Bull. Intern., Cracovie Akad. Umiejet, Classe sc. math. et nat. 705–724. (*43*)

1913 *1.* Sundman, K. F.: Mémoire sur le problème des trois corps. Acta Math. **36**, 105–179. (*19*)

1914 *1.* Hellinger, E.: Allgemeine Ansätze der Mechanik der Kontinua. Enclykl. Math. Wiss. IV (4), pp. 601–694. Leipzig: Teubner. (*16*)

2. Southwell, R. V.: On the general theory of elastic stability. Phil. Trans. Roy. Soc. London, Ser. A **213**, 187–244. (*47, 78*)

1922 *1.* Leibenson, L. S.: Ob adnom sposobie opredelenia ustoichivosti uprugovo ravnovesia. Azerb. Neft. choz-ve No. 3. Baku = Coll. Works **1** (1951). (*47*)

1923 *1.* Hadamard, J.: Lectures on Cauchy's Problem in Linear Partial Differential Equations. New Haven: Yale University Press. Reprinted at New York: Dover 1952. (*6, 8*)

2. Mises, R. v: Über die Stabilitätsprobleme der Elastizitätstheorie. Z. Angew. Math. Mech. **3**, 406–422. (*47*)

1924 *1.* Beliaev, N. M.: Stability of prismatic rods subject to variable longitudinal forces [in Russian]. Eng. Const. Struct. Mechs. Leningrad Put, pp. 149–167 (*64*)

2. Southwell, R. V., and S. W. Skan: On the stability under shearing forces of a flat elastic strip. Proc. Roy. Soc. London, Ser. A **105**, 582–607. (*45, 47*)

1925 *1.* Dean, W. R.: On the theory of elastic stability. Proc. Roy. Soc. London, Ser. A **107**, 734–760. (*47*)

2. Funk, P.: Über die Stabilität der beiderseits eingespannten Elastika und ähnliche Fragen. Z. Angew. Math. Mech. **5**, 467–472. (*35*)

3. Reissner, H.: Energiekriterium der Knicksicherheit. Z. Angew. Math. Mech. **5**, 474–478. (*47*)

1927 *1.* Birkhoff, G. D.: Dynamical Systems. Am. Math. Soc. Colloq. Publ. New York. (*1, 8, 19*)

1928 *1.* Biezeno, C. B., and H. Hencky: On the general theory of elastic stability. Proc. Koninkl. Ned. Akad. Wetenschap. **31**, 569–592. (*47, 78*)

2. — — Sur les équations générales de la stabilité élastique. Atti Congr. Intern. Mat. Bologna **6**, 233–237. (*47, 78*)

3. Nikolai, E. L.: On the stability of the rectilinear form of a compressed and twisted bar (in Russian). Izv. Leningrad Politekhn. in-ta. **31** = Studies in Mechanics. Gos. Izdat. Tekhniko-Teoreticheskoi Literatury, pp. 357–387. Moscow 1955. (*65*)

1929 *1.* Biezeno, C. B., and H. Hencky: On the general theory of elastic stability. Proc. Koninkl. Ned. Akad. Wetenschap. **32**, 444–456. (*47, 78*)

2. NIKOLAI, E. L.: On the problem of stability of a twisted bar [in Russian]. Vestn. prikl. mat. i mekh. **1**, = Studies in Mechanics. Gos. Izdat. Tekhniko-Teoreticheskoi Literatury, pp. 388–406. Moscow 1955. (*65*)

1930 *1.* NIKOLAI, E. L.: Über den Einfluß der Torsion auf die Stabilität rotierender Wellen. Proc. Third Internat. Congr. Appl. Mech. Stockholm, pp. 103–104. (*65*)

2. TREFFTZ, E.: Über die Ableitung der Stabilitätskriterien des elastischen Gleichgewichts aus der Elastizitätstheorie endlicher Deformationen. Proc. Third Internat. Congr. Appl. Mech. Stockholm **3**, 44–50. (*31, 47*)

1931 *1.* BOLZA, O.: Lectures on the Calculus of Variations. New York: Stechert (reprint of Edition 1904). 2nd Ed. reprinted at New York: Chelsea 1961. (*18*)

2. MARKOV, A. A.: Sur une propriété générale des ensembles minimaux de M. Birkhoff. Comp. Rend. **193**, 823–825. (*1*)

1932 *1.* WHITNEY, H.: Regular families of curves. Proc. Nat. Acad. Sci. U.S.A. **18**, 340–342. (*1*)

1933 *1.* TREFFTZ, E.: Zur Theorie der Stabilität des elastischen Gleichgewichts. Z. Angew. Math. Mech. **13**, 160–165. (*30, 31, 35, 47*)

1936 *1.* TIMOSHENKO, S.: Theory of Elastic Stability. New York: McGraw-Hill. (*47*)

1937 *1.* WHITTAKER, E. T.: A Treatise on the Analytical Dynamics of Particles and Rigid Bodies. 4th Ed. Cambridge: University Press. (*31*)

1938 *1.* CHETAEV, N. G.: On the instability of equilibrium when the force function is not a maximum [in Russian]. Uch. zap. Kazan University. (*31*)

2. MARGEURRE, K.: Über die Behandlung von Stabilitätsproblemen mit Hilfe der energetischen Methode. Z. Angew. Math. Mech. **18**, 57–73. (*47*)

3. — Über die Anwendung der energetischen Methode auf Stabilitätsprobleme. Jahrb. Dtsch. Luftfahrtforschung **1**, 433–443. English Translation NACA TM No. 1138. (*47*)

1939 *1.* KAPPUS, R.: Zur Elastizitätstheorie endlicher Verschiebungen, I und II.Z. Angew. Math. Mech. **19**, 271–285, 344–361. (*47*)

2. KARMAN, T. v., and H. S. TSIEN: Buckling of spherical shells by external pressure. J. Aero. Sci. **7**, 43–50. (*47*)

3. NIKOLAI, B. L.: On the stability criterion of elastic systems [in Russian]. Tr. Odessk. in-ta inzh. grazhd. i komm. str.-va. No. 1. (*81*)

4. REUT, V. I.: On the theory of elastic stability [in Russian]. Tr. Odessk. in-ta inzh. grazhd. i komm. str.-va. No. 1. (*81*)

1943 *1.* NEUBER, H.: Die Grundgleichungen der elastischen Stabilität in allgemeinen Koordinaten und ihre Integration. Z. Angew. Mech. **23**, 321–330. (*47*)

1945 *1.* KOITER, W. T.: Over de stabiliteit van het elastisch evenwicht. Thesis Delft: Amsterdam: H. J. Paris. Translated into English as NASA Report TT F-10, 833 (1967). (*18, 30, 31, 32, 47*)

1946 *1.* CATTANEO, C.: Su un teorema fondamentale nella teoria delle onde di discontinuità. Atti Accad. Sci. Lincei Rend., Classe Sci. Fis. Mat. Nat. (8) **1**, 66–72, 728–734. (*18*)

1947 *1.* FRIEDRICHS, K. O.: On the boundary value problems of the theory of elasticity and Korn's inequality. Ann. of Math. **48**, 441–471. (*35, 43*)

2. HOVE, L. VAN: Sur l'extension de la condition de Legendre du calcul des variations aux intégrales multiples à plusieurs fonctions inconnues. Proc. Koninkl. Ned. Akad. Wetenschap. A **50** (1), 18–23. (*18, 35*)

3. NEMYTSKII, V. V., and V. V. STEPANOV: Qualitative Theory of Differential Equations. 1st Ed. Moscow-Leningrad. 2nd Ed. 1949. English translation: Princeton Math. Ser. No. 22. Princeton: University Press 1960. (*1, 4, 10*)

4. PRAGER, W.: The general variational principle of the theory of structural stability. Quart. Appl. Math. **4**, 378–384. (*47*)

1948 *1.* BARBASHIN, E. A.: On the theory of generalised dynamical systems [in Russian]. Učenye Zapiski Moskov. Gos. Univ., No. 135. Matematika **2**, 110–133. (*1*)

2. RIVLIN, R. S.: Large elastic deformations of isotropic materials. II. Some uniqueness theorems for pure, homogeneous deformations. Phil. Trans. Roy. Soc. London, Ser. A **240**, 491–508. (*57*)

1949 *1.* HAMEL, G.: Theoretische Mechanik. Berlin-Göttingen-Heidelberg: Springer. (*16*)

2. MOISEEV, N. D.: An Outline of the Development of the Theory of Stability. Moscow [in Russian]. (*5, 15*)

1950 *1.* COURANT, R.: Dirichlet's Principle, Conformal Mapping, and Minimal Surfaces. New York: Interscience. (*18*)

1951 *1.* EYDUS, D. M.: On the mixed problem of the theory of elasticity [in Russian]. Dokl. Akad. Nauk. SSSR **76**, 181–184. (*43*)

2. MURNAGHAN, F. D.: Finite Deformation of an Elastic Solid. New York: Wiley. Reprinted at New York: Dover 1967. (*23*)

3. Ziegler, H.: Ein nichtkonservatives Stabilitätsproblem. Z. Angew. Math. Mech. **31**, 265–266. (*78, 80*)
4. — Stabilitätsprobleme bei geraden Stäben und Wellen. Z. Angew. Math. Phys. **2**, 265–289. (*80*)
1952 *1.* Beck, M.: Die Knicklast des einseitig eingespannten, tangential gedrückten Stabes. Z. Angew. Math. Phys. **3**, 225–228. (*65, 80*)
2. Dean, W. R.: The Green's function of an elastic plate. Proc. Cambridge Phil. Soc. **48**, 149–167. (*47*)
3. Malkin, I. G.: Theory of Stability of Motion. Translated from the Russian Edition of 1952. U.S. Atomic Energy Commission. Office of Technical Information 1956. (*31*)
4. Troesch, A.: Stabilitätsprobleme bei tordierten Stäben und Wellen. Ing.-Arch. **20**, 258–277. (*80*)
5. Ziegler, H.: Kritische Drehzahlen unter Torsion und Druck. Ing.-Arch. **20**, 377–390. (*80*)
6. — Knickung gerader Stäbe unter Torsion. Z. Angew. Math. Phys. **3**, 96–118. (*80*)
7. — Die Stabilitätskriterien der Elastomechanik. Ing.-Arch. **20**, 49–56. (*34, 80, 82*)
1953 *1.* Courant, R., and D. Hilbert: Methods of Mathematical Physics, Vol. I. New York: Interscience. (*6, 18*)
2. Feodos'ev, V. I.: Selected Problems and Questions in Strength of Materials [in Russian], pp. 38, 165. Gostekhizdat. (*80*)
3. Novozhilov, V. V.: Foundations of the Non-linear Theory of Elasticity. Translated from the first (1948) Russian edition by F. Bagemihl, H. Komm, W. Seidel. Rochester, New York: Graylock Press. (*47*)
4. Timoshenko, S.: History of Strength of Materials. New York: McGraw-Hill. (*47*)
5. Wittrick, W. H.: Stability of a bimetallic disk. Part I. Part II by W. H. Wittrick, D. M. Myers and W. R. Blunden. Quart. J. Appl. Math. Mech. **6**, 15–31. (*58*)
6. Ziegler, H.: Linear elastic stability. Z. Angew. Math. Phys. **4**, 89–121, 167–185. (*34*)
1954 *1.* Petrovsky, I. G.: Lectures on Partial Differential Equations. English Translation from the Russian edition by A. Shenitzer. Oxford: Pergamon. (*6, 8*)
1955 *1.* Biezeno, C. B., and R. Grammel: Engineering Dynamics, Vol. I. London: Blackie = English translation of Technische Dynamik. Berlin: Springer 1939. (*47, 78*)
2. Bottema, O.: On the stability of the equilibrium of a linear mechanical system. Z. Angew. Math. Phys. **6**, 97–104. (*82*)
3. Deineko, K. S., and M. Ia. Leonov: The dynamic method of investigating the stability of a bar in compression [in Russian]. Prikl. Mat. Mekh. **19**, 738–744. (*80*)
4. Gottschalk, W. H., and G. S. Hedlund: Topological dynamics. A.M.S. Colloquium Publications **36**, Providence, R. I. (*1*)
5. John, F.: A note on "improper" problems in partial differential equations. Comm. Pure Appl. Math. **8**, 591–594. (*8*)
6. Pucci, C.: Sui problemi di Cauchy non "ben posti". Rend. Acad. Naz. Lincei **18**, 437–477. (*48*)
7. Stoker, J. J.: On the stability of mechanical systems. Comm. Pure Appl. Math. **8**, 133–142. (*5, 15, 23*)
8. Wilkes, E. W.: On the stability of a circular tube under end thrust. Quart. J. Mech. Appl. Math. **8**, 88–100. (*47*)
9. Yoshizawa, T.: Note on the boundedness and the ultimate boundedness of solutions of $x' = F(t, x)$. Mem. Coll. Sci. Univ. Kyoto, Ser. A **29**, Math. (3), 275–291. (*10, 20*)
1956 *1.* Ericksen, J. L., and R. A. Toupin: Implications of Hadamard's conditions for elastic stability with respect to uniqueness theorems. Can. J. Math. **8**, 432–436. (*32, 45*)
2. Pearson, C. E.: General theory of elastic stability. Quart. Appl. Math. **14**, 133–144. (*30, 31, 32, 47, 67*)
3. Ziegler, H.: On the concepts of elastic stability. Advances in Applied Mechanics **4**, 351–403. New York: Academic Press. (*34, 65, 78*)
1957 *1.* Hill, R.: On uniqueness and stability in the theory of finite elastic strain. J. Mech. Phys. Solids **5**, 229–241. (*23, 31, 32, 45, 47, 55, 57*)
2. Lavrentiev, M. M.: On the problem of Cauchy for linear elliptic equations of the second order [in Russian]. Dokl. Akad. Nauk. SSSR **112**, 195–197. (*48*)
3. Lefschetz, S.: Differential Equations: Geometric Theory. New York: Interscience. (*6, 7, 9*)
1958 *1.* Kantorovich, L. V., and V. I. Krylov: Approximate Methods of Higher Analysis. Groningen: Nordhoff. (*18*)
2. Taylor, A. E.: Introduction to Functional Analysis. New York: Wiley. (*5*)
1959 *1.* Coleman, B. D., and W. Noll: On the thermostatics of continuous media. Arch. Rational Mech. Anal. **4**, 97–128. (*30*)

2. GREEN, A.E., and A.J.M. SPENCER: The stability of a circular cylinder under finite extension and torsion. J. Math. Phys. **37**, 316–338. (*47*)
3. HAHN, W.: Bemerkungen zu einer Arbeit von Herrn Vejvoda über Stabilitätsfragen. Math. Nachr. **20**, 21–24. (*15*)
4. — Theorie und Anwendung der direkten Methode von Ljapunov. Berlin: Springer = Theory and Application of Liapounov's Direct Method. Translated by H.H. HOSENTHIEN and S.H. LEHNIGK. Englewood Cliff, N. J.: Prentice-Hall 1963. (*7, 15*)
5. MOVCHAN, A.A.: The direct method of Liapounov in stability problems of elastic systems. J. Appl. Math. Mech. **23**, 686–670 = Prikl. Mat. Mekh. **23**, 483–493. (*6, 17*)
6. PEARSON, C.E.: Theoretical Elasticity. Harvard: University Press. (*47*)
7. SERRIN, J.: Mathematical principles of classical fluid mechanics. Handbuch der Physik, Vol. VIII/1, pp. 125–263. Berlin-Göttingen-Heidelberg: Springer. (*14*)
1960 *1.* BERNSTEIN, B., and R.A. TOUPIN: Korn inequalities for the sphere and for the circle. Arch. Rational Mech. Anal. **6**, 51–64. (*43*)
2. CORDUNEANU, C.: The application of differential inequalities to the theory of stability [in Russian]. An. Şti. Uni. "Al. I. Cuza" Iaşi. Sect. I.N.S. 6, 47–58. (*15*)
3. GREEN, A.E., and J.E. ADKINS: Large Elastic Deformations and Non-linear Continuum Mechanics. 1st Ed. Oxford: University Press. (*23, 54, 57*)
4. HELLWIG, G.: Partielle Differentialgleichungen. Stuttgart: Teubner 1960. English translation of the original German edition by E. GERLACH. New York: Blaisdell 1964. (*4*)
5. JOHN, F.: Continuous dependence on data for solutions of partial differential equations with a prescribed bound. Comm. Pure Appl. Math. **13**, 551–585. (*8, 14, 37*)
6. MOVCHAN, A.A.: On the stability of motion of continuous bodies. Lagrange's theorem and its converse [in Russian]. Inzh. Sb. **29**, 3–20. (*17*)
7. — Stability of processes with respect to two metrics. J. Appl. Math. Mech. **24**, 1506–1524 = Prikl. Mat. Mekh. **24**, 988–1001. (*6, 9, 12, 15, 17, 19*)
1961 *1.* BOLOTIN, V.V.: Non-conservative Problems of the Theory of Elastic Stability. Moscow. English Translation by T.K. LUSHER, edit. by G. HERRMANN. Oxford: Pergamon 1963. (Chap. *E*, Sects. *65, 78, 80, 81, 82*)
2. CHETAEV, N.G.: The Stability of Motion. Translated into English by M. NADLER. Oxford: Pergamon. (*19, 31, 82*)
3. ENGLAND, A.H., and A.E. GREEN: Steady-state thermoelasticity for initially stressed bodies. Phil. Trans. Roy. Soc. London, Ser. A **253**, 517–542. (*24*)
4. GOLDBERG, R.R.: Fourier transforms. Cambridge Tracts in Mathematics and Mathematical Physics, Vol. 52. (*35*)
5. LASALLE, J.P., and S. LEFSCHETZ: Stability by Liapounov's Direct Method with Applications. New York: Academic Press. (*7, 10, 15, 20*)
6. PAYNE, L.E., and H.F. WEINBERGER: On Korn's inequality. Arch. Rational Mech. Anal. **8**, 89–98. (*43*)
7. TIMOSHENKO, S.P., and J.M. GERE: Theory of Elastic Stability. 2nd Ed. New York: McGraw-Hill. (*34, 47*)
1962 *1.* ANTOSIEWICZ, H.A.: An inequality for approximate solutions of ordinary differential equations. Math. Z. **8**, 44–52. (*15*)
2. BRAMBLE, J.H., and L.E. PAYNE: Some inequalities for vector functions with applications in elasticity. Arch. Rational Mech. Anal. **11**, 16–28. (*41*)
3. COURANT, R., and D. HILBERT: Methods of Mathematical Physics, Vol. II. New York: Interscience. (*4*)
4. FUNK, P.: Variationsrechnung und ihre Anwendung in Physik und Technik. Berlin-Göttingen-Heidelberg: Springer. (*18*)
5. GOBERT, J.: Une inégalité fondamentale de la théorie de l'élasticité. Bull. Soc. Roy. Sci. Liège **31**, 182–191. (*43*)
6. GREEN, A.E.: Thermoelastic stresses in initially stressed bodies. Proc. Roy. Soc. London, Ser. A **266**, 1–19. (*24*)
7. HILL, R.: Uniqueness criteria and extremum principles in self-adjoint problems of continuum mechanics. J. Mech. Phys. Solids **10**, 185–194. (*66, 67*)
8. KERR, A.D.: On the instability of elastic solids. Proceedings of the Fourth U.S. National Congr. on Applied Mechanics, Vol. 1, pp 647–656. (*47*)
9. KOITER, W.T.: Stability of Equilibrium of Continuous Bodies. Technical Report No. 79. Division of Applied Mathematics. Brown University. (*18, 30*)
10. LAVRENTIEV, M.M.: On the Incorrect Problems of Mathematical Physics [in Russian]. Novosibirsk. (*6, 48*)
11. LEIPHOLZ, H.: Anwendung des Galerkinschen Verfahrens auf nichtkonservative Stabilitätsprobleme des elastischen Stabes. Z. Angew. Math. Phys. **13**, 359–372. (Chap. *E*, Sects. *80, 81*)

12. Leipholz, H.: Die Knicklast des einseitig eingespannten Stabes mit gleichmäßig verteilter, tangentialer Langbelastung. Z. Angew. Math. Phys. **13**, 581–589. (Chap. *E*)
13. Slobodkin, A.M.: On the stability of the equilibrium of conservative systems with an infinite number of degrees of freedon. J. Appl. Math. Mech. **26**, 513–517 = Prikl. Mat. Mekh. **26**, 356–358. (*17, 35*)
14. Wesołowski, Z.: Some stability problems of tension in the light of the theory of finite strains. Bull. Acad. Polon. Sci., Sér. Sci. Tech. **10**, 123–128. (*47*)
15. Wolmir, A.S.: Biegsame Platten und Schalen. Translated from the Russian Ed. by A. Duda. Berlin: VEB Verlag für Bauwesen. (*47*)
16. Zorski, H.: On the equations describing small deformations superposed on finite deformations. Proceedings of the International Symposium on Second-Order Effects in Elasticity, Plasticity and Fluid Dynamics. Haifa 1962. Oxford: Pergamon 1964. (*35*)
1963 *1.* Cesari, L.: Asymptotic Behaviour and Stability Problems in Ordinary Differential Equations. 2nd Ed. Berlin-Göttingen-Heidelberg: Springer. (*6, 7, 8, 10, 23*)
2. Evan-Iwanoski, R.M.: Parametric (dynamic) stability of elastic systems. Theoretical and Applied Mechanics **1**, 111–119 (1962). New York: Plenum Press. (*64*)
3. Fosdick, R.L., and R.T. Shield: Small bending of a circular bar superposed on finite extension or compression. Arch. Rational Mech. Anal. **12**, 223–248. (*47*)
4. Gelfond, I.M., and S.V. Fomin: Calculus of Variations. Translated by R.A. Silverman. Englewood Cliffs, N. J.: Prentice-Hall. (*18*)
5. Hahn, W.: The present state of Lyapounov's direct method. Proceedings of Symposium on Non-Linear Problems. Ed. by R.E. Langer, pp. 195–205. Madison: University of Wisconsin Press. (*15*)
6. Hale, J.K., and J.P. Lasalle: Differential equations: linearity versus nonlinearity. SIAM Review **5** (3), 249–272. (*15, 23*)
7. Koiter, W.T.: The concept of stability of equilibrium for continuous bodies. Proc. Koninkl. Ned. Akad. Wetenschap. B **66** (4), 173–177. (*6, 16, 31, 34*)
8. — Elastic stability and post-buckling behaviour. Proceedings of Symposium on Non-Linear Problems. Ed. by R.E. Langer, pp. 257–275. Madison: University of Wisconsin Press. (*18, 30, 31, 32, 34, 47*)
9. Kordas, Z., and M. Życzkowski: On the loss of stability of a rod under supertangential force. Arch. Mech. Stos. **15**, 7–31. (*81*)
10. Kozin, F.: On almost sure stability of linear systems with random coefficients. J. Math. Phys. **42**, 59–67. (*77*)
11. Krasovskii, N.N.: Stability of Motion. English translation by J.L. Brenner. Stanford: University Press. (*7, 15*)
12. Leipholz, H.: Über ein Kriterium für die Gültigkeit der statischen Methode zur Bestimmung der Knicklast von elastischen Stäben unter nichtkonservativer Belastung. Ing.-Arch. **32**, 286–296. (*80*)
13. — Über das statische Kriterium bei nicht konservativen Stabilitätsproblemen der Elastomechanik. Ing.-Arch. **32**, 214–220. (*80*)
14. — Über die Konvergenz des Galerkinschen Verfahrens bei nichtselbstadjungierten und nicht konservativen Eigenwertproblemen. Z. Angew. Math. Phys. **14**, 70–79. (Chap. *E*, Sect. *81*)
15. Leonov, M.Ia., and L.M. Zovii: Effect of friction on the critical load of a compressed rod. Soviet Phys. Doklady **7**, 611–613. (*82*)
16. Movchan, A.A.: On the stability of processes concerning deformation of solid bodies [in Russian]. Arch. Mech. Stos. **15**, 659–682. (*19, 30, 32, 44, 82*)
17. Serrin, J.: The initial value problem for the Navier-Stokes equations. Proceedings of Symposium in Non-Linear Problems. Edit. by R.E. Langer, pp. 69–98. Madison: University of Wisconsin Press. (*14*)
18. Shield, R.T., and A.E. Green: On certain methods in the stability theory of continuous systems. Arch. Rational Mech. Anal. **12**, 354–360. (*6, 25*)
19. Truesdell, C.: The meaning of Betti's reciprocal theorem. J. Res. Nat. Bur. Stand. B **67**, 85–86. (*23*)
20. Wesołowski, Z.: The axially symmetric problem of stability loss of an elastic bar subject to tension. Arch. Mech. Stos. **15**, 383–395. (*47*)
21. — The axial-symmetric problem of instability in the case of tension of the elastic rod. Bull. Acad. Polon. Sci., Sér. Sci. Tech. **11**, 53–58. (*47*)
22. Zhong-Heng, G., and W. Urbanowski: Stability of non-conservative systems in the theory of elasticity of finite deformations. Arch. Mech. Stos. **15**, 309–321. (*66, 80*)
1964 *1.* Eringen, A.C., and E.S. Suhubi: Nonlinear theory of simple micro-elastic solids I. Int. J. Eng. Sci. **2**, 189–203. (*38*)

2. GREEN, A.E., and R.S. RIVLIN: Simple forces and stress multipoles. Arch. Rational Mech. Anal. **16**, 325–352. (*38*)
3. — — Multipolar continuum mechanics. Arch. Rational Mech. Anal. **17**, 113–147. (*38*)
4. HERRMANN, G., and R.W. BUNGAY: On the stability of elastic systems subjected to nonconservative forces. J. Appl. Mech. **31**, 435–440. (*80*)
5. HILL, R.: Principles of Dynamics. Oxford: Pergamon. (*19*)
6. HOLDEN, J.T.: Estimation of critical loads in elastic stability theory. Arch. Rational Mech. Anal. **17**, 171–183. (*23, 31, 40, 43, 51, 53, 55, 57*)
7. JOHN, F.: Hyperbolic and parabolic equations. Partial Differential Equations, Ed. by L. BERS, F. JOHN and M. SCHECHTER, pp. 1–123. New York: Wiley. (*6*)
8. LEIPHOLZ, H.: Über den Einfluß der Dampfung bei nichtkonservativen Stabilitätsproblemen elastischer Stäbe. Ing.-Arch. **33**, 308–321. (*82*)
9. MIKHLIN, S.G.: Variational Methods in Mathematical Physics. Translated by T. BODDINGTON. Oxford: Pergamon. (Chap. *E*, Sects. *35*)
10. PFLÜGER, A.: Stabilitätsprobleme der Elastostatik. 2nd Ed. (1st Ed. 1950). Berlin-Göttingen-Heidelberg: Springer. (*9, 32, 47, 65, 78, 80*)
11. SHAPOVALA, L.A.: Thermal stability of plates and shells. Strength and Deformation in Non-Uniform Temperature Fields. Ed. by YA.B. FRIDMAN. Consultants Bureau, N.Y. (*58*)
12. SOBOLEV, S.L.: Partial Differential Equations of Mathematical Physics. Translated from the 3rd Russian edition by E.R. DAWSON and T.A.A. BROADBENT. Oxford: Pergamon. (*4, 6, 8*, Chap. *E*)
13. TOUPIN, R.A.: Theories of elasticity with couple-stress. Arch. Rational Mech. Anal. **17**, 85–112. (*38*)
14. VAINBERG, M.M.: Variational Methods for the Study of Nonlinear Operators with a Chapter on Newton's Method by L.V. KANTOROVICH and G.P. AKILOV. Translated and supplemented by A. FEINSTEIN. San Francisco: Holden-Day. (*18*)
15. WESOŁOWSKI, Z.: Stability of a full elastic sphere uniformly loaded on the surface. Arch. Mech. Stos. **16**, 1131–1151. (*66, 80*)
16. — The stability of an elastic orthotropic parallelepiped subject to finite elongation. Bull. Acad. Polon. Sci., Sér. Sci. Tech. **12**, 155–160. (*47*)
17. ZUBOV, V.I.: Methods of A.M. Lyapunov and Their Application. English Edition translated under the auspices of the U.S. Atomic Energy Commission and edit. by L.F. BORON. Groningen: Noordhoff. (*1, 2, 4, 6, 15, 19*)

1965 *1.* BEATTY, M.F.: Some static and dynamic implications of the general theory of elastic stability. Arch. Rational Mech. Anal. **19**, 167–168. (*31, 32, 40, 53*)
2. BHATIA, N.P., and V. LAKSHMIKANTHAM: An extension of Lyapounov's direct method. Mich. Math. J. **12**, 183–191. (*20*)
3. BIOT, M.A.: Mechanics of Incremental Deformations. New York: Wiley. (*47*)
4. BRUN, L.: Sur l'unicité en thermoélasticité dynamique et diverses expressions analogues à la formule de Clapeyron. Comp. Rend. **261**, 2584–2587. (*50, 61*)
5. BUFLER, H.: Zur Gleichgewichtsmethode in der Theorie der elastischen Stabilität. Acta Mech. **1**, 386–394. (*47*)
6. CAUGHEY, T.K., and A.H. GRAY, JR.: On the almost sure stability of linear dynamic systems with stochastic coefficients. J. Appl. Mech. **32**, 365–372. (*77*)
7. ECKHAUS, W.: Studies in Non-Linear Stability Theory. Springer Tracts in Natural Philosophy, Vol. 6. Berlin-Heidelberg-New York: Springer. (Chap. *E*)
8. FICHERA, G.: Linear elliptic differential systems and eigenvalue problems. Lecture Notes in Mathematics, Vol. 8. Berlin-Heidelberg-New York: Springer. (*35, 37*)
9. GOFFMAN, C.: Calculus of Several Variables. New York: Harper and Row. (*6*)
10. HERRMANN, G., and I-C. JONG: On the destabilizing effect of damping in nonconservative elastic systems. J. Appl. Mech. **32**, 592–597. (*82*)
11. INFANTE, E.F., and L. WEISS: On the stability of systems defined over a finite time interval. Proc. Nat. Acad. Sci. U.S.A. **54** (1), 44–48. (*7*)
12. KOITER, W.T.: The energy criterion of stability for continuous elastic bodies. I and II. Proc. Koninkl. Ned. Akad. Wetenschap. B **68**, 178–189, 190–202. (*6, 16, 18, 21, 30, 31, 32, 34, 35, 38, 82*)
13. — On the instability of equilibrium in the absence of a minimum of the potential energy. Proc. Koninkl. Ned. Akad. Wetenschap. B **68**, 107–113. (*19, 31, 82*)
14. LEIPHOLZ, H.: Über die Konvergenz des Verfahrens von Galerkin bei quasi-linearen Randwertproblemen. Acta Mech. **1**, 339–353. (*81*)
15. — Über die Zulässigkeit des Verfahrens von Galerkin bei linearen, nichtselbstadjungierten Eigenwertproblemen. Z. Angew. Math. Phys. **16**, 837–843. (*81*)
16. LYNDEN-BELL, D.: On the evolution of frictionless ellipsoids. Astrophys. J. **142**, 1648–1649. (*82*)

17. MIKHLIN, S. G.: The Problem of the Minimum of a Quadratic Functional. San Francisco: Holden-Day. (*15, 35, 43*)
18. NEUBER, H.: Theorie der elastischen Stabilität bei nichtlinearer Vervorformung. Acta Mech. **1**, 285–293. (*47*)
19. OGAWA, H.: Lower bounds for solutions of differential inequalities in Hilbert Space. Proc. Am. Math. Soc. **16**, 1241–1243. (*48*)
20. SEWELL, M. J.: On the calculation of potential functions defined on curved boundaries. Proc. Roy. Soc. London, Ser. A **286**, 402–411. (*66, 67*)
21. SHIELD, R. T.: On the stability of linear continuous systems. Z. Angew. Math. Phys. **16**, 649–686. (*6, 16, 17, 25, 34, 35, 37, 70, 73, 82*)
22. TRUESDELL, C., and W. NOLL: The non-linear field theories of mechanics. Handbuch der Physik, Vol. III/3. Berlin-Heidelberg-New York: Springer. (*18, 22, 23, 30, 31, 32, 40, 53*)
23. WANG, P. K. C.: Stability analysis of a simplified flexible vehicle via Lyapunov's direct method. Am. Inst. Aero. Astro. J. **3**, 1764–1766. (*77*)
1966 *1.* AGMON, L.: Unicité et convexité dans les problèmes différentiels. Sem. Math. Sup. 1965. Montreal: University Press. (*14, 48, 50*)
2. BÜRGERMEISTER, G., H. STEUP u. H. KRETZCHMAR: Stabilitätstheorie, Teil 1. 3rd Ed. Berlin: Springer. (*47*)
3. COMO, M.: Lateral buckling of a cantilever subjected to a transverse follower force. Int. J. Sols. Structs. **2**, 515–523. (*81*)
4. CONTRI, L.: On the stability of elastic systems under non-conservative follower forces. Meccanica **1**, 61–64. (*80*)
5. DOWELL, E. H.: Non-linear oscillations of a fluttering plate. Am. Inst. Aero. Astro. J. **4**, 1267–1275. (*81*)
6. DUGUNDJI, J.: Theoretical considerations of panel flutter at high supersonic Mach numbers. Am. Inst. Aero. Astro. J. **4**, 1257–1266. (*81*)
7. ERICKSEN, J. L.: A thermo-kinetic view of elastic stability theory. Int. J. Sols. Structs. **2**, 573–580. (*21, 25, 28, 30, 32*)
8. ERICKSEN, J. L.: Thermoelastic stability. Proc. 5th U.S. National Congr. of Appl. Mech. ASME, pp. 187–193. (*21, 25, 28, 30, 32, 34*)
9. HERBERT, R. E.: Stability of elastic bodies subjected to body and pressure forces with specialisation to thin shells. J. Inst. Math. Appl. **2**, 248–265. (*67, 68, 80*)
10. HERRMANN, G., and I.-C. JONG.: On non-conservative stability problems of elastic systems with slight damping. J. Appl. Mech. **33**, 125–133. (*82*)
11. — S. NEMAT-NASSER, and S. N. PRASAD: Models demonstrating instability of non-conservative mechanical systems. Technical Report No. 66-4. Structural Mechanics Laboratory, Department of Civil Engineering, The Technological Institute, Northwestern University. (*65, 78, 81*)
12. HSU, C. S.: On dynamic stability of elastic bodies on the prescribed initial conditions. Int. J. Eng. Sci. **4**, 1–21. (*6, 9, 21, 25, 47, 82, 83*)
13. — Dynamic stability of autonomous and continuous elastic systems. Proc. 5th U.S. National Congr. of Appl. Mech. ASME, p. 128. (*9, 21, 25, 47, 82, 83*)
14. JOSEPH, D. D.: Non-linear stability of the Boussinesq equations by the method of energy. Arch. Rational Mech. Anal. **22**, 163–184. (*14*)
15. KNOPS, R. J., and E. W. WILKES: On Movchan's theorems for stability of continuous systems. Int. J. Eng. Sci. **4**, 303–329. (*17, 21, 27, 35, 37*)
16. KOITER, W. T.: Purpose and archievements of research in elastic stability. Proc. 4th Technical Conference. Soc. for Eng. Sci. (N. Carolina). (*21, 22*)
17. KRALL, G.: Osservazioni sui principi varionazionali per la stabilità in elastostatica e loro applicazioni. Atti Accad. Naz. Lincei Mem. Classe Sci. Fis. Mat. Nat. **8**, 53–84 (1966–1967). (*31, 47*)
18. LEIPHOLZ, H.: Über die Anwendung der Methoden von Liapounov auf Stabilitätsprobleme der Elastostatik. Ing.-Arch. **35**, 181–191. (*81*)
19. LEVINSON, M.: Application of the Galerkin and Ritz methods to nonconservative problems of elastic stability. Z. Angew. Math. Phys. **17**, 431–442. (Chap. *E*, Sect. *81*)
20. LIAPOUNOV, A. M.: Stability of Motion. New York: Academic Press. (*15*)
21. MORREY, C. B., JR.: Multiple integrals in the Calculus of Variations. Berlin-Heidelberg-New York: Springer. (*18, 31*)
22. NEMAT-NASSER, S., and G. HERRMANN: On the stability of equilibrium of continuous systems. Ing.-Arch. **35**, 17–24. (*6, 63, 68, 81*)
23. — — Adjoint systems in nonconservative problems of elastic stability. Am. Inst. Aero. Astro. J. **4**, 2221–2222. (*81*)
24. — — Torsional instability of cantilevered bars subjected to nonconservative loading. J. Appl. Mech. **33**, 102–104. (*80*)

25. Nemat-Nasser, S., and G. Herrmann: Some general considerations concerning the destabilizing effect in non-conservative systems. Z. Angew. Math. Phys. **17**, 305–313. (*82*)
26. — S. N. Prasad, and G. Herrmann: Destabilizing effect of velocity-dependent forces in nonconservative continuous systems. Am. Inst. Aero. Astro. J. **4**, 1276–1280. (*82*)
27. Parks, P. C.: A stability criterion for panel flutter via the second method of Liapounov. Am. Inst. Aero. Astro. J. **4**, 175–177. (*79*)
28. Payne, L. E.: On some non-well posed problems for partial differential equations. Numerical Solutions of Non-Linear Differential Equations. Ed. by D. Greenspan, pp. 239–263. New York: Wiley. (*14, 48*)
29. Rivlin, R. S.: The fundamental equations of nonlinear continuum mechanics. Symposium on Dynamics of Fluids and Plasmas. Ed. by S. I. Pai, pp. 83–126. New York: Academic Press. (*22, 23*)
30. Rosenberg, R. M.: On geometric representations in dynamics. Acta Mech. **2**, 144–170. (*6*)
31. Takahashi, Y.: A short introduction into the theory of the state space concept. Parts 1 and 2. Regelungstechnik **14**, 449–455, 513–518. (*6*)
32. Wang, P. K. C.: Stability analysis of elastic and aeroelastic systems via Lyapounov's direct method. J. Franklin Inst. **281**, 51–72. (*1, 6*)

1967 *1.* Achenbach, J. D.: The propagation of stress discontinuities according to the coupled equations of thermoelasticity. Acta Mech. **3**, 342–351. (*6*)
2. Agmon, S., and L. Nirenberg: Lower bounds and uniqueness theorems of differential equations in a Hilbert space. Comm. Pure Appl. Math. **20**, 207–229 (*14, 48*)
3. Barbashin, E. A.: Introduction to the theory of stability [in Russian]. Nauka 223: Moscow. (*15*)
4. Beatty, M. F.: A theory of elastic stability for incompressible, hyperelastic bodies. Int. J. Sols. Structs. **3**, 23–37. (*23, 31, 32, 40, 53, 54, 55, 57*)
5. Bhatia, N. P., and G. P. Szegö: Dynamical systems: Stability theory and applications. Lecture Notes in Mathematics, Vol. 35. Berlin-Heidelberg-New York: Springer. (*1, 4, 5, 6, 7, 10, 15*)
6. Budiansky, B.: Dynamic buckling of elastic structures: criteria and estimates. Proc. of Int. Conference on Dynamic Stability of Structures. Northwestern University 1965. Ed. by G. Herrmann, pp. 83–106. Oxford: Pergamon. (*32*)
7. Carroll, R.: Some growth and convexity theorems for second order equations. J. Math. Anal. Appl. **17**, 508–517. (*14*)
8. Caughey, T. K., and J. R. Dickerson: Stability of linear dynamic systems with narrow-band parametric excitation. J. Appl. Mech. **34**, 709–713. (*77*)
9. Eckhaus, W.: On the perturbation method in the stability theory of continuous systems. Math. Inst. Tech. Hogesch. Delft. (Chap. *E*)
10. Gilbert, J. E., and R. J. Knops: Stability of general systems. Arch. Rational Mech. Anal. **25**, 271–284. (*3, 6, 7, 9, 10, 12, 15, 17, 19*)
11. Hahn, W.: Stability of Motion. English Translation by Arne P. Baartz. Berlin-Heidelberg-New.York: Springer. (*15*)
12. Herrmann, G.: Stability of equilibrium of elastic systems subjected to non-conservative forces. Appl. Mech. Rev. **20**, 103–108. (*64, 65, 78, 81, 82*)
13. —, and S. Nemat-Nasser: Instability modes of cantilevered bars induced by fluid flow through attached pipes. Int. J. Sols. Structs. **3**, 39–52. (*82*)
14. — — Energy considerations in the analysis of stability of non-conservative structural systems. Proc. of Int. Conf. Dynamic Stability of Structures. Northwestern University 1965. Ed. by G. Herrmann, pp. 299–308. Oxford: Pergamon. (*63, 82*)
15. Hill, R.: Eigenmodal deformations in elastic/plastic continua. J. Mech. Phys. Sols. **15**, 371–386. (*40, 51, 52, 53, 54, 55, 57*)
16. Hsu, C. S.: The effects of various parameters on the dynamic stability of a shallow arch. J. Appl. Mech. **34**, 349–358. (*47, 83*)
17. Infante, E. F., and L. Weiss: Finite time stability under perturbing forces and on product spaces. Int. Symp. Diff. Eqs. and Dynamical Systems. Univ. of Puerto Rico 1965. Ed. by J. K. Hale and J. P. LaSalle, pp. 341–350. New York: Academic Press. (*7*)
18. Koiter, W. T.: On the thermodynamic background of elastic stability theory. Report No. 360 Lab. of Eng. Mech., Dept. of Mech. Eng., Technological Univ. Delft. (*6, 16, 21, 22, 23, 28, 30, 32, 82*)
19. Kuzmin, G. A.: On the stability of motion in the case of two small positive roots. J. Appl. Math. Mech. **31**, 150–155 = Prikl. Math. Mech. **31**, 140–144. (*7*)

20. LASALLE, J.P.: An invariance principle in the theory of stability. Int. Symp. Diff. Eqs. and Dynamical Systems. Univ. of Puerto Rico 1965. Ed. by J.K. HALE and J.P. LASALLE, pp. 277–286. New York: Academic Press. (*3, 20*)
21. LAVRENTIEV, M.M.: Some improperly posed problems of mathematical physics (Translation revised by R. J. SACKER). Springer Tracts in Natural Philosophy, Vol. 11. Berlin-Heidelberg-New York: Springer 1962. (*6, 48*)
22. LEIPHOLZ, H.: Über die Wahl der Ansatzfunktionen bei der Durchführung des Verfahrens von GALERKIN. Acta Mech. **3**, 295–317. (*81*)
23. LIN, K.-H., S. NEMAT-NASSER, and G. HERRMANN: Stability of a bar under eccentric follower force. J. Eng. Mech. Div. Proc. Am. Soc. Civil Eng. **93**, 105–115. (*81*)
24. METTLER, E.: Stability and vibration problems of mechanical systems under harmonic excitation. Proc. of Int. Conf. Dynamic Stability of Structures. Northwestern Univ. 1965. Ed. by G. HERRMANN, pp. 169–188. Oxford: Pergamon. (*64*)
25. NEMAT-NASSER, S., and J. ROORDA: On the energy concepts in the theory of elastic stability. Acta Mech. **4**, 296–307. (*34, 63, 68*)
26. OGAWA, H.: Lower bounds for solutions of parabolic differential inequalities. Can. J. Math. **19**, 667–672. (*48*)
27. PARKS, P.C.: A stability criterion for a panel flutter problem via the second method of Liapounov. Int. Symp. Diff. Eqs. Dynamical Systems. Univ. of Puerto Rico 1965. Ed. by J.K. HALE and J.P. LASALLE, pp. 287–298. New York: Academic Press. (*79*)
28. PLAUT, R.H.: A study of the dynamic stability of continuous elastic systems by Liapounov's direct method. Coll. of Eng. Report No. AM-67-3. Univ. of California, Berkeley. (*68, 79*)
29. PRASAD, S.N., and G. HERRMANN: The usefulness of adjoint systems in solving non-conservative stability problems of elastic continua. Technical Report No. 67-5. Structural Mech. Lab. Tech. Inst. Northwestern Univ. (Chap. *E*, Sect. *81*)
30. PRASAD, S.N., and G. HERRMANN: Complex treatment of a class of nonconservative stability problems. Technical Report No. 67-1. Structural Mech. Lab. Tech. Inst. Northwestern Univ. (*82*)
31. SEWELL, M.J.: On configuration-dependent loading. Arch. Rational Mech. Anal. **23**, 327–351. (*65, 66, 67, 78*)
32. STOKER, J.J.: Stability of continuous systems. Proc. of Int. Conf. Dynamic Stability of Structures. Northwestern Univ. 1965. Ed. by G. HERRMANN, pp. 45–63. Oxford: Pergamon. (*5, 15, 23*)
33. WEBER, J.-D.: Sur une condition générale de stabilité d'un corps élastique non homogène. Comp. Rend., Sér. A **265**, 66–68. (*31*)
34. WILLKE, H.L., JR.: Stability in time-symmetric flows. J. Math. Phys. **46**, 151–163. (*7*)
35. WOLKOWISKY, J.H.: Existence of buckled states of circular plates. Comm. Pure Appl. Math. **20**, 549–560. (*47*)
1968 *1.* AUGUSTI, G.: Instability of struts subject to radial heat. Meccanica **3**,t167–176. (*58*)
2. BEATTY, M.F.: Stability of the undistorted states of an isotropic elas ic body. Int. J. Non-Linear Mech. **3**, 337–349. (*23, 31, 32, 40, 53, 54, 55, 56, 57*)
3. —, and D.E. HOOK: Some experiments on the stability of circular rubber bars under end thrust. Int. J. Sols. Structs. **4**, 623–625. (*57*)
4. CAUGHEY, T.K., and R.T. SHIELD: Instability and the energy criterion for continuous systems. Z. Angew. Math. Phys. **19**, 485–491. (*32, 44, 45, 82*)
5. COLEMAN, B.D., and E.H. DILL: On the stability of certain motions of incompressible materials with memory. Arch. Rational Mech. Anal. **30**, 197–224. (*29*)
6. DAFERMOS, C.M.: Some remarks on Korn's inequality. Z. Angew. Math. Phys. **19**, 913–920. (*43*)
7. — On the existence and the asymptotic stability of solutions to the equations of linear thermoelasticity. Arch. Rational Mech. Anal. **29**, 241–271. (*2, 23, 35, 58, 61, 62*)
8. DUPUIS, G.: Contribution à l'étude de la stabilité élastique. Thèse. Lausanne. (*31*)
9. GILL, K.F., J. SCHWARZENBACH, and G.E. HARLAND: Stability analysis for the practical engineer. Engineer **224**, 247–254. (*15*)
10. GREEN, A.E., and W. ZERNA: Theoretical Elasticity. 2nd Ed. Oxford: University Press. (*22, 78*)
11. HILL, R.: On constitutive inequalities for simple materials. J. Mech. Phys. Sols **16**, 229–242. (*30*)
12. HSU, C.S.: Equilibrium configurations of a shallow arch of arbitrary shape and their dynamic stability character. Int. J. Non-linear Mech. **3**, 113–136. (*34, 47, 83*)
13. — Stability of shallow arches against snap-through under time-wise step loads. J. Appl. Mech. **35**, 31–39. (*47, 82, 83*)

14. HSU, C. S., KUO, C. T., and S. S. LEE: On the final states of shallow arches on elastic foundations subjected to dynamical loads. J. Appl. Mech. **35**, 713–723. (*83*)
15. INFANTE, E. F.: On the stability of some linear nonautonomous random systems. J. Appl. Mech. **35**, 7–11. (*77*)
16. KNOPS, R. J., and L. E. PAYNE: On the stability of solutions of the Navier-Stokes equations backward in time. Arch. Rational Mech. Anal. **29**, 331–335. (*48*)
17. — — Stability of the traction boundary value problem in linear elastodynamics. Int. J. Eng. Sci. **6**, 351–357. (*41*)
18. — — Uniqueness in classical elastodynamics. Arch. Rational Mech. Anal. **27**, 349–355. (*23, 35, 50*)
19. — — Stability in linear elasticity. Int. J. Sols. Structs. **4**, 1233–1242. (*50*)
20. LEIPHOLZ, H.: Über die Konvergenz des Verfahrens von Grammel und seine mögliche Erweiterung. Z. Angew. Math. Phys. **10**, 94–113. (*81*)
21. LEVINSON, M.: Stability of a compressed neo-Hookean rectangular parallelepiped. J. Mech. Phys. Sols. **16**, 403–415. (*47*)
22. NEMAT-NASSER, S.: On local stability of a finitely deformed solid subjected to follower type loads. Quart. Appl. Math. **26**, 119–129. (*63, 67, 68, 78, 79*)
23. OGAWA, H.: On lower bounds and uniqueness for solutions of the Navier-Stokes equations. J. Math. Mech. **18**, 445–452. (*48*)
24. PRITCHARD, A. J.: Stability boundaries for a two-dimensional panel. Am. Inst. Aero. Astro. J. **6**, 1590–1592. (*79, 81*)
25. SHEN, M. K.: Note on the static criterion of elastic stability. Z. Angew. Math. Mech. **48**, 356–357. (*47*)
26. STOKER, J. J.: Nonlinear Elasticity. London: Nelson. (*47*)
27. WASHIZU, K.: Variational Methods in Elasticity and Plasticity. Oxford: Pergamon. (*18*)
28. WATHER, W. W., and LEVINSON, M.: Destabilisation of a nonconservative loaded elastic system due to rotary inertia. C. A. S. I. Trans. **1**, 91–93. (*82*)
29. ZIEGLER, H.: Principles of Structural Stability. Waltham: Blaisdell. (*34, 64*)
30. ZHINZHER, N. I.: The destabilising effect of friction on the stabilising of non-conservative elastic systems [in Russian]. Inzh. Zh. Mekh. Tver. Tcha. **3**, 44–47. (*82*)
1969 *1.* ARBOCZ, J., and C. D. BABCOCK: The effect of general imperfections on the buckling of cylindrical shells. J. Appl. Mech. **36**, 28–38. (*47*)
2. BRUN, L.: Méthodes énergétiques dans les systèmes évolutifs linéaires. Premier Partie: Séparation des énergies. Deuxième Partie: Théorèmes d'unicité. J. de Mech. **8**, 125–166, 167–192. (*50, 61*)
3. CAUGHEY, T. K., and J. R. DICKERSON: Stability of continuous dynamic systems with parametric excitation. J. Appl. Mech. **36**, 212–216. (*77*)
4. COTSAFTIS, M.: On the stability of motion. J. Inst. Math. Appl. **5**, 19–54. (Chap. *E*, Sect. *81*)
5. DAVIS, S. H.: On the principle of exchange of stabilities. Proc. Roy. Soc. London, Ser. A **310**, 341–358. (*46, 81*)
6. DIPRIMA, R. C., and G. J. HABETLER: Completeness theorem for nonself-adjoint eigenvalue problems in hydrodynamic stability. Arch. Rational Mech. Anal. **34**, 218–227. (Chap. *E*, Sect. *81*)
7. DUPUIS, G.: Stabilité élastique des structures unidimensionelles. Z. Angew. Math. Phys. **20**, 94–106. (*31*)
8. EVAN-IWANOWSKI, R. M.: Non-stationary vibrations of mechanical systems. Appl. Mech. Rev. **22**, 213–219. (*64*)
9. FELDT, W. T., S. NEMAT-NASSER, S. N. PRASAD, and G. HERRMANN: Instability of a mechanical system induced by an impinging fluid jet. J. Appl. Mech. **36**, 693–701. (*81*)
10. HALE, J. K.: Dynamical systems and stability. J. Math. Anal. Appl. **26**, 39–59. (*1, 2, 3 4, 19, 20*)
11. — Ordinary Differential Equations. New York: Wiley-Interscience. (*6, 7, 8, 31*)
12. HSU, C. S., C.-T. KUO, and R. H. PLAUT: Dynamic stability criteria for clamped shallow arches under time-wise step loads. Am. Inst. Aero. Astro. J. **7**, 1925–1931. (*83*)
13. KNOPS, R. J., and L. E. PAYNE: Continuous data dependence for the equations of classical elastodynamics. Proc. Cambridge Phil. Soc. **66**, 481–491. (*75, 76*)
14. —, and T. R. STEEL: On the stability of a mixture of two elastic solids. J. Comp. Mats. **3**, 652–662. (*48*)
15. LAKSHMIKANTHAM, V., and S. LEELA: Differential and Integral Inequalities. New York: Academic Press. (*15*)

16. Leipholz, H.: Application of Liapounov's direct method to the stability problem of rods subject to follower forces. Solid Mech. Div. Report No. 26. University of Waterloo. (*79*)
17. Nowinski, J. L.: Instability of a thick nonhomogeneous elastic layer under high initial stress. J. Appl. Mech. **36**, 639–640. (*47*)
18. —, and M. Shahinpoor: Stability of an elastic circular tube of arbitrary wall thickness subject to external pressure. Int. J. Non-Linear Mech. **4**, 143–158. (*47*)
19. Sagan, H.: Introduction to the Calculus of Variations. New York: McGraw-Hill. (*6, 18*)
20. Sewell, M. J.: A method of post-buckling analysis. J. Mech. Phys. Sols. **17**, 219–233. (*47*)
21. Slemrod, M.: Ph.D. Dissertation. Div. Appl. Math. Brown Univ. (*3*)
22. Thompson, J. M. T.: A general theory for the equilibrium and stability of discrete conservative systems. Z. Angew. Math. Phys. **20**, 797–846. (*31*)
23. Vahidi, B., and N. C. Huang: Thermal buckling of shallow bimetallic two-hinged arches. J. Appl. Mech. **36**, 768–774. (*58*)
24. Weber, J.-D.: Sur des conditions suffisantes de stabilité de corps élastiques et plastoélastiques. Comp. Rend., Sér. A **268**, 971–973. (*31*).
25. Wu, C.-H., and O. E. Widera: Stability of a thick rubber solid subject to pressure loads. Int. J. Sols. Structs. **5**, 1107–1117. (*47*)
1970 *1.* Aggarwala, B. D., and E. Saibel: Thermal stability of bimetallic shallow spherical shells. Int. J. Non-Linear Mech. **5**, 49–62. (*58*)
2. Beatty, M. F.: Stability of hyperelastic bodies subject to hydrostatic loads. Int. J. Non-Linear Mech. **5**, 367–383. (*40, 67, 68*)
3. — A theory of elastic stability for constrained, hyperelastic Cosserat continua. Arch. Mech. Stos. **12**, 585–606. (*31*)
4. —, and D. C. Leigh: A mathematical theory for the traction on a body in motion in a continuum. Part I. The foundation principles. Arch. Rational Mech. Anal. **38**, 81–106. (*65*)
5. Bhatia, N. P., and G. P. Szegö: Stability Theory of Dynamical Systems. Berlin-Heidelberg-New York: Springer. (*1, 4, 10*)
6. Bromberg, E.: Buckling of a very thick rectangular block. Comm. Pure Appl. Maths. **23**, 511–528. (*47*)
7. Brown, A. L., and A. Page: Elements of Functional Analysis. London: Van Nostrand Reinhold. (*6, 16*)
8. Coleman, B. D.: On the stability of equilibrium states of general fluids. Arch. Rational Mech. Anal. **36**, 1–32. (*28*)
9. Dickey, R. W.: Free vibrations and dynamic buckling of the extensible beam. J. Math. Anal. Appl. **29**, 443–454. (*81*)
10. Hagedorn, P.: On the destabilizing effect of non-linear damping in non-conservative systems with follower forces. J. Non-Linear Mech. **5**, 341–358. (*82*)
11. Hill, R.: Constitutive inequalities for isotropic elastic solids under finite strain. Proc. Roy. Soc. London, Ser. A **314**, 457–472. (*30*)
12. Hlaváček, I., and J. Nečas: On inequalities of Korn's type. Parts I and II. Arch. Rational Mech. Anal. **36**, 305–334. (*17, 43*)
13. Huang, N. C., and R. C. Shieh: Thermomechanical coupling effect on the stability of non-conservative elastic continuous systems. Int. J. Mech. Sci. **12**, 39–49. (*24, 58, 82*)
14. Huseyin, K.: Fundamental principles in the buckling of structures under combined loading. Int. J. Sols. Structs. **6**, 479–487. (*46*)
15. John, F.: Partial differential equations. Mathematics Applied to Physics. Ed. by E. Roubine, pp. 229–315. UNESCO Paris: Springer. (*6, 8*)
16. Kerr, A. D., and L. El-Bayoumy: On the nonunique equilibrium states of a shallow arch subjected to a uniform lateral load. Quart. Appl. Math. **28**, 399–409. (*31*)
17. Kershaw, R. J.: Dynamical systems and their stability. M. Sc. Dissertation, University of Newcastle upon Tyne. (*3*)
18. Kiusalaas, J., and H. E. Davies: On the stability of elastic systems under retarded follower forces. Int. J. Sols. Structs. **6**, 399–409. (*78*)
19. Knops, R. J., and L. E. Payne: On uniqueness and continuous dependence in dynamical problems of linear thermoelasticity. Int. J. Sols. Structs. **6**, 1173–1184. (*61*)
20. Leipholz, H.: Über die Anwendung von Liapounovs direkter Methode auf Stabilitätsprobleme kontinuierlicher, nicht konservativer Systeme. Ing.-Arch. **39**, 357–368. (*79*)
21. Levine, H. A.: Logarithmic convexity, first order differential inequalities and some applications. Trans. Am. Math. Soc. **152**, 299–320. (*14, 48, 50*)

22. Levine, H. A.: Logarithmic convexity and the Cauchy problem for some abstract second order differential inequalities. J. Diff. Eqs. **8**, 34–55. (*14, 48, 50*)
23. Matkowsky, B. J.: Nonlinear dynamic stability: a formal theory. SIAM J. Appl. Math. **18**, 872–883. (*82*)
24. Nemat-Nasser, S.: On thermomechanics of elastic stability. Z. Angew. Math. Phys. **21**, 538–552. (*21, 22, 28, 30, 63, 65, 68, 78*)
25. Ogawa, H.: On the maximum rate of decay of solutions of parabolic differential inequalities. Arch. Rational Mech. Anal. **38**, 173–177. (*48*)
26. Plaut, R.H., and E.F. Infante: The effect of external damping on the stability of Beck's problem. Int. J. Sols. Structs. **6**, 491–496. (*82*)
27. Rosenbrock, H.H., and C. Storey: Mathematics of Dynamical Systems. Studies in Dynamical Systems. London: Nelson. (*1*)
28. Sewell, M. J.: On the branching of equilibrium paths. Proc. Roy. Soc. London, Ser. A **315**, 499–518. (*31, 47*)
29. Tihonov, A.N., A.B. Vasil'eva, and V.M. Volosov: Ordinary differential equations. Mathematics applied to physics. Ed. by E. Roubine, pp. 162–228. UNESCO Paris: Springer. (*18*)
1971 *1.* Cotsaftis, M.: On general theorems for stability. IUTAM Symp. on Instability of Continuous Systems. Herrenalb 1969, pp. 204–214. Berlin-Heidelberg-New York: Springer. (Chap. *E*, Sect. *81*)
2. Dikman, M.: Stability of the Cosserat surface. IUTAM Symp. on Instability of Continuous Systems. Herrenalb 1969, pp. 188–193. Berlin-Heidelberg-New York: Springer. (*31*)
3. Dowell, E.H.: Aeroelastic stability of plates and shells. IUTAM Symp. on Instability of Continuous Systems. Herrenalb 1969, pp. 65–77. Berlin-Heidelberg-New York: Springer. (*78, 81*)
4. Horgan, C.O., and J.K. Knowles: Eigenvalue problems associated with Korn's inequalities. Arch. Rational Mech. Anal. **40**, 384–402. (*43*)
5. Hsu, C.S., and T.H. Lee: A stability study of continuous systems under parametric excitation via Liapounov's direct method. IUTAM Symp. on Instability of Continuous Systems. Herrenalb 1969, pp. 112–118. Berlin-Heidelberg-New York: Springer. (*64, 77*)
6. Jospeh, D.D.: On the place of energy methods in global theory of hydrodynamic stability. IUTAM Symp. on Instability of Continuous Systems. Herrenalb 1969, pp. 132–142. Berlin-Heidelberg-New York: Springer. (*14*)
7. — Hydrodynamic stability. Springer Tracts in Natural Philosophy. (In Press.) (*14*)
8. Knops, R. J., and L.E. Payne: Uniqueness theorems in linear elasticity. Springer Tracts in Natural Philosophy, Vol. 19. Berlin-Heidelberg-New York: Springer. (*46*)
9. — — Hölder stability and logarithmic convexity. IUTAM Symp. on Instability of Continuous Systems. Herrenalb 1969, pp. 248–255. Berlin-Heidelberg-New York: Springer. (*8, 14*)
10. — — Growth estimates for solutions of evolutionary equations in Hilbert space with applications in elastodynamics. Arch. Rational Mech. Anal. **41**, 363–398. (*14, 48, 50*)
11. Matkowsky, B. J., and E.L. Reiss: On the asymptotic theory of dissipative wave motion. Arch. Rational Mech. Anal. **42**, 194–212. (*82*)
12. Slemrod, M., and E.F. Infante: An invariance principle for dynamical systems on Banach space: application to the general problem of thermoelastic stability. IUTAM Symp. on Instability of Continuous Systems. Herrenalb 1969, pp. 215–221. Berlin-Heidelberg-New York: Springer. (*2, 3, 20, 23, 35, 58, 62*)

Growth and Decay of Waves in Solids.

By

Peter J. Chen.

With 1 Figure.

I. Introduction.

1. Nature of this article. It has always been recognized that knowledge of the manner in which waves propagate in material bodies is of great importance in that it furnishes us with the understanding of how they react to sudden disturbances. There are many differing methods in acquiring this knowledge. During the latter part of the nineteenth century, Hugoniot developed a concept of wave propagation. He conceived of a wave as a disturbance which is limited, rigorously, to a surface, but the disturbance itself may be arbitrary in magnitude. The consequence of this concept is far reaching in that the resulting theory of propagating singular surfaces is mathematically exact, free from any approximations or assumptions. This theory of Hugoniot was developed to its present level by Hadamard during the latter part of the nineteenth century and the first part of this century.

In this article, I shall utilize the method of the theory of propagating singular surfaces to examine the behavior of waves in various types of materials and shall show that definite, concrete results can be obtained without having to appeal to any explicit representation for the constitutive relations of the materials. Therefore, the results obtained are common to all the materials within each particular class. Also, I shall show that the condition of a material body does have definite influences on the behavior of waves propagating in it and that it is possible for waves propagating in bodies of different materials under different conditions to have the same behavior. This, of course, has great physical significance in that experimenters must exercise great care in classifying materials by observing wave behavior. In particular, I shall show that the behavior of shock waves or acceleration waves propagating in bodies of simple material with memory which are at rest in homogeneous configurations, in deformed elastic non-conductors of heat and in bodies of inhomogeneous elastic material at rest in non-homogeneous reference configurations, can under certain circumstances be qualitatively the same. To put it more directly, in these situations it is not possible to distinguish between the various classes of materials by observing the wave behavior.

These conclusions not only demonstrate the power of the method of the theory of singular surfaces, but also they illuminate its elegance. It is difficult to conceive of any other present-day methods of analysis that will yield similar conclusions of such generality. For instance, it would be very difficult to obtain a solution representing a wave propagating in a body which is initially in any condition other than being at rest in a homogeneous configuration. Also, the method of harmonic oscillations assumes from the outset that the body is at rest in a homogeneous configuration.

Rather than examine the behavior of waves within the most general theory of materials conceivable and then specialize it to various degenerate cases, I believe that it is more illuminating to examine their behavior for each class of materials. After dispensing with the preliminaries in Chap. II, I consider the behavior of acceleration waves in three dimensional elastic bodies in Chap. III; of course, I consider only the situations for which the existence of the waves can be established. In particular, I examine the behavior of plane longitudinal and plane transverse waves in anisotropic elastic bodies, principal waves and waves of arbitrary shape in isotropic elastic bodies. Brief accounts of the influences of thermodynamics on the behavior of these waves are also given. In Chaps. IV, V, VI and VII, I consider successively one dimensional shock waves and acceleration waves in bodies of material with memory, elastic bodies, elastic non-conductors of heat and inhomogeneous elastic bodies, and obtain specific results characteristic of the waves for each class of materials even though the waves share certain common features.

In writing this article, I have tried to present a treatment that is self-contained, so that the reader need not have the appropriate references on hand. I do presume, however, that the reader has some knowledge of the modern theories of constitutive relations.

2. General scheme of notation. In this article, I shall adopt the notations of "The Classical Field Theories" by TRUESDELL and TOUPIN [1960, *2*] and "The Non-Linear Field Theories of Mechanics" by TRUESDELL and NOLL [1965, *6*]. In cases when confusion would arise by adhering too closely to their scheme I shall simply introduce my own notations.

Acknowledgments. I wish to thank Professor C. TRUESDELL for giving me this opportunity and Dr. T. M. BURFORD and Dr. D. R. MORRISON for allowing me the leisure to write this article. I am also indebted to Professor M. E. GURTIN and Professor C. TRUESDELL for their most helpful criticisms and suggestions offered during the course of this study. Thanks are also due to Miss BONNIE VIGIL for her assistance in preparing the manuscript and for bearing with me throughout the many changes and to Mrs. JOSEPHINE EMERY for minimizing my efforts in having to check the many revisions.

This work was supported by the U.S. Atomic Energy Commission.

II. Preliminaries.

3. Basic kinematical concepts. Consider a three dimensional body $\mathscr{B}$, identified with a region in Euclidean space. The *motion* of the body is described by a function $\chi(\cdot, \cdot)$ giving the position

$$\boldsymbol{x} = \chi(\boldsymbol{X}, \tau) \tag{3.1}$$

at time τ of each material point of the body whose position in the reference configuration is $\boldsymbol{X}$. We shall, as is customary, identify each material point with its position in the reference configuration. A property of the motion is that, for each fixed τ, the function $\chi(\cdot, \tau)$ is invertible; the inverse function is denoted by $\chi^{-1}(\cdot, \tau)$, and

$$\boldsymbol{X} = \chi^{-1}(\boldsymbol{x}, \tau), \tag{3.2}$$

which gives the material point $\boldsymbol{X}$ whose position at time τ is $\boldsymbol{x}$.

The derivatives of the motion

$$\begin{aligned} \boldsymbol{F}(\boldsymbol{X}, \tau) &\equiv \nabla_{\boldsymbol{X}}\, \chi(\boldsymbol{X}, \tau), \\ \dot{\boldsymbol{x}}(\boldsymbol{X}, \tau) &\equiv \frac{\partial}{\partial \tau}\, \chi(\boldsymbol{X}, \tau), \\ \ddot{\boldsymbol{x}}(\boldsymbol{X}, \tau) &\equiv \frac{\partial^2}{\partial \tau^2}\, \chi(\boldsymbol{X}, \tau) \end{aligned} \tag{3.3}$$

are, respectively, the *deformation gradient*, the *velocity* and the *acceleration* of $\boldsymbol{X}$ at time τ. Let t denote the present time. The function $\boldsymbol{F}^t(\boldsymbol{X}, \cdot)$ with values

$$\boldsymbol{F}^t(\boldsymbol{X}, s) = \boldsymbol{F}(\boldsymbol{X}, t-s), \qquad 0 \leqq s < \infty \tag{3.4}$$

is called the *history of the deformation gradient* of $\boldsymbol{X}$ up to time t. The restriction $\boldsymbol{F}^t_r(\boldsymbol{X}, \cdot)$ of the history to the open interval $(0, \infty)$ is called the *past history* of the deformation gradient. Clearly, the present value $\boldsymbol{F}^t(\boldsymbol{X}, 0)$ of the history is given by

$$\boldsymbol{F}^t(\boldsymbol{X}, 0) = \boldsymbol{F}(\boldsymbol{X}, t). \tag{3.5}$$

In this article, we shall have occasions to consider one dimensional motions. In these instances, the body is represented by an interval of the real line. Then, the motion is described by the function $\chi(\cdot, \cdot)$;

$$x = \chi(X, \tau) \tag{3.6}$$

is the position at time τ of the material point whose position in the reference configuration is X. Further, the derivatives $F(X, \tau)$, $\dot{x}(X, \tau)$, $\ddot{x}(X, \tau)$ of the motion and the function $F^t(X, \cdot)$ defined analogously to (3.3) and (3.4) are, respectively, the deformation gradient, the velocity, the acceleration and the history of the deformation gradient of the one dimensional body.

We call the function $\varepsilon^t(X, \cdot)$, defined by

$$\begin{aligned} \varepsilon^t(X, s) &= \varepsilon(X, t-s) \\ &= F(X, t-s) - 1, \qquad 0 \leqq s < \infty, \end{aligned} \tag{3.7}$$

the *strain history* of X up to time t. Clearly, its present value $\varepsilon^t(X, 0)$ is given by

$$\begin{aligned} \varepsilon^t(X, 0) &= \varepsilon(X, t) \\ &= F(X, t) - 1. \end{aligned} \tag{3.8}$$

4. Theory of singular surfaces. Consider a surface $\mathfrak{S}(t)$ which divides the body $\mathscr{B}$ into the regions $\mathscr{B}^+$ and $\mathscr{B}^-$ and forms a common boundary between them. The unit normal $\boldsymbol{N}$ of the surface is directed towards $\mathscr{B}^+$. Let $\psi(\cdot, \cdot)$ be a function such that $\psi(\cdot, t)$ is continuous within the regions $\mathscr{B}^+$ and $\mathscr{B}^-$, and let $\psi(\boldsymbol{X}, t)$ have definite limits ψ^+ and ψ^- as $\boldsymbol{X}$ approaches a point on the surface from paths entirely within the regions $\mathscr{B}^+$ and $\mathscr{B}^-$, respectively. The surface $\mathfrak{S}(t)$ is said to be *singular* with respect to $\psi(\cdot, t)$ at time t if[1]

$$[\psi] \equiv \psi^- - \psi^+ \neq 0.$$

Obviously, for each fixed time t, the jump $[\psi]$ is a function of position on $\mathfrak{S}(t)$.

[1] A rather comprehensive treatment of the theory of singular surfaces has been presented, for instance, by TRUESDELL and TOUPIN [1960, *2*]. Here, we shall simply point out some of the salient features of the theory. The interested reader may refer to Sect. 173 through Sect. 194 of their article for additional details as well as the historical aspects of the theory. For convenience, all citations from their article will be prefixed by the letters CFT.

The surface $\mathfrak{S}(t)$ may be represented by the relation

$$\boldsymbol{X} = \boldsymbol{Y}(\boldsymbol{V}, t) \tag{4.1}$$

at each time t. $\boldsymbol{V}$, with components V^Γ, $\Gamma = 1, 2$, is a pair of surface parameters. $\boldsymbol{V} =$ constant identifies a point on the surface $\mathfrak{S}(t)$. The *velocity* $\boldsymbol{U}$ of a point on the surface is defined by the relation

$$\boldsymbol{U}(\boldsymbol{V}, t) = \frac{\partial}{\partial t} \boldsymbol{Y}(\boldsymbol{V}, t). \tag{4.2}$$

Further, the parametric equation (4.1) implies that there exists a function $\Sigma(\cdot, \cdot)$ such that the surface $\mathfrak{S}(t)$ may also be represented by the relation

$$\Sigma(\boldsymbol{X}, t) = 0 \tag{4.3}$$

at each time t. Thus the unit normal to the surface $\mathfrak{S}(t)$ is given by the well-known formula

$$\boldsymbol{N} = \frac{\nabla_{\boldsymbol{X}} \Sigma}{|\nabla_{\boldsymbol{X}} \Sigma|}. \tag{4.4}$$

We call the quantity U_N, defined by the relation

$$\begin{aligned} U_N &\equiv \boldsymbol{U} \cdot \boldsymbol{N} \\ &= -\frac{\frac{\partial \Sigma}{\partial t}}{|\nabla_{\boldsymbol{X}} \Sigma|}, \end{aligned} \tag{4.5}$$

the *speed of propagation* of the surface $\mathfrak{S}(t)$; it is *a measure of the speed with which the surface $\mathfrak{S}(t)$ traverses the material.* Clearly, the speed of propagation U_N is independent of the choice of parametrization of the surface $\mathfrak{S}(t)$, while the velocity $\boldsymbol{U}$ is not. In other words, the speed of propagation is an inherent property of the surface, and *all possible velocities of the surface have the same normal component.* A surface $\mathfrak{S}(t)$ that is singular with respect to some quantity and has non-zero speed of propagation, i.e.,

$$U_N(\boldsymbol{X}, t) \neq 0,$$

is said to be a *wave*.

Corresponding to the surface $\mathfrak{S}(t)$, we have the alternate representation of the surface $\mathfrak{s}(t)$ defined by

$$\sigma(\boldsymbol{x}, t) = 0 \tag{4.6}$$

at each time t, with

$$\sigma(\boldsymbol{x}, t) = \Sigma\left(\chi^{-1}(\boldsymbol{X}, t), t\right).$$

Thus, the two surfaces are the duals of each other. Formula (4.3), representing the surface $\mathfrak{S}(t)$, gives the initial positions of the material particles in the reference configuration which are on the surface $\mathfrak{s}(t)$, represented by (4.6), at time t. That is, the surfaces $\mathfrak{S}(t)$ and $\mathfrak{s}(t)$ are, respectively, the material representation and the spatial representation of the singular surface. Further, the surface $\mathfrak{s}(t)$ also has the parametric representation

$$\boldsymbol{x} = \boldsymbol{y}(\boldsymbol{v}, t), \tag{4.7}$$

where $\boldsymbol{v}$, with components v^γ, $\gamma = 1, 2$, is a pair of surface parameters.

The unit normal $\boldsymbol{n}$ to the surface $\mathfrak{s}(t)$, also called the *direction of propagation*, is given by

$$\boldsymbol{n} = \frac{\nabla_{\boldsymbol{x}} \sigma}{|\nabla_{\boldsymbol{x}} \sigma|}; \tag{4.8}$$

and the normal speed u_n of the surface $\mathfrak{s}(t)$, called the *speed of displacement*, is given by

$$u_n = -\frac{\frac{\partial \sigma}{\partial t}}{|\nabla_{\boldsymbol{x}} \sigma|}. \tag{4.9}$$

The speed of displacement u_n is *a measure of the speed with which the surface $\mathfrak{s}(t)$ traverses in space*; it is also independent of the choice of parametrization of the surface $\mathfrak{s}(t)$. It can be readily verified that the relations between the normals and the speeds are

$$\begin{aligned} \boldsymbol{N} &= \boldsymbol{F}^T \boldsymbol{n} \frac{|\nabla_{\boldsymbol{x}} \sigma|}{|\nabla_{\boldsymbol{X}} \Sigma|} = \frac{\boldsymbol{F}^T \boldsymbol{n}}{|\boldsymbol{F}^T \boldsymbol{n}|}, \\ \boldsymbol{n} &= \overset{-1}{\boldsymbol{F}}{}^T \boldsymbol{N} \frac{|\nabla_{\boldsymbol{X}} \Sigma|}{|\nabla_{\boldsymbol{x}} \sigma|} = \frac{\overset{-1}{\boldsymbol{F}}{}^T \boldsymbol{N}}{|\overset{-1}{\boldsymbol{F}}{}^T \boldsymbol{N}|}, \\ U_N &= (u_n - \dot{\boldsymbol{x}} \cdot \boldsymbol{n}) \frac{|\nabla_{\boldsymbol{x}} \sigma|}{|\nabla_{\boldsymbol{X}} \Sigma|}. \end{aligned} \tag{4.10}$$

We call the quantity

$$U = u_n - \dot{\boldsymbol{x}} \cdot \boldsymbol{n} \tag{4.11}$$

the *local speed of propagation*; it is *a measure of the normal speed of the surface $\mathfrak{s}(t)$ with respect to the material particles* that are instantaneously situated upon it.

Thus far we have defined singular surfaces and have discussed some of the properties of moving surfaces associated with the motion of the material body $\mathscr{B}$. However, there are certain conditions which must be satisfied across the singular surfaces. First, there is the geometrical condition of compatibility; it relates the jump of the derivative of $\psi(\cdot, t)$ to the jump of the normal derivative of $\psi(\cdot, t)$, the tangential derivative of the jump of $\psi(\cdot, t)$ and the geometry of the singular surface. Then, there is the kinematical condition of compatibility; it relates the jump of the derivative of $\psi(\boldsymbol{X}, \cdot)$ to the jump of the normal derivative of $\psi(\cdot, t)$, the time rate of change of the jump of $\psi(\cdot, t)$ and the speed of the singular surface. These conditions may be iterated to yield other conditions giving the jumps of the higher order derivatives of $\psi(\cdot, \cdot)$. THOMAS[2] and TRUESDELL and TOUPIN[3] have presented rather detailed derivations of these conditions of compatibility. Thus, instead of repeating their analyses, we shall simply list the results which are of interest to us.

Actually, we are interested in the dual forms of their results. Writing

$$\begin{aligned} A &\equiv [\psi], \\ B &\equiv [N^\alpha \psi_{,\alpha}], \\ C &\equiv [N^\alpha N^\beta \psi_{,\alpha\beta}], \end{aligned} \tag{4.12}$$

then, by the duals of CFT $(175.11)_2$ and CFT $(176.8)_2$, we have the *geometrical condition of compatibility*

$$[\psi_{,\alpha}] = B N_\alpha + \mathfrak{g}_{\alpha\beta}\, \mathfrak{a}^{\Delta\Gamma}\, Y^\beta{}_{;\Delta}\, [\psi]_{;\Gamma}, \tag{4.13}$$

and the *iterated geometrical condition of compatibility*

$$\begin{aligned} [\psi_{,\alpha\beta}] = {} & C N_\alpha N_\beta + 2 N_{(\alpha} Y_{\beta);}{}^{\Gamma} (B_{;\Gamma} + \mathfrak{b}_\Gamma^\Delta A_{;\Delta}) \\ & + Y_{(\alpha;}{}^{\Gamma} Y_{\beta);}{}^{\Delta} (A_{;(\Gamma\Delta)} - \mathfrak{b}_{\Gamma\Delta} B); \end{aligned} \tag{4.14}$$

---

[2] THOMAS [1957, *1*].

[3] TRUESDELL and TOUPIN [1960, *2*, Sects. 175, 176, 180, and 181].

and, by the duals of CFT (180.3) and CFT (181.8), we have the *kinematical condition of compatibility*

$$\frac{\delta_D}{\delta t}[\psi] = [\dot{\psi}] + U_N B, \tag{4.15}$$

and the *iterated kinematical conditions of compatibility*

$$\begin{aligned} [\dot{\overline{\psi_{,\alpha}}}] &= \left(-U_N C + \frac{\delta_D B}{\delta t}\right) N_\alpha - g_{\alpha\beta}\, a^{\Delta\Gamma}\, Y^\beta{}_{;\Delta} (U_N B)_{;\Gamma}, \\ [\ddot{\psi}] &= U_N^2 C - 2 U_N \frac{\delta_D B}{\delta t} - B \frac{\delta_D U_N}{\delta t}. \end{aligned} \tag{4.16}$$

In (4.13), (4.14) and $(4.16)_1$, $\mathfrak{g}$, $\mathfrak{a}$ and $\mathfrak{b}$ are, respectively, the *metric tensor*, the *surface metric* of the surface $\mathfrak{S}(t)$, and the *second fundamental form* of the surface $\mathfrak{S}(t)$. In (4.15) and (4.16), $\delta_D/\delta t$ is called the *displacement derivative* with respect to the surface $\mathfrak{S}(t)$;[4] it gives time rate of change of any quantity which is a function of position on $\mathfrak{S}(t)$ and time t. It is given by the dual of CFT (179.8); i.e., let $\psi = \psi(\boldsymbol{X}, t)$ with $\boldsymbol{X} = \boldsymbol{Y}(\boldsymbol{V}, t)$, then

$$\frac{\delta_D \psi}{\delta t} = \dot{\psi} + U_N \psi_{,\alpha} N^\alpha. \tag{4.17}$$

In the one dimensional context, the material representation $\mathscr{S}(t)$ of the singular surface is given by

$$X = Y(t), \tag{4.18}$$

with *speed of propagation*

$$V(t) = \frac{dY(t)}{dt}; \tag{4.19}$$

and the spatial representation $\mathfrak{s}(t)$ of the singular surface is given by

$$x = y(t), \quad y(t) \equiv \chi(Y(t), t), \tag{4.20}$$

with *speed of displacement*

$$u(t) = \frac{dy(t)}{dt}. \tag{4.21}$$

Analogous to our discussions in the three dimensional context, we call the number v, defined by the relation

$$v = u - \dot{x}, \tag{4.22}$$

the *local speed of propagation* of the singular surface. By (4.19), (4.20), (4.21), and (4.22), it follows that

$$v = FV. \tag{4.23}$$

In one dimension, a singular surface is said to be a *wave* if

$$V(t) \neq 0.$$

Let $f(\cdot,\cdot)$ be a scalar-valued function such that $f(\cdot, t)$ is a continuous function except at $X = Y(t)$; then the jump in $f(\cdot, t)$ across the singular surface at time t is given by

$$[f] = f^- - f^+,$$

[4] Cf. THOMAS [1957, *1*] and TRUESDELL and TOUPIN [1960, *2*, Sect. 179]. A geometrical interpretation of the displacement derivative is given by BOWEN and WANG [1971, *2*].

where

$$f^+ \equiv f\left(Y(t)^+, t\right) = \lim_{X \downarrow Y(t)} f(X, t),$$
$$f^- \equiv f\left(Y(t)^-, t\right) = \lim_{X \uparrow Y(t)} f(X, t).$$

Therefore, the jump $[f]$ is a function of $Y(t)$ and t.

In the present context, the *entire study of the behavior of waves in material bodies rests upon the one dimensional counterpart of the kinematical condition of compatibility* (4.5). This condition is expressed as follows: Suppose that the functions

$$f(\cdot,\cdot), \quad \dot{f}(\cdot,\cdot) \text{ and } \quad \frac{\partial}{\partial X} f(\cdot,\cdot)$$

have jump discontinuities across $\mathscr{S}(t)$ but are continuous functions everywhere else, then

$$\frac{\delta_1}{\delta t}[f] = [\dot{f}] + V\left[\frac{\partial f}{\partial X}\right], \tag{4.24}$$

where $\delta_1/\delta t$ is the *one dimensional displacement derivative.* That is, let $f = f(X, t)$ with $X = Y(t)$, then

$$\frac{\delta_1 f}{\delta t} = \dot{f} + V \frac{\partial f}{\partial X}. \tag{4.25}$$

Here, $\delta_1/\delta t$ is simply the total time derivative and may be written alternately as d/dt. However, we choose to retain the former notation in order to emphasize the fact we are considering the time rate of change of quantities which are only defined on the singular surface $\mathscr{S}(t)$.

It should be pointed out that the particular assumptions stated above in writing down the kinematical condition of compatibility (4.24) are those usually encountered in practice. A necessary and sufficient condition for its existence is given by CHEN and WICKE;[5] in their paper it is pointed out that these particular assumptions can be weakened. Their results are given in the Appendix.

5. Definition of shock waves and acceleration waves. In this article, we shall consider three dimensional acceleration waves. *A propagating singular surface is said to be an acceleration wave if the following condition holds:*

(A-I). $\chi(\cdot,\cdot)$, $\dot{\boldsymbol{x}}(\cdot,\cdot)$, $\boldsymbol{F}(\cdot,\cdot)$ *are continuous functions everywhere;*

$$\ddot{\boldsymbol{x}}(\cdot,\cdot), \nabla_{\boldsymbol{X}} \boldsymbol{F}(\cdot,\cdot), \dot{\boldsymbol{F}}(\cdot,\cdot)$$

and also

$$\nabla_{\boldsymbol{X}}^2 \boldsymbol{F}(\cdot,\cdot), \nabla_{\boldsymbol{X}} \dot{\boldsymbol{F}}(\cdot,\cdot), \ddot{\boldsymbol{F}}(\cdot,\cdot) \text{ and } \dddot{\boldsymbol{x}}(\cdot,\cdot)$$

have jump discontinuities across $\mathfrak{S}(t)$ *but are continuous functions everywhere else.*

That is, across an acceleration wave, the motion, the velocity and the deformation gradient are continuous, but the acceleration and the second and third derivatives of the motion are not.[6]

The jump in the acceleration $[\ddot{\boldsymbol{x}}]$ is called the *amplitude vector* of the wave. Since this jump is a function of $\left(\boldsymbol{Y}(\boldsymbol{V}, t), t\right)$, it may be given by a function $\mathbf{s}(\cdot,\cdot)$ of $(\boldsymbol{V}, t)$, i.e.,

$$\mathbf{s}(\boldsymbol{V}, t) = [\ddot{\boldsymbol{x}}]\left(\boldsymbol{Y}(\boldsymbol{V}, t), t\right). \tag{5.1}$$

---

[5] CHEN and WICKE [1971, *10*].

[6] Cf. TRUESDELL and TOUPIN [1960, *2*, Sect. 190].

Hence, (4.12), (4.14), (4.16) with $\psi(\cdot,\cdot)=\chi(\cdot,\cdot)$ and (5.1) imply that

$$
\begin{aligned}
[\dot{\boldsymbol{F}}] &= -\frac{1}{U_N}\,\boldsymbol{s}\otimes\boldsymbol{N},\\
[\nabla_{\boldsymbol{X}}\boldsymbol{F}] &= \frac{1}{U_N^2}\,\boldsymbol{s}\otimes\boldsymbol{N}\otimes\boldsymbol{N}.
\end{aligned}
\tag{5.2}
$$

By $(4.10)_{1,3}$ and (4.11), (5.2) may be written in the alternate forms

$$
\begin{aligned}
[\dot{\boldsymbol{F}}] &= -\frac{1}{U}\,\boldsymbol{s}\otimes\boldsymbol{F}^T\boldsymbol{n},\\
[\nabla_{\boldsymbol{X}}\boldsymbol{F}] &= \frac{1}{U^2}\,\boldsymbol{s}\otimes\boldsymbol{F}^T\boldsymbol{n}\otimes\boldsymbol{F}^T\boldsymbol{n}.
\end{aligned}
\tag{5.3}
$$

A wave whose amplitude vector is parallel to the direction of propagation, i.e.,

$$\boldsymbol{s}\wedge\boldsymbol{n}=\boldsymbol{0}, \tag{5.4}$$

is called a *longitudinal wave*; a wave whose amplitude vector is perpendicular to the direction of propagation, i.e.,

$$\boldsymbol{s}\cdot\boldsymbol{n}=0, \tag{5.5}$$

is called a *transverse wave*.

Henceforth, we shall assume that

$$U_N(\boldsymbol{X},t)>0$$

and

$$U(\boldsymbol{X},t)>0.$$

While these assumptions will allow us to express our results more explicitly, there is no loss in generality in adopting them. Further, in view of the definition of singular surfaces, given in Sect. 4, and the definition of the speed of propagation (4.5), these assumptions are equivalent to the assertion that the material region ahead of the wave is $\mathscr{B}^+$ and that behind the wave is $\mathscr{B}^-$.

Balance of mass implies that

$$[\varrho U]=0, \tag{5.6}$$

and

$$[\dot{\varrho}+\varrho\,\operatorname{div}\dot{\boldsymbol{x}}]=0, \tag{5.7}$$

where ϱ is the present density of the material. Formula (5.6) is simply a direct consequence of the definition of balance of mass; formula (5.7) follows from the fact that the local equation of balance of mass holds in the region $\mathscr{B}^+$ and $\mathscr{B}^-$.[7] By (4.11) and the hypothesis **(A-I)**,

$$[U]=0.$$

Hence (5.6) implies that

$$[\varrho]=0$$

across an acceleration wave. Consequently, (5.7) with $(5.3)_1$ reduces to

$$[\dot{\varrho}]=\frac{\varrho}{U}\,\boldsymbol{s}\cdot\boldsymbol{n}, \tag{5.8}$$

where the density ϱ is evaluated at the wave.

---

[7] Cf. Truesdell and Toupin [1960, *2*, Sects. 156, 192].

By (5.8) and definition (5.4), we have the following

Remark 5.1. At a longitudinal acceleration wave

$$[\dot{\varrho}] = \frac{\varrho}{U}\, s\,,$$

where s denotes the magnitude $|\mathbf{s}|$ *of the amplitude vector* $\mathbf{s}$.

The above remark also motivates adopting the following definition: A longitudinal wave for which

$$s(\boldsymbol{V}, t) > 0$$

shall be called a *compressive* wave; a longitudinal wave for which

$$s(\boldsymbol{V}, t) < 0$$

shall be called an *expansive* wave.

In other words, across a compressive longitudinal wave, the rate of change of density $\dot{\varrho}^-$ behind the wave is greater than the rate of change of density $\dot{\varrho}^+$ ahead of the wave; and across an expansive longitudinal wave, the rate of change of density $\dot{\varrho}^-$ behind the wave is less than the rate of change of density $\dot{\varrho}^+$ ahead of the wave. In particular, if the material region ahead of the wave is at rest, so that $\dot{\varrho}^+ = 0$, then a longitudinal wave is compressive or expansive according as $\dot{\varrho}^-$ is positive or negative.

By (5.8) and definition (5.5), we also have the following

Remark 5.2. At a transverse acceleration wave

$$[\dot{\varrho}] = 0.$$

In other words, the rate of change of density is also continuous across a transverse wave. In particular, if the material region ahead of the wave is at rest, so that $\dot{\varrho}^+ = 0$, then $\dot{\varrho}^-$ also vanishes.

In our analyses of the growth and decay of acceleration waves later on in this article, we shall need explicit expressions for the jumps in $\nabla_{\boldsymbol{X}} \dot{\boldsymbol{F}}(\cdot,\cdot)$ and $\ddot{\boldsymbol{x}}(\cdot,\cdot)$. Hence, by (4.14) and $(4.16)_2$ with $\psi(\cdot,\cdot) = \dot{\boldsymbol{x}}(\cdot,\cdot)$ and $(5.2)_1$, we have the component relations

$$\begin{aligned} [\dot{F}^k{}_{\alpha,\beta}] &= [\dot{x}^k{}_{,\alpha\beta}] = c^k N_\alpha N_\beta - 2 N_{(\alpha} Y_{\beta);}{}^{\Gamma} \left(\frac{s^k}{U_N}\right)_{;\Gamma} + \frac{1}{U_N} Y_{(\alpha;}{}^{\Gamma} Y_{\beta);}{}^{\Lambda} \mathfrak{b}_{\Gamma\Lambda} s^k, \\ [\dddot{x}^k] &= U_N^2 c^k + 2 \frac{\delta_D s^k}{\delta t} - \frac{s^k}{U_N} \frac{\delta_D U_N}{\delta t}, \end{aligned} \tag{5.9}$$

where the vector $\boldsymbol{c}$ is given by the component relation

$$c^k = [\dot{x}^k{}_{,\alpha\beta}]\, N^\alpha N^\beta. \tag{5.10}$$

The vector $\boldsymbol{c}$ is called the *induced discontinuity* associated with the acceleration wave. In certain of the circumstances considered in this article, we shall show that the behavior of certain components of $\boldsymbol{c}$ is actually related to the behavior of the primary wave.[8]

Now, we shall turn our attention to one dimensional considerations. In this article, we shall consider the growth and decay of both shock waves and acceleration waves. Accordingly, let us define what we mean by a shock wave and an acceleration wave and examine the immediate consequences.

In one dimension, *a propagating singular surface is said to be a shock wave if the following condition holds*:

[8] Cf. Sects. 6, 7 and 8 of this article.

(S-1). $\chi(\cdot,\cdot)$ *is a continuous function everywhere;*

$$\dot{x}(\cdot,\cdot),\ F(\cdot,\cdot)$$

and also

$$\ddot{x}(\cdot,\cdot),\ \dot{F}(\cdot,\cdot)\ \textit{and}\ \frac{\partial F}{\partial X}(\cdot,\cdot)$$

have jump discontinuities across $\mathcal{S}(t)$ *but are continuous functions everywhere else.*

That is, across a shock wave, the motion is continuous; but the velocity, deformation gradient, acceleration and the second derivatives of the motion are not.[9]

In our discussions on shock waves given in this article, we shall regard $\varepsilon(\cdot,\cdot)$ as the independent variable in the constitutive relations rather than $F(\cdot,\cdot)$. Hence, it follows from the hypothesis (S-1), (3.8) and the kinematical condition of compatibility (4.24) with $f(\cdot,\cdot)=\chi(\cdot,\cdot)$, $\dot{x}(\cdot,\cdot)$ and $F(\cdot,\cdot)$ that the following conditions hold at a shock wave:

$$\begin{aligned} V[\varepsilon] &= -[\dot{x}],\\ \frac{\delta_1[\varepsilon]}{\delta t} &= [\dot{\varepsilon}] + V\left[\frac{\partial \varepsilon}{\partial X}\right],\\ \frac{\delta_1[\dot{x}]}{\partial t} &= [\ddot{x}] + V[\dot{\varepsilon}]. \end{aligned} \tag{5.11}$$

Henceforth, we assume that the speed of propagation of the shock is strictly positive, i.e.,

$$V(t) > 0.$$

In one dimension, balance of mass may be expressed in the form

$$\frac{\varrho_R}{\varrho} = \frac{\partial x}{\partial X}, \tag{5.12}$$

where ϱ_R is the density of the material body in the reference configuration and ϱ is the present density. For homogeneous materials, homogeneous configurations are always taken as the reference, so that ϱ_R, for each of these materials, is independent of X. However, for an inhomogeneous material, ϱ_R is given by a function of X:

$$\varrho_R = \hat{\varrho}_R(X).$$

We shall always assume that $\hat{\varrho}_R(\cdot)$ *is of class* C^0.

By (3.8) and (5.12), we have

$$[\varrho] = -\frac{\varrho^- \varrho^+}{\varrho_R}[\varepsilon].$$

Hence, in the application of one dimensional theory to longitudinal motions a wave is said to be a *compression shock* if

$$[\varepsilon] < 0, \tag{5.13}$$

while a wave is said to be an *expansion shock* if

$$[\varepsilon] > 0. \tag{5.14}$$

[9] Cf. TRUESDELL and TOUPIN [1960, *2*, Sect. 189] for the three dimensional counterpart of this definition.

In other words, across a compression shock the density ϱ^- behind the wave is greater than the density ϱ^+ ahead of the wave; and, across an expansion shock the density ϱ^- behind the wave is less than the density ϱ^+ ahead of the wave. In particular, if the material region ahead of the wave is at rest and unstrained, so that $\varepsilon^+ = 0$, then a wave is a compression shock or an expansion shock according as ε^- is negative or positive. In either case, the jump $[\varepsilon]$ in strain is called the *amplitude* of the shock.

Balance of momentum asserts that for each part of the body bounded by the pair of points X_α, X_β and for all times t

$$\frac{d}{dt}\int_{X_\alpha}^{X_\beta} \hat{\varrho}_R(X)\,\dot{x}(X,t)\,dX = \int_{X_\alpha}^{X_\beta} \hat{\varrho}_R(X)\,b(X,t)\,dX + T(X_\beta,t) - T(X_\alpha,t),$$

where $T(\cdot,\cdot)$ and $b(\cdot,\cdot)$ are the *stress* and the *external body force* per unit mass, respectively. The momentum balance law together with the appropriate smoothness hypotheses implies that for $X \neq Y(t)$

$$\begin{aligned} \frac{\partial T}{\partial X} + \varrho_R\, b &= \varrho_R\, \ddot{x}, \\ \frac{\partial \dot{T}}{\partial X} + \varrho_R\, \dot{b} &= \varrho_R\, \dddot{x}, \end{aligned} \tag{5.15}$$

and[10] at $X = Y(t)$

$$\begin{aligned} [T] &= -\varrho_R V[\dot{x}], \\ \left[\frac{\partial T}{\partial X}\right] &= \varrho_R[\ddot{x}], \\ \left[\frac{\partial \dot{T}}{\partial X}\right] &= \varrho_R[\dddot{x}]. \end{aligned} \tag{5.16}$$

In deriving $(5.16)_{2,3}$, we have tacitly assumed that $b(\cdot,\cdot)$ and $\dot{b}(\cdot,\cdot)$ are continuous functions everywhere. Henceforth, unless otherwise noted, we shall always assume that the external body force is absent.

Formulae $(5.11)_1$ and $(5.16)_1$ imply the well-known result

$$\varrho_R V^2 = \frac{[T]}{[\varepsilon]} \tag{5.17}$$

for the speed of propagation of a shock. Further, by $(5.11)_{2,3}$ and $(5.16)_2$, we have

Remark 5.3. The amplitude $[\varepsilon]$ *of a shock wave must obey the equation*

$$2V\frac{\delta_1[\varepsilon]}{\delta t} + [\varepsilon]\frac{\delta_1 V}{\delta t} = V^2\left[\frac{\partial \varepsilon}{\partial X}\right] - \frac{1}{\varrho_R}\left[\frac{\partial T}{\partial X}\right]. \tag{5.18}$$

Formula (5.18) is a direct consequence of balance of momentum and the kinematical conditions for a shock; it does not depend on the constitutive relation for the stress.[11] It is important to point out that (5.18) is valid for inhomogeneous as well as homogeneous materials.

In one dimension, *a propagating singular surface is said to be an acceleration wave if the following condition holds*:[12]

---

[10] See, for example, TRUESDELL and TOUPIN [1960, *2*, Sects. 193, 205] for the derivation of the three dimensional counterpart of $(5.16)_1$.

[11] For homogeneous bodies, (5.18) is given independently by ACHENBACH and HERRMANN [1970, *1*] and by CHEN and GURTIN [1970, *6*].

[12] Of course, this definition is analogous to **(A-I)**, the three dimensional counterpart.

(A-1). *$\chi(\cdot,\cdot)$, $\dot{x}(\cdot,\cdot)$ and $F(\cdot,\cdot)$ are continuous functions everywhere;*

$$\ddot{x}(\cdot,\cdot),\ \frac{\partial F}{\partial X}(\cdot,\cdot),\ \dot{F}(\cdot,\cdot),$$

and also

$$\frac{\partial^2 F}{\partial X^2}(\cdot,\cdot),\ \frac{\partial \dot{F}}{\partial X}(\cdot,\cdot),\ \ddot{F}(\cdot,\cdot) \text{ and } \dddot{x}(\cdot,\cdot)$$

have jump discontinuities across $\mathscr{S}(t)$ but are continuous functions everywhere else.

The jump in acceleration is called the amplitude of the acceleration wave. Since this jump is a function of $Y(t)$ and t, the amplitude may be given by a function $a(\cdot)$ of t alone, i.e.,

$$a(t) = [\ddot{x}]\,(Y(t), t). \tag{5.19}$$

The hypothesis (A-1) and the compatibility condition (4.24) with $f(\cdot,\cdot) = \dot{x}(\cdot,\cdot)$ and $F(\cdot,\cdot)$ imply that

$$[\ddot{x}] = -V[\dot{F}] = V^2\left[\frac{\partial F}{\partial X}\right], \tag{5.20}$$

while (4.24) with $f(\cdot,\cdot) = \ddot{x}(\cdot,\cdot)$ yields the relation

$$\frac{\delta_1 a}{\delta t} = [\dddot{x}] + V[\ddot{F}]. \tag{5.21}$$

Henceforth, we shall assume that the speed of propagation of the acceleration wave is strictly positive, i.e.,

$$V(t) > 0.$$

Formula (5.12) and the hypothesis (A-1) imply that $[\varrho] = 0$ across an acceleration wave. Further, differentiating (5.12) and utilizing (5.19) and (5.20), we have

Remark 5.4. Across a one dimensional acceleration wave

$$[\dot{\varrho}] = \frac{\varrho^2}{\varrho_R V}\, a,$$

where the densities ϱ_R and ϱ are evaluated at the wave.

Hence, in the application of one dimensional theory to longitudinal motions an acceleration wave is said to be *compressive* if $a(t) > 0$; an acceleration wave is said to be *expansive* if $a(t) < 0$.[13]

Formula $(5.16)_1$ and the hypothesis (A-1) imply that

$$[T] = 0. \tag{5.22}$$

Hence, by $(5.16)_2$ and (4.24) with $f(\cdot,\cdot) = T(\cdot,\cdot)$, we have

Remark 5.5. Across a one dimensional acceleration wave

$$[\dot{T}] = -\varrho_R V[\ddot{x}]. \tag{5.23}$$

As we shall see, (5.23) is useful in determining the speed of propagation V of an acceleration wave when the constitutive relation for the stress of a particular material is known.[14]

---

[13] Of course, these definitions are analogous to those of the three dimensional counterpart.

[14] The three dimensional counterpart of (5.23) was given by CHEN [1965, *2*]. He showed that in terms of the Cauchy stress tensor $\boldsymbol{t}$

$$[\dot{\boldsymbol{t}}]\,\boldsymbol{n} = -\varrho\, U[\ddot{\boldsymbol{x}}],$$

and in terms of the Piola-Kirchhoff stress tensor $\boldsymbol{T}_R$

$$[\dot{\boldsymbol{T}}_R]\,\boldsymbol{N} = -\varrho_R\, U_N[\ddot{\boldsymbol{x}}].$$

The former result is also given by HILL [1962, *3*] without proof.

Formula $(5.16)_3$, (5.23) and (4.24) with $f(\cdot,\cdot)=\dot{T}(\cdot,\cdot)$ imply that

$$[\ddot{x}] = -\frac{1}{\varrho_R V}[\ddot{T}] - \frac{\delta_1 a}{\delta t} - \frac{a}{V}\frac{\delta_1 V}{\delta t}. \tag{5.24}$$

Combining (5.21) and (5.24), we have

Remark 5.6. The amplitude a of an acceleration wave must obey the equation

$$2\frac{\delta_1 a}{\delta t} = -\frac{1}{\varrho_R}\frac{\delta_1 \varrho_R}{\delta t}a - \frac{a}{V}\frac{\delta_1 V}{\delta t} - \frac{1}{\varrho_R V}[\ddot{T}] + V[\ddot{F}]. \tag{5.25}$$

Formula (5.25) is a direct consequence of the kinematical conditions of compatibility and balance of momentum; it does not depend on the constitutive relation for the stress.[15] Notice that in the applications of (5.25) to homogeneous materials, the first term of the right-hand member of (5.25) vanishes.

In this article, we shall also consider the growth and decay of one dimensional shock waves and acceleration waves in materials including thermodynamic influences. Balance of energy asserts that for each part of the body bounded by the pair of points X_α, X_β and for all times t

$$\begin{aligned}\frac{d}{dt}\int_{X_\alpha}^{X_\beta}\left(e(X,t)+\frac{1}{2}\hat{\varrho}_R(X)\,\dot{x}^2(X,t)\right)dX = \int_{X_\alpha}^{X_\beta}\hat{\varrho}_R(X)\left(\dot{x}(X,t)\,b(X,t)+r(X,t)\right)dX\\ +T(X_\beta,t)\,\dot{x}(X_\beta,t)-T(X_\alpha,t)\,\dot{x}(X_\alpha,t)\\ -q(X_\beta,t)+q(X_\alpha,t),\end{aligned}$$

where $e(\cdot,\cdot)$ is the *internal energy*, $q(\cdot,\cdot)$ the *heat flux* and $r(\cdot,\cdot)$ the *external radiation* per unit mass. The balance law together with the appropriate smoothness hypotheses, implies that for $X \neq Y(t)$

$$\begin{aligned}\dot{e} &= T\dot{F} - \frac{\partial q}{\partial X} + \varrho_R r\\ &= T\dot{\varepsilon} - \frac{\partial q}{\partial X} + \varrho_R r,\end{aligned} \tag{5.26}$$

and[16] at $X=Y(t)$

$$-V[e+\tfrac{1}{2}\varrho_R\,\dot{x}^2] = [T\dot{x}] - [q], \tag{5.27$_1$}$$

$$\begin{aligned}[\dot{e}] &= [T\dot{F}] - \left[\frac{\partial q}{\partial X}\right]\\ &= [T\dot{\varepsilon}] - \left[\frac{\partial q}{\partial X}\right].\end{aligned} \tag{5.27$_2$}$$

In deriving $(5.27)_2$, we have tacitly assumed that $r(\cdot,\cdot)$ is a continuous function everywhere. Henceforth, unless otherwise noted, we shall always assume that the external radiation is absent.

III. Acceleration waves in elastic bodies.

6. Longitudinal waves in anisotropic elastic bodies. Here, we consider the behavior of longitudinal acceleration waves in bodies of homogeneous elastic material which does not possess any symmetry.[17] Such a material is defined by

[15] A somewhat different form of (5.25) has been given by Coleman, Greenberg and Gurtin [1966, *3*].

[16] See, for example, Truesdell and Toupin [1960, *2*, Sects. 193, 241] for the derivation of the three dimensional counterpart of $(5.27)_1$.

[17] The results given in this section are due to Chen [1970, *4*].

the constitutive relation

$$\boldsymbol{T}_R(\boldsymbol{X}, t) = \boldsymbol{h}\big(\boldsymbol{F}(\boldsymbol{X}, t)\big), \tag{6.1}$$

where $\boldsymbol{T}_R$ is the Piola-Kirchhoff stress tensor and $\boldsymbol{h}(\cdot)$ the response function. Here, we have taken a homogeneous configuration to be the reference, so that $\boldsymbol{h}(\cdot)$ and the density ϱ_R in this configuration are independent of $\boldsymbol{X}$.

We assume that $\boldsymbol{h}(\cdot)$ is of class C^2. Thus, the *elasticity tensor* $\boldsymbol{A}(\boldsymbol{F})$, defined by the component relation

$$A_{k\ m}^{\ \alpha\ \beta} = \frac{\partial h_k^{\ \alpha}}{\partial F^m_{\ \beta}}, \tag{6.2}$$

is of class C^1; and the *second-order elasticity tensor* $\boldsymbol{B}(\boldsymbol{F})$, defined by the component relation

$$B_{k\ m\ n}^{\ \alpha\ \beta\ \gamma} = \frac{\partial^2 h_k^{\ \alpha}}{\partial F^m_{\ \beta}\,\partial F^n_{\ \gamma}} = \frac{\partial^2 h_k^{\ \alpha}}{\partial F^n_{\ \gamma}\,\partial F^m_{\ \beta}}, \tag{6.3}$$

is of class C^0.

The second-order elasticity tensor plays an important role in the analysis that follows. In fact, the behavior of the longitudinal waves is governed by certain properties of this tensor.

It can be readily verified that balance of linear momentum

$$\operatorname{Div} \boldsymbol{T}_R + \varrho_R\, \boldsymbol{b} = \varrho_R\, \ddot{\boldsymbol{x}}, \tag{6.4}$$

the constitutive relation (6.1) together with the hypothesis **(A-I)** on the properties of acceleration waves given in Sect. 5, (5.1) and $(5.2)_2$ imply that the amplitude vector $\boldsymbol{s}$ and the speed of propagation U_N of an acceleration wave with unit normal $\boldsymbol{n}$ obey the following *propagation condition*:[18]

$$\{\hat{\boldsymbol{Q}}(\boldsymbol{N}, \boldsymbol{F}) - \varrho_R\, U_N^2\, \boldsymbol{1}\}\, \boldsymbol{s} = \boldsymbol{0}, \tag{6.5}$$

where $\hat{\boldsymbol{Q}}(\boldsymbol{N}, \boldsymbol{F})$ is the *acoustic tensor* defined by the component relation

$$\hat{Q}_{km} = A_{k\ m}^{\ \alpha\ \beta}\, N_\alpha\, N_\beta. \tag{6.6}$$

The propagation condition (6.5) states that the amplitude vector $\boldsymbol{s}$ is a right proper vector of $\hat{\boldsymbol{Q}}(\boldsymbol{N}, \boldsymbol{F})$ corresponding to the proper value $\varrho_R U_N^2$. Thus, any real right proper vector $\boldsymbol{s}$ of $\hat{\boldsymbol{Q}}(\boldsymbol{N}, \boldsymbol{F})$ is a possible amplitude vector provided that the corresponding proper value is real and positive.

In general, the acoustic tensor $\hat{\boldsymbol{Q}}(\boldsymbol{N}, \boldsymbol{F})$ need not have real and positive proper values and real right proper vectors. However, there is a condition which suffices to insure that a longitudinal wave may exist and propagate at each point in the material body; i.e., it guarantees the existence of a wave for which its squared speed of propagation is real and positive and its amplitude vector is parallel to the direction of propagation. This result is due to TRUESDELL.[19]

Consider a fixed deformation gradient at a point in the material body. Thus the acoustic tensor depends on the normal $\boldsymbol{N}$ alone. By $(4.10)_1$, we observe that it may also depend on the direction of propagation $\boldsymbol{n}$ alone. That is,

$$\hat{\boldsymbol{Q}}(\boldsymbol{N}) = \overline{\boldsymbol{Q}}(\boldsymbol{n}) \tag{6.7}$$

---

[18] Cf. TRUESDELL [1961, *3*, Sect. 2].
[19] TRUESDELL [1966, *9*].

for each fixed $\boldsymbol{F}$. If $\boldsymbol{m}$ is a known unit amplitude vector, then by (6.5)

$$\varrho_R U_N^2 = \boldsymbol{m} \cdot \bar{\boldsymbol{Q}}(\boldsymbol{n})\, \boldsymbol{m}. \tag{6.8}$$

We say that the strained material has *positive longitudinal elasticity* if

$$\boldsymbol{n} \cdot \bar{\boldsymbol{Q}}(\boldsymbol{n})\, \boldsymbol{n} > 0 \tag{6.9}$$

for all unit vectors $\boldsymbol{n}$.

Theorem 6.1. *At a point of a deformed elastic body such that the material has positive longitudinal elasticity, there is at least one direction in which a longitudinal wave may exist and propagate.*

The proof of this theorem, which follows from (6.5), (6.8) and (6.9), will not be duplicated here; the interested reader may refer to TRUESDELL'S article for the details.[20]

The preceding theorem asserts that a longitudinal wave may exist and propagate at each point in the material body; of course, the direction of propagation will, in general, differ for different points, even though (6.9) may be satisfied in all instances. However, if we assume that the body is initially homogeneously deformed and at rest, and that (6.9) is satisfied, then the theorem implies that a plane longitudinal wave may exist and propagate in the same direction throughout the body. In the sequel, we shall consider the behavior of this longitudinal wave.

We assume that since time $t=0$, the longitudinal wave has been propagating into a region which is at rest in a fixed homogeneous configuration. Let the constant tensor $\boldsymbol{F}_0$ characterize the homogeneous configuration, and let the constant unit vector $\boldsymbol{n}_1$ denote the direction in which the longitudinal wave may exist and propagate. Thus, by $(4.10)_1$, the fixed unit normal $\boldsymbol{N}_1$ corresponding to $\boldsymbol{n}_1$ is given by

$$\boldsymbol{N}_1 = \frac{\boldsymbol{F}_0^T \boldsymbol{n}_1}{|\boldsymbol{F}_0^T \boldsymbol{n}_1|}. \tag{6.10}$$

In view of definition (5.4) and Remark 5.1, the amplitude vector of the longitudinal wave may be expressed in the form

$$\mathbf{s} = s\, \boldsymbol{n}_1, \tag{6.11}$$

and, by (6.5) and (6.10), its speed of propagation is given by

$$\varrho_R U_N^2 = |\hat{\boldsymbol{Q}}(\boldsymbol{N}_1, \boldsymbol{F}_0)\, \boldsymbol{n}_1|. \tag{6.12}$$

We assume that *the discontinuity across the wave is uniform*, so that the amplitude s depends on t alone, independent of the surface parameters $\boldsymbol{V}$.

By taking the material derivative of the balance of linear momentum (6.4), and evaluating the jump of the resulting relation with the aid of (5.2), (5.9), (6.1), and (6.11), we have

$$2\varrho_R \frac{\delta_D s}{\delta t} \boldsymbol{n}_1 + \frac{\boldsymbol{b}}{U_N^3} s^2 = \left(\hat{\boldsymbol{Q}}(\boldsymbol{N}_1, \boldsymbol{F}_0) - \varrho_R U_N^2 \mathbf{1}\right) \boldsymbol{c}, \tag{6.13}$$

[20] TRUESDELL [1966, *9*]. TRUESDELL attributes the mathematical arguments associated with the proof of this result to STIPPES [1965, *10*] who considered the existence of longitudinal waves within the infinitesimal theory of elasticity. KOLODNER [1966, *6*] extended the results of STIPPES; he proved that if the body is at rest in a natural state, then there are at least three distinct directions along which longitudinal waves may exist and propagate. In this regard, also refer to TRUESDELL [1968, *8*].

where the vector $\boldsymbol{c}$, the induced discontinuity, is defined by (5.10) with $\boldsymbol{N}=\boldsymbol{N}_1$, and the vector $\boldsymbol{b}$ is defined by the component relation

$$b_k = B_{k\,m\,n}^{\alpha\,\beta\,\gamma} N_{1\alpha} N_{1\beta} N_{1\gamma} n_1^m n_1^n, \tag{6.14}$$

with $B_{k\,m\,n}^{\alpha\,\beta\,\gamma}$ evaluated at $\boldsymbol{F}_0$. Further, in (6.13), U_N is given by (6.12).

Formula (6.13) is fundamental to the derivation of our main results. In general, it not only furnishes information regarding the behavior of the amplitude of the longitudinal wave, but, under certain circumstances, it also tells us a great deal about the behavior of certain components of the induced discontinuity $\boldsymbol{c}$ associated with the wave.

It is well-known that (6.13) has a unique solution for $\boldsymbol{c}$ if and only if the homogeneous equation

$$\{\hat{\boldsymbol{Q}}(\boldsymbol{N}_1, \boldsymbol{F}_0) - \varrho_R U_N^2 \boldsymbol{1}\}\boldsymbol{u} = \boldsymbol{0}$$

has just the trivial solution $\boldsymbol{u}=\boldsymbol{0}$. However, it has been shown that the homogeneous equation does have a non-trivial solution; i.e., the direction of propagation $\boldsymbol{n}_1$. Therefore, if a solution $\boldsymbol{c}$ of (6.13) exists, it cannot be unique.

The two cases of (6.13) which must be considered are:

(α) *The induced discontinuity* $\boldsymbol{c}$ *is a right proper vector of* $\hat{\boldsymbol{Q}}(\boldsymbol{N}_1, \boldsymbol{F}_0)$ *corresponding to the proper value* $\varrho_R U_N^2$.[21]

(β) *The induced discontinuity* $\boldsymbol{c}$ *is not a right proper vector of* $\hat{\boldsymbol{Q}}(\boldsymbol{N}_1, \boldsymbol{F}_0)$ *corresponding to the proper value* $\varrho_R U_N^2$.

Case (α).

In this case,[22] the right-hand member of (6.13) vanishes. Thus, taking the inner product of (6.13) with $\boldsymbol{n}_1$, we have

$$\frac{\delta_D s}{\delta t} = -\frac{\boldsymbol{b}\cdot\boldsymbol{n}_1}{2\varrho_R U_N^3} s^2. \tag{6.15}$$

Integrating (6.15), we have

Theorem 6.2. *Let Theorem 6.1 hold. Consider the longitudinal wave which since* $t=0$ *has been propagating into a region that is at rest in a homogeneous configuration with deformation gradient* $\boldsymbol{F}_0$. *Suppose that the induced discontinuity associated with the wave is parallel to an axis of the acoustic tensor corresponding to the proper value* $\varrho_R U_N^2$. *Then the amplitude s of the wave has the following explicit dependence on time*:

$$s(t) = \frac{1}{\left(\frac{1}{s(0)} + \frac{\boldsymbol{b}\cdot\boldsymbol{n}_1}{2\varrho_R U_N^3} t\right)}, \tag{6.16}$$

where U_N *and* $\boldsymbol{b}$ *are given by* (6.12) *and* (6.14), *respectively.*

Consequently, we have

Theorem 6.3. *Let Theorem 6.2 hold.*

(i) *The amplitude s of a compressive wave* (*i.e.*, $s(0)>0$) *will decay to zero monotonically in infinite time if and only if*

$$B_{k\,m\,n}^{\alpha\,\beta\,\gamma} N_{1\alpha} N_{1\beta} N_{1\gamma} n_1^k n_1^m n_1^n > 0.$$

[21] If $\varrho_R U_N^2$ is a *distinct* proper value of $\hat{\boldsymbol{Q}}(\boldsymbol{N}_1, \boldsymbol{F}_0)$, then the induced discontinuity is also *longitudinal.*

[22] GREEN [1965, 7] has shown that this situation actually does arise in his studies on principal waves in *isotropic* elastic materials.

(ii) *The amplitude s of an expansive wave (i.e., $s(0) < 0$) will decay to zero monotonically in infinite time if and only if*

$$B_{k\ m\ n}^{\alpha\ \beta\ \gamma} N_{1\alpha} N_{1\beta} N_{1\gamma} n_1^k n_1^m n_1^n < 0.$$

(iii) *In either case, the amplitude s will become infinite monotonically within a finite time t_∞, given by*

$$t_\infty = -\frac{2\varrho_R U_N^3}{s(0)\, \boldsymbol{b}\cdot\boldsymbol{n}_1},$$

if and only if

$$\operatorname{sgn} s = -\operatorname{sgn} B_{k\ m\ n}^{\alpha\ \beta\ \gamma} N_{1\alpha} N_{1\beta} N_{1\gamma} n_1^k n_1^m n_1^n.$$

Notice that the criteria governing the growth and decay of the amplitude depend only on the properties of the second-order elasticity tensor $\boldsymbol{B}(\boldsymbol{F})$, defined by (6.3).

Case (β).

For the moment, we assume that the material has a stored-energy function, i.e., the material is *hyperelastic*. For such a material $\hat{\boldsymbol{Q}}(\boldsymbol{N}_1, \boldsymbol{F}_0)$ is symmetric. Thus, taking the inner product of (6.13) with $\boldsymbol{n}_1$, we have

$$2\varrho_R \frac{\delta_D s}{\delta t} + \frac{\boldsymbol{b}\cdot\boldsymbol{n}_1}{U_N^3} s^2 = \boldsymbol{n}_1 \cdot \left(\hat{\boldsymbol{Q}}(\boldsymbol{N}_1, \boldsymbol{F}_0) - \varrho_R U_N^2 \boldsymbol{1}\right) \boldsymbol{c}. \tag{6.17}$$

But,

$$\boldsymbol{n}_1 \cdot \left(\hat{\boldsymbol{Q}}(\boldsymbol{N}_1, \boldsymbol{F}_0) - \varrho_R U_N^2 \boldsymbol{1}\right) \boldsymbol{c} = \boldsymbol{c} \cdot \left(\hat{\boldsymbol{Q}}(\boldsymbol{N}_1, \boldsymbol{F}_0) - \varrho_R U_N^2 \boldsymbol{1}\right) \boldsymbol{n}_1 = 0;$$

therefore (6.17) reduces to

$$\frac{\delta_D s}{\delta t} = -\frac{\boldsymbol{b}\cdot\boldsymbol{n}_1}{2\varrho_R U_N^3} s^2. \tag{6.18}$$

Notice that (6.18) is identical to (6.15). Thus, its solution is given by (6.16), and the criteria governing the growth and decay of the amplitude s are identical to those given in Theorem 6.3. However, for this case the material is hyperelastic; therefore the second-order elasticity tensor has the following symmetries:

$$B_{k\ m\ n}^{\alpha\ \beta\ \gamma} = B_{k\ n\ m}^{\alpha\ \gamma\ \beta} = B_{n\ m\ k}^{\gamma\ \beta\ \alpha} = B_{m\ k\ n}^{\beta\ \alpha\ \gamma}.$$

So far, nothing has been said about the induced discontinuity $\boldsymbol{c}$. However, for this case the explicit time dependence of certain components of $\boldsymbol{c}$ can actually be determined. Since $\hat{\boldsymbol{Q}}(\boldsymbol{N}_1, \boldsymbol{F}_0)$ is symmetric, it has two other mutually orthogonal principal axes, say $\boldsymbol{n}_2$ and $\boldsymbol{n}_3$, also orthogonal to $\boldsymbol{n}_1$. Thus, taking the inner product of (6.13) with $\boldsymbol{n}_\mathfrak{a}$, $\mathfrak{a} = 2, 3$, we have

$$\frac{\boldsymbol{b}\cdot\boldsymbol{n}_\mathfrak{a}}{U_N^3} s^2 = \boldsymbol{c} \cdot \left(\hat{\boldsymbol{Q}}(\boldsymbol{N}_1, \boldsymbol{F}_0)\, \boldsymbol{n}_\mathfrak{a} - \varrho_R U_N^2 \boldsymbol{n}_\mathfrak{a}\right). \tag{6.19}$$

However,

$$\hat{\boldsymbol{Q}}(\boldsymbol{N}_1, \boldsymbol{F}_0)\, \boldsymbol{n}_\mathfrak{a} = \lambda_\mathfrak{a} \boldsymbol{n}_\mathfrak{a},$$

where $\lambda_\mathfrak{a}$, $\mathfrak{a} = 2, 3$, are the proper values; therefore by (6.19)

$$\boldsymbol{c}(t) \cdot \boldsymbol{n}_\mathfrak{a} = \frac{\boldsymbol{b}\cdot\boldsymbol{n}_\mathfrak{a}\, s^2(t)}{U_N^3 (\lambda_\mathfrak{a} - \varrho_R U_N^2)}, \tag{6.20}$$

provided that $\lambda_\mathfrak{a} \neq \varrho_R U_N^2$, and where $s(t)$ is given by (6.16). Formula (6.20) states that *the component of the induced discontinuity* $\boldsymbol{c}$ *in the* $\boldsymbol{n}_\mathfrak{a}$ *direction will either grow or decay according as the amplitude s of the primary wave grows or decays.* Notice also that the behavior of these components of $\boldsymbol{c}$ is proportional to the square of the behavior of the primary wave. However, if $\lambda_2 = \lambda_3 = \lambda$, there exists a unit vector, say $\boldsymbol{t}$, in the principal plane corresponding to λ such that $\boldsymbol{c} \cdot (\boldsymbol{n}_1 \wedge \boldsymbol{t}) = 0$ and

$$\boldsymbol{c}(t) \cdot \boldsymbol{t} = \frac{\boldsymbol{b} \cdot \boldsymbol{t} s^2(t)}{U_N^3 (\lambda - \varrho_R U_N^2)}, \tag{6.21}$$

giving the explicit time dependence of the *transverse component* of the induced discontinuity.

Finally, we should point out that as long as $\lambda_\mathfrak{a} \neq \varrho_R U_N^2$ and $\lambda \neq \varrho_R U_N^2$ the components of $\boldsymbol{c}$ in the $\boldsymbol{n}_\mathfrak{a}$ and $\boldsymbol{t}$ directions are unique, even though $\boldsymbol{c}$ itself is not. Therefore, the results given in (6.20) and (6.21) are not empty.

Now, if the material is not hyperelastic, so that $\hat{\boldsymbol{Q}}(\boldsymbol{N}_1, \boldsymbol{F}_0)$ need not be symmetric, we may still obtain the solution of (6.13) for s and derive similar results for certain components of $\boldsymbol{c}$. Taking the inner product of (6.13) with $\boldsymbol{v}_1$, $\boldsymbol{v}_1$ being the left unit proper vector of $\hat{\boldsymbol{Q}}(\boldsymbol{N}_1, \boldsymbol{F}_0)$ corresponding to the proper value $\varrho_R U_N^2$, we have

$$\frac{\delta_D s}{\delta t} = -\frac{\boldsymbol{b} \cdot \boldsymbol{v}_1}{2 \varrho_R U_N^3 \boldsymbol{n}_1 \cdot \boldsymbol{v}_1} s^2. \tag{6.22}$$

Integrating (6.22), we have

Theorem 6.4. *Let Theorem 6.1 hold. Consider the longitudinal wave which since $t = 0$ has been propagating into a region that is at rest in a homogeneous configuration with deformation gradient $\boldsymbol{F}_0$. Then the amplitude s of the wave has the following explicit dependence on time*:

$$s(t) = \frac{1}{\left(\frac{1}{s(0)} + \frac{\boldsymbol{b} \cdot \boldsymbol{v}_1}{2 \varrho_R U_N^3 \boldsymbol{n}_1 \cdot \boldsymbol{v}_1} t\right)}, \tag{6.23}$$

where U_N and $\boldsymbol{b}$ are given by (6.12) *and* (6.14), *respectively.*

Consequently, we have

Theorem 6.5. *Let Theorem 6.4 hold.*

(i*) *The amplitude s of a compressive wave (i.e., $s(0) > 0$) will decay to zero monotonically in infinite time if and only if*

$$\frac{\boldsymbol{b} \cdot \boldsymbol{v}_1}{\boldsymbol{n}_1 \cdot \boldsymbol{v}_1} > 0.$$

(ii*) *The amplitude s of an expansive wave (i.e., $s(0) < 0$) will decay to zero monotonically in infinite time if and only if*

$$\frac{\boldsymbol{b} \cdot \boldsymbol{v}_1}{\boldsymbol{n}_1 \cdot \boldsymbol{v}_1} < 0.$$

(iii*) *In either case, the amplitude s will become infinite monotonically within a finite time t_∞, given by*

$$t_\infty = -\frac{2 \varrho_R U_N^3 \boldsymbol{n}_1 \cdot \boldsymbol{v}_1}{s(0)\, \boldsymbol{b} \cdot \boldsymbol{v}_1},$$

if and only if

$$\operatorname{sgn} s = -\operatorname{sgn}\left(\frac{\boldsymbol{b} \cdot \boldsymbol{v}_1}{\boldsymbol{n}_1 \cdot \boldsymbol{v}_1}\right).$$

Notice that the results of Theorem 6.5 are quite different from those of Theorem 6.3.

Suppose that $\lambda_\mathfrak{v}$, $\mathfrak{v}=2, 3$, are the two other proper values of $\hat{\mathbf{Q}}(\mathbf{N}_1, \mathbf{F}_0)$, and $\mathbf{v}_\mathfrak{v}$ are the corresponding left unit proper vectors, i.e.,

$$\hat{\mathbf{Q}}^T(\mathbf{N}_1, \mathbf{F}_0)\,\mathbf{v}_\mathfrak{v} = \lambda_\mathfrak{v}\,\mathbf{v}_\mathfrak{v}.$$

Further, it is known that each left proper vector of $\hat{\mathbf{Q}}(\mathbf{N}_1, \mathbf{F}_0)$ corresponding to a particular proper value is orthogonal to every right proper vector of $\hat{\mathbf{Q}}(\mathbf{N}_1, \mathbf{F}_0)$ corresponding to any different proper value. Therefore, if $\lambda_\mathfrak{v} \neq \varrho_R U_N^2$, $\mathbf{v}_\mathfrak{v}$ must be orthogonal to $\mathbf{n}_1$. Thus, taking the inner product of (6.13) with $\mathbf{v}_\mathfrak{v}$, $\mathfrak{v}=2, 3$, we have

$$\frac{\mathbf{b}\cdot\mathbf{v}_\mathfrak{v}}{U_N^3}\,s^2 = \mathbf{c}\cdot\left(\hat{\mathbf{Q}}^T(\mathbf{N}_1, \mathbf{F}_0)\,\mathbf{v}_\mathfrak{v} - \varrho_R U_N^2\,\mathbf{v}_\mathfrak{v}\right),$$

which in turn implies that

$$\mathbf{c}(t)\cdot\mathbf{v}_\mathfrak{v} = \frac{\mathbf{b}\cdot\mathbf{v}_\mathfrak{v}\,s^2(t)}{U_N^3(\lambda_\mathfrak{v} - \varrho_R U_N^2)}, \tag{6.24}$$

where $s(t)$ is given by (6.23). Formula (6.24) states that *the component of the induced discontinuity* $\mathbf{c}$ *in the* $\mathbf{v}_\mathfrak{v}$ *direction will either grow or decay according as the amplitude* s *of the primary wave grows or decays.* Again, the behavior of these components of $\mathbf{c}$ is proportional to the square of the behavior of the primary wave.

If $\lambda_2 = \lambda_3 = \bar{\lambda}$, then there are two distinct circumstances which must be considered:

(a) If there is only one left unit proper vector, say $\boldsymbol{\tau}$, of $\hat{\mathbf{Q}}(\mathbf{N}_1, \mathbf{F}_0)$ corresponding to $\bar{\lambda}$, then

$$\mathbf{c}(t)\cdot\boldsymbol{\tau} = \frac{\mathbf{b}\cdot\boldsymbol{\tau}\,s^2(t)}{U_N^3(\bar{\lambda} - \varrho_R U_N^2)}, \tag{6.25}$$

giving the explicit time dependence of the component of $\mathbf{c}$ in the $\boldsymbol{\tau}$ direction.

(b) If there are two linearly independent left unit proper vectors of $\hat{\mathbf{Q}}(\mathbf{N}_1, \mathbf{F}_0)$ corresponding to $\bar{\lambda}$, then there exists a unit vector, say $\hat{\boldsymbol{\tau}}$, in the principal plane corresponding to $\bar{\lambda}$ associated with $\hat{\mathbf{Q}}^T(\mathbf{N}_1, \mathbf{F}_0)$ such that $\mathbf{c}\cdot(\mathbf{n}_1 \wedge \hat{\boldsymbol{\tau}}) = 0$, and

$$\mathbf{c}(t)\cdot\hat{\boldsymbol{\tau}} = \frac{\mathbf{b}\cdot\hat{\boldsymbol{\tau}}\,s^2(t)}{U_N^3(\bar{\lambda} - \varrho_R U_N^2)}, \tag{6.26}$$

giving the explicit time dependence of the transverse component of the induced discontinuity.

Finally, since $\lambda_\mathfrak{v} \neq \varrho_R U_N^2$ and $\bar{\lambda} \neq \varrho_R U_N^2$, the components of $\mathbf{c}$ in the $\mathbf{v}_\mathfrak{v}$, $\boldsymbol{\tau}$ and $\hat{\boldsymbol{\tau}}$ directions are unique.

7. Transverse waves in anisotropic elastic bodies. While the condition of positive longitudinal elasticity (6.9), insures that at a point of the material body a longitudinal wave may exist and propagate, it does not guarantee the existence of transverse waves. However, if the acoustic tensor is symmetric, it has a triad of orthogonal principal axes. Thus if we assume that, for a fixed deformation gradient at a point of the material body, the elasticity tensor $\mathbf{A}(\mathbf{F})$ is strongly elliptic, in the sense that

$$\mathbf{m}\cdot\bar{\mathbf{Q}}(\mathbf{n})\,\mathbf{m} > 0 \tag{7.1}$$

for all unit vectors $\mathbf{m}$ and $\mathbf{n}$, then, by (6.8) and Theorem 6.1, we have[23]

[23] This theorem is due to TRUESDELL [1966, *9*].

Theorem 7.1. *At a point of a deformed hyperelastic body such that the material has strongly elliptic elasticity, there is at least one direction in which a longitudinal wave and two transverse waves with orthogonal amplitude vectors may exist and propagate.*

Here, we shall consider the behavior of the transverse waves.[24] We assume that since time $t=0$ each of the transverse waves has been propagating into a region which is at rest in a fixed homogeneous configuration with deformation gradient $\boldsymbol{F}_0$, and that the elasticity tensor $\boldsymbol{A}(\boldsymbol{F}_0)$ is strongly elliptic. Let $\boldsymbol{n}_1$ denote the fixed direction of propagation, and let $\boldsymbol{N}_1$, defined by (6.10), denote the fixed unit normal corresponding to $\boldsymbol{n}_1$. Thus, by Theorem 7.1, the transverse waves are necessarily plane and their amplitude vectors are parallel to $\boldsymbol{n}_\mathfrak{a}$, $\mathfrak{a}=2, 3$, the two other principal axes of $\hat{\boldsymbol{Q}}(\boldsymbol{N}_1, \boldsymbol{F}_0)$. In view of definition (5.5) and Remark 5.2, their amplitude vectors $\boldsymbol{s}_\mathfrak{a}$ are given by

$$\boldsymbol{s}_\mathfrak{a} = s_\mathfrak{a}\, \boldsymbol{n}_\mathfrak{a}, \qquad s_\mathfrak{a} > 0; \tag{7.2}$$

and their speeds of propagation $U_{N_\mathfrak{a}}$, $\mathfrak{a}=2, 3$, are given by

$$\varrho_R\, U_{N_\mathfrak{a}}^2 = \lambda_\mathfrak{a} = |\hat{\boldsymbol{Q}}(\boldsymbol{N}_1, \boldsymbol{F}_0)\, \boldsymbol{n}_\mathfrak{a}|, \tag{7.3}$$

where $\lambda_\mathfrak{a}$, $\mathfrak{a}=2, 3$, are the proper values of $\hat{\boldsymbol{Q}}(\boldsymbol{N}_1, \boldsymbol{F}_0)$. Here, the proper values of $\hat{\boldsymbol{Q}}(\boldsymbol{N}_1, \boldsymbol{F}_0)$, i.e., λ_1 and $\lambda_\mathfrak{a}$, $\mathfrak{a}=2, 3$, are regarded to be distinct. We also assume that *the discontinuities across the waves are uniform*, so that the scalars $s_\mathfrak{a}$, $\mathfrak{a}=2, 3$, depend on t alone, independent of the surface parameters $\boldsymbol{V}$.

Analogous to the derivation of (6.13), we can show that for each of the transverse waves

$$2\varrho_R \frac{\delta_D s_\mathfrak{a}}{\delta t}\, \boldsymbol{n}_\mathfrak{a} + \frac{\boldsymbol{b}_\mathfrak{a}}{U_{N_\mathfrak{a}}^3}\, s_\mathfrak{a}^2 = \left(\hat{\boldsymbol{Q}}(\boldsymbol{N}_1, \boldsymbol{F}_0) - \varrho_R\, U_{N_\mathfrak{a}}^2\, \mathbf{1}\right) \boldsymbol{c}, \tag{7.4}$$

where each of the vectors $\boldsymbol{b}_\mathfrak{a}$ is defined by the component relation

$$b_{\mathfrak{a}k} = B_{k\ m\ n}^{\ \alpha\ \beta\ \gamma}\, N_{1\alpha}\, N_{1\beta}\, N_{1\gamma}\, n_\mathfrak{a}^m\, n_\mathfrak{a}^n, \tag{7.5}$$

with $B_{k\ m\ n}^{\ \alpha\ \beta\ \gamma}$ evaluated at $\boldsymbol{F}_0$. In (7.4), the vector $\boldsymbol{c}$, defined by (5.10) with $\boldsymbol{N}=\boldsymbol{N}_1$, is the induced discontinuity associated with each of the transverse waves. Formula (7.4) is identical in form to (6.13). Here, the two cases of (7.4) which must be considered are:

(α) The induced discontinuity $\boldsymbol{c}$ is a proper vector of $\hat{\boldsymbol{Q}}(\boldsymbol{N}_1, \boldsymbol{F}_0)$ corresponding to each of the proper values $\varrho_R\, U_{N_\mathfrak{a}}^2$.

(β) The induced discontinuity $\boldsymbol{c}$ is not a proper vector of $\hat{\boldsymbol{Q}}(\boldsymbol{N}_1, \boldsymbol{F}_0)$ corresponding to each of the proper values $\varrho_R\, U_{N_\mathfrak{a}}^2$.

Since $\hat{\boldsymbol{Q}}(\boldsymbol{N}_1, \boldsymbol{F}_0)$ is symmetric, the solutions of (7.4) corresponding to case (α) and case (β) are the same. That is, we have

Theorem 7.2. *Let Theorem 7.1 hold. Consider each of the transverse waves which since $t=0$ has been propagating into a region that is at rest in a homogeneous configuration with deformation gradient $\boldsymbol{F}_0$. Then the amplitude $s_\mathfrak{a}$ of each of the transverse waves has the following explicit dependence on time:*

$$s_\mathfrak{a}(t) = \frac{1}{\left(\dfrac{1}{s_\mathfrak{a}(0)} + \dfrac{\boldsymbol{b}_\mathfrak{a}\cdot \boldsymbol{n}_\mathfrak{a}}{2\varrho_R\, U_{N_\mathfrak{a}}^3}\, t\right)}, \tag{7.6}$$

where $U_{N_\mathfrak{a}}$ and $\boldsymbol{b}_\mathfrak{a}$ are given by (7.3) and (7.5) respectively.

[24] The results given in this section are due to CHEN [1970, 5].

Consequently, we have

Theorem 7.3. *Let Theorem 7.2 hold.*

(i) *The amplitude of each of the transverse waves will decay to zero monotonically in infinite time if and only if*

$$B_{k\ m\ n}^{\alpha\ \beta\ \gamma} N_{1\alpha} N_{1\beta} N_{1\gamma} n_{\mathfrak{a}}^{k} n_{\mathfrak{a}}^{m} n_{\mathfrak{a}}^{n} > 0 .$$

(ii) *The amplitude of each of the transverse waves will become infinite monotonically within a finite* t_∞, *given by*

$$t_\infty = -\frac{2\varrho_R U_{N_\mathfrak{a}}^3}{s_\mathfrak{a}(0)\, \boldsymbol{b}_\mathfrak{a}\cdot \boldsymbol{n}_\mathfrak{a}},$$

if and only if

$$B_{k\ m\ n}^{\alpha\ \beta\ \gamma} N_{1\alpha} N_{1\beta} N_{1\gamma} n_{\mathfrak{a}}^{k} n_{\mathfrak{a}}^{m} n_{\mathfrak{a}}^{n} < 0 .$$

Here again, the criteria governing the growth and decay of the amplitude of each of the transverse waves depend only on the properties of the second-order elasticity tensor $\boldsymbol{B}(\boldsymbol{F})$, defined by (6.3).

For case (β), we may derive results exhibiting the behavior of certain components of the induced discontinuity $\boldsymbol{c}$ associated with each of the transverse waves. Taking the inner product of (7.4) with $\boldsymbol{n}_\mathfrak{v}$, $\mathfrak{v}=2, 3$, we have

$$\boldsymbol{c}(t)\cdot \boldsymbol{n}_\mathfrak{v} = \frac{\boldsymbol{b}_\mathfrak{a}\cdot \boldsymbol{n}_\mathfrak{v}\, s_\mathfrak{a}^2(t)}{\varrho_R U_{N_\mathfrak{a}}^3 (U_{N_\mathfrak{v}}^2 - U_{N_\mathfrak{a}}^2)}, \qquad \mathfrak{a}\neq \mathfrak{v}, \tag{7.7}$$

where $s_\mathfrak{a}(t)$ is given by (7.6). On the other hand, taking the inner product of (7.4) with $\boldsymbol{n}_1$, we have

$$\boldsymbol{c}(t)\cdot \boldsymbol{n}_1 = \frac{\boldsymbol{b}_\mathfrak{a}\cdot \boldsymbol{n}_1\, s_\mathfrak{a}^2(t)}{\varrho_R U_{N_\mathfrak{a}}^3 (U_N^2 - U_{N_\mathfrak{a}}^2)}, \tag{7.8}$$

giving the explicit time dependence of the *longitudinal component* of the induced discontinuity associated with each of the transverse waves.[25] In (7.8), U_N is the speed of the longitudinal wave propagating in the $\boldsymbol{n}_1$ direction; it is given by (6.12). Formulae (7.7) and (7.8) state that *the components in the* $\boldsymbol{n}_\mathfrak{v}$ *and* $\boldsymbol{n}_1$ *directions of the induced discontinuity associated with the transverse wave in the* $\boldsymbol{n}_\mathfrak{a}$ *direction will either grow or decay according as the amplitude* $s_\mathfrak{a}$ *of the primary wave grows or decays.*

Finally, it should be pointed out that, while it is not possible to obtain a unique solution of (7.4) for $\boldsymbol{c}$ corresponding to each $\lambda_\mathfrak{a}$, the components of $\boldsymbol{c}$ associated with each of the transverse waves in these directions are unique, i.e., the results (7.7) and (7.8) are not empty.

8. Thermodynamic influences on waves in anisotropic elastic bodies. In our discussions on the behavior of plane longitudinal and transverse waves, given in Sects. 6 and 7, we did not include the influence of thermodynamics. Here, we shall show that these results can be readily extended to waves propagating in elastic non-conductors of heat.[26]

[25] In his studies on principal waves in *isotropic* elastic bodies, GREEN [1965, 7] has shown that the induced discontinuity associated with each of the primary transverse waves is indeed longitudinal. Also, in considering the global solutions of transverse waves in isotropic elastic bodies, DAVISON [1966, 5] arrived at the same conclusion.

[26] The results given in this section are due to CHEN [1971, 4].

TRUESDELL showed that in the absence of external radiation the amplitude vector $\mathbf{s}$ and the speed of propagation U_N of an acceleration wave propagating in an elastic non-conductor of heat obey the propagation condition [27]

$$\{\hat{\boldsymbol{Q}}(\boldsymbol{N}, \boldsymbol{F}, \eta) - \varrho_R U_N^2 \mathbf{1}\} \mathbf{s} = \mathbf{0}, \tag{8.1}$$

provided that the entropy $\eta(\cdot, \cdot)$ is continuous everywhere. In (8.1), the acoustic tensor $\hat{\boldsymbol{Q}}(\boldsymbol{N}, \boldsymbol{F}, \eta)$ is defined by the component relation

$$\hat{Q}_{km} = \hat{A}_{k\ m}^{\ \alpha\ \beta} N_\alpha N_\beta, \tag{8.2}$$

where $\hat{\boldsymbol{A}}(\boldsymbol{F}, \eta)$ is the isentropic elasticity tensor

$$\hat{A}_{k\ m}^{\ \alpha\ \beta} = \frac{\partial \hat{h}_k^{\ \alpha}}{\partial F^m_{\ \beta}} = \frac{\partial^2 \hat{e}}{\partial F^m_{\ \beta}\, \partial F^k_{\ \alpha}}. \tag{8.3}$$

Here, $\boldsymbol{h}$ and $\hat{e}$ are, respectively, the response functions for the stress and the internal energy. We assume that $\hat{e}(\boldsymbol{F}, \eta)$ is three times continuously differentiable, so that $\hat{\boldsymbol{Q}}(\boldsymbol{N}, \boldsymbol{F}, \eta)$ is symmetric. Finally, the isentropic second-order elasticity tensor $\hat{\boldsymbol{B}}(\boldsymbol{F}, \eta)$ is given by the component relation

$$\hat{B}_{k\ m\ n}^{\ \alpha\ \beta\ \gamma} = \frac{\partial \hat{A}_{k\ m}^{\ \alpha\ \beta}}{\partial F^n_{\ \gamma}}, \tag{8.4}$$

with the obvious symmetry properties.

The propagation condition (8.1) is identical in form to (6.5), the propagation condition for the purely mechanical theory. Therefore, we can readily extend the results on the existence of waves, i.e., those corresponding to Theorem 6.1 and Theorem 7.1, to elastic non-conductors of heat. This extension is quite trivial. In fact, we have

Remark 8.1. At a point of a deformed homogeneous elastic non-conductor of heat, such that the isentropic elasticity tensor has positive longitudinal elasticity, in the sense analogous to (6.9), *there is at least one direction in which a longitudinal wave may exist and propagate.*

In general, even though the acoustic tensor is symmetric, there is no guarantee that the squared speeds corresponding to the possible transverse amplitude vectors are positive. However, if the isentropic elasticity tensor is strongly elliptic, in a sense analogous to (7.1), then there is at least one direction in which a longitudinal wave and two transverse waves with orthogonal amplitude vectors may exist and propagate.

In view of the preceding results, if we assume that the body is at rest in a homogeneous configuration with constant and uniform entropy, and that in this configuration the isentropic elasticity tensor is strongly elliptic, then a plane longitudinal and two plane transverse waves may exist and propagate in the same direction throughout the body and for all times. Thus, we may examine the behavior of the amplitudes of these waves.

In view of the preceding observation, we have the following correspondence:

Remark 8.2. Consider an anisotropic elastic non-conductor of heat which is at rest in a homogeneous configuration with constant and uniform entropy, characterized by, say, $(\boldsymbol{F}_0, \eta_0)$. *Suppose that in this configuration the isentropic elasticity tensor* $\hat{\boldsymbol{A}}(\boldsymbol{F}_0, \eta_0)$ *is strongly elliptic.*

---

[27] TRUESDELL [1961, *3*, Sect. 13]. TRUESDELL attributes this result to DUHEM [1903, *1*] and [1906, *1*, Chap. I, Sect. 2].

(i) *The amplitude s of the plane longitudinal wave propagating in, say, the $\boldsymbol{n}_1$ direction obeys the equation*

$$2\varrho_R \frac{\delta_D s}{\delta t} \boldsymbol{n}_1 + \frac{\hat{\boldsymbol{b}}}{U_N^3} s^2 = \left(\hat{\boldsymbol{Q}}(\boldsymbol{N}_1, \boldsymbol{F}_0, \eta_0) - \varrho_R U_N^2 \boldsymbol{1}\right) \boldsymbol{c}, \tag{8.5}$$

where $\boldsymbol{N}_1$ is the fixed unit normal corresponding $\boldsymbol{n}_1$, $\boldsymbol{c}$ is the induced discontinuity, defined by (5.10) *with $\boldsymbol{N}=\boldsymbol{N}_1$, and the vector $\hat{\boldsymbol{b}}$ is defined by the component relation*

$$\hat{b}_k = \hat{B}_{k\ m\ n}^{\ \alpha\ \beta\ \gamma} N_{1\alpha} N_{1\beta} N_{1\gamma} n_1^m n_1^n,$$

with $\hat{B}_{k\ m\ n}^{\ \alpha\ \beta\ \gamma}$ evaluated at $(\boldsymbol{F}_0, \eta_0)$. Here, U_N is the longitudinal wave speed, given by

$$\varrho_R U_N^2 = |\hat{\boldsymbol{Q}}(\boldsymbol{N}_1, \boldsymbol{F}_0, \eta_0)\, \boldsymbol{n}_1|.$$

(ii) *The amplitude $s_{\mathfrak{a}}$, $\mathfrak{a}=2, 3$, of each of the transverse waves propagating in, say, the $\boldsymbol{n}_1$ direction with its amplitude vector parallel to the $n_{\mathfrak{a}}$ principal direction, obeys the equation*

$$2\varrho_R \frac{\delta_D s_{\mathfrak{a}}}{\delta t} \boldsymbol{n}_{\mathfrak{a}} + \frac{\hat{\boldsymbol{b}}_{\mathfrak{a}}}{U_{N_{\mathfrak{a}}}^3} s_{\mathfrak{a}}^2 = \left(\hat{\boldsymbol{Q}}(\boldsymbol{N}_1, \boldsymbol{F}_0, \eta_0) - \varrho_R U_{N_{\mathfrak{a}}}^2 \boldsymbol{1}\right) \boldsymbol{c}, \tag{8.6}$$

where the vector $\boldsymbol{c}$ is the induced discontinuity associated with each of the transverse waves, and each of the vectors $\hat{\boldsymbol{b}}_{\mathfrak{a}}$ is defined by the component relation

$$\hat{b}_{\mathfrak{a}k} = \hat{B}_{k\ m\ n}^{\ \alpha\ \beta\ \gamma} N_{1\alpha} N_{1\beta} N_{1\gamma} n_{\mathfrak{a}}^m n_{\mathfrak{a}}^n,$$

with $\hat{B}_{k\ m\ n}^{\ \alpha\ \beta\ \gamma}$ evaluated at $(\boldsymbol{F}_0, \eta_0)$. Here, $U_{N_{\mathfrak{a}}}$ is the speed of each of the transverse waves, given by

$$\varrho_R U_{N_{\mathfrak{a}}}^2 = |\hat{\boldsymbol{Q}}(\boldsymbol{N}_1, \boldsymbol{F}_0, \eta_0)\, \boldsymbol{n}_{\mathfrak{a}}|.$$

This remark is a direct consequence of balance of linear momentum and balance of energy. For the sake of brevity, we omit the somewhat lengthy but direct derivations of these results.

Notice that (8.5) and (8.6) are identical in form to (6.13) and (7.4), respectively. Therefore, the amplitudes of acceleration waves propagating in elastic non-conductors of heat behave in exactly the same way as those in the purely mechanical theory. In other words, *the thermodynamic properties of the material do not affect the behavior of these waves.* Finally, we should point out that the results for the purely mechanical theory, given in Sects. 6 and 7, can be directly carried over to the present situation, provided that the isentropic elasticity and second-order elasticity tensors are taken at fixed entropy.

9. Waves in isotropic elastic bodies. In our considerations of the properties of acceleration waves in homogeneous anisotropic elastic bodies, given in Sects. 6 and 7, we assume that the material regions ahead of the waves are at rest in homogeneous configurations. This assumption, taken together with Theorem 6.1 and Theorem 7.1, insures that there is at least one direction in which plane longitudinal and transverse waves may exist and propagate. However, if the material is isotropic rather than anisotropic, and if the material region ahead of the waves is at rest in a homogeneous configuration, then plane longitudinal and transverse waves may travel down the principal axes of stress and strain provided that their squared speeds are positive and different from each other. That is, there are at least three directions in which plane longitudinal and transverse waves may exist and propagate. In the sequel, we shall examine the properties of these so called principal waves.[28]

A homogeneous isotropic elastic material is defined by the constitutive relation

$$\boldsymbol{t}(\boldsymbol{x}, t) = \boldsymbol{h}(\boldsymbol{B}(\boldsymbol{x}, t)), \tag{9.1}$$

where $\boldsymbol{t}$ is the Cauchy stress tensor, and $\boldsymbol{B}$ the left Cauchy-Green tensor. Further, the response function $\boldsymbol{h}(\cdot)$ has the explicit representation

$$\boldsymbol{h}(\boldsymbol{B}) = h_0 \boldsymbol{1} + h_1 \boldsymbol{B} + h_2 \boldsymbol{B}^2, \tag{9.2}$$

where h_0, h_1 and h_2 are functions of the principal invariants I, II and III of $\boldsymbol{B}$. Here, we have taken an undistorted homogeneous configuration as the reference, so that $\boldsymbol{h}(\boldsymbol{B})$ has the representation (9.2), and the density ϱ_R in this configuration is independent of $\boldsymbol{X}$. We assume that h_0, h_1 and h_2 are of class C^2.

Within the context of the present theory of isotropic elasticity, DAVISON[29] has exhibited explicit solutions of plane longitudinal acceleration and shock waves as well as those of plane transverse acceleration waves and plane propagating vortex sheets.

In the present context, the condition of balance of linear momentum assumes the form

$$\operatorname{div} \boldsymbol{t} + \varrho \boldsymbol{b} = \varrho \ddot{\boldsymbol{x}}, \tag{9.3}$$

where ϱ is the present density. Formula (9.3), the constitutive relation (9.1), (5.1), $(5.3)_2$ together with the hypothesis **(A-I)** on the properties of acceleration waves given in Sect. 5 imply that the amplitude vector $\boldsymbol{s}$ and the local speed of propagation U of an acceleration wave with unit normal $\boldsymbol{n}$ obey the following propagation condition:[30]

$$\{\tilde{\boldsymbol{Q}}(\boldsymbol{n}, \boldsymbol{B}) - \varrho U^2 \boldsymbol{1}\} \boldsymbol{s} = \boldsymbol{0}, \tag{9.4}$$

where the acoustic tensor $\tilde{\boldsymbol{Q}}(\boldsymbol{n}, \boldsymbol{B})$ is defined by the component relation

$$\tilde{Q}^k{}_m = 2 \frac{\varrho_R}{\varrho} \frac{\partial h^{kp}}{\partial B^{mq}} B^{qs} n_p n_s. \tag{9.5}$$

By (9.4), (9.5) and (9.2), we have the following theorem due to TRUESDELL:

Theorem 9.1. *For a wave travelling down a principal axis of stress and strain in an isotropic elastic body, the proper vectors of the acoustic tensor coincide with the principal axes of stress and strain.*

In other words, the principal axes $\boldsymbol{n}_\mathfrak{a}$, $\mathfrak{a} = 1, 2, 3$, of stress and strain are the proper vectors of $\tilde{\boldsymbol{Q}}(\boldsymbol{n}_\mathfrak{a}, \boldsymbol{B})$. Thus, we see that if, for the directions of propagation corresponding to the principal axes of stress and strain, the possible longitudinal and transverse wave speeds are unequal to each other, then TRUESDELL'S theorem implies that waves travelling down the principal axes of stress and strain are always either longitudinal or transverse. Such waves are called *principal waves.*

Let $v_\mathfrak{a}$, $\mathfrak{a} = 1, 2, 3$, denote the principal stretches. By (9.4), (9.5) and (9.2), we can show that, for the $\boldsymbol{n}_1$ principal direction, the speed U_{11} of the longitudinal wave is given by

$$\varrho U_{11}^2 = 2 v_1^2 \left\{ h_1 + 2 v_1^2 h_2 + \sum_{\Gamma=0}^{2} v_1^{2\Gamma} D_1 h_\Gamma \right\}, \tag{9.6}$$

[28] The results given in this section are due to GREEN [1964, *3*]. CHEN [1968, *5*] extended these results to include thermodynamic influences. Corresponding results for waves in inhomogeneous isotropic elastic bodies are given by BOWEN and WANG [1970, *3*].

[29] DAVISON [1966, *5*].

[30] Cf. TRUESDELL [1961, *3*, Sect. 7].

with

$$D_1 = \frac{\partial}{\partial \mathrm{I}} + (v_2^2 + v_3^2)\frac{\partial}{\partial \mathrm{II}} + v_2^2 v_3^2 \frac{\partial}{\partial \mathrm{III}}; \tag{9.7}$$

and the speed U_{12} of the transverse wave with amplitude vector in the $\boldsymbol{n}_2$ principal direction is given by

$$\varrho U_{12}^2 = v_1^2\{h_1 + (v_1^2 + v_2^2)\, h_2\}. \tag{9.8}$$

In terms of the principal stresses $t_\mathfrak{a}$, $\mathfrak{a}=1, 2, 3$, as functions of the principal stretches, (9.6) and (9.8) may be expressed in the alternate forms

$$\begin{aligned} \varrho U_{11}^2 &= 2 v_1^2 \frac{\partial t_1}{\partial v_1^2}, \\ \varrho U_{12}^2 &= v_1^2 \frac{t_1 - t_2}{v_1^2 - v_2^2}, \end{aligned} \tag{9.9}$$

where $t_\mathfrak{a} = h_0 + h_1 v_\mathfrak{a}^2 + h_2 v_\mathfrak{a}^4$. The assumption that the speeds U_{11} and U_{12} are real and non-zero is equivalent to the condition that the right hand members of (9.6) and (9.8) or of (9.9) are strictly positive.

As we have remarked earlier, we shall examine the properties of principal waves propagating, say, in the $\boldsymbol{n}_1$ principal direction. We suppose that since time $t=0$ the waves have been propagating into regions which are at rest in homogeneous configurations; i.e., we suppose that the stretches $v_\mathfrak{a}$, $\mathfrak{a}=1, 2, 3$, are constants. Hence not only are the principal waves either longitudinal or transverse, but they are necessarily plane waves. Further, we assume that *the discontinuity across each of the waves is uniform*, so that the amplitude vector $\boldsymbol{s}$ of each of the waves is given a function of t alone, independent of the surface parameters $\boldsymbol{V}$.

Balance of linear momentum (9.3), the constitutive relation (9.1) and (5.7) together with the hypothesis **(A-I)** imply that

$$\varrho[\ddot{x}^k] - \varrho[\dot{x}^r_{,r}][\ddot{x}^k] = \frac{\partial^2 h^{km}}{\partial B^{ab}\,\partial B^{jl}}[\dot{B}^{ab}][B^{jl}_{,m}] + \frac{\partial h^{km}}{\partial B^{jl}}[\overline{B^{jl}_{,m}}^{\boldsymbol{\cdot}}]. \tag{9.10}$$

Briefly, the differential equation governing the behavior of the amplitude of the principal longitudinal wave propagating in the $\boldsymbol{n}_1$ direction is derived by taking the inner product of (9.10) with $\boldsymbol{n}_1$; while that of the principal transverse wave propagating in the $\boldsymbol{n}_1$ direction with amplitude vector parallel to the $\boldsymbol{n}_2$ direction is derived by taking the inner product of (9.10) with $\boldsymbol{n}_2$.

Longitudinal principal waves. In view of definition (5.4) and Remark 5.1, the amplitude vector $\boldsymbol{s}$ of the longitudinal wave propagating in the $\boldsymbol{n}_1$ principal direction may be expressed in the form

$$\boldsymbol{s} = s\boldsymbol{n}_1. \tag{9.11}$$

Thus, taking the inner product of (9.10) with $\boldsymbol{n}_1$ and evaluating the resulting relation with the aid of (5.1), (5.3) and (5.9), we have the equation

$$\frac{\delta_D s}{\delta t} = -\frac{1}{2U_{11}}\left(1 + \frac{4v_1^4}{\varrho U_{11}^2}\frac{\partial^2 t_1}{\partial v_1^2\,\partial v_1^2}\right)s^2, \tag{9.12}$$

which the amplitude of the longitudinal principal wave must obey. Setting

$$Q_1 = \frac{1}{2U_{11}}\left(1 + \frac{4v_1^4}{\varrho U_{11}^2}\frac{\partial^2 t_1}{\partial v_1^2\,\partial v_1^2}\right) \tag{9.13}$$

and integrating (9.12), we have

Theorem 9.2. *Consider the longitudinal wave which since time $t=0$ has been propagating in the $\boldsymbol{n}_1$ principal direction in an isotropic elastic body that is at rest in a homogeneous configuration. Then the amplitude s has the following explicit dependence on time:*

$$s(t)=\frac{1}{\frac{1}{s(0)}+Q_1 t}. \tag{9.14}$$

Consequently, by $(9.9)_1$ and (9.13), we have

Theorem 9.3. *Let Theorem 9.2 hold.*

(i) *The amplitude of a compressive wave (i.e., $s(0)>0$) will decay to zero monotonically in infinite time if and only if*

$$\frac{\partial t_1}{\partial v_1^2}+2v_1^2\frac{\partial^2 t_1}{\partial v_1^2\,\partial v_1^2}>0.$$

(ii) *The amplitude of an expansive wave (i.e., $s(0)<0$) will decay to zero monotonically in infinite time if and only if*

$$\frac{\partial t_1}{\partial v_1^2}+2v_1^2\frac{\partial^2 t_1}{\partial v_1^2\,\partial v_1^2}<0.$$

(iii) *In either case, the amplitude s will become infinite monotonically within a finite time t_∞, given by*

$$t_\infty=-\frac{1}{s(0)\,Q_1},$$

if and only if

$$\operatorname{sgn} s=-\operatorname{sgn}\left(\frac{\partial t_1}{\partial v_1^2}+2v_1^2\frac{\partial^2 t_1}{\partial v_1^2\,\partial v_1^2}\right).$$

Transverse principal waves. In view of definition (5.5) and Remark 5.2, the amplitude vector $\mathbf{s}$ of the transverse wave propagating in the $\boldsymbol{n}_1$ principal direction with $\mathbf{s}$ parallel to, say, the $\boldsymbol{n}_2$ direction may be expressed in the form

$$\mathbf{s}=\hat{s}\,\boldsymbol{n}_2, \qquad \hat{s}>0. \tag{9.15}$$

Thus, taking the inner product of (9.10) with $\boldsymbol{n}_2$ and evaluating the resulting relation with the aid of (5.1), (5.3) and (5.9), we have the equation

$$\frac{\delta_D \hat{s}}{\delta t}=0, \tag{9.16}$$

which the amplitude of the principal transverse wave must obey. Consequently, we have

Theorem 9.4. *In an isotropic elastic body which is at rest in a homogeneous configuration, transverse principal waves propagate with constant amplitudes.*

10. Waves of arbitrary shape in isotropic elastic bodies. Thus far we have considered circumstances which insure that plane acceleration waves may exist and propagate in elastic bodies. In particular, in our considerations of the properties of acceleration waves in homogeneous anisotropic and isotropic elastic bodies, given in Sects. 6, 7 and 9, we assume that the regions ahead of the waves are at rest in homogeneous configurations. However, if, for an isotropic body, the material region ahead of the waves is subjected to constant hydrostatic stress, then acceleration waves of arbitrary shape may exist and propagate. Here, we shall examine the properties of these waves; in particular, we shall obtain explicit expressions for cylindrically expanding and spherically expanding waves.

Our observation of the existence of waves of arbitrary shape follows directly from Theorem 9.1. First of all, if the isotropic elastic body is subjected to hydrostatic stress, then every direction is a principal axis of stress and strain. Therefore, if the possible longitudinal and transverse wave speeds are unequal, then every wave is either longitudinal or transverse; and at a point all longitudinal waves have the same speed and all transverse waves have the same speed. Further, if the region ahead of the waves is at rest besides being in an undistorted configuration, then all longitudinal waves have the same constant speed and all transverse waves have the same constant speed throughout the isotropic body. Thus we may examine the properties of longitudinal and transverse waves of arbitrary shape.[31]

Let the constant stretch v characterize the undistorted configuration. By (9.6), (9.7) and (9.8), the longitudinal wave speed $U_\parallel$ and the transverse wave speed $U_\perp$ are given by

$$\begin{aligned} \varrho U_\parallel^2 &= 2v^2\left(h_1 + 2v^2 h_2 + \sum_{\Gamma=0}^{2} v^{2\Gamma} D h_\Gamma\right), \\ \varrho U_\perp^2 &= v^2 (h_1 + 2v^2 h_2), \end{aligned} \tag{10.1}$$

where

$$D = \frac{\partial}{\partial \mathrm{I}} + 2v^2 \frac{\partial}{\partial \mathrm{II}} + v^4 \frac{\partial}{\partial \mathrm{III}}. \tag{10.2}$$

In (10.1) h_1, h_2 and Dh_Γ, $\Gamma = 0, 1, 2$ are evaluated at $\mathrm{I} = 3v^2$, $\mathrm{II} = 3v^4$, $\mathrm{III} = v^6$. Formula $(10.1)_1$ may be expressed in two alternate ways. First, in terms of the principal stresses $t_\mathfrak{a}$, $\mathfrak{a} = 1, 2, 3$, as functions of the principal stretches $v_\mathfrak{a}$, $U_\parallel$ is given by

$$\varrho U_\parallel^2 = 2v_\mathfrak{a}^2 \left.\frac{\partial t_\mathfrak{a}}{\partial v_\mathfrak{a}^2}\right|_{v_\mathfrak{b} = v}, \qquad \mathfrak{b} = 1, 2, 3. \tag{10.3}$$

Second, since the stress is hydrostatic, the constitutive relation (9.1) reduces to $\boldsymbol{t} = -p\mathbf{1}$, with $p = p(\varrho)$. It therefore follows that $U_\parallel$ is also given by

$$\varrho U_\parallel^2 = \frac{4}{3} v^2 (h_1 + 2v^2 h_2) + \varrho \frac{dp}{d\varrho}. \tag{10.4}$$

For physical reasons, it is reasonable to assume that

$$\frac{dp}{d\varrho} > 0$$

in all circumstances. Further, if we assume that the squared speed of transverse waves is positive, i.e., $h_1 + 2v^2 h_2 > 0$, then by (10.4) the squared speed of longitudinal waves is also positive.

We assume that since time $t = 0$ the waves under consideration have been propagating into regions that are at rest in undistorted configurations. In the common frame and under the present conditions, the motion ahead of the waves

[31] The results presented in this section are due to Chen [1968, *4*]. These results generalize those given by Thomas [1957, *2*], [1961, *2*] for waves in ideal gases and in the linear elastic bodies. Degenerate cases of the present theory are those of Chen [1968, *3*], Juneja and Nariboli [1970, *8*] and Suhubi [1970, *10*]. Other results on waves of arbitrary shape are those by Becker and Schmitt [1968, *2*] and Doria and Bowen [1970, *7*] for waves in fluids with internal state variables, and those by Bowen and Wang [1970, *3*] for waves in inhomogeneous isotropic elastic bodies.

reduces to

$$\boldsymbol{x} = v\boldsymbol{X}, \tag{10.5}$$

so that, by the hypothesis **(A-I)**, (4.10) and (4.11),

$$\boldsymbol{n} = \boldsymbol{N}, \tag{10.6}$$

and

$$U = u_n = v\,U_N. \tag{10.7}$$

Thus, the normal distance $n - n_0$, traversed by the waves in the deformed material, is related to the normal distance $N - N_0$, traversed by the waves in the reference configuration, by the formula

$$n - n_0 = v(N - N_0), \tag{10.8}$$

with

$$n - n_0 = Ut, \qquad N - N_0 = U_N\,t. \tag{10.9}$$

We assume that the *discontinuity across each of the waves is uniform*, so that the amplitude of each of the waves is given by a function of t alone, independent of the surface parameters $\boldsymbol{V}$. Thus, by (10.7), (10.8), (10.9) and the dual of CFT (179.19), we see that it may be given by a function of n alone and

$$\frac{\delta_D \boldsymbol{s}}{\delta t} = U\frac{d\boldsymbol{s}}{dn}. \tag{10.10}$$

Longitudinal waves. Here, the amplitude vector $\boldsymbol{s}$ of a longitudinal wave may be expressed in the form

$$\boldsymbol{s} = s\boldsymbol{n}. \tag{10.11}$$

Thus, taking the inner product of (9.10) with $\boldsymbol{n}$ and evaluating the resulting relation with the aid of (5.1), (5.3) and (5.9), we have

$$\begin{aligned} 2\varrho\,U_{\|}^2\frac{ds}{dn} &= v^2(h_1 + 2v^2 h_2)\,(b_\delta^\delta s - s^i_{;\delta}\,y_{i;}{}^\delta) - 2v^2\sum_{\Gamma=0}^{2} v^{2\Gamma} D h_\Gamma\, s^i_{;\delta}\,y_{i;}{}^\delta \\ &\quad - \frac{1}{U_{\|}^2}\left\{\varrho\,U_{\|}^4 + 4v^4\left(2h_2 + 2D h_1 + 4v^2 D h_2 + \sum_{\Gamma=0}^{2} v^{2\Gamma} D^2 h_\Gamma\right)\right\} s^2, \end{aligned} \tag{10.12}$$

where $\boldsymbol{b}$ is the second fundamental form of the surface $\mathfrak{s}(t)$. By (10.12), $(10.1)_1$ and the fact that[32]

$$s^i_{;\gamma}\,y_{i;}{}^\delta = -b_\gamma^\delta\,s,$$

we have

Theorem 10.1. *The amplitude of a longitudinal acceleration wave propagating in an isotropic elastic body subjected to constant hydrostatic stress obeys the equation*

$$\frac{ds}{dn} = \frac{\overline{K}}{2}s - \frac{1}{2U_{\|}^2}\left(1 + \frac{\varphi(v)}{\varrho\,U_{\|}^2}\right)s^2, \tag{10.13}$$

[32] This relation follows from the well-known Gauss-Weingarten equation

$$y^k_{;\gamma\delta} = b_{\gamma\delta}\,n^k,$$

and the fact that for a longitudinal wave

$$s^i\,y_{i;}{}^\delta = 0.$$

where $\overline{K}=b^{\delta}_{\delta}$ *is the mean curvature of the wave shape,* $U_{\|}$ *is given by* $(10.1)_1$, *and*

$$\begin{aligned}\varphi(v) &= 4v^4\left(2h_2+2D\,h_1+4v^2\,Dh_2+\sum_{\Gamma=0}^{2} v^{2\Gamma}\,D^2\,h_{\Gamma}\right)\\ &= 4v_{\mathfrak{a}}^4\,\frac{\partial^2 t_{\mathfrak{a}}}{\partial v_{\mathfrak{a}}^2\,\partial v_{\mathfrak{a}}^2}\bigg|_{v_{\mathfrak{v}}=v}, \qquad \mathfrak{v}=1,2,3.\end{aligned} \tag{10.14}$$

For cylindrically expanding and spherically expanding waves the mean curvatures are

$$-\frac{1}{r}\quad\text{and}\quad-\frac{2}{r},$$

respectively. Setting $n=r$, and

$$Q(v)=\frac{1}{2U_{\|}^2}\left(1+\frac{\varphi(v)}{\varrho\,U_{\|}^2}\right), \tag{10.15}$$

we have, on integrating (10.13),

Theorem 10.2. *Let Theorem 10.1 hold. Then for a cylindrically expanding wave*

$$s(r)=\frac{1}{\left(\dfrac{1}{s(r_0)}+2Q(v)\,r_0^{\frac{1}{2}}\left(r^{\frac{1}{2}}-r_0^{\frac{1}{2}}\right)\right)}\left(\frac{r_0}{r}\right)^{\frac{1}{2}}; \tag{10.16}$$

and for a spherically expanding wave

$$s(r)=\frac{1}{\left(\dfrac{1}{s(r_0)}+Q(v)\,r_0\ln\dfrac{r}{r_0}\right)}\left(\frac{r_0}{r}\right). \tag{10.17}$$

Consequently, we have[33]

Theorem 10.3. *Let Theorem 10.2 hold.*

(i) *If* $\operatorname{sgn} s(r_0)=\operatorname{sgn} Q(v)$, *the amplitude of a compressive or an expansive cylindrical wave will decay to zero monotonically in infinite time. If* $\operatorname{sgn} s(r_0)=-\operatorname{sgn} Q(v)$, *the amplitude of a compressive or an expansive cylindrical wave will become infinite within a finite time* t_∞, *given by*

$$t_\infty=\frac{1}{4\,U_{\|}\,s^2(r_0)\,Q^2(v)\,r_0}-\frac{1}{U_{\|}\,s(r_0)\,Q(v)}.$$

(ii) *If* $\operatorname{sgn} s(r_0)=\operatorname{sgn} Q(v)$, *the amplitude of a compressive or an expansive spherical wave will decay to zero monotonically in infinite time. If* $\operatorname{sgn} s(r_0)=-\operatorname{sgn} Q(v)$, *the amplitude of a compressive or an expansive spherical wave will become infinite within a finite time* t_∞, *given by*

$$t_\infty=\frac{r_0}{U_{\|}}\left\{e^{-\frac{1}{s(r_0)Q(v)r_0}}-1\right\}.$$

Since the amplitude of a cylindrical wave will decay if and only if

$$-Q(v)\,s(r)<\frac{1}{2r},$$

or grow if and only if

$$-Q(v)\,s(r)>\frac{1}{2r},$$

[33] The results of Theorem 10.3 are stronger than those given previously by Chen [1968, *4*].

and since the amplitude of a spherical wave will decay if and only if

$$-Q(v)\, s(r) < \frac{1}{r},$$

or grow if and only if

$$-Q(v)\, s(r) > \frac{1}{r},$$

it follows that when $\operatorname{sgn} s(r_0) = -\operatorname{sgn} Q(v)$, the amplitudes of these waves may decay initially, but at some larger radii they will begin to grow and become infinite within finite times. An example, illustrating this observation, has been given by CHEN.[34]

Finally, by setting

$$r = r_0 + \Delta r, \qquad \Delta r > 0,$$

that is,

$$\frac{r}{r_0} = 1 + \frac{\Delta r}{r_0}, \qquad \frac{r_0}{r} = 1 - \frac{\Delta r}{r_0 + \Delta r},$$

it can be shown that the solutions for cylindrically and spherically expanding waves, (10.16) and (10.17), reduce to that for plane waves in the limit as $r_0 \to \infty$. This is what is expected in that *cylindrical and spherical waves may be regarded locally as plane waves when their radii are large.*

Transverse waves. Let $\hat{s}$, the absolute value of $|\mathbf{s}|$, denote the amplitude of a transverse wave. Taking the inner product of (9.10) with $\mathbf{s}$ and evaluating the resulting relation with the aid of (5.1), (5.3) and (5.9), we have

$$\begin{aligned} \varrho\, U_\perp^2 \frac{d\hat{s}^2}{dn} = {} & v^2 (h_1 + 2v^2 h_2)\, b_\delta^\delta\, \hat{s}^2 + v^2 (s^m s^n\, y_{m;}{}^\delta\, y_{n;}{}^\lambda\, b_{\delta\lambda} - s^m_{;\delta}\, n_m\, s^n\, y_{n;}{}^\delta) \\ & \cdot \left\{ h_1 + 2v^2 h_2 + 2 \sum_{\Gamma=0}^{2} v^{2\Gamma} D h_\Gamma \right\}, \end{aligned} \tag{10.18}$$

which together with $(10.1)_2$ and the fact that[35]

$$s^i_{;\gamma}\, n_i = s^i\, y_{i;}{}^\delta\, b_{\gamma\delta}$$

yields the following

Theorem 10.5. *The amplitude of a transverse wave propagating in an isotropic elastic body subjected to constant hydrostatic stress obeys the equation*

$$\frac{d\hat{s}^2}{dn} = \overline{K}\, \hat{s}^2, \tag{10.19}$$

where $\overline{K}$ is the mean curvature of the wave shape.

This theorem states that *in an isotropic elastic body subjected to constant hydrostatic stress the behavior of the amplitude of a transverse wave is purely geometrical, independent of the response function which defines a particular material.* In other words, *for a given stretch transverse waves of the same shape behave in exactly the same way in all isotropic elastic bodies.* Finally, we observe that cylindrically and spherically expanding transverse waves, with respective mean curva-

---

[34] CHEN [1968, *3*].

[35] This relation follows from the well-known Gauss-Weingarten equation

$$n^k_{;\gamma} = -b_\gamma^\delta\, y^k_{;\delta},$$

and the fact that for a transverse wave

$$\mathbf{s} \cdot \mathbf{n} = 0.$$

tures

$$-\frac{1}{r} \text{ and } -\frac{2}{r},$$

will always decay to zero monotonically in infinite time.

11. Thermodynamic influences on waves in isotropic elastic bodies. The results given in Sects. 9 and 10 concerning the behavior of plane principal waves and waves of arbitrary shape have been extended to include the influence of thermodynamics.[36] Here, we shall simply list the results for materials which do not conduct heat. The interested reader may refer to the original article for the details and other results.

The constitutive relation for the stress of an isotropic elastic material which does not conduct heat may be expressed in the form

$$\boldsymbol{t} = \tilde{h}_0 \, \mathbf{1} + \tilde{h}_1 \boldsymbol{B} + \tilde{h}_2 \boldsymbol{B}^2, \tag{11.1}$$

where $\tilde{h}_0$, $\tilde{h}_1$ and $\tilde{h}_2$ are functions of the principal invariants of $\boldsymbol{B}$ and of the entropy η. Hence, the principal stresses $\tilde{t}_\mathfrak{a}$, $\mathfrak{a} = 1, 2, 3$, are given by

$$\tilde{t}_\mathfrak{a} = \tilde{h}_0 + \tilde{h}_1 v_\mathfrak{a}^2 + \tilde{h}_2 v_\mathfrak{a}^4, \tag{11.2}$$

where $v_\mathfrak{a}$, $\mathfrak{a} = 1, 2, 3$, are the principal stretches.

Here, as in the case of the purely mechanical theory, every principal wave is either longitudinal or transverse, provided that the possible longitudinal and transverse wave speeds are unequal to each other. Thus we see that the results given in Sects. 9 and 10 can be readily extended to the present circumstances. In fact, we have the following:

Remark 11.1. Consider an isotropic elastic body which does not conduct heat.

(i) *Suppose that the body is at rest in a homogeneous configuration with constant and uniform entropy. Then the amplitude of a plane longitudinal wave propagating in the* $\boldsymbol{n}_1$ *principal direction obeys the equation*

$$\frac{\delta_D s}{\delta t} = -\frac{1}{2\,U_{11}}\left(1 + \frac{4\,v_1^4}{\varrho\,U_{11}^2}\,\frac{\partial^2 \tilde{t}_1}{\partial v_1^2\,\partial v_1^2}\right) s^2,$$

with

$$\varrho\,U_{11}^2 = 2\,v_1^2\,\frac{\partial \tilde{t}_1}{\partial v_1^2}.$$

(ii) *Suppose that the body is at rest in a homogeneous configuration with constant and uniform entropy. Then plane transverse principal waves propagate with constant amplitudes.*

(iii) *Suppose that the body is at rest in an undistorted configuration with constant and uniform entropy. Let the constant stretch v characterize this configuration. Then the amplitudes of longitudinal waves of arbitrary shape obey the equation*

$$\frac{ds}{dn} = \frac{\bar{K}}{2}\,s - \frac{1}{2\,U_\parallel^2}\left(1 + \frac{\varphi(v)}{\varrho\,U_\parallel^2}\right) s^2,$$

with

$$\varrho\,U_\parallel^2 = 2\,v_\mathfrak{a}^2\,\frac{\partial \tilde{t}_\mathfrak{a}}{\partial v_\mathfrak{a}^2}\bigg|_{v_\mathfrak{b} = v}, \qquad \mathfrak{b} = 1, 2, 3,$$

---

[36] This extension has been carried out by CHEN [1968, *5*]; including extensive treatments on the propagation properties of waves. CHEN's results have been extended by BOWEN and WANG [1971, *3*] to include material inhomogeneity. Also, CHADWICK and POWDRILL [1965, *1*] have considered the properties of waves within the infinitesimal theory of thermoelasticity.

and

$$\varphi(v) = 4 v_{\mathfrak{a}}^4 \frac{\partial^2 \tilde{t}_{\mathfrak{a}}}{\partial v_{\mathfrak{a}}^2 \partial v_{\mathfrak{a}}^2}\bigg|_{v_{\mathfrak{v}}=v}, \quad \mathfrak{v} = 1, 2, 3.$$

(iv) *Suppose that the body is at rest in an undistorted configuration with constant and uniform entropy. Let the constant stretch v characterize this configuration. Then the amplitudes of transverse waves of arbitrary shape obey the equation*

$$\frac{d\hat{s}^2}{dn} = \overline{K}\,\hat{s}^2.$$

Notice that the above equations are identical in form to the corresponding equations given in Sects. 9 and 10. Thus, for the situations considered, *the thermodynamic properties of the material do not affect the behavior of these waves*; and the results for the purely mechanical theory can be directly carried over to the present situation, provided that the stress is taken at fixed entropy.

IV. One dimensional waves in bodies of material with memory.

12. Acceleration waves in bodies of material with memory. Here, we shall consider the behavior of one dimensional acceleration waves propagating in a homogeneous body of simple material with memory.[37] Such a material is characterized by the response functional $\mathscr{F}(\cdot,\cdot)$; it gives the present value of the stress $T(X, t)$ at the material point X when the past history and the present value of the deformation gradient are known at X, i.e.,

$$T(X, t) = \mathscr{F}\big(F_r^t(X,\cdot), F(X, t)\big). \tag{12.1}$$

Here, we have taken a homogeneous configuration to be the reference, so that $\mathscr{F}(\cdot,\cdot)$ and the density ϱ_R in this configuration are independent of X.

Formula (12.1) says that the present value of the stress at the point X depends in an arbitrary way on the entire history of the deformation gradient at that point. Here, we introduce the assumption that the functional $\mathscr{F}(\cdot,\cdot)$ has fading memory in the sense considered by COLEMAN and NOLL.[38] To state this assumption precisely, let $h(\cdot)$ defined on $(0, \infty)$ be a given fixed influence function, i.e., a positive, continuous, monotone-decreasing function such that

$$\int_0^\infty h^2(s)\, ds < \infty;$$

let the norm $\|f\|$ of a real-valued function f on $(0, \infty)$ be defined by

$$\|f\|^2 = \int_0^\infty |f(s)|^2\, h^2(s)\, ds; \tag{12.2}$$

let $\mathscr{H}$ denote the space of real-valued functions f on $(0, \infty)$ such that $\|f\|$ is finite; and let $\mathscr{H}^+$ denote the cone in $\mathscr{H}$ of strictly positive functions. It is assumed that

[37] The results given in this section are due to COLEMAN and GURTIN [1965, *3*] and COLEMAN, GREENBERG and GURTIN [1966, *3*]. However, our results are somewhat different from theirs; nevertheless, correspondence between the two can be readily established. Experimental evidence of the existence of acceleration waves in polymethyl methacrylate is given, for example, by BARKER and HOLLENBACH [1970, *2*, Figs. 4 and 8].

[38] COLEMAN and NOLL [1960, *1*] and [1961, *1*].

the domain of definition of the functional $\mathscr{F}(\cdot,\cdot)$ is $\mathscr{D}=\mathscr{H}^+\times(0,\infty)$ and that $\mathscr{F}(\cdot,\cdot)$ is of class C^1.

Physically, the fading memory assumption says that the more recent past has more influence on the present value of the stress than does the more distant past. For histories whose present values are the same and whose past are close in the sense of the norm $\|\cdot\|$ (12.2), the present values of the stress are also close, even though the values of the past histories may be quite different for sufficiently large s.

In defining the response functional $\mathscr{F}(\cdot,\cdot)$, we have rendered explicit its dependence on the present value of the deformation gradient. This allows us to make precise the concept of *instantaneous elasticity*, in that we allow for changes in the present value of the deformation gradient while holding its past fixed. As we shall see, this concept is essential to the study of wave propagation.

Since $\mathscr{F}(\cdot,\cdot)$ is of class C^1, it has a continuous partial derivative

$$D_F\mathscr{F}(F_r^t,F)\in(-\infty,\infty) \tag{12.3}$$

with respect to F, and a continuous partial (Fréchet) derivative

$$\delta\mathscr{F}(F_r^t,F;\cdot):\mathscr{H}\to(-\infty,\infty) \tag{12.4}$$

with respect to F_r^t.

$D_F\mathscr{F}(F_r^t,F)$ is simply the ordinary partial derivative and is given by the formula

$$D_F\mathscr{F}(F_r^t,F)=\frac{\partial}{\partial F}\mathscr{F}(F_r^t,F);$$

the Fréchet derivative $\delta\mathscr{F}(F_r^t,F;\cdot)$ is given by

$$\delta\mathscr{F}(F_r^t,F;f)=\frac{d}{d\alpha}\mathscr{F}(F_r^t+\alpha f,F)\big|_{\alpha=0}$$

for all $f\in\mathscr{H}$, provided that $F_r^t+\alpha f\in\mathscr{H}^+$.

Since the functional $\delta\mathscr{F}(F_r^t,F;\cdot)$ is linear, it follows from the Riesz-Fréchet Theorem that there exists a unique function $k\in\mathscr{H}$ such that

$$\delta\mathscr{F}(F_r^t,F;f)=\int_0^\infty k(s)\,f(s)\,h^2(s)\,ds,$$

for all $f\in\mathscr{H}$.

Let $G(s)$ be the unique solution of

$$G_t'(s)=\frac{d}{ds}G_t(s)\equiv k(s)\,h^2(s),$$

with

$$G_t(0)=D_F\mathscr{F}(F_r^t,F). \tag{12.5}$$

Therefore

$$\delta\mathscr{F}(F_r^t,F;f)=\int_0^\infty G_t'(s)\,f(s)\,ds. \tag{12.6}$$

We assume that

$$\lim_{s\to\infty}G_t'(s)=0. \tag{12.7}$$

The function $G_t(\cdot)$ is called the *stress-relaxation function* corresponding to the history (F_r^t,F). Of course, $G_t(\cdot)$ depends on the history (F_r^t,F). This dependence can be made explicit by writing

$$G_t(s)=\mathfrak{G}(F_r^t,F;s). \tag{12.8}$$

We assume that $\mathfrak{G}(\cdot,\cdot;s)$, for each fixed s, is a class C^1 functional over $\mathcal{D}$, and that, for each fixed history (F_r^t, F) in $\mathcal{D}$, $G_t(\cdot) = \mathfrak{G}(F_r^t, F, \cdot)$ is a twice differentiable function on $(0, \infty)$. Putting

$$G_t''(s) = \frac{d}{ds} G_t'(s),$$

we further assume that the mapping $(F_r^t(X,\cdot), F(X,t)) \to G_t''(\cdot)\, h^{-2}(\cdot)$ carries $\mathcal{D}$ into $\mathcal{H}$ and is continuous.

Henceforth, we assume that the response functional $\mathcal{F}(\cdot,\cdot)$ is of class C^2. This in turn insures the existence and continuity of the following partial derivatives:

$$D_F^2 \mathcal{F}(F_r^t, F) \in (-\infty, \infty),$$

$$\delta^2 \mathcal{F}(F_r^t, F; \cdot, \cdot): \mathcal{H} \times \mathcal{H} \to (-\infty, \infty),$$

$$D_F\, \delta \mathcal{F}(F_r^t, F; \cdot): \mathcal{H} \to (-\infty, \infty).$$

We call the number

$$E_t = D_F \mathcal{F}(F_r^t, F) \tag{12.9}$$

the *instantaneous tangent modulus*; it is a measure of the initial slope of the stress-strain law for instantaneous response to strain impulses superimposed on the history (F_r^t, F) at time t. Notice that, by (12.5) and (12.9),

$$G_t(0) = E_t. \tag{12.10}$$

The number

$$\tilde{E}_t = D_F^2 \mathcal{F}(F_r^t, F) \tag{12.11}$$

is called the *instantaneous second-order tangent modulus*; it is a measure of the curvature of the instantaneous stress-strain law.

In addition to the hypothesis (A-1) on the properties of acceleration waves given in Sect. 5, we assume that

(A-2). *The mapping* $(X,t) \to F_r^t(X,\cdot)$ *is a smooth function with respect to the norm* $\|\cdot\|$ (12.2), *in the following sense*:

(i) *For each fixed* (X, t),

$$\lim_{\substack{\alpha\to 0\\ \beta\to 0}} \|F_r^{t+\alpha}(X+\beta,\cdot) - F_r^t(X,\cdot)\| = 0.$$

(ii) *There exist two functions in* $\mathcal{H}$, *denoted by*

$$\frac{\partial}{\partial X} F_r^t(X,\cdot) \quad \text{and} \quad \dot{F}_r^t(X,\cdot),$$

such that for each fixed (X, t)

$$\lim_{\beta\to 0} \left\| \frac{F_r^t(X+\beta,\cdot) - F_r^t(X,\cdot)}{\beta} - \frac{\partial}{\partial X} F_r^t(X,\cdot) \right\| = 0,$$

$$\lim_{\alpha\to 0} \left\| \frac{F_r^{t+\alpha}(X,\cdot) - F_r^t(X,\cdot)}{\alpha} - \dot{F}_r^t(X,\cdot) \right\| = 0.$$

(iii) *For each fixed* (X, t),

$$\lim_{\substack{\alpha\to 0\\ \beta\to 0}} \left\| \frac{\partial}{\partial X} F_r^{t+\alpha}(X+\beta,\cdot) - \frac{\partial}{\partial X} F_r^t(X,\cdot) \right\| = 0,$$

$$\lim_{\substack{\alpha\to 0\\ \beta\to 0}} \|\dot{F}_r^{t+\alpha}(X+\beta,\cdot) - \dot{F}_r^t(X,\cdot)\| = 0.$$

(A-3). *For each fixed* (X, t),

$$\limsup_{s\to\infty} F_r^t(X, s) < \infty .$$

The constitutive relation (12.1) together with the smoothness properties of the functional $\mathscr{F}(\cdot,\cdot)$, the hypotheses (A-1) and (A-2) implies that for $X \neq Y(t)$

$$\dot{T} = \delta\mathscr{F}(F_r^t, F; \dot{F}_r^t) + D_F\mathscr{F}(F_r^t, F)\,\dot{F}, \tag{12.12}$$

which together with (12.5), (12.10), $(5.20)_2$, the smoothness properties of the functional $\mathscr{F}(\cdot,\cdot)$, the hypotheses (A-1) and (A-2) in turn implies that across the wave

$$\begin{aligned}[\dot{T}] &= -\frac{1}{V} E_{t_X}[\ddot{x}]\\ &= -\frac{1}{V} G_{t_X}(0)\,[\ddot{x}],\end{aligned} \tag{12.13}$$

where t_X denotes the time at which the wave passes the material point X; i.e., the mapping $X \to t_X$ is the inverse of the mapping $t \to Y(t)$. In addition,

$$\begin{aligned} E_{t_X} &= D_F\mathscr{F}\big(F_r^{t_X}(X,\cdot), F(X, t_X)\big)\\ &= D_F\mathscr{F}\big(F_r^t(Y(t),\cdot), F(Y(t), t)\big),\\ G_{t_X}(0) &= \mathfrak{G}\big(F_r^{t_X}(X,\cdot), F(X, t_X); 0\big)\\ &= \mathfrak{G}\big(F_r^t(Y(t),\cdot), F(Y(t), t); 0\big).\end{aligned} \tag{12.14}$$

Thus, by (5.23) and (12.13), we have the following expression for the speed of propagation of the acceleration wave:

$$V^2 = \frac{E_{t_X}}{\varrho_R} = \frac{G_{t_X}(0)}{\varrho_R}. \tag{12.15}$$

Formula (12.15) implies that the speed of propagation is real if and only if

$$E_{t_X} = G_{t_X}(0) > 0. \tag{12.16}$$

Henceforth, we assume that (12.16) holds.

Further, differentiating (12.12) with respect to time, we obtain the result

$$\begin{aligned}\ddot{T} &= \delta^2\mathscr{F}(F_r^t, F; \dot{F}_r^t, \dot{F}_r^t) + 2D_F\,\delta\mathscr{F}(F_r^t, F; \dot{F}_r^t)\,\dot{F}\\ &\quad + \frac{\partial}{\partial\tau}\,\delta\mathscr{F}\left(F_r^t, F; \frac{\partial F_r^\tau}{\partial\tau}\right)\bigg|_{\tau=t} + D_F^2\mathscr{F}(F_r^t, F)\,\dot{F}^2 + D_F\mathscr{F}(F_r^t, F)\,\ddot{F},\end{aligned} \tag{12.17}$$

which is valid for $X \neq Y(t)$. Thus, by (12.9), 12.11), the hypotheses (A-1) and (A-2) and the smoothness properties of the functional $\mathscr{F}(\cdot,\cdot)$, formula (12.17) implies that at $X = Y(t)$

$$[\ddot{T}] = \left[\frac{\partial}{\partial\tau}\,\delta\mathscr{F}\left(F_r^t, F; \frac{\partial F_r^\tau}{\partial\tau}\right)\bigg|_{\tau=t}\right] + 2I_{t_X}[\dot{F}] + \tilde{E}_{t_X}[\dot{F}^2] + E_{t_X}[\ddot{F}], \tag{12.18}$$

where

$$\begin{aligned} I_{t_X} &= D_F\,\delta\mathscr{F}\big(F_r^{t_X}(X,\cdot), F(X, t_X); \dot{F}_r^{t_X}(X,\cdot)\big)\\ &= D_F\,\delta\mathscr{F}\big(F_r^t(Y(t),\cdot), F(Y(t), t); \dot{F}_r^t(Y(t),\cdot)\big),\\ \tilde{E}_{t_X} &= D_F^2\mathscr{F}\big(F_r^{t_X}(X,\cdot), F(X, t_X)\big)\\ &= D_F^2\mathscr{F}\big(F_r^t(Y(t),\cdot), F(Y(t), t)\big).\end{aligned} \tag{12.19}$$

In order to evaluate the first term of the right-hand member of (12.18), we note that (12.6) and the hypothesis (A-2) permit us to write

$$\frac{\partial}{\partial\tau}\,\delta\mathscr{F}\left(F_r^t, F; \frac{\partial F_r^\tau}{\partial\tau}\right)\bigg|_{\tau=t} = \frac{\partial}{\partial\tau}\int\limits_0^\infty G_t'(s)\,\frac{\partial}{\partial\tau}F(\tau-s)\,ds\bigg|_{\tau=t}, \tag{12.20}$$

for $X \neq Y(t)$. Now

$$\begin{aligned}\int\limits_0^\infty G_t'(s)\,\frac{\partial}{\partial\tau}F(\tau-s)\,ds &= -\int\limits_0^\infty G_t'(s)\,\frac{\partial}{\partial s}F(\tau-s)\,ds\\ &= -G_t'(s)\,F(\tau-s)\Big|_0^\infty + \int\limits_0^\infty G_t''(s)\,F(\tau-s)\,ds.\end{aligned} \tag{12.21}$$

Next, since the integral

$$\int\limits_0^\infty G_t'(s)\,F_r^\tau(s)\,ds$$

is finite and the function $G_t'(\cdot)\,F_r^\tau(\cdot)$ is continuous, it follows from the hypothesis (A-3) and the properties of $G_t'(\cdot)$ (12.7), that

$$\lim_{s\to\infty} G_t'(s)\,F_r^\tau(s) = 0. \tag{12.22}$$

Thus, (12.20) together with (12.21) and (12.22) reduces to

$$\frac{\partial}{\partial\tau}\,\delta\mathscr{F}\left(F_r^t, F; \frac{\partial F_r^\tau}{\partial\tau}\right)\bigg|_{\tau=t} = G_t'(0)\,\dot F + \int\limits_0^\infty G_t''(s)\,\dot F_r^t(s)\,ds. \tag{12.23}$$

Hence, (12.23) together with the assumed smoothness of $\mathfrak{G}(\cdot,\cdot,\cdot)$, the hypotheses (A-1) and (A-2) yields the desired result

$$\left[\frac{\partial}{\partial\tau}\,\delta\mathscr{F}\left(F_r^t, F; \frac{\partial F_r^\tau}{\partial\tau}\right)\bigg|_{\tau=t}\right] = G_{t_X}'(0)\,[\dot F], \tag{12.24}$$

where

$$\begin{aligned}G_{t_X}'(0) &= \mathfrak{G}'(F_r^{t_X}(X,\cdot), F(X,t_X); 0)\\ &= \mathfrak{G}'(F_r^t(Y(t),\cdot), F(Y(t),t); 0).\end{aligned} \tag{12.25}$$

Consequently, (12.18) becomes

$$[\ddot T] = (G_{t_X}'(0) + 2I_{t_X})\,[\dot F] + \widetilde{E}_{t_X}[\dot F^2] + E_{t_X}[\ddot F]. \tag{12.26}$$

Substitution of this result into (5.25) and utilizing $(5.20)_2$ and (12.15) completes the derivation of

Theorem 12.1. *The amplitude of an acceleration wave propagating in a body of simple material with fading memory obeys the equation*

$$\frac{\delta_1 a}{\delta t} = -\mu_{t_X}\,a + \beta_{t_X}\,a^2, \tag{12.27}$$

with

$$\mu_{t_X} \equiv \frac{1}{4G_{t_X}(0)}\,\frac{\delta_1 G_{t_X}(0)}{\delta t} - \frac{1}{2G_{t_X}(0)}\left(G_{t_X}'(0) + 2I_{t_X} + 2\dot F^+\widetilde{E}_{t_X}\right), \tag{12.28}$$

and

$$\beta_{t_X} \equiv -\frac{\widetilde{E}_{t_X}}{2G_{t_X}(0)\,V}. \tag{12.29}$$

In (12.28) *and* (12.29), $G_{tX}(0)$, $G'_{tX}(0)$, I_{tX} *and* $\widetilde{E}_{tX}$ *are given by* $(12.14)_2$, (12.25) *and* $(12.19)_{1,2}$, *respectively.*

Formula (12.27) is a differential equation of the Bernoulli type.[39] The coefficients μ_{tX} and β_{tX} are functions of time. In general, we would expect that

$$\mu_{tX}(t) \neq 0, \qquad \beta_{tX}(t) \neq 0. \tag{12.30}$$

Notice that the coefficient μ_{tX} depends on the dissipative as well as the instantaneous elastic properties of the material; it also depends on the deformation rate just ahead of the wave. On the other hand, in view of (12.10) and (12.15), the coefficient β_{tX} depends on the instantaneous elastic properties of the material alone. This latter observation permits us to say a great deal about the behavior of the amplitude a; these results are given in the following section.

Finally, on integrating (12.27), we have

Remark 12.1. The amplitude of an acceleration wave propagating in a body of simple material with fading memory is given by the explicit formula

$$a(t) = \frac{e^{-\int_0^t \mu_{tX}(\tau)\,d\tau}}{\dfrac{1}{a(0)} - \displaystyle\int_0^t \beta_{tX}(\tau)\, e^{-\int_0^\tau \mu_{tX}(s)\,ds}\, d\tau}, \qquad 0 \leqq t < \infty. \tag{12.31}$$

13. The local and global behavior of the amplitudes of acceleration waves. In the preceding section, we have shown that the amplitudes of acceleration waves propagating in bodies of simple material with fading memory obey the well-known Bernoulli equation. However, this is not the only circumstance for which this is true. We shall see that, in our subsequent considerations of the properties of acceleration waves in various other classes of materials, the Bernoulli equation is always the governing differential equation for the amplitudes. That is, the amplitudes always obey the equation[40]

$$\frac{\delta_1 a}{\delta t} = -\mu a + \beta a^2, \qquad t \in [0, \infty). \tag{13.1}$$

In general, the coefficients μ and β are functions of time t. A careful examination of these coefficients reveals that there are two common features. Briefly,

(i) the coefficient μ depends on the class of materials under consideration and the consequences of the physical assumptions introduced regarding the conditions of the material body ahead of the waves, and

(ii) the coefficient β depends only on the elastic response of the material. In view of this latter observation, we may examine the general behavior of the amplitudes of acceleration waves once and for all. Indeed, this study has been carried out by BAILEY and CHEN.[41] Their results, which are given in the remainder

[39] Notice that (12.27) with (12.28) and (12.29) is different from that given by COLEMAN and GURTIN [1965, *3*, Eq. (2.38)]. However, if we assume that the functional $\mathfrak{G}'(\cdot, \cdot; 0)$ defined on $\mathscr{D}$ is differentiable, then we can readily show that the two equations have the same reduced form.

[40] While the subsequent discussions of this article are restricted to the one dimensional context, this equation also governs the behavior of the amplitudes of waves in three dimensional situations; see, e.g., BOWEN and CHEN [1972, *2*, Eq. (4.13)] who considered the behavior of waves in bodies of anisotropic thermoelastic material with internal state variables, and BOWEN and WANG [1970, *3*, Eq. (6.11)] who considered the behavior of waves in inhomogeneous isotropic elastic bodies.

[41] BAILEY and CHEN [1971, *1*].

of this section, not only are applicable to the specific case considered in Sect. 12, but also they apply to all the other cases which we shall consider later on in this article.

Remark 13.1. Since the differential equation (13.1) *also governs the behavior of the amplitude of acceleration waves in three dimensional situations and since the coefficients* μ *and* β *also have the properties stated above, the results given in the remainder of this section, though here restricted to the one dimensional context, are also valid in the general setting. We simply need to interpret the terms accordingly.*

In the one dimensional context, the coefficient β of (13.1) always has the form

$$\beta = -\frac{\tilde{E}}{2EV}, \tag{13.2}$$

where E and $\tilde{E}$ are, respectively, the tangent modulus and the second-order tangent modulus of the elastic stress-strain law at the wave, and V denotes the acceleration wave speed. Since we always assume that

$$\left.\begin{array}{l} E(t) > 0 \\ \tilde{E}(t) \neq 0 \\ V(t) > 0 \end{array}\right\}, \quad 0 \leqq t < \infty, \tag{13.3}$$

we see that, by (13.2),

$$\operatorname{sgn} \beta(t) = -\operatorname{sgn} \tilde{E}(t), \quad 0 \leqq t < \infty. \tag{13.4}$$

As we shall see, formula (13.4) is crucial to the derivations of our results.

We assume that for all $t \in [0, \infty)$ either

$$\operatorname{sgn} \tilde{E}(t) = +1, \tag{13.5}$$

or

$$\operatorname{sgn} \tilde{E}(t) = -1. \tag{13.6}$$

Physically, condition (13.5) states that the elastic stress-strain law is always concave from above; condition (13.6) states that it is always concave from below.

By (13.2) and (13.3),

$$\beta(t) \neq 0, \quad 0 \leqq t < \infty; \tag{13.7}$$

and, if (13.5) holds,

$$\operatorname{sgn} \beta(t) = -1, \tag{13.8}$$

or if (13.6) holds,

$$\operatorname{sgn} \beta(t) = +1. \tag{13.9}$$

We assume that the coefficients

$$\mu \text{ and } \beta$$

are integrable on every finite sub-interval of $[0, \infty)$, and, as is physically reasonable, that

$$\liminf_{t\to\infty} |\beta(t)| \neq 0. \tag{13.10}$$

Formula (13.10) is equivalent to the assertion that β is bounded away from zero; it is only needed in the proofs of Theorem 13.2, part (V) of Theorem 13.3 and part (ii) of Theorem 13.4. It should be pointed out that part (ii) of Theorem 13.4 is valid also under the much weaker condition $\int_0^\infty |\beta(t)|\, dt = \infty$.

Finally, we observe that if a solution of (13.1) is zero at some time, then it is identically zero. Therefore, we shall only consider these solutions for which

$$a(t) \neq 0, \qquad 0 \leqq t < \infty.$$

In view of the physical assumptions (13.5) and (13.6), the differential equation (13.1) and the assumption (13.10) tell us a great deal about the local and global behavior of the amplitudes of acceleration waves. Before proceeding to state these results, let us define the quantity λ by the relation

$$\lambda = \frac{\mu}{\beta}. \tag{13.11}$$

For the *local behavior* of the amplitudes, we have the following:

Theorem 13.1. *Consider the differential equation* (13.1).

(i) *At any instant, if either* $\tilde{E}(t) < 0$ *and* $a(t) < \lambda(t)$, *or* $\tilde{E}(t) > 0$ *and* $a(t) > \lambda(t)$, *then*

$$\frac{\delta_1 |a(t)|}{\delta t} < 0.$$

(ii) *At any instant,* $a(t) = \lambda(t)$ *if and only if*

$$\frac{\delta_1 a(t)}{\delta t} = 0.$$

(iii) *At any instant, if either* $\tilde{E}(t) < 0$ *and* $a(t) > \lambda(t)$, *or* $\tilde{E}(t) > 0$ *and* $a(t) < \lambda(t)$, *then*

$$\frac{\delta_1 |a(t)|}{\delta t} > 0.$$

For the *global behavior* of the amplitude, we have the following three theorems:

Theorem 13.2. *Consider the differential equation* (13.1). *Let* $\operatorname{sgn} a(0) = \operatorname{sgn} \tilde{E}(t)$. *If* λ *is bounded* $\begin{Bmatrix} above \\ below \end{Bmatrix}$ *or tends to a* $\begin{Bmatrix} non\text{-}negative \\ non\text{-}positive \end{Bmatrix}$ *finite or infinite limit* L, *the same is true for any solution* $\begin{Bmatrix} a(t) > 0 \\ a(t) < 0 \end{Bmatrix}$.

Theorem 13.3. *Consider the differential equation* (13.1). *Let* μ and β *be integrable on every finite sub-interval of* $[0, \infty)$; *and let* $a_1(t)$, $a_2(t)$ *be any two solutions of* (13.1) *with* $\operatorname{sgn} a_1(0) = a_2(0) = \operatorname{sgn} \tilde{E}(t)$.

(i) *If* $\lim_{t\to\infty} |a_1(t)| = \infty$, *then* $\lim_{t\to\infty} |a_2(t)| = \infty$.

(ii) *If* $a_1(t)$ *is bounded, then* $a_2(t)$ *is also bounded.*

(iii) *If* $\liminf_{t\to\infty} |a_1(t)| = 0$, *then* $\liminf_{t\to\infty} |a_2(t)| = 0$, *and* $\liminf_{t\to\infty} |a_1(t) - a_2(t)| = 0$.

(iv) *If* $\lim_{t\to\infty} |a_1(t)| = 0$, *then* $\lim_{t\to\infty} |a_2(t)| = 0$, *and* $\lim_{t\to\infty} |a_1(t) - a_2(t)| = 0$.

(v) *If* $a_1(t)$ *is bounded and bounded away from zero, then* $a_2(t)$ *is also bounded and bounded away from zero, and* $\lim_{t\to\infty} |a_1(t) - a_2(t)| = 0$.

Theorem 13.4. *Consider the differential equation* (13.1). *Let* $\operatorname{sgn} a(0) = -\operatorname{sgn} \tilde{E}(t)$. *Let* μ *and* β *be integrable on every finite sub-interval of* $[0, \infty)$; *and let*

$$\alpha = \frac{1}{\int_0^\infty |\beta(t)|\, e^{-\int_0^t \mu(\tau)\, d\tau}\, dt}. \tag{13.12}$$

(i) *If* $|a(0)|>\alpha$, *there exists a unique finite time* $t_\infty>0$ *such that*

$$\int_0^{t_\infty} \beta(t)\, e^{-\int_0^t \mu(\tau)\,d\tau}\, dt = \frac{1}{a(0)}, \tag{13.13}$$

and

$$\lim_{t\to t_\infty} |a(t)| = \infty.$$

(ii) *If* $|a(0)|<\alpha$, *then*

$$\liminf_{t\to\infty} |a(t)| = 0.$$

The derivation of the local results, Theorem 13.1, follows directly from the differential equation (13.1) and (13.4); but the derivations of the global results, stated in Theorems 13.2, 13.3 and 13.4, are quite technical. The proofs of these theorems, however, are given in the Appendix.

Statements (i) and (iii) of Theorem 13.1 say, in particular, that when $\operatorname{sgn} a(t) = \operatorname{sgn} \lambda(t) = \operatorname{sgn} \tilde{E}(t)$, $|a(t)|$ is increasing or decreasing according as $|a(t)|$ is less than or greater than $|\lambda(t)|$; and when $\operatorname{sgn} a(t) = \operatorname{sgn} \lambda(t) = -\operatorname{sgn} \tilde{E}(t)$, $|a(t)|$ is increasing or decreasing according as $|a(t)|$ is greater than or less than $|\lambda(t)|$.

Theorem 13.2 says that if λ is well behaved, then the eventual behavior of the amplitudes is the same as that of λ. However, even though the behavior of λ is not known, Theorem 13.3 says that if the amplitude of a wave behaves one way, then after a sufficiently long time the amplitudes of all other waves behave the same way.

In view of Theorem 13.4, we may call the number α, defined by (13.12), the *critical initial amplitude* for acceleration waves. That is, if the initial amplitude of a wave is greater in absolute value than the critical amplitude, the amplitude of the wave will grow to infinity within a finite time;[42] or if the initial amplitude is less in absolute value than the critical amplitude, the amplitude of the wave will become arbitrarily small.[43] The former case, of course, suggests the formation of a shock.

Now, let us examine in more detail the asymptotic behavior of the solutions of (13.1) under the conditions of Theorem 13.4. These results are due to BAILEY and CHEN;[44] the proofs are given in the Appendix.

Theorem 13.5. *Let the conditions of Theorem 13.4 hold.*

(i) *If* $|a(0)|>\alpha$ *and* $\beta(t)$ *is continuous from below at* t_∞, *then*

$$a(t) = \frac{1}{\beta(t_\infty)(t-t_\infty)}\bigl(1+o(1)\bigr) \quad as \quad t\to t_\infty, \tag{13.14}$$

where $\beta(t_\infty) = \lim_{t\to t_\infty} \beta(t)$.

(ii) *If* $|a(0)|>\alpha$, $\beta(0)\neq 0$, *and* $\beta(t)$ *is differentiable with* $\left|\frac{\beta'(t)}{\beta(t)} - \mu(t)\right|$ *bounded near zero, then*

$$t_\infty = \frac{1}{a(0)\,\beta(0)}\bigl(1+o(1)\bigr) \quad as \quad |a(0)|\to\infty. \tag{13.15}$$

[42] While COLEMAN, GREENBERG and GURTIN [1966, *3*] showed that it is possible for the amplitude of a wave to become infinite within a finite length of time, they did not define the critical initial amplitude α in the sense of (13.12).

[43] The critical initial amplitude α, defined by (13.12), may sometimes be equal to zero; see, e.g., Sects. 21 and 25 of this article.

[44] BAILEY and CHEN [1972, *1*].

(iii) *If* $|a(0)| < \alpha$, *then*

$$a(t) = \frac{a(0)\, e^{-\int_0^t \mu(\tau)\, d\tau}}{1 - \frac{|a(0)|}{\alpha}} \left(1 + o(1)\right) \tag{13.16}$$

as $t \to \infty$ *or as* $a(0) \to 0$.

Our results on the asymptotic behavior of the solutions of (13.1) state that *if the initial amplitude of a wave is greater in absolute value than the critical amplitude then for times close to* t_∞ *the behavior of the wave is dominated by the value of the coefficient* β *at* t_∞, *and, in addition, if the initial amplitude is very large in absolute value, then* t_∞ *is dominated by the initial value of the coefficient* β. As we have remarked earlier, the coefficient β depends on the elastic response of the material alone. Thus, in these circumstances, it is *the elastic response of the materials that predominates*. In particular, we have the following corollary:

Corollary: *Let the conditions of parts* (i) *and* (ii) *of Theorem 13.5 hold simultaneously. Then*

$$a(t) = \frac{1}{\beta\left(\frac{1}{a(0)\,\beta(0)}\right)(t - t_\infty)} \left(1 + o(1)\right) \tag{13.17}$$

as $t \to t_\infty$ *and* $|a(0)| \to \infty$, *where* t_∞ *is given by* (13.15).

On the other hand, *if the initial amplitude of a wave is less in absolute value than the critical amplitude, then for sufficiently large times or for an initial amplitude which is sufficiently small in absolute value the behavior of the wave is dominated by the coefficient* μ. Here, we recall that the coefficient μ in (13.1) depends on the material under consideration and the consequences of the physical assumptions introduced regarding the condition of the body ahead of the wave.

In the special case when μ and β are finite non-zero constants, say μ_0 and β_0, it follows from (13.11) and (13.12) that

(i) if $\mu_0 > 0$, then

$$\alpha = \frac{\mu_0}{|\beta_0|} = |\lambda_0|, \qquad \lambda_0 \equiv \frac{\mu_0}{\beta_0}; \tag{13.18}$$

(ii) if $\mu_0 < 0$, then

$$\alpha = 0. \tag{13.19}$$

While (i) states that if $\mu_0 > 0$, there is indeed a *non-zero* critical amplitude,[45] (ii) states that if $\mu_0 < 0$, all waves with non-zero initial amplitudes satisfying the conditions of Theorem 13.4 will have "infinite amplitudes" within finite times. In either case, it follows from (13.13) that the time t_∞ it takes a wave of a certain initial amplitude, which satisfies the conditions of Theorem 13.4, to "blow up" in this sense is given by

$$t_\infty = -\frac{1}{\mu_0} \ln\left(1 - \frac{\lambda_0}{a(0)}\right). \tag{13.20}$$

Further, the asymptotic results (13.14), (13.15), (13,16) and (13.17) will have certain obvious reduced forms in this special case. In this regard, also refer to Sect. 14 of this article.

---

[45] The number $|\lambda_0|$, defined by (13.18), is what COLEMAN and GURTIN [1965, *3*] called critical amplitude in their considerations of the properties of acceleration waves in bodies of material with fading memory which are initially at rest in homogeneous configurations; cf. Sect. 14 of this article.

Finally, when $\operatorname{sgn} a(0) = \operatorname{sgn} \lambda(t) = -\operatorname{sgn} \tilde{E}(t)$, and if $|\lambda(0)| < |a(0)| < \alpha$, it follows from the results of Theorems 13.1 and 13.4 that $|a(t)|$ will increase initially, but eventually it will become arbitrarily small. In order to illustrate this observation, let

$$\mu(t) = (1+t),$$
$$\beta(t) = -1.$$

By (13.11),

$$\lambda(t) = -(1+t),$$

so that

$$\lambda(0) = -1;$$

and, by (13.12), the critical initial amplitude is approximately

$$\alpha \doteqdot 1.54.$$

The solutions of the differential equation (13.1) corresponding to three different initial values, i.e.,

$$|a_1(0)| < |\lambda(0)|,$$
$$|\lambda(0)| < |a_2(0)| < \alpha,$$
$$|a_3(0)| > \alpha,$$

are shown graphically in the figure.[46]

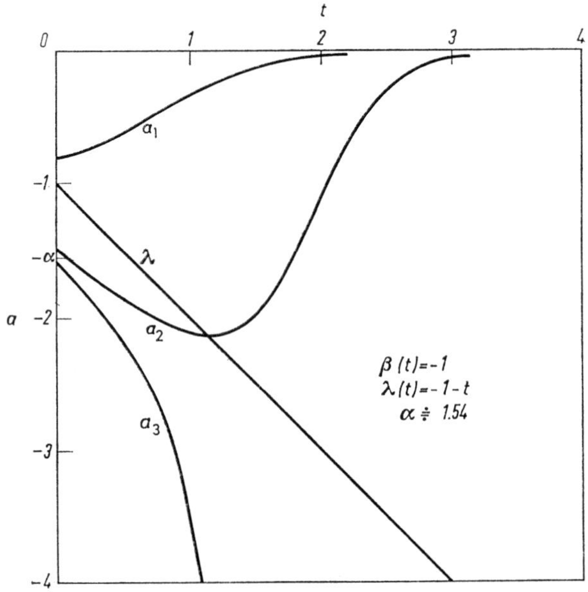

Fig. 1.

14. Acceleration waves entering homogeneously deformed bodies of material with memory. In Sect. 12 we have derived the differential equation governing the amplitudes of acceleration waves propagating in bodies of simple material with fading memory. The general behavior of the amplitudes as predicted by this equation

[46] This example is furnished by my colleague P. B. BAILEY.

has been examined in Sect. 13. Here, we shall consider a specific application of formula (12.27) and examine the consequences.

To be more precise, we shall assume that since $t=0$ an acceleration wave has been propagating into a region which previously has always been at rest in a fixed homogeneous configuration;[47] i.e., since $V(t)>0$, we suppose that for $t\geqq 0$ and $X\geqq Y(t)$ the motion reduces to

$$\chi(X,\tau)=F_0X+X_0, \qquad \tau\in(-\infty,t], \tag{14.1}$$

where F_0 and X_0 are constants. Let the function $F^\dagger(\cdot)$ with values

$$F_0^\dagger(s)=F_0, \qquad s\in(0,\infty),$$

denote the constant past history; then for either $X\geqq Y(0)$ and $t\leqq t_X$ or $t\geqq 0$ and $X\geqq Y(t)$, (14.1) implies that

$$\left(F_r^t(X,\cdot),\,F(X,t)\right)=\left(F_0^\dagger(\cdot),\,F_0\right). \tag{14.2}$$

Hence (14.2) together with the hypotheses (A-1) and (A-2), $(12.14)_2$, $(12.19)_2$ and (12.25) implies that

$$G_{t_X}(0)=G_0(0), \qquad G'_{t_X}(0)=G'_0(0), \qquad \tilde{E}_{t_X}=\tilde{E}_0, \quad \text{for} \quad t\geqq 0, \tag{14.3}$$

where

$$\begin{aligned} G_0(0)&=\mathfrak{G}(F_0^\dagger,F_0,0)=\text{const},\\ G'_0(0)&=\mathfrak{G}'(F_0^\dagger,F_0,0)=\text{const},\\ \tilde{E}_0&=D_F^2\mathscr{F}(F_0^\dagger,F_0)=\text{const}. \end{aligned} \tag{14.4}$$

Notice that, by (12.15), (14.3) and (14.4), the speed of propagation is now independent of time; it is given by

$$V_0^2=\frac{G_0(0)}{\varrho_R}. \tag{14.5}$$

Further, (14.1) implies that

$$\dot{F}(X,t-s)=0,$$

for $X\geqq Y(0)$ and $t<t_X$ or $t\geqq 0$ and $X\geqq Y(t)$, and $s\in(0,\infty)$. Hence by the hypothesis (A-2), $\dot{F}^{t_X}(X,\cdot)$ is a constant function with value zero; consequently, I_{t_X}, defined by $(12.19)_1$, reduces to

$$I_{t_X}\equiv 0, \quad \text{for} \quad t\geqq 0. \tag{14.6}$$

Finally, (14.2) also implies that

$$\dot{F}^+=0, \quad \text{for} \quad t\geqq 0. \tag{14.7}$$

Remark 14.1. The above assumption together with the hypothesis (A-1) *imply that the mapping* $(X,t)\to F_r^t(X,\cdot)$ *is a smooth function with respect to the norm* $\|\cdot\|$ (12.2).

This result, which is the hypothesis (A-2), is due to COLEMAN and GURTIN.[48] The interested reader is referred to their article for the details of the proof.

[47] The results given in this section are due to COLEMAN and GURTIN [1965, *3*].
[48] COLEMAN and GURTIN [1965, *3*].

It follows from the preceding results that the amplitude equation (12.27) reduces to[49]

$$\frac{\delta_1 a}{\delta t} = -\mu_0 a + \beta_0 a^2, \tag{14.8}$$

where, by (12.28), (12.29), (14.3) through (14.7),

$$\mu_0 = -\frac{G_0'(0)}{2G_0(0)} = \text{const}, \quad \beta_0 = -\frac{\tilde{E}_0}{2G_0(0)\,V_0} = \text{const}. \tag{14.9}$$

On integrating (14.8), we obtain

Theorem 14.1. *Consider an acceleration wave which since $t=0$ has been propagating into a region that has always been at rest in a homogeneous configuration with deformation gradient F_0. Then for $t \geqq 0$, its speed V_0 is a constant given by* (14.5), *and its amplitude a has the following explicit dependence on t:*

$$a(t) = \frac{\lambda_0}{\left(\frac{\lambda_0}{a(0)} - 1\right) e^{\mu_0 t} + 1}, \tag{14.10}$$

with

$$\lambda_0 \equiv \frac{\mu_0}{\beta_0} = \frac{G_0'(0)\,V_0}{\tilde{E}_0}. \tag{14.11}$$

For a given material μ_0 and λ_0 are constants depending only on F_0.

In the physical application of the preceding results, we expect to have[50]

$$G_0(0) > 0, \quad G_0'(0) \leqq 0, \quad \tilde{E}_0 \neq 0.$$

When $G_0'(0) = 0$, which is the case when there is no instantaneous mechanical dissipation, (14.8), with (14.5) and (14.9), reduces to

$$\frac{\delta_1 a}{\delta t} = -\frac{\tilde{E}_0}{2\varrho_R V_0^3} a^2,$$

which has the solution

$$a(t) = \frac{1}{\frac{1}{a(0)} + \frac{\tilde{E}_0}{2\varrho_R V_0^3} t}. \tag{14.12}$$

Notice that formula (14.12) is identical in form to formulae (6.16) and (9.13), those for plane longitudinal waves in anisotropic and isotropic elastic bodies.[51] By (14.12), we have

Remark 14.2. When there is no instantaneous mechanical dissipation, i.e., $G_0'(0) = 0$, the behavior of the amplitude of an acceleration wave propagating in a body of simple material with fading memory is the same as that of a wave propagating in an elastic body. Indeed, if $\operatorname{sgn} a(0) = \operatorname{sgn} \tilde{E}_0$, then $a(t) \to 0$ monotonically as $t \to \infty$; and if $\operatorname{sgn} a(0) = -\operatorname{sgn} \tilde{E}_0$, then $|a(t)| \to \infty$ monotonically within a finite time t_∞, given by

$$t_\infty = -\frac{2\varrho_R V_0^3}{a(0)\,\tilde{E}_0}.$$

[49] Within the theory of finite linear viscoelasticity, a relation of the form (14.8) is given by CHEN [1965, *2*]. Similar results for waves of arbitrary shape in bodies of finite linear viscoelastic material are given by CHEN [1966, *2*], and in bodies of viscoelastic material of the integral type at rest in natural states by VARLEY [1965, *11*].

[50] GURTIN and HERRERA [1965, *8*] have shown that a material with $G_0'(0) > 0$ would not be dissipative in the sense of their definition. In this regard, also refer to BREUER and ONAT [1962, *1*] and SHU and ONAT [1966, *8*].

[51] In this regard, also refer to Sect. 18 of this article.

When $G_0'(0) < 0$, $(14.9)_1$ and (14.11) imply that

$$\mu_0 > 0, \qquad \operatorname{sgn} \lambda_0 = -\operatorname{sgn} \tilde{E}_0. \tag{14.13}$$

Since $\mu_0 > 0$, it follows from (13.14) that there is indeed a *non-zero* critical initial amplitude,[52] given by $|\lambda_0|$. Hence, by $(14.13)_2$, we have[53]

Theorem 14.2. *Let Theorem 14.1 hold, and suppose that*

$$G_0(0) > 0, \qquad G_0'(0) < 0, \qquad \tilde{E}_0 \neq 0.$$

(i) *If either* $\operatorname{sgn} a(0) = \operatorname{sgn} \tilde{E}_0$, *or* $\operatorname{sgn} a(0) = -\operatorname{sgn} \tilde{E}_0$ *and* $|a(0)| < |\lambda_0|$, *then* $a(t) \to 0$ *monotonically as* $t \to \infty$.

(ii) *If* $a(0) = \lambda_0$, *then* $a(t) = a(0)$.

(iii) *If* $\operatorname{sgn} a(0) = -\operatorname{sgn} \tilde{E}_0$ *and* $|a(0)| > |\lambda_0|$, *then* $|a(t)| \to \infty$ *monotonically within a finite time* t_∞ *given by*

$$t_\infty = -\frac{1}{\mu_0} \ln\left(1 - \frac{\lambda_0}{a(0)}\right). \tag{14.14}$$

The results given in the preceding theorem are direct consequences of (14.13) and those on the general behavior of the amplitudes of acceleration waves given in Sect. 13. We omit the straightforward arguments leading to the statement of the above results.

In view of Theorem 14.2, we see that *the presence of instantaneous mechanical dissipation*, manifested by a strictly negative value for $G_0'(0)$, *does not imply that an acceleration wave propagating into a homogeneous region must always be damped out.* In other words, in the present theory including mechanical dissipation and non-linear instantaneous elastic response, either the damping effects of dissipation predominate or the reinforcing effects of non-linear elastic response predominate, depending on whether the initial amplitude of the wave is less than or greater than a certain critical value.[54]

Now, if Δ, defined by the relation

$$\Delta = \frac{\lambda_0}{a(0)},$$

is small in magnitude compared with unity, then, by (13.20), (14.5) and $(14.9)_2$, we see that

$$t_\infty \doteqdot -\frac{2\varrho_R V_0^3}{a(0)\tilde{E}_0}.$$

In view of this result and Remark 14.2, we have

Remark 14.3. Let Theorem 14.1 hold. When the initial amplitude of a wave satisfying the conditions of Theorem 14.2 is large compared with the critical initial amplitude, the time t_∞ *it takes the amplitude of the wave to approach* ∞ *depends on the instantaneous elastic response alone; it is the same as that predicted by elasticity theory, not including dissipative effects.*

Thus, *for waves whose initial amplitudes are large* compared with the critical initial amplitude *we shall not be able to ascertain the dissipative effects of the materials*

[52] As we shall see, the existence of a non-zero critical initial amplitude is not unique to the present situation, cf. Sects. 21 and 25 of this article.

[53] Within the theory of finite linear viscoelasticity, PIPKIN [1966, *7*] has exhibited explicit solutions in which steady acceleration waves occur.

[54] In this regard, also refer to TRUESDELL [1969, *2*, p. 77].

under consideration by observing the behavior of the waves; *the instantaneous elastic response of the materials suffices completely to predict their behavior.*

Finally, it is of interest to point out that if δ, defined by the relation

$$\delta = \frac{a(0)}{\lambda_0}, \tag{14.15}$$

is small in magnitude compared with unity, (14.10) yields

$$a(t) = a(0) \exp\left\{\frac{G_0'(0)}{2G_0(0)}\, t\right\} \left(1 + O(\delta)\right), \tag{14.16}$$

where $O(\delta)$ is uniform in t. To within an error of order $O(\delta)$, *acceleration waves of small amplitudes decay exponentially.*

15. Thermodynamic influences on acceleration waves in bodies of material with memory. In Sect. 12 we have derived the differential equation which the amplitude of an acceleration wave, propagating in a body of simple material with fading memory, must obey, and in which we neglected to consider the influence of thermodynamics. Thermodynamic influences can indeed be included by allowing the present value of the stress to depend not only on the history of the deformation gradient but also on the history of a thermodynamic variable such as the absolute temperature or the entropy. No restrictions on the influences of the histories of the deformation gradient and the thermodynamic variable on the stress, other than those which are consequences of the laws of thermodynamics and the principle of fading memory, need be imposed. That is, we only require that the materials obey COLEMAN's theory of thermodynamics of materials with fading memory.[55]

Within this general class of materials, we can consider two distinct types: The first, called conductors of heat, has the property that the derivative of the heat flux with respect to the present temperature gradient is always strictly negative; the second, called non-conductors of heat, consists of materials for which the heat flux is always zero. Of course, the heat flux depends also on the histories of the deformation gradient and the absolute temperature or the entropy.

This extension has been carried out by COLEMAN and GURTIN, in that they have derived the differential equations which the amplitudes of acceleration waves propagating in conductors and non-conductors of heat must obey.[56] Here, we shall not repeat their analysis but shall simply point out the general features of their results. The interested reader may refer to their article for the details.

First, they showed that the amplitudes of acceleration waves propagating in both a conductor and a non-conductor of heat again obey the Bernoulli equation, i.e., (13.1) with β given by (13.2). However, for conductors of heat, the coefficient β depends on the instantaneous isothermal elastic response of the material; and, for non-conductors of heat, the coefficient β depends on the instantaneous isentropic elastic response of the material. In either case, the coefficient μ depends on the dissipative properties, the thermodynamic properties and the instantaneous elastic properties of the material; it also depends on the properties of the motion just ahead of the waves. In view of this observation, we see that the results on the local and global behavior of the amplitudes of accelera-

[55] COLEMAN [1964, *1*].

[56] Cf. COLEMAN and GURTIN [1965, *4*, Eqs. (6.17) and (10.18)]. They have also presented extensive treatments on the propagation properties of waves; these results of theirs are beyond the scope of this article.

tion waves, given in Sect. 13, are applicable to these situations provided that we interpret the coefficients μ and β of the Bernoulli equation (13.1) accordingly.

Second, they showed that if since $t=0$ an acceleration wave has been propagating into a non-conductor of heat which has always been at rest in a fixed homogeneous configuration with constant and uniform entropy, the coefficients μ and β are constants. More importantly, μ now depends on the dissipative properties and the instantaneous elastic properties of the material alone, independent of its thermodynamic properties. That is, as in the special case of elastic materials, the thermodynamic properties of simple materials with fading memory which do not conduct heat have no effect on the behavior of the amplitudes of acceleration waves. Finally, we should point out that the results given in Sect. 14 for the purely mechanical case can be directly carried over to the present situation provided that the stress-relaxation function and the instantaneous second-order tangent modulus are taken at the fixed entropy history ahead of the waves.

16. Shock waves entering unstrained bodies of material with memory. In our studies on the behavior of the amplitude of an acceleration wave propagating in a homogeneous body of simple material with memory, we have shown that whenever the initial amplitude $a(0)$ of a wave and the instantaneous second-order tangent modulus have opposite signs and $|a(0)|$ is greater than a certain critical value, the amplitude of the wave will become infinite within a finite length of time. This suggests the formation of a shock.

Here, we shall examine the behavior of one dimensional shock waves propagating in homogeneous bodies of simple material with memory.[57] In the present situation, it is more convenient to presume that the present value of the stress $T(X,t)$ is determined by the strain history $\varepsilon^t(X,\cdot)$:

$$T(X,t)=\mathscr{S}\left(\varepsilon^t(X,\cdot)\right). \tag{16.1}$$

Here, as in Sect. 12, we have taken a homogeneous configuration to be the reference.

We shall restrict our discussion to waves propagating in regions that previously have been unstrained at all past times. Thus the values of strain histories $\varepsilon^t(s)$ we consider vanish for sufficiently large s, say $s>s_0$. Let $\mathscr{D}$ be the normed space of integrable functions f on $[0,\infty)$ that vanish on (s_0,∞) with norm[58]

$$\|f\|=|f(0)|+\int_0^\infty |f(s)|\,ds. \tag{16.2}$$

We assume that the domain $\mathscr{D}_{\mathscr{S}}$ of $\mathscr{S}(\cdot)$ is the subset in $\mathscr{D}$ of strain histories that satisfy

$$-1<\varepsilon^t(s)<\infty, \qquad 0\leqq s<\infty.$$

We assume that *the response functional $\mathscr{S}(\cdot)$ can be approximated by a linear viscoelastic relation for small relative strains.* More precisely, we assume that given any strain history $\varepsilon^t(\cdot)\in\mathscr{D}_{\mathscr{S}}$ there exists a smooth function $G_t(\cdot)$ on $[0,\infty)$

[57] The results given in this section are due to Chen and Gurtin [1970, *6*].

[58] Since every function in $\mathscr{D}$ vanishes on (s_0,∞), the norm (16.2) is equivalent to the norm

$$\|f\|=|f(0)|+\int_0^\infty |f(s)|\,h(s)\,ds$$

introduced by Coleman and Noll [1960, *1*] in their theory of fading memory.

with the following property: If $\hat{\varepsilon}^t(\cdot)\in\mathcal{D}_{\mathcal{S}}$ and $\gamma^t(\cdot)$, with

$$\gamma^t(s)=\hat{\varepsilon}^t(s)-\varepsilon^t(s),$$

is the corresponding relative strain history, then

$$\mathcal{S}(\hat{\varepsilon}^t)=\mathcal{S}(\varepsilon^t)+G_t(0)\,\gamma^t(0)+\int_0^\infty G_t'(s)\,\gamma^t(s)\,ds+o(\|\gamma^t\|) \tag{16.3}$$

as $\hat{\varepsilon}^t(\cdot)\to\varepsilon^t(\cdot)$ in the sense of the norm $\|\cdot\|$ (16.2).

The function $G_t(\cdot)$ is the *stress-relaxation function* corresponding to the strain history $\varepsilon^t(\cdot)$. In general,

$$G_t(s) \quad\text{and}\quad G_t'(s)=\frac{d}{ds}G_t(s)$$

will depend on the underlying strain history $\varepsilon^t(\cdot)$. To make explicit this dependence, we write

$$G_t(s)=\mathcal{G}(\varepsilon^t;s),\qquad G_t'(s)=\mathcal{G}'(\varepsilon^t;s);$$

we assume that $\mathcal{G}(\cdot;\cdot)$ and $\mathcal{G}'(\cdot;\cdot)$ are continuous on $\mathcal{D}_{\mathcal{S}}\times[0,\infty)$.

Remark 16.1. Superficially, our assumption (16.3) *may seem artificial. This, however, is not the case. Indeed, the constitutive relation* (12.1) *together with the smoothness properties on the response functional* $\mathcal{F}(\cdot,\cdot)$ *yields the relation* (16.3)[59] *with*

$$\|\gamma^t\|=\sup_s|\gamma^t(s)|.$$

Given a number $\varepsilon\in(-1,\infty)$, the history $\varepsilon^*(\cdot)\in\mathcal{D}_{\mathcal{S}}$, defined by

$$\varepsilon^*(s)=\begin{cases}\varepsilon, & s=0,\\ 0, & 0<s<\infty,\end{cases} \tag{16.4}$$

is the strain history resulting from a strain impulse of amount ε applied suddenly to a material point that has been unstrained at all past times. The function $T_I(\cdot)$, defined by

$$T_I(\varepsilon)=\mathcal{S}(\varepsilon^*), \tag{16.5}$$

is called the *instantaneous elastic response function* for the material. The hypothesis (16.3) laid down for $\mathcal{S}(\cdot)$ insures that $T_I(\cdot)$ is a smooth function of ε; in fact

$$\frac{d\,T_I(\varepsilon)}{d\varepsilon}=G(0), \tag{16.6}$$

where $G(0)$ is the initial value of the stress-relaxation function $G(\cdot)$ corresponding to the strain history $\varepsilon^*(\cdot)$, defined by (16.4). We assume that $T_I(\cdot)$ is of class C^2; we call the numbers

$$E_I(\varepsilon)=\frac{d\,T_I(\varepsilon)}{d\varepsilon}\quad\text{and}\quad \tilde{E}_I(\varepsilon)=\frac{d^2\,T_I(\varepsilon)}{d\varepsilon^2}, \tag{16.7}$$

respectively, the *instantaneous tangent modulus* and the *instantaneous second-order tangent modulus* corresponding to the strain ε.

We assume that

$$\left.\begin{array}{l}E_I(\varepsilon)>0\\ \tilde{E}_I(\varepsilon)\neq 0\end{array}\right\},\qquad \varepsilon\in(-1,\infty), \tag{16.8}$$

[59] This observation is due to COLEMAN and GURTIN [1965, *3*].

and that[60]

$$G'(0) \leqq 0 \tag{16.9}$$

whenever $G(\cdot)$ is the stress-relaxation function corresponding to a strain history of the form (16.4).

As we have remarked earlier, we suppose that the material ahead of the shock has been unstrained for all past times. That is, since

$$V(t) > 0,$$

we assume that for all $X \geqq Y(t)$

$$\varepsilon(X, t-s) = 0, \qquad 0 \leqq s < \infty. \tag{16.10}$$

Hence

$$[\varepsilon] = \varepsilon^-, \qquad \left[\frac{\partial \varepsilon}{\partial X}\right] = \left(\frac{\partial \varepsilon}{\partial X}\right)^-. \tag{16.11}$$

Further, it follows from the hypothesis (S-1) on the properties of shock waves given in Sect. 5 and the assumed smoothness of the functional $\mathscr{S}(\cdot)$ that

$$\left[\frac{\partial T}{\partial X}\right] = E_I^- \left(\frac{\partial \varepsilon}{\partial X}\right)^- - \frac{G'(0)}{V} \varepsilon^-. \tag{16.12}$$

The derivation of this relation, which is quite technical, is given in the Appendix. In (16.12),

$$E_I^- = E_I(\varepsilon^-(t)) \tag{16.13}$$

is the instantaneous tangent modulus corresponding to the strain ε^-, and $G'(0)$ is the initial slope of the stress-relaxation function $G(\cdot)$ corresponding to the strain history

$$\varepsilon^t(s) = \begin{cases} \varepsilon^-(t), & s = 0, \\ 0, & 0 < s < \infty. \end{cases} \tag{16.14}$$

Note that, in (16.13) and (16.14), $\varepsilon^-(t) = \varepsilon(Y(t)^-, t)$. Further, it follows from the continuity of $\mathscr{S}(\cdot)$ and the definition (16.5) of the instantaneous elastic response function that

$$[T] = T_I(\varepsilon^-) - T_I(0), \tag{16.15}$$

so that, by (5.17), the speed of propagation of the shock is given by[61]

$$V^2 = \frac{T_I(\varepsilon^-) - T_I(0)}{\varrho_R\, \varepsilon^-}. \tag{16.16}$$

On the other hand, the amplitude equation (5.18) with (16.11) and (16.12) yields the relation

$$2V \frac{\delta_1 \varepsilon^-}{\delta t} = \left(\frac{G'(0)}{\varrho_R V} - \frac{\delta_1 V}{\delta t}\right) \varepsilon^- - \left(\frac{E_I^-}{\varrho_R} - V^2\right)\left(\frac{\partial \varepsilon}{\partial X}\right)^-; \tag{16.17}$$

and, by $(16.7)_1$, (16.13) and (16.16),

$$\frac{\delta_1 V}{\delta t} = \frac{(E_I^- - \varrho_R V^2)}{2\varrho_R V \varepsilon^-} \frac{\delta \varepsilon^-}{\delta t}. \tag{16.18}$$

Substitution of (16.18) into (16.17) completes the derivation of[62]

---

[60] Cf. GURTIN and HERRERA [1965, *8*].

[61] Cf. COLEMAN, GURTIN and HERRERA [1965, *5*].

[62] For a non-linear Maxwell material a relation of the form (16.19) has been derived by DUVALL and ALVERSON [1963, *1*] and AHRENS and DUVALL [1966, *1*]. Within the linear theory of viscoelasticity LEE and KANTER [1953, *1*] and CHU [1962, *2*] have exhibited explicit solutions for shock waves; in their solutions the amplitude of a shock obeys a reduced form of (16.19), and it always decays exponentially; cf. CHEN and GURTIN [1970, *6*, Eq. (3.16)].

Theorem 16.1. *Consider a shock wave propagating in a body of simple material with memory, and assume that the region ahead of the wave has always been unstrained. Then the amplitude of the shock wave obeys the equation*

$$\frac{\delta_1 \varepsilon^-}{\delta t} = \frac{E_I^- - \varrho_R V^2}{2\varrho_R V\left(1 + \frac{E_I^- - \varrho_R V^2}{4\varrho_R V^2}\right)} \left\{\lambda^* - \left(\frac{\partial \varepsilon}{\partial X}\right)^-\right\}, \tag{16.19}$$

where

$$\lambda^* = \frac{G'(0)\, \varepsilon^-}{V(E_I^- - \varrho_R V^2)}. \tag{16.20}$$

Theorem 16.1 is valid for both compression shocks and expansion shocks. However, in considering a compression shock,[63] for which

$$\varepsilon^- < 0, \tag{16.21}$$

we assume that the instantaneous stress-strain law in compression is concave from below, i.e.,

$$\frac{d^2 T_I(\varepsilon)}{d\varepsilon^2} < 0 \quad \text{for} \quad \varepsilon \leqq 0; \tag{16.22}$$

in considering an expansion shock, for which

$$\varepsilon^- > 0, \tag{16.23}$$

we assume that the instantaneous stress-strain law in tension is concave from above; i.e.,

$$\frac{d^2 T_I(\varepsilon)}{d\varepsilon^2} > 0 \quad \text{for} \quad \varepsilon \geqq 0. \tag{16.24}$$

Remark 16.2. The assumptions (16.22) *and* (16.24) *are not artificial. In fact,* (16.22) *and* (16.24) *are, respectively, the necessary conditions for the existence of compression and expansion shocks in bodies of simple material with memory. They are also consistent with the conditions under which the amplitudes of compressive and expansive acceleration waves become infinite.*[64]

In either case, (16.22) or (16.24) with $(16.8)_1$, $(16.11)_1$, (16.13) and (16.16) implies that

$$E_I^- > \varrho_R V^2. \tag{16.25}$$

Hence, by (16.9), (16.21), (16.23) and (16.25), we have

Theorem 16.2. *Let Theorem 16.1 hold.*

(i) *Consider a compression shock. Suppose that the instantaneous stress-strain law in compression is concave from below. Then* $\lambda^* \geqq 0$ *and at any instant*

$$\left(\frac{\partial \varepsilon}{\partial X}\right)^- < \lambda^* \Leftrightarrow \frac{\delta_1 |\varepsilon^-|}{\delta t} < 0,$$

$$\left(\frac{\partial \varepsilon}{\partial X}\right)^- > \lambda^* \Leftrightarrow \frac{\delta_1 |\varepsilon^-|}{\delta t} > 0,$$

$$\left(\frac{\partial \varepsilon}{\partial X}\right)^- = \lambda^* \Leftrightarrow \frac{\delta_1 \varepsilon^-}{\delta t} = 0.$$

---

[63] Experimental evidence of the existence of shock waves in polymethyl methacrylate is given, for example, by HALPIN and GRAHAM [1965, *9*] and BARKER and HOLLENBACH [1970, *2*, Figs. 5 and 6]. Relevant also are the shock wave studies by DUVALL [1964, *2*] (Sioux quartzite), AHRENS and DUVALL [1966, *1*] (Arkansas novaculite, Sioux and Eureka quartzite), and JOHNSON and BARKER [1969, *1*] (aluminum) in which the materials exhibit instantaneous elasticity and gradual stress-relaxation.

[64] Cf. COLEMAN and GURTIN [1965, *3*].

(ii) *Consider an expansion shock. Suppose that the instantaneous stress-strain law in tension is concave from above. Then* $\lambda^* \leq 0$ *and at any instant*

$$\left(\frac{\partial \varepsilon}{\partial X}\right)^- < \lambda^* \Leftrightarrow \frac{\delta_1 \varepsilon^-}{\delta t} > 0,$$

$$\left(\frac{\partial \varepsilon}{\partial X}\right)^- > \lambda^* \Leftrightarrow \frac{\delta_1 \varepsilon^-}{\delta t} < 0,$$

$$\left(\frac{\partial \varepsilon}{\partial X}\right)^- = \lambda^* \Leftrightarrow \frac{\delta_1 \varepsilon^-}{\delta t} = 0.$$

Theorem 16.2 says, in particular, that whether the amplitude of a shock is increasing or decreasing depends on the relative magnitudes of the strain gradient $(\partial\varepsilon/\partial X)^-$ behind the wave and the number λ^*, defined by (16.20). Therefore, we call the number λ^* the *critical jump in strain gradient* for shock waves propagating in unstrained bodies of simple material with memory.[65] Notice that, in particular, λ^* depends on the dissipative properties as well as the instantaneous elastic properties of the material; it also depends on the strain ε^- behind the shock wave. In view of the definition of $G'(0)$, (16.13) and (16.16), it follows that λ^* is given by a function of the strain ε^- behind the shock wave.

In Theorem 16.2, we have also given a necessary condition for the existence of steady shocks.[66] For a steady shock, the compatibility conditions (5.11), (16.18) and Theorem 16.2 imply that the speed of propagation V is constant, and

$$\lambda^* = \left(\frac{\partial \varepsilon}{\partial X}\right)^- = \frac{1}{V^2}\ddot{x}^-. \tag{16.26}$$

Thus, $\lambda^*(\varepsilon^-)$ can be determined by measuring $(\partial\varepsilon/\partial X)^-$ or $\ddot{x}^-$ for steady shocks of various strengths.

Further, by (16.19) and (16.20), we have

Remark 16.3. When there is no instantaneous mechanical dissipation; i.e., $G_0'(0)=0$, *the critical jump in strain gradient* λ^* *vanishes. Thus Theorem 16.1 implies that the behavior of the amplitude of a shock depends solely on whether the strain gradient behind the shock* $(\partial\varepsilon/\partial X)^-$ *is positive or negative. This behavior is the same as that of a wave propagating in an elastic body.*[67]

Now, we shall determine the limit of $\lambda^*(\varepsilon^-)$ as $\varepsilon^- \to 0$. As we shall see, this limit is of special interest. By (16.13) and (16.15), we have

$$E_I^- = E_0 + \widetilde{E}_0\,\varepsilon^- + o(\varepsilon^-),$$

$$T_I(\varepsilon^-) - T_I(0) = E_0\,\varepsilon^- + \tfrac{1}{2}\widetilde{E}_0(\varepsilon^-)^2 + o\left((\varepsilon^-)^2\right),$$

where

$$E_0 = \left.\frac{d T_I}{d\varepsilon}\right|_{\varepsilon=0} \quad\text{and}\quad \widetilde{E}_0 = \left.\frac{d^2 T_I}{d\varepsilon^2}\right|_{\varepsilon=0}.$$

Thus, we conclude from (16.16) that

$$E_I^- - \varrho V^2 = \tfrac{1}{2}\widetilde{E}_0\,\varepsilon^- + o(\varepsilon^-). \tag{16.27}$$

and

$$V = V_0 + o(1), \tag{16.28}$$

[65] The existence of the critical jump in strain gradient is not unique to the present situation; cf. Sects. 22 and 26 of this article.

[66] Within the theory of finite linear viscoelasticity, Pipkin [1966, *7*] has exhibited explicit solutions for steady shock waves. Extending the work of Pipkin, Greenberg [1967, *1*] and [1968, *6*] proved that steady shock waves can exist in a large class of materials with memory.

[67] Cf. Remark 19.1 of this article.

where

$$V_0 = \left(\frac{E_0}{\varrho_R}\right)^{\frac{1}{2}}.$$

Since $G'(0)$ depends continuously on ε^-,

$$G'(0) = G_0'(0) + o(1), \tag{16.29}$$

where $G_0'(0)$ is the initial slope of the stress-relaxation function $G_0(\cdot)$ corresponding to the zero history; i.e., $\varepsilon^t(s) = 0$, $0 \leqq s < \infty$. Thus, (16.20) with (16.27) through (16.29) implies that

$$\lambda^*(\varepsilon^-) \to \frac{2G_0'(0)}{V_0 \tilde{E}_0} \quad \text{as} \quad \varepsilon^- \to 0. \tag{16.30}$$

By (3.7), (5.20) and (14.11), $|\lambda_0|$, the critical amplitude for acceleration waves in homogeneously deformed bodies of simple material with memory, expressed in terms of the strain gradient $(\partial\varepsilon/\partial X)^-$, reduces to

$$\left|\frac{G_0'(0)}{V_0 \tilde{E}_0}\right|.$$

Consequently, we have

Remark 16.4. As the amplitude of a shock wave tends to zero, the associated critical jump in strain gradient tends to a limit equal in absolute value to twice the critical amplitude for acceleration waves.

Superficially, it might be expected that this limit would be equal in absolute value to the critical amplitude for acceleration waves. However, it should be remembered that the quantity λ^* governs the behavior of the amplitude ε^- of a shock rather than the behavior of the strain gradient $(\partial\varepsilon/\partial X)^-$ behind the shock.

In the following section, we shall show that the results given in this section together with experimentally measurable quantities of steady shocks will allow us to predict the behavior of the amplitudes of acceleration waves.

17. Consequences of the existence of steady shock waves. In Sect. 16 we have shown that if a shock wave, propagating in a body of simple material with memory which is initially at rest in a homogeneous configuration, is steady, then the strain gradient $(\partial\varepsilon/\partial X)^-$ or the acceleration $\ddot{x}^-$ behind the wave must be equal to a certain critical value, which depends only on the strength ε^- of the shock. As the shock strength ε^- becomes vanishingly small, this critical value has a limit equal in absolute value to twice the critical amplitude for acceleration waves. Here, we shall show that these results taken together with experimentally measurable quantities of steady shocks of various strengths allow us to predict the behavior of acceleration waves in bodies of simple material with memory at rest in homogeneous configurations.[68]

Usually, experimenters measure directly or indirectly the speed of propagation V as well as the particle velocity $\dot{x}$ at a given material point[69] as functions of

[68] The results given in this section are due to CHEN and GURTIN [1972, *6*]. They were first announced in a talk entitled: "Wave Propagation in Non-linear Viscoelastic Materials" at the 11th Meeting of the Society for Natural Philosophy, University of Cincinnati, November 12–13, 1970.

[69] A laser interferometry technique has been developed by BARKER [1968, *1*]; it measures the velocity history of a reflecting surface buried in a transparent specimen. Of course, the reflecting surface is identified with the material particle situated at that location. Thus, the particle velocity history can be measured directly. It should be pointed out that the experimental accuracy of this technique is of the order nano-seconds.

time for each steady shock.[70] Knowing the graph of particle velocity $\dot{x}$, we can determine the velocity $\dot{x}^-$ and the acceleration $\ddot{x}^-$ immediately behind the shock. That is, we know the quantities

$$\dot{x}^-, \ \ddot{x}^-, \ \text{and} \ V$$

for each steady shock.

By $(5.11)_1$, we see that

$$\varepsilon^- = -\frac{1}{V}\dot{x}^-;$$

therefore we can determine ε^-, the strain behind each shock, which yields the speed $V(\varepsilon^-)$ as a function of ε^-. Hence, by (16.16), the instantaneous response function is known to within some constant; i.e., we can compute the values of

$$\overline{T}_I(\varepsilon) \equiv T_I(\varepsilon) - T_I(0).$$

Hence, we can determine the instantaneous tangent modulus

$$E_I(\varepsilon) = \frac{d\overline{T}_I(\varepsilon)}{d\varepsilon},$$

and the instantaneous second-order tangent modulus

$$\widetilde{E}_I(\varepsilon) = \frac{d^2\overline{T}_I(\varepsilon)}{d\varepsilon^2}.$$

Of course, $E_I(\varepsilon)$ is equal to $G(0)$, the initial value of the stress-relaxation function corresponding to the strain history $\varepsilon^*(\cdot)$, defined by (16.4).

It follows from (16.20) and (16.26) that for a steady shock

$$G'(0) = \frac{\ddot{x}^-(E_I^- - \varrho_R V^2)}{V\varepsilon^-};$$

therefore, we can also determine $G'(0)$, the initial slope of the stress-relaxation function corresponding to the strain history $\varepsilon^*(\cdot)$.

Extrapolating $G'(0)$, $G(0)$ and $\widetilde{E}_I(\varepsilon)$ to zero strain, we have $G_0'(0)$, $G_0(0)$ and $\widetilde{E}_0$. Thus, by (14.5), $(14.9)_1$ and (14.11), we can determine the constants μ_0 and λ_0. That is, we know the critical amplitude for acceleration waves $|\lambda_0|$. Finally, by (14.14), t_∞, the time it takes the amplitude of an acceleration wave of a certain initial amplitude to grow to ∞, is determined.

The preceding analysis shows that all of the material constants needed to predict the behavior of acceleration waves propagating in bodies of simple material with memory which are at rest in homogeneous configurations can be obtained from the experimental results of steady shocks. This provides a convenient method of verifying the theories of wave propagation discussed in Sects. 14 and 16.

Remark 17.1. We have arrived at this conclusion without specifying any particular explicit constitutive relation, and it is valid for all materials with instantaneous elastic response and fading memory. More importantly, agreement of experimental results with the theoretical results on acceleration waves cannot be used to justify the validity of a particular constitutive relation.

---

[70] Adopting BARKER's technique, SCHULER [1970, *9*] showed that his experimental results and those of BARKER and HOLLENBACH [1970, *2*] indicate that shocks of certain strengths in polymethyl methacrylate are indeed steady. Also, using these experimental data, SCHULER and WALSH [1971, *11*] have determined the critical jump in strain gradient λ^* as a function of the strain ε^- behind the shock.

As a check of the above procedure, the critical amplitude for acceleration waves $|\lambda_0|$ can be determined by a different method. By (16.20) and (16.26), we have the expected result

$$\ddot{x}^-(\varepsilon^-) = V^2(\varepsilon^-)\,\lambda^*(\varepsilon^-) \to \frac{2G_0'(0)\,V_0}{\tilde{E}_0} \quad \text{as} \quad \varepsilon^- \to 0;$$

i.e., $|\lambda_0|$ is simply one-half of the absolute value of this limit.

V. One dimensional waves in elastic bodies.

18. Acceleration waves in elastic bodies. It is clear that the governing differential equation of the amplitudes of acceleration waves in homogeneous elastic bodies[71] is a special case of that of waves in bodies of simple material with memory, and the general behavior of the amplitudes would be as we have predicted in Sect. 13. However, for elastic materials, the amplitudes of acceleration waves have certain interesting properties.

For a homogeneous elastic body, the stress $T(X,t)$ is determined by the deformation gradient $F(X,t)$:

$$T(X,t) = \hat{T}\big(F(X,t)\big). \tag{18.1}$$

Here, we have taken a homogeneous configuration to be the reference so that $\hat{T}(\cdot)$ and the density ϱ_R in this configuration are independent of X.

We assume that $\hat{T}(\cdot)$ is of class C^2. The *tangent modulus* E_e and the *second-order tangent modulus* $\tilde{E}_e$ are defined by the relations

$$E_e = \frac{d\hat{T}}{dF}, \quad \tilde{E}_e = \frac{d^2\hat{T}}{dF^2}, \tag{18.2}$$

and we assume that

$$E_e(F) > 0, \quad \tilde{E}_e(F) \neq 0. \tag{18.3}$$

By (5.20), (5.23), the hypothesis (A-1) given in Sect. 5 and the constitutive relation (18.1), it follows that the speed of propagation of an acceleration wave is given by

$$V^2 = \frac{E_e}{\varrho_R}, \tag{18.4}$$

where E_e is evaluated at $F\big(Y(t), t\big)$.

Differentiating (18.4), we have directly

$$\frac{\delta_1 V}{\delta t} = \frac{\dot{F}^+\tilde{E}_e}{2\varrho_R V} + \frac{\tilde{E}_e}{2\varrho_R}\left(\frac{\partial F}{\partial X}\right)^+. \tag{18.5}$$

Further, the constitutive relation (18.1), (18.2) and the hypothesis (A-1) imply that

$$[\ddot{T}] = 2\dot{F}^+\tilde{E}_e[\dot{F}] + \tilde{E}_e[\dot{F}]^2 + E_e[\ddot{F}]. \tag{18.6}$$

Finally, by (5.19), (5.20), (5.25), (18.4), (18.5) and (18.6), we have

Theorem 18.1. *The amplitude of an acceleration wave propagating in an elastic body obeys the differential equation*

$$\frac{\delta_1 a}{\delta t} = -\mu_e\, a + \beta_e\, a^2, \tag{18.7}$$

[71] Experimental evidence of the existence of acceleration waves in fused silica, which is elastic, is given, for example, by BARKER and HOLLENBACH [1970, *2*, Fig. 10]. In this regard, also refer to WACKERLE [1962, *4*].

where

$$\mu_e = \frac{1}{4}\left\{\frac{V\tilde{E}_e}{E_e}\left(\frac{\partial F}{\partial X}\right)^+ - 3\,\frac{\dot{F}^+\tilde{E}_e}{E_e}\right\}, \tag{18.8}$$

and

$$\beta_e = -\frac{\tilde{E}_e}{2E_e V}. \tag{18.9}$$

In (18.8) *and* (18.9), E_e *and* $\tilde{E}_e$ *are evaluated at* $F(Y(t), t)$.

Notice that (18.7) is of the same form as (13.1), so that the local and global behavior of the amplitude would be as we have predicted in Sect. 13. However, analogous to definition (13.11), we have

$$\begin{aligned}\lambda_e &= \frac{\mu_e}{\beta_e}\\ &= \frac{V^2}{2}\left\{\frac{3\dot{F}^+}{V} - \left(\frac{\partial F}{\partial X}\right)^+\right\}.\end{aligned} \tag{18.10}$$

We see that λ_e depends on the deformation rate and the gradient of the deformation ahead of the wave, and it depends on the properties of the material only through the dependence of the speed on the tangent modulus. Hence, in view of Theorems 13.1 and 13.2, we have the following observations:

Remark 18.1. Consider a particular elastic material. If $\operatorname{sgn}\tilde{E}_e(t)$ *is known, then the local behavior of the amplitude of an acceleration wave in a body of that material depends only on the properties of the motion ahead of the wave and the tangent modulus of the material; it is independent of the non-linearity of the material.*

Remark 18.2. Consider a particular elastic material. If λ_e *is well-behaved, in the sense of Theorem 13.2, then the global behavior of the amplitude of an acceleration wave with* $\operatorname{sgn} a(0) = \operatorname{sgn}\tilde{E}_e(t)$ *depends only on the properties of the motion ahead of the wave and the tangent modulus of the material; it is independent of the non-linearity of the material.*

Finally, we see that if the body ahead of the wave is at rest in a homogeneous configuration, then μ_e, defined by (18.8), vanishes, and β_e, defined by (18.9), is constant. In this situation, the behavior of the amplitude of an acceleration wave is well-known.[72]

19. Shock waves in elastic bodies. In our previous discussion concerning shock waves in bodies of material with memory, given in Sect. 16, we have assumed that the material regions ahead of the shocks are at rest in homogeneous configurations. Here, we consider the behavior of one dimensional shock waves propagating in homogeneous elastic bodies without adopting similar assumptions.[73] We simply assume that the motions of the material regions ahead of the shocks are known and that certain types of shocks are propagating into these regions.

In the present situation, it is more convenient to presume that the stress $T(X, t)$ is determined by the strain $\varepsilon(X, t)$:

$$T(X, t) = \tilde{T}(\varepsilon(X, t)). \tag{19.1}$$

Here, as in Sect. 18, we have taken a homogeneous configuration to be the reference.

[72] See, e.g., (14.12) and Remark 14.2.
[73] The results given in this section are due to Chen [1971, 7].

We assume that $\tilde{T}(\cdot)$ is of class C^2. Here, the tangent modulus E_e is given by

$$E_e = \frac{d\tilde{T}}{d\varepsilon} \tag{19.2}$$

and, as is customary, we assume that

$$E_e(\varepsilon) > 0. \tag{19.3}$$

By (19.1), (19.2) and the hypothesis (S-1) on the properties of shock waves given in Sect. 5, the shock amplitude equation (5.18) becomes

$$2V \frac{\delta_1[\varepsilon]}{\delta t} + [\varepsilon] \frac{\delta_1 V}{\delta t} = V^2 \left[\frac{\partial \varepsilon}{\partial X}\right] - \frac{1}{\varrho_R}\left[E_e \frac{\partial \varepsilon}{\partial X}\right], \tag{19.4}$$

where

$$E_e^- = E_e(\varepsilon^-), \quad E_e^+ = E_e(\varepsilon^+); \tag{19.5}$$

and, by (5.17) and (19.1), the speed of propagation is given by

$$V^2 = \frac{T^- - T^+}{\varrho_R [\varepsilon]}, \tag{19.6}$$

where

$$T^- = \tilde{T}(\varepsilon^-), \quad T^+ = \tilde{T}(\varepsilon^+). \tag{19.7}$$

The jump

$$\left[E_e \frac{\partial \varepsilon}{\partial X}\right]$$

occurring in the 2nd term of the right-hand member of (19.4) may be expressed in the alternate form

$$\left[E_e \frac{\partial \varepsilon}{\partial X}\right] = E_e^- \left[\frac{\partial \varepsilon}{\partial X}\right] + \left(\frac{\partial \varepsilon}{\partial X}\right)^+ [E_e]. \tag{19.8}$$

Differentiating (19.6), we have

$$2\varrho_R V \frac{\delta_1 V}{\delta t} = \frac{E_e^-}{[\varepsilon]} \frac{\delta_1 \varepsilon^-}{\delta t} - \frac{E_e^+}{[\varepsilon]} \frac{\delta_1 \varepsilon^+}{\delta t} - \frac{\varrho_R V^2}{[\varepsilon]} \frac{\delta_1 [\varepsilon]}{\delta t},$$

which may be written in the alternate form

$$2\varrho_R V \frac{\delta_1 V}{\delta t} = \frac{1}{[\varepsilon]} (E_e^- - \varrho_R V^2) \frac{\delta_1 [\varepsilon]}{\delta t} + \frac{[E_e]}{[\varepsilon]} \frac{\delta_1 \varepsilon^+}{\delta t}, \tag{19.9}$$

where $\delta_1 \varepsilon^+/\delta t$ is given by

$$\frac{\delta_1 \varepsilon^+}{\delta t} = \dot{\varepsilon}^+ + V\left(\frac{\partial \varepsilon}{\partial X}\right)^+. \tag{19.10}$$

Thus, by (19.4), (19.8), (19.9) and (19.10), we have[74]

Theorem 19.1. *The amplitude of a shock wave propagating in an elastic body obeys the equation*

$$\frac{\delta_1[\varepsilon]}{\delta t} = -\frac{E_e^- - \varrho_R V^2}{2\varrho_R V\left(1 + \frac{E_e^- - \varrho_R V^2}{4\varrho_R V^2}\right)} \left\{\frac{[E_e]}{2(E_e^- - \varrho_R V^2)} \left\{\frac{\dot{\varepsilon}^+}{V} + 3\left(\frac{\partial \varepsilon}{\partial X}\right)^+\right\} + \left[\frac{\partial \varepsilon}{\partial X}\right]\right\}. \tag{19.11}$$

In general, it is not possible to deduce any information from (19.11) regarding the behavior of the amplitude of a shock without adopting additional assumptions.

[74] HARRIS [1950, *1*] has derived the governing equations for plane, cylindrical and spherical shock waves in ideal gases which are initially at equilibrium.

Thus, in the sequel, we shall assume that the motion ahead of the shock is known and shall consider two distinct cases for which definite conclusions do follow from Theorem 19.1.

(α) *We consider a* $\begin{pmatrix}\textit{compression}\\ \textit{expansion}\end{pmatrix}$ *shock entering a body which is initially in* $\begin{pmatrix}\textit{compression}\\ \textit{tension}\end{pmatrix}$.

(β) *We consider a* $\begin{pmatrix}\textit{compression}\\ \textit{expansion}\end{pmatrix}$ *shock entering a body which is initially in* $\begin{pmatrix}\textit{tension}\\ \textit{compression}\end{pmatrix}$ *such that the body behind the wave remains in* $\begin{pmatrix}\textit{tension}\\ \textit{compression}\end{pmatrix}$.

Before proceeding to do this, let us rewrite (19.11) in the more suggestive form

$$\frac{\delta_1[\varepsilon]}{\delta t} = \frac{E_\varepsilon^- - \varrho_R V^2}{2\varrho_R V\left(1 + \dfrac{E_\varepsilon^- - \varrho_R V^2}{4\varrho_R V^2}\right)}\left\{\lambda_s - \left[\frac{\partial\varepsilon}{\partial X}\right]\right\}, \tag{19.12}$$

where

$$\lambda_s = -\frac{[E_\varepsilon]}{2(E_\varepsilon^- - \varrho_R V^2)}\left\{\frac{\dot\varepsilon^+}{V} + 3\left(\frac{\partial\varepsilon}{\partial X}\right)^+\right\}. \tag{19.13}$$

Case (α).

Here, in considering a compression shock, for which

$$\varepsilon^+ < 0 \quad \text{and} \quad [\varepsilon] < 0, \tag{19.14}$$

we assume that the stress-strain law in compression is concave from below; i.e.,

$$\frac{d^2\tilde T(\varepsilon)}{d\varepsilon^2} < 0 \quad \text{for} \quad \varepsilon \leqq 0; \tag{19.15}$$

in considering an expansion shock, for which

$$\varepsilon^+ > 0 \quad \text{and} \quad [\varepsilon] > 0, \tag{19.16}$$

we assume that the stress-strain law in tension is concave from above; i.e.,

$$\frac{d^2\tilde T(\varepsilon)}{d\varepsilon^2} > 0 \quad \text{for} \quad \varepsilon \geqq 0. \tag{19.17}$$

In either case,[75] (19.14) and (19.15) or (19.16) and (19.17) with (19.3), (19.5) and (19.6) imply

$$E_\varepsilon^- > \varrho_R V^2, \tag{19.18}$$

and

$$[E_\varepsilon] > 0. \tag{19.19}$$

Thus, by (19.12), (19.13), (19.18) and (19.19), we have

Theorem 19.2. *Let Theorem* 19.1 *hold.*

(i) *Consider a compression shock entering a body which is initially in compression. Suppose that the stress-strain law in compression is concave from below. Then at any instant*

$$\left[\frac{\partial\varepsilon}{\partial X}\right] < \lambda_s \Leftrightarrow \frac{\delta_1|[\varepsilon]|}{\delta t} < 0,$$

$$\left[\frac{\partial\varepsilon}{\partial X}\right] > \lambda_s \Leftrightarrow \frac{\delta_1|[\varepsilon]|}{\delta t} > 0,$$

$$\left[\frac{\partial\varepsilon}{\partial X}\right] = \lambda_s \Leftrightarrow \frac{\delta_1[\varepsilon]}{\delta t} = 0.$$

[75] Here, (19.15) and (19.17) are, respectively, the necessary conditions for the existence of the compression shock and the expansion shock.

(ii) *Consider an expansion shock entering a body which is initially in tension. Suppose that the stress-strain law in tension is concave from above. Then at any instant*

$$\left[\frac{\partial \varepsilon}{\partial X}\right] < \lambda_s \Leftrightarrow \frac{\delta_1 [\varepsilon]}{\delta t} > 0,$$

$$\left[\frac{\partial \varepsilon}{\partial X}\right] > \lambda_s \Leftrightarrow \frac{\delta_1 [\varepsilon]}{\delta t} < 0,$$

$$\left[\frac{\partial \varepsilon}{\partial X}\right] = \lambda_s \Leftrightarrow \frac{\delta_1 [\varepsilon]}{\delta t} = 0.$$

(iii) *For either the compression shock or the expansion shock*

$$\operatorname{sgn} \lambda_s = -\operatorname{sgn}\left\{\frac{\dot{\varepsilon}^+}{V} + 3\left(\frac{\partial \varepsilon}{\partial X}\right)^+\right\}.$$

Case (β).

Here, in considering a compression shock, for which

$$\varepsilon^- > 0, \quad \varepsilon^+ > 0 \quad \text{and} \quad [\varepsilon] < 0, \tag{19.20}$$

we assume that the stress-strain law in tension is convex from above; in considering an expansion shock, for which

$$\varepsilon^- < 0, \quad \varepsilon^+ < 0 \quad \text{and} \quad [\varepsilon] > 0, \tag{19.21}$$

we assume that the stress-strain law in compression is convex from below.[76] In either case, it can be shown that

$$E_e^- - \varrho_R V^2 > 0, \tag{19.22}$$

and

$$[E_e] > 0. \tag{19.23}$$

Thus, by (19.12), (19.13), (19.22) and (19.23), we have

Theorem 19.3. *Let Theorem 19.1 hold.*

(i*) *Consider a compression shock entering a body which is initially in tension such that the region behind the wave remains in tension. Suppose that the stress-strain law in tension is convex from above. Then at any instant*

$$\left[\frac{\partial \varepsilon}{\partial X}\right] < \lambda_s \Leftrightarrow \frac{\delta_1 |[\varepsilon]|}{\delta t} < 0,$$

$$\left[\frac{\partial \varepsilon}{\partial X}\right] > \lambda_s \Leftrightarrow \frac{\delta_1 |[\varepsilon]|}{\delta t} > 0,$$

$$\left[\frac{\partial \varepsilon}{\partial X}\right] = \lambda_s \Leftrightarrow \frac{\delta_1 [\varepsilon]}{\delta t} = 0.$$

(ii*) *Consider an expansion shock entering a body which is initially in compression such that the region behind the wave remains in compression. Suppose that the stress-strain law in compression is convex from below. Then at any instant*

$$\left[\frac{\partial \varepsilon}{\partial X}\right] < \lambda_s \Leftrightarrow \frac{\delta_1 [\varepsilon]}{\delta t} > 0,$$

$$\left[\frac{\partial \varepsilon}{\partial X}\right] > \lambda_s \Leftrightarrow \frac{\delta_1 [\varepsilon]}{\delta t} < 0,$$

$$\left[\frac{\partial \varepsilon}{\partial X}\right] = \lambda_s \Leftrightarrow \frac{\delta_1 [\varepsilon]}{\delta t} = 0.$$

[76] Notice that these conditions are in contrast to those for case (α).

(iii*) *For either the compression shock or the expansion shock*

$$\operatorname{sgn} \lambda_s = -\operatorname{sgn}\left\{\frac{\dot{\varepsilon}^+}{V} + 3\left(\frac{\partial \varepsilon}{\partial X}\right)^+\right\}.$$

It is of interest to point out that statements (i), (ii) and (iii) of Theorem 19.2 and statements (i*), (ii*) and (iii*) of Theorem 19.3 are the same in consequence.

The criteria governing the growth or decay of the amplitudes of shock waves, given in Theorems 19.2 and 19.3, are based on the relative magnitudes of the jump $[\partial \varepsilon/\partial X]$ in strain gradient and λ_s. Therefore, we call the number λ_s, defined by (19.13), the *critical jump in strain gradient* for shock waves in elastic bodies.[77]

Notice that if $(\partial \varepsilon/\partial X)^- = (\partial \varepsilon/\partial X)^+$, then the criteria reduce simply to the conditions of whether λ_s is positive or negative. By statements (iii) of Theorem 19.2 and (iii*) of Theorem 19.3,

$$\begin{aligned} \lambda_s > 0 &\Leftrightarrow \left(\frac{\partial \varepsilon}{\partial X}\right)^- < -\frac{\dot{\varepsilon}^+}{3V}, \\ \lambda_s < 0 &\Leftrightarrow \left(\frac{\partial \varepsilon}{\partial X}\right)^- > -\frac{\dot{\varepsilon}^+}{3V}, \end{aligned} \tag{19.24}$$

whenever $(\partial \varepsilon/\partial X)^- = (\partial \varepsilon/\partial X)^+$.

Further, by (19.12) and (19.13), we have

Remark 19.1. If the region ahead of a shock is unstrained, so that λ_s and the strain gradient ahead of the wave vanish, then the behavior of the shock depends solely on whether the strain gradient $(\partial \varepsilon/\partial X)^-$ behind the shock is positive or negative.

Now, we shall determine the limit of λ_s as $\varepsilon^- \to \varepsilon^+$. By (19.5) and (19.7)

$$\begin{aligned} E_e^- &= E_e^+ + \tilde{E}_e^+ [\varepsilon] + o([\varepsilon]), \\ T^- &= T^+ + E_e^+ [\varepsilon] + \tfrac{1}{2}\tilde{E}_e^+ [\varepsilon]^2 + o([\varepsilon]^2), \end{aligned} \tag{19.25}$$

where

$$\tilde{E}_e^+ = \left.\frac{d^2 \tilde{T}}{d\varepsilon^2}\right|_{\varepsilon=\varepsilon^+}.$$

Thus, by (19.6) and (19.25), we have

$$[E_e] = \tilde{E}_e^+ [\varepsilon] + o([\varepsilon]), \tag{19.26}$$

$$E_e^- - \varrho_R V^2 = \tfrac{1}{2}\tilde{E}_e^+ [\varepsilon] + o([\varepsilon]), \tag{19.27}$$

and

$$V = V^+ + o(1), \qquad V^+ = \left(\frac{E_e^+}{\varrho_R}\right)^{\frac{1}{2}}. \tag{19.28}$$

Therefore, (19.12), (19.26), (19.27) and (19.28) imply that

$$\lambda_s \to -\left\{\frac{\dot{\varepsilon}^+}{V^+} + 3\left(\frac{\partial \varepsilon}{\partial X}\right)^+\right\} \quad \text{as} \quad \varepsilon^- \to \varepsilon^+. \tag{19.29}$$

Remark 19.2. As the amplitude of a shock wave tends to zero, the associated critical jump in strain gradient has a limit which depends only on the properties of motion ahead of the wave and the tangent modulus of the material; it is independent of the non-linearity of the material.

---

[77] Cf. Sect. 16 of this article.

VI. One dimensional waves in elastic non-conductors of heat.

20. Acceleration waves in elastic non-conductors of heat. In this section, we shall derive the governing differential equation of an acceleration wave in a homogeneous elastic body which does not conduct heat.[78] For such a material, the internal energy $e(X, t)$, the stress $T(X, t)$ and the absolute temperature $\theta(X, t)$ are determined by the deformation gradient $F(X, t)$ and the entropy $\eta(X, t)$:

$$\begin{aligned} e(X, t) &= \hat{e}\big(F(X, t), \eta(X, t)\big), \\ T(X, t) &= \hat{T}\big(F(X, t), \eta(X, t)\big), \\ \theta(X, t) &= \hat{\theta}\big(F(X, t), \eta(X, t)\big), \end{aligned} \tag{20.1}$$

with

$$\hat{T}(F, \eta) = \frac{\partial \hat{e}(F, \eta)}{\partial F}, \quad \hat{\theta}(F, \eta) = \frac{\partial \hat{e}(F, \eta)}{\partial \eta}. \tag{20.2}$$

Here, we have taken a homogeneous configuration to be the reference, so that $\hat{e}(\cdot, \cdot)$, $\hat{T}(\cdot, \cdot)$, $\hat{\theta}(\cdot, \cdot)$ and the density ϱ_R in this configuration are independent of X.

We assume that $\hat{e}(\cdot, \cdot)$ is of class C^3; thus, by (20.2), $\hat{T}(\cdot, \cdot)$ and $\hat{\theta}(\cdot, \cdot)$ are of class C^2. We call the quantities

$$E_n = \frac{\partial \hat{T}}{\partial F}, \quad G_n = \frac{\partial \hat{T}}{\partial \eta}, \quad \tilde{E}_n = \frac{\partial^2 \hat{T}}{\partial F^2} \tag{20.3}$$

the *isentropic tangent modulus*, the *stress-entropy modulus*, and the *second-order isentropic tangent modulus*, respectively, and assume that

$$E_n(F, \eta) > 0, \quad G_n(F, \eta) \neq 0, \quad \tilde{E}_n(F, \eta) \neq 0. \tag{20.4}$$

In addition to the hypothesis (A-1) on the properties of acceleration waves given in Sect. 5, we assume that[79]

(A-4). *$\eta(\cdot, \cdot)$ is a continuous function everywhere.*

In the absence of external radiation, (20.1), (20.2) and balance of energy (5.26) imply that for $X \neq Y(t)$

$$\dot{\eta} = 0. \tag{20.5}$$

Consequently, (A-4), (20.5) and (4.24) with $f(\cdot, \cdot) = \eta(\cdot, \cdot)$ imply that

$$[\dot{\eta}] = \left[\frac{\partial \eta}{\partial X}\right] = 0, \tag{20.6}$$

i.e., an acceleration wave in an elastic non-conductor of heat is homentropic.

By (5.20), (5.23), the hypothesis (A-1), the constitutive relation $(20.1)_2$ and $(20.3)_1$, it follows that the speed of propagation of an acceleration wave is given by

$$V^2 = \frac{E_n}{\varrho_R}, \tag{20.7}$$

where E_n is evaluated at $\big(F(Y(t), t), \eta(Y(t), t)\big)$.

---

[78] Of course, this equation is a special case of that given by COLEMAN and GURTIN [1965, *4*, Eq. (10.18)]. However, in an application of this equation in the following section, we shall see that much *deeper* results on the behavior of the amplitude of the wave can be obtained, and that these results have important physical significance.

[79] Actually, the hypothesis (A-4) follows from (A-1), (5.22), $(5.27)_1$ and the fact that the absolute temperature is always strictly positive; cf. DORIA and BOWEN [1970, *7*].

Differentiating (20.7), we have upon using $(5.15)_1$, $(20.1)_2$, (20.3) and (20.6)

$$\frac{\delta_1 V}{\delta t} = \frac{\dot{F}^+ \tilde{E}_n}{2\varrho_R V} + \frac{1}{2}\frac{\partial E_n}{\partial \eta}\frac{\ddot{x}^+}{G_n} + \frac{1}{2\varrho_R}\left(\tilde{E}_n - \frac{\partial E_n}{\partial \eta}\frac{E_n}{G_n}\right)\left(\frac{\partial F}{\partial X}\right)^+ . \tag{20.8}$$

Further, the constitutive relation $(20.1)_2$, $(20.3)_{1,3}$ and the hypothesis (A-1) imply that

$$[\ddot{T}] = 2\dot{F}^+ \tilde{E}_n [\dot{F}] + \tilde{E}_n [\dot{F}]^2 + E_n [\ddot{F}] . \tag{20.9}$$

Finally, by (5.19), (5.20), (5.25), (20.7), (20.8) and (20.9), we have[80]

Theorem 20.1. *The amplitude of an acceleration wave propagating in an elastic body which does not conduct heat obeys the equation*

$$\frac{\delta_1 a}{\delta t} = -\mu_n a + \beta_n a^2, \tag{20.10}$$

where

$$\mu_n = \frac{1}{4}\left\{\frac{V}{E_n}\left(\tilde{E}_n - \frac{\partial E_n}{\partial \eta}\frac{E_n}{G_n}\right)\left(\frac{\partial F}{\partial X}\right)^+ - 3\frac{\dot{F}^+ \tilde{E}_n}{E_n} + \frac{1}{V}\frac{\partial E_n}{\partial \eta}\frac{\ddot{x}^+}{G_n}\right\}, \tag{20.11}$$

and

$$\beta_n = -\frac{\tilde{E}_n}{2E_n V} . \tag{20.12}$$

In (20.11) *and* (20.12), E_n, $\tilde{E}_n$, G_n *and* $\partial E_n/\partial \eta$ *are evaluated at* $(F(Y(t), t), \eta(Y(t), t))$.

Notice that (20.10) is of the same form as (13.1), so that the local and global behavior of the amplitude would be as we have predicted in Sect. 13. In the following section, we shall consider a specific application of (20.10) from which important consequences do follow.

21. Acceleration waves entering deformed elastic non-conductors. In the preceding section, we have derived the differential equation which governs the behavior of an acceleration wave propagating in an elastic non-conductor of heat. This equation has been obtained under the most general conditions, in that we did not adopt any assumptions regarding the nature of the deformation ahead of the wave. Here, we shall consider a specific application of this equation.[81]

We suppose that the body ahead of the wave is at rest in an equilibrium configuration, i.e., for all t and all $X > Y(t)$ the motion and the entropy are known and are given by

$$\chi(X, t) = \mathring{\chi}(X), \qquad \eta(X, t) = \mathring{\eta}(X), \tag{21.1}$$

and satisfying the equilibrium condition

$$\frac{d\mathring{\eta}}{dX} = -\frac{\mathring{E}_n}{\mathring{G}_n}\frac{d\mathring{F}}{dX}, \tag{21.2}$$

with

$$\mathring{F}(X) = \frac{d\mathring{\chi}(X)}{dX} . \tag{21.3}$$

Of course, (21.2) is the consequence of balance of momentum $(5.15)_1$, the constitutive relation $(20.1)_2$, $(20.3)_{1,2}$ and (21.1). Further, in (21.2), $\mathring{E}_n$ and $\mathring{G}_n$ are, respectively, the equilibrium isentropic tangent modulus and the equilibrium

---

[80] The results of Theorem 20.1 have been extended by Chen [1972, *4*] to include material inhomogeneity.

[81] The results presented in this section are due to Chen [1972, *3*].

stress-entropy modulus, i.e.,

$$\mathring{E}_n = E_n(\mathring{F}, \mathring{\eta}), \quad \mathring{G}_n = G_n(\mathring{F}, \mathring{\eta}) \tag{21.4}$$

for all t and all $X > Y(t)$.

Formula $(21.1)_1$ implies that

$$\dot{F}(X,t) = 0, \quad \ddot{x}(X,t) = 0$$

for all t and all $X > Y(t)$; hence taking the limit as $X \downarrow Y(t)$, we have

$$\dot{F}^+ = 0, \quad \ddot{x}^+ = 0. \tag{21.5}$$

Further, the hypotheses (A-1) and (A-4) together with $(21.1)_2$ and (21.3) imply that

$$F(Y(t),t) = \mathring{F}(Y(t)), \quad \eta(Y(t),t) = \mathring{\eta}(Y(t)), \quad \left(\frac{\partial F}{\partial X}\right)^+ = \left(\frac{d\mathring{F}}{dX}\right)^+; \tag{21.6}$$

and, by $(20.3)_3$ and $(21.4)_1$, we have

$$\frac{\partial \mathring{E}_n}{\partial \mathring{\eta}} = \frac{\partial E_n(\mathring{F}, \mathring{\eta})}{\partial \mathring{\eta}}, \tag{21.7$_1$}$$

$$\begin{aligned}\mathring{\tilde{E}}_n &= \tilde{E}_n(\mathring{F}, \mathring{\eta}) \\ &= \frac{\partial E_n(\mathring{F}, \mathring{\eta})}{\partial \mathring{F}}\end{aligned} \tag{21.7$_2$}$$

for all t and all $X > Y(t)$.

Consequently, by (20.10) and (21.4) through (21.7), we have

Theorem 21.1. *Let Theorem 20.1 hold. Suppose that the region ahead of the wave is at rest in an equilibrium configuration. Then the amplitude of an acceleration wave obeys the equation*

$$\frac{\delta_1 a}{\delta t} = -\mathring{\mu}_n a + \mathring{\beta}_n a^2, \tag{21.8}$$

where

$$\mathring{\mu}_n = \frac{\mathring{V}}{4\mathring{E}_n}\left(\mathring{\tilde{E}}_n - \frac{\partial \mathring{E}_n}{\partial \mathring{\eta}} \frac{\mathring{E}_n}{\mathring{G}_n}\right)\left(\frac{d\mathring{F}}{dX}\right)^+, \tag{21.9}$$

and

$$\mathring{\beta}_n = -\frac{\mathring{\tilde{E}}_n}{2\mathring{E}_n \mathring{V}}. \tag{21.10}$$

In (21.9) *and* (21.10), $\mathring{E}_n$, $\mathring{G}_n$, $\partial\mathring{E}_n/\partial\mathring{\eta}$ *and* $\mathring{\tilde{E}}_n$ *are evaluated at* $(\mathring{F}(Y(t)), \mathring{\eta}(Y(t)))$, *and*

$$\mathring{V} = \left(\frac{\mathring{E}_n}{\varrho_R}\right)^{\frac{1}{2}}. \tag{21.11}$$

Formula (21.8), which is a special case of (20.10), is also of the same form as (13.1), so that the local and global behavior of the amplitude would be as we have predicted in Sect. 13, and that nothing more specific can be said without further assumptions.

In Sects. 8 and 11, we showed that the thermodynamic properties of an elastic non-conductor of heat do not affect the behavior of an acceleration wave when the material region ahead of the wave is at rest in a homogeneous configuration.

However, notice that here the coefficient $\mathring{\mu}_n$ depends on the thermodynamic properties of the material as well as its isentropic elastic properties. In particular, it depends on

$$\frac{\partial \mathring{E}_n}{\partial \mathring{\eta}} \text{ and } \mathring{G}_n .$$

Therefore, we have the important observation

Remark 21.1. Consider an elastic body which does not conduct heat. Suppose that the body is initially at rest in an equilibrium configuration which is not homogeneous. Then the thermodynamic properties of the material do have a definite influence on the behavior of the amplitude of an acceleration wave.

Analogous to definition (13.11) and by (21.9) and (21.10), we have

$$\begin{aligned} \mathring{\lambda}_n &= \frac{\mathring{\mu}_n}{\mathring{\beta}_n} \\ &= \frac{\mathring{V}^2}{2} \left(\frac{\partial \mathring{E}_n}{\partial \mathring{\eta}} \frac{\mathring{E}_n}{\mathring{\tilde{E}}_n \mathring{G}_n} - 1 \right) \left(\frac{d\mathring{F}}{dX} \right)^+ . \end{aligned} \tag{21.12}$$

Further, by (20.8), (21.5), (21.6), (21.7) and (21.12), we see that,

$$\frac{\delta_1 \mathring{V}}{\delta t} = - \frac{\mathring{\tilde{E}}_n}{\mathring{E}_n} \mathring{\lambda}_n . \tag{21.13}$$

Thus, by $(20.4)_1$ and (21.13), we have

Remark 21.2. Let Theorem 21.1 hold. Then at any instant

$$\begin{aligned} \frac{\delta_1 \mathring{V}(t)}{\delta t} > 0 &\Leftrightarrow \operatorname{sgn} \mathring{\tilde{E}}_n(t) = - \operatorname{sgn} \mathring{\lambda}_n(t), \\ \frac{\delta_1 \mathring{V}(t)}{\delta t} < 0 &\Leftrightarrow \operatorname{sgn} \mathring{\tilde{E}}_n(t) = \operatorname{sgn} \mathring{\lambda}_n(t). \end{aligned} \tag{21.14}$$

That is, whether the speed of propagation is increasing or decreasing depends on the relative signs of the isentropic second-order tangent modulus $\mathring{\tilde{E}}_n$ and $\mathring{\lambda}_n$.

As we have remarked earlier, nothing more specific can be said about the behavior of the amplitude without adopting additional assumptions. Before proceeding to do this, let us consider the following degenerate case:

Remark 21.3. Let Theorem 21.1 hold. Suppose that

$$\mathring{\tilde{E}}_n(t) = \frac{\partial \mathring{E}_n(t)}{\partial \mathring{\eta}} \frac{\mathring{E}_n(t)}{\mathring{G}_n(t)} .$$

(i) *If* sgn $a(0) =$ sgn $\mathring{\tilde{E}}_n(t)$, *then* $a(t) \to 0$ *monotonically as* $t \to \infty$.

(ii) *If* sgn $a(0) = -$ sgn $\mathring{\tilde{E}}_n(t)$, *then* $|a(t)| \to \infty$ *monotonically within a finite time* t_∞ *given by*

$$\int_0^{t_\infty} \mathring{\beta}_n(\tau)\, d\tau = \frac{1}{a(0)},$$

where $\mathring{\beta}_n$ *is defined by* (21.10).

For the sake of brevity, we omit the somewhat tedious but straightforward derivations of the above results. In view of this remark and Remark 14.2, we see that the behavior of the amplitude of the wave is qualitatively the same as that of a wave in a body of material with memory when there is no instantaneous mechanical dissipation or as that of a wave in an elastic body.

Now, if we assume that

$$G_n(F,\eta)<0 \quad \text{and} \quad \frac{\partial E_n(F,\eta)}{\partial\eta}<0, \tag{21.15}$$

as is the case for most solids,[82] then, by $(21.4)_2$, $(21.7)_1$, (21.9), and (21.12), we observe that there are *three* distinct circumstances which may occur. That is,

(α) *If* $\mathring{\tilde{E}}_n(t)<0$, *then*

$$\operatorname{sgn}\mathring{\mu}_n(t) = -\operatorname{sgn}\left(\frac{d\mathring{F}}{dX}\right)^+(t)$$

and

$$\operatorname{sgn}\mathring{\lambda}_n(t) = -\operatorname{sgn}\left(\frac{dF}{dX}\right)^+(t).$$

(β) *If* $\mathring{\tilde{E}}_n(t)>0$, *then*

$$\operatorname{sgn}\mathring{\mu}_n(t) = \operatorname{sgn}\left(\frac{d\mathring{F}}{dX}\right)^+(t)$$

and

$$\operatorname{sgn}\mathring{\lambda}_n(t) = -\operatorname{sgn}\left(\frac{d\mathring{F}}{dX}\right)^+(t)$$

if and only if

$$\mathring{\tilde{E}}_n(t) > \frac{\partial\mathring{E}_n(t)}{\partial\eta}\,\frac{\mathring{E}_n(t)}{\mathring{G}_n(t)}.$$

(γ) *If* $\mathring{\tilde{E}}_n(t)>0$, *then*

$$\operatorname{sgn}\mathring{\mu}_n(t) = -\operatorname{sgn}\left(\frac{d\mathring{F}}{dX}\right)^+(t)$$

and

$$\operatorname{sgn}\mathring{\lambda}_n(t) = \operatorname{sgn}\left(\frac{d\mathring{F}}{dX}\right)^+(t)$$

if and only if

$$\mathring{\tilde{E}}_n(t) < \frac{\partial\mathring{E}_n(t)}{\partial\mathring{\eta}}\,\frac{\mathring{E}_n(t)}{\mathring{G}_n(t)}.$$

In the sequel, we shall consider the consequences of the general results on the local and global behavior of the amplitude, given in Sect. 13, in light of the above circumstances.[83] In particular, we shall consider the situation for which an acceleration is propagating into a region of *increasing* strain or a region of *decreasing* strain.

[82] For example, (21.15) is valid for sapphire and quartz; cf. ROHDE and JONES [1968, 7] and TOULOUKIAN [1967, 2].

[83] Of course, we can also establish the corresponding results for the situations when

$$\operatorname{sgn} G_n(F,\eta) \text{ and/or } \operatorname{sgn}\frac{\partial E_n(F,\eta)}{\partial\eta}$$

are not specified by (21.15).

Case (α).

(I) Suppose that the wave is propagating into a region of increasing strain, i.e.,

$$\left(\frac{d\overset{\circ}{F}}{dX}\right)^{+}(t) > 0.$$

Then

$$\overset{\circ}{\mu}_n(t) < 0, \qquad \overset{\circ}{\lambda}_n(t) < 0;$$

and, by (21.14),

$$\frac{\delta_1 \overset{\circ}{V}(t)}{\delta t} < 0.$$

Suppose that $\overset{\circ}{\mu}_n$ is bounded away from zero and $\overset{\circ}{\beta}_n$ is bound above. Then, by (13.12), it is not difficult to verify that the critical initial amplitude, denoted by $\overset{\circ}{\alpha}_n$, vanishes.

Summarizing the preceding results and those consequences of the general results given in Sect. 13, we have

(i) *The speed of propagation $\overset{\circ}{V}$ always decreases.*

(ii) *The critical initial amplitude $\overset{\circ}{\alpha}_n = 0$.*

(iii) *The amplitude of a compressive wave $(a(t) > 0)$ will always become infinite within a finite time.*

(iv) *The amplitude of an expansive wave $(a(t) < 0)$ will grow if $|a(0)| < |\overset{\circ}{\lambda}_n(0)|$ but will decay if $|a(0)| > |\overset{\circ}{\lambda}_n(0)|$.*

(II) Suppose that the wave is propagating into a region of decreasing strain, i.e.,

$$\left(\frac{d\overset{\circ}{F}}{dX}\right)^{+}(t) < 0.$$

Then

$$\overset{\circ}{\mu}_n(t) > 0, \qquad \overset{\circ}{\lambda}_n(t) > 0;$$

and, by (21.14),

$$\frac{\delta_1 \overset{\circ}{V}(t)}{\delta t} > 0.$$

Again, suppose that $\overset{\circ}{\mu}_n$ is bounded away from zero and $\overset{\circ}{\beta}_n$ is bounded above, then, by (13.12), we see that the critical initial amplitude $\overset{\circ}{\alpha}_n \neq 0$ and is finite. Therefore, we have the following summary:

(i*) *The speed of propagation $\overset{\circ}{V}$ always increases.*

(ii*) *The critical initial amplitude $\overset{\circ}{\alpha}_n \neq 0$ and is finite.*

(iii*) *The amplitude of a compressive wave will become infinite within a finite time if $a(0) > \overset{\circ}{\alpha}_n$ but will become arbitrarily small if $a(0) < \overset{\circ}{\alpha}_n$.*

(iv*) *The amplitude of an expansive wave always become arbitrarily small.*

Case (β).

(I*) Suppose that the wave is propagating into a region of increasing strain. Then

$$\overset{\circ}{\mu}_n(t) > 0, \qquad \overset{\circ}{\lambda}_n(t) < 0;$$

and we have the following:

(i) *The speed of propagation* $\mathring{V}$ *always increases.*

(ii) *The critical initial amplitude* $\mathring{\alpha}_n \neq 0$ *and is finite.*

(iii) *The amplitude of a compressive wave will always become arbitrarily small.*

(iv) *The amplitude of an expansive wave will become infinite within a finite time if* $|a(0)| > \mathring{\alpha}_n$ *but will become arbitrarily small if* $|a(0)| < \mathring{\alpha}_n$.

(II*) Suppose that the wave is propagating into a region of decreasing strain. Then

$$\mathring{\mu}_n(t) < 0, \quad \mathring{\lambda}_n(t) > 0;$$

and we have the following:

(i*) *The speed of propagation* $\mathring{V}$ *always decreases.*

(ii*) *The critical initial amplitude* $\mathring{\alpha}_n = 0$.

(iii*) *The amplitude of a compressive wave will grow if* $a(0) < \mathring{\lambda}_n(0)$ *but will decay if* $a(0) > \mathring{\lambda}_n(0)$.

(iv*) *The amplitude of an expansive wave will always become infinite within a finite time.*

Case (γ).

(I†) Suppose that the wave is propagating into a region of increasing strain. Then

$$\mathring{\mu}_n(t) < 0, \quad \mathring{\lambda}_n(t) > 0;$$

and we have the following:

(i) *The speed of propagation* $\mathring{V}$ *always decreases.*

(ii) *The critical initial amplitude* $\mathring{\alpha}_n = 0$.

(iii) *The amplitude of a compressive wave will grow if* $a(0) < \mathring{\lambda}_n(0)$ *but will decay if* $a(0) > \mathring{\lambda}_n(0)$.

(iv) *The amplitude of an expansive wave will always become infinite within a finite time.*

(II†) Suppose that the wave is propagating into a region of decreasing strain. Then

$$\mathring{\mu}_n(t) > 0, \quad \mathring{\lambda}_n(t) < 0;$$

and we have the following:

(i*) *The speed of propagation* $\mathring{V}$ *always increases.*

(ii*) *The critical initial amplitude* $\mathring{\alpha}_n \neq 0$ *and is finite.*

(iii*) *The amplitude of a compressive wave will always become arbitrarily small.*

(iv*) *The amplitude of an expansive wave will become infinite within a finite time if* $|a(0)| > \mathring{\alpha}_n$ *but will become arbitrarily small if* $|a(0)| < \mathring{\alpha}_n$.

We see that in some of the circumstances considered, there is indeed a *non-zero* critical initial amplitude. Therefore, in view of Theorem 14.2 and the preceding results, we have the following important observation:[84]

Remark 21.4. In the presence of initial strain, the behavior of the amplitudes of acceleration waves in elastic non-conductors of heat is in certain situations qualitatively the same as that of acceleration waves in bodies of simple material with memory which are initially at rest in homogeneous configurations.

[84] In this regard, also refer to Remark 25.1 of this article.

22. Shock waves in elastic non-conductors of heat. In this section, we shall examine the properties of a shock wave in a homogeneous elastic body which does not conduct heat.[85] Here, it is more convenient to presume that the internal energy $e(X, t)$, the stress $T(X, t)$ and the absolute temperature $\theta(X, t)$ are determined by the strain $\varepsilon(X, t)$ and the entropy $\eta(X, t)$:

$$\begin{aligned} e(X,t) &= \tilde{e}\big(\varepsilon(X,t), \eta(X,t)\big), \\ T(X,t) &= \tilde{T}\big(\varepsilon(X,t), \eta(X,t)\big), \\ \theta(X,t) &= \tilde{\theta}\big(\varepsilon(X,t), \eta(X,t)\big), \end{aligned} \tag{22.1}$$

with

$$\begin{aligned} \tilde{T}(\varepsilon, \eta) &= \frac{\partial \tilde{e}(\varepsilon, \eta)}{\partial \varepsilon}, \\ \tilde{\theta}(\varepsilon, \eta) &= \frac{\partial \tilde{e}(\varepsilon, \eta)}{\partial \eta}. \end{aligned} \tag{22.2}$$

Here, as in Sect. 20, we have taken a homogeneous configuration to be the reference.

We assume that $\tilde{e}(\cdot,\cdot)$ is of class C^3, so that, by (22.2), $\tilde{T}(\cdot,\cdot)$ and $\tilde{\theta}(\cdot,\cdot)$ are of class C^2. Here, the isentropic tangent modulus E_n and the stress-entropy modulus G_n are given by

$$E_n = \frac{\partial T}{\partial \varepsilon}, \qquad G_n = \frac{\partial \tilde{T}}{\partial \eta}, \tag{22.3}$$

and, as in Sect. 20, we assume that[86]

$$E_n(\varepsilon, \eta) > 0, \qquad G_n(\varepsilon, \eta) \neq 0. \tag{22.4}$$

By $(22.1)_2$, (22.3) and the hypothesis (S-1) on the properties of shock waves given in Sect. 5, the shock amplitude equation (5.18) becomes

$$\begin{aligned} 2V\frac{\delta_1[\varepsilon]}{\delta t} + [\varepsilon]\frac{\delta_1 V}{\delta t} &= \left(V^2 - \frac{E_n^-}{\varrho_R}\right)\left[\frac{\partial \varepsilon}{\partial X}\right] \\ &\quad - \frac{G_n^-}{\varrho_R}\left[\frac{\partial \eta}{\partial X}\right] - \frac{[E_n]}{\varrho_R}\left(\frac{\partial \varepsilon}{\partial X}\right)^+ - \frac{[G_n]}{\varrho_R}\left(\frac{\partial \eta}{\partial X}\right)^+, \end{aligned} \tag{22.5}$$

where

$$E_n^- = E_n(\varepsilon^-, \eta^-), \qquad E_n^+ = E_n(\varepsilon^+, \eta^+), \tag{22.6}$$

and

$$G_n^- = G_n(\varepsilon^-, \eta^-), \qquad G_n^+ = G_n(\varepsilon^+, \eta^+). \tag{22.7}$$

Further, by (5.17) and $(22.1)_2$, the speed of propagation of the shock is given by

$$V^2 = \frac{T^- - T^+}{\varrho_R[\varepsilon]}, \tag{22.8}$$

where

$$T^- = \tilde{T}(\varepsilon^-, \eta^-), \qquad T^+ = \tilde{T}(\varepsilon^+, \eta^+). \tag{22.9}$$

In order that we may simplify (22.5), we need to determine the expressions for

$$\left[\frac{\partial \eta}{\partial X}\right] \text{ and } \frac{\delta_1 V}{\delta t}.$$

[85] The results given in this section are due to Chen [1971, *6*]. These results generalize those given by Chen and Gurtin [1971, *9*] who considered the properties of shock waves in elastic non-conductors of heat which are initially unstrained.

[86] As we have pointed out earlier, $G_n(\varepsilon, \eta) < 0$ for most solids.

In the absence of external radiation, (22.1), (22.2) and balance of energy (5.26) imply that for $X \neq Y(t)$

$$\dot{\eta} = 0, \tag{22.10}$$

so that across a shock wave in an elastic non-conductor

$$[\dot{\eta}] = 0. \tag{22.11}$$

Balance of energy $(5.27)_1$ with $(5.11)_1$ and (22.8) reduces to

$$[e] - \tfrac{1}{2}(T^- + T^+)\,[\varepsilon] = 0. \tag{22.12}$$

On the other hand, balance of energy $(5.27)_2$ and (4.24) with $f(\cdot,\cdot) = \varepsilon(\cdot,\cdot)$ imply that

$$[\dot{e}] = T^- \frac{\delta_1[\varepsilon]}{\delta t} - T^- V\left[\frac{\partial \varepsilon}{\partial X}\right] + \dot{\varepsilon}^+[T]; \tag{22.13}$$

and, by (22.1) and (22.2), we have

$$\left[\frac{\partial e}{\partial X}\right] = T^-\left[\frac{\partial \varepsilon}{\partial X}\right] + \left(\frac{\partial \varepsilon}{\partial X}\right)^+[T] + \theta^-\left[\frac{\partial \eta}{\partial X}\right] + \left(\frac{\partial \eta}{\partial X}\right)^+[\theta] \tag{22.14}$$

where

$$\theta^- = \tilde{\theta}(\varepsilon^-, \eta^-), \qquad \theta^+ = \tilde{\theta}(\varepsilon^+, \eta^+). \tag{22.15}$$

Thus, differentiating (22.12) and using (22.13), (22.14) and (4.24) with $f(\cdot,\cdot) = e(\cdot,\cdot)$, we have

$$\begin{aligned} \theta^- V\left[\frac{\partial \eta}{\partial X}\right] &= -\frac{[T]}{2}\frac{\delta_1[\varepsilon]}{\delta t} - [T]\frac{\delta_1 \varepsilon^+}{\delta t} \\ &\quad - V\left(\frac{\partial \eta}{\partial X}\right)^+[\theta] + \frac{[\varepsilon]}{2}\left(\frac{\delta_1 T^-}{\delta t} + \frac{\delta_1 T^+}{\delta t}\right). \end{aligned} \tag{22.16}$$

On the other hand, differentiating (22.8), we find that

$$2\varrho_R V\frac{\delta_1 V}{\delta t} = \frac{1}{[\varepsilon]}\frac{\delta_1[T]}{\delta t} - \frac{\varrho_R V^2}{[\varepsilon]}\frac{\delta_1[\varepsilon]}{\delta t}. \tag{22.17}$$

Now, by (22.3), (22.9), (22.10), (22.11), and (4.24) with $f(\cdot,\cdot) = \eta(\cdot,\cdot)$,

$$\begin{aligned} \frac{\delta_1 T^-}{\delta t} &= E_n^- \frac{\delta_1 \varepsilon^-}{\delta t} + G_n^- \frac{\delta_1 \eta^-}{\delta t} \\ &= E_n \frac{\delta_1[\varepsilon]}{\delta t} + G_n^- V\left[\frac{\partial \eta}{\partial X}\right] + E_n^- \frac{\delta_1 \varepsilon^+}{\delta t} + G_n^- V\left(\frac{\partial \eta}{\partial X}\right)^+, \end{aligned} \tag{22.18}$$

and

$$\frac{\delta_1 T^+}{\delta t} = E_n^+ \frac{\delta_1 \varepsilon^+}{\delta t} + G_n^+ V\left(\frac{\partial \eta}{\partial X}\right)^+. \tag{22.19}$$

Substituting (22.18) and (22.19) into (22.16) and (22.17), we have the expressions[87]

$$\begin{aligned} \left[\frac{\partial \eta}{\partial X}\right] &= \frac{E_n^-(1-\xi)}{G_n^- V(2\tau - 1)}\frac{\delta_1[\varepsilon]}{\delta t} \\ &\quad + \left(\frac{2E_n^-(1-\xi)}{G_n^- V(2\tau-1)} - \frac{[E_n]}{G_n^- V(2\tau-1)}\right)\frac{\delta_1 \varepsilon^+}{\delta t} \\ &\quad + \left(\frac{G_n^- + G_n^+}{G_n^-(2\tau-1)} - \frac{2[\theta]}{G_n^-(2\tau-1)[\varepsilon]}\right)\left(\frac{\partial \eta}{\partial X}\right)^+, \end{aligned} \tag{22.20}$$

---

[87] Clearly, we may derive an expression for $\delta_1[\eta]/\delta t$ by noting that

$$\frac{\delta_1[\eta]}{\delta t} = V\left[\frac{\partial \eta}{\partial X}\right].$$

We shall examine the consequences of this observation later on in this section.

and

$$\begin{aligned}\frac{\delta_1 V}{\delta t} = {} & \frac{\tau E_n^-(1-\xi)}{\varrho_R V(2\tau-1)[\varepsilon]}\frac{\delta_1[\varepsilon]}{\delta t} \\ & + \left(\frac{E_n^-(1-\xi)}{\varrho_R V(2\tau-1)[\varepsilon]} + \frac{(\tau-1)[E_n]}{\varrho_R V(2\tau-1)[\varepsilon]}\right)\frac{\delta_1 \varepsilon^+}{\delta t} \\ & + \left(\frac{[G_n]}{2\varrho_R[\varepsilon]} + \frac{G_n^- + G_n^+}{2\varrho_R(2\tau-1)[\varepsilon]} - \frac{[\theta]}{\varrho_R(2\tau-1)[\varepsilon]^2}\right)\left(\frac{\partial \eta}{\partial X}\right)^+,\end{aligned} \tag{22.21}$$

where $\delta_1 \varepsilon^+/\delta t$ is given by

$$\frac{\delta_1 \varepsilon^+}{\delta t} = \dot{\varepsilon}^+ + V\left(\frac{\partial \varepsilon}{\partial X}\right)^+. \tag{22.22}$$

In (22.20) and (22.21),

$$\xi = \frac{\varrho_R V^2}{E_n^-}, \qquad \tau = \frac{\theta^-}{G_n^-[\varepsilon]}. \tag{22.23}$$

Thus, by (22.5), (22.20) and (22.21), we have[88]

Theorem 22.1. *The amplitude of a shock wave propagating in an elastic body which does not conduct heat obeys the equation*

$$\frac{\delta_1[\varepsilon]}{\delta t} = \frac{V(1-\xi)(2\tau-1)}{(3\xi+1)\tau-(3\xi-1)}\left\{\lambda_n^* - \left[\frac{\partial \varepsilon}{\partial X}\right]\right\}, \tag{22.24}$$

where

$$\begin{aligned}\lambda_n^* = {} & -\left(\frac{3}{V(2\tau-1)} + \frac{(\tau-2)[E_n]}{E_n^-(1-\xi)(2\tau-1)V}\right)\dot{\varepsilon}^+ \\ & - \left(\frac{3}{(2\tau-1)} + \frac{3(\tau-1)[E_n]}{E_n^-(1-\xi)(2\tau-1)}\right)\left(\frac{\partial \varepsilon}{\partial X}\right)^+ \\ & - \left(\frac{3[G_n]}{2E_n^-(1-\xi)} + \frac{3(G_n^- + G_n^+)}{2E_n^-(1-\xi)(2\tau-1)} - \frac{3[\theta]}{E_n^-(1-\xi)(2\tau-1)[\varepsilon]}\right)\left(\frac{\partial \eta}{\partial X}\right)^+.\end{aligned} \tag{22.25}$$

Formula (22.24) is quite complicated; and, as in the case of the purely mechanical theory considered in Sect. 19, it is not possible to deduce any information from (22.24) regarding the behavior of the amplitude of a shock without adopting additional assumptions. Here, we shall consider a specific application of Theorem 22.1. The interested reader may examine its consequences for other examples if the need arises.[89]

In the sequel, *we shall consider a* $\binom{\textit{compression}}{\textit{expansion}}$ *shock entering a body which is initially in* $\binom{\textit{compression}}{\textit{tension}}$.

Consider a particular instant of time. Then the strain ε^+ and the entropy η^+ just ahead of the shock are *fixed*. By $(22.1)_2$, $(22.3)_2$ and $(22.4)_2$, we have

$$\eta = \tilde{T}^{-1}(\varepsilon, T), \tag{22.26}$$

where $\tilde{T}^{-1}(\varepsilon, \cdot)$ denotes the inverse of $\tilde{T}(\varepsilon, \cdot)$.

Now, by $(22.1)_1$, (22.12) and (22.26), we define the function $H(\cdot,\cdot)$ by the relation

$$H(\varepsilon^-, T^-) = \tilde{e}\left(\varepsilon^-, \tilde{T}^{-1}(\varepsilon^-, T^-)\right) - \tilde{e}\left(\varepsilon^+, \tilde{T}^{-1}(\varepsilon^+, T^+)\right) - \tfrac{1}{2}(T^- + T^+)[\varepsilon] = 0. \tag{22.27}$$

[88] For a fluid with internal state variables which is initially at equilibrium, a relation of the form (22.24) has been given by CHEN and GURTIN [1971, 8].

[89] For example, the situation corresponding to case (β) considered in Sect. 19 of this article.

Formula (22.27) is the well-known Hugoniot relation; it gives all the possible thermodynamical states (ε^-, T^-) which can be reached across a shock from an initial state (ε^+, T^+). We assume that *in the* (ε^-, T^-)*-plane the Hugoniot relation* (22.27) *can be represented in the form* $T^- = T_H(\varepsilon^-)$.

With the preceding assumption, we can prove the following:

Remark 22.1. If in addition to (22.4) *we assume that*

$$\frac{\partial^2 \tilde{T}(\varepsilon, \eta)}{\partial \varepsilon^2} < 0$$

for all $\varepsilon \leqq 0$, *then*

(i) *the wave is a compression shock, i.e.,*

$$[\varepsilon] < 0;$$

(ii) *the speed of propagation is supersonic with respect to the front of the shock and subsonic with respect to the rear of the shock, i.e.,*

$$E_n^+ < \varrho_R V^2 < E_n^-,$$

where E_n^+ *and* E_n^- *are given by* (22.6), *and* $\varrho_R V^2$ *is given by* (22.8);

(iii) *along the whole Hugoniot curve the entropy increases with decreasing strain.*

Remark 22.2. If in addition to (22.4) *we assume that*

$$\frac{\partial^2 \tilde{T}(\varepsilon, \eta)}{\partial \varepsilon^2} > 0$$

for all $\varepsilon \geqq 0$, *then*

(i*) *the wave is an expansion shock, i.e.*

$$[\varepsilon] > 0;$$

(ii*) *the speed of propagation is supersonic with respect to the front of the shock and subsonic with respect to the rear of the shock, i.e.,*

$$E_n^+ < \varrho_R V^2 < E_n^-,$$

where E_n^+ *and* E_n^- *are given by* (22.6), *and* $\varrho_R V^2$ *is given by* (22.8);

(iii*) *along the whole Hugoniot curve the entropy increases with increasing strain.*

We omit giving the somewhat tedious derivations of the results stated in the preceding Remarks. In fact, they follow analogously from the known derivations of the corresponding results for compression shocks in elastic fluids given by BETHE and WEYL.[90] BETHE's assumptions differ from those adopted in this section.

Let us examine the consequences of Remarks 22.1 and 22.2; they will prove to be quite essential in determining the behavior of the shocks which we are considering.

First it follows from (ii) of Remark 22.1 and (ii*) of Remark 22.2 that in either case ξ, defined by $(22.23)_1$, has the property

$$0 < \xi < 1. \tag{22.28}$$

[90] Cf. BETHE [1942, *1*] and WEYL [1949, *2*]. In this regard, also refer to COURANT and FRIEDRICHS [1948, *1*, pp. 141–148]. Statements (i*) and (ii*) of Remark 22.2 are degenerate cases of those given by COLEMAN and GURTIN [1966, *4*] for materials with memory.

Next, (22.26) and our assumption on the properties of the Hugoniot relation insure the existence of a function $\eta_H(\cdot,\cdot,\cdot)$ such that

$$[\eta]=\eta_H([\varepsilon],\varepsilon^+,\eta^+); \tag{22.29}$$

and in view of (iii) of Remark 22.1 and (iii*) of Remark 22.2, we have

$$\frac{\partial\eta_H([\varepsilon],\varepsilon^+,\eta^+)}{\partial[\varepsilon]}<0 \quad \text{whenever} \quad \varepsilon^+<0, \quad [\varepsilon]<0, \tag{22.30}$$

and

$$\frac{\partial\eta_H([\varepsilon],\varepsilon^+,\eta^+)}{\partial[\varepsilon]}>0 \quad \text{whenever} \quad \varepsilon^+>0, \quad [\varepsilon]>0. \tag{22.31}$$

Now formula (22.12) with $(22.1)_{1,2}$ and (22.9) may be written in the form

$$\begin{aligned}&\tilde{e}(\varepsilon^++[\varepsilon],\eta^++\eta_H([\varepsilon],\varepsilon^+,\eta^+))-\tilde{e}(\varepsilon^+,\eta^+)\\&\quad-\tfrac{1}{2}\{\tilde{T}(\varepsilon^++[\varepsilon],\eta^++\eta_H([\varepsilon],\varepsilon^+,\eta^+))+\tilde{T}(\varepsilon^+,\eta^+)\}\,[\varepsilon]=0.\end{aligned} \tag{22.32}$$

Differentiating (22.32) with respect to $[\varepsilon]$ and utilizing (22.2) and (22.8) we have

$$\frac{\partial\eta_H}{\partial[\varepsilon]}=\frac{E_n^-(1-\xi)}{G_n^-(2\tau-1)}. \tag{22.33}$$

Formula (22.33) when taken together with the preceding results has great significance. Indeed, by $(22.23)_2$, (22.28), (22.30), (22.31), and (22.33), we have

Remark 22.3. For the compression shock under consideration

$$\begin{aligned}&G_n^-<0\Leftrightarrow\tau>\tfrac{1}{2},\\&G_n^->0\Leftrightarrow\tau<0.\end{aligned} \tag{22.34}$$

Remark 22.4. For the expansion shock under consideration

$$\begin{aligned}&G_n^-<0\Leftrightarrow\tau<0,\\&G_n^->0\Leftrightarrow\tau>\tfrac{1}{2}.\end{aligned} \tag{22.35}$$

It should be pointed out that $(22.34)_2$ and $(22.35)_1$ are direct consequences of the definition of τ given by $(22.23)_2$; the results of Remarks 22.1 and 22.2 are not essential insofar as their validity are concerned.

In view of (22.28) and the results of Remarks 22.3 and 22.4, we have

Theorem 22.2. *Let Theorem 22.1 hold.*

(I) *Consider a compression shock entering a body which is initially in compression. Suppose that the isentropic stress-strain law in compression is concave from below.*

(i) *If*

$$G_n^-<0 \quad \textit{or} \quad G_n^->0 \quad \textit{and} \quad \tau<\frac{3\xi-1}{3\xi+1},$$

then at any instant

$$\left[\frac{\partial\varepsilon}{\partial X}\right]<\lambda_n^*\Leftrightarrow\frac{\delta_1|[\varepsilon]|}{\delta t}<0,$$

$$\left[\frac{\partial\varepsilon}{\partial X}\right]>\lambda_n^*\Leftrightarrow\frac{\delta_1|[\varepsilon]|}{\delta t}>0,$$

$$\left[\frac{\partial\varepsilon}{\partial X}\right]=\lambda_n^*\Leftrightarrow\frac{\delta_1[\varepsilon]}{\delta t}=0.$$

(ii) *If*

$$G_n^- > 0 \quad and \quad \tau > \frac{3\xi - 1}{3\xi + 1},$$

then at any instant

$$\left[\frac{\partial \varepsilon}{\partial X}\right] < \lambda_n^* \Leftrightarrow \frac{\delta_1 |[\varepsilon]|}{\delta t} > 0,$$

$$\left[\frac{\partial \varepsilon}{\partial X}\right] > \lambda_n^* \Leftrightarrow \frac{\delta_1 |[\varepsilon]|}{\delta t} < 0,$$

$$\left[\frac{\partial \varepsilon}{\partial X}\right] = \lambda_n^* \Leftrightarrow \frac{\delta_1 [\varepsilon]}{\delta t} = 0.$$

(II) *Consider an expansion shock entering a body which is initially in tension. Suppose that the isentropic stress-strain law in tension is concave from above.*

(i*) *If*

$$G_n^- > 0 \quad or \quad G_n^- < 0 \quad and \quad \tau < \frac{3\xi - 1}{3\xi + 1},$$

then at any instant

$$\left[\frac{\partial \varepsilon}{\partial X}\right] < \lambda_n^* \Leftrightarrow \frac{\delta_1 [\varepsilon]}{\delta t} > 0,$$

$$\left[\frac{\partial \varepsilon}{\partial X}\right] > \lambda_n^* \Leftrightarrow \frac{\delta_1 [\varepsilon]}{\delta t} < 0.$$

$$\left[\frac{\partial \varepsilon}{\partial X}\right] = \lambda_n^* \Leftrightarrow \frac{\delta_1 [\varepsilon]}{\delta t} = 0.$$

(ii*) *If*

$$G_n^- < 0 \quad and \quad \tau > \frac{3\xi - 1}{3\xi + 1},$$

then at any instant

$$\left[\frac{\partial \varepsilon}{\partial X}\right] < \lambda_n^* \Leftrightarrow \frac{\delta_1 [\varepsilon]}{\delta t} < 0,$$

$$\left[\frac{\partial \varepsilon}{\partial X}\right] > \lambda_n^* \Leftrightarrow \frac{\delta_1 [\varepsilon]}{\delta t} > 0,$$

$$\left[\frac{\partial \varepsilon}{\partial X}\right] = \lambda_n^* \Leftrightarrow \frac{\delta_1 [\varepsilon]}{\delta t} = 0.$$

Since $G_n(\varepsilon, \eta) < 0$ for most solids, by Theorems 19.2 and 22.2, we have

Remark 22.5. In general, the behavior of compression shocks in elastic bodies which do not conduct heat is qualitatively the same as that of waves predicted by the purely mechanical theory, and that of expansion shocks is qualitatively the same as that of waves predicted by the purely mechanical theory when

$$\tau < \frac{3\xi - 1}{3\xi + 1}.$$

In view of the fact that the criteria governing the growth and decay of the amplitudes of shock waves given in Theorem 22.2 are based on the relative magnitudes of the jump $[\partial \varepsilon / \partial X]$ in strain gradient and λ_n^*, we call the number λ_n^* the *critical jump in strain gradient* for shock waves in bodies of elastic material which do not conduct heat.[91] Notice that λ_n^*, defined by (22.25), depends on the thermodynamic properties and the isentropic elastic properties of the material; it also depends on the properties of the motion and the entropy ahead of the wave.

[91] Cf. Sects. 16 and 19 of this article.

Before proceeding further to examine the properties of the jump $[\theta]$ in temperature across a shock, we assume that $\tilde{\theta}(\varepsilon, \cdot)$ is invertible and introduce the specific heat $\varkappa$, defined by the relation

$$\varkappa(\varepsilon, \eta) = \tilde{\theta}(\varepsilon, \eta) \left(\frac{\partial \tilde{\theta}(\varepsilon, \eta)}{\partial \eta} \right)^{-1}. \tag{22.36}$$

We assume that

$$\varkappa(\varepsilon, \eta) > 0 \tag{22.37}$$

for all (ε, η). Now, consider a particular instant of time, so that the strain ε^+ and the entropy η^+ just ahead of the shock are fixed. By $(22.1)_3$ and (22.29),

$$[\theta] = \tilde{\theta}(\varepsilon^+ + [\varepsilon], \eta^+ + \eta_H([\varepsilon], \varepsilon^+, \eta^+)) - \tilde{\theta}(\varepsilon^+, \eta^+). \tag{22.38}$$

Differentiating (22.38) with respect to $[\varepsilon]$, we have on utilizing (22.2), $(22.3)_2$ and (22.36)

$$\frac{\partial [\theta]}{\partial [\varepsilon]} = G_n^- + \frac{\theta^-}{\varkappa^-} \frac{\partial \eta_H}{\partial [\varepsilon]}. \tag{22.39}$$

In view of (22.30), (22.37) and (22.39), we have

Remark 22.6. Across a compression shock with $\varepsilon^+ < 0$.

(i) *If* $G_n^- < 0$, *then*

$$\frac{\partial [\theta]}{\partial [\varepsilon]} < 0.$$

(ii) *If* $G_n^- > 0$, *then*

$$\frac{\partial [\theta]}{\partial [\varepsilon]} < 0 \quad \textit{whenever} \quad G_n^- < -\frac{\theta^-}{\varkappa^-} \frac{\partial \eta_H([\varepsilon], \varepsilon^+, \eta^+)}{\partial [\varepsilon]};$$

or

$$\frac{\partial [\theta]}{\partial [\varepsilon]} > 0 \quad \textit{whenever} \quad G_n^- > -\frac{\theta^-}{\varkappa^-} \frac{\partial \eta_H([\varepsilon], \varepsilon^+, \eta^+)}{\partial [\varepsilon]}.$$

Similarly, in view of (22.31), (22.37) and (22.39), we have

Remark 22.7. Across an expansion shock with $\varepsilon^+ > 0$.

(i*) *If* $G_n^- < 0$, *then*

$$\frac{\partial [\theta]}{\partial [\varepsilon]} > 0 \quad \textit{whenever} \quad -G_n^- < \frac{\theta^-}{\varkappa^-} \frac{\partial \eta_H([\varepsilon], \varepsilon^+, \eta^+)}{\partial [\varepsilon]};$$

or

$$\frac{\partial [\theta]}{\partial [\varepsilon]} < 0 \quad \textit{whenever} \quad -G_n^- > \frac{\theta^-}{\varkappa^-} \frac{\partial \eta_H([\varepsilon], \varepsilon^+, \eta^+)}{\partial [\varepsilon]}.$$

(ii*) *If* $G_n^- > 0$, *then*

$$\frac{\partial [\theta]}{\partial [\varepsilon]} > 0.$$

Of course, the results given in Remarks 22.6 and 22.7 are of importance in that they give the conditions when there is an increase or decrease in temperature across a shock.[92] In particular, Remark 22.6 states that across a compression shock in most materials the temperature along the whole Hugoniot curve always increases with decreasing strain; and Remark 22.7 states that across an expansion shock in most materials the temperature along the whole Hugoniot curve may either increase or decrease with increasing strain. For a weak expansion shock, however, we know that the temperature jump always decreases in most materials.[93]

---

[92] The corresponding results for *weak* shock are, of course, known; see, for example, COLEMAN and GURTIN [1966, *4*].

[93] Cf. COLEMAN and GURTIN [1966, *4*].

Now, let us determine explicitly the derivatives $\partial\eta_H/\partial\varepsilon^+$ and $\partial\eta_H/\partial\eta^+$ of the function $\eta_H(\cdot,\cdot,\cdot)$ defined by (22.29). Taking the $\delta_1/\delta t$ derivative of (22.29) and upon using (22.10), (22.20), (22.33), and (4.24) with $f(\cdot,\cdot)=\eta(\cdot,\cdot)$, we have

$$\begin{aligned}\frac{\partial\eta_H}{\partial\varepsilon^+}&=\frac{\partial\eta_H}{\partial[\varepsilon]}\left(2-\frac{[E_n]}{E_n^-(1-\xi)}\right),\\ \frac{\partial\eta_H}{\partial\eta^+}&=\frac{1}{E_n^-(1-\xi)}\,\frac{\partial\eta_H}{\partial[\varepsilon]}\left\{G_n^-+G_n^+-2\frac{[\theta]}{[\varepsilon]}\right\}.\end{aligned}\tag{22.40}$$

That is, the derivatives of $\eta_H(\cdot,\cdot,\cdot)$ are related to one another.

Finally, let us assume that the region ahead of the shock is at rest in its reference configuration with constant entropy. By (22.24) and (22.25), we immediately have the following:[94]

Remark 22.8. If the region ahead of a shock is at rest in its reference configuration with constant entropy, so that λ_n^* and the strain gradient ahead of the wave vanish, then the behavior of the shock depends solely on whether the strain gradient $(\partial\varepsilon/\partial X)^-$ behind the shock is positive or negative.

Further, (22.11), (22.20) and (4.24) with $f(\cdot,\cdot)=\eta(\cdot,\cdot)$ imply that

$$\frac{\delta_1[\eta]}{\delta t}=\frac{E_n^-(1-\xi)}{G_n^-(2\tau-1)}\,\frac{\delta_1[\varepsilon]}{\delta t}.$$

Hence, by (22.28), Remarks 22.3 and 22.4, we have

Remark 22.9. Consider an elastic non-conductor of heat. Suppose that the body is initially at rest in its reference configuration with constant entropy.

(i) *For a compression shock*

$$\operatorname{sgn}\frac{\delta_1[\eta]}{\delta t}=\operatorname{sgn}\frac{\delta_1|[\varepsilon]|}{\delta t}.$$

(ii) *For an expansion shock*

$$\operatorname{sgn}\frac{\delta_1[\eta]}{\delta t}=\operatorname{sgn}\frac{\delta_1[\varepsilon]}{\delta t}.$$

While the above results are expected from (iii) of Remark 22.1 and (iii*) of Remark 22.2, they are in general not true.

Within the framework of the present theory, CHEN[95] examined the influence of discontinuous external body force b and discontinuous external radiation r on the behavior of a shock wave assuming that the body is initially unstrained. He showed that the amplitude of the shock again obeys an equation of the form (22.24) with λ_n^* replaced by

$$\lambda_d^*=\frac{\varrho_R}{E_n^-(1-\xi)}\left(\frac{G_n^-r^-}{\theta^-V}-b^-\right),$$

so that its behavior would be as predicted by Theorem 22.2 with $\varepsilon^+=0$. He also outlined a particular experiment for which there is no external body force but the external radiation is indeed discontinuous. A linearized version of CHEN's result had been given earlier by WALSH;[96] his governing equation for the amplitude of the shock is no longer of the same form as (22.24).

23. Shock waves entering deformed elastic non-conductors. In the preceding section, we have derived the differential equation which the amplitude of a shock wave propagating in an elastic non-conductor of heat must obey. We have shown that the behavior of the shock depends on the relative magnitudes of the jump

[94] This result was first obtained directly by CHEN and GURTIN [1971, *9*].
[95] CHEN [1971, *5*].
[96] WALSH [1967, *3*].

$[\partial\varepsilon/\partial X]$ in strain gradient and a number λ_n^* called the critical jump in strain gradient, and that this critical strain gradient vanishes if the region ahead of the shock is at rest in its reference configuration with constant entropy. Here, we shall assume that the region ahead of the wave is at rest in an equilibrium configuration and examine the consequences of this assumption.[97]

We suppose that the body is initially at rest in an equilibrium configuration in the sense made precise in Sect. 21. Let $\mathring{\varepsilon}(X)$ denote the strain field ahead of the wave, i.e.,

$$\mathring{\varepsilon}(X) = \mathring{F}(X) - 1, \tag{23.1}$$

where $\mathring{F}(X)$ is defined by (21.3). In view of the results given in Sect. 21, (22.8), $(22.9)_2$, and (23.1), the speed of propagation $\mathring{V}$ of the shock is given by

$$\mathring{V}^2 = \frac{T^- - \mathring{T}^+}{\varrho_R[\varepsilon]}, \tag{23.2}$$

where

$$\mathring{T}^+ = \tilde{T}(\mathring{\varepsilon}^+, \mathring{\eta}^+).$$

Further, by $(22.6)_2$, $(22.7)_2$ and $(22.15)_2$,

$$\begin{aligned} \mathring{E}_n^+ &= E_n(\mathring{\varepsilon}^+, \mathring{\eta}^+), \\ \mathring{G}_n^+ &= G_n(\mathring{\varepsilon}^+, \mathring{\eta}^+), \\ \mathring{\theta}^+ &= \tilde{\theta}(\mathring{\varepsilon}^+, \mathring{\eta}^+); \end{aligned} \tag{23.3}$$

and, by (22.23),

$$\mathring{\xi} = \frac{\varrho_R \mathring{V}^2}{E_n^-}. \tag{23.4}$$

In view of the equilibrium condition (21.2), (23.1) through (23.4), we have

Theorem 23.1. *Let Theorem 22.1 hold. Suppose that the region ahead of the wave is at rest in an equilibrium configuration. Then the amplitude of a shock wave obeys the equation*

$$\frac{\delta_1[\varepsilon]}{\delta t} = \frac{\mathring{V}(1-\mathring{\xi})(2\tau-1)}{(3\mathring{\xi}+1)\,\tau-(3\mathring{\xi}-1)} \left\{\mathring{\lambda}_n^* - \left[\frac{\partial\varepsilon}{\partial X}\right]\right\}, \tag{23.5}$$

where

$$\mathring{\lambda}_n^* = -\frac{3}{(2\tau-1)(1-\mathring{\xi})} \left\{\tau - \mathring{\xi} - \frac{\mathring{\theta}^+}{\mathring{G}_n^+[\varepsilon]} \frac{\mathring{E}_n^+}{E_n^-}\right\} \left(\frac{d\mathring{\varepsilon}}{dX}\right)^+. \tag{23.6}$$

Since (23.5) is of the same form as (22.24), the behavior of the amplitude of the shock would be the same as that of the more general case. In general, however, we do not expect that $\mathring{\lambda}_n^*$ will vanish. Therefore, in view of Theorems 16.2 and 22.2, we have the following important observation:[98]

Remark 23.1. In the presence of initial strain, the behavior of the amplitudes of compression shocks and that of expansion shocks with

$$\tau < \frac{3\mathring{\xi}-1}{3\mathring{\xi}+1}$$

in most elastic non-conductors of heat are qualitatively the same as that of shock waves in bodies of simple material with memory which are initially unstrained.

---

[97] The results given in this section are due to CHEN [1971, *6*].

[98] In this regard, also refer to Remark 27.1 of this article.

VII. One dimensional waves in inhomogeneous elastic bodies.

24. Acceleration waves in inhomogeneous elastic bodies. In all of our previous considerations of the properties of acceleration waves, our discussions have been confined to homogeneous bodies. This does not mean that material inhomogeneity has no influence on the behavior of waves. Here, we shall consider the properties of acceleration waves in inhomogeneous elastic bodies and show that material inhomogeneity does have a definite influence on the behavior of these waves.[99]

For an inhomogeneous elastic body, the present value $T(X,t)$ of the stress at the material point X depends on the present value $F(X,t)$ of the deformation gradient at X, and X:

$$T(X,t)=\hat{T}\big(F(X,t);X\big). \tag{24.1}$$

In other words, there is no reference configuration which renders $\hat{T}(\cdot\,;\cdot)$ independent of X; hence the density ϱ_R, in the chosen configuration, is also given by a function of X.

We assume that $\hat{T}(\cdot\,;X)$ is of class C^2. We call the quantities

$$E_X=\frac{\partial\hat{T}}{\partial F},\qquad \tilde{E}_X=\frac{\partial^2\hat{T}}{\partial F^2} \tag{24.2}$$

the *local tangent modulus*, and the *local second-order tangent modulus*, respectively, and assume that

$$E_X(F;X)>0,\qquad \tilde{E}_X(F;X)\neq 0. \tag{24.3}$$

We further assume that $\hat{T}(\cdot\,;\cdot)$, $E_X(\cdot\,;\cdot)$ and $\tilde{E}_X(\cdot\,;\cdot)$ are of class C^0, and that $\hat{\varrho}_R(\cdot)$, $\hat{T}(F;\cdot)$ and $E_X(F;\cdot)$ are differentiable.[100]

By (24.1) and $(24.2)_1$, we have for $X\neq Y(t)$

$$\dot{T}=E_X\dot{F}, \tag{24.4}$$

so that, by (5.20), (5.23) and the hypothesis (A-1) on the properties of acceleration waves given in Sect. 5, the speed of propagation given by

$$V^2=\frac{E_{Y_t}}{\varrho_R}, \tag{24.5}$$

where

$$E_{Y_t}=E_X\big(F(Y(t),t);Y(t)\big), \tag{24.6}$$

with Y_t being an abbreviation for $Y(t)$. Of course, in (24.5), ϱ_R is evaluated at $Y(t)$.

Further, differentiating (24.4) with respect to time, we have, upon using $(24.2)_2$, the result

$$\ddot{T}=\tilde{E}_X\dot{F}^2+E_X\ddot{F},$$

which is valid for $X\neq Y(t)$, so that, by the hypothesis (A-1),

$$[\ddot{T}]=2\dot{F}^+\tilde{E}_{Y_t}[\dot{F}]+\tilde{E}_{Y_t}[\dot{F}]^2+E_{Y_t}[\ddot{F}], \tag{24.7}$$

where

$$\tilde{E}_{Y_t}=\tilde{E}_X\big(F(Y(t),t);Y(t)\big). \tag{24.8}$$

---

[99] The results given in this section are due to COLEMAN, GREENBERG and GURTIN [1966, *3*]. These results have been extended by CHEN [1972, *4*] to include the influence of thermodynamics.

[100] COLEMAN, GREENBERG and GURTIN [1966, *3*] assumed that $\hat{\varrho}_R(\cdot)$ is of class C^1.

Differentiating (24.5), we have

$$\begin{aligned}\frac{\delta_1 V}{\delta t} &= -\frac{V}{2\varrho_R}\frac{\delta_1 \varrho_R}{\delta t} + \frac{1}{2\varrho_R V}\frac{\delta_1 E_{Y_t}}{\delta t} \\ &= -\frac{V}{2\varrho_R}\frac{\delta_1 \varrho_R}{\delta t} + \frac{\dot{F}^+ \tilde{E}_{Y_t}}{2\varrho_R V} + \frac{\tilde{E}_{Y_t}}{2\varrho_R}\left(\frac{\partial F}{\partial X}\right)^+ + \frac{1}{2\varrho_R}\left(\frac{\partial E_X}{\partial X}\right)^+ .\end{aligned} \tag{24.9}$$

Finally, by (5.19), (5.20), (5.25), (24.5), (24.7), and (24.9), we have

Theorem 24.1. *The amplitude of an acceleration wave propagating in an inhomogeneous elastic body obeys the equation*

$$\frac{\delta_1 a}{\delta t} = -\mu_X a + \beta_X a^2, \tag{24.10}$$

where

$$\begin{aligned}\mu_X &= \frac{1}{4}\left\{\frac{1}{\varrho_R}\frac{\delta_1 \varrho_R}{\delta t} + \frac{1}{E_{Y_t}}\frac{\delta_1 E_{Y_t}}{\delta t}\right\} - \frac{\dot{F}^+ \tilde{E}_{Y_t}}{E_{Y_t}} \\ &= \frac{1}{4}\left\{\frac{1}{\varrho_R}\frac{\delta_1 \varrho_R}{\delta t} - 3\frac{\dot{F}^+ \tilde{E}_{Y_t}}{E_{Y_t}} + \frac{\tilde{E}_{Y_t}}{\varrho_R V}\left(\frac{\partial F}{\partial X}\right)^+ + \frac{1}{\varrho_R V}\left(\frac{\partial E_X}{\partial X}\right)^+\right\},\end{aligned} \tag{24.11}$$

and

$$\beta_X = -\frac{\tilde{E}_{Y_t}}{2 E_{Y_t} V}. \tag{24.12}$$

In (24.11) *and* (24.12), ϱ_R *is evaluated at* $Y(t)$.

Notice that the coefficient μ_X, defined by (24.11), depends on the local elastic properties of the material as well as the inhomogeneity of the body; it also depends on the deformation rate and gradient of deformation ahead of the wave. On the other hand, the coefficient β_X, defined by (24.12), depends only on the local elastic properties of the material. In view of this observation we see that the local and global behavior of the amplitude of the wave would be as predicted in Sect. 13. In the following section, we shall consider a specific application of (24.10).

25. Acceleration waves in inhomogeneous elastic bodies at rest. In the preceding section, we have derived the differential equation which the amplitude of an acceleration wave propagating in an inhomogeneous elastic body must obey. Here, we shall consider a specific application of (24.10) from which important consequences do follow.[101]

We assume that the body is initially at rest in a non-homogeneous reference configuration; that is, for all t and all $X > Y(t)$ the motion ahead of the wave is known and is given by

$$\chi(X,t) = \mathring{\chi}(X), \tag{25.1}$$

and satisfying the equilibrium condition

$$\mathring{E}_X \frac{d\mathring{F}}{dX} + \frac{\partial \hat{T}}{\partial X} = 0, \tag{25.2}$$

with

$$\mathring{F}(X) = \frac{d\mathring{\chi}(X)}{dX}. \tag{25.3}$$

---

[101] The results given in this section are due to P. J. CHEN; they have not been published previously in the open literature.

Of course, (25.2) is the consequence of balance of momentum $(5.15)_1$, the constitutive relation (24.1), $(24.2)_1$ and (25.1). Further, in (25.2), $\mathring{E}_X$ is the equilibrium local tangent modulus, i.e.,

$$\mathring{E}_X = E_X(\mathring{F}; X) \tag{25.4}$$

for all t and all $X > Y(t)$.

Formula (25.1) implies that for all t and all $X > Y(t)$

$$\dot{F}(X, t) = 0,$$

hence taking the limit as $X \downarrow Y(t)$, we have

$$\dot{F}^+ = 0. \tag{25.5}$$

Further, the hypothesis (A-1) together with (25.1) and (25.3) implies that

$$\begin{aligned} F(Y(t); t) &= \mathring{F}(Y(t)), \\ \left(\frac{\partial F}{\partial X}\right)^+ &= \left(\frac{d\mathring{F}}{dX}\right)^+, \end{aligned} \tag{25.6}$$

and, by (25.4),

$$\mathring{\tilde{E}}_X = \tilde{E}_X(\mathring{F}; X) \tag{25.7}$$

for all t and all $X > Y(t)$. Consequently, by (24.10) and (25.4) through (25.7), we have

Theorem 25.1. *Let Theorem 24.1 hold. Suppose that the region ahead of the wave is at rest in a non-homogeneous reference configuration. Then the amplitude of an acceleration wave obeys the equation*

$$\frac{\delta_1 a}{\delta t} = -\mathring{\mu}_X\, a + \mathring{\beta}_X\, a^2, \tag{25.8}$$

where

$$\begin{aligned} \mathring{\mu}_X &= \frac{1}{4}\left\{\frac{\mathring{V}}{\varrho_R}\left(\frac{d\varrho_R}{dX}\right)^+ + \frac{\mathring{V}}{\mathring{E}_{Y_t}}\left(\frac{d\mathring{E}_X}{dX}\right)^+\right\} \\ &= \frac{1}{4}\left\{\frac{\mathring{V}}{\varrho_R}\left(\frac{d\varrho_R}{dX}\right)^+ + \frac{\mathring{V}\mathring{\tilde{E}}_{Y_t}}{\mathring{E}_{Y_t}}\left(\frac{d\mathring{F}}{dX}\right)^+ + \frac{\mathring{V}}{\mathring{E}_{Y_t}}\left(\frac{\partial \mathring{E}_X}{\partial X}\right)^+\right\}, \end{aligned} \tag{25.9}$$

and

$$\mathring{\beta}_X = -\frac{\mathring{\tilde{E}}_{Y_t}}{2\mathring{E}_{Y_t}\mathring{V}}. \tag{25.10}$$

In (25.9) *and* (25.10), ϱ_R *is evaluated at* $Y(t)$, *and*

$$\mathring{V} = \left(\frac{\mathring{E}_{Y_t}}{\varrho_R}\right)^{\frac{1}{2}}. \tag{25.11}$$

Since $\mathring{\mu}_X$ and $\mathring{\beta}_X$ are functions of time, it is clear that the local and global behavior of the amplitude of an acceleration wave would again be as predicted in Sect. 13. However, here we may be more specific about its behavior simply knowing certain properties of the material.

Before proceeding to do this, let us define the quantity $\mathring{\lambda}_X$ analogous to (13.11). Thus, by (25.9) and (25.10),

$$\mathring{\lambda}_X = -\frac{\mathring{V}^2}{2\mathring{\tilde{E}}_{Y_t}}\left\{\mathring{V}^2\left(\frac{d\varrho_R}{dX}\right)^+ + \left(\frac{d\mathring{E}_X}{dX}\right)^+\right\}. \tag{25.12}$$

In the sequel, we shall consider two specific physical circumstances:

(α) *An acceleration wave is propagating into a region of increasing density and increasing tangent modulus.*

(β) *An acceleration wave is propagating into a region of decreasing density and decreasing tangent modulus.*

Case (α).

If case (α) holds, i.e.,

$$\operatorname{sgn}\left(\frac{d\varrho_R}{dX}\right)^+ = \operatorname{sgn}\left(\frac{d\mathring{E}_X}{dX}\right)^+ = +1,$$

then, by (25.9) and (25.12), we see that

$$\mathring{\mu}_X(t) > 0,$$

and

$$\operatorname{sgn} \mathring{\lambda}_X(t) = -\operatorname{sgn} \mathring{\tilde{E}}_{Y_t}(t).$$

Suppose that $\mathring{\mu}_X$ is bounded away from zero and $|\mathring{\beta}_X|$ is bounded above. Then, by (13.12), we see that the critical initial amplitude, denoted by $\mathring{\alpha}_X$, is non-zero and is finite.

In view of the preceding results and the general results given in Sect. 13, we have the following summary:

(i) *The critical initial amplitude* $\mathring{\alpha}_X \neq 0$ *and is finite.*

(ii) *When* $\mathring{\tilde{E}}_{Y_t}(t) < 0$, *the amplitude of a compressive wave* $(a(t) > 0)$ *will either become infinite within a finite time if* $a(0) > \mathring{\alpha}_X$ *or become arbitrarily small if* $a(0) < \mathring{\alpha}_X$; *the amplitude of an expansive wave* $(a(t) < 0)$ *will always become arbitrarily small.*

(iii) *When* $\mathring{\tilde{E}}_{Y_t}(t) > 0$, *the amplitude of a compressive wave will always become arbitrarily small; the amplitude of an expansive wave will either become infinite within a finite time if* $|a(0)| > \mathring{\alpha}_X$ *or become arbitrarily small if* $|a(0)| < \mathring{\alpha}_X$.

Case (β).

If case (β) holds, i.e.,

$$\operatorname{sgn}\left(\frac{d\varrho_R}{dX}\right)^+ = \operatorname{sgn}\left(\frac{d\mathring{E}_X}{dX}\right)^+ = -1,$$

then, by (25.9) and (25.12),

$$\mathring{\mu}_X(t) < 0,$$

and

$$\operatorname{sgn} \mathring{\lambda}_X(t) = \operatorname{sgn} \mathring{\tilde{E}}_{Y_t}(t).$$

Again, suppose that $\mathring{\mu}_X$ is bounded away from zero and $|\mathring{\beta}_X|$ is bounded above. Then, by (13.12), we see that the critical initial amplitude $\mathring{\alpha}_X$ vanishes. Therefore, we have the following summary:

(i*) *The critical initial amplitude* $\mathring{\alpha}_X = 0$.

(ii*) *When* $\mathring{\tilde{E}}_{Y_t}(t) < 0$, *the amplitude of a compressive wave will become infinite within a finite time; the amplitude of an expansive wave will either grow if* $|a(0)| < |\mathring{\lambda}_X(0)|$ *or decay if* $|a(0)| > |\mathring{\lambda}_X(0)|$.

(iii*) *When* $\mathring{\tilde{E}}_{Y_t}(t) > 0$, *the amplitude of a compressive wave will either grow if* $a(0) < \mathring{\lambda}_X(0)$ *or decay if* $a(0) > \mathring{\lambda}_X(0)$; *the amplitude of an expansive wave will become infinite within a finite time.*

Notice that the results for case (α) are quite different from those for case (β). In view of the results for case (α) for which $\mathring{\alpha}_X \neq 0$ and those of Theorem 14.2, we have the following important observation:[102]

Remark 25.1. The behavior of the amplitude of an acceleration wave propagating into a region of increasing density and increasing local tangent modulus of an inhomogeneous elastic body is qualitatively the same as that of a wave propagating in a body of simple material with memory which is initially at rest in a homogeneous configuration.

26. Shock waves in inhomogeneous elastic bodies. In Sects. 24 and 25, we have shown that under certain conditions it is possible for the amplitude of an acceleration wave propagating in inhomogeneous elastic bodies to become infinite. This, of course, suggests the formation of a shock.

In this section, we shall examine the properties of shock waves in inhomogeneous elastic bodies.[103] Here, it is more convenient to presume that the stress $T(X, t)$ is determined by the strain $\varepsilon(X, t)$ at X, and X:

$$T(X,t) = \tilde{T}(\varepsilon(X,t); X). \tag{26.1}$$

We assume that $\tilde{T}(\cdot, X)$ is of class C^2. Here, the tangent modulus E_X and the second-order tangent modulus $\tilde{E}_X$ are given by

$$E_X = \frac{\partial \tilde{T}}{\partial \varepsilon}, \quad \tilde{E}_X = \frac{\partial^2 \tilde{T}}{\partial \varepsilon^2}; \tag{26.2}$$

and, as in Sect. 24, we assume that $\tilde{T}(\cdot;\cdot)$, $E_X(\cdot;\cdot)$ and $\tilde{E}_X(\cdot;\cdot)$ are of class C^0, $\hat{\varrho}_R(\cdot)$, $\tilde{T}(\varepsilon;\cdot)$ and $E_X(\varepsilon;\cdot)$ are differentiable, and

$$E_X(\varepsilon; X) > 0, \quad \tilde{E}_X(\varepsilon; X) \neq 0. \tag{26.3}$$

By (26.1), (26.2) and the hypothesis (S-1) on the properties of shock waves given in Sect. 5, the shock amplitude equation (5.18) becomes

$$2V\frac{\delta_1[\varepsilon]}{\delta t} + [\varepsilon]\frac{\delta_1 V}{\delta t} = \left(V^2 - \frac{E_{Y_t}^-}{\varrho_R}\right)\left[\frac{\partial \varepsilon}{\partial X}\right] - \frac{[E_{Y_t}]}{\varrho_R}\left(\frac{\partial \varepsilon}{\partial X}\right)^+ - \frac{1}{\varrho_R}\left[\frac{\partial \tilde{T}}{\partial X}\right], \tag{26.4}$$

where

$$E_{Y_t}^- = E_X(\varepsilon^-; Y(t)), \quad E_{Y_t}^+ = E_X(\varepsilon^+; Y(t)); \tag{26.5}$$

and, by (5.17) and (26.1), the speed of propagation is given by

$$V^2 = \frac{T^- - T^+}{\varrho_R[\varepsilon]}, \tag{26.6}$$

where

$$T^- = \tilde{T}(\varepsilon^-; Y(t)), \quad T^+ = \tilde{T}(\varepsilon^+; Y(t)). \tag{26.7}$$

Differentiating (26.6), we have

$$2\varrho_R V\frac{\delta_1 V}{\delta t} = -V^2\frac{\delta_1 \varrho_R}{\delta t} + \frac{E_{Y_t}^-}{[\varepsilon]}\frac{\delta_1 \varepsilon^-}{\delta t} + \frac{V}{[\varepsilon]}\left(\frac{\partial \tilde{T}}{\partial X}\right)^- - \frac{E_{Y_t}^+}{[\varepsilon]}\frac{\delta_1 \varepsilon^+}{\delta t} - \frac{V}{[\varepsilon]}\left(\frac{\partial \tilde{T}}{\partial X}\right)^+ - \frac{\varrho_R V^2}{[\varepsilon]}\frac{\delta_1[\varepsilon]}{\delta t},$$

102 In this regard, also refer to Remark 21.4 of this article.

103 The results given in this section are due to CHEN [1972, 5].

which may be rewritten in the form

$$2\varrho_R V \frac{\delta_1 V}{\delta t} = -V^2 \frac{\delta_1 \varrho_R}{\delta t} + \frac{(\overline{E_{Y_t}} - \varrho_R V^2)}{[\varepsilon]} \frac{\delta_1 [\varepsilon]}{\delta t} + \frac{[E_{Y_t}]}{[\varepsilon]} \frac{\delta_1 \varepsilon^+}{\delta t} + \frac{V}{[\varepsilon]} \left[\frac{\partial \tilde{T}}{\partial X}\right], \tag{26.8}$$

where $\delta_1 \varepsilon^+/\delta t$ is given by

$$\frac{\delta_1 \varepsilon^+}{\delta t} = \dot{\varepsilon}^+ + V\left(\frac{\partial \varepsilon}{\partial X}\right)^+. \tag{26.9}$$

Thus, by (26.4), (26.8) and (26.9), we have

Theorem 26.1. *The amplitude of a shock wave propagating in an inhomogeneous elastic body obeys the equation*

$$\frac{\delta_1 [\varepsilon]}{\delta t} = \frac{\overline{E_{Y_t}} - \varrho_R V^2}{2\varrho_R V \left(1 + \dfrac{\overline{E_Y} - \varrho_R V^2}{4\varrho_R V^2}\right)} \left\{\lambda_X^* - \left[\frac{\partial \varepsilon}{\partial X}\right]\right\}, \tag{26.10}$$

where

$$\lambda_X^* = -\frac{1}{2(\overline{E_{Y_t}} - \varrho_R V^2)} \left\{[E_{Y_t}]\left\{\frac{\dot{\varepsilon}^+}{V} + 3\left(\frac{\partial \varepsilon}{\partial X}\right)^+\right\} + 3\left[\frac{\partial \tilde{T}}{\partial X}\right] - [\varepsilon] V \frac{\delta_1 \varrho_R}{\delta t}\right\}. \tag{26.11}$$

Notice that (26.10) is of the same form as (19.11), that of a shock wave in a homogeneous elastic body. Therefore, it is clear that we may determine the behavior of the amplitudes of certain shock waves which correspond to those considered in Sect. 19. We simply need to assume that the local elastic response of the inhomogeneous elastic body has certain properties. Without duplicating the proofs, we have

Theorem 26.2. *Let Theorem 26.1 hold.*

(i) *Consider a compression shock entering a body which is initially in compression. Suppose that the local stress-strain law in compression is concave from below. Then at any instant*

$$\left[\frac{\partial \varepsilon}{\partial X}\right] < \lambda_X^* \Leftrightarrow \frac{\delta_1 |[\varepsilon]|}{\delta t} < 0,$$

$$\left[\frac{\partial \varepsilon}{\partial X}\right] > \lambda_X^* \Leftrightarrow \frac{\delta_1 |[\varepsilon]|}{\delta t} > 0,$$

$$\left[\frac{\partial \varepsilon}{\partial X}\right] = \lambda_X^* \Leftrightarrow \frac{\delta_1 [\varepsilon]}{\delta t} = 0.$$

(ii) *Consider an expansion shock entering a body which is initially in tension. Suppose that the local stress-strain law in tension is concave from above. Then at any instant*

$$\left[\frac{\partial \varepsilon}{\partial X}\right] < \lambda_X^* \Leftrightarrow \frac{\delta_1 [\varepsilon]}{\delta t} > 0,$$

$$\left[\frac{\partial \varepsilon}{\partial X}\right] > \lambda_X^* \Leftrightarrow \frac{\delta_1 [\varepsilon]}{\delta t} < 0,$$

$$\left[\frac{\partial \varepsilon}{\partial X}\right] = \lambda_X^* \Leftrightarrow \frac{\delta_1 [\varepsilon]}{\delta t} = 0.$$

Theorem 26.3. *Let Theorem 26.1 hold.*

(i*) *Consider a compression shock entering a body which is initially in tension such that the region behind the wave remains in tension. Suppose that the local stress-*

strain law in tension is convex from above. Then at any instant

$$\left[\frac{\partial \varepsilon}{\partial X}\right] < \lambda_X^* \Leftrightarrow \frac{\delta_1 |[\varepsilon]|}{\delta t} < 0,$$

$$\left[\frac{\partial \varepsilon}{\partial X}\right] > \lambda_X^* \Leftrightarrow \frac{\delta_1 |[\varepsilon]|}{\delta t} > 0,$$

$$\left[\frac{\partial \varepsilon}{\partial X}\right] = \lambda_X^* \Leftrightarrow \frac{\delta_1 [\varepsilon]}{\delta t} = 0.$$

(ii*) *Consider an expansion shock entering a body which is initially in compression such that the region behind the wave remains in compression. Suppose that the local stress-strain law in compression is convex from below. Then at any instant*

$$\left[\frac{\partial \varepsilon}{\partial X}\right] < \lambda_X^* \Leftrightarrow \frac{\delta_1 [\varepsilon]}{\delta t} > 0,$$

$$\left[\frac{\partial \varepsilon}{\partial X}\right] > \lambda_X^* \Leftrightarrow \frac{\delta_1 [\varepsilon]}{\delta t} < 0,$$

$$\left[\frac{\partial \varepsilon}{\partial X}\right] = \lambda_X^* \Leftrightarrow \frac{\delta_1 [\varepsilon]}{\delta t} = 0.$$

Of course, we call the number λ_X^*, defined by (26.11), the *critical jump in strain gradient* for shock waves in inhomogeneous elastic bodies.[104] Notice that, for a given material, λ_X^* depends on its local elastic properties as well as the material inhomogeneity; it also depends on the properties of the motion ahead of the shock.

27. Shock waves in inhomogeneous elastic bodies at rest. In the preceding section, we have derived the differential equation which the amplitude of a shock wave propagating in an inhomogeneous elastic body must obey. We showed that the behavior of the shock depends on the relative magnitudes of the jump $[\partial\varepsilon/\partial X]$ in strain gradient and a number λ_X^* called the critical jump in strain gradient. Here, we shall assume that the region ahead of the wave is at rest.[105]

We suppose that the material is initially at rest in a non-homogeneous reference configuration in the sense made precise in Sect. 25. Let $\mathring{\varepsilon}(X)$ denote the strain field ahead of the wave, i.e.,

$$\mathring{\varepsilon}(X) = \mathring{F}(X) - 1, \tag{27.1}$$

where $\mathring{F}(X)$ is defined by (25.3). In view of the results given in Sect. 25, (26.6), $(26.7)_2$, and (27.1) imply that the speed of propagation $\mathring{V}$ is given by

$$\mathring{V}^2 = \frac{T^- - \mathring{T}^+}{\varrho_R [\varepsilon]}, \tag{27.2}$$

where

$$\mathring{T}^+ = \tilde{T}(\mathring{\varepsilon}^+; Y(t)).$$

Further, by (26.1) and $(26.5)_2$,

$$\left(\frac{\partial \tilde{T}}{\partial X}\right)^+ = \lim_{X \downarrow Y(t)} \frac{\partial \tilde{T}(\mathring{\varepsilon}; X)}{\partial X}, \qquad E_{Y_t}^+ = E_X(\mathring{\varepsilon}^+; Y(t)). \tag{27.3}$$

In view of the equilibrium condition (25.2), (27.1), (27.2), and (27.3), we have

Theorem 27.1. *Let Theorem 26.1 hold. Suppose that the region ahead of the wave is at rest in a non-homogeneous reference configuration. Then the amplitude*

[104] Cf. Sects. 16, 19, 22 of this article.

[105] The results given in this section are due to CHEN [1972, *5*].

of a shock wave obeys the equation

$$\frac{\delta_1[\varepsilon]}{\delta t} = \frac{E\bar{Y}_t - \varrho_R \overset{\circ}{V}^2}{2\varrho_R \overset{\circ}{V}\left(1 + \frac{E\bar{Y}_t - \varrho_R \overset{\circ}{V}^2}{4\varrho_R \overset{\circ}{V}^2}\right)} \left\{\overset{\circ}{\lambda}{}_X^* - \left[\frac{\partial \varepsilon}{\partial X}\right]\right\}, \tag{27.4}$$

where

$$\overset{\circ}{\lambda}{}_X^* = -\frac{1}{2(E\bar{Y}_t - \varrho_R \overset{\circ}{V}^2)} \left\{3\left(\frac{\partial \tilde{T}}{\partial X}\right)^- - 3\frac{E^-}{E^+}\left(\frac{\partial \tilde{T}}{\partial X}\right)^+ - [\varepsilon]\, \overset{\circ}{V}^2 \left(\frac{d\varrho_R}{dX}\right)^+\right\}. \tag{27.5}$$

Since (27.4) is of the same form as (26.10), the behavior of the amplitude of the shock would be the same as that of the general cases considered in Sect. 26. More importantly, by Theorems 16.2, 26.2 and 26.3, we have[106]

Remark 27.1. Consider an inhomogeneous elastic body which is at rest in a non-homogeneous reference configuration. The behavior of the amplitude of a shock wave is qualitatively the same as that of a shock wave in a body of simple material with memory which is initially unstrained.

Appendix.

1. Existence of the one dimensional kinematical condition of compatibility. Here, we shall provide a rigorous derivation of the one dimensional kinematical condition of compatibility (4.24), in that we give a necessary and sufficient condition for its existence. Since this condition exists under much weaker assumptions than those encountered in practice, the treatment as well as the notation given here should be considered to be self-contained; and the smoothness hypotheses adopted previously in writing down (4.24) should be ignored.

Let $\mathscr{R}$ denote the set of real numbers, let $\mathscr{B}$ denote an interval (open, closed, or semi open; finite or infinite) in $\mathscr{R}$, and let $\overset{\circ}{\mathscr{B}}$ denote the interior of $\mathscr{B}$. For $x, y \in \mathscr{R}$, with $x < y$, $I(x, y)$ denotes $\{z \in \mathscr{R}: x < z < y\}$.

Let $Y: \mathscr{R} \to \overset{\circ}{\mathscr{B}}$, be a differentiable function. Let $\mathscr{D}$ denote

$$\{(X, t) \in \overset{\circ}{\mathscr{B}} \times \mathscr{R}: X \neq Y(t)\},$$

and let $f: \mathscr{D} \to \mathscr{R}$ be a continuous function which possesses partial derivatives f_1 and f_2 in $\mathscr{D}$. Suppose that $f_1: \mathscr{D} \to \mathscr{R}$ is continuous. (Alternately, we may assume that $f_2: \mathscr{D} \to \mathscr{R}$ is continuous; and, in addition, for each fixed $t \in \mathscr{R}$,

$$f_1(\cdot, t): \overset{\circ}{\mathscr{B}} \setminus \{Y(t)\} \to \mathscr{R}$$

is continuous.) Further, we assume that the following limits exist and we define functions as indicated for all $t \in \mathscr{R}$:

$$\begin{aligned} g_u(Y(t), t) &= \lim_{X \downarrow Y(t)} f(X, t), \\ h_u(Y(t), t) &= \lim_{X \downarrow Y(t)} f_1(X, t), \\ i_u(Y(t), t) &= \lim_{X \downarrow Y(t)} f_2(X, t), \end{aligned} \tag{1.1}$$

[106] In this regard, also refer to Remark 23.1 of this article.

and

$$g_l(Y(t), t) = \lim_{X \uparrow Y(t)} f(X, t),$$
$$h_l(Y(t), t) = \lim_{X \uparrow Y(t)} f_1(X, t), \tag{1.2}$$
$$i_l(Y(t), t) = \lim_{X \uparrow Y(t)} f_2(X, t).$$

Now, if we set

$$[g] = g_l(Y(t), t) - g_u(Y(t), t),$$
$$[h] = h_l(Y(t), t) - h_u(Y(t), t), \tag{1.3}$$
$$[i] = i_l(Y(t), t) - i_u(Y(t), t),$$

then the kinematical condition of compatibility states that

$$\frac{d}{dt}[g] = [h]\frac{dY(t)}{dt} + [i]. \tag{1.4}$$

The proof of the existence of (1.4) is based on the following simple lemma. The lemma will be stated only for the case of real valued functions defined on subsets of the plane, although more general statements are clearly valid.

Lemma. *Suppose $A \times B \subset \mathscr{R} \times \mathscr{R}$, φ is a real valued function whose domain includes $A \times B$, and (a, b) is a limit point of $A \times B$. If the limits*

$$\lim_{x \to a} \varphi(x, y) = \psi(y), \qquad \lim_{y \to b} \varphi(x, y) = \varrho(x), \qquad \lim_{x \to a} \lim_{y \to b} \varphi(x, y) = l$$

exist (where $(x, y) \in A \times B$), then a necessary and sufficient condition that

$$\lim_{y \to b} \lim_{x \to a} \varphi(x, y) = l$$

is that for every $\varepsilon > 0$ and $\delta > 0$ there exist $z \in I(a-\delta, a+\delta) \cap A$ and $\eta > 0$ such that $|\varphi(z, y) - \psi(y)| < \varepsilon$ for all $y \in I(b-\eta, b+\eta) \cap B$.

The statement of the lemma motivates the condition of the theorem below. A certain amount of care is required in its application with regard to the functions involved. In an attempt to clarify this, we introduce the following notation. For each $t \in \mathscr{R}$ consider the function F_t defined as follows: The domain of F_t is

$$\mathscr{D}_t = \{(\alpha, \beta) \in \mathscr{R} \times (R \setminus \{0\}): (\alpha + Y(t+\beta), t+\beta) \in \mathscr{D} \text{ and } (\alpha + Y(t), t) \in \mathscr{D}\},$$

and for $(\alpha, \beta) \in \mathscr{D}_t$

$$F_t(\alpha, \beta) = \frac{f(\alpha + Y(t+\beta), t+\beta) - f(\alpha + Y(t), t)}{\beta}.$$

Note that $\mathscr{D}_t$ is an open set, $(0, 0) \notin \mathscr{D}_t$, but is a point of accumulation of $\mathscr{D}_t$; and if $(\alpha, \beta) \in \mathscr{D}_t$, then $(\alpha, \beta') \in \mathscr{D}_t$ for all $\beta' \neq 0$ in some neighborhood of 0.

Also let G_u and G_l denote the functions on $\mathscr{R} \setminus \{0\}$ into $\mathscr{R}$ defined, for each $\beta \in \mathscr{R} \setminus \{0\}$, by

$$G_u(\beta) = \frac{g_u(Y(t+\beta), t+\beta) - g_u(Y(t), t)}{\beta},$$
$$G_l(\beta) = \frac{g_l(Y(t+\beta), t+\beta) - g_l(Y(t), t)}{\beta}.$$

Theorem. *Under the assumptions on f and Y, the condition of compatibility* (1.4) *is valid if and only if for each $t \in \mathscr{R}$ and $\varepsilon, \delta > 0$ there exist $\alpha_1 \in I(0, \delta)$, $\alpha_2 \in I(-\delta, 0)$,* *and* $\eta > 0$ *such that for all* $\beta \in I(-\eta, \eta)\setminus\{0\}$:

$$(\alpha_i, \beta) \in \mathscr{D}_t, \qquad i = 1, 2,$$
$$|F_t(\alpha_1, \beta) - G_u(\beta)| < \varepsilon, \qquad |F_t(\alpha_2, \beta) - G_l(\beta)| < \varepsilon.$$

Proof. *Sufficiency.* Of course, we presume that the validity of (1.4) is equivalent to the validity of both of the following for each $t \in \mathscr{R}$:

$$\frac{d}{dt} g_u(Y(t), t) = h_u(Y(t), t)\frac{dY(t)}{dt} + i_u(Y(t), t), \tag{1.5}$$

$$\frac{d}{dt} g_l(Y(t), t) = h_l(Y(t), t)\frac{dY(t)}{dt} + i_l(Y(t), t). \tag{1.6}$$

We will only establish (1.5), since (1.6) follows by a similar argument. For $t \in \mathscr{R}$, there exist open intervals $U = I(0, \alpha')$ and $V = I(-\beta', \beta')$, $\alpha', \beta' > 0$, such that $U \times (V\setminus\{0\}) \subset \mathscr{D}_t$. For $\alpha \in U$ we have, by the smoothness assumptions on f and Y,

$$\frac{d}{dt} f(\alpha + Y(t), t) = f_1(\alpha + Y(t), t)\frac{dY(t)}{dt} + f_2(\alpha + Y(t), t), \tag{1.7}$$

i.e., $\lim\limits_{\substack{\beta\to 0\\ \beta\in V\setminus\{0\}}} F_t(\alpha, \beta)$ exists and is equal to the right-hand member of (1.7). By definitions $(1.1)_{2,3}$ the right-hand limit of this expression exists:

$$\lim_{\alpha\downarrow 0}\lim_{\beta\to 0} F_t(\alpha, \beta) = h_u(Y(t), t)\frac{dY(t)}{dt} + i_u(Y(t), t).$$

Since $\lim\limits_{\alpha\downarrow 0} F_t(\alpha, \beta) = G_u(\beta)$ for $\beta \in V$, the hypothesis of the lemma is satisfied by F_t and G_u if we take $A = U$ and $B = V\setminus\{0\}$. The condition given in the theorem for F_t and G_u implies the condition of the lemma. Hence

$$\lim_{\beta\to 0} G_u(\beta) = \lim_{\beta\to 0}\lim_{\alpha\downarrow 0} F_t(\alpha, \beta) = \lim_{\alpha\downarrow 0}\lim_{\beta\to 0} F_t(\alpha, \beta), \tag{1.8}$$

which is (1.5).

Necessity. The validity of (1.4) implies the existence and equality of the iterated limits (1.8), and

$$\lim_{\beta\to 0}\lim_{\alpha\uparrow 0} F_t(\alpha, \beta) = \lim_{\alpha\uparrow 0}\lim_{\beta\to 0} F_t(\alpha, \beta).$$

By applying the lemma to F_t and G_u and then to F_t and G_l it may be seen that the condition of the theorem holds.

2. Proofs of Theorems 13.2, 13.3, 13.4 and 13.5. Here, we shall present the proofs of the theorems on the global behavior of acceleration waves stated in Sect. 13. The presentation given here is mathematically self-contained, and one need not refer to Sect. 13 for additional details.

The differential equation

$$\frac{da}{dt} = -\mu a + \beta a^2, \qquad t \in [0, \infty), \tag{2.1}$$

with $\operatorname{sgn}\beta(t) = -\operatorname{sgn}\widetilde{E}(t)$, can be readily solved in the usual way by introducing the change of variable

$$a(t) = \frac{1}{b(t)},$$

so that (2.1) reduces to

$$\frac{db}{dt} - \mu b = -\beta.$$

Let μ and β be integrable on every finite sub-interval of $[0, \infty)$; then the above equation has the solution

$$b(t) = e^{\int_0^t \mu(\tau)\,d\tau} \left\{ b(0) - \int_0^t \beta(\tau)\, e^{-\int_0^\tau \mu(s)\,ds}\, d\tau \right\}.$$

Therefore, (2.1) has the solution

$$a(t) = \frac{e^{-\int_0^t \mu(\tau)\,d\tau}}{\dfrac{1}{a(0)} - \displaystyle\int_0^t \beta(\tau)\, e^{-\int_0^\tau \mu(s)\,ds}\, d\tau}. \tag{2.2}$$

Theorem 13.2. *Consider the differential equation* (2.1). *Let* $\operatorname{sgn} a(0) = \operatorname{sgn} \tilde{E}(t)$ *and* $\liminf_{t\to\infty} |\beta(t)| \neq 0$. *If* λ *is bounded* $\left\{\begin{matrix}above\\below\end{matrix}\right\}$ *or tends to a* $\left\{\begin{matrix}non\text{-}negative\\non\text{-}positive\end{matrix}\right\}$ *finite or infinite limit* L, *the same is true for any solution* $\left\{\begin{matrix}a(t) > 0\\a(t) < 0\end{matrix}\right\}$.

Proof. Consider the case for which $a(0) > 0$. Since

$$\frac{da(t)}{dt} > 0$$

only if $a(t) < \lambda(t)$, therefore a can be unbounded only if λ is unbounded. Thus, if λ is bounded, then a is bounded.

If $\lim_{t\to\infty} \lambda(t) = 0$, then for any $\varepsilon > 0$, there exists a $T > 0$ such that

$$\lambda(t) < \varepsilon \quad \text{for all} \quad t \geqq T.$$

Then, for $t \geqq T$, either

$$a(t) < 2\varepsilon, \tag{2.3}$$

or

$$a(t) \geqq 2\varepsilon. \tag{2.4}$$

If (2.4) holds, then

$$a(t) - \lambda(t) \geqq \varepsilon$$

and, by (2.1),

$$\left|\frac{da(t)}{dt}\right| \geqq |\beta(t)| \cdot 2\varepsilon \cdot \varepsilon = 2\varepsilon^2 |\beta(t)|.$$

This means that either (2.3) holds or $da(t)/dt$ is negative and bounded away from zero. In the latter case, $a(t)$ will become less than 2ε within a finite length of time $T^* > T$. Since $\lambda(t) < \varepsilon$ for all $t \geqq T$, clearly $a(t)$ must remain less than 2ε for all $t \geqq T^*$. It follows that $\lim_{t\to\infty} a(t) = 0$.

If λ tends to a finite limit $L > 0$, a very similar argument works. Let ε be arbitrary except that $0 < \varepsilon < \frac{1}{10} L$. Let $T > 0$ be such that

$$|\lambda(t) - L| < \varepsilon \quad \text{for all} \quad t \geqq T.$$

Then, for $t \geqq T$, either

$$|a(t) - L| < 2\varepsilon \tag{2.5}$$

or

$$|a(t) - L| \geqq 2\varepsilon. \tag{2.6}$$

If (2.6) holds, either $a(t) \geqq L + 2\varepsilon$ and $a(t) - \lambda(t) \geqq \varepsilon$, so $a(t)$ is decreasing with

$$\left|\frac{d\,a(t)}{d\,t}\right| \geqq |\beta(t)| \cdot \varepsilon (L + 2\varepsilon);$$

or $a(t) \leqq L - 2\varepsilon$ and $\lambda(t) - a(t) \geqq \varepsilon$, so $a(t)$ is increasing with

$$\left|\frac{d\,a(t)}{d\,t}\right| \geqq |\beta(t)| \cdot a(T)(L - 2\varepsilon).$$

Thus, in both cases $d\,a(t)/d\,t$ is bounded away from zero, and after a finite length of time (2.5) holds, say, for all $t \geqq T^* > T$. Thus $\lim_{t\to\infty} a(t) = L$.

Finally, if $\lim_{t\to\infty} \lambda(t) = \infty$, then for $\varepsilon > 0$ there exists a $T > 0$ such that

$$\lambda(t) > \frac{1}{\varepsilon} \quad \text{for all} \quad t \geqq T.$$

Then, for $t \geqq T$, either

$$a(t) > \frac{1}{2\varepsilon}, \tag{2.7}$$

or

$$a(t) \leqq \frac{1}{2\varepsilon}. \tag{2.8}$$

If (2.8) holds, $a(t)$ is increasing with

$$\lambda(t) - a(t) \geqq \frac{1}{\varepsilon}$$

and so

$$\left|\frac{d\,a(t)}{d\,t}\right| \geqq |\beta(t)| \cdot a(T) \cdot \frac{1}{\varepsilon}.$$

Hence $d\,a(t)/d\,t$ is bounded away from zero, and after a finite length of time (2.7) holds, say, for all $t \geqq T^* > T$. Thus, $\lim_{t\to\infty} a(t) = \infty$.

The case for which $a(0) < 0$ can be treated in exactly the same way. ∎

Theorem 13.3. *Consider the differential equation* (2.1). *Let* μ *and* β *be integrable on every finite sub-interval of* $[0, \infty)$, *and let* $a_1(t)$, $a_2(t)$ *be any two solutions of* (2.1), *given by* (2.2), *with* $\operatorname{sgn} a_1(0) = \operatorname{sgn} a_2(0) = \operatorname{sgn} \tilde{E}(t)$.

(i) *If* $\lim_{t\to\infty} |a_1(t)| = \infty$, *then* $\lim_{t\to\infty} |a_2(t)| = \infty$.

(ii) *If* $a_1(t)$ *is bounded, then* $a_2(t)$ *is also bounded.*

(iii) *If* $\liminf_{t\to\infty} |a_1(t)| = 0$, *then* $\liminf_{t\to\infty} |a_2(t)| = 0$, *and* $\liminf_{t\to\infty} |a_1(t) - a_2(t)| = 0$.

(iv) *If* $\lim_{t\to\infty} |a_1(t)| = 0$, *then* $\lim_{t\to\infty} |a_2(t)| = 0$, *and* $\lim_{t\to\infty} |a_1(t) - a_2(t)| = 0$.

(v) *Let* $\liminf_{t\to\infty} |\beta(t)| \neq 0$. *If* $a_1(t)$ *is bounded and bounded away from zero, then* $a_2(t)$ *is also bounded and bounded away from zero, and*

$$\lim_{t\to\infty} |a_1(t) - a_2(t)| = 0.$$

Proof.

(i) If $a_1(t)$ and $a_2(t)$ are any two solutions of (2.1), then

$$\frac{1}{a_1(t)} - \frac{1}{a_2(t)} = \left\{\frac{1}{a_1(0)} - \frac{1}{a_2(0)}\right\} e^{\int_0^t \mu(\tau)\,d\tau}. \tag{2.9}$$

Since $a_1(0)$ and $\beta(t)$ have opposite signs,

$$\frac{1}{a_1(0)} - \int_0^t \beta(\tau)\, e^{-\int_0^\tau \mu(s)\,ds}\, d\tau$$

is bounded away from zero. So, if

$$\lim_{t\to\infty} |a_1(t)| = \infty,$$

then, by (2.2), we see that

$$\lim_{t\to\infty} e^{-\int_0^t \mu(\tau)\,d\tau} = \infty.$$

Hence, by (2.9),

$$\lim_{t\to\infty} \left| \frac{1}{a_1(t)} - \frac{1}{a_2(t)} \right| = 0,$$

which implies that

$$\lim_{t\to\infty} |a_2(t)| = \infty.$$

(ii) If $a_1(t)$ and $a_2(t)$ are any two solutions of (2.1), then

$$a_1(t) - a_2(t) = \left(\frac{1}{a_2(0)} - \frac{1}{a_1(0)}\right) \frac{a_1(t)}{\left\{\frac{1}{a_2(0)} - \int_0^t \beta(\tau)\, e^{-\int_0^\tau \mu(s)\,ds}\, d\tau\right\}}. \tag{2.10}$$

Since

$$\frac{1}{a_2(0)} - \int_0^t \beta(\tau)\, e^{-\int_0^\tau \mu(s)\,ds}\, d\tau$$

is bounded away from zero, we see that if $a_1(t)$ is bounded, then the right-hand member of (2.10) is bounded. Hence $a_1(t) - a_2(t)$ is bounded; therefore $a_2(t)$ is also bounded.

(iii) Here, we need to consider three distinct possibilities: First, if

$$\liminf_{t\to\infty} e^{-\int_0^t \mu(\tau)\,d\tau} = 0$$

then obviously

$$\liminf_{t\to\infty} |a_1(t)| = 0 \quad \text{and} \quad \liminf_{t\to\infty} |a_2(t)| = 0.$$

On the other hand, if

$$\liminf_{t\to\infty} |a_1(t)| = 0 \quad \text{and} \quad \liminf_{t\to\infty} e^{-\int_0^t \mu(\tau)\,d\tau} > 0$$

but is not ∞, then by (2.2) it must be that

$$\lim_{t\to\infty} \left| \frac{1}{a_1(0)} - \int_0^t \beta(\tau)\, e^{-\int_0^\tau \mu(s)\,ds}\, d\tau \right| = \infty. \tag{2.11}$$

Since

$$e^{-\int_0^t \mu(\tau)\,d\tau}$$

does not tend to ∞, there exists some finite constant $K > 0$ such that

$$\liminf_{t\to\infty} e^{-\int_0^t \mu(\tau)\,d\tau} < K. \tag{2.12}$$

Therefore, if

$$\liminf_{t\to\infty} |a_1(t)| = 0,$$

then, by (2.10), (2.11) and (2.12),

$$\liminf_{t\to\infty} |a_2(t)| = 0.$$

Finally, if

$$\liminf_{t\to\infty} e^{-\int_0^t \mu(\tau)\,d\tau} = \infty,$$

then

$$\limsup_{t\to\infty} e^{\int_0^t \mu(\tau)\,d\tau} = 0,$$

and

$$\limsup_{t\to\infty} \left|\frac{1}{a_1(t)}\right| = \limsup_{t\to\infty} \left|\frac{1}{a_1(t)} + \left(\frac{1}{a_2(0)} - \frac{1}{a_1(0)}\right) e^{\int_0^t \mu(\tau)\,d\tau}\right|. \tag{2.13}$$

Hence, by (2.9) and (2.13),

$$\limsup_{t\to\infty} \left|\frac{1}{a_1(t)}\right| = \limsup_{t\to\infty} \left|\frac{1}{a_2(t)}\right|.$$

Therefore, if

$$\liminf_{t\to\infty} |a_1(t)| = 0, \quad \text{then} \quad \liminf_{t\to\infty} |a_2(t)| = 0.$$

Since

$$\frac{1}{a_2(0)} - \int_0^t \beta(\tau)\, e^{-\int_0^\tau \mu(s)\,ds} d\tau$$

is bounded away from zero, and if

$$\liminf_{t\to\infty} |a_1(t)| = 0,$$

then, by (2.10),

$$\liminf_{t\to\infty} |a_1(t) - a_2(t)| = 0.$$

(iv) The proof of this part of the theorem is strictly analogous to the preceding proof of (iii); i.e., we simply need to replace

$$\liminf_{t\to\infty} \text{ and } \limsup_{t\to\infty}$$

wherever they occur in the proof of (iii) by

$$\lim_{t\to\infty}.$$

(v) If $|a_1(t)|$ is bounded and bounded away from zero, it follows directly from statements (ii) and (iii) of this theorem that the same is true for $|a_2(t)|$.

Since $|a_1(t)|$ is bounded away from zero, then

$$\liminf_{t\to\infty} e^{-\int_0^t \mu(\tau)\,d\tau} > 0,$$

and, since β is bounded away from zero,

$$\lim_{t\to\infty} \int_0^t |\beta(\tau)|\, e^{-\int_0^\tau \mu(s)\,ds} d\tau = \infty.$$

Therefore, since $|a_1(t)|$ is bounded above, it follows from (2.10) that

$$\lim_{t\to\infty} |a_1(t) - a_2(t)| = 0. \quad \blacksquare$$

Theorem 13.4. *Consider the differential equation* (2.1). *Let* sgn $a(0) = -\operatorname{sgn}\tilde{E}(t)$. *Let μ and β be integrable on every finite sub-interval of* $[0, \infty)$; *and let*

$$\alpha = \frac{1}{\int_0^\infty |\beta(t)|\, e^{-\int_0^t \mu(\tau)\,d\tau}\, dt}.$$

(i) *If* $|a(0)| > \alpha$, *there exists a unique finite time* $t_\infty > 0$ *such that*

$$\int_0^{t_\infty} \beta(t)\, e^{-\int_0^t \mu(\tau)\,d\tau}\, dt = \frac{1}{a(0)}, \tag{2.14}$$

and

$$\lim_{t\to t_\infty} |a(t)| = \infty.$$

(ii) *Let* $\liminf_{t\to\infty} |\beta(t)| \neq 0$. *If* $|a(0)| < \alpha$, *then* $\liminf_{t\to\infty} |a(t)| = 0$.

Proof.

(i) Since $a(0)$ and $\beta(t)$ have the same sign, and since the integral

$$\int_0^t \beta(\tau)\, e^{-\int_0^\tau \mu(s)\,ds}\, d\tau$$

is a continuous, monotone function of t, it follows that if $|a(0)| > \alpha$ then there is a unique value of time, say t_∞, such that (2.14) is satisfied and the denominator of (2.2) vanishes. Obviously,

$$\lim_{t\to t_\infty} |a(t)| = \infty.$$

(ii) In order that $|a(0)| < \alpha$ be possible it must be that $\alpha > 0$, which is to say the integral

$$\int_0^t \beta(\tau)\, e^{-\int_0^\tau \mu(s)\,ds}\, d\tau$$

tends to a finite limit as $t\to\infty$. Since $\beta(t)$ is of constant sign and is bounded away from zero, it must be that

$$\liminf_{t\to\infty} e^{-\int_0^t \mu(\tau)\,d\tau} = 0;$$

and since

$$\frac{1}{a(0)} - \int_0^t \beta(\tau)\, e^{-\int_0^\tau \mu(s)\,ds}\, d\tau$$

is bounded away from zero, it follows from (2.2) that

$$\liminf_{t\to\infty} |a(t)| = 0. \quad \blacksquare$$

Theorem 13.5. *Let the conditions of Theorem 13.4 hold.*

(i) *If* $|a(0)| > \alpha$ *and* $\beta(t)$ *is continuous from below at* t_∞, *then*

$$a(t) = \frac{1}{\beta(t_\infty)(t - t_\infty)}\left(1 + o(1)\right) \quad \text{as} \quad t \to t_\infty, \tag{2.15}$$

where $\beta(t_\infty) = \lim_{t\to t_\infty} \beta(t)$.

(ii) *If* $|a(0)|>\alpha$, $\beta(0)\neq 0$, *and* $\beta(t)$ *is differentiable with* $\left|\frac{\beta'(t)}{\beta(t)}-\mu(t)\right|$ *bounded near zero, then*

$$t_\infty=\frac{1}{a(0)\,\beta(0)}\big(1+o(1)\big) \quad \text{as} \quad |a(0)|\to\infty. \tag{2.16}$$

(iii) *If* $|a(0)|<\alpha$, *then*

$$a(t)=\frac{a(0)\,e^{-\int_0^t \mu(\tau)\,d\tau}}{1-\frac{|a(0)|}{\alpha}}\big(1+o(1)\big) \tag{2.17}$$

as $t\to\infty$ *or as* $a(0)\to 0$.

Proof.

(i) Let

$$F(t)\equiv e^{-\int_0 \mu(\tau)\,d\tau}.$$

By (2.2),

$$a(t)=\frac{F(t)}{(t_\infty-t)\,\beta(t_\infty)\,F(t_\infty)}\left\{1-\frac{\int_t^{t_\infty}[\beta(\tau)\,F(\tau)-\beta(t_\infty)\,F(t_\infty)]\,d\tau}{\int_t^{t_\infty}\beta(\tau)\,F(\tau)\,d\tau}\right\}.$$

Clearly, both integrals vanish when $t=t_\infty$, and the ratio of their derivatives is simply

$$\frac{\beta(t)F(t)-\beta(t_\infty)\,F(t_\infty)}{\beta(t)\,F(t)},$$

which has limit zero as $t\to t_\infty$. Hence, by L'Hospital's rule, it follows that

$$a(t)=\frac{F(t)}{(t_\infty-t)\,\beta(t_\infty)\,F(t_\infty)}\big(1+o(1)\big) \quad \text{as} \quad t\to t_\infty.$$

Since $\frac{F(t)}{F(t_\infty)}=1+o(1)$,

$$a(t)=\frac{1}{(t_\infty-t)\,\beta(t_\infty)}\big(1+o(1)\big) \quad \text{as} \quad t\to t_\infty.$$

(ii) Formula (2.14) defines t_∞ as a function of the variable $x=\frac{1}{a(0)}$, which tends to zero as $|a(0)|\to\infty$; i.e., $t_\infty(0)=0$. By Taylor's theorem

$$t_\infty(x)=t_\infty(0)+x\,t'_\infty(0)+\frac{x^2}{2}\,t''_\infty(\theta) \qquad \text{for some } \theta\in(0,x).$$

But

$$t'_\infty(x)=\frac{1}{\beta(t_\infty)\,F(t_\infty)},$$

$$\begin{aligned} t''_\infty(x)&=-\frac{1}{\beta(t_\infty)\,F(t_\infty)}\left\{\frac{\beta'(t_\infty)}{\beta(t_\infty)}+\frac{F'(t_\infty)}{F(t_\infty)}\right\}\\ &=-\frac{1}{\beta(t_\infty)\,F(t_\infty)}\left\{\frac{\beta'(t_\infty)}{\beta(t_\infty)}-\mu(t_\infty)\right\}. \end{aligned}$$

Hence

$$\begin{aligned} t_\infty\left(\frac{1}{a(0)}\right)&=\frac{1}{a(0)}\cdot\frac{1}{\beta(0)\,F(0)}-\frac{1}{2(a(0))^2}\cdot\frac{1}{\beta(t_\infty(\theta))\,F(t_\infty(\theta))}\left(\frac{\beta'(t_\infty(\theta))}{\beta(t_\infty(\theta))}+\frac{F'(t_\infty(\theta))}{F(t_\infty(\theta))}\right)\\ &=\frac{1}{a(0)\,\beta(0)}\left\{1-\frac{1}{2(a(0))}\cdot\frac{\beta(0)}{\beta(t_\infty(\theta))\,F(t_\infty(\theta))}\left(\frac{\beta'(t_\infty(\theta))}{\beta(t_\infty(\theta))}-\mu(t_\infty(\theta))\right)\right\}. \end{aligned}$$

Since β and F are continuous near zero, hence bounded there, and since

$$\frac{\beta'(t_\infty(\theta))}{\beta(t_\infty(\theta))} - \mu(t_\infty(\theta))$$

is bounded near zero by assumption, it follows that

$$t_\infty\left(\frac{1}{a(0)}\right) = \frac{1}{a(0)\,\beta(0)}\,(1+o(1)) \quad \text{as} \quad |a(0)| \to \infty .$$

(iii) By (2.2),

$$\begin{aligned} a(t) &= F(t)\left[\frac{1}{a(0)} - \int_0^\infty \beta(\tau)\,F(\tau)\,d\tau + \int_t^\infty \beta(\tau)\,F(\tau)\,d\tau\right]^{-1} \\ &= F(t)\left[\frac{1}{a(0)} - \int_0^\infty \beta(\tau)\,F(\tau)\,d\tau\right]^{-1} \left\{1 - \frac{\int_t^\infty \beta(\tau)\,F(\tau)\,d\tau}{\frac{1}{a(0)} - \int_0^t \beta(\tau)\,F(\tau)\,d\tau}\right\}. \end{aligned}$$

Since $\int_0^\infty \beta(\tau)\,F(\tau)\,d\tau$ is finite,

$$\int_t^\infty \beta(\tau)\,F(\tau)\,d\tau \to 0$$

as $t\to\infty$, and since

$$\frac{1}{a(0)} - \int_0^t \beta(\tau)\,F(\tau)\,d\tau$$

is bounded away from zero, it follows that

$$a(t) = a(0)\,F(t)\left[1 - a(0)\int_0^\infty \beta(\tau)\,F(\tau)\,d\tau\right]^{-1} (1+o(1)) \quad \text{as} \quad t\to\infty .$$

Clearly, the preceding result is also valid as $a(0)\to 0$ with $o(1)$ being uniform in t. In other words,

$$a(t) = \frac{a(0)\,F(t)}{1 - \frac{|a(0)|}{\alpha}}\,(1+o(1)) \quad \text{as} \quad t\to\infty \text{ or as } a(0)\to 0. \quad \blacksquare$$

Corollary. *Let the conditions of parts* (i) *and* (ii) *of Theorem 13.5 hold simultaneously. Then*

$$a(t) = \frac{1}{\beta\left(\frac{1}{a(0)\,\beta(0)}\right)(t-t_\infty)}\,(1+o(1)) \tag{2.18}$$

as $t\to t_\infty$ *and* $|a(0)|\to\infty$, *where* t_∞ *is given by* (2.16).

Proof.

$$\frac{1}{\beta(t_\infty)} = \frac{1}{\beta\left(\frac{1}{a(0)\,\beta(0)}\right)}\left\{1+\beta\left(\frac{1}{a(0)\,\beta(0)}\right)\left[\frac{1}{\beta(t_\infty)} - \frac{1}{\beta\left(\frac{1}{a(0)\,\beta(0)}\right)}\right]\right\}.$$

In view of (2.16) and the continuity of β from below at t_∞, the quantity in square brackets tends to zero as $|a(0)|\to\infty$. Hence

$$\frac{1}{\beta(t_\infty)}=\frac{1}{\beta\left(\frac{1}{a(0)\,\beta(0)}\right)}\big(1+o(1)\big) \quad \text{as} \quad |a(0)|\to\infty. \tag{2.19}$$

Since the $o(1)$ terms appearing in (2.15) and in (2.19) are completely independent of each other, the two estimates may be combined to give (2.18). ▮

A possibility not mentioned in Theorem 13.4 is that a solution of (2.1) could have its initial value exactly equal in absolute value to α, in which case its global behavior could be almost anything at all. This mathematical possibility is not of any practical interest, however, since all other solutions beginning arbitrarily nearby must either "blow up" in a finite length of time or else become arbitrarily small.

3. Derivation of (16.12). Since $V(t)>0$, the function $Y(\cdot)$ is invertible. We let $\hat{t}(\cdot)$ designate its inverse and, for convenience, we write

$$Y_t=Y(t), \qquad t_X=\hat{t}(X). \tag{3.1}$$

Choose $(Z,\tau)\in\mathcal{N}-\Sigma$ with $Z<Y_t$ and let

$$\Omega=\{(X,t):\ X_0<X<X_1,\ -\infty<t<t_0\},$$
$$\Omega^-=\{(X,t)\in\Omega:\ X<Y_t\},$$

with X_0, X_1, and t_0 chosen so that $(Z,\tau)\in\Omega$, $[X_0,X_1]$ is contained in the domain of $\hat{t}(\cdot)$, and

$$\Omega^-\subset\mathcal{N}.$$

For any $(X,t)\in\Omega$ let

$$\tilde{\varepsilon}(X,t)=\begin{cases}\varepsilon^-(t), & t\geqq t_X\\ 0, & t<t_X\end{cases} \tag{3.2}$$

and let

$$\hat{\varepsilon}(X,t)=\varepsilon(X,t)-\tilde{\varepsilon}(X,t). \tag{3.3}$$

Then $\hat{\varepsilon}$ is continuous on Ω and vanishes on $\Omega-\Omega^-$.

Now choose $h>0$ with $(Z+h,\tau)\in\Omega^-$ and let $\delta_h\in\mathcal{D}$ denote the function

$$\delta_h=\frac{1}{h}\left[\tilde{\varepsilon}^t(Z+h)-\tilde{\varepsilon}^t(Z)\right], \tag{3.4}$$

where, for any $(X,t)\in\Omega$, $\tilde{\varepsilon}^t(X)$ is the function with values $\tilde{\varepsilon}^t(X)(s)=\tilde{\varepsilon}(X,t-s)$. By (3.2) and (3.4)

$$\delta_h(s)=\begin{cases}0, & 0<s<\tau-t_{Z+h},\\ -\dfrac{\varepsilon^-(\tau-s)}{h}, & \tau-t_{Z+h}<s<\tau-t_Z,\\ 0, & \tau-t_Z<s<\infty.\end{cases} \tag{3.5}$$

In view of the constitutive relation (16.1), the assumed properties of the response function $\mathcal{S}(\cdot)$ (16.3), (3.2), (3.3), and (3.4),

$$\begin{aligned}\frac{T(Z+h,\tau)-T(Z,\tau)}{h}&=\frac{1}{h}\left\{\mathcal{S}\left(\varepsilon^\tau(Z+h)\right)-\mathcal{S}\left(\varepsilon^\tau(Z)\right)\right\}\\ &=I_1(h)+I_2(h)+I_3(h)+o(\Phi_h)\end{aligned} \tag{3.6}$$

as $h \downarrow 0$, where

$$\begin{aligned} I_1(h) &= G_\tau(Z, \tau; 0)\left\{\frac{\varepsilon(Z+h, \tau) - \varepsilon(Z, \tau)}{h}\right\}, \\ I_2(h) &= \int_0^{s_0} G'_\tau(Z, \tau; s)\, \delta_h(s)\, ds, \\ I_3(h) &= \int_0^{s_0} G'_\tau(Z, \tau; s)\left\{\frac{\hat{\varepsilon}(Z+h, \tau-s) - \hat{\varepsilon}(Z, \tau-s)}{h}\right\} ds, \\ \Phi_h &= \left|\frac{\varepsilon(Z+h, \tau) - \varepsilon(Z, \tau)}{h}\right| + \int_0^{s_0} |\delta_h(s)|\, ds + \int_0^{s_0}\left|\frac{\hat{\varepsilon}(Z+h, \tau-s) - \hat{\varepsilon}(Z, \tau-s)}{h}\right| ds, \end{aligned} \tag{3.7}$$

and

$$G_\tau(Z, \tau; 0) = \mathscr{G}(\varepsilon^\tau(Z); 0), \quad G'_\tau(Z, \tau; s) = \mathscr{G}'(\varepsilon^\tau(Z); s). \tag{3.8}$$

All of the upper limits are taken to be s_0 rather than ∞ since all of the integrands vanish on (s_0, ∞).

Since $(Z, \tau) \notin \Sigma$

$$I_1(h) \to G_\tau(Z, \tau; 0)\frac{\partial \varepsilon(Z, \tau)}{\partial Z} \tag{3.9}$$

as $h \downarrow 0$. Next, in view of (3.5),

$$I_2(h) = -\frac{1}{h}\int_{\tau - t_{Z+h}}^{\tau - t_Z} G'_\tau(Z, \tau; s)\, \varepsilon^-(\tau - s)\, ds, \tag{3.10}$$

and since the integrand is a continuous function of s

$$I_2(h) = -\left(\frac{t_{Z+h} - t_Z}{h}\right) G'_\tau(Z, \tau; s^*)\, \varepsilon^-(\tau - s^*), \tag{3.11}$$

where

$$\tau - t_{Z+h} \leqq s^* \leqq \tau - t_Z. \tag{3.12}$$

Further, as $\hat{t}(\cdot)$ is smooth,

$$t_{Z+h} - t_Z = \frac{h}{V(t_{Z^*})}, \tag{3.13}$$

where

$$Z \leqq Z^* \leqq Z + h, \tag{3.14}$$

and (3.11), (3.12) and (3.13) imply that

$$I_2(h) \to -\frac{1}{V(t_Z)} G'_\tau(Z, \tau; \tau - t_Z)\, \varepsilon^-(t_Z) \tag{3.15}$$

as $h \downarrow 0$.

Now let

$$M = \sup\left\{\left|\frac{\partial \varepsilon(X, t)}{\partial X}\right| : (X, t) \in \Omega^-\right\}.$$

It follows from the properties of shock waves that the strain ε is smooth on the closure of Ω^-; thus

$$M < \infty$$

and we have the inequalities

$$\begin{aligned} \left|\frac{\varepsilon(Z+h, \tau) - \varepsilon(Z, \tau)}{h}\right| &\leqq M, \\ \left|\frac{\hat{\varepsilon}(Z+h, \tau-s) - \hat{\varepsilon}(Z, \tau-s)}{h}\right| &\leqq M. \end{aligned} \tag{3.16}$$

For $Y_{\tau-s} \notin (X_0, X_1)$, the second inequality is obvious as the left-hand side vanishes. For $Y_{\tau-s} \in (X_0, X_1)$, this inequality is a consequence of the following observation: not only is the function $X \to \hat{\varepsilon}(X, \tau - s)$ continuous on $[X_0, X_1]$, but also its derivative is continuous on $[X_0, Y_{\tau-s})$, vanishes on $(Y_{\tau-s}, X_1]$, and suffers at most a jump discontinuity at $Y_{\tau-s}$. Further by (3.2) and (3.3),

$$\frac{\hat{\varepsilon}(Z+h, \tau-s) - \hat{\varepsilon}(Z, \tau-s)}{h} \to \frac{\partial \varepsilon(Z, \tau-s)}{\partial Z}$$

as $h \downarrow 0$ whenever $\tau - s \neq t_Z$. Therefore, since $G'_\tau(Z, \tau; \cdot)$ is bounded on $[0, s_0]$, $(3.16)_2$ and Lebesgue's dominated convergence theorem imply that

$$I_3(h) \to \int_0^{s_0} G'_\tau(Z, \tau; s) \frac{\partial \varepsilon(Z, \tau-s)}{\partial Z} ds \tag{3.17}$$

as $h \downarrow 0$.

Let

$$K = \sup\{|\varepsilon^-(t)| : t \in [t_{X_0}, t_{X_1}]\},$$

$$L = \sup\left\{\frac{1}{V(t)} : t \in [t_{X_0}, t_{X_1}]\right\}.$$

Since ε^- and V are continuous and $V(t) > 0$, K and L are finite. Thus it follows from (3.5) and (3.13) that

$$\int_0^{s_0} |\delta_h(s)| ds \leqq KL,$$

and we conclude from (3.16) that the function Φ_h, defined by $(3.7)_4$, remains bounded as $h \downarrow 0$. Therefore it follows from (3.6), (3.9), (3.15), and (3.17) that

$$\begin{aligned} \lim_{h \downarrow 0} \left\{\frac{T(Z+h, \tau) - T(Z, \tau)}{h}\right\} = {} & G_\tau(Z, \tau; 0) \frac{\partial \varepsilon(Z, \tau)}{\partial Z} \\ & - \frac{1}{V(t_Z)} G'_\tau(Z, \tau; \tau - t_Z)\, \varepsilon^-(t_Z) \\ & + \int_0^{s_0} G'_\tau(Z, \tau; s) \frac{\partial \tau(Z, \tau-s)}{\partial Z} ds. \end{aligned} \tag{3.18}$$

The same result also holds as $h \uparrow 0$. Thus the right-hand side of (3.18) is equal to $\partial T(Z, \tau)/\partial Z$. Now consider the mapping $(Z, \tau) \to \varepsilon^\tau(Z)$ as a function from Ω^- into the normed space $\mathscr{D}$. From the properties of the strain field ε it is not difficult to verify that this function is continuous right up to the boundary of Ω^-, part of which is a portion of Σ. In fact

$$\varepsilon^\tau(Z) \to \varepsilon^\dagger \quad \text{as} \quad Z \uparrow Y_\tau, \tag{3.19}$$

where $\varepsilon^\dagger \in \mathscr{D}$ is the function

$$\varepsilon^\dagger(s) = \begin{cases} \varepsilon^-(\tau), & s = 0, \\ 0, & 0 < s < \infty. \end{cases} \tag{3.20}$$

Thus it follows from (3.8) and the properties of the functionals $\mathscr{G}$ and $\mathscr{G}'$ that

$$\begin{aligned} G_\tau(Z, \tau; 0) &\to G(0) \\ G'_\tau(Z, \tau; s) &\to G'(s) \end{aligned} \tag{3.21}$$

as $Z \uparrow Y_\tau$, where $G(\cdot)$ is the stress-relaxation function corresponding to the history $\varepsilon^\dagger$, defined by (3.20). In addition it follows from the above remarks and the continuity of $\mathcal{G}'$ that $G'_\tau(Z, \tau; s)$ is bounded on $\Omega^- \times [0, s_0)$. Further, for $s \neq 0$

$$\frac{\partial \varepsilon(Z, \tau - s)}{\partial Z} \to 0 \tag{3.22}$$

as $Z \uparrow Y_\tau$ and is bounded by M. Therefore,

$$\left(\frac{\partial T}{\partial Z}\right)^-(\tau) = \lim_{Z \uparrow Y_\tau} \frac{\partial T(Z, \tau)}{\partial Z} = G(0)\left(\frac{\partial \varepsilon}{\partial Z}\right)^-(\tau) - \frac{1}{V(\tau)} G'(0)\, \varepsilon^-(\tau).$$

Next, since $\varepsilon = 0$ on $\Omega - \Omega^-$ it is clear that

$$\left(\frac{\partial T}{\partial Z}\right)^+(\tau) = 0.$$

Thus

$$\left[\frac{\partial T}{\partial Z}\right](\tau) = G(0)\left(\frac{\partial \varepsilon}{\partial Z}\right)^-(\tau) - \frac{1}{V(\tau)} G'(0)\, \varepsilon^-(\tau).$$

References.

List of works cited. Italic numbers in parentheses following the reference indicate the sections in which it has been cited.

1903 *1.* DUHEM, P.: Sur la propagation des ondes dans un milieu parfaitement élastique affecté de déformations finies. Compt. Rend. **136**, 1379–1381. (*8*)

1906 *1.* — Recherches sur l'élasticité. Quatrième Partie. Propriétés générales des ondes dans les milieux visqueux et non visqueux. Ann. Ecole Normale (3) **23**, 169.223. (*8*)

1942 *1.* BETHE, H.: Report on "The theory of shock waves for an arbitrary equation of state." Office of Scientific Research and Development, Division B, Report No. 545. (*22*)

1948 *1.* COURANT, R., and K. O. FRIEDRICHS: Supersonic flow and shock waves. New York: Interscience. (*22*)

1949 *1.* WEYL, H.: Shock waves in arbitrary fluids. Comm. Pure Appl. Math. **2**, 103–122. (*22*)

1950 *1.* HARRIS, A. J.: The decay of plane, cylindrical and spherical shock waves. In: Underwater explosion research, vol. 1, entitled: Shock waves. A compendium of British and American reports. Washington, D.C.: Office of Naval Research. (*19*)

1953 *1.* LEE, E. H., and I. KANTER: Wave propagation in finite rods of viscoelastic materials. J. Appl. Phys. **24**, 1115–1122. (*16*)

1957 *1.* THOMAS, T. Y.: Extended compatibility conditions for the study of surfaces of discontinuity in continuum mechanics. J. Math. Mech. **6**, 311–322. (*4*)

2. — The growth and decay of sonic discontinuities in ideal gases. J. Math. Mech. **6**, 455–469. (*10*)

1960 *1.* COLEMAN, B. D., and W. NOLL: An approximation theorem for functionals with applications in continuum mechanics. Arch. Rational Mech. Anal. **6**, 355–370. (*12, 16*)

2. TRUESDELL, C., and R. TOUPIN: The classical field theories. Handbuch der Physik, vol. III/1, p. 225–739. Berlin-Göttingen-Heidelberg: Springer. (*2, 4, 5*)

1961 *1.* COLEMAN, B. D., and W. NOLL: Foundations of linear viscoelasticity. Rev. Mod. Phys. **33**, 239–249; **36**, 1103. (*12*)

2. THOMAS, T. Y.: Plastic flow and fracture in solids. New York-London: Academic Press. (*10*)

3. TRUESDELL, C.: General and exact theory of waves in finite elastic strain. Arch. Rational Mech. Anal. **8**, 263–296. Reprinted in Continuum Mechanics IV. Problems of Non-Linear Elasticity. New York-London-Paris: Gordon and Breach, 1965, and in Wave Propagation in Dissipative Materials. New York: Springer 1965. (*6, 8, 9*)

1962 *1.* BREUER, S., and E. T. ONAT: On uniqueness in linear viscoelasticity. Quart. Appl. Math. **19**, 355–359. (*14*)

2. CHU, B. T.: Stress waves in isotropic linear viscoelastic materials. J. de Mécanique **1**, 439–462. (*16*)

3. HILL, R.: Acceleration waves in solids. J. Mech. Phys. Solids **10**, 1–16 (*5*).

4. WACKERLE, J.: Shock wave compression of quartz. J. Appl. Phys. **33**, 922–937. (*18*)

1963 *1.* Duvall, G. E., and R. C. Alverson: Fundamental research in support of Vela-uniform. Semiannual Technical Summary Report No. 4. Menlo Park, Calif.: Stanford Research Institute. (*16*)

1964 *1.* Coleman, B. D.: Thermodynamics of materials with memory. Arch. Rational Mech. Anal. **17**, 1–46. (*15*)

2. Duvall, G. E.: Propagation of plane waves in a stress-relaxing medium, p. 20–32. In: Stress waves in anelastic solids. Proc. IUTAM Symposium held at Brown Univ., Apr. 3–5. Berlin-Göttingen-Heidelberg: Springer. (*16*)

3. Green, W. A.: The growth of plane discontinuities propagating into a homogeneous deformed elastic material. Arch. Rational Mech. Anal. **16**, 79–88. (*9*)

1965 *1.* Chadwick, P., and B. Powdrill: Singular surfaces in linear thermoelasticity. Int. J. Eng. Sci. **3**, 561–595. (*11*)

2. Chen, P. J.: Acceleration waves in rheological materials. Thesis, University of Washington. (*5, 14*)

3. Coleman, B. D., and M. E. Gurtin: Waves in materials with memory II. On the growth and decay of one-dimensional acceleration waves. Arch. Rational Mech. Anal. **19**, 239–265. Reprinted in Wave Propagation in Dissipative Materials. New York: Springer 1965. (*12, 13, 14, 16*)

4. — — Waves in materials with memory III. Thermodynamic influences on the growth and decay of acceleration waves. Arch. Rational Mech. Anal. **19**, 266–298. Reprinted in Wave Propagation in Dissipative Materials. New York: Springer 1965. (*15, 20*)

5. — —, and I. Herrera: Waves in materials with memory I. The velocity of one-dimensional shock and acceleration waves. Arch. Rational Mech. Anal. **19**, 1–19. Reprinted in Wave Propagation in Dissipative Materials. New York: Springer 1965. (*16*)

6. Truesdell, C., and W. Noll: The non-linear field theories of mechanics. Handbuch der Physik, vol. III/3, p. 1–602. Berlin-Heidelberg-New York: Springer. (*2*)

7. Green, W. A.: The growth of plane discontinuities propagating into a homogeneously deformed elastic material, corrections and additional results. Arch. Rational Mech. Anal. **19**, 20–23. (*6, 7*)

8. Gurtin, M. E., and I. Herrera: On dissipation inequalities and linear viscoelasticity. Quart. Appl. Math. **23**, 235–245. (*14, 16*)

9. Halpin, W. J., and R. A. Graham: Shock wave compression of plexiglas from 3 to 20 kilobars. Fourth International Symposium on Detonation, U. S. Naval Ordnance Laboratory, White Oak, Oct. 12–15. (*16*)

10. Stippes, M.: Steady state waves in anisotropic media. Ann. Meeting Soc. Eng. Sci. (unpublished). (*6*)

11. Varley, E.: Acceleration fronts in viscoelastic materials. Arch. Rational Mech. Anal. **19**, 215–225. (*14*)

1966 *1.* Ahrens, T. J., and G. E. Duvall: Stress relaxation behind elastic shock waves in rocks. J. Geophys. Res. **71**, 4349–4360. (*16*)

2. Chen, P. J.: On the attenuation of acceleration waves of arbitrary form in isotropic finite linear viscoelastic materials. Fourth Tech. Meeting Soc. Eng. Sci., Oct. 31–Nov. 2, Raleigh, N. C. (unpublished). (*14*)

3. Coleman, B. D., J. M. Greenberg, and M. E. Gurtin: Waves in materials with memory V. On the amplitude of acceleration waves and mild discontinuities. Arch. Rational Mech. Anal. **22**, 333–354. (*5, 12, 13, 24*)

4. —, and M. E. Gurtin: Thermodynamics and one dimensional shock waves in materials with memory. Proc. Roy. Soc., A, **292**, 562–574. (*22*)

5. Davison, L. W.: Propagation of plane waves of finite amplitude in elastic solids. J. Mech. Phys. Solids **14**, 249–270. (*7, 9*)

6. Kolodner, I. I.: Existence of longitudinal waves in anisotropic media. J. Acoust. Soc. Am. **40**, 730–731. (*6*)

7. Pipkin, A. C.: Shock structure in a viscoelastic fluid. Quart. Appl. Math. **23**, 297–303. (*14, 16*)

8. Shu, L. S., and E. T. Onat: On anisotropic linear viscoelastic solids. In: Mechanics and chemistry of solid propellants. Oxford: Pergamon. (*14*)

9. Truesdell, C.: Existence of longitudinal waves. J. Acoust. Soc. Am. **40**, 729–730. (*6, 7*)

1967 *1.* Greenberg, J. M.: The existence of steady shock waves in non-linear materials with memory. Arch. Rational Mech. Anal. **24**, 1–21. (*16*)

2. Touloukian, Y. S.: Thermophysical properties of high temperature solid materials, vol. 4, part I. New York: MacMillan. (*21*)

3. WALSH, E. K.: Induced one-dimensional waves in elastic non-conductors. J. Appl. Mech. **34**, 937–941. (*22*)

1968 *1.* BARKER, L. M.: Fine structure of compressive and release wave shapes in aluminum measured by the velocity interferometer technique. Proceedings of the IUTAM symposium H.D.P., comportement des milieux denses sous hautes pressions dynamiques, Paris, 1967. New York: Gordon and Breach. (*17*)

2. BECKER, E., u. H. SCHMITT: Die Entstehung von ebenen, zylinder- und kugelsymmetrischen Verdichtungsstößen in relaxierenden Gasen. Ingenieur-Arch. **36**, 335–347. (*10*)

3. CHEN, P. J.: Growth of acceleration waves in isotropic elastic materials. J. Acoust. Soc. Am. **43**, 982–987. (*10*)

4. — The growth of acceleration waves of arbitrary form in homogeneously deformed elastic materials. Arch. Rational Mech. Anal. **30**, 81–89. (*10*)

5. — Thermodynamic influences on the propagation and the growth of acceleration waves in elastic materials. Arch. Rational Mech. Anal. **31**, 228–254; **32**, 400–401. (*9, 11*)

6. GREENBERG, J. M.: Existence of steady waves for a class of nonlinear dissipative materials. Quart. Appl. Math. **26**, 27–34. (*16*)

7. ROHDE, R. W., and O. E. JONES: Mechanical and piezoelectric properties of shock-loaded x-cut quartz at 573 °K. Rev. Sci. Instr. **39**, 313–316. (*21*)

8. TRUESDELL, C.: Comment on longitudinal waves. J. Acoust. Soc. Am. **43**, 170. (*6*)

1969 *1.* JOHNSON, J. N., and L. M. BARKER: Dislocation dynamics and steady plastic wave profiles in 6061-T6 aluminum. J. Appl. Phys. **40**, 4321–4333. (*16*)

2. TRUESDELL, C.: Rational thermodynamics. New York: McGraw-Hill. (*14*)

1970 *1.* ACHENBACH, J. D., and G. HERRMANN: Propagation of second-order thermomechanical disturbances in viscoelastic solids. Proceedings IUTAM Symposium East Kilbride, June 25–28, 1968, ed. by B. A. BOLEY. Vienna-New York: Springer. (*5*)

2. BARKER, L. M., and R. E. HOLLENBACH: Shock-wave studies of PMMA, fused silica, and sapphire. J. Appl. Phys. **41**, 4208–4226. (*12, 16, 17, 18*)

3. BOWEN, R. M., and C. C. WANG: Acceleration waves in inhomogeneous isotropic elastic bodies. Arch. Rational Mech. Anal. **38**, 13–45. (*9, 10, 13*)

4. CHEN, P. J.: On the growth of longitudinal waves in anisotropic elastic materials. Arch. Rational Mech. Anal. **36**, 381–389. (*6*)

5. — On the growth of transverse waves in anisotropic elastic materials. Z. Angew. Math. Phys. **21**, 846–850. (*7*)

6. —, and M. E. GURTIN: On the growth of one-dimensional shock waves in materials with memory. Arch. Rational Mech. Anal. **36**, 33–46. (*5, 16*)

7. DORIA, M. L., and R. M. BOWEN: Growth and decay of curved acceleration waves in chemically reacting fluids. Phys. Fluids **13**, 867–876. (*10, 20*)

8. JUNEJA, B. L., and G. A. NARIBOLI: Growth of acceleration waves in an unstrained non-linear isotropic elastic medium. Int. J. Non-Linear Mech. **5**, 513–524. (*10*)

9. SCHULER, K. W.: Propagation of steady shock waves in polymethyl methacrylate. J. Mech. Phys. Solids **18**, 277–293. (*17*)

10. SUHUBI, E. S.: The growth of acceleration waves of arbitrary form in deformed hyperelastic materials. Int. J. Eng. Sci. **8**, 699–710. (*10*)

1971 *1.* BAILEY, P. B., and P. J. CHEN: On the local and global behavior of acceleration waves. Arch. Rational Mech. Anal. **41**, 121–131. (*13*)

2. BOWEN, R. M., and C. C. WANG: On displacement derivatives. Quart. Appl. Math., **29**, 29–39. (*4*)

3. — — Thermodynamic influences on acceleration waves in inhomogeneous isotropic elastic bodies with internal state variables. Arch. Rational Mech. Anal. **41**, 287–318. (*11*)

4. CHEN, P. J.: A correspondence for acceleration waves in elastic non-conductors. Int. J. Non-Linear Mech., **6**, 701–705. (*8*)

5. — External influences on the growth and decay of one-dimensional shock waves in elastic non-conductors. Int. J. Solids and Struct., **7**, 1697–1703. (*22*)

6. — One dimensional shock waves in elastic non-conductors. Arch. Rational Mech. Anal. **43**, 350–362. (*22, 23*)

7. — One dimensional shock waves in elastic materials. Istituto Lombardo di Scienze, Rendiconti A **105**, 907–916 (*19*)

8. —, and M. E. GURTIN: Growth and decay of one-dimensional shock waves in fluids with internal state variables. Phys. Fluids **14**, 1091–1094. (*22*)

9. — — The growth of one-dimensional shock waves in elastic non-conductors. Int. J. Solids and Struct. **7**, 5–10. (*22*)

10. CHEN, P, J., and H. H. WICKE: Existence of the one-dimensional kinematical condition of compatibility. Istituto Lombardo di Scienze, Rendiconti A **105**, 322–328. (*4*)

11. SCHULER, K. W., and E. K. WALSH: Critical induced acceleration for shock propagation in polymethyl methacrylate. J. Appl. Mech., **38**, Ser. E, 641–645. (*17*)

1972 *1.* BAILEY, P. B., and P. J. CHEN: On the local and global behavior of acceleration waves: Addendum, asymptotic behavior. Arch. Rational Mech. Anal., **44**, 212–216. (*13*)

2. BOWEN, R. M., and P. J. CHEN: Acceleration waves in anisotropic thermoelastic materials with internal state variables. Acta Mechanica, to appear. (*13*)

3. CHEN, P. J.: On the behavior of acceleration waves in deformed elastic non-conductors, J. Appl. Mech., **39**, Ser. E, 114–118 (*21*)

4. — One dimensional acceleration waves in inhomogeneous elastic non-conductors. Acta Mechanica, to appear. (*20, 24*)

5. — One dimensional shock waves in inhomogeneous elastic materials, Int. J. Solids and Struct., **8**, 409–414 (*26, 27*)

6. —, and M. E. GURTIN: On the use of experimental results concerning steady shock waves to predict the acceleration wave response of nonlinear viscoelastic materials. J. Appl. Mech., **39**, Ser. E, 295–296 (*17*)

Additional references. The following additional papers are relevant to the subject matter of this article. Since the scope of the article cannot be all-encompassing, I did not have occasion to refer to them.

1870 a. RANKINE, W. J. M.: On the thermodynamic theory of waves of finite longitudinal disturbance. Phil. Trans. Roy. Soc. London **160**, 277–288.

1903 a. HADAMARD, J.: Leçons sur la propagation des ondes et les équations de l'hydrodynamique. Paris: Hermann.

1953 a. ERICKSEN, J. L.: On the propagation of waves in isotropic incompressible perfectly elastic materials. J. Rational Mech. Anal. **2**, 329–337.

1960 a. HAYES, W. D.: Gasdynamic discontinuities. Princeton: Princeton Univ. Press.

b. HUNTER, S. C.: Viscoelastic waves. Progress in solid mechanics, ed. by SNEDDON and HILL, vol. I, p. 3. Amsterdam: North-Holland Publishing Co.

1961 a. FLAVIN, J. N., and A. E. GREEN: Plane thermo-elastic waves in an initially stressed medium. J. Mech. Phys. Solids **8**, 179–190.

b. HILL, R.: Discontinuity relations in mechanics of solids. Progress in solid mechanics, ed. by SNEDDON and HILL, vol. II, p. 245. Amsterdam: North-Holland Publishing Co.

c. TOUPIN, R. A., and B. J. BERNSTEIN: Sound waves in deformed perfectly elastic materials. Acousto-elastic effect. J. Acoust. Soc. Am. **33**, 216–225.

1963 a. GREEN, A. E.: A note on wave propagation in initially deformed bodies. J. Mech. Phys. Solids **11**, 119–126.

b. NARIBOLI, G. A.: The propagation and growth of sonic discontinuities in magnetohydrodynamics. J. Math. Mech. **12**, 141–148.

1964 a. BLAND, D. R.: Dilational waves and shocks in large displacement isentropic dynamic elasticity. J. Mech. Phys. Solids **12**, 245–267.

b. CHU, B. T.: Finite amplitude waves in incompressible perfectly elastic materials. J. Mech. Phys. Solids **12**, 45–57.

c. KOLSKY, H.: Stress waves in solids. J. Sound and Vibration **1**, 88–110.

d. NARIBOLI, G. A.: Growth and propagation of waves in hypoelastic media. J. Math. Anal. Appl. **8**, 57–65.

1965 a. COLEMAN, B. D., and M. E. GURTIN: Waves in materials with memory IV. Thermodynamics and the velocity of general acceleration waves. Arch. Rational Mech. Anal. **19**, 317–338. Reprinted in Wave Propagation in Dissipative Materials. New York: Springer 1965.

b. DUNWOODY, J., and N. T. DUNWOODY: Acceleration waves in viscoelastic media of Maxwell-Zaremba type. Int. J. Eng. Sci. **3**, 417–427.

c. FISHER, G. M. C., and M. E. GURTIN: Wave propagation in the linear theory of viscoelasticity. Quart. Appl. Math. **23**, 257–263.

d. GRAHAM, R. A. F. W. NEILSON, and W. B. BENEDICK: Piezoelectric current from shock-loaded quartz—A submicrosecond stress gauge. J. Appl. Phys. **36**, 1775–1783.

e. HERRERA, I., and M. E. GURTIN: A correspondence principle for viscoelastic wave propagation. Quart. Appl. Math. **22**, 360–364.

f. MCCARTHY, M. F.: Propagation of plane acceleration discontinuities in hyperelastic dielectrics. Int. J. Eng. Sci. **3**, 603–623.

g. VARLEY, E., and E. CUMBERBATCH: Non-linear theory of wave-front propagation. J. Inst. Maths. Applics. **1**, 101–112.

h. VARLEY, E., and J. DUNWOODY: The effect of non-linearity at an acceleration wave. J. Mech. Phys. Solids **13**, 17–28.

1966 a. BÜRGER, W.: Zur Entstehung von Verdichtungsstößen beim „Kolbenversuch" in Gasen mit thermodynamischer Relaxation. Z. Angew. Math. Mech. **46**, 149–151.

b. — Entstehung von Verdichtungsstößen in Gasen mit thermodynamischer Relaxation. Z. Angew. Math. Mech. **46**, T187–T189.

c. MCCARTHY, M. F.: The growth of magneto-elastic waves in a Cauchy elastic material of finite electrical conductivity. Arch. Rational Mech. Anal. **23**, 191–217.

d. —, and W. A. GREEN: The growth of plane acceleration discontinuities propagating into a homogeneously deformed hyperelastic dielectric material in the presence of a magnetic field. Int. J. Eng. Sci. **4**, 403–422.

e. NARIBOLI, G. A.: Wave propagation in anisotropic elasticity. J. Math. Anal. Applics. **16**, 108–122.

1967 a. COLEMAN, B. D., and M. E. GURTIN: Growth and decay of discontinuities in fluids with internal state variables. Phys. Fluids **10**, 1454–1458.

1969 a. BOWEN, R. M.: Acceleration and higher order waves in a mixture of chemically reacting elastic materials. Arch. Rational Mech. Anal. **33**, 169–180.

b. MCCARTHY, M. F., and A. C. ERINGEN: Micropolar viscoelastic waves. Int. J. Eng. Sci. **7**, 447–458.

1970 a. BECKER, E.: Die Entstehung von Verdichtungsstößen in kompressiblen Medien Ing.-Arch. **5**, 302–316.

b. — Entstehung von Verdichtungsstößen. Z. Angew. Math. Mech. **50**,T167–T168

c. — Relaxation effects in gas dynamics. Aeron. J. Roy. Aero. Soc. **74**, 736–748.

Ideal Plasticity.

By

HILDA GEIRINGER.

With 50 Figures.

Introduction. This article is a revised version of Part Two: *The Ideal Plastic Body* of the paper by A. M. FREUDENTHAL and H. GEIRINGER "The mathematical theories of the inelastic continuum", published 1958 in Vol. VI of this Encyclopedia. On the whole, I have tried to conserve the general character of the original though carefully checking it, word by word and thought by thought. In addition, I have attempted to fill certain gaps, some of which were present in the original version while many more were due to the development over the intervening twelve years. The emphasis implicit in the choice of this added material is evidently an expression of my personal judgment. I am aware of the fact that this attempt to update the material remains sporadic and has not produced a thoroughly modern survey. The shorter and longer inserts have been worked into the text and the footnotes.

Missing is a study of plasticity under non-homogeneous conditions. I have abstained from attempting this because of the existence of the comprehensive article on the subject in the seventh volume of "Advances in Applied Mechanics", pp. 131–214, by W. OLZSAK, W. RYCHLEWSKI and W. URBANOWSKI, which is rich in content, clearly and authoritatively written and includes nine pages of references. Although this paper is from 1962, I could hardly have hoped to go in a worthwhile way beyond it. While, of course, many more subjects deserved some presentation, there were natural limits to this revision.

I hope that the easy readability which some readers liked in the original version has not been lost in this "revised edition".

A. The basic equations.

I. The three-dimensional problem.

1. Quadratic yield condition. I present in this section the basic equations of the three-dimensional perfectly plastic body under the assumption of the quadratic yield condition proposed (1913) by VON MISES and of the associated stress-strain relations of SAINT VENANT, LÉVY, and V. MISES. I refrain from an attempt to sketch the early historical development, which led to a complete statement of the concept of the ideal plastic body by V. MISES,[1] since this history has been set forth often and may be found e.g. in the books by PRAGER and HODGE [*23*] (1951), HILL [*6*] (1950), NADAI [*18*] (1950) and [*17*] (1931), SOKOLOVSKY [*26*] (1950), GEIRINGER [*3*] (1937); see in this connection also an article by PRAGER.[2]

[1] R. V. MISES: Göttinger Nachr. Math. phys. Kl. 582–592 (1913).

[2] W. PRAGER: James Clayton Lecture. Proc. Inst. Mech. Engrs. (London) **169**, 41–57 (1955).

The basic equations are, to begin with, the *continuity equation* and the *equation of motion*. In the first of these:

$$\varrho \operatorname{div} \boldsymbol{v} + \frac{d\varrho}{dt} = 0, \tag{1.1}$$

$\boldsymbol{v}$ denotes the vector of flow velocity, ϱ the density and d/dt the "material derivative". If ϱ = constant, as is assumed in general in the theory of perfectly plastic solids, $d\varrho/dt = 0$, and (1.1) reduces to

$$\operatorname{div} \boldsymbol{v} = 0 \quad \text{or} \quad \sum_{i=1}^{3} \frac{\partial v_i}{\partial x_i} = 0. \tag{1.1'}$$

In the equation of motion let $\boldsymbol{k}$ be the resultant of the external forces per unit of mass and σ_{ij} the components of the stress tensor Σ. The mean pressure p is then defined as the invariant expression

$$p = -\tfrac{1}{3}(\sigma_{11} + \sigma_{22} + \sigma_{33}) = -\tfrac{1}{3} J_1, \tag{1.2}$$

where J_1 is the first invariant of Σ, and the resultant pressure force per unit volume is grad p. Finally, there is the stress deviator force $\boldsymbol{d}$ (corresponding to the viscous force in viscous flow theory) with components

$$d_i = \sum_{j=1}^{3} \frac{\partial s_{ij}}{\partial x_j} \qquad (i = 1, 2, 3).$$

The s_{ij}, the components of the stress deviator tensor S, are,

$$s_{ij} = \sigma_{ij} + p\,\delta_{ij}, \tag{1.3}$$

and the equation of motion appears in the form

$$\varrho \frac{d\boldsymbol{v}}{dt} = \varrho\,\boldsymbol{k} - \operatorname{grad} p + \boldsymbol{d}, \tag{1.4}$$

or in component form, with X_i as components of $\varrho\boldsymbol{k}$,

$$\varrho \frac{dv_i}{dt} = X_i - \frac{\partial p}{\partial x_i} + \sum_{j=1}^{3} \frac{\partial s_{ij}}{\partial x_j}, \tag{1.4'}$$

or

$$\varrho \frac{dv_i}{dt} = X_i + \sum_{j=1}^{3} \frac{\partial \sigma_{ij}}{\partial x_j}. \tag{1.4''}$$

Next, we need a relation which specifies s_{ij} in terms of the other variables p, ϱ, v_i, and possibly also x_1, x_2, x_3, t. The Saint Venant-Lévy-v. Mises relations[3] do this in the form

$$s_{ij} = \mu \cdot \frac{1}{2}\left(\frac{\partial v_i}{\partial x_j} + \frac{\partial v_j}{\partial x_i}\right), \quad \mu > 0 \tag{1.5}$$

or

$$\frac{1}{2}\left(\frac{\partial v_i}{\partial x_j} + \frac{\partial v_j}{\partial x_i}\right) = \lambda s_{ij}, \quad \lambda \geqq 0.^{4} \tag{1.5'}$$

[3] A. J. C. B. DE SAINT VENANT: C. R. Acad. Sci., Paris **70**, 473–480 (1870). — M. LÉVY: C. R. Acad. Sci., Paris **70**, 1323–1325 (1870). — R. v. MISES, see Ref. 1 of this article.

[4] Explicit expressions of Eqs. (1.4') in both cylindrical and spherical coordinates may be found, for example, in the German edition of [*26*], pp. 59/60 and on pp. 63/64 are the expressions for the left sides of (1.5') in these coordinates.

The relation (1.5′) is a relation between two symmetric tensors, the tensor E'' of plastic flow velocity with components

$$\dot{\varepsilon}''_{ij} = \frac{1}{2}\left(\frac{\partial v_i}{\partial x_j} + \frac{\partial v_j}{\partial x_i}\right), \qquad i, j = 1, 2, 3$$

and the stress deviator tensor S with components (1.3). In ordinary x, y, z-space we use also the notations $x_1 = x$, $x_2 = y$, $x_3 = z$ and

$$\dot{\varepsilon}''_{11} = \dot{\varepsilon}''_x, \; \dot{\varepsilon}''_{xy} = \dot{\varepsilon}''_{yx} = \tfrac{1}{2}\dot{\gamma}''_z, \; \ldots, \; s_{11} = s_{xx} = s_x, \; s_{12} = s_{xy} = s_{yx} = \tau_z \ldots.$$

With these notations (1.5′) becomes:

$$\dot{\varepsilon}''_z = \lambda s_x, \; \ldots, \; \tfrac{1}{2}\dot{\gamma}''_x = \lambda \tau_x. \tag{1.5''}$$

In E'' and its components the dot stands for d/dt since we use flow *velocities*, and the double primes stand for "plastic" in contrast to "elastic" (see Sect. 6).

Note that in contrast to viscous flow theory the μ of Eqs. (1.5) is an unknown (positive) function of the coordinates and possibly also of t. Since $s_{11} + s_{22} + s_{33} = 0$, Eqs. (1.5) are compatible with (1.1′) and (1.5) stands for five independent scalar equations.

In addition to the $1+3+5=9$ Eqs. (1.1′), (1.4), (1.5′) for ten unknowns v_i, p, s_{ij}, μ one assumes a *yield condition*, which limits the admissible plastic stress tensors to ∞^5. We shall give it in this section in v. Mises' "quadratic" form.

If we consider problems of equilibrium, the acceleration force

$$\varrho \frac{d\boldsymbol{v}}{dt}$$

vanishes; usually, the body force $\varrho \boldsymbol{k}$ is also disregarded. The basic $1+3+5+1=10$ equations of the theory are then

$$\frac{\partial v_1}{\partial x_1} + \frac{\partial v_2}{\partial x_2} + \frac{\partial v_3}{\partial x_3} = 0, \tag{1.6}$$

$$-\frac{\partial p}{\partial x_i} + \sum_{j=1}^{3} \frac{\partial s_{ij}}{\partial x_j} = 0, \qquad i = 1, 2, 3, \tag{1.7}$$

$$\frac{1}{2}\left(\frac{\partial v_i}{\partial x_j} + \frac{\partial v_j}{\partial x_i}\right) = \lambda s_{ij}, \qquad \lambda \geqq 0, \qquad i, j = 1, 2, 3 \tag{1.8}$$

$$\sum_{i,j}^{1\ldots 3} s_{ij} s_{ij} = (s_x^2 + s_y^2 + s_z^2) + 2(\tau_x^2 + \tau_y^2 + \tau_z^2) = 2\tau_0^2 = 2k^2 \tag{1.9}$$

where τ_0 is the yield stress in simple shear.[5]

Another form of (1.9) is

$$(\sigma_x - \sigma_y)^2 + (\sigma_y - \sigma_z)^2 + (\sigma_z - \sigma_x)^2 + 6(\tau_x^2 + \tau_y^2 + \tau_z^2) = 6\tau_0^2 = 2\sigma_0^2, \tag{1.9'}$$

where σ_0 is the yield stress in simple tension. (Thus, in v. Mises' theory, $\sigma_0 = \sqrt{3}\,\tau_0$.) We may also easily see that

$$\begin{aligned} J_2 &\equiv -(s_x s_y + s_y s_z + s_z s_x) + (\tau_x^2 + \tau_y^2 + \tau_z^2) \\ &= \tfrac{1}{2}(s_x^2 + s_y^2 + s_z^2) + (\tau_x^2 + \tau_y^2 + \tau_z^2) = k^2, \end{aligned} \tag{1.9''}$$

where J_2 is the second invariant of the deviator S.

[5] Eq. (1.9) is the simplest differentiable isotropic incompressible yield condition.

Denoting by subscripts 1, 2, 3 the (common) *principal directions* of the tensors Σ and S, we obtain also

$$s_1^2+s_2^2+s_3^2=2k^2, \text{[6]} \tag{1.$\bar{9}$}$$

$$(\sigma_1-\sigma_2)^2+(\sigma_2-\sigma_3)^2+(\sigma_3-\sigma_1)^2=6k^2, \tag{1.$\bar{9}'$}$$

$$J_2\equiv-(s_1 s_2+s_2 s_3+s_3 s_1)=k^2, \text{[7]} \tag{1.$\bar{9}''$}$$

where σ_1, σ_2, σ_3 are the principal stresses. The *principal stresses* are determined as the roots of the determinantal equation $|\sigma_{ij}-\sigma\delta_{ij}|=0$, where δ_{ij} has the usual meaning. It is well known that this equation has only real solutions, which may be simple, double or triple. The greatest of these roots equals the greatest principal stress. We ordinarily assume these ordered so that $\sigma_1\geqq\sigma_2\geqq\sigma_3$. Also, introducing the principal shear stresses,

$$\tau_1=\tfrac{1}{2}(\sigma_2-\sigma_3)=\tfrac{1}{2}(s_2-s_3), \qquad \text{etc.}, \tag{1.10}$$

we find that

$$\tau_1^2+\tau_2^2+\tau_3^2=\tfrac{3}{2}k^2, \quad \text{where} \quad \tau_1+\tau_2+\tau_3=0 \tag{1.9'''}$$

results.[8] The criterion (1.9) has been generalized by v. MISES[9] and by F. SCHLEICHER[10] by replacing the constant right side in it by a function of p in accordance with experiments by W. LODE[11] (cf. also O. MOHR's form, see Ref. 21).

We rewrite Eq. (1.8) in the form

$$\dot\varepsilon_i''=\lambda s_i, \qquad i=1,2,3. \tag{1.8'}$$

Here λ can be eliminated by squaring the equations, summing over all subscripts and using (1.$\bar{9}$). We obtain

$$\lambda^2=\frac{1}{2k^2}\left[(\dot\varepsilon_1'')^2+(\dot\varepsilon_2'')^2+(\dot\varepsilon_3'')^2\right]. \tag{1.11}$$

Alternatively, we may multiply both sides of Eq. (1.8) by s_{ij} and sum over all subscripts. Then, using also (1.5″) and introducing

$$\dot W''=\sum_{i,j}\dot\varepsilon_{ij}'' s_{ij}=\dot\varepsilon_x'' s_x+\dot\varepsilon_y'' s_y+\dot\varepsilon_z'' s_z+\dot\gamma_x''\tau_x+\dot\gamma_y''\tau_y+\dot\gamma_z''\tau_z=\sum_i\dot\varepsilon_i'' s_i \tag{1.12}$$

we obtain

$$\lambda=\frac{1}{2k^2}\dot W''. \tag{1.13}$$

[6] In "stress space" (orthogonal coordinates in the principal stress directions) Eq. (1.$\bar{9}$) defines a circular cylinder of radius $k\sqrt{2}$, the axis of which coincides with the space diagonal and which intersects the plane $J_1=0$, which is normal to the space diagonal along a circle having the radius of the cylinder.

[7] Thus double subsripts 11 or 12 have from now on the same meaning as in the equations preceding (1.5"), while a single subscript 1 or 2 alone denotes a principal direction.

[8] If $\sigma_1\geqq\sigma_2\geqq\sigma_3$ then τ_1 and τ_3 are positive, and τ_2 negative and $-\tau_2=|\tau|_{\max}$, where $|\tau|_{\max}$ means the greatest absolute value of τ. We also note, SOKOLOVSKY [*26*], p. 31, that (1.9‴) is approximately equivalent to the condition

$$\left(\frac{1}{\sqrt{3}}+\frac{1}{2}\right)\cdot|\tau|_{\max}=k.$$

Cf. also HILL [*6*, p. 117]).

[9] R. v. MISES: Z. angew. Mech. **5**, 147–149 (1925).

[10] F. SCHLEICHER: Z. angew. Math. Mech. **5**, 478 (1925); **6** 199, (1926).

[11] W. LODE: Z. angew. Math. Mech. **5**, 142–144 (1925). — Z. Physik **36**, 913–939 (1926).

Note that on account of (1.1′) also:

$$\dot{W}'' = \sum_{i,j} \dot{\varepsilon}''_{ij}\, \sigma_{ij}. \tag{1.12′}$$

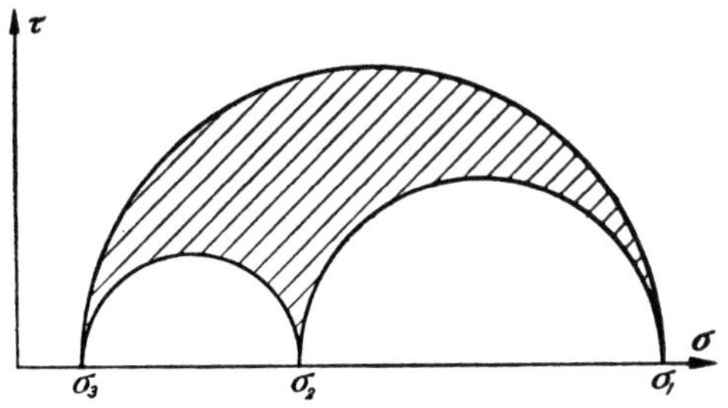

Fig. 1. Mohr circles.

Hence from (1.13) λ is proportional to the rate of *plastic work* $\dot{W}''$ *producing a change in shape*—the change in volume being zero by (1.1′). Comparing (1.11) and (1.13), we find, using (1.9̄), the equality

$$\sum_i s_i^2 \sum_i (\dot{\varepsilon}_i'')^2 = \Big(\sum_i s_i\, \dot{\varepsilon}_i''\Big)^2. \tag{1.14}$$

2. Some basic formulas. Mohr circles. The stress tensor Σ, like any tensor, associates with a direction ν a vector $\boldsymbol{t}_\nu$, the stress vector or traction, which may be resolved into a component σ in the ν-direction, the normal stress, and into a component τ perpendicular to it. From the formulas for σ and τ one deduces the well-known stress representation due to O. Mohr.[12] He has proved that *the points which correspond to all possible ∞^2 directions ν* (for a fixed point in the material) *lie in the shaded area between the three circles of Fig. 1, called circles of Mohr, which intercept the σ-axis in the three points with abscissae $\sigma_3, \sigma_2, \sigma_1$ respectively* (Fig. 1).

If a_1, a_2, a_3 are the direction cosines $\cos(\nu 1)$, $\cos(\nu 2)$, $\cos(\nu 3)$, then for $a_3 = 0$, $a_1^2 + a_2^2 = 1$, the σ, τ satisfy the equation

$$\left(\sigma - \frac{\sigma_1 + \sigma_2}{2}\right)^2 + \tau^2 = \left(\frac{\sigma_1 - \sigma_2}{2}\right)^2, \tag{2.1}$$

i.e., the equation of the half-circle to the right, and similarly for $a_2 = 0$ and for $a_1 = 0$, respectively. The *principal tangential stresses*

$$\tau_1 = \tfrac{1}{2}(\sigma_2 - \sigma_3), \qquad \tau_2 = \tfrac{1}{2}(\sigma_3 - \sigma_1), \qquad \tau_3 = \tfrac{1}{2}(\sigma_1 - \sigma_2) \tag{2.2}$$

are the radii of the circles of Mohr.

Nadai introduced and explained the octahedral shearing stress τ_{oct}^2:

$$\begin{aligned} \tau_{\text{oct}}^2 &= \tfrac{1}{9}\left[(\sigma_1 - \sigma_2)^2 + (\sigma_2 - \sigma_3)^2 + (\sigma_3 - \sigma_1)^2\right] = \tfrac{2}{3}\, k^2 = \tfrac{2}{3}\, J_2, \\ J_2 &= \tfrac{3}{2}\, \tau_{\text{oct}}^2. \end{aligned} \tag{2.3}$$

This last equation has been used as a physical interpretation of the invariant J_2. Another physical interpretation in terms of the energy associated with a change

[12] Otto Mohr: Abhandlungen aus dem Gebiete der technischen Mechanik, 2nd ed., pp. 192–235. Berlin 1914. See [*18*, p. 96], footnote, for more bibliographical details and p. 96seq. for details of proof and illustrations.

of shape was given by HENCKY.[13] PRAGER and HODGE [*23*] very rightly remark "v. Mises' yield condition derives its importance in the mathematical theory of plasticity not from the fact that the invariant J_2 appearing therein can be interpreted physically in this or that manner, but from the fact that it has the simplest form compatible with the general postulates which any yield condition must fulfill." Actually, it is also in very good agreement with experimental evidence, particularly for ductile metals (cf. e.g. [*6*, p. 22]).

3. Plastic potential. Denote the left side of (1.9') by $6g(\sigma_x, \sigma_y, \ldots, \tau_z)$. We see that

$$3\frac{\partial g}{\partial s_x} = 3\frac{\partial g}{\partial \sigma_x} = 2\sigma_x - \sigma_y - \sigma_z = 3\left(\sigma_x - \frac{\sigma_x+\sigma_y+\sigma_z}{3}\right) = 3\,s_x, \ldots$$
$$\frac{1}{2}\frac{\partial g}{\partial \tau_x} = \tau_x, \quad \text{etc.} \tag{3.1}$$

Accordingly we may replace Eqs. (1.8) by

$$\operatorname{Grad} g = \mu \dot{E}'', \tag{3.1'}$$

where $\dot{E}''$ denotes the (plastic) flow velocity tensor, and "Grad" denotes a symbolic tensor (in analogy to the symbolic vector "grad"), symmetric and with components

$$\frac{\partial g}{\partial \sigma_x}, \quad \frac{1}{2}\frac{\partial g}{\partial \tau_z}, \quad \text{etc.}$$

The tensor $\mu\dot{E}''$ has a *potential*. Or, if we use $\chi = (1/\mu)\,g$, we obtain

$$\operatorname{Grad} \chi = \dot{E}''. \tag{3.1''}$$

In (3.1) $g(\sigma_{ij})$ may denote the left side of a yield condition which need not be that of Sect. 1. This rule which associates to a stress Σ a plastic strain $\dot{E}''$, to within an arbitrary positive factor, is called *v. Mises' theory of plastic potential.*[14] A proof is needed to ensure that by the operation "Grad" the components of a tensor are obtained, i.e., scalar quantities which obey the transformation laws. This is proved in tensor calculus. An elementary proof limited to rectangular Cartesian coordinates has been given in [*3*, p. 20], see also GEIRINGER.[15]

Component forms of Eq. (3.1') are

$$\mu\dot{\varepsilon}_x'' = \frac{\partial g}{\partial \sigma_x}, \quad \mu\dot{\varepsilon}_y'' = \frac{\partial g}{\partial \sigma_y}, \quad \mu\dot{\varepsilon}_z'' = \frac{\partial g}{\partial \sigma_z}$$
$$\mu\dot{\gamma}_x'' = \frac{\partial g}{\partial \tau_x}, \quad \mu\dot{\gamma}_y'' = \frac{\partial g}{\partial \tau_y}, \quad \mu\dot{\gamma}_z'' = \frac{\partial g}{\partial \tau_z}. \tag{3.1'}$$

It follows from (1.1') that

$$\frac{\partial g}{\partial \sigma_x} + \frac{\partial g}{\partial \sigma_y} + \frac{\partial g}{\partial \sigma_z} = 0. \tag{3.1'''}$$

In recent years the concept of the plastic potential has gained importance in the theory of *limit analysis*, which uses this relation between yield condition and flow rule.[16]

[13] H. HENCKY: Proc. 1st Internat. Congr. Appl. Mech. Delft 1924, pp. 312–317.

[14] R. v. MISES: Z. angew. Math. Mech. **8**, 161–185 (1928), see p. 180seq.

[15] H. GEIRINGER: Some recent results in the theory of an ideal plastic body. In: Advances in Applied Mechanics, vol. III, pp. 197–294. New York 1953. We cite this paper henceforth as GEIRINGER: Advances.

[16] See for example, D. C. DRUCKER, W. PRAGER, and H. J. GREENBERG: Quart. Appl. Mech. **9**, 381–389 (1952). Regarding the plastic-potential rule see also R. HILL: Phil. Mag. (7) **40**, 971–983 (1949).

Setting $g(\sigma_x, \sigma_y, \sigma_z, \tau_x, \tau_y, \tau_z = G(\sigma_1, \sigma_2, \sigma_3)$ we obtain a geometric interpretation of (3.1) by writing it in the form

$$\dot{\varepsilon}_1'' : \dot{\varepsilon}_2'' : \dot{\varepsilon}_3'' = \frac{\partial G}{\partial \sigma_1} : \frac{\partial G}{\partial \sigma_2} : \frac{\partial G}{\partial \sigma_3}. \tag{3.2}$$

The sign of the yield function $G(\sigma_1, \sigma_2, \sigma_3)$ may be chosen in such a way that the exterior normal points in the direction of increasing G. Then (3.2) associates to every point $(\sigma_1, \sigma_2, \sigma_3)$ on the yield surface the vector direction $\dot{\varepsilon}_1'' : \dot{\varepsilon}_2'' : \dot{\varepsilon}_3''$ as that of the exterior normal to the yield surface. This leads to a unique direction $\dot{\varepsilon}_1'' : \dot{\varepsilon}_2'' : \dot{\varepsilon}_3''$ only at those points on the yield surface where there is a unique outward normal (cf. Sect. 4).

An *extremum principle* due to v. MISES (p. 184 of the paper cited as Ref. 14) is obtained as follows:

We write $\dot{W}''$, the plastic work of the stresses per unit volume in the form (1.12'), vary the stresses but not the strain velocities, and use (3.1). We obtain

$$\begin{aligned}\delta \dot{W}'' &= \dot{\varepsilon}_x'' \,\delta\sigma_x + \dot{\varepsilon}_y'' \,\delta\sigma_y + \dot{\varepsilon}_z'' \,\delta_{\sigma_z} + \dot{\gamma}_x'' \,\delta\tau_x + \dot{\gamma}_y'' \,\delta\tau_y + \dot{\gamma}_z'' \,\delta\tau_z \\ &= \lambda\left(\frac{\partial g}{\partial \sigma_x}\,\delta\sigma_x + \cdots + \frac{\partial g}{\partial \tau_x}\,\tau_x + \cdots\right) = \lambda\,\delta g.\end{aligned}$$

Here λ is the λ of Eq. (1.8), which depends, in general, on space and time. If the stresses are at the yield limit, $g = \text{constant}$, then $\delta g = 0$, and we obtain MISES' interpretation of (3.1): *For a perfectly plastic body the strain velocities do not perform additional work if the stress tensor Σ satisfies the yield condition and is varied along the yield surface.*

PRAGER[17] introducing generalized stresses Q_i and "corresponding" generalized strains q_i, considers a generalization of (1.12')

$$L = Q_1\, q_1 + Q_2\, q_2 + \cdots + Q_n\, q_n.$$

The term "generalized stress" indicates that the Q_i used to specify the state of stress need not be stresses proper; they may be loads or moments, etc.; they may be made dimensionless. When a set $Q_1, \ldots, Q_n$ has been chosen, the corresponding generalized strains $q_1, \ldots, q_n$ are defined by the condition that L is the work that the stresses perform on the strains. In terms of the Q_i the yield limit is to be given by a continuously differentiable function $F(Q_1, \ldots, Q_n)$ which is chosen so that $F < 0$ below the yield limit. This F is then used as a plastic potential [see Eqs. (3.5)]. It is easily seen that v. MISES' extremum principle still applies.

Several authors have tried to provide a basis for the postulate of the plastic potential by reducing it to other postulates (see DRUCKER, ZIEGLER).[18] W. T. KOITER[19] extends the idea of the plastic potential to *piecewise continuous functions* [for Tresca's yield this had been done by v. MISES (l. c.[14])]. In place of the one function F the yield limit may be specified by several yield functions

$$F_1(Q_1, \ldots, Q_n), \;\ldots,\; F_k(Q_1, \ldots, Q_n). \tag{3.3}$$

A state of stress $Q_1, \ldots, Q_n$ is below the yield limit if *all* these functions are negative. *At* the yield limit, at least one of the F_ν vanishes, and none is positive. The plastic-

[17] W. PRAGER: Proc. 8th Internat. Congr. Appl. Mech. (1952), Istanbul, 1955, pp. 65–72.

[18] H. ZIEGLER: Quart. Appl. Math. **19**, 39 (1961). He rejects previous attempts and merely shows that if the rule is used in MISES' original sense for an element of volume then PRAGER'S generalization holds likewise. — D. C. DRUCKER: Quart. Appl. Math. **14**, 35 (1956) gives an instructive example showing lack of uniqueness in a case where the rule is violated, in as much as Tresca's criterion is applied in connection with v. Mises' flow rule.

[19] W. T. KOITER: Biezeno Anniversary Volume. Haarlem 1953, pp. 232–251; also W. T. KOITER: Quart. Appl. Math. **77**, 350–354 (1953).

potential theory gives then, if, for example, the functions $F_1, \ldots, F_i$ vanish and all others are negative:

$$\dot{q}_1'' = \lambda_1 \frac{\partial F_1}{\partial Q_1} + \cdots + \lambda_i \frac{\partial F_i}{\partial Q_1}, \ldots, \dot{q}_n'' = \lambda_1 \frac{\partial F_1}{\partial Q_n} + \cdots + \lambda_i \frac{\partial F_i}{\partial Q_n} \tag{3.4}$$

where the λ_j are non-negative and at least one of them is positive. Eqs. (3.4) takes the place of the equations

$$\dot{q}_1'' = \lambda \frac{\partial F}{\partial Q_1}, \ldots, \dot{q}_n'' = \lambda \frac{\partial F}{\partial Q_n}, \tag{3.5}$$

which correspond to PRAGERS' generalized stresses and strains.

Returning to (3.1′) we note that a function $g(\sigma_{ij})$ which is to serve as a yield function and as a plastic potential is subject to certain obvious restrictions.[20] In the case of an isotropic material, g must not depend on the chosen coordinate system, but only, as in (2.2), on the three invariants J_1, J_2, J_3 of the stress tensor, or, in other words, only on the values of the principal stresses, $\sigma_1, \sigma_2, \sigma_3$, and not on the principal directions; this dependence must be symmetric. Since g is to serve as the potential of an incompressible perfectly plastic medium we have to assume that it does not depend on the first invariant. It follows that g remains unchanged if Σ is replaced by $S = \Sigma + pI$, where I is the unit tensor, viz,

$$g(\sigma_{ij}) = g(s_{ij}) = G(\sigma_1, \sigma_2, \sigma_3) = G(s_1, s_2, s_3),$$

where G is a symmetric function of its arguments. Consequently G depends only on the differences $\sigma_1 - \sigma_2$, etc., hence only on the τ_1, τ_2, τ_3. Thus, we may write

$$g(\sigma_{ij}) = G(\sigma_1, \sigma_2, \sigma_3) = K(\tau_1, \tau_2, \tau_3) = 0 \tag{3.6}$$

as yield condition for an isotropic, incompressible, perfectly plastic body. It follows from (3.1‴) that

$$\frac{\partial G}{\partial \sigma_1} + \frac{\partial G}{\partial \sigma_2} + \frac{\partial G}{\partial \sigma_3} = \frac{\partial g}{\partial \sigma_x} + \frac{\partial g}{\partial \sigma_y} + \frac{\partial g}{\partial \sigma_z} = 0. \tag{3.7}$$

In general, it is also assumed that the yield surface must be *convex* (that means: any line segment cuts it in at most two points or, exceptionally, lies in the surface). This has been shown, e.g., by HODGE [7, p. 60], to be a consequence of a requirement of DRUCKER[20] which is essentially a statement of plastic reversibility. We see that through

$$\dot{\varepsilon}_{ij}'' = \frac{\partial g}{\partial \sigma_{ij}}, \tag{3.8}$$

which follows from (3.1′), if $\mu \neq 0$ the incompressibility of the plastic strain rate tensor and the condition (3.7) are interdependent. Since

$$\frac{\partial g}{\partial \sigma_{ij}} = \frac{\partial g}{\partial s_{ij}}$$

we may also write

$$\dot{e}_{ij}'' = \lambda \frac{\partial g}{\partial s_{ij}}, \tag{3.8′}$$

where $\dot{e}_{ij}'' = \dot{\varepsilon}_{ij}'' - \delta_{ij}\dot{\varepsilon} = \dot{\varepsilon}_{ij}''$ is the strain rate deviator. The complete system of v. Mises' general equations consists then of Eqs. (1.1) or (1.7), (3.6)—which has the property (3.7)—and (3.8). These last are often called associated flow conditions, where "associated" relates to the yield condition $g = 0$.

[20] See v. MISES[14] and D. C. DRUCKER: Proc. 1st. Nat. Congr. Appl. Mech., Chicago 1950, pp. 487–491.

4. Tresca's yield criterion. "Singular" yield conditions. We consider now the yield condition due to TRESCA. It had been used in connection with the Lévy-v. Mises flow rule (1.8) by the early pioneers in the field, and often this is still done. This is in contrast to the idea of the plastic potential explained in Sect. 3, which associates (1.8) with v. Mises' yield criterion. Tresca's criterion is written as

$$|\tau|_{\max} = \text{constant} \tag{4.1}$$

and states that in perfectly plastic flow yielding occurs when the greatest shear stress reaches a certain limit value which is the same for all stress tensors throughout the material. If the stresses are ordered so that

$$\sigma_1 \geqq \sigma_2 \geqq \sigma_3, \tag{4.2}$$

the condition is

$$2|\tau|_{\max} = \sigma_1 - \sigma_3 = 2\tau_0 = 2k.^{21} \tag{4.1'}$$

In general we do not know which σ_i is the largest, which the smallest principal stress. All we can say is that one of the three differences has absolute value $2k$. It follows that *the yield surface is a regular hexagonal prism, inscribed in the cylinder of the v. Mises yield condition.* V. MISES (see Ref. 14), applied to this yield function the theory of plastic potential, in the form (3.2) which gives for (4.2)

$$\dot{\varepsilon}_1'' : \dot{\varepsilon}_2'' : \dot{\varepsilon}_3'' = 1 : 0 : -1, \tag{4.3}$$

and he showed that the corresponding transformation *is a plane slip in the directions of* $\tau_{\max}$.

We have seen how the theory of plastic potential applies to the faces of the Tresca prism, where there exists a uniquely defined normal. V. MISES, KOITER[22] and PRAGER[23] have complemented the rule for the edges of the prism, where the direction of the normal to the yield surface is not unique. E.g. for points on the edge formed by the adjacent faces $\sigma_1 > \sigma_3 > \sigma_2$ and $\sigma_1 > \sigma_2 > \sigma_3$ we obtain possible flow mechanisms by a linear combination of the flow mechanisms for these two faces in the form

$$\dot{\varepsilon}_1'' : \dot{\varepsilon}_2'' : \dot{\varepsilon}_3'' = -1 : r : 1 - r, \qquad 0 \leqq r \leqq 1. \tag{4.4}$$

This rule may be adapted to yield surfaces with a corner, etc.[24]

The geometric equivalent of Mohr's general condition is a body formed of six curved surfaces and curved edges.[25]

5. "Compatibility" relations. Consider the system (1.6)–(1.9) of ten equations with ten unknowns v_i, σ_{ij}, λ. Following V. MISES (Ref. 1), one may eliminate

[21] H. TRESCA: Mém. prés. par divers savants **18**, 733–799 (1868). The condition had also been stated by B. DE SAINT VENANT, J. Math. pures appl. II, **16**, 308 (1871). The condition can be regarded as a particular case of COULOMB's (1773) condition, basic in the theory of earth pressure, viz, $\sigma_1 - \sigma_3 = c_1 + c_2(\sigma_1 + \sigma_3)$. — J. J. GUEST: Phil. Mag., Ser. V, **50**, 69 (1900) proposed to use it for ductile metals with c_2 a constant small compared to c_1. More general is O. MOHR's form: $\sigma_1 - \sigma_3 = f(\sigma_1 + \sigma_3)$, which states that in yielding the maximum shear stress $(\sigma_1 - \sigma_3)/2$ depends only on the corresponding normal stress $(\sigma_1 + \sigma_3)/2$; no influence is attributed to the median stress σ_2.

[22] W. T. KOITER: see Ref. 19.

[23] W. PRAGER: J. Appl. Mech. **20**, 317 (1953).

[24] For general discussion of edges and corners see P. G. HODGE, JR.: J. Rational Mech. Anal. **5**, 917–938 (1956). HODGE developed in several papers a theory of piecewise linear yield surfaces designed to simplify the computations.

[25] C. TORRE: Oesterr. Ing. Arch. **1**, 36 and 316 (1948), studied such conditions applied to a thick-walled cylinder stressed by internal pressure. (His slip line field is to be compared to that of Fig. 9. The slip lines are however then no longer orthogonal.)

the σ_{ij} and obtain five equations for the five functions v_i, p and $\mu = 1/\lambda$. We simply replace in (1.7) the s_{ij} by means of (1.8) or (3.1), write (1.9′) as

$$(\dot{\varepsilon}_x'' - \dot{\varepsilon}_y'')^2 + \cdots + 6(\dot{\gamma}_x''^2 + \dot{\gamma}_y''^2 + \dot{\gamma}_z''^2) = 6\lambda^2 k^2, \tag{5.1}$$

and use as the fifth equation the incompressibility relation $\dot{\varepsilon}_x'' + \dot{\varepsilon}_y'' + \dot{\varepsilon}_z'' = 0$.

On the other hand, if we are only interested in the six stresses, we may consider the four Eqs. (1.7) and (1.9) (or a more general yield condition) and complement them by two *compatibility conditions* (relations between the stress components), derived by W. JENNE.[26] These relations simplify considerably in the case of plane deformation and in the axially symmetric case.

6. The flow equations of Prandtl[27] and Reuss[28]. The perfectly plastic body as defined by v. MISES is based on the following conception.[29] There exists a function g of the six stresses and a constant C such that at each point of the material $g \leqq C$. Wherever $g < C$ the laws of elasticity hold; wherever $g = C$ the flow law (3.8) holds for the plastic flow velocities with a (non-negative) proportionality factor λ varying in space and time. λ can assume any value between zero and infinity. It is permissible, but in no way necessary, to neglect the elastic displacements in either domain—if the problem warrants it. (See also Sect. 58.)

In the flow theory of PRANDTL and REUSS (PRANDTL dealt with the plane problem, REUSS with the general case), both plastic and elastic strains are considered simultaneously in the domain where $g = C$. Following REUSS we denote elastic strains by a prime, plastic strains by a double prime and total strains without a superscript, hence e.g., $E = E' + E''$ or, the dot denoting d/dt, $\dot{E} = \dot{E}' + \dot{E}''$ etc. Using

$$\sigma = -p = \frac{1}{3}(\sigma_x + \sigma_y + \sigma_z), \qquad \varepsilon = \frac{1}{3}(\varepsilon_x + \varepsilon_y + \varepsilon_z), \qquad \varepsilon'' = \frac{1}{3}(\varepsilon_x'' + \varepsilon_y'' + \varepsilon_z'')$$

$$\dot{\varepsilon}'' = \frac{1}{3}\left(\frac{\partial v_x}{\partial x} + \frac{\partial v_y}{\partial y} + \frac{\partial v_z}{\partial z}\right),$$

we need the strain deviator e_{ij} and likewise e_{ij}', e_{ij}'', $\dot{e}_{ij}$, etc. We put

$$\varepsilon_{ij} = e_{ij} + \varepsilon\,\delta_{ij}, \qquad \dot{\varepsilon}_{ij} = \dot{e}_{ij} + \dot{\varepsilon}\,\delta_{ij}. \tag{6.1}$$

As before, plastic incompressibility, $\dot{\varepsilon}'' = 0$, is assumed as well as $\varepsilon'' = 0$; hence $\varepsilon = \varepsilon'$, $\dot{\varepsilon} = \dot{\varepsilon}'$. Hooke's law is written in the form

$$e_{ij}' = \frac{1}{2G} s_{ij}, \qquad \varepsilon' = \varepsilon = \frac{1}{K}\sigma. \tag{6.2}$$

Here G and K are related to the elasticity modulus E and Poisson's ratio ν by

$$G = \frac{E}{2(1+\nu)}, \qquad K = \frac{E}{1-2\nu}. \tag{6.2'}$$

From (6.2) we obtain

$$\dot{e}_{ij}' = \frac{1}{2G}\dot{s}_{ij}, \qquad \dot{\varepsilon}' = \frac{1}{K}\dot{\sigma}. \tag{6.3}$$

For the plastic part of the deformation we assume, in addition to $\varepsilon'' = 0$, that (1.8) holds

$$\dot{\varepsilon}_{ij}'' = \lambda s_{ij}. \tag{6.4}$$

---

[26] W. JENNE: Z. angew. Math. Mech. **8**, 18–44 (1928). Cf. his Eqs. (18).

[27] L. PRANDTL: Proc. 1st Internat. Congr. Appl. Mech., Delft 1924, pp. 43–54.

[28] E. REUSS: Z. angew. Math. Mech. **10**, 266–274 (1930).

[29] See e.g. R. v. MISES: H. Reissner Anniversary Volume, Ann Arbor, Michigan 1949, pp. 415–429 (paper presented Feb. 1948 at the meeting on plasticity in Providence, R.I.).

Adding the first Eqs. (6.3) and (6.4), and observing that $\dot{\varepsilon}''_{ij} = \dot{e}''_{ij}$, we obtain

$$\dot{e}_{ij} = \frac{1}{2G}\,\dot{s}_{ij} + \lambda s_{ij}; \tag{6.5}$$

that is, using also the last Eq. (6.2) we obtain the system of six equations

$$\frac{d}{dt}(e_{ij}) = \frac{1}{2G}\,\frac{d}{dt}(s_{ij}) + \lambda s_{ij} \quad \text{and} \quad \frac{d\sigma}{dt} = K\,\frac{d\varepsilon}{dt}. \tag{6.5'}$$

These six equations together with three equilibrium equations and the yield condition are now again ten equations for the $10 = 5+1+3+1$ unknown quantities s_{ij}, σ, v_i, λ. Mathematically, Eq. (6.5) are much more complicated than (1.8) since in (6.5) the s_{ij} *and* their time derivatives both appear. The Eqs. (6.5) apply only during plastic flow, i.e., they apply for $J_2 = k^2$. If in a problem $\dot{s}_{ij} = 0$, then from (6.3) $\dot{e}'_{ij} = 0$, hence $\dot{e}_{ij} = \dot{e}''_{ij} = \dot{\varepsilon}''_{ij}$, and the Reuss equation (6.5) reduces to the v. Mises equation (1.8). Likewise, if the limit $G \to \infty$ is considered, $e'_{ij} \to 0$, in general, from (6.3) and therefore in this case again Eqs. (1.8) are obtained. It is, however, quite unjustified to take this—conversely—as a reason for identifying the v. Mises theory with a Reuss theory for $G \to \infty$. In fact, in the plastic domain of a body, D_{pl}, where $J_2 = k^2$, the v. Mises theory offers the theory of the *plastic* deformation, viz. (6.4), and it makes no specific statement concerning the elastic displacements in D_{pl}.[30] With respect to D_{el}, the region below the yield limit, v. MISES stated repeatedly and used in much of his work, that there the material is elastic (see as one of many examples the treatment reproduced in our Sect. 55). Of course, in appropriate cases the elasticity in D_{el} may be disregarded (see Sect. 58 seq.), as, for example in certain problems of unrestricted plastic flow; but, in problems of contained plastic deformation, this "neglecting" would blot out the difference between regions of contained plastic deformation and of elastic deformation since both regions would be considered rigid (see, however, footnote 255).

To eliminate λ, in (6.5) we multiply each Eq. (6.5) by s_{ij} and add all six equations:

$$\sum_{i,j} s_{ij}\,\dot{e}_{ij} = \frac{1}{2G}\sum_{i,j} s_{ij}\,\dot{s}_{ij} + \lambda \sum_{i,j} s_{ij}\,s_{ij}.$$

Now in the plastic range, according to Eq. (1.9′) the sum in the second term on the right equals $2\tau_0^2 = 2k^2$ and therefore the first term to the right is zero. Hence

$$\sum_{i,j} s_{ij}\,\dot{e}_{ij} = 2\lambda k^2,$$

and, writing as an abbreviation

$$\sum_{i,j} s_{ij}\,\dot{e}_{ij} = \dot{W}, \tag{6.6}$$

we obtain

$$\lambda = \frac{\dot{W}}{2k^2}. \tag{6.7}$$

If in (6.6) we write $\dot{e}_{ij} = \dot{e}'_{ij} + \dot{e}''_{ij}$, we have $\dot{W} = \dot{W}' + \dot{W}''$, where $\dot{W}'$ and $\dot{W}''$ denote the elastic and plastic contributions, respectively. Now, in the plastic

[30] The Reuss theory complements rather than contradicts the Mises theory. It is worth mentioning that in the discussion following v. MISES' general lecture, given at the Third International Congress of Applied Mechanics, Stockholm, 1930 (see R. v. MISES: Proc. 3rd Internat. Congr. Appl. Mech. II, pp. 3–13) the new Reuss theory, just published 1930 in the Z. angew. Math. Mech., is denoted by the participants in the discussion as v. Mises-Reuss theory.

domain, where

$$\sum_{i,j} s_{ij}\, s_{ij} = 2k^2,$$

by (6.3)

$$\dot{W}'' = \frac{1}{2G} \sum_{i,j} s_{ij}\, \dot{s}_{ij} = 0.$$

Hence, there $\dot{W} = \dot{W}''$, and (6.7) is replaced by

$$\lambda = \frac{\dot{W}''}{2k^2}, \tag{6.7'}$$

as in (1.13). Hence Eqs. (6.5) may then be written

$$\dot{e}_{ij} = \frac{1}{2G}\, \dot{s}_{ij} + \frac{\dot{W}''}{2k^2}\, s_{ij}. \tag{6.8}$$

Of these six equations only four are independent. First, the sum of the three equations for $\dot{e}_{11}$, $\dot{e}_{22}$ and $\dot{e}_{33}$ vanishes identically. In addition, the combination

$$\sum_{i,j} \dot{e}_{ij}\, s_{ij} = \frac{1}{2G} \sum_{i,j} \dot{s}_{ij}\, s_{ij} + \frac{\dot{W}''}{2k^2} \sum_{i,j} s_{ij}\, s_{ij}$$

is an identity since it stands for

$$\dot{W} = 0 + \frac{\dot{W}}{2k^2} \cdot 2k^2.$$

The balance of equations is therefore the same as before, since now there is one less unknown.

We note from (6.7), since λ is not negative, that the same holds for $\dot{W}$. *Hence Eqs.* (6.8) *hold at the yield limit,* $J_2 = k^2$ *and for* $\dot{W} \geqq 0$. If either the yield limit is not reached, $J_2 < k^2$, or $J_2 = k^2$ but $\dot{W} < 0$ (elastic unloading from a plastic state), the flow is elastic and (6.3) valid.

Again, for $G \to \infty$ we obtain from (6.8)

$$\dot{e}_{ij} = \frac{\dot{W}''}{2k^2}\, s_{ij},$$

i.e., Eqs. (1.8′) with λ replaced by (1.13), *if* in (6.3) as $G \to \infty$ the ratio $s_{ij}/2G \to 0$, hence $\dot{e}'_{ij} \to 0$, so that the $\dot{e}_{ij}$ on the left side in (6.8) become $\dot{e}''_{ij} = \dot{\varepsilon}''_{ij}$.

It is certainly physically satisfactory that in the Prandtl-Reuss equations (6.5) or (6.8) the elastic strains in the plastic domain are incorporated. On the other hand the Reuss equations introduce the mathematical difficulty of containing both the s_{ij} and $\dot{s}_{ij}$. Even with v. Mises' relations the complete system of equations is difficult to handle.

7. Further stress strain laws. Both the Lévy-v. Mises theory and the Prandtl-Reuss theory assume a sudden transition from the elastic to the plastic state. Hence two different sets of equations have to be used in the two domains, between which the boundary is in general not known but has to be determined as part of the problem.

In an attempt to overcome this difficulty PRAGER[31] has proposed stress-strain relations which reflect a continuous transition from the elastic to the plastic state.[32]

---

[31] W. PRAGER: Proc. 5th Internat. Congr. Appl. Mech., Cambridge 1938, pp. 234–237, W. PRAGER: Duke Math. J. 9, 228–233 (1942).

[32] We refer the reader to the original papers, in particular to the first one, Eqs. (5) and (6). Cf. also HILL [6, p. 49].

In a survey of proposed stress-strain laws PRAGER[33] introduces a useful terminology. He calls a stress-strain law of *flow type* or of *deformation type* depending on whether it links the stresses (stress deviations) and rates of stress (of stress deviation) *either to the rate of strain* (*and* maybe to the strain) *or* only to the *strain*. We have discussed here only flow type laws. The best known deformation-type law is due to W. HENCKY.[34] It may be considered as a development of the theory of A. HAAR and TH. VON KÁRMÁN[35] (cf. discussion in HILL [*6*, p. 45]. For work up to 1930 see the survey article on the mechanics of continua by v. MISES.[36]

8. Remarks on some three-dimensional problems. The problems discussed and solved in the theory of the perfectly plastic body are mainly problems of plane strain, of plane stress, problems with axial symmetry, with spherical symmetry, etc. Three-dimensional problems seem so far amenable to rigorous discussion only if (a) the plastic zone is narrow in one dimension or (b) some high symmetry prevails.

The fundamentals reported in the preceding sections relate to the general three-dimensional problem. In addition, the characteristics of the three-dimensional problem will be given presently. Here we still mention a few results of a more restricted importance.

W. JENNE[37] considers some generalizations of the problem of plane deformation. In plane deformation or plane strain the state of stress is the same in all planes perpendicular to one of the principal directions of stress, which is the same for all points of the body. JENNE studies the case of equal states of stress in planes perpendicular to an arbitrary space curve and further generalizations in this direction. The considerations are interesting but involved. We wish to report here an elegant auxiliary result, a particular case of which we shall use in Chap. C, Eqs. (34.4). Denote by u, v, w the principal axes of Σ. The change of the directions of this triad can be studied by means of a tensor of angular velocity. Consider, e.g., a point $P(x, y, z)$ and a neighboring point $P'(x+dx, y+dy, z+dz)$ at distance ds from P; then, in the transition from P to P' the triad of principal axes of Σ will rotate with the angular velocity $\boldsymbol{\omega}_x\, dx + \boldsymbol{\omega}_y\, dy + \boldsymbol{\omega}_z\, dz$. Denote by $\omega_{uu}, \omega_{u,v}\ldots$ the nine components of the tensor of angular velocity Ω with respect to the u, v, w-system. Then for the derivatives of the σ_i in the u, v, w-directions we have the formulas

$$\frac{\partial\sigma_1}{\partial u} = 2\tau_2\,\omega_{wv} + 2\tau_3\,\omega_{vw}, \qquad \frac{\partial\sigma_2}{\partial v} = 2\tau_3\,\omega_{uw} + 2\tau_1\,\omega_{wu}, \qquad \frac{\partial\sigma_3}{\partial\omega} = 2\tau_1\,\omega_{vu} + 2\tau_2\,\omega_{uv}. \tag{8.1}$$

In a paper of 1954 on three-dimensional plastic flow PRAGER[38] cites a paper by SIMONI,[39] which he shows to reveal itself at closer inspection as a plane problem, and a genuinely three-dimensional flow field studied by HILL[40] where the incipient plastic flow in a prismatic bar of plastic-rigid material subject to combined tension, torsion and bending, is described. PRAGER then investigates completely the general three-dimensional plastic flow possible under a uniform state of stress.[41]

---

[33] W. PRAGER: J. Appl. Mech. **14**, 226–233 (1948). It seems that the terminology had been used earlier by A. A. ILYUSHIN, Prikl. Mat. Mekh. **9**, 207–218 (1945).

[34] W. HENCKY: Proc. 1st Internat. Congr. Appl. Mech., Delft 1924, pp. 312–317. For a more detailed presentation see, for example [*27*, p. 95] or [*25*, p. 10].

[35] A. HAAR and TH. VON KÁRMÁN: Nachr. kgl. Ges. Wiss. Göttingen 1909, p. 204–218.

[36] R. v. MISES: Proc. 3rd Internat. Congr. Appl. Mech. II, pp. 3–13 (pp. 9–13 contain an interesting discussion of his paper).

[37] W. JENNE: see Ref. 26.

[38] W. PRAGER: Rev. Fac. Sci. Univ. Istanbul **19**, 23—27 (1954).

[39] F. DE SIMONI: Ist. Lombardo Sci. Lett., Rendic. Cl. Sci. Nat. (3) **15**, 623–634 (1951).

[40] R. HILL: Quart. Appl. Math. **1**, 18–28 (1948).

[41] See also L. FINZI: Torino R. Accad. Sci. **76**, 1–19 (1941), and L. FINZI: Ist. Lombardo Sci. Rendic. Cl. Sci. Mat. Nat. **90**, 528–535 (1956).

8bis. Remarks on uniqueness for rigid plastic solids. In 1951 R. HILL began studying this problem and followed it up in a series of papers.[42] He considers a rigid plastic work-hardening solid[43] and supposes the identity of the plastic potential and the yield condition. The yield surface is everywhere strictly convex and contains the origin. HILL[42] showed in 1951 that when over a certain part S_F of the surface of the body the traction $\boldsymbol{F}$ is given and over the remaining surface $S_v = S - S_F$ the velocity $\boldsymbol{v}$ is given, then the state of *stress*, Σ, is uniquely determined in a certain region[44] which forms part (or sometimes all) of the plastic region. But the mode of *deformation* may not be uniquely determined by these boundary conditions. (An example is the well-known problem of the indentation of a plane surface by a flat punch [*6*, p. 255] where there exists an infinity of such modes.)

In paper I (1956a) HILL made the assumption that it may be permitted to neglect positional changes and rotation of the material. If the mode of deformation is not uniquely determined by the above boundary conditions, a set of "virtual" modes ε^* can be found, compatible with $\boldsymbol{v}$ on S_v and with the existing Σ. However, only one of these modes is "actual" (see footnote[44]). It is defined if over S_F not only $\boldsymbol{F}$ but also $\dot{\boldsymbol{F}}$ is given and over S_v not only $\boldsymbol{v}$ but also $\dot{\boldsymbol{v}}$. Hence the indeterminacy disappears if the right boundary conditions are considered.[45]

In contribution II (1956b) these results are illustrated by application to the torsion of a prismatic bar.[46]

In III, the assumption (made in I) that changes in geometry can be disregarded is dropped, and a sufficient criterion for uniqueness is established for this more general situation.

In IV, the actual mode which satisfies the criterion derived in III is characterized by an extremum principle which appears as a generalization of an extremum principle which holds in I for the actual mode.[47]

II. Discontinuous solutions.[48]

a) Characteristics. Application to the three-dimensional problem of the perfectly plastic body.

9. Introduction. In many branches of mathematical physics concepts like "discontinuity surface", "characteristic surface", or "shock" play a great role.

[42] R. HILL: Phil. Mag. **42**, 868 (1951). J. Mech. and Phys. Solids **4**, 247 (1956a); **5**, 1 (1956b); **5**, 153 (1957a); **5**, 302 (1957b).

[43] Work hardening is not included in this article. For this reason we cannot go beyond a brief sketch of this remarkable work.

[44] A method of delimiting this zone has been given by BISHOP, GREEN, and HILL in the above journal **4**, 256 (1956). Basing their study on the work of HILL, 1951, they present a procedure for finding the greatest extent of a "deformable region" in a rigid plastic body yielding under given boundary conditions. As "deformable region" is denoted the domain which is occupied by the complete set of modes. (If there is more than one mode, the isolation of the above domain is only the first step towards selecting the mode that will actually operate — the "actual" mode.) — A generalization of their result has been given by R. M. HAYTHORNTHWAITE and R. T. SHIELD, J. Mech. and Phys. Solids **6**, 127 (1958).

[45] For a non-hardening material a uniqueness theorem cannot be derived directly. HILL recommends determining the actual mode under work-hardening and considering the non-hardening solid as a limit case if the rate of hardening becomes vanishingly small.

[46] HILL's work relates to regular yield conditions. R. M. HAYTHORNTHWAITE and W. PRAGER in J. Mech. and Phys. Solids **6**, 9 (1957) extend some of HILL's results to rigid work-hardening solids with singular yield condition.

[47] In the same Journal **9**, 114 (1961) and **10**, 185 (1962) HILL generalized his work.

[48] The considerations of this part II apply to three dimensions except for Sect. 24.

The mathematicians to whom we owe much of our present insight into these theories—here we only mention the names of B. RIEMANN, E. B. CHRISTOFFEL, H. HUGONIOT, J. HADAMARD, T. LEVI-CIVITA—developed the problems and concepts essentially with respect to the theory of fluids and elastic solids, in particular compressible fluid flow, and numerous investigations followed.

In recent years similar investigations have been attempted regarding perfectly plastic solids. This theory has not yet reached a clarity and completeness comparable to achievements in *compressible fluid* flow. We shall report on some of the results obtained so far; in doing so we think it appropriate to explain at least to a certain degree some of the mathematical foundations which might prove helpful to one or another research worker in our field, where considerations of this type are often neglected.

10. Examples. α) Denote by φ a function of x and y and consider one of the simplest partial differential equations of second order:

$$\frac{\partial^2 \varphi}{\partial x\, \partial y} = 0. \tag{10.1}$$

Along some *initial curve*, $y = \alpha(x)$, e.g., the straight line $y = x$, the values of φ and of $\partial\varphi/\partial x$ may be prescribed, *initial data*, or *Cauchy data*, as arbitrarily given functions of one variable,

$$\varphi(x, x) = f(x), \qquad \left(\frac{\partial\varphi}{\partial x}\right)_{y=x} = g(x). \tag{10.2}$$

We recognize immediately the truth of the well-known result that by these data a solution is uniquely given. For, introduce $G(x) = \int^x g(x)\, dx$, and put

$$\varphi(x, y) = G(x) + f(y) - G(y).$$

This function satisfies (10.1) and also (10.2), since

$$\varphi(x, x) = G(x) + f(x) - G(x) = f(x)$$

$$\left(\frac{\partial\varphi}{\partial x}\right)_{y=z} = G'(x) = g(x).$$

Uniqueness follows immediately.

Next, we choose as initial curve the line $x = c$ and prescribe along this initial line the values of φ and of $\partial\varphi/\partial x$:

$$\varphi(c, y) = f(y), \qquad \left(\frac{\partial\varphi}{\partial x}\right)_{x=c} = g(y). \tag{10.3}$$

It is known that the general form of the solution of (10.1) is $\varphi(x, y) = h(x) + k(y)$, where h and k are arbitrary functions of one variable; thus, from (10.3)

$$\varphi(c, y) = h(c) + k(y) = f(y), \qquad h'(c) = g(y).$$

We thus see that *it is not possible to prescribe* $g(y)$ *arbitrarily*; this function must reduce to a constant, say k, since it is to be equal to $h'(c)$. On the other hand, if $g(y)$ is a constant, then there are clearly *infinitely many solutions* which all satisfy the given initial conditions. Take, e.g., $k = 3$, $c = 1$, and

$$h(x) = x^3, \qquad h'(x) = 3x^2, \qquad h'(c) = 3 = k.$$

But for $h(x) = \frac{3}{10} x^{10}$, likewise $h'(1) = 3$ or for $h(x) = -\frac{6}{\pi} \cos \frac{\pi}{2} x$. We see that two completely different solutions

$$(1) \quad \varphi(x, y) = x^3 + f(y) - 1, \qquad \text{and} \qquad (2) \quad \varphi(x, y) = f(y) - \frac{6}{\pi} \cos \frac{\pi}{2} x$$

(and infinitely many others) both satisfy (10.1) and (10.3), viz,

$$\varphi(1, y) = f(y), \quad \left(\frac{\partial \varphi}{\partial x}\right)_{x=1} = 3. \tag{10.3'}$$

Along $x = c$, both solutions have the same

$$\frac{\partial \varphi}{\partial x}, \quad \frac{\partial \varphi}{\partial y}, \quad \frac{\partial^2 \varphi}{\partial x\, \partial y}, \quad \frac{\partial^2 \varphi}{\partial y^2}, \ldots$$

etc. But the values of $\partial^2 \varphi / \partial x^2$, etc. are different. We may assert the result also in the following way: Along the *exceptional* or *characteristic* line $x = c$, the two different solutions (1) and (2) can be patched together without violating (10.1) or (10.3). Derivatives of φ along the characteristic line are determined by (10.1) and (10.3), but derivatives across this line remain undetermined.

β) Consider next the more general partial differential equation[49]

$$A \frac{\partial^2 \varphi}{\partial x^2} + 2B \frac{\partial^2 \varphi}{\partial x\, \partial y} + C \frac{\partial^2 \varphi}{\partial y^2} = F, \tag{10.4}$$

where A, B, C, F may be functions of x, y, φ and of its first derivatives. Assume again, as before, initial conditions along the line $x = c$ namely:

$$\varphi(c, y) = f(y), \quad \left(\frac{\partial \varphi}{\partial x}\right)_{x=c} = g(y). \tag{10.5}$$

We now ask: To what extent does the differential equation (10.4) together with the initial conditions (10.5) determine a solution? Obviously we can compute from (10.5) for

$$x = c: \frac{\partial \varphi}{\partial y}, \quad \frac{\partial^2 \varphi}{\partial x\, \partial y}, \quad \frac{\partial^2 \varphi}{\partial y^2}, \quad \text{etc.}$$

In order to compute $\partial^2 \varphi / \partial x^2$ we need the differential equation (10.4) and find if $A \neq 0$ that

$$\left[\frac{\partial^2 \varphi}{\partial x^2}\right]_{x=c} = \left[\frac{F}{A} - \frac{C}{A} \frac{\partial^2 \varphi}{\partial y^2} - \frac{2B}{A} \frac{\partial^2 \varphi}{\partial x\, \partial y}\right]_{x=c}.$$

By using both (10.5) and (10.4)—which we may differentiate—we can thus compute for $x = c$ as many derivatives as we wish and set up a Taylor expansion which determines $\varphi(x, y)$ in a neighborhood of the initial curve $x = c$.

Or, we may conclude as follows—always if $A \neq 0$. Consider the neighboring line $x = x_1 = c + dc$. Then, approximately

$$\begin{aligned} \varphi(c + dc, y) &= \varphi(c, y) + \frac{\partial \varphi}{\partial x}(c, y)\, dc, \\ \frac{\partial \varphi}{\partial x}(c + dc, y) &= \frac{\partial \varphi}{\partial x}(c, y) + \frac{\partial^2 \varphi}{\partial x^2}(c, y)\, dc, \quad \text{etc.} \end{aligned} \tag{10.5'}$$

Thus, we known approximately φ and $\partial \varphi / \partial x$ along the line $x = x_1$. We then may proceed to a line $x = x_2$ adjacent to $x = x_1$ and we may use (10.5′) similarly as we have used (10.5). Likewise, we may proceed towards the left to $x = x_1' = c - dc$, etc.

These simple considerations show the role of the condition $A \neq 0$ for our present example. [In example (α), which corresponds to the case $A = 0$, the

[49] Cf. the presentation in R. v. MISES, Mathematical theory of compressible fluid flow. Completed by HILDA GEIRINGER and G. S. S. LUDFORD. New York: Academic Press 1958, in particular articles 9 and 10.

condition for a unique solution was that the initial curve should not be parallel to either of the two axes x or y.] Assume now that it is possible—as in example (α)—to find two different solutions of (10.4) and (10.5), say one, $\varphi=\varphi_1$, to the right of $x=c$, one, $\varphi=\varphi_2$, to the left, but such that along that line,

$$\varphi_1=\varphi_2, \quad \frac{\partial\varphi_1}{\partial x}=\frac{\partial\varphi_2}{\partial x}.$$

We may then call this combination of φ_1 and φ_2 a "solution" in the combined domain to the right *and* to the left. In fact, this combined solution satisfies the differential equation in both domains, including the line $x=c$, and the boundary conditions along $x=c$. Also along this vertical line

$$\frac{\partial^2\varphi_1}{\partial x\,\partial y}=\frac{\partial^2\varphi_2}{\partial x\,\partial y}, \quad \frac{\partial^2\varphi_1}{\partial y^2}=\frac{\partial^2\varphi_2}{\partial y^2}.$$

It is true that, in general,

$$\frac{\partial^2\varphi_1}{\partial x^2}\neq\frac{\partial^2\varphi_2}{\partial x^2};$$

this quantity, however, has now the coefficient $A=0$. Hence we see, as in (α) that *along $x=c$ two entirely different solutions can be patched together*, if $A=0$. In this case the y-direction is exceptional.

The preceding formulations are still dependent on the arbitrary coordinate system. We must obtain an *invariant interpretation* of a condition like $A\neq0$. Such an interpretation can be achieved by mathematical considerations or by physical conclusions.

γ) *Examples of invariant interpretations.* Consider the Eqs. (1.7) and (1.9) for a "plane problem". We shall consider this theory in detail in Chap. B. For the present we write (with $\tau_{xy}=\tau$):

$$\frac{\partial\sigma_x}{\partial x}+\frac{\partial\tau}{\partial y}=0, \quad \frac{\partial\tau}{\partial x}+\frac{\partial\sigma_y}{\partial y}=0, \quad (\sigma_x-\sigma_y)^2+4\tau^2=4k^2.$$

To integrate these we introduce a function $\varphi(x, y)$ such that

$$\frac{\partial\varphi}{\partial x}=-\tau, \quad \frac{\partial\varphi}{\partial y}=\sigma_x.$$

The first equation is then identically satisfied, and the second combined with the yield condition gives

$$(\sigma_y-\sigma_x)\frac{\partial^2\varphi}{\partial x^2}-4\tau\frac{\partial^2\varphi}{\partial x\,\partial y}+(\sigma_x-\sigma_y)\frac{\partial^2\varphi}{\partial y^2}=0,$$

an equation of type (10.4). Assume that along the y-axis τ and σ_x are given. We know from the previous example that this will determine a solution uniquely unless $A=0$, viz. unless $\sigma_y-\sigma_x=0$. If $A=0$, the y-direction is exceptional or *characteristic*. To obtain the condition $A=0$ in a meaningful form we consider the angle ϑ' of the first principal direction with the y-direction and find $\tan 2\vartheta'=2\tau/(\sigma_y-\sigma_x)$. If $\sigma_y-\sigma_x=0$, $2\vartheta'=90°$, or $\vartheta'=45°$. *Hence the characteristic bisects the angle of the principal directions*; and this is now a geometrical characterization of the exceptional direction which is independent of any coordinate system (see Sect. 29).

As a second example we establish an important geometrical property of a characteristic. We consider the general yield condition $g(\sigma_x, \sigma_y, \tau)=0$ and obtain, with φ defined as before, and using the abbreviation $\varphi_{xx}=\partial^2\varphi/\partial x^2$, etc.,

$$\frac{\partial g}{\partial\sigma_y}\varphi_{xx}-\frac{\partial g}{\partial\tau}\varphi_{xy}+\frac{\partial g}{\partial\sigma_x}\varphi_{yy}=0.$$

The y-direction will be characteristic if $A = \partial g/\partial\sigma_y = 0$. But according to (3.5), this gives $\dot{\varepsilon}_y = 0$. Hence *the rate of extension in a characteristic direction vanishes.* This is an invariant geometrical characterization valid for a general yield condition under the strain-stress law (3.5).

11. Systems of differential equations. Instead of the single partial differential equation (10.4) with two independent variables we now consider a system of m equations for m unknown functions $u_1, u_2, \ldots, u_m$ of $n+1$ independent variables $x_0, x_1, \ldots, x_n$. It is convenient for our purpose to consider a system of equations of order two rather than one of order one, although, theoretically, the former can be reduced to the latter.

We remember the condition, $A \neq 0$, which appeared in the discussion of (10.4), in connection with data given on the line $x = \text{const.} = c$. From the fact that

$$\varphi(c, y) \quad \text{and} \quad \frac{\partial\varphi}{\partial x}(c, y)$$

were given, we knew all derivatives with respect to y of $\varphi(c, y)$ or, as we may say, all *interior derivatives* along the line $x = c$. In addition our knowledge of the normal *exterior* derivative

$$\frac{\partial\varphi}{\partial x}(c, y)$$

led us beyond the line $x = c$. The meaning of $A \neq 0$ was that exterior derivatives like

$$\frac{\partial^2\varphi}{\partial x^2}(c, y)$$

which could not be found from the initial conditions could be computed from the differential equation since in this case, we could solve Eq. (10.4) with respect to $\partial^2\varphi/\partial x^2$. (If the equation is given in the explicit form

$$\frac{\partial^2\varphi}{\partial x^2} = \Phi\left(x, y, \frac{\partial\varphi}{\partial x}, \frac{\partial\varphi}{\partial y}, \frac{\partial^2\varphi}{\partial x\,\partial y}, \frac{\partial^2\varphi}{\partial y^2}\right)$$

then the condition $A \neq 0$ is clearly satisfied, and the line $x = c$ cannot be a characteristic.) These ideas can be generalized.

Consider the m equations

$$\sum_{k=1}^{m}\sum_{i,j}^{0\ldots n} a_{\mu k i j} + \frac{\partial^2 u_k}{\partial x_i\,\partial x_j} + b_\mu = 0 \qquad \mu = 1, 2, \ldots, m, \tag{11.1}$$

where the $a_{\mu k i j}$ and the b_μ depend on the x_i, on the u_k, and the first partial derivatives $\partial u_k/\partial x_i$ where always $i, j = 0, 1, \ldots, n$; $\mu, k = 1, \ldots, m$. Instead of the line $x = c$ we consider now the n-dimensional hyperplane, ω: $x_0 = c$ (an ordinary plane if $n+1 = 3$), and assume the *initial data*: all u_k are given on ω, as well as all $\partial u_k/\partial x_0$; or, more explicitly

$$u_k(c, x_1, \ldots, x_n) = f_k(x_1, \ldots, x_n), \qquad \frac{\partial u_k}{\partial x_0}(c, x_1, \ldots, x_n) = g_k(x_1, \ldots, x_n), \tag{11.2}$$

as generalization of (10.5). Note that from the knowledge of $u_k(c, x_1, x_2, \ldots, x_n)$ we can compute *all interior derivatives*

$$\frac{\partial u_k}{\partial x_\nu}(c, x_1, \ldots, x_n), \qquad \nu = 1, 2, \ldots, n$$

along ω; in addition the *exterior derivatives* $\partial u_k/\partial x_0$ are given. By further differentiation in ω we can compute higher interior derivatives. But, in order to be able to compute the exterior derivatives $\partial^2 u_k/\partial x_0^2$, it is necessary that the system (11.1) can be solved with respect to

$$\frac{\partial^2 u_1}{\partial x_0^2}, \quad \frac{\partial^2 u_2}{\partial x_0^2}, \quad \ldots, \quad \frac{\partial^2 u_m}{\partial x_0^2}.$$

A system solved with respect to these derivatives is briefly called *normal*. Such a system is of the form

$$\frac{\partial^2 u_\mu}{\partial x_0^2} = U_\mu, \qquad \mu = 1, 2, \ldots, m, \tag{11.3}$$

where U_μ may depend on the x_i, $i = 0, 1, \ldots, n$, the u_k, $k = 1, 2, \ldots, m$, the first and second derivatives of the u_k, except $\partial^2 u_\mu/\partial x_0^2$. One denotes as the *Cauchy problem the determination in a neighborhood of ω of solutions $u_1, u_2, \ldots, u_m$ of* (11.1) *which assume the values* (11.2) *on ω*. This determination is uniquely possible for a normal system under appropriate assumptions for the $a_{\mu k i j}$, the b_μ the f_k and g_k which appear in (11.1) and (11.2).[50] In order to be able to transform (11.1) into a normal system a certain determinant of coefficients must be different from zero. With $a_{\mu k 00} = a_{\mu k}$ and $\|a_{\mu k}\|$ denoting the $m \times m$ determinant of these coefficients,

$$\Omega = \|a_{\mu k}\| \neq 0 \tag{11.4}$$

is the generalization of the condition $A \neq 0$. In this case we can solve with respect to the $\partial^2 u_\mu/\partial x_0^2$ ($\mu = 1, 2, \ldots, m$), and by further differentiations of these equations, higher exterior derivatives of the u_μ can also be computed. It is then plausible (thinking in terms of Taylor expansions in the neighborhood of ω or of step by step numerical computation), and indeed true, that a solution of (11.1), which assumes the given values (11.2) on ω, exists in a certain neighborhood of ω.

On the other hand, we call ω, i.e. the plane $x_0 = c$, *exceptional* or *characteristic* if

$$\Omega = 0. \tag{11.5}$$

Clearly, the condition (11.4) depends on the coordinate system and must be *transformed into an invariant form* by means of physical or mathematical considerations (see Sect. 14).

If the given system is of first order

$$\sum_{k=1}^{m} \sum_{i=0}^{n} a_{\mu k i} \frac{\partial u_k}{\partial x_i} + b_\mu = 0, \qquad \mu = 1, 2, \ldots, m, \tag{11.6}$$

and the $a_{\mu k i}$ and b_μ are functions of the x_i and the u_k, the initial values or Cauchy data consist of values of the u_μ on ω. Analogous considerations as before lead to the condition

$$\Omega = \|a_{\mu k}\| \neq 0, \tag{11.7}$$

where $a_{\mu k} = a_{\mu k 0}$ is the coefficient of $\partial u_k/\partial x_0$ in the μ-th equation, and ω is called exceptional or characteristic if $\Omega = 0$. Again the system is called normal if it is solved for the $\partial u_k/\partial x_0$, $k = 1, 2, \ldots, m$. Since the $a_{\mu k}$ in (11.4) or (11.7) depend on the given Cauchy data, *the same geometrically specified plane may or may not be exceptional, depending on these given data.*

[50] The proof is due to CAUCHY and SONJA KOVALEVSKA (see, e.g. E. GOURSAT: Cours d'Analyse, vol. III, and many other sources).

12. Characteristics of the v. Mises plasticity equations. The following investigations are due to T. Y. THOMAS.[51] Particular cases had been studied before.[52]

From the basic equations (Sect. 1) we eliminate the five stress components s_{ij}, and using (1.5) we obtain the equation

$$\frac{\partial p}{\partial x} = \frac{\partial}{\partial x}(\mu \dot{\varepsilon}_{xx}) + \frac{\partial}{\partial y}(\mu \dot{\varepsilon}_{xy}) + \frac{\partial}{\partial z}(\mu \dot{\varepsilon}_{xz}) \tag{12.1}$$

and two similar ones;[53] then, as in (1.11),

$$\mu^2 = \frac{2\tau_0^2}{\dot{\varepsilon}_1^2 + \dot{\varepsilon}_2^2 + \dot{\varepsilon}_3^2}, \quad \text{or} \quad \mu = \frac{\sqrt{2}\,\tau_0}{A}, \quad \text{where} \quad A^2 = \sum_{i,j} \dot{\varepsilon}_{ij}^2, \tag{12.2}$$

and

$$\dot{\varepsilon}_1 + \dot{\varepsilon}_2 + \dot{\varepsilon}_3 = 0. \tag{12.3}$$

The Eq. (12.1), the second Eq. (12.2) and Eq. (12.3) form a system of five equations with three independent variables x, y, z and five unknowns p, v_x, v_y, v_z, μ. The system is non-linear and of second order since there are second order derivatives of the $v_x, \ldots$ and the coefficients of these derivatives contain unknowns. Compared with (11.1) we have $m=5$, $n=2$.

We now identify z with x_0 of the preceding section and try to solve our system with respect to the highest derivatives

$$\frac{\partial^2 v_x}{\partial z^2}, \quad \frac{\partial^2 v_y}{\partial z^2}, \quad \frac{\partial^2 v_z}{\partial z^2}, \quad \frac{\partial p}{\partial z}, \quad \frac{\partial \mu}{\partial z}.$$

From (12.1), it follows that for $i = 1, 2, 3$

$$\frac{\partial p}{\partial x_i} = \sum_j^{1\ldots 3} \frac{\partial}{\partial x_j}(\mu\,\dot{\varepsilon}_{ij}) = \sum_j \frac{\partial \mu}{\partial x_j}\dot{\varepsilon}_{ij} + \mu\frac{\partial \dot{\varepsilon}_{ij}}{\partial x_j}, \tag{12.4}$$

where x_1, x_2, x_3 stands for x, y, z; next

$$\frac{\partial \mu}{\partial z} = -\frac{\sqrt{2}\,\tau_0}{A^2}\frac{\partial A}{\partial z},$$

$$\frac{\partial A}{\partial z} = \frac{1}{A}\left(\dot{\varepsilon}_{xx}\frac{\partial^2 v_x}{\partial z^2} + \dot{\varepsilon}_{yz}\frac{\partial^2 v_y}{\partial z^2} + \dot{\varepsilon}_{zz}\frac{\partial^2 v_z}{\partial z^2} + \cdots\right),$$

$$\frac{\partial \mu}{\partial z} = -\frac{\sqrt{2}\,\tau_0}{A^3}\left(\dot{\varepsilon}_{xx}\frac{\partial^2 v_x}{\partial z^2} + \cdots\right).$$

If in the first Eq. (12.4) we write only the terms that matter, we have

$$\begin{aligned}\frac{\partial p}{\partial x} &= \frac{\sqrt{2}\,\tau_0}{2A}\frac{\partial^2 v_x}{\partial z^2} + \cdots - \frac{\sqrt{2}}{A^3}\tau_0\,\dot{\varepsilon}_{xz}\left(\dot{\varepsilon}_{xz}\frac{\partial^2 v_x}{\partial z^2} + \dot{\varepsilon}_{yz}\frac{\partial^2 v_y}{\partial z^2} + \dot{\varepsilon}_{zz}\frac{\partial^2 v_z}{\partial z^2}\right) + \cdots \\ &= \frac{\sqrt{2}\,\tau_0}{A}\left[\frac{\partial^2 v_x}{\partial z^2}\left(\frac{1}{2} - \frac{\dot{\varepsilon}_{xz}^2}{A^2}\right) - \frac{\partial^2 v_y}{\partial z^2}\frac{\dot{\varepsilon}_{xz}\dot{\varepsilon}_{yz}}{A^2} - \frac{\partial^2 v_z}{\partial z^2}\frac{\dot{\varepsilon}_{xz}\dot{\varepsilon}_{zz}}{A^2} + \cdots\right],\end{aligned}$$

and two similar equations. Differentiation with respect to z of Eq. (12.3) gives

$$\frac{\partial^2 v_z}{\partial z^2} + \cdots = 0.$$

[51] T. Y. THOMAS: J. Rational Mech. Anal. **1**, 343–357 (1952).

[52] The characteristics for the problem of plane deformation are well known. (See Chaps. B and C of this article.) P. S. SYMONDS: Quart. Appl. Math. **6**, 448–452 (1949) investigated the problem in the case of axial symmetry; cf. also R. HILL [*6*, p. 263]. This problem is essentially elliptic, which means there are in general no real characteristics.

[53] Here plastic strains are denoted by ε_{ij}, $\dot{\varepsilon}_{ij}$, etc., since no confusion with elastic strains is possible. The A of (12.2) has nothing to do with the notation A in Sect. 10.

Hence for the Ω in (11.5) we obtain the symmetric determinant

$$\begin{vmatrix} A^2-2\dot{\varepsilon}_{xz}^2 & -2\dot{\varepsilon}_{xz}\dot{\varepsilon}_{yz} & -2\dot{\varepsilon}_{xz}\dot{\varepsilon}_{zz} & 0 \\ -2\dot{\varepsilon}_{xz}\dot{\varepsilon}_{yz} & A^2-2\dot{\varepsilon}_{yz}^2 & -2\dot{\varepsilon}_{yz}\dot{\varepsilon}_{zz} & 0 \\ -2\dot{\varepsilon}_{xz}\dot{\varepsilon}_{zz} & -2\dot{\varepsilon}_{yz}\dot{\varepsilon}_{zz} & A^2-2\dot{\varepsilon}_{zz}^2 & 1 \\ 0 & 0 & 1 & 0 \end{vmatrix} = \begin{vmatrix} A^2-2\dot{\varepsilon}_{xz}^2 & -2\dot{\varepsilon}_{xz}\dot{\varepsilon}_{yz} \\ -2\dot{\varepsilon}_{xz}\dot{\varepsilon}_{yz} & A^2-2\dot{\varepsilon}_{yz} \end{vmatrix}. \tag{12.5}$$

Upon expanding we obtain

$$A^2-2\dot{\varepsilon}_{xz}^2-2\dot{\varepsilon}_{yz}^2=0. \tag{12.6}$$

It remains to find an invariant form of this last condition. For this purpose we write the left-hand side in (12.6) in terms of the stresses and replace

$$\sum_{i,j} s_{ij}^2 \quad \text{by} \quad 2\tau_0^2.$$

We find simply

$$\tau_{xz}^2+\tau_{yz}^2=\tau_0^2=k^2. \tag{12.7}$$

If this relation holds, the plane $z=$ constant is tangent to a characteristic surface element. Hence we have the result: *The shear stress corresponding to a characteristic surface element equals* $\pm k$. With σ_1 the largest, σ_3 the smallest principal stress, $\tau_{\max}=\frac{1}{2}(\sigma_1-\sigma_3)=\frac{1}{2}(s_1-s_3)$. Next we put $\tau_{xz}^2+\tau_{yz}^2=\tau^2$ and $\tau=\tau_{\max}-\delta$ where $\delta>0$. Then from (12.7) and $\sum s_{ij}^2=2k^2$ we see that

$$\big((s_1-s_3)/2-\delta\big)^2=k^2,$$

where $s_1^2+s_2^2+s_3^2=2k^2$. This is possible only if

$$\delta=0, \qquad s_2=0, \qquad s_1+s_3=0. \tag{12.8}$$

The last two conditions give

$$\sigma_2=(\sigma_1+\sigma_3)/2=-p, \tag{12.9}$$

and the yield condition reduces to

$$(\sigma_1-\sigma_3)^2=4k^2, \tag{12.10}$$

a well-known formula of plane strain theory.

Hence, *the condition* $s_2=0$ *is necessary and sufficient that there exist at a point two real characteristic surface elements which are identical in direction with the planes of maximum and minimum shearing stress. If* s_2 *does not vanish, real characteristic surface elements do not exist.* The direction cosines of the normals to these characteristic elements with respect to a coordinate system which has the three principal directions are therefore

$$\left(\frac{1}{\sqrt{2}},\, 0,\, \frac{1}{\sqrt{2}}\right) \quad \text{and} \quad \left(\frac{1}{\sqrt{2}},\, 0,\, -\frac{1}{\sqrt{2}}\right),$$

and with respect to an arbitrary x_1, x_2, x_3-system the direction cosines are[54]

$$\alpha_i=\frac{1}{\sqrt{2}}\left[\cos(x_i\,1)+\cos(x_i\,3)\right] \quad \text{and} \quad \frac{1}{\sqrt{2}}\left[\cos(x_i\,1)-\cos(x_i\,3)\right]. \tag{12.11}$$

[54] In the same paper THOMAS has also discussed the v. Mises flow equations in connection with Tresca's yield condition (thus abandoning the theory of plastic potential). The result is similar, except that the condition $s_2=0$ does not appear. There exist two and only two real characteristic directions at each point of the medium, and the characteristic surface elements, whose normals they are, are the surface elements of maximum and minimum shearing stress.

This is THOMAS' result. We see that the cone of normals to the characteristic surface elements (Sect. 14) is in the present case either fully imaginary, or, if $s_2=0$, and therefore $\dot{\varepsilon}_2 \neq 0$, $s_1=-s_3$, $\dot{\varepsilon}_1+\dot{\varepsilon}_3=0$, imaginary with the exception of two real directions. The problem reduces to that of "plane strain".[55]

13. Further results and comments. The above result has been discussed by W. PRAGER.[56] As particular cases he mentions the cases of plane strain, of plastic torsion and that of the plastic twisting of a circular ring studied by FREIBERGER[57] (Sect. 48). On the other hand, it is seen that in the vast majority of three-dimensional problems real characteristics will not arise if the quadratic yield condition with associated flow rule is used. The existence of real characteristic surfaces presents a great mathematical advantage, and (as PRAGER remarks) inasmuch as characteristic surfaces are slip surfaces, it seems almost a physical necessity. PRAGER[58] adds that it might perhaps be possible to apply some adjustment to the yield condition so as to make the problem hyperbolic. For a similar purpose v. MISES, 1948, introduced a modification of his quadratic yield condition in the case of plane stress (see Sect. 35).

THOMAS has also investigated the question whether the characteristic surface elements of the v. Mises plasticity equations, if they exist, join so as to form characteristic surfaces. Starting at some point of the material we consider the space curve which has at every point the direction—say $\boldsymbol{\alpha}$—of one of the two characteristic normals and obtain a two-dimensional family of space curves, a congruence of curves. We ask whether these curves admit a family of surfaces normal to them—a family of orthogonal trajectories. This is not obvious, in space.

If a family of surfaces $f(x, y, z)=$ constant are the orthogonal trajectories of a given vector field $\boldsymbol{\alpha}$, then the gradient of f must have the $\boldsymbol{\alpha}$-direction. Hence, λ denoting a non-vanishing scalar function,

$$\operatorname{grad} f = \lambda \boldsymbol{\alpha}$$

must hold. Since curl grad $f=0$, we obtain, using an easily verified formula of vector calculus,

$$0 = \operatorname{curl}(\lambda \boldsymbol{\alpha}) = \boldsymbol{\alpha} \times \operatorname{grad} \lambda - \lambda \operatorname{curl} \boldsymbol{\alpha},$$

or $\boldsymbol{\alpha} \times \operatorname{grad} \lambda = \lambda \operatorname{curl} \boldsymbol{\alpha}$. But $\boldsymbol{\alpha} \cdot (\boldsymbol{\alpha} \times \operatorname{grad} \lambda)=0$. Hence $\lambda(\boldsymbol{\alpha} \cdot \operatorname{curl} \boldsymbol{\alpha})=0$. Since we may perform these steps also in reverse, we see that

$$\boldsymbol{\alpha} \cdot \operatorname{curl} \boldsymbol{\alpha} = 0 \tag{13.1}$$

is a necessary and sufficient condition for a vector field in space to admit orthogonal trajectories. (In the plane, this is identically satisfied.) In our problem the $\boldsymbol{\alpha}$ are given by (12.11), varying as the directions 1, 2, 3 vary.

THOMAS has also studied[59] the characteristics of the general three-dimensional problem, where instead of v. Mises' flow equations the Prandtl-Reuss equations (Sect. 6) are used and total incompressibility (in contrast to plastic incompressibility) is assumed. Derivations and results are rather involved, and we refer the reader to the original paper and the references cited there. See also [27, p. 123].

b) Continuation.

14. Characteristic surfaces. Characteristic condition. We return to systems as considered in Sect. 11. We shall obtain the characteristic condition in a general

[55] THOMAS' result was reestablished by him in 1953, T. Y. THOMAS: J. Rational Mech. Anal. **2**, 339–381 (1953), and by J. L. ERICKSEN: J. Math. Phys. **34**, 74–79 (1955). W. PRAGER presented the same result in a seminar lecture (1954) at Brown University.

[56] W. PRAGER: Proc. 2nd U. S. Nat. Congr. Appl. Mech., Michigan 1954, pp. 21–32, see p. 25. This paper contains extensive material on discontinuous solutions in plasticity.

[57] W. FREIBERGER: Aeron. Res. Lab. Australia, Report SM 213, (1953).

[58] W. PRAGER: James Clayton Lecture. Proc. Inst. Mech. Engrs (London) **169**, 41–57 (1955), see particularly p. 52.

[59] T. Y. THOMAS: J. Rational Mech. Anal. **5**, 251–262 (1956).

and invariant form.[60] Consider the system of first order (11.6) of m equations for m unknown functions $u_k(x_0, x_1, \ldots, x_n) = u_k(x)$, $k=1, \ldots, m$. Instead of the hyperplane $x_0 = c$ of Sect. 11, we consider a hypersurface S

$$z(x_0, x_1, \ldots, x_n) = a_0, \quad \text{briefly} \quad z(x) = a_0, \tag{14.1}$$

assuming that by an appropriate change of the variables $x_0, x_1, \ldots, x_n$ into new variables $z, z_1, \ldots, z_n$ the plane $x_0 = c$ is transformed into S. Then, the families $z(x) = \text{const.}$, $z_1(x) = \text{const.}, \ldots, z_n(x) = \text{const.}$ form the new coordinate surfaces. *Cauchy data on S, i.e. values of the $u_1, \ldots, u_m$ in terms of the $z_1, \ldots, z_n$* will determine a solution in the neighborhood of S if the system (11.6) can be transformed into a normal system with respect to these variables (see Sect. 11).

Let us compute this condition. With $p_i = \partial z/\partial x_i$ we have for $k=1, 2, \ldots, m$:

$$\frac{\partial u_k}{\partial x_i} = \frac{\partial u_k}{\partial z}\frac{\partial z}{\partial x_i} + \frac{\partial u_k}{\partial z_1}\frac{\partial z_1}{\partial x_i} + \cdots + \frac{\partial u_k}{\partial z_n}\frac{\partial z_n}{\partial x_i} = \frac{\partial u_k}{\partial z} p_i + \cdots, \tag{14.2}$$

where the dots at the end of (14.2) indicate that we are only interested in the first term

$$\frac{\partial u_k}{\partial z} p_i.$$

If we substitute the $\partial u_k/\partial x_i$ into Eqs. (11.6), we obtain for the left-hand sides

$$\sum_{k=1}^{m} \frac{\partial u_k}{\partial z} \sum_{i=0}^{n} a_{\mu k i} p_i + \cdots, \qquad \mu = 1, 2, \ldots, m.$$

The coefficient of $\partial u_k/\partial z$ in the μ-th equation is thus seen to be

$$\sum_{i=0}^{n} a_{\mu k i} p_i.$$

The p_i are proportional to the direction cosines α_i of the normal α to S, so that we may use these α_i instead of the p_i. The coefficients

$$\omega_{\mu k} = \sum_{i=0}^{n} a_{\mu k i} \alpha_i \tag{14.3}$$

generalize the $a_{\mu k}$ of Eq. (11.7). The transformed system can be solved for the

$$\frac{\partial u_1}{\partial z}, \ldots, \frac{\partial u_m}{\partial z}$$

if

$$\Omega = \|\omega_{\mu k}\| \neq 0. \tag{14.4}$$

The $\omega_{\mu k}$ are linear forms in the α_i and Ω is homogeneous of degree m in the α_i.

The equation

$$\Omega = \|\omega_{\mu k}\| = \|\textstyle\sum a_{\mu k i} \alpha_i\| = 0, \tag{14.5}$$

singles out directions α^* as directions to the normals to a surface S^* which *in connection with values u_k^* on S^* becomes exceptional or characteristic* (see also Sect. 15). For such a characteristic surface the Cauchy problem cannot be solved in general.

[60] Attention is called to the beautiful presentation in T. Levi-Civita: Caractéristiques des systèmes différentielles et propagation des ondes. Paris 1932.

In the *linear case* where the $a_{\mu k i}$ and the b_μ in Eq. (11.6) depend only on the x_i but not on the u_k, the characteristic surfaces are determined once and for all for a given system. In the more general case which holds in plasticity—as well as in gas dynamics—the exceptional character depends on the surface S and on the given values on it. Eq. (14.5) is called *characteristic condition* or *direction condition*. If Eq. (14.5) is expressed in terms of the α_i, the result is a homogeneous algebraic equation of degree m with coefficients depending on the coefficients of the given equations at the point P under consideration. Thus the endpoint S of the $\boldsymbol{\alpha}$ which satisfy condition (14.5) lie on a "cone" of order m in n-space with vertex at P. This cone need not be real and may degenerate. For the generalization of (11.4) (which refers to a system of second order) see the end of Sect. 20.

15. Compatibility conditions. In a similar way as before, we denote at any point of S the derivatives of the u_k with respect to the $z_1, z_2, \ldots, z_n$ as *interior derivatives* and $\partial u_k/\partial z$ $(k=1, 2, \ldots, m)$ as *exterior derivatives*. We have seen that the m original Eqs. (11.6) fail to determine the m exterior derivatives if and only if Eq. (14.5) holds for some direction $\boldsymbol{\alpha}^*$, or, equivalently, at least one combination (of the original equations) exists—say $s \geqq 1$ such combinations—which do not contain any derivative in the z-direction. This shows again that on an exceptional surface S^* the u_k^* are not "arbitrary", since there necessarily exist $s \geqq 1$ relations between them. *These s relations between the interior derivatives on S^* are called compatibility relations* since they restrict the arbitrariness of the u_k^* on S^*. *We call the S^* with compatible values u_k^* on it exceptional or characteristic.* An analytic formulation of the compatibility relations follows in Sect. 21. In addition to these s compatibility relations *there remain $(m-s)$ equations each of which contains at least one exterior derivative, i.e. derivative in the $\boldsymbol{\alpha}^*$-direction.* These $(m-s)$ equations cannot determine m exterior derivatives. Hence in no neighborhood of S^* is a solution uniquely determined. Along S^* two different solutions can be patched together.

16. Discontinuous solutions. In the light of the preceding facts, following HADAMARD, LEVI-CIVITA and particularly v. MISES,[61] we define *discontinuous solutions of a system* (11.6) *across a surface S^** as follows:

(1) On both sides of S^* all differential equations are satisfied. *On S^** the s compatibility relations hold. These do not contain a cross derivative of any of the u_k.

(2) Across S^* at least one of the u_k or its derivatives has a jump.

A finite jump of u_k across S^* causes the derivative of u_k in the direction normal to S^* to become infinite. Then, if for example u_1 has such a jump, condition (1) can still hold if the normal derivative of u_1 does not appear in the $(m-s)$ "remaining" equations (see the end of Sect. 15)—or, if the value $+\infty$ or $-\infty$ for $\partial u_1/\partial z$ does not contradict these $(m-s)$ equations.[62]

One may ask whether such a "separation surface" S^* is necessarily characteristic. This can be answered affirmatively if we know that in a domain which includes S^* the given system is hyperbolic. It is impossible then that two different solutions meet along a surface S for which (14.4) holds. For, in this case (under certain formal restrictions which are not severe in a hyperbolic region), the Cauchy problem would admit a unique solution in the neighborhood of S and neither a u_k nor a derivative of u_k could change abruptly across S.

[61] R. v. MISES: Proc. 1st Nat. Congr. Appl. Mech., Chicago 1950, pp. 667–671.
[62] Thus the "remaining equations" forbid certain jumps.

The preceding definition of discontinuous solutions differs from usual ones, since we include the case when a u_k itself may jump across S^*—*absolute discontinuity* in the terminology of HADAMARD.

The above distinctions and considerations help to clarify the important question of *which variables may undergo abrupt changes across characteristics.* Both HADAMARD and LEVI-CIVITA use physical reasoning to show why in flow of a compressible fluid, for example, the pressure must not jump or why certain velocity components cannot change abruptly. From our present point of view the answer is simple: *A variable whose finite exterior derivative appears in the* $(m-s)$ *"remaining" equations cannot jump.* Other variables may undergo sudden changes subject to the conditions (1) and (2) of this section.

17. Preliminary comments on discontinuous solutions in plasticity. For the general space problem the discontinuities found by THOMAS present an important and general result. The investigation was based on the Eqs. (12.1)–(12.3), the first of which are equations of second order in the velocities. Following HADAMARD we call a surface S (a line S) *discontinuous of order n* with respect to a magnitude if all derivatives up to the order $(n-1)$ are continuous across S but the n-th derivative is discontinuous across S $(n=1, 2, \ldots)$.[63] A jump in the magnitude itself may be denoted as a discontinuity of order zero but, preferably, as an *absolute dicontinuity.*

Another classification is the following: a discontinuity has been called *weak* by PRAGER if its order is not lower than the order of the highest derivative of the respective quantity in the equations under consideration,[64] and hence one may distinguish between weak and strong discontinuities. With HADAMARD's notation the characteristic surface elements found by THOMAS are discontinuous of order two in the velocities and of order one with respect to the pressure.

Next, ERICKSEN[65] has shown that *across the same surfaces also a discontinuity of order one is possible* for tangential components of the velocity.[66] *The stresses,* and consequently *p, are always continuous* across these characteristics.

We shall see that in the *plane problems* of "plane strain" as well as of "plane stress" an *absolute discontinuity of a tangential velocity component* is possible across a characteristic. This will be in line with the principle pronounced at the end of Sect. 16. We shall see that stresses cannot jump across characteristics.[67]

[63] This definition must be used with care since the dependent variables may be, for example, the deformation-velocity components or the components of the deformation.

[64] "Weak" discontinuities, which for Eq. (11.6) would be discontinuities of first order derivatives, are sometimes identified with discontinuities across characteristics. Actually, both "weak" and "strong" discontinuities can take place across characteristics. E.g. in a general compressible fluid flow, surfaces composed of streamlines are, in general, characteristic and may exhibit different values of the density ϱ and (or) of the tangential velocity-components on both sides.

The "shocks" of compressible fluid flow are, in general, not across characteristics. As pointed out repeatedly by v. MISES (see footnote 61) for mathematical as well as physical reasons, these shocks cannot be considered as discontinuous solutions (in the sense of the present section) of the ideal flow equations.

[65] J. L. ERICKSEN: J. Math. Phys. **34**, 74–79 (1955).

[66] Note that here a discontinuity of order one in the velocities is possible although Eqs. (12.1) contain second-order derivatives in the velocities. We see that a discontinuity across a characteristic is not necessarily weak. (More details in Sect. 23).

[67] In a later paper, T. Y. THOMAS, J. Math. Mech. **6**, 67–85 (1957) investigates discontinuities of order one for the problem of plane stress (see Sect. 27); "order one" means here that the stress components and velocity components are continuous across the discontinuity surface S while at least one of their first derivatives with respect to a space coordinate is discontinuous across S.

A particular type of *absolute discontinuity of the stresses*, first investigated by PRAGER, will be discussed later. Across such a discontinuity line the velocities cannot be discontinuous (see Sects. 24, 33, 47). We shall return to most of these questions.

c) Hadamard's theory.

18. Moving surfaces. The theory of discontinuity surfaces or singular surfaces can be approached from a point of view best known through the work of HADAMARD.[68] These suggestive concepts and ideas have occasionally been used by authors on plasticity, for example ERICKSEN, PRAGER, HILL, MANDEL, THOMAS.[69] A perfect treatise on the subject is TRUESDELL and TOUPIN'S article in this Encyclopedia, vol. III/1, p. 226. See in particular Chap. C (see also Chap. VI of the forthcoming treatise by WANG and TRUESDELL, *Introduction to Rational Elasticity*). The work is rich in conceptual and mathematical results. Like all the works of TRUESDELL it gives a wealth of historical information. Unfortunately, it does not consider plasticity. The problems are general and the derivations rigorous so that the Handbuch contains a general and correct exposition of the theory. The prevalent consideration of the duality between spatial and material presentation clarifies the exposition (cf. pp. 506—508).[70]

We present HADAMARD'S ideas briefly. As one application we shall use them to derive the mathematical form of the general compatibility conditions. (Sect. 15, Sect. 21.)

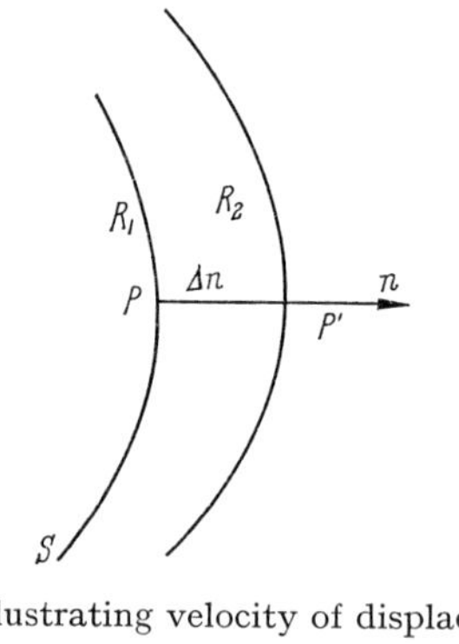

Fig. 2. Illustrating velocity of displacement.

Let x stand for x_1, x_2, x_3. Consider a region $R(t)$ in ordinary space (Fig. 2) varying in time which is divided into two parts R_1 and R_2 by a surface $S(t)$ of equation $z(t|x)=0$. $z(t|x)$ has continuous first partial derivatives with respect to the x_i and to t. Consider on S a point P which at time t has coordinates x and a normal n to S at P positive in direction from R_1 to R_2. The surface S moves and at time $t+\Delta t$ it intersects n again at the point P'. Let Δn be the algebraic measure of PP'; we call

$$\lim_{\Delta t\to 0}\frac{\Delta n}{\Delta t}=\frac{dn}{dt}$$

[68] J. HADAMARD: Leçons sur la propagation des ondes et les équations de l'hydrodynamique. Paris 1903. HADAMARD'S work was preceded and influenced by the work of RIEMANN (1860), CHRISTOFFEL (1877) and HUGONIOT (1885).

[69] See an introduction to these ideas in T. Y. THOMAS, J. Rational Mech. Anal. **2**, 339–381 (1953), Sects. 1–8 and a sequence of papers in 1956, 1957, 1958, 1959 in the same journal (see Ref. 71).

[70] TRUESDELL feels that the material standpoint adopted by HUGONIOT and HADAMARD lends itself to more complete and clear results and more rigorous derivations than the spatial viewpoint used by THOMAS and HILL.

the *velocity of displacement*, T. Let $p_i = \partial z/\partial x_i$, $i = 1, 2, 3$, and

$$g^2 = \sum_1^3 p_i^2$$

and $v_i = p_i/g$ the direction cosines of PP'. If x_i and $x_i + \Delta x_i = x_i + v_i \Delta n$ are the coordinates of P and P', at times t and $t + \Delta t$, respectively and $z(t|x) = 0$, $z(t + \Delta t|x + \Delta x) = 0$ and $\partial z/\partial t = p_0$ one has:

$$0 = \Delta n \sum_1^3 p_i v_i + p_0 \Delta t = \Delta n \cdot g + p_0 \Delta t \tag{18.1}$$

to within first order of Δn. Hence we obtain for T, the *displacement velocity* of the surface,

$$T = \lim_{\Delta t \to 0} \frac{\Delta n}{\Delta t} = -\frac{p_0}{g} = -\frac{\partial z}{\partial t}/g. \tag{18.2}$$

The result is also obtained as follows. Differentiation of $z = 0$ gives

$$\frac{\partial z}{\partial x_i} \frac{dx_i}{dt} + \frac{\partial z}{\partial t} = 0 \quad \text{or} \quad \operatorname{grad} z \cdot \boldsymbol{u} + \frac{\partial z}{\partial t} = 0, \tag{18.1'}$$

where $\boldsymbol{u}$ is the velocity with components dx_i/dt. Denoting as before by v_i the components of the unit normal $\boldsymbol{n}$ to S we obtain $T = \boldsymbol{u}\boldsymbol{n} = u_n$, the velocity or speed of displacement

$$u_n = T = \boldsymbol{u}\boldsymbol{n} = \boldsymbol{u} \operatorname{grad} z/|\operatorname{grad} z| = -p_0/|\operatorname{grad} z| = -p_0/g. \tag{18.2'}$$

T is counted as positive if the motion is from R_1 to R_2.

Let v_n be the *normal velocity of the medium* (we assume here that this velocity does not jump in the transition); then to v_n belongs a *speed of propagation* $\theta = T - v_n$, i.e., speed of S with respect to the medium:

$$\theta = T - v_n. \tag{18.3}$$

If $\theta = 0$ the discontinuity is called *material*. If $T = 0$ the discontinuity is *stationary* (see Sect. 58).

Following up (18.1′) we consider an important instance of an absolute velocity discontinuity. Let x^1 and x^2 be situated at time t both at the same point of S, but let x^1 belong to R_1 and x^2 to R_2. Write (18.1′) once for x^1, once for x^2 and subtract. Denote by $\boldsymbol{u}$ the velocity vector, as in (18.1′). Then

$$\frac{\partial z}{\partial x_i} \left[\frac{dx_i}{dt}\right] = 0 \quad \text{or} \quad \operatorname{grad} z \cdot [\boldsymbol{u}] = 0. \tag{18.4}$$

A jump in velocity is tangential to S. If $[\boldsymbol{u}] = 0$, a similar computation shows that a jump in acceleration is tangential to S.

19. Geometrical and kinematical discontinuity conditions. Consider a differentiable function $f(x, t)$ in $R(t)$. The partial derivatives of f are continuous in $R_1 + S$ and in $R_2 + S$, but across S jumps of derivatives may occur. Discontinuity of order n ($n = 1, 2, \ldots$) of S, with respect to f has been defined in Sect. 17. An absolute discontinuity (discontinuity of f itself across S) is not considered in the present context. We shall use subscripts, or superscripts, 1 and 2, for values in the regions R_1, R_2 on both sides of S and denote in a customary manner $f_2 - f$ as $[f]$ etc., and occasionally we shall for brevity write $t = x_0$. *A discontinuity of order one* is then expressed by

$$[f] = 0, \quad \text{at least one} \quad \left[\frac{\partial f}{\partial x_i}\right] \neq 0, \qquad i = 0, 1, 2, 3.$$

Consider a point P on S. We have from the first of these equations: $f_P^1 = f_P^2$ and for a neighboring point Q on S: $f_Q^1 = f_Q^2$, hence $f_Q^1 - f_P^1 = f_Q^2 - f_P^2$. If we take Q very close to P and go to the limit, then $df_P^1 = df_P^2$, or

$$\sum_{i=0}^{3} \frac{\partial f^1}{\partial x_i} dx_i = \sum_{i=0}^{3} \frac{\partial f^2}{\partial x_i} dx_i, \quad \text{or} \quad \sum_{i=0}^{3} \left[\frac{\partial f}{\partial x_i}\right] dx_i = 0. \tag{19.1}$$

Since the dx_i are in S, we have also

$$dz = \sum_{i=0}^{3} p_i\, dx_i = 0, \qquad \sum_{i=0}^{3} \frac{p_i}{g}\, dx_i = 0. \tag{19.2}$$

The coexistence of (19.1) and (19.2) is expressed by writing with an arbitrary factor λ:

$$\sum_{i=0}^{3}\left(\left[\frac{\partial f}{\partial x_i}\right]-\lambda\frac{p_i}{g}\right)dx_i=0. \tag{19.3}$$

Assume $p_0 \neq 0$ and determine λ from

$$\left[\frac{\partial f}{\partial x_0}\right]-\lambda\frac{p_0}{g}=0.$$

Since the dx_i are arbitrary, this leads to

$$\left[\frac{\partial f}{\partial x_i}\right]-\lambda\frac{p_i}{g}=0, \qquad i=1,2,3.$$

Hence with

$$\frac{p_0}{g}=-T, \quad \frac{p_i}{g}=\nu_i:$$

$$\left[\frac{\partial f}{dx_i}\right]=\lambda\nu_i, \quad i=1,2,3, \quad \left[\frac{\partial f}{\partial t}\right]=-\lambda T, \tag{19.4}$$

where the ν_i are the direction cosines of the normal to S.

We have the first result: *if f itself is continuous, the $(n+1)$ jumps of the first derivatives must obey* (19.4).

For a discontinuity of order two we obtain:

$$\left[\frac{\partial^2 f}{\partial x_i\,\partial x_j}\right]=\lambda\nu_i\nu_j, \quad \frac{\partial^2 f}{\partial x_i\,\partial t}=-\lambda T\nu_i, \quad \frac{\partial^2 f}{\partial t^2}=\lambda T^2, \quad i=1,2,3. \tag{19.5}$$

Conditions (19.4), (19.5) etc. have been denoted by HADAMARD as *geometrical and kinematical conditions of compatibility*. More precisely: the first three conditions (19.4) and the first (19.5) are geometrical conditions, the last (19.4) and the last (19.5) are kinematical and the middle conditions (19.5) are of a "mixed" nature. The right hand sides of (19.5) depend on λ and on T.[71]

Now consider Eqs. (19.4) applied to some *vector* $\boldsymbol{w}$. Then:

$$\left[\frac{\partial^2 \boldsymbol{u}}{\partial x_i}\right]=\boldsymbol{\lambda}\nu_i, \quad \left[\frac{\partial \boldsymbol{u}}{\partial t}\right]=\boldsymbol{\lambda}T, \quad \left[\frac{\partial^2 \boldsymbol{w}}{\partial t^2}\right]=\boldsymbol{\lambda}T^2, \quad \text{etc.} \tag{19.6}$$

HADAMARD calls $\boldsymbol{\lambda}$ the *characteristic segment*. If $\boldsymbol{\lambda}$ is normal to S he calls the discontinuity *longitudinal*, and if it is tangent to S he calls it *transversal*.

19bis. Continuation. HADAMARD considered also the more general case where—in a discontinuity of order one—f was not continuous over S. This and other generalizations of the content of Sect. 19 are denoted as *extended compatibility conditions*, T. Y. THOMAS considered many such situations.[72] He followed up this work in a book of 1961 (see [*27*]) and an article of 1963.[73] We consider a few instances.

Assume $[f]$ continuous over S and continuously differentiable. We consider a jump in the first space derivative over an S which may be moving or not. A spatial condition of first order is then for example:

$$\left[\frac{\partial f}{\partial x_i}\right]=\lambda\nu_i+g_{ij}\,g^{\alpha\beta}\,\frac{\partial x_j}{\partial u_\alpha}\,\frac{\partial[f]}{\partial u_\beta}, \tag{19bis.1}$$

where $\lambda=[\partial f/\partial\nu]$. Clearly this reduces to the first Eq. (19.4) if $[f]=0$.

If $[f]\neq 0$, a generalization of the last Eq. (19.4) is

$$\left[\frac{\partial f}{\partial t}\right]=-\lambda T+\frac{\delta[f]}{\delta t}, \tag{19bis.2}$$

where $\delta/\delta t$ is called the displacement derivative.

[71] It is assumed that in (19.4) and (19.5) the discontinuities are of orders one and two respectively.

[72] T. Y. THOMAS: J. Rational Mech. Anal. **2**, 339 (1953); **5**, 251 (1956); **6**, 67 (1957); **6**, 311 (1957a); **6**, 455 (1957b); **7**, 141 (1958a); **7**, 291 (1958b); **7**, 893 (1958c); **8**, 1 (1959).

[73] T. Y. THOMAS: Intern. J. Eng. Sciences **4**, 207 (1966).

Compatibility conditions of second order concern jumps in second derivatives in dependence on $[f]$ and $[\partial f/\partial v]$. Second derivatives are supposed existing in R_1 and in R_2 but they tend to different limits on the two sides of S. These and many more such conditions have been computed by T. Y. THOMAS.[74]

In 1961 appeared a review article by R. HILL on discontinuity relations in the mechanics of solids.[75] His systematic analysis achieves some simplified derivations of previous results and some new ones. He studies in particular discontinuities in stress rates (under either continuous or discontinuous stress). He also studies various jumps (in velocity, in stress, in strain rate) for elastic solids and recovers a result by LAMPARIELLO [Rend. Accad. nation. Lincei (1931) presented by LEVI-CIVITA, citation 60 of our text.] He finally considers discontinuities in rigid plastic solids giving, in particular, a general proof of the repeatedly mentioned fact that (in case of a convex yield surface) the strain rate tensor must vanish on both sides of a stress discontinuity surface.

We mention still a few papers on our present subject. In a valuable paper J. MANDEL[76] studies moving discontinuity surfaces. He distinguishes four types according to the state of affairs on the two sides of the moving surface: $E\to E$, $E\to P$, $P\to E$, and $P\to P$ where P means plastic and E elastic.

H. ZIEGLER[77] extends the main theorems of limit analysis to three-dimensional isotropic rigid plastic bodies which may contain discontinuities of stresses or (and) velocities. He allows various—convex—yield conditions.

We finally mention an interesting Russian paper[78] on stress discontinuities (see our Sect. 22). The three-dimensional body under consideration is rigid-plastic; the yield condition convex but otherwise arbitrary.

20. Application to a system of equations. So far we have not considered any differential equations. Now we combine the conditions (19.4) or (19.5) with a system of differential equations. Consider Eqs. (11.6) where the $a_{\mu k i}$ and the b_μ are continuous across S. Therefore jumps of $\partial u_k/\partial x_i$ must satisfy the conditions

$$\sum_{k=1}^{m}\sum_{i=1}^{n} a_{\mu k i}\left[\frac{\partial u_k}{\partial x_i}\right]=0, \qquad \mu=1,\ldots,m.$$

and from (19.4) writing $p_0/g=v_0$

$$\left[\frac{\partial u_k}{\partial x_i}\right]=\lambda_k v_i, \qquad k=1,\ldots,m,\ i=0,\ldots,n, \tag{20.1}$$

so that

$$\sum_{k=1}^{m}\sum_{i=0}^{n} a_{\mu k i}\,\lambda_k v_i=0, \qquad \mu=1,\ldots,m. \tag{20.2}$$

Hence with the $\omega_{\mu k}$ of (14.3), since $g\neq 0$, we obtain

$$\sum_{k=1}^{m}\omega_{\mu k}\,\lambda_k=0. \tag{20.3}$$

These are m homogeneous equations for the m unknowns $\lambda_1,\ldots,\lambda_m$. The λ_k are parameters which characterize the discontinuities of the first derivatives of the m functions $u_1,\ldots,u_m$. Eqs. (20.3) are termed by HADAMARD *dynamical conditions* (of compatibility).[79] This system has a solution not identically zero [i.e., a solution such that, according to (20.1), not all derivatives are continuous across S] *if and only if* $\|\omega_{\mu k}\|=0$, i.e. if *condition* (14.5) *holds*. We see that here *the characteristic condition* (14.5) *is recovered by equating to zero the coefficient determinant of the dynamical conditions* (20.3). As mentioned before, Eq. (14.5) is that of a

[74] His is obviously the priority regarding much of the work on extended compatibility conditions. I feel, however, uneasy about some applications he makes of his results to rupture, fracture, Lueders' Bands, etc. Competent experts should form an opinion.

[75] R. HILL, Progress in Solid Mechanics, vol. II, Chap. VI.: North Holland Publishing Co. 1961.

[76] J. MANDEL: J. Mécan. **1**, 3 (1962).

[77] H. ZIEGLER: Z. angew. Math. u. Phys. **20**, 81 (1969).

[78] G. I. BYKOVTSERV, D. D. IVLEV, and M. MIASNIANKIN: P.M.M. **32**, 472 (1968).

[79] It is in relation to our terminology confusing to speak of kinematical or dynamical conditions "*of compatibility*". The term "*compatibility conditions*" should in this context be reserved to the relations between the interior derivatives, as in Sects. 15 and 21.

cone of order m in $(n+1)$-dimensional (here four-dimensional) space. Each plane normal to a solution ν^* of (14.5) is tangent to a characteristic surface element.[80]

Analogous considerations may be applied to a system like (11.1), which has second-order discontinuities; in correspondence to Eqs. (20.1), (14.3), (14.5) we obtain

$$\left[\frac{\partial^2 u_k}{\partial x_i\,\partial x_j}\right]=\varrho\,\nu_i\,\nu_j,\qquad k=1,\ldots,m,\ i,j=0,\ldots,n,\tag{20.4}$$

$$\omega_{\mu k}=\sum_{i,j=0}^{n} a_{\mu k i j}\,\nu_i\,\nu_j,\qquad k,\mu=1,\ldots,m,\tag{20.5}$$

$$\sum_{k=1}^{m}\omega_{\mu k}\,\varrho_k=0,\qquad \mu=1,\ldots,m,\qquad \text{dynamical conditions},\tag{20.6}$$

$$\|\omega_{\mu k}\|=0,\qquad \text{characteristic condition}.\tag{20.7}$$

(20.7), the *characteristic condition*, generalizes (11.4). If $m=n=1$, the last determinant reduces to a single element, and we obtain instead of Eq. (10.4)

$$A\nu_1^2+2B\,\nu_1\nu_2+C\nu_2^2=0,\qquad \tan\varphi=\frac{1}{A}\left(-B\pm\sqrt{B^2-AC}\right),\tag{20.8}$$

where φ is the angle of a characteristic curve with the x-axis. For $\varphi=\pi/2$, this was our starting point in Sect. 10.

21. Compatibility conditions. We can now derive in a very simple way the analytic form of the compatibility conditions explained in Sect. 15. Let ν^* be a solution of Eq. (14.5) and put

$$\omega_{\mu k}^*=\sum_i a_{\mu k i}\,\nu_i^*$$

as in (14.3). Then according to (20.1)–(20.3)

$$\sum_{k=1}^{m}\omega_{\mu k}^*\,\lambda_k^*=0,\qquad \mu=1,\ldots,m\tag{21.1}$$

defines the jumps of the derivatives of the u_k. Consider the transposed system (21.1),

$$\sum_{\mu=1}^{m}\omega_{\mu k}^*\,\gamma_\mu^*=0.\tag{21.2}$$

We shall prove that these γ_μ^* determine *those linear combinations of the original equations,* explained in Sect. 15, *for which only interior derivatives,* i.e. derivatives perpendicular to $\boldsymbol{\nu}^*$, *appear*. In fact, consider the linear combinations of Eqs. (11.6).

$$\sum_{\mu=1}^{m}\gamma_\mu^*\sum_{k=1}^{m}\sum_{i=0}^{n}a_{\mu k i}\frac{\partial u_k}{\partial x_i}=-\sum_\mu \gamma_\mu^*\,b_\mu=B^*.\tag{21.3}$$

If vectors $\boldsymbol{A}_k^*$ with components

$$A_{ki}^*=\sum_{\mu=1}^{m}\gamma_\mu^*\,a_{\mu k i},\qquad k=1,\ldots,m,\ i=0,\ldots,n,\tag{21.4}$$

are introduced, Eq. (21.3) become

$$\sum_{k=1}^{m}\sum_{i=0}^{n}A_{ki}^*\frac{\partial u_k}{\partial x_i}=B^*.\tag{21.5}$$

Now, using (21.2) and (21.4), we have for all k:

$$\sum_\mu\left(\sum_i a_{\mu k i}\,\nu_i^*\right)\gamma_\mu^*=\sum_i\left(\sum_\mu \gamma_\mu^* a_{\mu k i}\right)\nu_i^*=\sum_i A_{ki}^*\,\nu_i=0,\tag{21.6}$$

[80] HADAMARD's theory has been applied by THOMAS and ERICKSEN to the v. Mises equations of plasticity, cf. T. Y. THOMAS: J. Rational Mech. Anal. **2**, 339–381 (1953), and Ref. 67; see ERICKSEN, Ref. 65.

and we see that the m vectors $\boldsymbol{A}_k^*$ defined by (21.4) are normal to $\boldsymbol{\nu}^*$. Therefore the left side of (21.5) is seen to contain *only differentiations normal to* $\boldsymbol{\nu}^*$ and hence only "interior" derivatives (Sect. 15), and therefore (21.5) *is the compatibility condition corresponding to the direction* $\boldsymbol{\nu}^*$, the normal to S^*.

d) Shock conditions. Stress discontinuities.

22. "Shock conditions". In our definition of discontinuous solutions it was not implied that discontinuities need occur only across characteristics. We ask whether one might define some analogue to the shocks of compressible fluid theory[81] and to the shock conditions which hold in this case.

We consider Eqs. (1.1) and (1.4). The method outlined in MISES' work (see footnote 81) may be used in the derivation of shock conditions.[82] Since our final result will be very obvious, we do not give complete derivations. Denote by S the surface which separates the two regions 1 and 2 under consideration, by T the normal velocity of the moving discontinuity surface S in the direction n leading from the side 1 to the side 2 of S. Then, the result corresponding to the continuity Eq. (1.1) is the condition

$$\varrho_1(v_{1n}-T)=\varrho_2(v_{2n}-T), \quad \text{or} \quad [\varrho(v_n-T)]=0, \tag{22.1}$$

where $T-v_n$ is the velocity of propagation (Sect. 18).

Next, denote by $\boldsymbol{t}_n$ the stress vector corresponding to the n-direction [viz. $\boldsymbol{t}_n = \boldsymbol{t}_x\cos(nx)+\boldsymbol{t}_y\cos(ny)+\boldsymbol{t}_z\cos(nz)$, where σ_x, τ_{xy}, τ_{xz} are the components of $\boldsymbol{t}_x$, etc.]. Then the result derived from (1.4) is

$$[\varrho(v_n-T)\,\boldsymbol{v}]=[\boldsymbol{t}_n], \tag{22.2}$$

or, in components,

$$[\varrho(v_n-T)\,v_i]=[\sigma_{ni}], \qquad i=1,2,3, \tag{22.3}$$

where σ_{ni} is the component of $\boldsymbol{t}_n$ in the i-direction.

However, in our present problem these conditions simplify very much. Since we assume $\varrho=$ constant, Eq. (22.1) reduces to $v_{1n}=v_{2n}$. If, next, we consider instead of Eqs. (1.4) merely the equilibrium conditions (1.7), the left side of (22.3) is seen to vanish and our results are simply

$$[v_n]=0, \qquad [\sigma_{ni}]=0. \tag{22.4}$$

If, for example, a discontinuity surface has the normal in the x-direction at the point under consideration, then v_x must remain continuous, and likewise σ_x, τ_{xy}, τ_{xz}, whereas σ_y, σ_z, τ_{yz} may change rapidly across the element. [This follows in this case also from consideration of our basic system. In fact in (1.7) the only stresses which are not differentiated in the x-direction are σ_y, σ_z, τ_{yz}. A jump of a stress which *is* differentiated in the x-direction would imply an infinite value of the respective derivative.]

The general conditions (22.1), (22.2) may be used in "dynamic problems" where the above simplifying assumptions are not made. Of course, these equations, or (22.4) in the simplified case, are only necessary conditions. The possible jumps must be compatible with the basic equations of Sect. 1.

[81] We mentioned before (footnote 64) that these shocks cannot be considered as a phenomenon of ideal fluid flow. One has to assume a "transition zone" D_λ, where the fluid is viscous while outside viscosity is negligible. See R. v. MISES, Mathematical theory of compressible fluid flow. New York 1958, Arts. 11 and 14.

[82] See also derivations in T. Y. THOMAS: J. Rational Mech. Anal. **2**, 339–381 (1953) and Math. Mag. **22**, 169–189 (1949).

23. On the classification of discontinuities. The definitions of Sect. 16 are general. They apply to discontinuities across any surface (line), characteristic or otherwise. HADAMARD'S definitions and geometrical and kinematical conditions (Sects. 18, 19) are even independent of any system of differential equations, but they do not refer to absolute discontinuities.

PRAGER[83] classifies discontinuities as weak or strong (see footnote 64), say with respect to a system of first order differential equations, like our Eqs. (11.6): The discontinuity of any u_k is weak if it is of order one, or higher, but not of order zero; in the latter case the discontinuity is called strong. One must, however keep in mind that the same physical system can be written in various forms depending on whether any and which quantities are eliminated, etc. At any rate one must remember not to identify weak discontinuities with discontinuities across characteristics. In fact, in compressible flow as well as in our problem, weak and strong discontinuities are possible across characteristics. (See Sects. 16 and 17.)

Discontinuities for which the propagation velocity $\vartheta = T - v_n = 0$ [see Eq. (18.3)], which we called material discontinuities, are often called "contact discontinuities".

In addition to the obvious distinction between the various physical quantities affected by a discontinuity we shall here simply distinguish between discontinuities across characteristics and discontinuities across non-characteristic surfaces, characteristics being defined by the vanishing of the respective characteristic determinants like (11.4), (14.5), (20.7). In each case the order of the discontinuity with respect to a chosen system of equations may also be considered.

The characteristics of the three-dimensional problem of Sect. 1 have been studied (Sects. 12, 13), other discontinuities have been briefly indicated (Sect. 17), and details will follow. However, a survey of discontinuities in plasticity would be incomplete without the description of an important and frequent type of absolute stress discontinuity introduced and studied by PRAGER.[84] We shall give here briefly the description in the case of the *plane* problem and shall return to it when we know more about this problem.

24. Stress discontinuities.[85] In the problem of *plane strain* it is assumed that one of the three principal directions of the stress tensor Σ [and therefore, according to (1.8), also of the strain rate tensor $\dot{E}$] is, at any point, parallel to one and the same direction. We take it as the z-direction. Then, $\dot{\gamma}_x = \dot{\gamma}_y = 0$, $\tau_x = \tau_y = 0$. In addition, it is assumed that all stresses and all strains are independent of z and that $v_z =$ constant. Therefore

$$\begin{aligned} &\dot{\varepsilon}_z = \dot{\varepsilon}_3 = 0, \qquad \dot{\varepsilon}_x + \dot{\varepsilon}_y = 0 \\ &\tau_x = \tau_y = 0, \qquad s_z = \sigma_z + p = 0, \qquad p = -\tfrac{1}{2}(\sigma_x + \sigma_y) = -\sigma_z, \end{aligned} \tag{24.1}$$

and the Eqs. (1.6)–(1.9) take on the simple form, with $\tau_{xy} = \tau$:

$$\frac{\partial \sigma_x}{\partial x} + \frac{\partial \tau}{\partial y} = 0, \qquad \frac{\partial \tau}{\partial x} + \frac{\partial \sigma_y}{\partial y} = 0, \tag{24.2}$$

$$(\sigma_x - \sigma_y)^2 + 4\tau^2 = \tfrac{4}{3}\sigma_0^2 = 4\tau_0^2 = 4k^2, \tag{24.3}$$

$$\dot{\varepsilon}_x + \dot{\varepsilon}_y = 0, \qquad \frac{\dot{\varepsilon}_x - \dot{\varepsilon}_y}{\dot{\gamma}_z} = \frac{\sigma_x - \sigma_y}{2\tau} \tag{24.4}$$

for the unknowns $\sigma_x, \sigma_y, \tau, v_x, v_y$.

[83] W. PRAGER: Proc. 2nd U.S. Nat. Congr. Appl. Mech., Michigan 1954, pp. 21–32.

[84] W. PRAGER: R. Courant Anniversary Volume, pp. 289–299. New York 1948.

[85] Since we are now dealing with plastic strains only, we omit the double primes, writing E or $\dot{E}$ instead of E'' or $\dot{E}''$.

Take the direction of a curve of discontinuity at a typical point P as the y-direction. Then, from (22.4), v_x as well as σ_x and τ must be continuous across that curve at P. Also, since $\partial v_y/\partial x$ appears in (24.4), v_y must be continuous across the curve at P if infinite values of τ are excluded.[86] Consider Eq. (19.4) with $f = v_y$. We find (since the n-direction is the x-direction) for

$$x_i = y, \quad \alpha_i = \cos(ny) = \cos(xy) = 0 \quad \text{that} \quad [\partial v_y/\partial y] = 0, \quad \text{or} \quad [\dot{\varepsilon}_y] = 0,$$ [87]

and from the first Eq. (24.4) that $[\dot{\varepsilon}_x] = 0$.

We now write n and t instead of x and y and have from (24.3), since only σ_t may jump,

$$\sigma_t = \sigma_n \pm 2/\sqrt{k^2 - \tau^2}. \tag{24.5}$$

Thus, the jump amounts to $4\sqrt{k^2 - \tau^2}$. The corresponding jump in pressure $p = \frac{1}{2}(\sigma_n + \sigma_t)$, equals $2\sqrt{k^2 - \tau^2}$. We think of the discontinuity line as of a narrow transition zone through which σ_n and τ remain unchanged while σ_t changes rapidly from $\sigma_t^{(1)}$ to $\sigma_t^{(2)}$, and $\sigma_t^{(2)} \leqq \sigma_t \leqq \sigma_t^{(1)}$ (Fig. 3). *In* the transition zone neither of the

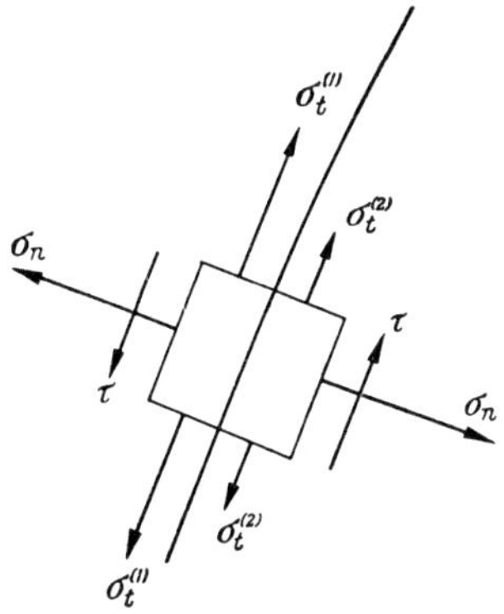

Fig. 3. Stress discontinuity.

equality signs holds, since otherwise the respective part of the transition zone would simply add to the continuous stress region. Hence, there

$$(\sigma_t - \sigma_n)^2 + 4\tau^2 < 4k^2.$$

Now consider the velocities and strain rates. From (24.5) we conclude (using for simplicity again x, y for n, t)

$$\sigma_y^{(1)} - \sigma_x = -(\sigma_y^{(2)} - \sigma_x). \tag{24.6}$$

On the other hand, according to (24.6), with $\dot{E} = \lambda S$

$$\dot{\varepsilon}_y^{(1)} = \lambda^{(1)} \frac{\sigma_y^{(1)} - \sigma_x}{2}, \qquad \dot{\varepsilon}_y^{(2)} = \lambda^{(2)} \frac{\sigma_y^{(2)} - \sigma_x}{2} = -\lambda^{(2)} \frac{\sigma_y^{(1)} - \sigma_x}{2}.$$

---

[86] If $\tau \to \infty$ the y-direction is characteristic, since it will be seen by (29.5) that

$$\tan 2\varphi = (\sigma_y - \sigma_x)/2\tau,$$

where φ is the angle of a characteristic with the x-axis. We shall see, however, that across a characteristic the stresses must be continuous (Sect. 33). That the velocity is continuous across a stress discontinuity has been shown for a more difficult plane problem by R. Hill: J. Mech. Phys. Solids **1**, 19–30 (1952).

[87] This had been *assumed*, for physical reasons, in H. Geiringer: Advances, p. 276.

We have proved before that $\dot\varepsilon_y^{(1)} = \dot\varepsilon_y^{(2)}$ (which is physically obvious), also $\sigma_y^{(1)} - \sigma_x$ cannot vanish since otherwise $\sigma_y^{(1)} = \sigma_x = \sigma_y^{(2)}$ and there is then no discontinuity. Hence $\lambda^{(1)} = -\lambda^{(2)}$; since a negative λ-value is excluded, it follows that *at every discontinuity point*

$$\lambda = 0, \quad \dot\varepsilon_x = \dot\varepsilon_y = \dot\gamma = 0, \quad \text{or} \quad \dot E = 0, \tag{24.7}$$

where $\dot E$ is the strain-rate tensor.[88]

The last result, together with the inequality which follows Eq. (24.5), suggests the previously mentioned consideration of the transition region as a thin, inextensible, elastic membrane between two plastic regions. This has been proposed by PRAGER, LEE, and HILL for physical reasons.[89] We shall consider stress discontinuities repeatedly in the following. So far the main general results are: (a) The stress vector $\boldsymbol{t}_n$ associated to the direction normal to the discontinuity line is continuous across this line. (b) The tangential component σ_t may jump subject to the condition $\sigma_t^{(1)} + \sigma_t^{(2)} = 2\sigma_n$. (c) The velocity vector $\boldsymbol{v}$ is continuous across the line. (d) The strain-rate tensor $\dot E$ vanishes for each point of the line. In my opinion these properties sufficiently describe the discontinuity. Other properties and interpretations follow from them.

Most of the results and consideration in Chap. A.II were concerned with the general three-dimensional problem. This last section, exceptionally, has dealt with a plane problem. We shall now give the theory of the plane problem of a perfectly plastic body.

B. Plane problems.[90]

I. Plane strain, plane stress, and generalizations.

25. Plane strain with v. Mises' or with Tresca's yield condition derived from three-dimensional problem. The assumptions given at the beginning of Sect. 24 lead to

$$\dot\gamma_x = 0, \quad \dot\gamma_y = 0, \quad v_z = \text{constant} \tag{25.1}$$

and to Eqs. (24.2)–(24.4).

Next, we consider Tresca's condition: Since $\sigma_3 = \frac{1}{2}(\sigma_1 + \sigma_2)$, the three principal τ-values (1.10) are $\tau_1 = \frac{1}{4}(\sigma_2 - \sigma_1)$, $\tau_2 = \frac{1}{4}(\sigma_2 - \sigma_1)$, $\tau_3 = \frac{1}{2}(\sigma_1 - \sigma_2)$. Hence τ_3 is the greatest, and Tresca's condition takes the form:

$$(\sigma_1 - \sigma_2)^2 = (\sigma_x - \sigma_y)^2 + 4\tau^2 = 3\tau_0^2 = 3k^2. \tag{25.2}$$

It is seen that the conditions (24.2) and (25.3) are now of the same form.

---

[88] See also HILL [*6*, p. 160] and his paper in J. Mech. Phys. Solids **1**, 19–30 (1952), and p. 26 in E. H. LEE, Proc. 3rd Sympos. Appl. Math., New York, 1950, p. 213–228. See GEIRINGER, Advances, p. 276/277. See the careful discussion in R. HILL, Progress in Solid Mechanics, vol. II, Chap. VI, p. 273, in particular the equation after Eq. (41). It follows from HILL's and GEIRINGER's considerations that $\dot E = 0$ holds also in the three-dimensional case if the yield surface is convex.

[89] See W. PRAGER: R. Courant Anniversary Volume, New York 1948, pp. 289–299, and particularly E. H. LEE: Proc. 3rd Symp. Appl. Math., New York 1950, pp. 213–228; R. HILL [*6*], pp. 157–160. See also J. L. ERICKSEN: J. Math. Phys. **34**, 74–79 (1955).

[90] In Chaps. B and C the strains and strain rates are plastic, and a distinction between elastic and plastic strains does not arise. Hence the strains and strain rates are denoted by ε_{ij} or $\dot\varepsilon_{ij}$ etc. rather than by $\dot\varepsilon_{ij}''$ etc. This applied also to Sect. 24.

26. Plane strain under general yield condition. We now consider a general yield function as in Eq. (3.3), and take it as the plastic potential. For such a function g of the stresses *it can be shown that the tensor* $\partial g/\partial\sigma_{ij}$ *is co-axial with the stress tensor* σ_{ij}.[91] From Eqs. (3.8), together with (24.1), it follows that

$$\frac{\partial g}{\partial \tau_x}=0, \quad \frac{\partial g}{\partial \tau_y}=0, \quad \frac{\partial g}{\partial \sigma_z}=0, \tag{26.1}$$

or, with the previous notation (3.6)

$$\frac{\partial G}{\partial \sigma_3}=0. \tag{26.2}$$

Using the above property of $\partial g/\partial\sigma_{ij}$ we conclude that $\tau_x=\tau_y=0$ and $\sigma_z=\sigma_3$; we then compute σ_3 in terms of σ_1 and σ_2 and obtain

$$\tau_x=\tau_y=0, \qquad \sigma_3=m(\sigma_1,\sigma_2)=l(\sigma_x,\sigma_y,\tau)=\sigma_z, \tag{26.3}$$

where m is some symmetric function of σ_1, σ_2. Hence with

$$G\big(\sigma_1,\sigma_2,m(\sigma_1,\sigma_2)\big)=F(\sigma_1,\sigma_2)$$

the plane yield condition (with a symmetric F) is

$$f(\sigma_x,\sigma_y,\tau)=F(\sigma_1,\sigma_2)=0. \tag{26.4}$$

We show that

$$\frac{\partial f}{\partial \sigma_x}+\frac{\partial f}{\partial \sigma_y}=\frac{\partial F}{\partial \sigma_1}+\frac{\partial F}{\partial \sigma_2}=0. \tag{26.5}$$

Denote by the subscript m that, after differentiation, we put $\sigma_3=m(\sigma_1,\sigma_2)$. Then

$$\frac{\partial F}{\partial \sigma_1}+\frac{\partial F}{\partial \sigma_2}=\left[\frac{\partial G}{\partial \sigma_1}+\frac{\partial G}{\partial \sigma_2}+\frac{\partial G}{\partial \sigma_3}\left(\frac{\partial \sigma_3}{\partial \sigma_1}+\frac{\partial \sigma_3}{\partial \sigma_2}\right)\right]_m=\left[\frac{\partial G}{\partial \sigma_1}+\frac{\partial G}{\partial \sigma_2}\right]_m=0,$$

on account of (26.2) and (3.7).

The three remaining relations (3.8) are, with $\tau_{xy}=\tau_{yx}=\tau$,

$$\dot{\varepsilon}_x=\lambda\frac{\partial f}{\partial \sigma_x}, \quad \dot{\varepsilon}_y=\lambda\frac{\partial f}{\partial \sigma_y}, \quad \dot{\gamma}_z=\lambda\frac{\partial f}{\partial \tau_z}, \tag{26.6}$$

where

$$\dot{\varepsilon}_x+\dot{\varepsilon}_y=0, \qquad \frac{\partial f}{\partial \sigma_x}+\frac{\partial f}{\partial \sigma_y}=0. \tag{26.7}$$

The present set of equations thus consists of the two Eqs. (24.2), of the yield condition (26.4) with the restriction (26.5), and two independent Eqs. (26.6), hence of six equations for $\sigma_x, \sigma_y, \tau, v_x, v_y, \lambda$.

On account of

$$\frac{\partial F}{\partial \sigma_1}+\frac{\partial F}{\partial \sigma_2}=0,$$

the yield function $F(\sigma_1,\sigma_2)$ only depends on the difference $\sigma_1-\sigma_2$; hence

$$F(\sigma_1,\sigma_2)=A\,[(\sigma_1-\sigma_2)],$$

where A is a function of one variable. Finally, eliminating λ from (26.6), we obtain for plane strain under a general yield condition the system

$$\begin{gathered}\frac{\partial \sigma_x}{\partial x}+\frac{\partial \tau}{\partial y}=0, \quad \frac{\partial \tau}{\partial x}+\frac{\partial \sigma_y}{\partial y}=0,\\ f(\sigma_x,\sigma_y,\tau)=A\,[(\sigma_1-\sigma_2)]=0,\\ \dot{\varepsilon}_x+\dot{\varepsilon}_y=0, \quad (\dot{\varepsilon}_x-\dot{\varepsilon}_y)/\dot{\gamma}=\left(\frac{\partial f}{\partial \sigma_x}-\frac{\partial f}{\partial \sigma_y}\right)\Big/\frac{\partial f}{\partial \tau}.\end{gathered} \tag{26.8}$$

[91] See H. Geiringer: Advances, p. 206.

In addition: $\tau_{yz}=\tau_x=0$, $\tau_{xz}=\tau_y=0$, $\sigma_3=m(\sigma_1,\sigma_2)$ and $v_z=$ constant. Here we have, indeed, a particular solution of the general problem of Sect. 3. However, we see that in plane strain, the "general" yield condition is not much more general than the quadratic condition of the preceding section.[92]

27. Plane stress with quadratic yield condition. The main assumption is again that one of the principal directions of Σ be the same everywhere. This direction being chosen as the z-direction, the essential additional assumption is now that $\sigma_z=0$ everywhere, and hence

$$\tau_x=\tau_y=\sigma_z=0 \tag{27.1}$$

throughout the body. In addition we assume that the remaining stresses are independent of z.

The third equilibrium condition is identically satisfied, as before. In plane strain, however, the relations (25.1) lead only to the restriction that v_x and v_y do not depend on z, whereas here (27.1) eliminates three unknowns. Thus, there remain $2+1+6=9$ equations (two equilibrium, one yield, six strain rate equations) for $3+3+1=7$ unknowns $\sigma_x, \sigma_y, \tau, v_x, v_y, v_z, \lambda$. As a consequence of (27.1) we obtain from corresponding stress-strain relations

$$\frac{\partial v_y}{\partial z}+\frac{\partial v_z}{\partial y}=0, \quad \frac{\partial v_x}{\partial z}+\frac{\partial v_z}{\partial x}=0, \quad 3\frac{\partial v_z}{\partial z}+\lambda(\sigma_x+\sigma_y)=0. \tag{27.2}$$

It is not impossible that velocity distributions $\boldsymbol{v}(x, y, z)$ can be found which satisfy all conditions, though this is not presently known.

In *the usual problem of plane stress* we wish to determine the stresses and the velocities v_x and v_y (and consequently λ) *as functions of* x, y, from the six Eqs. (27.3), which are analogous to the six equations of plane strain. This problem, as we shall see later, cannot be considered as a particular case of the three-dimensional problem.[93] With v. Mises' yield function as the plastic potential we formulate the system

$$\begin{gathered}\frac{\partial\sigma_x}{\partial x}+\frac{\partial\tau}{\partial y}=0, \quad \frac{\partial\tau}{\partial x}+\frac{\partial\sigma_y}{\partial y}=0,\\ \sigma_x^2+\sigma_y^2-\sigma_x\sigma_y+3\tau^2=\sigma_1^2+\sigma_2^2-\sigma_1\sigma_2=3\tau_0^2=\sigma_0^2,\\ \dot\varepsilon_x=\lambda(2\sigma_x-\sigma_y), \quad \dot\varepsilon_y=\lambda(2\sigma_y-\sigma_x), \quad \dot\gamma=\lambda\cdot 6\tau.\end{gathered} \tag{27.3}$$

Here the yield condition follows from (1.9′) by putting $\sigma_z=0$.[94] The last three equations are derived from the corresponding ones in (1.8), using $\sigma_z=0$, $p=-\frac{1}{3}(\sigma_x+\sigma_y)$.[95] We note that in (27.3) $\dot\varepsilon_x+\dot\varepsilon_y$ is not zero since $\dot\varepsilon_z\neq 0$. The problem (27.3) must be considered on its own merit as a two-dimensional problem.

Generalizing both problems, (26.8) and (27.3), we define a *general complete plane problem* with a general yield function $F(\sigma_1,\sigma_2)=f(\sigma_x,\sigma_y,\tau)$, F being sym-

[92] Cf. H. ZIEGLER, Z. angew. Math. Phys. **20**, Sect. 10 (1961).

[93] The same is true for the plane stress problem of elasticity theory and for various other important problems in mechanics. In GEIRINGER: Advances I tried—erroneously— to derive the problem of plane stress as a particular case of the three-dimensional problem (see pp. 202, 203, and 209). That this is indeed not possible will be seen later (Sect. 36) since its characteristics cannot be recovered as a special case of the general result of Sect. 12. Here we note that some difficulties arise immediately if Eqs. (27.2) and (27.3), are considered simultaneously.

[94] We mention also the form $(\sigma_1+\sigma_2)^2+3(\sigma_1-\sigma_2)^2=12\tau_0^2$, which shows dependence of the yield condition on $(\sigma_1+\sigma_2)$ in contrast to (25.3). See R. v. MISES: Z. angew. Math. Mech. **5**, 147–149 (1925). — F. SCHLEICHER: Z. angew. Math. Mech. **6**, 199–216 (1926), footnote 10.

[95] The first general study of statically determined problems of plane stress, governed by Eqs. (27.3), is due to SOKOLOVSKY, who studied the problem in several papers of 1946. His work is presented in [*26*, pp. 266–307].

metric in σ_1 and σ_2 (compare the "restrictions" discussed in Sect. 3):

$$\frac{\partial\sigma_x}{\partial x}+\frac{\partial\tau}{\partial y}=0,\qquad \frac{\partial\tau}{\partial x}+\frac{\partial\sigma_y}{\partial y}=0,$$
$$f(\sigma_x,\sigma_y,\tau)=F(\sigma_1,\sigma_2)=0, \tag{27.4}$$
$$\lambda\frac{\partial f}{\partial\sigma_x}=\dot\varepsilon_x,\qquad \lambda\frac{\partial f}{\partial\sigma_y}=\dot\varepsilon_y,\qquad \lambda\frac{\partial f}{\partial\tau}=\dot\gamma.$$

These reduce to (26.8) if

$$\frac{\partial f}{\partial\sigma_x}+\frac{\partial f}{\partial\sigma_y}=0,$$

and to (27.3) if $f(\sigma_x,\sigma_y,\tau)$ is chosen as in the second line of (27.3); other "plane stress" yield functions will be considered[96] in Sect. 35.

27 bis. Plane stress. Continued. PRAGER uses the term *yield mechanism* to denote a plastic strain increment whose components are specified to within a positive factor. It then follows (PRAGER, [*24*, Chap. I]) that under v. Mises' yield condition and the theory of the plastic potential *there is a one-to-one correspondence between yield mechanism and state of plane stress at the yield limit.* In fact: from (27.3) we have $\sigma_1^2+\sigma_2^2-\sigma_1\sigma_2-4k^2=0$, $\dot\varepsilon_1=\lambda(2\sigma_1-\sigma_2)$, $\dot\varepsilon_2=\lambda(2\sigma_2-\sigma_1)$. Hence, for a state of stress at the yield limit *the last two equations determine a unique yield mechanism.*

To show the converse solve the above system with respect to σ_1 and σ_2. The result is, with $e=(\dot\varepsilon_1^2+\dot\varepsilon_2^2+\dot\varepsilon_3^2)^{\frac12}$, and $4k^2=\sigma_0^2$:

$$\sigma_1=\frac{\sigma_0}{\sqrt3}\,\frac{2\dot\varepsilon_1+\dot\varepsilon_2}{e},\qquad \sigma_2=\frac{\sigma_0}{\sqrt3}\,\frac{2\dot\varepsilon_2+\dot\varepsilon_1}{e}. \tag{27bis}$$

The right-hand sides of these equations are homogeneous in the $\dot\varepsilon_1$, $\dot\varepsilon_2$ and thus depend only on the yield mechanism, which therefore is *seen to determine uniquely the state of stress.*

Consider next plane stress, the Tresca yield and KOITER's generalization (3.4). The principal stresses are σ_1, σ_2, 0 and at least one of the differences between two different stresses has the absolute value $2k$. Clearly, along a vertical or horizontal side of the Tresca hexagon the same yield mechanism corresponds to infinitely many states of stress, while at a corner infinitely many yield mechanisms correspond to one state of stress (Fig. 12a). In formulas: there are 6 yield functions

$$\begin{aligned}&F_1=\sigma_1-2k, && F_2=\sigma_2-2k, && F_3=\sigma_2-\sigma_1-2k,\\ &F_4=-\sigma_1-2k, && F_5=-\sigma_2-2k, && F_6=-\sigma_2+\sigma_1-2k.\end{aligned} \tag{27ter}$$

Along the vertical side $\sigma_1=2k$, $0<\sigma_2<\sigma_3$ the function F_1 vanishes, all other F_i are negative, and the rule of the generalized potential gives $\dot\varepsilon_1=\lambda_1$, $\dot\varepsilon_2=0$ where $\lambda_1>0$ if any plastic flow is to occur. Hence to the single yield mechanism $\dot\varepsilon_1=\lambda_1$, $\dot\varepsilon_2=0$ belong *an infinity of states of stress.* If, on the other hand, $\sigma_1=0$, $\sigma_2=2k$, then F_2 and F_3 vanish and all other F_i are negative and Eq. (3.4) give $\dot\varepsilon_1=-\lambda_3$, $\dot\varepsilon_2=\lambda_2+\lambda_3$. Hence one single state of stress is associated *with infinitely many yield mechanisms.* With Tresca's yield condition and the rule of the plastic potential *in plane stress there is no longer a one-to-one correspondence between yield mechanism and state of stress at the yield limit.*

[96] The classification adopted in this paper differs from that used by several authors: We consider in the present Chap. B the problem of plane strain with dependence of the yield function on $(\sigma_1-\sigma_2)$ only, and (27.7) satisfied. All other yield conditions which depend also on $(\sigma_1+\sigma_2)$ are treated in Chap. C, among them those of "plane stress" under v. Mises' and Tresca's yield condition as studied first by SOKOLOVSKY [see Eqs. (35.1), (35.5), etc.]. Other authors, including SOKOLOVSKY, HILL, MANDEL consider separately the problem of plane strain (our Chap. B), that of plane stress, and finally other yield conditions which depend also on $(\sigma_1+\sigma_2)$, considered as generalizations in the sense of O. MOHR, SCHLEICHER and MISES (see footnotes 10, 11). In this connection the concept of the envelope of Mohr circles is then taken as starting point, and, for example, SOKOLOVSKY and MANDEL (not HILL) limit the problem to the case of real contact with the envelope (hyperbolic problem). — From a physical point of view the separate consideration of the problem of plane stress is certainly justified, particularly if this is considered in the general form of Sect. 28; from a mathematical point of view the first classification seems more logical.

28. Generalized plane stress. A thin plate loaded only in its plane is approximately in a state of plane stress. In a more general approach the thickness h of the plate is not neglected. In formulating this problem R. HILL [*6*, p. 300] uses the yield criterion as in (27.3). However, his considerations are not restricted to this assumption.[97]

Consider a plate of small thickness h under conditions of plane stress (viz. $\sigma_z=\tau_x=\tau_y=0$). Forces are applied along the edge of the plate. The thickness h need not remain uniform during the deformation; but if $\partial h/\partial s$ (ds parallel to the surface of the plate) is small compared to unity, the state may still be considered as approximately "plane". We assume that at a certain instant t_0 the thickness $h(x, y)$ is a given function of x, y. Denote by σ_x, σ_y, τ the stress components averaged over the thickness of the plate and use these in the yield criterion. By v_x, v_y we denote the averaged velocity components, and the strain rates are defined as usual in terms of these components. Then, the equations of this problem are

$$\frac{\partial}{\partial x}(h\sigma_x)+\frac{\partial}{\partial y}(h\tau)=0,\qquad \frac{\partial}{\partial x}(h\tau)+\frac{\partial}{\partial y}(h\sigma_y)=0,\qquad f(\sigma_x,\sigma_y,\tau)=0$$
$$\dot\varepsilon_x:\dot\varepsilon_y:\dot\gamma=\frac{\partial f}{\partial\sigma_x}:\frac{\partial f}{\partial\sigma_y}:\frac{\partial f}{\partial\tau}. \tag{28.1}$$

The unknown functions are v_x, v_y, σ_z, σ_y, τ.

After v_x and v_y have been found, the equation of continuity may serve to compute *the change of* h at the instant t_0. If d/dt denotes the material derivative, the rate of strain $\dot\varepsilon_z$ averaged through the thickness equals $(1/h)\,(dh/dt)$, or

$$\dot\varepsilon_z=\frac{1}{h}\frac{dh}{dt}=\frac{1}{h}\left(\frac{\partial h}{\partial t}+v_x\frac{\partial h}{\partial x}+v_y\frac{\partial h}{\partial y}\right),$$

and the continuity equation (1.6) takes the form

$$\frac{1}{h}\frac{dh}{dt}+\dot\varepsilon_x+\dot\varepsilon_y=0,$$

or

$$\frac{\partial h}{\partial t}+\frac{\partial}{\partial x}(hv_x)+\frac{\partial}{\partial y}(hv_y)=0. \tag{28.2}$$

If $h(x, y)$ is known at $t=t_0$ and v_x, v_y have been determined, Eq. (28.2) serves to determine $[\partial h/\partial t]_{t=t_0}$. Mathematically, this problem (28.1), together with (28.2), does not differ much from (27.4).

On the other hand HODGE[98] considers the system (28.1), (28.2) with $(\partial h/\partial t)_{t=t_0}=0$ and $h(x, y)$ as an unknown function. This is then a system of six equations with the additional unknown function h. Mathematically, this problem is much more difficult than (27.4) since we have to consider the six equations simultaneously. In this formulation a uniform h is excluded since from (28.2) it would amount to $\dot\varepsilon_x+\dot\varepsilon_y=0$—plane strain rather than plane stress.

II. The theory of plane strain.

This is the best developed branch of the mathematical theory of plasticity, and it has found a wide field of applications; probably the two facts are interdependent. The three first Eqs. (27.4) with

$$\frac{\partial f}{\partial\sigma_x}+\frac{\partial f}{\partial\sigma_y}=0$$

form a non-linear "reducible" or "quasi-linear" system, and several integration procedures have been developed for them.

a) Differential relations.

29. Basic equations. Our system with Mises' yield condition, consists of the Eqs. (24.2)–(24.4) for the functions $\sigma_x(x, y)$, $\sigma_y(x, y)$, $\tau(x, y)$, $v_x(x, y)$, $v_y(x, y)$.

[97] See R. HILL: J. Mech. Phys. Solids **1**, 19 (1952).
[98] G. P. HODGE: Quart. Appl. Math. **8**, 381–386 (1951).

As long as we do not consider boundary conditions, Eqs. (24.2) and (24.3) may be considered as three quasilinear equations for three unknown functions. The three equations may be reduced to one single equation by means of the *function of* AIRY,[99] $F = F(x, y)$, where

$$\sigma_x = \frac{\partial^2 F}{\partial y^2}, \qquad \tau = -\frac{\partial^2 F}{\partial x\, \partial y}, \qquad \sigma_y = \frac{\partial^2 F}{\partial x^2}. \tag{29.1}$$

The equilibrium Eqs. (24.2) are thus satisfied, and (24.3) gives the second-order equation

$$\left(\frac{\partial^2 F}{\partial x^2} - \frac{\partial^2 F}{\partial y^2}\right)^2 + \left(2\frac{\partial^2 F}{\partial x\, \partial y}\right)^2 = 4k^2. \tag{29.2}$$

This equation—which has not been much used in modern presentations—can be reduced to a Monge-Ampère equation.[100] First introduce the new independent variables u, v by:

$$u = xi + y, \qquad v = x + yi,$$

and obtain by a simple computation

$$\frac{\partial^2 F}{\partial u^2}\frac{\partial^2 F}{\partial v^2} + h^2 = 0, \qquad h = \frac{k}{2}, \tag{29.2'}$$

which, with usual notation for second derivatives, is $rt + h^2 = 0$, or $r = -h^2/t$. Differentiating this last equation with respect to v, we obtain

$$\left(\frac{\partial z}{\partial v}\right)^2 \frac{\partial^2 z}{\partial u^2} - h^2 \frac{\partial^2 z}{\partial v^2} = 0, \quad \text{where} \quad z = \frac{\partial F}{\partial v}. \tag{29.2''}$$

This equation, linear in the second derivatives, is a Monge-Ampère equation and may be used as the basis for further discussion. Many of the results on our problem can be reached by this approach, which, however, we shall not follow here.[101]

Denote now by ψ the angle which an arbitrary direction ν makes with the positive x-direction, by $\sigma_\nu = \sigma$ the component in the ν-direction of the stress vector $\boldsymbol{t}_\nu$ associated with that direction, by $\tau_\nu = \tau$ the component normal to ν (see Sect. 2). Then, with $\tau_{xy} = \tau_z$:

$$\begin{aligned} \sigma &= \sigma_x \cos^2\psi + 2\tau_z \cos\psi \sin\psi + \sigma_y \sin^2\psi \\ &= \frac{\sigma_x + \sigma_y}{2} + \frac{\sigma_x - \sigma_y}{2}\cos 2\psi + \tau_z \sin 2\psi, \\ \tau &= -\frac{\sigma_x - \sigma_y}{2}\sin 2\psi + \tau_z \cos 2\psi. \end{aligned} \tag{29.3}$$

The *normal stress* σ reaches extremum values, $(d\sigma/d\psi = 0)$, for $\psi = \vartheta$ and $\psi = \vartheta + 90°$, where

$$\tan 2\vartheta = \frac{2\tau_z}{\sigma_x - \sigma_y}. \tag{29.4}$$

For these *principal directions*, the u-direction and the v-direction, we see from (29.3) that τ vanishes and the extreme values σ_1 and σ_2 of σ are equal to

$$\sigma_{1,2} = \frac{1}{2}(\sigma_x + \sigma_y) \pm \left[\left(\frac{\sigma_x - \sigma_y}{2}\right)^2 + \tau_z^2\right]^{\frac{1}{2}}.$$

---

[99] G. B. AIRY: British Ass. Reps. 1862.

[100] Cf. E. STORCHI: Istituto Lombardo di Scienze e Lettere. Rendiconti Classe di Scienze Matematiche e Naturali **68**, 694–713 (1953).

[101] The non-linear partial differential equation (29.2) has been successfully studied by A. GALIN, Prikl. Mat. Mehk. **10**, 367–368 (1946).

On the other hand, $d\tau/d\psi = 0$ defines two directions $\psi = \varphi$ and $\psi = \varphi + 90°$ where

$$\tan 2\varphi = \frac{\sigma_y - \sigma_x}{2\tau_z} \tag{29.4'}$$

for which the *shear stress* τ takes on extremum values. They are seen to bisect the angles of the principal directions and are called directions of *maximum shear stress*. The corresponding values are: $\tau = \pm\frac{1}{2}(\sigma_1 - \sigma_2)$; for each of these two directions $\sigma = \frac{1}{2}(\sigma_1 + \sigma_2) = -p$. All this holds for any symmetric tensor Σ.

Next, in the theory of the plastic potential, the strain-rate tensor $\dot{E}$ is coaxial to Grad g, and hence to Σ (see proof in Advances, p. 206). Hence the *shear strain* $\dot{\gamma}$—derived from $\dot{E}$, as τ was from Σ—has its absolute maximum value for the same directions (29.4'). *The directions of extremum shear strain are called slip directions or directions of the slip lines.* The corresponding *normal strain rate* is $\dot{\varepsilon} = \frac{1}{2}(\dot{\varepsilon}_1 + \dot{\varepsilon}_2)$; it vanishes in plane strain theory.

Now, we compute the *characteristic directions*. We take from Sect. 10 γ) the result that the *stress characteristics*, i.e. the characteristics of the system (24.2), (24.3) make the angles $\pm 45°$ with the first principal direction. It remains to compute from (24.4) the directions of the *velocity characteristics*. Eqs. (24.4), viz.

$$\frac{\partial v_x}{\partial x} + \frac{\partial v_y}{\partial y} = 0,$$

$$\frac{\partial v_x}{\partial x} - \left(\frac{\partial v_x}{\partial y} + \frac{\partial v_y}{\partial x}\right)\cot 2\vartheta - \frac{\partial v_y}{\partial y} = 0$$

are two linear first-order equations, for v_x, v_y. Denote by m the slope of a characteristic direction and apply (14.5) for $k = 2$. We obtain

$$m^2 - 2m\tan\vartheta - 1 = 0, \tag{29.5}$$

and

$$m = \frac{dy}{dx} = \frac{\sin 2\vartheta \pm 1}{\cos 2\vartheta} = \tan(\vartheta \pm 45°). \tag{29.5'}$$

Hence, we have the following results (Fig. 4): *The directions of maximum and minimum shear stress, which coincide with the directions of maximum and minimum shear strain, make angles $\vartheta \pm 45°$ with the positive x-axis,*[102] *ϑ being the angle of the first principal direction with this axis. In plane strain, these directions are also the directions of the stress- and of the velocity characteristics, and along each of them the normal strain rate is zero.*

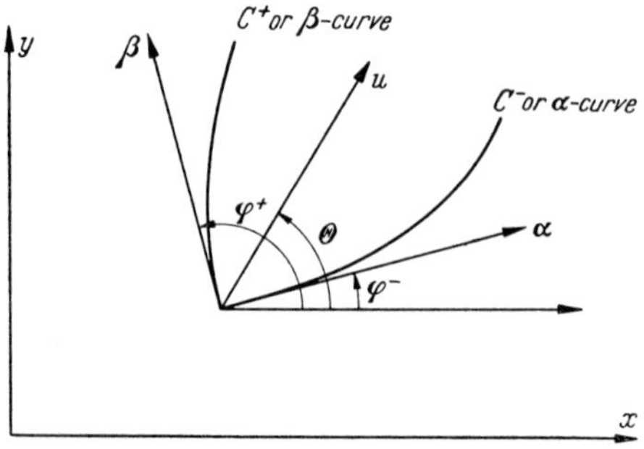

Fig. 4. Characteristic directions in plain strain.

[102] This holds also for the general problem (27.4).

We denote the curves which have at each point these distinguished directions, as slip lines, or also as characteristics C^- and C^+. They form a curvilinear net $x=x(\alpha,\beta)$, $y=y(\alpha,\beta)$ where α,β are the parameters of the net. A curve $\beta=$constant, a C^-, is called an α-curve, a curve $\alpha=$constant, a C^+, is called a β-curve.

Next, consider the well-known formulas, valid for any symmetric tensor,

$$\begin{aligned}
\sigma_x &= \sigma_1\cos^2\vartheta+\sigma_2\sin^2\vartheta=\frac{\sigma_1+\sigma_2}{2}+\frac{\sigma_1-\sigma_2}{2}\cos 2\vartheta,\\
\sigma_y &= \sigma_1\sin^2\vartheta+\sigma_2\cos^2\vartheta=\frac{\sigma_1+\sigma_2}{2}-\frac{\sigma_1-\sigma_2}{2}\cos 2\vartheta,\\
\tau_z &= \sigma_1\cos\vartheta\sin\vartheta-\sigma_2\sin\vartheta\cos\vartheta=\frac{\sigma_1-\sigma_2}{2}\sin 2\vartheta.
\end{aligned} \tag{29.6}$$

In our problem, where $(\sigma_1+\sigma_2)/2=(\sigma_x+\sigma_y)/2=-p$, $(\sigma_1-\sigma_2)/2=k$, they take on the forms[103]

$$\sigma_x=-p+k\cos 2\vartheta,\qquad \sigma_y=-p-k\cos 2\vartheta,\qquad \tau_z=k\sin 2\vartheta,$$

or

$$s_z=-s_y=k\cos 2\vartheta,\qquad \tau_z=k\sin 2\vartheta,\qquad s_x^2+\tau_z^2=k^2. \tag{29.6'}$$

These equations form a representation of the stresses, in terms of p, k, ϑ, for which the yield condition holds.

Now compute

$$\frac{\partial\sigma_x}{\partial x}+\frac{\partial\tau_z}{\partial y}$$

and likewise

$$\frac{\partial\tau_z}{\partial x}+\frac{\partial\sigma_y}{\partial y}$$

by means of (29.6'). Since both sums are zero, we obtain

$$\frac{\partial}{\partial x}\left(\frac{p}{2k}\right)=-\frac{\partial\vartheta}{\partial x}\sin 2\vartheta+\frac{\partial\vartheta}{\partial y}\cos 2\vartheta,\qquad \frac{\partial}{\partial y}\left(\frac{p}{2k}\right)=\frac{\partial\vartheta}{\partial x}\cos 2\vartheta+\frac{\partial\vartheta}{\partial y}\sin 2\vartheta.$$

The integration theory of the present Chap. B, which we are going to exhibit, as well as that of the more general problem of Chap. C, is based on the exchange of dependent and independent variables, used for "plane strain" as early as 1923 by several authors (see note 105). In Chap. C, Sect. 34 we obtain, following v. MISES,[104] by an exchange of variables the Eqs. (34.10). SOKOLOVSKY [*26*, p. 176seq.] uses the same variables, (34.9), in the particular case of plane strain and obtains the equations to which (34.10) reduce in this particular case. However, in this case, where the angle of the characteristics with the first principal direction is constant, one gets more elegant results by using differentiation with respect to characteristic variables. These are subsequently introduced by SOKOLOVSKY in connection with an approximate method due to CHRISTIANOWITSCH [see his Eqs. (5.47), (5.48)]. Eqs. (30.12) or (31.14) are, however, simpler, without being approximate (cf. also footnote 125).

Denote by $\partial/\partial u$, $\partial/\partial v$ directional derivatives in the first and second principal directions, viz.

$$\frac{\partial}{\partial u}=\cos\vartheta\frac{\partial}{\partial x}+\sin\vartheta\frac{\partial}{\partial y},\qquad \text{etc.}$$

We obtain the intermediate result (which will become of interest in the case of Sect. 34)

$$\frac{\partial}{\partial u}\left(\frac{p}{2k}\right)=\frac{\partial\vartheta}{\partial v},\qquad \frac{\partial}{\partial v}\left(\frac{p}{2k}\right)=\frac{\partial\vartheta}{\partial u}.$$

---

[103] The magnitude $p/2k$, which plays an important role in all that follows, is often denoted by χ.

[104] R. v. MISES: p. 415–429, Reissner Anniversary Volume. Ann Arbor, Michigan 1949.

Using these formulas, we compute the directional derivative $\partial/\partial l^+$ of $p/2k$ in the β-direction, viz.

$$\frac{\partial}{\partial l^+}=\frac{1}{\sqrt{2}}\left(\frac{\partial}{\partial u}+\frac{\partial}{\partial v}\right),$$

and similarly $\partial/\partial l^-$ for the α-direction. We obtain

$$\frac{\partial}{\partial l^-}\left(\frac{p}{2k}\right)=-\frac{\partial\vartheta}{\partial l^-},\qquad \frac{\partial}{\partial l^+}\left(\frac{p}{2k}\right)=\frac{\partial\vartheta}{\partial l^+}. \tag{29.7}$$

These are equivalent to

$$\frac{\partial}{\partial \alpha}\left(\frac{p}{2k}\right)=-\frac{\partial\vartheta}{\partial \alpha},\qquad \frac{\partial}{\partial \beta}\left(\frac{p}{2k}\right)=\frac{\partial\vartheta}{\partial \beta}. \tag{29.7'}$$

Eqs. (29.7) or (29.7') are *compatibility equations*, along the slip lines C^- or C^+, respectively, since each contains only differentiations along one characteristic direction. They can be integrated and we obtain,[105] using for convenience the angle $\vartheta-45°=\varphi$ rather than ϑ:

$$\begin{aligned}\frac{p}{2k}+\varphi&=\text{constant along each } C^-,\\ \frac{p}{2k}-\varphi&=\text{constant along each } C^+.\end{aligned} \tag{29.8}$$

Of course the values of the constants vary from one slip line to another. The first constant must be a function of β alone (since $\beta=$ constant along each C^-) and the second constant a function of α alone. Thus we have the equations equivalent to (29.8):

$$\frac{p}{2k}=g(\beta)+f(\alpha),\qquad \varphi=g(\beta)-f(\alpha). \tag{29.9}$$

From (29.9) the equations of the slip lines have the form

$$F_1(x,y,\alpha)=0,\qquad F_2(x,y,\beta)=0,$$

or, if solved with respect to x and y

$$x=x(\alpha,\beta),\qquad y=y(\alpha,\beta) \tag{29.9'}$$

or, equivalently $\boldsymbol{r}=\boldsymbol{r}(\alpha,\beta)$ where $d\boldsymbol{r}/d\alpha$ has the direction of the positive α, and $d\boldsymbol{r}/d\beta$ that of the positive β.

Also, as a consequence of (29.7') or (29.9), we have the integrability conditions

$$\frac{\partial^2}{\partial\alpha\,\partial\beta}\left(\frac{p}{2k}\right)=0,\qquad \frac{\partial^2\varphi}{\partial\alpha\,\partial\beta}=0. \tag{29.10}$$

We finally want the *velocity compatibility relations*. They must consist of formulas expressing the previous result that the rate of extension along a slip line is zero: $\dot\varepsilon_\alpha=0$, $\dot\varepsilon_\beta=0$. The rate of extension in the α-direction is, with $\varphi^-=\vartheta-45°=\varphi$,

$$\begin{aligned}\dot\varepsilon_\alpha&=\dot\varepsilon_x\cos^2\varphi+\dot\gamma_z\sin\varphi\cos\varphi+\dot\varepsilon_y\sin^2\varphi\\ &=\cos\varphi\left(\frac{\partial v_x}{\partial x}\cos\varphi+\frac{\partial v_x}{\partial y}\sin\varphi\right)+\sin\varphi\left(\frac{\partial v_y}{\partial x}\cos\varphi+\frac{\partial v_y}{\partial y}\sin\varphi\right)\\ &=\frac{\partial v_x}{\partial l^-}\cos\varphi+\frac{\partial v_y}{\partial l^-}\sin\varphi=0,\end{aligned}$$

$$\dot\varepsilon_\beta=\frac{\partial v_x}{\partial l^+}\cos\varphi^++\frac{\partial v_y}{\partial l^+}\sin\varphi^+=0.$$

[105] Cf. H. HENCKY: Z. angew. Math. Mech. 3, 241–251 (1923). — C. CARATHÉODORY and E. SCHMIDT: Z. angew. Math. Mech. 3, 468–475 (1923).

A geometric interpretation of this result will be discussed presently. We now express $\dot{\varepsilon}_\alpha$ and $\dot{\varepsilon}_\beta$ in terms of the components v_1, v_2 of $\boldsymbol{v}$ in the α- and β-directions.[106] Using $v_1 = v_x \cos\varphi^- + v_y \sin\varphi^-$, $v_2 = v_x \cos\varphi^+ + v_y \sin\varphi^+$ and the definitions of $\partial/\partial l^-$, $\partial/\partial l^+$, we find by an elementary computation

$$\frac{\partial v_1}{\partial l^-} = \left(\frac{\partial v_x}{\partial l^-}\cos\varphi^- + \frac{\partial v_y}{\partial l^-}\sin\varphi^-\right) + v_2\frac{\partial\varphi^-}{\partial l^-},$$

$$\frac{\partial v_2}{\partial l^+} = \left(\frac{\partial v_x}{\partial l^+}\cos\varphi^+ + \frac{\partial v_y}{\partial l^+}\sin\varphi^+\right) = v_1\frac{\partial\varphi^+}{\partial l^+}.$$

The expressions to the right, in parentheses, are zero, and we have the result:

$$\dot{\varepsilon}_\alpha = \frac{\partial v_1}{\partial l^-} - v_2\frac{\partial\varphi^-}{\partial l^-} = 0, \qquad \dot{\varepsilon}_\beta = \frac{\partial v_2}{\partial l^+} + v_1\frac{\partial\varphi^+}{\partial l^+} = 0 \tag{29.11}$$

or, since $d\varphi^- = d\varphi^+ = d\vartheta$,

$$dv_1 - v_2\, d\vartheta = 0 \quad \text{along a } C^-; \qquad dv_2 + v_1 d\vartheta = 0 \quad \text{along a } C^+. \tag{29.11'}$$

These are the compatibility relations in terms of v_1 and v_2.[107] In terms of v_x, v_y they read

$$\frac{\partial v_y}{\partial l^-} = -\frac{\partial v_x}{\partial l^-}\cot\varphi^-, \qquad \frac{\partial v_y}{\partial l^+} = -\frac{\partial v_x}{\partial l^+}\cot\varphi^+. \tag{29.12}$$

These are less useful than (29.11′) since $\cot\varphi^-$ and $\cot\varphi^+$ depend on the stresses.

Consider a *velocity plane*, i.e. a plane with rectangular coordinates v_x, v_y, so that to a point P with coordinates x, y in the *physical plane* corresponds the point $\bar{P}$ with coordinates v_x, v_y. The images of the characteristics C^+, C^- at P are curves $\bar{C}^+, \bar{C}^-$ at $\bar{P}$; these are at the same time the *characteristics in the velocity plane*. Eqs. (29.12) state *that the velocity characteristic $\bar{C}^+$ at $\bar{P}$ is normal to the C^+ at P, and similarly for the C^- and $\bar{C}^-$*[108] (cf. Fig. 5).

More important than the velocity plane is the *stress plane*, or stress graph—introduced by NEUBER, V. MISES, and SAUER[109]—since the stress equations are non-linear while the velocity equations are linear. In the stress plane the independent variables are two stresses, e.g. σ_x and σ_y, or p and ϑ. With any of these as independent variables (and some physical variables as dependent variables)[110] thus *by means of an interchange*[110] *of independent and dependent variables*, the stress equations become linear. The images of the C^+, C^- in the stress plane with coordinates p, ϑ, say, may be called Γ^+, Γ^- and Eqs. (19.8) hold for the Γ^+ and Γ^-, respectively.

30. Continuation. Slip line field. The equations of the orthogonal slip lines, the characteristics, may be written as

$$x = x(\alpha, \beta), \qquad y = y(\alpha, \beta).$$

Then, with $A > 0$, $B > 0$:

$$(dr)^2 = dx^2 + dy^2 = A^2\, d\alpha^2 + B^2\, d\beta^2, \qquad A = \frac{dl^+}{d\alpha}, \qquad B = \frac{dl^+}{d\beta}. \tag{30.1}$$

[106] Note that we use $\partial/\partial u$, $\partial/\partial v$ as directional derivatives in the two principal directions.

[107] H. GEIRINGER: Proc. 3rd Internat. Congr. Appl. Mech. II, 1930, pp. 185–190.

[108] H. GEIRINGER: Proc. Nat. Acad. Sci. U.S. **37**, 214–220 (1951); there the theorem is proved for the general problem (27.4). Cf. also W. PRAGER: Trans. Roy. Inst. Technol. Stockholm **65**, 1–26 (1953), where the orthogonality for the present problem, viz. plane strain, is proved.

[109] R. v. MISES: Reissner Anniversary Volume, pp. 415–429, Ann. Arbor, Michigan 1949 (see footnote 29). R. SAUER: Z. angew. Math. Mech. **29**, 274–279 (1949). H. NEUBER: Z. angew. Math. Mech. **28**, 253–257 (1948).

[110] This interchange requires that the appropriate Jacobian, e.g. $\partial(p, \vartheta)/\partial(\alpha, \beta)$ be different from zero.

Writing

$$\varphi^- = \varphi, \qquad \varphi^+ = 90^\circ + \varphi,$$

we have

$$\begin{aligned} \frac{\partial x}{\partial \alpha} &= A\cos\varphi, & \frac{\partial x}{\partial \beta} &= -B\sin\varphi, \\ \frac{\partial y}{\partial \alpha} &= A\sin\varphi, & \frac{\partial y}{\partial \beta} &= B\cos\varphi. \end{aligned} \tag{30.2}$$

From

$$\frac{\partial^2 x}{\partial\alpha\,\partial\beta} = \frac{\partial^2 x}{\partial\beta\,\partial\alpha} \quad\text{and}\quad \frac{\partial^2 y}{\partial\alpha\,\partial\beta} = \frac{\partial^2 y}{\partial\beta\,\partial\alpha},$$

we obtain

$$\cos\varphi\left(\frac{\partial A}{\partial\beta} + B\frac{\partial\varphi}{\partial\alpha}\right) + \sin\varphi\left(\frac{\partial B}{\partial\alpha} - A\frac{\partial\varphi}{\partial\beta}\right) = 0,$$

$$\sin\varphi\left(\frac{\partial A}{\partial\beta} + B\frac{\partial\varphi}{\partial\alpha}\right) - \cos\varphi\left(\frac{\partial B}{\partial\alpha} - A\frac{\partial\varphi}{\partial\beta}\right) = 0.$$

Therefore

$$\frac{\partial A}{\partial\beta} + B\frac{\partial\varphi}{\partial\alpha} = 0, \qquad \frac{\partial B}{\partial\alpha} - A\frac{\partial\varphi}{\partial\beta} = 0. \tag{30.3}$$

Denote by R^+ and R^- the radii of curvature of a C^+ or C^- chosing the sign positive if the center of curvature of an α-line (β-line) is in direction of increasing β (of increasing α). Put, accordingly,

$$\frac{1}{R^-} = \frac{d\varphi^-}{dl^-} = \frac{\frac{\partial\varphi}{\partial\alpha}}{\frac{dl^-}{d\alpha}} = \frac{1}{A}\frac{\partial\varphi}{\partial\alpha}, \qquad \frac{1}{R^+} = -\frac{d\varphi^+}{dl^+} = -\frac{\frac{\partial\varphi}{\partial\beta}}{\frac{dl^+}{d\beta}} = -\frac{1}{B}\frac{\partial\varphi}{\partial\beta}. \tag{30.4}$$

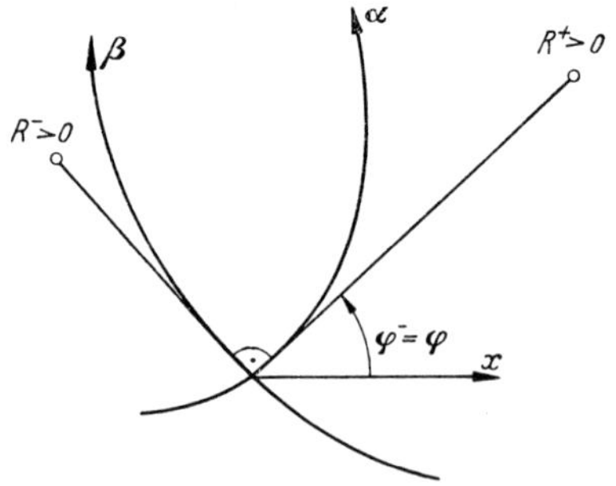

Fig. 5. Signs of the radii of curvature in slip-line field.

[In the second Eq. (30.4) the negative sign must be chosen: Indeed (see Fig. 5), while α is increasing with increasing x, β is increasing with decreasing x. Hence, if as in the figure $R^- > 0$, $R^+ > 0$, by our definition, the $\varphi^- = \varphi$ is indeed increasing with increasing α along the α-curve, but φ is decreasing along the β-curve with increasing β.] Now we compute $\partial R^+/\partial\alpha$, using (30.3) and (30.1), and in the same way we find $\partial R^-/\partial\beta$. The result is

$$\frac{\partial R^+}{\partial\alpha} + \frac{dl^-}{d\alpha} = \frac{B}{\left(\frac{\partial\varphi}{\partial\beta}\right)^2}\frac{\partial^2\varphi}{\partial\alpha\,\partial\beta}, \qquad \frac{\partial R^-}{\partial\beta} + \frac{dl^+}{d\beta} = -\frac{A}{\left(\frac{\partial\varphi}{\partial\alpha}\right)^2}\frac{\partial^2\varphi}{\partial\alpha\,\partial\beta}. \tag{30.5}$$

All these formulas hold *for any orthogonal net.*

Now, however, consider the second Eqs. (29.9) and (29.10), viz.

$$\varphi = g(\beta) - f(\alpha), \qquad \frac{\partial^2 \varphi}{\partial\alpha\,\partial\beta} = 0. \tag{30.6}$$

We then obtain from (30.5)

$$\frac{\partial R^+}{\partial\alpha} + \frac{dl^-}{d\alpha} = 0, \qquad \frac{\partial R^-}{\partial\beta} + \frac{dl^+}{d\beta} = 0. \tag{30.7}$$

We actually see that from each of the four Eqs. (30.6), (30.7) the other three follow.

Eqs. (30.6), (30.7) form the analytic expression of two well-known geometric properties of the slip lines. For formulations and figures see GEIRINGER [*3*, pp. 35 and 39], HILL [*6*, p. 139], HODGE and PRAGER [*23*, pp. 130–134]. The equivalent of the first slip line property is Eq. (30.6), that of the second property are the two Eqs. (30.7), each for one family of curves, and we have shown, following CARATHÉODORY and E. SCHMIDT,[111] that it is sufficient to *assume ones of these properties for one family in order to obtain both properties for both families.*[112]

We consider now a region where $f'(\alpha) \neq 0$, $g'(\beta) \neq 0$. Since, from (30.4) and (30.6),

$$\frac{1}{R^-} = -\frac{f'(\alpha)}{A}, \qquad \frac{1}{R^+} = -\frac{g'(\beta)}{B}, \tag{30.4'}$$

the last assumption implies that in the considered region the curvatures of the curves of the net are different from zero. (The opposite case, which arises for solutions denoted as "degenerate solution", "simple wave", "fan", will be discussed in connection with the general plane problem.[113]) In order to take care of the signs of the R^+, R^- we use two constants a, b, where $|a| = |b| = 1$, and an additive constant c for convenience and according to (29.9) we introduce the *characteristic coordinates* ξ, η by

$$g(\beta) = b\eta + c, \qquad f(\alpha) = -a\xi, \tag{30.8}$$

$$\varphi = a\xi + b\eta + c, \qquad \frac{p}{2k} = -a\xi + b\eta + c. \tag{30.9}$$

Then [see (30.4)] e.g., $a = -1$, $b = +1$ corresponds to $R^- < 0$, $R^+ < 0$. Denote by h_1 and h_2 the values of A and B for these new characteristic parameters. The simplified Eqs. (30.4), (30.7), (30.3) are then

$$h_1 = aR^-, \qquad h_2 = -bR^+. \tag{30.10}$$

$$\begin{aligned} &\frac{\partial R^+}{\partial\xi} + aR^- = 0, \qquad \frac{\partial R^-}{\partial\eta} - bR^+ = 0, \\ &\frac{\partial h_2}{\partial\xi} - bh_1 = 0, \qquad \frac{\partial h_1}{\partial\eta} + a h_2 = 0. \end{aligned} \tag{30.11}$$

Since also $\partial\vartheta/\partial\xi = a$, $\partial\vartheta/\partial\eta = b$, Eqs. (30.11) may also be written

$$dR^+ + R^-\,d\vartheta = 0, \qquad dR^- - R^+\,d\vartheta = 0, \tag{30.11'}$$

[111] CARATHÉODORY and E. SCHMIDT: Z. angew. Math. Mech. **3**, 468–475 (1923).

[112] A further geometric property has been found by I. KAPUANO: Rev. Fac. Sci. Univ. Istanbul A, 36–39 (1941).

[113] A misleading because too general name is used in the translation [*26*]. The "fans" are denoted as "integrals of the plasticity equations".

along a C^- or C^+, respectively, in obvious analogy to (29.11′).[114] From (30.11) we conclude that

$$\frac{\partial^2 R^+}{\partial\xi\,\partial\eta}+abR^+=0,\qquad \frac{\partial^2 R^-}{\partial\xi\,\partial\eta}+abR^-=0. \tag{30.12}$$

Finally, Eqs. (30.2) take the form[115]

$$\begin{aligned}\frac{\partial x}{\partial\xi}&=aR^-\cos\varphi, & \frac{\partial x}{\partial\eta}&=bR^+\sin\varphi,\\ \frac{\partial y}{\partial\xi}&=aR^-\sin\varphi, & \frac{\partial y}{\partial\eta}&=-bR^+\cos\varphi.\end{aligned} \tag{30.13}$$

$$\frac{\partial y}{\partial\xi}=\tan\varphi\,\frac{\partial x}{\partial\xi},\qquad \frac{\partial y}{\partial\eta}=-\operatorname{ctg}\varphi\,\frac{\partial x}{\partial\eta}. \tag{30.13′}$$

Our main result, as far as the stresses are concerned, is the Eqs. (30.12) and (30.13). If we know R^+, R^- from (30.12) [or from (30.11)] the Eqs. (30.13) give $x(\xi,\eta)$, $y(\xi,\eta)$. Then (30.9) give φ and $p/2k$, and, using also (29.5) and the yield condition, we know the stress tensor.

Variables different from R^+, R^- may be used in a similar way,[116] namely the "moving coordinates"

$$X=x\cos\varphi+y\sin\varphi,\qquad Y=-x\sin\varphi+y\cos\varphi.$$

These X and Y are the components of the radius vector $\boldsymbol{r}$ in the characteristic directions. Then:

$$\frac{\partial Y}{\partial\xi}=\left(-\frac{\partial x}{\partial\xi}\sin\varphi+\frac{\partial y}{\partial\xi}\cos\varphi\right)+(x\cos\varphi+y\sin\varphi).$$

Since the expression in the first parenthesis vanishes along a ξ-line, we obtain, with a similar computation for $\partial X/\partial\eta$:

$$\frac{\partial Y}{\partial\xi}=X,\qquad \frac{\partial X}{\partial\eta}=Y,\qquad \frac{\partial^2 X}{\partial\xi\,\partial\eta}=X,\qquad \frac{\partial^2 Y}{\partial\xi\,\partial\eta}=Y, \tag{30.14}$$

in complete analogy to (30.11) and (30.12).

We have thus various effective linearizations, and the *method of superposition* might be applied to find new solutions as linear combinations of given ones. The superposition can be performed analytically or geometrically. If a slip-line field is drawn for equidistant ξ-, η-values then from $\vartheta=\eta-\xi$, $p/2k=\eta+\xi$, the diagonal curves are the isoclines and isobares. Consider two slip line nets such that points with the same ξ,η (hence also isoclines and isobars) correspond to each other then the vector addition $\overrightarrow{OP_1}+\overrightarrow{OP_2}$ where P_1, P_2 are corresponding points (O arbitrary) gives again a slip line net: $\overrightarrow{OP_1}+\overrightarrow{OP_2}=\overrightarrow{OP}$. If P_1, and therefore P_2, describes a slip line the same holds for P. The components of $\overrightarrow{OP_1}$ and of $\overrightarrow{OP_2}$ satisfy the same Eqs. (30.13′), and so do their sum and their difference.[117]

[114] More specifically, R^+ is replaced by v_1, and R^- by $-v_2$.

[115] From these equations follows $x=a\int R^-\cos\varphi\,d\varphi$, $y=a\int R^-\sin\varphi\,d\varphi$ along a ξ-line, with similar relations along an η-line.

[116] H. GEIRINGER: Mémoires sur la mécanique des fluides, offerts à M. D. RIABOUCHINSKY, Paris, 1955.

[117] Independently of early remarks by R. v. MISES, Z. angew. Math. Mech. **5**, 147 (1925), and H. GEIRINGER [*3*, p. 55], R. HILL, J. Mech. Phys. Solids **15**, 255 (1967), studied the superposition of the slip line nets and developed practical methods. HILL showed also that every net can be generated by superposition of centered fans. His work was followed up and applied to the problems of compression and extrusion by I. F. COLLINS, J. Mech. Phys. Solids **16**, 137 (1968) and Proc. Roy. Soc. Lond., Ser. A **303**, 317.

Consider finally the velocity Eqs. (29.11). If ξ, η are introduced they become

$$\frac{\partial v_1}{\partial \xi} - a v_2 = 0, \qquad \frac{\partial v_2}{\partial \eta} + b v_1 = 0, \tag{30.15}$$

and from these

$$\frac{\partial^2 v_1}{\partial \xi\, \partial \eta} + a b v_1 = 0, \qquad \frac{\partial^2 v_2}{\partial \xi\, \partial \eta} + a b v_2 = 0, \tag{30.16}$$

equations of the same form as (30.12).

From (29.12) we obtain

$$\frac{\partial v_x}{\partial \xi} + \frac{\partial v_y}{\partial \xi} \tan\varphi = 0, \qquad \frac{\partial v_x}{\partial \eta} - \frac{\partial v_y}{\partial \eta} \cot\varphi = 0. \tag{30.17}$$

Comparing the two groups (30.15), (30.17) with (30.11), (10.13') we see that the functions v_1, v_2, v_x, v_y satisfy the same equations as R^+, $-R^-$, y, $-x$. Hence we can apply Riemann's method (Sect. 31) in both cases in the same way.

b) Integration. Particular solutions.

31. Integration. The actual solution of a concrete mechanical problem often meets with great difficulties, some of which we shall discuss at the end of this section. However, the theory contained in the last few sections presents in the ξ,η-plane considerable advantages: (1) The equations are hyperbolic for the stresses as well as for the velocities. (2) The equations for both, the stresses and the velocities, are simple, and their *Riemann function* is known.

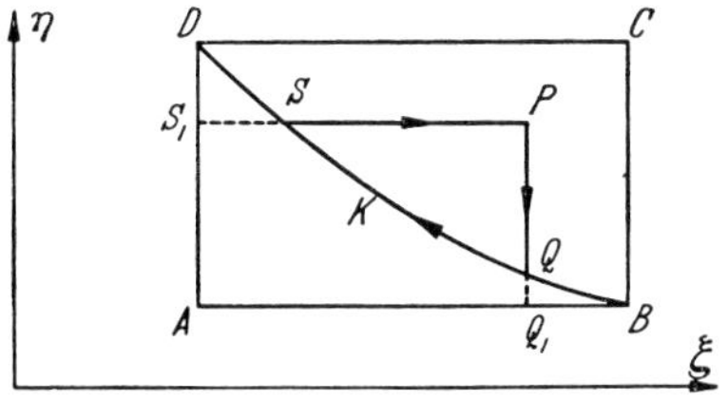

Fig. 6. Illustrating Riemann's integration method.

The Riemann method applies to the linearized Eqs. (30.11), (30.12) and pro vides explicit solutions for certain fundamental initial-value problems, if in the ξ η-plane appropriate initial values are given. But, two additional questions arise: (a) How do we find initial values in the ξ, η-plane from meaningful data in the physical plane? (b) If, by means of the Riemann method and of appropriate initial data, $R^+(\xi,\eta)$ and $R^-(\xi,\eta)$ have been found in a certain domain, how do we find the corresponding stress distribution in the physical plane? For the velocities these two questions do not arise since, after the stress problem has been solved, the velocity problem is linear and can be solved in the x, y-plane.

The exact explicit solutions are often cumbersome, and therefore numerical as well as graphical approximation methods have been devised. We shall speak of these methods also.

Now consider question (a). An arc $\mathscr{K}$ is given in the x, y-plane by $x = x(t)$, $y = y(t)$. Along this arc we know e.g. the stresses: $p/k = -s$ and ϑ [equivalent to giving σ_x, σ_y, τ, on account of Eq. (29.6')]. The ϑ given along $\mathscr{K}$ must be such that the actual inclination of $\mathscr{K}$ is at no point equal to $\vartheta + 45°$ or $\vartheta - 45°$, that means that $\mathscr{K}$ has nowhere a characteristic direction. Then, from (30.9) (since

$\varphi = \vartheta - 45°$) we find ξ and η at every point of $\mathcal{K}$, say $\xi = \xi(t)$, $\eta = \eta(t)$, and by (30.13), R^- and R^+ follow along $\mathcal{K}$.

To the arc $\mathcal{K}$ in the x, y-plane there corresponds K in the ξ, η-plane, and we know R^-, R^+ along it. The preceding Riemann formula then gives R^- or R^+ in $ABCD$; the other of these functions follows by (30.11).

Finally (question b), if $R^-(\xi, \eta)$, $R^+(\xi, \eta)$ have been found in $ABCD$, Eqs. (30.13) give $\partial x/\partial\xi$, $\partial x/\partial\eta$, $\partial y/\partial\xi$, $\partial y/\partial\eta$ in $ABCD$, and $x = x(\xi, \eta)$, $y = y(\xi, \eta)$ are found by quadratures.[118] Thus the slip-line field in the physical plane is known. The slip-line field will have singularities at points where

$$J = \frac{\partial x}{\partial \xi}\frac{\partial y}{\partial \eta} - \frac{\partial x}{\partial \eta}\frac{\partial y}{\partial \xi} = 0.$$

(Since, by (30.13), $J = h_1 h_2 = R^- R^+$, we see that along such lines, say along $h_1(\xi\,\eta) = 0$, the ξ-curves will have cusps.[119]) In a region D where $J \neq 0$ we can invert the above formulas and find $\xi = \xi(x, y)$, $\eta = \eta(x, y)$ and therefore the stresses at every point (x, y) in D.[120]

Cauchy data for the linear equation[121]

$$\frac{\partial^2 z}{\partial \xi\,\partial \eta} = z$$

consist in giving z and $\partial z/\partial\xi$ along a curve K in the ξ, η-plane whose tangent is nowhere parallel to either axis. If then the Riemann function is known, the solution is determined in the characteristic rectangle whose horizontal and vertical sides pass through the two end-points of K (Fig. 6):

$$z(P) = z(S) + \int\limits_{QS} \left(\omega \frac{\partial z}{\partial \xi}\, d\xi + z \frac{\partial \omega}{\partial \eta}\, d\eta\right),$$

where $\int\limits_{QS}$ is the line integral along K. The Riemann function $\omega(\xi, \eta; \xi_0\,\eta_0)$ for our equation is the Bessel function

$$\omega = \sum_{n=0}^{\infty} \frac{(\xi - \xi_0)^n (\eta - \eta_0)^n}{n!\, n!} = I[(\xi - \xi_0)(\eta - \eta_0)].$$

We do not enter into details regarding this well-known method.[122] The main point is that R^+ (and R^-) are explicitly determined in terms of ξ, η if R^+ and $\partial R^+/\partial\xi$ are known along K. (A similar formula for $z(P)$ holds if R^+ and $\partial R^+/\partial\eta$ are given along K.)

In the second or *characteristic initial-value problem* of $\frac{\partial^2 z}{\partial \xi\,\partial \eta} = z$ we need the value of z along AB and AD. Denote by Q_1 and S_1 the points of intersection of the vertical and horizontal straight line through P with AB and AD (see Fig. 6). The Riemann formula is

$$z(P) = z(S_1) + \int\limits_{S_1}^{A} z \frac{\partial \omega}{\partial \eta}\, d\eta + \int\limits_{A}^{Q_1} \omega \frac{\partial z}{\partial \xi}\, d\xi,$$

---

118 Compare the formulas in footnote 115.

119 More on these singularities in Sect. 40.

120 The stress tensor is determined at every point by (30.9) and the yield condition.

121 To fix the ideas let us take here $ab = -1$.

122 See e.g. GEIRINGER [3, pp. 44seq.] where the method is explained in relation to the present problem.

where ω is the same Bessel function as before. The initial data in the x, y-plane now simply consist in giving two curves intersecting at a right angle, and designated as C^-, C^+, respectively. In other words, we wish to determine a system of slip lines which contains these two curves as a ξ-curve and as an η-curve.[123] We take the x- and y-axis tangent to the C^- and C^+ at their point of intersection, O. Then, denoting by φ_0 and s_0 the values of φ and $s=-p/k$ at O and taking $b=1$, $a=-1$, $c=0$ in (30.9), we have $\varphi_0=0=\eta_0-\xi_0$, $\frac{1}{2}s_0=-\eta_0-\xi_0=-2\eta_0$, where η_0 must be given. Then, along the ξ-curve $\xi=\eta_0-\varphi$, while along the other $\eta=\varphi^+ +\xi_0-90°=\varphi^+ +\eta_0-90°$. On the other hand, we know by inspection the radii of curvature, R^- along the ξ-curve, R^+ along the η-curve. To find $R^+(\xi,\eta)$ in the rectangle we use the Riemann formula just given where we need R^+ along S_1A, and $\partial R^+/\partial\xi$ along the horizontal piece. We find $\partial R^+/\partial\xi$ by (30.11) since it equals R^-. (We proceed similarly if we wish to find R^-, using a second formula similar to the above, for $z(P)$.) When we have found, say, R^+ in the rectangle $ABCD$, the problem continues as before.[124]

Consider, now, for the *velocities*, say, the Cauchy problem. If stresses are given along $\mathscr{K}$ in the x, y-plane we know ξ, η there; if then the velocity components v_1, v_2 are given along $\mathscr{K}$, we know there $v_1(\xi,\eta)$, $v_2(\xi,\eta)$, and the Riemann method gives these functions in $ABCD$. If, finally $\xi=\xi(x,y)$, $\eta=\eta(x,y)$ have been found (stress problem solved) in $ABCD$, we know there

$$v_1(\xi(x,y),\eta(x,y))=v_1(x,y)$$

and, similarly, $v_2(x,y)$.

We review the results:

Cauchy problem. Along a curve which has nowhere characteristic direction the tensor of stresses and the vector of velocity are given. The stresses and velocities are then determined in the corresponding characteristic quadrangle.

Characteristic value problem. In the physical plane, two intersecting slip lines are given and along each of them one of the components v_1, v_2 of the velocity vector. The velocities as well as the stresses are then determined in the corresponding characteristic quadrangle. (See details in [*3*, p. 64 and p. 48], see also p. 76.)

We therefore have the remarkable result that we possess *explicit solutions* of important initial-value problems.[125] The Riemann method can be adapted to the "third" or "mixed" initial-value problem (see e.g. GEIRINGER [*3*, p. 46]; see also the solutions in Sects. 59 to 61 of the present article[126]).

All these are exact analytic solutions. In addition, rapidly coverging approximate solutions are available. A geometric method based on Eqs. (29.7) and (29.11′) has been worked out by PRAGER,[127] and other approximation methods are described by HILL [*6*, pp. 141–151] (see the original publication[128]). Some more will be added in connection with the general case.

---

[123] The problem has been carried through by C. CARATHÉODORY and E. SCHMIDT, Z. angew. Math. Mech. **3**, 468–475 (1923).

[124] As a recent instance of the direct application of Riemann's method we mention M. SAYIR'S [Z. angew. Math. Phys. **20**, 298 (1969)] continuation of PRANDTL'S solution of the indentation problem (1920), treated later by HILL [*6*, p. 254]. See also: M. SAYIR and H. ZIEGLER, Ingr. Arch. **36**, 294 (1968).

[125] A similar situation exists in the problem of unsteady parallel flow (x, t-problem) of a compressible fluid. This problem is also everywhere hyperbolic, and the Riemann function is known, as well as explicit solutions of the two initial problems. Such is not the case for the steady plane problem of a compressible fluid and for the general problem of plasticity, (27.4).

[126] See papers by W. HAACK and G. HELLWIG: Math. Z. **53**, 244–266 (1950), and 340–356 (1950), and H. BECKERT: Ber. Verh. Sächs. Akad. Wiss. Leipzig **97**, 68 (1950).

[127] W. PRAGER: Trans. Roy. Inst. Technol., Stockholm **65**, 1–26 (1953). See also Sect. 46.

[128] R. HILL, E. H. LEE and S. J. TUPPER: J. Appl. Mechanics **8**, 46–52 (1951).

Unfortunately, the concrete problems of plasticity theory do not reduce often to a succession of initial-value problems as described in this section. Fig. 7 indicates the following concrete problem. A plane plate of cross-section $ABCD$

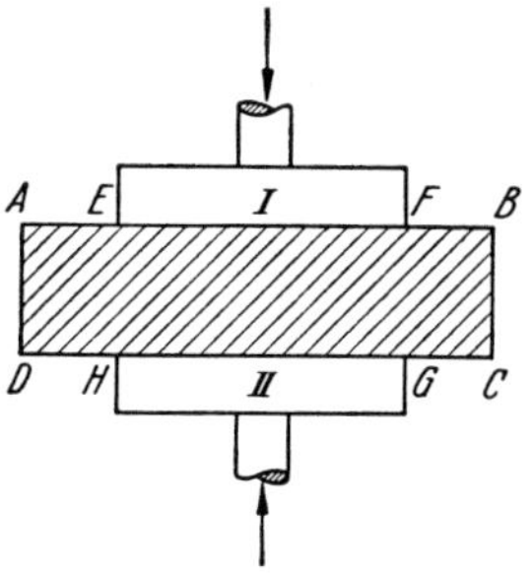

Fig. 7. Plastic material between two plates.

is pressed between two pistons moved toward each other by the vertical forces P. Boundary conditions are:

1. Along BC and AD: $\sigma_x = \tau = 0$.
   Along FB, CG, HD, AE: $\sigma_y = \tau = 0$.
2. Along EF and HG:
   a) The vertical velocity components equals $+c$ or $-c$,
   b) if there is no friction: $\tau = 0$;
   if the coefficient of friction equals ψ: $\tau/\sigma = \psi$.

Hence along each piece of the boundary we know two scalar functions. The mathematical question is whether for such a physically meaningful problem a solution exists.[129]

32. Examples of exact particular solutions. The theory of the equation

$$\frac{\partial^2 u}{\partial \xi \, \partial \eta} = u \quad \text{(telegraph equation)}$$

is well known. Corresponding to any solution $u(\xi, \eta)$ of this equation, there is a slip-line field. An especially simple particular solution is the following:

1. Put here and in the following example $b = 1$, $a = -1$ and consider

$$R^+ = A\, e^{c\xi + \frac{1}{c}\eta}, \qquad R^- = A\, c\, e^{c\xi + \frac{1}{c}\eta}, \tag{32.1}$$

$$\begin{aligned} x &= -A \frac{c}{1+c^2} e^{c\xi + \frac{1}{c}\eta} (c \cos\varphi - \sin\varphi) = -\frac{A\,c}{\sqrt{1+c^2}} e^{c\xi + \frac{1}{c}\eta} \cos(\varphi + \gamma), \\ y &= -A \frac{c}{1+c^2} e^{c\xi + \frac{1}{c}\eta} (\cos\varphi + c \sin\varphi) = -\frac{A\,c}{\sqrt{1+c^2}} e^{c\xi + \frac{1}{c}\eta} \sin(\varphi + \gamma), \end{aligned} \tag{32.2}$$

where $\sin\gamma = (1 + c^2)^{-\frac{1}{2}}$, $\cot\gamma = c$. The slip lines form two families of logarithmic spirals of opposite sense and making angles γ and $90 - \gamma$ with the radius vector. The origin is a singular point. Through each other point passes one spiral of each family.

[129] As early as 1925, R. v. MISES: Z. angew. Math. Mech. **5**, 147–149 (1925), called attention to this situation and illustrated it. See also GEIRINGER and PRAGER: Ergeb. exakt. Naturw. **13**, 310–363 (1934), especially p. 343. Other contributions in this direction are due to E. H. LEE: J. Appl. Mech. **19**, 97–103 (1952).

Obviously, we can write in (32.1), (32.2) c_r and γ_r for c and γ and *consider the respective sums* (finite or infinite) since the principle of superposition holds.

2. Consider with α a constant,

$$\begin{aligned} R^+ &= A\cos\left(d\xi-\frac{\eta}{d}\right)-B\sin\left(d\xi-\frac{\eta}{d}\right),\\ R^- &= -Bd\cos\left(d\xi-\frac{\eta}{d}\right)-Ad\sin\left(d\xi-\frac{\eta}{d}\right), \end{aligned} \tag{32.3}$$

and, with $\varphi=\eta-\xi$, $\psi=d\xi-\frac{\eta}{d}$,

$$\begin{aligned} x &= \frac{2Ad}{1-d^2}(d\cos\psi\cos\varphi-\sin\psi\sin\varphi)+\frac{2Bd}{1-d^2}(\cos\psi\sin\varphi+d\sin\psi\cos\varphi),\\ y &= \frac{2Ad}{1-d^2}(d\cos\psi\sin\varphi-\sin\psi\cos\varphi)-\frac{2Bd}{1-d^2}(\cos\psi\cos\varphi-d\sin\psi\sin\varphi). \end{aligned} \tag{32.4}$$

The slip lines are epicycloids and hypocycloids. For $d=1$, $R^-=A\sin\varphi-B\cos\varphi$, $R^+=A\cos\varphi+B\sin\varphi$, the slip lines are cycloids. Again, superposition gives new solutions.

3. Next, consider any solution $u=h_1$ of

$$\frac{\partial^2 h_1}{\partial\xi\,\partial\eta}=h_1$$

and form, see (30.11), $\partial h_1/\partial\eta=h_2$, from which $\partial h_2/\partial\xi=h_1$. A slip-line field is found by forming

$$\begin{aligned} x(\xi,\eta) &= h_1\cos(\eta-\xi)-h_2\sin(\eta-\xi),\\ y(\xi,\eta) &= h_1\sin(\eta-\xi)+h_2\cos(\eta-\xi). \end{aligned} \tag{32.5}$$

We compute $\partial x/\partial\xi$, $\partial x/\partial\eta$, etc. and find, with $\varphi=\eta-\xi$,

$$\begin{aligned} \frac{\partial x}{\partial\xi} &= \left(\frac{\partial h_1}{\partial\xi}+h_2\right)\cos\varphi, & \frac{\partial x}{\partial\eta} &= -\left(h_1+\frac{\partial h_2}{\partial\eta}\right)\sin\varphi,\\ \frac{\partial y}{\partial\xi} &= \left(\frac{\partial h_1}{\partial\xi}+h_2\right)\sin\varphi, & \frac{\partial y}{\partial\eta} &= \left(h_1+\frac{\partial h_2}{\partial\eta}\right)\cos\varphi. \end{aligned} \tag{32.6}$$

Thus we see using (30.13) that the radii of curvature R^-, R^+ of the slip-line field (32.5) are

$$R^-=-\left(\frac{\partial h_1}{\partial\xi}+h_2\right)$$

and

$$R^+=-\left(h_1+\frac{\partial h_2}{\partial\eta}\right).$$

The field (32.5) is the sum of two fields, one with $k_1=\partial h_1/\partial\xi$, $k_2=h_1$, the other with $l_1=h_2$, $l_2=\partial h_2/\partial\eta$.

4. The Bessel function $I(\xi\cdot\eta)$, where

$$I(z)=1+\frac{z}{1!}+\frac{z^2}{2!\,2!}+\frac{z^3}{3!\,3!}+\cdots \tag{32.7}$$

satisfies our equation. Since this is also true for the partial derivatives $\partial^n I/\partial\xi^n$, $\partial^m I/\partial\eta^m$, we may set up a general solution in the form

$$u(\xi,\eta)=a\,I(\xi\cdot\eta)+a_1\frac{\partial I}{\partial\xi}+a_2\frac{\partial^2 I}{\partial\xi^2}+\cdots+b_1\frac{\partial I}{\partial\eta}+b_2\frac{\partial^2 I}{\partial\eta^2}+\cdots. \tag{32.8}$$

One might ask whether a solution of the form (32.8) is enough general to satisfy general boundary conditions (e.g. those of the Cauchy problem or the second problem.) The following theorem holds: *If two polynomials are given, viz.*

$$f(\alpha)=A_0+A_1\alpha+\cdots+A_n\alpha^n \quad \textit{in an interval } \overline{OA}$$

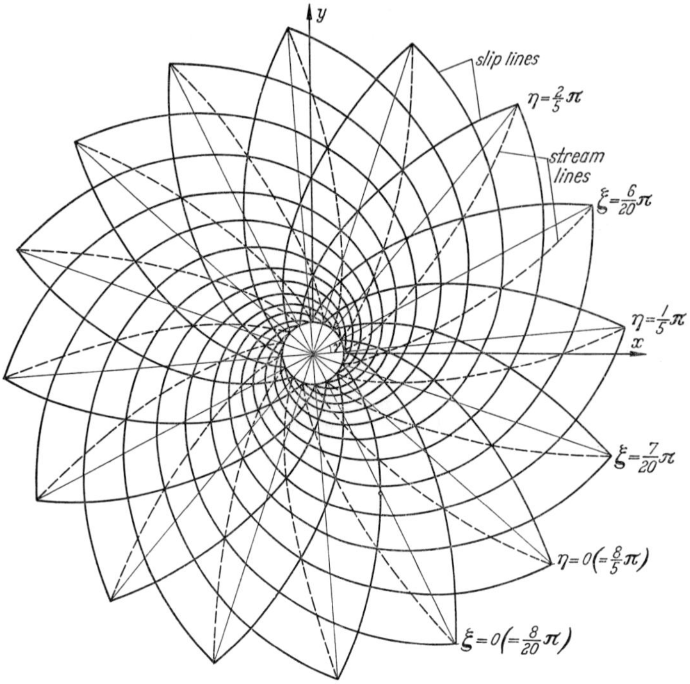

Fig. 8. Logarithmic spirals as slip-lines (full) and as streamlines (dashed).

and

$$g(\beta)=A_0+B_1\beta+\cdots+B_m\beta^m \quad \textit{in an interval } \overline{OB}$$

then the function

$$u(\alpha,\beta)=A_0+\sum_{\mu=1}^{m}B_\mu\,\mu!\,\frac{\partial^\mu I}{\partial\alpha^\mu}+\sum_{\gamma=1}^{n}A_\nu\,\nu!\,\frac{\partial^\nu I}{\partial\beta^\nu} \tag{32.8'}$$

satisfies the equation

$$\frac{\partial^2 u}{\partial\alpha\,\partial\beta}=u$$

in the characteristic quadrangle defined by $\overline{OA}$ *and* $\overline{OB}$ *and reduces to* $f(\alpha)$ *on* $\overline{OA}$ *and to* $g(\beta)$ *on* $\overline{OB}$. The simple proof is in [*3*, p. 57].

Since a continuous function can be approximated by a polynomial, the theorem allows us to solve the above-named characteristic boundary value problems in rather general cases (using tables of the Bessel function and its derivatives).

We do not go into considerations concerning convergence, completeness, etc.

We now consider examples of velocities. Obviously, for the velocity components v_1, v_2 we can use the same particular solutions just considered. Thus, for example:

$$v_1=\sum_r A_r\,e^{c_r\xi+\frac{1}{c_r}\eta},\qquad v_2=-\sum_r A_r\,c_r\,e^{c_r\xi+\frac{1}{c_r}\eta},$$

$$\left.\begin{aligned} v_x&=\sum_r A_r\sqrt{1+c_r^2}\,e^{c_r\xi+\frac{1}{c_r}\eta}\cos(\gamma_r-\varphi), &\quad \sin\gamma_r&=\frac{c_r}{\sqrt{1+c_r^2}},\\ v_y&=\sum_r A_r\sqrt{1+c_r^2}\,e^{c_r\xi+\frac{1}{c_r}\eta}\sin(\gamma_r-\varphi), &\quad \cot\gamma_r&=c_r. \end{aligned}\right\} \tag{32.9}$$

If in a flow $v_1 = v_2$, the flow direction at every point bisects the angle of the characteristics and has therefore a principal direction. An example of such a *diagonal flow* is obtained from (32.9) by puttng $c = -1$:

$$v_1 = v_2 = e^{-(\xi+\eta)}.\text{[130]} \tag{32.10}$$

We may combine it with the slip-line field (32.2). The differential equations of the streamlines being $dy/dx = \tan\vartheta$, $\vartheta = \varphi + 45°$, we easily find

$$\tan\varphi = \frac{cy - x}{cx + y}, \qquad \tan\vartheta = \frac{1+\tan\varphi}{1-\tan\varphi} = \frac{y(1+c) - x(1-c)}{x(1+c) + y(1-c)}. \tag{32.11}$$

These streamlines are again logarithmic spirals; they intersect the radius vector at the constant angle

$$\gamma = \operatorname{arc\,tan}\frac{1-c}{1+c}.$$

In Fig. 8 the full lines are the two families of slip lines, the dashed lines are the diagonal streamlines.

For $c = 1$, the slip-line spirals are symmetric to the radii, and the radii are diagonal streamlines (Fig. 9). This pattern is approximately realized if a punch

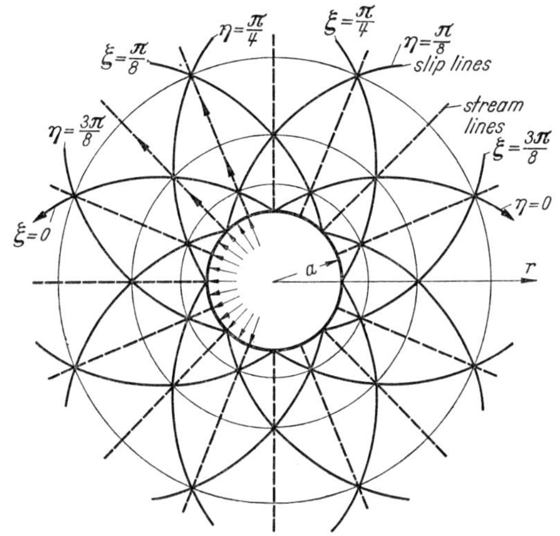

Fig. 9. Logarithmic spirals as slip-lines and radial streamlines for thick-walled tube under internal pressure.

of circular cross-section is forced into a rigid plastic material. It has been observed in wrought iron plates and in the cross-section of thick-walled tubes (see NADAI [*17*, p. 228]). Uniform normal pressure p acts along the circle of radius $r = a$. Hence at $r = a$, $\sigma_r = -p$, $\tau_{r\vartheta} = 0$. Thus, by the yield condition, $\sigma_\vartheta = -p + 2k$. The inner circle plays the role of the curve $\mathcal{K}$ of a Cauchy problem. The solution of the equations

$$\frac{d}{dr}(r\,\sigma_r) - \sigma_\vartheta = 0, \quad \sigma_\vartheta - \sigma_r = 2k,$$

[130] Cf. [*3*, pp. 65–69].

with the above boundary conditions, is due to PRANDTL:

$$\sigma_r = -p + 2k \log \frac{r}{a}, \qquad \sigma_\vartheta = -k + 2k\left(1 + \log \frac{r}{a}\right), \qquad \tau_{r\vartheta} = 0.$$

The characteristics are the two families of orthogonal spirals

$$\vartheta \pm \log \frac{r}{a} = \text{constant}.$$

The problem where along $r = a$ the more general stress distribution holds:

$$\sigma_r = -p, \qquad \tau_{r\vartheta} = -t, \qquad p \geqq 0$$

has been considered by SOKOLOVSKY and by MICHLIN.[131] It has likewise a simple explicit solution for the stresses. The characteristics are two families of spirals which approach logarithmic spirals as the distance from the circular hole increases.

A paper of R. HILL[132] called my attention to an interesting instance of diagonal flow. "In 1962 RICHMOND and DEVENPECK indicated a die profile that produces purely diagonal flow in a strip drawn through it, assuming perfect lubrication and no hardening. Calculations show that transverse lines marked on the strip would still be exactly transverse after drawing so that the final reduction in thickness is homogeneous." HILL reformulated the concept of diagonal flow, deriving some of its elementary properties, and applying them to the result of RICHMOND and DEVENPECK.

CHRISTIANOWITSCH, in a well-known and somewhat controversial paper[133] considered instead of the circular hole one of arbitrary continuous smooth boundary (cf. [*26*, pp. 184–192]) for the case of a normal pressure at the hole (see also pp. 192–195).

As another type of example consider a particular case of (32.3). But here in order to obtain positive curvature of the slip lines we take $a = 1$, $b = -1$, $\varphi^- = \xi - \eta - 45°$, $\varphi^+ = \xi - \eta + 45°$. Then:

$$\begin{aligned} &R^- = 4A\cos(\xi - \eta - 45°), \qquad R^+ = -4A\sin(\xi - \eta - 45°),\\ &x = 2A(\xi + \eta) - A\sin 2(\xi - \eta + 45°),\\ &y = A\cos 2(\xi - \eta + 45°). \end{aligned} \tag{32.12}$$

The slip lines form two families of orthogonal cycloids. PRANDTL, who first considered this slip-line field, found that it originates if a plastic mass is compressed between two rigid, rough plates. However, the streamlines which correspond to this problem cannot be found by guessing, as in the case of the thick-walled tube—a boundary problem must be solved (see Sects. 60, 61).

33. Discontinuities. We return to our discussion of discontinuities (Sects. 16, 22, 24). Consider Eqs. (29.7′). For the characteristic C^- the first of Eqs. (29.7′) is the compatibility equation, while the second one is the "remaining equation" (see Sects. 15, 16). Since in the "remaining equation" both variables ϑ and $p/2k$ are differentiated across C^-, none of these variables can jump. *The stresses must be continuous across a C^-. The same holds likewise for a C^+.* Consider, next, the velocities and Eqs. (30.15) or (29.11). For a C^- the first Eq. (30.15) is the compatibility equation and the second the "remaining equation". In this second equation v_2 is differentiated across the characteristic, but v_1 is not. Therefore *the tangential velocity of the C^- may jump, while the normal component must remain continuous*, and, of course, the same holds for the C^+.

[131] S. G. MICHLIN: The mathematical theory of plasticity. Publ. Acad. Sci. USSR. 1938.

[132] R. HILL: J. Mech. Phys. Solids **14**, 245 (1966). O. RICHMOND and M. L. DEVENPECK: Proc. 4th U.S. Natl. Congr. Appl. Mech. (1962), p. 1053.

[133] S. A. CHRISTIANOWITSCH: Mat. Sbornik **1**, 511–543 (1936).

Hence, stress discontinuities can arise only across non-characteristic lines We have seen, Sect. 24, that across such a line σ_n, τ, and $\boldsymbol{v}$ must be continuous and only σ_t may jump as in (24.5). Also, $\dot{E}=0$ at the discontinuity line.

We can now add some more information. Let us express the jump conditions in terms of $s=-p/k$ and ϑ. Denote by λ the angle of the normal to the discontinuity line with the first principal direction. Then Eqs. (29.6) hold with σ_n, σ_t, τ on the left, and λ (rather then ϑ) on the right-hand side, or, with

$$k=\frac{1}{2}(\sigma_1-\sigma_2), \qquad s=\frac{1}{2k}(\sigma_1+\sigma_2),$$

$$\sigma_n=ks+k\cos 2\lambda, \qquad \sigma_t=ks-k\cos 2\lambda, \qquad \tau=-k\sin 2\lambda. \tag{33.1}$$

Since σ_n and τ remain continuous, we obtain

$$\begin{aligned} s_2-s_1+\cos 2\lambda_2-\cos 2\lambda_1&=0,\\ \sin 2\lambda_2-\sin 2\lambda_1&=0, \end{aligned} \tag{33.2}$$

or

$$\lambda_1+\lambda_2=90^\circ, \qquad s_2=s_1+2\cos 2\lambda_1. \tag{33.3}$$

Or, with $\varphi^+=\vartheta+45^\circ$ and γ the angle of the normal and the x-axis:

$$\begin{aligned} \varphi_2^+ +\varphi_1^+&=2\gamma,\\ s_2-s_1&=2\sin 2(\varphi_1^+-\gamma). \end{aligned} \tag{33.3'}$$

At a point of the discontinuity line there are now four slip-line directions C_1^+, C_2^+ and C_1^-, C_2^- where $C_1^+\perp C_1^-$, $C_2^+\perp C_2^-$. The above equations show that the discontinuity line bisects both the angles of $C_1^-\,C_2^-$ and of $C_1^+\,C_2^+$ (Fig. 10).

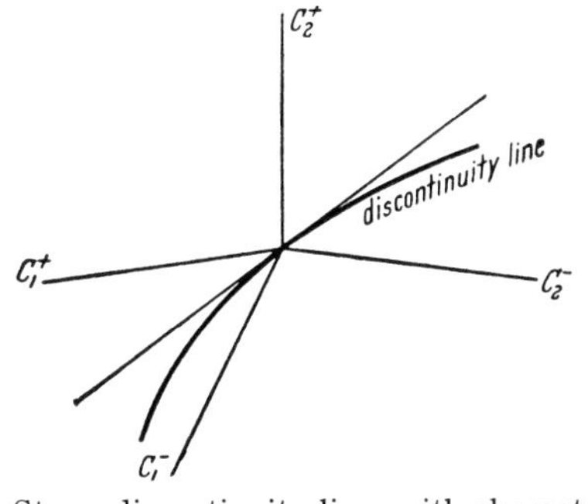

Fig. 10. Stress discontinuity lines with characteristics.

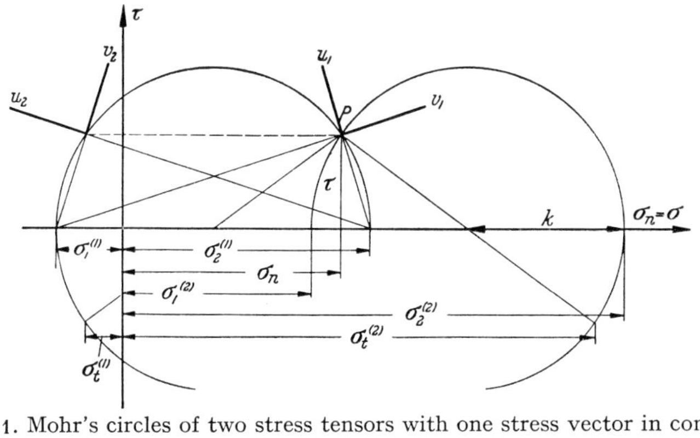

Fig. 11. Mohr's circles of two stress tensors with one stress vector in common.

Fig. 11 shows the Mohr circles of two stresses $\Sigma^{(1)}$, $\Sigma^{(2)}$ on both sides of the discontinuity. The σ_n and τ are the same for both tensors. The figure shows the directions u_1, v_1 and u_2, v_2 of two pairs of principal directions compatible with a common stress vector $\boldsymbol{t}_n$, which in both tensors corresponds to the direction $\boldsymbol{n}$, normal to the discontinuity line. From the figure we read, with $\sigma_n = \sigma$:

$$\begin{aligned} 4\sigma &= (\sigma_1^{(1)} + \sigma_2^{(1)}) + (\sigma_1^{(2)} + \sigma_2^{(2)}), \\ 2\sigma &= \sigma_2^{(1)} + \sigma_1^{(2)} = \sigma_1^{(1)} + \sigma_2^{(2)}; \end{aligned} \tag{33.4}$$

$$\begin{aligned} \sigma_1^{(1)} + \sigma_1^{(2)} &= 2\sigma - 2k, \\ \sigma_2^{(1)} + \sigma_2^{(2)} &= 2\sigma + 2k. \end{aligned} \tag{33.5}$$

It can be seen that the curvatures $1/R^+$, $1/R^-$ may change across a discontinuity line (see PRAGER,[134] HILL [*6*, p. 159]). Of the contributions concerned with stress discontinuities we mention a few in addition to those discussed in the preceding chapter. LEE[135] has given the velocity distribution for a problem with a discontinuity line. CARRIER and WINZER[136] have studied the interaction of discontinuities meeting at a point. HILL[137] has carried out the investigation of discontinuities for the problem of plane stress.

C. The general plane problem.

I. Basic theory.

a) The equations.

34. Linearization. In this chapter we shall consider the theory of the problem (27.4) which contains the problems of plane strain and of plane stress as particular cases.[138] Consider the first three Eqs. (27.4)

$$\frac{\partial \sigma_x}{\partial x} + \frac{\partial \tau}{\partial y} = 0, \qquad \frac{\partial \tau}{\partial x} + \frac{\partial \sigma_y}{\partial y} = 0, \tag{34.1}$$

$$f(\sigma_x, \sigma_y, \tau) = F(\sigma_1, \sigma_2) = 0. \tag{34.2}$$

They are nonlinear but can be transformed into a linear system in many ways. Following v. MISES[139] we use a parametric representation of the yield condition

$$F(\sigma_1, \sigma_2) = 0, \qquad \sigma_1 = \sigma_1(s), \qquad \sigma_2 = \sigma_2(s), \tag{34.3}$$

where s is an adequate parameter, e.g. $s = -p/k = (\sigma_1 + \sigma_2)/2k$ with k a constant. Denote by ϑ the angle between the x-axis and the first principal direction, which we denote as the u-direction, the second being the v-direction. We assume in the usual

[134] W. PRAGER: R. Courant Anniversary Volume, pp. 289–299. New York 1948.

[135] E. H. LEE: Proc. 3rd Symp. Appl. Math., pp. 213–228. New York 1950.

[136] A. WINZER and G. F. CARRIER: J. Appl. Mechanics, 213–264 (1948).

[137] R. HILL: J. Math. Phys. Solids **1**, 19–30 (1952). See also Appendix.

[138] Cf. H. GEIRINGER: Advances. We mention particularly the valuable monograph, H. MANDEL: Sur les équilibres par tranches parallèles des milieux plastiques à la limite d'écoulement. Paris: Louis Jean 1942. This article is concerned only with the stress distributions, not with velocities; hence in our terminology MANDEL does not consider the complete problem. MANDEL's starting point is the envelope of Mohr circles (see Sect. 37)—which determines a yield condition—and all points of contact with the circles are assumed to be real. This limits the study to hyperbolic states of stress. The same approach is that of SOKOLOVSKY [*26*, Chap. X] (based on a paper of 1948). Cf. also footnote 98.

[139] R. v. MISES: Reissner Anniversary Volume. Ann Arbor, Michigan 1949.

way $\sigma_1 \geqq \sigma_2$.[140] Using Eqs. (29.6), we compute

$$\frac{\partial \sigma_x}{\partial x} + \frac{\partial \tau}{\partial y}$$

as well as

$$\frac{\partial \tau}{\partial x} + \frac{\partial \sigma_y}{\partial y}.$$

With directional derivatives, as in Sect. 29, we obtain

$$\frac{\partial \sigma_1}{\partial u} \cos \vartheta + \frac{\partial \sigma_2}{\partial v} \sin \vartheta + (\sigma_2 - \sigma_1) \left(\frac{\partial \vartheta}{\partial u} \sin \vartheta - \frac{\partial \vartheta}{\partial v} \cos \vartheta \right) = 0,$$

$$\frac{\partial \sigma_1}{\partial u} \sin \vartheta + \frac{\partial \sigma_2}{\partial v} \cos \vartheta - (\sigma_2 - \sigma_1) \left(\frac{\partial \vartheta}{\partial u} \cos \vartheta + \frac{\partial \vartheta}{\partial v} \sin \vartheta \right) = 0.$$

If we multiply the first of these equations by $\cos \vartheta$, the second by $\sin \vartheta$, and add, we obtain Jenne's equations (see Sect. 5), a particular case of (8.1):

$$\frac{\partial \sigma_1}{\partial u} = (\sigma_2 - \sigma_1) \frac{\partial \vartheta}{\partial v}, \qquad \frac{\partial \sigma_2}{\partial v} = (\sigma_2 - \sigma_1) \frac{\partial \vartheta}{\partial u}. \tag{34.4}$$

Using (34.3) and putting

$$\frac{d\sigma_1}{ds} = \sigma_1', \qquad \frac{d\sigma_2}{ds} = \sigma_2', \qquad \frac{\sigma_2 - \sigma_1}{\sigma_1'} = g(s), \qquad \frac{\sigma_2 - \sigma_1}{\sigma_2'} = h(s), \tag{34.5}$$

we obtain the "reducible" equations[141]

$$\frac{\partial s}{\partial u} = g(s) \frac{\partial \vartheta}{\partial v}, \qquad \frac{\partial s}{\partial v} = h(s) \frac{\partial \vartheta}{\partial u}, \tag{34.6}$$

which generalize the equations preceding (29.7).

Together with (34.3) these determine σ_1, σ_2 and ϑ, hence Σ. The plane where s and ϑ are polar coordinates has been termed by v. MISES the stress graph (see end of Sect. 29).

The dependent and independent variables may now be interchanged if

$$d = \frac{\partial (s, \vartheta)}{\partial (u, v)} \neq 0.$$

With

$$\frac{\partial s}{\partial u} = d \frac{\partial v}{\partial \vartheta}, \qquad \frac{\partial s}{\partial v} = -d \frac{\partial u}{\partial \vartheta}, \qquad \frac{\partial \vartheta}{\partial u} = -d \frac{\partial v}{\partial s}, \qquad \frac{\partial \vartheta}{\partial v} = d \frac{\partial u}{\partial s}, \tag{34.7}$$

we find v. MISES' linear equations

$$\frac{\partial v}{\partial \vartheta} = g(s) \frac{\partial u}{\partial s}, \qquad \frac{\partial u}{\partial \vartheta} = h(s) \frac{\partial v}{\partial s}. \tag{34.8}$$

Note that in these equations u and v are not coordinates.

The basic equations of compressible fluid flow

$$\frac{\partial q}{\partial s} = \frac{q}{M^2 - 1} \frac{\partial \vartheta}{\partial n}, \qquad \frac{\partial q}{\partial n} = q \frac{\partial \vartheta}{\partial s},$$

[140] In Advances, I assumed $\sigma_1 \leqq \sigma_2$, which simplifies some formulas in the present problem; it is, however, unusual, and with this assumption the formulas would not directly reduce to those of Chap. B.

[141] By that we mean equations, linear and homogeneous in the first order derivatives with coefficients which are functions of the dependent variables only.

with q and ϑ the polar coordinates of the velocity vector, and M the Mach number, are of the same mathematical form as Eqs. (34.6); if φ and ψ denote potential and stream functions, the basic linearized equations for these functions,

$$\frac{\partial\varphi}{\partial\vartheta}=\frac{q}{\varrho}\frac{\partial\psi}{\partial q}, \qquad \frac{\partial\psi}{\partial\vartheta}=-\frac{\varrho q}{1-M^2}\frac{\partial\varphi}{\partial q},$$

are similar to (34.8).[142] The stress graph corresponds to the hodograph, our s and ϑ to q and ϑ, the yield condition to the adiabatic condition, principal stress trajectories to stream lines and potential lines, and characteristics to characteristics. Mathematical insight and heuristic ideas regarding our subject can be gained from the more developed field of gas dynamics.[143]

We may now derive in various ways equations for quantities in the physical plane. Consider, for example the "moving coordinates", end of Sect. 30.[144]

$$X=x\cos\vartheta+y\sin\vartheta, \qquad Y=y\cos\vartheta-x\sin\vartheta, \tag{34.9}$$

and, according to the definition of u- and v-directions,

$$du=dx\cos\vartheta+dy\sin\vartheta, \qquad dv=dy\cos\vartheta-dx\sin\vartheta. \tag{34.9'}$$

Hence from (34.9) and (34.9')

$$dX=du+Y\,d\vartheta, \qquad dY=dv-X\,d\vartheta$$

whence

$$\frac{\partial X}{\partial\vartheta}=\frac{\partial u}{\partial\vartheta}+Y, \quad \frac{\partial X}{\partial s}=\frac{\partial u}{\partial s}, \quad \frac{\partial Y}{\partial\vartheta}=\frac{\partial v}{\partial\vartheta}-X, \quad \frac{\partial Y}{\partial s}=\frac{\partial v}{\partial s}. \tag{34.9''}$$

Then, using (34.8) we obtain

$$\frac{\partial X}{\partial\vartheta}=h(s)\frac{\partial Y}{\partial s}+Y, \qquad \frac{\partial Y}{\partial\vartheta}=g(s)\frac{\partial X}{\partial s}-X \tag{34.10}$$

and

$$\begin{aligned}\frac{\partial^2 X}{\partial\vartheta^2}-hg\frac{\partial^2 X}{\partial s^2}&=-X+\frac{\partial X}{\partial s}(g'h+g-h),\\ \frac{\partial^2 Y}{\partial\vartheta^2}-hg\frac{\partial^2 Y}{\partial s^2}&=-Y+\frac{\partial Y}{\partial s}(h'g+g-h).\end{aligned} \tag{34.11}$$

We see that the problem is hyperbolic, elliptic, or parabolic according as

$$hg>0, \quad hg<0, \quad hg=0 \text{ or } \infty. \tag{34.12}$$

We note that $hg=(\sigma_2-\sigma_1)^2/\sigma_1'\sigma_2'=(4-s^2)^2/(3-s^2)$ if $s=-p/k$. In plane strain, $\sigma_1=k(s+1)$, $\sigma_2=k(s-1)$ is a parametric presentation of $(\sigma_1-\sigma_2)^2-4k^2=0$. Hence $g=h=-2$, $gh=4$; in this case (34.11) reduces to $\partial^2X/\partial\vartheta^2-4\partial^2X/\partial s^2+X=0$, which again is the telegraph equation.[145] Again, solutions may be superposed.

[142] If a scale factor is given, M and ϱ depend on q only; in (34.6), (34.8), $g(s)$ and $h(s)$ depend on s only.

[143] Essentially, the same analogy has been applied by NOBUO INOUE to some exact solutions in plane strain and soil mechanics. NOBUO INOUE: J. Phys. Soc. Japan 7, 518–523, 604–609, 610–618 (1952), where further literature is cited. Cf. also R. HILL: J. Mech. Phys. of Solids 2, 110–116 (1954), who discusses and in some instances develops INOUE's gas dynamical analogy.

[144] Cf. R. v. MISES: Reissner Anniversary Volume. Ann Arbor, Michigan 1949.

[145] Other essentially equivalent equations have been given in GEIRINGER: Advances, pp. 224–227.

Another linear equation, valid also in the case of non-isotropy is due to NEUBER[146] and to SAUER.[147] The first Eq. (34.1) is satisfied by means of a stress potential φ:

$$\sigma_x = \frac{\partial \varphi}{\partial y}, \qquad \tau = -\frac{\partial \varphi}{\partial x}.$$

Assume that the yield condition can be solved with respect to σ_y, say:

$$\sigma_y = k(\sigma_x, \tau), \qquad \frac{\partial k}{\partial \sigma_x} = -k_1, \qquad \frac{\partial k}{\partial \tau} = k_2. \tag{34.13}$$

Then the second Eq. (34.1) gives

$$\frac{\partial^2 \varphi}{\partial x^2} + k_2 \frac{\partial^2 \varphi}{\partial x\, \partial y} + k_1 \frac{\partial^2 \varphi}{\partial y^2} = 0, \tag{34.14}$$

a non-linear equation whose coefficients depend on $\partial\varphi/\partial x$ and $\partial\varphi/\partial y$. This transforms into a linear equation by means of the Legendre transformation

$$\begin{gathered} \Phi = -\tau x + \sigma_x y - \varphi, \qquad d\Phi = -x\, d\tau + y\, d\sigma_x - d\varphi, \\ x = -\frac{\partial \Phi}{\partial \tau}, \qquad y = \frac{\partial \Phi}{\partial \sigma_x}, \end{gathered} \tag{34.15}$$

and, if

$$\frac{\partial(\sigma_x, \tau)}{\partial(x, y)} \neq 0,$$

we obtain the linear equation

$$\frac{\partial^2 \Phi}{\partial \sigma_x^2} + k_2 \frac{\partial^2 \Phi}{\partial \sigma_x\, \partial \tau} + k_1 \frac{\partial^2 \Phi}{\partial \tau^2} = 0. \tag{34.16}$$

On account of the linearity of (34.16) a new $\phi_3 = c_1 \phi_1 + c_2 \phi_2$ follows from two solutions ϕ_1 and ϕ_2 and by (34.15) also $x_3 = c_1 x_1 + c_2 x_2$, $y_3 = c_1 y_1 + c_2 y_2$. Hence a new field follows from two given ones by division of the segment $P_1(x_1, y_1)$, $P_2(x_2, y_2)$ in a fixed proportion.

In the case of isotropy (symmetry), Eqs. (34.11) seem preferable to (34.16), since the coefficients depend on one variable only.[148] Eqs. (34.11) simplify by the use of the new independent variable μ, or λ, defined by

$$\frac{d\mu}{ds} = (hg)^{-\frac{1}{2}} \quad \text{or} \quad \frac{d\lambda}{ds} = (-hg)^{-\frac{1}{2}}$$

in the hyperbolic and elliptic cases, respectively. A more detailed study of these equations depends on the particular yield condition.

35. Various yield conditions. α) Assume *v. Mises' quadratic condition* (Sect. 1) and $\sigma_3 = 0$. The yield condition is the equation of an ellipse in the σ_1, σ_2-plane, with minor axis $2k\sqrt{2}/\sqrt{3}$, major axis $2k\sqrt{2}$. The principal axes of the ellipse bisect the angles of the 1, 2-directions. We use the following parametric representation

---

146 H. NEUBER: Z. angew. Math. Mech. **28**, 253–257 (1948).

147 R. SAUER: Z. angew. Math. Mech. **29**, 274–279 (1949).

148 A different derivation is in GEIRINGER: Advances, p. 224.

and formulas [g and h defined in (34.5)]:

$$\begin{aligned}
&\sigma_1^2+\sigma_2^2-\sigma_1\sigma_2-4k^2=0,\\
&s=\frac{1}{2k}(\sigma_1+\sigma_2), \qquad (-2\leqq s\leqq 2),\\
&\sigma_1=k\left(s+\sqrt{\frac{4-s^2}{3}}\right), \qquad \sigma_2=k\left(s-\sqrt{\frac{4-s^2}{3}}\right),\\
&\sigma_1'=k\left(1-\frac{s}{\sqrt{12-3s^2}}\right), \qquad \sigma_2'=k\left(1+\frac{s}{\sqrt{12-3s^2}}\right),\\
&g=\frac{-2(4-s^2)}{\sqrt{12-3s^2}-s}, \qquad h=\frac{-2(4-s^2)}{\sqrt{12-3s^2}+s},\\
&gh=\frac{(\sigma_1-\sigma_2)^2}{\sigma_1'\sigma_2'}=\frac{(4-s^2)^2}{3-s^2}.
\end{aligned} \tag{35.1}$$

Another useful parametric presentation is due to NADAI. Let an angle δ be defined by

$$\sin\delta=\frac{s}{2}, \qquad \tan\delta=\frac{s}{\sqrt{4-s^2}}, \qquad -90°\leqq\delta\leqq 90°. \tag{35.2}$$

Then, we see that

$$\sigma_1=\frac{4k}{\sqrt{3}}\sin(\delta+30°), \qquad \sigma_2=\frac{4k}{\sqrt{3}}\sin(\delta-30°). \tag{35.3}$$

(Practically the same presentation has been used by SOKOLOVSKY [*26*, p. 268 seq.]; his parameter is $w=90°-\delta$.) Using (34.2) and the abbreviation $F_i=\partial F/\partial\sigma_i$, $i=1, 2$, and anticipating the formula of the next section, $(F_1+F_2)/(F_1-F_2)=-\cos 2\alpha$ [see Eq. (36.3) etc.], we see that in the hyperbolic part

$$\tan\delta=-\sqrt{3}\cos 2\alpha. \tag{35.4}$$

It is easy to express in terms of δ the various statements which we shall make in terms of s.

β) *Tresca's condition*, with $\sigma_3=0$, leads to the intersection of the spatial hexagonal prism (Sect. 4) with the σ_1, σ_2-plane.[149] Hence, with the same s:

$$\begin{aligned}
\sigma_1-\sigma_2&=2k, &&\text{if } \sigma_1\sigma_2\leqq 0,\\
&=4k-|\sigma_1+\sigma_2|, &&\text{if } \sigma_1\sigma_2\geqq 0,\\
\sigma_1=k(s+1),\quad \sigma_2&=k(s-1),\quad g=h=-2,\quad gh=4, &&|s|\leqq 1\\
\sigma_1=k[2+s-|s|],\quad \sigma_2&=k[-2+s+|s|], &&1\leqq|s|\leqq 2.
\end{aligned} \tag{35.5}$$

The first and the third line give the 45° line (to the right) of the hexagon. The second and fourth line are the equations of the vertical and the horizontal line of the hexagon. The 45°-lines correspond to a hyperbolic, the others to a parabolic problem.[150]

γ) The quadratic limit (35.1) is hyperbolic for values $-\sqrt{3}<s<+\sqrt{3}$, while for $|s|$ between $\pm\sqrt{3}$ and ± 2 the problem is elliptic. To avoid this considerable difficulty, v. MISES proposed a comparatively small adjustment of his yield condition. He introduced a *parabola limit* for which the problem is hyperbolic throughout. In the σ_1, σ_2-plane this limit is represented by two branches of parabolas

[149] A "generalized Tresca condition" was introduced by P. G. HODGE Jr.: J. Math. Phys. **29**, 38–48 (1950).

[150] For both v. Mises' and Tresca's yield condition for plane stress, SOKOLOVSKY was first to investigate the stress problem: V. V. SOKOLOVSKY: C. R. Acad. Sci. USSR. **51**, 175–178 (1946) and same journal, **51**, 421–424 (1946), cf. [*26*, p. 272 seq.].

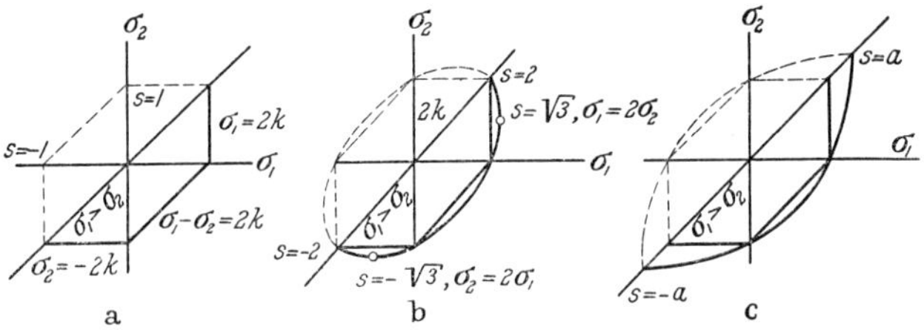

Fig. 12a–c. Yield conditions in plane stress: (a) hexagonal condition; (b) quadratic condition; (c) parabola condition.

passing through four corners of the Tresca hexagon. Their equation is, with $1+\sqrt{2}=a$, where $a^2-1=2a$

$$\frac{\sigma_1-\sigma_2}{k}=\pm\frac{1}{a}\left[a^2-\left(\frac{\sigma_1+\sigma_2}{2k}\right)^2\right],\quad \text{upper sign for right branch,}$$

$$\sigma_1=ks+\frac{k}{2a}(a^2-s^2),\qquad \sigma_2=ks-\frac{k}{2a}(a^2-s^2),\qquad |s|\leqq a,\qquad g=-(a+s),\qquad h=-(a-s). \tag{35.6}$$

The three conditions are shown in Fig. 12.[151]

δ) Coulomb's condition used in soil mechanics (see Ref. 21), written in the above variables, is as follows:

$$\pm\frac{\sigma_2-\sigma_1}{2}=c\cos\Phi+\frac{\sigma_1+\sigma_2}{2}\sin\Phi,\qquad s\geqq-\cot\Phi,$$
$$\sigma_1=c(s+s\sin\Phi+\cos\Phi),\qquad \sigma_2=c(s-s\sin\Phi-\cos\Phi), \tag{35.7}$$
$$\sigma_1'=c(1+\sin\Phi),\qquad \sigma_2'=c(1-\sin\Phi),\qquad \sigma_1'\sigma_2'=c^2\cos^2\Phi.$$

Here c is the cohesion and Φ, a constant, the angle of internal friction (see Fig. 13a). The problem is hyperbolic.[152]

Better known is the geometric interpretation of the Coulomb condition in a σ, τ-coordinate system. It is assumed that the envelope of the Mohr circles consists of two straight lines symmetric to the σ-axis, making the angle $+\phi$, $-\phi$ with this axis and intersecting the τ-axis at $\tau=c$. (In soil mechanics, which is the main field of application of this condition, compression is taken positive, in contrast to the usage in mechanics of metals.) The equation of the condition is then:

$$\pm\tau=c+\sigma\tan\phi. \tag{35.8}$$

[151] C. CARATHÉODORY and E. SCHMIDT: Z. angew. Math. Mech. **3**, 468–475 (1923), discussed the possibility of isometric nets of principal stress trajectories of the plane strain yield condition, $\sigma_1-\sigma_2=$ constant. P. F. NEMÉNYI and A. VAN TUYL: Quart. J. Mech. Appl. Math. **5**, 1–11 (1952), considered a general yield condition and showed that only for three distinct families of yield conditions do there exist isometric nets (beyond the trivial nets consisting of concentric circles and radial straight lines). Among these, one is the "parabola condition".

[152] C. A. COULOMB: Mém. Sav. Etr. **7**, 343 (1773); O. MOHR: Z. Arch. u. Ing. Verein **17**, 344 (1871); **18**, 67, 245 (1872).

Some more recent papers on the subject are: D. C. DRUCKER and W. PRAGER: Quart. J. Appl. Math. **10**, 157–165 (1952). — D. C. DRUCKER: J. Mech. Phys. Solids **1**, 217–226 (1953). — R. T. SHIELD: Quart. Appl. Math. **11**, 61–75 (1953). — R. T. SHIELD: J. Math. Phys. **33**, 144–156 (1954). — R. T. SHIELD: J. Mech. Phys. Solids **4**, 10 (1955). — R. M. HAYTHORNTHWAITE: Prager Anniversary Volume, p. 235. New York 1963. — A very interesting paper on the kinematics of soil mechanics is by A. J. M. SPENCER: J. Mech. Phys. Solids **12**, 337 (1964). SPENCER is particularly interested in the velocity field. Regarding the somehow related problem of the mechanics of snow: see H. ZIEGLER, Z. angew. Math. Phys. **14**, 113 (1963).

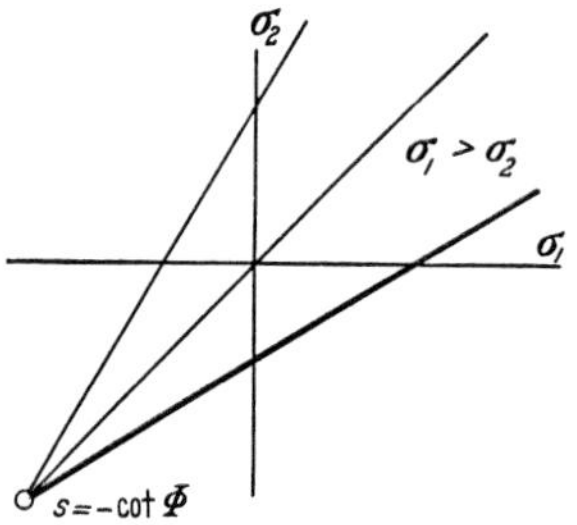

Fig. 13a. Coulomb's yield condition.

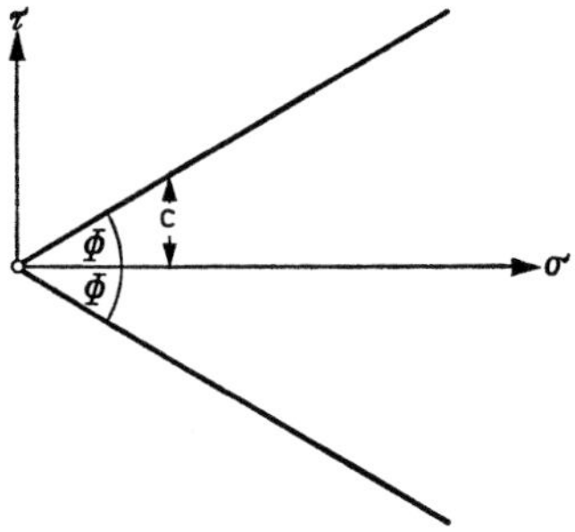

Fig. 13b. Coulomb's yield condition.

The basis of the use of the Coulomb condition in soil mechanics is as follows: by analogy with the properties of Coulomb friction between separate bodies it is assumed that the soil will yield if on any plane the magnitude of the shearing stress τ acting on that plane reaches or exceeds $c+\sigma\tan\phi$, where σ is the normal stress on the plane.

More in Sect. 38.

b) Characteristics of the complete plane problem.

36. Characteristic directions and compatibility relations. The first three Eqs. (27.4) have been replaced by Eqs. (34.6). Hence the complete problem is equivalent to Eqs. (34.6) and those in the last line of (27.4), which we write in the form

$$\frac{\dot{\varepsilon}_x-\dot{\varepsilon}_y}{\dot{\gamma}}=\frac{\dfrac{\partial f}{\partial\sigma_x}-\dfrac{\partial f}{\partial\sigma_y}}{\dfrac{\partial f}{\partial\tau}},\qquad \frac{\dot{\varepsilon}_x+\dot{\varepsilon}_y}{\dot{\gamma}}=\frac{\dfrac{\partial f}{\partial\sigma_x}+\dfrac{\partial f}{\partial\sigma_y}}{\dfrac{\partial f}{\partial\tau}}. \tag{36.1}$$

This is a system of four equations with the dependent variable s, ϑ, v_x, v_y and two independent variables x, y. The first two equations do not contain velocities, the last two contain no derivatives of the stresses. It is readily seen[153] that the characteristic determinant of order four resolves into the product of two determinants of order two which define separately the characteristics of the stress problem and those of the velocity problem.

Eqs. (34.6) relate to the principal directions as axes. Denote by α the angle of a characteristic direction with the first principal direction (the u-direction), The characteristic determinant equated to zero is:

$$\begin{vmatrix}\sin\alpha & g\cos\alpha\\ \cos\alpha & h\sin\alpha\end{vmatrix}=0. \tag{36.2}$$

Hence, using (34.5) and writing with the notation of (27.4), and in analogy to (3.1), Grad $f=\mu\dot{E}$, we obtain, with $F_i=\partial F/\partial\sigma_i$

$$\begin{aligned}\tan^2\alpha&=\frac{g}{h}=\frac{\sigma_2'}{\sigma_1'}=-\frac{F_1}{F_2}=-\frac{\dot{\varepsilon}_1}{\dot{\varepsilon}_2},\\ \tan\alpha&=\frac{dv}{du}=\pm\sqrt{\frac{d\sigma_2}{d\sigma_1}}.\end{aligned} \tag{36.3}$$

[153] H. GEIRINGER: Advances, p. 227.

We see again that hyperbolic points are those where g and h or σ_1' and σ_2' have the same sign and where $\dot{\varepsilon}_1, \dot{\varepsilon}_2$ or F_1, F_2 have opposite signs. The characteristics make angles $\vartheta \pm \alpha$ with the x-axis.

In the first Eq. (36.1) the right side equals $\cot 2\vartheta$ since Grad f and Σ have the same axes and Eqs. (29.4) hold for any symmetric tensor. In the second Eq. (36.1) the right side equals $(F_1+F_2)/(F_1-F_2) \sin 2\vartheta = -\cos 2\alpha/\sin 2\vartheta$, and the velocity equations become, after a simple computation,

$$\frac{\dot{\varepsilon}_x - \dot{\varepsilon}_y}{\dot{\gamma}} = \cot 2\vartheta, \qquad \frac{\dot{\varepsilon}_x - \dot{\varepsilon}_y}{\dot{\gamma}} = -\frac{\cos 2\alpha}{\sin 2\vartheta}, \tag{36.1'}$$

two linear differential equations for v_x, v_y. Let $m = dy/dx$ denote the slope of a velocity characteristic. The characteristic determinant equated to zero gives:

$$\begin{gathered} m^2(\cos 2\alpha + \cos 2\vartheta) - 2m \sin 2\vartheta + (\cos 2\alpha - \cos 2\vartheta) = 0, \\ m = \frac{dy}{dx} = \frac{\sin 2\vartheta \pm \sin 2\alpha}{\cos 2\vartheta + \cos 2\alpha} = \tan(\vartheta \pm \alpha). \end{gathered} \tag{36.4}$$

Hence, also in this general problem *the directions of stress characteristics and of velocity characteristics coincide under the theory of the plastic potential.*

Consider finally the directions for which the normal strain vanishes. If such a direction makes the angle ψ with the first principal direction, then the corresponding strain rate is

$$\dot{\varepsilon} = \dot{\varepsilon}_1 \cos^2\psi + \dot{\varepsilon}_2 \sin^2\psi = 0, \quad \text{or} \quad \tan^2\psi = -\dot{\varepsilon}_1/\dot{\varepsilon}_2 = \tan^2\alpha.$$

Hence, we have the result (Fig. 14a):

At a hyperbolic point, i.e. at a point where $F_1 F_2 < 0$, there are two real characteristics C^+, C^-. These, each being of multiplicity two, represent the four characteristics of the system (34.6), (36.1). *They form the angles $\vartheta + \alpha$, $\vartheta - \alpha$ with the x-axis, where α is given by* (36.3). *The normal strain in a characteristic direction vanishes. The directions of maximum shear stress, which coincide with the directions of maximum shear strain and bisect the angle of the principal directions, are in general not characteristic. This happens only if $F_1 + F_2 = 0$, $\dot{\varepsilon}_1 + \dot{\varepsilon}_2 = 0$ as in (generalized) plane strain.*

The stress equations have been studied by many authors, the complete problem much less frequently.[154]

To derive one pair of *compatibility equations* we multiply the first Eq. (34.6) by $\sin\alpha$ the second by $\sin^2\alpha/\cos\alpha = g\cos\alpha/h$, and add. The right side then equals

$$g\,\frac{\partial\vartheta}{\partial l^+},$$

the left side is

$$\frac{\partial s}{\partial u}\sin\alpha + \frac{\partial s}{\partial v}\,\frac{\sin^2\alpha}{\cos\alpha} = \tan\alpha\left(\frac{\partial s}{\partial u}\cos\alpha + \frac{\partial s}{\partial v}\sin\alpha\right) = \tan\alpha\,\frac{\partial s}{\partial l^+},$$

so that we obtain

$$\frac{\partial s}{\partial l^+} = g\cot\alpha\,\frac{\partial\vartheta}{\partial l^+},$$

and

$$\frac{\partial s}{\partial l^+} = h\tan\alpha\,\frac{\partial\vartheta}{\partial l^+}.$$

If, as before,

[154] SOKOLOVSKY, v. MISES, NEUBER, SAUER, GEIRINGER, HODGE, and HILL studied the stress problem. HODGE [7, p. 325], GEIRINGER [3] and various papers, and HILL [6, p. 305] dealt with the complete problem.

$$\tan\alpha = +\sqrt{\frac{g}{h}} = +\sqrt{\frac{\sigma_2'}{\sigma_1'}}$$

then

$$h\tan\alpha = \frac{\sigma_2-\sigma_1}{\sigma_2'}\left(+\sqrt{\frac{\sigma_2'}{\sigma_1'}}\right) = \frac{\sigma_2-\sigma_1}{+\sqrt{\sigma_1'\sigma_2'}} < 0 \quad \text{if } \sigma_1 > \sigma_2 .$$

Hence, $g\cot\alpha = h\tan\alpha = -\sqrt{gh}$, if $\sigma_1 > \sigma_2$, and[155]

$$\frac{\partial s}{\partial l^+} = -\sqrt{gh}\,\frac{\partial\vartheta}{\partial l^+}, \qquad \frac{\partial s}{\partial l^-} = +\sqrt{gh}\,\frac{\partial\vartheta}{\partial l^-} \tag{36.5}$$

are the compatibility relations which generalize Eqs. (29.7).[156] The C^-, C^+ which make the respective angles $\varphi^- = \vartheta - \alpha$, $\varphi^+ = \vartheta + \alpha$ with the x-axis form a net which is, in general, not orthogonal.

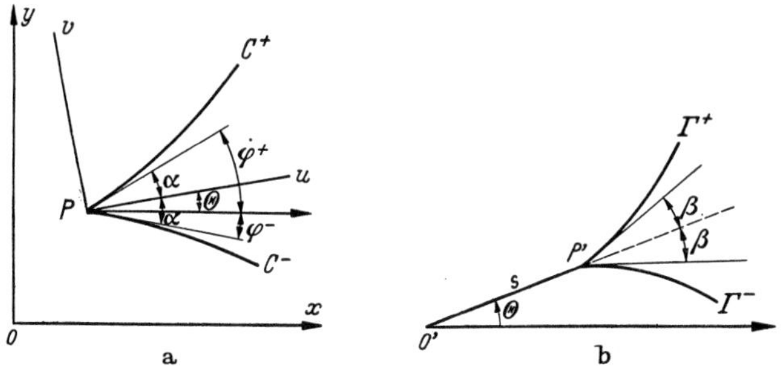

Fig. 14a and b. Characteristics in physical plane and stress graph.

The C^-, C^+ are the images (Fig. 14) of the fixed charcteristics Γ^-, Γ^+ in the stress graph which have the equations (upper sign for Γ^+):

$$\frac{d\vartheta}{ds} = \mp\frac{1}{\sqrt{gh}} \tag{36.6}$$

and make equal angles β with the radius vector in the stress graph, where s and ϑ are polar coordinates and $\tan\beta = s\,d\vartheta/ds$. These Γ-curves can be found once

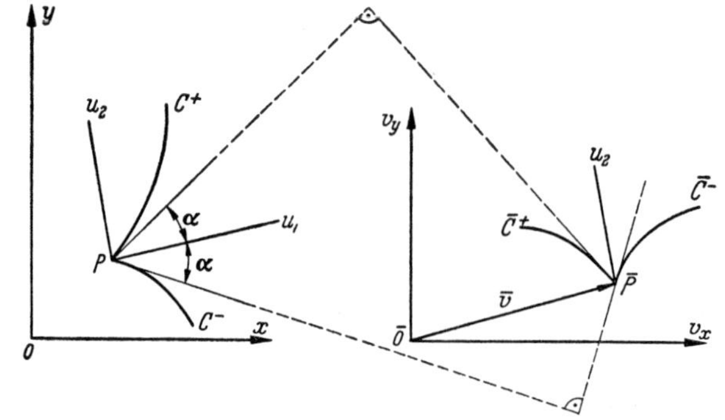

Fig. 15. Physical plane and velocity plane.

[155] This involved derivation was made to avoid arbitrariness of signs of roots, which might lead to materially wrong formulas.

[156] Note that in "plane strain", from (35.5), $g = h = -2$, $+\sqrt{gh} = 2$; then Eqs. (36.5) agree with (29.7), since $p/k = -s$.

and for all for specific yield conditions. With

$$S(s) = -\int^{s} \frac{ds}{\sqrt{gh}} = \int^{s} \frac{\sqrt{\sigma_1' \sigma_2'}}{\sigma_2 - \sigma_1} \, ds, \tag{36.7}$$

we obtain for Γ^-, Γ^+, respectively,

$$S(s) + \vartheta = \text{constant}, \qquad S(s) - \vartheta = \text{constant}, \tag{36.8}$$

which generalize Eqs. (29.8). In plane strain

$$S = -\frac{s}{2} = \frac{p}{2k}.$$

There is a *second pair of compatibility conditions* since the characteristics are of multiplicity two. The fact that along a C^- and a C^+ the respective rate of extension is zero gives, exactly as in (29.12),

$$\frac{\partial v_x}{\partial l^-} \cos\varphi^- + \frac{\partial v_y}{\partial l^-} \sin\varphi^- = 0, \qquad \frac{\partial v_x}{\partial l^+} \cos\varphi^+ + \frac{\partial v_y}{\partial l^+} \sin\varphi^+ = 0. \tag{36.9}$$

The interpretation in terms of the *velocity plane* (Fig. 15) is the same as in Sect. 29 (cf. also footnote 108). (In the figure, u_1, u_2 are used instead of u, v in order to avoid confusion which velocity components.) In the velocity plane there are two families of characteristics $\bar{C}^-$, $\bar{C}^+$ the images of the C^-, C^+ in the physical plane and $\bar{C}^+ \perp C^+$, $\bar{C}^- \perp C^-$ the C^+, C^- make equal angles with the second principal direction.

Next, we want equations analogous to (29.11). Denote by v_1, v_2, v_3, v_4 components of $\boldsymbol{v}$ in the direction of C^-, in the direction of C^+, in the direction normal to C^-, and in the direction normal to C^+ (Fig. 16a). Equivalently,

$$v_1 = v_x \cos\varphi^- + v_y \sin\varphi^-, \qquad v_3 = v_y \cos\varphi^- - v_x \sin\varphi^-, \qquad \text{etc.}$$

By straigntforward computation, analogous to that in Sect. 29, we obtain

$$\frac{\partial v_1}{\partial l^-} = \left(\frac{\partial v_x}{\partial l^-} \cos\varphi^- + \frac{\partial v_y}{\partial l^-} \sin\varphi^-\right) + (v_y \cos\varphi^- - v_x \sin\varphi^-) \frac{\partial \varphi^-}{\partial l^-},$$

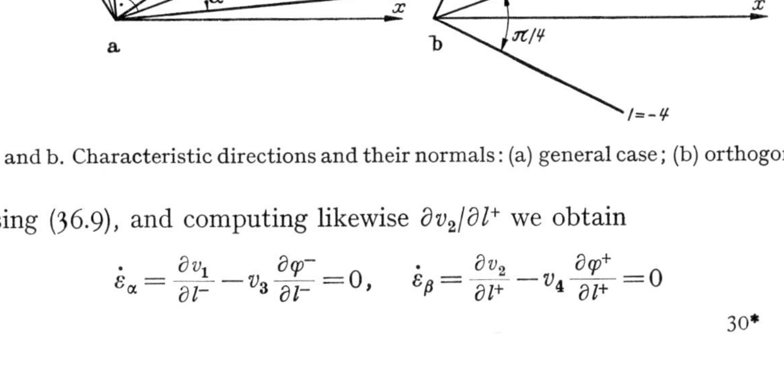

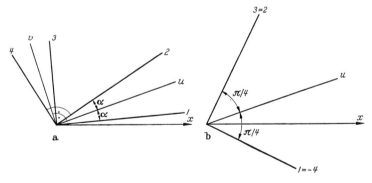

Fig. 16a and b. Characteristic directions and their normals: (a) general case; (b) orthogonal case.

and, using (36.9), and computing likewise $\partial v_2/\partial l^+$ we obtain

$$\dot{\varepsilon}_\alpha = \frac{\partial v_1}{\partial l^-} - v_3 \frac{\partial \varphi^-}{\partial l^-} = 0, \qquad \dot{\varepsilon}_\beta = \frac{\partial v_2}{\partial l^+} - v_4 \frac{\partial \varphi^+}{\partial l^+} = 0 \tag{36.10}$$

or, briefly,

$$dv_1 - v_3\, d\varphi^- = 0 \quad \text{along a } C^-, \qquad dv_2 - v_4\, d\varphi^+ = 0 \quad \text{along a } C^+. \tag{36.10'}$$

In plane strain, Fig. 16b, $v_2 = v_3$, $v_4 = -v_1$, $d\varphi^- = d\varphi^+ = d\vartheta$ and we recover (29.11'). Finally, we express v_3 and v_4 in terms of v_1, v_2, the components in the characteristic directions, and obtain

$$\frac{\partial v_1}{\partial \varphi^-} = \frac{v_2 - v_1 \cos 2\alpha}{\sin 2\alpha}, \qquad \frac{\partial v_2}{\partial \varphi^+} = \frac{v_2 \cos 2\alpha - v_1}{\sin 2\alpha}, \tag{36.11}$$

along a C^- and C^+, respectively. Or, introducing R^-, R^+, the radii of curvature of C^-, C^+ respectively, we obtain

$$\frac{\partial v_2}{\partial l^+} = \frac{v_2 \cos 2\alpha - v_1}{R^+ \sin 2\alpha}, \qquad \frac{\partial v_1}{\partial l^-} = \frac{v_2 - v_1 \cos 2\alpha}{R^- \sin 2\alpha}. \tag{36.11'}$$

37. Continuation. Relation to O. Mohr's theory. Differential equations in characteristic coordinates. In a σ, τ-coordinate system (see Sect. 2) the equation of a Mohr circle is (Fig. 17b)

$$\left(\sigma - \frac{\sigma_1 + \sigma_2}{2}\right)^2 + \tau^2 = \left(\frac{\sigma_1 - \sigma_2}{2}\right)^2. \tag{37.1}$$

Here $OA = \sigma$, $AP = \tau$.[157] From the figure we find:

$$\begin{aligned} \sigma &= \frac{\sigma_1 + \sigma_2}{2} - \frac{\sigma_1 - \sigma_2}{2} \cos 2\varepsilon = \sigma_1 \sin^2 \varepsilon + \sigma_2 \cos^2 \varepsilon, \\ \tau &= \frac{\sigma_1 - \sigma_2}{2} \sin 2\varepsilon = (\sigma_1 - \sigma_2) \sin \varepsilon \cos \varepsilon. \end{aligned} \tag{37.2}$$

From Eqs. (37.2) or from Fig. 17b we obtain

$$\frac{\sigma_1 + \sigma_2}{2} = \sigma + \tau \cot 2\varepsilon, \qquad \frac{\sigma_1 - \sigma_2}{2} = \frac{\tau}{\sin 2\varepsilon}. \tag{37.3}$$

We now consider the one-dimensional family of Mohr circles, whose σ_1, σ_2 satisfy a given equation $F(\sigma_1, \sigma_2) = 0$, which we identify with the yield condition of a plasticity problem. These circles have an envelope whose point of contact

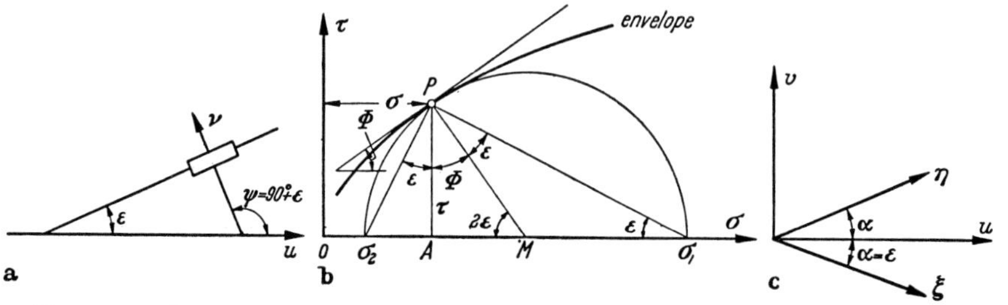

Fig. 17a–c. Characteristic directions defined by means of envelope of Mohr circles.

with any circle may be real or imaginary. Consider a point P of real contact with the envelope. To each value of the parameter s corresponds a Mohr circle with $\sigma_1 = \sigma_1(s)$, $\sigma_2 = \sigma_2(s)$, which at the point of contact with the envelope has

[157] The main value of a Mohr circle (see Fig. 17) consists in its furnishing directly the angle ε which corresponds to each point on the circumference of the circle, i.e., to each stress vector.

coordinates $\sigma=\sigma(s)$, $\tau=\tau(s)$, and an angle $\varepsilon=\varepsilon(s)$. Call Φ the angle which the tangent to the envelope makes with the positive σ-axis; then we have along the envelope

$$\frac{d\tau}{d\sigma}=\tan\Phi;^{158} \tag{37.4}$$

also $\Phi=90°-2\varepsilon$, and Eqs. (37.3) become

$$\frac{\sigma_1+\sigma_2}{2}=\sigma+\tau\tan\Phi, \qquad \frac{\sigma_1-\sigma_2}{2}=\frac{\tau}{\cos\Phi}. \tag{37.3'}$$

Now, differentiating and using (37.4), we obtain:

$$\begin{aligned}\frac{\sigma_1'+\sigma_2'}{2}&=\sigma'+\tau'\tan\Phi+\frac{\tau}{\cos^2\Phi}\Phi'=\frac{\sigma'+\tau\Phi'}{\cos^2\Phi},\\ \frac{\sigma_1'-\sigma_2'}{2}&=\frac{\sin\Phi}{\cos^2\Phi}(\sigma'+\tau\Phi');\end{aligned} \tag{37.5}$$

dividing, we find

$$\frac{\sigma_1'-\sigma_2'}{\sigma_1'+\sigma_2'}=\sin\Phi=\cos 2\varepsilon. \tag{37.5'}$$

Using (36.3), we find that $(\sigma_1'-\sigma_2')/(\sigma_1'+\sigma_2')=\cos 2\alpha$. Thus we see (cf. Figs. 17a, c) that *the angle* ε *in* (37.2) *which corresponds on any Mohr circle to a point of real contact with the envelope, is equal to the angle* α *which a characteristic makes with the first principal direction.*[159] The points with real contact with the envelope of Mohr circles are the hyperbolic points. Hence, *we have for* σ *and* τ:

$$\sigma=\sigma_1\sin^2\alpha+\sigma_2\cos^2\alpha, \qquad \tau=(\sigma_1-\sigma_2)\sin\alpha\cos\alpha. \tag{37.2'}$$

From these formulas we obtain a new expression for the S' of (36.7). From (37.5) and (37.3'):

$$\frac{\sigma'+\tau\Phi'}{2\tau}=\frac{\sigma_1'+\sigma_2'}{2(\sigma_1-\sigma_2)}\cos\Phi.$$

But

$$\cos\Phi=\sin 2\alpha=2\frac{\sqrt{\sigma_1'\sigma_2'}}{\sigma_1'+\sigma_2'}, \quad \text{if} \quad \frac{\sigma_2'}{\sigma_1'}=\tan^2\alpha$$

is used. Hence

$$\frac{\sigma'+\tau\Phi'}{2\tau}=\frac{\sqrt{\sigma_1'\sigma_2'}}{\sigma_1-\sigma_2}=-S'(s), \tag{37.6}$$

or, in terms of

$$\alpha=45°-\frac{\Phi}{2},$$

$$S'=\alpha'-\frac{\sigma'}{2\tau}. \tag{37.6'}$$

Characteristic coordinates. On account of Eqs. (36.8), and in analogy to Eqs. (30.9), we put[160]

$$S(s)=\eta+\xi, \qquad \vartheta=\eta-\xi. \tag{37.7}$$

---

[158] This contains the particular case (35.8).

[159] The first proof of this interesting relation is due to H. MANDEL: Sur les équilibres par tranches parallèles des milieux plastiques à la limite d'écoulement. Paris: Louis Jean 1942, pp. 38, 39. The proof given here which is simpler than that of H. GEIRINGER in Advances; 1953, p. 236, is independent of, and different from, a proof in HILL [*6*, p. 295]. In several papers C. TORRE see e.g. Z. angew. Math. Mech. **31**, 275 (1951) has proved a closely related result.

[160] A theorem like HENCKY's first theorem follows readily from Eqs. (37.7).

These characteristic coordinates,[161]

$$S+\vartheta=2\eta, \quad S-\vartheta=2\xi, \tag{37.8}$$

do not apply in a region with rectilinear characteristics.

Now introduce, as in Sect. 30,

$$h_1=\frac{d s_1}{d\xi}, \quad h_2=\frac{d s_2}{d\eta}, \tag{37.9}$$

where $d s_1$ and $d s_2$ denote line elements of a C^- and C^+, respectively. Then the following equations [corresponding to (30.13)] will hold:

$$\begin{aligned} \frac{\partial x}{\partial \xi}&=h_1\cos\varphi^-, & \frac{\partial x}{\partial \eta}&=h_2\cos\varphi^+,\\ \frac{\partial y}{\partial \xi}&=h_1\sin\varphi^-, & \frac{\partial y}{\partial \eta}&=h_2\sin\varphi^+ \end{aligned} \tag{37.10}$$

and

$$\frac{\partial y}{\partial \xi}=\frac{\partial x}{\partial \xi}\tan(\vartheta-\alpha), \quad \frac{\partial y}{\partial \eta}=\frac{\partial x}{\partial \eta}\tan(\vartheta+\alpha). \tag{37.11}$$

Next,

$$\frac{\partial\varphi^+}{\partial\eta}=\frac{\partial\vartheta}{\partial\eta}+\frac{\partial\alpha}{\partial\eta}=\frac{\partial\vartheta}{\partial\eta}+\frac{\partial\alpha}{\partial S}\,\frac{\partial S}{\partial\eta}=1+\alpha^{`}=1+\frac{\alpha'}{S'}, \tag{37.12}$$

where the two different accents denote differentiation with respect to $s\,(')$ and with respect to $S(`)$, respectively. There are three more such formulas, and we obtain

$$\frac{\partial\varphi^+}{\partial\eta}=-\frac{\partial\varphi^-}{\partial\xi}=1+\alpha^{`}, \quad \frac{\partial\varphi^-}{\partial\eta}=-\frac{\partial\varphi^+}{\partial\xi}=1-\alpha^{`}, \tag{37.13}$$

and introducing the radii of curvature with the same sign conventions as in (30.4), but writing R_1, R_2 rather than R^-, R^+, we have, corresponding to the relations (30.10),

$$h_1=-R_1(1+\alpha^{`}), \quad h_2=-R_2(1+\alpha^{`}). \tag{37.14}$$

Now using the conditions

$$\frac{\partial^2 x}{\partial\xi\,\partial\eta}=\frac{\partial^2 x}{\partial\eta\,\partial\xi}$$

and

$$\frac{\partial^2 y}{\partial\xi\,\partial\eta}=\frac{\partial^2 y}{\partial\eta\,\partial\xi}$$

we obtain from Eqs. (37.13) the relations[162]

$$\begin{aligned} \frac{\partial h_1}{\partial\eta}\sin 2\alpha-\left(1-\frac{\alpha'}{S'}\right)(h_1\cos 2\alpha+h_2)&=0,\\ \frac{\partial h_2}{\partial\xi}\sin 2\alpha-\left(1-\frac{\alpha'}{S'}\right)(h_1+h_2\cos 2\alpha)&=0, \end{aligned} \tag{37.15}$$

[161] The choice of (37.8) corresponds to $b=+1$, $a=-1$, $c=0$ in (30.9).

[162] Just as in plane strain there is a parallelism between the velocity equations (36.11) and the Eqs. (37.16). If characteristic coordinates are used, the latter become [use (37.14)]

$$\sin 2\alpha\,\frac{\partial v_1}{\partial\xi}=\left(1+\frac{\alpha'}{S'}\right)(v_1\cos 2\alpha-v_2), \quad \sin 2\alpha\,\frac{\partial v_2}{\partial\eta}=\left(1+\frac{\alpha'}{S'}\right)(v_2\cos 2\alpha-v_1)$$

which reduce to (30.15) for $\alpha=45°$, $b=1$, $a=-1$. However, these velocity equations are not the same equations as Eqs. (37.16).

which reduce for $\alpha=45°$ to (30.11) with $b=1$, $a=-1$. With the abbreviation

$$A(s)=\sin 2\alpha\Big/\left(1-\frac{\alpha'}{S'}\right)$$

we may write

$$A\frac{\partial h_1}{\partial\eta}=h_1\cos 2\alpha+h_2,\qquad A\frac{\partial h_2}{\partial\xi}=h_1+h_2\cos 2\alpha, \tag{37.16}$$

which reduce to (30.11) for $\phi=0$, $\alpha=45°$.

These equations may be simplified further by introducing

$$h_1\sqrt{\tau}=m_1,\qquad h_2\sqrt{\tau}=m_2. \tag{37.17}$$

Then

$$\frac{\partial h_1}{\partial\eta}=\frac{1}{\sqrt{\tau}}\frac{\partial m_1}{\partial\eta}-\frac{m_1}{2\tau\sqrt{\tau}}\frac{\partial\tau}{\partial\eta},$$

and, using (37.6′) and $\tau'/\sigma'=\cot 2\alpha$, we find that

$$\begin{aligned}A\frac{\partial h_1}{\partial\eta}&=\frac{A}{\sqrt{\tau}}\frac{\partial m_1}{\partial\eta}+\frac{m_1}{2\tau\sqrt{\tau}}\frac{\tau}{S'}\frac{\sin 2\alpha}{\sigma'}\cdot 2\tau' S'\\&=\frac{A}{\sqrt{\tau}}\frac{\partial m_1}{\partial\eta}+\frac{m_1}{\sqrt{\tau}}\cos 2\alpha.\end{aligned}$$

The second term on the right is $h_1\cos 2\alpha$; hence, on account of (37.16),

$$A\frac{\partial h_1}{\partial\eta}=\frac{A}{\sqrt{\tau}}\frac{\partial m_1}{\partial\eta}+h_1\cos 2\alpha=h_1\cos 2\alpha+h_2=h_1\cos 2\alpha+\frac{m_2}{\sqrt{\tau}}.$$

We obtain

$$A\frac{\partial m_1}{\partial\eta}=m_2,\qquad A\frac{\partial m_2}{\partial\xi}=m_1. \tag{37.18}$$

These equations, due to MANDEL, are probably as simple as any obtainable. (For plane strain, $A=1$ and $m_1=h_1\sqrt{c}$, $m_2=h_2\sqrt{c}$, Eqs. (30.11) result.] Cross differentiation of Eqs. (37.18) leads to

$$\begin{aligned}A^2\frac{\partial^2 m_1}{\partial\xi\,\partial\eta}+A A^{\backprime}\frac{\partial m_1}{\partial\eta}-m_1&=0,\\A^2\frac{\partial^2 m_2}{\partial\xi\,\partial\eta}+A A^{\backprime}\frac{\partial m_2}{\partial\xi}-m_2&=0.\end{aligned} \tag{37.19}$$

Here $A^{\backprime}=dA/dS$.

38. Examples for Sects. 36 and 37.

α) *Quadratic condition.* Using (35.1), we obtain from (36.6), (36.7)

$$\begin{aligned}&\frac{\partial\vartheta}{ds}=\pm\frac{\sqrt{3-s^2}}{4-s^2},\qquad S(s)=-\int^{s}\frac{\sqrt{3-s^2}}{4-s^2}\,ds,\\&\pm\vartheta=\arctan\frac{s}{\sqrt{3-s^2}}-\frac{1}{2}\arctan\frac{s}{2\sqrt{3-s^2}}+\text{constant},\end{aligned} \tag{38.1}$$

as the equations of the fixed characteristics in the stress plane and at the same time the compatibility conditions in the physical plane. The upper sign holds along a ξ-line. Also

$$\tan^2\alpha=\frac{\sigma_2'}{\sigma_1'}=(\sqrt{12-3s^2}+s)/(\sqrt{12-3s^2}-s),$$

and

$$\tan\alpha=\pm\frac{\sqrt{12-3s^2}+s}{2\sqrt{3-s^2}}. \tag{38.2}$$

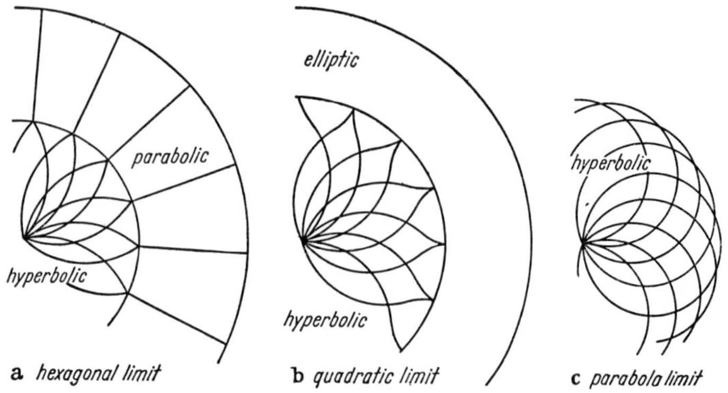

Fig. 18a–c. Fixed characteristics for the three yield conditions of Fig. 12; (a) hexagonal limit; (b) quadratic limit; (c) parabola limit.

The points (see Fig. 12) $s = \pm\sqrt{3}$ are parabolic: for $s = +\sqrt{3}$ we have $\sigma_1 = 2\sigma_2$ and $\alpha = 90°$; for $s = -\sqrt{3}$, $\sigma_2 = 2\sigma_1$ and $\alpha = 0°$; for both, $\beta = 0$, where β is the angle of Γ^+ or Γ^- with the radius vector in the stress graph. The hyperbolic part extends from $s = -\sqrt{3}$ to $s = +\sqrt{3}$, the elliptic part through $\sqrt{3} < |s| < 2$. Fig. 18b shows the Γ-curves in polar coordinates after v. MISES.[163]

Next we compute σ and τ using (37.2′) and (35.1):

$$\sigma = \sigma_1 \sin^2\alpha + \sigma_2 \cos^2\alpha = \frac{1}{2k}(\sigma_1 \sigma_2' + \sigma_2 \sigma_1')$$

$$= \frac{1}{2k}(\sigma_1 \sigma_2)' = \frac{2k}{3}(s^2 - 1)' = \frac{4k}{3}s,$$

$$\tau = \frac{2k}{3}\sqrt{3 - s^2}.$$

Hence

$$\sigma = \frac{4k}{3}s, \quad \tau = \frac{2k}{3}\sqrt{3 - s^2}, \quad \sigma^2 + 4\tau^2 = \frac{16k^2}{3}; \tag{38.3}$$

thus the yield condition appears as an ellipse in the σ, τ-plane.[164]

β) *Hexagonal condition* (TRESCA, ST. VENANT). The problem is hyperbolic for

$$|s| < 1; \quad \text{then} \quad S(s) = -\frac{s}{2}.$$

The Γ-characteristics in the stress graph are ordinary spirals

$$\pm\vartheta = \frac{s}{2} + \text{constant},$$

or straight lines, in polar or rectangular coordinates s, ϑ, respectively. In the parabolic region the unique set of characteristics consists of radii through the center (Fig. 18a). In the hyperbolic region $\alpha = 45°$, $\varphi^{\pm} = \vartheta \pm 45°$ and the characteristics are orthogonal.[165]

[163] R. v. MISES: Reissner Anniversary Volume. Ann Arbor, Michigan 1949. In rectangular coordinates s, ϑ a Γ-line is shown in H. GEIRINGER, Advances, p. 238.

[164] A yield condition in the σ, τ-plane, considered as envelope of Mohr circles, is called by MANDEL "courbe intrinsèque".

[165] We have seen that the assumptions of isotropy and orthogonality lead to $\dot{\varepsilon}_1 + \dot{\varepsilon}_2 = 0$, and the yield condition is of the form as in (26.8). The most general plane yield condition with orthogonal characteristics is of the form $\tau = f[(\sigma_x - \sigma_y)]$, where f is a function of one variable, as pointed out by R. SAUER, Z. angew. Math. Mech. **29**, 274–279 (1949).

γ) *Parabola condition* (Fig. 18c). The problem is hyperbolic everywhere, with the exception of the two points $s=\pm a$, where $hg=0$ [see (35.6)]. We find

$$\begin{gathered}\frac{d\vartheta}{ds}=\pm(a^2-s^2)^{-\frac{1}{2}},\qquad \vartheta=\pm\ \text{arc sin}\,\frac{s}{a}+\text{constant},\\ -S=\text{arc sin}\,\frac{s}{a},\qquad \sin S=-\frac{s}{a},\\ \tan\alpha=\sqrt{\frac{a+s}{a-s}},\qquad \cos 2\alpha=-\frac{s}{a},\qquad S=90^\circ-2\alpha,\\ 1-\frac{\alpha'}{S'}=\frac{3}{2},\qquad A(S)=\frac{\sin 2\alpha}{1-a'/S'}=\frac{2}{3a}\sqrt{a^2-s^2}=\frac{2}{3}\cos S=\frac{2}{3}\cos(\xi+\eta).\end{gathered}\tag{38.4}$$

We see that Eqs. (37.18), (37.19) become here comparatively simple.

δ) SOKOLOVSKY[166] introduced the yield condition

$$\frac{\sigma_1-\sigma_2}{2K}=\sin\left(\frac{\sigma_0}{K}=\frac{\sigma_1+\sigma_2}{2K}\right),\tag{38.5}$$

where σ_0 and K are constants, and

$$2\sigma_0-\pi K\leqq(\sigma_1+\sigma_2)\leqq 2\sigma_0.\tag{38.6}$$

It is easily seen that for this yield condition $F_1 F_2\leqq 0$: the stress problem is hyperbolic. In the σ, τ-plane this condition represents a cycloid; certain mathematical simplifications result which allow explicit solutions (see [*26*, p. 311 seq.]). Such hyperbolic yield conditions, which depend on $(\sigma_1+\sigma_2)$, may be considered as a generalization of the plane strain yield condition in the sense of SCHLEICHER[167] and v. MISES;[168] the parabola condition considered in this and previous sections, which is also of this type, has the expressed purpose of approximating the quadratic condition for plane stress $\sigma_1^2+\sigma_2^2-\sigma_1\sigma_2=$ constant.

ε) *Coulomb condition.* The problem is hyperbolic since, as seen in (35.7) $\sigma_1'\sigma_2'=k^2\cos^2\Phi>0$, if $\varphi<90^\circ$. From Sect. 37: $\alpha=45^\circ-\Phi/2$, and

$$\begin{gathered}\varphi^-=\vartheta-45^\circ+\Phi/2,\qquad \varphi^+=\vartheta+45^\circ-\Phi/2,\\ \sqrt{gh}=2(1+s\tan\Phi),\\ S=-\int\frac{ds}{\sqrt{gh}}=-\frac{1}{2}\cot\Phi\log(1+s\tan\Phi).\end{gathered}\tag{38.7}$$

Also we saw that

$$\pm\tau=k+\sigma\tan\Phi\tag{38.8}$$

is the equation of the yield locus in the σ, τ-system. Along the stress characteristics, C^+, C^- $\vartheta\pm S$ are constant. These lines[169] are in general identified with the *failure lines*. The Γ-curves in the stress graph are logarithmic spirals.

Since $\alpha'=0$, $A(S)=\cos\Phi=$ constant, we see from (37.14) that $h_1=-R_1$, $h_2=-R_2$. The two Eqs. (37.19) become

$$\cos^2\Phi\frac{\partial^2 m_i}{\partial\xi\,\partial\eta}-m_i=0,\qquad i=1,2,\tag{38.9}$$

and the integration problem is similar to that in plane strain, if $\xi/\cos\Phi$ and $\eta/\cos\Phi$ are now considered as independent variables.

[166] V. V. SOKOLOVSKY: J. Appl. Math. Mech. **13** (1949).
[167] F. SCHLEICHER: Z. angew. Math. Mech. **6**, 199–216 (1926).
[168] R. v. MISES: Z. angew. Math. Mech. **5**, 147–149 (1925).
[169] References given by R. T. SHIELD: Ref. 152.

Thus e.g.

$$m_1 = c\, e^{c\frac{\xi}{\cos\Phi}+\frac{1}{c}\frac{\eta}{\cos\Phi}}, \qquad m_2 = e^{c\frac{\xi}{\cos\Phi}+\frac{1}{c}\frac{\eta}{\cos\Phi}}$$

is a particular solution of (38.9), and new solutions can be derived by addition, integration with respect to a parameter, etc. The stress problem consisting of the usual equilibrium equations plus the Coulomb yield condition seems to be non-controversial.

Next, if as proposed by DRUCKER and PRAGER[170] the theory of the plastic potential is used in the velocity problem, we obtain from (36.1′)

$$\begin{aligned} \dot\varepsilon_x : \dot\varepsilon_y : \dot\gamma &= (\cos 2\alpha - \cos 2\vartheta) : (\cos 2\alpha + \cos 2\vartheta) : -\sin 2\vartheta \\ &= (\sin\Phi - \cos 2\vartheta) : (\sin\Phi + \cos 2\vartheta) : -\sin 2\vartheta . \end{aligned} \tag{38.10}$$

The problem is almost as simple as that of plane strain, since Φ has a constant value. Consider the Eqs. (36.11); now, $d\varphi^- = d\varphi^+ = d\vartheta$, $2\alpha = 90° - \Phi$, and

$$\begin{aligned} d v_1 &= (v_2 \sec\Phi - v_1 \tan\Phi)\, d\vartheta , \\ d v_2 &= (-v_1 \sec\Phi + v_2 \tan\Phi)\, d\vartheta \end{aligned} \tag{38.11}$$

along a C^- and C^+, respectively.

However, a consequence of the above assumption (rule of the plastic potential) is that the *rate of dilatation*

$$\dot\varepsilon_x + \dot\varepsilon_y = \sin\phi\,[(\dot\varepsilon_x - \dot\varepsilon_y)^2 + \dot\gamma_z^2]^{\frac{1}{2}} \tag{38.12}$$

is positive if $\phi > 0$: the plastic deformation is accompanied by an *increase in volume*, and such an increase has not been verified experimentally. This is the difficulty connected with taking the Coulomb condition as plastic potential for the soil. SPENCER (see footnote 152 points out the impossibility of reconciling the following desirable postulates: that of incompressibility; that of the associated flow rule; of SAINT VENANT'S hypothesis (that the principal axes of stress and of strain rate coincide); and the usual requirement that the *observed* failure lines should coincide with the stress characteristics. SPENCER'S hypothesis is that the soil deforms in plane strain by shear along the stress characteristics (see Sect. 3 of his paper); in addition isotropy[171] and incompressibility are assumed. He presents on this basis a theory which is very clear and coherent and reduces when $\phi = 0$ to plane strain of a rigid plastic material. However, J. MANDEL,[172] an expert in the mathematics and physics of the subject, while appreciating SPENCER'S mathematical theory, sees physical difficulties "de base", to explain which here would lead too far.

c) Remarks on integration. Examples.

39. On integration.[173] If for a specific yield condition a solution $m_1(\xi,\eta), m_2(\xi,\eta)$ of (37.17) has been found, then $h_1 = m_1/\sqrt{\tau}$, $h_2 = m_2/\sqrt{\tau}$, where $\tau = \frac{1}{2}(\sigma_1 - \sigma_2)\sin 2\alpha$

[170] D. C. DRUCKER and W. PRAGER: Quart. J. Appl. Math. **10**, 157–165 (1952).

[171] In SPENCER'S paper a distinction is pointed out between isotropic material and material for which Saint Venant's hypothesis holds. To the great number of relevant references given by SPENCER we add: H. ZIEGLER, Z. angew. Math. Phys. **20**, 659 (1969), who discusses in detail the question of the plastic potential in soil mechanics examined by SPENCER.

[172] J. MANDEL: J. Mech. Phys. Solids **14**, 303 (1966).

[173] The great mathematical difficulties which result from the combination of a *mixed problem* with *non-linearity* are well known in the theory of flows of compressible fluids. In plasticity theory, these difficulties are the same.

is a known function of $(\xi+\eta)$. Then, along a ξ-line

$$dx = h_1 \cos(\vartheta-\alpha)\, d\xi, \qquad dy = h_1 \sin(\vartheta-\alpha)\, d\xi$$

with $\vartheta=\eta-\xi$, and α a known function of $(\eta+\xi)$; similar relations hold along an η-line. Hence

$$\begin{aligned} dx &= \frac{1}{\sqrt{\tau}}\left[m_1 \cos(\vartheta-\alpha)\, d\xi + m_2 \cos(\vartheta+\alpha)\, d\eta\right], \\ dy &= \frac{1}{\sqrt{\tau}}\left[m_1 \sin(\vartheta-\alpha)\, d\xi + m_2 \sin(\vartheta+\alpha)\, d\eta\right], \end{aligned} \tag{39.1}$$

and $x=x(\xi,\eta)$, $y=y(\xi,\eta)$ can be found by quadratures.

General integration procedures (as discussed for "plane strain") for finding solutions of (37.18) are available in the case of the Coulomb condition and to a certain degree in the case of the parabola yield condition. We add a few remarks, due to MANDEL (Ref. 138) regarding the integration of (37.18).

Assume that we know two particular solutions of (37.18), say m_1, m_2 and γ_1, γ_2. *Then* $m_1\gamma_1\, d\xi + m_2\gamma_2\, d\eta$ *is a total differential,* as can be verified immediately. Now, conversely, consider the expressions

$$m_1\,\delta_1\, d\xi + m_2\,\delta_2\, d\eta, \qquad \gamma_1\,\delta_1\, d\xi + \gamma_2\,\delta_2\, d\eta,$$

each of which is assumed to be a total differential. If then m_1, m_2 *and* γ_1, γ_2 *are independent solutions,* it is easily seen *that* δ_1, δ_2 *are likewise solutions.* We know e.g. from (39.1) that

$$\frac{m_1}{\sqrt{\tau}}\cos(\vartheta-\alpha)\, d\xi + \frac{m_2}{\sqrt{\tau}}\cos(\vartheta+\alpha)\, d\eta$$

is a complete differential for any m_1, m_2. It follows that

$$\frac{1}{\sqrt{\tau}}\cos(\vartheta-\alpha)=\delta_1, \qquad \frac{1}{\sqrt{\tau}}\cos(\vartheta+\alpha)=\delta_2 \tag{39.2}$$

is a solution of (37.18). We shall consider this example below.

Next set

$$\Sigma=\int\frac{dS}{A(S)} \quad \text{where} \quad A=\frac{\sin 2\alpha}{1-\dfrac{\alpha'}{S'}}. \tag{39.3}$$

It is then easily seen that

$$du = e^{\Sigma}(m_1\, d\xi + m_2\, d\eta) \quad \text{and} \quad dv = e^{-\Sigma}(-m_1\, d\xi + m_2\, d\eta) \tag{39.4}$$

are total differentials and that $dv=0$ $(du=0)$ *is the differential equation of the u-lines (of the v-lines).*

It is also seen immediately that

$$m_1=m_2=e^{\Sigma} \quad \text{as well as} \quad m_1=-m_2=e^{-\Sigma} \tag{39.5}$$

are particular solutions of (37.18).

As an *example* we consider now the following particular solution of (37.18):

$$m_1=\frac{1}{\sqrt{\tau}}\cos(\vartheta-\alpha), \qquad m_2=\frac{1}{\sqrt{\tau}}\cos(\vartheta+\alpha). \tag{39.2'}$$

Carrying out calculations explained at the beginning of this section, we find

$$\begin{aligned} x &= \int\frac{dS}{2\tau}+\frac{1}{4\tau}\sin 2\alpha\cos 2\vartheta, \\ y &= \frac{1}{4\tau}\sin 2\alpha\sin 2\vartheta, \end{aligned} \tag{39.6}$$

which gives x and y in terms of s and ϑ or rather S and ϑ.

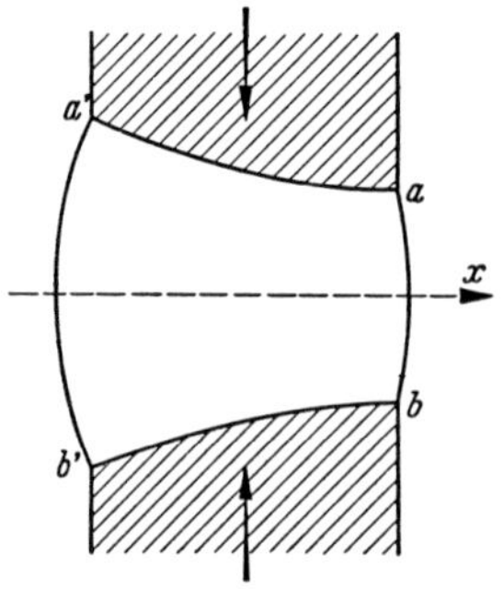

Fig. 19. Plastic mass between curved plates (MANDEL).

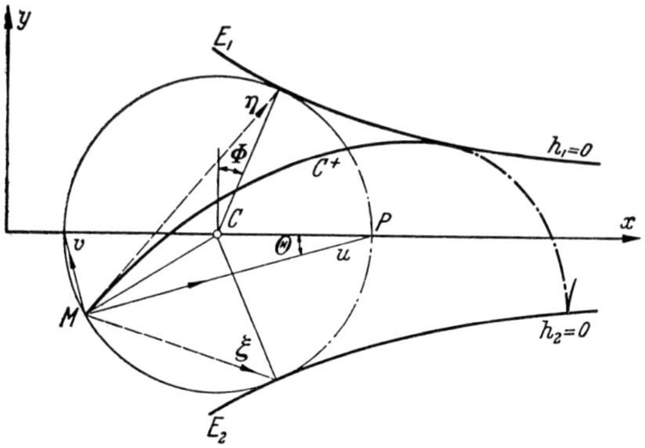

Fig. 20. Compression of a plastic mass between curved plates (MANDEL).

The lengthy computation which leads to the result (39.6) can be omitted if we use the following simple remark, also due to MANDEL: If a total differential is of the form

$$d\Pi = A(S)\,B(\vartheta)\,dS + C(S)\,D(\vartheta)\,d\vartheta$$

then Π will be obtained, ω being a constant, as

$$\Pi = \omega\, C(S)\,B(\vartheta) + \text{constant},$$

as may be verified immediately. Now for the m_1, m_2 of (39.2) we have indeed, using (39.1),

$$\begin{aligned} dx &= \frac{m_1}{\sqrt{\tau}}\cos\varphi^-\,d\xi + \frac{m_2}{\sqrt{\tau}}\cos\varphi^+\,d\eta \\ &= \frac{1}{\tau}\left[\cos^2(\vartheta-\alpha)\,\frac{dS-d\vartheta}{2} + \cos^2(\vartheta+\alpha)\,\frac{dS+d\vartheta}{2}\right] \\ &= \frac{1}{2\tau}\left[dS + \cos 2\alpha\cos 2\vartheta\,dS - \sin 2\alpha\sin 2\vartheta\,d\vartheta\right], \\ dy &= \frac{1}{2\tau}\left[\cos 2\alpha\sin 2\vartheta\,dS + \sin 2\alpha\cos 2\vartheta\,d\vartheta\right]. \end{aligned}$$

Here for dx, with $A=\cos 2\alpha$, $B=\cos 2\vartheta$, $C=\sin 2\alpha$, $D=-\sin 2\vartheta$, the above situation obtains, with $\omega=\frac{1}{2}$, and we find the first Eq. (39.6) and similarly, the second one.

In this example the lines of constant S (or s), the *isobars*, are the circles

$$(x-a)^2 + y^2 = \frac{1}{16\tau^2}\sin^2\alpha, \quad \text{where} \quad a = \int \frac{dS}{2\tau}. \tag{39.7}$$

The radius equals $\sin 2\alpha/4\tau = 1/4R$, where $R = \tau/\sin 2\alpha$ is the radius of Mohr's circle.

The lines aa' and bb' (Fig. 19) which separate the plastic and the rigid material must have characteristic directions. They are (Fig. 20) actually envelopes of one family of characteristics and loci of cusps of the other family. Such lines, *limit lines* E_1 and E_2 (see general theory, Sect. 40 seq.), are the images in the physical plane of the lines $m_1 = 0$ and $m_2 = 0$. In our example $m_1 = \tau^{-\frac{1}{2}} \cos(\vartheta - \alpha) = 0$ for $\vartheta - \alpha = 90°$ and $m_2 = 0$ for $\vartheta + \alpha = 90°$. The first one is the envelope of characteristics C^+, and the C^- have cusps there forming the constant angle 90° with the envelope; the second is the envelope of the C^- and locus of the cusps of the C^+ (cf. Sect. 40_1). These same curves E_1 and E_2 are also envelopes of the *isobars* (lines of constant s) and of the *isoclines* (lines of constant ϑ), while the principal stress lines have cusps at the limit lines. The solution corresponds to the physical problem of a plastic mass pressed between two completely rough plates, gliding in the direction of divergence; the traces which represent the walls must be similar to the envelopes. More details on this example may be found in MANDEL's book, p. 123 seq.

In the particular case α = constant the two envelopes E_1 and E_2 are straight lines which make the angle 2Φ with each other (Sect. 37). For $\phi = 0°$ they are parallel lines and the characteristics are cycloids.

II. Singular solutions and various remarks.[174]

a) Limit line singularities and branch line singularities.

40. Limit line singularities. We have noticed before that the transition from a solution $x = x(s, \vartheta)$, $y = y(s, \vartheta)$ to the inverse $s = s(x, y)$, $\vartheta = \vartheta(x, y)$ is impossible at points or along lines where the Jacobian $\partial(x, y)/\partial(s, \vartheta)$ vanishes. Such a line will be called a *limit line.* We found such lines in the preceding example.

We have to consider singularities of two kinds: *limit type singularities* and *branch type singularities.* The first type is characterized by the vanishing of a determinant like $D = \partial(x, y)/\partial(s, \vartheta)$ with stress graph coordinates in the denominator, physical plane coordinates in the numerator; the second by $d = 0$, where $d = D^{-1}$. (Many other Jacobians are equivalent to D in this respect, e.g. $\partial(u, v)/\partial(s, \vartheta)$, or in a hyperbolic problem $\partial(x, y)/\partial(\xi, \eta)$ etc.; for instance

$$\frac{\partial(u, v)}{\partial(s, \vartheta)} = \frac{\partial(u, v)}{\partial(x, y)} \frac{\partial(x, y)}{\partial(s, \vartheta)} = \frac{\partial(x, y)}{\partial(s, \vartheta)}, \quad \text{or} \quad \frac{\partial(x, y)}{\partial(\xi, \eta)} = \frac{2}{S'} \frac{\partial(x, y)}{\partial(s, \vartheta)}.$$

These last two are equivalent to each other wherever S' is neither zero nor infinite.) Roughly speaking, the first type of singularity prevents the transition from the stress graph to the physical plane, the second that from the physical plane to the stress graph. Using Eqs. (34.8), (36.7) and (37.7), we find by a straightforward computation that

$$D = \frac{\partial(u, v)}{\partial(s, \vartheta)} = \frac{1}{h} \frac{\partial u}{\partial \xi} \frac{\partial u}{\partial \eta} = -\frac{1}{g} \frac{\partial v}{\partial \xi} \frac{\partial v}{\partial \eta}. \tag{40.1}$$

174 Here, in a way, the theory precedes the applications (in contrast to gas dynamics, where the development of the theory was prompted by the knowledge of examples). At this moment, I cannot mention examples in our field which exhibit the various singularities. However, the presentation and classification of limit line and branch line singularities may help other workers to clarify their problems. (E.g. an eminent author like SOKOLOVSKY fuses together limit lines and lines of stress discontinuity.)

We can deal briefly with the case, not important in the present connection, of an *elliptic region*,[175] where $gh<0$. Using Eqs. (34.6), we obtain

$$d=\frac{\partial(s,\vartheta)}{\partial(x,y)}=\frac{\partial(s,\vartheta)}{\partial(u,v)}=g\left(\frac{\partial\vartheta}{\partial v}\right)^2-h\left(\frac{\partial\vartheta}{\partial u}\right)^2. \tag{40.2}$$

Here both terms have the same signs, since $gh<0$, and therefore d can vanish only if

$$\frac{\partial\vartheta}{\partial u}=\frac{\partial\vartheta}{\partial v}=0.$$

Then, also

$$\frac{\partial s}{\partial u}=\frac{\partial s}{\partial v}=0$$

from (34.6). Clearly, such a singularity must be isolated.

In a similar way consider

$$D=\frac{\partial(u,v)}{\partial(s,\vartheta)}=g\left(\frac{\partial u}{\partial s}\right)^2-h\left(\frac{\partial v}{\partial s}\right)^2.$$

These two terms can vanish only [see (34.9″) and (34.8)] if

$$\frac{\partial X}{\partial s}=0,\quad \frac{\partial Y}{\partial s}=0,\quad \frac{\partial X}{\partial\vartheta}-Y=0,\quad \frac{\partial Y}{\partial\vartheta}+X=0.$$

A brief computation shows that these imply that

$$\frac{\partial x}{\partial s}=0,\quad \frac{\partial y}{\partial s}=0,\quad \frac{\partial x}{\partial\vartheta}=0,\quad \frac{\partial y}{\partial\vartheta}=0,$$

hence again an isolated singularity.

We now consider the main case, that of a *hyperbolic region*, and, using (37.11),

$$J=\frac{\partial(x,y)}{\partial(\xi,\eta)}=h_1 h_2\sin 2\alpha. \tag{40.3}$$

Assume that $\alpha\neq0°$, $\alpha\neq90°$; since $\tan\alpha=\sqrt{g/h}$, this can be excluded if neither g nor h is zero or infinite.

Consider now in the ξ,η-plane the locus $h_1(\xi,\eta)=0$ for a given stress distribution, and in particular the mapping onto the x,y-plane in the neighborhood of this line.[176] Consider a point p with coordinates ξ_0,η_0, where $h_1(\xi_0,\eta_0)=0$, $(\partial h_1/\partial\eta)_{\xi_0,\eta_0}\neq0$. Then, by the implicit function theorem there is a curve $\eta=g(\xi)$, with $\eta_0=g(\xi_0)$, on which $h_1(\xi,\eta)=0$. We call it the *critical curve* and denote it by l_1, and we call its image L_1 in the physical plane the *limit line*. A point P at which $h_1=0$ is called a *limit point*.

Consider a point m on l_1 where *both* $\partial h_1/\partial\xi$ and $\partial h_1/\partial\eta$ are *different from zero*. Hence l_1 does not have the ξ-direction at m, and, using (37.16) and $\alpha\neq0°$, $\neq90°$,

[175] The case where $gh=0$ or $gh\to\infty$ will be partly considered in Sect. 41. An s-value for which g or h both vanish will in general designate an exceptional point of the yield limit. E.g., in the quadratic limit $g=h=0$ for $s=\pm2$. This is the absolute maximum of s, where $\sigma_1=\sigma_2$. The circle $s=2$ plays the role of the maximum circle $q=q_{\max}$ in the flow of a compressible fluid; it delimits the region of possible plastic flow (for $|s|>2$ the σ_i become imaginary while the yield condition is formally satisfied). This boundary $|s|=2$ is not of great interest.

[176] Our presentation reproduces to some extent the excellent study of MANDEL, l.c., Chap. V. However, we discuss several features not considered by MANDEL. We use also R. v. MISES, Mathematical theory of compressible fluid flow. New York: Academic Press 1958, particularly Art. 19 (written by GEIRINGER). SOKOLOVSKY considers envelopes of characteristics; he calls them "lines of rupture"; this same notation is used by him for what we call stress-discontinuity lines.

we see that $h_2 \neq 0$ at m. Then (37.10) shows that at M (image of m) L_1 has the η-direction, i.e. L_1 is tangent to the C^+ at M. The same conclusion holds for any curve C through M whose image c does not have the ξ-direction at m; all such curves C are tangent to L_1 at M.

For example, the images in the ξ, η-plane of the lines of constant s, the isobars, or the lines of constant ϑ, the isoclines, do not have the *exceptional direction*, the ξ-direction, at m. In fact, e.g. for the latter,

$$d\vartheta = 0 \quad \text{and} \quad d\eta = \tfrac{1}{2}(dS + d\vartheta) = \tfrac{1}{2} dS \neq 0 \quad \text{if} \quad dS = -ds/\sqrt{gh} \neq 0;$$

similar conclusions hold for the isobars.

If, however, a *curve has the ξ-direction* at m, this means $d\eta/d\xi = f'(\xi) = 0$, we can no longer conclude that its image has the η-direction at M. *In this case it has a cusp at M.* In fact,

$$\frac{dx}{d\xi} = \frac{\partial x}{\partial \xi} + f'(\xi)\frac{\partial x}{\partial \eta}, \qquad \frac{dy}{d\xi} = \frac{\partial y}{\partial \xi} + f'(\xi)\frac{\partial y}{\partial \eta},$$

and since both $\partial x/\partial\xi$ and $\partial y/\partial\xi$ vanish because $h_1 = 0$, and since $f'(\xi) = 0$, both $dx/d\xi$ and $dy/d\xi$ are zero at M; it can also easily be shown that both second derivatives $d^2x/d\xi^2$, $d^2y/d\xi^2$ cannot vanish. It follows that the C^- has a cusp at M, its tangent making the angle 2α with the direction of L_1. Since the principal trajectories bisect the angles of the C^-- and the C^+-directions, neither of them is tangent to L_1 at M; hence both their images must have the ξ-direction at m (see Fig. 21) and therefore they both must have cusps at M. All these results have their counterparts for points of L_2, along which $h_2 = 0$. We review: *Consider in the ξ, η-plane the locus $h_1(\xi, \eta) = 0$, the critical curve l_1, and points of l_1 at which $\partial h_1/\partial\xi \neq 0$; $\partial h_1/\partial\eta \neq 0$. Its image L_1, the limit line, is the envelope of the C^+, of the isobars, and of the isoclines; it is the locus of cusps of the C^- and of the principal stress trajectories* (Fig. 21).

We call *streamlines* those lines which have at every point the direction of $\boldsymbol{v}$. This direction will be known only after the whole problem has been solved. In general [see Eqs. (36.11)] a streamline will not coincide with (part of) a characteristic.

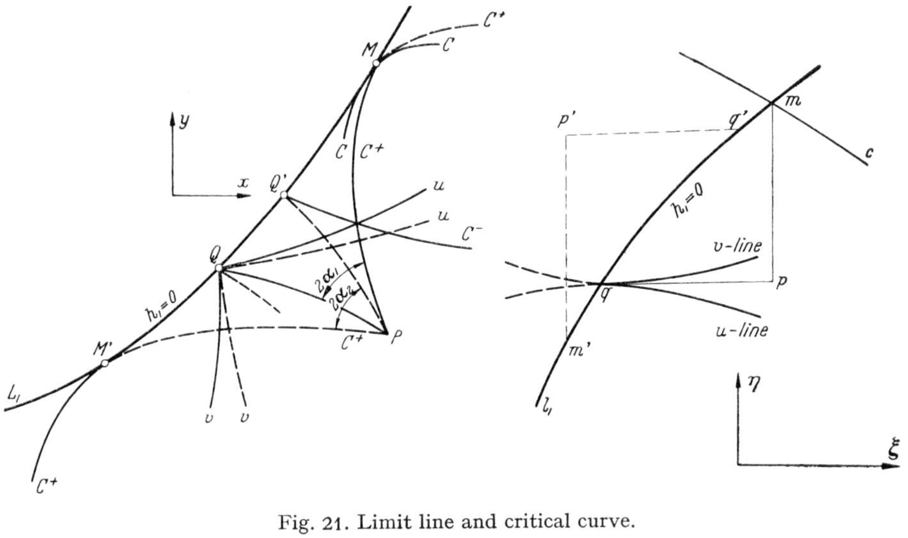

Fig. 21. Limit line and critical curve.

At a point M of L_1, where a streamline of course does not have the C^+-direction, its image will have the ξ-direction at m, and therefore the streamline has a cusp at L_1.

In general, two solutions meet at the limit line: A point P in the neighborhood of L_1 in the x, y-plane is the image of two points p, p' in the ξ, η-plane. In fact, through P pass two C^+-lines which touch the L_1 at M and M', respectively, with corresponding m and m' on l_1. On the η-line through m lies p; on that through m' lies p'. The points p, p' are on opposite sides of the l_1 since η varies in the opposite sense along PM and along PM'. Likewise two C^--lines PQ and PQ' pass through P making angles $2\alpha_1$ and $2\alpha_2$ with the C^+-characteristics PM and PM', respectively. The point q corresponding to Q lies on the ξ-line through p and on l_1, while q' lies on the ξ-line through p' and on l_1. This correspondence may be discussed in more detail.

We compute $\partial s/\partial u$, $\partial s/\partial v$, viz. the changes of pressure in a principal direction. We now write ω rather than s to avoid confusion with line elements and $\partial/\partial s_u$ for $\partial/\partial u$, $\partial/\partial s_v$ for $\partial/\partial v$. Then

$$\frac{\partial\omega}{\partial s_1} = \frac{\partial\omega}{\partial s_u}\cos\alpha - \frac{\partial\omega}{\partial s_v}\sin\alpha, \qquad \frac{\partial\omega}{\partial s_2} = \frac{\partial\omega}{\partial s_u}\cos\alpha + \frac{\partial\omega}{\partial s_v}\sin\alpha.$$

Hence

$$\frac{\partial\omega}{\partial s_u} = \frac{1}{2\cos\alpha}\left(\frac{\partial\omega}{\partial s_1} + \frac{\partial\omega}{\partial s_2}\right), \qquad \frac{\partial\omega}{\partial s_1} = \frac{\partial\omega}{\partial\xi}\,\frac{\partial\xi}{d s_1} = \frac{1}{h_1}\,\frac{d\omega}{dS} = \frac{1}{h_1 S'},$$

and therefore

$$\frac{\partial\omega}{\partial s_u} = \frac{1}{2\cos\alpha\cdot S'}\cdot\left(\frac{1}{h_1} + \frac{1}{h_2}\right), \qquad \frac{\partial\omega}{\partial s_u} = \frac{1}{2\sin\alpha\cdot S'}\left(-\frac{1}{h_1} + \frac{1}{h_2}\right). \tag{40.4}$$

Thus grad ω = grad $s \to \infty$ *at a limit line* L_1 *or* L_2 unless $S' \to \infty$, *and the same holds for* grad ϑ, in agreement with the fact that the curvatures of the stress trajectories were seen to have cusps at L_1 and L_2.

41. Limit line singularities. Continuation. So far we assumed that at the point(s) m, where $h_1 = 0$, both $\partial h_1/\partial\xi \neq 0$, $\partial h_1/\partial\eta \neq 0$. If $h_1 = 0$ and $\partial h/\partial\eta = 0$, it follows from (37.15) that $h_2 = 0$. Such a point, which appears as a point of intersection of an L_1 and an L_2, is called a *double limit point*. This is an isolated point, which we do not discuss further.

Next, assume $h_1 = 0$, $\partial h_1/\partial\eta \neq 0$, $\partial h_1/\partial\xi = 0$, $\partial^2 h_1/\partial\xi^2 \neq 0$. Then the l_1 has the ξ-direction and the Jacobian J changes sign there. Writing $h_1(\xi, \eta) = 0$ in the form $\eta = g(\xi)$, we find

$$g'(\xi) = -\frac{\partial h_1}{\partial\xi}\Big/\frac{\partial h_1}{\partial\eta} = 0$$

at m while $g''(\xi) \neq 0$. At M, the image of m,

$$\frac{dx}{d\xi} = \frac{\partial x}{d\xi} + \frac{\partial x}{\partial\eta}\,g'(\xi) = 0,$$

and likewise $dy/d\xi = 0$, and it can be seen that $d^2x/d\xi^2$, $d^2y/d\xi^2$ cannot both vanish if $h_2 \neq 0$. Hence it is seen that the L_1 has a cusp at M (Fig. 22). Further study again shows that the C^+ touches the L_1 at M; this means it has the cusp tangent there, but it does not have itself a cusp; the same holds for any curve through M whose image does not have the ξ-direction at m, e.g. for the isoclines and isobars. On the other hand, the image of the C^- has, of course, the ξ-direction at m, and the same is true of the images of both principal trajectories, since neither has the L_1-direction at M. The image of both trajectories contacts the l_1, which has an extremum at m.

It remains to describe briefly the limit type singularities along a "sonic" line. (In Sect. 56 there is an example.) We mean by that a line s = constant along which gh changes sign. *We consider the two cases* $g \to \infty$, h finite, and $h \to \infty$, g finite. (Cf. footnote 175 for some remarks on $g = 0$, $h = 0$.) In the quadratic con-

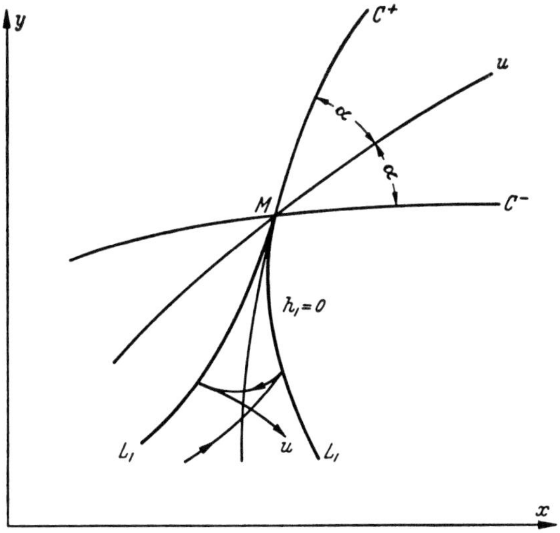

Fig. 22. Limit line with cusp.

dition $S' = \sqrt{3-s^2}/(4-s^2) \to 0$ for $s = \pm\sqrt{3}$; for $s = +\sqrt{3}$, $g \to \infty$, h finite $\alpha = 90°$ while for $s = -\sqrt{3}$, g finite, $h \to \infty$, $\alpha = 0°$. Of course, a sonic line, where *necessarily* $S' = 0$ on account of $S =$ constant, is in general not a limit line.[177]

We consider now the case $g \to \infty$, h finite, hence $S' = 0$, and $\alpha = 90°$, from (36.3). With a view to Eq. (40.1) we consider $\partial u/\partial\xi$ and $\partial u/\partial\eta$. Using (34.8) we find, as before

$$\frac{\partial u}{\partial \xi} = \frac{\partial u}{\partial s}\frac{\partial s}{\partial \xi} + \frac{\partial u}{\partial \vartheta}\frac{\partial \vartheta}{\partial \xi} = \frac{1}{g}\frac{\partial v}{\partial \vartheta}\cdot\frac{1}{S'} - h\frac{\partial v}{\partial s},$$

$$\frac{\partial u}{\partial \eta} = \frac{\partial u}{\partial s}\frac{\partial s}{\partial \eta} + \frac{\partial u}{\partial \vartheta}\frac{\partial \vartheta}{\partial \eta} = \frac{1}{g}\frac{\partial v}{\partial \vartheta}\cdot\frac{1}{S'} + h\frac{\partial v}{\partial s}.$$

Now,

$$\frac{1}{g S'} = -\sqrt{\frac{h}{g}}$$

tends to zero as $g \to \infty$ (and hence

$$\frac{1}{g S'}\frac{\partial v}{\partial \vartheta} \to 0$$

unless $dv/\partial\vartheta$ tends strongly towards infinity, which we exclude). Hence at a parabolic point ($g \to \infty$, h finite)

$$\frac{\partial u}{\partial \xi} = -h\frac{\partial v}{\partial s}, \quad \frac{\partial u}{\partial \eta} = h\frac{\partial v}{\partial s}.$$

Laying aside the case when $D \to \infty$, which corresponds to branch type singularities, wc define as *ordinary sonic point* (in contrast to limit point and branch point) a point where $g \to \infty$, h is finite, $\partial v/\partial s \neq 0$; hence both $\partial u/\partial\xi$, $\partial u/\partial\eta$ are different from zero, and $D \neq 0$. Likewise the sonic point where $h \to \infty$, g is finite is *ordinary* if $\partial u/\partial s \neq 0$ and consequently $D \neq 0$ by (40.1).

[177] It seems to me that MANDEL assumes that his "isobare limite"—our sonic line—is always a limit line. This is not so.

A new type of limit point and limit line (different from a sonic point of an L_1 or of an L_2), which we shall call *sonic limit point*, is characterized by $\alpha=90°$ (viz. $g\to\infty$, h finite) and $\partial v/\partial s=0$. At such a point either $\partial v/\partial s=0$, $\partial v/\partial\vartheta\neq 0$, or $\partial v/\partial s=0$, $\partial v/\partial\vartheta=0$. The first case can clearly happen along a whole arc of a curve which we then denote as *sonic limit line*, L_t. In the second case the point is isolated[178] (or it may present some more complicated singularity, like the intersection of an L_t and an L_1, etc.).

Since $\partial u/\partial\xi=0$, $\partial u/\partial\eta=0$, *the* L_t *under consideration is a v-line.* Hence at every point of the L_t the u-direction is normal to the L_t, and since $\alpha=90°$ it follows that both the C^+ and the C^- are enveloped by the L_t (see Fig. 23). Since at the L_t,

$$\frac{\partial v}{\partial s}=0 \quad \frac{\partial v}{\partial\vartheta}\neq 0, \quad dv=\frac{\partial v}{\partial s}ds+\frac{\partial v}{\partial\vartheta}d\vartheta=\frac{\partial v}{\partial\vartheta}d\vartheta,$$

we see that if $dv=0$ (u-line), also $d\vartheta=0$, and vice versa, hence the line $\vartheta=$ constant, the isocline, is likewise normal to the L_t. More generally, and in analogy to our study of the L_1 and L_2, since on the L_t we have $du=0$, $\partial v/\partial s=0$ $\partial v/\partial\vartheta\neq 0$, we conclude that

$$dx=\frac{\partial x}{\partial v}dv=\frac{\partial x}{\partial v}\frac{\partial v}{\partial\vartheta}d\vartheta=-\sin\vartheta\frac{\partial v}{\partial\vartheta}d\vartheta,$$

$$dy=\frac{\partial y}{\partial v}dv=\frac{\partial y}{\partial v}\frac{\partial v}{\partial\vartheta}d\vartheta=\cos\vartheta\frac{\partial v}{\partial\vartheta}d\vartheta.$$

Thence

$$\frac{dy}{dx}=-\cot\vartheta, \quad \text{if} \quad d\vartheta\neq 0. \tag{41.1}$$

Therefore (Fig. 23) *an element of any curve in the stress graph on which* $d\vartheta\neq 0$ *maps onto an element in the x, y-plane which has the second principal direction.* (This is true for the sonic limit line itself, *which therefore points in the v-direction and is itself a* v-line.) The characteristics make the angle $\alpha=90°$ with the u-direction, by which they are separated; hence they are tangent to the v-direction, and one is the continuation of the other.

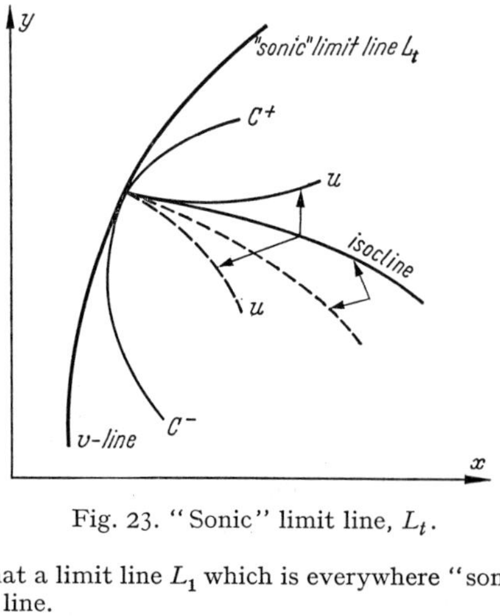

Fig. 23. "Sonic" limit line, L_t.

[178] It can be shown that a limit line L_1 which is everywhere "sonic" is not possible. Hence L_t is the only sonic limit line.

Since the sonic limit line is a v-line and since along it $s=$ constant, it follows that $\partial s/\partial v=0$ along the L_t and $\partial s/\partial u\to\infty$ across the L_t as may be seen from the first Eq. (40.4) with $h_1=h_2$, $S'\to 0$.

If, for an element in the stress graph, $d\vartheta=0$ (exceptional direction), its map in the physical plane does not have the v-direction, and it can be shown that it has a cusp at the L_t. This holds true for the u-line and for the isocline. For the latter, clearly, $d\vartheta=0$, and the former, being normal to the v-line in the x, y-plane must have the exceptional (radial) direction in the stress graph.

In the second case, where $h\to\infty$, g finite, $\alpha=0°$, Eq. (41.5) is replaced by $dy/dx=\tan\vartheta$, if $d\vartheta\neq 0$. The roles of first and second principal directions are interchanged.

We review: A sonic limit line L_t, i.e., a sonic line along which $\partial v/\partial s=0$ is (piecewise) a v-line. At a point of it at which $\partial v/\partial\vartheta\neq 0$ it is envelope of both families of characteristics and, actually, of all curves whose images in the stress graph do not have there the exceptional direction ($d\vartheta=0$); the curves whose images have this direction have cusps at the L_t.

42. Branch line singularities. Here $J=h_1 h_2\sin 2\alpha$ cannot become infinite unless either h_1 or h_2 becomes infinite. The loci $h_1(x,y)=\infty$ and $h_2(x,y)=\infty$ are called branch lines B_1 and B_2; there is no analogue to the sonic limit line. We may achieve formal similarity with preceding considerations by putting

$$k_1=(h_1\sin 2\alpha)^{-1},\qquad k_2=(h_2\sin 2\alpha)^{-1},\qquad \alpha\neq 0°,\ 90°. \tag{42.1}$$

Interchanging (x,y) with (ξ,η) in Eqs. (37.10) we obtain

$$\begin{aligned}
\frac{\partial\xi}{\partial x}&=k_1\sin\varphi^+, &\qquad \frac{\partial\eta}{\partial x}&=-k_2\sin\varphi^-,\\
\frac{\partial\xi}{\partial y}&=-k_1\cos\varphi^+, &\qquad \frac{\partial\eta}{\partial y}&=k_2\cos\varphi^-,\\
j&=(h_1h_2\sin 2\alpha)^{-1}=k_1k_2\sin 2\alpha. & &
\end{aligned}\tag{42.2}$$

The image in the ξ, η-plane, or in the s, ϑ-plane, of the branch line B_1 of the x, y-plane is called *the edge*, b_1 (Fig. 24). If $k_1=0$, then $d\xi=0$, hence $\xi=$ constant. *Therefore b_1 is an η-line in the ξ, η-plane.* It follows, just as before, that all lines

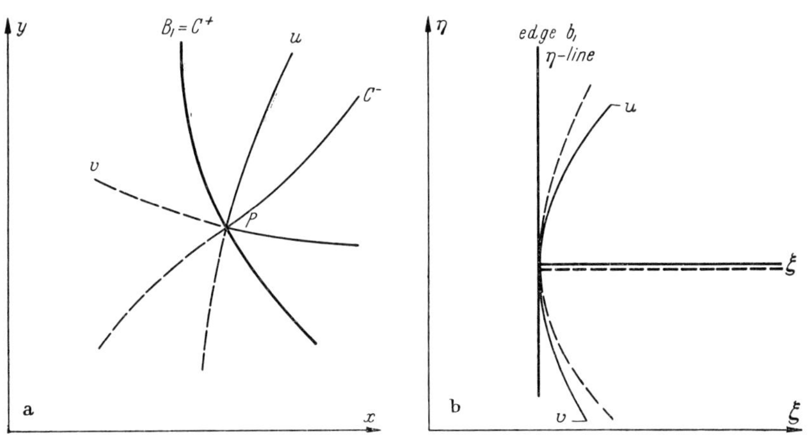

Fig. 24a and b. Branch line in (a) physical plane and (b) characteristic plane.

in the x, y-plane, with the exception of such lines in the x, y-plane as have the C^--direction (the exceptional direction) at the point of intersection with the B_1, appear in the ξ, η-plane as tangents to the straight vertical line, b_1, the edge. Among these are *the principal trajectories, and b_1 is an envelope of them.* It can be concluded, as before, that a line which has ξ-direction at the intersection with the b_1, and in particular a ξ-line, must have a cusp there. Since, however, *the ξ-lines* are straight horizontal lines, they *return at the edge.* For a C^--line,

$$d(\vartheta - \alpha) = \frac{\partial(\vartheta - \alpha)}{\partial \xi} d\xi = -\left(1 + \frac{\alpha'}{S'}\right) d\xi,$$

and since $d\xi$ changes sign at an intersection with b_1, the same must be true for $d(\vartheta - \alpha)$ $\left(\text{unless,}^{179} \text{ quite exceptionally, } 1 + \frac{\alpha'}{S'} = 0\right)$; hence *each C^- has an inflection point at its intersection with the B_1.* Since the isobars and isoclines are the lines $\eta \pm \xi =$ constant, they intersect the vertical edge at $\pm 45°$ and are not tangent to it. Therefore, in the flow x, y-plane *they have the exceptional, the C^--direction, at B_1.* Hence, at each point of B_1, *the C^-, the isobar and the isocline touch each other, while the trajectories bisect the angles between B_1 and the inflection tangent of the C^-* and cross the B_1 without singularity. The direction of the vector grad s at B_1 is perpendicular to the C^--direction, and the same holds for grad ϑ.

A branch line has physical reality. In fact, its characteristic property of dividing two plastic regions in which the same stress occurs at different points is nothing out of the ordinary. However, one and the same single-valued stress-graph solution cannot represent such a distribution. Hence a representation of such a distribution, e.g. in form of a series, must break down at the edge. To save space we do not review the above properties of branch lines.

We repeat: A limit line L_1 is the image of $h_1(\xi, \eta) = 0$, where a single-valued solution of the linearized stress-graph equations (or characteristic-plane equations) is assumed; $J(\xi, \eta) = 0$ along the L_1; a separate discussion is needed at "sonic" points. A branch line B_1, defined by $k_1(x, y) = 0$, corresponds to a single-valued solution of the original non-linear physical plane equations, and along it $j(x, y) = 0$, $j = J^{-1}$.

In the next sections we shall study *simple waves,* which may be defined by the condition that $j(x, y) = 0$ not along a line only but in a two-dimensional region, and where the transition to the stress graph, which in Sect. 34 formed our starting point, is no longer possible.

b) Simple waves.

43. Definition. Simple waves is the name of an important class of solutions of the basic equations. These particular solutions play a very useful role in building up solutions to boundary-value problems in plasticity.[180] They may be introduced in various essentially equivalent ways. We ask, e.g., for a stress

[179] For the Coulomb condition

$$1 + \frac{\alpha'}{S'} = 1;$$

for the parabola condition

$$1 + \frac{\alpha'}{S'} = \frac{1}{2}.$$

[180] In gas dynamics the term "Prandtl-Meyer solution" is used; in plasticity, the terms "fan" and "lost solution" are likewise in use.

distribution in which the lines $s = \text{constant}$, the isobars, coincide with the isoclines; we call such lines w-lines. Each w-line is thus mapped, by definition, onto one point (s, ϑ) of the stress graph. We assume that these points do not all coincide, i.e. that the solution does not merely represent a region of constant state. Hence the whole set of w-lines, or the whole region R in the x, y-plane covered by w-lines, is mapped onto one line Λ of the stress graph.

The existence of the line Λ in the stress graph implies the existence of a relation between s and ϑ, and, as a consequence, the vanishing throughout R of the Jacobian d of (40.2).

We conclude, just as in Sect. 40, that if $gh < 0$, the Jacobian d can only vanish at isolated points, and we consider the case where there are real characteristics.

Among the lines crossing the w-lines there must be at least one set of characteristics C, and the image of each of these must be on Λ. In fact, each point of such a line C must map onto a point of Λ. It follows that Λ is a Γ^+ or a Γ^- and that each characteristic of the other set—each C^- if Λ is a Γ^+—is mapped onto a single point of Λ. Thus, the w-lines form this second set of characteristics. Since on each of them both ϑ and s, and therefore α as a function of s, are constant, it follows that $\vartheta \mp \alpha$, that is, the slope of each w-line, is constant. Hence, the w-lines are straight.

Since the whole region R covered by the w-lines is mapped onto one characteristic, Γ^+ or Γ^-, the equation of this characteristic, $S \mp \vartheta = \text{constant}$, is valid throughout R. Hence the definition: *A plane stress distribution is a simple wave solution if one set of characteristics consists of straight lines on each of which the stress tensor Σ is constant. The image of this region in the stress graph is an arc of a Γ-characteristic. If it is a Γ^+, the C^- are straight and $S - \vartheta$ has a constant value throughout R, and correspondingly in the case of a Γ^-.*

Throughout the simple wave region we have $d = \partial(s, \vartheta)/\partial(x, y) = 0$. Therefore, the interchange of variables (x, y) and (s, ϑ), so useful in general, is not possible here, and in this sense simple waves are "lost solutions" since they cannot be found as solutions of the linearized equations in the stress graph.

The simple wave pattern forms the transition between the general distribution, where a region of the x, y-plane is mapped onto an area of the stress graph, and the extremely degenerate "constant state", where an area of the x, y-plane maps onto a single point.

This is expressed in the theorem: *The region adjacent to a domain of constant state is either another domain of constant state or a simple wave region.* In other words, a region of constant state, which maps into a single stress graph point, cannot be directly adjacent to a "general" state of stress. A simple wave must form the link between them.[181]

A simple wave can connect any uniform hyperbolic state Σ_1 with another uniform hyperbolic state Σ_2, provided either $S + \vartheta$ or $S - \vartheta$ has the same value in both states. By combining a Γ^+-wave and a Γ^--wave and inserting a uniform state between the two, a given final state Σ_2 can be reached, in general, and in many cases in two ways.

Simple waves, denoted in our Chap. B as "degenerate solutions", have already been considered by the pioneers in our field. In the present, more general, case they have been studied independently by MANDEL[182] and GEIRINGER.[183] Cf. also [*26*, p. 323].

[181] See proof in H. GEIRINGER: Advances, p. 258.
[182] H. MANDEL, reference 138.
[183] GEIRINGER: Advances, pp. 257–270.

44. Simple waves. Continuation. An individual wave may be specified in several ways, e.g. by giving *a certain characteristic* Γ_0' *as the image of the whole "minus" wave*; and, in addition, giving in the x, y-plane *a family of straight lines to represent the* C^+. If these C^+ have a point in common, we speak of a *centered wave*. The stress distribution for the wave follows immediately, if merely φ, the angle which a straight C^+ (a straight C^-) makes with the positive x-axis, is given for each straight characteristic. Then for any specific yield condition the two relations hold

$$S(s) \mp \vartheta = \text{constant}, \qquad \vartheta \mp \alpha = \varphi. \tag{44.1}$$

The relations (44.1) do not depend on the type of the family of straight lines. For example it makes no difference whether the wave is centered or not. As in all similar formulas, the upper (lower) sign holds for a Γ^+ (a Γ^-) wave, and the "constant", as well as φ, are known. Hence along each single C^- (C^+), s and ϑ are determined by the two Eqs. (44.1).

For the computation of the *principal trajectories* and of the *cross characteristics*, i.e. the other set of characteristics, we need the equation of the particular family of lines which form the straight characteristics. An adequate way to specify this set of straight lines is to give e.g. *one* cross characteristic or *one* principal trajectory. In fact, knowing one trajectory—say a u-line—in the x, y-plane we know ϑ at each point of this line; then, in case of a minus wave, $S(s) = \text{constant} - \vartheta$ determines s and, therefore, α; hence, at each point of the given u-line, the direction $\vartheta + \alpha$ of the C^+ through this point is known. Similarly, if one C^- is given, $\vartheta - \alpha$ is known along it and $S(s) + \alpha(s) = \text{constant} - (\vartheta - \alpha)$ provides s; then α, and finally $\alpha + \vartheta$ follow at all points of the C^-. To present the computations consider the first case, where one u-trajectory is given: $x = a(t)$, $y = b(t)$, $db/da = \tan\vartheta$; t is a parameter. Consider a minus wave with the C^+ as straight characteristics. For a point $P(x, y)$ on the C^+ passing through the point P_0 on the given initial trajectory: $x = a + r\cos\varphi^+$ and $y = b + r\sin\varphi^+$, where $r = P_0P$ (Fig. 25). If we consider, a, b, r and $\varphi^+ = \varphi$ as functions of t, the equation of the

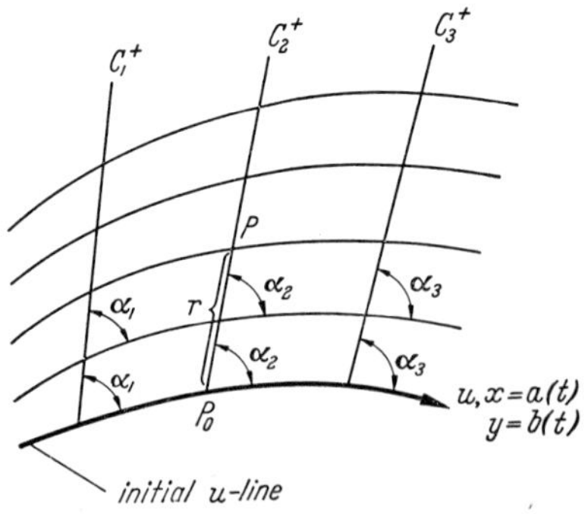

Fig. 25. Straight characteristics and cross characteristics in (backward) simple wave.

trajectory through P is

$$(db + r\cos\varphi\, d\varphi + dr\sin\varphi)\cos\vartheta - (da - r\sin\varphi\, d\varphi + dr\cos\varphi)\sin\vartheta = 0,$$

and since

$$db\cos\vartheta - da\sin\vartheta = 0,$$

it follows, with $\varphi - \vartheta = \alpha$, that

$$r\, d\varphi\cos\alpha + dr\sin\alpha = 0,$$

or

$$\frac{r\,d\varphi}{dr} = -\tan\alpha. \tag{44.2}$$

We note that this is the same equation which can be written down immediately for a centered wave, with r, φ as polar coordinates. For the cross characteristics, always assuming a Γ_0^--wave, we find in the same way

$$\frac{r\,d\varphi}{dr} = \tan 2\alpha. \tag{44.3}$$

From (44.1), writing $2\eta_0$ for "constant", we have

$$\alpha = \varphi - \vartheta = \varphi - 2\eta_0 + S(s), \quad \text{or} \quad \alpha(s) - S(s) = \varphi - 2\eta_0.$$

Thus φ is determined in terms of s; likewise s and $\alpha(s)$ are determined in terms of φ. Examples follow in the next section.

We proceed with the integration of (44.3). In $dr/r = -\cot 2\alpha\, d\varphi$ we substitute $d\varphi = d\vartheta + d\alpha$ and $d\vartheta = -dS$. Then

$$\frac{dr}{r} = \cot 2\alpha (dS - d\alpha).$$

Using (37.6′) and (37.4) we obtain

$$\frac{dr}{r} = -\cot 2\alpha \frac{d\sigma}{2\tau} = -\frac{d\tau}{d\sigma}\frac{d\sigma}{2\tau} = -\frac{d\tau}{2\tau}, \tag{44.4}$$

which gives

$$r\sqrt{\tau} = \text{constant}. \tag{44.5}$$

To obtain the above equation in r, φ we need τ in terms of φ for the specific yield condition. The same relation (44.5) results in case of a Γ^+-wave. The cross characteristics have cusps on the envelope of the rectilinear characteristics.

In the case of a centered wave, the cross characteristics are similar to each other and r in (44.5) is the radius vector. If $\alpha = \text{constant}$ they are logarithmic spirals (see Sect. 32 for $\alpha = 45°$).

45. Simple waves for particular yield conditions. α) *Quadratic yield condition.* Consider the plus wave

$$\vartheta = S(s) \tag{45.1}$$

where

$$S(s) = -\arctan\frac{s}{\sqrt{3-s^2}} + \frac{1}{2}\arctan\frac{s}{2\sqrt{3-s^2}}. \tag{45.1′}$$

The straight C^- makes the angle $\varphi^- = \varphi = \vartheta - \alpha$ with the x-axis.

Using (38.1), (38.2), we find that

$$\varphi = \varphi^- = \vartheta - \alpha = -\arctan\frac{s}{\sqrt{3-s^2}} + \frac{1}{2}\arctan\frac{s}{2\sqrt{3-s^2}} - \arctan\frac{\sqrt{12-3s^2}+s}{2\sqrt{3-s^2}}.$$

To simplify we introduce

$$t = \frac{s}{\sqrt{3-s^2}}, \quad s^2 = \frac{3t^2}{1+t^2}. \tag{45.2}$$

Then

$$\varphi^- = -\arctan t + \frac{1}{2}\arctan\frac{t}{2} - \arctan\left(\frac{1}{2}\sqrt{4+t^2} + \frac{t}{2}\right). \tag{45.3}$$

It may be verified that

$$\arctan\frac{\sqrt{4+t^2}+t}{2} - \frac{1}{2}\arctan\frac{t}{2} = 45°.$$

Therefore

$$\varphi = -\operatorname{arc\,tan} t - 45^\circ, \qquad t = -\tan(\varphi + 45^\circ),$$
$$s = -\sqrt{3}\sin(\varphi + 45^\circ), \qquad \vartheta = S(s). \tag{45.4}$$

As s goes from $-\sqrt{3}$ to zero, to $+\sqrt{3}$, t goes from $-\infty$, to zero, to $+\infty$, ϑ from 45°, to zero, to $-45°$, and φ from 45°, to $-45°$, to $-135°$. Hence, in a *complete wave*, φ rotates through 180° (Fig. 26).

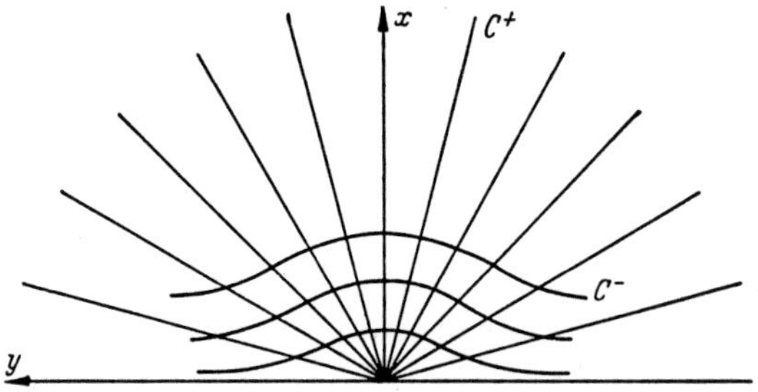

Fig. 26. Quadratic limit. Complete centered wave. Straight characteristics and cross characteristics.

The image of the Γ_0^-:

$$\vartheta = -S(s) - 45^\circ = \operatorname{arc\,tan} t - \frac{1}{2}\operatorname{arc\,tan}\frac{t}{2} - 45^\circ, \tag{45.5}$$

gives with $\varphi^+ = \varphi = \vartheta + \alpha$,

$$\varphi = -45^\circ + \operatorname{arc\,tan} t - \frac{1}{2}\operatorname{arc\,tan}\frac{t}{2} + \operatorname{arc\,tan}\left[\frac{1}{2}\sqrt{4+t^2} + \frac{t}{2}\right] = \operatorname{arc\,tan} t. \tag{45.5'}$$

Thus

$$\tan\varphi = t, \qquad \varphi = \operatorname{arc\,tan} t, \qquad s = \sqrt{3}\sin\varphi. \tag{45.6}$$

As s runs from $-\sqrt{3}$ to $+\sqrt{3}$, φ increases from $-90°$ to $+90°$, and ϑ from zero to 90°. We compute the *cross characteristics* for this last wave using (44.5). To express τ in terms of φ we use (38.3) and (45.6) and find

$$\sigma = \frac{4k}{3}s = \frac{4k}{\sqrt{3}}\sin\varphi, \qquad \tau = \frac{2k}{\sqrt{3}}\cos\varphi, \tag{45.7}$$

which satisfy the last Eq. (38.3). From (44.5)

$$r^2\cos\varphi = \text{constant}. \tag{45.8}$$

β) *Parabola limit.* With $t = s/a$ we consider the Γ^- wave, using (38.4)

$$\vartheta = -S(s) = \arcsin t, \quad \text{thus} \quad t = \sin\vartheta,$$
$$\tan\alpha = \sqrt{\frac{1+t}{1-t}} = \tan\left(45^\circ + \frac{\vartheta}{2}\right). \tag{45.9}$$

$$\alpha = \frac{\vartheta}{2} + 45^\circ, \qquad \varphi^+ = \vartheta + \alpha = \frac{3\vartheta}{2} + 45^\circ. \tag{45.10}$$

Hence, with $\varphi^+ = \varphi$,

$$\vartheta = \tfrac{2}{3}(\varphi - 45^\circ), \qquad s = a\sin\vartheta. \tag{45.11}$$

Thus we know s and ϑ along each straight C^+ which makes the angle φ with the x-axis. Here the range of ϑ is 180° (from $-90°$ to $+90°$); that for φ is 270°, from $-90°$ to $+180°$.

For the *cross characteristics* (Fig. 27a) we use (44.3):

$$\frac{r\,d\varphi}{dr} = -\tan 2\alpha = -\tan(\vartheta + 90°) = -\tan\frac{2\varphi + 180°}{3}, \tag{45.12}$$

$$r = r_0\left[\sin\frac{2\varphi + 180°}{3}\right]^{-\frac{3}{2}}. \tag{45.13}$$

The *principal stress* lines may likewise be found (Fig. 27b).

Velocities. We first ask whether *velocity distributions* exist *such that* $\boldsymbol{v}$ = constant *along each straight characteristic.*

In other words, is it possible to prescribe initial values v_x, v_y along a non-characteristic curve K such that the above holds? Obviously, then the initial distribution along K is subject to a condition. In fact, consider the image $\bar{K}$ of K in the velocity plane. Since to each C^+ corresponds only one point of $\bar{K}$ (because of v_x = constant, v_y = constant on C^+), the $\bar{K}$ is the image of the whole simple wave region, hence also of all C^-. Therefore [see (36.9)], along $\bar{K}$ the relation

$$\frac{dv_y}{dv_x} = -\cot(\vartheta - \alpha) \tag{45.14}$$

must hold. More explicitly: Through every point of K passes a straight C^+ associated with constant values s, ϑ and an angle $(\vartheta - \alpha)$ which is given in terms of s or of ϑ, say

$$\cot(\vartheta - \alpha) = H(s).$$

Suppose K given as $x = x(s)$, $y = y(s)$ and along it $v_x = v_x(s)$, $v_y = v_y(s)$; then $v_y'/v_x' = -H(s)$ must hold along K. It can be shown that if this condition holds along K, and consequently in the whole region, there will exist a corresponding velocity distribution.

Now, consider the general case. Let us have a coordinate system consisting of the characteristics with the straight C^+ as η-lines, the cross characteristics

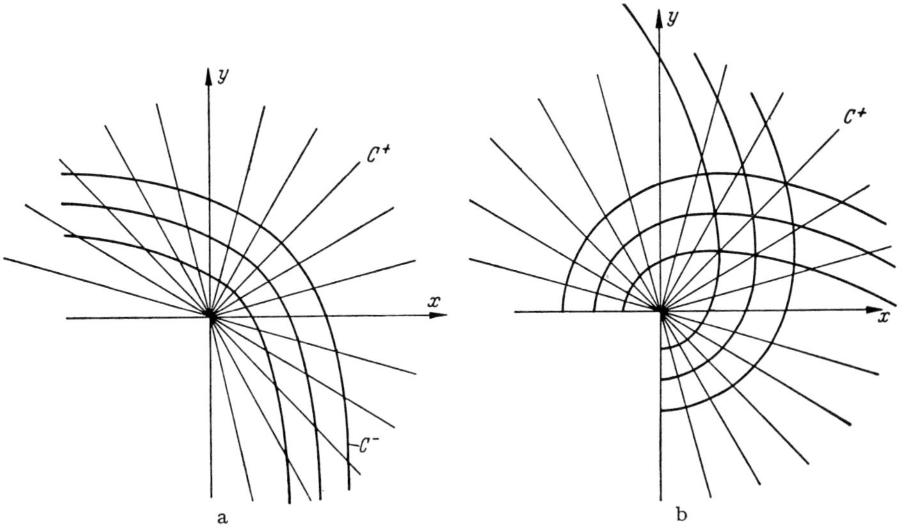

Fig. 27a and b. Parabola limit. Complete entered wave. (a) Straight characteristics and cross characteristics. (b) Lines of principal stress.

as ξ-lines. Since the angles φ^-, φ^+, α are all constant along the C^+, any of these may serve as ξ-coordinate. Choose $\xi = \varphi^+ = \varphi$. To define η we choose a certain C^+ with $\varphi = \varphi_0$, and on it a point O, and take for the η of an arbitrary point P the distance OP' where P' is the point of intersection of the C^- through P with the fixed C^+ (Fig. 28). (Of course, the characteristic coordinates of the previous sections cannot be used here.)

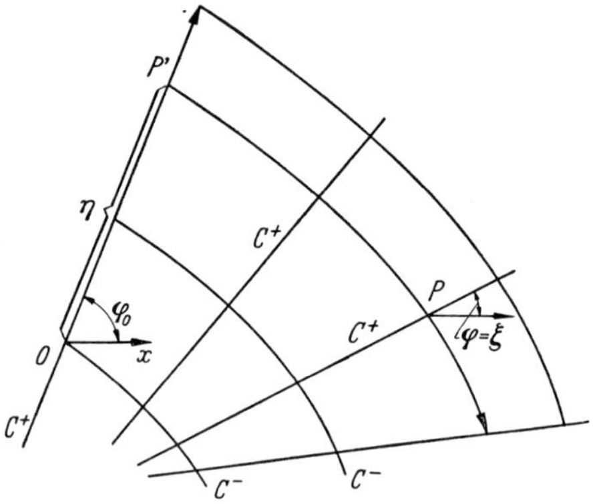

Fig. 28. Characteristic coordinates in simple wave.

As before, v_1 and v_2 are the components of $\boldsymbol{v}$ in the directions of C^-, C^+ respectively. Consider Eqs. (36.10). The second gives $dv_2/dl^+=0$. *The component of* $\boldsymbol{v}$ *in the direction of a straight* C^+ *is constant along it,*

$$\frac{\partial v_2}{\partial \eta}=0. \tag{45.15}$$

The first Eq. (36.11) gives

$$\frac{\partial v_1}{\partial \varphi^-}+v_1 \cot 2\alpha=\frac{v_2}{\sin 2\alpha}. \tag{45.16}$$

For $\alpha=45°$ this yields

$$\frac{\partial v_1}{\partial \xi}-v_2=0.$$

In general, since v_2 and α do not change along a C^+, we see that *not only* v_2, *but also*

$$\frac{\partial v_1}{\partial \varphi^-}+v_1 \cot 2\alpha,$$

remains constant along a C^+. (The constant value of this expression is, of course, $v_2/\sin 2\alpha$.)

To integrate (45.16) we observe that if v_2 is given along a curve which intersects the C^+, then v_2 is known everywhere. We use $\xi=\varphi^+=\varphi$ as independent variable and have $\varphi^-=\varphi^-(\xi)$, $\alpha=\alpha(\xi)$, $v_2=v_2(\xi)$. Putting

$$\cot 2\alpha \frac{\partial \varphi^-}{\partial \xi}=a(\xi), \qquad \frac{v_2}{\sin 2\alpha}\frac{\partial \varphi^-}{\partial \xi}=b(\xi),$$

we reduce Eq. (45.16) to the linear differential equation

$$\frac{\partial v_1}{\partial \xi}+a(\xi)\, v_1-b(\xi)=0,$$

the integral of which is

$$v_1(\xi,\eta)\, e^{\int_{\xi_0}^{\xi} a\, dt}=\int_{\xi_0}^{\xi} b\, e^{\int_{\xi_0}^{\xi} a\, dt}\, d\xi+\psi(\eta), \tag{45.17}$$

where

$$\psi(\eta)=v_1(\xi_0,\eta).$$

Consider a *Cauchy problem*. Along a non-characteristic curve K with equation $\eta = \eta(\xi)$ both v_1 and v_2 are given: $v_1 = g(\xi)$, $v_2 = h(\xi)$. We know v_2 along each C^+ intersecting K, and we know v_1 from Eq. (45.17).

In a characteristic initial-value problem we assume e.g. that v_2 is given on an arc OC of the C^-: $\eta = \eta_0$, while v_1 is given on a segment OB of the straight C^+: $\xi = \xi_0$. We then know v_2 on each C^+ intersecting OC, we know v_1 from (45.17), and $\psi(\eta) = v_1(\xi_0, \eta)$. Further problems of this type are equally easy.

Thus, in a simple wave region we can determine explicitly both stresses and velocities.

c) Various remarks.

46. Remarks on the approximate solution of initial-value problems. (See Sect. 31.) In the *Cauchy problem, along an arc K given by $x = x(t)$, $y = y(t)$, with suitable regularity assumptions, values of s and ϑ are given in such a way that K nowhere has the characteristic direction*; that means that the slope of K nowhere is equal to either $\tan(\vartheta + \alpha(s))$ or to $\tan(\vartheta - \alpha(s))$. These data determine a solution on both sides in a neighborhood of K which is contained in the corresponding characteristic quadrangle. *If, in addition, two components of $\boldsymbol{v}$ are given along K, also the velocity* is uniquely determined in the characteristic quadrangle.

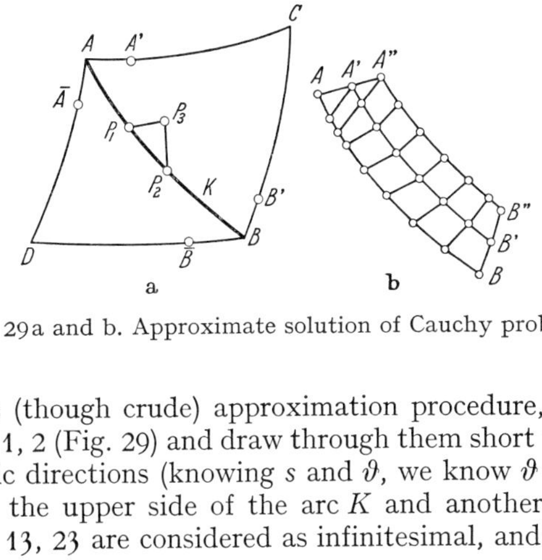

Fig. 29a and b. Approximate solution of Cauchy problem.

For an obvious (though crude) approximation procedure, consider on K the neighboring points 1, 2 (Fig. 29) and draw through them short rectilinear segments in the characteristic directions (knowing s and ϑ, we know $\vartheta \pm \alpha$). They have an intersection, 3, on the upper side of the arc K and another on the lower side. If all distances 12, 13, 23 are considered as infinitesimal, and if we neglect terms of higher order, this amounts to

$$s_3 - s_1 = \pm (\sqrt{g\,h})_1 \cdot (\vartheta_3 - \vartheta_1), \qquad s_3 - s_2 = \mp (\sqrt{g\,h})_2 \cdot (\vartheta_3 - \vartheta_2), \tag{46.1}$$

in an obvious notation. In these linear equations for s_3, ϑ_3 either both upper signs hold or both lower ones. The determinant $\pm[(\sqrt{gh})_1 + (\sqrt{gh})_2]$ is different from zero, since $\sqrt{gh}$ cannot change sign between 1 and 2, and both s_3, ϑ_3 can be evaluated. In this way, starting from a sequence of points on K, say between A and B, one can derive from the given values s, ϑ on AB the values along a second row of points, $A'B'$. Continuing in the same manner, one eventually finds s- and ϑ-values for all lattice points within a curvilinear triangle ABC, where AC and BC are characteristics of two different kinds—provided the procedure does not break down earlier, which may happen if the direction of a cross line such as $A'B'$, $A''B''$, ... somewhere approaches a characteristic direction. However, due to the non-characteristic nature of AB and the continuity of all functions involved, this can happen only at a finite distance from AB. All these conclusions apply, of course, also to the triangle ABD on the lower side of AB.

Thus, while such a step procedure does not constitute an existence proof, it shows that a solution exists in a neighborhood of AB, and, insofar as a solution exists in $ABCD$, it is uniquely determined there, and we can approximate it by this step procedure, unless the procedure breaks down in the manner explained above.

As a second case (Fig. 30), *consider data given along two intersecting lines AB, AC, one of them a characteristic.* Assume that AB is a minus characteristic and that the non-characteristic arc AC lies in the angular space between the minus characteristic AB and the positively directed plus characteristic through A. Values of s and ϑ along the C^- must be given in such a way that at each point

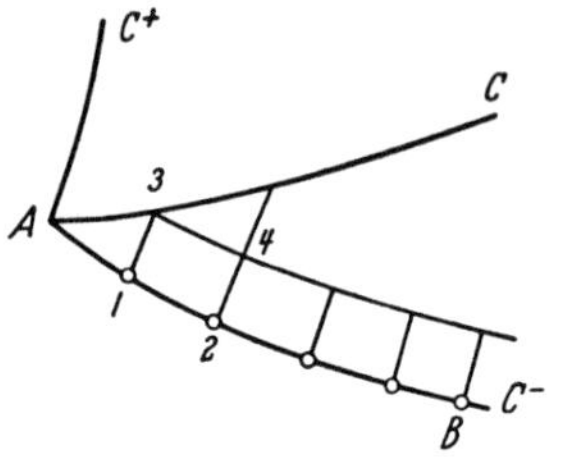

Fig. 30. Approximate solution of third or mixed problem.

its angle φ^- with the x-axis equals $\vartheta - \alpha(s)$ and such that the second Eq. (36.5) holds. It follows that, if the geometric shape of AB is given, we may prescribe only the value of either s or ϑ at *one* point of AB; then s, ϑ follow along AB.

We further suppose that either ϑ or s or a combination of both is given along the non-characteristic arc AC ("*third*" or "mixed" problem). From the data along AB, the initial elements of the plus characteristics at all points of AB can be derived, and we assume that they are plotted in the direction towards AC. If the point 1 is adjacent to A, the characteristic element through 1 will intersect the line AC in some point 3. From a given value at 3 and from $s_3 - s_1 = (\sqrt{gh})_1(\vartheta_3 - \vartheta_1)$, the quantities ϑ_3, s_3 can be derived. This, then, enables us to find the beginning of the minus characteristic 34 through 3, and the above used compatibility relations, applied to the segments 34 and 24, give the values s_4, ϑ_4. In this way, step by step, a triangle ABC where C is the intersection of the C^+-characteristic through B with the curve C can be filled by a net of points at which s and ϑ are known.

A slight modification of the procedure takes place if AC is likewise a characteristic, here a C^+ (*second*, or *characteristic problem*). If we know that of two geometrically given intersecting arcs one is a C^-, the other a C^+, then neither ϑ nor s can be arbitrarily prescribed anywhere along these curves. Their values follow from the two characteristic conditions and the two compatibility conditions, with both s and ϑ at A being determined by the two characteristic conditions. A stepwise construction of the net inside the characteristic quadrangle $ABDC$ is analogous to the preceding one. Point 3 is the intersection of a small C^+-segment 13 and a small C^--segment 1'3. The existence of a solution, however, is guaranteed only in a neighborhood of point A and, similarly, for the mixed problem.

In each case, the complete problem also calls for the determination of the velocities. Knowing $s = s(x, y)$, $\vartheta = \vartheta(x, y)$ at a point, we know $\alpha(x, y)$. We then may use the finite difference equivalent of the velocity Eqs. (36.11), or perhaps (36.9), the choice dependent on what exactly are the initial data.

Returning to the stress problem, we recall that the knowledge of the net of characteristics in a region is equivalent to knowing the stress tensor. To construct this net we used the compatibility relations (36.5) in the form of finite difference equations and the characteristic conditions (36.3). This most direct but also most crude procedure, essentially due to MASSAU,[184] becomes a practical approximation procedure if the compatibility relations are used in the form (36.8), together with a tabulation of $S(s)$ and $\alpha(s)$ for a chosen yield condition. A first approximation thus found can then be improved by iteration.

We may arrange the approximation procedure also in the following not essentially different way. Consider, e.g. the characteristic initial-value problem where we know that two arcs OA, OB in the x, y-plane are parts of a ξ-line and an η-line, respectively. We assume that neither of them is straight. (If e.g. OB is straight, all C^+ characteristics in the simple wave region must be straight.) Consider a subdivision $\xi_0, \xi_1, \ldots, \xi_m, \ldots$ on OA and $\eta_0, \eta_1, \ldots, \eta_n, \ldots$ on OB and a corresponding lattice where a general nodal point has the coordinates (ξ_m, η_n) or, briefly, (m, n). We saw that, if we know that OA and OB are characteristics, we know both ϑ and s along them, as well as the constants ξ_0, η_0, where $\xi = \xi_0$, on OC, $\eta = \eta_0$ on OB. Then from Eqs. (37.7):

$$\vartheta_{mn} - \vartheta_{m0} - \vartheta_{0n} + \vartheta_{00} = 0, \tag{46.2}$$

and similarly,

$$S_{mn} - S_{m0} - S_{0n} + S_{00} = 0. \tag{46.3}$$

Hence, we know ϑ and S, i.e. ϑ and s, at all nodal points.

We still need the coordinates x_{mn}, y_{mn} in the x, y-plane to which the s_{mn}, ϑ_{mn} belong. To find these coordinates, our original independent variables, we need a step-by-step procedure. In the method explained at the beginning of this section, this procedure consisted in the actual drawing of characteristic segments. In the present method, it consists in setting down the equations

$$\left.\begin{aligned} y_{mn} - y_{m-1,n} &= \tan \frac{\overset{+}{\varphi}_{m-1,n} + \overset{+}{\varphi}_{m,n}}{2} \cdot (x_{mn} - x_{m-1,n}), \\ y_{mn} - y_{m,n-1} &= \tan \frac{\overset{-}{\varphi}_{m,n-1} + \overset{-}{\varphi}_{m,n}}{2} \cdot (x_{mn} - x_{m,n-1}). \end{aligned}\right\} \tag{46.4}$$

These are two simultaneous linear equations for x_{mn}, y_{mn}, if $x_{m,n-1}, \ldots, y_{m-1,n}$ are known. The $x_{m,0}, y_{m,0}, x_{0,n}, y_{0,n}$ are the known coordinates of the original subdivision points along OA and OB. The φ^+, φ^- are known at all nodal points since we know $s_{m,n}, \vartheta_{m,n}$. Again the procedure can be improved by iteration.

This and other more refined procedures have been described by HILL [*6*, p. 140] for plane strain. They can be generalized, to a certain extent, to the case of other yield conditions. A more geometric arrangement, using stress graph and velocity plane is also useful.[185]

A graphical method for the solution of boundary-value problems based on the simultaneous use of physical plane, hodograph and stress plane has been sug-

---

[184] I. MASSAU: Mémoire sur l'intégration des équations aux dérivées partielles, Gand, Meya van Loos, 1899; reprinted as Edition du Centénaire, Mons, 1952.

[185] H. GEIRINGER: Advances, p. 250seq.

gested independently by A. P. GREEN[186] and W. PRAGER. A more recent paper by A. M. SOKOLOVSKY[187] is along similar lines. In a brief note GREEN extends the method to the case of plane stress.[188]

47. Summary remarks on some further problems. α) *Discontinuities*. The same principles as used in Sect. 33 may be applied to the general problem of this Chapter. Consider Eqs. (36.5). For a characteristic C^- the second Eq. (36.5) is the compatibility relation and the first the "remaining equation" (see Sect. 16). Since in this first equation both variables s and ϑ are differentiated across the C^-, none of these variables can jump. *The stresses must be continuous across a characteristic.* Consider next the velocities and Eqs. (36.10), (36.11). For a C^- the first Eq. (36.11) is the compatibility equation and the second the "remaining equation". In this second equation v_2 is differentiated across the C^-, but v_1 is not. *Hence the tangential velocity v_1 may jump across C^- while the normal component must remain continuous,* and, of course, similarly for the C^+.

Hence stress discontinuities can arise only across non-characteristic lines. As in plane strain, it follows that *across such a discontinuity line d,* which is assumed to have the y-direction at the point under consideration, *σ_x, τ and $\boldsymbol{v}$ remain continuous, while σ_y changes abruptly.* It also follows, as in Sect. 24, that $\dot{E}=0$ at every point of d. In fact

$$\dot{\varepsilon}_y^{(1)}=\lambda^{(1)}\left(\frac{\partial f}{\partial\sigma_y}\right)_1,\qquad \dot{\varepsilon}_y^{(2)}=\lambda^{(2)}\left(\frac{\partial f}{\partial\sigma_y}\right)_2,$$

where f is the yield function, used as plastic potential, and the subscript of $\partial f/\partial\sigma_y$ means that $\sigma_y^{(1)}$ or $\sigma_y^{(2)}$ are used in $\partial f(\sigma_x,\sigma_y,\tau)/\partial\sigma_y$. These two values must be opposite in sign, just as in Sect. 24, and the conclusion remains the same.

In the case of Coulomb's condition [Eqs. (35.1) and (38.7)] the stress discontinuities have been studied by SHIELD.[189] He applies his results to the wedge, which was also PRAGER'S[190] first example. The discontinuities for plane stress, for v. Mises' yield condition, have been investigated by HILL.[191]

β) *Plastic rigid boundary.* It has been shown by HILL [*6*, p. 150] and by LEE[192] that in plane strain the boundary between plastic and rigid material must consist of slip lines. (It can also be an envelope of slip lines, a limit line.) The idea is simple (see also [*23*, p. 145]). Assume a non-characteristic curve C to be the boundary between a rigid (or contained plastic) domain, which we assume at rest, and a free flowing plastic domain. At a point P of the rigid domain adjacent to C the velocity vector $\boldsymbol{v}$ vanishes. Hence $\boldsymbol{v}=0$ at every point of C, and if the velocity is continuous across C, this vector is also zero on C adjacent to the plastic domain. We have then Cauchy data, namely $\boldsymbol{v}=0$ along C and therefore in a triangle in the plastic domain $\boldsymbol{v}=0$ throughout, in contrast to the assumption that C be the plastic-rigid boundary. The assumption that $\boldsymbol{v}$ be continuous across C is justified since we have seen that a velocity discontinuity is possible only across a characteristic. It follows that C has characteristic direc-

186 A. P. GREEN: Phil. Mag. **42**, 900 (1951). W. PRAGER presented the method in a course of lectures at Imperial College, London in 1952 and published it in Trans. Roy. Inst. Technol. Stockholm, **65** (1953). J. ALEXANDER used this method very skillfully in papers of 1955, 1961, 1962, 1967, Intern. J. Mech. Sci. **9** (1967).

187 SOKOLOVSKY: J. Mech. Phys. Solids **10**, 353 (1962).

188 A. P. GREEN: J. Mech. Phys. Solids **2**, 296 (1954).

189 R. T. SHIELD: J. Math. Phys. **33**, 144–156 (1954).

190 W. PRAGER: Courant Anniversary Volume, p. 289–299. New York 1948.

191 R. HILL: J. Mech. Phys. of Solids **1**, 19–30 (1952).

192 E. H. LEE: J. Appl. Mechanics **19**, 97–103 (1952).

tion everywhere. The tangential component of $\boldsymbol{v}$ can then change across C from zero to a non-zero value while the normal component is zero.

In the general case where the slip lines and characteristics are not identical a more elaborate discussion is necessary.[193] In this case the separation line is characteristic.

D. Boundary-value problems.

We consider in this last chapter a small number of classical problems in some detail with the aim of pointing out the approach, the methods, and the difficulties. More space has been given to elastic-plastic than to rigid-plastic problems since they are physically more significant.

I. Some elastic-plastic problems.[194]

a) The torsion problem.

48. Fully elastic and fully plastic torsion. We consider the equilibrium of a cylindrical bar under axial moments, in other words, a bar which is twisted about an axis parallel to the generators by equal and opposite couples. We take the z-axis in the direction of the generators.

If, then, we treat the stress distribution for a bar with concave corners as an elastic problem, infinite stresses are obtained at the corners. We conclude that at these corners the yield limit has been reached and Hooke's law is no longer valid. Outside of a neighborhood of the corners the material will be elastic (Fig. 31). The general case is that, for any cross section, the material will be plastic in some region and elastic in the remainder of the cross section. A main problem is to determine in a given case the boundary between the plastic and elastic regions; in addition, we wish to evaluate the stresses, displacements, and velocities in the respective regions. In our problem of plane shear, the only non-vanishing stresses are $\tau_{xz} = \tau_x$, $\tau_{yz} = \tau_y$, the components of the shear vector τ (the notation is different from that in Chap. A). Thus $\sigma_x = \sigma_y = \sigma_z = \tau_{xy} = 0$, and

$$\frac{\partial \tau_x}{\partial x} + \frac{\partial \tau_y}{\partial y} = 0 \tag{48.1}$$

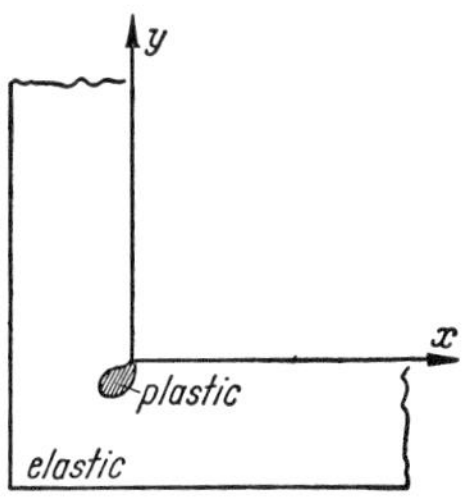

Fig. 31. Plastic region attached to concave corner of L-beam (TREFFTZ).

[193] H. GEIRINGER: Advances p. 253 seq. In a letter of Fall 1953, Professor HILL informed me that he had reached the result already in J. Mech. Appl. Math. **2** (1949); and he pointed out that the discussion in his work [*6*, pp. 150/151] is general.

[194] For this chapter cf. also HOFFMAN and SACHS [*8*], but, in particular, the new article by TING, following in this volume of the Encyclopedia.

is the only equilibrium equation which is not satisfied trivially. It is solved by introducing a *stress function* $\psi(x, y)$, where

$$\frac{\partial\psi}{\partial y}=\tau_x, \qquad \frac{\partial\psi}{\partial x}=-\tau_y. \tag{48.2}$$

It follows that the *contour lines*, $\psi=$constant (the "streamlines" if we think in terms of an isochoric flow) *have everywhere the direction of* τ, and hence we call them *stress lines, while the normals to the contour lines* (which correspond to the potential lines) *have the direction of the vector* grad ψ. Thus, if we denote by $\partial/\partial s$ and $\partial/\partial n$ differentiation in direction of and normal to the contour lines, respectively,

$$\frac{\partial\psi}{\partial s}=0, \qquad \frac{\partial\psi}{\partial n}=\tau, \qquad \left(\frac{\partial\psi}{\partial x}\right)^2+\left(\frac{\partial\psi}{\partial y}\right)^2=\tau^2. \tag{48.3}$$

Since the lateral surface of the bar is free of stress, the stress vector τ must be tangent to the boundary of the cross section. Hence this boundary is a stress line, $\psi=$constant. This last condition is general no matter whether the problem is completely elastic or completely plastic or partly elastic, partly plastic. Thus

$$\psi=0 \quad \text{on the boundary.} \tag{48.4}$$

In *the elastic region*[195] Hooke's law shows that $\varepsilon_x=\varepsilon_y=\varepsilon_z=\gamma_{xy}=0$ and $\gamma_{xz}=\gamma_x, \gamma_{yz}=\gamma_y$ are the only non-vanishing strains. If u_x, u_y, u_z are the small displacements of a point (x, y, z), the deformation is described by

$$u_x=-\alpha y z, \qquad u_y=\alpha x z, \qquad u_z=\alpha w(x, y), \tag{48.5}$$

where α is the twist per unit length and $w(x, y)$ is the *warping function* (see SAINT VENANT[196] and NADAI,[197] as well as textbooks on elasticity theory). Hence by Hooke's law

$$\tau_x=G\alpha\left(-y+\frac{\partial\omega}{\partial x}\right), \qquad \tau_y=G\alpha\left(x+\frac{\partial\omega}{\partial y}\right), \tag{48.6}$$

and by elimination of w the relation

$$\operatorname{curl}\tau=\frac{\partial\tau_y}{\partial x}-\frac{\partial\tau_x}{\partial y}=2G\alpha \tag{48.7}$$

results. By use of (48.2) this becomes

$$\Delta\psi=-2\alpha G. \tag{48.8}$$

The boundary-value problem (48.8), (48.4) determines ψ uniquely if the boundary is given.

Denote by A the area of the cross section. The twisting moment or *torque* required to produce the *twist per unit length* α is

$$T=\iint_A(x\tau_y-y\tau_x)\,dx\,dy=-\iint_A\left(x\frac{\partial\psi}{\partial x}+y\frac{\partial\psi}{\partial y}\right)dx\,dy=2\iint_A\psi\,dx\,dy. \tag{48.9}$$

The last equality follows by integrating by parts and using $\psi=0$ on the boundary. All this holds true for elastic as well as plastic torsion.

---

[195] In this chapter the distinction between ε', ε'', and ε (see Sect. 6) is made only where it seems desirable for clarity.

[196] B. DE SAINT VENANT: Mém. prés. par div. sav. Acad. Sci., math. et phys. **14**, 233–560 (1856).

[197] A. NADAI: Z. angew. Math. Mech. **3**, 442–454 (1923), see p. 448.

PRANDTL[198] has given a well-known intuitive interpretation of this problem. Suppose a thin membrane (soap film) is fastened along the contour C of a fully elastic cross section and loaded by a uniform surface pressure, p, proportional to α. If S is the surface tension of the membrane, its vertical displacement, u, satisfies the equation

$$\frac{\partial^2 u}{\partial x^2} + \frac{\partial^2 u}{\partial y^2} = -\frac{p}{S},$$

and $u=0$ along C. For $p=2\alpha G S$ the right side of the above equation is $-2\alpha G$ and comparison with (48.9) shows that we may put $\psi=u$: the membrane surface reproduces the *elastic stress surface*. According to (48.3) the resultant shearing stress $|\tau|$ is equal, at any point (x, y) to the greatest slope of the membrane, and its direction is that of the contour line through this point.

Next assume that over part of the cross section *the material is plastic*. We then obtain for both v. Mises' and Tresca's conditions

$$\tau_x^2+\tau_y^2=\tau_0^2=k^2, \quad \text{or} \quad \tau=\sqrt{\tau_x^2+\tau_y^2}=k, \tag{48.10}$$

and at the boundary (48.4) is valid. By (48.2)

$$|\operatorname{grad}\psi| = \left[\left(\frac{\partial\psi}{\partial x}\right)^2+\left(\frac{\partial\psi}{\partial y}\right)^2\right]^{\frac{1}{2}} = \frac{\partial\vartheta}{\partial x} = k. \tag{48.11}$$

If ϑ denotes the angle of τ with the x-axis, Eq. (48.10) is satisfied by

$$\tau_x=k\cos\vartheta, \qquad \tau_y=k\sin\vartheta; \tag{48.12}$$

if this is substituted into (48.1)

$$\sin\vartheta\frac{\partial\vartheta}{\partial y} - \cos\vartheta\frac{\partial\vartheta}{\partial y}=0 \tag{48.13}$$

results. Since by (48.11) *the gradient of the plastic stress surface ψ is constant* this surface is a *surface of constant slope* (Böschungsfläche). Such surfaces have well-known interesting properties (see e.g. NADAI, l.c., p. 445). Denote by y' the slope of a stress line and by $\partial/\partial s$, $\partial/\partial n$ the same as in (48.3). Then

$$\frac{\partial\psi}{\partial s}=0, \quad y'=-\frac{\partial\psi}{\partial x}\bigg/\frac{\partial\psi}{\partial x}=\tan\vartheta, \quad \frac{\partial\psi}{\partial n}=\tau=k \tag{48.14}$$

results, which is the same as in (48.3), except that now $\tau=k$. The contour lines have the direction of τ everywhere, and, since $\partial\psi/\partial n=k$, they are *equidistant curves*. Next, using Eq. (48.13), we see that *the orthogonal trajectories of the stress lines are straight lines* of slope $-\cot\vartheta$ where ϑ corresponds to the point of intersection with C. We call those straight trajectories *the normals*; along each of them τ remains the same; they are the lines of steepest descent, having the direction of the vector $\operatorname{grad}\psi$; the stress lines have the direction of $\operatorname{grad}\vartheta$.

From the first-order partial differential equation (48.13) we see that *the characteristics* have the slope $-\cot\vartheta$; hence they *coincide with the orthogonal trajectories*.

A physical construction of a surface of constant slope Σ belonging to a given cross section has been indicated by NADAI.[199] If the contour of the cross-section is cut out (physically), laid down horizontally and covered with a fine

[198] L. PRANDTL: Phys. Z. 4, 758–759 (1903).
[199] A. NADAI: Z. angew. Math. Mech. 3, 442–454 (1923).

powder, the natural surface of constant slope will form above the cross-section *The shape of Σ is independent of α.*

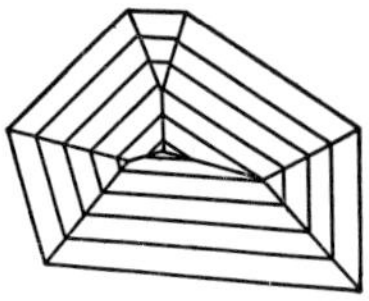

Fig. 32. Sandhill over polygon (NADAI).

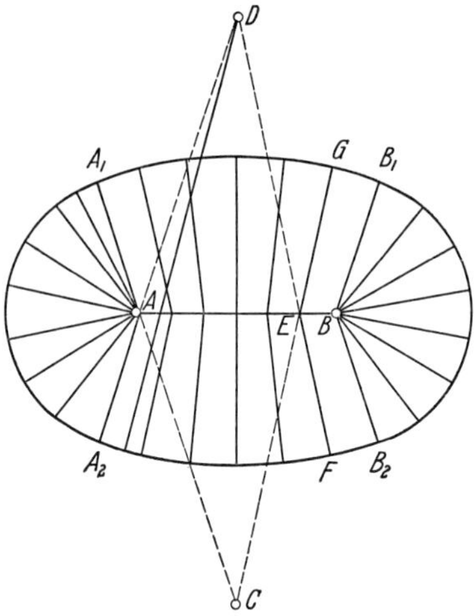

Fig. 33. Fully plastic oval with characteristics and stress discontinuity line (PRAGER, HODGE).

Fig. 32 shows the construction of the projection of Σ having a polygon as contour.[200] In case C consists of straight segments and arcs of circles, Σ consists of planes and conic surfaces.

Even if the contour C of the cross section has no corners, edges like those in Fig. 32 will in general appear. An instructive example is due to PRAGER and HODGE.[201] Fig. 33 shows an oval contour consisting of four circular arcs with centers A, B, C and D. The orthogonal trajectories, the characteristics, are therefore straight lines through these centers. All characteristics through points of the arc $\widehat{A_1 A_2}$ intersect at A, all through $\widehat{B_1 B_2}$ at B; each other point of AB is a point of intersection of two characteristics. Since at any point of any characteristic the stress vector τ is perpendicular to it, it follows that the direction of τ changes abruptly at a typical point E of AB. Thus AB is a *stress discontinuity line*,[202] as studied in Sects. 24, 33, etc. In general, the characteristics intersect on a line of stress discontinuity. Since the normal component of τ cannot jump, the discontinuity line must bisect the angle formed by the characteristics meeting at that line.

[200] See photos of "sandhills" in NADAI [*17*, p. 134seq.]. For the case of concave corners, cf. also PRAGER and HODGE [*23*, p. 66].

[201] W. PRAGER and P. G. HODGE [*23*, p. 64], and W. PRAGER, Proc. 2nd U.S. Nat. Congr. Appl. Mech., Ann Arbor, Michigan 1954, pp. 21–32. Cf. also the fully plastic distribution in the oval cross-section similar to an ellipse in [*26*, pp. 126/127] and photographs p. 129.

[202] See footnote 204.

The strain rate $\dot{E}''$ in a fully plastic problem is given by v. MISES' relations

$$\dot{\varepsilon}_x'' = \dot{\varepsilon}_y'' = \dot{\varepsilon}_z'' = \dot{\gamma}_{xy}'' = 0, \quad \dot{\gamma}_{xz}'' = \lambda \tau_{xz}, \quad \dot{\gamma}_{yz}'' = \lambda \tau_{yz}. \tag{48.15}$$

As in Sect. 24, *the strain-rate tensor* $\dot{E}''$ *vanishes* at the discontinuity line. The characteristics drawn through an arc of the contour C will in general have an envelope, and beyond this limit line they cannot be continued. In PRAGER and HODGE's example the limit line shrinks to the four points A, B, C, D. It may, however, be a curve inside the contour C.

W. FREIBERGER[203] has studied the fully plastic torsion of circular ring sectors. This problem is fairly general since the body is part of a torus rather than of a cylinder (Fig. 34). The characteristics are still the orthogonal trajectories of the stress lines in the x, y-plane and they intersect on a *line of stress discontinuity* which at every point bisects the angle of the characteristics. In this problem the characteristics as well as the discontinuity line are curved; $\dot{E}''$ vanishes at the discontinuity line.[204]

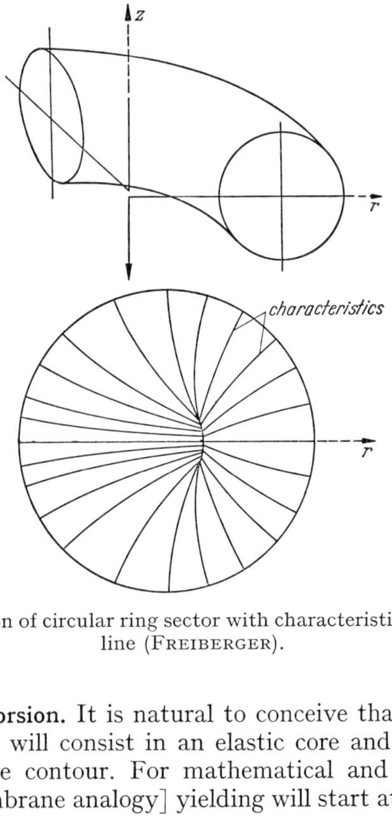

Fig. 34. Fully plastic torsion of circular ring sector with characteristics and stress discontinuity line (FREIBERGER).

49. Elastic-plastic torsion. It is natural to conceive that in general a solution of the torsion problem will consist in an elastic core and one or several plastic pieces, attached to the contour. For mathematical and physical reasons [see Eq. (48.8) and the membrane analogy] yielding will start at the contour C. As the

[203] W. FREIBERGER: Quart. Appl. Math. **14**, 259–264 (1956) and Commonwealth of Australia, Dept. of Supply, Aer. Res. Labor., Rept. SM 213 (1953). The fully plastic twisting of a circular ring sector of arbitrary cross-section is also considered by A. J. WANG and W. PRAGER: J. Mech. Phys. Solids **3**, 169–175 (1955). With respect to the velocity problem the approach differs from FREIBERGER's.

[204] FREIBERGER points out that this discontinuous fully plastic solution is an *exact* solution; the discontinuity line may be regarded as the limit of an elastic core. If, however, the material is considered rigid perfectly plastic, a fully plastic torsion exists.

twisting moment is increased, plastic zones spread out. The plastic-elastic boundary Γ is unknown. If α is small enough that small changes in the contour C can be neglected, we obtain the following model.

Above the given contour C the surface of constant slope, the "roof", has been erected—materially, in cardboard—corresponding to the given contour and the given slope k; above the base of the roof a stretched membrane is loaded by a pressure p, proportional to α. If this pressure reaches a certain value, parts of the membrane will touch the surface of the roof from within; that means that along the lines of contact the slope of the membrane (which can never exceed k) has reached the slope k of the roof. The stress surface ψ is formed by the free parts of the membrane and by those parts which rest on the roof. The horizontal projection of those parts of the membrane which touch the roof gives the plastic region of the cross-section; below the free parts of the membrane the material remains elastic. If the pressure on the membrane is increased more and more, an increasing part will rest on the roof, but the part under the edges of the roof will still remain free, corresponding to the elastic neighborhood of the lines of discontinuity. This combined model is due to NADAI[205] and has been used by TREFFTZ[206] and by later authors.[207]

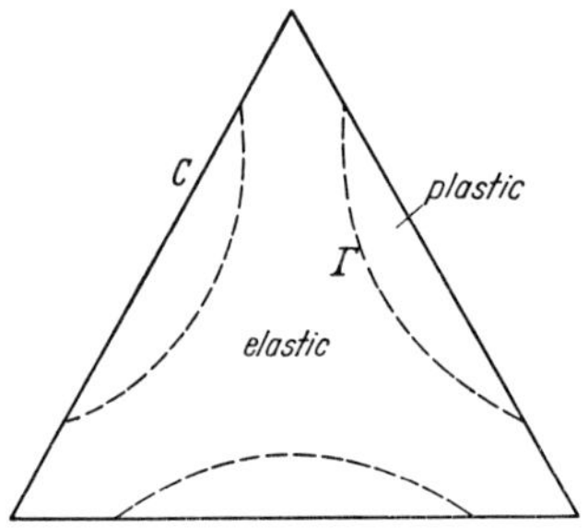

Fig. 35. Scheme of elastic-plastic triangle in torsion.

This interpretation suggests that the corresponding mathematical problem has a solution. The mathematical problem is the following: A closed contour C is given (Fig. 35), enclosing a region A. We ask for a vector field $\tau(x, y)$, having a given constant curl, viz

$$\frac{\partial \tau_y}{\partial x} - \frac{\partial \tau_x}{\partial y} = 2\alpha G$$

in some inner region, a given magnitude τ ($\tau = k$) in the remaining "outer" part, including the (unknown) boundary Γ between the two parts, and a given direction along C. Also, τ should be continuous across Γ; this appears physically plausible (see NADAI, l.c.) and has been proved by PRAGER and HODGE [23, p. 70]. The boundary Γ separating "inner" and "outer" regions is unknown. In terms of ψ the problem is as follows: We seek, in A, a continuous function ψ, with continuous $\partial\psi/\partial x$, $\partial\psi/\partial y$, with given constant $\Delta\psi$ in some unknown inner

[205] A. NADAI: Z. angew. Math. Mech. **3**, 442–454 (1923).

[206] E. TREFFTZ: Z. angew. Math. Mech. **5**, 64–73 (1925).

[207] In 1951 PRAGER and HODGE [23] pointed out that the Prandtl-Nadai analogy and the elastic plastic torsion problem are identical only if certain restrictions are imposed on the shape of the cross section. In 1963 (Prager Anniversary Volume, New York 1963, pp. 251–259) HODGE showed that these restrictions are automatically satisfied for cylinders with simply connected cross section, but that for multiply connected cross section the Prandtl-Nadai analogy may fail.

part of A, given $|\operatorname{grad}\psi|$ in the remaining outer part, and given constant value on the closed contour C. As pointed out by TREFFTZ and v. MISES, the problem is of the type of the well-known free-boundary problems or jet problems of hydromechanics, where the separation line between the regions in which ψ is subject to different conditions is likewise unknown. v. MISES[208] remarked that this analogy, as well as the membrane analogy, leads to the assumption that the problem is determinate.[209] *This seems indeed to be so*; see my remarks end of Sect. 50, regarding TING'S article.

The *strains in the elastic region* are determined by Hooke's law:

$$\gamma_x/\tau_x = \gamma_y/\tau_y. \tag{49.1}$$

In the *plastic region* the elastic strains are assumed to be so small that time-changes in the contour C and displacements of elements can be neglected—as in elasticity.[210] Once the element has become plastic, the stress $\boldsymbol{\tau}$ at a fixed element does not change, since its magnitude τ equals k and its direction is tangent to the contour line. Hence with the notation of Sect. 6 $\dot{\tau}_x = 0$, $\dot{\tau}_y = 0$, and from (6.3), $\dot{e}'_{ij} = 0$, $\dot{e}_{ij} = \dot{e}''_{ij}$, the Reuss equations reduce to the Lévy-v. Mises stress-strain equations and $\dot{\gamma}_x/\tau_x = \dot{\gamma}_y/\tau_y$. Since τ_x, τ_y are constant (from an instant t_0 on) we can even integrate these last equations from t_0 to t, and since

$$\gamma_x(t_0)/\tau_x - \gamma_y(t_0)/\tau_y = 0,$$

by (49.1), we see that (49.1) holds also for the strains in the plastic region.

We can no longer assume that the warping is proportional to the twist α; hence we replace the last Eq. (48.5) by $u_z = w(x, y, \alpha)$ and obtain in the elastic region, instead of (48.6),

$$\frac{\partial \omega}{\partial x} - \alpha y = \frac{\tau_x}{G}, \qquad \frac{\partial w}{\partial y} + \alpha x = \frac{\tau_y}{G}. \tag{49.2}$$

Introducing the strains from (49.1), we have

$$\frac{\gamma_x}{\gamma_y} = \left(\frac{\partial w}{\partial x} - \alpha y\right) \Big/ \left(\frac{\partial w}{\partial y} + \alpha x\right) = \frac{\tau_x}{\tau_y} \tag{49.3}$$

in both the plastic and elastic regions. Once the stresses have been evaluated, this equation determines the only unknown function, $w(x, y, \alpha)$. Since $\tau_x/\tau_y = \cot\vartheta$, (49.3) becomes

$$\frac{\partial w}{\partial x} \sin\vartheta - \frac{\partial w}{\partial y} \cos\vartheta = \alpha(x\cos\vartheta + y\sin\vartheta). \tag{49.4}$$

The characteristics of this linear first-order partial differential equation, defined by

$$dx : dy : dw = \sin\vartheta : -\cos\vartheta : \alpha(x\cos\vartheta + y\sin\vartheta), \tag{49.5}$$

are again the normals to the contour. To determine w we write (49.4) as

$$\frac{\partial w}{\partial n} = -\alpha d, \tag{49.6}$$

where d is the distance from the normal through the point (x, y) to the origin. This can be integrated[211] and gives (Fig. 36),

$$w(x, y) - w(\xi, \eta) = -\alpha l d = \alpha(x\eta - y\xi), \tag{49.7}$$

[208] R. v. MISES: Reissner Anniversary Volume, p. 415–429. Ann Arbor, Mich. 1949.
[209] See G. P. CHEREPANOV: J. Appl. Math. Mech. **28**, 162 (1964).
[210] See also HILL [*6*, p. 88].
[211] Cf. P. G. HODGE Jr.: J. Appl. Mechanics **16**, 399–406 (1949), see p. 400.

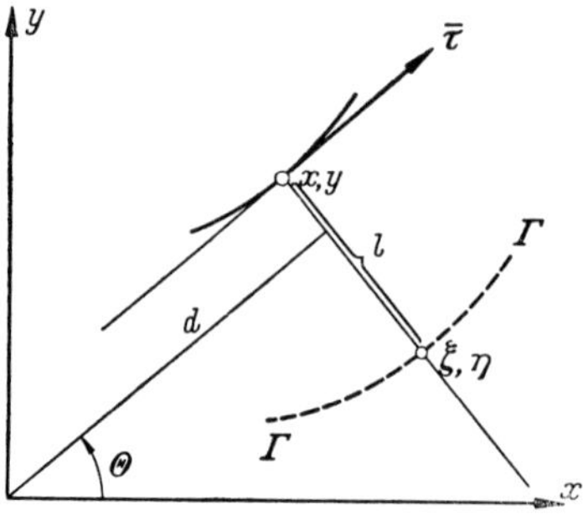

Fig. 36. Illustrating computation of warping in torsion.

where (ξ, η) are the coordinates of the point of intersection of the normal through (x, y) and the boundary Γ, and l is the distance between (x, y) and (ξ, η). Hence w is known in the plastic region after it has been determined in the elastic region. There, however, it is known from Eqs. (48.6), which are of the form

$$\partial w/\partial x = f(x, y), \qquad \partial w/\partial y = g(x, y),$$

with f and g known. Thus the strains can be found.[212]

50. Examples, further problems and concluding remarks. An exact direct solution of the stress boundary value problem is possible when the shape of Γ can be estimated from the beginning. This is e.g. possible for the *bar of circular section*, radius a, where Γ is obviously a concentric circle of radius $\varrho < a$. The elastic stress function valid inside this circle is

$$\psi = \tfrac{1}{2} G\alpha(\varrho^2 - x^2 - y^2), \qquad \tau_x = -G\alpha y, \qquad \tau_y = G\alpha x, \qquad \tau = G\alpha r. \tag{50.1}$$

The value $\tau = k$ is reached for $\alpha = k/G\varrho$ and, correspondingly, $\varrho = k/G\alpha$. Hence, for

$$\varrho \leqq r \leqq a: \tau_x = -k y/r, \qquad \tau_y = k x/r, \qquad \tau_y/\tau_x = -\frac{x}{y} = \tan\vartheta,$$

and the torque is therefore

$$T = 4\frac{\pi}{2}\int_0^a r^2 \tau\, dr = 2\pi k\left[\int_0^\varrho \frac{r^3}{\varrho}\, dr + \int_\varrho^a r^2\, dr\right] = 2\pi k\left(\frac{a^3}{3} - \frac{\varrho^3}{12}\right), \tag{50.2}$$

where $\varrho = k/G\alpha$ may be substituted. The axial displacement $u_z = w$ is zero in the elastic as well as in the plastic zone. A "fully plastic solution" corresponds to $\alpha \to \infty$; then

$$T = \frac{2\pi}{3} k a^3.$$

The above example was first treated by SAINT VENANT.

[212] Arguing against MISES' flow equations, PRAGER [*23*, pp. 81–84], who insists on the identity of v. MISES' theory with that of a "rigid-plastic" material (see our Sect. 58), states that in v. MISES' theory necessarily $w = 0$, in both the elastic and (contained) plastic region, as a consequence of $\alpha \to 0$ in both regions which in turn follows from $G \to \infty$. It would indeed be strange to study torsion with $\alpha = 0$. But this is in no way implied by or connected with v. MISES' theory (see also p. 413, line 9seq. of our article). Actually, in the torsion problem the Reuss equations *reduce* to v. MISES' equations on account of $\dot{s}_{ij} = 0$ (and not on account of $G \to \infty$); the wrong result $\alpha = 0$, $w = 0$, follows if torsion is considered as a problem of a rigid plastic material which, indeed, would be quite inadequate. (Cf. v. MISES' treatment of elastic plastic torsion problems in Reissner Anniversary Volume, l.c. and our Sects. 54, 55, and 58.)

A direct problem has been solved by TREFFTZ,[213] who computed the stress distribution for an L-shaped beam (the two legs of the L having equal widths), in particular in the vicinity of the re-entrant corner. The method is based in a natural way on complex function theory, and is approximate, but the errors are small.

Practically all known examples use inverse methods, where the boundary C is derived from an assumed solution and an assumed separation line Γ. Essentially the same indirect method has been described by SOKOLOVSKY[214] and by v. MISES.[215] We illustrate the method for an oval cross-section, as considered by SOKOLOVSKY.

We start with the solution of the elastic problem (48.8):

$$\frac{\psi}{\alpha G} = -\frac{b^2 x^2 + a^2 y^2}{a^2 + b^2} \tag{50.3}$$

and with the assumption of the elliptic boundary Γ

$$\frac{\xi^2}{a^4} + \frac{\eta^2}{b^4} = 1. \tag{50.4}$$

From (50.3)

$$\frac{\partial \psi}{\partial x} = -2\alpha G \frac{b^2}{a^2+b^2} x = -\tau_y, \qquad \frac{\partial \psi}{\partial y} = -2\alpha G \frac{a^2}{a^2+b^2} y = \tau_z. \tag{50.5}$$

Hence, on Γ, from (48.10)

$$k^2 = \frac{4\alpha^2 G^2}{(a^2+b^2)^2} (b^4 \xi^2 + a^4 \eta^2) = \frac{4\alpha^2 G^2}{(a^2+b^2)^2} a^4 b^4, \tag{50.6}$$

and we obtain

$$2\alpha G = k \frac{a^2+b^2}{a^2 b^2} \tag{50.7}$$

as the relation valid for the twist at which yielding occurs along (50.4). If the equation of Γ is written in parametric form, using a', b' for a^2, b^2 viz. $\xi = -a' \sin\vartheta$, $\eta = b' \cos\vartheta$, the equation of a normal through (ξ, η) is

$$y - \eta = -\cot\vartheta\,(x - \xi), \quad \text{or} \quad y + x\cot\vartheta = (b' - a')\cos\vartheta. \tag{50.8}$$

The differential equation of the orthogonal trajectories to these normals, i.e. of the stress lines, is

$$\frac{dy}{dx} = \tan\vartheta = -\frac{x-\xi}{y-\eta} = -\frac{x + a'\sin\vartheta}{y - b'\cos\vartheta}, \tag{50.9}$$

and by integration, with c as an integration constant, we obtain

$$\begin{aligned} x &= -\sin\vartheta\left[\frac{a'}{2} + c + \frac{a'-b'}{2}\cos^2\vartheta\right], \\ y &= \cos\vartheta\left[\frac{b'}{2} + c - \frac{a'-b'}{2}\sin^2\vartheta\right]. \end{aligned} \tag{50.10}$$

These curves are ovals of double symmetry and differ very little from ellipses with semi-axes

$$\frac{a'}{2} + c \quad \text{and} \quad \frac{b'}{2} + c.$$

---

[213] E. TREFFTZ: Z. angew. Math. Mech. 5, 64–73 (1925).
[214] V. V. SOKOLOVSKY: J. Appl. Math. Mech. 6, 241–246 (1942); see [*26*, p. 132].
[215] R. v. MISES: Reissner Anniversary Volume, l.c.

We have thus a solution of the elastic-plastic problem for an approximately elliptic contour C (see Fig. 37).

On the other hand, if we assume the oval contour as given by the "axes"

$$A = \frac{a'}{2} + c, \quad B = \frac{b'}{2} + c,$$

we can evaluate a' and b', the axes of the ellipse Γ, from the two equations

$$\frac{2\alpha G}{k} = \frac{a' + b'}{a' b'}, \quad 2(A - B) = a' - b,$$

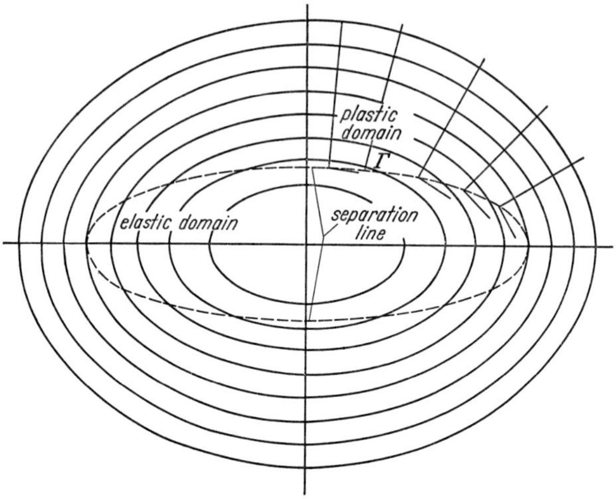

Fig. 37. Elastic-plastic torsion of approximately elliptical oval (SOKOLOVSKY).

where the first equations is (50.7) again, and we find, with $d = A - B$:

$$a' = \frac{k}{2G\alpha} + d + \left[d^2 + \left(\frac{k}{2G\alpha}\right)^2\right]^{\frac{1}{2}}, \quad b' = \frac{k}{2G\alpha} - d + \left[d^2 + \left(\frac{k}{2G\alpha}\right)^2\right]^{\frac{1}{2}}. \qquad (50.11)$$

The expressions for a' and b' depend only on the difference $d = A - B$. Our solution makes sense only if Γ lies entirely in C, i.e. if $a' < A$, $b' < B$, which leads to the condition

$$\alpha > \frac{k}{G} \frac{B}{A(2B - A)}. \qquad (50.12)$$

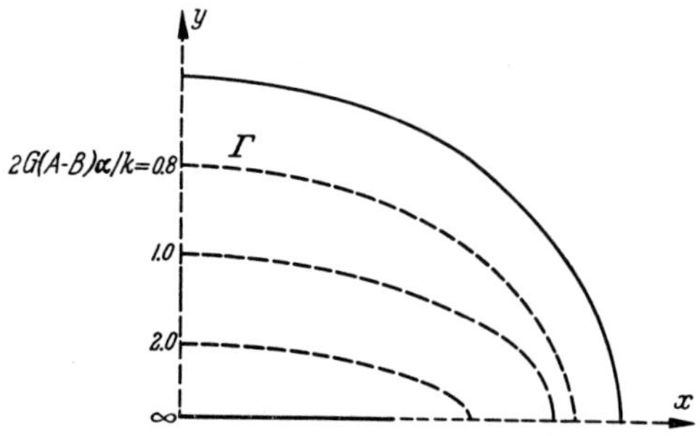

Fig. 38. Spreading of separation line Γ for increasing twist (PRAGER).

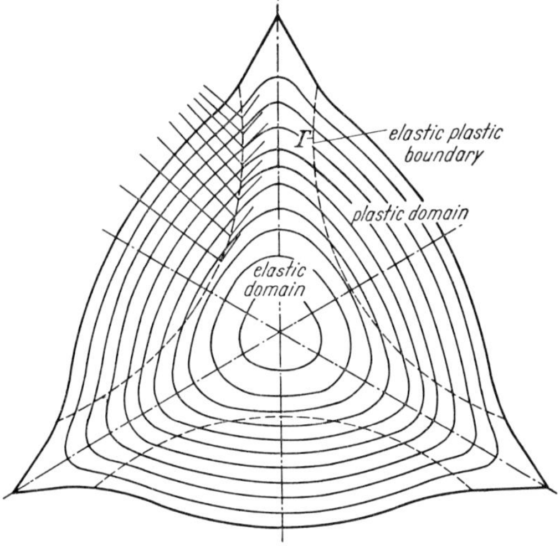

Fig. 39. Triangular solution for elastic-plastic torsion (v. MISES).

Fig. 38 (cf. [23, p. 80]) shows the spreading of the separation line Γ with increasing twist α.

The warping in the elastic region is found from (48.6) and (50.5) as

$$w = -xy\frac{a'-b'}{a'+b'}$$

[in this combined elastic-plastic problem $w(x, y, \alpha)$ is not proportional to α, not even in the elastic section]; on Γ we have

$$w = \frac{a'-b'}{a'+b'}\, a'\, b' \sin\vartheta \cos\vartheta\,, \tag{50.13}$$

and the warping in the plastic region is determined by (49.5) (cf. [26, p. 133). The moment T can likewise be found ([26, p. 134]).

v. MISES (l.c.) has applied (Fig. 39) the same indirect method (combined with graphical integration) to find an approximate solution for an equilateral triangular section. Elastic-plastic torsion for particular cross sections has also been studied by GALIN.[216]

A survey on general theorems for elastic-plastic solids has been written by W. T. KOITER, Progress in Solid Mechanics, vol. I, chap. IV, 1960.

The last decade has brought great progress regarding the torsion problem (see part D of the article by T. W. TING, following in this volume of the Encyclopedia), particularly in regard to existence, uniqueness, and qualitative theory of solutions. (See in TING's bibliography the papers [26, 47, 49, 61, 63]: see the interesting general paper [30] and in particular TING's own contributions [41, 51].)

An approximate formula for warping under large twist has been given by HODGE.[217] In most actual solutions numerical or graphical methods are needed to carry out the steps. As a typical example we mention the numerical solution of the elastic-plastic torsion problem of a shaft of rotational symmetry and varying

---

[216] L. A. GALIN: J. Appl. Math. Mechanics **13** (1949).

[217] P. G. HODGE Jr.: J. Appl. Mechanics **16**, 399–406 (1949). He also considers in this paper the problem of repeated loading.

diameter.[218] In this problem[219] the characteristics are no longer straight lines (although still normal to the contour lines); they may have an envelope, i.e. the solution may have a limit line, which may, however, be outside the contour). Non-rectilinear characteristics also appear in the elastic-plastic solution of the twisted ring sector.[220] Here, the envelope of the characteristics which, for the fully plastic solution is a limit line inside the contour, does not materialize in the plastic part of the elastic-plastic solution.

b) The thick walled tube.

In Sect. 32 we briefly commented on the fully plastic radial yielding in a thick tube subjected to high internal pressure considered as a problem of plane strain. Here we study the elastic-plastic problem, where the tube yields only in part, so that there results an inner plastic region surrounded by an elastic region; also $\varepsilon_z = 0$ is no longer a necessary assumption.

51. Expansion of a cylindrical tube. Let the inner radius of the circular cylindrical ring be a, the outer b, and let the generators lie in the z-direction; the tube is expanded by an internal pressure p and loaded at both ends by some equal and opposite longitudinal forces of magnitude L. The tube is long enough to justify the assumption that the stress and strain distribution is the same in normal cross-sections sufficiently far from the end. It follows by symmetry that any originally plane normal cross-section remains plane and that ε_z is the same all over a representative cross-section and remains the same through the expansion.[221]

Three main *end conditions* are usually considered. The tube may be closed at both ends by plugs firmly attached to the ends: *closed-end* conditions; or it is closed by floating pistons: *open-end* conditions; a third condition is that of *zero extension or plane strain.*

The principal directions are radial, axial and circumferential and, with respect to such a coordinate system, $\tau_{r\vartheta} = \tau_{z\vartheta} = \tau_{rz} = 0$. Our problem is independent of ϑ and is assumed also independent of z. Thus, two of the three equilibrium conditions are identically satisfied, and the remaining one reduces to

$$\frac{\partial \sigma_r}{\partial r} + \frac{\sigma_r - \sigma_\vartheta}{r} = 0. \tag{51.1}$$

The non-vanishing strains are

$$\varepsilon_r = \frac{\partial u_r}{\partial r}, \qquad \varepsilon_\vartheta = \frac{u_r}{r}, \qquad \varepsilon_z = \frac{\partial u_z}{\partial z} = \text{constant}, \tag{51.2}$$

where $u_r = u$ is the radial, u_z the axial displacement. There therefore remain altogether four unknowns; σ_r, σ_ϑ, σ_z, and u.

We first consider *elastic expansion*. The radial displacement u is of the form

$$u = A\,r + \frac{B}{r}. \tag{51.3}$$

[218] R. P. EDDY and F. S. SHAW: J. Appl. Mechanics **16**, 139–148 (1949), and further citations of numerical methods in HILL [*6*, p. 94]. See also the presentation in [*26*, p. 140seq.].

[219] See the presentation in HILL [*6*, pp. 94–96], in P. G. HODGE JR.: An introduction to the mathematical theory of perfectly plastic solids. Providence: Brown University 1950, pp. 107–118, also SOKOLOVSKY: J. Appl. Mech. **1**, 343–346 (1945).

[220] See footnote 203.

[221] The presentation in this section follows, HILL [*6*, p. 106seq.], cf. also SOKOLOVSKY [*26*, Chap. 3]. He takes strain hardening into consideration; his investigation uses Hencky's stress-strain law.

If E is Young's modulus, ν Poisson's ratio, we have from Hooke's law

$$\begin{aligned} E\,\varepsilon_r &= E\,\frac{\partial u}{\partial r} = E\left(A - \frac{B}{r^2}\right) = \sigma_r - \nu\,(\sigma_\vartheta + \sigma_z)\,, \\ E\,\varepsilon_\vartheta &= E\,\frac{u}{r} \quad = E\left(A + \frac{B}{r^2}\right) = \sigma_\vartheta - \nu\,(\sigma_z + \sigma_r)\,, \\ E\,\varepsilon_z &= \qquad\qquad\qquad\qquad\; = \sigma_z - \nu\,(\sigma_r + \sigma_\vartheta)\,. \end{aligned} \tag{51.4}$$

Solving these equations and using the abbreviation

$$\alpha = (1+\nu)\,(1-2\nu)/E\,,$$

we obtain

$$\begin{aligned} \alpha\,\sigma_z &= A - (1-2\nu)\,\frac{B}{r^2} + \nu\,e_z \\ \alpha\,\sigma_\vartheta &= A + (1-2\nu)\,\frac{B}{r^2} + \nu\,\varepsilon_z \\ \alpha\,\sigma_z &= 2\nu\,A + (1-\nu)\,\varepsilon_z\,. \end{aligned} \tag{51.5}$$

It is seen that σ_z as well as $\sigma_r + \sigma_\vartheta$, and therefore $\sigma_r + \sigma_\vartheta + \sigma_z = 3\,\sigma$, are independent of r.

If the flow is elastic throughout, the boundary conditions are

$$\begin{aligned} \sigma_r &= -p \quad \text{for} \quad r = a \\ &= 0 \qquad \text{for} \quad r = b\,. \end{aligned} \tag{51.6}$$

Substituting these into (51.5), we find the values of A and B, and, using the abbreviation $p' = a^2\,p/(b^2 - a^2)$, we obtain

$$\begin{aligned} \sigma_r &= -p'\left(\frac{b^2}{r^2} - 1\right), \qquad \sigma_\vartheta = p'\left(\frac{b^2}{r^2} + 1\right), \\ \sigma_z &= E\,\varepsilon_z + 2\nu\,p'\,, \qquad \sigma = \frac{1}{3}\,[2\,(1+\nu)\,p' + E\,\varepsilon_z] \end{aligned} \tag{51.7}$$

and

$$u = -\nu\,\varepsilon_z\,r + \frac{1+\nu}{E}\,p'\left[(1-2\nu)\,r + \frac{b^2}{r}\right]. \tag{51.8}$$

In general, not ε_z but the axial force L is prescribed, and, from this, ε_z may be computed for the different end conditions. The axial force L is here

$$L = \pi\,(b^2 - a^2)\,\sigma_z\,.$$

Using the third Eq. (51.7), we find the axial strain ε_z, corresponding to given load L and pressure p,

$$\varepsilon_z = \frac{\sigma_z}{E} - \frac{2\nu}{E}\,p' = \frac{a^2}{(b^2 - a^2)\,E}\left(\frac{L}{\pi\,a^2} - 2\nu\,p\right). \tag{51.9}$$

In case of closed-end conditions, L is a tension equal to the force $\pi a^2 p$ exerted by the internal pressure on each plug. Hence from (51.9)

$$\varepsilon_z = \frac{1-2\nu}{E}\,p'\,, \qquad \sigma_z = p' = \frac{1}{2}\,(\sigma_r + \sigma_\vartheta) \qquad \text{(closed end)}\,. \tag{51.10}$$

When both ends are open, $L = 0$, and

$$\varepsilon_z = -\frac{2\nu}{E}\,p'\,, \qquad \sigma_z = 0 \qquad \text{(open end)}\,; \tag{51.11}$$

and under plane strain

$$\varepsilon_z = 0, \quad \sigma_z = 2\nu p' = \nu(\sigma_r + \sigma_\vartheta) \quad \text{(plane strain)}. \tag{51.12}$$

This case is identical with the first for $\nu = \frac{1}{2}$, which means (elastic) incompressibility, $\varepsilon_r + \varepsilon_\vartheta + \varepsilon_z = 0$, as may be verified from (51.4). We note that σ_r and σ_ϑ are not influenced by the end conditions while σ_z is different in the three cases and in all is intermediate between σ_r and σ_ϑ.

The above formulas are valid throughout the tube for values of p such that the yield function remains below the yield limit. We consider for v. Mises' limit the region where

$$J_M \equiv (\sigma_r - \sigma_\vartheta)^2 + (\sigma_\vartheta - \sigma_z)^2 + (\sigma_z - \sigma_r)^2 < 6k^2, \tag{51.13}$$

and from (51.7)

$$J_M = 6p'^2 \frac{b^4}{r^4} + 2(\sigma_z - p')^2. \tag{51.13'}$$

Yielding occurs first at the inner surface $r = a$; hence the pressure p^* at which $J_M = 6k^2$ is determined by

$$3\left(\frac{b^2}{b^2 - a^2}\right)^2 p^{*2} + \left[\sigma_z - \frac{a^2}{b^2 - a^2} p^*\right]^2 = 3k^2. \tag{51.14}$$

Inserting for σ_z the values from Eqs. (51.10)–(51.12), we obtain

$$\begin{aligned} p^* &= k\left(1 - \frac{a^2}{b^2}\right) && \text{(closed end)},\\ &= k\left(1 - \frac{a^2}{b^2}\right)\Big/\sqrt{1 + \frac{a^4}{3b^4}} && \text{(open end)},\\ &= k\left(1 - \frac{a^2}{b^2}\right)\Big/\sqrt{1 + (1 - 2\nu)^2 \frac{a^4}{3b^4}} && \text{(plane strain)}. \end{aligned} \tag{51.15}$$

We see that of the three yield pressures that for "open end" is the smallest, that for "closed end" the largest.

With Tresca's yield condition we have to consider

$$\sigma_\vartheta - \sigma_r = 2p' \frac{b^2}{r^2} = \frac{2p}{r^2} \frac{a^2 b^2}{b^2 - a^2}.$$

Putting $r = a$, we obtain the yield pressure p_0:

$$p_0 = \frac{k\sqrt{3}}{2}\left(1 - \frac{a^2}{b^2}\right), \tag{51.16}$$

the same for all end conditions.

52. Partly plastic tube. α) *Determination of σ_ϑ and σ_r for Tresca's yield.* For a pressure somewhat exceeding the yield pressure, the tube will become partly plastic. On account of symmetry the elastic-plastic boundary Γ is a circle with radius ϱ. Consider first Tresca's yield, where there is independence of the end conditions. We assume that σ_z remains the intermediate principal stress. Then, using (51.7), we find that

$$\sigma_\vartheta - \sigma_r = \left[\frac{2p}{r^2} \frac{a^2 b^2}{b^2 - a^2}\right]_{r=\varrho} = k\sqrt{3}, \tag{51.1}$$

and substituting this value of p into Eqs. (51.7) and (51.8) we obtain the *solution for the partly plastic tube in the elastic part, in the case of Tresca's yield conditions*;

with $p''=\sqrt{3}\,k\,\varrho^2/2b^2$:

$$\begin{aligned}\sigma_r &= -p''\left(\frac{b^2}{r^2}-1\right), & \sigma_\vartheta &= p''\left(\frac{b^2}{r^2}+1\right),\\ \sigma_z &= E\,\varepsilon_z+2\nu\,p'', & u &= -\nu\,\varepsilon_z r+\frac{1+\nu}{E}\,p''\left[(1-2\nu)\,r+\frac{b^2}{r}\right],\end{aligned} \tag{52.2}$$

where the p' of (51.7), (51.8) is now replaced by p''.

In the *plastic region* $a \leqq r \leqq \varrho$, both the yield condition

$$\sigma_\vartheta-\sigma_r = k\sqrt{3} \tag{52.3}$$

and the equilibrium condition hold. By means of these two equations the two stresses σ_ϑ, σ_r can be found, and for all end conditions, without solving the complete problem. Using (51.1) and (52.3), we obtain the simple differential equation

$$\frac{\partial\sigma_r}{\partial r} = k\sqrt{3}\cdot\frac{1}{r}.$$

Observing that σ_r and σ_ϑ at $r=\varrho$ must coincide with the values from (52.2), we find for $a \leqq r \leqq \varrho$ that

$$\sigma_r=\sqrt{3}\,k\left[\frac{1}{2}\left(\frac{\varrho^2}{b^2}-1\right)-\log\frac{\varrho}{r}\right],\qquad \sigma_\vartheta=\sqrt{3}\,k\left[\frac{1}{2}\left(\frac{\varrho^2}{b^2}+1\right)-\log\frac{\varrho}{r}\right], \tag{52.4}$$

$$p=-(\sigma_r)_{r=a}=\sqrt{3}\,k\left[\frac{1}{2}\left(1-\frac{\varrho^3}{b^2}\right)+\log\frac{\varrho}{a}\right]. \tag{52.5}$$ [222]

We may eliminate the unknown ϱ from (52.4) by means of (52.5) and obtain

$$\sigma_r=k\sqrt{3}\,\log\frac{r}{a}-p,\qquad \sigma_\vartheta=k\sqrt{3}\left(1+\log\frac{r}{a}\right)-p. \tag{52.6}$$

Thus, if a, b, k, p are given σ_r, σ_ϑ follow from (52.6). Hence the problem is partly statically determined, and this for all end conditions.[223]

β) Continuation under assumption of plane strain. Tresca's yield. In contrast to the above determination of σ_r, σ_ϑ the determination of σ_z and u requires the use of the Prandtl-Reuss equations and this leads to a more difficult problem.[224] To simplify the work we restrict ourselves to plane strain, $\varepsilon_z=0$. From (51.4) we obtain by addition, with $\sigma_r+\sigma_\vartheta+\sigma_z=3\sigma$:

$$\frac{\partial u}{\partial r}+\frac{u}{r}=3\,\frac{1-2\nu}{E}\,\sigma. \tag{52.7}$$

We have to consider both plastic and elastic strains in the region of plastic flow which is restricted by the surrounding elastic ring. From the Reuss equations, with $s_z=\sigma_z-\sigma$ and $\sigma_r+\sigma_\vartheta=A$, we have $2\sigma=A+s_z$, $2\sigma_z=A+3s_z$ where, from (52.4),

$$A=2k\sqrt{3}\left(\frac{\varrho^2}{2b^2}-\log\frac{\varrho}{r}\right).$$

[222] According to HILL this formula for p was given by L. B. TURNER: Trans. Cambridge Phil. Soc. **21**, 377 (1909), and Engineering **92**, 115 (1911). The solution for σ_ϑ, σ_r, p agrees with NADAI's; A. NADAI: Trans. Am. Soc. Mech. Engrs **52**, 193 (1930), and [*17*, p. 196].

[223] See HILL [*6*, Fig. 14, p. 111], setting $\sqrt{3}k$ for Y and ϱ for c; likewise [*23*, Figs. 24, 25], and [*26*, Fig. 27].

[224] R. HILL, E. H. LEE and S. J. TUPPER: Proc. Roy. Soc. Lond., Ser. A **191**, 278–303 (1947). (Cf. also HILL [*6*, p. 114].) A numerical solution is included. See also R. HILL, E. H. LEE, and S. J. TUPPER: Proc. 1st U.S. Nat. Congr. Mech., Chicago, p. 561, 1951.

Following HILL, we take

$$q = \frac{\sqrt{3}}{2k} s_z$$

as parameter. Then

$$\sigma_z = k\sqrt{3}\left(q + \frac{\varrho^2}{2b^2} - \log\frac{\varrho}{r}\right), \qquad \sigma = k\sqrt{3}\left(\frac{q}{3} + \frac{\varrho^2}{2b^2} - \log\frac{\varrho}{r}\right),$$

and (52.7) takes the form

$$\frac{\partial u}{\partial r} + \frac{u}{r} = \frac{1-2\nu}{E} k\sqrt{3}\left(q - 3\log\frac{\varrho}{r} + \frac{3\varrho^2}{2b^2}\right). \tag{52.8}$$

We write the Reuss equations in the form

$$\dot{\varepsilon}_r = \frac{1}{E}\left[\dot{\sigma}_r(1+\nu) - 3\nu\dot{\sigma}\right] + \lambda(s_r)$$

(where we have used

$$\dot{\sigma}_r - \nu\dot{\sigma}_\vartheta - \nu\dot{\sigma}_z = \dot{\sigma}_r(1+\nu) - 3\nu\dot{\sigma} \text{ and } \sigma_r - \sigma = s_r)$$

and two similar ones, and we eliminate λ from the equations for $\dot{\varepsilon}_\vartheta$ and $\dot{\varepsilon}_z$. In these equations the dot (which means differentiation with respect to time) may be interpreted to mean differentiation with respect to some magnitude which increases monotonically with time. We choose ϱ as such a parameter, as long as $\varrho < b$, and obtain

$$s_z\left[E\dot{\varepsilon}_\vartheta - (1+\nu)\dot{\sigma}_\vartheta + 3\nu\dot{\sigma}\right] + s_\vartheta\left[(1+\nu)\dot{\sigma}_z - 3\nu\dot{\sigma}\right] = 0, \tag{52.9}$$

where

$$\dot{\varepsilon}_\vartheta = \frac{1}{r}\frac{\partial u}{\partial\varrho}, \qquad s_\vartheta = \sigma_\vartheta - \sigma,$$

with σ_ϑ from (52.6), $\sigma = \frac{1}{2}(A + s_z)$ where $A = \sigma_r + \sigma_\vartheta$, $s_z = \frac{2k}{\sqrt{3}} q$. Carrying out differentiations and collecting terms, we obtain by a laborious computation

$$\frac{E}{k\sqrt{3}}\frac{4q}{3r}\frac{\partial u}{\partial\varrho} + \left[1 - \frac{2}{3}(1-2\nu)q\right]\frac{\partial q}{\partial\varrho} + (1-2\nu)\frac{2q-1}{\varrho}\left(1 - \frac{\varrho^2}{b^2}\right) = 0. \tag{52.10}$$

The boundary conditions are supplied by the continuity of both q and u across Γ, hence, from (52.2), at $r = \varrho$, with $\varepsilon_z = 0$ we find that

$$\left.\begin{aligned} u &= k\sqrt{3}\,\frac{1+\nu}{2E}\,\frac{\varrho^2}{b^2}\left[(1-2\nu)\varrho + \frac{b^2}{\varrho}\right], \\ q &= (2\nu - 1)\frac{\varrho^2}{2b^2}. \end{aligned}\right\} \tag{52.11}$$

Eqs. (52.8) and (52.10) are two non-linear partial differential equations of first order for the unknowns u, q and the independent variables r, ϱ; they are of the type

$$\frac{\partial u}{\partial x} = c, \qquad a\frac{\partial u}{\partial y} + b\frac{\partial v}{\partial y} = d,$$

where a, b, c, d may depend on all four variables. The equations are hyperbolic with fixed characteristics, parallel to the axes in the r, ϱ-plane. The boundary conditions are Cauchy data, u and q given along the non-characteristic line $\varrho = r$ (Fig. 40). A step-by-step method, as previously explained, can be used to find u and q in the shaded characteristic triangle of Fig. 40. It is obvious from the figure

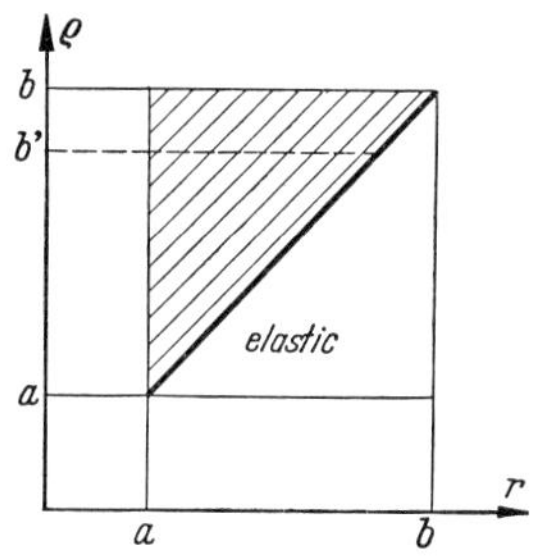

Fig. 40. r, ϱ-plane in problem of thick-walled tube.

that if the problem has been solved for a certain ratio b/a, its solution is known for the smaller ratio b'/a. After q has been found, the longitudinal force L can be evaluated

$$L = 2\pi \left[\int_a^\varrho \sigma_z\, r\, dr + \int_\varrho^b \sigma_z\, r\, dr \right], \tag{52.12}$$

with the plastic value of σ_z in the first, the elastic in the second integral. A graph of σ_z as a function of r/a for various values of ϱ/a may be found in [*6*, p. 116].

γ) *Complete problem for plane strain. v. Mises yield.*[225] Instead of (52.1) we now consider $J_M = 6k^2$, for $r = \varrho$. Using (51.13′) with $\sigma_z = 2\nu p'$ from (51.7), we obtain

$$p'^2 \left[\frac{b^4}{\varrho^4} + \frac{1}{3} (2\nu - 1)^2 \right] = k^2, \tag{52.13}$$

which is satisfied for a pressure

$$p'' = k \left[\frac{b^4}{\varrho^4} + \frac{1}{3} (2\nu - 1)^2 \right]^{-\frac{1}{2}}. \tag{52.13′}$$

Thus, we obtain the solution for *the elastic part of the partly plastic tube under v. Mises' yield*, if the p' of (51.7), (51.8) is replaced by the p'' of (52.13′) [similarly as in the result (52.2)].

In the *plastic region*, $a \leqq r \leqq \varrho$, we now have no (partially) statically determinate problem. The unknown functions are three stresses and two strains, since $\varepsilon_z = 0$. The equations are the equilibrium condition, yield condition, two independent Prandtl-Reuss equations, and a "compatibility condition" for the strains ε_r and ε_ϑ. This problem has been studied by Hodge and White.[225]

53. Further solutions. Comments.

(a) If *the assumption* $\varepsilon_z = 0$ *is dropped*, then in (52.8), on the left side, ε_z must be added, and in (52.9) $-E\dot{\varepsilon}_z$ appears in the second bracket. The equation for u is the same as in the second line of (52.2). The modified Eq. (52.10) then contains a term with $\dot{\varepsilon}_z$. A third finite relation between the three unknowns u, q, ε_z is supplied by the end condition (see [*6*, p. 114] for details).

(b) The above problem has been studied by Hill [*6*, pp. 106–114] for *strains* of any magnitude.[226] The unknown u is then replaced by $v = dr/d\varrho$; one family

---

225 P. G. Hodge Jr., and G. N. White Jr.: J. Appl. Mechanics **17**, 180–184 (1950). See also [*23*, p. 95]. See R. Hill, E. H. Lee and S. J. Tupper: Proc. Roy. Soc. Lond., Ser. A **191**, 278 (1947).

226 Cf. also C. W. MacGregor, L. F. Coffin Jr., and J. C. Fisher: J. Appl. Phys. **19**, 291–297 (1948). — The same authors, in J. Franklin Inst. **245**, 135–158 (1948), computed the closed-end problem for v. Mises' yield condition and Hencky's stress-strain law. For comparison see, Hodge and White, l.c. p. 183.

of characteristics is still straight, $\varrho=$constant, but the second consists in curves $dr - v\, d\varrho = 0$, which are not known in advance.

(c) HILL, LEE, and TUPPER have also obtained a solution (including numerical results) of the problem of Sect. 52α), β) for *closed-end conditions.*[227]

(d) The equations in Sect. 52α), β) may be written in *non-dimensional form* with the variables σ_r/k, σ_ϑ/k, q and Eu/k. It can be seen [6, p. 115] that these are functions of r/a, b/a, ν but not of E. Thus, scaling of a solution obtained for a given b/a and ν supplies solutions for the same b/a and ν, but varying k and E.

(e) Various assumptions have been made in order to simplify the procedures of Sect. 52, which necessitate numerical treatment. The best known is that of total incompressibility in the plastic region.[228] (We always assume $\varepsilon''=0$, but the elastic part of the total strain is not *a priori* isochoric.) In order to avoid an unrealistic discontinuity at $r=\varrho$, some authors also assume incompressibility in the elastic region.[229] This makes, however, no essential difference.

We make again the assumption of plane strain, $\varepsilon_z=0$; if in addition, $\varepsilon=0$, we obtain

$$\varepsilon_r+\varepsilon_\vartheta=\frac{\partial u}{\partial r}+\frac{u}{r}=0, \tag{53.1}$$

and this equation is satisfied by

$$u(r,\varrho)=\frac{D(\varrho)}{r}, \qquad a\leqq r\leqq b. \tag{53.1'}$$

Hence, independently of the stresses,

$$\varepsilon_r=-\frac{D}{r^2}, \qquad \varepsilon_\vartheta=\frac{D}{r^2}, \qquad \varepsilon_z=0, \qquad \varepsilon=0, \quad \text{when} \quad a\leqq r\leqq b. \tag{53.2}$$

Eqs. (53.2) hold for all stages of the deformation. From (51.4) we see that elastic incompressibility, $\varepsilon'=\varepsilon=0$, leads to $\nu=\frac{1}{2}$, unless $\sigma=0$. To obtain the stresses in the elastic region we may thus use (52.13'), with $2\nu-1=0$, and observe the two lines that follow this equation. Then, with $p''=k\varrho^2/b^2$, we obtain

$$\left.\begin{aligned} &\sigma_r=p''\left(1-\frac{b^2}{r^2}\right), && \sigma_\vartheta=p''\left(1+\frac{b^2}{r^2}\right),\\ &\sigma_z=p''=\sigma, && u=k\varrho^2/2G\,r. \end{aligned}\right\}\quad r\geqq\varrho \tag{53.3}$$

where $2G=\dfrac{E}{1+\nu}$. The Reuss equations are:

$$\dot{s}_r:\dot{s}_\vartheta:\dot{s}_z=(2G\,\dot{e}_r-\lambda s_r):(2G\,\dot{e}_\vartheta-\lambda s_\vartheta):(2G\,\dot{e}_z-\lambda s_z). \tag{53.4}$$

From Eqs. (53.2) we see that $\dot{e}_r:\dot{e}_\vartheta:\dot{e}_z=1:-1:0$, from (53.3), that $s_r:s_\vartheta:s_z=1:-1:0$ in the elastic region; using this and the continuity across Γ we see that also *in the plastic region*

$$s_r:s_\vartheta:s_r=\dot{s}_r:\dot{s}_\vartheta:\dot{s}_z=1:-1:0. \tag{53.5}$$

Substituting these relations into the v. Mises yield condition $s_r^2+s_\vartheta^2+s_r\,s_\vartheta=k^2$, we find for $r\leqq\varrho$

$$\sigma_r=\sigma-k, \qquad \sigma_\vartheta=\sigma+k, \qquad \sigma_z=\sigma, \tag{53.6}$$

[227] R. HILL, E. H. LEE and S. J. TUPPER: Ministry of Supply Armament Res. Dept. Theoret. Res. Report 11/46.

[228] A. NADAI: Trans. Am. Soc. Mech. Engrs **52**, 193 (1930).

[229] W. PRAGER: Theory of Plasticity (mimeographed lecture notes) Brown University Providence, R. I. 1942. Cf. also the presentation in HODGE, Ref. 219, and in [*23*, p. 100].

and, using the equation of equilibrium, we obtain

$$\sigma = 2k\left(\frac{\varrho^2}{2b^2} + \log\frac{r}{\varrho}\right). \tag{53.7}$$

Thus, in the present approach in the elastic as well as plastic domain, $\sigma_r + \sigma_\vartheta = 2\sigma_z$. This relation holds in the plastic region in NADAI's original work, but in the elastic region he has, as from (52.2), $(\sigma_r + \sigma_\vartheta)\,\nu = \sigma_z$, and these coincide only for $\nu = \frac{1}{2}$; in general this amounts at $r = \varrho$ for σ_z to a percentage discontinuity of $(1-2\nu)/2\nu$, viz. $67\% \geqq$ for $\nu = 0.3$.[230] However, there is no real advantage in avoiding this discontinuity. HODGE and WHITE, for $b/a = 2$, $\varrho/a = 0.5$, $\nu = 0.3$, find hardly any difference between the present and their exact theory as far as σ_r, σ_ϑ are concerned; however, for σ_z and u there is a similar discrepancy as in NADAI's original work[231] (Fig. 41). The authors suggest that by using correction factors for σ_z and u the simplified solution gives a good approximation to the exact one in the case of plane strain.[232] This "adjusted" solution is not recommended for closed-end tubes. Careful discussion and comparison of results found under various assumptions is given by HODGE and WHITE.[233]

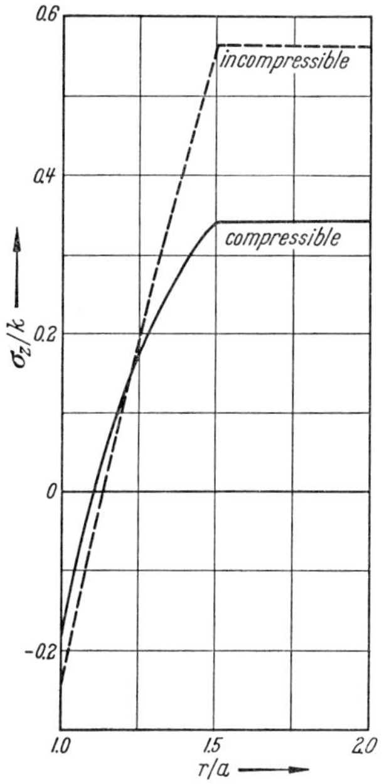

Fig. 41. Distribution of axial stress for compressible (full line) and incompressible (dashed line) material (HODGE, WHITE).

(f) HODGE (pp. 147–155 of the book cited in footnote 219) and PRAGER and HODGE [*23*, p. 110ff.] consider also the problem of the variation of stresses and displacements for *unloading* and for *repeated loading*. For simplification, incompressibility is assumed.[234] The stress field after unloading is obtained

[230] Cf. NADAI [*17*, p. 197, Eqs. (46), (47)], and the discussion in the paper of HODGE and WHITE, Ref. 225.

[231] SOKOLOVSKY's procedure is similar to NADAI's, see, e.g. [*26*, p. 95].

[232] It should be remembered that for $\sigma_z = \frac{1}{2}(\sigma_r + \sigma_\vartheta)$, as in (53.6), the form of v. Mises' and Tresca's criteria become the same viz. $\sigma_\vartheta - \sigma_r = 2k$ (v. MISES), and $\sigma_\vartheta - \sigma_r = \sqrt{3}\,k$ (TRESCA), so that there is only *one* incompressibility-approximation theory.

[233] D. N. DE G. ALLEN and D. G. SOPWITH: Proc. Roy. Soc. Lond., Ser. A **205**, 69 (1951) obtain a solution in closed form by using Hencky's stress-strain relations (Sect. 7). For the present problem the results based on these "deformation type" relations are acceptable. Cf. HODGE and WHITE, l.c.

[234] Cf. also RUTH MOUFANG: Z. angew. Math. Mech. **28**, 33–42 (1948).

by superposition of an elastic stress field, producing the stress-free inner surface by the applied pressure $-p$.

(g) We show finally, following a noteworthy idea of KOITER,[235] that considerable simplification results if Tresca's yield condition is used together with its associated flow rule (see Sect. 4). We *assume* that σ_z is the intermediate stress, $\sigma_r \leqq \sigma_z \leqq \sigma_\vartheta$, in the elastic as well as the plastic part. The stress distribution in the elastic part is the same as in (52.2). In the plastic part σ_r and σ_ϑ are given by (52.4) and the pressure by (52.5). So far, this is the same as above.

Now, in the plastic region, we write the components of total strain as sums of elastic and plastic strains:

$$\varepsilon_r = \varepsilon_r' + \varepsilon_r'', \qquad \varepsilon_\vartheta = \varepsilon_\vartheta' + \varepsilon_\vartheta'', \qquad \varepsilon_z = \varepsilon_z' + \varepsilon_z'', \tag{53.8}$$

where Hooke's law (51.4) holds for the elastic strains, and v. Mises' law (4.3) for the plastic strain rates.[236] Thus, here

$$\dot{\varepsilon}_\vartheta'' : \dot{\varepsilon}_z'' : \dot{\varepsilon}_r'' = 1 : 0 : -1. \tag{53.9}$$

It follows—this is the decisive simplification—that here

$$\dot{\varepsilon}_z'' = 0, \qquad \varepsilon_z'' = 0, \qquad \varepsilon_z = \varepsilon_z'. \tag{53.10}$$

Therefore, the axial stress σ_z can be computed from the third Eq. (51.4) (which holds for the elastic strain ε_z', which here equals ε_z) together with (52.4):

$$\sigma_z = E\,\varepsilon_z + \nu(\sigma_r + \sigma_\vartheta) = E\,\varepsilon_z + \sqrt{3}\,k\,\nu\left(2\log\frac{r}{\varrho} + \frac{\varrho^2}{b^2}\right). \tag{53.11}$$

The radial displacement u follows, as in (52.7) from the dilatation,

$$\varepsilon = \frac{1-2\nu}{E}\cdot 3\sigma,$$

where ε is purely elastic, since $\varepsilon'' = 0$. Using this and (51.2) we obtain the differential equation for u

$$\frac{\partial u}{\partial r} + \frac{u}{r} = \frac{1}{E}(1-2\nu)(1+\nu)\,k\sqrt{3}\left[2\log\frac{r}{\varrho} + \frac{\varrho^2}{b^2}\right] - 2\nu\,\varepsilon_z. \tag{53.12}$$

This equation can be integrated. If the continuity of u for $r = \varrho$ is used, we obtain

$$u = \frac{1-\nu^2}{E}\,k\sqrt{3}\,\frac{\varrho^2}{r} + \frac{(1+\nu)(1-2\nu)}{E}\,k\sqrt{3}\,r\left[\log\frac{r}{\varrho} + \frac{1}{2}\left(\frac{\varrho^2}{b^2} - 1\right)\right] - \nu r\,\varepsilon_z. \tag{53.13}$$

For plane strain, $\varepsilon_z = 0$, the solution is thus completed. For tubes with open ends or closed ends, ε_z must be expressed in terms of the axial load L.[237]

KOITER then investigates his assumption that σ_z is the intermediate stress: At the boundary, for $r = \varrho$, this holds true. For $r < \varrho$ the results are: In plane strain, the assumption is always satisfied if $b/a \leqq 5.75$. For open-end tubes and closed-end tubes the inequality depends on ν. For example, for $\nu = 0.3$ (for which some of the previously reported investigations were made) the assumption is justified for:

$$b/a \leqq 5.75 \text{ (plane strain)}, \quad \leqq 6.19 \text{ (open end)}, \quad \leqq 5.43 \text{ (closed end)}. \tag{53.14}$$

[235] W. T. KOITER: Ref. 19.

[236] There is some major confusion in terminology: KOITER denotes v. MISES' (1913) Eqs. (1.8): $\dot{\varepsilon}_{ij}'' = \lambda s_i$ (also rightly called Lévy-v. Mises relations) as the "well-known Prandtl-Reuss equations" and with respect to (53.9) (v. MISES, 1928) he quotes a passage of [*6*], 1950.

[237] See KOITER: l.c., p. 240.

B. CROSSLAND and R. HILL study a combination of the problems considered here, namely the plastic behavior of thick tubes under combined torsion and internal pressure.[238]

c) Flat ring and flat sheet in plane stress. Further elastic-plastic problems.

54. Flat ring radially stressed as a problem of plastic-elastic equilibrium. Consider a flat ring with inner radius $r=a$ and outer radius $r=b$ stressed by a pressure p, uniformly distributed over the circumference of the inner circle and acting in the plane of the sheet, while the outer circumference is free of stress. The conditions of "plane stress" are assumed satisfied, and in the plastic domain the radial equation of equilibrium

$$\sigma_\vartheta = \frac{d}{dr}(r\,\sigma_r) \tag{54.1}$$

and the Mises' yield condition [see (35.1)]

$$\sigma_r^2 + \sigma_\vartheta^2 - \sigma_r\,\sigma_\vartheta = 4k^2 \tag{54.2}$$

both hold. In general, in a plasticity problem the flow equations determine plastic flow velocities $\boldsymbol{v}$. In certain problems, however, a complete solution with $\boldsymbol{v}=0$ exists for a certain range of stresses [as a consequence of $\lambda=0$ in (1.8)], i.e. a solution of plastic equilibrium. In the present problem, which we study as representative for this circumstance, this equilibrium is characterized by a one-to-one relation between the dimension of the ring, defined by b/a, and the applied pressure p (or rather $p/2k$). The combinations of plastic and elastic equilibrium and the characterization of the situation for plastic *flow* ($\boldsymbol{v}\neq 0$) have been fully investigated.[239]

Plastic equilibrium. With the dimensionless variables $t=r/b$, $u=\sigma_r/2k$, $v=\sigma_\vartheta/2k$, Eqs. (54.1), (54.2) read (the prime denoting d/dr), since $v=ru'+u$ from (54.1):

$$v^2 - uv + u^2 = 1, \quad \text{or} \quad v = \frac{u}{2} \pm \frac{1}{2}\sqrt{4-3u^2} = ru' + u.$$

Putting, for abbreviation,

$$z = +\sqrt{4-3u^2}, \qquad \text{hence } v = \tfrac{1}{2}(u+z), \tag{54.3}$$

we obtain the differential equation

$$dr/r = 2\,du/(z-u) \tag{54.4}$$

or

$$\log r = 2\int \frac{du}{z-u} = -\frac{1}{2}\log(z-u) - \sqrt{3}\arctan\frac{z+2}{u\sqrt{3}} + \text{const.} \tag{54.5}$$

or, since $u=0$ at $r=b$,

$$\frac{r^2}{b^2} = \frac{2}{z-u}\,e^{2\sqrt{3}\vartheta}, \quad \text{where} \quad \tan\vartheta = \frac{u\sqrt{3}}{z+2}, \quad -\frac{\pi}{4} \leqq \vartheta \leqq 0. \tag{54.6}$$

The last relation between u, z and ϑ can be written in terms of the parameter ϑ as $u=\frac{2}{\sqrt{3}}\sin 2\vartheta$, $z=2\cos 2\vartheta$; using (54.3), we obtain

$$\begin{aligned} u &= \frac{2}{\sqrt{3}}\sin 2\vartheta, & v &= \frac{2}{\sqrt{3}}\sin\left(2\vartheta + \frac{\pi}{3}\right), \\ \frac{p}{2k} &= -\frac{2}{\sqrt{3}}\sin 2\vartheta, & \frac{b^2}{r^2} &= \frac{2}{\sqrt{3}}\sin\left(\frac{\pi}{3} - 2\vartheta\right)\mathrm{e}^{-2\sqrt{3}\vartheta}. \end{aligned} \tag{54.7}$$

[238] B. CROSSLAND and R. HILL: J. Mech. Phys. Solids **2**, 27–39 (1953).

[239] R. v. MISES: Reissner Anniversary Volume, p. 415–429. Ann Arbor, Michigan 1949. — See NADAI [*18*, p. 472], [*17*, p. 191].

It is seen that for $\vartheta=0$, $u=0$, $p/2k=0$, $r=b$, while $p/2k$ reaches its maximum, viz.

$$\frac{2}{\sqrt{3}} \quad \text{for} \quad \vartheta=-\frac{\pi}{4}.$$

Thus, collecting the results, we have

$$\vartheta=0, \qquad u=0, \qquad v=1, \qquad \frac{b^2}{r^2}=1,$$

$$\vartheta=-\frac{\pi}{4}, \quad u=-\frac{2}{\sqrt{3}}, \quad v=-\frac{1}{\sqrt{3}}, \quad \frac{b^2}{r^2}=\frac{1}{\sqrt{3}}\,e^{\sqrt{3}\frac{\pi}{2}}.$$ [240]

By differentiation we see that, for

$$\vartheta=-\frac{\pi}{4},$$

the right side of the last Eq. (54.7) reaches its maximum; hence

$$\left(\frac{b}{r}\right)_{\max}=2.964, \qquad \left(\frac{r}{b}\right)_{\min}=0.338. \tag{54.7'}$$

Thus, if for a flat ring $b/a>2.964$, overall yielding cannot arise merely by the application of an internal pressure. An outer portion of the ring must remain elastically strained. On the other hand, if $b/a\leqq 2.964$, then, corresponding to the specific b/a-value of this ring there exists a unique pressure

$$p=-\frac{4k}{\sqrt{3}}\sin 2\vartheta_1,$$

where ϑ_1 follows from the last Eq. (54.7), with $r=a$ such that the whole ring is in plastic equilibrium; and vice versa, to any pressure between 0 and $4k/\sqrt{3}$ corresponds the dimension b/a of a corresponding ring in fully plastic equilibrium. We now wish to determine the state of the body if this specific relation between b/a and $p/2k$ does not hold hence for partial yielding.

55. Continuation: Plastic-elastic equilibrium. *If the whole ring is elastic*, we obtain, as in (51.7)

$$\sigma_r=-p'\left(\frac{b^2}{r^2}-1\right), \qquad \sigma_\vartheta=p'\left(\frac{b^2}{r^2}+1\right), \qquad p'=\frac{p\,a^2}{b^2-a^2}, \tag{55.1}$$

or, in terms of u, v and using $t=r/b$, $t_1=a/b$, $u_1=u_{r=a}$, etc.

$$u=\varkappa\left(1-\frac{1}{t^2}\right), \qquad v=\varkappa\left(1+\frac{1}{t^2}\right), \qquad \varkappa=\frac{p}{2k}\,\frac{t_1^2}{1-t_1^2}; \tag{55.2}$$

or

$$u=u_1\left(1-\frac{1}{t^2}\right)\Big/\left(1-\frac{1}{t_1^2}\right), \qquad v=u_1\left(1+\frac{1}{t^2}\right)\Big/\left(1-\frac{1}{t_1^2}\right). \tag{55.2'}$$

The condition for this fully elastic equilibrium is that $u_1^2+v_1^2-u_1v_1\leqq 1$, even for $r=a$. Using (55.2) we obtain this condition in the form:

$$\frac{p}{2k}\leqq\frac{1-t_1^2}{\sqrt{3+t_1^4}}, \quad \text{and} \quad \left(\frac{p}{2k}\right)_{\max}=\frac{1}{\sqrt{3}}. \tag{55.3}$$

---

[240] The relation between NADAI's parameter δ, Sect. 35, and v. MISES' ϑ is:

$$2\vartheta=\delta-\frac{\pi}{6}.$$

In the present problem δ varies from

$$\frac{\pi}{6} \text{ to } -\frac{\pi}{3}, \qquad \vartheta \text{ from } 0 \text{ to } -\frac{\pi}{4}.$$

But if

$$\frac{p}{2k} > \frac{1-t_1^2}{\sqrt{3+t_1^4}},$$

smaller, however, than the $p/2k$ value which corresponds in (54.7) to the particular t_1 (>0.338) under consideration, then the ring will be neither fully elastic nor fully plastic. For example, for a ring with $b=2a$, $t_1=\frac{1}{2}$, the right side of (55.3) equals $\frac{3}{7}\sim 0.43$. If the applied $p/2k$ is less than 0.43, the ring is fully elastic; on the other hand, the value of $p/2k$ which in (54.7) corresponds to $b/a=2$ is approximately 0.77 and for this $p/2k$ the whole ring is in plastic equilibrium. For each value of $p/2k$ between 0.43 and 0.77 the ring will be plastic in an inner annulus between $r=a$ and a specific $r=\varrho$, the plastic-elastic boundary Γ, and elastic between $r=\varrho$ and $r=b$; and, conversely, to each ϱ between $r=a$ and $r=2a$ there corresponds a $p/2k$ which generates this particular boundary. (We shall see that, for example, to $p/2k=0.5$ corresponds $\varrho=1.58a$.)

We denote the values corresponding to ϱ by $\bar{t}, \bar{u}, \bar{v}$, etc. The formula corresponding to (55.2′) is then

$$u=\bar{u}\left(1-\frac{1}{t^2}\right)\Big/\left(1-\frac{1}{\bar{t}^2}\right), \qquad v=\bar{u}\left(1+\frac{1}{t^2}\right)\Big/\left(1-\frac{1}{\bar{t}^2}\right). \tag{55.4}$$

We may eliminate $\bar{t}$ and, using (54.4), we obtain (55.4) in the form

$$u=\frac{1}{4}(\bar{z}+3\bar{u})-\frac{1}{4t^2}(\bar{z}-\bar{u}), \tag{55.5}$$

as the relation between u and $\bar{u}, \bar{v}$ at Γ [remember that $v=\frac{1}{2}(u+z)$]. Since from (55.2) $u+v=2\varkappa=\bar{u}+\bar{v}$, we have for v in terms of u

$$v=\bar{u}+\bar{v}-u=\tfrac{1}{2}(\bar{z}+3\bar{u})-u. \tag{55.6}$$

In Fig. 42a the line AB on the left gives the relation between u and $t=r/b$ for fully plastic equilibrium [Eqs. (54.6) or (54.7)], b being the abscissa of the point B; the curve CD gives the corresponding v, t-relation. These relations hold only for t-values between 0.338 ($=1/2.964$) and one. On the u-line we choose now an arbitrary point $\bar{A}$ with corresponding $\bar{u}$; and the point $\bar{C}$ above has the corresponding $\bar{v}$ as ordinate. The line $\bar{A}B_2$ is then plotted from (55.5) until its intersection with the t-axis, where $u=0$; and $\bar{C}D_2$ is the corresponding elastic v-line. [We see easily that the two u-lines, the elastic and the plastic one, are tangent at $u=\bar{u}$. In fact, with $x=\log r$ we have from (55.2)

$$u=\varkappa-\varkappa b^2 e^{-2x}, \qquad \left(\frac{du}{dx}\right)_{\text{elast}}=2\varkappa\frac{b^2}{r^2}=2(\varkappa-u),$$

while for the u in the plastic region, from (54.5),

$$\frac{du}{dx}=\frac{z-u}{2};$$

at the transition

$$\left(\frac{du}{dx}\right)_{\text{elast}}=2\varkappa-2\bar{u}=\bar{u}+\bar{v}-2\bar{u}=\bar{v}-\bar{u}=\left(\frac{du}{dx}\right)_{\text{pl}}.]$$

We thus obtain the "mixed" lines $A\bar{A}B_2$ for u, and $C\bar{C}D_2$ for v. This has the following meaning: B_2 (with abscissa ~ 1.35 in the figure) corresponds to a ring with

$$\frac{b}{a}\sim 1.35/0.338\sim 4\,(>2.964).$$

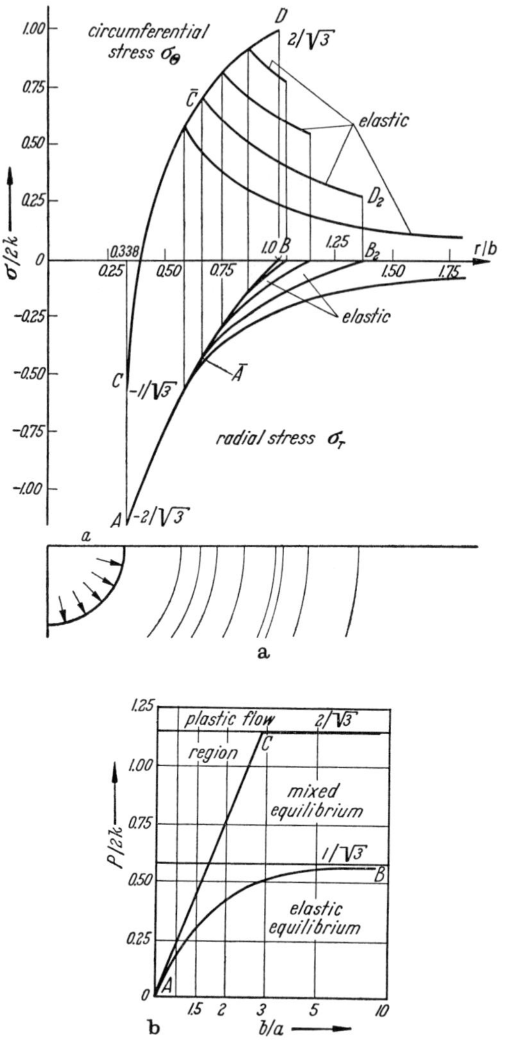

Fig. 42a and b. Flat ring under internal pressure as elastic-plastic problem: (a) radial stress and circumferential stress; (b) elastic, plastic, and mixed equilibrium (v. MISES).

The point $\bar{A}$ has an abscissa $\varrho \sim 2a$; hence between $r=a$ and $r=2a$ the ring is in plastic, and between $r=2a$ and $r=4a$ in elastic equilibrium. The $p/2k$ which leads to this solution follows from (54.7); corresponding to the ratio 2 it is approximately 0.77; hence in a ring with $b=4a$ the above particular plastic-elastic distribution results for $p/2k=0.77$. For a smaller (larger) pressure the plastic annulus will be smaller (larger) than in the example, but its maximum width (for $p/2k=2/\sqrt{3}$) ranges from a to $2.964a$.

The limits of possible states of equilibrium are indicated in Fig. 42b. The line AC shows Eq. (55.3), with equality sign, and independent variable $1/t_1$. If, for *any* ring ratio, the pressure is below or on this line, the ring is fully elastic. If $p \leqq 4k/\sqrt{3}$, any point in the "mixed region" defines a ring in mixed equilibrium; e.g. the point with abscissa 4 and ordinate 0.77 corresponds to the above-discussed example. To all points with the same $p/2k$ there belongs one and the same intersection of this horizontal with AC. The abscissa of this point of intersection defines the outer boundary Γ of the fully plastic annulus, and the remainder is elastic. If, finally, to a ring of $b/a < 2.964$ we apply a pressure higher than the "corresponding" pressure on AC, or to a ring with $b/a \geqq 2.964$ a pressure $> 4k/\sqrt{3}$, then *free plastic flow* sets in; if, for example, to a ring with $b/a = 1.5$ the pressure $p/2k = 1/\sqrt{3} = 0.575$ is applied, the ring starts to flow plastically, while under a slightly smaller over-pressure (compare Fig. 42b) it would be entirely in *contained plastic equilibrium*. Such flow will cause a thickening around the edge of the hole. This thickening has been studied by G. I. TAYLOR and R. HILL,[241] in the slightly simpler situation where $b \to \infty$, under the assumption of a plastic rigid material (see Sect. 58).

56. Expansion of a circular hole in an infinite sheet. This problem may be treated as a particular case of the preceding one, as $b \to \infty$, $t \to 0$. A fully plastic state, as studied in Sect. 54 for the ring, is now, of course, impossible.

The sheet is overall elastic if

$$\frac{p}{2k} \leqq \frac{1}{\sqrt{3}}, \tag{56.1}$$

as seen from (55.3) and Fig. 42b. The corresponding stresses, from (55.1), are

$$\sigma_r = -p\,\frac{a^2}{r^2}, \qquad \sigma_\vartheta = p\,\frac{a^2}{r^2}. \tag{56.2}$$

If $2k/\sqrt{3} \leqq p \leqq 4k/\sqrt{3}$, or, from (54.7),

$$-\frac{\pi}{4} \leqq \vartheta_1 \leqq -\frac{\pi}{12}.$$

there is an annulus in plastic equilibrium, which reaches its maximum width if

$$p = 4k/\sqrt{3}, \qquad \vartheta_1 = -\frac{\pi}{4}.$$

The value of b^2/r^2 from (54.7), for

$$\vartheta_1 = -\frac{\pi}{4}, \quad \text{is} \quad \frac{1}{\sqrt{3}}\, e^{\sqrt{3}\frac{\pi}{2}},$$

that for

$$\vartheta_1 = -\frac{\pi}{12} \quad \text{is} \quad \frac{2}{\sqrt{3}}\, e^{\sqrt{3}\frac{\pi}{6}};$$

the ratio is

$$\tfrac{1}{2}\, e^{\sqrt{3}\frac{\pi}{3}} \sim (1.751)^2.$$

In this simpler case we easily compute directly and find that if $p/2k > 1/\sqrt{3}$, the elastic stresses in the region $r \geqq \varrho$ are from (56.2),

$$u = -\frac{1}{\sqrt{3}}\,\frac{\varrho^2}{r^2}, \qquad v = \frac{1}{\sqrt{3}}\,\frac{\varrho^2}{r^2}. \tag{56.3}$$

---

[241] Cf. [*18*, p. 477], and [*26*, p. 287 seq.], with generalization for a non-circular hole. In the presentation of HILL [*6*, p. 307], where work of G. I. TAYLOR, Quart. J. Mech. Appl. Math. **1**, 103–124 (1948), is incorporated, the deformation is studied which arises around the edge of the hole as soon as *free* plastic flow sets in. The material is considered as "rigid-plastic" (see Sect. 58).

We find the first three Eqs. (54.7) as before; for the last we write

$$\frac{C^2}{r^2} = \frac{2}{\sqrt{3}} \sin\left(\frac{\pi}{3} - 2\vartheta\right) e^{-2\sqrt{3}\vartheta}.$$

At $r = \varrho$, from (56.3),

$$u = -\frac{1}{\sqrt{3}},$$

which arises in (54.7) for

$$2\vartheta = -\frac{\pi}{6},$$

and this gives

$$\frac{C^2}{\varrho^2} = e^{\sqrt{3}\frac{\pi}{6}}.$$

Hence the last Eq. (54.7) is to be replaced by

$$\frac{\varrho^2}{r^2} = \frac{2}{\sqrt{3}} e^{-\sqrt{3}\left(2\vartheta + \frac{\pi}{6}\right)} \sin\left(\frac{\pi}{3} - 2\vartheta\right) = \frac{2}{\sqrt{3}} e^{-\sqrt{3}\delta} \cos\delta, \qquad -\frac{\pi}{3} \leqq \delta \leqq 0, \quad a \leqq r \leqq \varrho, \tag{56.4}$$

where δ is Nadai's parameter (footnote 137), which simplifies the formula. The discussion is simple. If, at $r = a$, a pressure $p/2k$, between $1/\sqrt{3}$ and $2/\sqrt{3}$ is applied, the corresponding δ_1 is between

$$0 \text{ and } -\frac{\pi}{3}, \quad \text{and} \quad \varrho^2 = a^2 e^{-\sqrt{3}\delta_1} \cos\delta_1;$$

the δ for the stresses in this ring is between 0 and δ_1, and beyond this annulus the sheet is elastic. This plastic equilibrium is hyperbolic corresponding to the range $0 \geqq s \geqq -\sqrt{3}$, in terms of the parameter s of Chap. C; in the problem of Sect. 54, s was between $+1$ and $-\sqrt{3}$ [remember that the "hyperbolic range" is $(-\sqrt{3}, +\sqrt{3})$]. For $\delta = \pi/3$, $s = -\sqrt{3}$ the problem becomes parabolic (see Sect. 35), the angle α between the first principal direction—that of the circumferential stress σ_ϑ and the characteristics—is zero and the coinciding characteristics envelop the edge of the hole. If p becomes greater than $4k/\sqrt{3}$, the radius ϱ may increase further to $\varrho' > \varrho = 1.751a$ (at least at the beginning), but the ratio between the outer and the inner radius defined by (56.4) remains the same so that the inner radius of this annulus now equals $\varrho'/1.75 = 0.57\varrho' = a' > a$. At this circle, $r = a'$, the equations become parabolic; if we consider the constrained material in the ring between $r = a'$ and $r = \varrho'$ as rigid (see Sect. 58), the circle $r = a'$ forms the "plastic-rigid" boundary-enveloped by characteristics—between the "rigid" material and that which flows freely (see Sect. 47β and the sketch Fig. 43). This flow will cause a deformation in the domain from $r = a$ to $r = a'$ where the material near the hole piles up into a thickened crater. The mechanics of this deformation has been clarified by TAYLOR[242] and by HILL, using the theory of Sect. 28, and under Tresca's yield condition.

In the above problem, Eqs. (56.3) show that $u \to 0$, $v \to 0$ as $r \to \infty$. NADAI [*18*, p. 481] and SOKOLOVSKY[243] has pointed out that the plastic equilibrium of an infinite disk with a circular hole (with or without pressure at $r = a$), stressed uniformly in its plane by a stress at infinity, can be treated in a similar way.[243]

[242] G. I. TAYLOR: Quart. J. Mech. Appl. Math. **1**, 103 (1948). See also [*6*, 307].
[243] See also SOKOLOVSKY [*26*, 283].

In a domain corresponding to $\pi/3 \leqq \delta \leqq \pi/2$ (or s between $\sqrt{3}$ and 2) the stress distribution will be elliptic.

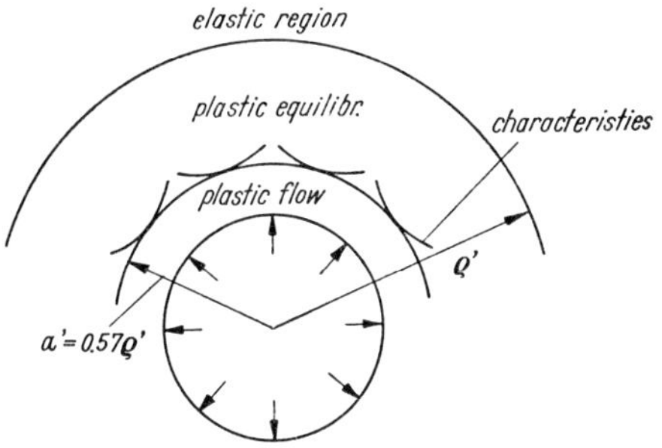

Fig. 43. Schematic distribution of states of stress in expansion of circular hole (HILL).

57. A few further elastic-plastic problems. We mention a few examples. (a) In problems of cylindrical or spherical symmetry the determination of the plastic-elastic interface is facilitated by the symmetry of the problem, which requires the boundary to be spherical or cylindrical. The expansion of a *spherical shell* by internal pressure is the problem originally studied by REUSS,[244] which led him to propose his stress-strain relations. The problem may be found in detail in HILL [*6*, p. 97] for finite deformation.[245] The thorough analysis is straightforward.

(b) The problems of *rotating cylinders* and *rotating disks* are somewhat similar to those of thick-walled tubes and flat rings (see NADAI [*18*, p. 482], HODGE (see footnote 219) p. 158seq., SOKOLOVSKY [*26*, p. 100seq.]). A related problem is that of a rotating disc with a hole.[246]

(c) A problem which, in contrast to the preceding ones, does not depend on r but only on the polar angle ϑ is that of a wedge under a uniform pressure acting on one face; it has been considered by many authors, cf. extensive study in SOKOLOVSKY [*26*, Chap. XI, particularly p. 342seq.] (see also our Sect. 59).

(d) NADAI[247] indicated a problem, later solved rigorously by TREFFTZ,[248] viz. that of stresses and displacements in pure shear in the neighborhood of a semi-circular grove. Around the hole a plastic domain will form, limited by an unknown elastic-plastic boundary, Γ. The problem is a planc one; $\tau_{xz} = \tau_x$, $\tau_{yz} = \tau_y$ are the only non-zero stresses, γ_{xz}, γ_{yz} the only non-zero strains. TREFFTZ determined the elastic-plastic boundary Γ and the complete solution by means of conformal mapping.

(e) In the *elastic-plastic problems of plane strain* a classical analogy plays a role similar to the sand hill and soap film analogies in elastic-plastic torsion: the so-called *plate analogy*. To fix the ideas, we consider an infinite plane region with a circular hole of radius a, free from loads along the circumference C of the hole. At infinity where the directions x, y are chosen as principal directions there acts a general uniform stress:

$$\sigma_x = A, \quad \sigma_y = B, \quad \tau_{x,y} = 0.$$

[244] E. REUSS: Z. angew. Math. Mech. **10**, 266–274 (1930).

[245] R. HILL: J. Appl. Mechanics **16**, 295 (1949). — SOKOLOVSKY [*26*, p. 95–100], presenting his paper in J. Appl. Math. Mech. **8** (1944), uses Hencky's stress-strain relations.

[246] Cf. H. I. WEISS and W. PRAGER: J. Aer. Sci. **21**, 196 (1954).

[247] A. NADAI: Z. angew. Math. Mech. **3**, 442–454 (1923), see p. 449.

[248] E. TREFFTZ: Z. angew. Math. Mech. **5**, 64–73 (1925).

We use Airy's stress function F, Eq. (29.1), by means of which we satisfy the two plane equilibrium equations identically. At infinity

$$F_\infty = \tfrac{1}{2}(Bx^2 + Ay^2) + Cx + Dy + E, \tag{57.1}$$

with C, D, E arbitrary constants. Since the hole is free of surface traction,

$$\sigma_n = \frac{\partial^2 F}{\partial s^2} = 0, \qquad \tau_{sn} = -\frac{\partial^2 F}{\partial s\,\partial n} = 0,$$

we can assume that

$$F = 0, \quad \frac{\partial F}{\partial n} = 0 \quad \text{on } C. \tag{57.2}$$

As long as the applied forces at infinity are such that the whole domain is elastic, the stress function F satisfies the biharmonic equation

$$\Delta_2 F = \frac{\partial^4 F}{\partial x^4} + 2\frac{\partial^4 F}{\partial x^2\,\partial y^2} + \frac{\partial^4 F}{\partial y^4} = 0, \tag{57.3}$$

and the boundary conditions (57.1), (57.2). This stress function can be visualized by means of an infinite elastic plate exterior to C, clamped along C and bent by appropriate moments at infinity. The deflection w of this plate satisfies the biharmonic equation, and the moments at infinity may be such that (57.1) holds for w at infinity, while along the clamped contour conditions (57.2) hold for w.[249] If, however, the loads are increased until local yielding occurs around the hole up to a boundary Γ, then, outside Γ elastic and inside plastic conditions prevail. In the plastic domain F satisfies the yield condition (29.2), say, and the boundary conditions (57.2), while outside Γ, (57.3) and the conditions at infinity must hold. Also, F is continuous across Γ, together with its first and second derivatives. The generalization of the plate analogy, found by PRAGER[250] and by GALIN,[251] works under certain assumptions regarding the elastic-plastic boundary Γ. (See [*23*, p. 202], cf. [*26*, p. 200 seq.].)

L. A. GALIN[251] solved analytically the more general problem where a normal internal pressure p is applied to the hole in addition to the biaxial tension $\sigma_x = A$, $\sigma_y = B$ at infinity. The stresses in the plastic neighborhood of the cavity are given by (52.6) if Tresca's yield condition is used. Since the present F is independent of ϑ, the definition of F in polar coordinates,

$$\sigma_r = \frac{1}{r}\frac{\partial F}{\partial r} + \frac{1}{r^2}\frac{\partial^2 F}{\partial r^2}, \qquad \sigma_\vartheta = \frac{\partial^2 F}{\partial r^2}, \qquad \tau_{r\vartheta} = \frac{1}{r^2}\frac{\partial F}{\partial \vartheta} - \frac{1}{r}\frac{\partial^2 F}{\partial r\,\partial\vartheta},$$

becomes here

$$\sigma_r = \frac{1}{r}\frac{\partial F}{\partial r}, \qquad \sigma_\vartheta = \frac{\partial^2 F}{\partial r^2}, \qquad \tau_{r\vartheta} = 0, \tag{57.4}$$

and F can be found explicitly, using (52.6):

$$F = \frac{k\sqrt{3}}{2}\left[r^2 \log\frac{r}{a} - \frac{r^2}{2}\left(1 + \frac{2p}{k\sqrt{3}}\right)\right]. \tag{57.5}$$

Since this is a (degenerate) solution of the biharmonic equation, F in this problem satisfies the same equation as the stress function in the elastic region. GALIN took this as the starting point for his solution of this particular problem by

249 K. WIEGHARDT: Mitt. Geb. Ingenieurwes. **49**, 15–30 (1908).

250 W. PRAGER: Theory of Plasticity. Mimeographed lecture notes. Brown University, Providence, R. I. 1942.

251 L. A. GALIN: J. Appl. Math. Mech. **10**, 365–386 (1946), Translation: GDAM, Brown University 1947, and L. A. GALIN: J. Appl. Math. Mech. **12**, 757–760 (1948).

means of complex function theory. (See HODGE (footnote 219), p. 348, PRAGER and HODGE [*23*, p. 203], HILL [*6*, p. 253].)[252]

(f) In continuation of his investigation on the uniqueness problem of rigid plastic solids (see Sect. 45 bis) R. HILL[253] examined in 1958 the problem of uniqueness and stability for elastic plastic bodies. For a comprehensive class of such solids he established a result which corresponds to the results in his contributions III and IV for the rigid-plastic case.

II. Some plastic-rigid problems.

We present only a few typical plastic-rigid problems, and these not in great detail, referring the reader to the presentations in the literature cited. Some attention is given to the *concept* of the plastic-rigid body.

a) Various remarks.

58. The plastic-rigid body. α) The perfectly plastic body as defined 1913 by v. MISES contains, in general, an elastic and a plastic part, according as $F<C$ or $F=C$, where F is an appropriate yield function and C a constant. If $F=C$, the plastic strain rate $\dot{E}''$ is proportional to a tensor of stress, in general derived from F according to the theory of the plastic potential (Sect. 3); we obtain $\dot{E}''=\lambda S$, where S is the stress deviator, if F is v. Mises' yield function, and we find Tresca's "associated flow rule" (Sect. 4) if F is Tresca's yield function. The positive proportionality factor λ varies in space and time and can assume any value between zero and infinity. In a particular problem, the elastic displacements in either part may or may not be considered as negligible. It is not implied in the theory that the elastic displacements in regions below the yield limit or in regions of contained plastic deformation are to be neglected.

In the problem of the thick-walled tube (Sects. 51–53) the stresses in the plastic part have reached the yield limit, but the deformation is restrained by the surrounding elastic material. It is a natural approach to use Hooke's law in the elastic part, the Prandtl-Reuss equations in the plastic one. In the problem of elastic-plastic torsion (Sects. 48–50) the elastic strains in the elastic part are of course relevant; in the plastic part, $\dot{S}=0$ holds, so that the Reuss and the v. Mises equations become identical (see Sect. 6), and the elastic strains in the plastic region are automatically "neglected". In the elastic-plastic analysis of the problem of a flat ring under internal pressure (Sects. 54 and 55) let it be assumed for the moment that $b/a>2.964$ (so that fully plastic equilibrium is excluded) and $p/2k>1/\sqrt{3}$ (so that the ring is not fully elastic). Then, as long as $p/2k<2/\sqrt{3}$, there is an annulus of definite width in "plastic equilibrium", surrounded by an elastic region. This plastic annulus is in a state of contained plastic deformation. This appears mathematically as follows: If there is a complete solution of a problem which satisfies all differential equations (of plasticity) and all boundary conditions, with identically vanishing proportionality factor λ (in the flow equations), then the plastic deformation velocities *are zero* (no matter whether we think of v. Mises' or Reuss' relations), and we have a case of plastic equilibrium. If then a pressure $p>4k/\sqrt{3}$ is applied, *free plastic flow* sets in.

In the problem of the thick-walled tube, as well as in that of elastic-plastic torsion, the regions of contained deformation show the same basic character.

[252] A generalization of the above problem has been considered and solved by conformal mapping by G. P. CHEREPANOV, J. Appl. Math. Mech. **27**, 644 (1963).

[253] R. HILL: J. Mech. Phys. Solids **6**, 236 (1958).

In each of these problems a complete solution is found, satisfying all equations and boundary conditions, with plastic deformation velocities zero. The warping function w in the torsion problem and the radial displacement u in the case of the tube are *strains*; at a fixed time t_0—for a fixed radius $\varrho=\varrho_0$ of the elastic-plastic boundary—the corresponding strain *rates* are zero. The problem of the ring as well as that of torsion show the situation particularly clearly because there we can compute, to begin with, all stresses in the domain of plastic equilibrium, whereas in the problem of the tube [see the treatment of Sect. 52β)] the stress σ_z and the displacement u appeared as simultaneous "unknowns". The contained deformation in the three cases is, however, of the same type.

It seems instructive, in this connection, to look at KOITER's skillful treatment of the thick-walled tube problem (end of Sect. 53). He does not "neglect" elastic strains in the plastic region, but he sets up separately Hooke's law for the elastic strains and v. Mises' for the plastic strain rates. A similar approach might be helpful in other problems,[254] though probably not as simple as in the case of the Tresca yield condition and associated flow rule.

In the so called *rigid-plastic* analysis, strains which are of the order of elastic strains are neglected throughout: the material below the yield limit—which otherwise would be considered as elastic—is treated as rigid; in a domain of unrestricted plastic flow the total strain (or strain rate) is identified with the plastic one since this part is overwhelmingly large; regions of contained plastic flow are also considered as rigid, since there all strains are of elastic order of magnitude. There is obviously no reason to use the Reuss equations in this case.[255]

We shall consider, in the following, a few examples of this type of analysis.

β) The state of a material is completely described if the stress Σ and the velocity $\boldsymbol{v}$ are known as functions of the radius vector $\boldsymbol{r}$ and the time t: $\boldsymbol{v}=\boldsymbol{v}(\boldsymbol{r}, t)$, $\Sigma=\Sigma(\boldsymbol{r}, t)$. Another way of description is the "material" one, where stress and velocity of each individual particle are given functions of time. It is known that, correspondingly, two types of differentiation are distinguished: the local differentiation $\partial/\partial t$, at fixed $\boldsymbol{r}$ and the material differentiation, d/dt, or $\delta/\delta t$ for a fixed particle, where

$$\frac{d}{dt}=\frac{\partial}{\partial t}+v_x\frac{\partial}{\partial x}+v_y\frac{\partial}{\partial y}+v_z\frac{\partial}{\partial z}=\frac{\partial}{\partial t}+v\frac{\partial}{\partial s}, \tag{58.1}$$

and $\partial/\partial s$ means differentiation in the direction of $\boldsymbol{v}$. Streamlines are lines which, at a fixed time t_0, have at each point the direction of $\boldsymbol{v}$, principal trajectories are lines which at $t=t_0$ have everywhere the first (second) principal stress direction etc. In general, the one-dimensional infinity of streamlines (in the plane) or of trajectories *change in time*; so do the characteristics, the slip-line pattern, etc.

We call a state *steady* or *stationary* if $\partial/\partial t=0$ for all magnitudes under consideration, i.e. at a fixed place $\boldsymbol{r}$ the same $\boldsymbol{v}$ and Σ hold for all time. In this case there

[254] The addition of the plastic and elastic strain rates [Eqs. (6.3)–(6.5)] as in the Reuss equations, while leading to elegant equations, contributes in some cases to mathematical complication. This addition is not essential for the consideration of elastic strains in the plastic region.

[255] Historically, the "plastic-rigid analysis" was not a conception of v. MISES, who emphasized the basically elastic-plastic character of the perfectly plastic body. However, this analysis was accepted by him whenever the problem warrants neglecting the elastic contributions. (See, as one example, the problem of Sect. 60). Rigid-plastic analysis is applied to the flat ring or sheet by G. I. TAYLOR and by HILL (footnote 242) because they are mainly interested in the mechanics of the distortion, which appears as a thickening near the hole for $p>4k/\sqrt{3}$ when free plastic flow has set in; on the other hand, v. MISES, who was interested in the elastic-plastic state for $p<4k/\sqrt{3}$, and in particular in the contained plastic deformation, used an elastic-plastic analysis (Sects. 54, 55).

is *one* family of streamlines, of trajectories, of slip lines; the pattern does not change in time. Of course, the same spatial region is not filled at different moments by the same particles; but the new particle has again the same $\boldsymbol{v}$ and Σ at a fixed place as had the particle which left the place.

In an unsteady problem the theory gives the solution for a particular instant t_0. If desired, we may then use this solution to formulate boundary conditions for the next time interval, and so on. This is extremely laborious and rarely done. The study for the fixed moment t_0 is termed the problem of *incipient* plastic flow. HILL, LEE and TUPPER [Proc. Roy. Soc. Lond., A **188**, 273 (1947)] were the first to draw attention to so-called problems of *pseudosteady* state, where the slip line pattern, while not remaining fixed, changes only in scale; its image in the stress plane and hodograph plane remains fixed.

Numerous interesting and practically important problems of steady and pseudosteady flow have been worked out by A. P. GREEN, HILL, HODGE, LEE, PRAGER and other authors; these are problems of sheet drawing, of extrusion, piercing, strip rolling and others.[256] A. P. GREEN, for example, investigated among various problems *unsymmetrical* extrusion in plane strain;[257] the plastic yielding of deep-notched bars (1953) and of shallow-notched bars (1956).

In [*24*, p. 99] PRAGER calls attention to the fact that in technological forming problems such as most of the above-mentioned ones, plastic deformation cannot be assumed small. However, under the theory of rigid-perfectly-plastic solids this fact (which, for example in the theory of elasticity entails very great complications) does not introduce difficulties. The reason is simply that in "flow"-type deformation theories—like the Saint Venant-v. Mises equations—the stress is related to the strain *rate* [just as in the theory of viscous fluids (see p. 405)] and not to the strain as in elasticity. In strain rate theories the finiteness of deformation does not influence the differential equations.

We mention briefly a problem of a type not considered so far (on which we might have reported at the end of Part C) namely that of a plane ideal-plastic-rigid solid[258] where inertia forces cannot be considered negligible. The equations are the usual ones of plane strain except that the equilibrium conditions are replaced by

$$\frac{\partial \sigma_x}{\partial x} + \frac{\partial \tau}{\partial y} = \varrho\, a_x, \qquad \frac{\partial \tau}{\partial x} + \frac{\partial \sigma_y}{\partial y} = \varrho\, a_y,$$

where ϱ is the constant density and a_x, a_y the components of the vector of particle acceleration. There are three independent variables, x, y, and the time t and four dependent ones, two stresses and two velocity components.

The characteristic condition is investigated, and it is seen that derivatives with respect to t do not appear. Each characteristic has multiplicity 2; stress and velocity characteristics coincide. These α- and β-lines are intersections of characteristic surfaces in x, y, t-space with planes $t=$constant. With respect to the moving coordinate system α, β, t the equations of the problem simplify considerably to give generalized Hencky equations for the stress field and generalized Geiringer equations for the velocity field. To this system SPENCER applies successfully his perturbation method (see next section).

58bis. Axial symmetry. A few remarks. Problems exhibiting *axial symmetry* have also been considered under a plastic rigid analysis (see [*6*, p. 262seq.]). The

[256] See explanations and literature in W. PRAGER: James Clayton Lecture, Proc. Inst. Mech. Engrs **169**, 41–57 (1955), particularly p. 51, compare HILL [*6*, Chap. VII]. See also [*24*, Chap. IV] and [*26*], etc.

[257] A. P. GREEN: J. Mech. Phys. Solids **3**, 189 (1955); — Quart. J. Mech. Appl. Math. **6**, 223 (1953); — J. Mech. Phys. Solids **4**, 259 (1956).

[258] A. J. M. SPENCER, above Journal **8**, 262 (1960).

independent variables are r and z (and not r alone as in problems considered in Chap. D, I). It is assumed that stresses and strains are independent of ϑ. The non-zero stresses and strains are $\sigma_r, \sigma_\vartheta, \sigma_z, \tau_{rz}$ and $\varepsilon_r, \varepsilon_\vartheta, \varepsilon_z, \gamma_{rz}$. In 1955 R. T. SHIELD[259] surveyed the problem. He assumed plastic rigid material and Tresca's yield condition.

The four strains correspond to two velocity components. Hence there are seven unknowns: four stresses, two velocities, and the λ of the flow equations, and there are seven equations, two equilibrium equations, one yield condition and four flow equations. The problem is, in general, not statically determinate. P. S. SYMONDS has shown that the problem is not hyperbolic.

The governing stress equations become statically determined if a well-known hypothesis of HAAR and v. KÁRMÁN[260] is assumed satisfied which, in this case, takes the form that the circumferential stress σ_ϑ is equal to one of the two principal stresses in the r, z-plane. The legitimacy of the assumption remains questionable. (See HILL [*6*, p. 280].) SHIELD studied several problems under this assumption.

D. H. PARSON[261] considers perfectly plastic flow under axial symmetry and v. Mises' yield criterion and shows how by a suitable choice of new variables the number of dependent variables can be reduced to four. In 1957 PARSON showed that most of the results of his earlier paper are not restricted to assuming v. Mises' flow criterion.

A. J. M. SPENCER[262] applied to problems of axially symmetric rigid plastic flow a perturbation method invented by him 1960 and used in a variety of problems. In each case the slip line field is the basic element to be perturbed. H. LIPPMANN,[263] criticizing the Haar-v. Kármán principle assumes Tresca's yield and the three principal stresses *pairwise different*; he adds the assumption $\dot{\varepsilon}_{\text{med.}} = 0$, where this is the strain rate corresponding to the intermediate principal stress, and in this way obtains a "kinematically determined" system of three equations for three kinematic variables u, w, ψ, where u is the radial, w the axial velocity and ψ the angle between the direction of the axis and one of the two other (mutually orthogonal) principal directions in the plane normal to z. This system has real characteristics, which he showed (1965) to be identical with the stress characteristics.

We pass to the study of a few complete plane rigid-plastic problems, which we shall present in some detail: the problem of the *unilaterally loaded wedge* as a problem of incipient plastic flow, which also exemplifies a *stress discontinuity*, and the pseudosteady problem of a *plastic mass pressed between two rough rigid plates*.

b) Wedge with pressure on one face.

59. General discussion and velocity distribution. Consider a wedge of perfectly plastic material with vertex angle $2\beta_0 \geqq 90°$. There is a continuous stress solution due to PRANDTL.[264] In Fig. 44a the triangles ABC and BDE are regions of

[259] R. T. SHIELD: Proc. Roy. Soc. Lond., Ser. A **233**, 267–287 (1955); also J. Mech. Phys. Solids **3**, 246 (1955).

[260] A. HAAR and TH. v. KÁRMÁN: Nachr. Ges. Wiss. Göttingen, math.-phys. Kl. S. 204–218 (1909).

[261] H. D. PARSON: Proc. London Math. Soc. (3) **6**, 610 (1956), also J. London Math. Soc. **32**, 233 (1957).

[262] A. J. M. SPENCER: J. Mech. Phys. Solids **12**, 231 (1964). Same Journal: 1960, 1961, 1962.

[263] H. LIPPMANN: J. Mech. Phys. Solids **10**, 111 (1962). Same Journal **13**, 29 (1965).

[264] L. PRANDTL: Göttinger Nachr. S. 74–89 (1920), and L. PRANDTL: Z. angew. Math. Mech. **1**, 15–20 (1921).

constant state. In ABC the principal stresses are $-p_0$, normal to AB, and $2k-p_0$, parallel to AB. In BDE they are zero normal to BE and $-2k$ parallel to it. The region BCD is filled by a centered "fan" where the slip lines are radii and arcs of circles. The value of p_0 that produces the plastic flow is

$$p_0 = 2k(1 + 2\beta_0 - 90°). \tag{59.1}$$

On the other hand, we may consider the discontinuous solution of Fig. 44b;[265] the corresponding pressure value is easily found to be

$$p_0' = 2k(1 - \cos 2\beta_0). \tag{59.2}$$

Hence $p_0' < p_0$ for $2\beta_0 > 90°$, while $p_0 = p_0' = 2k$ for $2\beta_0 = 90°$. HENCKY[266] and PRANDTL[267] have suggested that, in the case of two alternate solutions, the correct one is that requiring the smaller force. However, it is seen that only the continuous solution gives a consistent velocity distribution. Indeed, in Fig. 44b the rigid region below $A'AHCKEE'$ is at rest, hence the particles in the plastic region of constant state, ACB and BCE, can move only parallel to AC and CE, respectively. The discontinuity line BC is inextensible (see Sect. 24); its particles can only move perpendicular to this line or not at all. These requirements are not compatible (unless all velocities are zero) and therefore no complete solution of this type exists. On the other hand, one easily sees that a uniquely determined velocity field exists in the case of Fig. 44a if the velocities are e.g. given along AB.

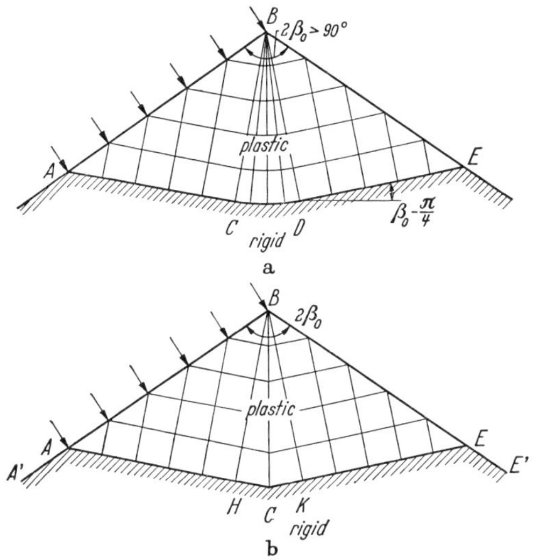

Fig. 44a and b. Wedge, $2\beta_0 > 90°$, under unilateral pressure: (a) continuous and (b) discontinuous slip-line field (PRANDTL).

We now turn to the more interesting case, $2\beta_0 < 90°$. Fig. 45a shows clearly that, without a discontinuity line, an (unacceptable) multivalued stress field would result. PRAGER[268] and, in particular, LEE[269] have determined a complete

[265] For details and additional references cf., for example, H. GEIRINGER, Advances, p. 278 seq.
[266] H. HENCKY: Z. angew. Math. Mech. 3, 241–251 (1923).
[267] L. PRANDTL: Z. angew. Math. Mech. 3, 401–406 (1923).
[268] W. PRAGER: R. Courant Anniversary Volume, pp. 289–299. New York 1948.
[269] E. H. LEE: Proc. 3rd Symp. Appl. Math., pp. 213–228. New York 1950.

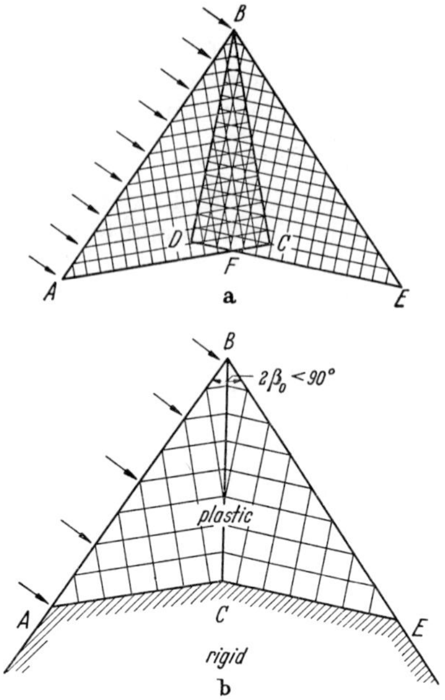

Fig. 45a and b. Wedge, $2\beta_0 \leqq 90°$, under unilateral pressure: (a) continuous and (b) discontinuous slip-line field (PRAGER).

solution with a stress discontinuity. The slip lines are shown in the figure. We shall now determine the velocities.

The plastic-rigid boundary (Fig. 46) is a velocity characteristic; since stress and velocity characteristics coincide, the characteristics ACE form this boundary. Along this boundary the normal velocity is zero. Along the discontinuity line BC, $\dot{\varepsilon}_y = 0$ (Sect. 33), $v_y = \text{constant} = 0$, since $v_y = 0$ at C. Along AB we prescribe one component of velocity, say the normal component; then along each of the lines BA, BC, AC, CE one velocity component is known. The two velocity equations are two linear differential equations of first order. In ACD we have a "mixed" boundary-value problem (Sect. 46) since we know one unknown along the characteristic AC, one along AD; in CDF we again have a mixed problem, with data on CD and CF, and so on through smaller and smaller triangles. If we carry this out analytically, denoting by v_1 and v_2 the components of $\boldsymbol{v}$ in the characteristic directions, the expressions for v_1 and v_2 are seen to be sums of an increasing number of terms; as we approach the point B these infinite sums approach a limit. Actually, the velocity at B turns out to be normal to BC and of magnitude of $f_B/\cos\beta_0$ (as we had to expect, since the component of $\boldsymbol{v}$ in direction of BC is zero). We can verify that the condition $\dot{E} = 0$ at the discontinuity line holds true.

On the right side of the wedge the computation is easy, since now we know both components of velocity on BC. This determines the velocities in BCG; finally, from data along the characteristics CG and CE, the velocities in GCE follow. It can be seen that the velocities on the right are symmetric to those on the left—see the "typical streamline" in Fig. 46. (This is not obvious since the original data are not symmetric.) For details and comments, particularly regarding

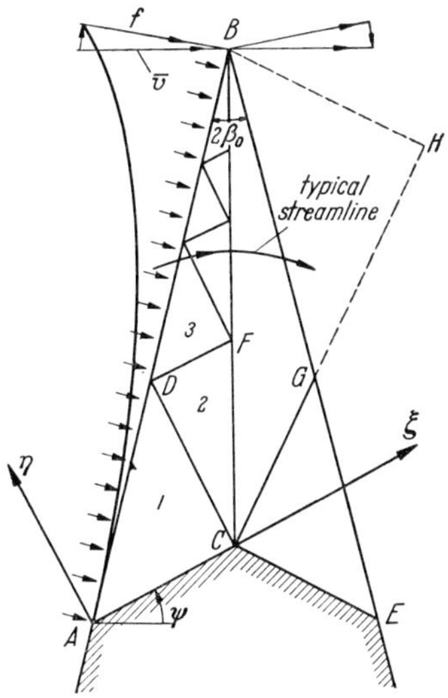

Fig. 46. Computation of velocities for wedge under unilateral pressure (LEE).

the given function f, which must have a positive second derivative, see LEE'S original paper[270], and GEIRINGER, Advances, pp. 278–287.

c) Plastic mass between rough rigid plates.

60. Infinite slab. An infinitely long slab of height $2h$ is pressed between two rough rigid plates which are forced to remain parallel, while they move at a uniform downward (upward) velocity c. The origin is at 0, and the x-axis is parallel to the trace of the plates. While the mass becomes plastic, below each plate, symmetric to the y-axis, a rigid kernel will form, bounded by characteristics (Fig. 47). Since material on the right side of this axis flows to the right, that on the left to the left, a discontinuity would arise if the slab were in plastic flow throughout. The stress field is that of (32.12) with $2A = h$, as given by PRANDTL.[271] The characteristics are two families of orthogonal cycloids with the horizontal lines $y = \pm h$ as envelopes.

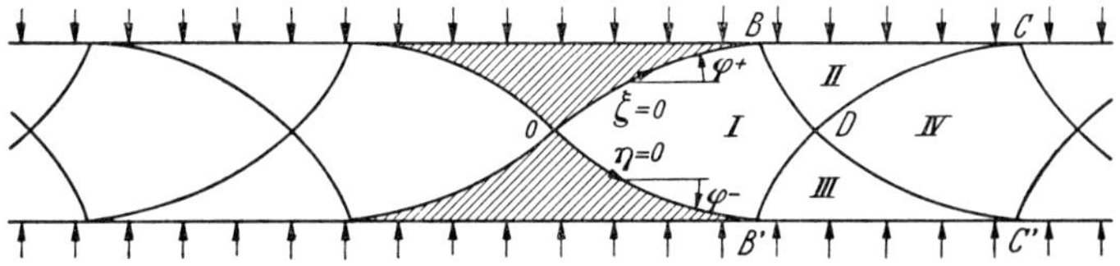

Fig. 47. Plastic mass between infinitely long rough plates.

To find the velocities, a boundary-value problem must be solved.[272] The boundary conditions are that everywhere where rigid material is adjacent to plastic material the normal velocities coincide. This boundary consists of parts of the

[270] E. H. LEE, preceding footnote.
[271] L. PRANDTL: Proc. 1st Internat. Congr. Appl. Mech., Delft 1924, pp. 43–54.
[272] H. GEIRINGER: Proc. 3rd Internat. Congr. Appl. Mech. II, 1930, pp. 185–190. See [3] for details.

slip lines $\xi=0$ and $\eta=0$ adjacent to the rigid kernel and of the horizontal envelope of the cycloids adjacent to the plates, beginning from the points of contact with these slip lines. The domain I is limited by the four cycloids $\xi=0$, $\eta=0$, $\xi=\pi/4$, $\eta=\pi/4$. Along $\xi=0$ the normal component $c\cos\varphi'$ of the pressing rigid material equals the normal component v_1 of the plastic mass (v_1, v_2 are the components of $\boldsymbol{v}$ in the ξ- and η-directions) and similarly for $\eta=0$; hence, writing φ' and φ for the φ^+ and φ^- of Sect. 32:

$$\begin{aligned}&\text{along } \xi=0\colon\ v_1=c\cos\varphi'=c\cos\left(\frac{\pi}{4}-\eta\right)=\frac{c}{\sqrt{2}}(\sin\eta+\cos\eta)=g(\eta),\\&\text{along } \eta=0\colon\ v_2=c\cos\varphi=c\cos\left(-\frac{\pi}{4}+\xi\right)=\frac{c}{\sqrt{2}}(\cos\xi+\sin\xi)=g(\xi).\end{aligned}\tag{60.1}$$

By integration of (30.15), v_2 and v_1 are found along $\xi=0$ and $\eta=0$, respectively, where the integration constants are such that the same value of v_1 and v_2 results at $\xi=\eta=0$. Thus, with $f(x)=\frac{c}{\sqrt{2}}(\sin x-\cos x+2)$, $g(x)=\frac{c}{\sqrt{2}}(\sin x+\cos x)$

$$\begin{aligned}&\text{along } \xi=0\colon\ v_1=g(\eta),\ v_2=f(\eta),\\&\text{along } \eta=0\colon\ v_1=f(\xi),\ v_2=g(\xi).\end{aligned}\tag{60.2}$$

We want a function $v_1(\xi,\eta)$ which satisfies Eq. (30.16) in I and takes on the values $g(\eta)$ and $f(\xi)$ on $\xi=0$ and $\eta=0$, respectively. Next there follows, by symmetry,

$$v_2(\xi,\eta)=v_1(\eta,\xi).\tag{60.3}$$

The above characteristic boundary-value problem for v_1 can be solved explicitly by means of Riemann's method (Sect. 31) and it reduces to quadratures. Then, in the domain II (Fig. 47), there is a mixed boundary-value problem since, in addition to knowing v_1 on the slip line between I and II, we know $v_1=c$ on the

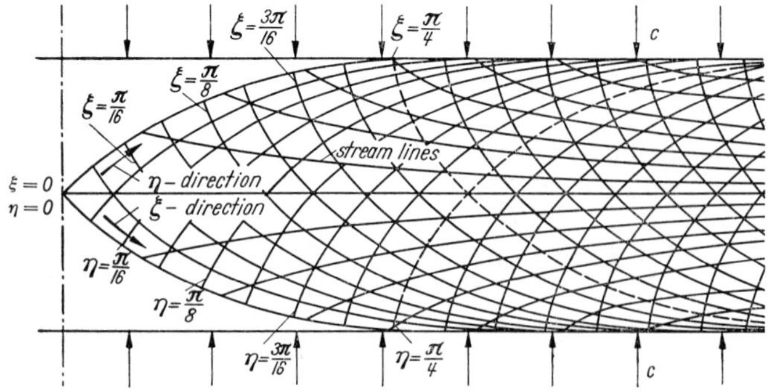

Fig. 48. Streamlines for plastic mass between infinitely long rough plates (GEIRINGER).

horizontal straight line. To solve this problem, we consider instead of v_1 the function

$$w=v_1-c\cos\left(\frac{\pi}{4}+\xi-\eta\right).$$

For w the same differential equation holds but with $w=0$ on BC.[273] This problem can be treated by substituting for it a characteristic-value problem with such

[273] The use of w simplifies the formulas also in region I.

an asymmetry in the data that $w=0$ on BC. Thus v_1, and consequently v_2 are found in II, and similarly in III. In IV we then have again a characteristic-value problem, etc. It should be noted that we obtain non-vanishing tangential velocities along $\xi=0$ and $\eta=0$ which are not equal to those of the pressing material. This tangential discontinuity across a characteristic is admissible, however. Eventually, after v_1 and v_2 are found, the stream lines can be plotted (Fig. 48).

Thus a complete solution has been found. However, it is based on the assumption of plastic flow throughout (except for the rigid kernel at the center), of an infinite mass between infinite plates; these assumptions simplify the problem but may not be realistic. In the next section we shall consider the same problem in a physically more realistic formulation.

61. Slab of material with overhanging ends between rough plates. We consider the compression of a slab of plastic material between parallel rigid rough plates of initial distance $2h$ and such that the slab overhangs them.[274] It is then clear that the material to the right of some curve ABA', where A is the right upper corner of the upper plate, will not reach the yield limit; it is considered rigid and AB and $A'B$ must be shear lines (Fig. 49). If ABA' is known, the stress field can in principle, be determined (Fig. 50) step-by-step in I, II, III, IV, ... by the second differential equation (29.10) and the fact that we know φ along ABA', along AE, and symmetrically. In fact, the ξ-lines (η-lines) meet the upper (lower) plate orthogonally on account of its complete roughness. The horizontal line AE is the envelope of η-lines, a limit line, just as in the preceding problem.

A suitable choice of AB is[274] to take it as a straight line of slope one, and $A'B$ symmetrically (Fig. 50). The singular points A and A' are centers of centered simple waves; the upper wave is between AB and AC, with concentric circular arcs as cross characteristics. For this choice of ABA', the material on the right of it is able to move as a rigid body. In fact, along AB the velocity v_2 in the direction

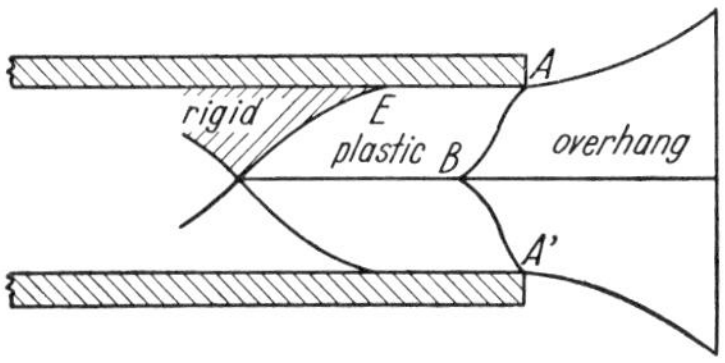

Fig. 49. Slab with overhanging ends between rough plates (HILL, LEE and TUPPER).

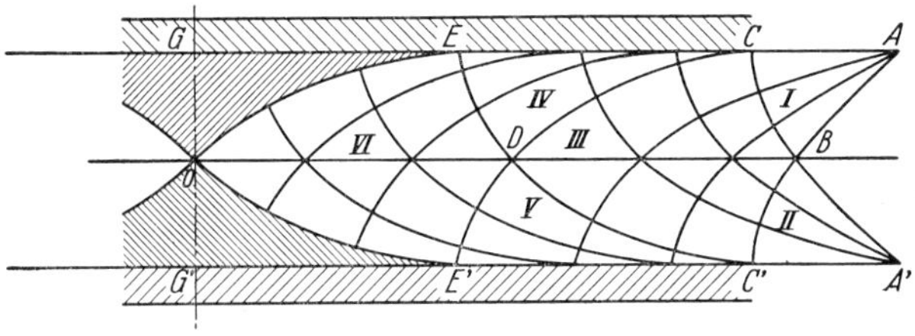

Fig. 50. Slip-line field for slab with overhanging ends between rough plates (HILL, LEE and TUPPER).

[274] See L. PRANDTL, Proc. 1st Internat. Congr. Appl. Mech., Delft 1924, pp. 43–54, who considers the problem of the overhanging plates. See R. HILL, Dissertation, Cambridge, 1948, Ministry of Supply Survey 1–48, also SOKOLOVSKY [26, p. 243].

of AB is constant by the second Eq. (29.11′), and since along this radius $v_2\,d\vartheta$ is constant, the first Eq. (29.11′) shows that dv_1 is constant along it. Therefore, if in this fan the component v_1 normal to one radius is constant, the same holds for each radius; but along AC the normal component *is* constant. Hence the velocity along AB is constant.

The slip-line field in I (i.e., in ABC) and II is thus known. For the further computation Eqs. (30.11), (30.12) are used. The arrangement from here on is exactly the same as in the problem of the preceding section since the Eqs. (30.11) and (30.15) are of the same type. First, in III we have in $CBC'D$ a characteristic-value problem for either R^- or R^+ (which is explicitly solved by Bessel functions), and $R^+(\xi, \eta) = R^-(\eta, \xi)$, as in (60.3). In IV there is a mixed problem with R^- given along CD and $R^- = 0$ along the limit line EC. By the same trick as in the preceding section, we solve instead a characteristic problem with such data that $R^- = 0$ on CD. Then V and VI follow. The length-to-height ratio is now, of course, not arbitrary but, depends on the stage of compression. In the work of HILL, LEE and TUPPER[275] (see Fig. 50) it is 6.72. The curves in Fig. 50 towards the center are similar in form to cycloids and approach them for increasing length-to-height ratio; as the characteristics approach the cycloidal form the stresses approach PRANDTL'S stress distribution in the problem in Sect. 60. The value of p follows at each point from (29.9). The initial pressure at B equals k from equilibrium considerations, hence $p = k$ all along AB; the pressure along AC is easily found as

$$k\left(1 + \frac{\pi}{2}\right),$$

and so on. The pressure distribution approaches that of PRANDTL.

From the solution indicated in the figure we can obtain solutions for smaller l/h ratios, the smallest being 3.64, corresponding to a solution progressing to point D only (from the right). Cutting off the field at various points between D and 0 corresponds to ratios between 3.64 and 6.72. By continuing the computations beyond field VI of the figure we obtain larger ratios; i.e., as the compression continues we deal with a sequence of blocks of increasing l/h ratio. HILL (see [*6*, p. 230]) has also considered the case where $l/h < 3.64$, but greater than unity. There exists a solution where the central rigid kernel extends over the whole pressing plate.

After the slip-line field has been found (the problem is statically determinate like the preceding problems), the velocity distribution can be determined in exactly the same way as in the preceding section, of course based on the present slip-line field: we start in the middle and finish with the distribution across ABA', which must turn out constant. Again there is a tangential discontinuity along the limit line (and symmetrically).[276]

Other non-steady motion problems include those of *indentation*, started in 1920 by PRANDTL (see [*6*, p. 254 seq.], [*26*, p. 206 seq.]). See also the interesting critical discussion by LEE.[277]

---

[275] R. HILL, E. H. LEE and S. J. TUPPER: J. Appl. Mechanics **8**, 46–52 (1951).

[276] The problem of the block pressed between overhanging plates, which had been considered by PRANDTL, has been studied further in HILL'S dissertation, Cambridge 1948, see also SOKOLOVSKY [*26*, p. 243].

[277] E. H. LEE: J. Appl. Mechanics **19**, 97 (1952). Discussed in GEIRINGER, Advances, p. 289.

Some reference books.

[1] COLONETTI, G.: L'équilibre des corps déformables. Paris: Dunod 1955.
[2] FREUDENTHAL, A. M.: Inelastic behavior of engineering materials and structures. New York: Wiley 1950. German transl. Berlin: VEB 1955.
[3] GEIRINGER, H.: Fondéments mathématiques de la théorie des corps plastiques isotropes. Paris 1937.
[4] GOLDENBLATT, I. I.: Problems of the mechanics of deformable bodies. Moscow 1955.
[5] HANSEN, BENT: A theory of plasticity for ideal frictionless materials. Kopenhagen 1965.
[6] HILL, R.: Mathematical theory of plasticity. Oxford: University Press 1950.
[7] HODGE, P. G., Jr.: The mathematical theory of plasticity. Surveys in applied mathematics, vol. I. New York: Wiley 1958.
[8] HOFFMAN, O., and G. SACHS: Theory of plasticity for engineers. New York 1953.
[9] ILYUSHIN, A. A.: Plasticity. Moscow 1948. French transl. Paris: Eyrolles 1956.
[10] — Plasticity. Izdat Akad. Nauk, SSSR. Moscow 1963.
[11] KACHANOV, L. M.: Theory of plasticity. Moscow 1956.
[12] KALISKI, S.: Dynamic boundary value problems in the theory of elastic and inelastic bodies. Warsaw 1957.
[13] LEBENSON, L. S.: Elements of the mathematical theory of plasticity. Moscow 1943.
[14] LIPPMANN, H., and O. MAHRENHOLTZ: Plasto-mechanik, Berlin-Heidelberg-New York: Springer 1967.
[15] LOHR, E.: Mechanik der Festkörper. Berlin: W. de Gruyter 1952.
[16] MANDEL, J.: Equilibres par tranches planes des solides à la limite d'écoulement. Paris: Louis Jean 1942.
[17] NADAI, A.: Plasticity. New York: McGraw-Hill 1931.
[18] — Theory of flow and fracture of solids, Vol. 1. New York: McGraw-Hill 1950.
[19] NEAL, B. G.: Plastic methods of structural analysis. New York: Wiley 1956.
[20] OLSZAK, W.: Non homogeneity in elasticity and plasticity. Proc. of the I.U.T.A.M Symposium. London: Pergamon Press 1958.
[21] — J. RYCHLEWSKI, and W. URBANOWSKI: Plasticity under non-homogeneous conditions. Advances in applied mechanics, vol. VII, 1962, p. 131–214. New York: Academic Press 1962.
[22] PHILLIPS, A.: Introduction to plasticity. New York: Ronald 1956.
[23] PRAGER, W., and P. G. HODGE: Theory of perfectly plastic solids. New York: Wiley 1951.
[24] — Probleme der Plastizitätstheorie. Basel: Birkhäuser 1955. English: An Introduction to plasticity. Boston: Addison Wesley 1959.
[25] RECKLING, K. A.: Plastizitätstheorie und ihre Anwendung auf Festigkeitsprobleme. Berlin-Heidelberg-New York: Springer 1967.
[26] SOKOLOVSKY, V. V.: Theory of plasticity. Moscow 1946, 1950. Revised German ed. Berlin; VEB 1955.
[27] THOMAS, T. Y.: Plastic flow and fracture in solids. New York: Academic Press 1961.
[28] WESTERGAARD, H. M.: Theory of elasticity and plasticity. Cambridge: Harvard University Press 1952.

Topics in the Mathematical Theory of Plasticity.

By

TSUAN WU TING.

With 2 Figures.

A. Introduction.

Three main topics are taken up in this article: foundation of plasticity theory, general existence theorems and the elastic-plastic torsion problem in the Prandtl-Reuss theory.

In discussing the most recent development for the foundation of the theory, I have considered elastic-plastic deformations as thermodynamic processes. It should be said that the splendid idea of elastic continuations is due to OWEN [*55, 56*]. Both he and Professor TRUESDELL read an earlier version of the part of this article that concerns the foundation and made various comments.

Up to now a complete existence theory for elastic-plastic deformations remains to be developed. However, under certain restrictions on the yielding functionals LANGENBACH'S minimum principle [*9*] answers basic questions for Hencky materials. I present that principle here in its original form with some slight modifications. With the motivation of processes of manufacture such as extrusion, wire drawing and sheet or strip rolling, etc., I establish the existence of steady plastic flows with "small" boundary data.

The St. Venant torsion problem for Prandtl-Reuss materials became an active topic of research in the last few years. The details of the presentation below are based on my own work since 1963. For simply connected cylindrical bars, the results given here are the most comprehensive and quantitative yet published. In my search for a relatively complete answer I have been constantly encouraged by Professors GARABEDIAN, THOMAS and TRUESDELL. Also, the article of VON MISES [*1*] and GEIRINGER'S article of 1958 on the ideal plastic body [*7*] gave me much inspiration. Readers interested in the results but not in the details of proof should turn to p. 578 for the main conclusions about elastic-plastic torsion according to the Prandtl-Reuss and Hencky theories.

Clearly this article concerns only certain topics. The references listed are only those related to the subjects discussed here. For a survey of the classical theory and for full references to it, the reader should consult the article by GEIRINGER, just preceding in this volume. The preparation of the manuscript has been supported in part by the U.S. National Science Foundation.

B. Foundation of the theory.

In this part, a general mathematical theory for elastic-plastic materials is derived by applying basic principles for thermo-mechanical processes and certain ideas of OWEN. The Prandtl-Reuss theory is then deduced as a special case within this framework. In addition, derivations of the St. Venant-Lévy-v. Mises theory and of the Hencky theory have been provided, along with some comments.

I. Thermodynamics for elastic-plastic materials.

In this section general stress relations, a dissipation inequality and a formula for "plastic power" will be derived from the principles of balance of energy, linear momentum and moment of momentum and from the principle of production of entropy by considering an elastic-plastic deformation as a thermodynamic process.

1. Motion and deformation history. Consider a moving body B. For the present theory there is no loss of generality if we assume B connected. Let E_n be a real n-dimensional Euclidean space and let R_τ in E_3 be the region (open connected set) occupied by B at the time τ. Let X and x be two coordinate systems covering E_3. The position in E_3 of a material point P in B at time $\tau=0$ will be denoted by $X=(X^1, X^2, X^3)$ and that at all other times by $x=(x^1, x^2, x^3)$. The principle of impenetrability [*11*, p. 244] requires that at each instant τ the correspondences, $X \leftrightarrow P \leftrightarrow x$ be one to one. Hence the spatial (Eulerian) coordinates of a material point P are uniquely determined by 4-tuple $(X^1, X^2, X^4; \tau)$ of real numbers:

$$x^i=\chi^i(X^1, X^2, X^3; \tau), \quad i=1, 2, 3, \ 0 \leqq \tau \leqq t, \ X \in R_0. \tag{1.1}$$

In what follows, the function $\chi(X; \tau)$ will be assumed to be of class $C^3(R_0)$ for each fixed τ and continuous and piecewise smooth in τ for each X in R_0, so that the Jacobian determinant with respect to X is always positive:

$$\det\left(\nabla_X \chi(X; \tau)\right) \equiv \det\left(\partial x^i/\partial X^\alpha\right) > 0 \quad \text{in } R_0 \times [0, t]. \tag{1.2}$$

For a given motion (1.1) the *deformation gradient* $F(X; \tau)$ of a material point P at the time τ relative to the configuration R_0 is defined by the formula

$$F(X; \tau) = \nabla_X \chi(X; \tau), \quad 0 \leqq \tau \leqq t, \ X \in R_0. \tag{1.3}$$

Then $F(X; \tau)$ is of class $C^2(R_0)$ and is continuous and piecewise smooth in $\tau \in [0, t]$. Conversely, if $F(X; \tau)$ is given on the cylinder $R_0 \times [0, t]$ with the regularity properties just stated, then a theorem of DARBOUX states that the Cauchy problem,

$$\partial\chi^i/\partial X^\alpha = F^i_\alpha(X^1, X^2, X^3; \tau), \quad \chi^i(X^1_0, X^2_0, X^3_0; \tau) = x^i_0(\tau), \tag{1.4}$$

$i, \alpha, \beta = 1, 2, 3$, $X \in R_0$, always possesses a unique solution $\chi(X; \tau)$ provided the following integrability conditions,

$$\partial F^i_\alpha/\partial X^\beta = \partial F^i_\beta/\partial X^\alpha, \quad i, \alpha, \beta = 1, 2, 3, \tag{1.5}$$

are satisfied everywhere in $R_0 \times [0, t]$. Further, the solution $\chi(X; \tau)$ is of class $C^3(R_0)$ and is continuous and piecewise smooth in $\tau \in [0, t]$.

In the above description, a material point has been identified by its place X (Lagrangian coordinates) at $\tau=0$. Let $T(x; \tau)$ denote the Cauchy stress, and let $\varrho(X; \tau)$, $\theta(X; \tau)$, $\varepsilon(X; \tau)$, $\eta(X; \tau)$ and $\psi(X; \tau)$ denote respectively the mass density, temperature, specific internal energy, specific entropy and the specific free energy associated with a material point as functions of the Lagrangian coordinates X and the time τ. Let $q(X; \tau)$ be the heat flux of P at time τ and let its components $q^\alpha(X;\tau)$, $\alpha=1,2,3$, relative to the X-system be functions of X and τ. Let $g(X;\tau)$ be the spatial gradient of the temperature $\theta(X;\tau)$. Then

$$g(X;\tau)=\operatorname{grad}_x\theta\left(\chi^{-1}(x,\tau);\tau\right)\quad\text{or}\quad F^T g=\nabla_X\theta(X;\tau),\tag{1.6}$$

where F^T stands for the transpose of F. Similar notations will also be used for other second-order tensors. Let

$$\Gamma(X;\tau)\equiv\left(F(X;\tau),\theta(X;\tau)\right),\qquad X\in R_0,\ 0\leqq\tau\leqq t,\tag{1.7}$$

be given. If $\Gamma(X;\tau)$ is of class $C^2(R_0)$ and is continuous and piecewise smooth in τ and such that the integrability conditions in (1.5) and the impenetrability conditions (1.2) are satisfied everywhere in $R_0\times[0,t]$, then $\Gamma(X;\tau)$ will be called a *deformation history of B relative to the configuration R_0 from the time $\tau=0$ to the time $\tau=t$.* Restricting $\Gamma(X;\tau)$ to an arbitrarily fixed interior point X_0 in R_0, we write

$$f^t(\tau)\equiv\Gamma(X_0;\tau),\qquad 0\leqq\tau\leqq t,\tag{1.8}$$

where the superior index t over f indicates the domain of definition of the function f^t. The function $f^t(\tau)$, $0\leqq\tau\leqq t$, will be referred to as the deformation history of the material point $(X_0,0)$ from $\tau=0$ to $\tau=t$.

2. Constitutive functionals. From now on, the interior material point P_0 of B will be fixed arbitrarily. Thus, to simplify notations, the variable X_0 or x_0 will not be written. The point of departure is to regard the motion of an elastic-plastic body as a thermodynamic process [*55*]. Further it is a basic postulate that *for elastic-plastic materials the quantities $T(t)$, $q(t)$, $\eta(t)$ and $\psi(t)$ at $(X_0;t)$ are uniquely determined by the history f^t and the present temperature gradient $g(t)$.* As was observed by Coleman [*20*, § 5], to do this it suffices to assume that $\Sigma(t)$, $\psi(t)$ and $q(t)$ are uniquely determined by f^t and $g(t)$, where $\Sigma(t)$ is the stress-entropy vector, which is defined by

$$\Sigma\equiv\left(T(F^{-1})^T/\varrho,\,-\eta\right).\tag{2.1}$$

Further, it was shown by Coleman [*20*, § 6] that as a consequence of the principles of balance of energy, linear momentum, moment of momentum and of the second law of thermodynamics, the values of $\psi(t)$ and $T(t)$, are actually independent of $g(t)$. Thus the basic postulates thus amount to

$$\Sigma=\Sigma(f^t),\qquad\psi=\psi(f^t),\qquad q=q\left(f^t;g(t)\right),\tag{2.2}$$

where for convenience the distinction between a functional and its values has not been made. It may be noted that the symbols on the right-hand sides of the first two equations in (2.2) indicate that the quantities Σ and ψ associated with P_0 at time $\tau=t$ are uniquely determined by the history f^t. Also, they may depend upon the entire history, $f^t(\tau)$, $0\leqq\tau\leqq t$, not just the present value $f^t(t)$. Of course, given $f^t(\tau)$, $0\leqq\tau\leqq t$, $\Sigma(f^\tau)$ and $\psi(f^\tau)$ are well-defined *domain* functionals on $[0,\tau]$. On the other hand, for fixed t, Σ and ψ are just ordinary functionals defined on the class of histories with the same domain $[0,t]$. As for the heat flux, q, it is a functional of the history f^t and a function of the present value of the temperature gradient, $g(t)$.

In order to derive physically meaningful results from the above postulates, it is, of course, necessary that the three functionals in (2.2) possess certain regularity properties. For the present purpose, it suffices to require that for every history $f^t(\tau)$, $0\leqq\tau\leqq t$, the functions

$$\Sigma(f^\tau),\,\psi(f^\tau),\,q\left(f^\tau;g(\tau)\right),\qquad 0\leqq\tau\leqq t,$$

are continuous and piecewise smooth in τ and twice continuously differentiable in the unwritten parameter X_0.

3. Elastic-plastic resolution of deformations. With the motivation of a one-dimensional "stress-strain" curve obtained by interpolations between experimental data for loadings and unloadings, an elastic-plastic material will be idealized as follows: Let the motion of a body B be described by the equations in (1.1). At each instant τ, $0 \leqq \tau \leqq t$, if the stress in B is gradually released while the temperature field in B is maintained, then the body B will not recover completely from the deformed state R_τ but will remain in a permanently deformed state $p(R_\tau)$ specified by the equations,

$$Y^I = Y^I(X; \tau), \quad 0 \leqq \tau \leqq t, \; X \in R_0, \; I = 1, 2, 3 \tag{3.1}$$

relative to the undeformed state R_0. The coordinate system Y in (3.1) can be arbitrary but (Y, τ), (x, τ) and $(X, 0)$ identify the same material point in B. Conversely, the body B will return from the permanently deformed state $p(R_\tau)$ to the state R_τ if the stress field in B is restored and the temperature distribution in B is maintained. Thus, the deformation from $p(R_\tau)$ to R_τ,

$$x^i = x^i(Y; \tau), \quad 0 \leqq \tau \leqq t, \; Y \in p(R_\tau), \tag{3.2}$$

is purely elastic in the sense of complete recovery. By combining (1.1), (3.1), and (3.2), we have

$$x^i = x^i\big(Y(X; \tau); \tau\big) = \chi^i(X, \tau), \quad 0 \leqq \tau \leqq t, \; X \in R_0. \tag{3.3}$$

It should be mentioned that the time required for the deformation from $p(R_\tau)$ to R_τ or from R_τ to $p(R_\tau)$ has been deliberately neglected in (3.2) and (3.1). Hence, the motions (3.1) and (3.2) are both fictitious. They are mathematical artifices, suggested by experimental evidence, for precise description of the theory.

We shall restrict the functions $Y^I(X; \tau)$, $x^i(Y; \tau)$ in (3.1) and (3.2) to be of class C^3 in the space variables and to be continuous and piecewise smooth in τ. Accordingly, the principle of impenetrability implies that

$$\det\left(\partial Y^I/\partial X^\alpha\right) > 0, \quad \det\left(\partial x^i/\partial Y^I\right) > 0 \tag{3.4}$$

for $0 \leqq \tau \leqq t$ throughout the body B. Further, upon an application of the chain rule for differentiations, we obtain from (3.3) the *polar decomposition*,

$$F^i_\alpha = \partial x^i/\partial X^\alpha = \left(\partial x^i/\partial Y^I\right)\left(\partial Y^I/\partial X^\alpha\right), \tag{3.5}$$

for the deformation gradient throughout B for all time τ. For simplicity, we write

$$p(f^\tau) \equiv \left(\partial Y^I/\partial X^\alpha\right), \quad e(f^\tau) \equiv \left(\partial x^i/\partial Y^I\right), \quad 0 \leqq \tau \leqq t \tag{3.6}$$

and call them the *permanent* and *elastic deformation gradients* respectively.

Kinematically, there are infinitely many admissible decompositions like that in (3.5). However, we shall assume that for a given material the decomposition (3.5) is uniquely determined by the deformation history so that both $e(f^t)$ and $p(f^t)$ are functionals of the deformation history $f^t(\tau)$, $0 \leqq \tau \leqq t$. From equations in (1.7), (1.8), (3.5) and (3.6) we now have

$$f^t(\tau) = \big(e(f^\tau)\, p(f^\tau), \theta(\tau)\big), \quad 0 \leqq \tau \leqq t. \tag{3.7}$$

To simplify notations we shall write

$$e(\tau) \equiv e(f^\tau), \quad p(\tau) \equiv p(f^\tau), \quad 0 \leqq \tau \leqq t. \tag{3.8}$$

Also I shall restrict $e(\tau)$, $p(\tau)$ to be continuous and piecewise smooth in τ.

To summarize what has been discussed I lay down the following assumption:

A1. *For every deformation history f^t of a given elastic-plastic material there is a unique decomposition (3.7) with $e(f^t)$ and $p(f^t)$ being functionals of f^t such that $p(f^\tau)$, $0 \leqq \tau \leqq t$, is a permanent deformation gradient history corresponding to f^t.*

Since each deformation history can be resolved as a composition of elastic and plastic ones, it is natural to assume that the free energy functional $\psi(f^t)$ is actually a functional of the elastic deformation history alone. This leads to the following somewhat stronger assumption,

A2. *Corresponding to a given elastic-plastic material there is a unique continuously differentiable function ψ^* defined on E_{10} such that for every history $f^t(\tau)$, $0 \leqq \tau \leqq t$,*

$$\psi(f^\tau) = \psi^*(e(\tau), \theta(\tau)), \quad 0 \leqq \tau \leqq t. \tag{3.9}$$

Again, following the motivation of the one-dimensional "stress-strain" curve obtained by interpolating between experimental data for loadings and unloadings, I assume that all deformations relative to a permanently deformed state can be completely recovered provided they are sufficiently small. In other words, the elastic property relative to a permanently deformed state is similar to that with respect to an undeformed state. For convenience in later applications, I state this simple ideas as two assumptions:

A3. *For a given elastic-plastic material, there is a positive number ε depending only on the history f^t such that for every positive number $s \leqq \varepsilon$ it is possible to continue the history f^t to a history f^{t+s} with the following properties:*

(i) $f^{t+s}(\tau)$ *is continuously differentiable on* $[t, t+s]$,

(ii) $f^{t+s}(t+s) = f^t(t-s)$,

(iii) $p(f^{t+u}) = p(f^t)$ for all $u \in [0, s]$.

If f^{t+s} is a continuation of f^t, satisfying (i) and (iii) in A_3, then f^{t+s} will be called an *elastic continuation* of f^t.

A4. *If f^{t+s} is an elastic continuation of f^t, then there are two positive numbers δ and μ depending only on f^{t+s} such that for every positive $r < \mu$ every continuation f^{t+s+r} of f^{t+s} is elastic provided*

$$\max_{0 \leqq s' \leqq r} |f(t+s+s') - f(t+s)| \leqq \delta,$$

where $|\;|$ stands for the usual norm on V_{10} and V_n denotes a vector space of dimension n over the reals.

4. Consequences of the Clausius-Duhem inequality. Assuming the symmetry of the Cauchy stress T and applying the principles of balance of energy and linear momentum, one can show [*20*, § 6] that the principle of production of entropy leads to the Clausius-Duhem inequality.

$$\dot{\psi}(f^\tau) - \Sigma(f^\tau) \cdot \dot{f}^\tau + q(f^\tau; g(\tau)) \cdot g(\tau)/\varrho(\tau)\,\theta(\tau) \leqq 0 \tag{4.1}$$

for $0 \leqq \tau \leqq t$, where the symbol dot " · " stands for the inner product in V_{10} or V_3 and where $\dot{\psi}$, $\dot{f}$ stand for one-side material derivatives.

Since the specific free energy functional ψ is given through the function ψ^* in (3.9) and ψ^* is continuously differentiable over E_{10}, an application of the chain rule for differentiations leads to

$$\dot{\psi}(f^\tau) = \nabla\psi^*(e(\tau), \theta(\tau)) \cdot (\dot{e}(\tau), \dot{\theta}(\tau)), \quad 0 \leqq \tau \leqq t. \tag{4.2}$$

According to assumption A3, it is possible for all sufficiently small s to extend the history f^t to a history f^{t+s} such as to fulfill all the requirements (i)–(iii) in A3. Define the *history gradient* $\delta\psi$ of the free energy ψ at the instant t by the following rule,

$$\delta\psi(f^t) = \lim_{s\downarrow 0}[\psi(f^{t+s}) - \psi(f^{t-s})]/s. \tag{4.3}$$

To see that the limit of the difference quotient in (4.3) really exists, we note from the conditions in A3 that

$$\begin{aligned}\psi(f^{t+s}) &= \psi^*(e(t+s), \theta(t+s))\\ &= \psi^*(F(t+s)\,p^{-1}(t+s), \theta(t+s))\\ &= \psi^*(F(t-s)\,p^{-1}(t), \theta(t-s))\end{aligned} \tag{4.4a}$$

$$\psi(f^{t-s}) = \psi^*(F(t-s)\,p^{-1}(t-s), \theta(t-s)). \tag{4.4b}$$

By substituting into (4.3) the expressions in (4.4) for $\psi^*(f^{t+s})$ and $\psi^*(f^{t-s})$ and then letting $s \to 0$, we conclude from the differentiability of ψ^* on E_{10} and from the piecewise smoothness of $p(\tau)$ in τ that

$$\delta\psi(f^t) = \nabla_e\psi^*(e(t), \theta(t)) \cdot F(t)\, dp^{-1}(t)/dt, \tag{4.5}$$

where dp^{-1}/dt is the left material derivative of $p^{-1}(t)$ at time t. It is easy to check that the left material derivative of ψ can be written in the form

$$\dot{\psi}(f^\tau) = \nabla\psi^*(e(\tau), \theta(\tau)) \cdot (\dot{F}(\tau)\, p^{-1}(\tau), \dot{\theta}(\tau)) + \delta\psi(f^\tau) \tag{4.6}$$

for $0 \leqq \tau \leqq t$. Substitution of the expression (4.6) into the Clausius-Duhem inequality (4.1) yields

$$\begin{aligned}&\nabla\psi^*(e(\tau), \theta(\tau)) \cdot (\dot{F}(\tau)\, p^{-1}(\tau), \theta(\tau)) + \delta\psi(f^\tau)\\ &\qquad - \Sigma(f^\tau) \cdot \dot{f}^\tau + q(f^\tau; g(\tau)) \cdot g(\tau)/\varrho(\tau)\,\theta(\tau) \leqq 0\end{aligned} \tag{4.1a}$$

for $0 \leqq \tau \leqq t$, where the dots over F or f stand for their left material derivatives.

From (3.1)–(3.3) we see that

$$\nabla_e\psi^*(e(\tau), \theta(\tau)) \cdot F(\tau)\, p^{-1}(\tau) = \nabla_e\psi^*(e(\tau), \theta(\tau))\,(p^{-1}(\tau))^T \cdot \dot{F}(\tau),$$

or, in component form,

$$\frac{\partial\psi^*}{\partial x^i_I}\,\dot{x}^i{}_\alpha X^\alpha_I = \frac{\partial\psi^*}{\partial x^i_I}\,X^\alpha_I\,\dot{x}^i_\alpha.$$

Using the above identity, one can now write (4.1) as follows:

$$\begin{aligned}&\{(\nabla_e\psi^*(e(\tau), \theta(\tau))\,(p^{-1}(\tau)^T), \psi^*_\theta) - \Sigma(f^\tau)\} \cdot \dot{f}^\tau\\ &\qquad + \delta\psi(f^\tau) + q(f^\tau; g(\tau))/\varrho(\tau)\,\theta(\tau) \leqq 0\end{aligned} \tag{4.1b}$$

for all τ in $[0, t]$, where $\dot{f}$ is the left material derivative of f at τ.

Let $f^{t+\varepsilon}$ be an elastic continuation of f^t. Let h be an arbitrary element in V_{10}. Then according to Assumption A4, for all sufficiently small δ the function,

$$f^{t+\varepsilon+\delta}(\tau) \equiv \begin{cases} f^{t+\varepsilon}(\tau), & \tau \text{ in } [0, t+\varepsilon],\\ f^{t+\varepsilon}(t+\varepsilon) + (t+\varepsilon+\delta-\tau)\,h, & \tau \text{ in } [\tau+\varepsilon, t+\varepsilon+\delta],\end{cases} \tag{4.7}$$

is a deformation history from $\tau=0$ to $\tau=t+\varepsilon+\delta$, and it is an elastic continuation of $f^{t+\varepsilon}$. Now, for the history $f^{t+\varepsilon}$,

$$\delta\psi(f^{t+\varepsilon})=0 \quad \text{at the instant} \quad t+\varepsilon. \tag{4.8}$$

Moreover, the right material derivative $\dot{f}^{t+\varepsilon+\delta}$ of $f^{t+\varepsilon+\delta}$ at the instant $t+\varepsilon$ is equal to h, i.e.,

$$\dot{f}^{t+\varepsilon+\delta}(t+\varepsilon)=h\equiv\dot{f}^{t+\varepsilon}(t+\varepsilon). \tag{4.9}$$

This shows that the right material derivative of $f^{t+\varepsilon}$ at $t+\varepsilon$ can be assigned arbitrarily. By applying (4.1) to the history $f^{t+\varepsilon}$ at the instant $t+\varepsilon$ and using the relations in (4.8), it is easy to derive, by reasoning similar to that used in deriving (4.1a), that for all $\varepsilon>0$

$$\begin{aligned}\{(\nabla_e\psi^*(e(t+\varepsilon),\theta(t+\varepsilon))(p^{-1}(t+\varepsilon)),\psi_\theta^*)-\Sigma(f^{t+\varepsilon})\}\cdot\dot{f}_r^{t+\varepsilon}(t+\varepsilon)\\+q(f^{t+\varepsilon};g(t+\varepsilon))\cdot g(t+\varepsilon)/\varrho(t+\varepsilon)\,\theta(t+\varepsilon)\leqq 0\end{aligned} \tag{4.1a'}$$

where $\dot{f}_r^{t+\varepsilon}(t+\varepsilon)$ is defined in (4.9). Since $\dot{f}_r^{t+\varepsilon}(t+\varepsilon)$ can be any element h in V_{10}, while the coefficients of it and of the other terms in (4.1a') are all uniquely determined by $f^{t+\varepsilon}$ and $g(t+\varepsilon)$, for (4.1a') to hold it is necessary that for all $\varepsilon>0$

$$\nabla_e\psi^*(e(t+\varepsilon),\theta(t+\varepsilon))\,(p^{-1}(t+\varepsilon))^T-\psi_\theta^*=\Sigma(f^{t+\varepsilon}). \tag{4.10}$$

By letting $\varepsilon\to 0$ in (4.10), it follows simply from the continuity in time that the *generalized stress relations*,

$$\begin{aligned}\partial_\theta\psi^*(e(t),\theta(t))&=-\eta(t),\\ T(f^t)&=\varrho(t)\,\nabla_e\psi^*(e(t),\theta(t))\,(p^{-1}(t))^T(F(t))^T,\end{aligned} \tag{4.11}$$

or

$$t_i^j=\varrho(t)\,(\partial\psi^*/\partial x_I^i)\,X_I^\alpha\,x_\alpha^j,$$

must hold. Combining (4.11) and (4.1b) yields the *dissipation inequality*,

$$\delta\psi(f^t)+q(f^t;g(t))\cdot g(t)/\varrho(t)\,\theta(t)\leqq 0. \tag{4.12}$$

Since (4.12) holds for all values of g, we may set $g(t)$ equal to zero to conclude that

$$\delta\psi(f^t)=\nabla_e\psi^*(e(t),\theta(t))\cdot F(t)\,\frac{d}{dt}(p^{-1}(t))\leqq 0. \tag{4.13}$$

Since $pp^{-1}=I$, we have $dp^{-1}/dt=-p^{-1}(t)\,\dot{p}(t)\,p^{-1}(t)$. Hence inequality (4.13) can be written as

$$\delta\psi(f^t)=-\nabla_e\psi^*(e(t),\theta(t))\cdot F(t)\,p^{-1}(t)\,\dot{p}(t)\,p^{-1}(t)\leqq 0, \tag{4.14}$$

or, in component form,

$$\frac{\partial\psi^*}{\partial x_I^i}\,x_J^i\,\dot{Y}_\beta^J\,X_I^\beta\geqq 0.$$

Denote by $L_p\equiv\dot{p}(t)\,p^{-1}(t)$ the *velocity gradient* of the permanent motion (3.1). Then the *permanent stretching tensor* is given by

$$D_p(t)\equiv\tfrac{1}{2}(L_p(t)+(L_p(t))^T). \tag{4.15}$$

Thus (4.14) can also be written as

$$-\delta\psi(f^t)=e^T(t)\,\nabla_e\psi^*(e(t),\theta(t))\cdot L_p(t)\geqq 0. \tag{4.16}$$

It is clear from (4.16) that

$$-\delta\psi(f^t) = \left(\nabla_e \psi^*(e(t), \theta(t))^T e(t)\right) \cdot L_p(t) \geqq 0. \tag{4.17}$$

Moreover, using the symmetry of the Cauchy stress, i.e.,

$$\varrho(\nabla_e \psi^*)\, e^T = (t^i_j) = (t^j_i) = \varrho\, e(\nabla_e \psi^*)^T,$$

one can verify that

$$e^T \nabla_e \psi^* = (\nabla_e \psi^*)^T e. \tag{4.18}$$

Hence, by adding the corresponding sides of (4.16) and (4.17) and by applying the symmetry relations (4.18), we find that

$$-\delta\psi(f^t) = e^T(t)\, \nabla_e \psi^*\left(e(t), \theta(t)\right) \cdot D_p(t) \geqq 0. \tag{4.19}$$

To see the physical meaning of (4.19) we note that

$$T = (t^j_i) = \varrho(\nabla_e \psi^*)\, e^T$$

and hence, to within a factor of proportionality,

$$e^T \nabla_e \psi^* = e^T \frac{1}{\varrho}\left[\varrho(\nabla_e \psi^*)\, e^T\right](e^T)^{-1}$$

are the components of the second Piola-Kirchhoff stress [35, § 43A] relative to the permanently deformed state. Thus (4.19) means that *the rate of work in producing permanent motion is always non-negative.* Such an inequality of the monotone type is not only physically meaningful but also narrows down the possible relations between the Cauchy stress and the permanent stretching tensor.

Similarly by combining (4.10) and (4.1a′) and then letting $\varepsilon \to 0$ in the resulting equation, we get

$$q\left(f^t; g(t)\right) \cdot g(t) \leqq 0. \tag{4.20}$$

This inequality means that *the heat always flows from a hotter place to a colder place.*

Finally we note that the arguments also applies to the history f^τ, $0 \leqq \tau \leqq t$. Hence the conclusions (4.11), (4.12), (4.19), and (4.20) hold for all τ in $[0, t]$.

5. Rate independent materials. Consider the motion of a body B,

$$x = \chi(X, \tau), \quad 0 \leqq \tau \leqq t, \; X \text{ in } R_0. \tag{5.1}$$

Let $\phi(t)$ be a continuous monotone piecewise smooth function defined on $[0, t]$. Then the equations

$$x = \hat{\chi}\left(X, \phi(\tau)\right) \equiv \chi(X, \tau), \quad 0 \leqq \tau \leqq t, \; X \text{ in } R_0, \tag{5.2}$$

define another motion of B. The function $\phi(t)$ with the specified properties will be called a *rescaling function.*

Let $f^t(\tau) \equiv \left(F(\tau), \theta(\tau)\right)$, $0 \leqq \tau \leqq t$, be a history, $F(\tau)$ being derived from (5.1). Changing the time scale through a rescaling function ϕ gives another deformation history:

$$\hat{f}^{\phi(t)}\left(\phi(\tau)\right) = \left(\hat{F}(\phi(\tau)), \hat{\theta}(\phi(\tau))\right), \quad 0 \leqq \tau \leqq t,$$

with $\hat{F}(\phi(\tau))$ being derived from (5.2). An elastic-plastic material will be said to be *rate independent* if the range of values of the specific energy functional remains unchanged under all possible changes of time scale. More precisely, for all rescaling function $\phi(\tau)$,

$$\psi(f^{\tau}) = \psi(\hat{f}^{\phi(\tau)}), \quad 0 \leqq \tau \leqq t. \tag{5.3}$$

This definition of rate independence was first given by TRUESDELL and NOLL [*35*, § 99]. So far no such a restriction has been made on the materials considered above. Accordingly, all the preceeding results holds for rate dependent materials. We wish to show that if *the material is rate independent in the sense just defined, then the stress is also rate independent.*

Let $\hat{e}$ and $\hat{p}$ be respectively the elastic and plastic resolution of $\hat{f}$. Then for $0 \leqq \tau \leqq t$

$$\hat{e}(\phi(\tau)\,\hat{p}(\phi(\tau))) = \hat{F}(\phi(\tau)) = F(\tau) = e(\tau)\,p(\tau). \tag{5.4}$$

In terms of the function ψ^*, (5.3) can be written as follows,

$$\psi^*(\hat{e}(\phi(\tau)), \hat{\theta}(\phi(\tau))) = \psi^*(e(\tau), \theta(\tau)), \quad 0 \leqq \tau \leqq t. \tag{5.5}$$

According to Assumption A3, there is an $\varepsilon > 0$ such that it is possible to extend f^t elastically to $f^{t+\varepsilon}$. According to Assumption A4, for all sufficiently small $\mu > 0$, $f^{t+\varepsilon}$ can be extended elastically and linearly in time to f^{t+s}, $s = \varepsilon + \mu$. Again by A3, there is $\nu > 0$ such that f^{t+s}, $s = \varepsilon + \mu$, can be further extended elastically to f^{t+r}, $r = \varepsilon + \mu + \nu$, in such a way that

$$f^{t+r}(t+r) = f^{t+s}(t+s-\nu), \quad \nu = r - s > 0. \tag{5.6}$$

Now, we extend $\phi(\tau)$, $0 \leqq \tau \leqq t$, to the interval $[0, t+r]$ by setting

$$\phi(t+\tau) = \phi(t) + \tau, \quad 0 \leqq \tau \leqq r.$$

The function ϕ so extended is again a rescaling function, and hence a history $\hat{f}^{\phi(t)+r}$, $r = \varepsilon + \mu + \nu$, can be defined through the extended rescaling function ϕ, which is, of course, a continuation of $\hat{f}^{\phi(t)}$. Moreover, the relation

$$f^{\phi(t)+r}(\phi(t)+r) = f^{t+r}(t+\tau), \quad 0 \leqq \tau \leqq r,$$

implies that for all $0 \leqq \tau \leqq r$

$$\left(\frac{d}{dt}\hat{F}(\phi(t)+\tau), \frac{d}{dt}\hat{\theta}(\phi(t)+\tau)\right) = (\dot{F}(t+\tau), \dot{\theta}(t+\tau)), \tag{5.7}$$

where the dots denote differentiations with respect to τ.

Since the conditions in (5.6) holds for all $v > 0$, it follows that

$$\delta\psi(f^{t+s}) = \nabla_e\psi^*(e(t+s), \theta(t+s)) \cdot \dot{p}(t+s) = 0. \tag{5.8}$$

Also, from the condition of rate independence,

$$\psi(f^{\phi(t)+\tau}) = \psi(f^{t+\tau}), \quad 0 \leqq \tau \leqq r,$$

it follows that

$$\delta\psi(f^{\phi(t)+\tau}) = \nabla_{\hat{e}}\psi^*(\hat{e}(\phi(t)+\tau), \hat{\theta}(\phi(t)+\tau)) \cdot \frac{d}{ds}\hat{p}^{-1}(\phi(t)+s) = 0. \tag{5.9}$$

On the other hand, differentiating the equation

$$\psi^*(\hat{e}(\phi(t)+\tau), \hat{\theta}(\tau(t)+\tau)) = \psi^*(e(t+\tau), \theta(t+\tau)), \quad 0 \leqq \tau \leqq r,$$

with respect to τ, and then evaluating the resulting equation at $\tau = s$ yields

$$\begin{aligned} \nabla\psi^*\big(\hat{e}(\phi(t)+s), \hat{\theta}(\phi(t)+s)\big) \cdot \left(\frac{d}{ds}\hat{e}(\phi(t)+s), \frac{d}{ds}\hat{\theta}(\phi(t)+s)\right) \\ = \nabla\psi^*\big(e(t+s), \theta(t+s)\big) \cdot \big(\dot{e}(t+s), \dot{\theta}(t+s)\big). \end{aligned} \tag{5.10}$$

Using the relations in (5.7)–(5.9) and the fact that

$$\dot{e}(t+s) = \dot{F}(t+s)\, p^{-1}(t+s) + F(t+s)\, \dot{p}^{-1}(t+s),$$

$$\frac{d}{ds}\hat{e}(t+s) = \left[\frac{d}{ds}\hat{F}(\phi(t)+s)\right]\hat{p}^{-1}(\phi(t)+s) + \hat{F}(\phi(t)+s)\,\frac{d}{ds}\hat{p}^{-1}(\phi(t)+s),$$

we can write (5.10) in a more useful form:

$$\begin{aligned} &[\nabla_e\psi^*\big(\hat{e}(\phi(t)+s), \hat{\theta}(\phi(t)+s)\big)\big(\hat{p}^{-1}(\phi(t)+s)\big)^T \\ &\quad -\nabla_e\psi^*\big(e(t+s), \theta(t+s)\big)\big(p^{-1}(t+s)\big)^T] \cdot \dot{F}(t+s) \\ &\quad +[\partial_{\hat{\theta}}\psi^*\big(\hat{e}(\phi(t)+s), \hat{\theta}(\phi(t)+s)\big) - \partial_\theta\psi^*\big(e(t+s), \theta(t+s)\big)]\,\dot{\theta}(t+s) = 0. \end{aligned} \tag{5.11}$$

Since $\dot{f}(t+s) = \big(\dot{F}(t+s), \dot{\theta}(t+s)\big)$ can be chosen arbitrarily, for (5.11) to hold the coefficient of $\dot{f}(t+s)$ must vanish. Furthermore, by letting $\mu \to 0$ we see from (5.11) that for all $\varepsilon > 0$

$$\begin{aligned} \nabla_{\hat{e}}\psi^*\big(\hat{e}(\phi(t)+\varepsilon), \hat{\theta}(\phi(t)+s)\big)\big(\hat{p}^{-1}(\phi(t)+\varepsilon)\big)^T \\ = \nabla_e\psi^*\big(e(t+\varepsilon), \theta(t+\varepsilon)\big)\big(p^{-1}(t+\varepsilon)\big)^T, \\ \partial_{\hat{\theta}}\psi^*\big(\hat{e}(\phi(t)+\varepsilon), \hat{\theta}(\phi(t)+\varepsilon)\big) = \partial_\theta\psi^*\big(e(t+\varepsilon), \theta(t+\varepsilon)\big). \end{aligned}$$

Using the continuity argument and letting $\varepsilon \to 0$ in the above equations, we arrive at *the rate independence of the generalized stress.*

$$\begin{aligned} \partial_e\psi^*\big(e(\phi(t)), \hat{\theta}(\phi(t))\big)\big(\hat{p}^{-1}(\phi(t))\big)^T = \nabla_e\psi^*\big(e(t), \theta(t)\big)\big(p^{-1}(t)\big)^T \\ \partial_{\hat{\theta}}\psi^*\big(\hat{e}(\phi(t)), \hat{\theta}(\phi(t))\big) = \partial_\theta\psi^*\big(e(t), \theta(t)\big). \end{aligned} \tag{5.12}$$

Reviewing the derivation of (5.12), we see that it also holds for all $\tau \in [0, t]$.

A natural question arises. Does the rate independence of the free energy functional ψ imply that of the elastic-strain functional $e(f^t)$? The answer is easily seen to be affirmative if the correspondence between the stress and the elastic strain is invertible. Otherwise, the question is open, and probably the answer is no.

6. Isotropy and principle of frame indifference. Let $f^t(\tau)$, $0 \leq \tau \leq t$, be a given deformation history. For a given material there is a unique resolution,

$$F(\tau) = e(\tau)\, p(\tau), \qquad 0 \leq \tau \leq t,$$

with $e(\tau)$ and $p(\tau)$ being, respectively, the elastic and the permanent deformations. Since the elastic-permanent resolution of the motion characterizes the material, it is natural to require that the functionals $e(\tau) \equiv e\big(f^\tau(s)\big)$, $p(\tau) \equiv p\big(f^\tau(s)\big)$ satisfy *the principle of frame indifference* [35, § 19]. More precisely, if $Q(\tau)$ is a one-parameter family of orthogonal transformations of the spatial coordinates x, then for every history $f^t(\tau)$, $0 \leq \tau \leq t$,

$$\begin{aligned} Q(\tau)\, e(\tau) = e\big(Q(s)\, f^\tau(s)\big), \qquad 0 \leq s \leq \tau, \\ p(\tau) = p\big(Q(s)\, f^\tau(s)\big), \qquad 0 \leq s \leq \tau. \end{aligned} \tag{6.1}$$

Thus the elastic deformation history of the history $Q(\tau)\, f^t(\tau)$, $0 \leqq \tau \leqq t$, is simply $Q(\tau)\, e(\tau)$.

Since the free energy functional ψ can be expressed as a function of the elastic deformation gradients, i.e.,

$$\psi(f^t(\tau)) = \psi^*(e(\tau), \theta(\tau)), \qquad 0 \leqq \tau \leqq t, \tag{6.2}$$

an application of the principle of frame indifference leads to

$$\psi^*(e(\tau), \theta(\tau)) = \psi(f^\tau(s)) = \psi(Q(s)\, F^\tau(s)) = \psi^*(Q(\tau)\, e(\tau), \theta(\tau))$$

for $0 \leqq \tau \leqq t$. Thus, the function ψ^* has the property that

$$\psi^*(Q(\tau)\, e(\tau), \theta(\tau)) = \psi^*(e(\tau), \theta(\tau)) \tag{6.3}$$

for all orthogonal transformation Q.

From the generalized stress in (4.11) and the frame indifference relations in (6.1) and (6.3) we see that

$$\begin{aligned} T(Q(s)\, f^\tau(s)) &= p(\tau)\, \nabla_{Qe}\psi^*(Q(\tau)\, e(\tau), \theta(\tau))\, (p^{-1}(\tau))^T\, (Q(\tau)\, F(\tau))^T \\ &= Q(\tau)\, p(\tau)\, \nabla_e\psi^*(e(\tau), \theta(\tau))\, (p^{-1}(\tau))^T\, F^T(\tau)\, Q^T(\tau) \\ &= Q(\tau)\, T(f^\tau)\, Q^T(\tau), \qquad 0 \leqq \tau \leqq t. \end{aligned} \tag{6.4}$$

This shows that *the frame indifference of the stress tensor is a consequence of the frame indifference of* ψ and $e(f^t)$.

Consider now isotropic elastic-plastic materials. By this I mean that for every history f and for every orthogonal transformation Q the stress functional $T(f^t)$ satisfies the identity [*35*, § 31],

$$T(Q\, f^\tau\, Q^T) = Q\, T(f^\tau)\, Q^T, \qquad 0 \leqq \tau \leqq t. \tag{6.5}$$

Since for every $0 \leqq \tau \leqq t$,

$$Q\, f^\tau\, Q^T = (Q\, e(\tau)\, Q^T)\, (Q\, p(\tau)\, Q^T),$$

application of the principle of frame indifference shows that $Q\, e(\tau)\, Q^T$ and $Q\, p(\tau)\, Q^T$ are, respectively, the elastic and the permanent deformations of the history $Q f^\tau Q^T$ at the instant τ. Accordingly, from the generalized stress relations in (4.11) we conclude that

$$\begin{aligned} \varrho(\tau)\, \nabla_e\psi^*(Q\, e(\tau)\, Q^T, \theta(\tau))\, (Q\, e\, Q^T)^T &= T(Q\, f^\tau\, Q^T) = Q\, T(f^\tau)\, Q^T \\ &= Q\, \varrho(\tau)\, \nabla_e\psi^*(e(\tau), \theta(\tau))\, e^T\, Q^T. \end{aligned}$$

It follows immediately that the tensor function $\nabla_e\psi^*$ satisfies the equations

$$\nabla_e\psi^*(Q\, e(\tau)\, Q^T, \theta(\tau)) = Q\, \nabla_e\psi^*(e(\tau), \theta(\tau))\, Q^T, \tag{6.6}$$

identically in Q and e. That is, $\nabla_e\psi^*$ *is an isotropic function of the tensor* e. This proves that *the isotropy condition* (6.6) *is a consequence of the isotropy condition* (6.5) *and the principle of frame indifference*. It is now immediate from (6.6) and (4.11) that the Cauchy stress (t_i^j) is an isotropic tensor function of the elastic deformation e.

II. Prandtl-Reuss theory.

In this section the Prandtl-Reuss equations will be derived for isotropic elastic-plastic materials under the assumptions that *the elastic deformation is always small* and that the permanent stretching tensor, if it does not vanish, is proportional to the stress.

7. Flow rule. The generalized stress relations in (4.11) ensure that all thermomechanical behavior of an elastic-plastic material is completely characterized by a single free-energy functional ψ of the deformation history f^t. Under the assumption that $\psi(f^t(\tau)) = \psi^*(e(\tau), \theta(\tau))$, we need to specify a function ψ^* and a tensor-valued functional $e(f^t(\tau)) \equiv e(\tau)$ so as to define the functional $\psi(f^t(\tau))$. Since $e(\tau)\,p(\tau) = F(\tau)$, we may specify the functional $p(f^t(\tau))$ instead of $e(f^t(\tau))$. However, since an elastic continuation of a history is assumed to be always possible, there is no functional relationship between the Cauchy stress and the permanent deformation gradient. But it is permissible to consider the permanent deformation p as a functional of the history of the Cauchy stress, which itself is a functional of the deformation history. Accordingly, it is consistent with the previous theory to assume that

A5. *Whenever the motion is not elastic, i.e., whenever the permanent stretching tensor does not vanish*:

$$D_p \equiv \tfrac{1}{2}\left[(\dot{p}\,p^{-1}) + (\dot{p}\,p^{-1})^T\right] \neq 0,$$

then

$$\left(e^{-1}(\tau)\right)^T D_p(\tau)\, e^T(\tau) = \Phi\left(T^*(f^t(\tau)), \theta(\tau)\right), \qquad 0 \leqq \tau \leqq t, \tag{7.1}$$

where Φ, a sufficiently smooth function defined on E_7, characterizes the plastic behavior of the material. The tensor T^* is the *deviator* of the Cauchy stress T and is defined by $T^* \equiv T - \frac{1}{3}\operatorname{tr}(T)\, I$. The factors $(e^{-1})^T$ and e^T appearing on the left-hand side of (7.1) are introduced so as to express the resulting tensor in terms of its components relative to the Eulerian coordinate system x. The choice of T^* instead of T was suggested by the experimental evidence that the permanent motion is nearly isochoric, i.e., $\operatorname{tr}(D_p) = 0$.

Now we restrict our attention to the isotropic materials. Then for every orthogonal transformation Q

$$T^*(Q\, f^t\, Q^T) = Q\, T^*(f^t)\, Q^T, \qquad 0 \leqq \tau \leqq t.$$

Hence, under every orthogonal transformation Q, Eq. (7.1) becomes

$$Q\,\Phi(T^*)\,Q^T = Q(e^{-1})^T D_p\, e^T Q = \Phi\left(T^*(Q f^\tau Q^T)\right) = \Phi(Q T^* Q^T),$$

which means that $\Phi(T^*)$ is an isotropic tensor function of T^*. By the well-known representation theorem [35, § 12] for isotropic tensor functions,

$$(e^{-1})^T D_p\, e^T = \alpha_0 I + \alpha_1 T^* + \alpha_2 (T^*)^2, \tag{7.2}$$

where $\alpha_0, \alpha_1, \alpha_2$ are scalar functions of the invariants of T^*. Since the permanent motion is assumed to be isochoric, i.e., $\operatorname{tr}(D_p) = 0$, for all time τ, upon taking the trace of the matrix equation (7.2) we get

$$3\alpha_0 + \alpha_2 \operatorname{tr}\left((T^*)^2\right) = 0.$$

Hence (7.2) reduces to

$$(e^{-1})^T D_p\, e^T = \alpha_1 T^* + \alpha_2\left((T^*)^2 - \tfrac{1}{3}\operatorname{tr}(T^{*2})\, I\right), \tag{7.3}$$

whenever $D_p \neq 0$. Although without sufficient physical reason, for mathematical simplicity we assume that $\alpha_2 = 0$ identically. Then

$$(e^{-1})^T D_p\, e^T = \alpha_1 T^* \quad \text{whenever} \quad D_p \neq 0. \tag{7.4}$$

However, two things may be mentioned with regard to the vanishing of α_2. First, the trace of the both sides of (7.4) vanishes, and secondly (7.4) ensures that the *plastic power* [see Eq. (4.19)] is always non-negative.

Assume now that the elastic deformation is always very small, i.e.,

$$e(\tau) - I = 0(\varepsilon) \quad \text{for} \quad 0 \leqq \tau \leqq t. \tag{7.5}$$

It follows immediately from (7.4) that whenever $D_p \neq 0$.

$$D_p = \alpha_1 T^* \tag{7.4a}$$

with error of order $0(\varepsilon)$. This set of equations is known as the *flow rule*. Of course, for (7.4a) to be meaningful it is necessary to identify the coordinates x and y. In passing, we note from the exact relations in (7.3) that the type of materials considered here is rate dependent.

8. Prandtl-Reuss equations. According to the results derived in Chap. B. I, the Cauchy stress is a tensor function of the elastic deformation e and the temperature θ, i.e.,

$$T(f^t(\tau)) = \hat{T}(e(\tau), \theta(\tau)), \quad 0 \leqq \tau \leqq t. \tag{8.1}$$

If we assume that $\hat{T}$ is a smooth tensor function defined on E_{10}, material differentiation of (8.1) leads to

$$\dot{T}(f^t(\tau)) = \nabla_e \hat{T}(e(\tau), \theta(\tau))\, \dot{e}(\tau) + \partial_\theta \hat{T}(e(\tau), \theta(\tau))\, \dot{\theta}, \quad 0 \leqq \tau \leqq t. \tag{8.2}$$

To calculate $\dot{e}$, we note that $e = F p^{-1}$, $p p^{-1} = I$. Hence

$$\dot{e}(\tau) = \dot{F}(\tau)\, p^{-1}(\tau) - F(\tau)\, p^{-1}(\tau)\, \dot{p}(\tau)\, p^{-1}(\tau), \quad 0 \leqq \tau \leqq t. \tag{8.3}$$

Upon combining (8.2) and (8.3), we find that

$$\dot{T}(f^t(\tau)) = \nabla_e \hat{T}(e(\tau), \theta(\tau))\,(L(\tau)\, e(\tau) - e(\tau)\, L_p(\tau)) + (\partial_\theta \hat{T}(e, \theta))\, \dot{\theta}(\tau), \tag{8.4}$$

$$L \equiv \dot{F}\, F^{-1} \quad \text{and} \quad L_p \equiv \dot{p}\, p^{-1}.$$

In view of the assumption (8.5), then to within errors of the order $0(\varepsilon)$

$$e = I, \qquad Le = L, \qquad e L_p = L_p,$$

provided we identify the coordinates x and y. Hence (8.4) can be written with error of order $0(\varepsilon)$ as

$$\dot{T}(f^t(\tau)) = \nabla_e T(I, \theta)\,(L(\tau) - L_p(\tau)) + (\partial_\theta T(I, \theta))\, \dot{\theta} \tag{8.4a}$$

provided T is a smooth tensor function. For isotropic materials, the function $\hat{T}(e, \theta)$ is an isotropic tensor function of the tensor e. By using the explicit representation for the gradient of an isotropic tensor function [*35*, Eq. 45.12, 43 A.11, 45.1, 4.52 and 47.20] and by neglecting the distinction between the coordinates x and y, we can derive the formula

$$\nabla_e \hat{T}(I, \theta)\, U = \lambda_0 (\operatorname{tr} U)\, I + (\lambda_1 + 2\lambda_2)\,(U + U^T) \tag{8.5}$$

for all tensors U, where λ_0, λ_1 and λ_2 are scalar functions of the temperature $\theta(\tau)$. Of course, if the material is not homogeneous, then λ_0, λ_1 and λ_2 are functions of the material point. By substituting (8.5) into (8.4a) for $\nabla_e \hat{T}(I, \theta)$ we obtain

$$\dot{T}^* = \lambda_0 (\operatorname{tr} D(\tau))\, I + 2(\lambda_1 + 2\lambda_2)\,(D(\tau) - D_p(\tau)) + (\partial_\theta T)\, \dot{\theta}, \tag{8.6}$$

from which it follows that to within an error of order $0(\varepsilon)$

$$\dot{T}^* = 2(\lambda_1 + 2\lambda_2)\,(D^*(\tau) - D_p^*(\tau)) + (\partial_\theta T^*)\, \dot{\theta}. \tag{8.6a}$$

Now substituting the non-trivial relations in (7.4a) into (8.6a) for $D_p^* = D_p$, we obtain *the Prandtl-Reuss equations,*

$$\dot{T}^* = 2(\lambda_1 + 2\lambda_2)(D^* - \alpha_1 T^*) + (\partial_\theta T^*)\,\dot{\theta}, \tag{8.7}$$

whenever $D_p \neq 0$. This set of equations establishes *relations between the deviators of the stress and of the stretching tensor.* It is a merit of these equations that they do not involve the unknown permanent stretching tensor. However, they only hold when $D_p \neq 0$, and hence the real difficulty has not been completely removed.

If the deformation process is such that thermal effects can be neglected, then the term $(\partial_\theta T^*)\,\dot{\theta}$ drops out. In general, there are only four independent relations in (8.7). To this we can add the equation

$$\mathrm{tr}(\dot{T}) = [3\lambda_0 + 2(\lambda_1 + 2\lambda_2)]\;\mathrm{tr}(D) + \mathrm{tr}\,(\partial_\theta T)\,\dot{\theta},$$

obtained by taking the trace of (8.6). In addition, it is assumed that the stress satisfies a single functional equation known as *yield condition.* But then we have six independent relations among the components of the stress and stretching tensors.

III. St. Venant-Lévy-v. Mises theory.

This is the earliest theory intended to serve as a model for metals under purely permanent deformations without any elastic effect. Although a highly idealized one, yet it is a perfect theory in the sense that every term involved is well-defined, no approximation argument is used and all the basic principles in continuum mechanics are satisfied.

The Eulerian description will be used throughout this section. Denote by v the velocity, by D the stretching tensor and by T the Cauchy stress. We shall consider only those motions for which thermodynamic effects can be neglected. Following the elegant work of THOMAS [*13*, pp. 71–78], I characterize the motion of isotropic perfectly plastic materials by the following three assumptions:

A1. The flow of the plastic material is isochoric, i.e., $\mathrm{div}(v) = 0$.

A2. The deviator of the Cauchy stress is proportional to the deviator of the stretching tensor, i.e.

$$T^* = \phi(D^*)\,D^*, \qquad \phi(D^*) > 0. \tag{1}$$

A3. The correspondence between T^* and D^* fails to be one-to-one.

Since the material is isotropic, we have for every orthogonal transformation Q

$$Q T^* Q^T = \phi(Q D^* Q^T)\,Q D^* Q^T = Q\,\phi(Q D^* Q^T)\,D^* Q^T,$$

from which it follows that

$$\phi(D^*)\,D^* = T^* = \phi(Q D^* Q^T)\,D^*.$$

Accordingly, for all orthogonal transformations Q,

$$\phi(Q D^* Q^T) = \phi(D^*).$$

Thus, by the representation theorem for scalar invariant functions [*35*, § 12], we conclude that *ϕ is a function of the invariants of D^*.* Let $\hat{\xi}, \hat{\zeta}$ and ξ, ζ be, repectively, the second and the third invariants of T^* and D^*, i.e.,

$$\hat{\xi} = \mathrm{tr}(T^* T^{*T}), \qquad \hat{\zeta} = \mathrm{tr}\left((T^*)^2\, T^{*T}\right), \qquad \text{etc.}$$

Then it follows from A2 and the results just proved that

$$\hat{\xi} = \phi^2(\xi, \zeta)\, \xi, \qquad \hat{\zeta} = \phi^3(\xi, \zeta)\, \zeta. \tag{2}$$

According to A3, the system of functional equations in (2) cannot be solved for ξ and ζ as functions of $\hat{\xi}$ and $\hat{\zeta}$. By the implicit function theorem, this is so only when the Jacobian of the above system vanishes for all values of ξ and ζ, i.e.,

$$2\xi \frac{\partial \phi}{\partial \xi} + 3\zeta \frac{\partial \phi}{\partial \zeta} + \phi = 0, \tag{3}$$

provided $\phi(\xi, \zeta)$ is a smooth function. The general solution of this first-order linear partial differential Eq. (3) is given by

$$\phi = (\xi)^{-\frac{1}{2}} M(\zeta^{\frac{1}{3}}/\xi^{\frac{1}{2}}). \tag{4}$$

Hence, we have from A2 the following relations

$$T^* = (\xi)^{-\frac{1}{2}} M(\zeta^{\frac{1}{3}}/\xi^{\frac{1}{2}})\, D^*. \tag{5}$$

It is easy to see that there are, in general, four independent equations in (5). However, a combination of equations in (2) gives

$$\hat{\zeta}^{\frac{1}{3}}/\hat{\xi}^{\frac{1}{2}} = \zeta^{\frac{1}{3}}/\xi^{\frac{1}{2}}.$$

Using this relation, we show from the system of equations in (5) that

$$\operatorname{tr}(T^* T^{*T}) = \hat{\xi} = M(\hat{\zeta}^{\frac{1}{3}}/\hat{\xi}^{\frac{1}{2}}). \tag{6}$$

It should be noted that this equation is, in general, independent of those in (1) because in its derivation we have used the equations (4), which are consequences of A3. Thus, the function M completely characterizes the mechanical behavior of the material under considerations.

The five independent Eqs. (5) *and* (6) *together with the isochoric condition* A1, *the equation for conservation of mass and the three equations for balance of linear momentum constitute ten independent equations for determining the ten unknowns* T, v *and the density* ϱ.

If we take the material function M in (6) to be a constant $2k^2$, then

$$\operatorname{tr}(T^* T^{*T}) = 2k^2, \tag{7}$$

which is the well-known *v. Mises yield condition.* It is interesting to note that the general yield condition (6) is a consequence of A2 and A3 rather than an independent hypothesis. Under v. Mises' yield condition (7), the relations in (5) become

$$T^* = 2kD^*/\operatorname{tr}(D^* D^{*T}),$$

which is known as the St. Venant-Lévy-v. Mises relation between the stress and the rate of strain.

IV. Theory of Hencky.

In the preceeding discussion, the non-elastic behavior of materials is described by relations between the stress and the rate of strain. Accordingly, these theories are called *flow theories.* On the other hand, Hencky described plastic behavior by means of relations between stress and strain. His theory is often called a *deformation theory.* In what follows an outline of a derivation of that theory will be presented for *isotropic* materials. It goes without saying that all thermodynamic effects will be neglected.

Consider a connected body B occupying a region R_0 in E_3 in its undeformed state. Suppose that its deformed state is given by the equations

$$x^i = \chi^i(X), \quad X \equiv (X^1, X^2, X^3) \text{ in } R_0, \quad i = 1, 2, 3. \tag{1}$$

We shall say that the deformation (1) is elastic-plastic if and only if the following assumption holds:

A 1. *There is a unique decomposition.*

$$x^i = \hat{\chi}^i(Y), \quad Y \equiv (Y^1, Y^2, Y^3), \quad i = 1, 2, 3, \tag{2}$$

$$Y^I = Y^I(X), \quad X \in R_0, \quad I = 1, 2, 3, \tag{3}$$

such that the deformation (1) *is a composition of* (2) *and* (3):

$$x^i = \hat{\chi}^i(Y(X)) = \chi^i(X), \quad X \in R_0, \quad i = 1, 2, 3. \tag{4}$$

Further, if the applied forces are released then the body B will not, in general, recover completely from its deformed state but will remain in its permanently deformed state (3).

As usual, the deformations (2) and (3) will be called elastic and plastic deformations respectively. Consider now the Green-St. Venant tensor,

$$E \equiv \tfrac{1}{2}(F^T F - I), \quad F \equiv (\partial x^i / \partial X^\alpha). \tag{5}$$

It follows at once that

$$E = E_p + E_e \quad \text{with} \quad E_p \equiv \tfrac{1}{2}(p^T p - I), \quad E_e = \tfrac{1}{2} p^T (e^T e - I)\, p, \tag{6}$$

where e and p stand for the elastic and plastic deformation gradients, i.e., $e = (\partial x^i / \partial Y^I)$, $p \equiv (\partial Y^I / \partial X^\alpha)$. We shall call E_e and E_p, respectively, the elastic and the plastic part of the strain tensor E. However, both E_e and E_p are referred to the undeformed state, and they are non-linear in e and p. A severe restriction, though it simplifies things mathematically in this theory, is to assume that

A 2. *Both the elastic deformation gradient e and the plastic deformation gradient p are infinitesimal.*

Under this assumption one can apply the rules for neglect of infinitesimals of higher order and identify the coordinates x, Y and X which specify the positions of the same material point at different deformed states. Although Assumption A 2 is quite restrictive, it is not necessary for developing a theory. The essential hypotheses in the present theory are

A 3. *The elastic deformation* (2) *obeys the usual law for perfectly elastic solids with respect to the permanently deformed state Y.*

A 4. *The permanent deformation* (3) *is always volume preserving, and the deviator E_p^* of the permanent deformation tensor E_p is a function of the stress deviator T^*; i.e., there is a tensor function Φ defined on E_9 such that*

$$(e^{-1})^T E_p^* e^T = \Phi(T^*). \tag{7}$$

Under Assumption A 3, it follows [*35*, Eq. (94.1)] that

$$\tilde{T} = \lambda \operatorname{tr}(E_e)\, I + 2\mu\, E_e,$$

where $\widetilde{T} \equiv J e^{-1} T (e^{-1})^T$, $J \equiv |\det(e)|$ and E_e is the elastic deformation tensor defined in (6). Hence an application of A2 yields

$$t_{ij} = \lambda\, e_{kk}\, \delta_{ij} + 2\mu\, e_{ij}, \qquad i = 1, 2, 3, \tag{8}$$

where t_{ij} stands for the components of Cauchy stress, $(e_{ij}) = E_e$ is the linearized elastic strain tensor, $e_{kk} = \operatorname{tr}(e_{ij})$ and λ and μ are the Lamé constants.

By the reasoning used to derive Eq. (7.4), it follows from (7) that

$$(e^{-1})^T E_p^* \, e^T = \alpha T^*, \tag{9}$$

where α is a scalar function of the first three invariants of T^*. According to A2, $e - I = 0(\varepsilon)$ $p - I = 0(\varepsilon)$ when ε is arbitrarily small, so (9) reduces to

$$E_p^* = \alpha T^* \tag{10}$$

with error of the order $0(\varepsilon)$. By combining (10) and (8) we obtain the five independent equations

$$\operatorname{tr}(T) = (3\lambda + 2\mu) \operatorname{tr}(E), \qquad T^* = 2\mu (E^* - \alpha T^*), \tag{11}$$

which relate the components of Cauchy stress to the components of Green-St. Venant strain, which is usually linearized according to A2.

The five independent equations in (11) together with some kind of yield condition, equations of equilibrium (or motions) and the equation for conservation of mass constitute ten equations for ten unknowns.

C. General theorems.

In this section, a variational principle for solving equilibrium problems of the Hencky material will be presented. Also certain existence theorems for steady flows of the St. Venant-Lévy-v. Mises material will be established. Apparently, there is as yet no general existence theory for elastic-plastic materials.

I. Intrinsic formulation of the variational principle.

This variational principle due to Langenbach can now be phrased in terms of the monotonicity method [27]. Since it is a minimum principle and hence leads to various methods for finding approximate solutions, I shall present it here, giving it in its original form [9] with some simplifications.

Let Ω be a bounded domain in R_3. Let $C^m(\Omega)$ be the space of m-times continuously differentiable functions in Ω. For f in $C^m(\Omega)$ we denote by $\|f\|_m$ the m-fold Dirichlet norm of f:

$$\|f\|_m^2 = \sum_{|\alpha| \leq m} \int_\Omega |D^\alpha f|^2 \, dx, \tag{1}$$

where $\alpha \equiv (\alpha_1, \alpha_2, \alpha_3)$, $|\alpha| = \alpha_1 + \alpha_2 + \alpha_3$, $D^\alpha f \equiv \partial^\alpha f / \partial x_1^{\alpha_1} \partial x_2^{\alpha_2} \partial x_3^{\alpha_3}$. We denote by $H^m(\Omega)$ the Hilbert space which is the completion of $C^m(\Omega)$ with respect to $\| \; \|_m$-norm. The subspace of $H^m(\Omega)$, which is the completion of the space of infinitely differentiable functions with compact support in Ω with respect to $\| \; \|_m$-norm, we denote by $H_0^m(\Omega)$.

Let u be a real-valued function in $H_0^{2m}(\Omega)$. Consider a non-linear partial differential operator of order $2m$ in the divergence form,

$$P(u) = \sum_{|\alpha| \leq m} (-1)^{|\alpha|} D^\alpha P_\alpha(u), \tag{2}$$

where each $P_\alpha(u)$ is a continuous function of $D^\alpha u$. It should be said that this restriction is imposed only for simplicity and that the case when $P_\alpha(u)$ depends on x can also be handled. Our problem is to find a function u in $H_0^m(\Omega)$ of the non-homogeneous equation $Pu=f$ in the sense that for all ϕ in $C_0^\infty(\Omega)$

$$\sum_{|\alpha|\leqq m}(P_\alpha(u), D^\alpha \phi)_0=(f, \phi)_0, \tag{3}$$

where f is a given function in $L_2(\Omega)$ and $(f, \phi)_0$ stands for the L_2 inner product of f and ϕ. Needless to say, the expression on the left-hand side is meaningful because of the restriction on each $P_\alpha(u)$. Further, we assume that $P_\alpha(u)$ satisfies the following conditions:

(i) for each α, the weak differential defined by

$$P'_\alpha(u)\, v \equiv \lim_{t\to 0}\frac{1}{t}[P_\alpha(u+tv)-P_\alpha(u)], \quad |\alpha|\leqq m, \tag{4}$$

exists as an element in $L_2(\Omega)$ for all u, v in $H_0^m(\Omega)$ and it maps $H_0^m(\Omega)\times H_0^m(\Omega)$ continuously into $L_2(\Omega)$.

(ii) $\sum_{|\alpha|\leqq m}(P'_\alpha(u)\, v_1, D^\alpha v_2)_0=\sum_{|\alpha|\leqq m}(P'_\alpha(u)\, v_2, D^\alpha v_1)_0$ for all u, v_1, v_2 in $H_0^m(\Omega)$.

(iii) There is a positive constant c, depending only on P and Ω, such that for all u, v in $H_0^m(\Omega)$

$$\sum_{|\alpha|\leqq m}(P'_\alpha(u)\, v, D^\alpha v)_0\geqq c\|v\|_m^2.$$

Theorem 1. The problem defined by (3) has at most one solution in $H_0^m(\Omega)$. If u in $H_0^m(\Omega)$ solves that problem, then it minimizes the functional

$$J[u]\equiv\int_0^1\sum_{|\alpha|\leqq m}(P_\alpha(tu), D^\alpha u)_0\, dt-(f, u)_0, \tag{5}$$

among all functions in $H_0^m(\Omega)$. Conversely, if u in $H_0^m(\Omega)$ minimizes $J[u]$ over $H_0^m(\Omega)$, then it solves problem (3).

Proof. Let u_1, u_2 be two solutions in $H_0^m(\Omega)$ of the problem defined by (3). Then

$$\sum_{|\alpha|}(P_\alpha(u_1)-P_\alpha(u_2), D^\alpha\phi)_0=0 \qquad \text{for all } \phi \text{ in } C_0^\infty(\Omega).$$

As a consequence, we have

$$\sum_{|\alpha|}(P_\alpha(u_1)-P_\alpha(u_2), D^\alpha(u_1-u_2))_0=0.$$

Hence

$$\begin{aligned} 0&=\sum_{|\alpha|\leqq m}(P_\alpha(u_1)-P_\alpha(u_2), D^\alpha(u_1-u_2))_0\\ &=\sum_{|\alpha|}\left(\int_0^1 P'_\alpha(\xi_t)(u_1-u_2)\, dt, D^\alpha(u_1-u_2)\right)_0\\ &=\int_0^1\sum_{|\alpha|}(P'_\alpha(\xi_t)(u_1-u_2), D^\alpha(u_1-u_2))_0\, dt\geqq c\|u_1-u_2\|_m^2, \end{aligned}$$

where $\xi_t\equiv tu_1+(1-t)\, u_2$ and where the last inequality follows from Assumption (iii). This proves the uniqueness of the solution.

Let u in $H_0^m(\Omega)$ be a solution of problem (3). For all v in $H_0^m(\Omega)$,

$$J[u+v]-J[u]=\int_0^1 \sum_{|\alpha|\leq m} (P_\alpha(t(u+v)), D^\alpha v)_0\, dt + \int_0^1 \sum_{|\alpha|} (P_\alpha(t(u+v)) - P(tu), D^\alpha u)_0\, dt - (f, v)_0. \tag{6}$$

For the second term on the right-hand side we have

$$\begin{aligned}\int_0^1 \sum_{|\alpha|} (P_\alpha(t(u+v)) - P_\alpha(tu), D^\alpha u)_0\, dt &= \int_0^1 \sum_{|\alpha|} \left(\int_0^t \frac{d}{ds} P_\alpha(tu+sv)\, ds, D^\alpha u\right)_0 dt \\ &= \int_0^1 ds \int_s^1 \sum_{|\alpha|} (P'_\alpha(tu+sv)\, v, D^\alpha u)_0\, dt \\ &= \int_0^1 ds \int_s^1 \sum_{|\alpha|} (P'_\alpha(tu+sv)\, u, D^\alpha v)_0\, dt \\ &= \int_0^1 \sum_{|\alpha|} (P_\alpha(u+tv) - P_\alpha(t(u+v)), D^\alpha v)_0\, dt.\end{aligned}$$

Thus (6) can be written as follows:

$$J[u+v]-J[u]=\int_0^1 \sum_{|\alpha|} (P_\alpha(u+tv), D^\alpha v)_0\, dt - (f, v)_0. \tag{7}$$

Accordingly, for all v in $H_0^m(\Omega)$,

$$J[u+v]-J[u]=\int_0^1 \frac{dt}{t} \int_0^1 \sum_{|\alpha|} (P'_\alpha(u+stv)\, tv, D^\alpha tv)_0\, ds \geqq 0,$$

which shows the minimizing property of u.

Suppose now that u is a minimizing extremal of $J[u]$ over $H_0^m(\Omega)$. Then for all ϕ in $C_0^\infty(\Omega)$

$$J[u+t\phi]-J[u]=\int_0^1 \sum_{|\alpha|} (P_\alpha(u+st\phi), D^\alpha t\phi)_0\, ds - (f, t\phi)_0.$$

Dividing both sides by t and then letting $t\to 0$ yields Eq. (3). The proof is now complete.

Theorem 2. The minimum problem (5) possesses a solution u in $H_0^m(\Omega)$.

Proof. First, we check that $J[u]$ is bounded from below on $H_0^m(\Omega)$. Indeed, for every u in $H_0^m(\Omega)$,

$$\begin{aligned}J[u] &= \int_0^1 \sum_{|\alpha|} (P(tu), D^\alpha u)_0\, dt = \int_0^1 \sum_{|\alpha|} (P_\alpha(tu), D^\alpha tu)_0 \frac{dt}{t} - (f, u)_0 \\ &\geqq \int_0^1 \frac{dt}{t} \sum_{|\alpha|} \left(\int_0^1 \frac{d}{ds} P_\alpha(stu)\, ds, D^\alpha tu\right)_0 - |(f, u)_0| \\ &\geqq \int_0^1 \frac{dt}{t} \int_0^1 \sum_{|\alpha|} (P'_\alpha(stu)\, tu, D^\alpha tu)_0\, ds - |(f, u)_0| \\ &\geqq c\|u\|_m^2 - \|f\|_0 \|u\|_0 \geqq -\frac{1}{c}\|f\|_0^2.\end{aligned}$$

Thus, if a minimizing extremal cannot be found, there is a minimizing sequence $\{u_n\}$ in $H_0^m(\Omega)$ such that

$$\lim_{n\to\infty} J[u_n] = \inf J[u] \quad \text{over} \quad H_0^m(\Omega) \equiv d.$$

We wish to show that the minimizing sequence is actually a Cauchy sequence in $H_0^m(\Omega)$. To this end, we define

$$\varrho(u_j, u_k) \equiv \tfrac{1}{2} J[u_j] + \tfrac{1}{2} J[u_k] - \tfrac{1}{2} J[(u_j + u_k)/2] \tag{8}$$

for all u_j, u_k in $H_0^m(\Omega)$. It is easy to see that

$$\varrho(u_j, u_k) = \varrho(u_k, u_j).$$

Upon setting $u = u_j$, $u + v = u_k$, we see from (7) that

$$\begin{aligned}
\varrho(u_j, u_k) &= \frac{1}{2}\{J[u] - J[u + v/2]\} + \frac{1}{2}\{J[u+v] - J[u+v/2]\} \\
&= \frac{1}{2}\int_0^1 \sum_{|\alpha|} \left(P_\alpha\left(u + \frac{v}{2} + \frac{s}{2} v\right) - P\left(u + \frac{s}{2} v\right), \frac{v}{2}\right)_0 ds \\
&= \frac{1}{2}\int_0^1 ds \sum_{|\alpha|} \left(\int_0^1 \frac{d}{dt}\left(P_\alpha\left(u + \frac{s+t}{2} v\right) dt, D^\alpha \frac{v}{2}\right)\right)_0 \\
&\geqq \frac{1}{2}\int_0^1 ds \int_0^1 \frac{c}{4} \|v\|_m^2 \, dt = \frac{1}{4} c \|v\|_m^2.
\end{aligned}$$

Using this estimate, we easily verify that

$$\varrho(u, v) + \varrho(v, w) \geqq \varrho(u, w)$$

for all u, v, w in $H_0^m(\Omega)$. Thus ϱ, defined in (8), induces an equivalent norm on $H_0^m(\Omega)$.

Consider now the minimizing sequence $\{u_n\}$. For every $\varepsilon > 0$ we can choose j and k so large that

$$J[u_j] \leqq d + \varepsilon, \qquad J[u_k] \leqq d + \varepsilon.$$

Hence, for all large j and k

$$\begin{aligned}
\varrho(u_j, u_k) &= \tfrac{1}{2}\{J[u_j] + J[u_k]\} - J[\tfrac{1}{2}(u_j + u_k)] \\
&\leqq \tfrac{1}{2}(d + \varepsilon) + \tfrac{1}{2}(d + \varepsilon) - d = \varepsilon.
\end{aligned}$$

This proves that $\{u_n\}$ is a Cauchy sequence in $H_0^m(\Omega)$. The completeness of $H_0^m(\Omega)$ assures the existence of a function u in $H_0^m(\Omega)$ such that $\|u - u_n\|_m \to 0$ as $n \to \infty$. It follows from the continuity of $P_\alpha(u)$ in $D^\alpha u$ that

$$J[u] = \lim_{n\to\infty} J[u_n] = d.$$

II. Examples.

We list a few examples for which the above intrinsic theory is realized, omitting the detailed derivations, which are given in [9, pp. 150–152].

9. Elastic-plastic torsion in the sense of Hencky. According to HENCKY, the stress function $u(x, y)$ must satisfy the equations

$$\sum_{|\alpha|=1} D^\alpha P_\alpha(u) = \frac{\partial}{\partial x}[f(T^2)\, u_x] + \frac{\partial}{\partial y}[f(T^2)\, u_y] = -\omega \quad \text{in } \Omega,$$
$$u = 0 \quad \text{on } \partial\Omega,$$

where $T^2 \equiv |\text{grad } u|^2$ and ω is a positive constant. Assume that the material function $f(\xi^2)$ is continuously differentiable. We see that the assumptions in (i) and (ii) are satisfied. Further, if

$$f(T^2) > f_0 > 0, \qquad f_0 = \text{constant}$$

and

$$f(T^2) + 2f'(T^2)\, T^2 \geqq k > 0,$$

then the assumption (iii) holds. Indeed, in this case

$$-\sum_{|\alpha|} (P'_\alpha(u)\, v, D^\alpha v)_0 = \int_\Omega [f(T^2)(v_x^2 + v_y^2) + 4f'(T^2)(u_x v_x + u_y v_y)^2]\, dx.$$

10. Stationary creep deformation of a plate. For a simply supported plate, according to the Hencky theory the displacement of the middle surface $u(x, y)$ must satisfy the equation

$$\sum_{|\alpha|\leqq 2} (-1)^{|\alpha|} D^\alpha P_\alpha(u) = [g(H^2)(u_{xx} + \tfrac{1}{2} u_{yy})]_{xx} + [g(H^2)(u_{yy} + \tfrac{1}{2} u_{xx})]_{yy}$$
$$+ [g(H^2)\, u_{xy}]_{xy} = f(x, y)$$

in a plane region Ω and vanish on $\partial\Omega$, where

$$H^2(u) \equiv u_{xx}^2 + u_{yy}^2 + u_{xy}^2 + u_{xx} u_{yy}$$

and g is a given material function. Clearly the assumptions in (i) and (ii) are satisfied. Simple calculation shows that

$$\sum_{|\alpha|\leqq 2} (P'_\alpha(u)\, v, D^\alpha v)_0 = \int_\Omega \{g(H^2(u))\, H^2(v) + 2g'(H^2(u))\, (H^2(u, v))^2\}\, dx\, dy,$$

where

$$H^2(u, v) \equiv u_{xx} v_{xx} + u_{yy} v_{yy} + u_{xy} v_{xy} + \tfrac{1}{2}(u_{xx} v_{yy} + u_{yy} v_{xx}).$$

Accordingly, if g is sufficiently smooth and if

$$g(\xi^2) \geqq g_0 > 0, \qquad g_0 = \text{constant}, \qquad g(\xi^2) + 2g'(\xi^2)\, \xi^2 \geqq k > 0,$$

then one easily deduces that for all v in $H_0^m(\Omega)$,

$$\sum_{|\alpha|\leqq 2} (P'_\alpha(u)\, v, D^\alpha v)_0 \geqq \text{const.}\, \|v\|_2^2.$$

11. Plane stress and plane deformation problems. LANGENBACH also made detailed computations and verified that for these problems the assumptions (i)–(iii) are all satisfied provided the material function (or yield condition) fulfills certain requirements. We refer to his article for these details.

Remark 1. The essential hypotheses in the above minimum principle are the monotonicity, Assumption (i), and the semi-boundedness, Assumption (iii). These assumptions are realized in the examples listed above because of the restrictions upon the yield condition. As we shall see, under the v. Mises yield condition both the strict monotonicity and coerciveness conditions are lost, and hence the problem becomes non-standard.

Remark 2. For simplicity we have restricted our discussions to homogeneous boundary conditions. However, under the assumptions (i)–(iii), the method can be extended to handle non-homogeneous boundary value problems. Also, it can be extended to handle a system of non-linear equations [*9*, pp. 154–156].

III. Existence of certain steady plastic flows.

Although the basic existence theory has been well developed for the Navier-Stokes equations [*12*, *19*], apparently, there is yet no such a theory for elastic-plastic materials. Of course, the non-linearity of unconventional nature in the constitutive equations causes difficulties. Most difficult of all is that the body changes its shape in the course of motion or deformation. Accordingly, in almost all realistic problems the determination of the shape of the body at every instant is a part of the problem. In what follows, an existence theorem for certain steady plastic flows for St. Venant-Lévy-v. Mises materials will be established.

The problem setup for consideration is motivated by continuous production processes such as extrusion, wire drawing and metal rolling. For all these fabrication process, the motion of the material may be regarded being stationary after a short transient period. Also, in these processes part of the boundary of the flow region consists of idealized rigid walls, while the remainder of the boundary consists of imagined partition surfaces which separate the flow region under consideration from a bigger one. Accordingly, the region of plastic flow under consideration is fixed for all time. In other words, we shall have a boundary-value problem in a fixed, smoothly bounded region Ω in R_3.

According to the classical theory [*13*, p. 111], the mechanical behavior of perfectly plastic materials are governed by the following system of equations

$$\sigma_{ij}^* = k\left(\frac{d_{ab}\,d_{ab}}{2}\right)^{-\frac{1}{2}} d_{ij} \qquad \text{(stress-stretching relations)}, \tag{1}$$

$$d\varrho/dt + \varrho v_{i,i} = 0 \qquad \text{(equation of continuity)}, \tag{2}$$

$$dv_i/dt = \sigma_{ij,j} + f_i \qquad \text{(balance of linear momentum)}, \tag{3}$$

$$v_{i,i} = 0 \qquad \text{(incompressibility)}. \tag{4}$$

Here the Eulerian description, Cartesian tensor notation and the summation convention have been used. In Eqs. (1)–(4), σ_{ij} are the components of the Cauchy stress; $\sigma_{ij}^* \equiv \sigma_{ij} - p\,\delta_{ij}$, the components of the stress deviator, $p \equiv -\sigma_{ii}/3$ the pressure; v_i are the velocity components; $d_{ij} \equiv (v_{i,j} + v_{j,i})/2$ the stretching tensor; f_i are the body force components; and ϱ is the mass density.

12. Formulation of a weak problem. Substituting the expression $\sigma_{ij} = \sigma_{ij}^* + \sigma\delta_{ij}$ in the system of equations in (3) and making use of the constitutive relations (1), we find

$$\varrho\,(\partial v^i/\partial t + v_{i,j}\, v_j) - [k\,(d_{ab}\, d_{ab}/2)^{-\frac{1}{2}}\, d_{ij}]_{,j} = p_i + f_i. \tag{12.1}$$

The Eqs. (2) and (4) imply that $d\varrho/dt = 0$: the density ϱ of a material point remains constant following the motion. This condition will be satisfied, in particular, if the density is a constant, independent of the position x and time t. For simplicity, it will be assumed that ϱ is a constant in the following discussion. Under this assumption, the system of the three equations in (12.1) and the Eq. (4) constitute four equations for determining the four unknowns v_1, v_2, v_3 and p.

Let Ω be a sufficiently smoothly bounded region in R_3. Suppose that body force is absent. We seek time independent functions $v_i(x)$, $i=1, 2, 3$, and $p(x)$ that satisfy the Eqs. (4) and (12.1) in the domain Ω and take on prescribed values v^0 on $\partial\Omega$, i.e.,

$$v_i(x) = v_i^0(x) \quad \text{on } \partial\Omega, \qquad i=1, 2, 3. \tag{12.2}$$

We shall solve the problem (4), (12.1), and (12.2) only in a weak sense, which will be defined now. To this end, we assume for the moment that a strict solution v to this problem exists. Then for every solenoidal vector $\varphi(x)$ in $C_0^\infty(\Omega)$, we have from (4), upon an integration by parts,

$$\int_\Omega [k(2/|d(v)|^2)^{\frac{1}{2}} d_{ij}(v)\, \varphi_{i,j} + v_{i,j} v_j \varphi_i]\, dx = \int_\Omega p_i\, \varphi_i\, dx, \tag{12.3}$$

where $d \equiv (d_{ij})$, $|d(v)| \equiv (d_{ij}(v)\, d_{ij}(v))^{\frac{1}{2}}$ and $d_{ij}(v)$ denotes the components of the stretching tensor calculated from the velocity field v. Note that

$$-\int_\Omega p_i\, \varphi_i\, dx = \int_\Omega p\, \varphi_{i,i}\, dx = 0, \qquad -\int_\Omega v_{i,j} v_j\, \varphi\, dx = \int_\Omega \varphi_{i,j} v_i v_j\, dx,$$

because $\varphi_{i,i} = 0$ and $v_{i,i} = 0$. Hence (12.3) can be written as

$$\int_\Omega [k(2/|d(v)|^2)^{\frac{1}{2}} d_{ij}(v)\, \varphi_{i,j} - \varphi_{i,j} v_i v_j]\, dx = 0. \tag{12.4}$$

Let $\hat{H}_0^1(\Omega)$ be the pre-Hilbert space consisting of *solenoidal* vectors u in $C_0^1(\Omega)$. For u in $\hat{H}_0^1(\Omega)$, we denote by $\|u\|_1$, $\|u\|_0$ respectively the Dirichlet and the L_2-norm of u. Let $H_0^1(\Omega)$ be the closure of $\hat{H}_0^1(\Omega)$ under $\|\ \|_1$-norm. Then $H_0^1(\Omega)$ is a separable Hilbert space.

It is well-known [*12*, p. 206] that if $\partial\Omega$ is sufficiently smooth and if $v_i^0(x)$, $i=1, 2, 3$ are sufficiently smooth functions on $\partial\Omega$, then for every $\varepsilon > 0$ there is a twice continuously differentiable solenoidal vector $a(x)$ defined in Ω such that

$$a_i(x) = v_i^0(x) \quad \text{on } \partial\Omega, \qquad \int_\Omega a_{i,j} v_i v_j\, dx \leqq \varepsilon \int_\Omega v_{i,j} v_{i,j}\, dx \tag{12.5}$$

holds for all v in $H_0^1(\Omega)$. If $v(x)$ is a strict solution of (4), (12.1), and (12.2), then $v(x) - a(x) \equiv u(x)$ belongs to $\hat{H}_0^1(\Omega)$. Under these circumstances we can write (12.4) as follows:

$$\begin{aligned}\int_\Omega [k(2/|d(u+a)|^2)^{\frac{1}{2}} d_{ij}(u+a)\, d_{ij}(\varphi) - \varphi_{i,j}(u_i u_j + a_i u_j + u_i a_j)]\, dx \\ = \int_\Omega \varphi_{i,j}\, a_i a_j\, dx,\end{aligned} \tag{12.6}$$

where we have used the fact that $d_{ij} = d_{ji}$ for $i, j = 1, 2, 3$. Clearly, (12.6) holds for all φ in $\hat{H}_0^1(\Omega)$, and hence by continuity it holds also for all φ in $H_0^1(\Omega)$. We shall say that *a vector $v(x)$ is a weak solution of the problem* (4), (12.1), and (12.2) *if* (i) *$u = v - a$ belongs to the Hilbert space $H_0^1(\Omega)$ and* (ii) *it satisfies the Eq.* (12.6) *for all φ in $H_0^1(\Omega)$.*

13. Existence of solution of certain weak problems. To solve the problem in a weak sense, we shall apply SHINBROT's version of a fixed-point theorem in a separable Hilbert space [*25*, p. 257].

Theorem (SHINBROT). Let A be a weakly continuous operator mapping a separable Hilbert space H into itself. That is, $x_n \to x$ weakly on H implies that

$A x_n \to A x$ weakly on H. Let y be an element in H. If there is a positive number r such that either

$$Re(Ax - y, x) \geqq 0 \quad \text{for all } x \text{ on } S_r$$

or

$$Re(Ax - y, x) \leqq 0 \quad \text{for all } x \text{ on } S_r,$$

then y is in the range of A, where S_r stands for the sphere with radius r and centered at the origin.

Consider the non-linear functional $J[u, \varphi]$ defined on $H_0^1(\Omega) \times H_0^1(\Omega)$ by the expression on the left-hand side in (12.6). Clearly, it is, for fixed u in $H_0^1(\Omega)$, a linear function of φ in $H_0^1(\Omega)$. We assert that it is a bounded linear functional of φ. Indeed, for every u in $H_0^1(\Omega)$,

$$|d_{ij}(u+a)/|d(u+a)|| = 1 \quad \text{a.e. in } \Omega. \tag{13.1}$$

It follows immediately that for all φ in $H_0^1(\Omega)$

$$\Big|\int_\Omega k(2/|d(u+a)|^2)^{\frac{1}{2}} d_{ij}(u+a)\, d_{ij}(\varphi)\, dx\Big| \leqq \text{const.}\ \|\varphi\|_1 .$$

Moreover, Sobolev's inequality assures that u in $H_0^1(\Omega)$ also belongs to $L_4(\Omega)$. Accordingly,

$$\int_\Omega \varphi_{i,j}\, u_i\, u_j\, dx \leqq \text{const.}\ \|\varphi\|_1 .$$

The other two terms in $J[u, \varphi]$ are clearly bounded linear functionals of φ. Thus our assertion is proved. Now, Ritz's representation theorem assures that for every u in $H_0^1(\Omega)$ there is an element $A u$ in $H_0^1(\Omega)$ such that

$$J[u, \varphi] = (A u, \varphi), \quad \text{for all } \varphi \text{ in } H_0^1(\Omega), \tag{13.2}$$

where

$$(u, \varphi)_1 \equiv \int_\Omega (u_i\, \varphi_i + u_{i,j}\, \varphi_{i,j})\, dx .$$

Consider now the right-hand side in (12.6). It is clear that for fixed choice of the field $a(x)$ it is a bounded linear functional of φ. Hence there is an element α in $H_0^1(\Omega)$, depending only on the choice of $a(x)$, such that

$$\int_\Omega \varphi_{i,j}\, a_i\, a_j\, dx = (\alpha, \varphi)_1 \quad \text{for all } \varphi \text{ in } H_0^1(\Omega).$$

In order to apply SHINBROT's fixed-point theorem we show that the operator A in (13.2) which maps $H_0^1(\Omega)$ into itself is weakly continuous. Suppose that $u_n \to u$ weakly in $H_0^1(\Omega)$. Then $\{u_n\}$ stays in a bounded set in $H_0^1(\Omega)$. Hence we can apply the Rellich selection principle to conclude that there is a subsequence $\{u_n\}$ such that $u_n \to u$ a.e. in Ω. Accordingly, as $n \to \infty$,

$$\int_\Omega \varphi_{ij}(a_i\, u_{nj} + u_{ni}\, a_j)\, dx \to \int_\Omega \varphi_{ij}(a_i\, u_j + u_i\, a_j)\, dx ,$$

for all φ in $H_0^1(\Omega)$. Moreover, the Sobolev imbedding theorem insures that the boundedness of $\{u_n\}$ in $H_0^1(\Omega)$ implies the boundedness of $\{u_n\}$ in $L_4(\Omega)$. Hence, by choosing a subsequence if necessary, we conclude that for all φ in $H_0^1(\Omega)$

$$\int_\Omega \varphi_{ij}\, u_{ni}\, u_{nj}\, dx \to \int_\Omega \varphi_{ij}\, u_i\, u_j\, dx \quad \text{as } n \to \infty .$$

Finally, we consider the term

$$\int_\Omega k(2/|d(u+a)|^2)^{\frac{1}{2}} d_{ij}(u+a)\, \varphi_{i,j}\, dx$$

in $(Au, \varphi)_1$. Again equality holds in (13.1) for all u_n. We conclude from the weak compactness of $L_2(\Omega)$ that it converges for all φ in $H_0^1(\Omega)$ as $n\to\infty$. The weak continuity of the operator A on $H_0^1(\Omega)$ is now established.

Consider now the values of $(Au, u)_1$. We have

$$\begin{aligned}(Au, u)_1 &= \int_\Omega [k(2/|d(u+a)|^2)^{\frac{1}{2}} d_{ij}(u+a)\, d_{ij}(u) + a_{i,j} u_i u_j]\, dx \\ &= \int \left\{k\left[|d(u+a)| - \frac{\sqrt{2}}{|d(u+a)|} d_{ij}(u+a)\, d_{ij}(a)\right] + a_{i,j} u_i u_j\right\} dx,\end{aligned} \tag{13.3}$$

where the term cubic in u as well as the terms $u_{i,j} u_i a_j$ have been dropped out because of the solenoidal character of u. By Schwarz's inequality

$$|\sqrt{2}\, k |d(u+a)|^{-1} d_{ij}(u+a)\, d_{ij}(a)| \leq \sqrt{2}\, k |d(a)|. \tag{13.4}$$

By the choice of the field $a(x)$ in (12.5) we have

$$\int_\Omega a_{ij} u_i u_j\, dx \leq \varepsilon \int_\Omega u_{i,j} u_{i,j}\, dx \leq \varepsilon \|u\|_1^2. \tag{13.5}$$

According to the well-known Korn estimates, there is a $c_1 > 0$ such that

$$\left(\int_\Omega |d(u+a)|\, dx\right)^2 \geq \int_\Omega |d(u+a)|^2\, dx \geq c_1^2 \|u+a\|_1^2 \geq c_1^2 (\|u\|_1 - \|a\|_1)^2.$$

Hence

$$\int_\Omega |d(u+a)|\, dx \geq c_1 (\|u\|_1 - \|a\|_1). \tag{13.6}$$

Using the estimates in (13.4)–(13.6), we deduce from (13.3) that

$$\begin{aligned}(Au, u)_1 &\geq \sqrt{2}\, k(\|u\|_1 - 2\|a\|_1) - \varepsilon \|u\|_1^2 \\ &= (\sqrt{2}\, c_1 k - \varepsilon \|u\|_1)\, \|u\|_1 - 2\sqrt{2}\, c_1 k \|a\|_1.\end{aligned} \tag{13.7}$$

On the other hand, for all u in $H_0^1(\Omega)$,

$$|(\alpha, u)_1| = \left|\int_\Omega u_{i,j} a_i a_j\, dx\right| \leq \|a\|_0\, \|u\|_1. \tag{13.8}$$

The above estimates indicate that the operator A is not strictly monotone. Also it is not coercive. Thus neither the monotonicity method nor SHINBROT'S fixed-point theorem can be applied directly. This difficulty is caused by the yield condition. If this difficulty cannot be overcome, perhaps a new technique is needed. However, we can consider the boundary data $v^* \equiv \delta v^0$ with δ being a small number. For the small boundary data v^* we have $a^*(x) = \delta a(x)$, which is a sufficiently smooth continuation of v^* into the whole domain Ω. Thus, for the boundary data v^*, we have, instead of (13.7) and (13.8),

$$(Au, u)_1 \geq (\sqrt{2}\, c_1 k - \varepsilon \|u\|_1) - 2\sqrt{2}\, c_1 k \|a\|_1 \delta, \tag{13.7*}$$

$$(\alpha, \varphi)_1 \leq \|a\|_0 \|u\|_1 \delta. \tag{13.8*}$$

Fixing the choice of ε and the field $a(x)$, we restrict the norm of u to be such that

$$\|u\|_1 \leq \frac{1}{2\varepsilon} \sqrt{2}\, c_1 k. \tag{13.9}$$

Next, we choose δ to be so small that

$$\tfrac{1}{2} \sqrt{2}\, c_1 k \|u\|_1 \geq (2\sqrt{2}\, c_1 k \|a\|_1 + \|a\|_0 \|u\|_1)\, \delta.$$

That is,

$$\|u\|_1 \geqq 2\sqrt{2}\, c_1\, k \|a\|_1\, \delta / (\tfrac{1}{2}\sqrt{2}\, c_1\, k - \delta \|a\|_0). \tag{13.10}$$

It is easy to check that for (13.9) and (13.10) to hold simultaneously we need only to require that

$$\delta \leqq \frac{\sqrt{2}\, c_1\, k}{4\varepsilon \|a\|_1 + \|a\|_0} \tag{13.11}$$

Thus, for the boundary data $v^* = \delta v$ with δ subject to the inequality (13.11), we have for all u on S_r with

$$\tfrac{1}{2}\sqrt{2}\, c_1\, k \geqq r \geqq 2\sqrt{2}\, c_1\, k \|a\|_1\, \delta / (\tfrac{1}{2}\sqrt{2}\, c_1\, k - \delta \|a\|_0),$$

the inequality

$$(A\,u - \alpha,\, u)_1 \geqq 0.$$

By SHINBROT's theorem there is an element u^* in the ball B_r enclosed by S_r such that $(A\,u^*\,\alpha,\, \varphi) = 0$ for φ in $H_0^1(\Omega)$. Clearly, $u^* + a^*$ is a weak solution of the weak problem (12.6) with boundary data v^*. We state the above results as

Theorem 3. If the boundary data of the weak problem (12.6) is sufficiently small, then it possesses a solution in the weak sense.

D. Torsion problems.

Consider a cylindrical bar twisted by terminal couples. For a small angle of twist per unit length, within the linear theory of elasticity ST. VENANT formulated this problem by means of a semi-inverse method, so reducing it to a Neumann problem [*0*, pp. 323–333, 346–350]. With the help of potential theory, this problem has been studied extensively and is well understood. However, for pure torsion beyond the linearly elastic range, the formulation of the problem varies with the theory for the materials under consideration. As far as the theory for ideal elastic-plastic bodies is concerned, much progress has been achieved only in recent years.

According to the classical theory for elastic-plastic solids, the twisted bar will remain elastic if the applied torque is sufficiently small. As the applied torque is increased to a certain critical value, some portion of the cylindrical surface will become plastic. Furthermore, the plastic portion of the bar will continue to grow as the applied torque increases. However, for completely plastic torsion, i.e., for the entire bar to become plastic, an infinitely large angle of twist per unit length is required in this theory. Accordingly, completely plastic torsion will never occur in practice, and hence all large torsions will be, according to the classical theory for ideal elastic-plastic solids, of the elastic-plastic type. In what follows, a comprehensive result for elastic-plastic torsion will be presented. In particular, the phenomena just described will be shown to be mathematical consequences of the classical theory for ideal plastic-solids. The reader is referred to Chap. D of the preceding article by GEIRINGER for numerical results and physical interpretations.

I. Completely plastic torsion.

Although completely plastic torsion can never be realized, it may be regarded as the limiting case of a sequence of elastic-plastic torsions when the angle of twist per unit length increases beyond all bounds. As we shall see, the torque for completely plastic torsion is, indeed, the least upper bound of those for all large torsions. The main reason for investigating plastic torsion completely is that a thorough understanding of it is needed for treating the difficult elastic-plastic torsion problem. As we shall see also, to determine the existence and properties of the stress function ψ for elastic-plastic torsion it is necessary to know the corresponding function for completely plastic torsion.

14. Variational formulation of the problem. In what follows, I always denote by G the cross section of a cylindrical bar. For definiteness and simplicity, I shall restrict G to be a Jordan domain with the property that ∂G, boundary of G, possesses continuously varying non-zero radius of curvature except at a finite number of corners. By completely plastic torsion I mean the following: *find a non-negative function $\Psi(x, y)$ such as to satisfy the following conditions*:

(i) *Ψ is continuous in $\bar{G}$ (closure of G), piecewise smooth in G and zero on ∂G.*

(ii) $|\mathrm{grad}\,\Psi|^2 = k^2 > 0$ *in $\bar{G}$,*

(iii) *Ψ maximizes the integral*

$$\iint\limits_G \Phi(x, y)\,dx\,dy$$

in the class of all functions Φ that satisfy the conditions (i) *and* (ii).

The reason for allowing the discontinuities in grad Ψ is that as a partial differential equation of the first order, the equation in (ii) does not have a solution continuous in $G + \partial G$, smooth in the entire domain G and zero on ∂G. On the other hand, without the restriction (iii) the problem would lose its uniqueness of solution, as one can show by constructing simple examples. Physically, the function Ψ is known as a stress function, and the introduction of it is suggested by the equilibrium requirements; the vanishing of Ψ on ∂G implies that the lateral surface of the cylindrical bar is free from any applied force; the condition (ii) is the well known yield criterion, and the condition (iii) states that the solution function Ψ is the one which gives the maximum resisting torque [*41*, p. 17].

For completely plastic torsion the conditions (i) and (ii) are well known, and the natural condition (iii) was first stated explicitly in [*41*]. It may be noted that in the formulation of the problem only the conditions of equilibrium, the yield condition and the boundary conditions on the lateral surface have been used. Thus, no constitutive relation is needed for determining the stresses unless one needs to compute the displacement.

15. Solution of the problem. Although the problem has been formulated as a maximum problem with a differential equation as a side constraint, yet the method of characteristics for solving first-order partial differential equations leads one to suspect that the solution of our problem is still given by the formula:

$$\Psi(q) = k\varrho(q, \partial G), \qquad q \text{ in } \bar{G}, \tag{15.1}$$

where q stands for a point in $\bar{G}$, ϱ the distance from the point q to ∂G and k is the positive constant appearing in (ii). Indeed, we can verify that the function Ψ defined in (15.1) fulfills all the requirements (i), (ii) and (iii).

Clearly, the function Ψ defined in (15.1) vanishes on ∂G. Also, one can easily establish its continuity in $\bar{G}$ by applying the triangle inequalities for the distance function ϱ [*41*]. However, it is not obvious that Ψ is piecewise smooth in G. The difficulty lies in confirming the topological fact that the set of points where grad Ψ is discontinuous is not only of 2-dimensional measure zero but also forms a piecewise smooth curve in certain sense. Since information about the set of discontinuities of grad Ψ is essential in the present treatment of the elastic-plastic torsion problem, I state some important properties of the ridge of a Jordan domain.

By the ridge of a Jordan domain G I mean the set of points γ in G such that if γ is in G and s on ∂G with $\varrho(\gamma, s) = \varrho(\gamma, \partial G)$ then the open disk $D(\gamma, \varrho(\gamma, s))$ is the maximum open disk contained in G. This disk touches ∂G at s. From this definition for Γ it is easy to show that if G is a convex polygon then Γ consists of line segments each of which bisects some interior angle of G. In general, we have the following two theorems:

Theorem 1. By distinguishing the two sides of a curve, the ridge Γ of a Jordan domain G can be represented as a joint union of the continuous images of the subarcs of ∂G between two neighboring corners and the continuous images of the interior angles of the reentrant corners of G.

Theorem 2. If the curvature of ∂G is continuous and monotone in a generalized sense between any two consecutive corners of G, then the ridge Γ of G possesses continuously turning tangents except at a finite number of points in Γ.

The first theorem ensures the existence of a parametric representation for the ridge Γ, and hence it is a curve in the usual sense if we distinguish its two sides. The second theorem ensures that if the two sides of Γ are distinguished, then Γ is a piecewise smooth curve. Although the smoothness of Γ will not be needed here, it will play a very important role in solving the elastic-plastic torsion problem. Since the proofs for these two theorems, while elementary, are rather complicated, the reader is referred to the original article [*41*] for the details.

From the definition of Γ and the formula (15.1) for defining Ψ it is easy to show that the set of points of discontinuity of grad Ψ in G is precisely the ridge Γ of G and that in the open set $G - \bar{\Gamma}$, Ψ is continuously differentiable with $|\text{grad}\,\Psi| = k$. This proves that the conditions in (i) and (ii) are satisfied by Ψ. Also, the formula (15.1) and the properties of Γ ensure that the function Ψ does maximize the integral in (iii) and that it is the unique solution of the completely plastic torsion problem.

16. St. Venant's conjecture and its extension. By examining the numerical results in his study of purely elastic torsion, ST. VENANT conjectured that *among all solid cylindrical bars with the same cross-sectional area the circular one possesses the maximum torsional rigidity.* This profound conjecture was first confirmed by PÓLYA *et al.* [2, § 5.17]. Later, PÓLYA and WEINSTEIN extended the conjecture as the following theorem [Note A, 2]. *Among all multiply connected cylindrical bars with the same cross-sectional area and with the same joint area of the holes, the cylindrical pipe possesses the maximum torsional rigidity.* For completely plastic torsion of a solid cylindrical bar, ST. VENANT'S conjecture is also a theorem, first proved by LEAVITT and UNGAR [*14*]. Interested readers are referred to the original articles for the proofs.

II. Elastic-plastic torsion.

As was already mentioned, the large torsion problem varies with the constitutive equations governing the materials under consideration. As far as plasticity theory is concerned, GAJEWSKI and LANGENBACH used Hencky's deformation theory to formulate a class of variational problems, and they answered certain questions of existence. In particular, large torsion problems have also been attacked by them [*29*]. By applying the linear theory of elasticity and the

v. Mises-Prandtl theory of plasticity, the problem of elastic-plastic torsion was stated explicitly a long time ago [*1*]. The difficulty in solving it lies in the fact that different branches of the constitutive relation should be applied, according to the state of deformation. It is a part of the problem to determine how the cylindricial bar is partitioned into elastic "regions" over which the linearly elastic relation applies and into plastic "regions" over which the completely plastic relation applies. Moreover, across the "surface" which separates the elastic and plastic "regions", the dynamical conditions of compatibility [*13*, p. 55] require that the stress components be continuous. Because of these difficulties, existence and uniqueness of solution in the strict sense has been established only recently and for certain special cases [*51*]. However, by using the variational formulation, the existence of a weak solution has been established recently in complete generality [*49*]. Thanks to the regularity results for the solutions of certain variational inequalities [63], this problem can now be solved in the strict sense with relatively complete generality.

17. Formal statement of the problem. Let G be the cross-section of a solid cylindrical bar with the properties listed in Sect. 14. In addition, G is required to satisfy the cone condition, so that the concept of weak and strong derivatives of a function over G will be identical. In the present case, the cone condition is equivalent to forbidding that any corner of G shall be a cusp. The elastic-plastic torsion problem is to *find a function* $\psi(x, y)$ *continuous in* $\bar{G}$, *continuously differentiable in* G *and vanishing on* ∂G, *such that* (i) $|\text{grad}\,\Psi|$ *is always less than or equal to a positive constant* k *(the yield constant) in* G *and* (ii) *wherever* $|\text{grad}\,\psi|$ *is strictly less than* k, ψ *must be twice continuously differentiable and satisfy the Poisson equation* $\Delta\psi = -2\mu\,\theta_0$. Here θ_0 stands for the angle of twist per unit length and μ stands for the (positive) shear modulus. For physical derivation of the problem and for more details, see Chap. D of the preceding article by GEIRINGER.

18. Variational formulation of the problem. The problem as stated above has never been attacked directly. Also, the uniqueness of its solution remains an open question. In what follows, it will be formulated as a variational minimum problem with an inequality as a side constraint. It will be shown that the extremal will solve the elastic-plastic problem as just stated. Moreover, much information about the problem will be derived from this variational formulation.

According to the results in Sect. 14, the solution to the corresponding completely plastic torsion problem is given by the formula,

$$\Psi(q) = k\,\varrho(q, \partial G), \qquad q \text{ in } \bar{G}. \tag{18.1}$$

Let $C_0^\infty(G)$ be the class of infinitely differentiable functions with compact support in G. For every f in $C_0^\infty(G)$, we denote by $\|f\|_1$ the Dirichlet norm of f:

$$\|f\|_1^2 = \iint_G [(\text{grad}\,\psi)^2 + f^2]\,dx\,dy. \tag{18.2}$$

Let $H_0^1(G)$ be the completion of $C_0^\infty(G)$ under $\|\ \|_1$-norm. Then $H_0^1(G)$ is a complete Hilbert space. Also, $H_0^0(G) = H^0(G)$ consists of square integrable functions over G. Let F be the family of functions u such that

(i) u belongs to $H_0^1(G)$,

(ii) $u \leqq \Psi$ almost everywhere in G,

where Ψ is the function defined in (18.1). Now I state the elastic-plastic torsion problem as follows: *to find a function* $\psi(x, y)$ *such as to minimize the integral*

$$J[u] \equiv \iint_G [(\text{grad } u)^2 - 4\mu\,\theta_0\, u]\, dx\, dy \tag{18.3}$$

among all functions u *in* F, *and to show that the extremal* ψ *solves the elastic-plastic problem in the strict sense.*

The present formulation differs from the usual one [61] in that instead of using the yield condition in the mean, the admissible functions are majorized by the distance function Ψ defined in (18.1). In this way the class of admissible functions has been enlarged so that more information about the extremal ψ can be derived by means of variations. However, it turns out that the extremal ψ is smooth in G and satisfies the yield condition pointwise.

Remark 1. The values of the expression in (18.3) are bounded from below. For if u^* is the solution of the Dirichlet problem,

$$\Delta u^* = -2\mu\,\theta_0 \quad \text{in } G, \qquad u^* = 0 \quad \text{on } \partial G, \tag{18.4}$$

then for every admissible u, $J[u] \geqq J[u^*]$. Further, from the integral representation for the solution u^* of (18.4),

$$u^*(x, y) = 2\mu\,\theta_0 \iint_G G(x, y; \xi, \eta)\, d\xi\, d\eta,$$

with $G(x, y; \xi, \eta)$ being Green's function of (18.4) in G, we see that $u^* > \Psi$ somewhere in G provided that $\mu\theta_0$ is sufficiently large. Therefore, the admissibility condition (ii) is essential.

Remark 2. If $\mu\theta_0$ is so small that the solution u^* of (18.4) is less than or equal to Ψ in $\bar{G}$, then it is the solution of our minimum problem. Thus the present formulation of the elastic-plastic torsion problem does include a purely elastic torsion problem as a special case.

Remark 3. Consider the integral,

$$J[u]/4\mu\,\theta_0 = \iint_G [(\text{grad } u)^2/4\mu\,\theta_0 - u]\, dx\, dy.$$

If we let $\theta_0 \to \infty$, i.e., the angle of twist per unit length increases beyond all bounds, then the minimum problem is that of maximizing the integral

$$\iint_G u(x, y)\, dx\, dy$$

subject to the side condition $u \leqq \Psi$ in $\bar{G}$. Accordingly, the present formulation also includes completely plastic torsion as a limiting case.

Remark 4. The minimum problem (18.3) may be regarded as a principle of minimizing the stress energy

$$\iint_G (\text{grad } u)^2\, dx\, fy$$

under the additional isoperimetric condition that the value of the integral is equal to a given constant. That is, the applied torque is given. Indeed, the two minimum problems are equivalent if we identify the constant $-4\mu\,\theta_0$ as a Lagrange multiplier.

19. Existence and uniqueness of the extremal. Suppose that an extremal cannot be found. Then there is a minimizing sequence $\{\psi_n\}$ in F such that

$$J[\psi_n] \to \inf_{u \in F} J[u] = d \quad \text{as } n \to \infty. \tag{19.1}$$

Using the parallelogram law,

$$\|\tfrac{1}{2}(\psi_m+\psi_n)\|_1^2+\|\tfrac{1}{2}(\psi_m-\psi_n)\|_1^2=\tfrac{1}{2}\{\|\psi_m\|_1^2+\|\psi_n\|_1^2\},$$

and the well known Poincaré inequality,

$$\iint_G \psi_n^2\,dx\,dy \leqq \text{const.}\iint_G (\operatorname{grad}\psi_n)^2\,dx\,dy,$$

one can verify that the minimizing sequence $\{\psi_n\}$ in (19.1) is also a Cauchy sequence in the complete Hilbert space $\mathring{H}^1_0(G)$. Hence there is a function ψ in $\mathring{H}^1_0(G)$ such that $\|\psi_n-\psi\|_1\to 0$ as $n\to\infty$. Next, one easily checks that the admissible family F is a closed subset of $\mathring{H}^1_0(G)$, and hence the limit function ψ also belongs to F. Finally one can apply the triangle inequalities in $L_2(G)$ to verify that

$$J[\psi]=\lim_{n\to\infty} J[\psi_n]=d.$$

This establishes the existence of an extremal ψ.

For the uniqueness, let ψ,ψ^* be two minimizing extremals. Then $\hat{\psi}\equiv(\psi+\varphi^*)/2$ also belongs to F. Accordingly, $J[\hat{\psi}]\geqq J[\psi]=J[\psi^*]$. On the other hand, the convexity of the functional J implies that

$$J[\hat{\psi}]\leqq\tfrac{1}{2}\{J[\psi]+J[\psi^*]\}.$$

Consequently, $J[\frac{1}{2}(\psi+\psi^*)]=\frac{1}{2}\{J[\psi]+J[\psi^*]\}$. This is so if and only if

$$\iint_G (\operatorname{grad}(\psi-\psi^*))^2\,dx\,dy=0.$$

But then there is the Poincaré inequality

$$\iint_G (\psi-\psi^*)^2\,dx\,dy\leqq\text{const.}\iint_G (\operatorname{grad}(\psi-\psi^*))^2\,dx\,dy=0,$$

and hence $\|\psi-\psi^*\|_1=0$. The uniqueness of the extremal is established.

20. Hölder continuity of the extremal. As before, we denote by Γ the ridge of the domain G and proceed to show that the extremal ψ can be identified by a function Hölder continuous in $G-\bar{\Gamma}$. The proof consists in establishing a Dirichlet growth condition for ψ and then applying a theorem of Morrey.

Let s_i, s_{i+1} be two consecutive regular (non-reentrant) corners of G and let $G_{i,i+1}$ be the simply connected domain bounded by the subarc $s_i s_{i+1}$ of ∂G and a portion of Γ, [*41*]. Let

$$x=f(s),\qquad y=g(s),\qquad s_i\leqq s\leqq s_{i+1}, \tag{20.1}$$

be the parametric representation of the arc $s_i s_{i+1}$. Then, for every point (x, y) in $G_{i,i+1}$, we can write

$$x=f(s)+t n_x(s),\qquad y=g(s)+t n_y(s), \tag{20.2}$$

where $n_x(s)$, $n_y(s)$ are respectively the x- and y-components of the unit inward normal to ∂G at s and where t stands for the distance from the point (x, y) to ∂G. As was shown in [*41*], the equations in (20.2) define a continuously differentiable coordinate transformation, $(s,t)\leftrightarrow(x,y)$, with positive Jacobian,

$$\frac{\partial(x,y)}{\partial(s,t)}=1-t\varkappa(s)>0\qquad\text{in } G-\bar{\Gamma}, \tag{20.3}$$

where $\varkappa(s)$ is the curvature of ∂G at s.

For every point 0 in $G_{i,i+1}$, we can choose R so small that the disk $D(0, R)$ in the (s, t)-plane with center at 0 and radius,

$$R^2 \equiv (s-s_0)^2 + (t-t_0)^2,$$

is contained in $G_{i,i+1}$, where (s_0, t_0) are curvilinear coordinates of the point 0. Of course, $D(0, R)$ is, in general, not a disk in the (x, y)-plane.

For every function u in F with $u=\psi$ in $H_0^1(G-D(0,R))$, $0<r\leqq R$, we have for $0<r\leqq R$,

$$\iint\limits_{D(0,r)} [(\operatorname{grad}\psi)^2 - 4\mu\,\theta_0\,\psi]\,\frac{\partial(x,y)}{\partial(s,t)}\,ds\,dt \leqq \iint\limits_{D(0,r)} [(\operatorname{grad} u)^2 - 4\mu\,\theta_0\,u]\,\frac{\partial(x,y)}{\partial(s,t)}\,ds\,dt.$$

In view of the fact that $u, \psi \leqq \Psi$ in $\bar{G}$ we can rewrite this relation as follows:

$$\iint\limits_{D(0,r)} (\operatorname{grad}\psi)^2\,\frac{\partial(x,y)}{\partial(s,t)}\,ds\,dt \leqq \iint\limits_{D(0,r)} (\operatorname{grad} u)^2\,\frac{\partial(x,y)}{\partial(0,r)}\,ds\,dt + \text{const.}\, r^2, \tag{20.4}$$

for $0<r\leqq R$. Since the chain rule for differentiations still holds for functions in $H^1(G_{i,i+1})$ under smooth coordinate transformations, $(x, y)\leftrightarrow(s, t)$, and since

$$s_x \equiv \frac{\partial s}{\partial x} = n_y(s)/(1-t\varkappa(s)), \qquad s_y \equiv \frac{\partial s}{\partial y} = -n_x(s)/(1-t\varkappa(s)),$$
$$t_x \equiv \frac{\partial t}{\partial x} = n_x(s), \qquad t_y \equiv \frac{\partial t}{\partial y} = n_y(s),$$

relative to the curvilinear coordinate system (s, t),

$$\begin{aligned}(\operatorname{grad}\psi)^2 &= (\psi_s\,s_x + \psi_t\,t_x)^2 + (\psi_s\,s_y + \psi_t\,t_y)^2\\ &= \psi_s^2/(1-t\varkappa(s))^2 + \psi_t^2.\end{aligned}$$

Also, it follows from continuity and from the inequality in (20.3) that there are two constants, $0<c<1$, $C>1$ such that

$$0<c\leqq 1-t\varkappa(s)\leqq C \quad \text{in } D(0,R).$$

Accordingly, for $0<r\leqq R$,

$$\begin{aligned}&\iint\limits_{D(0,r)} (\operatorname{grad}\psi)^2\,\frac{\partial(x,y)}{\partial(s,t)}\,ds\,dt \geqq \frac{c}{C^2}\iint\limits_{D(0,r)} (\psi_s^2+\psi_t^2)\,ds\,dt,\\ &\iint\limits_{D(0,r)} (\operatorname{grad} u)^2\,\frac{\partial(x,y)}{\partial(s,t)}\,ds\,dt \leqq \frac{C}{c^2}\iint\limits_{D(0,r)} (u_s^2+u_t^2)\,ds\,dt.\end{aligned} \tag{20.5}$$

Combining the inequalities in (20.4) and (20.5) yields

$$\iint\limits_{D(0,r)} (\psi_s^2+\psi_t^2)\,ds\,dt \leqq c_1 \iint\limits_{D(0,r)} (u_s^2+u_t^2)\,ds\,dt + c_2\,r^2, \tag{20.6}$$

where the constants are independent of r.

For almost all values of r, ψ belongs to $H^1(D(0,r))$ and is therefore essentially continuous there. For such a value of r we can define

$$u(\varrho,\phi) = \bar{\psi} + \frac{\varrho}{r}\,[\psi(r,\phi) - \bar{\psi}], \quad 0\leqq\varrho\leqq r, \quad 0\leqq\phi\leqq 2\pi, \tag{20.7}$$

where (ϱ, ϕ) are polar coordinates with origin at 0 and where

$$\bar{\psi} \equiv \frac{1}{2\pi} \int_0^{2\pi} \psi(r, \phi)\, d\phi, \qquad r^2 \equiv (s - s_0)^2 + (t - t_0)^2.$$

Similarly, we use $\bar{\Psi}$ to stand for the average of Ψ over $\partial D(0, r)$. Note that in the (s, t)-plane, $\Psi(s, t) = kt$. Hence Ψ is harmonic in $G_{i,i+1}$, i.e.,

$$\frac{\partial^2 \Psi}{\partial s^2} + \frac{\partial^2 \Psi}{\partial t^2} = 0 \quad \text{there.}$$

By the mean-value theorem, $\bar{\Psi} = \Psi(0, \phi)$. Thus,

$$\Psi(\varrho, \phi) = \bar{\Psi} + \frac{\varrho}{r} [\Psi(r, \phi) - \bar{\Psi}]. \tag{20.8}$$

By taking the difference of the corresponding sides of (20.7) and (20.8) we find that

$$\Psi(\varrho, \phi) - u(\varrho, \phi) = (\bar{\Psi} - \bar{\psi}) \left(1 - \frac{\varrho}{r}\right) + \frac{\varrho}{r} [\Psi(r, \phi) - \psi(r, \phi)] \geqq 0$$

almost everywhere in $D(0, r)$. This proves that if u is defined by (20.7) in $D(0, r)$ and $u = \psi$ in $H^1(G - D(0, r))$, then it belongs to F and hence the inequality (20.6) holds for such a function u.

Now, in the disk $D(0, r)$ in the (s, t)-plane,

$$u_s^2 + u_t^2 = \left[\frac{1}{r} (\psi(r, \phi) - \bar{\psi})\right]^2 + \left[\frac{1}{r} \psi_\phi(r, \phi)\right]^2.$$

Hence, from (20.6) and (20.7) we see that for $0 < r \leqq R$,

$$\begin{aligned} \iint\limits_{D(0,r)} (\psi_s^2 + \psi_t^2)\, ds\, dt &\leqq c_1 \iint\limits_{D(0,r)} \left\{\left[\frac{\psi(r, \phi) - \bar{\psi}}{r}\right]^2 + \left[\frac{\psi_\phi(r, \phi)}{r}\right]^2\right\} \varrho\, d\varrho\, d\phi + c_2 r^2 \\ &\leqq \frac{1}{2} c_1 \int_0^{2\pi} \{[\psi(r, \phi) - \bar{\psi}]^2 + [\psi\phi(r, \phi)]^2\}\, d\phi + c_2 r^2. \end{aligned} \tag{20.8'}$$

Since $\psi(r, \phi)$ is essentially continuous in ϕ for almost all values of r, we have the Poincaré inequality

$$\int_0^{2\pi} [\psi(r, \phi) - \bar{\psi}]^2\, d\phi \leqq 4\pi^2 \int_0^{2\pi} [\psi_\phi(r, \phi)]^2\, d\phi. \tag{20.9}$$

By combining (20.8′) and (20.9) we find

$$\begin{aligned} \iint\limits_{D(0,r)} (\psi_s^2 + \psi_t^2)\, ds\, dt &\leqq c_3 \int_0^{2\pi} [\psi_\phi(r, \phi)]^2\, d\phi + c_2 r^2 \\ &\leqq c_3 r^2 \int_0^{2\pi} \left\{[\psi_r(r, \phi)]^2 + \left[\frac{1}{r} \psi_\phi(r, \phi)\right]^2\right\} d\phi + c_2 r^2 \\ &\leqq c_3 r \frac{d}{dr} \int_0^r \int_0^{2\pi} \left\{[\psi_\varrho(\varrho, \phi)]^2 + \left[\frac{1}{\varrho} \psi_\phi(\varrho, \phi)\right]^2\right\} \varrho\, d\varrho\, d\phi + c_2 r^2 \\ &= c_3 r \frac{d}{dr} \iint\limits_{D(0,r)} (\psi_s^2 + \psi_t^2)\, ds\, dt + c_2 r^2, \end{aligned} \tag{20.10}$$

where c_2, c_3 are constants independent of r. Upon solving the above inequality we find the Dirichlet growth condition,

$$\iint\limits_{D(0,r)} (\psi_s^2+\psi_t^2)\, ds\, dt \leqq \text{const.}\, r^\lambda \quad \text{for} \quad r\leqq R, \quad 0<\lambda<1,$$

$$r^2 \equiv (s-s_0)^2+(t-t_0)^2,$$

for the function $\psi(s,t)$ in the domain $G_{i,i+1}$. Hence, a theorem of MORREY asserts that $\psi(s,t)$ is Hölder continuous in the variables s and t. Accordingly, the function

$$\psi(x,y) \equiv \psi\big(s(x,y),\, t(x,y)\big),$$

with $(x,y) \leftrightarrow (s,t)$ being given by (20.2) is Hölder continuous in the variables x and y for all (x,y) in $G_{i,i+1}$.

Consider a domain G_j bounded by the two extreme inward normals at a reentrant corner s_j and a portion of Γ. Relative to plane polar coordinates (ϱ, ϕ) with origin at s_j,

$$\Psi(\varrho,\phi) = k\varrho \quad \text{in } \bar{G}_j, \qquad \frac{\partial^2\Psi}{\partial\varrho^2}+\frac{\partial^2\Psi}{\partial\phi^2}=0 \quad \text{in } G_j.$$

Consequently, by reasoning completely similar to that given above, we conclude also that the extremal ψ is Hölder continuous in G_j.

Since the unions of such $\bar{G}_j, \bar{G}_{i,i+1}$ completely cover G, it follows that ψ is Hölder continuous in $G-\bar{\Gamma}$ with possible exceptions along the extreme inward normals at the reentrant corners. However, in a neighborhood of an extreme inward normal at a reentrant corner s_j, we have the coordinate transformation (20.2) on one side of the inward normal, and we may employ the following coordinate transformation:

$$x=f(s_j)+t n_x(s;s_j), \qquad y=g(s_j)+t n_y(s;s_j), \tag{20.2'}$$

where n_x and n_y stand for the x- and y-components of the unit inward normal at s_j making a non-obtuse angle φ with the extreme inward normal at s_j and where $s\equiv\varkappa(s_j)\phi$ with $\varkappa$ being the one-side limit of the curvature of ∂G at s_j. It is easy to see that the coordinate curves $s=$constant and $t=$constant of (20.2) and (20.2') join smoothly along the extreme inward normal, i.e., x_s, x_t, y_s, y_t are all continuous across the extreme inward normal at s_j. Hence the same reasoning as was given above ensures that ψ is also Hölder continuous along the extreme inward normals at each of the reentrant corners. This completes a proof that ψ can be identified with a function Hölder continuous in $G-\bar{\Gamma}$.

21. The existence of an elastic core. By an *elastic core* we mean an open neighborhood of the ridge Γ, in which the extremal ψ is strictly less than Ψ. Further, ψ is twice continuously differentiable, and, in the case it satisfies the Poisson equation $\Delta\psi=-2\mu\theta_0$. To this end, we recall that the ridge Γ is an open connected set contained in G and possessing a continuously turning tangent except at a finite number of points. First, we consider the general case.

Case (1). γ_0 on Γ is a point where a unique tangent to Γ is defined. In this case there are two and only two points such that

$$\varrho(\gamma_0, s_0^*) = \varrho(\gamma_0, s_0^{**}) = \varrho(\gamma_0, \partial G). \tag{21.1}$$

Further, the tangent to Γ at γ_0 bisects the angle between the segments $\gamma_0 s_0^*$ and $\gamma_0 s_0^{**}$ [*41*]. Thus, in a neighborhood of γ_0, Γ has the following two parametric representations:

$$x=f(s^*)+\xi(s^*)\, n_x(s^*), \qquad y=g(s^*)+\xi(s^*)\, n_y(s^*), \tag{21.2a}$$

where s^* varies in some open interval containing s_0^* and

$$x = f(s^{**}) + \xi(s^{**})\, n_x(s^{**}), \qquad y = g(s^{**}) + \xi(s^{**})\, n_y(s^{**}) \tag{21.2b}$$

where s^{**} varies in some open interval containing s_0^{**}. Moreover,

$$|\varkappa(s_0^*)| - \xi(s_0^*) > 0, \qquad |\varkappa(s_0^{**})| - \xi(s_0^{**}) > 0. \tag{21.3}$$

By the continuity of $\varkappa(s)$ and $\xi(s)$, the strict inequalities in (21.3) remain valid in some neighborhoods of s_0^* and s_0^{**}, respectively. Since the subarc of Γ defined by (21.2a) and (21.2b) is a smooth Jordan curve, it is well known that for all sufficiently small ε the disk $D(\gamma_0, \varepsilon)$ with center at γ_0 and radius ε intersects Γ in a single arc $\gamma_1\gamma_2$ which partitions $D(\gamma_0, \varepsilon)$ into two parts say D^* and D^{**}. In what follows we shall choose the positive ε as small as we need. In particular, it will be chosen so small that $D(\gamma_0, \varepsilon)$ is contained in the domain G.

Denote by Ψ^* and Ψ^{**} the restrictions of Ψ to D^* and D^{**}, respectively. In view of the strict inequalities in (21.3), we can choose ε so small that both Ψ^* and Ψ^{**} can be extended to the closed disk $\bar{D}(\gamma_0, \varepsilon)$ by the following rules:

$$\begin{aligned} \Psi^*(q^{**}) &= k\varrho(q^{**}, s^*) \quad \text{for } q^{**} \text{ in } D^{**}, \\ \Psi^{**}(q^*) &= k\varrho(q^*, s^{**}) \quad \text{for } q^* \text{ in } D^*, \end{aligned} \tag{21.4}$$

where the segments s^*q^{**} and $s^{**}q^*$ are normal to ∂G and s^* and s^{**} lie in some small interval containing s_0^* and s_0^{**} respectively (see Fig. 1). From the equations (18.1) and (21.4), we see that the functions Ψ^* and Ψ^{**} so extended are twice

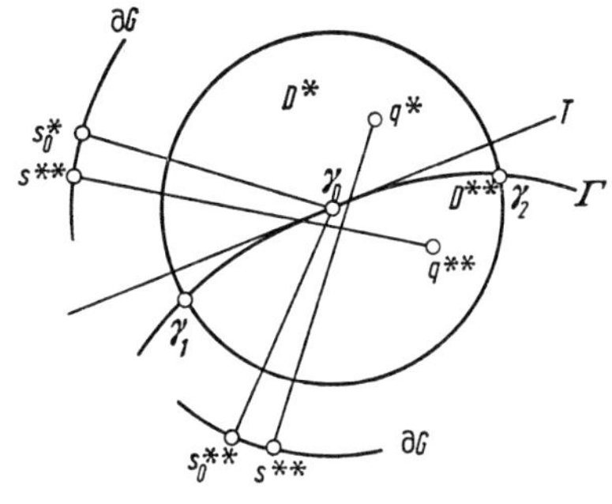

Fig. 1.

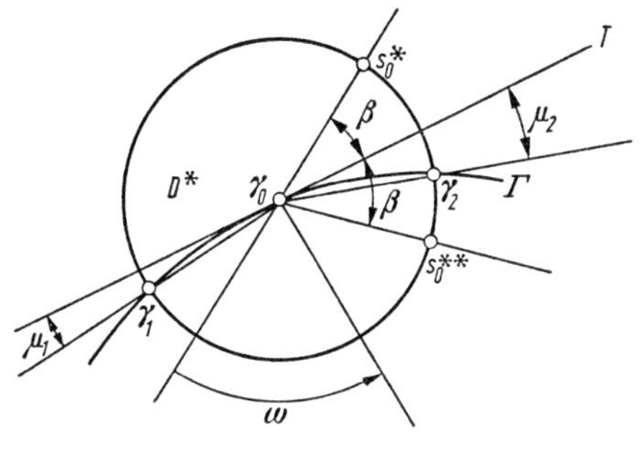

Fig. 2.

continuously differentiable in the closed disk $\bar{D}(\gamma_0, \varepsilon)$. Indeed, direct computation based on (18.1) and (21.4) gives

$$\begin{aligned} \Delta\Psi^* &= \Psi^*_{xx} + \Psi^*_{yy} = -k\varkappa(s^*)/[1 - t\varkappa(s^*)] \quad \text{in } \bar{D}(\gamma_0, \varepsilon), \\ \Delta\Psi^{**} &= \Psi^{**}_{xx} + \Psi^{**}_{yy} = -k\varkappa(s^{**})/[1 - t\varkappa(s^{**})] \quad \text{in } \bar{D}(\gamma_0, \varepsilon), \end{aligned} \tag{21.5}$$

where (s^*, t) and (s^{**}, t) are the curvilinear coordinates used in the discussion of Hölder continuity of ψ.

We proceed to show that if ε is sufficiently small then the solution ψ_ε of the Dirichlet problem,

$$\Delta\psi_\varepsilon = -2\mu\,\theta_0 \quad \text{in } D(\gamma_0, \varepsilon), \qquad \psi_\varepsilon = \psi \quad \text{on } \partial D(\gamma_0, \varepsilon), \tag{21.6}$$

is strictly less than Ψ in D. Indeed, ψ is Hölder continuous on $\partial D(\gamma_0, \varepsilon)$ with possible exceptions at the points γ_1, γ_2. Accordingly, there is a unique solution of

(21.6), which is of class $C^2(D(\gamma_0, \varepsilon))$ and of class $C^0(\bar{D}(\gamma_0, \varepsilon))$ with possible exceptions at γ_1 and γ_2. Suppose that the above assertion has been proved. Then the function,

$$\psi^* \equiv \begin{cases} \psi_\varepsilon \text{ in } D(\gamma_0, \varepsilon) \\ \psi \text{ in } \bar{G} - D(\gamma_0, \varepsilon), \end{cases}$$

belongs to the admissible class F. Since ψ is the minimizing extremal, it is necessary that

$$\begin{aligned} &\iint_G [(\operatorname{grad}\psi)^2 - 4\mu\,\theta_0\,\psi]\,dx\,dy - \iint_G [(\operatorname{grad}\psi^*)^2 - 4\mu\,\theta_0\,\psi^*]\,dx\,dy \\ &\quad = \iint_{D(\gamma_0,\varepsilon)} [(\operatorname{grad}\psi)^2 - 4\mu\,\theta_0\,\psi]\,dx\,dy - \iint_{D(\gamma_0,\varepsilon)} [\operatorname{grad}\psi^* - 4\mu\,\theta_0\,\psi^*]\,dx\,dy \leqq 0. \end{aligned} \tag{21.7}$$

On the other hand, ψ_ε is the solution of (21.6). It also follows that

$$\iint_{D(\gamma_0,\varepsilon)} [(\operatorname{grad}\psi^*)^2 - 4\mu\,\theta_0\,\psi^*]\,dx\,dy - \iint_{D(\gamma_0,\varepsilon)} [(\operatorname{grad}\psi)^2 - 4\mu\,\theta_0\,\psi]\,dx\,dy \leqq 0. \tag{21.8}$$

Thus, the equality sign in (21.7) and (21.8) must hold. Consequently, we conclude from the continuity and the uniqueness of the solution of problem (21.6) that $\psi = \psi_\varepsilon$ identically in $D(\gamma_0, \varepsilon)$. Hence, to show that $\psi < \Psi$ along the smooth part of Γ it suffices to show that for every such a regular point γ_0 on Γ, $\psi_\varepsilon < \Psi$ in $D(\gamma_0, \varepsilon)$ provided ε is sufficiently small.

Instead of the function ψ_ε, we now consider the function φ_ε that is the solution of the Dirichlet problem:

$$\Delta\varphi_\varepsilon = -2\mu\,\theta_0 \quad \text{in } D(\gamma_0, \varepsilon), \qquad \varphi_\varepsilon = \Psi \quad \text{on } \partial D(\gamma_0, \varepsilon). \tag{21.9}$$

If we can show that $\varphi_\varepsilon < \Psi$ in $D(\gamma_0, \varepsilon)$, then $\psi_\varepsilon < \Psi$ in $D(\gamma_0, \varepsilon)$. To see this we note that

$$\Delta(\varphi_\varepsilon - \psi_\varepsilon) = 0 \quad \text{in } D(\gamma_0, \varepsilon), \qquad \varphi_\varepsilon - \psi_\varepsilon = \Psi - \psi \geqq 0 \quad \text{on } \partial D(\gamma_0, \varepsilon). \tag{21.10}$$

Although $\Psi - \psi$ might have discontinuities on $\partial D(\gamma_0, \varepsilon)$ at the points γ_1, γ_2 where it intersects Γ, the unique solution of (21.9) is still given by the Poisson integral formula,

$$(\varphi_\varepsilon - \psi_\varepsilon)(r, \theta) = \int_0^{2\pi} \frac{(\Psi - \psi)(r, \omega)\,d\omega}{1 - 2\left(\frac{r}{\varepsilon}\right)\cos(\theta - \omega) + \left(\frac{r}{\varepsilon}\right)^2}, \qquad 0 \leqq r < \varepsilon,$$

which is clearly non-negative in $D(\gamma_0, \varepsilon)$. Thus, we need show only that if Γ is smooth at γ_0 then there is an $\varepsilon_0 > 0$ such that

$$\varphi_\varepsilon < \Psi \quad \text{in } D(\gamma_0, \varepsilon) \quad \text{for all} \quad 0 < \varepsilon \leqq \varepsilon_0. \tag{21.11}$$

To establish the inequality (21.11), we consider the function,

$$\chi^* \equiv \varphi_\varepsilon - \Psi^* \quad \text{in } D(\gamma_0, \varepsilon),$$

where Ψ^* is the extension of the restriction of Ψ to D^* according to the first equation in (21.4). Then it follows from (21.5) and (21.9) that

$$\Delta\chi^* = -2\mu\,\theta_0 + k\varkappa(s^*)/(1 - t\varkappa(s^*)) \geqq -2c^+ \quad \text{in } D(\gamma_0, \varepsilon), \tag{21.12}$$

where c^+ is an non-negative constant independent of the small number ε. Also, it follows from (21.9) and (21.4) that

$$\chi^* = 0 \quad \text{on } \partial D^* - D(\gamma_0, \varepsilon), \qquad \chi^* = \Psi - \Psi^* < 0 \quad \text{on } \partial D^{**} - \bar{D}(\gamma_0, \varepsilon). \tag{21.13}$$

Now, we write $\chi^* = \chi_1^* + \chi_2^*$ such that

$$\Delta\chi_1^* = 0 \quad \text{in } D(\gamma_0, \varepsilon), \qquad \chi_1^* = \Psi - \Psi^* \quad \text{on } \partial D(\gamma_0, \varepsilon), \tag{21.14}$$

and

$$\begin{aligned} \Delta\chi_2^* &= -2\mu\,\theta_0 + k\varkappa(s^*)/[1 - t\varkappa(s^*)] \quad \text{in } D(\gamma_0, \varepsilon) \\ \chi_2^* &= 0 \quad \text{on } \partial D(\gamma_0, \varepsilon). \end{aligned} \tag{21.15}$$

Then the unique solution of problem (21.15) is subject to the estimate

$$\chi_2^* < c^+(\varepsilon^2 - r^2), \qquad 0 \leqq r \leqq \varepsilon, \tag{21.16}$$

where c^+ is the constant (non-negative) in (21.12) and where $r^2 \equiv x^2 + y^2$ with y_0 being the origin of the coordinate system. By the mean-value theorem for the harmonic function χ_1^*,

$$\chi_1^*(0, \theta) = \frac{1}{2\pi} \int_0^{2\pi} [\Psi(\varepsilon, \omega) - \Psi^*(\varepsilon, \omega)]\, d\omega. \tag{21.17}$$

The inequality (21.16) shows that the values of χ_2^* in $D(\gamma_0, \varepsilon)$ is of the order $0(\varepsilon^2)$. On the basis of the exact relation (21.17) we are going to show that the value of χ_1^* in $D(\gamma_0, \varepsilon)$ is a constant multiple of ε with error of the order $0(\varepsilon^2)$. Here, $0(\varepsilon^n) \equiv \text{const.}\ \varepsilon^n$, $n = 1, 2, 3$, where the constant is independent of ε. To this end, we first obtain certain estimates for the values of $\Psi - \Psi^*$ on $\partial D(\gamma_0, \varepsilon)$.

Denote by T the unique tangent to Γ at γ_0 and by β the non-obtuse angle between the tangent T and the segment $\gamma_0\, s_0^*$, which is equal to that between T and the segment $\gamma_0\, s_0^{**}$. Let γ_0 be the origin of a plane polar coordinate system with the initial line coinciding with segment $\gamma_0\, s_0^*$ (see Fig. 2). By the mean-value theorem we have

$$\Psi^*(\varepsilon, \omega) = \Psi^*(0, \omega) + \varepsilon \operatorname{grad} \Psi(r_0, \omega) \cdot u_r, \qquad 0 < r_0 < \varepsilon, \tag{21.18}$$

where u_r stands for a unit vector along the radial line $\omega = \text{constant}$. By the very method for constructing Ψ^*,

$$|\operatorname{grad} \Psi^*(r, \omega)| = k \qquad \text{in } \bar{D}(\gamma_0, \varepsilon),$$

and Ψ^* is twice continuously differentiable in $\bar{D}(\gamma_0, \varepsilon)$. Hence, the direction of grad Ψ^* at every point in $D(\gamma_0, \varepsilon)$ will differ from that at γ_0 by an amount

$$\lambda^*(\varepsilon) = 0(\varepsilon) \qquad \text{as } \varepsilon \to 0. \tag{21.19}$$

Let $\mu_1(\varepsilon)$ and $\mu_2(\varepsilon)$ be the non-obtuse angles between T and the segments $\gamma_0\,\gamma_1$ and $\gamma_0\,\gamma_2$, respectively. Then

$$\mu_1(\varepsilon), \mu_2(\varepsilon) = 0(\varepsilon) \qquad \text{as } \varepsilon \to 0. \tag{21.20}$$

In terms of the quantities $\lambda^*(\varepsilon)$, $\mu_1(\varepsilon)$ and $\mu_2(\varepsilon)$, we can write (21.18) as follows:

$$\Psi^*(\varepsilon, \omega) = \Psi(0, \omega) + k\,\varepsilon \cos(\omega + \lambda^*), \qquad -\beta + \psi_1 \leqq \omega \leqq \pi - \beta - \mu_2. \tag{21.21}$$

By taking (21.19) into account, we see from (21.21) that

$$\Psi^*(\varepsilon, \omega) = \Psi(0, \omega) + k\,\varepsilon \cos\omega + 0(\varepsilon^2) \tag{21.22}$$

for $-\beta + \mu_1(\varepsilon) \leqq \omega \leqq \pi - \beta - \mu_2(\varepsilon)$. Completely similar reasoning shows that

$$\Psi^{**}(\varepsilon, \omega) = \Psi(0, \omega) + k\,\varepsilon \cos(2\beta + \omega) + 0(\varepsilon^2) \tag{21.23}$$

for $-\beta+\mu_1(\varepsilon)\leqq\omega\leqq\pi-\beta-\mu_2(\varepsilon)$. It follows from (21.22) and (21.23) and the definitions for Ψ^*, Ψ^{**} that

$$\begin{aligned}\Psi(\varepsilon,\omega)-\Psi^*(\varepsilon,\omega)&=\Psi^{**}(\varepsilon,\omega)-\Psi^*(\varepsilon,\omega)\\&=k\,\varepsilon\left[\cos(2\beta+\omega)-\cos\omega\right]+0(\varepsilon^2),\end{aligned}\tag{21.24}$$

for $-\beta+\mu_1(\varepsilon)\leqq\omega\leqq\pi-\beta-\mu_2(\varepsilon)$ and that the left-hand side is equal to zero on the remaining part of $\partial D(\gamma_0,\varepsilon)$.

By combining the estimates in (21.24) and Eq. (21.17) we have

$$\begin{aligned}\chi_1^*(0,\theta)&=\frac{1}{2\pi}\int\limits_{-\beta+\mu_1(\varepsilon)}^{\pi-\beta-\mu_2(\varepsilon)}k\,\varepsilon\left[\cos(2\beta+\omega)-\cos\omega+0(\varepsilon)\right]d\omega\\&=-\frac{k\,\varepsilon}{\pi}\left[\sin\beta\cos\mu_2(\varepsilon)+\sin\beta\cos\mu_1(\varepsilon)+0(\varepsilon)\right]\\&=-\frac{2k\,\varepsilon}{\pi}\sin\beta+0(\varepsilon^2),\qquad\text{as }\varepsilon\to0,\end{aligned}$$

where the relations in (21.20) have been used in deriving the last equality. This proves that $\chi_1^*(0,\theta)$ is negative and is of the order $0(\varepsilon)$ as $\varepsilon\to0$.

To derive an estimate for $\chi_1^*(r,\theta)$ in $D(\gamma_0,\varepsilon)$, we note that HOPF's maximum principle ensures that $\chi_1^*(r,\theta)<0$ in $D(\gamma_0,\varepsilon)$. Consequently, we have the following Harnack inequality [23] for χ_1^* at every point (r,θ) in $\bar{D}(\gamma_0,\varepsilon)$:

$$\frac{\varepsilon+r}{\varepsilon-r}\chi_1^*(0,\theta)\leqq\chi_1^*(r,\theta)\leqq\frac{\varepsilon-r}{\varepsilon+r}\chi_1^*(0,\theta)=-\frac{\varepsilon-r}{\varepsilon+r}\left[\frac{2k\,\varepsilon}{\pi}\sin\beta+0(\varepsilon^2)\right].\tag{21.25}$$

Finally, by combining the estimates in (21.25) and (21.16) we find for every point (r,θ) in $\bar{D}(\gamma_0,\varepsilon)$

$$\begin{aligned}\chi^*(r,\theta)&=\chi_1^*(r,\theta)+\chi^*(r,\theta)\\&\leqq-\frac{\varepsilon-r}{\varepsilon+r}\left[\frac{2k\,\varepsilon}{\pi}\sin\beta+0(\varepsilon^2)\right]+c^+(\varepsilon^2-r^2)\\&=-(\varepsilon-r)\left[\frac{k}{\pi}\sin\beta-c^+(\varepsilon+r)+0(\varepsilon^2)\right]\end{aligned}\tag{21.26}$$

as $\varepsilon\to0$. Since $k>0$, $0<\beta\leqq\pi/2$ are both fixed, c^+ is a non-negative constant independent of r and ε, and we can choose a positive ε so small that the value of the expression within the square bracket is positive. This proves that for all sufficiently small ε,

$$\chi^*\equiv\varphi_\varepsilon-\Psi^*=\varphi_\varepsilon-\Psi<0\quad\text{in }D^*(\gamma_0,\varepsilon).$$

Completely parallel analysis shows that for all small ε,

$$\chi^{**}\equiv\varphi_\varepsilon-\Psi^{**}=\varphi_\varepsilon-\Psi<0\quad\text{in }D^{**}(\gamma_0,\varepsilon).$$

The above two inequalities combined imply that

$$\varphi_\varepsilon<\Psi\quad\text{in }D(\gamma_0,\varepsilon),$$

which is what we set out to prove.

Case (2). Let γ_0 be a point on Γ where no unique tangent to Γ can be defined. There are only two subcases.

Subcase (i). There is one and only one point s_0 on ∂G such that

$$\varrho(\gamma_0,s_0)=\varrho(\gamma_0,\partial G).$$

Under this circumstance, $\varrho(\gamma_0, s_0)$ is equal to the radius of curvature of ∂G at s_0 and $\varkappa(s_0)$ is a proper local maximum [41]. For example, γ_0 is an end point of the ridge of the domain enclosed by an ellipse. Consider the parametric equations for Γ in a neighborhood of s_0,

$$x = f(s) + \xi(s)\, n_x(s), \quad y = g(s) + \xi(s)\, n_y(s), \tag{21.27}$$

where $(f(s), g(s))$ are coordinates of the point s on ∂G and $\xi(s)$ stands for the distance between $(f(s), g(s))$ and the point (x, y) on Γ. These two equations define a Jordan arc with γ_0 as an end point. More specifically, the Jordan arcs described by (21.27) for $s > s_0$ and $s < s_0$ coincide, because s_0 is the only point on ∂G such that $\varrho(\gamma_0, s_0) = \varrho(\gamma_0, \partial G)$. It follows immediately that for all sufficiently small ε the disk $D(\gamma_0, \varepsilon)$ with center at γ_0 and radius ε intersects $D(\gamma_0, \varepsilon)$ in a Jordan arc with one end at γ_0 and with the other end γ_1 on $\partial\big(D(\gamma_0, \varepsilon)\big)$. This means that the disk $D(\gamma_0, \varepsilon)$ can be completely covered by the inward normals of ∂G in a neighborhood of s_0. Accordingly, Ψ is twice continuously differentiable in $D(\gamma_0, \varepsilon)$ except along the arc $\gamma_0\gamma_1$ of Γ, and

$$\Delta\Psi = -\frac{k\varkappa(s)}{1 - t\varkappa(s)} \quad \text{in } D(\gamma_0, \varepsilon) - \gamma_0\gamma_1.$$

Since $\xi(s_0) = 1/\varkappa(s_0)$, $\Delta\Psi$ becomes negatively infinite as the point γ_0 is approached. By the continuity of $\varkappa(s)$,

$$-\Delta\Psi - 2\mu\,\theta_0 > \text{constant } c > 0 \quad \text{in } D(\gamma_0, \varepsilon) - \gamma_0\gamma_1.$$

Let ψ_ε be the function of class $C^2(D - \gamma_0\gamma_1) \cap C^0(\overline{D - \gamma_0\gamma_1})$ such that

$$\Delta\psi_\varepsilon = -2\mu\,\theta_0 \quad \text{in } D - \gamma_0\gamma_1, \quad \psi_\varepsilon = \psi \quad \text{on } \partial\big(D(\gamma_0, \varepsilon) - \gamma_0\gamma_1\big).$$

Then the function $\chi \equiv \psi_\varepsilon - \Psi$ satisfies the equations,

$$\Delta\chi > c > 0 \quad \text{in } D(\gamma_0, \varepsilon) - \gamma_0\gamma_1, \quad \chi \leqq 0 \quad \text{on } \partial D + \gamma_0\gamma_1.$$

Hence, an application of maximum principles shows that $\chi < 0$ in $D(\gamma_0, \varepsilon) - \gamma_0\gamma_1$. By the same reasoning as before, we conclude that $\psi_\varepsilon = \psi$ in $D(\gamma_0, \varepsilon)$. Since ψ is known to be less than Ψ along $\gamma_1\gamma_0$ except at γ_0, and since ψ satisfies the equation $\Delta\psi = -2\mu\,\theta_0$ in a neighborhood of the arc $\gamma_0\gamma_1$, by the results for case (1) it follows that

$$\Delta\psi = -2\mu\,\theta_0 \quad \text{in } D(\gamma_0, \varepsilon) - \gamma_0, \quad \psi < \Psi \quad \text{in } D(\gamma_0, \varepsilon) - \gamma_0.$$

Furthermore, by the theorem on removal of singularities for harmonic functions, we know that ψ can be so defined at γ_0 that

$$\Delta\psi = -2\mu\,\theta_0 \quad \text{in } D(\gamma_0, \varepsilon), \quad \psi(\gamma_0) \leqq \Psi(\gamma_0).$$

The fact that $\psi(\gamma_0) < \Psi(\gamma_0)$ can now be proved by a slight modification of ∂G at s_0, but also it can be seen readily after we have established the smoothness of the extremal ψ in the entire domain G.

Subcase (ii). There is more than one point s_0 on ∂G such that

$$\varrho(\gamma_0, s_0) = \varrho(\gamma_0, \partial G).$$

However, Γ can have only a finite number of branches at γ_0, and hence the circumference of a small disk centered at γ_0 intersects Γ at a finite number of points, say $\gamma_1, \ldots, \gamma_j$. Let $s_1, s_2, \ldots, s_j$ be points on ∂G such that

$$\varrho(\gamma_i, s_i) = \varrho(\gamma_i, \partial G), \quad i = 1, 2, \ldots, j.$$

Consider two neighboring arcs $\gamma_0\gamma_i, \gamma_0\gamma_{i+1}$. If $s_i = s_{i+1}$, then s_i is a reentrant corner of G. Under this circumstance, the arc $\gamma_i\gamma_0\gamma_{i+1}$ of Γ is defined parametrically by

$$x = f(s_i) + \xi(\theta)\cos\theta, \quad y = g(s_i) + \xi(\theta)\sin\theta, \quad \theta_i \leqq \theta \leqq \theta_{i+1},$$

where $|\theta_{i+1} - \theta_i|$ is the angle between the two extreme inward normals at $s_i = s_{i+1}$. By the same analysis as that for case (1), one concludes that $\psi < \Psi$ in the closure of the sector $\gamma_i\gamma_0\gamma_{i+1}$ and $\Delta\psi = -2\mu\,\theta_0$ in its interior.

Suppose that $s_i \neq s_{i+1}$. Without loss of generality, we may assume that $s_i s_{i+1}$ is a smooth arc of ∂G. Thus, the two neighboring arc $\gamma_0\gamma_i$ and $\gamma_0\gamma_{i+1}$ are defined by the parametric equations

$$x = f(s) + \xi(s)\, n_x(s), \quad y = g(s) + \xi(s)\, n_x(s), \quad s_i \leqq s \leqq s_{i+1}.$$

If $1/\varkappa(s) > \xi(s)$ for $s_i \leqq s \leqq s_{i+1}$ then for all sufficiently small ε the disk $D(\gamma_0, \varepsilon)$ centered at γ_0 with radius ε can be completely covered by the inward normals to the arc $s_i s_{i+1}$ of ∂G. Under this circumstance one can apply the same analysis as that for the case (1) to conclude that $\psi < \Psi$ in the closure of the sector $\gamma_i\gamma_0\gamma_{i+1}$ and $\Delta\psi = -2\mu\,\theta_0$ in its interior. If the arc $s_i s_{i+1}$ of ∂G contains a circular arc $s_i' s_{i+1}'$ with its center at γ_0, let γ_i', γ_{i+1}' be the intersection points of $\partial D(\gamma_0, \varepsilon)$ with the segments $\gamma_0 s_i'$ and $\gamma_0 s_{i+1}'$, respectively. Then by the same reasoning as for subcase (i) one concludes that $\psi < \Psi$ and $\Delta\psi = -2\mu\,\theta_0$ in the interior of the sector $\gamma_i'\gamma_0\gamma_{i+1}'$. By slight modification of the arcs $s_i s_i', s_{i+1} s_{i+1}'$ of ∂G near the points s_i' and s_{i+1}' one concludes, by the same reasoning as that for case (1), that $\psi < \Psi$ and $\Delta\psi = -2\mu\,\theta_0$ in the interior of the sectors $\gamma_i\gamma_0\gamma_i'$ and $\gamma_{i+1}\gamma_0\gamma_{i+1}'$. The case when the subarc $s_i s_{i+1}$ of ∂G contains a finite number of disjoint circular arcs with their center at γ_0 can be treated in a similar way. Thus we have proved that $\psi < \Psi$ in a complete neighborhood of Γ, with possible exceptions at those isolated points γ on Γ for each of which there is one and only one point s on ∂G such that $\varrho(\gamma, s) = \varrho(\gamma, \partial G)$. Our proof also shows that in a complete neighborhood of Γ the extremal ψ is twice continuously differentiable and satisfies the Poisson equation $\Delta\psi = -2\mu\,\theta_0$ there.

22. Continuity of stress. Suppose that the extremal ψ is a strict solution of the elastic-plastic torsion problem. Then the components of stress that do not vanish identically are partial derivatives of ψ. Accordingly, we need to establish the smoothness of ψ in G so as to assure the continuity of stress components.

In [*63*] LEWY and STAMPACCHIA established profound theorems of regularity for the solutions of elliptic inequalities under the assumption that the majorant function belongs to $H_p^2(G)$ with p being an integer greater than the dimensions of the space. Their result cannot be applied directly here, because grad Ψ is discontinuous across the ridge Γ and hence does not belong to $H_p^2(G)$ unless G is a circular disk. However, the existence of an elastic core $N(\Gamma)$ has been established, and the extremal ψ is actually analytic there. Accordingly, we need only to show that ψ is smooth in $G - N(\Gamma)$.

To this end we first note that the ridge Γ partitions G into a finite number of simply connected subdomains, say $G_1, \ldots, G_n$. Each of the subdomains G_j is bounded by a portion Γ_j of the ridge Γ and a portion of ∂G between two consecutive regular (non-reentrant) corners, say s_j and s_{j+1}. Further, at each regular corner s_j, there is a number $\varepsilon_j > 0$ such that in the intersection of G with the disk $D(s_j, \varepsilon_j)$ with radius ε_j and centered at s_j, the extremal ψ is strictly less than Ψ. For by looking back at the estimate in (21.26) we see that near the corner s_j the angle β is bounded away from zero and $\pi/2$ while the other terms are bounded by a constant multiple of ε. Thus, there is a positive ε_j such that for all points on Γ near the corner s_j, the estimate in (21.26) holds for all $\varepsilon \leqq \varepsilon_j$.

We say that the boundary ∂D of a domain D is of class $C_{2+\alpha}$ if ∂D can be covered by a finite number of disks such that in each of these disks ∂D has a twice continuously differentiable representation and that all derivatives up to and including to the second order are Hölder continuous with exponent α, $0<\alpha<1$. Since the ridge Γ is contained in an elastic core $N(\Gamma)$ and since the intersections of G_j with small disks at the corners s_j and s_{j+1} are also elastic regions, i.e., $\psi<\Psi$ there, we can draw a curve Γ_j^* of class $C_{2+\alpha}$ such that

(i) Γ_j^* and the portion of ∂G between s_j and s_{j+1} enclose a simply connected subdomain G_j^* of G_j,

(ii) ∂G_j^* is of class $C_{2+\alpha}$, $0<\alpha<1$,

(iii) Γ_j^* lies in the intersection of the elastic core and G_j, i.e. $\psi<\Psi$ along the curve Γ_j^*. For example, we may first map G_j conformally onto a unit disk and then draw a circular arc lying in the image of the elastic core in G_j and then connect the circular arc in a sufficiently smooth way to the image of ∂G between s_j and s_{j+1} so as to form a closed curve of class $C_{2+\alpha}$. The desired curve Γ_j^* is simply the inverse image of the circular arc with smoothly filled ends.

Consider now the restriction of ψ on ∂G_j^*. ψ vanishes identically on $\partial G \cap \partial G_j^*$. Since Γ_j^* lies in the elastic core, ψ is analytic along Γ_j^*. Let s_j^* and s_{j+1}^* be the points where Γ_j^* meets ∂G. Since ∂G_j^* is of class $C_{2+\alpha}$ and s_j and s_{j+1} are boundary points of elastic core, it follows from Kellogg's theorem that ψ is of class $C_{2+\alpha}$ at the points s_j and s_{j+1}. Again Kellogg's theorem ensures that the solution of the Dirichlet problem:

$$\Delta u=0 \quad \text{in } G_j^*, \qquad u=\psi \quad \text{on } \partial G_j^*, \tag{22.1}$$

is of class $C_{2+\alpha}(\bar{G}_j^*)$ i.e., the derivatives of u up to and including the second order are Hölder continuous in $\bar{G}_j^*$ with exponent α. This shows that the restriction of ψ of ∂G_j^* can be extended to the whole domain G_j^* so that it belongs to $C_{2+\alpha}(\bar{G}_j^*)$.

Since the subdomain G_j^* does not touch Γ, Ψ is of class $C_2(\bar{G}_j^*)$ if there is no reentrant corner between s_j and s_{j+1}. In any case, Ψ is continuously differentiable and piecewise twice continuously differentiable in $\bar{G}_j^*$. Hence, Ψ belongs to $H_p^2(G_j^*)$ for all $p\geqq 2$, where $H_p^2(G_j^*)$ stands for the Banach space of functions whose distribution derivatives up to and including the second order are in $L_p(G_j^*)$. We shall denote by $\|u\|_{2,p}$ the norm of u in $H_p^2(G_j^*)$:

$$\|u\|_{2,p}=\left\{\iint\limits_{G_j^*}\left[\sum_{i,j=1}^{2}\left|\frac{\partial^2 u}{\partial x_i\,\partial x_j}\right|^p+\sum_{i=1}^{2}\left|\frac{\partial u}{\partial x_i}\right|^p+|u|^p\right]dx_1\,dx_2\right\}^{1/p}.$$

We are now ready to show that the extremal ψ possesses Hölder continuous first derivatives in G. Our proof is a slight variation of the one given by Lewy and Stampacchia to construct a sequence of functions $\{\psi_\varepsilon\}$ in $H_p^2(G_j^*)$ which converge together with their first derivatives uniformly to the extremal ψ in G_j^*. To this end, we consider for each $\varepsilon>0$ the nonlinear Dirichlet problem.

$$\varepsilon\,\Delta\psi_\varepsilon=(\psi_\varepsilon-\psi)^{\frac{1}{p-1}}+\varepsilon\,\Delta\Psi \quad \text{in } G_j^*, \qquad \psi_\varepsilon=\psi \quad \text{on } \partial G_j^*, \tag{22.2}$$

where $p>2$ is an *even* integer. To assure that this problem has a solution in $H_p^2(G_j^*)$ for every $\varepsilon>0$ we consider the linear problem

$$\varepsilon\,\Delta u=(w-\psi)^{\frac{1}{p-1}}+\varepsilon\,\Delta\Psi \quad \text{in } G_j^*, \qquad u=\psi \quad \text{on } \partial G_j^*, \tag{22.3}$$

where w is any given function in $H_p^1(G_j^*)$. Under the present circumstances, it is known that problem (22.3) has a unique solution $H_p^2(G_j^*)$ and that it is subject to the L_p-estimates,

$$\|u\|_{2,p} \leqq c(p, G_j^*)\left\{\left\|(w-\psi)^{\frac{1}{p-1}}\right\|_{0,p} + \|\varepsilon\, \Delta\Psi\|_{0,p} + \|\psi\|_{2-\frac{1}{p},p}\right\}, \tag{22.4}$$

where

$$\|\varphi\|_{0,p} \equiv \Big[\iint\limits_{G_j^*} \varphi^p\, dx\, dy\Big]^{1/p}, \qquad \|\psi\|_{2-\frac{1}{p},p} \equiv \text{g.l.b}\ \|v\|_{2,p}$$

and the g.l.b. is taken over all functions v in $H_p^2(G_j^*)$ which equal ψ on ∂G_j^*. Note that $\|\psi\|_{2-\frac{1}{p},p}$ is indeed finite.

Since the problem (22.3) always has a unique solution u for arbitrarily assigned w in $H_p^1(G_j^*)$, it defines a mapping T of $H_p^1(G_j^*)$ into $H_p^2(G_j^*)$, i.e.,

$$u = T(w) \quad \text{in } H_p^2(G_j^*) \quad \text{for all } w \text{ in } H_p^1(G_j^*).$$

We assert that as a mapping of $H_p^1(G_j^*)$ into itself T is compact. Indeed, the L_p-estimate in (22.5) indicates that T carries a bounded set in $H_p^1(G_j^*)$ into a bounded set in $H_p^2(G_j^*)$, which is, according to the Rellich selection principle, a compact set in $H_p^1(G_j^*)$.

Next we assert that T maps $H_p^1(G_j^*)$ *continuously* into itself.

Suppose that $w_m \to w$ in $H_p^1(G_j^*)$. Let $u_m = T(w_m)$, $u = T(w)$ and $v = u_m - u$. It follows from (22.3) that

$$\varepsilon\, \Delta v = (w_m - \psi)^{\frac{1}{p-1}} - (w-\psi)^{\frac{1}{p-1}} \quad \text{in } G_j^*, \qquad v = 0 \quad \text{on } \partial G_j^*. \tag{22.5}$$

Applying the L_p-estimates to the solution v of (22.5), we find that

$$\|v\|_{2,p} \equiv c(p, G_j^*)\left\{\left\|(w_m-\psi)^{\frac{1}{p-1}} - (w-\psi)^{\frac{1}{p-1}}\right\|_{0,p}\right\}. \tag{22.6}$$

It is easy to check that for any positive numbers σ_1, σ_2, $0 \leqq \mu \leqq 1$, the following inequalities hold:

$$\frac{\sigma_1^\mu + \sigma_2^\mu}{(\sigma_1+\sigma_2)^\mu} \leqq 2^{1-\mu}, \qquad \frac{\sigma_1^\mu - \sigma_2^\mu}{|\sigma_1-\sigma_2|^\mu} \leqq 1 \quad \text{with} \quad \sigma_1 \neq \sigma_2.$$

Hence if $(w_m - \psi)$ and $(w-\psi)$ are of the same sign, then

$$\left|(w_m-\psi)^{\frac{1}{p-1}} - (w-\psi)^{\frac{1}{p-1}}\right| \leqq |w_m - w|^{\frac{1}{p-1}};$$

and if $(w_m-\psi)$ and $(w-\psi)$ are of the opposite sign, then

$$\left|(w_m-\psi)^{\frac{1}{p-1}} - (w-\psi)^{\frac{1}{p-1}}\right| \leqq 2^{1-\frac{1}{p-1}} |w_m - w|^{\frac{1}{p-1}} \leqq 2|w_m-w|^{\frac{1}{p-1}}.$$

Consequently, we have

$$\begin{aligned}
\left\|(w_m-\psi)^{\frac{1}{p-1}} - (w-\psi)^{\frac{1}{p-1}}\right\|_{0,p} &= \Big\{\iint\limits_{G_j^*} \left|(w_m-\psi)^{\frac{1}{p-1}} - (w-\psi)^{\frac{1}{p-1}}\right|^p dx\, dy\Big\}^{\frac{1}{p}} \\
&\leqq \Big\{\iint\limits_{G_j^*} |w_m - w|^{\frac{p}{p-1}} dx\, dy\Big\}^{\frac{1}{p}} \\
&\leqq \text{const.}\, [\|w_m - w\|_{0,p}]^{\frac{1}{p-1}}.
\end{aligned}$$

The continuity of T now follows from this estimate and that in (22.6).

From the estimate in (22.4) we have

$$\begin{aligned}\|u\|_{2,p} &\leqq c\left\{c_1(\|w-\psi\|_{0,p})^{\frac{1}{p-1}}+\|\varepsilon\,\Delta\Psi\|_{0,p}+\|\psi\|_{2-\frac{1}{p},p}\right\}\\ &\leqq c\left\{c_1(\|w\|_{0,p})^{\frac{1}{p-1}}+c_1(\|\psi\|_{0,p})^{\frac{1}{p-1}}+\|\varepsilon\,\Delta\Psi\|_{0,p}+\|\psi\|_{2-\frac{1}{p},p}\right\},\end{aligned} \tag{22.7}$$

where c and c_1 are constants depending on p and G_j^*. Now let

$$m=\max\left\{1,\, c_1(\|\psi\|_{0,p})^{\frac{1}{p-1}}+\|\varepsilon\,\Delta\Psi\|_{0,p}+\|\psi\|_{2-\frac{1}{p},p}\right\}.$$

Let B be the bounded set in $H_p^1(G_j^*)$ such that for w in B

$$\|w\|_{1,p}\leqq c\,N m,$$

where c is the same constant in (22.7) and N is a number to be fixed. For w in $B\subset H_p^1(G^*)$ and for $u=T(w)$ we conclude from (22.7) that

$$\|u\|_{2,p}\leqq c\left\{c_1(N m)^{\frac{1}{p-1}}+m\right\}.$$

Accordingly, $\|u\|_{2,p}\leqq c\,Nm$ if N is chosen so large that

$$\left[c_1(c)^{\frac{1}{p-1}}/N^{\frac{p-2}{p-1}}\right]+\frac{1}{N}\leqq 1. \tag{22.8}$$

Now we fix B by choosing N to satisfy (22.8). Then $T(B)\subset B$. Thus T maps B into itself and is continuous and compact. By Schauder's fixed-point theorem, T has a fixed point. The existence of a solution to (22.2) for every $\varepsilon>0$ is now established.

Consider now a solution of (22.2). We assert that $\psi_\varepsilon\leqq\Psi$ in G_j^*. Indeed, $(\psi_\varepsilon-\Psi)$ solves the problem,

$$\varepsilon\,\Delta(\psi_\varepsilon-\Psi)=(\psi_\varepsilon-\psi)^{\frac{1}{p-1}}\quad\text{in } G_j^*,\qquad \psi_\varepsilon-\Psi\leqq 0\quad\text{on } \partial G_j^*. \tag{22.9}$$

Since $\psi_\varepsilon-\psi$ is Hölder continuous in G_j^* and the $\psi_\varepsilon-\Psi$ is continuous on ∂G_j^*, it follows that $\psi_\varepsilon-\Psi$ is twice continuously differentiable in G_j^*. If it should achieve an interior positive maximum at some point q in G_j^*, then (22.9) implies that at the point q

$$0>\varepsilon\,\Delta(\psi_\varepsilon-\Psi)=(\psi_\varepsilon-\psi)^{\frac{1}{p-1}}>(\Psi-\psi)^{\frac{1}{p-1}}\geqq 0$$

which is a contradiction. Our assertion is thereby proved.

Since $\psi_\varepsilon\leqq\Psi$ in G_j^* and $\psi_\varepsilon=\psi$ on ∂G_j^*, if we extend ψ_ε to the whole domain G by setting $\psi_\varepsilon=\psi$ in $G-G_j^*$, then it belongs to the admissible family F. Using this fact we proceed to show that ψ_ε stays in a bounded set of $H_p^2(G_j^*)$. For every λ, $0\leqq\lambda\leqq 1$, the function, $(1-\lambda)\,\psi+\lambda\,\psi_\varepsilon$, belongs to F if $\psi_\varepsilon=\psi$ in $G-G_j^*$. Accordingly, we have the variational condition

$$\frac{d}{d\lambda}\,J[(1-\lambda)\,\psi+\lambda\,\psi_\varepsilon]\geqq 0\quad\text{at}\quad\lambda=0,$$

which leads to the inequality

$$\begin{aligned}&\iint\limits_{G_j^*}[-\operatorname{grad}(\psi_\varepsilon-\psi)\cdot\operatorname{grad}\psi_\varepsilon+2\mu\,\theta_0(\psi_\varepsilon-\psi)]\,dx\,dy\\ &\qquad\leqq-\iint\limits_{G_j^*}[\operatorname{grad}(\psi_\varepsilon-\psi)]^2\,dx\,dy\leqq 0.\end{aligned} \tag{22.10}$$

An application of the divergence theorem to (22.10) yields

$$\iint_{G_j^*} (\Delta\psi_\varepsilon)(\psi_\varepsilon - \psi)\, dx\, dy \leqq -2\mu\,\theta_0 \iint_{G_j^*} (\psi_\varepsilon - \psi)\, dx\, dy. \tag{22.11}$$

By substituting (22.2) into (22.11) we find

$$\iint_{G_j^*} (\Delta\psi_\varepsilon)(\Delta\psi_\varepsilon - \Delta\Psi)^{p-1}\, dx\, dy \leqq -2\mu\,\theta_0 \iint_{G_j^*} (\Delta\psi_\varepsilon - \Delta\Psi)^{p-1}\, dx\, dy.$$

Thus, repeated application of the Hölder inequality to the above estimate leads to

$$\begin{aligned}(\|\Delta\psi_\varepsilon\|_{0,p})^p &\leqq \iint_{G_j^*}\left[\sum_{j=0}^{p-1} b_j |\Delta\psi_\varepsilon|^{p-1-j}\,|\Delta\Psi|^j + \sum_{j=1}^{p-1} c_j |\Delta\psi_\varepsilon|^{p-j}\,|\Delta\psi|^j\right] dx\, dy\\ &\leqq \sum_{j=1}^{p} d_j (\|\Delta\psi_\varepsilon\|_{0,p})^{p-j},\end{aligned}$$

where b_j, c_j and d_j are constants independent of ε. Clearly, for the last inequality to hold it is necessary that for all $\varepsilon > 0$

$$\|\Delta\psi_\varepsilon\|_{0,p} \leqq \text{fixed constant}. \tag{22.12}$$

Since the domain G_j^* is of class $C_{2+\alpha}$, we conclude from the L_p-estimate that for all $\varepsilon > 0$

$$\|\Delta\psi_\varepsilon\|_{2,p} \leqq \text{fixed constant}.$$

We recall now that, by a well known lemma of SOBOLEV and by MORREY's theorem on the Dirichlet growth condition, a function v of $H_p^2(G_j^*)$ with $p > 2$, has Hölder continuous first derivatives with exponent $1-2/p$. More precisely,

$$[\psi_{\varepsilon x}]_{0,1-\frac{2}{p}},\ [\psi_{\varepsilon y}]_{0,1-\frac{2}{p}} \leqq c' \|\psi_\varepsilon\|_{2,p}$$

where c' is independent of ε and

$$[\varphi]_{0,\alpha} \equiv \sup_{q,q' \text{ in } G_j^*} \frac{|\varphi(q) - \varphi(q')|}{|q - q'|^\alpha}.$$

Thus, the first derivatives of $\{\psi_\varepsilon\}$ are all Hölder continuous with the same exponent $1-2/p$ and with the same Hölder constant. Accordingly, ψ_ε together with its first derivatives converges uniformly to a limit function having Hölder continuous first derivatives. From (22.2) and (22.12) we see that ψ_ε actually converges to the extremal ψ in G_j^*. Hence ψ possesses Hölder continuous derivatives in G_j^*. Since ψ is known to be analytic in the elastic core $N(\Gamma)$, it follows that ψ belongs to $C_{1+\alpha}(G_j)$. Further, since $G = \cup G_j$ and the same analysis is applicable to each of the subdomains G_j, it follows that ψ belongs to $C_{1+\alpha}(G)$. The proof is now complete.

23. Natural partition of the cross section. We proceed to partition the cross section G of the cylindrical bar into an elastic part E and a plastic part P, according to which branch of the constitutive relation holds in each. Since both the majorant function Ψ and the extremal ψ are continuous in $\bar{G}$, we can define

$$E = \{q \mid E \text{ in } \bar{G},\ \psi(q) < \Psi(q)\}, \qquad P = \{q \mid q \text{ in } \bar{G},\ \psi(q) = \Psi(q)\}. \tag{23.1}$$

Then E is open and P is closed in $\bar{G}$, and they constitute a partition of $\bar{G}$. It will be shown that E is the elastic "region" and P is the plastic part of G in the usual sense.

(i) *Linear elasticity holds in E.* We show that the material in E obeys the linear theory of elasticity by proving that ψ satisfies the Poisson equation $\Delta\psi = -2\mu\,\theta_0$ in E in the strict sense. In fact, ψ is strictly less than Ψ in E, and E is open. Hence localized small variations can be chosen arbitrarily in E. Because of this fact one can derive an integral representation for the extremal ψ from the variational condition [*42*, § 6]. The integral representation for ψ will then show that ψ is of class $C^2(E)$ and is a solution of the Poisson equation there.

(ii) *The yield point is not reached in E.* We wish to show that the material in E stays below the yield point. To this end we recall that for each of the isolated points γ on Γ, for which there is one and only one point s on ∂G with

$$\varrho(\gamma, s) = \varrho(\gamma, \partial G),$$

the extremal ψ satisfies the Poisson equation in a disk centered at γ and that $\psi < \Psi$ in the disk punctured at γ. Denote by Γ_1 the finite set of these points on Γ and let $E^* = E + \Gamma_1$. Then E^* is also an open set, and each point γ_1 in Γ_1 is an interior point of E^*. For each γ_1 in Γ_1 is the center of some disk which is contained in $E^* = E + \Gamma_1$.

Now ψ satisfies the Poisson equation in E^*, and hence it is analytic there. Consequently, we can perform computation to check that

$$\Delta(\operatorname{grad}\psi)^2 = 2(\psi_{xx}^2 + 2\psi_{xy} + \psi_{yy}^2) \geqq 0 \quad \text{in } E^*. \tag{23.2}$$

On the other hand, ψ is smooth in G, $\psi \leqq \Psi$ in $\bar{G}$ and $|\operatorname{grad}\Psi| = k$ in $\bar{G}$, so we conclude that

$$(\operatorname{grad}\psi)^2 \leqq k^2 \quad \text{on } \partial E^*. \tag{23.3}$$

By the strong maximum principle we conclude from (23.2) and (23.3) that

$$(\operatorname{grad}\psi)^2 < k \quad \text{in } E^*.$$

That is, the material in E has not yielded. The results in (i) and (ii) now justify the statement that E is, indeed, an elastic region.

(iii) *Plasticity of P.* If $P - \partial G$ is non-empty, it is clear from (23.1) and (18.1) that

$$|\operatorname{grad}\psi|^2 = |\operatorname{grad}\Psi|^2 = k^2 \quad \text{in } P - \partial G,$$

which shows that P is plastic in the usual sense.

24. Some properties of E and P. (i) *An intersection property of P.* First we note that the set $E^* = E$, i.e., $\psi < \Psi$ at each point γ for which there is one and only one point s on ∂G with $\varrho(\gamma, s) = \varrho(\gamma, \partial G)$. For if this were not the case, then $\Psi(\gamma) = \psi(\gamma)$. But there is a point q on the segment γs such that $\psi(q) < \Psi(q)$. Accordingly, the smoothness of ψ in G ensures that there is a point q^* on the segment γq with the property that $|\operatorname{grad}\psi(q^*)| > k$, which is a contradiction. This also finishes our proof of the existence of an elastic core. The same reasoning also shows that *each inward normal of ∂G intersects the plastic part P in a single segment.* In particular, each of the inward normals at a reentrant corner of G intersects P in a single segment.

The above intersection property of P shows that P always adheres to ∂G. Since the torsion is always elastic provided the applied torque is small, we conclude that *yielding always starts from the lateral surface of the bar.* The last statement has been conjectured in the literature [*1*].

(ii) *Non-existence of completely plastic torsion.* We have shown that the elastic region E defined in (23.1) is the elastic region in the usual sense. Consequently, the existence of an elastic core for all values of θ_0 implies that the cross section G never becomes completely plastic no matter how large the value of θ_0 may be. Hence, completely plastic torsion cannot be realized in practice.

(iii) *Simple connectivity of* E. We have shown that the ridge Γ of G is properly contained in E. For every point q in E, let s be the point on ∂G such that

$$\varrho(q, s) = \varrho(q, \partial G).$$

Let γ be the intersection point of the extension of the segment sq with Γ. According to the intersection property of P, the segment $q\gamma$ is contained in E. Since Γ is known to be connected [*41*], it follows that the set E is simply connected. Our proof also describes qualitatively the relative position of E in G as well as the extent of E.

25. Elastic-plastic boundary. According to the intersection property of the plastic part P of G, an inward normal to ∂G at s intersects P in a single segment sq which may degenerate into a point on ∂G. Since the ridge Γ of G is contained in the elastic region E, the end point q of the segment sq must be such that $\varrho(q, s) = \varrho(q, \partial G)$. Consequently, if $\varrho(q, s) > 0$, then q lies on the elastic-plastic boundary,

$$l = \partial E \cap G = \partial P \cap \partial G.$$

For every point s on ∂G, we define the function $R(s)$ by

$$R(s) = \varrho(q, s), \qquad s \text{ on } \partial G, \tag{25.1}$$

with sq being the intersection of P with the inward normal of ∂G at s. For the moment let us represent the free boundary l formally as follows:

$$x = f(s) + R(s)\, n_x(s), \qquad y = g(s) + R(s)\, n_y(s), \tag{25.2}$$

where (x, y) are rectangular coordinates of the point q on l, $(f(s), g(s))$ are those of the point s on ∂G and n_x, n_y are respectively the x- and y-components of the unit inward normal of ∂G at s.

By using the intersection property of P and the fact that $\Delta\psi = -2\mu\,\theta_0$ in E, it has been proved [*67*, § *7*] that $R(s)$ *is actually a continuous function of* s. Since the functions f, g, n_x and n_y in (25.2) are all continuously differentiable in s, it follows that the *equations* (25.2) *indeed serve as a parametric representation for the elastic-plastic boundary* l, *if* $R(s) > s$.

It may be noted that *the continuity of* $R(s)$ *implies that the free boundary* l *does not contain any line segment perpendicular to* ∂G. Although the derivation of the representation (25.2) for l is not so trivial, we wish to know whether l consists of rectifiable curves so as to check the dynamical compatibility conditions across the elastic-plastic boundary. However, in the present general case, the rectifiability of the free boundary is an open question.

LEWY and STAMPACCHIA established the analyticity of the free boundary under the restriction that the majorant function be analytic and *strictly concave* [*63*]. If we restrict ∂G to be piecewise analytic, then the majorant function Ψ is analytic in $G - \Gamma$, and so is the modified majorant function

$$\Phi \equiv \Psi + \tfrac{1}{2}\mu\,\theta_0 (x^2 + y^2).$$

It is easy to show that $\Delta\Phi > 0$ in the interior of the plastic part P. However, it is not known whether $\Delta\Phi > 0$ everywhere along the free boundary l. Because of this open question, the results of LEWY and STAMPACCHIA cannot be applied directly. In the special case when ∂G consists of line segments only, $\Delta\Phi = 2\mu\,\theta_0$ everywhere in $G - \Gamma$. In this case, we may apply the theorem of LEWY and STAMPACCHIA to conclude that the *free boundary l consists of analytic arcs*. If G is a convex polygon, the analyticity of the free boundary has also been established in [*66*, § 8].

26. Dependence upon angle of twist. We shall determine how the extremal ψ depends upon the parameters μ and θ_0. Let ψ_1, ψ_2 be two extremals corresponding respectively to the angles of twist per unit length θ_1 and θ_2. Suppose that

$$\theta_2 > \theta_1 > 0. \tag{26.1}$$

We first show that

$$\psi_2(x, y) \geqq \psi_1(x, y) \quad \text{everywhere in } \bar{G}. \tag{26.2}$$

To this end, let E_1, P_1 and E_2, P_2 be, respectively, the elastic-plastic partitions of G by the extremals ψ_1 and ψ_2. Then

$$\begin{aligned} G &= E_1 + P_1 = E_2 + P_2, \\ G &= E_1 \cap E_2 + E_1 \cap P_2 + P_1 \cap E_2 + P_1 \cap P_2. \end{aligned} \tag{26.3}$$

According to the results established in Sect. 23, (ii),

$$\begin{aligned} &|\text{grad}\,\psi_1|^2 < k^2 \quad \text{in } E_1, \qquad |\text{grad}\,\psi_1|^2 = k^2 \quad \text{in } P_1, \\ &|\text{grad}\,\psi_2|^2 < k^2 \quad \text{in } E_2, \qquad |\text{grad}\,\psi_2|^2 = k^2 \quad \text{in } P_2. \end{aligned} \tag{26.4}$$

Now consider the function $\hat{\psi} \equiv \psi_2 - \psi_1$. It belongs to $C_0(\bar{G}) \cap C_{1+\alpha}(G)$. We assert that ψ is *nonnegative* in $\bar{G}$. For if such were not the case, ψ would have to achieve a *negative* interior minimum at some point q in G. Clearly, q does not belong to $P_1 \cap P_2$ because $\hat{\psi}$ vanishes identically there. If q belongs to the set $E_1 \cap P_2$, then we have the absurd inequality

$$\Psi(q) = \psi_2(q) < \psi_1(q) \leqq \Psi(q);$$

if q belongs to the set $E_2 \cap P_1$, then it is an interior point of E_2, but then we have, in view of (26.4), the absurd inequality

$$k^2 > |\text{grad}\,\psi_2(q)|^2 = |\text{grad}\,\psi_1(q)|^2 = k^2.$$

Thus q can belong only to the open set $E_1 \cap E_2$. However, for $\theta_2 > \theta_1$,

$$\Delta\psi = -2\mu(\theta_2 - \theta_1) < 0 \quad \text{in } E_1 \cap E_2.$$

Therefore, q cannot belong to the set $E_1 \cap E_2$ either. For if q were in $E_1 \cap E_2$, then necessarily $\Delta\hat{\psi}(q) \geqq 0$. Since there is no place for q in G as listed in (26.3), we conclude that $\hat{\psi}$ possesses no negative interior minimum in G. That is, the inequality (26.2) must hold. We proceed to refine this inequality and to explore its physical significance.

Consider the set $P_1 \cap E_2$. We assert that it is empty: For if there were a point q in this set, then it would follow from the very definitions of P_1 and E_2 that

$$\psi_1(q) = \Psi(q) > \psi_2(q).$$

But this contradicts the inequality (26.2), and the assertion follows. Now from the first equation in (26.3) we have

$$G \cap P_1 = E_1 \cap P_1 + P_1 \cap P_1 = E_2 \cap P_1 + P_2 \cap P_1.$$

Since both the set $E_1 \cap P_1$ and the set $E_2 \cap P_1$ are empty, this equation simply becomes $P_1 = P_2 \cap P_1$. It means that

$$P_1 < P_2, \quad E_1 > E_2. \tag{26.5}$$

This says that *the plastic portion of $\bar{G}$ grows as the angle of twist per unit length increases.*

Having established the growth property of the plastic portion P as was shown in (26.5), we can now derive more quantitative comparisons of the two extremals ψ_2 and ψ_1, namely,

$$\psi_2(x, y) > \psi_1(x, y) \quad \text{in } E_1. \tag{26.6}$$

This estimate is the best possible one in the sense that $\psi_2 = \psi_1 = \Psi$ in $\bar{G} - E_1 = P_1$. According to the existence theorem in Sect. 21 for the elastic core, the elastic region E_2 corresponding to the extremal ψ_2 has positive area for every finite angle of twist per unit length θ_2. Hence, it is meaningful to consider the restriction of $\hat{\psi} = \psi_2 - \psi_1$ in E_2. Then $\hat{\psi}$ belongs to $C_0(\bar{E}_2) \cap C_2(E_2)$, and it solves the Dirichlet problem

$$\Delta \hat{\psi} = -2\mu(\theta_2 - \theta_1) < 0 \quad \text{in } E_2, \qquad \hat{\psi} = \Psi - \psi_1 \geqq 0 \quad \text{on } \partial E_2. \tag{26.7}$$

If $\hat{\psi}$ assumed non-positive values in E_2, then it would achieve also a non-positive interior minimum at some point q in E_2. But then $\Delta \hat{\psi}(q) \geqq 0$, which contradicts the first inequality in (26.7). Thus we conclude that ψ_2 is strictly greater than ψ_1 in the elastic region E_2. Since in the open set $E_1 \cap P_2$ we have from the definitions for E_1 and P_2 that $\psi_2 = \Psi > \psi_1$ there, the strict inequality (26.6) is now established.

Since the elastic region E_1 has positive area, the estimate in (26.6) yields the mathematically trivial but physically significant inequality

$$\iint_G \hat{\psi}(x, y)\, dx\, dy = \iint_{E_1} \hat{\psi}(x, y)\, dx\, dy > 0, \tag{26.8}$$

which means that *the applied torque is a strictly monotone increasing function of the angle of twist per unit length, and conversely.*

To see how the plastic portion P of G grows as the angle of twist increases, we restrict our attention to the cases when the free boundary is twice continuously differentiable. For example, if G is a polygon, then the free boundary consists of analytic arcs. Let l_1 and l_2 be the elastic-plastic boundaries corresponding to the angles of twist θ_1 and θ_2 respectively. We refine the results in (26.5) by proving that *l_1 and l_2 have no point of contact* in $G = \bar{G} - \partial G$. For if l_1 and l_2 should have a point q in common which lies in G, then the inclusion relations in (26.5) would imply that the sufficiently smooth arcs l_1 and l_2 had a common tangent at the point q and that there was a disk contained in E_2 such as to touch l_1 and l_2 at the point q. Since the function $\hat{\psi}$ solves the Dirichlet problem (26.7), it cannot achieve its minimum in the region E_2, and hence $\hat{\psi}(q) = 0$ is the minimum of $\hat{\psi}$ in $\bar{E}_2$. Hence, an application of the strong maximum principle for elliptic differential inequalities shows that the inward normal derivative of $\hat{\psi}$ at q is greater than zero, i.e.,

$$\frac{d\hat{\psi}}{dn} = \frac{d\psi_2}{dn} - \frac{d\psi_1}{dn} > 0 \quad \text{at } q.$$

But this contradicts the fact that

$$\frac{d}{dn}\psi_2(q) = \frac{d}{dn}\psi_1(q) = \frac{d}{dn}\Psi(q),$$

which is necessary for ψ_1 and ψ_2 to be smooth in G. Thus our assumption that l_1 and l_2 have a common point in G is false, and hence the contention is proved. What has been proved means that *if the free boundary is sufficiently smooth, and if* $\theta_2 > \theta_1$, *then around each interior point* q *of* l_2 *we can draw a disk contained in* E_1.

Now consider the dependence of the values of the functional J upon the parameter θ_0. We have

$$\begin{aligned} J[\psi_1] &\equiv \iint_G [(\operatorname{grad}\psi_1)^2 - 4\mu\,\theta_1\,\psi_1]\,dx\,dy \\ &> \iint_G [(\operatorname{grad}\psi_1)^2 - 4\mu\,\theta_2\,\psi_1]\,dx\,dy \\ &> \iint_G [(\operatorname{grad}\psi_2)^2 - 4\mu\,\theta_2\,\psi_2]\,dx\,dy = J[\psi_2]. \end{aligned} \tag{26.9}$$

The first strict inequality sign in (26.9) follows from the fact that $\psi_1 > 0$ in G and from the assumption that $\theta_2 > \theta_1$. The second strict inequality in (26.9) holds because ψ_1 lies in the domain of the functional

$$J[u] \equiv \iint_G [(\operatorname{grad} u)^2 - 4\mu\,\theta_2\,u]\,dx\,dy$$

and because ψ_1 is not identical with the minimizing extremal ψ_2. Hence we conclude that *the value of the functional* J *is a strictly decreasing function of the angle of twist per unit length.*

Finally, we consider the dependence of the *complementary energy*,

$$\iint_G (\operatorname{grad}\psi)^2\,dx\,dy,$$

upon the angle of twist per unit length. For $\theta_2 > \theta_1$, from the inclusion relations in (26.6) we see that

$$\iint_G [(\operatorname{grad}\psi_2)^2 - (\operatorname{grad}\psi_1)^2]\,dx\,dy = \iint_{E_1} [(\operatorname{grad}\psi_2)^2 - (\operatorname{grad}\psi_1)^2]\,dx\,dy, \tag{26.10}$$

because $\psi_2 = \psi_1 = \Psi$ in P_1. Since the extremal ψ_1 solves the Dirichlet problem

$$\Delta\psi_1 = -2\mu\,\theta_1 \quad \text{in } E_1, \qquad \psi_1 = \Psi \quad \text{on } \partial E_1,$$

it minimizes the functional

$$J[u] \equiv \iint_{E_1} [(\operatorname{grad} u)^2 - 4\mu\,\theta_1\,u]\,dx\,dy \tag{26.11}$$

among all functions u in $H^1(E_1)$ which have the same boundary values Ψ on ∂E_1. Consequently,

$$\iint_{E_1} [(\nabla\psi_1)^2 - 4\mu\,\theta_1\,\psi_1]\,dx\,dy < \iint_{E_1} [(\nabla\psi_2)^2 - 4\mu\,\theta_1\,\psi_2]\,dx\,dy. \tag{26.12}$$

This is because the extremal ψ_2 lies in $H^1(E_1)$ and equals Ψ on ∂E_1 and hence is an admissible function for the minimum problem (26.11). Moreover, $\psi_2 > \psi_1$ in E_1, so the strict inequality in (26.12) must hold. By combining (26.10) and (26.12) we find that

$$\iint_G [(\nabla\psi_2)^2 - (\nabla\psi_1)^2]\,dx\,dy > \iint_{E_1} 4\mu\,\theta_1(\psi_2 - \psi_1)\,dx\,dy > 0.$$

This proves the *the complementary energy is a strictly monotone increasing function of the angle of twist per unit length.*

27. Dependence upon the yield strength. In this section the dependence of the stress function and the complementary energy upon the yield strength k will be derived. Let $k_2 > k_1 > 0$ be two yield strengths, and let Ψ_2 and Ψ_1 the stress functions for completely plastic torsion corresponding to k_2 and k_1, respectively. Then

$$\Psi_1(q) = k_1\,\varrho(q, \partial G), \quad \Psi_2(q) = k_2\,\varrho(q, \partial G), \quad q \text{ in } \bar{G}. \tag{27.1}$$

Let ψ_1 and ψ_2 be, respectively, the stress functions for elastic-plastic torsion corresponding to the two different yield strengths k_1 and k_2 but with the same angle of twist per unit length θ. Let E_1 and P_1 be the elastic-plastic partition of G by ψ_1, and let E_2 and P_2 be that by ψ_2. Let F_1 and F_2 be, respectively, the admissible families to which ψ_1 and ψ_2 belong. In what follows, we shall always assume that the plastic portion P_2 has *positive* area.

Let $\alpha = k_1/k_2 < 1$. Consider the function $\alpha\psi_2$. It is equal to $\alpha\Psi_2 = \Psi_1$ in P_2, satisfies the Poisson equation, $\Delta\alpha\psi_2 = -2\mu\,\theta\alpha$ in E_2, and is smooth in the entire domain G. We assert that $\alpha\psi_2$ *minimizes the functional*

$$J[u] \equiv \iint_G [(\nabla u)^2 - 4\mu\,\theta\alpha u]\,dx\,dy \tag{27.2}$$

among all functions u in F_1. To see this, let φ be the unique minimizing extremal of the problem (27.2). Let E_2^*, P_2^* be the elastic plastic partition of G by φ. Then

$$G = E_2 \cap E_2^* + E_2 \cap P_2^* + P_2 \cap E_2^* + P_2 \cap P_2^*. \tag{27.3}$$

We proceed to show that $\varphi - \alpha\psi_2$ vanishes identically in $\bar{G}$. Suppose that it achieves its positive interior maximum at some point q in G. Clearly, q does not belong to $P_2 \cap P_2^*$ because $\varphi = \alpha\psi_2 = \alpha\Psi_2$ there. Also, q does not belong to $P_2 \cap E_2^*$. For if it were in $P_2 \cap E_2^*$, then we should have the absurd inequality

$$\alpha\Psi_2(q) = \alpha\psi_2(q) < \varphi(q) < \Psi_1(q) = \alpha\Psi_2(q).$$

Similarly, q does not belong to $E_1 \cap P_2^*$. For if this were not this case, then we should have the absurd inequality

$$\alpha\,k_2 > |\text{grad}\,\alpha\psi_2(q)| = |\text{grad}\,\varphi(q)| = \alpha\,k_2 = k_1.$$

Accordingly, q can belong only to $E_2 \cap E_2^*$. But $\Delta(\varphi - \alpha\psi_2) = 0$ in $E_2 \cap E_2^*$, and $\partial(E_2 \cap E_2^*)$ is contained in $\partial G \cup (P_2 \cap P_2^*)$, as is seen easily from (27.3). So $\varphi - \alpha\psi_2 = 0$ identically on $\partial(E_2 \cap E_2^*)$. Hence q does not belong to $E_2 \cap E_2^*$ either. Since there is no place for q in G as listed in (27.3), we conclude that $\varphi - \alpha\psi_2$ has no positive interior maximum. Similarly, it possesses no interior negative minimum. Hence our assertion that $\varphi - \alpha\psi_2$ vanishes in G is now proved.

Since $\alpha\psi_2$ minimizes the functional (27.2) among all functions u in F_1, we conclude from the dependence of the extremal upon the parameter $\mu\theta$ that for $k_2 > k_1 > 0$,

$$P_1 \supset P_2, \quad E_2 \supset E_1, \tag{27.4}$$

$$\alpha\psi_2(x, y) \leqq \psi_1(x, y) \quad \text{in } G, \text{ and equality holds only in } P_2, \tag{27.5}$$

$$\iint_G (\nabla\alpha\psi_2)^2\,dx\,dy < \iint_G (\nabla\psi_1)^2\,dx\,dy, \tag{27.6}$$

$$\begin{aligned} \iint_G [\nabla\alpha\psi_2)^2 - 4\mu\,\theta\,\alpha^2\psi_2]\,dx\,dy &> \iint_G [(\nabla\psi_1)^2 - 4\mu\,\theta\,\psi_1]\,dx\,dy \\ &> \iint_G [(\nabla\psi_2)^2 - 4\mu\,\theta\,\psi_2]\,dx\,dy. \end{aligned} \tag{27.7}$$

The above interesting inequalities are not those one might infer by physical reasoning. We should expect rather that the torsional rigidity as well as the complementary energy should increase with yield strength of the material. As we are going to see, these facts are far from obvious. As a first step toward solving this problem we show that *for* $k_2 > k_1 > 0$ *and* $|P_2| > 0$,

$$\psi(x, y; \mu\theta, k_2) \geqq \psi(x, y; \mu\theta, k_1) \quad \text{in } G, \tag{27.8}$$

where the strict inequality sign holds in P_2.

Consider the function $\psi_2 - \psi_1$ in the simply connected region E_2. Then

$$\Delta(\psi_2 - \psi_1) = \begin{cases} 0 & \text{in } E_1 \\ -2\mu\theta - \Delta\Psi_1 & \text{in } E_2 \cap P_1. \end{cases} \tag{27.9}$$

Note that if $k_2 > k_1$ and if $|P_2| > 0$ then $E_2 \cap P_1$ has positive area. We wish to show that $\Delta(\psi_2 - \psi_1) \leqq 0$ in $E_2 \cap P_1$ with possible exceptions along the extreme inward normals at the reentrant corners of G, where $\Delta(\psi_2 - \psi_1)$ is not uniquely defined.

Suppose that $\Delta(\psi_2 - \psi_1) > 0$ at some point q in $E_2 \cap P_1$ and that q is off the extreme inward normals at the reentrant corner of G. It follows from continuity that there exists a positive ε such that the disk $D(q, \varepsilon)$ centered at q with radius ε is contained in $E_2 \cap P_1$ and does not intersect the extreme inward normals at the reentrant corners and that $\Delta(\psi_2 - \psi_1) > 0$ throughout $D(q, \varepsilon)$. Let u be the function defined by

$$\Delta u = -2\mu\,\theta \quad \text{in } D(q, \varepsilon), \qquad u = \Psi_1 \quad \text{on } \partial D(q, \varepsilon).$$

Then

$$\Delta(u - \Psi_1) = -2\mu\,\theta - \Delta\Psi_1 > 0 \quad \text{in } D(q, \varepsilon), \qquad u - \Psi_1 = 0 \quad \text{on } \partial D(q, \varepsilon).$$

Hence by the maximum principle $u < \Psi_1$ in $D(q, \varepsilon)$. Now the function ψ_1^* defined by

$$\psi_1^* = \begin{cases} u & \text{in } D(q, \varepsilon) \\ \psi_1 & \text{in } \bar{G} - D(q, \varepsilon), \end{cases}$$

has the property that

$$J[\psi_1^*, G] - J[\psi_1, G] = J[\psi_1^*, D] - J[\psi_1, D] < 0,$$

because ψ_1^* minimizes the functional

$$J[u, D] \equiv \iint_{D(q,\varepsilon)} [(\operatorname{grad} u)^2 - 4\mu\,\theta_0\, u]\, dx\, dy$$

among all functions u in $H^1(D)$ which are equal to Ψ_1 on $\partial D(q, \varepsilon)$. This contradicts the minimizing property of ψ_1 and the fact that $\psi_1 = \Psi_1$ in P_1. Thus $\Delta(\psi_2 - \psi_1) \leqq 0$ in $E_2 \cap P_1$ except along the extreme inward normals at the reentrant corners of G.

For the moment, we restrict our attention to the cases for which the free boundary $l_1 = \partial E_1 \cap G$ consists of rectifiable curves. For example, if G is a polygon, then l_1 consists of analytic arcs. Under this circumstance we multiply (27.9) by a function $\zeta(x, y)$ which is of class $C_0^1(E_2)$ and non-negative in E_2 and then integrate the resulting equation so as to derive

$$\iint_{E_2} \operatorname{grad}(\psi_2 - \psi_1) \operatorname{grad} \zeta\, dx\, dy = -\iint_{E_2} \Delta(\psi_2 - \psi_1) \cdot \zeta\, dx\, dy \geqq 0. \tag{27.10}$$

Indeed, the fact that $\Delta(\psi_2 - \psi_1)$ has only finite jump discontinuities along l_1 and along the extreme inward normals at the reentrant corners and that l_1 is smooth allows us to apply the divergence theorem to derive the expression on the left in

(27.10). Now choose a function $\lambda(t)$ in $C_1(-\infty, \infty)$ such that

$$\lambda(t)=0 \quad \text{for } t\geqq 0, \qquad \lambda'(t)<0 \quad \text{for } t<0,$$

and let $\zeta(x, y)=\lambda(\psi_2(x, y)-\psi_1(x, y))$ in (27.10). Then the inequality (27.10) leads to

$$0\geqq \iint_{(\psi_2-\psi_1<0)} \lambda'(\psi_2-\psi_1)\,\mathrm{grad}(\psi_2-\psi_1)\cdot\mathrm{grad}(\psi_2-\psi_1)\,dx\,dy\geqq 0.$$

Thus the set where $\psi_2-\psi_1<0$ is of measure zero, and hence we conclude from the continuity of $(\psi_2-\psi_1)$ that $\psi_2\geqq\psi_1$ in E_2. But then

$$\Delta(\psi_2-\psi_1)\leqq 0 \quad \text{in } E_2\cap P_1, \qquad \psi_2-\psi_1\geqq 0 \quad \text{on } l_1 \quad \text{and} \quad \psi_2-\psi_1>0 \quad \text{on } l_2,$$

where $l_2\equiv\partial E_2\cap G$. Of course $\psi_2-\psi_1=0$ on ∂G. Consequently, an application of the maximum principle ensures that $\psi_2>\psi_1$ in the set $E_2\cap P_1$. Thus, when the free boundary consists of rectifiable curves, we have proved a little more than was stated in (27.8).

In the general case, we approximate G by inscribed and circumscribed polygons G'_n and G''_n such that

(i) $G'_n<G'_{n+1}$, $G''_n>G''_{n+1}$, $n=1, 2, \ldots$, respectively.

(ii) the ridge of G'_n is properly contained in that of G''_n, i.e., G'_n and G''_{n+1} are similar with parallel corresponding sides,

(iii) $G'_n\to G$ and $G''_n\to G$ as $n\to\infty$.

Let $\psi'_n(x, y; \mu\theta, k)$, $\psi''_n(x, y; \mu\theta, k)$ be, respectively, the solutions of our variational minimum problems with domains G'_n and G''_n. Then, by the same analysis as was given in [67, § 2], we conclude that both ψ'_n and ψ''_n converge uniformly in G to the solution ψ of our variational minimum problem with domain G. Since G'_n and G''_n are polygons, and the inequality (27.8) holds both for ψ'_n and ψ''_n, it holds in the limit as $n\to\infty$. That is, the extremal ψ with domain G must satisfy (27.8). The proof of the desired inequality (27.8) is now complete. As we have seen, in the limiting process some information has been lost in the set $E_2\cap P_1$.

By combining (27.5) and (27.8) *we have the Harnack inequalities*

$$\frac{k_1}{k_2}\psi(x, y; \mu\theta, k_2)\leqq\psi(x, y; \mu\theta, k_1)\leqq\psi(x, y; \mu\theta, k_2), \tag{27.11}$$

provided $k_2>k_1>0$ *and* $|P_2|>0$, *where the first strict inequality sign holds in* E_2 *and the second strict inequality holds in* P_2.

Of course, the second inequality in (27.11) ensures that *the torsional rigidity of a simply connected cylindrical bar increases strictly with the yield strength of the material provided the torsion is not purely elastic.*

Naturally, the question arises whether a similar statement can be made for the complementary energy. This question turns out to be difficult. We shall restrict our attention to the case when G is *convex*. First, we consider the case when G is a convex polygon, for which, as we know, the free boundaries consist in analytic arcs.

Consider the function $\Delta(\psi_2-\psi_1)$ in a convex polygon G. We have

$$\begin{aligned}\Delta(\psi_2+\psi_1)&=-4\mu\,\theta \quad \text{in } E_1,\\ &=-2\mu\,\theta+\Delta\Psi_1=-2\mu\,\theta \quad \text{in } E_2\cap P_1,\\ &=\Delta\Psi_2+\Delta\Psi_1=0 \quad \text{in } P_2.\end{aligned} \tag{27.12}$$

Upon multiplying the both sides by $(\psi_2-\psi_1)$ and then integrating the resulting equation over G, we obtain, after applying the divergence theorem,

$$\iint_G [(\nabla\psi_2)^2-(\nabla\psi_1)^2]\,dx\,dy = 4\mu\,\theta \iint_{E_1} (\psi_2-\psi_1)\,dx\,dy + \iint_{E_2\cap P_1} 2\mu\,\theta\,(\psi_2-\psi_1)\,dx\,dy. \tag{27.13}$$

Since $\psi_2 \geqq \psi_1$ in E_1, $\psi_2 > \psi_1$ in $E_2 \cap P_1$ and $E_2 \cap P_1$ has positive area, we conclude that

$$\iint_G (\nabla\psi_2)^2\,dx\,dy > \iint_G (\nabla\psi_1)^2\,dx\,dy \tag{27.14}$$

if G is a convex polygon.

When G is a general convex domain, then in $G-\Gamma$

$$\Delta\Psi_2 = -\frac{k_2\varkappa(s)}{1-tK(s)} < 0, \qquad \Delta\Psi_1 = -\frac{k_1\varkappa(s)}{1-tK(s)} < 0,$$

where (t, s) are the same curvilinear coordinates as were used in Sect. 20, and where $\varkappa(s)$ stands for the curvature of ∂G and is positive. The difficulty in deriving an identity like (27.13) is that we are not sure whether the divergence theorem is applicable, because the free boundaries l_1 and l_2 are not known to be rectifiable. Otherwise, we could immediately get the inequality (27.14). To overcome this difficulty, we approximate a *general convex domain G* by a sequence of inscribed convex polygons $\{G_n\}$ such that

(i) $G_{n+1} \supset G_n$ for $n=1, 2, \ldots$,

(ii) $G_n \to G$ as $n \to \infty$.

Let $\{\psi_n\}$ be the sequence of the minimizing extremals of our variational problems with convex polygons G_n as their domains. Then $\{\psi_n\}$, being equibounded and Lipschitz continuous with the same Lipschitz constant k, converges uniformly over G to a Lipschitz continuous limit. As was shown in [67], $\psi_n \to \psi$ in the space $H_0^1(G)$. Since the strict inequality holds for each ψ_n, i.e.,

$$\iint_G (\nabla\psi_{n2})^2\,dx\,dy > \iint_G (\nabla\psi_{n1})^2\,dx\,dy,$$

by letting $n\to\infty$ we conclude that

$$\iint_G (\nabla\psi_2)^2\,dx\,dy \geqq \iint_G (\nabla\psi_1)^2\,dx\,dy. \tag{27.15}$$

What has been lost in passing to the limit is the strict inequality in (27.14), which is replaced by "$\geqq$" in (27.15), which asserts that *for a cylindrical bar with convex normal section, the complementary energy is a non-decreasing function of the yield strength.*

By combining the inequalities (27.6) and (27.15), for convex domains G we have the Harnack inequalities for the complementary energy, namely,

$$\left(\frac{k_1}{k_2}\right)^2 \iint_G (\nabla\psi_2)^2\,dx\,dy < \iint_G (\nabla\psi_1)^2\,dx\,dy \leqq \iint_G (\nabla\psi_2)^2\,dx\,dy. \tag{27.16}$$

Remark 5. If we restrict the yield strength k to a compact subset on the real line, then the corresponding stress function $\psi(x, y; \mu\theta, k)$ is bounded by a fixed constant. This fact together with the Harnack inequalities (27.11) implies that $\psi(x, h; \mu\theta, k)$ is Lipschitz continuous in the parameter k and consequently is differentiable with respect to k for almost all values of k. Similarly, if G is convex and if the values of k are restricted to a compact subset of the real line, then the Harnack inequalities in (27.16) imply that

$$\iint_G (\nabla\psi(x, y; \mu\theta, k)\, dx\, dy$$

is Lipschitz continuous in k and hence differentiable in k for almost all values of k.

28. Dependence upon the angle of twist, continued. We adopt the same hypotheses and use the same notations as in Sect. 26. Consider the functions, $\varphi_2=\alpha\,\psi_2$, $\Phi_2=\alpha\Psi$, $\alpha\equiv\theta_1/\theta_2$. Then by the same reasoning as was given in this section we conclude that φ_2 minimizes the functional

$$J[u]\equiv\iint_G [(\nabla u)^2-4\mu\,\theta_1\,u]\,dx\,dy$$

among all functions in the admissible family F_α:

$$F_\alpha=\{u\,|\,u \text{ in } H_0^1(G),\ u\leqq\alpha\Psi=\Phi_2\ a,e, \text{ in } \bar{G}\}.$$

Accordingly, $\varphi_2(x, y)$ solves the elastic-plastic torsion problem with angle of twist θ_1 per unit length and with yield constant equal to αk. That is, ψ_1 and φ_2 correspond to two different yield constants but to the same angle of twist per unit length.

From the results in (27.11) we easily conclude that

$$\psi(x, y; \mu\,\theta_1, k)\leqq\psi(x, y; \mu\,\theta_2, k)\leqq\frac{\theta_2}{\theta_1}\psi(x, y; \mu\,\theta_1, k), \tag{28.1}$$

where the first equality sign only holds in P_1 *and the second strict inequality sign holds in* P_2. Similarly, if G is convex, then we conclude from (27.15) that

$$\iint_G (\nabla\psi_2)^2\,dx\,dy>\iint_G (\nabla\psi_1)^2\,dx\,dy\geqq\iint_G \left(\frac{\theta_1}{\theta_2}\right)^2(\nabla\psi_2)^2\,dx\,dy. \tag{28.2}$$

Remark 6. Clearly, what has been said about the parameter θ applies equally to the parameter μ, the shear modulus. Also, if θ is restricted to a compact subset of the real line, then the Harnack inequalities (28.1) imply the Lipschitz continuity of $\psi(x, y; \mu\,\theta, k)$ in θ. The same statement can be made for the dependence of the complementary energy upon θ if the domain G is convex.

References.

[0] ST. VENANT, A. J. C. B. DE: Mémoire sur la torsion des prismes. Mém. Divers Savants Acad. Sci. Paris **14**, 233–560 (1855).

[1] MISES, R. VON: Three remarks on the theory of the ideal plastic body. Reissner Anniversary Volume, p. 415–419. Ann Arbor, Michigan: Edwards 1949.

[2] PÓLYA, G., and G. SZEGÖ: Isoperimetric inequalities in mathematical physics. Ann. Math. Studies, No. **27**. Princeton, N. J.: Princeton Univ. Press 1951.

[3] LI, JAMES C. M., and T. W. TING: Thermodynamics of elastic solids in the electrostatic field. J. Chem. Physics **27**, 693–700 (1957).

[4] TING, T. W., and JAMES C. M. LI: Thermodynamics of elastic solids. Phys. Rev. **106**, 1165–1167 (1957).

[5] DRUCKER, D. C.: Variational principles in the mathematical theory of plasticity. Proc. Symp. Appl. Math. **8**, 7–22 (1958).

[6] NAGHDI, P. M.: On plane stress solution of an elastic perfectly plastic wedge. J. Appl. Mech. **25**, 407–410 (1958). A compressible elastic perfectly plastic wedge. J. Appl. Mech. **25**, 239–242 (1958).

[7] GEIRINGER, H.: The ideal plastic body. In: this Encyclopedia, vol. 6, pp. 322–433. Berlin-Göttingen-Heidelberg: Springer 1958.

[8] DRUCKER, D. C.: A definition of stable inelastic materials. J. Appl. Mech. **26**, 101–106 (1959).
[9] LANGENBACH, A.: Variationsmethoden in der nichtlinearen Elastizitäts- und Plastizitätstheorie. Wiss. Z. Humboldt-Univ. Berlin, Math.-Naturw. Reihe **9**, 145–164 (1959).
[10] KOITER, W. T.: General theorems for elastic-plastic solids. In: Progress in Solid Mechanics, I., pp. 167–221. Amsterdam: North-Holland 1960.
[11] TRUESDELL, C., and R. A. TOUPIN: The classical field theories. In: this Encyclopedia, vol. III/1, pp. 226–793. Berlin-Göttingen-Heidelberg: Springer 1960.
[12] FINN, ROBERT: On the steady state solution of the Navier-Stokes equations III. Acta Math. **105**, 197–244 (1961).
[13] THOMAS, T. Y.: Plastic Flow and Fracture of Solids. New York: Academic Press 1961.
[14] LEAVITT, J., and P. UNGAR: Circle supports the largest sandpile. Comm. Pure. App. Math. **15**, 35–37 (1962).
[15] SEDOV, L. I.: Introduction to mechanics of a continuous medium. Moscow: State Phys. Math. Press 1962 (English translation, Reading, Mass.: Addison-Wesley 1965).
[16] HODGE, P. G., JR.: On the soap-film sand-hill analogy for elastic-plastic torsion. In: Progress in Appl. Mech., pp. 251–259. New York: MacMillan Co. 1963.
[17] LADYZHENSKAIA, O.: The Mathematical Theory of Viscous Incompressible Fluids. New York: Gordon and Breach 1963.
[18] NORDGREN, R. P., and P. M. NAGHDI: Finite twisting and expansion of a hole in rigid-plastic plate. J. App. Mech. **30**, 605–612 (1963).
[19] SERRIN, JAMES: The initial-value problem for the Navier-Stokes equations. In: Nonlinear Problems, pp. 69–98. Madison: Univ. Wisconsin Press 1963.
[20] COLEMAN, B. D.: Thermodynamics of materials with memory. Arch. Rat. Mech. Anal. **17**, 1–46 (1964).
[21] —, and W. NOLL: Material symmetry and thermostatic inequalities in finite elastic deformations. Arch. Rat. Mech. Anal. **15**, 87–111 (1964).
[22] CHEREPANOV, G. P.: On the solution of certain problems with an unknown boundary in the theory of elasticity and plasticity. J. Appl. Math. Mech. (PMM) **28**, 162–167 (1964).
[23] GARABEDIAN, P. R.: Partial Differential Equations. New York: John Wiley & Sons 1964.
[24] LANGENBACH, A.: Verallgemeinerte und exakte Lösungen des Problems der elastisch-plastischen Torsion von Stäben. Math. Nachr. **28**, 219–234 (1964).
[25] SHINBROT, M.: A fixed point theorem and some applications. Arch. Rat. Mech. Anal. **17**, 255–271 (1964).
[26] ANNIN, A. D.: Existence and uniqueness of the solution of the elastic-plastic torsion problem for a cylindrical bar of oval cross-section. J. Math. Mech. (PMM) **29**, 1038–1047 (1965).
[27] BROWDER, F. E.: Remarks on the direct method of the calculus of variations. Arch. Rat. Mech. Anal. **20**, 251–258 (1965).
[28] FICHERA, G.: Linear elliptic differential systems and eigenvalue problems. In: Lecture Notes in Mathematics, vol. 8. Berlin-Heidelberg-New York: Springer 1965.
[29] GAJEWSKI, H., u. A. LANGENBACH: Zur Konstruktion von Minimalfolgen für das Funktional des ebenen elastisch-plastischen Spannungszustandes. Math. Nachr. **30**, 165–180 (1965).
[30] GREEN, A. E., and P. M. NAGHDI: A general theory of an elastic-plastic continuum. Arch. Rat. Mech. Anal. **18**, 251–281 (1965).
[31] KANIEL, S.: Quasi-compact non-linear operators in Banach spaces and applications. Arch. Rat. Mech. Anal. **20**, 259–278 (1965).
[32] PIPKIN, A. C., and R. S. RIVLIN: Mechanics of rate independent materials. Z. Angew. Math. Phys. **16**, 313–327 (1965).
[33] THOMAS, T. Y., Concepts from Tensor Analysis and Differential Geometry. New York: Academic Press 1965.
[34] TRUESDELL, C.: Rational mechanics of deformation and flow (Bingham Medal Address), Proc. Intl. Congr. Rheol. 1963, **2**, 3–50 (1965).
[35] —, and W. NOLL: The non-linear field theories of mechanics. In: this Encyclopedia, vol. III/3. Berlin-Heidelberg-New York: Springer 1965.
[36] AMOSOV, G. JA.: On the torsion of prismatic bars in the case of elastic-plastic deformation. Vestn. Mosk. Univ. Ser. I. Math. Mech. **21**, 98–106 (1966).
[37] GREEN, A. E., and P. M. NAGHDI: A thermodynamic development of elastic-plastic continua. Proc. IUTAM Symposia, pp. 118–131 (1966).
[38] MROZ, A.: On the form of constitutive laws for elastic-plastic solids. Mech. Stov. **18**, 3–35 (1966).
[39] PONTER, A. R. S.: On plastic torsion. Internat. J. Mech. Sci. **8**, 227–235 (1966).
[40] SHOEMAKER, E. M.: A non-linear theory of plasticity for plane strain. Quart. Appl. Math. **24**, 19–27 (1966).

[41] TING, T. W.: The ridge of a Jordan domain and completely plastic torsion. J. Math. Mech. **15**, 15–48 (1966).
[42] — Elastic-plastic torsion of a square bar. Trans. A.M.S. **123**, 369–401 (1966).
[43] TRUESDELL, C.: Thermodynamics for beginners. Proceedings IUTAM Symposia, Vienna, pp. 273–289 (1966).
[44] WANG, C.-C., and R. M. BOWEN: On the thermodynamics of non-linear materials with quasi-elastic response. Arch. Rat. Mech. Anal. **22**, 79–99 (1966).
[45] COLEMAN, B. D., and V. J. MIZEL: A general theory of dissipation in materials with memory. Arch. Rat. Mech. Anal. **27**, 255–274 (1967).
[46] FRANCIS, P. H., and K. RIM: A method for solving certain plane problem of contained elasto-plasticity. Siam J. App. Math. **15**, 842–855 (1967).
[47] GAJEWSKI, H.: Ein konstruktiver Existenz- und Einzigkeitsnachweis der Lösung des elastisch-plastischen Torsionsproblems für prismatische Stäbe. Math. Nachr. **35**, 153–169 (1967).
[48] HODGE, P. G., JR.: Elastic-plastic torsion as a problem in non-linear programming. Intl. J. Solid. Structures **3**, 989–999 (1967).
[49] LANCHON, H., et G. DUVAUT: Sur la solution du problème de torsion élasto-plastique d'une barre cylindrique de section quelconque. Compt. Rend. **264**, 520–523 (1967).
[50] NOTROT, R., and R. TIMMAN: A general method for solving the plane elasto-plastic problem. J. Eng. Math. **1**, 19–36 (1967).
[51] TING, T. W.: Elastic-plastic torsion problem II. Arch. Rat. Mech. Anal. **25**, 342–366 (1967).
[52] TRUESDELL, C.: La thermodynamique de la déformation. Proc. Canadian Congr. Appl. Mech. **3**, 207–231 (1967).
[53] COLEMAN, B. D., and M. E. GURTIN: Thermodynamics and wave propagation in non-linear materials with memory. Proc. IUTAM Symposia, Vienna, 1966, pp. 54–76 (1968).
[54] —, and V. J. MIZEL: On the general theory of fading memory. Arch. Rat. Mech. Anal. **29**, 18–31 (1968).
[55] OWEN, D. R.: Thermodynamics of materials with elastic range. Arch. Rat. Mech. Anal. **31**, 91–112 (1968).
[56] —, and W. O. WILLIAMS: On the concepts of rate independence. Quart. Appl. Math. **26**, 321–329 (1968).
[57] PETROSKI, H. J., and D. E. CARLSON: Controllable states of rigid heat conductors. Z. Angew. Math. Phys. **19**, 372–376 (1968).
[58] — — Controllable states of elastic heat conductors. Arch. Rat. Mech. Anal. **31**, 127–150 (1968).
[59] THOMAS, T. Y.: Stress-strain relations for crystals containing plastic deformations. Proc. Acad. Sci. **60**, 1102–1104 (1968).
[60] — Velocity of dislocations in crystals. J. Math. Mech. **18**, 571–584 (1968).
[61] BREZIS, H. R., et G. STAMPACCHIA: Sur la régularité de la solution d'inéquations elliptiques. Bull. Soc. Math. France **96**, 153–180 (1969).
[62] LANCHON, H.: Solution du problème de torsion élasto-plastique d'une barre cylindrique de section quelconque. Compt. Rend. **269**, 791–794 (1969).
[63] LEWY, H., and G. STAMPACCHIA: On the regularity of the solution of a variational inequality. Comm. Pure Appl. Math. **22**, 153–188 (1969).
[64] OWEN, D. R., and W. O. WILLIAMS: On the time derivatives of equilibrated response functions. Arch. Rat. Mech. Anal. **33**, 287–306 (1969).
[65] — Properties of stress functionals with elastic range with applications to classical theories of elastic-plastic materials (to appear).
[66] TING, T. W.: Elastic-plastic torsion problem III. Arch. Rat. Mech. Anal. **34**, 228–244 (1969).
[67] — Elastic-plastic torsion of convex cylindrical bars. J. Math. Mech. **19**, 531–551 (1969).
[68] TRUESDELL, C.: Rational thermodynamics. McGraw-Hill series in Modern Appl. Math., New York 1969.
[69] — A precise upper limit for the correctness of the Navier-Stokes theory with respect to the kinetic theory. J. Statistical Phys. **1**, 313–318 (1969).

Added in proof.

[71] BREZIS, H. R., and M. SIBONY: Equivalence de deux inéquations variationelles et applications. Arch. Rat. Mech. Anal. **41**, 254–265 (1971).
[72] DINCA, G.: Opérateurs monotones dans la théorie de la plasticité. Ann. Scuola. Norm. Sup. Pisa, Ser. 3, 24 (1970), pp. 357–399.
[73] TING, T. W.: Elastic-plastic torsion of simply connected cylindrical bars. Indiana Univ. Math. J. **20**. 1047–1076 (1971).

Namenverzeichnis. — Author Index.

Achenbach, J.D. 119, 120, 139, 298, 313, 400.
Adkins, J.E. 180, 181, 244, 248, 294.
Aggarwala, B.D. 250, 301.
Agmon, L. 153, 228, 231, 297, 298.
Agmon, S. 62, 63, 121.
Ahrens, T.J. 351, 352, 399.
Airy, G.B. 441.
Akilov, G.P. 296.
Alexander, J. 494.
Alfrey, T. 68, 70, 72, 73, 115.
Al-Khozaie, S. 77–79, 81–86, 119, 121.
Allen, D.N. de G. 513.
Alverson, R.C. 351, 399.
Amosov, G. Ja. 589.
Annin, A.D. 589.
Antosiewicz, H.A. 157, 294.
Appleby, E.J. 122.
Arbocz, J. 228, 300.
Archimedes 196.
Atock, D. 118.
Augusti, G. 250, 299.
Axelrad, D.R. 122.

Babcock, C.D. 228, 300.
Babuška, I. 58, 59, 61, 63, 122.
Bailey, P.B. 339, 342, 344, 400.
Barbashin, E.A. 126, 153, 292, 298.
Barberán, J. 96, 104, 105, 122, 123.
Barker, L.M. 334, 352, 354–356, 400.
Bartenev, G.M. 123.
Basov, V.P. 155.
Beatty, M.F. 181, 183, 198, 199, 201, 205, 216, 217, 243, 244, 246, 248, 249, 257, 261–263, 296, 298, 299, 301.
Beck, M. 283, 285, 293.
Becker, E. 329, 400, 402.
Beckert, H. 451.
Beliaev, N.M. 257, 291.
Bendixson, I. 126, 291.
Benedick, W.B. 401.
Bergen, J.T. 118.
Bernstein, B. 221, 294.
Bernstein, B.J. 401.
Berry, D.S. 97, 116, 117.
Bethe, H. 372, 398.
Bhatia, N.P. 126, 131, 133, 140, 144, 149, 155, 157, 169, 296, 298, 301.
Biezeno, C.B. 198, 226, 279, 291, 293.
Biot, M.A. 3, 93, 116, 117, 121, 227, 228, 296.
Birkhoff, G.D. 126, 146, 167, 291.
Bishop, J.F.W. 416.
Bland, D.R. 39, 117, 118.
Blunden, W.R. 250, 293.
Boley, B.A. 35, 46, 118.
Bolotin, V.V. 170, 256, 258, 279, 280, 283–286, 294.
Boltzmann, L. 2, 10, 12, 13, 115.
Bolza, O. 165, 292.
Born, M. 208, 291.
Borodacov, M.M. 117.
Bottema, O. 286, 293.
Boussinesq, J. 85, 115, 154, 290.
Bowen, R.M. 308, 326, 329, 333, 339, 362, 400, 402, 590.
Bramble, J.H. 219, 294.
Breuer, S. 3, 53, 58, 96, 119–121, 123, 346, 398.
Brezis, H.R. 564, 590.
Bromberg, E. 228, 301.
Browder, F.E. 551, 589.
Brown, A.L. 141, 159, 301.
Brun, L. 231, 254, 296, 300.
Bryan, G.H. 154, 196, 197, 199, 226, 227, 290.
Budiansky, B. 202, 298.
Bürger, W. 402.
Bürgermeister, G. 227, 297.
Bufler, H. 227, 228, 296.
Bungay, R.W. 284, 296.
Bykovtserv, G.I. 431.

Carathéodory, C. 444, 447, 451, 463.
Carleman, T. 228.
Carlson, D.E. 590.
Carrier, G.F. 458.
Carroll, R. 153, 298.
Cattaneo, C. 166, 198, 292.
Cauchy, A.-L. 421.
Caughey, T.K. 200, 202, 222, 224, 228, 275, 278, 286, 296, 298–300.
Cesari, L. 134, 142–146, 149, 178, 295.
Chadwick, P. 333, 399.
Chao, C.C. 119, 120.
Chen, P.J. 96, 121, 303, 313–315, 322, 323, 326, 329, 331–333, 339, 342, 346, 351, 354, 357, 363, 369, 371, 376–379, 382, 384, 399–401.
Cherepanov, G.P. 501, 523, 589.
Chetaev, N.G. 167, 168, 196, 286, 292, 294.
Christensen, R.M. 119, 121.
Christianowitsch, S.A. 443, 456.
Christoffel, E.B. 417, 428.
Chu, B.T. 97, 118, 119, 351, 398, 401.
Clebsch, A. 147, 154, 196, 197, 200, 223, 289.
Coffin, L.F., Jr. 511.
Coleman, B.D. 1–4, 11, 14, 20, 22, 97, 100, 110, 118–123, 189, 190, 192, 195, 293, 299, 301, 315, 334, 339, 342, 343, 345, 348–352, 372, 375, 378, 398, 399, 401, 402, 537, 539, 589, 590.
Collins, I.F. 448.
Colonnetti, G. 533.
Como, M. 285, 297.
Contri, L. 283, 284, 297.
Cooper, H.F. 121.
Cooper, L. 118.
Corduneanu, C. 157, 294.
Corneliussen, A.H. 119.
Cosserat, E. and F. 199.
Cotsaftis, M. 170, 284, 300, 302.
Cotter, B.A. 116.
Coulomb, C.A. 411, 463.
Courant, R. 76, 119, 130, 134, 164, 292–294, 372, 398.
Courtney-Pratt, J.S. 1.
Crossland, B. 515.
Cumberbatch, E. 401.

Dafermos, C.M. 3, 20, 105, 106, 123, 178, 207, 221, 250, 254, 255, 299.
D'Alembert, J.L. 162.

Darboux, G. 536.
Davies, H. E. 280, 301.
Davis, S. H. 226, 284, 300.
Davison, L. W. 323, 326, 399.
Day, W. A. 3, 23, 123.
Dean, W. R. 228, 291, 293.
Deineko, K. S. 283, 293.
Devenpeck, M. L. 456.
Dickerson, J. R. 275, 278, 298, 300.
Dickey, R. W. 284, 301.
Dikman, M. 199, 302.
Dill, E. H. 192, 299.
Dinca, G. 590.
Diprima, R. C. 170, 285, 300.
Dirichlet, G. Lejeune 131, 132, 154, 178, 195–197, 203, 204, 289.
Distéfano, J. N. 117.
Doria, M. L. 329, 362, 400.
Doty, P. 115.
Dowel, E. H. 280, 285, 297, 302.
Drucker, D. C. 408–410, 463, 474, 588, 589.
Duffin, R. J. 74, 116.
Dugundji, J. 285, 297.
Duhem, P. 97, 118, 134, 158, 166, 195, 196, 198, 200, 201, 222, 223, 261, 290, 291, 324, 398.
Dunwoody, J. 401, 402.
Dunwoody, N. T. 401.
Dupuis, G. 199, 299, 300.
Duvall, G. E. 351, 399.
Duvaut, G. 563, 590.

Eckhaus, W. 170, 296, 298.
Eddy, R. P. 506.
Edelstein, W. 59, 63, 87, 88, 91–94, 96, 120–123.
Efimov, A. B. 122.
El-Bayoumy, L. 199, 301.
Elder, A. S. 74, 75, 118, 120.
England, A. H. 183, 185, 294.
Ericksen, J. L. 119, 173, 186, 190–192, 194, 200, 202, 205, 224, 293, 297, 401, 424, 427, 428, 432, 436.
Eringen, A. C. 213, 295, 402.
Eubanks, R. A. 77–79, 81, 84, 116.
Euler, L. 144, 226, 289.
Evan-Iwanowski, R. M. 256, 295, 300.
Ewing, W. M. 116.
Eydus, D. M. 221, 292.

Feldt, W. T. 285, 300.
Feodos'ev, V. I. 283, 293.
Ferry, J. D. 2, 3, 107, 123.
Fichera, G. 59, 62, 63, 121, 209, 211, 296, 589.
Finn, R. 556, 557, 589.
Finzi, L. 415.
Fisher, G. M. C. 1, 98, 100–102, 104, 111, 112, 121, 122, 401.
Fisher, J. C. 511.
Flavin, J. N. 401.
Fomin, S. V. 164, 295.
Fosdick, R. L. 96, 123, 228, 295.
Fourier, J. B. 189.
Francis, P. H. 590.
Freiberger, W. 424, 499.
Freudenthal, A. M. 116, 117, 403, 533.
Friedrichs, K. O. 74, 115, 208, 221, 292, 372, 398.
Froehlich, F. E. 1.
Funk, P. 164, 208, 291, 294.

Gajewski, H. 562, 589, 590.
Galin, A. 441, 505, 522.
Garabedian, P. R. 535, 572, 589.
Gårding, L. 209.
Geiringer, H. 117, 403, 408, 418, 435–438, 445, 448, 450, 452, 458, 460, 461, 464, 465, 469, 472, 478, 485, 493, 495, 527, 530, 532, 533, 535, 560, 588.
Gelfand, I. M. 164, 295.
Gere, J. M. 206, 228, 294.
Geyling, F. T. 1.
Gilbert, J. E. 128, 133, 141, 147, 149, 151, 155, 160, 298.
Gill, K. F. 153, 299.
Gobert, J. 221, 294.
Goffman, C. 141, 296.
Goldberg, R. R. 209, 294.
Goldenblatt, I. I. 533.
Gottenberg, W. G. 119.
Gottschalk, W. H. 126, 293.
Goursat, E. 421.
Graffi, D. 68, 72, 73, 116.
Graham, G. A. C. 122.
Graham, R. A. 352, 399, 401.
Grammel, R. 226, 279, 293.
Gratch, S. 121.
Gray, A. H., Jr. 275, 296.
Green, A. E. 1, 2, 116–118, 139, 160, 164, 173, 180, 181, 183–186, 204, 206, 210, 211, 213, 228, 244, 248, 279, 294–296, 299, 401, 589.
Green, A. P. 416, 494, 525.
Green, G. 147, 154, 196, 197, 289.
Green, W. A. 318, 323, 326, 399, 402.
Greenberg, H. J. 408.
Greenberg, J. M. 315, 334, 342, 353, 378, 399, 400.
Greenhill, A. G. 226, 227, 290.
Gross, B. 28, 34, 115–117.
Guest, J. J. 411.
Gurtin, M. E. 1–3, 7, 10–12, 15–19, 22–25, 27, 29–31, 34–37, 39, 44–50, 53–58, 65–68, 70, 72, 74–77, 79–81, 84, 85, 87, 91–99, 101, 110, 112, 119–123, 304, 313, 315, 334, 339, 342, 343, 345, 346, 348–352, 354, 362, 369, 371, 372, 375, 376, 378, 399–402, 590.

Haak, W. 451.
Haar, A. 415, 526.
Habetler, G. J. 170, 285, 300.
Hadamard, J. 97, 118, 134, 136, 145, 164, 166, 196–198, 204, 208, 211, 291, 303, 401, 417, 426–428, 430–432, 434.
Hagedorn, P. 286, 301.
Hahn, W. 143, 153, 294, 295, 298.
Hale, J. K. 126–131, 134, 142, 145, 146, 154, 158, 168, 169, 172, 178, 191, 196, 199, 250, 287–289, 295, 298, 300.
Halphen, G. H. 226, 290.
Halpin, W. J. 352, 399.
Hamel, G. 160, 164, 204, 206, 292.
Hamming, R. W. 28, 117.
Hansen, Bent 533.
Harland, G. E. 153, 299.
Harris, A. J. 358, 398.
Hayes, W. D. 401.
Haythornthwaite, R. M. 416, 463.
Hedlund, G. S. 126, 293.
Hellinger, E. 160, 164, 204, 206, 291.
Hellwig, G. 130, 294, 451.
Hencky, W. H. 198, 226, 279, 291, 408, 415, 444, 506, 513, 527, 549.
Herbert, R. E. 261–263, 283, 297.
Herrera, I. 22, 96–98, 104, 105, 121–123, 346, 351, 399, 401.
Herrmann, G. 139, 170, 256, 258, 263, 279, 280, 283–286, 296–300, 313, 400.
Herrmann, L. R. 122.
Hertelendy, P. 119.
Hilbert, D. 76, 119, 130, 134, 164, 293, 294.
Hill, R. 118, 167, 181, 195,

198, 199, 202, 216, 224, 227, 239, 242–244, 248, 249, 259, 260, 293, 294, 296, 298, 299, 301, 314, 398, 401, 403, 406, 408, 414–416, 422, 428, 431, 436, 439, 440, 447, 448, 451, 456, 458, 460, 465, 469, 494, 495, 501, 506, 509, 511, 512, 515, 519–521, 523–526, 531–533.
Hills, R. N. 125.
Hlavaček, I. 58, 59, 61, 63, 120, 122, 163, 221, 301.
Hodge, P. G., Jr. 403, 408, 410, 440, 447, 462, 465, 498–501, 505, 506, 511, 513, 521, 523, 525, 533, 589, 590.
Hoffman, O. 495, 533.
Holden, J. T. 181, 198, 217, 221, 239, 240, 296.
Hölder, O. 171.
Hollenbach, R. E. 334, 352, 355, 356, 400.
Hook, D. E. 249, 299.
Hopf, E. 572.
Hopkins, I. L. 28, 117.
Horgan, C.O. 221, 302.
Hsu, C. S. 139, 173, 186, 228, 257, 275, 287–289, 297–300, 302.
Hu, H. 65, 116.
Huang, N. C. 120, 122, 185, 250, 287, 301.
Hugoniot, H. 303, 417, 428.
Hunter, S. C. 68, 106, 107, 116, 118, 401.
Huseyin, K. 226, 301.

Ignaczak, J. 90, 91, 120.
Ilyushin, A. A. 533.
Infante, E. F. 128, 129, 143, 144, 169, 178, 207, 255, 275, 286, 296, 298, 300, 302.
Inoue, N. 460.
Ivlev, D. D. 431.

Jacobi, C. G. 222, 289.
Jardetsky, W. S. 116.
Jeffries, H. 68, 115.
Jenne, W. 412, 415.
John, F. 136, 145, 146, 153, 210, 293, 294, 296, 301.
Johnson, J. N. 352, 400.
Jones, O. E. 366, 400.
Jong, I. C. 286, 296, 297.
Joseph, D. D. 122, 153, 297, 302.
Juneja, B. L. 329, 400.
Jung, H. 117.

Kachanov, L. M. 533.
Kaliski, S. 120, 533.
Kamenkov 143, 298.
Kaniel, S. 589.
Kanter, I. 97, 116, 351, 398.
Kantorovich, L. V. 164, 293, 296.
Kappus, R. 227, 292.
Kapuano, I. 447.
Kármán, T. v. 228, 292, 415, 526.
Kelvin, Lord (W. Thomson) 39, 77, 115, 144, 147, 154, 167, 196, 197, 199, 286, 289, 290.
Kerr, A. D. 199, 227, 228, 294, 301.
Kershaw, R. J. 129, 301.
Kirchhoff, G. 125, 147, 154, 196, 197, 199, 289, 290
Kiusalaas, J. 280, 301.
Klein, F. 132, 147, 178, 290.
Kneser, A. 208, 291.
Knops, R. J. 125, 128, 133, 141, 146, 147, 149, 151, 153, 155, 160, 162, 173, 178, 189, 207–209, 211, 217, 226, 228, 229, 231, 232, 234–236, 252–254, 273, 275, 297, 298, 300–302.
Knowles, J. K. 88, 122, 123, 221, 302.
Koiter, W. T. 139, 160, 164–167, 173, 179, 189, 192, 194, 195, 198–200, 202, 206, 208, 213, 227, 228, 286, 292, 294–298, 409, 411, 439, 505, 514, 524, 589.
Kolodner, I. I. 317, 399.
Kolsky, H. 2, 40, 107, 116–120, 123, 401.
König, H. 2, 11, 12, 15, 117.
Kordas, Z. 285, 295.
Korn, A. 218, 221, 241, 291.
Kovalevska, S. 421.
Kozin, F. 275, 295.
Krall, W. T. 198, 226, 297.
Krasovskii, N. N. 143, 153, 295.
Kretzchmar, H. 227, 297.
Krylov, V. I. 164, 293.
Kuo, C. T. 287–289, 300.
Kuzmin, G. A. 143, 298.

Ladyzhenskaia, O. 589.
Lagrange, J. L. 131–133, 141, 147, 148, 153, 154, 163, 167, 171, 192, 195–199, 204, 226, 289.
Lakshmikantham, V. 157, 169, 296, 300.
Lampariello, G. 431.
Lanchon, H. 563, 590.
Langenbach, A. 535, 554–556, 562, 589.
Laplace, P. S. de 148.
Lasalle, J. P. 129, 143, 149, 154, 157, 169, 178, 294, 295, 298, 299.
Laurent, H. 154, 196, 290.
Lavrentiev, M. M. 136, 228, 293, 294, 299.
Leaderman, H. 3, 117.
Leavitt, J. 562, 589.
Lebenson, L. S. 533.
Lee, E. H. 3, 68, 70, 97, 116–123, 351, 398, 436, 451, 452, 458, 494, 509, 511, 512, 525, 527, 529, 531, 532.
Lee, S. S. 287–289, 300.
Lee, T. H. 257, 275, 302.
Lee, T. M. 121.
Leela, S. 157, 300.
Lefschetz, S. 140, 142, 143, 147, 149, 157, 169, 293, 294.
Leibenson, L. S. 227, 291.
Leigh, D. C. 257, 301.
Leipholz, H. 170, 280, 282–286, 294–297, 299–301.
Leitman, M. J. 1, 11, 15, 91, 111–114, 122, 123.
Leonov, M. Ia. 283, 286, 293, 295.
Levi-Civita, T. 417, 425–427, 431.
Levine, H. A. 153, 228, 231, 301, 302.
Levinson, M. 170, 228, 285, 287, 297, 300.
Lévy, M. 226, 290, 403, 404.
Lewy, H. 563, 574, 575, 580, 590.
Li, J. C. M. 588.
Lianis, G. 121.
Liapounov, A. M. 125, 126, 133, 143, 147, 153, 154, 156, 158, 160, 164, 167, 168, 171, 172, 187, 192, 196, 200, 204, 206, 208, 250, 275, 279, 281, 290, 297.
Lin, K.-H. 285, 299.
Lippmann, H. 526, 533.
Lipschitz, R. 147, 167, 196, 197, 200, 222, 223, 290.
Lockett, F. J. 118, 119.
Lode, W. 406.
Lohr, E. 533.
Lohwater, A. J. 1.
Lorsch, H. G. 116.
Love, A. E. H. 75, 77, 85, 115.
Love, E. R. 116.
Lubliner, J. 119.
Ludford, G. S. S. 418.
Lynden-Bell, D. 287, 296.

MacGregor, C. W. 511.
Mahrenholtz, O. 533.

Malkin, I. G. 196, 293.
Man, F. 119.
Mandel, J. 117, 428, 431, 439, 458, 469, 471, 474–476, 478, 481, 485, 533.
Marguerre, K. 227, 292.
Markov, A. A. 126, 292.
Martin, J. B. 3, 123.
Massau, I. 493.
Matkowsky, B. J. 287, 302.
Maxwell, J. C. 39, 115, 132, 289.
May, W. D. 118.
McCarthy, M. F. 401, 402.
Meixner, J. 2, 11, 12, 15, 117.
Mettler, E. 256, 299.
Meyer, M. L. 121.
Meyer, O. E. 39, 115.
Meyers, D. M. 250, 293.
Miasniankin, M. 431.
Mikhlin, S. G. (Michlin) 157, 170, 208, 296, 297, 456.
Miklowitz, J. 121.
Mikusinski, J. 9, 27, 118.
Minchin, G. M. 154, 196, 290.
Mindlin, R. D. 68, 115.
Mindling, F. 154, 289.
v. Mises, R. 85, 86, 115, 198, 227, 291, 403, 404, 406, 408–413, 415, 418, 424, 426, 433, 438, 439, 443, 445, 448, 452, 458, 459, 460, 462, 465, 472, 473, 478, 494, 501, 503, 505, 513–515, 518, 523, 524, 535, 579, 588.
Mişicu, M. 120.
Miyake, A. 116, 117.
Mizel, V. J. 1–4, 11, 14, 15, 20, 22, 122, 123, 590.
Mohr, O. 406, 407, 411, 439, 463.
Moiseev, N. D. 132, 153, 154, 292.
Morland, L. W. 3, 118, 120, 123.
Morrey, C. B., Jr. 164, 166, 197, 297, 565, 568, 578.
Morris, E. L. 118.
Morrison, J. A. 116.
Moufang, R. 513.
Movchan, A. A. 118, 133, 134, 136, 139, 147, 151, 155, 156, 157, 160, 164, 168, 195, 197, 200, 206, 223, 286, 294, 295.
Mroz, A. 589.
Muki, R. 3, 118.
Murnaghan, F. D. 181, 292.

Nadai, A. 403, 496–498, 500, 509, 512, 513, 515, 516, 520, 521, 533.
Naghdi, P. M. 588.
Nariboli, G. A. 329, 400–402.
Neal, B. G. 533.
Nečas, J. 163, 221, 301.
Neilson, F. W. 401.
Neis, V. V. 123.
Nemat-Nasser, S. 139, 173, 189, 190, 192, 193, 204, 205, 256–258, 262, 263, 279, 280, 283, 285, 286, 297–300, 302.
Neményi, P. F. 463.
Nemytskii, V. V. 126, 130, 149, 292.
Neuber, H. 227, 228, 292, 297, 445, 461, 465.
Nikolai, B. L. 285, 292.
Nikolai, E. L. 258, 279, 283, 291, 292.
Nirenberg, L. 153, 228, 298.
Noll, W. 1–4, 11, 12, 14, 16–18, 117, 120, 165, 166, 173, 180, 183, 194, 195, 197–199, 202, 217, 243, 293, 297, 304, 334, 349, 398, 399, 542–545, 550, 589.
Nordgren, R. P. 589.
Notrot, R. 590.
Novozhilov, V. V. 227, 293.
Nowacki, W. 118–120, 122.
Nowinski, J. L. 228, 301.

Odeh, F. 96, 122.
Odqvist, F. K. G. 123.
Oestreicher, H. L. 116.
Ogawa, H. 228, 297, 299, 300, 302.
Oldroyd, J. G. 115.
Olszak, W. 118, 403, 533.
Onat, E. T. 3, 22, 53, 58, 96, 119–122, 346, 398, 399.
Onsager, L. 23, 115.
Orr, W. McF. 154, 291.
Osgood, W. F. 208, 291.
Owen, D. R. 536, 590.

Page, A. 141, 159, 301.
Parks, J. R. 118.
Parks, P. C. 281, 298, 299.
Parson, D. H. 526.
Payne, L. E. 146, 153, 178, 207, 217, 219, 221, 226, 228, 229, 231, 232, 234–236, 252–254, 267, 273, 275, 294, 298, 300–302.
Pearson, C. E. 192, 198, 199, 202, 227, 261, 262, 293, 294.
Perzyna, P. 118.
Petrof, R. C. 121.
Petroski, H. J. 590.
Petrovsky, I. G. 137, 145, 146, 293.
Pflüger, A. 147, 202, 227, 258, 280, 283, 285, 286, 296.
Phillips, A. 533.
Pipkin, A. C. 23, 118, 120–122, 347, 353, 399, 589.
Plaut, R. H. 263, 281, 286, 287, 289, 299, 300, 302.
Poincaré, H. 126, 144, 153, 226, 290.
Poisson, S.-D. 46.
Pólya, G. 562.
Ponter, A. R. S. 3, 123, 589.
Powdrill, B. 333, 399.
Prager, W. 227, 292, 403, 408–411, 414–416, 424, 427, 428, 434, 436, 439, 445, 447, 451, 452, 458, 463, 474, 494, 498–500, 504, 512, 513, 521–523, 525, 527, 528, 533.
Prandtl, L. 412, 456, 497, 526, 527, 531, 532.
Prasad, S. N. 170, 258, 279, 285, 286, 297–300.
Predeléanu, M. 58, 59, 63, 120, 122.
Press, F. 116.
Pritchard, A. J. 281, 284, 300.
Pucci, C. 228, 293.

Radok, J. R. M. 68, 70, 117–119.
Rankine, W. J. M. 401.
Read, W. T. 68, 115.
Reckling, K. A. 533.
Reiss, E. L. 119, 121, 287, 302.
Reissner, H. 227, 291.
Reuss, E. 412, 521.
Reut, V. I. 285, 292.
Reynolds, O. 154, 290.
Riabouchinsky, M. D. 448.
Richmond, O. 456.
Riemann, B. 417, 428.
Riesz, F. 25, 116.
Rim, K. 590.
Rivlin, R. S. 1, 2, 116–118, 122, 173, 180, 213, 248, 292, 296, 298, 589.
Rogers, T. G. 23, 120, 121.
Rohde, R. W. 366, 400.
Roorda, J. 204, 205, 256, 263, 299.
Rosenberg, R. M. 140, 298.
Rosenbrock, H. H. 126, 302.
Routh, E. J. 132, 134, 143, 154, 196, 290.
Rychlewski, W. 403, 533.

Sachs, G. 495, 533.
Sackman, J. L. 119, 123.
Sagan, H. 134, 165, 166, 301.
Saibel, E. 250, 301.

Saint-Venant, A.-J.-C. B. de 85, 115, 403, 404, 411, 472, 474, 496, 560, 562, 588.
Sakar, S. K. 121.
Saks, S. 10, 121.
Sauer, R. 445, 461, 465, 472.
Sayir, M. 451.
Schapery, R. A. 3, 119–122.
Schleicher, F. 406, 438, 439, 473.
Schmidt, E. 444, 447, 451, 463.
Schmitt, H. 329, 400.
Scholte, J. G. 115.
Schreiner, R. N. 121.
Schuler, K. W. 355, 400.
Schwarzenbach, J. 153, 299.
Schwarzl, F. 3, 116.
Sedov, L. I. 589.
Semyakin, E. I. 116.
Serrin, J. 153, 294, 295, 589.
Sewell, M. J. 228, 257, 259–262, 279, 297, 299, 301, 302.
Shahinpoor, M. 228, 301.
Shapovala, L. A. 250, 296.
Shaw, F. S. 506.
Shen, M. K. 227, 300.
Shi, Y. Y. 40, 117.
Shieh, R. C. 185, 250, 287, 301.
Shield, R. T. 137, 139, 148, 160–164, 186, 200, 202, 204, 206, 209–213, 222, 224, 228, 236, 265, 269, 286, 295, 297, 299, 416, 463, 473, 494, 526.
Shinbrot, M. 557–559, 589.
Shinozuka, M. 120.
Shoemaker, E. M. 589.
Shu, L. S. 22, 122, 346, 399.
Sibony, M. 590.
Simha, R. 115.
Simoni, F. de 415.
Sips, R. 97, 116.
Skan, S. W. 224, 227, 291.
Slemrod, M. 128, 129, 169, 178, 207, 250, 254, 255, 301, 302.
Slobodkin, A. M. 160, 162, 208, 295.
Smith, G. F. 117.
Sneddon, I. N. 118.
Sobolev, S. L. 130–133, 137, 145, 146, 170, 296, 578.
Sokolovsky, A. M. 494.
Sokolovsky, V. V. 403, 406, 438, 439, 443, 456, 458, 462, 465, 473, 477, 478, 503, 506, 513, 520, 521, 532, 533.
Somigliana, C. 93, 115.
Sommerfeld, A. 132, 147, 178, 290.
Sopwith, D. G. 513.
Southwell, R. V. 198, 224, 226, 227, 279, 291.
Spencer, A. J. M. 117, 118, 228, 294, 463, 474, 525, 526.
Stampacchia, G. 564, 574, 575, 580, 590.
Staverman, A. J. 3, 116.
Steele, T. R. 228, 300.
Stepanov, V. V. 126, 130, 149, 292.
Sternberg, E. 2, 3, 10–12, 15, 16, 23–25, 27, 30, 34–37, 39, 44, 45, 48–50, 54–58, 70, 74–79, 81–86, 88, 116, 117, 119–121.
Steup, H. 227, 297.
Stippes, M. 317, 399.
Stoker, J. J. 132, 154, 228, 293, 299, 300.
Storchi, E. 441.
Storey, C. 126, 302.
Suhubi, E. S. 213, 295, 329, 400.
Sundman, K. F. 167, 291.
Symonds, P. S. 422, 526.
Szegö, G. P. 126, 131, 133, 140, 144, 149, 155, 157, 298, 301, 588.
Sz-Nagy, B. 25, 116.

Tadjbakhsh, I. 96, 122.
Tait, P. G. 77, 115, 144, 154, 167, 196, 197, 290.
Takahashi, Y. 140, 298.
Tao, L. N. 120.
Taylor, A. E. 133, 293.
Taylor, G. I. 519, 520, 524.
Taylor, R. L. 121.
Ter-Haar, D. 116.
Terry, N. B. 117.
Thomas, T. Y. 101, 119, 307, 308, 329, 398, 422–424, 427, 428, 430–433, 533, 535, 548, 563, 589, 590.
Thompson, J. M. T. 199, 301.
Thomson, W. (Lord Kelvin) 39, 77, 115, 144, 147, 154, 167, 196, 197, 199, 286, 289, 290.
Thurston, G. B. 118.
Thurston, R. N. 1, 3, 123.
Tihonov, A. N. (Tikhonov) 164, 302.
Timman, R. 590.
Timoshenko, S. 206, 226, 228, 292–294.
Ting, T. W. 123, 495, 501, 505, 535, 561–563, 565, 573, 579–581, 586–588, 590.
Torre, C. 411, 469.
Touloukian, Y. S. 366, 399.
Toupin, R. A. 18, 87, 88, 96, 97, 99, 100, 110, 118, 122, 202, 213, 221, 224, 293, 294, 296, 304, 305, 307–310, 312, 313, 315, 398, 401, 428, 589.
Trefftz, E. 194, 196, 198, 208, 227, 292, 500, 501, 503, 521.
Tresca, H. 411, 472, 513.
Troesch, A. 283, 293.
Truesdell, C. 1, 11, 12, 14, 16–18, 96, 97, 99, 100, 110, 118, 122, 165, 166, 173, 180, 183, 194, 195, 197–199, 202, 217, 243, 295, 297, 304, 305, 307–310, 312, 313, 315–317, 321, 324, 326, 347, 398–400, 428, 535, 542–545, 550, 589, 590.
Tsai, Y. M. 123.
Tsien, H. S. 68, 73, 88, 116, 228, 292.
Tupper, S. J. 451, 509, 511, 512, 525, 531, 532.
Turner, L. B. 509.

Ungar, P. 562, 589.
Urbanowski, W. 260, 283, 295, 403, 533.

Vahidi, B. 250, 301.
Vainberg, M. M. 165, 296.
Valanis, K. C. 3, 96, 97, 100, 121–123.
van Hove, L. 167, 209, 292.
van Tuyl, A. 463.
Varley, E. 96, 97, 100, 122, 346, 399, 401, 402.
Vasil'eva, A. B. 164, 302.
Voigt, W. 39, 115.
Volosov, V. M. 164, 302.
Volterra, E. 116.
Volterra, V. 2, 10, 11, 27, 53, 56–58, 118, 154, 160, 291.

Wackerle, J. 356, 398.
Walsh, E. K. 355, 376, 400.
Wang, A. J. 499.
Wang, C.-C. 308, 326, 329, 333, 339, 400, 428, 590.
Wang, P. K. C. 126, 136, 275, 278, 297, 298.
Ward, I. M. 120.
Washizu, K. 65, 116, 165, 300.
Wather, W. W. 287, 300.
Weber, J.-D. 198, 299, 301.
Weierstrass, K. 164, 292, 297.
Weinberger, H. F. 221, 294.
Weiner, J. H. 35, 46, 118.
Weinstein, A. 562.
Weiss, H. I. 521.

Weiss, L. 143, 144, 296, 298.
Wesolowski, Z. 228, 260, 283, 295, 296.
Westergaard, H. M. 533.
Weyl, H. 372, 398.
White, G. N., Jr. 511, 513.
Whitney, H. 126, 292.
Whittaker, E. T. 196, 292.
Widera, O. E. 228, 301.
Wieghardt, K. 522.
Wilkes, E. W. 125, 143, 160, 162, 173, 208, 209, 211, 228, 293, 297.
Williams, W. O. 590.
Willke, H. L., Jr. 143, 299.
Winzer, A. 458.
Wittrick, W. H. 250, 293.
Wolkowisky, J. H. 228, 299.
Wolmir, A. S. 228, 295.
Woodward, W. B. 68, 117.
Wu, C.-H. 228, 301.

Yoshizawa, T. 148, 149, 169, 293.

Zerna, W. 173, 279, 299.
Zhinzher, N. J. 286, 300.
Zhong-Heng Guo 260, 283.
Zhukovskii, N. E. 196, 290.
Ziegler, H. 204, 206, 256–258, 279, 280, 283, 286, 293, 300, 409, 431, 438, 451, 463.
Zorski, H. 208, 295.
Zovii, L. M. 286, 295.
Zubov, V. I. 126, 127, 130, 140, 153, 155, 156, 160, 164, 168, 197, 206, 296.
Ẓuyev, Y. S. 123.
Życzkowski, M. 285, 295.

Sachverzeichnis.

(Deutsch-Englisch.)

Bei gleicher Schreibweise in beiden Sprachen sind die Stichwörter nur einmal aufgeführt.

Abbildung T, *map T* 133.
— T_τ 130, 133, 140, 148, 149.
abgeschlossene Geschichte eines Prozesses, *past history of a process* 4, 5.
abgeschlossenes plastisches Gebiet, *contained plastic domain* 494, 519, 523, 524.
Abklingen oder Wachstum der Wellenintensität, elastisch, *decay or growth of the strength of waves, elastic* 320–323, 328, 330–333.
im allgemeinen, *in general* 341–344.
in elastischen Nichtleitern, *in elastic nonconductors* 362–376.
in inhomogenen elastischen Körpern, *in inhomogeneous elastic bodies* 378–385.
in Materialien mit Gedächtnis, *in materials with memory* 346–348, 352–354.
thermoelastisch, *thermoelastic* 333, 334.
viscoelastisch, *viscoelastic* 110.
Ablation der Grenzfläche, *ablating boundary* 3.
absolute Spannungsunstetigkeit, *absolute stress discontinuity* 434.
— Temperatur, *temperature* 174.
— Unstetigkeit, *discontinuity* 427.
achtflächige Scherspannung, *octahedral shearing stress* 407.
Adiabatenbedingung, *adiabatic condition* 460.
adjungiert (s. auch selbstadjungiert und nicht selbstadjungiert), *adjoint (see also self-adjoint and non-self adjoint)*.
adjungierter Operator, *adjoint operator* 170, 276, 280, 284, 285.
adjungiertes Gleichgewicht, *adjoint equilibrium* 263.
— Problem 285.
Airysche Spannungsfunktion, *Airy's stress function* 441, 496, 522.
akustischer Tensor, *acoustic tensor* 98, 316, 324, 326.
Alfrey-Graffi-Satz über elastisch-viscoelastische Korrespondenz, *Alfrey-Graffi theorem on elastic-viscoelastic correspondence* 72, 73.
Amplitude (s. auch Wellenstärke) einer Stoßwelle, *amplitude (see also strength) of shock* 313.
— von einem Prozeß oder einer Geschichte, *of a process or history* 7.
Amplitudenvektor, *amplitude vector* 309.
anfänglich, *incipient* 525.
anfängliche Antwort, *initial response* 13, 49.
— elastische Nachgiebigkeit, *elastic compliance* 24.
— Elastizität, *elasticity* 18.
— Isotropie der Relaxationsfunktion, Relation zur Wellenausbreitung, *isotropy of the relaxation function, relation to wave propagation* 99.
— mechanische Antwort, *mechanical response* 49.
— Spannung, *stress* 197, 227–240.
— Störung, *perturbation* 207.
— vergangene Geschichte, Definition, *past-history, definition* 50.
— — —, konsistente, *consistent* 51, 89.
— — —, Problem 61, 63, 64.
— — —, verschwindende, *vanishing* 52, 63.
anfänglicher quasistatischer viscoelastischer Zustand, *initial quasi-static viscoelastic state* 49.
anfängliches plastisches Fließen, *initial incipient plastic flow* 525.
— Randwertproblem, *boundary value problem* 151.
Anfangsbedingungen, *initial conditions or data* 35, 90, 95, 112, 130, 178, 179, 186, 211, 232, 234, 235, 237, 264, 271, 274, 275, 417, 420.
Anfangswert, *initial value* 246.
Anfangswertproblem, *initial value problem* 104.
anisotrop, *anisotropic* 18.
Anisotropie, *anisotropy* 461.
Antwortfunktion oder Antwortfunktional, *response function or functional* 11–13.
Äquibeschränktheit einer Nullösung, *equiboundedness of null solution* 149.
Äquivalenzprinzip für Dämpfung, gedämpfte, sich ausbreitende Wellen unendlicher Frequenz und sich ausbreitende singuläre Flächen, *equivalence principle for attenuation, infinite frequency damped progressive waves and propagating singular surfaces* 110.
— für Wellengeschwindigkeiten und Wellenachsen, gedämpfte, sich ausbreitende Wellen unendlicher Frequenz und sich ausbreitende singuläre Flächen, *for wave speeds and wave axes, infinite frequency damped progressive waves and propagating singular surfaces* 110.

Äquivalenzprinzip, Relationen, *equivalence principle, relations* 139, 157.
—, Semiäquivalenz 139.
Arbeit (s. auch Energie, Leistung), *work (see also energy, power)* 3, 23.
asymptotische Stabilität, *asymptotic stability* 32.
— —, Definition 142.
— — der Nullösung mit dissipativen Belastungen, *of null solution with dissipative loads* 287.
— —, erwähnt, *remarked upon* 129.
— —, Gebiet der, *region of* 142, 158, 169.
— —, Halesches Invarianzprinzip, *Hale's invariance principle* 169, 250, 254, 255.
— —, hinreichende Bedingung, *sufficiency theorem* 152.
— —, lineare Thermoelastizität, *linear thermoelasticity* 249, 254, 255.
— —, notwendige Bedingungen für, *necessary conditions for* 200 (Fußnote, *footnote*).
— — ohne Eigenfunktionszerlegung, *without eigenfunction decomposition* 275.
— —, Sätze von Hsu und Lee, *theorems of Hsu and Lee* 277.
— —, Satz (Korollar), *theorem (corollary)* 158, 169.
— — und Beschränktheit, *and boundedness* 169.
— — und Umklappen, *and snap-through* 288
asymptotisches Verhalten, Satz über, *asymptotic behavior, theorem on* 32.
augenblickliche elastische Antwortfunktion, *instantaneous elastic response function* 350.
— Elastizität, *elasticity* 335.
augenblicklicher Tangentenmodul, *instantaneous tangent modulus* 336, 350.
— — zweiter Ordnung, *second-order* 336, 350.
Ausbreitungsbedingungen (s. auch Welle), *propagation condition (see also wave)* 98, 105, 316, 324, 326.
Ausdehnung, Ausdehnungswelle, *expansion, expansive wave* 51.
—, —, Beschleunigungswelle, *acceleration* 311, 314.
—, —, Stoßwelle, *shock* 312.
ausgeglichene Oberflächenbelastungen, *equilibrated surface tractions* 63.
Auspressung, *extrusion* 525.
außergewöhnliche Linie, *exceptional line* 418.
— Richtung, *direction* 479, 483, 484.
äußere Ableitung, *exterior derivative* 420, 426.
äußeres Gebiet, *exterior region* 81.
axialähnliche Wellen, *axially similar waves* 101.
axiale Kraft, *axial force* 507.
— Symmetrie, *symmetry* 422, 525.
axialunähnliche Wellen, *axially dissimilar waves* 102.
Babuška und Hlavačeksches Theorem, *Babuška and Hlavaček's theorem* 61.
Bahn, *orbit* 128.
Bahnstabilität, *orbital stability* 144, 172.
Balken, *bars* 279.
Beattysches Kriterium, *Beatty's criterion* 201, 205.
bedingte Stabilität, allgemein, *conditional stability in general* 147, 284.
— —, Beispiele, *examples* 147, 217, 219, 220.
Bedingung für geschlossene Enden, *closed-end condition* 506, 508.
Bedingungen für offene Enden, *open-end conditions* 506, 508.
begrenzte Geschichten, *restricted histories* 5.
behinderte Deformation, *restrained deformation* 523.
Belastungen, abhängig von der Geschwindigkeit und der Zeit, *loads, velocity and time-varying* 257, 278.
—, feste, *dead* 202, 215, 222, 226, 244, 249, 260, 272, 273, 286.
—, Folgekräfte, *follower forces* 257, 258, 261, 278, 279, 285.
—, hydrostatische, *hydrostatic* 262, 263.
—, konservative, *conservative* 201, 202, 204, 226, 284.
—, nicht-feste, *non-dead* 256–258.
—, schwach konservative, *weakly conservative* 193, 194, 257–264, 286.
—, — —, Definitionen, *definitions* 259.
—, wegabhängige, *path-dependent* 193.
Belastungsparameter, *load parameter* 225, 226, 284, 289.
Beltrami-Donati-Michell-Gleichungen, verallgemeinerte, *generalized Beltrami-Donati-Michell equations* 45, 46.
benachbartes Gleichgewicht, Methode des, Diskussion, *adjacent quilibrium, method of, discussion* 224–226.
— —, — —, Folgekräfte, *follower forces* 282–284.
— —, — —, Geschichte und Anwendung, *history and application* 226–228.
— —, — —, Kritik, *criticism* 202, 203–206.
— —, — —, schwach konservative Kräfte, *weakly conservative forces* 263.
Bernoulli-Gleichung, *Bernoulli equation* 339.
Beschleunigungswellen, allgemein, *acceleration waves, in general* 97, 309, 313, 428–432.
—, Definition 309, 313.
— in elastischen Körpern, *in elastic bodies* 356.
— in inhomogenen elastischen Körpern, *in inhomogeneous elastic bodies* 378.
— in Materialien mit Gedächtnis, *in bodies of materials with memory* 334.
— in Nichtleitern, *in non-conductors* 362.
— in plastischen Körpern, *in plasticity* 434, 435.
beschränkte Geschichten, *bounded histories* 5.

Beschränktheit, allgemein, *boundedness, in general* 131, 156, 170.
—, Bedingung, *condition* 271.
— einer Abbildung, Definition, *of map, definition* 148.
— einer Lösung, *of solution* 148.
— einer Nullösung, Äquibeschränktheit, *of null solution, equiboundedness* 149.
—, endgültige, *ultimate* 149.
— und asymptotische Stabilität, *and asymptotic stability* 169.
— und Halesches Invarianzprinzip, *and Hale's invariance principle* 169.
—, uniforme, *uniform* 149, 231.
beständige plastische Strömung, *steady plastic flow* 524.
— Stoßwelle, *shock waves* 354.
beständige Stabilität, *persistent stability* 187, 256, 257.
Bettisches Theorem in der Viscoelastizität, *Betti's theorem in viscoelasticity* 55.
bewegte Flächen, *moving surfaces* 428.
— Koordinaten, *coordinates* 448.
Bewegung, Definition, *motion, definition* 304.
— im dynamischen System, *in dynamical system* 127, 128, 131.
Bewegungsgleichung, *equation of motion* 42, 89, 91, 176–181, 404.
Bezeichnung, *notation* 3.
Bezugskonfiguration, *reference configuration* 173.
Bezugszustand, *reference state* 182.
biharmonische Gleichung, *biharmonic equation* 75.
Bilanz der Energie, *balance of energy* 174, 175.
— des Impulses und Drehimpulses, *linear and angular momentum* 42.
Bilanzgleichungen, Energie, *laws of balance, energy* 174, 175.
—, Impuls und Drehimpuls, *linear and angular momentum* 42.
bleibende Deformation, *permanent deformation* 538, 542, 550.
— Deformationsgradienten, *deformation gradients* 538.
— Geschichte des Deformationsgradienten, *deformation gradient history* 539.
bleibender symmetrisierter Geschwindigkeitsgradient, *permanent stretching tensor* 541, 545.
Boggioscher Satz, viscoelastisches Gegenstück, *Boggio's theorem, viscoelastic counterpart* 94.
Boltzmann-Gesetze, *Boltzmann laws* 12, 18.
—, Charakterisierung, *characterization of* 14, 15.
— mit (starker) Antwortfunktion, *laws with (strong) response function* 13.
—, Kriechgesetz, *creep law* 43, 89.
—, Reduzierung auf Differentialoperator-Gesetz, *reduction to differential operator law* 36.
—, Relaxationsgesetz, *relaxation law* 43, 89, 91.
Boltzmann-Gesetze, zusätzliche Eigenschaften, *Boltzmann laws, additional properties* 31.
Boltzmann-Operator, -Kern, *Boltzmann operator, kernel* 10.
Boussinesq-Papkovich-Neuber-Lösung, verallgemeinerte, *Boussinesq-Papkovich-Neuber solution, generalized* 75–78.
Boussinesq-Papkovich-Neubersche Spannungsprozesse, *Boussinesq-Papkovich-Neuber stress processes* 78.
Boussinesq-Somigliana-Galerkin-Lösung, verallgemeinerte, *Boussinesq-Somigliana-Galerkin solution, generalized* 75–77.

Cauchy-elastische Körper, *Cauchy elastic bodies* 173.
Cauchys Lösung des Anfangswertproblems, *Cauchy's solution of initial value problem* 105, 106.
Cauchy-Poissonsches Theorem, *Cauchy-Poisson theorem* 42.
Cauchy-Problem 104, 421, 491.
Cauchysche Bedingungen, *Cauchy data* 417, 425, 450.
— Spannung, *stress* 175, 180, 183, 185, 326, 536.
Cauchysches Anfangswertproblem, *Cauchy initial-value problem* 105.
Charakteristiken der v. Mises-Plastizitätsgleichung, *characteristics of the v. Mises plasticity equations* 422.
— des vollständigen ebenen Problems, *of the complete plane problem* 464.
—, feste, *fixed* 471.
charakteristische Bedingung, *characteristic condition* 424, 426.
— Fläche, *surface* 416, 424.
— Flächenelemente, *surface elements* 423.
— Koordinaten, *coordinates* 447, 468, 469, 490.
— Linie, *line* 418, 419, 443, 485.
— Richtung, *direction* 451.
charakteristisches Anfangswertproblem, *characteristic initial-value problem* 450, 491.
— Randwertproblem, *boundary-value problem* 530.
— Rechteck, *quadrangle* 491.
— Segment 430.
Clausius-Duhemsche Ungleichung, *Clausius-Duhem inequality* 537, 539, 540.
Colemansche kanonische freie Energie, *Coleman's canonical free energy* 190.
Coriolische Kraft, *Coriolis forces* 286.
Coulombsche Bedingung, *Coulomb's condition* 463, 464, 473.

Dach, *roof* 500.
Dämpfer, *dashpots* 2.
Dämpfung von sich ausbreitenden singulären Flächen, *attentuation of damped, propagating singular surfaces* 106, 107.
— von Wellen, *of waves* 1.
Dämpfungsmatrix, reduzierte, *attentuation matrix, reduced* 102, 103.

Dämpfungsoperator, *attentuation operator* 110.
Dämpfungstensor, *attentuation tensor* 101.
Dafermosscher Satz, *Dafermos' theorem* 106.
Deformationsform, *mode of deformation* 416.
Deformationsgeschichte, *deformation history* 537.
Deformationsgradient, *deformation gradient* 305, 536.
Deformationstheorie, *deformation theory* 549.
Deformationstyp, *deformation type* 415.
deformierbares Gebiet, *deformable region* 416.
Dehnungsenergie, *strain energy* 87, 173, 225.
Dehnungsfeldprozeß, *strain field process* 16.
Dehnungsfunktion, *strain function* 225.
Dehnungsgeschichte, *strain history* 305.
Dehnungsrate, Tensor der, *strain-rate tensor* 499.
Dehnungs-Verschiebungsgleichung, *strain displacement equation* 43, 71, 89.
Determiniertheit, *determinism* 12.
Deviator der Dehnung, *deviator of strain* 29, 71.
— der Spannung, *of stress* 29, 71, 546.
— des symmetrisierten Geschwindigkeitsgradienten, *of stretching tensor* 548.
— des Tensors der permanenten Deformation, *of permanent deformation tensor* 550.
diagonaler Fluß, *diagonal flow* 455, 456.
Dichte, *density* 42.
dickwandige Röhre, *thick-walled tube* 506.
Differentialmodelle für viscoelastisches Verhalten, *differential models for viscoelastic behavior* 2.
Differentialoperator, Beziehung zur Relaxationsfunktion, *differential operator, relation to relaxation function* 34.
— vom Kriechtyp, *of creep type* 34.
— vom Relaxationstyp, *of relaxation type* 34.
Differentialoperatorgesetze, allgemeine, *differential operator laws, general* 34, 35.
—, entartete Fälle, *degenerate cases* 37.
—, normale, *standard* 39.
—, spezielle, *special* 39.
—, tensorielle, *tensor* 37.
— und Anfangsbedingungen, *and initial conditions* 44.
Dirichlet-Probleme in elliptischen Gleichungen, *Dirichlet problems in elliptic equations* 211.
Dirichletsche Wachstumsbedingungen, *Dirichlet growth condition* 565.
diskrete Systeme (s. auch Lagrange-Dirichletsche Stabilitätsvorstellung), *discrete systems (see also Lagrange-Dirichlet stability concept)* 131.
— —, Energiekriterium, *energy citerion* 203, 204.
diskrete Systeme, konservative, Diskussion von Lagranges Werk, *discrete systems, conservative, discussion of Lagrange's work* 153, 154, 163.
— —, Lagrange-Dirichletsches Kriterium, *Lagrange-Dirichlet criterion* 153.
— —, — und Energiekriterium, *and energy criterion* 164 (Fußnote, *footnote*), 195, 196, 203.
— —, Liapounovsche Sätze, *Liapounov theorems* 155, 156.
— —, Maße, *measures in* 141.
— —, Stabilität, Vergleich mit kontinuierlichen Systemen, *stability, comparison with continuous systems* 140, 159.
— —, Stabilitätssatz, *stability theorem* 155.
— —, —, Beweis, *proof* 159.
Dissipation, allgemein, *dissipation, general* 173, 249, 263, 275, 285–289.
—, Ungleichung, *inequality* 541.
dissipative Kraft, *dissipative force* 258, 285.
dissipatives Material, *dissipative material* 22.
Divergenz, Instabilität durch, *divergence, instability* 282–284.
Divergenzsatz, *divergence theorem* 585.
DONATI s. BELTRAMI
doppelt gemischte Randbedingungen, *mixed-mixed boundary data* 47, 57.
doppelter Grenzpunkt, *double limit point* 480.
Drehimpulsbilanz, *rotational balance* 42, 216, 242, 243.
Drehmoment, *torque, twisting moment* 496, 564, 582.
dreidimensionale Wellengleichung, *three-dimensional wave equation* 137, 162.
dreidimensionaler euklidischer Punktraum, *three-dimensional Euclidean point-space* 4.
— innerer Produktraum, *inner product space* 4.
— Unterraum schiefsymmetrischer Tensoren, *subspace of skew-symmetric tensors* 4.
dreidimensionales Problem der Plastizität, *three-dimensional problem of plasticity* 403.
Dreiecksungleichungen, *triangle inequalities* 561, 565.
Druck, *pressure* 404, 480.
duales Problem, *dual problem* 264, 266–269, 271, 274.
Dualität der materiellen und räumlichen Beschreibung, *duality of material and spatial descriptions* 428.
dynamische Kompatibilitätsbedingungen, *dynamic conditions of compatibility* 431, 563.
— lineare Viscoelastizität, *linear viscoelasticity* 89.
— Probleme, andere Formulierungen, *problems, other formulations* 90.
— Randwertprobleme, *boundary value problem* 89.
— Stabilität, *stability* 32.

dynamische Systeme, allgemeine Züge, *dynamical systems, general features* 127, 131.
— —, Beispiele, *examples* 127, 128, 160.
— —, Definition 128.
— —, — (T, X) 129.
— —, Differentialgleichungen für, *differential equations* 127, 129, 157, 172.
— —, Einflüsse auf Stabilität der, *stability of, influences on* 135.
— —, — — —, Beispiele, *examples* 136–139.
— —, — — —, dissipative Systeme, *dissipative systems* 288.
— —, — — —, Elastizitätstheorie, *elasticity* 171, 207, 211, 250, 254, 255.
— —, Gleichgewichtslösung, Eindeutigkeit der, *equilibrium solution, uniqueness of* 130.
— —, Stabilität, Definition, *stability of, definition* 134.
— — von HALE, of *Hale* 130, 131, 168, 169, 255.
— — von ZUBOV, *of Zubov* 130.
— Variationsprinzipien, *dynamic variational principles* 112.
— Viscosität η, *viscosity* η 34.
dynamischer Modul $\boldsymbol{G}$, *dynamic modulus* $\boldsymbol{G}$ 33.
— viscoelastischer Prozeß für $\mathcal{B}$ entsprechend $\boldsymbol{G}$, $\boldsymbol{b}$ und ϱ: DVP($\boldsymbol{G}$, $\boldsymbol{b}$, ϱ), *viscoelastic process for* $\mathcal{B}$ *corresponding to* $\boldsymbol{G}$, $\boldsymbol{b}$, *and* ϱ: *DVP*($\boldsymbol{G}$, $\boldsymbol{b}$, ϱ) 89.
— — — für $\mathcal{B}$ entsprechend $\boldsymbol{J}$, $\boldsymbol{b}$ und ϱ: DVP($\boldsymbol{J}$, $\boldsymbol{b}$, ϱ), *for* $\mathcal{B}$ *corresponding to* $\boldsymbol{J}$, $\boldsymbol{b}$, *and* ϱ: *DVP*($\boldsymbol{J}$, $\boldsymbol{b}$, ϱ) 89.

ebene Dehnung unter allgemeinen Bedingungen für plastisches Nachgeben, *plane strain under general yield conditions* 437.
— Spannung, *stress* 424, 427, 436.
— — mit quadratischen Bedingungen für plastisches Nachgeben, *with quadratic yield conditions* 438.
— Welle, *plane wave* 100.
— —, Ausbreitung, *propagation* 100.
— — der Ordnung N, Ausbreitung in einem homogenen, isotropen, viscoelastischen Material, *of order N propagating in a homogeneous, isotropic, viscoelastic material* 103.
— —, fortschreitende, *travelling* 107, 109.
— —, longitudinale, *longitudinal* 103.
— —, transversale, *transverse* 103.
Eigenfunktionsentwicklung, *eigenfunction expansions* 170, 258, 279, 280, 284, 285.
Eigenschwingung, *eigenmode* 225, 226.
Eigenvektor (s. auch Wellenachsen), *eigenvector (see also wave axis)* 98.
Eigenwerte, Spektrum eines Differentialoperatorgesetzes, *eigenvalues, spectrum of differential operator law* 38, 41, 42.
— und Wellengeschwindigkeiten, *and wave speeds* 98, 101–104.
einachsige Spannung, *uniaxial tension* 249.
Eindeutigkeit, allgemein, *uniqueness in general* 131.
—, Äquivalenz zur Liapounov-Stabilität, *equivalence with Liapounov stability* 151.
—, Cauchysches Anfangswertproblem, *Cauchy initial value problem* 105.
— für starre plastische Festkörper, *for rigid plastic solids* 416, 523.
— in der linearen Elastizitätstheorie, *in linear elasticity* 152, 231.
—, inkompressible Körper, *incompressible bodies* 72.
—, isotrope Körper, *isotropic bodies* 57.
—, Nichteindeutigkeit in Folgekraftproblemen, Divergenz, *non-uniqueness in follower force problems, divergence* 282.
—, — und Liapounov-Instabilität, *and Liapounov instability* 151, 152, 167, 224.
—, Problem mit anfänglicher vergangener Geschichte, *initial past-history problem* 57.
— quasistatischer viscoelastischer Prozesse, *of quasi-static viscoelastic processes* 55.
—, Satz, Formulierung mit Laplace-Transformierter, *theorem, Laplace transform formulation* 96.
—, — für elastische Zustände, *for elastic states* 49.
—, — für lineare Elastizität, Thermoelastizität, Anfangs- und Randwertproblem, *for linear elasticity, thermoelasticity, initial boundary value problem* 231, 254.
—, — für Prozesse mit unendlich langen Geschichten, *for processes with infinite histories* 59.
—, — für quasistatische viscoelastische Prozesse, *for quasi-static viscoelastic processes* 56.
—, —, gemischtes Problem, *mixed problem* 96.
—, —, in dynamischer Viscoelastizität, *in dynamic viscoelasticity* 95, 96.
—, —, Verschiebung als Randwerte, *displacement boundary data* 96.
— (und Existenz) viscoelastischer Prozesse, *(and existence) of viscoelastic processes* 58.
— und Stabilität für elastisch plastische Körper, *and stability for elastic-plastic bodies* 523.
eindimensionale Wellengleichung, *one-dimensional wave equation* 160.
eindimensionales Gegenstück der kinematischen Kompatibilitätsbedingung, *one-dimensional counterpart of kinematical condition of compatibility* 309.
einfach unterstützte elastische Platte, *simply supported elastic plate* 136.
— zusammenhängend, *connected* 45.
einfache Welle, *simple wave* 447, 484, 485.
einfaches Material, *simple material* 17.
— Vektorfeld, *vector field* 73, 112.

einseitig belasteter Keil, *unilaterally loaded wedge* 526.
elastische Antwort, anfängliche, augenblickliche, *elastic response, initial, instantaneous* 18.
— — im Gleichgewicht, *equilibrium* 18.
— Deformation 545, 550.
— Deformationsgeschichte, *deformation history* 539.
— Deformationsgradienten, *deformation gradients* 538, 550.
— Fortsetzung, *continuation* 539, 540.
— Konstanten zweiter Ordnung, *second-order elastic constants* 181.
— Mischungen, *mixtures* 286.
— Nachgiebigkeit, *compliance* 42.
— Spannungsfläche, *stress surface* 497.
elastischer Kern, *elastic core* 568, 575.
— Körper, elastisches Material, *body, elastic material* 19, 175.
— Zustand, *state* 49.
elastisches Differentialoperatorgesetz, *elastic differential operator law* 39.
— Gebiet, *region* 501, 563, 578, 580.
elastisch-plastisch, *elastic-plastic* 495.
elastisch-plastische Randfläche, *elastic-plastic boundary* 580.
— Torsion 499, 562–564.
elastisch-viscoelastische Korrespondenz, *elastic-viscoelastic correspondence* 68.
Elastizitätskoeffizienten, Elastizität (s. auch Tangentenmodul) *elasticities, elasticity (see also tangent modulus)*
— A_{iAkB} 178, 182.
— c_{ijkl} 180, 215, 218, 244.
— d_{ijkl} 180, 206, 209, 211, 213, 216, 217, 239, 255, 264.
—, Modul, *modulus* 412.
—, positive longitudinale, *positive longitudinal* 317.
—, Tensor 316.
Elastizitätstensor zweiter Ordnung, *second-order elasticity tensor* 316.
elastodynamische Spannungsfunktionen von Lamé und Somigliana, *elastodynamic Lamé and Somigliana stress functions* 93, 94.
elektromagnetische Effekte, *electromagnetic effects* 286.
elliptisch, Elliptizität, *elliptic, ellipticity* 211, 213, 460, 462, 478.
—, stark, *strong* 64, 99.
endliche Deformationen, *finite deformations* 3.
— Geschichtsfunktionen, *histories* 5.
Energie, Gesamtenergie, *energy, total* 137, 138, 156, 160, 162, 163, 172, 207, 211, 222, 224, 231–236, 250, 265, 267, 269, 287, 288.
—, gesamte, erster Ordnung, *total, first-order* 211
—, innere, *internal* 362.
Energiebilanz, *energy balance* 174–176.
Energiekriterium für Stabilität (s. auch Lagrange-Dirichletsches Stabilitätskonzept und diskrete Systeme, konservative), *energy criterion for stability (see also Lagrange-Dirichlet stability concept and discrete systems, conservative)*
— — —, Beattysches Kriterium, *Beatty's criterion* 201, 205.
— — —, Beziehung zur Liapounovschen Stabilität, *relation to Liapounov stability* 164 (Fußnote, *footnote*).
hinreichende Bedingungen, *sufficiency conditions* 203, 206.
notwendige Bedingungen, *necessity conditions* 200–202.
— — —, Formulierung, *statement of* 196.
— — —, Geschichte, *history* 196–199.
— — —, Hadamard-Stabilität, *Hadamard stability* 197.
— — —, Kriterium von Truesdell und Noll, *criterion of Truesdell and Noll* 202.
— — —, Kritik, *criticisms* 203–206.
— — —, Lagrange-Dirichletsches Kriterium, *Lagrange-Dirichlet criterion* 154, 159, 195, 203.
— — —, Pearsonsches Kriterium, *Pearson's criterion* 202.
Energiemethode, *energy method* 153.
Energien höherer Ordnung, *higher order energies* 211, 212, 236–238.
Energietest, *energy test* 154.
Entartungen, *degeneracies* 37, 447.
Entropieproduktion, *production of entropy* 539, 540.
Entropierestungleichung, *entropy, residual inequality* 183, 188.
Entropieungleichung (s. auch Clausius-Duhemsche Ungleichung, Spannungsentropie), *entropy inequality (see also Clausius-Duhem inequality, stress-entropy)* 174.
Enveloppe von Charakteristiken, *envelope of characteristics* 477, 479, 483.
— von Mohrschen Kreisen, *of Mohr circles* 439, 458.
Epizykloiden, *epicycloids* 453.
Erhaltungssatz, allgemein, *conservation law, general* 179.
— der Energien, *of energies* 179, 207, 211, 236.
— der Masse, *of mass* 174.
Erstbewegung, *primary motion* 179, 186, 256.
Erstspannung, *primary stress* 198, 215, 239, 241.
Erstzustand, *primary state* 176–182, 184, 185, 188, 206, 215, 217, 219, 240, 256.
erzeugende Gleichung für eine viscoelastische Relaxationsfunktion, *generating equation for a viscoelastic relaxation function* 36.
Erzeugung von Stoßwellen, *generation of shock waves* 97.
Euklidischer Punktraum, *Euclidean point space* 4.

Eulersche Koordinaten, *Eulerian coordinates* 428, 536.
— kritische Last, *Euler critical load* 277.
Existenz, Eindeutigkeit und Satz zur asymptotischen Stabilität für dynamische Probleme der Viscoelastizitätstheorie, *existence, uniqueness and asymptotic stability, theorem for dynamic problems in viscoelasticity* 104–106.
— eines elastischen Kerns, *of an elastic core* 568.
— quasistatischer viscoelastischer Prozesse, *of quasi-static viscoelastic processes* 59.
— viscoelastischer Wellen, *of viscoelastic waves* 97.

Faltung, Definition, *convolution, definition* 8.
—, Eigenschaften, *properties* 9.
— für Vektor- und Tensorprozesse, *for vectors and tensor processes* 9.
— und Heaviside-Prozesse, *and Heaviside processes* 28.
Fehlstellenfeld, *slip field* 445, 450.
Fehlstellenflächen, *slip surfaces* 424.
Fehlstellenlinien an der Grenze zum plastischen Gebiet, *slip lines, plastic boundary* 494.
—, Definition 442, 443.
—, Eigenschaften, *properties* 447.
Feldgleichungen der Viscoelastizität, dynamisch, *field equations of viscoelasticity, dynamic* 91.
— — —, quasistatisch, *quasi-static* 42.
Fixpunktsatz, Shinbrotscher, *Shinbrot's fixed-point theorem* 558.
Flattern, *flutter* 283, 284.
Fließen, plastisches, Art, *flow (plastic), type* 415.
—, —, Geschwindigkeiten, *velocities* 405.
—, —, Gleichungen von Prandtl und Reuss, *equations of Prandtl and Reuss* 412.
—, —, Regel, *rule* 547.
—, —, Theorien, *theories* 549.
Flüssigkeiten, *fluids* 2.
Fokussierung von Wellen im Ursprung, *focusing of waves at origin* 138.
Folgekraft, *follower force* 257, 258, 261, 278–285.
Fourier-cos- bzw. -sin-Transformierte, *half-range Fourier transforms* 21, 33.
Fourier- und Laplace-Transformierte, *Fourier and Laplace transforms* 28.
Fourierscher Integralsatz, *Fourier integral theorem* 33.
Fréchetsche Ableitung, *Fréchet derivative* 165, 194, 195, 203, 205.
freie Energie, *free energy* 3, 179, 180, 183, 191, 192.
— —, Funktional, *functional* 539.
— Randflächen, *boundaries* 501, 580, 581, 583, 585, 586.
— Schwingungen, *vibrations* 106, 111.
freies Fließen, *free flowing* 494.
freitragender Stab, *cantilevered rod* 282.
Fresnel-Hadamardsche Ausbreitungsbedingung, *Fresnel-Hadamard propagation condition* 98, 105, 316, 324, 326.
Funktionaldeterminante, *Jacobian* 445, 477.
Funktionen der Lage und der Zeit, *functions of position and time* 7.

G gehört zum Paar P, Q, **G** *belongs to the pair* P, Q 35.
Galerkin s. Boussinesq.
Galerkinsche Methode, *Galerkin procedure* 285.
Gateausche Ableitung, *Gateau derivative* 165, 194, 198, 203, 205.
Gebiet eines Boltzmannschen Gesetzes, *domain of a Boltzmann law* 14.
Gedächtnis, Materialien mit, *memory, materials with* 172, 334–339, 344–356.
Gegenbeispiele von Hellinger, Hamel, Shield und Green, *counterexamples of Hellinger, Hamel, Shield and Green* 139, 160, 204, 210.
gegenseitig umkehrend, *mutually inverse* 26.
gegenwärtiger Deformationszustand und gegenwärtige Spannung, *present state of deformation and stress* 1.
— Wert, *value* 5.
gemischtes Gebiet, *mixed region* 518.
— Problem, nichtlinear, *problem, nonlinear* 474.
geometrische Dispersion, *geometric dispersion* 107.
— Kompatibilitätsbedingung, *geometrical condition of compatibility* 307, 430.
— und kinematische Unstetigkeitsbedingungen, *and kinematical discontinuity conditions* 429.
Geschichte, beschränkte, *history, bounded, restricted* 5.
—, Definition 4.
— des Deformationsgradienten, *of the deformation gradient* 305.
—, endliche, *finite* 5.
—, integrierbare, *integrable* 5.
—, unendliche, *infinite* 5.
Geschichtsabhängigkeit der Lösungen des viscoelastischen Problems, *history dependence of the solutions to the viscoelastic problem* 61.
Geschichtsgradient, *history gradient* 540.
Geschwindigkeit, Charakteristiken, *velocity, speed, characteristics* 445.
—, Definition 8, 173, 305.
— der Ausbreitung (s. auch Welle), *of propagation (see also wave)* 96, 97, 306, 308, 429.
— der Verschiebung, *of displacement* 307, 308, 429.
— des Mediums, *of the medium* 429.
—, Gleichungen, *equations* 449.
—, Kompatibilitätsbedingungen, *compatibility relations* 444.
—, konstant auf Charakteristiken, *constant on characteristics* 489.

Geschwindigkeitsebene, *velocity plane* 445, 467.
Geschwindigkeitsfeldprozeß, *velocity field process* 8.
Geschwindigkeitsgradient, *velocity gradient* 541.
Geschwindigkeitsverteilungen, *velocity distributions* 438.
gespeicherte Energie, *stored energy* 173.
gestörte Bewegung, Erstbewegung, *perturbed motion, primary motion* 179, 186, 256.
— —, Erstzustand, *primary state* 176, 178–180, 182, 184, 188, 189.
— —, Leistungsinkrement, Gleichung für, *incremental rate of work equation* 178, 180.
— —, Temperaturinkrement, *incremental temperature variable* 183.
— —, Verschiebungsinkrement, *incremental displacement* 176, 183.
— —, Zustandsinkrement, *incremental state* 187–189.
— —, Zweitbewegung, *secondary motion* 178, 179, 186, 187.
— —, Zweitzustand, *secondary state* 176, 177.
— —, Grundgleichungen, Elastizität, linear isotherm, *basic equations, elasticity, linear isothermal* 177–183.
anisotrop, inkompressibel, *anisotropic incompressible* 181–183.
isotrop, homogen, *isotropic homogeneous* 180, 181.
isotrop, homogen, inkompressibel, *isotropic homogeneous incompressible* 181–183.
— —, —, —, linear mit Temperaturänderung, *linear thermal* 183–185.
isotrop, homogen, *isotropic homogeneous* 185.
isotrop, homogen, inkompressibel, *isotropic homogeneous incompressible* 185.
— —, nicht-linear isotherm, *non linear isothermal* 176, 177.
getrennte Oberflächenelemente, *separate surface elements* 46.
gewöhnliche Spiralen, *ordinary spirals* 472.
gewöhnlicher Schallpunkt, *ordinary sonic point* 481.
Gleichgewicht, Achse des, *equilibrium, axis of* 216, 217, 219, 229, 239, 242.
—, Lagrangesche Instabilität und Existenz des, *Lagrange instability and existence* 151.
—, Liapounovsche Instabilität des, *Liapounov instability of* 200.
—, Stabilität, Divergenz, Nichteindeutigkeit in Folgelast-Problemen, *stability of, divergence, non-uniqueness in follower force problems* 282, 283.
—, —, freitragender Stab, *cantilevered rod* 282.
—, —, gekrümmte Platte in der Windströmung, *curved panel in windstream* 281.
Gleichgewicht, Stabilität, Knickform, *equlibrium, stability, buckling mode* 287, 288.
—, —, Konvexität, *convexity* 234, 273.
—, —, Nichteindeutigkeit und Instabilität, *non-uniqueness and instability* 224.
—, —, — und neutrales Gleichgewicht, *non-uniqueness and neutral equilibrium* 226.
—, — unter zeit- und ortsabhängigen Bedingungen, *under time- and position-dependent data* 268, 269, 274.
—, — unter fester Last in der Klasse kleiner Verschiebungsänderungen, bewegliche Oberfläche, *under dead load in class of small incremental displacements, movable surface* 215.
feste Oberfläche, *fixed surface* 206.
Satz, *theorem* 219.
Gleichgewichtsachse, *axis of equilibrium* 216, 217, 219, 229, 239, 242.
Gleichgewichtsantwortfunktion, *equilibrium response* 13.
Gleichgewichtsbedingungen für Instabilität, hinreichende, *equilibrium, sufficient conditions of instability* 199, 200.
— für Stabilität, *conditions for stability* 192, 194.
— — —, hinreichende, *sufficient* 203.
— — — in der Aussage des Energiekriteriums, *in statement of energy criterion* 196.
— — —, notwendige, *necessary* 200–202.
Gleichgewichtsgleichung, *equation of equilibrium* 42, 43.
Gleichgewichtslösung, *equilibrium solution* 132.
—, Definition 129.
—, Eindeutigkeit und dynamisches Problem, *uniqueness and dynamic problem* 130.
—, Nichteindeutigkeit, *non-uniqueness* 144.
—, Randwertproblem, *boundary-value problem* 151.
Gleichgewichtsmodul, *equilibrium modulus* 18.
Gleichgewichtspunkte, *equilibrium points* 131.
gleichzeitige und separierbare Randwerte und Volumkräfte, *synchronous and separable boundary data and body forces* 48, 55.
„Grad" 408, 465.
GRAFFI s. ALFREY.
graphische Methode, *graphical method* 493.
Greensche Funktion, *function* 8, 564.
— Prozesse erster Art, *processes of the first kind* 82.
— — zweiter Art, *processes of the second kind* 83.
Greenscher Deformationstensor, *Green's deformation tensor* 180.
— Prozeß, *process* 81.
— —, Integrallösungen, *integral solutions* 81
Green-St.Venantscher Tensor, *Green-St.Venant tensor* 550.

Grenzfläche in Wellenausbreitung, *limit boundary in wave propagation* 3.
Grenzfolgen, *limit sets* 126.
Grenzlinie, *limit line* 477, 478, 480.
Grenzpunkt, *limit point* 478.
Grenztyp, Singularitäten, *limit type singularities* 477.
Grenzwertbetrachtung, *limit analysis* 408, 431.
größte invariante Menge, *largest invariant set* 169.
Grundgleichungen der Elastizitätstheorie, *basic equations of elasticity* 173–176.
— der gestörten Bewegung (s. gestörte Bewegung), *of perturbed motion (see perturbed motion)*
— der Plastizität, *of plasticity* 403.
Grundlagen der linearen Theorie der Viscoelastizität, *foundations of linear theory of viscoelasticity* 11.
gyroskopische Kraft, *gyroscopic force* 257, 260.

Haar und v. Kármánsche Hypothese, *Haar and v. Kármán's hypothesis* 526.
Hadamardsche Stabilität, *Hadamard stability* 198, 199, 204, 224, 246.
— —, hinreichende Bedingungen, *sufficient conditions* 246, 248.
— Superstabilität, *superstability* 224.
— Theorie singulärer Flächen, *theory of singular surfaces* 428.
Hadamardsches Stabilitätskriterium, *criterion of stability* 197.
halbkreisförmige Rille, *semi-circular groove* 521.
Halesche notwendige Bedingungen für Stabilität, *necessary conditions for stability* 199.
Halescher Instabilitätssatz, *instability theorem* 169, 199.
— Invarianzprinzip, *invariance principle* 158, 169, 191, 288, 289.
— — und asymptotische Stabilität, *and asymptotic stability* 254.
Halesches dynamisches System, *Hale's dynamical system* 168, 169.
harmonische Gleichung, *harmonic equation* 75.
Harnacksche Ungleichung, *Harnack inequalities* 587.
Hauptachsen der mittleren Spannung, *principal axes of mean stress* 216, 240–243.
Hauptrichtungen, *principal directions* 406, 441.
Hauptspannungen, *principal stresses*
Haupttangentialspannungen, *principal tangential stresses* 407.
Haupttrajektoren, *principal trajectories* 484, 486.
Hauptwellen, *principal waves* 326, 328.
Heaviside-Einheitsstufenprozeß, *Heaviside unit step process* 6.
Heaviside-Klassen, *Heaviside classes* 6.
Heaviside-Prozesse, Näherung für, *Heaviside processes, approximation of* 16.
Heaviside-Typ, *Heaviside type* 5.
Helmholtzsche freie Energie, *Helmholtz free energy* 175.
— — — für isotherme Deformationen, *for isothermal deformations* 191.
Henckysche Theorie der Plastizität, *Hencky theory of plasticity* 549.
hexagonale Bedingung, *hexagonal condition* 472.
hexagonales Prisma, *hexagonal prism* 411.
Hilbert-Raum, *Hilbert space* 551.
Hilfsfunktionen (s. Liapounov-Funktionen), *auxiliary functions (see Liapounov functions)* 153.
Hilfslösung, *auxiliary solution* 61.
Hochfrequenzgrenze der Dämpfung, *high frequency limit of the attentuation* 110.
Hochfrequenzgrenzen oscillatorischer viscoelastischer Prozesse, *high frequency limits of oscillatory viscoelastic processes* 106.
Hochfrequenzverhalten viscoelastischer Materialien, *high frequency behavior of a viscoelastic material* 22.
höhere Ordnung, Energien, *higher order, energies* 211.
— —, Kelvin-Prozesse, *Kelvin processes* 81.
Hölderches Invarianzprinzip und asymptotische Stabilität, *Hölder invariance principle and asymptotic stability* 254.
Höldersche Stabilität, *Hölder stability*
— —, Definition 146.
— —, elastische Körper, *elastic bodies* 171.
— —, hinreichende Bedingung für, *sufficient condition for* 153.
— —, kontinuierliche Abhängigkeit, *continuous dependence* 253.
— —, logarithmische Konvexität, *logarithmic convexity* 231.
— —, Nullösung in der nichtlinearen Thermoelastizität, *null solution in nonlinear thermoelasticity* 252.
— —, Satz 253.
— Stetigkeit, *continuity* 146, 274.
homogene Rand- und Anfangsbedingungen, *homogeneous boundary conditions and initial data* 270.
homogener, isotroper, inkompressibler, viscoelastischer Körper, *homogeneous isotropic incompressible viscoelastic body* 71.
— — viscoelastischer Körper vom Relaxationstyp, *viscoelastic body of relaxation type* 46.
— Körper, *body* 17.
Homomorphismus, *homomorphism* 130.
Hookesches Gesetz, *Hooke's law* 496.
Hsusches Kriterium, *Hsu's criterion* 288, 289.
Hugoniotsche Beziehung, *Hugoniot relation* 372.
hydrostatische Last, *hydrostatic load* 244–249, 262, 263.
hydrostatischer Druck, *hydrostatic pressure* 261.

hyperbolische Punkte, *hyperbolic points* 465.
hyperbolisches Gebiet, *hyperbolic region* 478.
— Problem 439, 460, 462, 526.
hyperelastisch, *hyperelastic* 173.
Hypozykloiden, *hypocycloids* 453.

ideal plastischer Körper, *ideal plastic body* 403.
induzierte Abbildung, *induced map* 131.
— Unstetigkeit, *discontinuity* 311.
induzierter Sprung einer ebenen Welle, *induced jump of a plane wave* 100.
infinitesimale Deformation, *infinitesimal deformation* 3.
— linearisierte Theorie der Viscoelastizität, *linearized theory of viscoelasticity* 1.
— starre Bewegung, *rigid motion* 17.
infinitesimaler Dehnungsprozeß, *infinitesimal stain process* 17.
— starrer Rotationsprozeß, *infinitesimal rigid rotation process* 17.
inhomogener Körper, *inhomogeneous body* 17.
inkompressibel, *incompressible* 2, 71, 176, 183, 248, 264, 512.
Inkrement der Leistung, Gleichung für, *incremental rate of work equation* 178, 180, 182, 183.
— der Temperaturvariablen, *temperature variable* 183.
— der Verschiebung, *displacement* 176, 183.
— der Volumkraft, *body force* 259, 286.
— des Spannungsvektors, *stress vector* 259–261, 286.
—, Größen mit, *quantities* 176.
innere Ableitungen, *interior derivatives* 420, 426.
— Energie, *internal energy* 174.
— Reibung, *friction* 463.
— Variable, *hidden variables* 3.
— Zustandsvariablen, *internal state variables* 3.
inneres Produkt, *inner product* 4, 539.
Instabilität, allgemein, *instability, general* 135.
—, Definitionen, asymptotische, *definition, asymptotic* 150.
—, —, exponentiell asymptotische, *exponentially asymptotic* 150.
—, —, Lagrange 150.
—, —, latente (Methode des angrenzenden Gleichgewichts), *latent (adjacent equilibrium technique)* 144.
—, —, Liapounov 135, 147.
—, Sätze aus der Nichteindeutigkeit, *theorems from non-uniqueness* 151, 167, 224.
—, —, diskrete konservative Systeme, *discrete conservative systems* 167.
—, —, Diskussion, *discussion* 168.
—, —, Hale 169, 199.
Instabilität, Sätze, Liapounov, *instability, theorems* 168.
—, —, —, hinreichende Bedingungen für Gleichgewichtslösungen, *conditions sufficient for equlibrium solution* 199, 200.
Integrabilitätsbedingungen, *integrability conditions* 444.
Integraldarstellung für die Lösung von gemischten Randwertproblemen, *integral representation for the solution of mixed boundary value problem* 84.
— — — — — Spannungs-Randwertproblemen, *traction boundary value problem* 84.
— — — — — Verschiebungs-Randwertproblemen, *displacement boundary value problem* 82.
Integralform der Energiebilanz, *integral form of energy balance equation* 176.
Integralformeln, *integral formulae* 81.
Integralgesetze, *integral laws* 34.
Integralgleichung vom Nachwirkungstyp, *hereditary integral equation* 61, 64.
Integralsätze, *integral theorems* 53.
integrierbare Geschichten, *integrable histories* 5.
invariante Interpretationen, *invariant interpretations* 419.
— Menge, *set* 129, 153, 169, 288.
Invarianten I_1, I_2, I_3, *invariants* I_1, I_2, I_3 180.
Invarianzprinzipien, *invariance principles* 129, 169, 191, 255, 288, 289.
isentroper Elastizitätstensor, *isentropic elasticity tensor* 324.
— — zweiter Ordnung, *second-order elasticity tensor* 324.
— Tangentenmodul, *tangent modulus* 362, 369.
— — zweiter Ordnung, *second-order* 362.
— Zustand, *states* 187–189.
Isobaren, *isobars* 476, 477, 484.
isochor, *isochoric* 548.
Isoklinen, *isoclines* 477, 484.
isoliert, allgemein, *isolated, general* 478, 482.
isolierter Punkt, *isolated point* 480.
isoperimetrische Bedingung, *isoperimetric condition* 564.
isothermes Problem, *isothermal problems* 178.
isotrope Funktion, *isotropic function* 545.
— Materialien, *materials* 2, 18, 29, 180, 185, 244–249, 544, 545, 550.
— Tensorfunktionen, *tensor functions* 546.
iterierte Kompatibilitätsbedingungen, *iterated conditions of compatibility* 307.

Jennesche Gleichung, *Jenne's equations* 459.
Jordan-Gebiet, *Jordan domain* 561.

Kante, *edge* 483.
Kausalität, *causality* 12.
Keil, *wedge* 521, 526.

Kelvin-Dublettprozeß, Eigenschaften, *Kelvin doublet process, properties* 79, 80.
— mit oder ohne Moment, *with moment or without moment* 80.
—, normalisiert, *normalized* 79.
Kelvin-Elemente, *Kelvin elements* 40.
Kelvin-Operatoren, *Kelvin operators* 39.
Kelvin-Problem in der Elastostatik, viscoelastisches Gegenstück, *Kelvin problem in elastostatics, viscoelastic counterpart* 77.
Kelvin-Prozeß, *Kelvin process* 78, 79.
Kelvinsche Prozesse höherer Ordnung, *higher-order Kelvin processes* 81.
Kelvin-Voigt-Elemente, *Kelvin-Voigt elements* 40.
— in Reihe, *in series* 41.
Kelvin-Voigt-Operatoren, *Kelvin-Voigt operators* 39.
kinematisch bestimmt, *kinematically determined* 526.
kinematische Kompatibilitätsbedingung, *kinematical condition of compatibility* 308, 309, 429, 430.
kinetische Energie, Definition, *kinetic energy, definition* 95.
— — n-ter Ordnung, *n-th order* 236.
Klassen C^N und C^∞, *classes C^N and C^∞* 5.
— H^N und H^∞, *H^N and H^∞* 6.
—, Heaviside 6.
—, H^{ac} 6.
Klassifikation von Unstetigkeiten, *classification of discontinuities* 434.
klassische Lösungen, *classical solutions* 138, 171, 173, 178–180, 211, 224, 250, 258.
klassisches Elastizitätsgesetz für isotrope Materialien, *classical elastic law for isotropic materials* 29.
kleine Deformation aus einem spannungsfreien isotropen Bezugszustand, *small deformation from a stress free isotropic reference state* 191.
Knicken, Verhalten nach, *post-buckling behavior* 125, 195, 202.
Knickform, *buckling mode* 226, 287, 288.
Knicklast, *buckling load* 225.
Koeffizienten b_{ijkl}, *coefficients b_{ijkl}* 182, 183.
koerzive Ungleichung, *coercive inequality* 225.
Kohäsion, *cohesion* 463.
Koitersche Verallgemeinerung, *Koiter's generalization* 439.
Kompatibilitätsbedingungen, *compatibility conditions* 426, 432, 467, 471.
—, erweiterte, *extended* 430.
—, geometrische, *geometrical* 307, 429, 430.
—, —, iterierte, *interated* 307.
—, kinematische, *kinematical* 308, 429, 430.
—, —, eindimensionale, *one-dimensional* 309.
—, —, iterierte, *iterated* 308.
Kompatibilitätsgleichung eines spurfrei-symmetrischen Tensorprozesses, *compatibility equation for traceless symmetric tensor process* 71.
Kompatibilitätsgleichungen, *compatibility equations* 45, 456.
Kompatibilitätsrelationen, *compatibility relations* 411, 466, 494.
Kompatibilitätssatz, *compatibility theorem* 45.
Komplementärenergie, *complementary energy* 583, 584, 586, 587.
komplexe Wellenzahl, *complex wave number* 107.
komplexer akustischer Tensor, *complex acoustic tensor* 107.
— Modul, *modulus* 106.
kompressible Flüssigkeit, Strömung einer, *compressible fluid flow* 417, 459.
Kompressionsmodul, *compression modulus* 30.
konforme Abbildung, *conformal mapping* 521.
Konservenbüchseneffekt, *tin-canning* 287.
konsistente anfängliche vergangene Geschichte, *consistent initial past-history* 51, 89.
konstanter Prozeß, *constant process* 5.
Kontaktprobleme, *contact problems* 3.
Kontaktunstetigkeiten, *contact discontinuities* 434.
Kontinuitätsgleichung, *continuity equation* 404.
Konturlinien, *contour lines* 496, 497.
konvex, *convex* 410.
konzentrierte Lasten, *concentrated load* 77, 78.
Koordinaten, *coordinates* 536.
Kornsche Konstante, *Korn's constant* 218, 221, 240, 241.
— Ungleichung, *inequality* 218, 221, 241.
Körper $\mathscr{B}$, Definition, *body $\mathscr{B}$, definition* 16.
— mit abgeschlossener Hülle $\bar{\mathscr{B}}$ und Rand $\partial\mathscr{B}$, *with closure $\bar{\mathscr{B}}$ and boundary $\partial\mathscr{B}$* 42.
Korrespondenz, elastisch-viscoelastisch, *elastic-viscoelastic correspondence* 68, 69.
Korrespondenzprinzip, *correspondence principle* 69.
—, Differentialoperator-Gesetze, *differential operator laws* 69.
— für einfache freie Schwingungen, *for simple free vibrations* 111, 112.
— von Alfrey-Graffi, *of Alfrey-Graffi* 72.
Kraftmultipole, *force multipoles* 213.
kreisförmige Röhre, *circular tube* 192.
kreisförmiger Ring, *circular ring* 499.
kreisförmiges Loch, *circular hole* 519.
Kriechgesetz, *creep law* 23.
Kriechnachgiebigkeit **J**, *creep compliance* **J** 24.
— für Relaxationsfunktion **G**, *for relaxation function* **G** 26.
— und Relaxationsfunktion, *and relaxation function* 28.
Kriechprobleme, Variationsprinzip, *creep problems, variational principle* 66, 67, 114.

Kriechtyp, Differentialoperatoren, *creep type order of pair of differential operators* 34.
kritische Anfangsamplitude, *critical initial amplitude* 342, 367, 368, 381.
— Last, *load* 146, 277, 284, 285, 287.
kritischer Punkt, *critical point* 288, 289.
— Sprung im Dehnungsgradienten, *jump in strain gradient* 353, 361, 374, 384.
— Wert, *value* 225, 226, 278, 283, 284.
Krümmungsradien, *radii of curvature* 446.
Kugelschale, *spherical shell* 192, 521.

L_2 inneres Produkt, L_2 *inner product* 552.
$\overset{*}{\mathcal{L}}$ ist ein Kriechgesetz für $\mathcal{L}$, $\overset{*}{\mathcal{L}}$ *is a creep law for* $\mathcal{L}$ 23.
L-Träger, *L-beam* 495.
labil, *labile* 147.
Lage, Ort, *position* 7.
Lagrangesche Instabilität, asymptotische, *Lagrange instability, asymptotic* 150.
— —, Beispiel in der Elastizitätstheorie, *example in elasticity* 202, 223.
— —, Beziehung zu Liapounov, *relation to Liapounov* 150.
— —, — zur Beschränktheit, *to boundedness* 150.
— —, Definition 150.
— —, Diskussion, *discussion* 150, 151.
— —, hinreichende Bedingungen, *sufficient conditions* 170.
— — in elastischen Körpern, *elastic bodies* 171.
— —, logarithmische Konvexität, *logarithmic convexity* 239.
— —, Wahl der Definition in der Elastizitätstheorie, *choice of definition in elasticity* 171.
Lagrange-Multiplikator, *Lagrange multiplier* 564.
Lagrange-Stabilität, Beziehung zur Beschränktheit, *Lagrange stability, relation to boundedness* 149.
—, Definition 149.
—, Liapounov-Stabilität, *Liapounov stability* 156.
Lagrange-Dirichletsches Stabilitätskonzept (s. auch diskrete konservative Systeme), Beweis, *Lagrange-Dirichlet stability concept (see also discrete conservative systems), proof* 159.
— —, Definition 131, 134.
— —, —, Beobachtungen, *observations* 134.
— —, —, Bestandteile, *ingredients* 132.
— —, Diskussion, *discussion* 153, 154, 163.
— —, Energiekriterium, *energy criterion* 203, 204 (Fußnote, *footnote*).
— —, Kriterium oder Energietest, *criterion or energy test* 154.
— —, Satz 155.
Lagrangesche Koordinaten, *Lagrangian coordinates* 428, 524, 536.
Lamé-Konstanten, *Lamé constants* 45, 181.
Lamé-Modulen, *Lamé moduli* 29.
Lamé-Spannungsfunktionen, *Lamé stress functions* 93.
Laplace-Gleichung, *Laplace equation* 136.
Laplace-Transformierte, *Laplace transform* 28, 68.
Legendre-Hadamardsche Bedingung, *Legendre-Hadamard condition* 166.
Legendre-Transformation 461.
Leistung, *rate at which work is done* 95.
—, aufgebrachte, Satz der, *power expended, theorem of* 95.
— und Energie, *and energy* 95.
— pro Umdrehung, aufgebrachte, *expended per cycle* 34.
—, Spannungsleistung, *stress* 95.
— von Oberflächenspannungen und Volumkraft, *rate of working of surface tractions and body force* 54.
Leistungsidentität, *power identity* 54.
Leitfähigkeitstensor, *conductivity tensor* 250, 251, 254, 255.
Liapounovsche Funktionen (Hilfsfunktionen), endliche Thermoelastizität aus der Energiebilanz, *Liapounov functions (auxiliary functions), finite thermoelasticity, from energy balance* V_1 188, V_2 189.
— —, Colemans kanonische freie Energie V_3, *Coleman's canonical free energy* V_3 189, 190.
— —, Entropieungleichung, *entropy production inequality* 189, V_3 190, V_4 190, 222.
— —, $F_{\tau,t}$ 154, 157, 163, 164, 168, 169.
— —, Folgekräfte, *follower forces* 280–282.
— —, Gesamtenergie, *total energy* 172, 272.
— — im Sinne von Lasalle, *in sense of Lasalle* 168, 169.
— — in der Elastizitätstheorie, Erzeugung oder Wahl, *elasticity, generation or choice* 171, 172, 177, 193 (Fußnote, *footnote*), 207, 212, 223 (Fußnote, *footnote*), 250, 255, 259.
— —, isothermer Ausgangszustand, thermische Störungen, *isothermal primary state, thermal perturbations* 188.
— —, Movchansche Funktion, *Movchan's function* 195.
— —, Movchanscher Satz, *Movchan's theorem* 156.
— —, potentielle Energie V, *potential energy* V 205.
— — von Hsu und Lee für asymptotische Stabilität, *of Hsu and Lee for asymptotic stability* 275, 276.
Liapounovsche Instabilität, Beziehung zur Lagrangeschen Instabilität, *Liapounov instability, relation to Lagrange instability* 150.
— —, Definition 147.
— —, Diskussion eines Beispiels, *discussion of example* 139.
— — durch Divergenz, *by divergence* 282.
— —, hinreichende Bedingungen, *sufficient conditions* 199, 200.

Liapounovsche Instabilität, Sätze, *Liapounov instability, theorems* 168, 169.
— — und Nichteindeutigkeit, *and non-uniqueness* 151, 152.
— —, Wahl der Definition in der Elastizitätstheorie, *choice of definition in elasticity* 171.
— — in der Elastizitätstheorie, dissipative Kräfte, *in elasticity, dissipative forces* 286.
— — — — —, feste Belastungen für lineare thermoelastische Verschiebungen, Nullösung, *dead loads for linear thermoelastic displacements, null solution* 251, 252.
— — — — —, Folgekräfte, *follower forces* 282–285.
— — — — —, Gleichgewichtslösung unter fester Belastung für lineare Verschiebungsänderung, *equilibrium solution under dead loads for linear incremental displacements* 222.
— — — — —, inkompressible elastische Körper, *incompressible elastic media* 248, 249.
— — — — —, negativ definite Gesamtenergie V_4, *negative-definite total energy* V_4 222.
— — — — —, schwach konservative Kräfte, *weakly conservative forces* 259.
— — — — —, Spannungs-Randbedingungen, hinreichende Bedingungen für mittlere Spannungen, *traction boundary conditions, sufficient conditions on mean stresses* 242.
— — — — —, spezielle Spannungs-Randbedingungen, Nullösung, *special traction boundary conditions, null solution* 246–248.
— — — — — und Nichteindeutigkeit, *and non-uniqueness* 224.
— — — — —, zeit- und ortsabhängige Bedingungen, *time- and position-dependent data* 266.
Liapounovsche Stabilität, *Liapounov stability* 186.
äquivalente Definitionen, *equivalent definitions* 140.
Äquivalenz zur Eindeutigkeit, *equivalence with uniqueness* 151.
Beispiele, *examples* 160–163.
Beziehung zu Lagrange-Dirichlet, *relation to Lagrange-Dirichlet* 153, 154.
— zur stetigen Abhängigkeit, *to continuous dependence* 146.
Definition 133.
dynamisches System, *dynamical system* 134.
Eindeutigkeit der Gleichgewichtslösung, *uniqueness of equilibrium solution* 144.
geometrische Darstellungen, *geometrical representations* 140.
Liapounovsche Stabilität in der Elastizitätstheorie, *Liapounov stability in elasticity*
bezogen auf lineare Probleme, *related to linear problems* 194.
der Nullösung, *of null solution*
Gebrauch des Liapounovschen Satzes, *use of Liapounov theorem* 212.
in inkompressiblen Medien, *in incompressible media* 213, 214.
mit bewegter Oberfläche, *with movable surface* 215.
mit zeitabhängigen Elastizitäten d_{ijkl}, *with time-dependent elasticities* d_{ijkl} 209.
Satz, *theorem* 219.
unter fester Belastung in der Klasse kleiner Verschiebungsänderungen (feste Oberfläche), *under dead loads in class of small incremental displacements (fixed surface)* 206.
Wahl der Maße, *choice of measures* 210–213.
des Anfangszustandes, *of primary state* 177.
des natürlichen Zustandes, *of natural state* 195.
Diskussion und Geschichte, *discussion and history* 195, 196, 199.
dynamische Systeme, *dynamical system* 171.
in der Klasse der nichtlinearen Störungen, *in class of non-linear perturbations* 191.
mit nicht-festen Belastungen, *with non-dead loads* 256–257.
beständige Stabilität, *persistent stability* 256.
mit bewegungsabhängigen Bedingungen, *with motion-dependent data* 257.
mit schwach konservativen Belastungen, *with weakly conservative loads* 258–264.
linear 263.
Stabilitätsrechnungen, *stability analyses* 263, 264.
mit zeit- und ortsabhängigen Bedingungen, *with time- and position-dependent data* 264–278.
asymptotische Stabilität der Nullösung, *asymptotic stability of null solution* 275.
Liapounovsche Funktion von Hsu und Lee, *Liapounov function of Hsu and Lee* 276.
Satz, *theorem* 257.
nichtverschwindende Anfangswerte, *nonzero initial data* 266.
Nullösung, *null solution* 266.
unter vorgeschriebenen Oberflächenspannungen, linearen Verschiebungsänderungen und verschwindenden Anfangswerten, *under prescribed surface tractions and linear incremental displacements and zero initial data* 266.

Liapounovsche Stabilität in der Elastizitätstheory, *Liapounov stability in elasticity*
mit zeit- und ortsabhängigen Bedingungen, *with time- and position-dependent data*
unter vorgeschriebenen Oberflächenverschiebungen und linearen Verschiebungsänderungen, *under prescribed surface displacements and linear incremental displacements*
Gleichgewichtslösung, *equilibrium solution* 268, 269.
nichtverschwindende Anfangswerte, *non-zero initial data* 266.
Nullösung, *null solution* 266.
unter vorgeschriebener Volumkraft und fester Belastung, *under prescribed body force and dead loads*
Änderung der Elastizitäten mit linearen Verschiebungsänderungen, *variation in elasticities with linear incremental displacements* 271, 272.
Änderungen der Anfangswerte, *changes in initial data* 272, 273.
Argumente zur Konvexität, *convexity arguments* 273.
Gleichgewichtslösung, *equilibrium solution* 269–271.
Nullösung des gemischten Problems mit linearen Verschiebungsänderungen, *null solution of mixed problem with linear* 273. *incremental displacements*
weitere Argumente, *further arguments* 275.
Movchansche Bedingungen, *Movchan's conditions* 195.
Rechnung für Spannungs-Randbedingungen, *analysis for traction boundary problem* 239–242.
für inkompressible Medien, *for incompressible media* 244.
hinreichende Bedingungen ohne Gleichgewichtsachse, *sufficient conditions without axis of equilibrium* 239–241.
mit Achse, *with axis* 242.
spezielle Spannungs-Randwertprobleme, *special traction boundary value problems* 244.
für inkompressible elastische Materialien, *for incompressible elastic materials* 248.
isotropes kompressibles Material unter hydrostatischer Spannung, *isotropic compressible material under hydrostatic stress* 244–248.
notwendige Bedingungen, Nullösung, *necessary conditions, null solution* 246.

Liapounovsche Stabilität in der Elastizitätstheorie, *Liapounov stability in elasticity*
spezielle Spannungs-Randwertprobleme, *special traction boundary value problems*
Satz über Eindeutigkeit der Gleichgewichtslösung, *theorem on uniqueness of equilibrium solution* 224.
Sätze über hinreichende Bedingungen in der nichtlinearen Thermoelastizität, *sufficiency theorems in non-linear thermoelasticity* 192.
Sätze über notwendige Bedingungen, *necessity theorems*
Beziehung zu hinreichenden Bedingungen, *relation to sufficiency conditions* 203.
für Gleichgewichtslösungen in der Thermoelastizität, *for equilibrium solution in non-linear thermoelasticity* 200–202.
und Energiekriterium (Kritik), *and energy criterion (criticism)* 203–206.
und Variationsrechnung, *and calculus of variations* 195, 205.
uniforme Stabilität, *uniform stability* 207.
unter fester Belastung in der Klasse der linearen thermoelastischen Verschiebungen, *under dead loads in class of linear thermoelastic displacements* 249–255.
Liapounovsche Funktion, *Liapounov function* 250.
Nullösung, *null solution* 251.
unter Folgekräften, *under follower forces* 278–285.
Beispiele, Nullösung, *examples, null solution* 280, 281.
Gleichgewichtslösung, *equilibrium solutions* 281, 282.
mit dissipativen Kräften, *with dissipative forces* 285–289.
Liapounovsche Stabilität, Maße, *Liapounov stability, measures* 135.
Nichteindeutigkeit der Gleichgewichtslösung, *non-uniqueness of equilibrium solution* 144.
Sätze, *theorems* 149, 150, 152, 155, 157.
Abänderungen, *variations* 155.
diskrete Systeme, *discrete systems* 155, 159.
Diskussion, *discussion* 156–160.
schwach asymptotisch, *weakly asymptotic* 158.
uniforme, *uniform* 157.
Versionen von Zubov und Movchan, *versions of Zubov and Movchan* 156.
und Beschränktheit, *and boundedness* 148–150.
und Lagrange-Stabilität, *and Lagrange stability* 148, 149, 156.
und Maximalprinzipien, *and maximum principles* 152.
und Variationsrechnung, *and calculus of variations* 158, 163–167.

Limes, Analyse, *limit, analysis* 408, 431.
— Mengen, *sets* 126, 202, 225, 246.
lineares elastisches Element (Feder), *linearly elastic element (spring)* 40.
— viscoses Element (Dämpfer), *viscous element (dashpot)* 40.
lineare, linearisierte, *linear, linearized*
Elastizitätstheorie, *theory of elasticity* 1, 176–183.
Halesches Invarianzprinzip, *Hale's invariance principle* 255.
Integralgleichungen, *integral equations* 77.
Nachwirkungsgesetze, *hereditary laws* 11.
Näherung, *approximation* 2.
polarisierte Welle, *polarized waves* 108.
schwache (verallgemeinerte) Lösungen, *weak (generalized) solutions* 254.
Superposition 2, 12.
Systeme, *systems* 178.
Theorie, basierend auf endlichen Deformationen, *theory based on finite deformations* 3.
— der Viscoelastizität, *of viscoelasticity* 1.
Thermoelastizität, *thermoelasticity*
—, asymptotische Stabilität, *asymptotic stability* 254, 255.
—, Grundgleichungen, *basic equations* 183–185, 249, 250.
—, Halesches Invarianzprinzip, *Hale's invariance principle* 255.
—, Höldersche Stabilität, *Hölder stability* 252–254.
—, Instabilität, *instability* 251, 252.
—, Liapounovsche Stabilität, *Liapounov stability* 250, 251.
—, schwache (verallgemeinerte) Lösungen, *weak (generalized) solutions* 254, 255.
Transformationen, *transformations* 3, 4.
Linearisierung, basierend auf infinitesimalen Deformationen, *linearization based upon infinitesimal deformations* 3.
Linearität, *linearity* 12.
linear-viscoelastischer Körper vom Kriechtyp, *linearly viscoelastic body of creep type* 23.
— — vom Relaxationstyp, *of relaxation type* 16, 17.
linear-viscoelastisches Material, *linearly viscoelastic material* 1, 16.
linear-viscoses Element (Dämpfe), *linearly viscous element (dashpot)* 40.
Linienintegral für den Verschiebungsprozeß, *line integral for the displacement process* 46.
linke Umkehrung, *left inverse* 23.
linker Cauchy-Greenscher Tensor, *left Cauchy-Green tensor* 326.
Lipschitzsche Konstante, *Lipschitz constant* 587.
Locher, *punch* 455.
logarithmische Konvexität, *logarithmic convexity* 228–239, 253, 273.
— — der Funktion F $(t;\alpha, t_0)$, *of function* $F(t;\alpha, t_0)$ 229.
— — und Höldersche Stabilität, *and Hölder stability* 231.
— Spiralen, *spirals* 452.
— Stetigkeit, *continuity* 146.
lokal integrierbar, *locally integrable* 5.
lokale Ausbreitungsgeschwindigkeit, *local speed of propagation* 307, 308.
— Differentiation 524.
lokaler Tangentenmodul, *local tangent modulus* 378, 382.
— — zweiter Ordnung, *second-order tangent modulus* 378, 382.
longitudinale Elastizität, *longitudinal elasticity* 317.
— Unstetigkeit, *discontinuity* 430.
— Wellen, definition, *waves, definition* 99, 310.
— — in elastischen Körpern, anisotrop, *in elastic bodies, anisotropic* 315–321.
in Hauptrichtung, *principal* 327.
isotrop, *isotropic* 327, 328.
Lösung des Cauchy-Anfangswertproblems für dynamische Viscoelastizität, *solution, Cauchy initial value problem for dynamic viscoelasticity* 105.
— des dynamischen viscoelastischen Randwertproblems, *dynamic viscoelastic boundary problem* 89.
— des Hilfsproblems für $\boldsymbol{E}^*$, *auxiliary problem für* $\boldsymbol{E}^*$ 61.
— des Problems mit anfänglicher vergangener Geschichte, ausgedrückt durch Verschiebungen, *of initial past history problem in terms of displacements* 52.
— des zugeordneten elastischen Problems, *of associated elastic problem* 60.
— in einem dynamischen System, *in dynamical system* 128, 131.
—, quasistatisch, ausgedrückt durch Spannungen, *quasi-static, in terms of stresses* 48.
—, —, ausgedrückt durch Verschiebungen, Relaxationsfunktion oder Kriechnachgiebigkeit, *in terms of displacements, relaxation function, or creep compliance* 47.
—, viscoelastisches Problem, *viscoelastic problem* 60.
—, — —, mit anfänglicher vergangener Geschichte, *initial past-history problem* 51.

m-faltige Dirichlet-Norm, *m-fold Dirichlet norm* 551.
Machzahl, *Mach number* 281.
Maße, Äquivalenzbeziehungen (Semi-Äquivalenz), *measures, equivalence relations (semi-equivalence)* 139.
— als positiv definite Funktionen in $X \times X$, *as positive-definite functions on* $X \times X$ 135.

Maße, Auswahl, *measures choice of* 134–139, 163.
—, — in der Elastizitätstheorie, *in elasticity* 171, 172, 210–213.
—, Beispiele, *examples* 133.
—, — in der Elastizitätstheorie, *in elasticity* 192–195, 207, 209, 210, 212, 219, 220, 222, 241, 245, 251, 254, 266, 268, 270, 271, 273, 280.
—, Definition einer positiv definiten Funktion, *definition of positiv-definite function* 132.
—, Diskussion, *discussion* 135, 158.
—, Einfluß auf Stabilität, Beispiele, *influence on stability, examples* 135, 136, 138.
— in diskreten Systemen, *in discrete systems* 141.
— und Beschränktheit und Existenz, *and boundedness and existence* 135.
— — — in der Elastizitätstheorie, *in elasticity* 171, 172, 210–213.
Massendichte, *mass density* 536.
Massenmittelpunkt, *mass centre* 217, 218, 247.
Material, elastisches, *material, elastic* 19, 175, 315.
—, —, einem viscoelastischen Material zugeordnet, *related to viscoelastic* 19.
—, elastisch-plastisches, *elastic-plastic* 537.
— mit Gedächtnis, *with memory* 172.
—, plastisches, *plastic* 403, 412–415.
—, thermoelastisches, *thermoelastic* 183.
Materialgleichungen oder -relationen für Elastizitätstheorie, *constitutive equations or relations for elasticity* 19, 175.
— — — für Thermoelastizität, *for thermoelasticity* 175.
— — — für Viscoelastizität, *for viscoelasticity* 16, 18.
Materialien mit Nebenbedingungen, *constrained materials* 2, 71.
materielle Ableitung, *material derivative* 404, 440, 540.
— Symmetrie, *symmetry* 2, 18, 29.
— Unstetigkeit, *discontinuities* 429, 434.
materieller Gesichtspunkt, *material standpoint* 428, 524.
— Punkt, *point* 16.
mathematische Bezeichnung, *mathematical notation* 125.
— Modelle, *models* 131, 150, 213.
Mathieu-Hillsche Gleichung, *Mathieu-Hill equation* 257.
maximale Scherung, *maximum shear* 442.
— und minimale Scherspannung, *and minimum shearing stress* 423, 465.
Maximalprinzipien, *maximum principles* 152.
Maxwell-Element, *Maxwell element* 40.
Maxwell-Elemente in Reihe, *Maxwell elements in parallel* 41.
Maxwell-Operatoren, *Maxwell operators* 39.
Maxwellscher Satz, *Maxwell's theorem* 98, 101.
mechanische Anregung, *mechanical forcing* 31.
Membran, *membrane* 208, 436, 497.
Menge aller linearen Transformationen, *set of all linear transformations* 4.
— $\mathscr{B}(T, X)$ 127, 128.
— X 127, 128, 130.
— von Anfangswerten, *of initial data* 130.
Michell s. Beltrami
minimalisierende Folge, *minimizing sequence* 564, 565.
mittlere Normalspannung und Normaldehnung, *mean normal stress and strain* 71.
— Spannung, *mean stress* 216, 240–243, *intermediate stress* 514.
Mischungen, *mixtures* 172.
Mohrsche Kreise, *Mohr circles* 407, 468, 477.
Momentenbilanz, *moments, balance of* 42, 216, 242, 243.
Monge-Ampèresche Gleichung, *Monge-Ampère equation* 441.
multipolare Elastizitätstheorie, *multipolar elasticity* 172, 213.
Murnaghansche elastische Konstanten zweiter Ordnung, *Murnaghan's second-order elastic constants* 181.

Nachbarschaft, *neighbourhood* 132, 133.
Nachwirkung, *hereditary response* 1.
Nachwirkungs-Wellenoperator ($\Box_K$), *hereditary wave operator* ($\Box_K$) 92.
Näherungslösung, *approximate solution* 491.
natürlicher Zustand, *natural state* 195.
Netzwerk linear-elastischer und viscoser Elemente (Federn und Dämpfer), *networks of linearly elastic and viscous elements (springs and dashpots)* 34.
neutrales Gleichgewicht, *neutral equilibrium* 144, 170, 202, 204, 205, 226, 227.
Newtonsches Potential, *Newtonian potential* 76.
Nicht-Eindeutigkeit, *non-uniqueness* 151, 157, 224.
nicht-einfache Bewegungen, *nonsimple motions* 112.
Nicht-Homogenität, *non-homogeneous condition* 403.
nicht hyperbolisch, *not hyperbolic* 526.
nichtkonservative Probleme, *non-conservative problems* 258, 279.
nichtlineare Liapounovsche Funktionen, *nonlinear Liapounov functions* 187–191.
— — Stabilität, Bedingungen, *stability, conditions* 192.
— — — in bezug auf lineare Probleme, *related to linear problems* 194.
— Theorien der Mechanik, *theories of mechanics* 1, 11.
— Thermoelastizität, Gleichungen der gestörten Bewegung, *thermoelasticity, equations of perturbed motion* 176, 177.
— —, Grundgleichungen, *basic equations* 173–176.

nichtlinearer natürlicher Zustand, Bedingungen für Stabilität, *nonlinear natural state, conditions for stability* 195.
nicht-selbstadjungierte Gleichungen, *non-self-adjoint equations* 258, 279.
— Probleme, *problems* 170, 264, 280.
nicht-singuläre Anfangs- und Gleichgewichtsreaktion, *non-singular initial and equilibrium response* 26.
Normaldehnungszuwachs, *normal strain rate* 442.
Normale, äußere, *normal, outward* 47.
Normalen, *normals* 497.
Normalenkegel, *cone of normals* 424.
normalisierter Kelvinprozeß, *normalized Kelvin process* 78.
Normalspannung, *normal stress* 441.
Normalsystem, *normal system* 421.
Nullösung, *null solution* 129, 134, 149, 150, 207, 209, 211, 224, 231, 239, 246–248, 251, 266, 268, 273, 277, 280–284.
Nullmoment, *zero moment* 217.
numerische Verfahren, *numerical techniques* 3.

Oberfläche konstanter Neigung, *surface of constant slope* 497.
Oberflächenbelastung, *surface loads* 85, 278.
Oberflächenkraft, *surface force* 174, 278.
Oberflächenpotentialdichte, *surface potential density* 259, 260.
Oberflächenspannungszug, *surface traction* 42, 85, 178, 259.
ohne Seitenausdehnung, *zero extension* 506.
ONSAGER, Prinzip von, *Onsager's principle* 23.
Operatorenkalkül, *operational calculus* 68.
Ordnung eines Operatorpaars, *order of a pair of operators* 34.
orthogonale Zykloiden, *orthogonal cycloids* 456.
ortsabhängige Werte, *position-dependent data* 264–278.
oscillatorische Verschiebungsprozesse, *oscillatory displacement processes* 106.
ovaler Querschnitt, *oval cross-section* 503.

Paar $< P, Q >$ gehört zur Relaxationsfunktion **G**, *pair* $< P, Q >$ *belongs to the relaxation function* **G** 34.
Parabelbedingung, *parabola condition* 473.
Parabelgrenzwert, *parabola limit* 462, 488.
parabolischer Punkt, *parabolic point* 481.
parabolisches Problem, *parabolic problem* 460.
parametrische Anregung, *parametric excitation* 256, 278.
— Stabilität, *stability* 187.
Pearsonsche Formulierung, *Pearson's formulation* 202.
Periode, *period* 129.
periodische Bewegungen, *periodic motions* 129.
— Spannungs- und Dehnungsprozesse, *stress and strain processes* 21, 31, 32.
Phasenabbildung, *phase map* 131.
Phasendispersion, *phase dispersion* 107, 108.
Phasengeschwindigkeit, *phase velocity* 107.
Phasenverschiebung, *phase lag* 22.
physikalische Ebene, *physical plane* 445, 471, 477.
Piola-Kirchhoffsche Spannung, *Piola-Kirchhoff stress* 174, 181, 183, 185, 316, 542.
plastische Arbeit, *plastic work* 407.
— Kraft, *power* 546.
— Masse, zwischen zwei rauhe starre Platten gepreßt, *mass pressed between two rough rigid plates* 526.
— Torsion 424.
plastischer Deformationsgradient, *plastic deformation gradient* 550.
— Körper, *body* 403.
plastisches Gleichgewicht, *plastic equilibrium* 523.
— Nachgeben, Bedingung oder Kriterium, *yield condition or criterion* 403, 405, 411, 461, 463, 464, 471, 473, 487, 548, 549, 561.
— —, Funktion, *function* 410.
— —, Mechanismus, *mechanism* 439.
— —, Spannung bei, in einfachem Zug, *stress in simple tension* 405.
— —, Stärke, *strength* 584, 586.
— —, überall, *overall* 516.
— Potential 408, 465.
plastisch-elastische Randfläche, *plastic-elastic boundary* 500, 517.
plastisch-starre Probleme, *plastic-rigid problems* 523.
— Randfläche, *boundary* 494, 520, 528.
plastisch-starres Gebiet, *plastic-rigid region* 501.
Platte, allgemein, *plate, general* 136, 137, 172, 278, 279, 281.
Plattenanalogie, *plate analogy* 521.
plötzlicher Übergang, *sudden transition* 414.
Poissonsche Zahl, *Poisson's ratio* 412.
Polarisationsvektor einer ebenen fortschreitenden Welle, *polarization vector of a plane travelling wave* 107.
— — Unstetigkeitsfläche, *of a singular surface* 97.
positiv-definite Elastizität, *positive-definite elasticity* 62.
— — Funktionen, *functions* 134, 135, 168.
— — — als Maße, *as measures* 135.
— — —, Beispiele, *examples* 133.
— — —, Definitionsgebiet, *sets of definition* 133.
— — —, Diskussion, *discussion* 186.
— — —, $F_{\tau,t}$ 154, 155, 157, 163, 164, 168, 169, 207, 212.
— — — in der Variationsrechnung, *in calculus of variations* 164.
— — — und Mengen, Definition, *and sets, definition* 133.
— — — und stetige Abhängigkeit, *and continuous dependence* 145.
positive longitudinale Elastizität, *positive longitudinal elasticity* 317.

Potential, plastisches, *plastic potential* 408.
Potentialdichte, *potential density* 259, 261.
Potentialfunktion, *potential function* 259, 260, 287.
Potentiallinien, *potential lines* 460.
Prandtl-Reußsche Gleichungen, *Prandtl-Reuss equations* 424, 548.
Prinzip der Entropieproduktion, *principle of production of entropy* 539.
— — Systemunabhängigkeit, *frame-indifference* 544, 545.
— — Undurchdringlichkeit, *impenetrability* 536, 538.
Problem der abgeschlossenen Geschichte, *past history problem* 50–53.
— mit stationären Werten, *stationary value problem* 66.
Prozeß, Prozesse, Definition, *process, processes, definition* 4.
— der Klasse C^N, *of class* C^N 5.
— — — C^∞ 5.
—, dynamische, viscoelastische, *dynamic viscoelastic* 89.
—, eingeschränkter, *restricted* 5.
—, konstanter, *constant* 5.
—, lokal integrierbarer, *locally integrable* 5.
—, quasistatischer, viscoelastischer, *quasistatic viscoelastic* 42–44.
—, starker, *strong* 6.
—, stetiger, *continuous* 5.
—, —, im Unendlichen, *continuous at infinity* 5.
—, unstetiger, *discontinuous* 6.
—, —, Approximation unstetiger durch stetige, *approximation by continuous* 16, 35.
— vom Heaviside-Typ, *of Heaviside type* 5, 16.
pseudostationär, *pseudosteady* 525.
Pulsfront, *pulse front* 96.
Punkte, *points* 4.
Punktfeld, *point field* 7.

quadratische Bedingung für plastisches Nachgeben, *quadratic yield condition* 403, 405, 461, 471, 487, 549.
quasilinear, *quasi-linear* 440.
quasistatische Annahme, *quasi-static assumption* 42.
— lineare Viscoelastizität, *linear viscoelasticity* 42.
— Theorie, *theory* 2.
— Variationsprinzipien, *variational principles* 64.
quasistatischer viscoelastischer Prozeß, entsprechend der Kriechfunktion $\boldsymbol{J}$ und der Volumkraft $\boldsymbol{b}$, *quasi-static viscoelastic process, corresponding to creep function* $\boldsymbol{J}$ *and body force* $\boldsymbol{b}$ 43.
— — —, entsprechend der Relaxationsfunktion $\boldsymbol{G}$ und der Volumkraft $\boldsymbol{b}$, *corresponding to relaxation function* $\boldsymbol{G}$ *and body force* $\boldsymbol{b}$ 43, 77.
quasistatisches Randwertproblem, *quasi-static boundary value problem* 47, 69.
quellenfreies Vektorfeld, *solenoidal vector field* 73, 557.

Randbedingungen, allgemein, *boundary conditions or data, general* 46, 47.
— für Spannung, *stress* 175, 215, 258, 268.
— für Temperatur, *temperature* 184, 188, 250.
—, gemischte, *mixed* 180, 184.
—, homogene, *homogeneous* 234.
— mit separierbaren und gleichzeitigen Volumkräften, *separable and synchronous body forces* 73.
Randpunkte, *boundary points* 47.
Randwertproblem, *boundary-value problem* 3, 47, 69, 95, 495.
Rate der Ausdehnung, *rate of dilatation* 474.
— — — in einer Richtung, *of extension* 420.
Raum $C_0(\mathscr{R}^+;\mathscr{X})$ von stetigen $(\mathscr{X})$-Geschichtsfunktionen, die im Unendlichen verschwinden, *space* $C_0(\mathscr{R}^+;\mathscr{X})$ *of continuous* $(\mathscr{X})$*-histories which vanish at infinity* 7.
—, euklidischer Punkt, *Euclidean punkt* 4.
— $L^1(\mathscr{R}^+;\mathscr{X})$ von integrierbaren $(\mathscr{X})$-Geschichtsfunktionen, $L^1(\mathscr{R}^+;\mathscr{X})$ *of integrable* $(\mathscr{X})$*-histories* 7.
— mit skalarem Produkt, *inner product* 4.
—, Translation 4.
—, Vektor, *vector* 4.
— von linearen Abbildungen in $\mathscr{T}$ (dem Raum von Tensoren vierter Stufe): $[\mathscr{T}]$, *of linear maps in* $\mathscr{T}$ *(the space of fourth-order tensors):* $[\mathscr{T}]$ 4.
— von Tensoren zweiter Stufe $[\mathscr{V}]\mathscr{T}$, *of (second-order) tensors* $[\mathscr{V}]\mathscr{T}$ 4.
räumliche Darstellung, *spatial description* 428, 536.
rechte materielle Ableitung, *right material derivative* 541.
— Umkehrung, *inverse* 23.
reduzierbare Gleichungen, *reducible equations* 459.
reduzierbares System, *reducible system* 440.
Regularitätserhaltung, *regularity, preservation of* 14.
Regularitätssätze, *regularity theorems* 74.
Reibungskoeffizient, *coefficient of friction* 452.
reine Ausdehnung, *pure dilatation* 30.
— Scherbewegung, *shearing motion* 30.
— Scherspannung, *shear stress* 30.
Relaxationsfunktion $\boldsymbol{G}$, *relaxation function* $\boldsymbol{G}$ 18.
— für Kriechnachgiebigkeit $\boldsymbol{J}$, *for creep function (compliance)* $\boldsymbol{J}$ 26.
— für Scherwirkung, *for shear response* 71.
— und Kriechnachgiebigkeit, *and creep compliance* 28.
Relaxationsfunktional, *relaxation functional* 17.
Relaxationsfunktionalgesetz für den Körper $\mathscr{B}$, *relaxation law for body* $\mathscr{B}$ 17.
— in $\boldsymbol{x}:\mathscr{L}\boldsymbol{x}$, *at* $\boldsymbol{x}:\mathscr{L}\boldsymbol{x}$ 17.

Relaxationsprobleme, Variationsprinzipien, *relaxation problems, variational principles* 65, 66, 113.
Relaxationsverhalten und Kriechgesetz, *relaxation and creep laws* 23.
Relaxationszeit für ein Maxwell-Element, *relaxation time for a Maxwell element* 41.
Relaxationszeiten und Differentialoperator, *relaxation times and differential operator* 37, 38.
Rellichsches Auswahlprinzip, *Rellich selection principle* 576.
Restentropieungleichung, *residual entropy production inequality* 175, 183, 188.
Restgleichung der Energiebilanz, *residual equation of energy balance* 174, 175, 183.
Restverschiebungen, *residual displacements* 219.
Retardierungszeiten, *retardation times* 38, 42.
Reziprozitätstheorem, *reciprocal theorem* 55.
Richtung, *direction* 484.
— der Ausbreitung, *of propagation* 97, 306.
Richtungsableitung, *directional derivative* 444.
Richtungsbedingung, *direction condition* 426.
Riemann-Darstellung, *Riemann representation* 104.
Riemann-Funktion, *Riemann function* 449, 450.
Riemann-Methode, *Riemann method* 449.
Riemann-Lebesgue-Lemma 21.
Ring 424, 515.
Ritzsches Darstellungstheorem, *Ritz's representation theorem* 558.
rotierende Zylinder und Scheiben, *rotating cylinders and disks* 521.
rückwärts, *backward* 486.

Schale, *shell* 172, 287.
Schallgrenzlinie, *sonic limit line* 482.
Schallgrenzpunkt, *sonic limit point* 482.
Schallinie, *sonic line* 480.
Schaudersches Fixpunkttheorem, *Schauder's fixed-point theorem* 577.
Scheibe, *sheet* 515.
Scheitelpunkte, *locus of cusps* 477, 479, 483.
Scherdehnung, *shear strain* 442.
Schermodul, *shear modulus* 30.
Scherspannung, *shear stress* 442.
schiefsymmetrische Tensoren, Unterraum der, *skew tensors, subspace of* 4.
schiefsymmetrischer (infinitesimaler) Rotationsprozeß **W**, *skew (infinitesimal) rotation field process* **W** 17.
schließlich monoton, *eventually monotone* 20.
Schnitteigenschaft des plastischen Teils, *intersection property of the plastic part* 580.
Schnüre, *strings* 278.
schwach-asymptotische Stabilität, Definition, *weakly asymptotic stability, definition* 142.
schwach-asymptotische Stabilität, Satz, *weakly asymptotic stability, theorem* 158.
— schwach gekrümmte Tafel, *shallow curved panel* 281.
— konservative Kräfte, *weakly conservative forces* 193, 258–264, 286.
— konservativer Spannungsvektor, *conservative stress vector* 263.
schwache Kompaktheit, *weak compactness* 559.
— Lösungen, *solutions* 162, 171, 173, 178, 179, 211, 215, 224, 229, 258.
schwacher quasistatischer viscoelastischer Prozeß, *weak quasi-static viscoelastic process* 44.
schwaches Differential, *weak differential* 552.
Schwarzsche Ungleichung, *Schwarz's inequality* 559.
schwindendes Gedächtnis, *fading memory* 2, 334.
selbstadjungiert, *self-adjoint* 264.
Semiäquivalenz, *semiequivalence* 139.
semi-inverse Methode, *semi-inverse method* 560.
Separation von Variablen, *separation of variables* 255, 256, 284.
separierbar, *separable* 170, 557.
Shieldscher Satz, *Shield's theorem* 213.
Shinbrotscher Fixpunktsatz, *Shinbrot's fixed-point theorem* 558.
singuläre Bedingungen plastischen Nachgebens, *singular yield conditions* 411.
— Flächen (s. auch Beschleunigungswellen, Stoßwellen), *surfaces (see also accleration wave, shock wave)* 96, 97, 305.
— — in Relation zu hochfrequenten Wellen, *related to high frequency waves* 109.
— Lösungen, *solutions* 77.
Singularität, Zweiglinien, *singularity, branch lines* 483.
Singularität, Zweigtyp, *singularity, branch type* 477.
skalares Feld, *scalar field* 7.
Skalenverschiebung, Funktion der, *rescaling function* 542, 543.
Sobolev-Einbettungssätze, *Sobolev embedding theorems* 140, 212.
— -Lemma 578.
— -Norm, *norms* 224.
— -Raum, *space* 212.
Sokolovskysche Bedingung, *Sokolovsky condition* 473.
SOMIGLIANA s. BOUSSINESQ.
Somiglianasche Spannungsfunktionen, viscoelastische, *Somigliana stress functions, viscoelastic* 94.
Spannung, CAUCHY, *stress, Cauchy* 175, 180, 183, 185, 325, 536.
—, PIOLA-KIRCHHOFF 174, 181, 183, 185, 316, 542.
Spannungsdeviatorkraft, *stress deviator force* 404.
Spannungsdeviatortensor, *stress deviator tensor* 404.

Spannungsebene, *stress plane* 445, 471.
Spannungsentropiemodul, *stress entropy modulus* 362, 369.
Spannungsentropievektor, *stress entropy vector* 537.
Spannungsfeldprozeß, *stress field process* 16.
Spannungsfunktionen, *stress functions* 496, 522, 561, 584.
—, viscoelastische, *viscoelastic* 73, 93.
Spannungsfunktions-Lösungen, *completeness of stress function solutions* 76, 93.
Spannungsgleichung für Gleichgewicht, *stress equation of equilibrium* 46.
— für Kompatibilität, *of compatibility* 45, 46.
Spannungsleistung, *stress power* 95, 174.
Spannungslinien, *stress lines* 496.
Spannungsmultipole, *stress multipoles* 213.
Spannungs- oder Kraftrandwerte, *stress or traction boundary data* 47.
Spannungs- oder Kraftrandwertproblem, *stress or traction boundary value problem* 47.
Spannungsprozeß, *stress process* 42.
Spannungsraum, *stress space* 406.
Spannungsrelaxation, *stress relaxation* 1, 18, 20, 32.
Spannungs-Relaxationseigenschaft, *stress relaxation property* 20.
Spannungs-Relaxationsfunktion, *stress relaxation function* 335, 350, 351.
Spannungsschaubild, *stress graph* 445, 477.
Spannungstensor, *stress tensor* 404.
Spannungsunstetigkeit, *stress discontinuities* 433, 457.
Spannungsunstetigkeitslinie, *stress discontinuity line* 498.
Spannungsvektor, *stress vector* 258–260, 261.
Speichermodul, *storage modulus* 33.
Spektrum von Differentialoperatorgesetzen, *spectrum of differential operator law* 38.
spezifische Entropie, *specific entropy* 536.
— freie Energie, *free energy* 536.
— innere Energie, *internal energy* 536.
— Wärme, *heat* 193, 201.
Spiralen, *spirals* 456.
Sprung der Geschwindigkeit, *jump of the velocity* 494.
— — Spannung, *stress* 435.
— einer Funktion, *of a function* 97.
Sprungunstetigkeit, *jump discontinuity* 14, 96, 97.
— einer ebenen Welle, *for plane waves* 97.
spurfrei-symmetrischer Tensorprozeß, *traceless symmetric tensor process* 71.
Stab, *rod* 172, 279, 282, 283, 285.
Stabilität, *stability* 3.
—, Definitionen, asymptotische, *definitions, asymptotic* 142.
—, —, —, der Bahnbewegung, *asymptotically orbital* 144.
—, —, bedingte, *conditional* 147.
—, —, bei Störung, *perturbation* 146 (Fußnote, *footnote*).
Stabilität, Definitionen, beständige, *stability, definitions, persistent* 187.
—, — der Bahnbewegung, *orbital* 144.
—, — des Gleichgewichts, *of equilibrium* 131.
—, — einer gegebenen Bewegung, *of given motion* 134.
—, — eines dynamischen Systems, *of dynamical system* 134.
—, —, endliche Zeit (praktische oder Lasalle und Lefschetz), *finite-time (practical or Lasalle and Lefschetz)* 143.
—, —, exponentiell asymptotische, *exponentially asymptotic* 142.
—, —, frühe, *early* 132 (Fußnote, *footnote*).
—, —, global asymptotische, *globally asymptotic* 142.
—, —, Hadamard 198.
—, —, Hölder 146.
—, —, infinitesimale Überstabilität, *infinitesimal superstability* 202.
—, —, kontraktive, *contractive* 144.
—, —, Lagrange 149.
—, —, Lagrange-Dirichlet 131.
—, —, —, wesentliche Bestandteile, *essential ingredients* 132.
—, —, Liapounov 133.
—, —, —, äquivalente Definitionen, *equivalent definitions* 132, 140.
—, —, —, Beobachtungen, *observations* 134.
—, —, lokale (Krasovskii), *local (Krasovskii)* 143.
—, —, neutrale, *neutral* 144.
—, —, parametrische, *parametric* 187.
—, —, quasi-kontraktive, *quasi-contractive* 143.
—, —, schwach asymptotische, *weakly asymptotic* 142.
—, —, Stabilitätsgebiet, *region of stability* 142.
—, —, trigonometrische, *trigonometric* 146 (Fußnote, *footnote*).
—, —, Überstabilität, *superstability* 203.
—, —, uniform asymptotische, *uniformly asymptotic* 142.
—, —, uniforme, *uniform* 141.
—, Theoreme, *theorems*
—, —, Diskussion der, *discussion of* 156–160.
—, —, — —, Beispiele, *examples* 160–163.
—, —, Energiekriterium, *energy criterion* 164 (Fußnote, *footnote*), 195.
—, —, hinreichende Bedingungen in nichtlinearer Thermoelastizität, *sufficiency in non-linear thermoelasticity* 192–194.
—, —, — —, natürlicher Zustand, *natural state* 195.
—, —, Liapounov, Nullösung, *further Liapounov, null solution* 209, 212, 219.
—, —, —, asymptotisch, *asymptotic* 152, 158.
—, —, — für die Elastizitätstheorie, *for elasticity* 172.
—, —, — für diskrete Systeme, *for discrete systems* 155, 159.

Stabilität, Definitionen, Liapounov, schwach asymptotisch, *stability, definitions, Liapounov, weakly asymptotic* 158.
—, —, —, uniform 152, 157.
—, —, Liapounov-Stabilität mit Maximalprinzipien, *Liapounov stability with maximum principles* 152.
—, —, — und Beschränktheit, *and boundedness* 149, 150.
—, —, — und Eindeutigkeit, *and uniqueness* 151.
—, —, Notwendigkeit, für Gleichgewichtslösung, *necessity, for equilibrium solution* 200–202.
—, —, Shieldscher Satz, *Shield's theorem* 213.
—, —, Veränderungen von LIAPOUNOV, ZUBOV-MOVCHAN, *variations of Liapounov, Zubov-Movchan* 155, 156.
—, — und Variationsrechnung, *and calculus of variations* 163–167, 195.
— und Knicken, *and buckling* 130.
—, Wahl in der Elastizitätstheorie, *choice in elasticity* 171.
Stabilitätsgrenze, *limit of stability* 202, 225, 244.
Standard-Differentialoperatorgesetz, *standard differential operator law* 39.
Standardmodell, *standard models* 40.
Stärke einer ebenen Welle (s. auch Amplitude), *strength of a plane wave (see also amplitude)* 100.
Stark dissipatives Material, *strongly dissipative material* 22.
starke Antwortfunktion, *strong response function* 13.
— Elliptizität, *ellipticity* 64, 99, 211, 213, 321.
— Unstetigkeiten, *discontinuities (see also shock)* 427, 434.
starker quasistatischer viscoelastischer Prozeß, *strong quasi-static viscoelastic process* 44.
starr, starre Bewegungen, *rigid, rigid-body motions* 17, 18, 23, 216, 246, 262, 266, 274.
starre Rotationen, *rigid rotations* 215, 216, 219, 239, 240, 242, 244, 246.
— Translationen, *translations* 215, 217, 219, 220, 239.
— translatorische Geschwindigkeiten, *translational velocities* 218, 220.
starr-plastische Rechnung, *rigid-plastic analysis* 524.
starr-plastisches Problem, *rigid-plastic problem* 495.
stationärer Zustand, *stationary state* 524.
statisch bestimmt, *statically determinate* 526.
stetig im Unendlichen, *continuous at infinity* 5.
stetige Abhängigkeit der Lösung von den Bedingungen, Beziehung zur Liapounov-Stabilität, *continuous dependence of the solution on its data, relation with Liapounov stability* 134, 145, 146.
stetige Abhängigkeit der Lösung von den Bedingungen, Definition, *continuous dependence of the solution on its data, definition* 145.
— — — — — — — lineare Thermoelastizität, *linear thermoelasticity* 253.
— — — — — — — mit zeit- und ortsabhängigen Bedingungen, *with time-dependent, positiondependent data* 273–275.
— — der Lösungen von den Anfangswerten, *of the solutions on the initial data* 134, 253.
— Prozesse, *processes* 35.
stetiger Übergang, *continuous transition* 414.
Stetigkeit, Definition, *continuity, definition* 12.
— einer Abbildung, *of map* 133, 140.
—, Kontinuitätsgleichung, *equation* 404.
Störungen, *perturbations* 171.
—, Methode, *method* 526.
—, verursacht durch Anfangswerte, *causes of, initial data* 186.
—, — — Belastung, *loading* 186.
—, — — materielle Eigenschaften, *material properties* 187.
Stoßbedingungen, *shock conditions* 433.
Stoßwelle, Definition, *shock waves, definition* 97, 311.
— in der Theorie der Plastizität, *in plasticity* 416.
— in elastischen Körpern, *in elastic bodies* 357–361.
— in inhomogenen elastischen Körpern, *in inhomogeneous elastic bodies* 382–385.
— in Körpern aus Materialien mit Gedächtnis, *in bodies of material with memory* 349–356.
— in Nichtleitern, *in non-conductors* 369–377.
— in Verbindung mit schwachen Lösungen, *in connection with weak solutions* 171.
—, Verdichtung oder Ausdehnung, *compression or expansion* 312.
Stromlinien, *streamlines* 455, 460, 479, 531.
Struttsches Diagramm, *Strutt's diagram* 257.
St. Venants Lösung, *St. Venant's solution* 87.
— Prinzip, *principle* 85.
— —, Abnahme von Spannungen, *decay of stresses* 87.
— —, bevorzugte Richtung, *preferred direction* 87.
— — der Elastostatik, *of elastostatics* 88.
— — für eine Faser, *for a filament* 88.
— — für viscoelastische Körper, *for viscoelastic bodies* 86.
— —, Körper beliebiger Form, *bodies of general shape* 87.
— —, punktweise Schätzung, *pointwise estimate* 88.
— — und Potentialtheorie, *and potential theory* 87.
— Vorschlag, *conjecture* 562.

Superposition, Eigenschaft, *superposition, property* 12.
—, Methode, *method* 448.
Symmetrie der Cauchyschen Spannung, *symmetry of Cauchy stress* 542.
— des akustischen Tensors, *of acoustic tensor* 99.
— von Greenschen Prozessen der ersten Art, *of Green's processes of the first kind* 82.
— — — — der zweiten Art, *of the second kind* 84.
Symmetriegruppe, *symmetry group* 18.
Symmetrietransformation, *symmetry transformation* 18.
symmetrische Anfangselastizität, *symmetric initial elasticity* 62.
— Tensoren, Unterraum der ($\mathcal{T}_{\text{sym}}$), *tensors, subspace of* ($\mathcal{T}_{\text{sym}}$) 4.
symmetrischer infinitesimaler Dehnungsfeldprozeß, *symmetric infinitesimal strain field process* 17.
symmetrisierter Geschwindigkeitsgradient, *stretching tensor* 548.
Systeme in der Kontinuummechanik, *systems in continuum mechanics* 172.
— von Differentialgleichungen, *of differential equations* 420.
Systemunabhängigkeit, *frame-indifference* 544, 545.

Tangentenmodul, *tangent modulus* 356, 358.
— zweiter Ordnung, *second-order* 356.
tatsächliche Deformationsform, *actual mode* 416.
Temperatur, *temperature* 174, 536.
Tensor, Definition 3, 4.
— der Winkelgeschwindigkeit, *of angular velocity* 415.
Tensorfeld, *tensor field* 7.
Tensorgesetze vom Differentialtypus, *tensor laws of differential type* 37.
Thermodynamik der Viscoelastizität, *thermodynamics of viscoelasticity* 3.
—, zweiter Hauptsatz, *second law* 537.
Thermoelastizität, *thermoelasticity* 173–176, 183–185, 187–195, 249–255.
thermorheologisch einfache Materialien, *thermorheologically simple materials* 3.
Titchmarshscher Satz, *Titchmarsh's theorem* 74.
topologische Form der Stabilität, *topological nature of stability* 139, 141.
Torsion 495.
Torsionssteifheit, *torsional rigidity* 585, 586.
Träger, *beam* 278, 503.
Trägheitskräfte, *inertia forces* 525.
Trajektore, *trajectory* 128, 151.
Trajektorie/Bahn, *trajectory/orbit* 129.
Translationsraum, *translation space* 4.
transversale Hauptwelle, transverse, *transversal, principal waves* 328.
— Unstetigkeit, *discontinuity* 430.
— Welle, *wave* 99, 310.
— — in elastischen Körpern, anisotropen, *in elastic bodies, anisotropic* 321–323.
transversale Unstetigkeit in elastischen Körpern, isotropen, *transversal, discontinuity in elastic bodies, isotropic* 328.
Trennfläche, *surface of separation* 426.
Trescasches Kriterium für plastisches Nachgeben, *Tresca's yield criterion* 411, 423, 462.
Truesdell und Nollsches Kriterium, *Truesdell and Noll's criterion* 202.

Übergangsabbildung, *transition map* 131.
Übergangsgebiet, *transition zone* 433, 435.
überhängende Enden, *overhanging ends* 531.
Überschall-Luftströmung, *supersonic airstream* 281.
Umkehrung von Boltzmannschen Gesetzen, *inversion of Boltzmann laws* 23.
Umkehrungssatz für Antwortfunktionen, *inversion theorem for response functions* 27, 64.
— für starke Antwortfunktionen, *for strong response functions* 25, 30.
Umkehrpunkt, *inflection* 484.
Umklappen, *snap-through* 148, 287–289.
unbegrenzte Körper, *unbounded bodies* 77.
undeformierter Zustand, *undeformed state* 538.
Undurchdringbarkeit, *impenetrability* 536, 538.
unendliche Geschichten, *infinite histories* 5.
— Platte, *slab* 529.
uniforme hydrostatische Belastung, *uniform hydrostatic loads* 262, 263.
— Stabilität, Definition, *stability, definition* 141.
— —, Satz (Korollar), *theorem (corollary)* 157.
— —, Satz für hinreichende Bedingungen, *sufficiency theorem* 152.
uniformer Druck, *uniform pressure* 30.
unstetige Lösungen, *discontinuous solutions* 416, 426.
— Prozesse, *processes* 35.
Unstetigkeit (s. auch Wellen), allgemeine, *discontinuity (see also waves), general* 429, 456.
— der Ordnung *n*, *of order n* 427.
— der Ordnung eins, *of order one* 429.
— der Ordnung zwei, *of order two* 430.
— der Spannung, *of stress* 457.
—, Klassifikation, *classification* 434.
—, Kompatibilitätsbedingung, *compatibility* 494.
—, stationäre, *stationary* 429.
Unstetigkeitsfläche, *discontinuity, surface* 416, 433.
unsymmetrisches Auspressen, *unsymmetrical extrusion* 525.
Unterräume symmetrischer und schiefsymmetrischer Tensoren $\mathcal{T}_{\text{sym}}$ und $\mathcal{T}_{\text{skw}}$, *subspaces of symmetric and skewsymmetric tensor* $\mathcal{T}_{\text{sym}}$ *and* $\mathcal{T}_{\text{skw}}$ 4.

Variation, *Definition* 64, 65.
— eines Funtionals, *of a functional* 112.
— in der Amplitude, zeitlich, *in amplitude with time* 96.
variationsmäßige Charakterisierung von Lösungen des quasistatischen Randwertproblems mit vorgeschriebener anfänglicher vergangener Geschichte, *variational characterization of solutions to the quasi-static boundary value problem with prescribed initial past history* 64.
Variationsprinzip, erstes, für Kriechprobleme, *variational principle, first, for creep problems* 66.
—, —, für Relaxationsprobleme, *for relaxation problems* 65, 68.
— für dynamische Kriechprobleme, *dynamic creep problems* 114.
— — — Relaxationsprobleme, *relaxation problems* 113.
— zweites, für Kriechprobleme, *second, for creep problems* 67.
—, —, für Relaxationsprobleme, *for relaxation problems* 66.
Variationsrechnung, Legendre-Hadamardsche Bedingung, *calculus of variations, Legendre-Hadamard condition* 166.
—, Minimumproblem, *minimum problem* 164.
— und Stabilität, *and stability* 158, 163–167, 195, 205.
—, van Hovescher Satz, *van Hove's theorem* 167.
—, wirkliches Minimum, *proper minimum* 205.
—, Wahl von positiv definiter Funktion, *choice of positive-definite function* 164 (Fußnote, *footnote*), 211.
Vektoren, *vectors* 3.
Vektorfeld, *vector field* 7.
verallgemeinerte Dehnungen, *generalized strains* 409.
— ebene Spannung, *plane stress* 440.
— Lösungen, *solutions* 128, 138, 254, 255.
— Somigliana-Lösung, *Somigliana solution* 94.
— Spannungen, *stresses* 409.
verallgemeinertes Potential, *generalized Potential* 439.
„verbleibende" Gleichungen der Plastizitätstheorie, *"remaining" equations of plasticity* 426, 456, 494.
— — für Scherwirkung, *for shear response* 71.
Verdichtungs-Beschleunigungswelle, *compressive acceleration wave* 311, 314.
Verdichtungsstoßwelle, *compression shock* 312.
Verdickung, *thickening* 519.
Verdrehung, *twist* 496, 582, 583.
Verdrehungswinkel pro Längeneinheit, *angle of twist per unit length* 496, 582, 583.
vergangene Geschichte eines Prozesses, *past history of a process* 4, 5.
— —, Problem 50–53.
verlorene Lösungen, *lost solutions* 485.
Verlustmodul, *loss modulus* 33.
Vernünftigkeit, *reasonableness* 146.
Verschiebung, Ableitung, *displacement, derivative* 308, 430.
—, allgemein, *general* 8.
— einer Funktion, *translate of a function* 128.
—, Geschwindigkeit der, *speed of* 307, 308, 429.
—, Gleichungen der Bewegung, *displacement equations of motion* 91, 92.
—, — im Gleichgewicht, *of equilibrium* 44.
—, — — — für isotrope und homogene elastische Körper, *for isotropic and homogenous elastic body* 45.
—, — — — für isotrope viscoelastische Körper, *for isotropic viscoelastic bodies* 44.
—, Randwerte, *boundary data* 47.
—, —, ausgedrückt durch Spannungen, *in terms of stresses* 48.
—, Randwertproblem, *boundary value problem* 47, 83.
Verschiebungsbeziehung, *displacement relation* 91.
Verschiebungs-Feldprozesse in einem Körper, *displacement field processes for a body* 8, 16, 42.
Verschiebungs-Gradientenprozesse, *displacement gradient processes* 8.
Verschiebungsproblem, charakterisiert durch Verschiebungen, *displacement problem, characterization in terms of displacements* 51.
— inkompressibler viscoelastischer Körper, *of incompressible viscoelasticity* 72.
Verschiebungsraum, *translation space* 4.
Verschiebungsstetigkeit, *translation continuity* 15.
verschobener Prozeß, *translate process* 12.
verschwindende Randwerte, *null boundary data* 56.
Vertauschung von abhängigen und unabhängigen Variablen, *exchange of dependent and independent variables* 443.
— — Stabilitäten, *stabilities* 226.
Verteilung parallelgeschalteter Maxwellscher Elemente, *distributed system of Maxwell element in parallel* 41.
verträgliche Werte, *compatible values* 426.
Verzerren, elastisches, *warping, elastic* 496.
—, plastisches, *plastic* 501, 505.
Verzweigung, *bifurcation* 125.
virtuelle Deformationsformen, *virtual modes* 416.
— Verschiebungen, *displacement* 197, 198, 202, 203, 261.
viscoelastische Analoga der Arbeits-Energie-Identitäten der Elastostatik, *viscoelastic analogs of work-energy identities of elastostatics* 53.
— Feldgleichungen (dynamisch), *field equations (dynamic)* 42, 91.

viscoelastische Grundlagen der linearen Theorie, *viscoelastic foundations of linear theory* 11.
— Lösung, *solution* 60.
— Potentialfunktionen, *potential function* 3.
— Spannungsanalysis, *stress analysis* 68.
— Spannungsfunktionen, *stress functions* 93, 94.
— Verallgemeinerung von Boussinesq-Somigliana-Galerkin und Boussinesq-Papkovitch-Neuber-Lösungen, *generalization of Boussinesq-Somigliana-Galerkin and Boussinesq-Papkovitch-Neuber solutions* 74.
viscoelastischer Kelvin-Prozeß, *viscoelastic Kelvin process* 78.
— Körper, *body* 17.
— Prozeß, der einem Differentialoperator-Gesetz entspricht, *process corresponding to differential operator law* 57.
viscoelastisches Analogon zum Boggioschen Theorem, *viscoelastic analogue to Boggio's theorem* 94.
— Gegenstück zu den Beltrami-Donati-Michellschen Gleichungen, *counterpart of Beltrami-Donati-Michell equations* 46.
— gemischtes Randwertproblem, *mixed boundary-value problem* 59.
— Verschiebungs-Randwertproblem, *displacement boundary value problem* 81.
— Zentrum der Rotation, *center of rotation* 81.
— — — Verdichtung, *compression* 80.
viscose Strömung, *viscous flow* 405.
Voigt s. Kelvin
vollständig elastisch oder plastisch, *fully elastic or plastic* 495.
vollständige Lösung des Problems des plastischen Keils, *complete solution of plastic wedge problem* 527.
— plastische Torsion, *plastic torsion* 560, 564.
— Welle, *wave* 488.
vollständiges Problem für die Ebene, *complete problem for plane* 438.
— — — ebene Dehnung, *plane strain* 511.
Vollständigkeit der dynamische Verschiebung erzeugenden Funktionen, *completeness of dynamic displacement generating functions* 92.
— — Spannungsfunktions-Lösungen, *stress function solutions* 76, 93.
Volterra-Gleichung, *Volterra's equation* 62.
— Lemma 56.
Volterras Eindeutigkeitssatz, *Volterra's uniqueness theorem* 56.
volumerhaltend, *volume preserving* 550.
Volumenkraft, *body force* 42, 174, 178, 263, 269–271, 273, 278, 283.
Volum-Potentialdichte, *volume potential density* 259.
v. Mises' allgemeine Gleichungen, *v. Mises' general equations* 410.
v. Mises' Bedingung für elastisches Nachgeben, *v. Mises yield condition* 403, 405, 461, 471, 487, 549.
— Extremalprinzip, *extremum principle* 409.
— lineare Gleichungen, *linear equations* 459.
— Theorie des plastischen Potentials, *theory of plastic potential* 408.
v. Mises-Reuss-Theorie, *v. Mises-Reuss theory* 413.
von selbst im Gleichgewicht, Kelvin-Prozeß, *self-equilibrated Kelvin process* 81.
— — — —, Oberflächenkräfte, *surface tractions* 83.
vorgegebene Werte entlang zweier sich schneidender Linien, *data given along two intersecting lines* 492.
Vorspannung, *pre-stress* 219.

w-Linien, *w-lines* 485.
Wachstum (s. Abklingen), *growth (see decay)*
Wärmefluß, *heat flux* 174, 175.
Wärmeleitungsgleichung, *heat conduction equation* 184, 185.
Wechselwirkung von Unstetigkeiten, *interaction of discontinuities* 458.
Welle, *shaft* 505.
— (s. auch Unstetigkeit), Achsen, *wave (see also discontinuity) axes* 97, 98.
—, Amplitude (s. auch zeitliche Veränderung der Amplitude), *amplitude (see also variation of amplitude)* 309–311, 315.
—, axial ähnliche, *axially similar* 101.
—, — unähnliche, *dissimilar* 102.
—, Beschleunigungswelle, *acceleration* 97, 309, 313, 334, 356, 362, 428–432, 434, 435.
—, Definition 96, 306.
—, Geschwindigkeit, *speed* 92, 96, 97, 197, 306–339, 345–347, 351–372, 378–380, 382–385.
—, Gleichung, *equation* 137, 160, 162.
—, Hauptwelle, *principal* 326, 328.
—, hochfrequente, in Beziehung zur Abnahme, *high frequency, related to decay* 109.
— in Körpern aus Material mit Gedächtnis, *in bodies of material with memory* 334–339, 344–356.
— — —, elastischen, *elastic* 315–334, 356–385.
— — —, — inhomogenen, *inhomogeneous* 378–385.
— — —, — isotropischen, *isotropic* 325–334.
— — —, —, Nichtleitern, *non-conductors* 362–377.
— — —, plastischen, *plastic* 416, 434, 435.
— — —, viscoelastischen, *viscoelastic* 1, 92, 96–104, 106–112.
—, longitudinale 99, 310, 315–321, 327, 328.
—, Stoßwelle, *shock* 97, 171, 311, 313, 349–361, 369–377, 382–385, 416.
—, transversale, *transverse* 99, 321–323, 328, 332, 334.

Welle, zeitliche Veränderung der Amplitude (Dämpfung, Abklingen, Anwachsen), *wave, variation of amplitude with time (attenuation, decay, growth)* 1, 96, 98, 106, 107, 110, 320–323, 328, 330–333, 341–348, 353, 364–376, 378–385.
—, zentrierte, *centered* 486.
Wellenachsen, *wave axes* 97, 98.
Wellenstärke, *strength of a wave* 100.
Wert von einem Prozeß zur Zeit *t*, *value of a process at time t* 4.
wiedergewinnbare Arbeit, *recoverable work* 3.
wirbelfreies Vektorfeld, *irrotational vector field* 73.
wirkliches Minimum von *V*, *proper minimum of V* 205.
Wirkung, Erwähnung, *action, remarked upon* 140.
Wirkungsabbildung, *action maps* 131.
wohlformuliert, *well-posed* 136, 146.

($\mathscr{X}$)-abgeschlossene Geschichte, *($\mathscr{X}$)-past-history* 4.
($\mathscr{X}$)-Feld, *($\mathscr{X}$)-field* 7.
($\mathscr{X}$)-Feldprozeß für $\mathscr{D}$, *($\mathscr{X}$)-field process for $\mathscr{D}$* 7.
($\mathscr{X}$)-Geschichte, *($\mathscr{X}$)-history* 4.
($\mathscr{X}$)-Prozeß, *($\mathscr{X}$)-process* 4.

Zeit, Rolle der gegenwärtigen, *time, role played by present* 13.
— *t* 4, 7.
Zeit-Translationsinvarianz, *time translation invariance* 12.
zeitabhängige Bedingungen, *time-dependent data* 264.
Zeitskala, Materialien unabhängig von, *rate independent materials* 542.
Zeitskalenunabhängigkeit, *rate independence* 543.
— der verallgemeinerten Spannung, *of the generalized stress* 544.
Zeitumkehrung, *time reversal* 23.
zentrierte Welle, *centered wave* 486.
Zubov, allgemeines System von, *Zubov, general system of* 130.
zugeordnete elastische Lösung, *associated elastic solution* 60.
zugeordneter elastischer Körper, *associated elastic body* 60.
zugeordnetes elastisches Material, *associated elastic material* 112.
— — Problem 60, 111.
Zugspannung (s. auch Spannung), *traction (see also stress)*
Zugspannungsproblem für inkompressible Medien, *traction problem for incompressible media* 72.
Zugspannungs-(s. auch Spannungs-)Randwertproblem, *traction (see also stress) boundary value problem* 83.
zulässiger Prozeß für einen Körper $\mathscr{B}[\boldsymbol{u}, \boldsymbol{E}, \boldsymbol{S}]$, *admissible process for a body* $\mathscr{B}[\boldsymbol{u}, \boldsymbol{E}, \boldsymbol{S}]$ 42.
— Zustand für einen Körper $\mathscr{B}[\hat{\boldsymbol{u}}, \hat{\boldsymbol{E}}, \hat{\boldsymbol{S}}]$, *state for a body* $\mathscr{B}[\hat{\boldsymbol{u}}, \hat{\boldsymbol{E}}, \hat{\boldsymbol{S}}]$ 42.
Zusammentreffen zweier Lösungen, *meeting of two solutions* 480.
Zustandsinkrement, *incremental state* 188, 189.
Zweiglinien, *branch lines* 483.
Zweiglinien-Singularitäten, *branch type singularities* 477.
zweite Piola-Kirchhoffsche Spannung, *second Piola-Kirchhoff stress* 542.
zweiter Hauptsatz der Thermodynamik, *second law of thermodynamics* 537, 539, 540.
Zweitbewegung, *secondary motion* 178, 179, 186, 187, 279.
Zweitzustand, *secondary state* 176, 177, 256.
zweites oder charakteristisches Problem, *second, or characteristic problem* 492.
Zykloiden, *cycloids* 453, 456, 529, 532.
Zylinder, *cylinder* 411.

Subject Index.

(English-German.)

Where English and German spellings of a word are identical, the German version is omitted.

ablating boundaries, *Ablation der Grenzfläche* 3.
absolute discontinuity, *absolute Unstetigkeit* 427.
— stress discontinuity, *Spannungsunstetigkeit* 434.
— temperature, *Temperatur* 174.
acceleration waves, definition, *Beschleunigungswellen, Definition* 309, 313.
— — in bodies of materials with memory, *in Materialien mit Gedächtnis* 334.
— — in elastic bodies, *in elastischen Körpern* 356.
— — in general, *allgemein* 97, 309, 313, 428–432.
— — in inhomogeneous elastic bodies, *in inhomogenen elastischen Körpern* 378.
— — in non-conductors, *in Nichtleitern* 362.
— — in plasticity, *in plastischen Körpern* 434, 435.
acoustic tensor, *akustischer Tensor* 98, 316, 324, 326.
action maps, *Wirkungsabbildung* 131.
—, remarked upon, *Wirkung, erwähnt* 140.
actual mode, *tatsächliche Deformationsform* 416.
adiabatic condition, *Adiabatenbedingung* 460.
adjacent equilibrium, method of, criticism, *benachbartes Gleichgewicht, Methode des, Kritik* 202, 203–206.
— —, — —, discussion, *Diskussion* 224–226.
— —, — —, follower forces, *Folgekräfte* 282–284.
— —, — —, history and application, *Geschichte und Anwendung* 226–228.
— —, — —, weakly conservative forces, *schwach konservative Kräfte* 263.
adjoint (see also self-adjoint and nonself-adjoint), *adjungiert (s. auch selbstadjungiert und nicht-selbstadjungiert)*
— operator, *adjungierter Operator* 170, 276, 280, 284, 285.
— problem, *adjungiertes Problem* 285.
admissible process for a body: $\mathscr{B}$ [$\boldsymbol{u}$, $\boldsymbol{E}$, $\boldsymbol{S}$], *zulässiger Prozeß für einen Körper:* $\mathscr{B}$ [$\boldsymbol{u}$, $\boldsymbol{E}$, $\boldsymbol{S}$] 42.
— state for a body: $\mathscr{B}$ [$\hat{\boldsymbol{u}}$, $\hat{\boldsymbol{E}}$, $\hat{\boldsymbol{S}}$], *Zustand für einen Körper:* $\mathscr{B}$ [$\hat{\boldsymbol{u}}$, $\hat{\boldsymbol{E}}$, $\hat{\boldsymbol{S}}$] 42.
AIRY's stress function, *Airysche Spannungsfunktion* 441, 496, 522.
ALFREY-GRAFFI theorem on elastic-viscoelastic correspondence, *Alfrey-Graffi-Satz über elastisch-viscoelastische Korrespondenz* 72, 73.
amplitude (see also strength) of a process or history, *Amplitude (s. auch Wellenstärke) von einem Prozeß oder einer Geschichte* 7.
— of shock, *einer Stoßwelle* 313.
— vector, *Amplitudenvektor* 309.
angle of twist per unit length, *Verdrehungswinkel pro Längeneinheit* 496, 582, 583.
anisotropic, *anisotrop* 18.
approximate solution, *Näherungslösung* 491.
associated elastic body, *zugeordneter elastischer Körper* 60.
— — material, *zugeordnetes elastisches Material* 112.
— — problem 60, 111.
— — solution, *zugeordnete elastische Lösung* 60.
asymptotic behavior, theorem on, *asymptotisches Verhalten, Satz über* 32.
— stability, *asymptotische Stabilität* 32.
— — and boundedness, *und Beschränktheit* 169.
— — and snap-through, *und Umklappen* 288.
— —, definition 142.
— —, HALE's invariance principle, *Halesches Invarianzprinzip* 169, 250, 254, 255.
— —, linear thermoelasticity, *lineare Thermoelastizität* 249, 254, 255.
— —, necessary conditions for, *notwendige Bedingungen für* 200 (footnote, *Fußnote*).
— —, of null solution with dissipative loads, *der Nullösung mit dissipativen Belastungen* 287.
— —, region of, *Gebiet der* 142, 158, 169.
— —, remarked upon, *erwähnt* 129.
— —, sufficiency theorem, *hinreichende Bedingung* 152.
— —, theorem (corollary), *Satz (Korollar)* 158, 169.
— —, — of HSU and LEE, *Sätze von Hsu und Lee* 277.
— — without eigenfunction decomposition, *ohne Eigenfunktionszerlegung* 275.
attentuation matrix, reduced, *Dämpfungsmatrix, reduzierte* 102, 103.

attentuation of damped, propagating singular surfaces, *Dämpfung von sich ausbreitenden singulären Flächen* 106, 107.
— of waves, *von Wellen* 1.
— operator, *Dämpfungsoperator* 110.
— tensor, *Dämpfungstensor* 101.
auxiliary functions (see Liapounov functions), *Hilfsfunktionen (s. Liapounov-Funktionen)* 153.
— solution, *Hilfslösung* 61.
axes, principal mean stress, *Hauptachsen der mittleren Spannung* 216, 240–243.
—, wave, *Wellenachsen* 97, 98.
axial force, *axiale Kraft* 507.
— symmetry, *Symmetrie* 422, 525.
axially dissimilar waves, *axialunähnliche Wellen* 102.
— similar waves, *axialähnliche Wellen* 101.
axis of equilibrium, *Gleichgewichtsachse* 216, 217, 219, 229, 239, 242.

Babuška and Hlavaček's theorem, *Babuška und Hlavačeksches Theorem* 61.
backward, *rückwärts* 486.
balance of energy, *Bilanz der Energie* 174, 175.
— — linear and angular momentum, *des Impulses und Drehimpulses* 42.
bars, *Balken* 279.
basic equations of elasticity, *Grundgleichungen der Elastizitätstheorie* 173–176.
— — of perturbed motion (see perturbed motion), *der gestörten Bewegung (s. gestörte Bewegung)*
— — of plasticity, *der Plastizität* 403.
beam, *Träger* 278, 503.
Beatty's criterion, *Beattysches Kriterium* 201, 205.
Beltrami-Donati-Michell equations, generalized, *verallgemeinerte Beltrami-Donati-Michell-Gleichungen* 45, 46.
Bernoulli equation, *Bernoulli-Gleichung* 339.
Betti's theorem in viscoelasticity, *Bettisches Theorem in der Viscoelastizität* 55.
bifurcation, *Verzweigung* 125.
biharmonic equation, *biharmonische Gleichung* 75.
body $\mathscr{B}$, definition, *Körper $\mathscr{B}$, Definition* 16.
— with closure $\bar{\mathscr{B}}$ and boundary $\partial\mathscr{B}$, *mit abgeschlossener Hülle $\bar{\mathscr{B}}$ und Rand $\partial\mathscr{B}$* 42.
body force, *Volumenkraft* 42, 174, 178, 263, 269–271, 273, 278, 283.
Boggio's theorem, viscoelastic counterpart, *Boggioscher Satz, viscoelastisches Gegenstück* 94.
Boltzmann laws, *Boltzmann-Gesetz* 12, 18.
— —, additional properties, *zusätzliche Eigenschaften* 31.
— —, characterization of, *Charakterisierung* 14, 15.
— —, creep, *Kriechgesetz* 43, 89.
Boltzmann laws, reduction of differential operator law, *Boltzmann-Gesetz, Reduzierung auf Differentialoperator-Gesetz* 36.
— —, relaxation, *Relaxationsgesetz* 43, 89, 91.
— —, with (strong) response function, *mit (starker) Antwortfunktion* 13.
— operator, kernel, *Boltzmann-Operator, -Kern* 10.
boundary, ablating, *Ablation der Grenzfläche* 3.
— conditions or data, general, *Randbedingungen, allgemein* 46, 47.
— — — —, homogeneous, *homogene* 234.
— — — —, mixed, *gemischte* 180, 184.
— — — —, separable and synchronous body forces, *mit separierbaren und gleichzeitigen Volumkräften* 73.
— — — —, stress, *für Spannung* 175, 215, 258, 268.
— — — —, temperature, *für Temperatur* 184, 188, 250.
— in wave propagation, *Grenzfläche in Wellenausbreitung* 3.
— points, *Randpunkte* 47.
boundary-value problem, *Randwertproblem* 3, 47, 69, 95, 495.
bounded histories, *beschränkte Geschichten* 5.
boundedness and asymptotic stability, *Beschränktheit und asymptotische Stabilität* 169.
— and Hale's invariance principle, *und Halesches Invarianzprinzip* 169.
— of map, definition, *einer Abbildung, Definition* 148.
— of null solution, equiboundedness, *einer Nullösung, Äquibeschränktheit* 149.
— of solution, *der Lösung* 148.
—, ultimate, *endgültige* 149.
—, uniform, *uniforme* 149, 231.
Boussinesq-Papkovich-Neuber solution, generalized, *verallgemeinerte Boussinesq-Papkovich-Neuber-Lösung* 75–78.
— stress processes, *Spannungsprozesse* 78.
Boussinesq-Somigliana-Galerkin solution, generalized, *verallgemeinerte Boussinesq-Somigliana-Galerkin-Lösung* 75–77.
branch lines, *Zweiglinien* 483.
— type singularities, *Zweiglinien-Singularitäten* 477.
buckling load, *Knicklast* 225.
— mode, *Knickform* 226, 287, 288.

calculus of variations and stability, *Variationsrechnung und Stabilität* 158, 163–167, 195, 205.
— — —, choice of positive-definite function, *Wahl von positiv definiter Funktion* 164 (footnote, *Fußnote*), 211.
— — —, Legendre-Hadamard condition, *Legendre-Hadamardsche Bedingung* 166.
— — —, minimum problem, *Minimumproblem* 164.

calculus of variations, proper minimum, *Variationsrechnung, wirkliches Minimum* 205.
— — —, van Hove's theorem, *v. Hovescher Satz* 167.
cantilevered rod, *freitragender Stab* 282.
Cauchy data, *Cauchysche Bedingungen* 417, 425, 450.
— elastic bodies, *elastische Körper* 173.
— initial-value problem, *Cauchysches Anfangswertproblem* 105.
— problem 104, 421, 491.
— stress, *Cauchysche Spannung* 175, 180, 183, 185, 326, 536.
—, solution of initial value problem, *Lösung des Anfangswertproblems* 105, 106.
Cauchy-Poisson theorem, *Cauchy-Poissonsches Theorem* 42.
causality, *Kausalität* 12.
centered wave, *zentrierte Welle* 486.
characteristic boundary-value problem, *charakteristisches Randwertproblem* 530.
— condition, *charakteristische Bedingung* 424, 426.
— coordinates, *Koordinaten* 447, 468, 469, 490.
— direction, *Richtung* 451.
— initial-value problem, *charakteristisches Anfangswertproblem* 450, 491.
— line, *charakteristische Linie* 418, 419, 443, 485.
— quadrangle, *charakteristisches Rechteck* 491.
— segment 430.
— surface, *charakteristische Fläche* 416, 424.
— — elements, *Flächenelemente* 423.
characteristics, fixed, *Charakteristiken, feste* 471.
— of the complete plane problem, *des vollständigen ebenen Problems* 464.
— of the v. Mises plasticity equations, *der v. Mises-Plastizitätsgleichung* 422.
circular hole, *kreisförmiges Loch* 519.
— ring, *kreisförmiger Ring* 499.
— tube, *kreisförmige Röhre* 192.
classes C^N and C^∞, *Klassen* C^N *und* C^∞ 5.
— H^{ac} 6.
— H^N and H^∞, *Klassen* H^N *und* H^∞ 6.
—, Heaviside, *Heaviside-Klasse* 6.
classical elastic law for isotropic materials, *klassisches Elastizitätsgesetz für isotrope Materialien* 29.
— solutions, *klassische Lösungen* 138, 171, 173, 178–180, 211, 224, 231, 250, 258.
classification of discontinuities, *Klassifikation von Unstetigkeiten* 434.
Clausius-Duhem inequality, *Clausius-Duhemsche Ungleichung* 537, 539, 540.
closed-end condition, *Bedingung für geschlossene Enden* 506, 508.
coefficient of friction, *Reibungskoeffizient* 452.
coefficients b_{ijkl}, *Koeffizienten* b_{ijkl} 182, 183.
coercive inequality, *koerzive Ungleichung* 225.
cohesion, *Kohäsion* 463.
Coleman's canonical free energy, *Colemansche kanonische freie Energie* 190.
compatibility conditions, *Kompatibilitätsbedingungen* 426, 432, 467, 471.
— —, extended, *erweiterte* 430.
— —, geometrical, *geometrische* 307, 429, 430.
— —, —, iterated, *iterierte* 307.
— —, kinematical, *kinematische* 308, 429, 430.
— —, —, iterated, *iterierte* 308.
— —, —, one-dimensional, *eindimensionale* 309.
— equation for traceless symmetric tensor process, *Kompatibilitätsgleichung eines spurfrei-symmetrischen Tensorprozesses* 71.
— equations, *Kompatibilitätsgleichungen* 45, 456.
— relations, *Kompatibilitätsrelationen* 411, 466, 494.
— theorem, *Kompatibilitätssatz* 45.
compatible values, *verträgliche Werte* 426.
complementary energy, *Komplementärenergie* 583, 584, 586, 587.
complete problem for plane, *vollständiges Problem für die Ebene* 438.
— — — — strain, *für ebene Dehnung* 511.
— solution of plastic wedge problem, *vollständige Lösung des Problems des plastischen Keils* 527.
— wave, *vollständige Welle* 488.
completely plastic torsion, *vollständig plastische Torsion* 560, 564.
completeness of dynamic displacement generating functions, *Vollständigkeit der dynamische Verschiebung erzeugenden Funktion* 92.
— — stress function solutions, *Spannungsfunktions-Lösungen* 76, 93.
complex acoustic tensor, *komplexer akustischer Tensor* 107.
— modulus, *Modul* 106.
— wave number, *komplexe Wellenzahl* 107.
compressible fluid flow, *kompressible Flüssigkeit, Strömung einer* 417, 459.
compression shock, *Verdichtungsstoßwelle* 312.
— modulus, *Kompressionsmodul* 30.
compressive acceleration wave, *Verdichtungs-Beschleunigungswelle* 311, 314.
concentrated load, *konzentrierte Lasten* 77, 78.
conditional stability, examples, *bedingte Stabilität, Beispiele* 147, 217, 219, 220.
— — in general, *allgemein* 147, 284.
conductivity tensor, *Leitfähigkeitstensor* 250, 251, 254, 255.
cone of normals, *Normalenkegel* 424.

conformal mapping, *konforme Abbildung* 521.
conservation law of energies, *Erhaltungssatz der Energien* 179, 207, 211, 236.
— — of mass, *der Masse* 174.
consistent initial past history, *konsistente anfängliche vergangene Geschichte* 51, 89.
constant process, *konstanter Prozeß* 5.
constitutive equations or relations for elasticity, *Materialgleichungen oder -relationen für Elastizitätstheorie* 19, 175.
— — — — for thermoelasticity, *für Thermoelastizität* 175.
— — — — for viscoelasticity, *für Viscoelastizität* 16, 18.
constrained materials, *Materialien mit Nebenbedingungen* 2, 71.
contact discontinuities, *Kontaktunstetigkeiten* 434.
— problems, *Kontaktprobleme* 3.
contained plastic domain, *abgeschlossenes plastisches Gebiet* 494, 519, 523, 524.
continuity, definition, *Stetigkeit, Definition* 12.
— equation, *Kontinuitätsgleichung* 404.
— of map, *Stetigkeit einer Abbildung* 133, 140.
continuous at infinity, *stetig im Unendlichen* 5.
— dependence of the solution on its data, definition, *stetige Abhängigkeit der Lösung von den Bedingungen, Definition* 145.
— — of the solution on its data, linear thermoelasticity, *der Lösung von den Bedingungen, lineare Thermoelastizität* 523.
— — of the solution on its data, relation with Liapounov stability, *der Lösung von den Bedingungen, Beziehung zur Liapounov-Stabilität* 134, 145, 146.
— — of the solution on its data with time-dependent, position-dependent data, *der Lösung von den Bedingungen, mit zeit- und ortsabhängigen Bedingungen* 273–275.
— — of the solutions on the initial data, *der Lösungen von den Anfangswerten* 134, 253.
— processes, *Prozesse* 35.
— transition, *stetiger Übergang* 414.
contour lines, *Konturlinien* 496, 497.
convex, *konvex* 410.
convolution and Heaviside processes, *Faltung und Heaviside-Prozesse* 28.
—, definition 8.
— for vectors and tensor processes, *für Vektor- und Tensorprozesse* 9.
—, properties, *Eigenschaften* 9.
coordinates, *Koordinaten* 536.
Coriolis forces, *Coriolische Kraft* 286.
correspondence, elastic-viscoelastic, *Korrespondenz, elastisch-viscoelastisch* 68, 69.
— principle, *Korrespondenzprinzip* 69.
— —, differential operator laws, *Differentialoperator-Gesetz* 69.
— — for simple free vibrations, *für einfache freie Schwingungen* 111, 112.
correspondence principle of ALFREY-GRAFFI, *Korrespondenzprinzip von Alfrey-Graffi* 72.
COULOMB's condition, *Coulombsche Bedingung* 463, 464, 473.
counter-examples of HELLINGER, HAMEL, SHIELD and GREEN, *Gegenbeispiele von Hellinger, Hamel, Shield und Green* 139, 160, 204, 210.
creep compliance **J**, *Kriechnachgiebigkeit* **J** 24.
— — and relaxation function, *und Relaxationsfunktion* 28.
— — for relaxation function **G**, *für Relaxationsfunktion* **G** 26.
— law, *Kriechgesetz* 23.
— problems, variational principle, *Kriechprobleme, Variationsprinzip* 66, 67, 114.
— type order of pair of differential operators, *Kriechtyp, Differentialoperatoren vom* 34.
critical initial amplitude, *kritische Anfangsamplitude* 342, 367, 368, 381.
— jump in strain gradient, *kritischer Sprung im Dehnungsgradienten* 353, 361, 374, 384.
— load, *kritische Last* 146, 277, 284, 285, 287.
— point, *kritischer Punkt* 288, 289.
— value, *Wert* 225, 226, 278, 283, 284.
cusps, *Scheitelpunkte* 477, 479, 483.
cycloids, *Zykloiden* 453, 456, 529, 532.
cylinder, *Zylinder* 411.

DAFERMOS' theorem, *Dafermosscher Satz* 106.
dashpots, *Dämpfer* 2.
data given along two intersecting lines, *vorgegebene Werte entlang zweier sich schneidender Linien* 492.
decay or growth of the strength of waves, elastic, *Abklingen oder Wachstum der Wellenintensität, elastisch* 320–323, 328, 330–333.
— — — of the strength of waves in elastic non-conductors, *der Wellenintensität in elastischen Nichtleitern* 362–376.
— — — of the strength of waves in general, *der Wellenintensität im allgemeinen* 341–344.
— — — of the strength of waves in inhomogeneous elastic bodies, *der Wellenintensität in inhomogenen elastischen Körpern* 378–385.
— — — of the strength of waves in materials with memory, *der Wellenintensität in Materialien mit Gedächtnis* 346–348, 352, 354.
— — — of the strength of waves, thermoelastic, *der Wellenintensität, thermoelastisch* 333, 334.
— — — of the strength of waves, viscoelastic, *der Wellenintensität, viscoelastisch* 110.
deformable region, *deformierbares Gebiet* 416.
deformation gradient, *Deformationsgradient* 305, 536.

deformation history, *Deformationsgeschichte* 537.
— theory, *Deformationstheorie* 549.
— type, *Deformationstyp* 415.
degeneracies, *Entartungen* 37, 447.
density, *Dichte* 42.
determinism, *Determiniertheit* 12.
deviator of permanent deformation tensor, *Deviator des Tensors der permanenten Deformation* 550.
— of strain, *der Dehnung* 29, 71.
— of stress, *der Spannung* 29, 71, 546.
— of stretching tensor, *des symmetrisierten Geschwindigkeitsgradienten* 548.
diagonal flow, *diagonaler Fluß* 455, 456.
differential models for viscoelastic behavior, *Differentialmodelle für viscoelastisches Verhalten* 2.
— operator laws and initial conditions, *Differentialoperatorgesetze und Anfangsbedingungen* 44.
— — —, degenerate cases, *entartete Fälle* 37.
— — —, general, *allgemeine* 34, 35.
— — —, special, *spezielle* 39.
— — —, standard, *normale* 39.
— — —, tensor, *tensorielle* 37.
— — of creep type, *Differentialoperator vom Kriechtyp* 34.
— — of relaxation type, *vom Relaxationstyp* 34.
— —, relation to relaxation function, *Beziehung zur Relaxationsfunktion* 34.
direction, *Richtung* 484.
—, condition, *Richtungsbedingung* 426.
— of propagation, *Richtung der Ausbreitung* 97, 306.
directional derivative, *Richtungsableitung* 444.
Dirichlet growth condition, *Dirichletsche Wachstumsbedingungen* 565.
— problems in elliptic equations, *Probleme in elliptischen Gleichungen* 211.
discontinuity (see also waves), classification, *Unstetigkeit (s. auch Wellen), Klassifikation* 434.
—, compatibility, *Kompatibilitätsbedingung* 494.
—, general, *allgemeine* 429, 456.
— of order *n*, *der Ordnung n* 427.
— of order one, *der ersten Ordnung* 429.
— of order two, *der zweiten Ordnung* 430.
— of stress, *der Spannung* 457.
—, stationary, *stationäre* 429.
—, surface, *Unstetigkeitsfläche* 416, 433.
discontinuous processes, *unstetige Prozesse* 35.
— solutions, *Lösungen* 416, 426.
discrete systems (see also Lagrange-Dirichlet stability concept), *diskrete Systeme (s. auch Lagrange-Dirichletsche Stabilitätsvorstellung)* 131.
— —, conservative, discussion of LAGRANGE's work, *konservative, Diskussion von Lagranges Werk* 153, 154, 163.
discrete systems, energy criterion, *diskrete Systeme, Energiekriterium* 203, 204.
— —, Lagrange-Dirichlet criterion, *Lagrange-Dirichletsches Kriterium* 153.
— —, — — and energy criterion, *und Energiekriterium* 164 (footnote, *Fußnote*), 195, 196, 203.
— —, Liapounov theorems, *Liapounovsche Sätze* 155, 156.
— —, measures in, *Maß* 141.
— —, stability, comparison with continuous systems, *Stabilität, Vergleich mit kontinuierlichen Systemen* 140, 159.
— —, — theorem, *Stabilitätssatz* 155.
— —, — —, proof, *Beweis* 159.
displacement, boundary data, *Verschiebung, Randwerte* 47.
—, — — in terms of stresses, *ausgedrückt durch Spannungen* 48.
—, — value problem, *Randwertproblem* 47, 83.
— derivative, *Ableitung* 308, 430.
— equations of equilibrium, *Gleichungen im Gleichgewicht* 44.
— — — — for isotropic and homogeneous elastic body, *für isotrope und homogene elastische Körper* 45.
— — — — for isotropic viscoelastic bodies, *für isotrope viscoelastische Körper* 44.
— — of motion, *der Bewegung* 91, 92.
— field processes for a body, *Verschiebungs-Feldprozesse in einem Körper* 8, 16, 42.
—, general, *allgemein* 8.
— gradient processes, *Verschiebungs-Gradientenprozesse* 8.
— problem, characterization in terms of displacements, *Verschiebungsproblem, charakterisiert durch Verschiebungen* 51.
— — of incompressible viscoelasticity, *inkompressibler viscoelastischer Körper* 72.
— relation, *Verschiebungsbeziehung* 91.
dissipation, general, *Dissipation, allgemein* 173, 249, 257, 263, 275, 285–289.
—, inequality, *Ungleichung* 541.
dissipative force, *dissipative Kraft* 258, 285.
— material, *dissipatives Material* 22.
distributed system of Maxwell elements in parallel, *Verteilung parallelgeschalteter Maxwellscher Elemente* 41.
divergence, instability, *Divergenz, Instabilität durch* 282–284.
— theorem, *Divergenzsatz* 585.
domain of a Boltzmann law, *Gebiet eines Boltzmannschen Gesetzes* 14.
DONATI, see BELTRAMI
double limit point, *doppelter Grenzpunkt* 480.
dual problem, *duales Problem* 264, 266–269, 271, 274.
duality of material and spatial descriptions, *Dualität der materiellen und räumlichen Beschreibung* 428.

dynamic boundary value problem, *dynamische Randwertprobleme* 89.
— conditions of compatibility, *Kompatibilitätsbedingungen* 431, 563.
— linear viscoelasticity, *lineare Viscoelastizität* 89.
— modulus **G**, *dynamischer Modul* **G** 33.
— problems, other formulations, *dynamische Probleme, andere Formulierungen* 90.
— stability, *dynamische Stabilität* 32.
— variational principles, *Variationsprinzipien* 112.
— viscoelastic process for $\mathscr{B}$ corresponding to **G**, **b**, and ϱ: DVP (**G**, **b**, ϱ), *dynamischer viscoelastischer Prozeß für* $\mathscr{B}$ *entsprechend* **G**, **b** *und* ϱ: *DVP* (**G**, **b**, ϱ) 89.
— — — for $\mathscr{B}$ corresponding to **J**, **b**, and ϱ: DVP (**J**, **b**, ϱ), *für* $\mathscr{B}$ *entsprechend* **J**, **b**, ϱ: *DVP* (**J**, **b**, ϱ) 89.
— viscosity η, *Viskosität* η 34.
dynamical systems, definition of, *dynamische Systeme, Definition* 128.
— —, differential equations, *Differentialgleichungen für* 127, 129, 157, 172.
— —, equilibrium solution, uniqueness of, *Gleichgewichtslösung, Eindeutigkeit der* 130.
— —, examples, *Beispiele* 127, 128, 160.
— —, general features, *allgemeine Züge* 127, 131.
— — of Hale, *von Hale* 130, 168, 169, 255.
— — of Zubov, *von Zubov* 130.
— —, stability of, definition, *Stabilität, Definition* 134.
— —, — —, influences on, *Einflüsse auf Stabilität* 135.
— —, — —, influences on, dissipative systems, *Einflüsse auf, dissipative Systeme* 288.
— —, — —, influences on, elasticity, *Einflüsse auf, Elastizitätstheorie* 171, 207, 211, 250, 254, 255.
— —, — —, influences on, examples, *Einflüsse auf, Beispiele* 136–139.

edge, *Kante* 483.
eigenfunction expansions, *Eigenfunktionsentwicklung* 170, 258, 279, 280, 284, 285.
eigenmode, *Eigenschwingung* 225, 226.
eigenvalues, spectrum of differential operator law, *Eigenwerte, Spektrum eines Differentialoperatorgesetzes* 38, 41, 42.
— and wave speeds, *und Wellengeschwindigkeiten* 98, 101–104.
eigenvector (see also wave axis), *Eigenvektor (s. auch Wellenachsen)* 98.
elastic body, material, *elastischer Körper, elastisches Material* 19, 175.
— compliance, *elastische Nachgiebigkeit* 42.
— continuation, *Fortsetzung* 539, 540.
— core, *elastischer Kern* 568, 575.
elastic deformation, *elastische Deformation* 545, 550.
— — gradients, *Deformationsgradienten* 538, 550.
— — history, *Deformationsgeschichte* 539.
— differential operator law, *elastisches Differentialoperatorgesetz* 39.
— mixtures, *elastische Mischungen* 286.
— region, *elastisches Gebiet* 501, 563, 578, 580.
— response, initial, instantaneous, *elastische Antwort, anfängliche, augenblickliche* 18.
— —, equilibrium, *im Gleichgewicht* 18.
— stress surface, *Spannungsfläche* 497.
— state, *elastischer Zustand* 49.
elastic-plastic, *elastisch-plastisch* 495.
— boundary, *elastisch-plastische Randfläche* 580.
— torsion 499, 562–564.
elastic-viscoelastic correspondence, *elastisch-viscoelastische Korrespondenz* 68.
elasticities, elasticity (see also tangent modulus), *Elastizitätskoeffizienten, Elastizität (s. auch Tangentenmodul)*
— A_{iAkB} 178, 182.
— c_{ijkl} 180, 215, 218, 244.
— d_{ijkl} 180, 206, 209, 211, 213, 216, 217, 239, 255, 264.
—, modulus, *Modul* 412.
—, positive longitudinal, *positiv longitudinal* 317.
—, tensor 316.
elastodynamic Lamé and Somigliana stress functions, *elastodynamische Spannungsfunktionen von Lamé und Somigliana* 93, 94.
electromagnetic effects, *elektromagnetische Effekte* 286.
elliptic, ellipticity, *elliptisch, Elliptizität* 211, 213, 460, 462, 478.
— strong, *stark* 64, 99.
energy balance, *Energiebilanz* 174–176.
— criterion for stability (see also Lagrange-Dirichlet stability concept and discrete systems, conservative), *Energiekriterium für Stabilität (s. auch Lagrange-Dirichletsches Stabilitätskonzept und diskrete Systeme, konservative)*
— — — —, Beatty's criterion, *Beattysches Kriterium* 201, 205.
— — — —, criterion of Truesdell and Noll, *Kriterium von Truesdell und Noll* 202.
— — — —, criticisms, *Kritik* 203–206.
— — — —, Hadamard stability, *Hadamard-Stabilität* 197.
— — — —, history, *Geschichte* 196–199.
— — — —, Lagrange-Dirichlet criterion, *Lagrange-Dirichletsches Kriterium* 154, 159, 195, 203.
— — — —, Pearson's condition, *Pearsonsches Kriterium* 202.
— — — —, relation to Liapounov stability, *Beziehung zur Liapounovschen Stabilität* 164 (footnote, *Fußnote*).

energy criterion for stability, relation to Liapounov stability, necessity conditions, *Energiekriterium für Stabilität, Beziehung zur Liapounovschen Stabilität, notwendige Bedingungen* 200–202.
— — — —, relation to Liapounov stability, sufficiency conditions, *Beziehung zur Liapounovschen Stabilität, hinreichende Bedingungen* 203, 206.
— — — —, statement of, *Formulierung* 196.
—, higher order energies, *Energien höherer Ordnung* 211, 212, 236–238.
—, internal, *innere* 362.
— method, *Energiemethode* 153.
— test, *Energietest* 154.
—, total, *Gesamtenergie* 137, 138, 156, 160, 162, 163, 172, 207, 211, 222, 224, 231–236, 250, 265, 267, 269, 287, 288.
entropy inequality (see also Clausius-Duhem inequality, stress-entropy), *Entropieungleichung (s. auch Clausius-Duhemsche Ungleichung, Spannungsentropie)* 174.
—, production of, *Entropieproduktion* 539, 540.
—, residual inequality, *Entropierestungleichung* 183, 188.
envelope of characteristics, *Enveloppe von Charakteristiken* 477, 479, 483.
— of Mohr circles, *von Mohrschen Kreisen* 439, 458.
epicycloids, *Epizykloiden* 453.
equilibrated surface tractions, *ausgeglichene Oberflächenbelastungen* 63.
equilibrium, axis of, *Gleichgewicht, Achse des* 216, 217, 219, 229, 239, 242.
— equation, *Gleichgewichtsgleichung* 42, 43.
—, conditions for stability of, *Gleichgewicht, Bedingungen für Stabilität des* 192, 194.
—, — in statement of energy criterion, *in der Aussage des Energiekriteriums* 196.
—, —, necessary, *notwendige* 200–202.
—, —, sufficient, *hinreichende* 203.
—, —, for instability, *Bedingungen für Instabilität, hinreichende* 199, 200.
—, Lagrange instability and existence, *Lagrangesche Instabilität und Existenz des* 151.
—, Liapounov instability of, *Liapounovsche Instabilität des* 200.
— modulus, *Gleichgewichtsmodul* 18.
— points, *Gleichgewichtspunkte* 131.
— response, *Gleichgewichts-Antwortfunktion* 13.
— solution, *Gleichgewichtslösung* 132.
— —, boundary-value problem, *Randwertproblem* 151.
— —, definition 129.
— —, non-uniqueness, *Nichteindeutigkeit* 144.
— —, uniqueness and dynamic problem, *Eindeutigkeit und dynamisches Problem* 130.
equilibrium, stability of, buckling mode, *Gleichgewicht, Stabilität, Knickform* 287, 288.
—, — —, cantilevered rod, *freitragender Stab* 282.
—, — —, convexity, *Konvexität* 234, 273.
—, — —, curved panel in windstream, *gekrümmte Platte in der Windströmung* 281.
—, — —, divergence, non-uniqueness in follower force problems, *Divergenz, Nichteindeutigkeit in Folgelastproblemen* 282, 283.
—, — —, non-uniqueness and instability, *Nichteindeutigkeit und Instabilität* 224.
—, — —, non-uniqueness and neutral equilibrium, *Nichteindeutigkeit und neutrales Gleichgewicht* 226.
—, — —, under dead load in class of small incremental displacements, fixed surface, *unter fester Last in der Klasse kleiner Verschiebungsänderungen, feste Oberfläche* 206.
—, — —, under dead load in class of small incremental displacements, movable surface, *unter fester Last in der Klasse kleiner Verschiebungsänderungen, bewegliche Oberfläche* 215.
—, — —, under dead load in class of small incremental displacements, theorem, *unter fester Last in der Klasse kleiner Verschiebungsänderungen, Satz* 219.
—, — —, under time- and position-dependent data, *unter zeit- und ortsabhängigen Bedingungen* 268, 269, 274.
equivalence principle for attenuation, infinite frequency damped progressive waves and propagating singular surfaces, *Äquivalenzprinzip für Dämpfung, gedämpfte sich ausbreitende Wellen unendlicher Frequenz und sich ausbreitende singuläre Flächen* 110.
— — for wave speeds and wave axes, infinite frequency damped progressive waves and propagating singular surfaces, *für Wellengeschwindigkeiten und Wellenachsen, gedämpfte sich ausbreitende Wellen unendlicher Frequenz und sich ausbreitende singuläre Flächen* 110.
— relations, *Äquivalenzrelationen* 139, 157.
—, semiequivalence, *Semi-Äquivalenz* 139.
Euclidean point space, *Euklidischer Punktraum* 4.
Euler critical load, *Eulersche kritische Last* 277.
Eulerian coordinates, *Eulersche Koordinaten* 428, 536.
eventually monotone, *schließlich monoton* 20.
exceptional direction, *außergewöhnliche Richtung* 479, 483, 484.
— line, *Linie* 418.
exchange of dependent and independent variables, *Vertauschung von abhängigen und unabhängigen Variablen* 443.

exchange of stabilities, *Vertauschung von Stabilitäten* 226.
existence of an elastic core, *Existenz eines elastischen Kerns* 568.
— of quasi-static viscoelastic processes, *quasistatischer viscoelastischer Prozesse* 59.
— of viscoelastic waves, *viscoelastischer Wellen* 97.
—, uniqueness, and asymptotic stability theorem for dynamic problems in viscoelasticity, *Eindeutigkeit, und Satz zur asymptotischen Stabilität für dynamische Probleme der Viscoelastizitätstheorie* 104–106.
expansion, expansive wave, *Ausdehnung, Ausdehnungswelle* 51.
—, — —, acceleration, *Beschleunigungswelle* 311, 314.
—, — —, shock, *Stoßwelle* 312.
exterior derivative, *äußere Ableitung* 420, 426.
— region, *äußeres Gebiet* 81.
extrusion, *Auspressung* 525.

fading memory, *schwindendes Gedächtnis* 2, 334.
field equations of viscoelasticity, dynamic, *Feldgleichungen der Viscoelastizität, dynamisch* 91.
— — — —, quasi-static, *quasistatisch* 42.
finite deformations, *endliche Deformationen* 3.
— histories, *endliche Geschichtsfunktionen* 5.
fixed-point theorem, SHINBROT'S, *Shinbrotscher Fixpunktsatz* 558.
flow (plastic), equations of PRANDTL and REUSS, *Fließen (plastisches), Gleichungen von Prandtl und Reuss* 412.
—, rule, *Regel* 547.
—, theories, *Theorien* 549.
—, type, *Art* 415.
—, velocities, *Geschwindigkeiten* 405.
fluids, *Flüssigkeiten* 2.
flutter, *Flattern* 283, 284.
focusing of waves at origin, *Fokussierung von Wellen im Ursprung* 138.
follower force, *Folgekraft* 257, 258, 261, 278–285.
force multipoles, *Kraftmultipole* 213.
foundations of linear theory of viscoelasticity, *Grundlagen der linearen Theorie der Viscoelastizität* 11.
Fourier and Laplace transforms, *Fourier- und Laplace-Transformierte* 28.
— integral theorem, *Fourierscher Integralsatz* 33.
frame-indifference, *Systemunabhängigkeit* 544, 545.
Fréchet derivative, *Fréchetsche Ableitung* 165, 194, 195, 203, 205.
free boundaries, *freie Randflächen* 501, 580, 581, 583, 585, 586.
— energy, *Energie* 3, 179, 180, 183, 191, 192.
— — functional, *Funktional* 539.
free flowing, *freies Fließen* 494.
— vibrations, *freie Schwingungen* 106, 111.
Fresnel-Hadamard propagation condition, *Fresnel-Hadamardsche Ausbreitungsbedingung* 98, 105, 316, 324, 326.
fully elastic or plastic, *vollständig elastisch oder plastisch* 495.
functions of position and time, *Funktionen der Lage und der Zeit* 7.

G belongs to the pair P, Q, G *gehört zum Paar* P, Q 35.
GALERKIN, see BOUSSINESQ.
Galerkin procedure, *Galerkinsche Methode* 285.
Gateau derivative, *Gateausche Ableitung* 165, 194, 198, 203, 205.
generalized plane stress, *verallgemeinerte ebene Spannung* 440.
— potential, *verallgemeinertes Potential* 439.
— solutions, *Lösungen* 128, 138, 254, 255.
— Somigliana solution, *Somiglianasche Lösung* 94.
— strains, *Dehnungen* 409.
— stresses, *Spannungen* 409.
generating equation for a viscoelastic relaxation function, *erzeugende Gleichung für eine viscoelastische Relaxationsfunktion* 36.
generation of shock waves, *Erzeugung von Stoßwellen* 97.
geometric dispersion, *geometrische Dispersion* 107.
geometrical and kinematical discontinuity conditions, *geometrische und kinematische Unstetigkeitsbedingungen* 429.
— condition of compatibility, *Kompatibilitätsbedingungen* 307, 430.
"Grad" 408, 465.
GRAFFI, see ALFREY.
graphical method, *graphische Methode* 493.
GREEN'S deformation tensor, *Greenscher Deformationstensor* 180.
— function, *Greensche Funktion* 8, 564.
— process, *Greenscher Prozeß* 81.
— — integral solutions, *Integrallösungen* 81.
— processes of the first kind, *Greensche Prozesse erster Art* 82.
— — — — second kind, *zweiter Art* 83.
Green-St. Venant tensor, *Green-St. Venantscher Tensor* 550.
growth (see decay), *Wachstum (s. Abklingen)*
gyroscopic force, *gyroskopische Kraft* 257, 260.

HAAR and v. KÁRMÁN'S hypothesis, *Haar- und v. Kármánsche Hypothese* 526.
HADAMARD stability, *Hadamardsche Stabilität* 198, 199, 204, 224, 246.
— superstability, *Superstabilität* 224.
HADAMARD'S criterion of stability, *Hadamardsches Stabilitätskriterium* 197.
— —, sufficient conditions, *hinreichende Bedingungen* 246, 248.

HADAMARD's theory of singular surfaces, *Hadamardsche Theorie singulärer Flächen* 428.
HALE's dynamical system, *Halesches dynamisches System* 168, 169.
— instability theorem, *Instabilitätssatz* 169, 199.
— invariance principle, *Invarianzprinzip* 158, 169, 191, 255, 288, 289.
— — — and asymptotic stability, *und asymptotische Stabilität* 254.
— necessary conditions for stability, *notwendige Bedingungen für Stabilität* 199.
half-range Fourier transforms, *Fourier-cos- bzw. -sin-Transformierte* 21, 33.
harmonic equation, *harmonische Gleichung* 75.
Harnack inequalities, *Harnacksche Ungleichung* 587.
heat conduction equation, *Wärmeleitungsgleichung* 184, 185.
— flux, *Wärmefluß* 174, 175.
Heaviside classes, *Heaviside-Klassen* 6.
— processes, approximation of, *Heaviside-Prozesse, Näherung für* 16.
— type, *Heaviside-Typ* 5.
— unit step process, *Heaviside-Einheitsstufenprozeß* 6.
Helmholtz free energy, *Helmholtzsche freie Energie* 175.
— — — for isothermal deformations, *für isotherme Deformationen* 191.
Hencky theory of plasticity, *Henckysche Theorie der Plastizität* 549.
hereditary integral equation, *Integralgleichung vom Nachwirkungstyp* 61, 64.
— response, *Nachwirkung* 1.
— wave operator ($\Box_K$), *Nachwirkungs-Wellenoperator* ($\Box_K$) 92.
hexagonal condition, *hexagonale Bedingung* 472.
— prism, *hexagonales Prisma* 411.
hidden variables, *innere Variablen* 3.
high frequency behavior of a viscoelastic material, *Hochfrequenzverhalten viscoelastischer Materialien* 22.
— — limit of the attenuation, *Hochfrequenzgrenze der Dämpfung* 110.
— — limits of oscillatory viscoelastic processes, *Hochfrequenzgrenzen oscillatorischer viscoelastischer Prozesse* 106.
higher order energies, *Energien höherer Ordnung* 211, 212, 236–238.
— — Kelvin processes, *Kelvinsche Prozesse höherer Ordnung* 81.
Hilbert space, *Hilbert-Raum* 551.
history, bounded, *Geschichte, beschränkte* 5.
— dependence of the solutions to the viscoelastic problem, *Geschichtsabhängigkeit der Lösungen des viscoelastischen Problems* 61.
—, definition 4.
—, finite, *endliche* 5.
— gradient, *Geschichtsgradient* 540.
history, infinite, *Geschichte, unendliche* 5.
—, integrable, *integrierbare* 5.
— of the deformation gradient, *des Deformationsgradienten* 305.
—, restricted, *beschränkte* 5.
Hölder continuity, *Höldersche Stetigkeit* 146, 274.
— stability, continuous dependence, *Stabilität, kontinuierliche Abhängigkeit* 253.
— —, definition 146.
— —, elastic bodies, *elastische Körper* 171.
— —, logarithmic convexity *logarithmische Konvexität* 231.
— —, null solution in linear thermoelasticity, *Nullösung in der linearen Thermoelastizität* 252.
— —, sufficient condition for, *hinreichende Bedingung für* 153.
— —, theorem, *Satz* 253.
homogeneous body, *homogener Körper* 17.
— boundary conditions and initial data, *homogene Rand- und Anfangsbedingungen* 270.
— isotropic incompressible viscoelastic body, *isotroper, inkompressibler, viscoelastischer Körper* 71.
— — viscoelastic body of relaxation type, *viscoelastischer Körper vom Relaxationstyp* 46.
homomorphism, *Homomorphismus* 130.
HOOKE's law, *Hookesches Gesetz* 496.
HSU's criterion, *Hsusches Kriterium* 288, 289.
Hugoniot relation, *Hugoniotsche Beziehung* 372.
hydrostatic load, *hydrostatische Last* 244–249, 262, 263.
— pressure, *hydrostatischer Druck* 261.
hyperbolic points, *hyperbolische Punkte* 465.
— problem, *hyperbolisches Problem* 439, 460, 462, 526.
— region, *Gebiet* 478.
hyperelastic, *hyperelastisch* 173.
hypocycloids, *Hypozykloiden* 453.

ideal plastic body, *ideal plastischer Körper* 403.
impenetrability, *Undurchdringbarkeit* 536, 538.
incipient, *anfänglich* 525.
incompressible, *inkompressibel* 2, 71, 176, 183, 248, 264, 512.
incremental body force, *Inkrement der Volumkraft* 259, 286.
— displacement, *der Verschiebung* 176, 183.
— quantities, *Größen mit Inkrement* 176.
— rate of work equation, *Inkrement der Leistung, Gleichung für* 178, 180, 182, 183.
— state, *Zustandsinkrement* 188, 189.
— stress vector, *Inkrement des Spannungsvektors* 259–261, 286.
— temperature variable, *der Temperaturvariablen* 183.

induced discontinuity, *induzierte Unstetigkeit* 311.
— jump of a plane wave, *induzierter Sprung einer ebenen Welle* 100.
— map, *induzierte Abbildung* 131.
inertia forces, *Trägheitskräfte* 525.
infinite histories, *unendliche Geschichten* 5.
— slab, *Platte* 529.
infinitesimal deformation, *infinitesimale Deformation* 3.
— linearized theory of viscoelasticity, *linearisierte Theorie der Viscoelastizität* 1.
— rigid motion, *starre Bewegung* 17.
— rotation process, *infinitesimaler starrer Rotationsprozeß* 17.
— strain process, *Dehnungsprozeß* 17.
inflection, *Umkehrpunkt* 484.
inhomogeneous body, *inhomogener Körper* 17.
initial boundary value problem, *anfängliches Randwertproblem* 151.
— conditions or data, *Anfangsbedingungen* 35, 90, 95, 112, 130, 178, 179, 186, 211, 232, 234, 235, 237, 264, 271, 274, 275, 417, 420.
— elastic compliance, *anfängliche elastische Nachgiebigkeit* 24.
— elasticity, *Elastizität* 18.
— isotropy of the relaxation function, relation to wave propagation, *Isotropie der Relaxationsfunktion, Relation zur Wellenausbreitung* 99.
— mechanical response, *mechanische Antwort* 49.
— past-history, consistent, *vergangene Geschichte, konsistente* 51, 89.
— —, definition 50.
— —, problem 61, 63, 64.
— —, vanishing, *verschwindende* 52, 63.
— perturbation, *Störung* 207.
— quasi-static viscoelastic state, *anfänglicher quasistatischer viscoelastischer Zustand* 49.
— response, *anfängliche Antwort* 13, 49.
— stress, *Spannung* 197, 227–240.
— value, *Anfangswert* 246.
— — problem, *Anfangswertproblem* 104.
inner product, *inneres Produkt* 4, 539.
instability, definitions, asymptotic, *Instabilität, Definitionen, asymptotische* 150.
—, —, exponentially asymptotic, *exponentiell asymptotische* 150.
—, —, Lagrange 150.
—, —, latent (adjacent equilibrium technique), *latente Methode des angrenzenden Gleichgewichts* 144.
—, —, Liapounov 135, 147.
—, general, *allgemein* 135.
—, theorems, discrete conservative systems, *Sätze, diskrete konservative Systeme* 167.
—, —, discussion, *Diskussion* 168.
—, — from non-uniqueness, *aus der Nichteindeutigkeit* 151, 167, 224.
instability theorems, Hale, *Instabilitätssätze, Hale* 169, 199.
— —, Liapounov 168.
— —, —, conditions sufficient for equilibrium solution, *hinreichende Bedingungen für Gleichgewichtslösungen* 199, 200.
instantaneous elastic response function, *augenblickliche elastische Antwortfunktion* 350.
— elasticity, *Elastizität* 335.
— second-order tangent modulus, *augenblicklicher Tangentenmodul zweiter Ordnung* 336, 350.
— tangent modulus, *Tangentenmodul* 336, 350.
integrability conditions, *Integrabilitätsbedingungen* 444.
integrable histories, *integrierbare Geschichten* 5.
integral form of energy balance equation, *Integralform der Energiebilanz* 176.
— formulae, *Integralformeln* 81.
— laws, *Integralgesetze* 34.
— representation for the solution of displacement boundary value problem, *Integraldarstellung für die Lösung von Verschiebungs-Randwertproblemen* 82.
— — for the solution of mixed boundary value problem, *für die Lösung von gemischten Randwertproblemen* 84.
— — — the solution of traction boundary value problem, *Spannungs-Randwertproblemen* 84.
— theorems, *Integralsätze* 53.
interaction of discontinuities, *Wechselwirkung von Unstetigkeiten* 458.
interior derivatives, *innere Ableitungen* 420, 426.
intermediate stress, *mittlere Spannung* 514.
internal energy, *innere Energie* 174.
— friction, *Reibung* 463.
— state variables, *Zustandsvariablen* 3.
intersection property of the plastic part, *Schnitteigenschaft des plastischen Teils* 580.
invariance principles, *Invarianzprinzipien* 129, 169, 191, 255, 288, 289.
invariant interpretations, *invariante Interpretationen* 419.
— set, *Menge* 129, 153, 169, 288.
invariants I_1, I_2, I_3, *Invarianten* I_1, I_2, I_3 180.
inversion of Boltzmann laws, *Umkehrung von Boltzmannschen Gesetzen* 23.
— theorem for response functions, *Umkehrungssatz für Antwortfunktionen* 27, 64.
— — for strong response functions, *für starke Antwortfunktionen* 25, 30.
irrotational vector field, *wirbelfreies Vektorfeld* 73.
isentropic elasticity tensor, *isentroper Elastizitätstensor* 324.

isentropic second-order elasticity tensor, *isentroper Elastizitätstensor zweiter Ordnung* 324.
— states, *Zustand* 187–189.
— tangent modulus, *Tangentenmodul* 362, 369.
isobars, *Isobaren* 476, 477, 484.
isochoric, *isochor* 548.
isoclines, *Isoklinen* 477, 484.
isolated, general, *isoliert, allgemein* 478, 482.
— point, *isolierter Punkt* 480.
isoperimetric condition, *isoperimetrische Bedingung* 564.
isothermal problems, *isothermes Problem* 178.
isotropic function, *isotrope Funktion* 545.
— materials, *Materialien* 2, 18, 29, 180, 185, 244–249, 544, 545, 550.
— tensor functions, *Tensorfunktionen* 546.
iterated conditions of compatibility, *iterierte Kompatibilitätsbedingungen* 307.

Jacobian, *Funktionaldeterminante* 445, 477.
JENNE's equations, *Jennesche Gleichung* 459.
Jordan domain, *Jordan-Gebiet* 561.
jump discontinuity, *Sprungunstetigkeit* 14, 96, 97.
— — for plane waves, *einer ebenen Welle* 97.
— of a function, *einer Funktion* 97.
— of the stress, *der Spannung* 435.
— of the velocity, *der Geschwindigkeit* 494.

Kelvin doublet process, normalized, *Kelvin-Dublettprozeß, normalisiert* 79.
— — —, properties of, *Eigenschaften* 79, 80.
— — — with moment or without moment, *mit oder ohne Moment* 80.
— elements, *Kelvin-Elemente* 40.
— operators, *Kelvin-Operatoren* 39.
— problem in elastostatics, viscoelastic counterpart, *Kelvin-Problem in der Elastostatik, viscoelastisches Gegenstück* 77.
— process, *Kelvin-Prozeß* 78, 79.
Kelvin-Voigt elements, *Kelvin-Voigt-Elemente* 40.
— — in series, *in Reihe* 41.
— operators, *Operatoren* 39.
kinematical condition of compatibility, *kinematische Kompatibilitätsbedingung* 308, 309, 429, 430.
kinematically determined, *kinematisch bestimmt* 526.
kinetic energy, definition, *kinetische Energie, Definition* 95.
— —, n-th order, *n-ter Ordnung* 236.
KOITER's generalization, *Koitersche Verallgemeinerung* 439.
KORN's constant, *Kornsche Konstante* 218, 221, 240, 241.
— inequality, *Ungleichung* 218, 221, 241.

$\boldsymbol{L_2}$ inner product, L_2 *inneres Produkt* 552.
$\overset{*}{\mathscr{L}}$ is a creep law for $\mathscr{L}$, $\overset{*}{\mathscr{L}}$ *ist ein Kriechgesetz für* $\mathscr{L}$ 23.
labil, *labile* 147.
L-beam, *L-Träger* 495.
Lagrange instability, asymptotic, *Lagrangesche Instabilität, asymptotische* 150.
— —, choice of definition in elasticity, *Wahl der Definition in der Elastizitätstheorie* 171.
— —, definition 150.
— —, discussion, *Diskussion* 150, 151.
— —, elastic bodies, *in elastischen Körpern* 171.
— —, example in elasticity, *Beispiel in der Elastizitätstheorie* 202, 223.
— —, logarithmic convexity, *logarithmische Konvexität* 239.
— —, relation to boundedness, *Beziehung zur Beschränktheit* 150.
— —, — to Liapounov, *Beziehung zu Liapounov* 150.
— —, sufficient conditions, *hinreichende Bedingungen* 170.
— multiplier, *Lagrange-Multiplikator* 564.
— stability, definition, *Lagrange-Stabilität, Definition* 149.
— —, Liapounov stability, *Liapounov-Stabilität* 156.
— —, relation to boundedness, *Beziehung zur Beschränktheit* 149.
Lagrange-Dirichlet stability concept (see also discrete conservative systems), criterion or energy test, *Lagrange-Dirichlet-Stabilitätskonzept (s. auch diskrete konservative Systeme), Kriterium oder Energietest* 154.
— — —, definition 131, 134.
— — —, —, ingredients, *Bestandteile* 132.
— — —, —, observations, *Beobachtungen* 134.
— — —, discussion, *Diskussion* 153, 154, 163.
— — —, energy criterion, *Energiekriterium* 203, 204 (footnote, *Fußnote*).
— — —, theorem, *Satz* 155.
— — —, —, proof, *Beweis* 159.
Lagrangian coordinates, *Lagrangesche Koordinaten* 428, 524, 536.
Lamé constants, *Lamé-Konstanten* 45, 181.
— moduli, *Lamé-Modulen* 29.
— stress functions, *Lamé-Spannungsfunktionen* 93.
Laplace equation, *Laplace-Gleichung* 136.
— transform, *Laplace-Transformierte* 28, 68.
largest invariant set, *größte invariante Menge* 169.
laws of balance, energy, *Bilanzgleichungen, Energie* 174, 175.
— — —, linear and angular momentum, *Impuls und Drehimpuls* 42.

left Cauchy-Green tensor, *linker Cauchy-Greenscher Tensor* 326.
— inverse, *linke Umkehrung* 23.
Legendre transformation, *Legendre-Transformation* 461.
Legendre-Hadamard condition, *Legendre-Hadamardsche Bedingung* 166.
Liapounov functions (auxiliary functions), COLEMAN's canonical free energy V_3, *Liapounovsche Funktionen (Hilfsfunktionen), Colemans kanonische freie Energie* V_3 189, 190.
— —, elasticity, generation or choice, *in der Elastizitätstheorie, Erzeugung oder Wahl* 171, 172, 177, 193 (footnote, *Fußnote*), 207, 212, 223 (footnote, *Fußnote*), 250, 255, 259.
— —, entropy production inequality, *Entropieungleichung* 189, V_3 190, V_4 190, 222.
— — $F_{\tau,t}$ 154, 157, 163, 164, 168, 169.
— —, finite thermoelasticity, from energy balance, *endliche Thermoelastizität, aus der Energiebilanz* V_1 188, V_2 189.
— —, follower forces, *Folgekräfte* 280–282.
— — in sense of LASALLE, *im Sinne von Lasalle* 168, 169.
— —, isothermal primary state, thermal perturbations, *isothermer Ausgangszustand, thermische Störungen* 188.
— —, MOVCHAN's function, *Movchansche Funktion* 195.
— —, — theorem, *Movchanscher Satz* 156.
— of HSU and LEE for asymptotic stability, *von Hsu und Lee für asymptotische Stabilität* 275, 276.
— —, potential energy V, *potentielle Energie* V 205.
— —, total energy, *Gesamtenergie* 172, 272.
Liapounov instability and non-uniqueness, *Liapounovsche Instabilität und Nichteindeutigkeit* 151, 152.
— — by divergence, *durch Divergenz* 282.
— —, choice of definition in elasticity, *Wahl der Definition in der Elastizitätstheorie* 171.
— —, definition 147.
— —, discussion of example, *Diskussion eines Beispiels* 139.
— —, relation to Lagrange instability, *Beziehung zur Lagrangeschen Instabilität* 150.
— —, sufficient conditions, *hinreichende Bedingungen* 199, 200.
— —, theorems, *Sätze* 168, 169.
— — in elasticity and nonuniqueness, *in der Elastizitätstheorie und Nichteindeutigkeit* 224.
— — — —, dead loads for linear thermoelastic displacements, null solution, *feste Belastungen für lineare thermoelastische Verschiebungen, Nullösung* 251, 252.
Liapounov instability in elasticity, dissipative forces, *Liapounovsche Instabilität in der Elastizitätstheorie, dissipative Kräfte* 286.
— — — —, equilibrium solution under dead loads for linear incremental displacements, *Gleichgewichtslösung unter fester Belastung für lineare Verschiebungsänderung* 222.
— — — —, follower forces, *Folgekräfte* 282–285.
— — — —, incompressible elastic media, *inkompressible elastische Körper* 248, 249.
— — — —, negative-definite total energy V_4, *negativ definite Gesamtenergie* V_4 222.
— — — —, special traction boundary conditions, null solution, *spezielle Spannungs-Randbedingungen, Nullösung* 246–248.
— — — —, time- and position-dependent data, *zeit- und ortsabhängige Bedingungen* 266.
— — — —, traction boundary conditions, sufficient conditions on mean stresses, *Spannungs-Randbedingungen, hinreichende Bedingungen für mittlere Spannungen* 242.
— — — —, weakly conservative forces, *schwach konservative Kräfte* 259.
Liapounov stability and boundedness, *Liapounovsche Stabilität und Beschränktheit* 148–150.
— — and calculus of variations, *und Variationsrechnung* 158, 163–167.
— — and Lagrange stability, *und Lagrangesche Stabilität* 148, 149, 156.
— — — maximum principles, *und Maximalprinzipien* 152.
— —, definition 133.
— —, dynamical system, *dynamisches System* 134.
— —, equivalence with uniqueness, *Äquivalenz zur Eindeutigkeit* 151.
— —, equivalent definitions, *äquivalente Definitionen* 140.
— —, examples, *Beispiele* 160–163.
— —, geometrical representations, *geometrische Darstellungen* 140.
— —, measures, *Maße* 135.
— —, non-uniqueness of equilibrium solution, *Nichteindeutigkeit der Gleichgewichtslösung* 144.
— —, relation to continuous dependence, *Beziehung zur stetigen Abhängigkeit* 146.
— —, — — LAGRANGE-DIRICHLET, *Lagrange-Dirichlet* 153, 154.
— —, theorems, *Sätze* 149, 150, 152, 155, 157.
— — —, discrete systems, *diskrete Systeme* 155, 159.
— — —, discussion, *Diskussion* 156–160.
— — —, uniform, *uniforme* 157.
— — —, variations, *Abänderungen* 155.
— — —, versions of ZUBOV and MOVCHAN, *Versionen von Zubov und Movchan* 156.
— — —, weakly asymptotic, *schwach asymptotisch* 158.

Liapounov stability in elasticity, *Liapounovsche Stabilität in der Elastizitätstheorie*
analysis for traction boundary problem, *Rechnung für Spannungs-Randbedingungen* 239–242.
for incompressible media, *für inkompressible Medien* 244.
sufficient conditions without axis of equilibrium, *hinreichende Bedingungen ohne Gleichgewichtsachse* 239–241.
with axis, *mit Achse* 242.
and calculus of variations, *und Variationsrechnung* 195, 205.
and energy criterion (criticism), *und Energiekriterium (Kritik)* 203–206.
discussion and history, *Diskussion und Geschichte* 195, 196, 199.
dynamical system, *dynamische Systeme* 171.
in class of non-linear perturbations, *in der Klasse der nichtlinearen Störungen* 191.
MOVCHAN's conditions, *Movchansche Bedingungen* 195.
necessity theorems, *Sätze über notwendige Bedingungen*
for equilibrium solution in non-linear thermoelasticity, *für Gleichgewichtslösungen in der nichtlinearen Thermoelastizität* 200–202.
relation to sufficiency conditions, *Beziehung zu hinreichenden Bedingungen* 203.
of natural state, *des natürlichen Zustands* 195.
of null solution, *der Nullösung*
choice of measures, *Wahl der Maße* 210–213.
in incompressible media, *in inkompressiblen Medien* 213, 214.
theorem, *Satz* 219.
under dead loads in class of small incremental displacements (fixed surface), *unter fester Belastung in der Klasse kleiner Verschiebungsänderungen (feste Oberfläche)* 206.
use of Liapounov theorem, *Gebrauch des Liapounovschen Satzes* 212.
with movable surface, *mit bewegter Oberfläche* 215.
with time-dependent elasticities d_{ijkl}, *mit zeitabhängigen Elastizitäten* d_{ijkl} 209.
of primary state, *des Anfangszustandes* 177.
related to linear problems, *bezogen auf lineare Probleme* 194.
special traction boundary value problems, *spezielle Spannungs-Randwertprobleme* 244.
for incompressible elastic materials, *für inkompressible elastische Materialien* 248.

Liapounov stability in elasticity, special traction boundary value problems, *Liapounovsche Stabilität in der Elastizitätstheorie, spezielle Spannungs-Randwertprobleme*
isotropic compressible material under hydrostatic stress, *isotropes kompressibles Material unter hydrostatischer Spannung* 244–248.
necessary conditions, null solution, *notwendige Bedingungen, Nullösung* 246.
sufficiency theorems in non-linear thermoelasticity, *Sätze über hinreichende Bedingungen in der nichtlinearen Thermoelastizität* 192.
theorem on uniqueness of equilibrium solution, *Satz über Eindeutigkeit der Gleichgewichtslösung* 224.
under dead loads in class of linear thermoelastic displacements, *unter fester Belastung in der Klasse der linearen thermoelastischen Verschiebungen* 249–255.
Liapounov function, *Liapounovsche Funktion* 250.
null solution, *Nullösung* 251.
under follower forces, *unter Folgekräften* 278–285.
equilibrium solutions, *Gleichgewichtslösung* 281, 282.
examples, null solution, *Beispiele, Nullösung* 280, 281.
with dissipative forces, *mit dissipativen Kräften* 285–289.
uniform stability, *uniforme Stabilität* 207.
with non-dead loads, *mit nicht-festen Belastungen* 256–257.
persistent stability, *beständige Stabilität* 256.
with motion-dependent data, *mit bewegungsabhängigen Bedingungen* 257.
with time- and position-dependent data, *mit zeit- und ortsabhängigen Bedingungen* 264–278.
asymptotic stability of null solution, *asymptotische Stabilität der Nullösung* 275.
Liapounov function of HSU and LEE, *Liapounovsche Funktion von Hsu und Lee* 276.
theorem, *Satz* 257.
non-zero initial data, *nichtverschwindende Anfangswerte* 266.
null solution, *Nullösung* 266.
under prescribed body force and dead loads, *unter vorgeschriebener Volumkraft und fester Belastung*
changes in initial data, *Änderungen der Anfangswerte* 272, 273.
convexity arguments, *Argumente zur Konvexität* 273.
equilibrium solution, *Gleichgewichtslösung* 269–271.

Liapounov stability in elasticity, with time- and position-dependent data, under prescribed body force and dead loads, *Liapounovsche Stabilität in der Elastizitätstheorie, mit zeit- und ortsabhängigen Bedingungen, unter vorgeschriebener Volumkraft und fester Belastung*
further arguments, *weitere Argumente* 275.
null solution of mixed problem with linear incremental diplacements, *Nullösung des gemischten Problems mit linearen Verschiebungsänderungen* 273.
variation in elasticities, with linear incremental displacements, *Änderung der Elastizitäten mit linearen Verschiebungsänderungen* 271, 272.
under prescribed surface displacements and linear incremental displacements, *unter vorgeschriebenen Oberflächenverschiebungen und linearen Verschiebungsänderungen*
equilibrium solution, *Gleichgewichtslösung* 268, 269.
non-zero initial data, *nicht verschwindende Anfangswerte* 266.
null solution, *Nullösung* 266.
under prescribed surface tractions and linear incremental displacements and zero initial data, *unter vorgeschriebenen Oberflächenspannungen, linearen Verschiebungsänderungen und verschwindenden Anfangswerten* 266.
with weakly conservative loads, *mit schwach konservativen Belastungen* 258–264.
linear 263.
stability analyses, *Stabilitätsrechnungen* 263, 264.
limit analysis, *Grenzwertbetrachtung* 408, 431.
— line, *Grenzlinie* 477, 478, 480.
— of stability, *Stabilitätsgrenze* 202, 225, 244.
— point, *Grenzpunkt* 478.
— sets, *Grenzfolgen* 126.
— type singularities, *Grenztyp, Singularitäten* 477.
line integral for the displacement process, *Linienintegral für den Verschiebungsprozeß* 46.
linear, linearized approximation, *lineare, linearisierte Näherung* 2.
— hereditary laws, *Nachwirkungsgesetze* 11.
— integral equations, *Integralgleichungen* 77.
— superposition 2, 12.
— systems, *Systeme* 178.
— theory based upon finite deformations, *Theorie basierend auf endlichen Deformationen* 3.
linear, linearized, theory of elasticity, *lineare, linearisierte* 1, 176–183.
— — — viscoelasticity, *Theorie der Viscoelastizität* 1.
thermoelasticity, *Thermoelastizität*
— —, asymptotic stability, *asymptotische Stabilität* 254, 255.
— —, basic equations, *Grundgleichungen* 183–185, 249, 250.
— —, HALE's invariance principle, *Halesches Invarianzprinzip* 255.
— —, Hölder stability, *Höldersche Stabilität* 252–254.
— —, instability, *Instabilität* 251, 252.
— —, Liapounov stability, *Liapounovsche Stabilität* 250, 251.
— —, weak (generalized) solutions, *schwache (verallgemeinerte) Lösungen* 254, 255.
— transformations, *Transformationen* 3, 4.
linearity, *Linearität* 12.
linearization based upon infinitesimal deformations, *Linearisierung basierend auf infinitesimalen Deformationen* 3.
linearly elastic element (spring), *lineares elastisches Element (Feder)* 40.
— polarized waves, *lineare polarisierte Welle* 108.
— viscoelastic body of creep type, *linearviscoelastischer Körper vom Kriechtyp* 23.
— — — of relaxation type, *vom Relaxationstyp* 16, 17.
— — material, *linear-viscoelastisches Material* 1, 16.
— viscous element (dashpot), *viskoses Element (Dämpfer)* 40.
Lipschitz constant, *Lipschitzsche Konstante* 587.
load parameter, *Belastungsparameter* 225, 226, 284, 289.
loads, conservative, *Belastungen, konservative* 201, 202, 204, 226, 284.
—, dead, *feste* 202, 215, 222, 226, 244, 249, 260, 272, 273, 286.
—, non-dead, *nicht-feste* 256–258.
—, follower forces, *Folgekräfte* 257, 258, 261, 278, 279, 285.
—, hydrostatic, *hydrostatische* 262, 263.
—, path-dependent, *wegabhängige* 193.
—, velocity and time-varying, *abhängig von der Geschwindigkeit und der Zeit* 257, 278.
—, weakly conservative, *schwach konservative* 193, 194, 257–264, 286.
—, — —, definitions, *Definitionen* 259.
local differentiation, *lokale Differentiation* 524.
— second- order tangent modulus, *lokaler Tangentenmodul zweiter Ordnung* 378, 382.
— speed of propagation, *lokale Ausbreitungsgeschwindigkeit* 307, 308.
— tangent modulus, *lokaler Tangentenmodul* 378, 382.
locally integrable, *lokal integrierbar* 5.

logarithmic continuity, *logarithmische Stetigkeit* 146.
— convexity, *Konvexität* 228–239, 253, 273.
— — and Hölder stability, *und Höldersche Stabilität* 231.
— — of function $F(t; \alpha, t_0)$, *der Funktion* $F(t; \alpha, t_0)$ 229.
— spirals, *Spiralen* 452.
longitudinal discontinuity, *longitudinale Unstetigkeit* 430.
— elasticity, *Elastizität* 317.
— waves, definition, *Wellen, Definition* 99, 310.
— — in elastic bodies, anisotropic, *in elastischen Körpern, anisotrop* 315–321.
— — — — —, isotropic, *isotrop* 327, 328.
— — — — —, principal, *in Hauptrichtung* 327.
loss modulus, *Verlustmodul* 33.
lost solutions, *verlorene Lösungen* 485.

m-fold Dirichlet norm, *m-faltige Dirichlet-Norm* 551.
Mach number, *Machzahl* 281.
map T_τ, *Abbildung* T_τ 130, 133, 140, 148, 149.
mass centre, *Massenmittelpunkt* 217, 218, 247.
— density, *Massendichte* 536.
material derivative, *materielle Ableitung* 404, 440, 540.
— discontinuities, *Unstetigkeit* 429, 434.
—, elastic, *Material, elastisches* 19, 175, 315.
—, elastic-plastic, *elastisch-plastisches* 537.
—, elastic, related to viscoelastic, *elastisches Material, einem viscoelastischem Material zugeordnet* 19.
— plastic, *plastisches* 403, 412–415.
— point, *materieller Punkt* 16.
— standpoint, *Gesichtspunkt* 428, 524.
— symmetry, *materielle Symmetrie* 2, 18, 29.
—, thermoelastic, *Material, thermoelastisches* 183.
— with memory, *mit Gedächtnis* 172.
mathematical models, *mathematische Modelle* 131, 150, 213.
— notation, *Bezeichnung* 125.
Mathieu-Hill equation, *Mathieu-Hillsche Gleichung* 257.
maximum and minimum shearing stress, *maximale und minimale Scherspannung* 423, 465.
— principles, *Maximalprinzipien* 152.
— shear, *maximale Scherung* 442.
Maxwell element, *Maxwell-Element* 40.
— elements in parallel, *-Elemente in Reihe* 41.
— operators, *-Operatoren* 39.
— theorem, *Maxwellscher Satz* 98, 101.
mean normal stress and strain, *mittlere Normalspannung und Normaldehnung* 71.
— stress, *Spannung* 216, 240–243.
measures and boundedness and existence, *Maße und Beschränktheit und Existenz* 135.
— as positive-definite functions on $X \times X$, *als positiv definite Funktionen in* $X \times X$ 135.
—, choice, *Auswahl* 134–139, 163.
—, — in elasticity, *in der Elastizitätstheorie* 171, 172, 210–213.
—, definition of positive-definite function, *Definition einer positiv definiten Funktion* 132.
—, discussion, *Diskussion* 135, 158.
—, equivalence relations (semi-equivalence), *Äquivalenzbeziehungen (Semi-Äquivalenz)* 139.
—, examples, *Beispiele* 133.
—, — in elasticity, *Beispiele in der Elastizitätstheorie* 192–195, 207, 209, 210, 212, 219, 220, 222, 241, 245, 251, 254, 266, 268, 270, 271, 273, 280.
— in discrete systems, *in diskreten Systemen* 141.
—, influence on stability, examples, *Einfluß auf Stabilität, Beispiele* 135, 136, 138.
mechanical forcing, *mechanische Anregung* 31.
meeting of two solutions, *Zusammentreffen zweier Lösungen* 480
membrane, *Membran* 208, 436, 497.
memory, materials with, *Gedächtnis, Materialien mit* 172, 334–339, 344–356.
MICHELL, see BELTRAMI.
minimizing sequence, *minimalisierende Folge* 564, 565.
mixed problem, nonlinear, *gemischtes Problem, nichtlinear* 474.
— region, *Gebiet* 518.
mixed-mixed boundary data, *doppelt gemischte Randbedingungen* 47, 57.
mixtures, *Mischungen* 172.
mode of deformation, *Deformationsform* 416.
Mohr circles, *Mohrsche Kreise* 407, 468, 477.
moments, balance of, *Momentenbilanz* 42, 216, 242, 243.
Monge-Ampère equation, *Monge-Ampèresche Gleichung* 441.
motion, definition, *Bewegung, Definition* 304.
—, equation of, *Bewegungsgleichung* 42, 89, 91, 176–181, 404.
— in dynamical system, *im dynamischen System* 127, 128, 131.
moving coordinates, *bewegte Koordinaten* 448.
— surfaces, *Flächen* 428.
multipolar elasticity, *multipolare Elastizitätstheorie* 172, 213.
MURNAGHAN's second order elastic constants, *Murnaghansche elastische Konstanten zweiter Ordnung* 181.
mutually inverse, *gegenseitig umkehrend* 26.

natural state, *natürlicher Zustand* 195.
neighbourhood, *Nachbarschaft* 132, 133.
networks of linearly elastic and viscous elements (springs and dashpots), *Netzwerk linear-elastischer und viscoser Elemente (Federn und Dämpfer)* 34.
neutral equilibrium, *neutrales Gleichgewicht* 144, 170, 202, 204, 205, 226, 227.
Newtonian potential, *Newtonsches Potential* 76.
non-conservative problems, *nichtkonservative Probleme* 258, 279.
non-homogeneous condition, *Nicht-Homogenität* 403.
non-isotropy, *Anisotropie* 461.
nonlinear Liapounov functions, *nichtlineare Liapounovsche Funktionen* 187–191.
— — stability, conditions, *Stabilität, Bedingungen* 192.
— — —, related to linear problems, *in Beziehung auf lineare Probleme* 194.
— natural state, conditions for stability, *nicht-linearer natürlicher Zustand, Bedingungen für Stabilität* 195.
— theories of mechanics, *nicht-lineare Theorien der Mechanik* 1, 11.
— thermoelasticity, basic equations, *Thermoelastizität, Grundgleichungen* 173–176.
— —, equations of perturbed motion, *Gleichungen der gestörten Bewegung* 176, 177.
non-self adjoint equations, *nicht-selbst-adjungierte Gleichungen* 258, 279.
— — problems, *Probleme* 170, 264, 280.
nonsimple motions, *nicht-einfache Bewegungen* 112.
non-singular initial and equilibrium response, *nicht-singuläre Anfangs- und Gleichgewichtsreaktion* 26.
non-uniqueness, *Nicht-Eindeutigkeit* 151, 157, 224.
normal, outward, *Normale, äußere* 47.
— strain rate, *Normaldehnungszuwachs* 442.
— stress, *Normalspannung* 441.
— system, *Normalsystem* 421.
normalized Kelvin process, *normalisierter Kelvin-Prozeß* 78.
normals, *Normalen* 497.
notation, *Bezeichnung* 3.
not hyperbolic, *nicht hyperbolisch* 526.
null boundary data, *verschwindende Randwerte* 56.
— solution, *Nullösung* 129, 134, 149, 150, 207, 209, 211, 224, 231, 239, 246–248, 251, 266, 268, 273, 277, 280–284.
numerical techniques, *numerische Verfahren* 3.

octahedral shearing stress, *achtflächige Scherspannung* 407.
one-dimensional counterpart of kinematical condition of compatibility, *eindimensionales Gegenstück der kinematischen Kompatibilitätsbedingung* 309.
one-dimensional wave equation, *eindimensionale Wellengleichung* 160.
ONSAGER's principle, *Onsager, Prinzip von* 23.
open-end conditions, *Bedingungen für offene Enden* 506, 508.
operational calculus, *Operatorenkalkül* 68.
orbit, *Bahn* 128.
orbital stability, *Bahnstabilität* 144, 172.
order of a pair of operators, *Ordnung eines Operatorpaars* 34.
orthogonal cycloids, *orthogonale Zykloiden* 456.
ordinary sonic point, *gewöhnlicher Schallpunkt* 481.
— spirals, *gewöhnliche Spiralen* 472.
oscillatory displacement processes, *oscillatorische Verschiebungsprozesse* 106.
oval cross-section, *ovaler Querschnitt* 503.
overall yielding, *plastisches Nachgeben, überall* 516.
overhanging ends, *überhängende Enden* 531.

pair $\langle P, Q\rangle$ belongs to the relaxation function **G**, *Paar* $\langle P, Q\rangle$ *gehört zur Relaxationsfunktion* **G** 34.
parabola condition, *Parabelbedingung* 473.
— limit, *Parabelgrenzwert* 462, 488.
parabolic point, *parabolischer Punkt* 481.
— problem, *parabolisches Problem* 460.
parametric excitation, *parametrische Anregung* 256, 278.
— stability, *Stabilität* 187.
past history of a process, *abgeschlossene Geschichte eines Prozesses* 4, 5.
— — problem, *Problem der abgeschlossenen Geschichte* 50–53.
PEARSON's formulation, *Pearsonsche Formulierung* 202.
period, *Periode* 129.
periodic motions, *periodische Bewegungen* 129.
— stress and strain processes, *Spannungs- und Dehnungsprozesse* 21, 31, 32.
permanent deformation, *bleibende Deformation* 538, 542, 550.
— —, gradient history, *Geschichte des Deformationsgradienten* 539.
— — gradients, *Deformationsgradienten* 538.
— stretching tensor, *bleibender symmetrisierter Geschwindigkeitsgradient* 541, 545.
persistent stability, *beständige Stabilität* 187, 256, 257.
perturbations, *Störungen* 171.
—, causes of, initial data, *verursacht durch Anfangswerte* 186.
—, — —, loading, *Belastung* 186.
—, — —, material properties, *materielle Eigenschaften* 187.
—, method, *Methode* 526.
perturbed motion, basic equations, elasticity, linear isothermal, *gestörte Bewegung, Grundgleichungen, Elastizität, linear isotherm* 177–183.

perturbed motion, basic equations, elasticity, linear isothermal, isotropic homogeneous, *gestörte Bewegung, Grundgleichungen, Elastizität, linear isotherm, isotrop, homogen* 180, 181.
— —, — —, —, linear isothermal, isotropic homogeneous incompressible, *linear isotherm, isotrop, homogen, inkompressibel* 181–183.
— —, — —, —, linear thermal, *linear mit Temperaturänderung* 183–185.
— —, — —, —, linear thermal, isotropic homogeneous, *linear mit Temperaturänderung, isotrop homogen* 185.
— —, — —, —, linear thermal, isotropic homogeneous incompressible, *linear mit Temperaturänderung, isotrop, homogen, inkompressibel* 185.
— —, — —, —, non-linear isothermal, *nicht-linear isotherm* 176, 177.
— —, incremental displacement, *Verschiebungsinkrement* 176, 183.
— —, — rate of work equation, *Leistungsinkrement, Gleichung für* 178, 180.
— —, —, state, *Zustandsinkrement* 187–189.
— —, — temperature variable, *Temperaturinkrement* 183.
— —, primary motion, *Erstbewegung* 179, 186, 256.
— —, — state, *Erstzustand* 176, 178–180, 182, 184, 188, 189.
— —, secondary motion, *Zweitbewegung* 178, 179, 186, 187.
— —, — state, *Zweitzustand* 176, 177.
phase lag, *Phasenverschiebung* 22.
— map, *Phasenabbildung* 131.
— velocity, *Phasengeschwindigkeit* 107.
— dispersion 107, 108.
physical plane, *physikalische Ebene* 445, 471, 477.
Piola-Kirchhoff stress, *Piola-Kirchhoffsche Spannung* 174, 181, 183, 185, 316, 542.
plane strain under general yield conditions, *Dehnung unter allgemeinen Bedingungen für plastisches Nachgeben* 437.
— stress, *ebene Spannung* 424, 427, 436.
— — with quadratic yield conditions, *mit quadratischen Bedingungen für plastisches Nachgeben* 438.
— wave, *ebene Welle* 100.
— —, longitudinal, *longitudinale* 103.
— — of order N propagating in a homogeneous, isotropic, viscoelastic material, *der Ordnung N, Ausbreitung in einem homogenen, isotropen, viscoelastischen Material* 103.
— — propagation, *Ausbreitung* 100.
— —, transverse, *transversale* 103.
— —, travelling, *fortschreitende* 107, 109.
plastic body, *plastischer Körper* 403.
— deformation gradient, *Deformationsgradient* 550.
plastic equilibrium, *plastisches Gleichgewicht* 523.
— mass pressed between two rough rigid plates, *plastische Masse, zwischen zwei rauhe starre Platten gepreßt* 526.
— potential, *plastisches Potential* 408, 465.
— power, *plastische Kraft* 546.
— torsion 424.
— work, *Arbeit* 407.
plastic-elastic boundary, *plastisch-elastische Randfläche* 500, 517.
plastic-rigid boundary, *plastisch-starre Randfläche* 494, 520, 528.
— problems, *plastisch-starre Probleme* 523.
— region, *plastisch-starres Gebiet* 501.
plate analogy, *Plattenanalogie* 521.
—, general, *Platte, allgemein* 136, 137, 172, 278, 279, 281.
point field, *Punktfeld* 7.
points, *Punkte* 4.
POISSON's ratio, *Poissonsche Zahl* 412.
polarization vector of a plane travelling wave, *Polarisationsvektor einer ebenen fortschreitenden Welle* 107.
— — of a singular surface, *einer Unstetigkeitsfläche* 97.
position, *Lage, Ort* 7.
position-dependent data, *ortsabhängige Werte* 264–278.
positive longitudinal elasticity, *positive longitudinale Elastizität* 317.
positive-definite elasticity, *positiv-definite Elastizität* 62.
positive-definite functions, *positiv-definite Funktionen* 134, 135, 168.
— — and continuous dependence, *und stetige Abhängigkeit* 145.
— — as measures, *als Maße* 135.
— —, discussion, *Diskussion* 186.
— —, examples, *Beispiele* 133.
— — $F_{\tau,t}$, 154, 155, 157, 163, 164, 168, 169, 207, 212.
— — in calculus of variations, *in der Variationsrechnung* 164.
— — and sets, definition, *und Mengen, Definition* 133.
post-buckling behavior, *Knicken, Verhalten nach* 125, 195, 202.
potential density, *Potentialdichte* 259, 261.
— function, *Potentialfunktion* 259, 260, 287.
— lines, *Potentiallinien* 460.
—, plastic, *Potential, plastisches* 408.
power and energy, *Leistung und Energie* 95.
—, expended per cycle, *pro Umdrehung aufgebrachte* 34.
—, —, theorem of, *aufgebrachte, Satz der* 95.
— identity, *Leistungsidentität* 54.
—, stress, *Spannungsleistung* 95, 174.
Prandtl-Reuss equations, *Prandtl-Reußsche Gleichungen* 424, 548.
pre-stress, *Vorspannung* 219.

present state of deformation and stress, *gegenwärtiger Deformationszustand und gegenwärtige Spannung* 1.
— value, *Wert* 5.
pressure, *Druck* 404, 480.
primary motion, *Erstbewegung* 179, 186, 256.
— state, *Erstzustand* 176–182, 184, 185, 188, 206, 215, 217, 219, 240, 256.
— stress, *Erstspannung* 198, 215, 239, 241.
principal axes of mean stress, *Hauptachsen der mittleren Spannung* 216, 240–243.
— directions, *Hauptrichtungen* 406, 441.
— stresses, *Hauptspannungen* 406.
— tangential stresses, *Haupttangentialspannungen* 407.
— trajectories, *Haupttrajektoren* 484, 486.
— waves, *Hauptwellen* 326, 328.
principle of frame-indifference, *Prinzip der Systemunabhängigkeit* 544, 545.
— — impenetrability, *Undurchdringlichkeit* 536, 538.
— — production of entropy, *Entropieproduktion* 539.
process, processes, *Prozeß, Prozesse*
 class C^N, *Klasse* C^N 5.
 — C^∞ 5.
 constant, *konstant* 5.
 continuous, *stetig* 5.
 continuous at infinity, *stetig im Unendlichen* 5.
 definition 4.
 discontinuous, *unstetig* 6.
 — approximation by continuous, *Approximation unstetiger Prozesse durch stetige* 16, 35.
 dynamic viscoelastic, *dynamische viscoelastische* 89.
 Heaviside type, *Heaviside-Typ* 5, 16.
 locally integrable, *lokal integrierbar* 5.
 quasistatic viscoelastic, *quasistatisch viscoelastisch* 42–44.
 restricted, *eingeschränkt* 5.
 strong, *stark* 6.
propagation condition (see also wave), *Ausbreitungsbedingungen (s. auch Welle)* 98, 105, 316, 324, 326.
proper minimum of V, *wirkliches Minimum von* V 205.
pseudosteady, *pseudostationär* 525.
pulse front, *Pulsfront* 96.
punch, *Locher* 455.
pure dilatation, *reine Ausdehnung* 30.
— shear stress, *Scherspannung* 30.
— shearing motion, *Scherbewegung* 30.

quadratic yield condition, *quadratische Bedingung für plastisches Nachgeben* 403, 405, 461, 471, 487, 549.
quasi-linear, *quasilinear* 440.
quasi-static assumption, *quasistatische Annahme* 42.
— boundary value problem, *quasistatisches Randwertproblem* 47, 69.
quasi-static linear viscoelasticity, *quasistatische lineare Viscoelastizität* 42.
— theory, *Theorie* 2.
— variational principles, *Variationsprinzipien* 64.
— viscoelastic process corresponding to creep function $\boldsymbol{J}$ and body force $\boldsymbol{b}$, *quasistatischer viscoelastischer Prozeß entsprechend der Kriechfunktion* $\boldsymbol{J}$ *und der Volumkraft* $\boldsymbol{b}$ 43.
— — — corresponding to relaxation function $\boldsymbol{G}$ and body force $\boldsymbol{b}$, *entsprechend der Relaxationsfunktion* $\boldsymbol{G}$ *und der Volumkraft* $\boldsymbol{b}$ 43, 77.

radii of curvature, *Krümmungsradien* 446.
rate at which work is done, *Leistung* 95.
— independence, *Zeitskalenunabhängigkeit* 543.
— — of the generalized stress, *der verallgemeinerten Spannung* 544.
— independent materials, *Zeitskala, Materialien unabhängig von* 542.
— of dilatation, *Rate der Ausdehnung* 474.
— of extension, *Rate der Ausdehnung in einer Richtung* 420.
— of working of surface tractions and body force, *Leistung von Oberflächenspannungen und Volumkraft* 54.
reasonableness, *Vernünftigkeit* 146.
reciprocal theorem, *Reziprozitätstheorem* 55.
recoverable work, *wiedergewinnbare Arbeit* 3.
reducible equations, *reduzierbare Gleichungen* 459.
— system, *reduzierbares System* 440.
reference configuration, *Bezugskonfiguration* 173.
— state, *Bezugszustand* 182.
regularity, preservation of, *Regularitätserhaltung* 14.
—, theorems, *Regularitätssätze* 74.
relaxation and creep laws, *Relaxationsverhalten und Kriechgesetze* 23.
— function $\boldsymbol{G}$, *Relaxationsfunktion* $\boldsymbol{G}$ 18.
— — and creep compliance, *und Kriechnachgiebigkeit* 28.
— — for creep function (compliance) $\boldsymbol{J}$, *für Kriechnachgiebigkeit* $\boldsymbol{J}$ 26.
— functional, *Relaxationsfunktional* 17.
— law at $\boldsymbol{x}:\mathscr{L}\boldsymbol{x}$, *Relaxationsfunktionalgesetz in* $\boldsymbol{x}:\mathscr{L}\boldsymbol{x}$ 17.
— — for body $\mathscr{B}$, *für den Körper* $\mathscr{B}$ 17.
— problems, variational principles, *Relaxationsprobleme, Variationsprinzipien* 65, 66, 113.
— time for a Maxwell element, *Relaxationszeit für ein Maxwell-Element* 41.
— times and differential operator, *Relaxationszeiten und Differentialoperator* 37, 38.
Rellich selection principle, *Rellichsches Auswahlprinzip* 576.
"remaining" equations of plasticity, *„verbleibende" Gleichungen der Plastizitätstheorie* 426, 456, 494.

"remaining" equations for shear response, „verbleibende" *Gleichungen für Scherwirkung* 71.
rescaling function, *Skalenverschiebung, Funktion der* 542, 543.
residual displacements, *Restverschiebungen* 219.
— entropy production inequality, *Restentropieungleichung* 175, 183, 188.
— equation of energy balance, *Restgleichung der Energiebilanz* 174, 175, 183.
response function or functional, *Antwortfunktion oder Antwortfunktional* 11–13.
restrained deformation, *behinderte Deformation* 523.
restricted histories, *begrenzte Geschichten* 5.
retardation times, *Retardierungszeiten* 38, 42.
Riemann function, *Riemann-Funktion* 449, 450.
— method, *Riemann-Methode* 449.
— representation, *Riemann-Darstellung* 104.
Riemann-Lebesgue lemma 21.
right inverse, *rechte Umkehrung* 23.
— material derivative, *materielle Ableitung* 541.
rigid, rigid-body motions, *starre Bewegung* 17, 18, 23, 216, 246, 262, 266, 274.
— rotations, *Rotationen* 215, 216, 219, 239, 240, 242, 244, 246.
— translational velocities, *translatorische Geschwindigkeiten* 218, 220.
— translations, *Translationen* 215, 217, 219, 220, 239.
rigid-plastic analysis, *starr-plastische Rechnung* 524.
— problem, *starr-plastisches Problem* 495.
ring 424, 515.
RITZ's representation theorem, *Ritzsches Darstellungstheorem* 558.
rod, *Stab* 172, 279, 282, 283, 285.
roof, *Dach* 500.
rotating cylinders and disks, *rotierende Zylinder und Scheiben* 521.
rotational balance, *Drehimpulsbilanz* 42, 216, 242, 243.

scalar field, *skalares Feld* 7.
SCHAUDER's fixed-point theorem, *Schaudersches Fixpunkttheorem* 577.
SCHWARZ's inequality, *Schwarzsche Ungleichung* 559.
second law of thermodynamics, *zweiter Hauptsatz der Thermodynamik* 537, 539, 540.
—, or characteristic problem, *zweites oder charakteristisches Problem* 492.
second-order elastic constants, *elastische Konstanten zweiter Ordnung* 181.
— elasticity tensor, *Elastizitätstensor zweiter Ordnung* 316.
second-order isentropic tangent modulus, *isentroper Tangentenmodul zweiter Ordnung* 362.
— tangent modulus, *Tangentenmodul zweiter Ordnung* 356.
second Piola-Kirchhoff stress, *zweite Piola-Kirchhoffsche Spannung* 542.
secondary motion, *Zweitbewegung* 178, 179, 186, 187, 279.
— state, *Zweitzustand* 176, 177, 256.
self-adjoint, *selbstadjungiert* 264.
self-equilibrated Kelvin process, *von selbst im Gleichgewicht, Kelvin-Prozeß* 81.
— surface tractions, *Oberflächenkräfte* 83.
semi-circular groove, *halbkreisförmige Rille* 521.
semi-inverse method, *semiinverse Methode* 560.
separable, *separierbar* 170, 557.
separate surface elements, *getrennte Oberflächenelemente* 46.
separation of variables, *Separation von Variablen* 255, 256, 284.
—, surface of, *Trennfläche* 426.
set $\mathscr{B}(T, X)$, *Menge* $\mathscr{B}(T, X)$ 127, 128.
— of all linear transformations, *aller linearen Transformationen* 4.
— of initial data, *von Anfangswerten* 130.
— X 127, 128, 130.
shaft, *Welle* 505.
shallow curved panel, *schwach gekrümmte Tafel* 281.
shear modulus, *Schermodul* 30.
— strain, *Scherdehnung* 442.
— stress, *Scherspannung* 442.
sheet, *Scheibe* 515.
shell, *Schale* 172, 287.
SHIELD's theorem, *Shieldscher Satz* 213.
SHINBROT's fixed-point theorem, *Shinbrotscher Fixpunktsatz* 558.
shock conditions, *Stoßbedingungen* 433.
— waves, compression or expansion, *Stoßwelle, Verdichtung oder Ausdehnung* 312.
— —, definition 97, 311.
— — in bodies of material with memory, *in Körpern aus Materialien mit Gedächtnis* 349–356.
— — in connection with weak solutions, *in Verbindung mit schwachen Lösungen* 171.
— — in elastic bodies, *in elastischen Körpern* 357–361.
— — in inhomogeneous elastic bodies, *in inhomogenen elastischen Körpern* 382–385.
— — in non-conductors, *in Nichtleitern* 369–377.
— — in plasticity, *in der Theorie der Plastizität* 416
simple material, *einfaches Material* 17.
— vector field, *Vektorfeld* 73, 112.
— wave, *Welle* 447, 484, 485.

simply connected, *einfach zusammenhängend* 45.
— supported elastic plate, *unterstützte elastische Platte* 136.
singular solutions, *singuläre Lösungen* 77.
— surfaces (see also acceleration wave, shock wave), *Flächen (s. auch Beschleunigungswellen, Stoßwellen)* 96, 97, 305.
— —, related to high frequency waves, *in Relation zu hochfrequenten Wellen* 109.
— yield conditions, *Bedingungen plastischen Nachgebens* 411.
singularity, branch lines, *Singularität, Zweiglinien* 483.
—, — type, *Zweigtyp* 477.
skew (infinitesimal) rotation field process **W**, *schiefsymmetrischer (infinitesimaler) Rotationsprozeß* **W** 17.
— tensors, subspace of, *schiefsymmetrische Tensoren, Unterraum der* 4.
slip field, *Fehlstellenfeld* 445, 450.
— lines, definition, *Fehlstellenlinien, Definition* 442, 443.
— —, plastic boundary, *an der Grenze zum plastischen Gebiet* 494.
— —, properties of, *Eigenschaften* 447.
— surfaces, *Fehlstellenflächen* 424.
small deformation from a stress free isotropic reference state, *kleine Deformation aus einem spannungsfreien isotropen Bezugszustand* 191.
snap-through, *Umklappen* 148, 287–289.
Sobolev embedding theorems, *Sobolevsche Einbettungssätze* 140, 212.
— lemma 578.
— norms, *Norm* 224.
— space, *Raum* 212.
Sokolovsky condition, *Sokolovskysche Bedingung* 473.
solenoidal vector field, *quellenfreies Vektorfeld* 73, 557.
solution, auxiliary problem for $\boldsymbol{E}^*$, *Lösung des Hilfsproblems für* $\boldsymbol{E}^*$ 61.
—, Cauchy initial value problem for dynamic viscoelasticity, *des Cauchy-Anfangswertproblems für dynamische Viscoelastizität* 105.
—, dynamic viscoelastic boundary value problem, *Lösung des dynamischen viscoelastischen Randwertproblems* 89.
— in dynamical system, *in einem dynamischen System* 128, 131.
— of associated elastic problem, *des zugeordneten elastischen Problems* 60.
— of initial past history problem in terms of displacements, *des Problems mit anfänglicher vergangener Geschichte, ausgedrückt durch Verschiebungen* 52.
—, quasi-static, in terms of displacements, relaxation function, or creep compliance, *quasistatisch, ausgedrückt durch Verschiebungen, Relaxationsfunktion oder Kriechnachgiebigkeit* 47.
—, —, in terms of stresses, *ausgedrückt durch Spannungen* 48.
solution, viscoelastic, initial past history problem, *Lösung, viscoelastisch, Problem mit anfänglicher vergangener Geschichte* 51.
—, — problem 60.
SOMIGLIANA, see BOUSSINESQ.
Somigliana stress functions, viscoelastic, *Somiglianasche Spannungsfunktionen, viscoelastisch* 94.
sonic limit line, *Schallgrenzlinie* 482.
— — point, *Schallgrenzpunkt* 482.
— line, *Schallinie* 480.
space $\mathscr{C}_0(\mathscr{R}^+; \mathscr{X})$ of continuous $(\mathscr{X})$-histories which vanish at infinity, *Raum* $\mathscr{C}_0(\mathscr{R}^+; \mathscr{X})$ *von stetigen* $(\mathscr{X})$-*Geschichtsfunktionen, die im Unendlichen verschwinden* 7.
— $\mathscr{L}^1(\mathscr{R}^+; \mathscr{X})$ of integrable $(\mathscr{X})$-histories, $\mathscr{L}^1(\mathscr{R}^+; \mathscr{X})$ *von integrierbaren* $(\mathscr{X})$-*Geschichtsfunktionen* 7.
—, Euclidean point, *euklidischer Punkt* 4.
—, inner product, *mit skalarem Produkt* 4.
— of linear maps in $\mathscr{T}$ (the space of fourth-order tensors): $[\mathscr{T}]$, *von linearen Abbildungen in* $\mathscr{T}$ *(dem Raum von Tensoren vierter Stufe)*: $[\mathscr{T}]$ 4.
— of (second-order) tensors $[\mathscr{V}] \equiv \mathscr{T}$, *von Tensoren zweiter Stufe* $[\mathscr{V}] \equiv \mathscr{T}$ 4.
— translation 4.
— vector, *Vektor-* 4.
spatial description, *räumliche Darstellung* 428, 536.
specific entropy, *spezifische Entropie* 536.
— free energy, *freie Energie* 536.
— heat, *Wärme* 193, 201.
— internal energy, *innere Energie* 536.
spectrum of differential operator law, *Spektrum von Differentialoperatorgesetzen* 38.
speed of displacement, *Geschwindigkeit der Verschiebung* 307, 308, 429.
— of propagation (see also wave), *der Ausbreitung (s. auch Welle)* 96, 97, 306, 308, 429.
spherical shell, *Kugelschale* 192, 521.
spirals, *Spiralen* 456.
stability, *Stabilität* 3.
—, and buckling, *und Knicken* 130.
—, choice in elasticity, *Wahl in der Elastizitätstheorie* 171.
—, definitions, asymptotic, *Definitionen, asymptotische* 142.
—, —, asymptotically orbital, *asymptotische, der Bahnbewegung* 144.
—, —, conditional, *bedingte* 147.
—, —, contractive, *kontraktive* 144.
—, —, early, *frühe* 132 (footnote, *Fußnote*)
—, —, exponentially asymptotic, *exponentiell asymptotische* 142.
—, —, finite-time, (practical or Lasalle and Lefschetz), *endliche Zeit (praktische oder Lasalle und Lefschetz)* 143.
—, —, globally asymptotic (completely), *global asymptotische (vollständig)* 142.
—, —, Hadamard 198.

stability, definitions, Hölder, *Stabilität, Definitionen* 146.
—, —, infinitesimal superstability, *infinitesimale Überstabilität* 202.
—, —, Lagrange 149.
—, —, Lagrange-Dirichlet 131.
—, —, —, essential ingredients, *wesentliche Bestandteile* 132.
—, —, Liapounov 133.
—, —, —, equivalent definitions, *äquivalente Definitionen* 132, 140
—, —, —, observations, *Beobachtungen* 134.
—, —, local (Krasovskii), *lokale* 143.
—, —, neutral, *neutrale* 144.
—, —, of dynamical system, *eines dynamischen Systems* 134.
—, —, of equilibrium, *des Gleichgewichts* 131.
—, —, of a given motion, *einer gegebenen Bewegung* 134.
—, —, orbital, *der Bahnbewegung* 144.
—, —, parametric, *parametrische* 187.
—, —, persistent, *beständige* 187.
—, —, perturbative, *bei Störung* 146 (footnote, *Fußnote*)
—, —, quasi-contractive, *quasi-kontraktive* 143.
—, —, region of stability, *Stabilitätsgebiet* 142.
—, —, superstability, *Überstabilität* 202.
—, —, trigonometric, *trigonometrische* 146 (footnote, *Fußnote*).
—, —, uniform, *uniforme* 141.
—, —, uniformly asymptotic, *uniform asymptotische* 142.
—, —, weakly asymptotic, *schwach asymptotische* 142.
— theorems, *Theoreme*
— — and calculus of variations, *und Variationsrechnung* 163–167, 195
— —, discussion of, *Diskussion der* 156–160.
— —, — —, examples, *Beispiele* 160–163
— —, energy criterion, *Energiekriterium* 164 (footnote, *Fußnote*), 195.
— —, further Liapounov, null solution, *Liapounov, Nullösung* 209, 212, 219.
— —, necessity, for equilibrium solution, *Notwendigkeit, für Gleichgewichtslösung* 200–202.
— —, sufficiency, natural state, *hinreichende Bedingungen, natürlicher Zustand* 195.
— —, — in non-linear thermo-elasticity, *in nicht-linearer Thermoelastizität* 192–194.
— —, SHIELD's theorem, *Shieldsches Theorem* 213.
standard differential operator law, *Standard-Differentialoperatorgesetz* 39.
— models, *Standardmodell* 40.
statically determinate, *statisch bestimmt* 526.
stationary state, *stationärer Zustand* 524.
— value problem, *Problem mit stationären Werten* 66.
steady plastic flow, *beständige plastische Strömung* 524.
— shock waves, *Stoßwelle* 354.
storage modulus, *Speichermodul* 33.
stored-energy, *gespeicherte Energie* 173.
strain displacement equation, *Dehnungsverschiebungsgleichung* 43, 71, 89.
— energy, *Dehnungsenergie* 87, 173, 225.
— field process, *Dehnungsfeldprozeß* 16.
— history, *Dehnungsgeschichte* 305.
strain-rate tensor, *Dehnungsrate, Tensor der* 499.
streamlines, *Stromlinien* 455, 460, 479, 531.
strength of a plane wave (see also amplitude), *Stärke einer ebenen Welle (s. auch Amplitude)* 100.
stress, Cauchy, *Spannung, Cauchy* 175, 180, 183, 185, 325, 536.
— deviator force, *Spannungsdeviatorkraft* 404.
— — tensor, *Spannungsdeviatortensor* 404.
— discontinuities, *Spannungsunstetigkeit* 433, 457.
— discontinuity line, *Spannungsunstetigkeitslinie* 498.
— entropy modulus, *Spannungsentropiemodul* 362, 369.
— — vector, *Spannungsentropievektor* 537.
— equation of compatibility, *Spannungsgleichung für Kompatibilität* 45, 46.
— — of equilibrium, *für Gleichgewicht* 46.
— field process, *Spannungsfeldprozeß* 16.
— functions, *Spannungsfunktionen* 496, 522, 561, 584.
— —, viscoelastic, *Spannungsfunktionen, viscoelastische* 73, 93.
— graph, *Spannungsschaubild* 445, 477.
— lines, *Spannungslinien* 496.
— multipoles, *Spannungsmultipole* 213.
— or traction boundary data, *Spannungs- oder Kraft-Randwerte* 47.
— — — — value problem, *Randwertproblem* 47.
—, Piola-Kirchhoff 174, 181, 183, 185, 316, 542.
— plane, *Spannungsebene* 445, 471.
— power, *Spannungsleistung* 95, 174.
— process, *Spannungsprozeß* 42.
— relaxation, *Spannungsrelaxation* 1, 18, 20, 32.
— — function, *Spannungs-Relaxationsfunktion* 335, 350, 351.
— — property, *Spannungs-Relaxationseigenschaft* 20.
— space, *Spannungsraum* 406.
— tensor, *Spannungstensor* 404.
— vector, *Spannungsvektor* 258–260, 261.
stretching tensor, *symmetrisierter Geschwindigkeitsgradient* 548.
strings, *Schnüre* 278.
strong discontinuities (see also shock), *starke Unstetigkeiten* 427, 434.

strong ellipticity, *starke Elliptizität* 64, 99, 211, 213, 321.
— quasi-static viscoelastic process, *starker quasistatischer viscoelastischer Prozeß* 44.
— response function, *starke Antwortfunktion* 13.
strongly dissipative material, *stark dissipatives Material* 22.
STRUTT's diagram, *Struttsches Diagramm* 257.
ST. VENANT's conjecture, *St. Venants Vorschlag* 562.
— principle, *Prinzip* 85.
— — and potential theory, *und Potentialtheorie* 87.
— —, bodies of general shape, *Körper beliebiger Form* 87.
— —, decay of stresses, *Abnahme von Spannungen* 87.
— — for a filament, *für eine Faser* 88.
— — for viscoelastic bodies, *für viscoelastische Körper* 86.
— — of elastostatics, *der Elastostatik* 88.
— —, pointwise estimate, *punktweise Schätzung* 88.
— —, preferred direction, *bevorzugte Richtung* 87.
— solution, *Lösung* 87.
subspaces of symmetric and skewsymmetric tensor $\mathcal{T}_{\mathrm{sym}}$ and $\mathcal{T}_{\mathrm{skw}}$, *Unterräume symmetrischer und schiefsymmetrischer Tensoren* $\mathcal{T}_{\mathrm{sym}}$ *und* $\mathcal{T}_{\mathrm{skw}}$ 4.
sudden transition, *plötzlicher Übergang* 414.
supersonic air-stream, *Überschall-Luftströmung* 281.
superposition, method, *Superposition, Methode* 448.
—, property, *Eigenschaft* 12.
surface force, *Oberflächenkraft* 174, 278.
— loads, *Oberflächenbelastung* 85, 278.
— of constant slope, *Oberfläche konstanter Neigung* 497.
— potential density, *Oberflächenpotentialdichte* 259, 260.
— traction, *Oberflächenspannungszug* 42, 85, 178, 259.
symmetric infinitesimal strain field process, *symmetrischer infinitesimaler Dehnungsfeldprozeß* 17.
— initial elasticity, *symmetrische Anfangselastizität* 62.
— tensors, subspace of ($\mathcal{T}_{\mathrm{sym}}$), *symmetrische Tensoren, Unterraum der* ($\mathcal{T}_{\mathrm{sym}}$) 4.
symmetry group, *Symmetriegruppe* 18.
— of acoustic tensor, *Symmetrie des akustischen Tensors* 99.
— of Cauchy stress, *der Cauchyschen Spannung* 542.
— of GREEN's processes of the first kind, *von Greenschen Prozessen der ersten Art* 82.
— — — — of the second kind, *der zweiten Art* 84.
symmetry transformation, *Symmetrietransformation* 18.
synchronous and separable boundary data and body forces, *gleichzeitige und separierbare Randwerte und Volumkräfte* 48, 55.
systems of differential equations, *Systeme von Differentialgleichungen* 420.

tangent modulus, *Tangentenmodul* 356, 358.
temperature, *Temperatur* 174, 536.
tensor, definition 3, 4.
— field, *Tensorfeld* 7.
— laws of differential type, *Tensorgesetze vom Differential-Typus* 37.
— of angular velocity, *Tensor der Winkelgeschwindigkeit* 415.
thermodynamics of viscoelasticity, *Thermodynamik der Viscoelastizität* 3.
—, second law, *zweiter Hauptsatz* 537.
thermoelasticity, *Thermoelastizität* 173–176, 183–185, 187–195, 249–255.
thermorheologically simple materials, *thermorheologisch einfache Materialien* 3.
thick-walled tube, *dickwandige Röhre* 506.
thickening, *Verdickung* 519.
three-dimensional Euclidean point-space, *dreidimensionaler euklidischer Punktraum* 4.
— inner product space, *innerer Produktraum* 4.
— problem of plasticity, *dreidimensionales Problem der Plastizität* 403.
— subspace of skew-symmetric tensors, *dreidimensionaler Unterraum schiefsymmetrischer Tensoren* 4.
— wave equation, *dreidimensionale Wellengleichung* 137, 162.
time t, *Zeit* t 4, 7.
— reversal, *Zeitumkehrung* 23.
—, role played by present, *Rolle der gegenwärtigen Zeit* 13.
— translation invariance, *Zeit-Translationsinvarianz* 12.
time-dependent data, *zeitabhängige Bedingungen* 264.
tin-canning, *Konservenbüchseneffekt* 287.
TITCHMARSH's theorem, *Titchmarshscher Satz* 74.
topological nature of stability, *topologische Form der Stabilität* 139, 141.
torque, *Drehmoment* 496, 564, 582.
torsion 495.
torsional rigidity, *Torsionssteifheit* 585, 586.
traceless symmetric tensor process, *spurfrei-symmetrischer Tensorprozeß* 71.
traction (see also stress) boundary value problem, *Zugspannungs- (s. auch Spannungs-) Randwertproblem* 83.
— problem for incompressible media, *-Problem für inkompressible Medien* 72.
trajectory, *Trajektore* 128, 151.
trajectory/orbit, *Trajektorie/Bahn* 129.

transition map, *Übergangsabbildung* 131.
— zone, *Übergangsgebiet* 433, 435.
translate of a function, *Verschiebung einer Funktion* 128.
— process, *verschobener Prozeß* 12.
translation continuity, *Verschiebungsstetigkeit* 15.
— space, *Translationsraum* 4.
transverse, transversal, discontinuity, *transversale Unstetigkeit* 430.
— principal waves, *Hauptwelle* 328.
— wave, *Welle* 99, 310.
— — in elastic bodies, anisotropic, *in elastischen Körpern, anisotropisch* 321–323.
— — in elastic bodies, isotropic, *isotropisch* 328.
TRESCA's yield criterion, *Trescasches Kriterium für plastisches Nachgeben* 411, 423, 462.
triangle inequalities, *Dreiecksungleichungen* 561, 565.
TRUESDELL and NOLL's criterion, *Truesdell und Nollsches Kriterium* 202.
twist, *Verdrehung* 496, 582, 583.
twisting moment, *Drehmoment* 496.

unbounded bodies, *unbegrenzte Körper* 77.
undeformed state, *undeformierter Zustand* 538.
uniaxial tension, *einachsige Spannung* 249.
uniform hydrostatic loads, *uniforme hydrostatische Belastung* 262, 263.
— pressure, *uniformer Druck* 30.
— stability, definition, *uniforme Stabilität, Definition* 141.
— —, sufficiency theorem, *Satz für hinreichende Bedingungen* 152.
— —, theorem (corollary), *Satz (Korollar)* 157.
unilaterally loaded wedge, *einseitig belasteter Keil* 526.
uniqueness and stability for elastic-plastic bodies, *Eindeutigkeit und Stabilität für elastisch-plastische Körper* 523.
—, Cauchy initial value problem, *Cauchysches Anfangswertproblem* 105.
—, equivalence with Liapounov stability, *Äquivalenz zur Liapounov-Stabilität* 151.
— for rigid plastic solids, *für starre plastische Festkörper* 416, 523.
— in general, *allgemein* 131.
— in linear elasticity, *in der linearen Elastizitätstheorie* 152, 231.
—, incompressible bodies, *inkompressible Körper* 72.
—, initial past-history problem, *Problem mit anfänglicher vergangener Geschichte* 57.
—, isotropic bodies, *isotrope Körper* 57.
—, non-uniqueness and Liapounov instability, *Nichteindeutigkeit und Liapounov-Instabilität* 151, 152, 167, 224.
—, — in follower force problems, divergence, *Nichteindeutigkeit in Folgekraftproblemen, Divergenz* 282.
uniqueness of quasi-static viscoelastic processes, *Eindeutigkeit quasistatischer viscoelastischer Prozesse* 55.
— (and existence) of viscoelastic processes, *(und Existenz) viscoelastischer Prozesse* 58.
—, theorem, displacement boundary data, *Satz, Verschiebung als Randwerte* 96.
—, — for elastic states, *für elastische Zustände* 49.
—, — for linear elasticity, thermoelasticity, initial boundary value problem, *für lineare Elastizität, Thermoelastizität, Anfangs- und Randwertproblem* 231, 254.
—, — for processes with infinite histories, *für Prozesse mit unendlich langen Geschichten* 59.
—, — for quasi-static viscoelastic processes, *für quasistatische viscoelastische Prozesse* 56.
—, — in dynamic viscoelasticity, *in dynamischer Viscoelastizität* 95, 96.
—, —, Laplace transform formulation, *Formulierung mit Laplace-Transformierter* 96.
—, —, mixed problem, *gemischtes Problem* 96.
unsymmetrical extrusion, *unsymmetrisches Auspressen* 525.

value of a process at time *t*, *Wert von einem Prozeß zur Zeit t* 4.
variation, definition 64, 65.
— in amplitude with time, *in der Amplitude, zeitliche* 96.
— of a functional, *eines Funktionals* 112.
variational characterization of solutions to the quasi-static boundary value problem with prescribed initial past history, *variationsmäßige Charakterisierung von Lösungen des quasistatischen Randwertproblems mit vorgeschriebener anfänglicher vergangener Geschichte* 64.
— principle, dynamic creep problems, *Variationsprinzip für dynamische Kriechprobleme* 114.
— —, — relaxation problems, *für dynamische Relaxationsprobleme* 113.
— —, first, for creep problems, *erstes, für Kriechprobleme* 66.
— —, —, for relaxation problems, *für Relaxationsprobleme* 65, 68.
— —, second, for creep problems, *zweites, für Kriechprobleme* 67.
— —, —, for relaxation problems, *für Relaxationsprobleme* 66.
vector field, *Vektorfeld* 7.
vectors, *Vektoren* 3.
velocity, characteristics, *Geschwindigkeit, Charakteristiken der* 445.
—, compatibility relations, *Kompatibilitätsbedingungen* 444.
—, constant on characteristics, *konstant auf Charakteristiken* 489.
—, definition 8, 173, 305.

velocity distributions, *Geschwindigkeitsverteilungen* 438.
— equations, *Gleichungen* 449.
— field process, *Geschwindigkeitsfeldprozeß* 8.
— gradient, *Geschwindigkeitsgradient* 541.
— (speed) of displacement, *der Verschiebung* 307, 308, 429.
— of the medium, *des Mediums* 429.
— plane, *Geschwindigkeitsebene* 445, 467.
virtual displacement, *virtuelle Verschiebungen* 197, 198, 202, 203, 261.
— modes, *Deformationsformen* 416.
viscoelastic analogs of work-energy identities of elastostatics, *viscoelastische Analoga der Arbeits-Energie-Identitäten der Elastostatik* 53.
— analogue to Boggio's theorem, *viscoelastisches Analogon zum Boggioschen Theorem* 94.
— body, *viscoelastischer Körper* 17.
— center of compression, *viscoelastisches Zentrum der Verdichtung* 80.
— — — rotation 81.
— counterpart of Beltrami-Donati-Michell equations, *Gegenstück zu den Beltrami-Donati-Michellschen Gleichungen* 46.
— displacement boundary value problem, *Verschiebungs-Randwertproblem* 81.
— field equations (dynamic), *viscoelastische Feldgleichungen (dynamisch)* 42, 91.
— foundations of linear theory, *Grundlagen der linearen Theorie* 11.
— generalization of Boussinesq-Somigliana-Galerkin and Boussinesq-Papkovitch-Neuber solutions, *Verallgemeinerung von Boussinesq-Somigliana-Galerkin- und Boussinesq-Papkovitch-Neuber-Lösungen* 74.
— Kelvin process, *viscoelastischer Kelvin-Prozeß* 78.
— mixed boundary-value problem, *viscoelastisches gemischtes Randwertproblem* 59.
— potential function, *Potentialfunktionen* 3.
— process corresponding to differential operator law, *viscoelastischer Prozeß, der einem Differentialoperatorgesetz entspricht* 57.
— solution, *viscoelastische Lösung* 60.
— stress analysis, *Spannungsanalysis* 68.
— — functions, *Spannungsfunktionen* 93, 94.
viscous flow, *viskose Strömung* 405.
Voigt, see Kelvin
v. Mises extremum principle, *v. Mises-Extremalprinzip* 409.
— general equations, *v. Misessche allgemeine Gleichungen* 410.
— linear equations, *lineare Gleichungen* 459.
— theory of plastic potential, *Theorie des plastischen Potentials* 408.
— yield condition, *Bedingung für elastisches Nachgeben* 403, 405, 461, 471, 487, 549.
v. Mises-Reuss theory, *v. Mises-Reuss-Theorie* 413.
Volterra's equation, *Volterra-Gleichung* 62.
— lemma, *Volterra-Lemma* 56.
— uniqueness theorem, *Volterras Eindeutigkeitssatz* 56.
volume potential density, *Volum-Potentialdichte* 259.
— preserving, *volumerhaltend* 550.

w-lines, *w-Linien* 485.
warping, elastic, *Verzerren, elastisches* 496.
—, plastic, *plastisches* 501, 505.
wave (see also discontinuity), acceleration, *Welle (s. auch Unstetigkeit), Beschleunigungswelle* 97, 309, 313, 334, 356, 362, 428–432, 434, 435.
— amplitude (see also variation of amplitude), *Amplitude (s. auch zeitliche Veränderung der Amplitude)* 309–311, 315.
— axes, *Achsen* 97, 98.
—, axially dissimilar, *axial unähnliche* 102.
—, — similar, *axial ähnliche* 101.
—, centered, *zentrierte* 486.
—, definition 96, 306.
— equation, *Gleichung* 137, 160, 162.
—, high frequency, related to decay, *hochfrequente, in Beziehung zur Abnahme* 109.
— in bodies, elastic, *in Körpern, elastischen* 315–334, 356–385.
— — —, — inhomogeneous, *inhomogenen* 378–385.
— — —, — isotropic, *isotropischen* 325–334.
— — —, — non-conductors, *Nichtleitern* 362–377.
— — — of material with memory, *aus Material mit Gedächtnis* 334–339, 344–356.
— — —, plastic, *plastischen* 416, 434, 435.
— in bodies, viscoelastic, *in Körpern, viscoelastischen* 1, 92, 96–104, 106–112.
—, longitudinal, *longitudinale* 99, 310, 315–321, 327, 328.
—, principal, *Hauptwelle* 326, 328.
— shock, *Stoßwelle* 97, 171, 311, 313, 349–361, 369–377, 382–385, 416.
— speed, *Geschwindigkeit* 92, 96, 97, 197, 306–339, 345–347, 351–372, 378–380, 382–385.
— strength, *Wellenstärke* 100.
—, transverse, *transversale* 99, 310, 321–323, 328, 332, 334.
—, variation of amplitude with time (attenuation, decay, growth), *zeitliche Veränderung der Amplitude (Dämpfung, Abklingen, Anwachsen)* 1, 96, 98, 106, 107, 110, 320–323, 328, 330–333, 341–348, 353, 364–376, 378–385.
weak compactness, *schwache Kompaktheit* 559.
— differential, *schwaches Differential* 552.

weak quasi-static viscoelastic process, *schwacher quasistatischer viscoelastischer Prozeß* 44.
— solutions, *schwache Lösungen* 162, 171, 173, 178, 179, 211, 215, 224, 229, 258.
weakly asymptotic stability, definition, *schwach-asymptotische Stabilität, Definition* 142.
— — —, theorem, *Satz* 158.
— conservative forces, *konservative Kräfte* 193, 258–264, 286.
— — stress vector, *konservativer Spannungsvektor* 263.
wedge, *Keil* 521, 526.
well-posed, *wohlformuliert* 136, 146.
work (see also energy, power), *Arbeit (s. auch Energie, Leistung)* 3, 23.

($\mathcal{X}$)-field, ($\mathcal{X}$)-*Feld* 7.
($\mathcal{X}$)-field process for $\mathcal{D}$, ($\mathcal{X}$)-*Feldprozeß für* $\mathcal{D}$ 7.
($\mathcal{X}$)-history, ($\mathcal{X}$)-*Geschichte* 4.
($\mathcal{X}$)-past-history, ($\mathcal{X}$)-*abgeschlossene Geschichte* 4.
($\mathcal{X}$)-process, ($\mathcal{X}$)-*Prozeß* 4.

yield condition or criterion, *plastisches Nachgeben, Bedingung oder Kriterium* 403, 405, 411, 461, 463, 464, 471, 473, 487, 548, 549, 561.
— function, *Funktion* 410.
— mechanism, *Mechanismus* 439.
— overall, *überall* 516.
— strength, *Stärke* 584, 586.
— stress in simple tension, *Spannung bei, in einfachem Zug* 405.

zero extension, *ohne Seitenausdehnung* 506.
— moment, *Nullmoment* 217.
Zubov, general system of, *Zubov, allgemeines System von* 130.